INSECT
BIODIVERSITY

INSECT BIODIVERSITY

SCIENCE AND SOCIETY

Edited by Robert G. Foottit and Peter H. Adler

WILEY-BLACKWELL

A John Wiley & Sons, Ltd., Publication

This edition first published 2009, © 2009 by Blackwell Publishing Ltd

Blackwell Publishing was acquired by John Wiley & Sons in February 2007. Blackwell's publishing program has been merged with Wiley's global Scientific, Technical and Medical business to form Wiley-Blackwell.

Registered office: John Wiley & Sons Ltd, The Atrium, Southern Gate, Chichester, West Sussex, PO19 8SQ, UK

Editorial Offices: 9600 Garsington Road, Oxford, OX4 2DQ, UK
The Atrium, Southern Gate, Chichester, West Sussex, PO19 8SQ, UK
111 River Street, Hoboken, NJ 07030-5774, USA

For details of our global editorial offices, for customer services and for information about how to apply for permission to reuse the copyright material in this books please see our website at www.wiley.com/wiley-blackwell

Library of Congress Cataloging-in-Publication Data

Insect biodiversity : science and society / edited by Robert G. Foottit and Peter H. Adler.
     p. cm.
  Includes bibliographical references and index.
  ISBN 978-1-4051-5142-9 (hardcover : alk. paper) 1. Insects–Variation. 2. Insects–Evolution. I. Foottit, R. (Robert G.)
II. Adler, Peter H. (Peter Holdridge)
  QL463.I65 2009   7014507
  595.717–dc22
                                                            2008042545
ISBN: 9781405151429

A catalogue record for this book is available from the British Library.

Set in 9/11 Photina by Laserwords Pvt Ltd, Chennai, India

Printed and bound in the UK

1   2009

CONTENTS

DETAILED CONTENTS

CONTRIBUTORS

PETER H. ADLER *Department of Entomology, Soils & Plant Sciences, Clemson University, Box 340315, 114 Long Hall, Clemson, South Carolina 29634-0315 USA* (padler@clemson.edu)

MAY R. BERENBAUM *Department of Entomology, University of Illinois, 505 S. Goodwin Ave., Urbana, Illinois 61801-3707 USA* (maybe@uiuc.edu)

PATRICE BOUCHARD *Environmental Health National Program (Biodiversity), Agriculture and Agri-Food Canada, K. W. Neatby Bldg., 960 Carling Ave., Ottawa, Ontario K1A 0C6 CANADA* (patrice.bouchard@agr.gc.ca)

MICHAEL F. CLARIDGE *84 The Hollies, Quakers Yard, Treharris, Wales CF46 5PP UK* (claridge@cardiff.ac.uk)

GREGORY W. COURTNEY *Department of Entomology, 432 Science II, Iowa State University, Ames, Iowa 50011-3222 USA* (gwcourt@iastate.edu)

PETER S. CRANSTON *Department of Entomology, University of California at Davis, Davis, California 95616 USA* (pscranston@ucdavis.edu)

HUGH V. DANKS *Biological Survey of Canada, Canadian Museum of Nature, P.O. Box 3443 Station D., Ottawa, Ontario K1P 6P4 CANADA* (hughdanks@yahoo.ca)

HUME DOUGLAS *Canadian Food Inspection Agency, Canadian National Collection of Insects, Arachnids and Nematodes, 960 Carling Avenue, Ottawa, Ontario K1A 0C6 CANADA* (douglash@inspection.gc.ca)

TERRY L. ERWIN *Smithsonian Institution, Department of Entomology, National Museum of Natural History, Mrc-187, Washington, DC 20560-0001 USA* (ErwinT@si.edu)

ROBIN M. FLOYD *Biodiversity Institute of Ontario and Department of Integrative Biology, University of Guelph, Guelph, Ontario N1G 2W1 CANADA* (rfloyd@uoguelph.ca)

ROBERT G. FOOTTIT *Environmental Health National Program (Biodiversity), Agriculture and Agri-Food Canada, K. W. Neatby Bldg., 960 Carling Ave., Ottawa, Ontario K1A 0C6 CANADA* (robert.foottit@agr.gc.ca)

CHRISTY J. GERACI *Smithsonian Institution, Department of Entomology, National Museum of Natural History, Mrc-187, Washington, DC 20560-0001 USA* (GeraciC@si.edu)

VASILY V. GREBENNIKOV *Canadian Food Inspection Agency, Canadian National Collection of Insects, Arachnids and Nematodes, 960 Carling Avenue, Ottawa, Ontario K1A 0C6 CANADA* (GrebennikovV@inspection.gc.ca)

PAUL D. N. HEBERT *Biodiversity Institute of Ontario and Department of Integrative Biology, University of Guelph, Guelph, Ontario N1G 2W1 CANADA* (phebert@uoguelph.ca)

THOMAS J. HENRY *Systematic Entomology Laboratory, PSI, Agriculture Research Service, U. S. Department of Agriculture, c/o Smithsonian Institution, Washington DC 20013-7012 USA* (Thomas.henry@ars.usda.gov)

JOHN M. HERATY *Department of Entomology, University of California, Riverside, California 92521 USA* (john.heraty@ucr.edu)

E. RICHARD HOEBEKE *Department of Entomology, Cornell University, Ithaca, New York 14853-2601 USA (erh2@cornell.edu)*

JOHN T. HUBER *Canadian Forestry Service, K. W. Neatby Bldg., 960 Carling Ave., Ottawa, Ontario K1A 0C6 CANADA (john.huber@agr.gc.ca)*

NORMAN F. JOHNSON *Department of Entomology, Ohio State University, 1315 Kinnear Road, Columbus, Ohio 43212 USA (johnson.2@osu.edu)*

KE CHUNG KIM *Frost Entomological Museum, Department of Entomology, Pennsylvania State University, University Park, Pennsylvania 16802 USA (kck@psu.edu)*

ALEXANDER S. KONSTANTINOV *Systematic Entomology Laboratory, PSI, Agricultural Research Service, U. S. Department of Agriculture, Smithsonian Institution, P. O. Box 37012, National Museum of Natural History, Rm. CE-709, MRC–168, Washington, DC 20560-0168 USA (alex.konstantinov@ars.usda.gov)*

BORIS A. KOROTYAEV *Zoological Institute, Russian Academy of Sciences, Universitetskaya nab. 1, St. Petersburg 199034 RUSSIA (baris@zin.ru)*

MERVYN W. MANSELL *ARC – Plant Protection Research Institute, Biosystematics Division, Private Bag X134, Pretoria 0001 SOUTH AFRICA (vrehmwm@plant5.agric.za)*

GARY L. MILLER *Systematic Entomology Laboratory, PSI, Agricultural Research Service, U.S. Department of Agriculture, Bldg. 005, BARC-West, Beltsville, Maryland 20705 USA (Gary.Miller@ars.usda.gov)*

JOHN C. MORSE *Department of Entomology, Soils & Plant Sciences, Clemson University, Box 340315, 114 Long Hall, Clemson, South Carolina 29634-0315 USA (jmorse@clemson.edu)*

THOMAS PAPE *Natural History Museum of Denmark, Zoological Museum, Universitetsparken 15, DK-2100 Copenhagen, DENMARK (tpape@snm.ku.dk)*

MICHAEL G. POGUE *Systematic Entomology Laboratory, PSI, Agricultural Research Service, U. S. Department of Agriculture, Smithsonian Institution,* P. O. Box 37012, National Museum of Natural History Museum, MRC-168, Washington, DC 20560-0168 USA (michael.pogue@ars.usda.gov)

AMANDA D. ROE *Department of Biological Sciences, CW 405a Biological Sciences Centre, University of Alberta, Edmonton, Alberta T6G 2E9 CANADA (adroe@hotmail.com)*

MICHAEL J. SAMWAYS *Department of Entomology and Centre for Agricultural Biodiversity, University of Stellenbosch, Private Bag X1, 7602 Matieland, SOUTH AFRICA (samways@sun.ac.za)*

CLARKE H. SCHOLTZ *Department of Zoology and Entomology, University of Pretoria, Pretoria 002, SOUTH AFRICA (chscholtz@zoology.up.ac.za)*

GEOFFREY G. E. SCUDDER *Department of Zoology, University of British Columbia, 7270 University Boulevard, Vancouver, British Columbia V6T 1Z4 CANADA (scudder@zoology.ubc.ca)*

BRADLEY J. SINCLAIR *Entomology - Ontario Plant Laboratories, Canadian Food Inspection Agency, K.W. Neatby Building, 960 Carling Avenue, Ottawa, Ontario K1A 0C6 CANADA (sinclairb@inspection.gc.ca)*

JEFFREY H. SKEVINGTON *Agriculture and Agri-Food Canada (AAFC), Canadian National Collection of Insects, Arachnids and Nematodes, K.W. Neatby Building, 960 Carling Avenue, Ottawa, Ontario K1A 0C6 CANADA (jeffrey.skevington@agr.gc.ca)*

ANDREW B. T. SMITH *Research Division, Canadian Museum of Nature, P.O. Box 3443, Station 'D', Ottawa, Ontario K1P 6P4 CANADA (asmith@mus-nature.ca)*

FELIX A. H. SPERLING *Department of Biological Sciences, CW 405a Biological Sciences Centre, University of Alberta, Edmonton, Alberta T6G 2E9 CANADA (felix.sperling@ualberta.ca)*

MARK G. VOLKOVITSH *Zoological Institute, Russian Academy of Sciences, Universitetskaya nab. 1, St. Petersburg 199034 RUSSIA (polycest@zin.ru)*

ALFRED G. WHEELER, JR. *Department of Entomology, Soils & Plant Sciences, Clemson University,*

Box 340315, 114 Long Hall, Clemson, South Carolina 29634-0315 USA (awhlr@clemson.edu)

QUENTIN D. WHEELER *International Institute for Species Exploration, Arizona State University, P.O. Box 876505, Tempe, AZ 85287-6505 USA* (Quentin.Wheeler@asu.edu)

JOHN J. WILSON *Biodiversity Institute of Ontario and Department of Integrative Biology, University of Guelph, Guelph, Ontario, N1G 2W1 CANADA* (jwilso04@uoguelph.ca)

PREFACE

Insects are the world's most diverse group of animals, making up more than 58% of the known global biodiversity. They inhabit all habitat types and play major roles in the function and stability of terrestrial and aquatic ecosystems. Insects are closely associated with our lives and affect the welfare of humanity in diverse ways. At the same time, large numbers of insect species, including those not known to science, continue to become extinct or extirpated from local habitats worldwide. Our knowledge of insect biodiversity is far from complete; for example, barely 65% of the North American insect fauna has been described. Only a relatively few species of insects have been studied in depth. We urgently need to explore and describe insect biodiversity and to better understand the biology and ecology of insects if ecosystems are to be managed sustainably and if the effect of global environment change is to be mitigated.

The scientific study of insect biodiversity is at a precarious point. Resources for the support of taxonomy are tenuous worldwide. The number of taxonomists is declining and the output of taxonomic research has slowed. Many taxonomists are reaching retirement age and will not be replaced with trained scientists, which will result in a lack of taxonomic expertise for many groups of insects. These trends contrast with an increasing need for taxonomic information and services in our society, particularly for biodiversity assessment, ecosystem management, conservation, sustainable development, management of climate-change effects, and pest management. In light of these contrasting trends, the scientific community and its leadership must increase their understanding of the science of insect biodiversity and taxonomy and ensure that policy makers are informed of the importance of biodiversity for a sustainable future for humanity.

We have attended and contributed to many scientific meetings and management and policy gatherings where the future, the resource needs, and importance of insect taxonomy and biodiversity have been debated.

In fact, discussion of the future of taxonomy is a favorite pastime of taxonomists; there is no shortage of "taxonomic opinion." Considerable discussion has focused on the daunting task of describing the diversity of insect life and how many undescribed species are out there. However, we felt that there was a need for an up-to-date, quantitative assessment of what insect biodiversity entails, and to connect what we know and do not know about insect biodiversity with its impact on human society.

Our approach was to ask authors to develop accounts of biodiversity in certain orders of insects and geographic regions and along selected subject lines. In all categories, we were limited by the availability of willing contributors and their time and resources. Many insect groups, geographic regions, and scientific and societal issues could not be treated in a single volume. It also was apparent to us, sometimes painfully so, that many taxonomists are wildly over-committed. This situation can be seen as part of the so-called taxonomic impediment – the lack of available taxonomic expertise is compounded by an overburdened community of present-day taxonomists with too much work and perhaps too much unrealistic enthusiasm.

In Chapter 1, we introduce the ongoing challenge to document insect biodiversity and develop its services. Chapter 2 provides a comprehensive overview of the importance and value of insects to humans. The next two sections deal with regional treatments and ordinal-level accounts of insect biodiversity. These approaches were a serious challenge to the contributors who had to compile information from a wide array of sources or, alternatively, deal with situations in which accurate information simply is insufficient. In Part III, we document some of the tools and approaches to the science of taxonomy and its applications. Perspective is provided on the past, present, and future of the science of insect taxonomy and the all-important influence of species concepts and their operational treatment on taxonomic science and insect biodiversity. Contributions on the increasing role of informatics and molecular approaches are provided, areas with ongoing controversy and differences of opinion. These chapters are followed by contributions on the applications of taxonomic science for which biodiversity information is fundamental, including the increasing impact of adventive insects, pest detection and management, human medical concerns, and the management and conservation of biodiversity. The book ends with an historical view of the continuing attempts to document the extent of world insect biodiversity.

Robert G. Foottit
Ottawa, Ontario

Peter H. Adler
Clemson, South Carolina

FOREWORD

Insects are the most exuberant manifestation of Earth's many and varied life forms. To me, one of their greatest fascinations is how a rather simple basic unifying body plan has become modified and adapted to produce an enormous variety of species, able to exploit virtually all terrestrial and freshwater environments on the planet while, as a paradox debated extensively a few decades ago, not becoming equally predominant in the seas and oceans. Features such as possession of wings and the complete metamorphosis of many species have been cited frequently as fostering this massive diversity. However, the 'success' of insects can be measured by many parameters: their long-term persistence and stability of their basic patterns, the variety of higher groups (with almost 30 orders commonly recognized) and, as emphasized in this book, the wealth of species and similar entities. Each of these species has its individual biological peculiarities, ecological role, distribution, and interactions within the local community. And each may differ in habit and appearance both from its closest relatives and also across its range to reflect local influences and conditions. Every species is thus a mosaic of physical variety and genetic constitution that can lead to both taxonomic and ecological ambiguity in interpreting its integrity and the ways in which it may evolve and persist.

Entomologists will continue to debate the number of insect 'species' that exist and the levels of past and likely future extinctions. Documenting and cataloging insect biodiversity as a major component of Earth's life is a natural quest of human inquiry but is not an end in itself and, importantly, is not synonymous with conserving insects or a necessary prerequisite to assuring their well-being. Despite many ambiguities in projecting the actual numbers of insect species, no one would query that there are a lot and that the various ecological processes that sustain ecosystems depend heavily on insect activity. Indeed, 'ecological services' such as pollination, recycling of materials, and the economically important activities of predators and parasitoids are signaled increasingly as part of the rationale for insect conservation because these values can be appreciated easily through direct economic impacts. All these themes are dealt with in this book, centered on questions related to our ignorance of the fundamental matters of 'how many are there?' and 'how important are they?' to which the broad answers of 'millions' and 'massive' may incorporate considerable uncertainty; this uncertainty, however, is reduced by many of the chapters here.

In any investigations of insect biodiversity, the role of inventory tends to be emphasized. Documenting numbers of species (however they are delimited or defined) gives us foci for conservation advocacy and is pivotal in helping to elucidate patterns of evolution and distribution. Recognizing and naming species allow us to transfer information, but high proportions of undescribed or unrecognizable species necessitate the use of terms such as 'morphospecies' in much ecological interpretation of diversity. Nevertheless, other than in some temperate regions, particularly in the Northern Hemisphere, many estimates of insect species richness and naming the species present are highly incomplete. Much of the tropics, for example, harbors few resident entomologists other than those involved with pressing problems of human welfare, and more basic and sustained documentation almost inevitably depends on assistance from elsewhere. Some insects, of course, have been explored much more comprehensively than others, so that selected taxonomic groups (such as butterflies, larger beetles, and dragonflies) and ecological groups ('pests') have received much more attention than many less charismatic or less economically important groups. Indeed, when collecting Psocoptera in parts of the tropics, I have occasionally been asked by local people why I am not collecting birdwing butterflies, stag beetles, or other 'popular' (or commercially desirable!) insects, and my responses have done little to change their opinions of my insanity! In short, many gaps in knowledge of insect diversity persist, and seem unlikely to be redressed effectively in the near future, other than by 'guesstimates' extrapolating from sometimes rather dubious foundations. However, sufficient knowledge does exist to endorse the practical need to protect natural habitats effectively from continued despoliation and, as far as practicable, from the effects of climate changes. Citations of impressively large numbers of insect species can become valuable advocacy in helping to conserve areas with largely unheralded wealth of biodiversity. Presence of unusual lineages of insects, of narrow range endemics, as well as highly localized radiations and distributional idiosyncrasies (such as isolated populations beyond the main range of the taxon) are all commonplace scenarios and may in various ways help us to designate priorities for allocating the limited conservation resources available. Many such examples from selected insect groups are revealed in this book – but evaluating the richness and ecological importance of the so-called meek inheritors, that vast majority of insects that do not intrude notably on human intelligence and welfare, remains a major challenge. Many such taxa receive attention from only a handful of entomologists at any time, and some are essentially 'orphaned' for considerable periods. Progress with their documentation is inevitably slow and sporadic. In addition, some hyperdiverse orders and families of insects exhibit daunting complexity of form and biology, as 'black hole groups' whose elucidation is among the major challenges that face us.

Insect conservation has drawn heavily on issues relevant to biodiversity and appreciation of the vast richness of insects – not only of easily recognizable 'species' but also of the occurrence of subspecies and other infraspecific variants such as significant populations, collectively 'evolutionarily significant units'. This more complex dimension of insect biodiversity is receiving considerable attention as new molecular tools (such as DNA analysis) enable us to probe characters in ways undreamed of only a decade or so ago to augment the perspectives provided by morphological interpretation. The vast arrays of cryptic species gradually being revealed suggest that even our most up-to-date estimates of species numbers based on morphological data may be woefully inadequate. Insect diversity equates to 'variety', but the subtleties of interpopulation variations in genetic constitution and ecological performance are difficult to appraise and to categorize formally – and perhaps even more difficult to communicate to non-entomologists whose powers may determine the future of the systems in which those insects participate. Education and communication, based on the soundest available information, are essential components of insect conservation. This book is a significant contribution to this endeavor, through indicating how we may come to interpret and understand insect biodiversity more effectively. In addition to providing a range of opinions and facts on insect richness in a variety of taxonomic, geographical, and methodological contexts, it helps to emphasize the importance of accurate species recognition. Failure to recognize adventive alien species may have dire economic or ecological consequences, for example, or confusion between biotypes or cryptic species may invalidate expensive management programs for their suppression or conservation.

A new generation of skilled insect systematists – whose visions encompass the wider ramifications of insect biodiversity, its importance in understanding the natural world and the accelerating impacts of humans upon it – is an urgent need. They enter an exciting and challenging field of endeavor, and the perspectives included in this volume are essential background to their future contributions. This book is thus both a foundation and a stimulating working tool toward that end, and I expect many of the chapters to be key references as we progressively refine and enlarge the bases of our understanding of insects and their activities in the modern world.

T.R. New,
Department of Zoology,
La Trobe University,
Victoria 3086, Australia

ACKNOWLEDGMENTS

We asked external reviewers to give us perspective on each chapter, and we are grateful for their efforts and appreciative of the time they took from their busy schedules. We would like to thank the following individuals who reviewed one or more chapters: P. Bouchard, C. E. Carlton, M. F. Claridge, P. S. Cranston, T. L. Erwin, C. J. Geraci, D. R. Gillespie, P. W. Hall, R. E. Harbach, J. D. Lafontaine, J. D. Lozier, P. G. Mason, H. E. L. Maw, J. C. Morse, L. A. Mound, G. R. Mullen, T. R. New, J. E. O'Hara, V. H. Resh, M. D. Schwartz, G. G. E. Scudder, D. S. Simberloff, A. Smetana, A. B. T. Smith, J. Sóberon, L. Speers, F. A. H. Sperling, I. C. Stocks, M. W. Turnbull, C. D. von Dohlen, D. L. Wagner, G. Watson, A. G. Wheeler, Jr., Q. D. Wheeler, B. M. Wiegmann, D. K. Yeates, and P. Zwick. We extend our gratitude to Eric Maw for his tremendous efforts in generating the taxonomic indices.

Finally, we acknowledge the encouragement and support, both moral and technical, of the staff at Wiley-Blackwell, particularly Ward Cooper, Delia Sandford, and Rosie Hayden.

Chapter 1

INTRODUCTION

Peter H. Adler[1] and Robert G. Foottit[2]

[1]Department of Entomology, Soils, and Plant Sciences, Clemson University, Clemson, South Carolina 29634-0315 USA

[2]Agriculture and Agri-Food Canada, Canadian National Collection of Insects, Ottawa, ON, K1A 0C6 Canada

Insect Biodiversity: Science and Society, 1st edition. Edited by R. Foottit and P. Adler

© 2009 Blackwell Publishing, ISBN 978-1-4051-5142-9

Every so often, a technical term born in the biological community enters the popular vocabulary, usually because of its timeliness, political implications, media hype, and euphonious ability to capture the essence of an issue. 'Biotechnology', 'human genome', and 'stem cells' are terms as common in public discourse as they are in scientific circles. 'Biodiversity' is another recent example. Introduced in its portmanteau form in the mid-1980s by Warren G. Rosen (Wilson 1988), the term has grown steadily in popularity. By March 2008, the keyword 'biodiversity' generated 12 million hits on Google Search. Three months later, the number of hits, using the same keyword search, had shot to more than 17 million.

Although the word 'biodiversity' might be familiar to many, its definition is often subject to individual interpretation. Abraham Lincoln grappled with a similar concern over the word 'liberty'. In an 1864 speech, Lincoln opined, 'The world has never had a good definition of the word liberty, and the American people, just now, are much in want of one ... but in using the same *word* we do not all mean the same *thing*' (Simpson 1998). To the layperson, 'biodiversity' might conjure a forest, a box of beetles, or perhaps the entire fabric of life. Among scientists, the word has been defined, explicitly and implicitly, *ad nauseum*, producing a range of variants (e.g., Gaston 1996). In its original context, the term 'biodiversity' encompassed a broad range of topics (Wilson 1988), and we embrace that perspective. Biodiversity, then, is big biology, describing a holistic view of life. It is 'the variety of all forms of life, from genes to species, through to the broad scale of ecosystems' (Faith 2007). The fundamental units of biodiversity – species – serve as focal points for studying the full panoply of life, allowing workers to zoom in and out along a scale from molecule to ecosystem. The species-centered view also provides a vital focus for conserving life forms and understanding the causes of declining biodiversity.

Despite disagreements over issues ranging from definitions of biodiversity to phylogenetic approaches, biologists can agree on four major points. (1) The world supports a great number of insects. (2) We do not know how many species of insects occupy our planet. (3) The value of insects to humanity is enormous. (4) Too few specialists exist to inventory the world's entomofauna.

By virtue of the sheer numbers of individuals and species, insects, more than any other life form, command the attention of biologists. The number of individual insects on earth at any given moment has been calculated at one quintillion (10^{18}) (Williams 1964), an unimaginably large number on par with the number of copepods in the ocean (Schubel and Butman 1998) and roughly equivalent to the number of sand grains along a few kilometers of beach (Ray 1996). The total number of insect species similarly bankrupts the mind. Estimates offered over the past four centuries have increased steadily from 10,000 species, proposed by John Ray in 1691 (Berenbaum, this volume), to as many as 80 million (Erwin 2004). Today's total of 1,004,898 described living species (Table 1.1) is more than 100 times the 1691 estimate. Based on a figure of 1.50–1.74 million described eukaryotic species in the world (May 1998), insects represent 58–67% of the total.

The members of the class Insecta are arranged in 29 orders (Grimaldi and Engel 2005, Arillo and Engel 2006). Four of these orders – the Coleoptera, Diptera, Hymenoptera, and Lepidoptera – account for 81% of all the described species of living insects. The beetles are far in front, leading the next largest order, the Lepidoptera, by a factor of about 2.3 (Table 1.1). A growing number of world checklists and catalogs are available online for various families and orders. Outfitted with search functions, they provide another tool for handling the taxonomic juggernaut of new species and nomenclatural changes. We can foresee a global registry of species in the near future that is updated with each new species or synonym, allowing real-time counts for any taxon.

The greatest concentration of insect species lies in tropical areas of the globe. One hectare of Amazonian rainforest contains more than 100,000 species of arthropods (Erwin 2004), of which roughly 85% are insects (May 1998). This value is more than 90% of the total described species of insects in the entire Nearctic Region. Yet, this tropical skew is based partly on a view of species as structurally distinct from one another. Morphologically similar, if not indistinguishable, species (i.e., sibling species) typically do not figure in estimates of the number of insect species. If organisms as large as elephants and giraffes are composites of multiple species (Brown et al. 2007), a leap of faith is not required to realize that smaller earthlings also consist of additional, reproductively isolated units of biodiversity. When long-recognized nominal species of insects, from black flies to butterflies, are probed more deeply, the repetitive result is an increase, often manyfold, in the number of species (Hebert et al. 2004, Post et al. 2007). We do not yet have a clear indication

Table 1.1 World totals of described, living species in the 29 orders of the class Insecta.

Order[1]	Described Species[2]	References
Archaeognatha	504	Mendes 2002, Zoological Record 2002–2008
Zygentoma	527	Mendes 2002, Zoological Record 2002–2008
Ephemeroptera	3046	Barber-James et al. 2008
Odonata	5680	Kalkman et al. 2008
Dermaptera	1967	Steinmann 1989, Zoological Record 1989–2008
Notoptera	39	Vrsansky et al. 2001, Engel and Grimaldi 2004
Plecoptera	3497	Fochetti and de Figueroa 2008
Embiodea	458	Ross 2001, Zoological Record 2002–2008
Zoraptera	34	Hubbard 2004, Zoological Record 2004–2008
Phasmatodea	2853	Brock 2008
Orthoptera	23,616	Eades and Otte 2008
Mantodea	2384	Ehrmann 2002, Zoological Record 2002–2008
Blattaria	4565	Beccaloni 2007
Isoptera	2864	Constantino 2008
Psocoptera	5574	New and Lienhard 2007, Zoological Record 2008
Phthiraptera	5024	Durden and Musser 1994, Price et al. 2003, L. Durden personal communication
Thysanoptera	5749	Mound 2005, personal communication
Hemiptera	100,428	Duffels and van der Laan 1985, Zoological Record 1981–2008 (Cicadidae); Remaudière and Remaudière 1997, G. L. Miller personal communication (Aphidoidea); McKamey 1998, 2007, personal communication (Cercopidae, Cicadellidae, Membracidae); Hollis 2002 (Psylloidea); Ben-Dov et al. 2006 (Coccoidea); Bourgoin 2005, McKamey personal communication (Fulgoroidea); Martin and Mound 2007 (Aleyrodidae); Henry, this volume (Heteroptera)
Coleoptera	359,891	Bouchard et al., this volume
Raphidioptera	225	Aspöck 2002, Oswald 2007, J. D. Oswald personal communication
Megaloptera	337	Cover and Resh 2008, J. D. Oswald personal communication
Neuroptera	5704	Oswald 2007, J. D. Oswald personal communication
Hymenoptera	144,695	Huber, this volume
Mecoptera	681	Penny 1997, Zoological Record 1998–2008
Siphonaptera	2048	Lewis 1998, Zoological Record 1998–2008
Strepsiptera	603	Proffitt 2005, Zoological Record 2005–2008
Diptera	152,244	Courtney et al., this volume
Trichoptera	12,868	Morse 2008
Lepidoptera	156,793	Pogue, this volume; Zoological Record 2007–2008
TOTAL	**1,004,898**	

[1]We follow the ordinal classification of Grimaldi and Engel (2005) for the class Insecta, updated to recognize Notoptera (i.e., Grylloblattodea + Mantophasmatodea; Arillo and Engel 2006). Thus, the three orders of the class Entognatha – the Collembola, Diplura, and Protura – are not included. These three orders would add roughly another 11,000 species to the total number in Table 1.1.

[2]Species were tallied in spring 2008, with the exception of Hymenoptera (Huber, this volume), which were counted primarily in 2006–2007, with earlier counts for some families, and Coleoptera (Bouchard et al., this volume), which were counted, with few exceptions, within the past decade.

across sufficient taxa to know whether a regional bias in sibling species of insects might exist or even vary among taxonomic groups.

The precise number of species, however, is not what we, as a global society, desperately need. Rather, we require a comprehensive, fully accessible library of all volumes (i.e., species) – a colossal compendium of names, descriptions, distributions, and biological information that ultimately can be transformed into a yellow pages of services. Insects hold a vast wealth of behavior, chemistry, form, and function that conservatively translates into an estimated $57 billion per annum in ecological services to the United States (Losey and Vaughan 2006), a value that does not include services provided by domesticated insects (e.g., honey bees) or their products (e.g., honey and shellac) or mass-reared biological control agents. To harvest the full range of benefits from insects, taxonomists and systematists must first reveal the earth's species and organize them with collateral information that can be retrieved with ease.

Biodiversity science must keep pace with the changing face of the planet, particularly species extinctions and reshufflings driven largely by human activities such as commerce, land conversion, and pollution. By 2007, for example, 1321 introduced species had been documented on the Galapagos Islands, of which at least 37% are insects (Anonymous 2007). As species of insects are being redistributed, others are disappearing, particularly in the tropics, though the data are murky. We are forced into an intractable bind, for we cannot know all that we are losing if we do not know all that we have. We do know, however, that extinction is an inevitable consequence of planetary abuse. The Brazilian government, for instance, announced that deforestation rates had increased in its portion of the Amazon, with a loss of 3235 sq km in the last 5 months of 2007. Using Erwin's (2004) figure of 3×10^{10} individual terrestrial arthropods per hectare of tropical rainforest, we lost habitat for more than 30 trillion arthropods in that one point in space and time.

The urgency to inventory the world's insect fauna is gaining some balance through the current revolution in technology. Coupled with powerful electronic capabilities, the explosion of biodiversity information, much of it now derived from the genomic level, can be networked worldwide to facilitate not only communication and information storage and retrieval but also taxonomy itself – cybertaxonomy (*sensu* Wheeler 2007). Efforts to apply new approaches and bioinformatics on a global

scale are now underway (e.g., Barcode of Life Data Systems 2008, Encyclopedia of Life 2008). We can imagine that in our lifetimes, automated complete-genome sequencing will be available to identify specimens as routinely as biologists today use identification keys. The futuristic handheld gadget that can read a specimen's genome and provide immediate identification, with access to all that is known about the organism (Janzen 2004), is no longer strictly science fiction. Yet, each new technique for revealing and organizing the elements of biodiversity comes with its own set of limitations, some of which we do not yet know. DNA-sequence readers, for instance, will do little to identify fossil organisms. An integrated methodology, mustering information from molecules to morphology, will continue to prove its merit, although it is the most difficult approach for the individual worker to master. Given the vast number of insect species, however, today's themes are likely to remain the same well beyond the advent of handheld, reveal-all devices: an unknown number of species, too few specialists, and too little appreciation of the value of insect biodiversity.

Those who study insect biodiversity do so largely out of a fascination for insects; no economic incentive is needed. But for most people, from a land developer to a hardscrabble farmer, a personal, typically economic reason is required to appreciate the value of insect biodiversity. This value, therefore, must be translated into economic gain. Today's biologists place a great deal of emphasis on discovering species, cataloging them, and inferring their evolutionary relationships. Rightly so. But these activities will not, in themselves, curry favor with the majority. We believe that, now, equal emphasis must be placed on developing the services of insects. We envision a new era, one of entrepreneurial biodiversity that crosses disciplinary boundaries and links the expertise of insect systematists with that of biotechnologists, chemists, economists, engineers, marketers, pharmacologists, and others. Only then can we expect to tap the magic well of benefits derivable from insects and broadly applicable to society, while ensuring a sustainable environment and conserving its biodiversity. And, this enterprise just might reinvigorate interest in biodiversity among the youth and aspiring professionals.

The chapters in this volume are written by biologists who share a passion for insect biodiversity. The text moves from a scene-setting overview of the value of insects through examples of regional biodiversity, taxon biodiversity, tools and approaches, and management and conservation to a historical view of the quest for the true number of insect species. The case is made throughout these pages that real progress has been achieved in discovering and organizing insect biodiversity and revealing the myriad ways, positive and negative, that insects influence human welfare. While the job remains unfinished, we can be assured that the number of insect-derived benefits yet to be realized is far greater than the number of species yet to be discovered.

ACKNOWLEDGMENTS

We thank those who answered queries and helped us tally species in their areas of expertise: G. W. Beccaloni, P. D. Brock, L. A. Durden, D. C. Eades, F. Haas, C. Lienhard, S. McKamey, G. L. Miller, L. A. Mound, and J. D. Oswald. We appreciate the review of the manuscript by A. G. Wheeler, Jr.

REFERENCES

Anonymous. 2007. CDF supports Galapagos in danger decision. *Galapagos News* Fall/Winter 2007: 2.

Arillo, A. and M. S. Engel. 2006. Rock crawlers in Baltic amber (Notoptera: Mantophasmatodea). *American Museum Novitates* 3539: 1–10.

Aspöck, H. 2002. The biology of Raphidioptera: a review of present knowledge. *Acta Zoologica Academiae Scientiarum Hungaricae* 48 (Supplement 2): 35–50.

Barber-James, H. M., J.-L. Gattolliat, M. Sartori, and M. D. Hubbard. 2008. Global diversity of mayflies (Ephemeroptera; Insecta) in freshwater. *Hydrobiologia* 595: 339–350.

Barcode of Life Data Systems. 2008. http://www.barcodinglife.org/views/login.php [Accessed 20 January 2008].

Beccaloni, G. W. 2007. Blattodea species file online. Version 1.2/3.3. http://Blattodea.SpeciesFile.org [Accessed 20 May 2008].

Ben-Dov, Y., D. R. Miller, and G. A. P. Gibson. 2006. ScaleNet. http://www.sel.barc.usda.gov/SCALENET/scalenet.htm [Accessed 21 May 2008].

Bourgoin, T. 2005. FLOW: Fulgomorpha lists on the web. http://flow.snv.jussieu.fr/cgi-bin/entomosite.pl [Accessed 22 May 2008].

Brock, P. D. 2008. Phasmida species file online. Version 2.1/3.3. http://Phasmida.SpeciesFile.org [Accessed 22 May 2008].

Brown, D. M., R. A. Brenneman, K.-P. Koepfli, J. P. Pollinger, B. Milá, N. J. Georgiadis, E. E. Louis, Jr., G. F. Grether, D. K. Jacobs, and R. K. Wayne. 2007. Extensive population genetic structure in the giraffe. *BMC Biology* 5: 57.

Constantino, R. 2008. On-line termite database. http://www.unb.br/ib/zoo/docente/constant/catal/catnew.html [Accessed 22 May 2008].

Cover, M. R. and V. H. Resh. 2008. Global diversity of dobsonflies, fishflies, and alderflies (Megaloptera; Insecta) and spongillaflies, nevrorthids, and osmylids (Neuroptera; Insecta) in freshwater. *Hydrobiologia* 595: 409–417.

Duffels, J. P. and P. A. van der Laan. 1985. *Catalogue of the Cicadoidea (Homoptera, Auchenorrhyncha) 1956–1980.* Series Entomologica. Volume 34. W. Junk, Dordrecht. xiv + 414 pp.

Durden, L. A. and G. G. Musser. 1994. The sucking lice (Insecta, Anoplura) of the world: a taxonomic checklist with records of mammalian hosts and geographical distributions. *Bulletin of the American Museum of Natural History* 218: 1–90.

Eades, D. C. and D. Otte. 2008. Orthoptera species file online. Version 2.0/3.3. http://Orthoptera.SpeciesFile.org [Accessed 21 May 2008].

Ehrmann, R. 2002. *Mantodea: Gottesanbeterinnen der Welt.* Natur und Tier – Verlag, Münster, Germany. 519 pp.

Encyclopedia of Life. 2008. http://www.eol.org/ [Accessed 30 May 2008].

Engel, M. S. and D. A. Grimaldi. 2004. A new rock crawler in Baltic amber, with comments on the order (Mantophasmatodea: Mantophasmatidae). *American Museum Novitates* 3431: 1–11.

Erwin, T. L. 2004. The biodiversity question: how many species of terrestrial arthropods are there? Pp. 259–269. *In* M. D. Lowman and H. B. Rinker (eds). *Forest Canopies,* Second Edition. Elsevier Academic Press, Burlington, Massachusetts.

Faith, D. P. 2007. Biodiversity. *In* E. N. Zalta, U. Nodelman, and C. Allen (eds). *Stanford Encyclopedia of Philosophy.* Metaphysics Research Lab, Stanford University, Stanford, California. http://plato.stanford.edu/entries/biodiversity/ [Accessed 11 May 2008].

Fochetti, R. and J. M. T. de Figueroa. 2008. Global diversity of stoneflies (Plecoptera; Insecta) in freshwater. *Hydrobiologia* 595: 365–377.

Gaston, K. J. (ed). 1996. *Biodiversity: A Biology of Numbers and Difference.* Blackwell Science, Oxford. 396 pp.

Grimaldi, D. A. and M. Engel. 2005. *The Evolution of Insects.* Cambridge University Press, Cambridge. 755 pp.

Hebert, P. D. N., E. H. Penton, J. M. Burns, D. H. Janzen, and W. Hallwachs. 2004. Ten species in one: DNA barcoding reveals cryptic species in the neotropical skipper butterfly *Astraptes fulgerator.* Proceedings of the National Academy of Sciences USA 101: 14812–14817.

Hollis, D. 2002. Psylloidea. http://www.environment.gov.au/cgi-bin/abrs/fauna/details.pl?pstrVol=PSYLLOIDEA;pstrTaxa=1;pstrChecklistMode=1 [Accessed 21 May 2008].

Hubbard, M. D. 2004. The Zoraptera database: catalog of the order Zoraptera. http://www.famu.org/zoraptera/catalog.html [Accessed 18 May 2008].

Janzen, D. H. 2004. Now is the time. *Philosophical Transactions of the Royal Society of London* B 359: 731–732.

Kalkman, V. J., V. Crausnitzer, K.-D. B. Dijkstra, A. G. Orr, D. R. Paulson, and J. van Tol. 2008. Global diversity of dragonflies (Odonata) in freshwater. *Hydrobiologia* 595: 351–363.

Lewis, R. E. 1998. Résumé of the Siphonaptera (Insecta) of the world. *Journal of Medical Entomology* 35: 377–389.

Losey, J. E. and M. Vaughan. 2006. The economic value of ecological services provided by insects. *BioScience* 56: 311–323.

Martin, J. H. and L. A. Mound. 2007. An annotated check list of the world's whiteflies (Insecta: Hemiptera: Aleyrodidae). *Zootaxa* 1492: 1–84.

May, R. M. 1998. The dimensions of life on Earth. Pp. 30–45. In P. H. Raven (ed). *Nature and Human Society: The Quest for a Sustainable World*. National Academy Press, Washington, DC.

McKamey, S. H. 1998. Taxonomic catalogue of the Membracoidea (exclusive of leafhoppers): second supplement to fascicle 1 – Membracidae of the general catalogue of the Hemiptera. *Memoirs of the American Entomological Institute* 60: 1–377.

McKamey, S. H. 2007. Taxonomic catalogue of the leafhoppers (Membracoidea). Part 1. Cicadellinae. *Memoirs of the American Entomological Institute* 78: 1–394.

Mendes, L. F. 2002. Taxonomy of Zygentoma and Microcoryphia: historical overview, present status and goals for the new millennium. *Pedobiologia* 46: 225–233.

Morse, J. C. 2008. Welcome to the world Trichoptera checklist. http://entweb.clemson.edu/database/trichopt/index.htm [Accessed 21 May 2008].

Mound, L. 2005. Thysanoptera (thrips) of the world – a checklist. http://www.ento.csiro.au/thysanoptera/worldthrips.html [Accessed 22 May 2008].

New, T. R. and C. Lienhard. 2007. The Psocoptera of tropical South-east Asia. *Fauna Malesiana Handbooks* 6. ix + 290 pp.

Oswald, J. D. 2007. Neuropterida species of the world: a catalogue of the species-group names of the extant and fossil Neuropterida (Insecta: Neuroptera, Megaloptera and Raphidioptera) of the world. Version 2.00. http://lacewing.tamu.edu/Species-Catalogue/ [Accessed 22 May 2008].

Penny, N. D. 1997. World checklist of extant Mecoptera species. http://research.calacademy.org/research/entomology/Entomology_Resources/mecoptera/index.htm [Accessed 18 May 2008].

Post, R. J., M. Mustapha, and A. Krueger. 2007. Taxonomy and inventory of the cytospecies and cytotypes of the *Simulium damnosum* complex (Diptera: Simuliidae) in relation to onchocerciasis. *Tropical Medicine and International Health* 12: 1342–1353.

Price, R. D., R. A. Hellenthal, R. L. Palma, K. P. Johnson, and D. H. Clayton. 2003. The chewing lice: world checklist and biological overview. *Illinois Natural History Survey Special Publication* 24: 1–501.

Proffitt, F. 2005. Twisted parasites from 'outer space' perplex biologists. *Science* 307: 343.

Ray, C. C. 1996. *Stars and sand*. New York Times, 5 March 1996.

Remaudière, G. and M. Remaudière. 1997. Catalogue des Aphididae du monde (Homoptera Aphidoidea). INRA Editions, Paris. 473 pp.

Ross, E. S. 2001. World list of extant and fossil Embiidina (=Embioptera). http://research.calacademy.org/research/entomology/Entomology_Resources/embiilist/embiilist.html [Accessed 18 May 2008].

Schubel, J. R. and A. Butman. 1998. Keeping a finger on the pulse of marine biodiversity: how healthy is it? Pp. 84–103. In P. H. Raven (ed). *Nature and Human Society: The Quest for a Sustainable World*. National Academy Press, Washington, DC.

Simpson, B. D. 1998. *Think Anew, Act Anew: Abraham Lincoln on Slavery, Freedom, and Union*. Harlan Davidson, Wheeling, Illinois. 205 pp.

Steinmann, H. 1989. *World Catalogue of Dermaptera*. Series Entomologica. Volume 43. Kluwer Academic Publishers. 934 pp.

Vrsansky, P., S. Y. Storozhenko, C. C. Labandeira, and P. Ihringova. 2001. *Galloisiana olgae* sp. nov. (Grylloblattodea: Grylloblattidae) and the paleobiology of a relict order of insects. *Annals of the Entomological Society of America* 94: 179–184.

Wheeler, Q. D. 2007. Invertebrate systematics or spineless taxonomy? *Zootaxa* 1668: 11–18.

Williams, C. B. 1964. *Patterns in the Balance of Nature and Related Problems in Quantitative Biology*. Academic Press, London. 324 pp.

Wilson, E. O. (ed). 1988. *Biodiversity*. National Academy of Sciences/Smithsonian Institution, Washington, DC. 538 pp.

THE IMPORTANCE OF INSECTS

Geoffrey G. E. Scudder

Department of Zoology, University of British Columbia,
Vancouver, BC V6T 1Z4 Canada

Insects nurture and protect us, sicken us, kill us. They bring both joy and sorrow. They drive us from fear to hate, then to tolerance. At times they bring us up short to a realization of the way the world really is, and what we have to do to improve it. Their importance to human welfare transcends the grand battles we fight against them to manage them for our own ends. Most of us hate them, but some of us love them. Indeed at times they even inspire us.

— *McKelvey (1975)*

Insect Biodiversity: Science and Society, 1st edition. Edited by R. Foottit and P. Adler
© 2009 Blackwell Publishing, ISBN 978-1-4051-5142-9

Insects are important because of their diversity, ecological role, and influence on agriculture, human health, and natural resources. They have been used in landmark studies in biomechanics, climate change, developmental biology, ecology, evolution, genetics, paleolimnology, and physiology. Because of their many roles, they are familiar to the general public. However, their conservation is a challenge. The goal of this chapter is to document the dominance of insects in the living world and to show how they have been central to many advances in science.

DIVERSITY

Considerable debate continues over how many species of insects are in the world. Estimates range from 2 to 50 million (Stork 1993). The lower figure is from Hodkinson and Casson (1991). The higher figure of up to 50 million is from Erwin (1988, 1993), and like an earlier estimate of 30 million (Erwin 1982, 1983), is based on numbers obtained from canopy fogging in the tropical forests of the Americas. These high estimates have been questioned, however, because of the assumptions made and the lack of real evidence for vast numbers of undescribed species (Stork 1993). Other methods of estimation have been used by May (1988), Stork and Gaston (1990), and Gaston (1991), and from these other data, Stork (1993) concluded that a global total of 5–15 million is more reasonable. Gaston (1991) gave a figure of about 5 million, and this estimate was accepted by Grimaldi and Engel (2005), although Hammond (1992) gave an estimate of 12.5 million species.

The number of insects described at present is estimated to be 925,000 (Grimaldi and Engel 2005) (updated to 1,004,898; this volume), in a total biota described to date of 1.4 to 1.8 million (Stork 1988, 1993, May 1990, Hammond 1992). Using the 925,000 species described, versus the estimate of 5 million total, Grimaldi and Engel (2005) suggested that only about 20% of the insects are named.

A majority of the species on earth are insects. They have invaded every niche, except the oceanic benthic zone (Grimaldi and Engel 2005). Hammond (1992) calculated that arthropods constitute 65% of the total known biodiversity, and Grimaldi and Engel (2005) put the figure at about 58%, while Samways (1993) noted that they constitute 81.3% of described animal species, excluding the Protozoa. Thus, from a modest beginning some 400 mya, insects have become the dominant component of the known diversity on earth, with 100 million species having ever lived (Grimaldi and Engel 2005).

Wheeler (1990) in his 'species scape', pictorially illustrated the current dominance of insects, and Samways (1993) noted that if all insect species on earth were described, the beetle representing the proportion of insect species in the world might have to be drawn up to 10 times larger. Wheeler (1990) used a beetle to depict the arthropods in his species scape because the Coleoptera are the dominant insect group, constituting 40% of the estimated total number of insects (Nielsen and Mound 2000). Dominance of the Coleoptera was said to have led J. B. S. Haldane, when asked what he could infer about the work of the Creator, to respond that the Creator must have had 'an inordinate fondness for beetles', although there is some doubt about the provenance of this phrase (Fisher 1988). The success of the order Coleoptera is claimed to have been enabled by the rise of flowering plants (Farrell 1998).

Although Wheeler's (1990) species scape is based on the described world biota, a similar species scape could depict most terrestrial communities and ecosystems. Asquith et al. (1990) calculated the species richness in old-growth Douglas-fir forest in Oregon, showing that in the H. J. Andrews Experimental Forest near Eugene, arthropods are dominant, constituting 84.9% of the richness, with vascular plants comprising 11.5%, and vertebrates only 3.6%. Asquith et al. (1990) remarked that in such terrestrial ecosystems, animal diversity is virtually synonymous with arthropod diversity. They noted, however, that this vast arthropod diversity is to a large extent an invisible diversity. Yet, it is the glue that holds diversity together (Janzen 1987).

Hexapods not only dominate in number of species, but also in number of individuals. Collembola can occur at densities of 10^4 to 10^5 per m^2 in most terrestrial ecosystems (Petersen and Luxton 1982). Such statistics led Fisher (1998) to state that 'whether measured in terms of their biomass or their numerical or ecological dominance, insects are a major constituent of terrestrial ecosystems and should be a critical component of conservation research and management programs'. In terms of biomass and their interactions with other terrestrial organisms, insects are the most important group of terrestrial animals (Grimaldi and Engel 2005), so important that if all were to disappear, humanity probably could not last more than a few months (Wilson 1992). On land, insects reign (Grimaldi and Engel 2005) and are the chief competitors with humans for the domination of this planet (Wigglesworth 1976).

ECOLOGICAL ROLE

Insects create the biological foundation for all terrestrial ecosystems. They cycle nutrients, pollinate plants, disperse seeds, maintain soil structure and fertility, control populations of other organisms, and provide a major food source for other taxa (Majer 1987). Virtually any depiction of a food web in a terrestrial or freshwater ecosystem will show insects as a key component, although food-web architectures in these two ecosystems are quite different (Shurin et al. 2005).

Insects are of great importance as a source of food for diverse predators (Carpenter 1928). Aquatic insect larvae serve as food for fishes, and many stream fish appear to be limited by the availability or abundance of such prey, at least on a seasonal basis (Richardson 1993). Adult mayflies are devoured in myriads at the season of their emergence by trout (Carpenter 1928), and this phenomenon forms the basis of the fly-fishing sport (McCafferty 1981). Insects provide the major food supply of many lizards. Many amphibia are carnivorous, especially after they reach maturity, and insects form the bulk of their animal food (Brues 1946).

Birds of many families take insects as their staple food, at least during part of the year (Carpenter 1928), with martins, swallows, and swifts virtually dependent on flying insects for survival. For the yellow-headed blackbird in the Cariboo region of British Columbia, success in rearing young is linked to the emergence of damselflies (Orians 1966).

Mammals such as the American anteater, sloth bears, sun bears, and the African and Oriental pangolins are especially tied to ant and termite colonies, and a number of mammalian predators use insects as food. The British badger often digs out wasp nests to feed on the grubs (Carpenter 1928), and in North America, black bears in north-central Minnesota feed on ants in the spring for quick sources of protein and to obtain essential amino acids and other trace elements unavailable in other spring foods (Noyce et al. 1997). Aggregations of the alpine army cutworm moth

Euxoa auxiliaris (Grote) are an important, high-quality, preferred summer and early-fall food for grizzly bears in Glacier National Park, Montana (White et al. 1998).

Insects are an important supplementary human food source of calories and protein in many regions of the world (Bodenheimer 1951, DeFoliart 1989, 1992, 1999), with some 500 species in more than 260 genera and 70 families of insects known to be consumed (DeFoliart 1989, Groombridge 1992). Insects of most major orders are eaten, but the most widely used species are those, such as termites, that habitually occur in large numbers in one place or that periodically swarm, such as locusts, or large species such as saturniid moth larvae. The seasonal abundance at certain times of the year makes them especially important when other food resources may be lacking (Groombridge 1992).

No accurate estimates are available for the total number of insect natural enemies of other insects, but probably as many, or perhaps more, entomophagous insects exist as do prey or hosts (DeBach 1974). The habit of feeding upon other insects is found in all major insect orders (Clausen 1940). Included here are predators and parasitoids, both of which are involved in natural and practical control of insects (Koul and Dhaliwal 2003). The control of the cottony-cushion scale *Icerya purchasi* Maskell in California by the predatory vedalia beetle *Rodolia cardinalis* (Mulsant) imported from Australia established the biological control method in 1888–1889 (DeBach 1974, Caltagirone 1981, Caltagirone and Doutt 1989).

Conservatively, some 400,000 species of known insects are plant feeders (New 1988). Thus, phytophagous insects make up approximately 25% of all living species on earth (Strong et al. 1984). The members of many orders of insects are almost entirely phytophagous (Brues 1946), conspicuous orders being the Hemiptera, Lepidoptera, and Orthoptera. The influence of insects, as plant-feeding organisms, exceeds that of all other animals (Grimaldi and Engel 2005).

Under natural conditions, insects are a prime factor in regulating the abundance of all plants, particularly the flowering plants, as the latter are especially prone to insect attack (Brues 1946). More thoroughly than any other animals, insects have exploited their food supply and profited wonderfully thereby (Brues 1946). This ability was employed when the moth *Cactoblastis cactorum* Berg was used to control the prickly pear cactus in Australia in 1920–1925 (DeBach 1974). Among the flowering plants are a number of truly insectivorous forms that belong to several diverse groups (Brues 1946).

Food webs involving insects can be quite complex (Elkinton et al. 1996, Liebhold et al. 2000) and relevant to human health in unexpected ways. In oak forests (*Quercus* spp.) of the eastern USA, defoliation by gypsy moths (*Lymantria dispar* L.) and the risk of Lyme disease are determined by interactions among acorns, white-footed mice (*Peromyscus leucopus* (Rafinesque)), gypsy moths, white-tailed deer (*Odocoileus virginianus* Zimmermann), and black-legged ticks (*Ixodes scapularis* Say) (Jones et al. 1998). Experimental removal of mice, which eat gypsy moth pupae, demonstrated that moth outbreaks are caused by reductions in mouse density that occur when there are no acorns. Experimental acorn addition increased mouse and tick density and attracted deer, which are key tick hosts. Mice are primarily responsible for infecting ticks with the Lyme disease agent, the spirochete bacterium *Borrelia burgdorferi*. Lyme disease risk and human health are thus connected to insects indirectly.

Miller (1993) has categorized how insects interact with other organisms as providers, eliminators, and facilitators. Insects serve as providers in communities and ecosystems by serving as food or as hosts for carnivorous plants, parasites, and predatory animals. They also produce byproducts, such as honeydew, frass, and cadavers that sustain other species. As eliminators, insects remove waste products and dead organisms (decomposers and detritivores), consume and recycle live plant material (herbivores), and eat other animals (carnivores).

Many insect taxa are coprophagous. The subfamilies Aphodiinae, Coprinae, and Geotrupinae of scarab beetles (Scarabaeidae) are well-known dung feeders (Ritcher 1958, Hanski and Cambefort 1991), with adults of some species provisioning larval burrows with balls of dung. The dung-beetle community in North America is dominated by accidentally or intentionally introduced species, with aphodiines dominant in northern localities and scarabaeines dominant in southern areas (Lobo 2000). Australia has imported coprophagous scarabs from South Africa and the Mediterranean Region for the control of cattle dung (Waterhouse 1974). African species also have been introduced into North America to improve yield of pasture land through effective removal of dung and to limit the proliferation of flies and nematodes that inhabit the dung (Fincher 1986). Dung beetles in tropical forests also play an important role in secondary seed dispersal because they bury seeds in dung, protecting them from rodent predators (Shepherd and Chapman 1998).

Leaf-cutter ants, not large herbivores, are the principal plant feeders in Neotropical forests (Wilson 1987). Insects, not birds or rodents, are the most important consumers in temperate old fields (Odum et al. 1962). Spittlebugs, for example, ingest more than do mice or sparrows (Wiegert and Evans 1967).

Insect herbivory can affect nutrient cycling through food–web interactions (Wardle 2002, Weisser and Siemann 2004). Insect herbivores influence competitive interactions in the plant community, affecting plant-species composition (Weisser and Siemann 2004). Tree-infesting insects are capable of changing the composition of forest stands (Swaine 1933), and insects can influence the floristic composition of grasslands (Fox 1957). Soil animals, many of which are insects, ultimately regulate decomposition and soil function (Moore and Walter 1988) through both trophic interactions and biophysical mechanisms, which influence microhabitat architecture (McGill and Spence 1985). Soil insects are essential for litter breakdown and provide a fast return of nutrients to primary producers (Wardle 2002). Ants and termites are fine-scale ecosystem engineers (Jones et al. 1994, Lavelle 2002, Hastings et al. 2006). The attine ants are the chief agents for introducing organic matter into the soil in tropical rain forests (Weber 1966). Overall, termites are perhaps the most impressive decomposers in the insect world (Hartley and Jones 2004) and major regulators of the dynamics of litter and soil organic matter in many ecosystems (Lavelle 1997).

Insects serve as facilitators for interspecific interactions through phoresy, transmission of pathogenic organisms, pollination, seed dispersal, and alteration of microhabitat structure by tunneling and nesting (Miller 1993). The process of insect pollination is believed to be the basis for the evolutionary history of flowering plants, spanning at least 135 million years (Crepet 1979, 1983), although the origin of insect pollination, which is an integrating factor of biocenoses (Vogel and Westerkamp 1991), is still being debated (Pellmyr 1992, Kato and Inoue 1994).

Approximately 85% of angiosperms are pollinated by insects (Grimaldi and Engel 2005). Yucca moths (*Tegeticula* spp.) exhibit an extraordinary adaptation for flower visitation, and the yuccas depend on these insects for pollination (Frost 1959, Aker and Udovic 1981, Addicott et al. 1990, Powell 1992). Similarly, figs and chalcid wasps have a remarkable association (Frost 1959, Baker 1961, Galil 1977, Janzen 1979, Wiebes 1979). Orchid species have developed floral color, form, and fragrance that allow these flowers to interject themselves into the life cycle of their pollinators to accomplish their fertilization (Dodson 1975).

EFFECTS ON NATURAL RESOURCES, AGRICULTURE, AND HUMAN HEALTH

Less than 1–2% of phytophagous insects that are potential pests ever achieve the status of even minor pests (DeBach 1974). However, those that do become major pests can have a devastating effect.

Insect defoliators have major effects on the growth (Mott et al. 1957) and survival of forest trees (Morris 1951), and can alter forest-ecosystem function (Naiman 1988, Carson et al. 2004). The native mountain pine beetle *Dendroctonus ponderosae* Hopkins, whose primary host is lodgepole pine (*Pinus contorta* var. *latifolia* Engel.), has devastated pine stands in British Columbia over the last decade. By 2002, the current outbreak, which began during the 1990s, had infested 4.5 million hectares (Taylor and Carroll 2004), and by 2006 had infested more than 8.7 million hectares. It still has not reached its peak in south-central parts of the province and could well spread into the whole of the boreal forest and sweep across Canada. As it does so, it also could negatively affect the stability of wildlife populations (Martin et al. 2006). In such circumstances, the beetle acts as a keystone species, causing strong top-down effects on the community (Carson et al. 2004).

Few would argue that one of the world's most destructive insects is the brown planthopper *Nilaparvata lugens* (Stål) (Nault 1994). Each year, it causes more than $1.23 billion in losses to rice in Southeast Asia (Herdt 1987). These losses are caused by damage from feeding injury and by plant viruses transmitted by this planthopper (Nault 1994).

Desert locusts are well known for their devastating effect on crops in Africa (Baron 1972), and almost any book on applied entomology will list innumerable pests. Pfadt (1962), for example, considers pests of corn, cotton, fruits, households, legumes, livestock, poultry, small grains, stored products, and vegetable crops.

Most major insect pests in agriculture are non-native species, introduced into a new ecosystem, usually without their natural biological control agents (Pimentel 2002). Introduced insects in Australia are responsible for as much as $5–8 billion in annual damage and control costs (Pimentel 2002).

Transmission of plant-disease agents by insects has been known for a long time (Leach 1940). Insect vectors of diseases probably have affected humans more than have any other eukaryotic animals (Grimaldi and Engel 2005). Their epidemics have profoundly shaped human culture, military campaigns, and history (Zinsser 1934, McNeill 1976, R. K. D. Peterson 1995). Enormous effort has been made over the years to control insect-borne diseases (Busvine 1993). Tens of millions of people throughout the world have died in historical times as a result of just six major insect-borne diseases: epidemic typhus (a spirochete carried by *Pediculus* lice), Chagas disease (a trypanosome carried by triatomine bugs), plague (a bacterium carried by *Pulex* and *Xenopsylla* fleas), sleeping sickness (a trypanosome carried by tsetse), malaria (*Plasmodium* carried by *Anopheles* mosquitoes), and yellow fever (a virus carried by *Aedes* mosquitoes) (Grimaldi and Engel 2005). Mosquitoes also are involved in the transmission of West Nile virus, now a major concern in North America (Enserink 2000). From the fifteenth century to the present, successive waves of invasion of mosquitoes have been facilitated by worldwide transport (Lounibos 2002).

The story of the struggle in Africa to overcome tsetse and the disease agents they transmit is one of the major epics in human history (McKelvey 1973). Worldwide, arthropod-borne pathogens still take an enormous toll in human mortality, morbidity, and loss of productivity (Aultman et al. 2000).

Of all the ills that affect humankind, few have taken a higher toll than malaria (Alvardo and Bruce-Chwatt 1962). More than 400 million people fall ill each year with malaria, and 1–3 million die, mostly children younger than 5 years old, and most of them in Africa (Marshall 2000). Public health experts also believe the toll has been increasing in recent years (Marshall 2000). As a result, malaria still casts a deadly shadow over Africa (Miller and Greenwood 2002).

Certain insects can be beneficial. Losey and Vaughan (2006) estimate that the annual value of ecological services provided by insects in the United States is at least $57 billion. Insect pollination, for example, is of great economic value in the fruit-growing industry, the greenhouse industry, and in the growing of forage crops such as alfalfa (Free 1993, Proctor et al. 1996).

The annual benefit of honeybees to US agricultural consumers is on the order of $1.6–8.3 billion (Southwick and Southwick 1992).

Ample evidence now shows, however, that pollinator diversity and crop pollination services are at risk as a result of the use of pesticides; habitat alteration, fragmentation, and destruction; introduction of alien species; and diseases (Johansen 1977, Kevan et al. 1990, Royce and Rossignol 1990, Watanabe 1994, Raloff 1996, Kearns and Inouye 1997, Kearns et al. 1998, Kremen et al. 2002, Steffan-Dewenter et al. 2005). The innumerable insect predators and parasites are invaluable for natural biological control. Natural pest-control services maintain the stability of agricultural systems worldwide and are crucial for food security, rural household incomes, and national incomes in many countries (Naylor and Ehrlich 1997). Enough cases of highly effective natural biological control have been studied to indicate that 99% or more of potential pest insects are under such natural control (DeBach 1974). Natural pest controls represent an important ecosystem service, with an annual replacement value estimated to be $54 billion (Naylor and Ehrlich 1997).

The first and perhaps the most spectacular success of applied biological control was the introduction of the vedalia beetle from Australia into California in 1888 to control the cottony-cushion scale (Hagen and Franz 1973, Caltagirone and Doutt 1989). This outstanding success has led to many introductions of parasitoids and predaceous insects for biological control and the improvement of biological control techniques (Huffaker 1971, DeBach 1974, Caltagirone 1981, DeBach and Rosen 1991, van Driesche and Bellows 1996). Although after many reckless insect importations, biological control is no longer recognized as a panacea (Hagen and Franz 1973); it even can pose threats to nontarget species (Samways 1997, Louda et al. 1997, 2003, Boettner et al. 2000, Follett and Duan 2000) and elevate threats to human health (Pearson and Callaway 2006). Biological control, thus, can be a double-edged sword (Louda and Stiling 2004).

INSECTS AND ADVANCES IN SCIENCE

One by one, the natural sciences have found insects ideal for study (Wigglesworth 1976). Their study has produced major advances in our understanding

of biomechanics, climate change, developmental biology, ecology, evolution, genetics, paleolimnology, and physiology. A few examples suffice to illustrate how the study of insects has advanced these areas of science, especially where insects 'have done it first' (Akre et al. 1992).

Biomechanics

Insects have evolved unique features in the animal world that are a surprise to experts in biomechanics and bioengineering because many are recent inventions of humans. For example, insect cuticle, with its plywood-like structure, is a laminated composite material, now well known in engineering, and used where high strength and stiffness to weight ratios are required (Barth 1973). Manufactured plywood is fairly new, but insect cuticle has been around for some 400 million years (Grimaldi and Engel 2005). Furthermore, in the insect exoskeleton, associated with the flight system in particular, are areas of resilin protein (Weis-Fogh 1960), 'the most perfect rubber known' (Neville 1975), with a high compliance (deforms easily) and low tensile strength (Bennet-Clark 1976).

Although the origin of insect flight is still debated (Wootton and Ellington 1991), insects certainly were the first animals to evolve wings, evidently during the Late Devonian or Early Carboniferous (Grimaldi and Engel 2005). However, if *Rhyniognatha hirsti* Tillyard from the Early Devonian (Pragian) chert of the Old Red Sandstone of Scotland is a pterygote insect, wings might have evolved 80 my earlier (Engel and Grimaldi 2004). Thus, insect wings and flight capacity developed about 90 my prior to the earliest winged vertebrates (Grimaldi and Engel 2005), or perhaps even 170 my earlier (Engel and Grimaldi 2004).

Not only did insects evolve active flight first, they remain unsurpassed in many aspects of aerodynamic performance and maneuverability (Dickinson et al. 1999). Although nobody knows how the smallest insects fly (Nachtigall 1989), the aerodynamic properties and design of certain insect wings have been perfected to such an extent that they are superior to the design of human-made fixed-wing aircraft in a number of respects (Nachtigall 1974). The structure of these organs and the way they are used are the envy of variable-wing plane designers (Scudder 1976). Insect wings typically produce two to three times more lift than can be accounted for by conventional aerodynamics (Ellington 1999). Most insects rely on a leading-edge vortex created by dynamic stall during the downstroke to provide high lift forces (Ellington et al. 1996, Ellington 1999).

The wings of archaic Odonatoidea from the Middle Carboniferous, about 320 mya, show features analogous to the 'smart' mechanics of modern dragonflies (Wootton et al. 1998). These mechanisms act automatically in flight to depress the trailing edge of the wing and facilitate wing twisting in response to aerodynamic loading, and suggest that these early insects already were becoming adapted for high-performance flight in association with a predatory habit.

Furthermore, modern dragonflies use 'unsteady aerodynamics', a mode of flight not previously recognized as feasible (Somps and Luttges 1985). In such flight, the forewings generate a small vortex, which the hindwing can then capture to provide added lift. Hovering insects do not rely on quasi-steady aerodynamics, but use rotational lift mechanisms, involving concentrated vortex shedding from the leading edge during wing rotation (Ellington 1984). These discoveries have opened new possibilities in flight technology and have applications not only to the design of planes, but also to features of turbine blades and racing cars.

Collaborative research between engineers and specialists in insect flight is resulting in the development of micro-air vehicles capable of industrial fault location in enclosed situations (Wootton 2000). Although many types of walking machines exist, engineers have not yet determined how best to make these machines handle unfamiliar situations. However, biologists with a detailed knowledge of insect walking and its control have linked with computer scientists to develop better robot designs, using computer models of insect locomotion (Pennisi 1991).

Genetics

Drosophila melanogaster Meigen, a tramp species of fruit fly, arguably is the best-known eukaryotic organism (Grimaldi and Engel 2005). The Columbia Group of geneticists (T. H. Morgan, C. B. Bridges, H. J. Müller, and A. H. Sturtevant) in the 'Fly Room' (Sturtevant 1965, Roberts 2006) helped launch the field of modern genetics, with their pioneering use of *D. melanogaster*, by clarifying or discovering fundamental concepts such as crossing over, linkage, mutation, sex-linked inheritance, and the linear arrangement of genes on chromosomes (DeSalle and Grimaldi 1991). Almost all significant basic concepts in transmission genetics were either first developed by *Drosophila* workers or conspicuously verified by them (Brown 1973).

Drosophila played a major role in the investigation of the nature and action of genes (Glass 1957), and the role of genes in the determination of sex was first deduced from the study of sex-linked inheritance in the Lepidoptera by Doncaster and Raynor (1906). The laws of heredity, as worked out with these insects, provide one of the main pillars that support the science of genetics (Wigglesworth 1976).

During the era of what Haldane (1932, 1964) called 'bean-bag genetics' came the demonstration that every feature governing the life of an organism, if inherited, was under genetic control. Such research has made *Drosophila melanogaster* a favorite species in laboratory research from cell biology, to behavior, ecology, evolution, and physiology (Grimaldi and Engel 2005). This species was the preeminent model organism from 1970 to 1980 (Roberts 2006).

Most of the early data on the genetic characteristics of central and marginal populations came from studies of chromosomal polymorphisms in various species of *Drosophila* (Brussard 1984). The genetic constitution of colonizing individuals used to establish insect populations for biological control is relevant to their success, and depends on whether populations of colonizers were drawn from the center or periphery of their range (Force 1967, Remington 1968).

Research with houseflies (Bryant et al. 1986) demonstrated the possibility of increased genetic variance after population bottlenecks (Carson 1990). Until this research, population bottlenecks were usually considered to reduce genetic variance (Nei et al. 1975).

After the discovery of transposable elements in corn (*Zea mays* L.) by Barbara McClintock in the 1940s (Sherratt 1995), research on *Drosophila* first demonstrated the presence of transposable elements in animals (Cohen and Shapiro 1980). Of all the eukaryotic transposable elements, the most heavily exploited has been the *Drosophila* P element (Kaiser et al. 1995). Use of transposable elements for insect pest control is now being investigated (Grigliatti et al. 2007) in an effort to add to the current techniques of genetic control of insect pests (Davidson 1974).

Developmental biology

Insects probably contain the greatest variety of developmental forms of any class of organisms (Kause 1960, Boswell and Mahowald 1985, Sander et al. 1985). Eggs of insects are either indeterminate or determinate (Seidel 1924), and differ in their mechanism of pattern formations (Sander et al. 1985). Some of the earliest research on fate maps in determinate eggs was done by either small local mechanical or ultraviolet-radiation ablation experiments (Schubiger and Newman 1982) or by analyzing genetic mosaics in insects (Janning 1978).

Determination, which is the process that results in cells being committed to specific developmental fates, has been well studied in holometabolous insects in their imaginal discs, first described by Lyonet (1762) in Lepidoptera, and later recognized to have developmental significance by Weismann (1864). Largely through the pioneering efforts of such scientists as D. Bodenstein, B. Ephrussi, E. Hadorn, and C. Stern, the developmental biology of imaginal discs has become an active field of investigation and a unique and most favorable system for the study of numerous problems in metazoan development (Oberlander 1985). Among these problems are pattern formation, positional information, transdetermination, and programmed cell death.

In insects, morphogenesis gradients are present in the embryo, where they influence the polarity and quality of body segments (Sander et al. 1985) and their appendages. The elucidation of the segmental polarity genes, which control the primary segmental pattern in *Drosophila*, made a significant contribution to the development of the polar coordinate model of animal development (Bryant 1993, Roberts 2006).

Homeosis, defined by Bateson (1894) as the replacement of the body part of one segment with the homologous body part of another segment, was pioneered by Bateson with several examples drawn from insects. As with many other phenomena in insect development, homeotic mutants have been studied most extensively in *Drosophila melanogaster* (Gehring and Nöthiger 1973, Ouweneel 1975).

Rapid advances have been made in understanding the genetic basis of development and pattern formation in animals (Patel 1994) as a result of pioneer studies in *Drosophila*. Homologous genes are now known to serve similar developmental functions in a number of diverse organisms, with conservation of the homeobox sequence in evolution (Patel 1994).

The homeobox genes, first identified by homeotic mutations in *Drosophila* (Scott and Weiner 1984), act as markers of position, defining different fates along the anterior–posterior axis of animal embryos (Akam 1995). They are thus important regulators of embryonic development, and have provided key indicators of the mysteries of evolution (Marx 1992). They indicate that extremely complex organizational changes, both functional and structural, can arise from few genetic events (Hunkapiller et al. 1982). The realization that nothing is lost in evolution, but instead is just not developed, has been supported by studies in *Drosophila melanogaster* on the segmental organization of the tail region in insects (Jürgens 1987).

Although the strongest evidence for an intrinsic death program in animal cells originally came from genetic studies of the nematode *Caenorhabditis elegans* (Maupas) (Horvitz et al. 1982), recent research on *Drosophila* has substantially extended our understanding of how programmed cell death in development is executed and regulated (Raff 1994, White et al. 1994). Defect mutants and homeotic mutants of *Drosophila* have been examined to determine to what extent morphological changes have been initiated by cell death and been brought about by subsequent modification (Lockshin 1985).

Transdetermination, a phenomenon unique to insects and first documented by Hadorn (1968) with imaginal discs of *Drosophila* cultured *in vivo*, reveals that there must be an underlying regulatory system for switching between alternate states of development (Shearn 1985). Further research with insects on this topic could provide a major step toward understanding the genetic programming of development.

Sir John Lubbock (Lord Avebury), the famous banker, practical sociologist, and amateur entomologist, pointed out (Lubbock 1873) the significance of metamorphosis in insects and paved the way for our current understanding, emphasizing the difference between development and adaptive changes (Wigglesworth 1976). We now realize that metamorphosis in holometabolous insects serves to free certain ectodermal cells from the task of forming functional cuticle, so that they can prepare for the formation of the specialized structures of the future adult (Wigglesworth 1985). Epidermal cells, thus, have a unique triple capacity to form larval, pupal, or adult characters, and metamorphosis is no longer regarded as a reversion to embryonic development (Wigglesworth 1985).

Evolution

Charles Darwin was fond of insects and referred to them extensively in his work and ideas on natural selection (Smith 1987). Insects, specifically the moth *Biston betularia* (L.), finally provided what Kettlewell (1959) called Darwin's missing evidence, namely natural selection in action in the form of industrial melanism (Kettlewell 1961, 1973).

The study of speciation was brought into the modern age and onto a firmer, more genetic basis by drosophilists during the 1930s and 1940s (Mallet 2006). A major event was the discovery of two populations of *Drosophila* that could not be differentiated by systematists, but which did not interbreed with each other, although they were fully fertile within each population (Ross 1973). Dobzhansky's (1937) reproductive isolation species concept, later incorporated into Mayr's (1942) biological species concept, was based in large part on the discovery of pairs of sibling species in *Drosophila*. Sibling species pairs have since been discovered in other groups of insects, such as black flies and green lacewings (Bickham 1983) and provide fertile ground for speciation models.

These sibling pairs, genetically close and almost morphologically identical, typically show strong reproductive isolation in the form of hybrid sterility, hybrid inviability, and assortative mating (Mallet 2006). Recent advances in molecular and genetic understanding of *Drosophila* speciation genes have now begun to open up our understanding of the genetic basis of hybrid inviability and sterility (Mallet 2006).

The notion that prezygotic reproductive isolation can be reinforced when allopatric taxa become sympatric, and that no single isolating mechanism is the 'stuff of speciation', was the result of studies on *Drosophila* (Coyne and Orr 1989). *Drosophila* research also shows that there is no evidence for extensive reorganization of gene pools in speciation (Throckmorton 1977), and research on insects, together with that in developmental and molecular biology, now suggests that neither the genome nor the gene pool of species is highly coadapted (Bush and Howard 1986).

Insects have been instrumental in demonstrating that sympatric speciation can occur (Scudder 1974), with host–plant separation (Bush 1969, Huettel and Bush 1972), habitat specialization (Rice 1987, Rice and Salt 1988), or seasonal diversification (Tauber and Tauber 1981) often being important components. The power of disruptive selection was first investigated in *Drosophila* (Thoday 1972), and the analysis of hybrid zones (Barton and Hewitt 1985) has been facilitated by studies of insects (Hewitt 1988, 1990).

Allochronic speciation has been documented in crickets (Alexander 1968), and stasipatric speciation has been claimed from a study of morabine grasshoppers (Key 1968), but these speciation examples pose problems with parapatry and species interactions (Bull 1991). Founder-flush and transilience theories of speciation were inspired largely by the endemic nature of Hawaiian *Drosophila* species (Carson 1970, 1975; Templeton 1981; Carson and Templeton 1984), and the molecular drive model of species evolution (Dover 1982) was derived from research on *Drosophila* genetics. Chromosomal mechanisms of speciation have been best documented in insects (White 1973), and insects are represented in examples of gynogenesis (Moore et al. 1956, Sanderson 1960), a form of asexual reproduction similar to parthenogenesis but requiring stimulation by sperm, without contributing genetic material to the offspring.

Studies of insects have been behind the conclusion that many different kinds of species exist as a result of different kinds of speciation processes (Scudder 1974, Foottit 1997). They provide a case for pluralism in species concepts (Mishler and Donoghue 1982).

For evolution above the species level, or what Simpson (1953) called the major features of evolution, insects have not made a major contribution recently. Early work by the famous Dutch naturalist Jan Swammerdam in about 1669, however, first gave a physiological description of insect metamorphosis, and a concept of 'preformation' was foremost in evolutionary ideas in the eighteenth century, although as noted by Wigglesworth (1976), this latter notion was much abused.

With respect to phylogeny, according to Wheeler et al. (2001), an epistemological revolution was brought about by the publication of Hennig's (1966) book, which is an English translation and revision of an earlier publication by the German entomologist (Hennig 1950). Hennig (1966) used mostly insect examples in the explanation of his phylogenetic systematics, having earlier (Hennig 1965) briefly explained his methodology. Brundin (1966) was one of the first biologists to adopt this new methodology in his consideration of transantarctic relationships of chironomid midges. Later, Hennig (1969, 1981) applied the method in his consideration of insect phylogeny. It has now been generally adopted in discussions of the phylogeny of other animals and plants (Wiley 1981), and prevails in more recent considerations of insect phylogeny (Kristensen 1981) and both morphological and molecular data on the phylogeny of arthropods as a whole (Wheeler et al. 1993, 2001, Boore et al. 1995, Friedrich and Tautz 1995, Giribet et al. 2001, Nardi et al. 2003).

Physiology

Insects have been used for the study of the fundamental problems of physiology (Wigglesworth 1976). Today, everybody knows about cytochromes, but there are probably many biochemists who do not realize that the discovery of cytochromes was a product of the study of insect physiology (Wigglesworth 1976). Keilin (1925), while following the fate of hemoglobin beyond the endoparasitic larval stage in the fly *Gasterophilus intestinalis* (DeGeer), was led to the discovery of cytochromes in flight muscles of the adult free-living insect (Kayser 1985).

Wigglesworth (1976) noted that insect Malpighian tubules afford exceptional opportunity for the study of the physiology of excretion, owing to the ability to isolate individual tubules and have them function *in vitro*, as first demonstrated by Ramsay (1955) in the stick insect *Carausius morosus* (Sinety). Bradley (1985) reviewed the structural diversity of Malpighian tubules in insects and showed them to have quite a diversity of function. The Malpighian tubules of the larvae of the saline water-tolerant mosquito *Aedes campestris* Dyar and Knab can actively transport sulfate ions, an unusual function in animals (Maddrell and Phillips 1975). This sulfate transport in the Malpighian tubules of larvae of *A. taeniorhynchus* (Wiedemann) is inducible and suggests that sulfate stress results in the synthesis and insertion of additional transport pumps into the Malpighian tubule membranes (Maddrell and Phillips 1978). Insect Malpighian tubules, in comparison with vertebrate glomerula or other invertebrate tubules, are highly impermeable to organic molecules (Bradley 1985), although several insects can excrete nicotine independent of ion movement. However, transport systems for organic bases might not be universal in insects (Maddrell and Gardner 1976). The Malpighian tubules of the grasshopper *Zonocerus variegatus* (L.) have an inducible transport mechanism that actively removes the cardiac glycoside ouabain from the hemolymph (Rafaeli-Bernstein and Mordue 1978). The isolated tubules of the large milkweed bug *Oncopeltus fasciatus* (Dallas) can excrete ouabain (Meredith et al. 1984). This excretory function is important for these insects, which feed on plants containing cardenolides. *Oncopeltus fasciatus* also has ouabain-resistant Na-, K-ATPases (Moore and Scudder 1985).

Weight for weight, the asynchronous flight muscles of insects generate more energy than any other tissue in the animal kingdom (Smith 1965). A unique feature of most insect flight muscles is that the fine ramifications of the tracheal system, the tracheoles, penetrate deeply into the muscle fibers (Beenakkers et al. 1985). These flight muscles are among the most active tissues known. They have highly efficient fuel management and a remarkably high metabolic rate (Beenakkers et al. 1984). In most insects, carbohydrate is the most important substance for flight, but many insects, particularly the Lepidoptera and Orthoptera, possess the capacity to use lipids for flight, and in some insect species proline is used (Beenakkers et al. 1985).

Insects and other terrestrial arthropods use several techniques to adapt to alpine and polar environments (Downes 1965, Ring 1981). The study of insects at low temperature (Lee and Denlinger 1991) has largely elucidated the phenomena of cold hardiness, freeze-tolerance, supercooling, and the role of antifreeze proteins and cryoprotectants. The polyhydric alcohol glycerol and other polyols such as mannitol, sorbitol, and threitol are responsible for most increases in supercooling in insects (Mullins 1985). Antifreeze proteins from insects are far more potent than those from fish (Bower 1997). Some insect antifreeze protein (AFP) cDNAs (Graham et al. 1997, Tyshenko et al. 1997) and genes (Guo et al. 2005, Qin and Walker 2006) have been isolated, and potential transfer to other hosts is now being investigated. However, neither transfer of fish AFP genes (Duncker et al. 1995) nor transfer of hyperactive spruce budworm AFP genes (Tyshenko and Walker 2004) to transgenic *Drosophila melanogaster* has conferred cold tolerance to the recipient.

Ecology

Insects provide some of the best material for the ecologist (Wigglesworth 1976). The long-term research by Schwerdtfeger (1941) on German forest insects provided an example of the constancy of animal numbers, an essential concept in the development of Darwin's theory of natural selection. Tamarin (1978) reviewed the major concepts and debates on population regulation in ecology, and showed that insects have played a major role. Many aspects of population dynamics; population regulation by predators, parasites, or parasitoids; and the understanding of density-dependent factors, density-independent factors, and key factor analysis (Clark et al. 1967, Varley et al. 1973) were developed through research on insects (Andrewartha and Birch 1973).

Crombie (1945) confirmed that Gause's experimental results on interspecific competition with *Paramecium* also held true in the Metazoa, by using a number of species of insects that lived in stored products. Chapman (1928) introduced the beetle *Tribolium* as a laboratory animal, and Park (1948), working with *T. castaneum* (Herbst) and *T. confusum* Jacquelin du Val, investigated some of the complicating factors in interspecific competition, although the early results were compromised by the discovery that the pathogenic coccidian parasite *Adelina tribolii* Bhatia was a third party in the conflict. Park (1954) later showed that the results of competition in *Adeline*-free cultures could be predicted from a comparison of the carrying capacity of the two species in single-species cultures, depending on the temperature and humidity conditions, and that changing such conditions frequently could lead to indefinite coexistence. Comparable results could occur in many species pairs, particularly for short-generation species such as insects (Hutchinson 1953). Nevertheless, Price (1984) noted that many elusive aspects of species competition were investigated in subsequent experiments with insects, many of which were reviewed by DeBach (1966) and Reitz and Trumble (2002). The term 'ecological character displacement', introduced by Brown and Wilson (1956), became a controversial theme in ecology and evolutionary biology, and continues to be a focus of much exciting research (Dayan and Simberloff 2005).

Insects have provided examples in the ecological literature for the existence of enemy-free space (Atsatt 1981, Jeffries and Lawton 1984, Denno et al. 1990), and research on water mite parasitism of waterboatmen (Scudder 1983, Smith 1988, Bennett and Scudder 1998) has shown how parasitism can be a cryptic determinant of animal-community structure (Minchella and Scott 1991). Furthermore, an invasive leafhopper pest, *Homalodisca coagulata*

(Say), might be engineering enemy-free space in French Polynesia (Suttle and Hoddle 2006).

Research on semiochemicals, whether pheromones concerned with interspecific communication or allochemicals (allomones or kairomones) (Brown et al. 1970) involved with interspecific communication, was pioneered using insects (Karlson and Butenandt 1959, Brown et al. 1970, Whittaker and Feeny 1971, Duffey 1977, Price 1981, Rutowski 1981, Mayer and McLaughlin 1990). The term 'pheromone' was proposed originally by Karlson and Lüscher (1959) for the first sex pheromone bombykol, identified by Butenandt et al. (1959) from the silkworm moth *Bombyx mori* (L.). Since the earliest observation of mate finding by Fabré (1911) and others, the power of female insects to lure males has astonished biologists (Phelan 1992). Probably no mate-communication system is better studied than that in the Lepidoptera (Phelan 1992). Sex-pheromone components for about 80 compounds from more than 120 lepidopterous species have been discovered (Tamaki 1985), and they are known from many other groups of insects (Jacobson 1972). Yet, this sexual-communication database has remained largely untapped by evolutionary biologists, outside the field of insect pheromones (Phelan 1992).

The observation by B. Hüber in 1914 that alarm behavior in honeybee workers could be triggered by volatile sting-derived components has been followed by discovery and research on alarm pheromones in many other insects (Blum 1985). Aggregation pheromones occur in six insect orders, although the majority have been discovered and studied in the Coleoptera, particularly the bark and timber beetles (Curculionidae: Scolytinae) (Borden 1985). Other insect semiochemicals function as trail pheromones, spacing (epideictic) pheromones, and courtship pheromones, as well as in various interspecific communication situations (Haynes and Birch 1985, Roitberg and Mangel 1988).

Sociality is the most striking and sophisticated innovation of the insects (Grimaldi and Engel 2005), and semiochemicals are the glue that holds social insect colonies together (Winston 1992). The detailed structure and function of these complex societies have amazed scientists and posed major problems in biology, such as the question of caste determination and kin selection. Caste determination and its control in ants, bees, termites, and wasps have been covered by Hardie and Lees (1985), and the biology of these insects is well known (Wilson 1971). Kin selection is still the subject of much debate (Benson 1971, Eberhard 1975, Guilford 1985, Malcolm 1986).

Insects have provided classic examples of commensalism, endosymbiosis, mimicry, mutualism, and phoresy, although many examples, especially those of mimicry (Punnett 1915), have been based on subjective natural history observations rather than on experimentation (Malcolm 1990). When experiments are carried out, some of the classic examples have had to be reassessed (Ritland and Brower 1991).

Ants and their relationship with *Acacia* (Janzen 1966) provide a good example of mutualism, as do the insects that cultivate fungus gardens (Batra and Batra 1967). The well-known fungus-growing leaf-cutter ants (Weber 1966, Martin 1970, Cherrett et al. 1989), in an association that is some 50 million years old (Mueller et al. 1998), involve a third mutualist in the system, namely the antibiotic-producing *Streptomyces* bacteria (Currie et al. 1999). These leaf-cutting ants (*Atta* spp.) evidently play an important ecological role through their long-distance transport redistribution and concentration of critical nutrients for plants growing near their nests (Sternberg et al. 2007). The food-for-protection association between ants and honey-producing hemipterans is one of the most familiar examples of mutualism; these keystone interactions can have strong and pervasive effects on the communities in which they are embedded (Styrsky and Eubanks 2007).

Insects are particularly prone to endosymbiotic associations (Henry 1967, Baumann et al. 1997), and have provided insights into the evolutionary biology, genetics, and physiology of this intimate association. Many organisms also live ectosymbiotically with insects (Henry 1967).

Phoresy among entomophagous insects is well documented (Clausen 1976). And the occurrence of midge and mosquito larvae in the pool of water held by the leaves of the carnivorous pitcher plant, where they feed on decaying invertebrate carcasses, is a classic example of commensalism (Heard 1994).

Because of their extreme mobility, insects have challenged many of the prevailing concepts in zoogeography (Johnson and Bowden 1973), showing that species adapted to staying put are among the most successful travelers. Research on insect dispersal and migration (Johnson 1969) has shown the complex interrelations with agriculture, meteorology, medicine, physiology, and many other areas of general science.

The dispersal and movement of insect pests is a growing concern (Stinner et al. 1983).

Movement of organisms from one habitat to another can have profound effects on the structure and dynamics of food webs (Polis et al. 2004). Energy subsidies can either stabilize or destabilize food webs, depending on the nature of the subsidy and what components of the food web are subsidized. Dispersal influences food-web structure, and studies of insect dispersal from this viewpoint, could add to our understanding of community ecology (Thompson 2006).

Insects have played a major role in the development of ecological and island biogeography theory (Hafernik 1992). Much has been learned about the effects of fragmentation on insect populations and the movement of individuals between patches (Hunter 2002). Species living in highly fragmented landscapes often occur as metapopulations, and much of the literature on metapopulation dynamics has been the result of studies on insects, particularly butterflies (Boughton 1999, Hanski 1999, Hanski and Singer 2001, Wahlberg et al. 2002, Hanski et al. 2006). Different population structures can have markedly different evolutionary outcomes (Barton and Whitlock 1997). The value of habitat corridors also has been evaluated (Haddad 1999a, 1999b, Haddad and Baum 1999, Collinge 2000). Their function depends on environmental variation, landscape context, patch size, and species characteristics (Collinge 2000). Studies on insect-plant food webs have shown that habitat fragmentation can affect trophic processes in highly complex food webs involving hundreds of species (Valladares et al. 2006). Research with butterflies also has shown that the surrounding matrix can significantly influence the effective isolation of habitat patches (Ricketts 2001). Because butterflies respond to often subtle habitat changes, they have been suggested as ecological indicators of endangered habitats (Arnold 1983), for which they could serve as umbrella species (Murphy et al. 1990).

Paleolimnology and climate change

Quaternary insect fossils are proving to be sensitive indicators of past environments and climates (Elias 1994). Insect exoskeletons are found chiefly in anoxic sediments that contain abundant organic detritus. Lakes, ponds, and kettleholes serve as reservoirs that collect insects, and sediments that accumulate in these waters act rapidly to cover their remains, preventing oxidation. Insect remains, thus, have played a vital role in paleolimnological studies aimed at reconstructing and interpreting past environmental conditions (Smol and Glew 1992).

Because insects in particular respond readily to changes in temperature more promptly and with greater intensity than other components of the terrestrial biota, they are providing evidence that major climatic changes in the past took place with unexpected suddenness, moving from glacial cold to interglacial warmth in decades rather than in millennia (Elias 1994). Such evidence is vital for current decision making with respect to the management of ecosystems (Smol 1992), and gives an indication of what might occur with climate change in the near future.

Crozier (2004), studying the sachem skipper *Atalopedes campestris* (Boisduval) in the Pacific Northwest, was the first to provide unequivocal data consistent with the hypothesis that current winter warming will drive butterfly range expansion in North America. Ample evidence now shows that insects are one of the first groups of living organisms to respond to ongoing global warming (Franco et al. 2006, Wilson et al. 2007). Predicting insect response is now an active area of investigation (Williams and Liebhold 1997, 2002). Understanding insect strategies for survival under these circumstances, such as how they cope with new food sources (Thomas et al. 2001, Braschler and Hill 2007) and adjust to more acute temperature and humidity fluctuations is still a challenge (Philogene 2006). Research on butterflies in Britain has shown that many species fail to track recent climate warming because of a lack of suitable habitat (Hill et al. 2002), leading to local extinctions at low-latitude range boundaries of species (Franco et al. 2006).

INSECTS AND THE PUBLIC

Insects have had a long connection with humankind. Primitive humans learned to obtain honey by robbing the nests of bees in hollow trees or rock crevices by about 7000 BC (Townsend and Crane 1973). Archaeological evidence shows that the cultivation of the silkworm *Bombyx mori* was begun before 4700 BC, and sericulture occupied an important part of peasant life in China between 4000 and 3000 BC (Konishi and Ito 1973). The notion that metamorphosis of the sacred scarab *Scarabaeus sacer* L. is a symbol of the resurrection of the dead, according to the Egyptians, might be of recent origin (Bodenheimer 1960), but the supposed health properties of this beetle were identified by at least 1550 BC (Harpaz 1973). Fumigation by burning toxic plants to kill insect pests dates from about 1200 BC (Konishi and Ito 1973).

Insects have been in competition with humans for the products of our labor ever since cultivation of soil began (Wigglesworth 1976). Many members of the public – agriculturalists, healthcare professionals, homeowners, and natural resource personnel – no doubt regard insects as perfect pests (Scudder 1976).

Some medical professions, however, find certain insects beneficial, even in modern medicine. Certain fly maggots, for example, are a valuable, cost-effective tool for treating wounds and ulcers unresponsive to conventional treatment and surgical intervention (Mumcuoglu et al. 1999). The saliva of hematophagous deerflies (*Chrysops*) contains a potent inhibitor of platelet aggregation (Grevelink et al. 1993) previously unreported from arthropods and of potential use in medical therapeutics.

Insect evidence can be paramount in establishing the postmortem interval for a decedent, as well as in providing additional information to investigators capable of deciphering the entomological clues (Byrd and Castner 2001). Forensic entomologists, thus, have found insects to be useful in criminal investigations (Catts and Goff 1992), and some criminal elements in our society have used insects, especially rare butterflies, in illegal trade.

Educators and student participants in science fairs have found insects useful in simple experiments, although this practice is now discouraged by some animal-rights activists. Wigglesworth (1976) pointed out that insects present desirable properties as objects for experimentation. They are tolerant of operation; they are so varied in form and habit that some species suited to the problem at hand surely can be found; and their small size makes it possible for the observer to be constantly aware of the whole, while focusing attention on the part (Wigglesworth 1976). As a result, insects are used in numerous studies of the living world (Kalmus 1948, Cummins et al. 1965), although in many jurisdictions, researchers and even field biologists working on insects must have animal-care certificates.

Fishing enthusiasts have a special interest in insects. Mayflies constitute the primary basis for the sport and technique of fly-fishing and fly-tying (McCafferty 1981). Natural insects also can be used (Petersen 1956).

Well-known insect products used by humans include lac from the lac insect (*Laccifer lacca* Kerr.) and cochineal from *Dactylopius coccus* Costa (Bishopp 1952). Real silk from the silk moth *Bombyx mori* must now compete with synthetic fibers (Bishopp 1952).

In the past, species of the butterfly genus *Morpho* with their precisely spaced, overlapping projections on the ridges of their wing scales that give rise to their striking iridescent blue (I. Peterson 1995) have been collected for jewelry to such an extent that they are now in danger of extinction (Scudder 1976). More recent efforts have promoted live jewelry, using showy insects, mostly beetles.

Humans are involved in food chains, and insects have been used as human food for centuries in many cultures (Bodenheimer 1951, DeFoliart 1989, 1992). Some insects also are eaten by gourmands and treated as conversation pieces.

Honeybees, mainly *Apis mellifera* L., remain the most economically valuable pollinators of crop monocultures worldwide (Klein et al. 2007). When wild bees do not visit agricultural fields, managed honeybees are often the only solution for farmers to ensure crop pollination, because they are cheap, convenient, and versatile, although they are not the most effective pollinators on a per flower basis in some crops (Klein et al. 2007). Farmers in the USA pay beekeepers more than $30 million annually for the use of honeybees to pollinate crops valued at $9 billion (Weiss 1989). Honey has valuable nutritional and other benefits (Buchmann and Repplier 2005, Simon et al. 2006). Beeswax has many modern uses in industry and the arts (Bishopp 1952), and even royal jelly has been promoted for its supposed health benefits (Crane 2003). The total value of crops resulting from pollinator activity in 1980 approached $20 billion, compared to the approximately $140 million worth of honey and beeswax produced in the USA (Levin 1983). These figures indicate that the activity of bee pollination is worth 143 times as much as the value of honey and beeswax, on which most beekeepers must make their living.

That insects, especially honeybees, are valuable in pollination (Vansell and Griggs 1952) is well known to the general public. The decline of bee populations in Europe and North America (Allen-Wardell et al. 1998, Goddard and Taron 2001, Biesmeijer et al. 2006) has raised alarm about pollinators that fertilize crops essential for the human food supply. Articles in the popular press (Bueckert 2007) point out that threats to pollinators include climate change, habitat destruction, human indifference, invasive species, and pesticides. As a result, many gardeners are now interested in growing plants that attract butterflies and other insects, and even some municipal governments have ventured into this sphere in their parks and green spaces. Farmers and others are told that they should be less zealous about eliminating weeds around the boundaries of farmland, in ditches, or along utility rights of way (Bueckert 2007).

The local gardener and nursery owner, as well as agriculturalists and foresters, now realize that insects, both parasitoids and predators (Clausen 1952), can be used in natural biological control. Ladybird beetles are cultured for this purpose, are readily available, and seem to be released everywhere in the world (Caltagirone and Doutt 1989). However, users are not always aware that some of these ladybird beetles can be a major threat to native species (Staines et al. 1990, Howarth 1991, Elliott et al. 1996, Brown and Miller 1998, Cottrell and Yeargan 1999, Turnock et al. 2003). In intraguild predation, larger species are favored unless protected chemically (Sato and Dixon 2004). Gardeners also now know that carabid ground beetles are useful predators that can be affected negatively by indiscriminate use of insecticides. Insect books especially directed to gardeners are now available (Cranshaw 2004).

Butterfly farms and insectaries are used as public attractions, and insects can be used in wildlife education centers to educate the public on the living world and the need for biodiversity conservation. The inclusion of insects now under various endangered species legislation also has increased public awareness of this group of animals and the need for habitat conservation (Hafernik 1992).

Butterflies in particular first induced many youth to become interested in collecting insects, leading many of them to become professional entomologists, interested in the outdoors and involved in the study of systematic entomology, ecology, and behavior (Michener 2007).

Although many members of the public find insects abhorrent, others, using helpful and readily available field guides, such as those on butterflies (Glassberg 2001) and dragonflies (Dunkle 2000), have found insects of interest as a vocation. Insects are now becoming almost as popular as birds in this regard, and even a birder's bug book is now available (Waldbauer 1998).

Insects are now so popular with the public that publishers seem to be flooding the market with books on these animals. These books vary from encyclopedias (Resh and Cardé 2003) and solid accounts of biology (Wigglesworth 1964, Berenbaum 1995) to more popular books with varying slants (Hutchins 1966, Blaney 1976, d'Entrèves and Zunino 1976, Waldbauer 1996, 2003, Eisner 2003). The books by Berenbaum (1995), in particular, emphasize insects and their influence on human affairs. Butterflies are one of the few insect groups with a positive image among the average citizen (Hafernik 1992). The migratory monarch butterfly *Danaus plexippus* (L.) is perhaps the most well-known and widely recognized butterfly, at least in North America (Nagano and Sakai 1989), and its annual migration is considered one of the epic phenomena of the animal kingdom. Monarch wintering colonies have been a tremendous attraction for tourists, have improved local economies (Nagano and Sakaii 1989), and have engendered public interest in biodiversity conservation.

Insect conservation is still in its infancy (Pyle et al. 1981), but the conservation of insects is of increasing public concern. The task for entomologists is to decide how best this can be accomplished. Despite their ecological importance, their conservation has received little attention (Hafernik 1992). However, with recent reviews (Samways 1994, 2005, 2007a, 2007b, New 1995), this status might change soon.

REFERENCES

Addicott, J. F., J. Bronstein, and F. Kjellberg. 1990. Evolution of mutualistic life-cycles: yucca moths and fig wasps. Pp. 143–161. *In* F. Gilbert (ed). *Genetics, Evolution and Coordination of Insect Life Cycles*. Springer, London.

Akam, M. 1995. Hox genes and the evolution of diverse body plans. *Philosophical Transactions of the Royal Society. B. Biological Sciences* 349: 313–319.

Aker, C. L. and D. Udovic. 1981. Oviposition and pollination of the yucca moth, *Tegeticula maculatae* (Lepidoptera: Prodoxidae), and its relation to the reproductive biology of *Yucca whipplei* (Agavaceae). *Oecologia* 49: 96–101.

Akre, R. D., G. S. Paulson, and E. P. Catts. 1992. *Insects Did It First*. Ye Galleon Press, Fairfield, Washington. 160 pp.

Alexander, R. D. 1968. Life cycle origins, speciation, and related phenomena in crickets. *Quarterly Review of Biology* 43: 1–41.

Allen-Wardell, G., P. Bernhardt, R. Bitner, A. Burquez, S. Buchmann, J. Cane, P. A. Cox, V. Dalton, P. Feinsinger, M. Ingram, D. Inouye, C. E. Jones, K. Kennedy, P. Kevan, H. Koopowitz, R. Medellin, S. Medellin-Morales, G. P. Nabhan, B. Pavik, V. Tepedino, P. Torchio, and S. Walker. 1998. The potential consequences of pollinator declines on the conservation of biodiversity and stability of food crop yields. *Conservation Biology* 12: 8–17.

Andrewartha, H. G. and L. C. Birch. 1973. The history of insect ecology. Pp. 229–266. *In* R. F. Smith, T. E. Mittler, and C. N. Smith (eds). *History of Entomology*. Annual Reviews, Palo Alto, California.

Arnold, R. A. 1983. Ecological studies of six endangered butterflies (Lepidoptera: Lycaenidae): island biogeography, patch dynamics, and the design of habitat preserves. *University of California Publications in Entomology* 99: 1–161.

Asquith, A., J. D. Lattin, and A. R. Moldenke. 1990. Arthropods: the invisible diversity. *Northwest Environmental Journal* 6: 404–405.

Atsatt, P. R. 1981. Lycaenid butterflies and ants: selection for enemy-free space. *American Naturalist* 118: 638–654.

Aultman, K. S., E. D. Walker, F. Gifford, D. W. Severson, C. B. Beard, and T. W. Scott. 2000. Managing risk of arthropod vector research. *Science* 288: 2321–2322.

Baker, H. G. 1961. *Ficus* and *Blastophaga*. *Evolution* 15: 378–379.

Baron, S. 1972. *The Desert Locust*. Charles Scribner's Sons, New York. 228 pp.

Barth, F. G. 1973. Microfiber reinforcement of an arthropod cuticle: laminated composite material in biology. *Zeitschrift für Zellforschung und Mikroskopische Anatomie* 144: 409–433.

Barton, N. H. and G. M. Hewitt. 1985. Analysis of hybrid zones. *Annual Review of Ecology and Systematics* 16: 113–148.

Barton, N. H. and M. C. Whitlock. 1997. The evolution of metapopulations. Pp. 183–210. *In* I. Hanski and M. E. Gilpin (eds). *Metapopulation Biology: Ecology, Genetics, and Evolution*. Academic Press, San Diego, California.

Bateson, W. 1894. *Materials for the Study of Variation Treated with Especial Regard to Discontinuity in the Origin of Species*. MacMillan, New York. 598 pp.

Batra, S. W. T. and L. R. Batra. 1967. The fungus garden of insects. *Scientific American* 217 (11): 112–120.

Baumann, P., N. A. Moran, and L. Baumann. 1997. The evolution and genetics of aphid endosymbiosis. *BioScience* 47: 12–20.

Beenakkers, A. M. T., D. J. van der Horst, and W. J. A. van Marrewijk. 1984. Insect flight muscle metabolism. *Insect Biochemistry* 14: 243–260.

Beenakkers, A. M. T., D. J. van der Horst, and W. J. A. van Marrewijk. 1985. Biochemical processes directed to flight muscle metabolism. Pp. 451–486. *In* G. A. Kerkut and L. I. Gilbert (eds). *Comprehensive Insect Physiology, Biochemistry and Pharmacology*. Volume 10. Biochemistry. Pergamon Press, Oxford.

Bennet-Clark, H. C. 1976. Energy storage in jumping insects. Pp. 421–443. *In* H. R. Hepburn (ed). *The Insect Integument.* Elsevier Scientific Publishing, Amsterdam.

Bennett, A. M. R. and G. G. E. Scudder. 1998. Differences in attachment of water mites on water boatmen: further evidence of differential parasitism and possible exclusion of a host from part of its potential range. *Canadian Journal of Zoology* 76: 824–834.

Benson, W. W. 1971. Evidence for the evolution of unpalatability through kin selection in Heliconiinae. *American Naturalist* 105: 213–226.

Berenbaum, M. R. 1995. *Bugs in the System. Insects and their Impact on Human Affairs.* Addison-Wesley Publishing, Reading, Massachusetts. 377 pp.

Bickham, J. W. 1983. Sibling species. Pp. 96–106. *In* C. M. Schonewald-Cox (ed). *Genetics and Conservation: a Reference for Managing Wild Animals and Plant Populations.* Benjamin/Cummings London.

Biesmeijer, J. C., S. P. M. Roberts, M. Reemer, R. Ohlemüller, M. Edwards, T. Peeters, A. P. Schaffers, S. G. Potts, R. Kleukers, C. D. Thomas, J. Settele, and W. E. Kunin. 2006. Parallel declines in pollinators and insect-pollinated plants in Britain and the Netherlands. *Science* 313: 351–354.

Bishopp, F. C. 1952. Insect friends of man. Pp. 79–87. *In Insects. The Yearbook of Agriculture 1952.* United States Department of Agriculture, Washington, DC.

Blaney, W. M. 1976. *How Insects Live.* Elsevier-Phaidon, London. 160 pp.

Blum, M. S. 1985. Alarm pheromones. Pp. 193–234. *In* G. A. Kerkut and L. I. Gilbert (eds). *Comprehensive Insect Physiology, Biochemistry and Pharmacology.* Volume 9. Behaviour. Pergamon Press, Oxford.

Bodenheimer, F. 1951. *Insects as Human Food: a Chapter of the Ecology of Man.* Junk, The Hague. 352 pp.

Bodenheimer, F. S. 1960. *Animals and Man in Bible Lands.* Brill, Leiden. 232 pp.

Boettner, G. H., J. S. Elkinton, and C. J. Boettner. 2000. Effects of a biological control introduction in three nontarget native species of saturniid moths. *Conservation Biology* 14: 1798–1806.

Boore, J. L., T. M. Collins, D. Stanton, L. L. Daehler, and W. M. Brown. 1995. Deducing the pattern of arthropod phylogeny from mitochondrial DNA rearrangement. *Nature* 376: 163–165.

Borden, J. H. 1985. Aggregation pheromones. Pp. 257–285. *In* G. A. Kerkut and L. I. Gilbert (eds). *Comprehensive Insect Physiology, Biochemistry and Pharmacology.* Volume 9. Behaviour. Pergamon Press, Oxford.

Boswell, R. E. and A. P. Mahowald. 1985. Cytoplasmic determinants in embryogenesis. Pp. 387–405. *In* G. A. Kerkut and L. I. Gilbert (eds). *Comprehensive Insect Physiology, Biochemistry and Pharmacology.* Volume 1. Embryogenesis and Reproduction. Pergamon Press, Oxford.

Boughton, D. S. 1999. Empirical evidence for complex source-sink dynamics with alternative states in a butterfly population. *Ecology* 80: 2727–2739.

Bower, B. 1997. Freeze! Insect proteins halt ice growth. *Science News* 152: 135.

Bradley, T. J. 1985. The excretory system: structure and physiology. Pp. 421–465. *In* G. A. Kerkut and L. I. Gilbert (eds). *Comprehensive Insect Physiology, Biochemistry and Pharmacology.* Volume 4. Regulation: Digestion, Nutrition, Excretion. Pergamon Press, Oxford.

Braschler, B. and J. K. Hill. 2007. Role of larval host plants in the climate-driven range expansion of the butterfly *Polygonia c-album. Journal of Animal Ecology* 76: 415–423.

Brown, M. W. and S. S. Miller. 1998. Cocinellidae (Coleoptera) in apple orchards of eastern West Virginia and the impact of invasion by *Harmonia axyridis. Entomological News* 109: 136–142.

Brown, S. W. 1973. Genetics – the long story. Pp. 407–432. *In* R. F. Smith, T. E. Mittler, and C. N. Smith (eds). *History of Entomology.* Annual Reviews, Palo Alto, California.

Brown, W. L., T. Eisner, and R. H. Whittaker. 1970. Allomones and kairomones: transspecific chemical messengers. *BioScience* 20: 21–22.

Brown, W. L. and E. O. Wilson. 1956. Character displacement. *Systematic Zoology* 5: 49–64.

Brues, C. T. 1946. *Insect Dietary. An Account of the Food Habits of Insects.* Harvard University Press, Cambridge, Massachusetts. 466 pp.

Brundin, L. 1966. Transantarctic relationships and their significance, as evidenced by chironomid midges. *Kungliga Svenska Vetenskapsakademiens Handlinger* 11: 1–472.

Brussard, P. F. 1984. Geographic patterns and environmental gradients: the central-marginal model in *Drosophila* revisited. *Annual Review of Ecology and Systematics* 15: 25–64.

Bryant, E. H., S. A. McCommas, and L. M. Combs. 1986. The effect of an experimental bottleneck upon quantitative variation in the housefly. *Genetics* 114: 1191–1211.

Bryant, P. J. 1993. The polar coordinate model goes molecular. *Science* 259: 471–472.

Buchmann, S. and B. Repplier. 2005. *Letters from the Hive. An Intimate History of Bees, Honey and Humankind.* Bantam Books, New York. 275 pp.

Bueckert, D. 2007. Decline in bee numbers raises alarm for future. Experts call for measures to protect all pollinators. *The Vancouver Sun,* January 20, 2007.

Bull, C. M. 1991. Ecology of parapatric distributions. *Annual Review of Ecology and Systematics* 22: 19–36.

Bush, G. L. 1969. Sympatric host race formation and speciation in frugivorous flies of the genus *Rhagoletis* (Diptera, Tephritidae). *Evolution* 23: 237–251.

Bush, G. L. and D. J. Howard. 1986. Allopatric and non-allopatric speciation: assumptions and evidence. Pp. 411–438. *In* S. Karl and E. Nevo (eds). *Evolutionary Processes and Theory.* Academic Press, New York.

Busvine, J. R. 1993. *Disease Transmission by Insects: Its Discovery and 90 Years of Effort to Prevent It.* Springer-Verlag, Berlin. 361 pp.

Butenandt, A., R. Beckman, D. Stamm, and E. Hecker. 1959. Uber den Sexuallockstoff der Seidenspinner *Bombyx mori*, Reidarstellung und Konstitution. *Zeitschrift für Naturforschung* B14: 283–284.

Byrd, J. H. and J. L. Castner (eds). 2001. *Forensic Entomology. The Utility of Arthropods in Legal Investigations.* CRC Press, Boca Raton, Florida. 418 pp.

Caltagirone, L. E. 1981. Landmark examples in classical biological control. *Annual Review of Entomology* 26: 213–232.

Caltagirone, L. E. and R. L. Doutt. 1989. The history of the vedalia beetle importation to California and its impact on the development of biological control. *Annual Review of Entomology* 34: 1–16.

Carpenter, G. H. 1928. *The Biology of Insects.* Sidgwick and Jackson, London. 473 pp.

Carson, H. L. 1970. Chromosomal traces of the origin of species. Some Hawaiian *Drosophila* species have arisen from single founder individuals in less than a million years. *Science* 168: 1414–1418.

Carson, H. L. 1975. The genetics of speciation at the diploid level. *American Naturalist* 109: 83–92.

Carson, H. L. 1990. Increased genetic variance after a population bottleneck. *Trends in Ecology and Evolution* 5: 228–230.

Carson, H. L. and A. R. Templeton. 1984. Genetic revolutions in relation to speciation phenomena: the founding of new populations. *Annual Review of Ecology and Systematics* 15: 97–131.

Carson, W. P., J. P. Cronin, and Z. T. Long. 2004. A general rule for predicting when insects will have strong top-down effects on plant communities: on the relationship between insect outbreaks and host concentration. Pp. 193–211. *In* W. W. Weisser and E. Sieman (eds). *Insects and Ecosystem Function.* Springer-Verlag, Berlin Heidelberg.

Catts, E. P. and M. L. Goff. 1992. Forensic entomology in criminal investigations. *Annual Review of Entomology* 37: 253–272.

Chapman, R. N. 1928. The quantitative analysis of environmental factors. *Ecology* 9: 111–122.

Cherrett, J. M., R. J. Powell, and D. J. Stradling. 1989. The mutualism between leaf-cutting ants and their fungus. Pp. 93–120. *In* N. Wilding, N. M. Collins, P. M. Hammons, and J. F. Webber (eds). *Insect–Fungus Interactions.* Academic Press, London.

Clark, L. R., P. W. Geier, R. D. Hughes, and R. F. Morris. 1967. *The Ecology of Insect Populations in Theory and Practice.* Methuen, London. 232 pp.

Clausen, C. P. 1940. *Entomophagous Insects.* McGraw-Hill Book Co., New York. 688 pp.

Clausen, C. P. 1952. Parasites and predators. Pp. 380–388. *In Insects. The Yearbook of Agriculture 1952.* United States Department of Agriculture, Washington, DC.

Clausen, C. P. 1976. Phoresy among entomophagous insects. *Annual Review of Entomology* 21: 343–368.

Cohen, S. N. and J. A. Shapiro. 1980. Transposable genetic elements. *Scientific American* 242 (2): 40–49.

Collinge, S. K. 2000. Effects of grassland fragmentation on insect species loss, colonization and movement patterns. *Ecology* 81: 2211–2226.

Cottrell, T. E. and K. V. Yeargan. 1999. Intraguild predation between an introduced lady beetle, *Harmonia axyridis* (Coleoptera: Coccinellidae), and a native lady beetle, *Coleomegilla maculata* (Coleoptera: Coccinellidae). *Journal of the Kansas Entomological Society* 71: 159–163.

Coyne, J. A. and H. A. Orr. 1989. Pattern of speciation in *Drosophila. Evolution* 43: 362–381.

Crane, E. 2003. Royal jelly. Pp. 1009–1010. *In* V. H. Resh and R. T. Cardé (eds). *Encyclopedia of Insects.* Academic Press, San Diego, California.

Cranshaw, W. 2004. *Garden Insects of North America: The Ultimate Guide to Backyard Bugs.* Princeton University Press, Princeton, New Jersey. 656 pp.

Crepet, W. L. 1979. Insect pollination: a paleontological perspective. *BioScience* 29: 102–108.

Crepet, W. L. 1983. The role of insect pollination in the evolution of angiosperms. Pp. 31–50. *In* L. Real (ed). *Pollination Biology.* Academic Press, New York.

Crombie, A. C. 1945. On competition between different species of graminivorous insects. *Proceedings of the Royal Society. B. Biological Sciences* 132: 362–395.

Crozier, L. 2004. Warmer winters drive butterfly range expansion by increasing survivorship. *Ecology* 85: 231–241.

Cummins, K. W., L. D. Miller, N. A. Smith, and R. M. Fox. 1965. *Experimental Entomology.* Reinhold Publishing, New York. 176 pp.

Currie, C. R., J. A. Scott, R. C. Summerbell, and D. Malloch. 1999. Fungus-growing ants use antibiotic-producing bacteria to control garden parasites. *Nature* 398: 701–704.

Davidson, G. 1974. *Genetic Control of Insect Pests.* Academic Press, London. 158 pp.

Dayan, T. and D. Simberloff. 2005. Ecological and community-wide character displacement: the next generation. *Ecology Letters* 8: 875–894.

DeBach, P. 1966. The competitive displacement and coexistence principle. *Annual Review of Entomology* 11: 183–212.

DeBach, P. 1974. *Biological Control by Natural Enemies.* Cambridge University Press, London. 323 pp.

DeBach, P. and O. Rosen. 1991. *Biological Control by Natural Enemies*, Second Edition. Cambridge University Press, Cambridge. 440 pp.

DeFoliart, G. R. 1989. The human use of insects as food and as animal feed. *Bulletin of the Entomological Society of America* 35: 22–35.

DeFoliart, G. R. 1992. Insects as human food. *Crop Protection* 11: 395–399.

DeFoliart, G. R. 1999. Insects as food and why the Western attitude is important. *Annual Review of Entomology* 44: 21–50.

Denno, R. F., S. Larsson, and K. L. Olmstead. 1990. Role of enemy-free space and plant quality in host-plant selection by willow beetles. *Ecology* 71: 124–137.

d'Entrèves, P. P. and M. Zunino. 1976. *The Secret Life of Insects.* Orbis Publishing, London. 384 pp.

Desalle, R. and D. A. Grimaldi. 1991. Morphological and molecular systematics of Drosophilidae. *Annual Review of Ecology and Systematics* 22: 447–475.

Dickinson, M. H., F. O. Lehmann, and S. P. Sane. 1999. Wing rotation and the aerodynamic basis of insect flight. *Science* 284: 1954–1960.

Dobzhansky, T. 1937. *Genetics and the Origin of Species.* Columbia University Press, New York. 364 pp.

Dodson, C. H. 1975. Coevolution of orchids and bees. Pp. 91–99. *In* L. E. Gilbert and P. H. Raven (eds). *Coevolution of Animals and Plants.* University of Texas Press, Austin.

Doncaster, L. and G. H. Raynor. 1906. Breeding experiments with Lepidoptera. *Proceedings of the Zoological Society of London* 1: 125–133.

Dover, G. 1982. Molecular drive: a cohesive mode of species evolution. *Nature* 299: 111–117.

Downes, J. A. 1965. Adaptations of insects in the arctic. *Annual Review of Entomology* 10: 257–274.

Duffey, S. S. 1977. Arthropod allomones: chemical effronteries and antagonists. *Proceedings of XV International Congress of Entomology Washington, DC* (1976) 15: 323–394.

Duncker, B. P., C.-P. Chen, P. L. Davies, and V. K. Walker. 1995. Antifreeze protein does not confer cold tolerance to transgenic *Drosophila melanogaster. Cryobiology* 32: 521–527.

Dunkle, S. W. 2000. *Dragonflies through Binoculars. A Field Guide to Dragonflies of North America.* Oxford University Press, New York. 274 pp.

Eberhard, M. J. W. 1975. The evolution of social behavior by kin selection. *Quarterly Review of Biology* 50: 1–34.

Eisner, T. 2003. *For Love of Insects.* Belknap Press, Cambridge, Massachusetts. 448 pp.

Elias, S. A. 1994. *Quaternary Insects and Their Environments.* Smithsonian Institution Press, Washington. 284 pp.

Elkinton, J. S., W. M. Healy, J. P. Buonaccorsi, G. H. Buettner, A. M. Hazzard, H. R. Smith, and A. M. Liebhold. 1996. Interactions among gypsy moths, white-footed mice, and acorns. *Ecology* 77: 2332–2342.

Ellington, C. P. 1984. The aerodynamics of hovering insect flight. VI. Lift and power requirements. *Philosophical Transactions of the Royal Society. B. Biological Sciences.* 305: 145–181.

Ellington, C. P. 1999. The novel aerodynamics of insect flight: applications to micro-air vehicles. *Journal of Experimental Biology* 202: 3439–3448.

Ellington, C. P., C. van den Berg, A. P. Willmott, and A. L. R. Thomas. 1996. Leading-edge vortices in insect flight. *Nature* 384: 626–630.

Elliott, N., R. Kieckhager, and W. Kauffman. 1996. Effects of an invading coccinellid on native coccinellids in an agricultural landscape. *Oecologia* 105: 537–544.

Engel, M. A. and D. A. Grimaldi. 2004. New light shed on the oldest insect. *Nature* 427: 627–630.

Enserink, M. 2000. The enigma of West Nile. *Science* 290: 1482–1484.

Erwin, T. L. [R]. 1982. Tropical forests: their richness in Coleoptera and other arthropod species. *Coleopterists Bulletin* 36: 74–75.

Erwin, T. R. 1983. Tropical forest canopies, the last biotic frontier. *Bulletin of the Entomological Society of America* 29: 14–19.

Erwin, T. R. 1988. The tropical forest canopy: the heart of biotic diversity. Pp. 123–129. *In* E. O. Wilson and F. M. Peter (eds). *Biodiversity.* National Academy Press, Washington, DC.

Erwin, T. L. [R]. 1993. Biodiversity at its utmost: tropical forest beetles. Pp. 27–68. *In* M. L. Reaka-Kudla, D. E. Wilson, and E. O. Wilson (eds). *Biodiversity II. Understanding and Protecting our Biological Resources.* Joseph Henry Press, Washington, DC.

Fabré, J. H. 1911. *Social Life in the Insect World.* Ernest Benn, London. 327 pp.

Farrell, B. D. 1998. ''Inordinate fondness'' explained: why are there so many beetles. *Science* 281: 555–559.

Fincher, G. T. 1986. Importation, colorization, and release of dung-burying scarabs. *Miscellaneous Publications of the Entomological Society of America* 62: 69–76.

Fisher, B. L. 1998. Insect behavior and ecology in conservation: preserving functional species interactions. *Annals of the Entomological Society of America* 91: 155–158.

Fisher, R. C. 1988. An inordinate fondness for beetles. *Biological Journal of the Linnean Society* 35: 313–319.

Follett, P. A. and J. J. Duan (eds). 2000. *Nontarget Effects of Biological Control.* Kluwer Academic Publishers, Boston, 316 pp.

Foottit, R. G. 1997. Recognition of parthenogenetic insect species. Pp. 291–307. *In* M. F. Claridge, H. A. Dawah, and M. R. Wilson (eds). *Species: The Units of Biodiversity.* Chapman and Hall, London.

Force, D. C. 1967. Genetics of the colorization of natural enemies for biological control. *Annals of the Entomological Society of America* 60: 723–729.

Fox, C. J. S. 1957. Remarks on influence of insects on the floristic composition of grassland. *Rapport de la Société de Québec pour la Protection des Plantes* 39: 49–51.

Franco, A. M. A., J. K. Hill, C. Kitschke, Y. C. Collingham, D. B. Roy, R. Fox, B. Huntley, and C. D. Thomas. 2006. Impacts of climate warming and habitat loss on extinctions at species low-latitude range boundaries. *Global Change Biology* 12: 1545–1553.

Free, J. B. 1993. *Insect Pollination of Crops*. Second Edition. Academic Press, London. 768 pp.

Friedrich, M. and D. Tautz. 1995. Ribosomal DNA phylogeny of the major extant arthropod classes and the evolution of myriapods. *Nature* 376: 165–167.

Frost, S. W. 1959. *Insect Life and Insect Natural History*. Dover Publications, New York. 526 pp.

Galil, J. 1977. Fig biology. *Endeavour (n. s.)* 1 (2): 52–56.

Gaston, K. J. 1991. The magnitude of global insect species richness. *Conservation Biology* 5: 283–296.

Gehring, W. J. and R. Nöthiger. 1973. The imaginal discs of *Drosophila melanogaster*. Pp. 212–290. *In* S. J. Counce and C. H. Waddington (eds). *Developmental System: Insects*. Volume 2. Academic Press, New York.

Giribet, G., G. D. Edgecombe, and W. C. Wheeler. 2001. Arthropod phylogeny based on eight molecular loci and morphology. *Nature* 413: 157–161.

Glass, B. 1957. In pursuit of a gene. *Science* 126: 683–689.

Glassberg, J. 2001. *Butterflies through Binoculars. A Field Guide to the Butterflies of Western North America*. Oxford University Press, New York. 384 pp.

Goddard, M. and D. Taron. 2001. Neglected flowers: the decline of honey bees in North America. *The NEB Transcript* 11 (1): 8–9.

Graham, L. A., Y.-C. Liou, V. K. Walker, and P. L. Davies. 1997. Hyperactive antifreeze protein from beetles. *Nature* 388: 727–728.

Grevelink, S. A., D. E. Youssef, J. Loscalzo, and E. A. Lerner. 1993. Salivary gland extracts from the deerfly contain a potent inhibitor of platelet aggregation. *Proceedings of the National Academy of Sciences USA* 90: 9155–9158.

Grigliatti, T. A., G. Meiser, and T. Pfeifer. 2007. TAC-TICS: Transposon-based biological pest management systems. Pp. 327–352. *In* M. Vurro and J. Gressel. (eds). *Novel Biotechnologies for Biocontrol Enhancement and Management*. NATO Science Series. IOS Press, Amsterdam.

Grimaldi, D. and M. S. Engel. 2005. *Evolution of the Insects*. Cambridge University Press, New York. 755 pp.

Groombridge, B. (ed). 1992. *Global Biodiversity. Status of the Earth's Living Resources. A Report Compiled by the World Conservation Monitoring Centre*. Chapman and Hall, London. 585 pp.

Guilford, T. 1985. Is kin selection involved in the evolution of warning coloration? *Oikos* 45: 31–36.

Guo, L.-Q., J.-F. Lin, S. Xiong, and S.-C. Chen. 2005. Transformation of *Volvariella volvacea* with a thermal hysteresis protein gene by particle bombardment. *Weishengwu Xuebao* 45: 39–43. [In Chinese].

Haddad, N. M. 1999a. Corridor and distance effects on interpatch movements: a landscape experiment with butterflies. *Ecological Applications* 9: 612–622.

Haddad, N. M. 1999b. Corridor use predicted from behaviors at habitat boundaries. *American Naturalist* 153: 215–227.

Haddad, N. M. and K. A. Baum. 1999. An experimental test of corridor effects on butterfly densities. *Ecological Applications* 9: 623–633.

Hadorn, E. 1968. Transdetermination in cells. *Scientific American* 219 (5): 110–120.

Hafernik, J. E., Jr. 1992. Threats to invertebrate biodiversity: implications for conservation strategies. Pp. 171–195. *In* P. L. Fiedler and S. K. Jain (eds). *Conservation Biology. The Theory and Practice of Nature Conservation, Preservation and Management*. Chapman and Hall, New York.

Hagen, K. S. and J. M. Franz. 1973. A history of biological control. Pp. 433–476. *In* R. F. Smith, T. E. Mittler, and C. N. Smith (eds). *History of Entomology*. Annual Reviews, Palo Alto, California.

Haldane, J. B. S. 1932. *The Causes of Evolution*. Harper and Brothers, London. 222 pp.

Haldane, J. B. S. 1964. A defense of bean bag genetics. *Perspectives in Biology and Medicine* 7: 343–359.

Hammond, P. 1992. Species inventory. Pp. 17–39. *In* B. Groombridge (ed). *Global Biodiversity. Status of the Earth's Living Resources. A Report Compiled by the World Conservation Monitoring Centre*. Chapman and Hall, London.

Hanski, I. 1999. *Metapopulation Ecology*. Oxford University Press, Oxford. 324 pp.

Hanski, I. and Y. Cambefort (eds). 1991. *Dung Beetle Ecology*. Princeton University Press, Princeton, New Jersey. 520 pp.

Hanski, I., M. Saastamoinen, and O. Ovaskainen. 2006. Dispersal-related life-history trade-offs in a butterfly metapopulation. *Journal of Animal Ecology* 75: 91–100.

Hanski, I. and M. C. Singer. 2001. Extinction-colorization dynamics and host-plant choice in butterfly metapopulations. *American Naturalist* 158: 341–353.

Hardie, J. and A. D. Lees. 1985. Endocrine control of polymorphism and polyphenism. Pp. 441–490. *In* G. A. Kerkut and L. I. Gilbert (eds) *Comprehensive Insect Physiology, Biochemistry and Pharmacology*. Volume 8. Endocrinology II. Pergamon Press, Oxford.

Harpaz, I. 1973. Early entomology in the Middle East. Pp. 21–36. *In* R. F. Smith, T. E. Mittler, and C. N. Smith (eds). *History of Entomology*. Annual Reviews, Palo Alto, California.

Hartley, S. E. and T. H. Jones. 2004. Insect herbivores, nutrient cycling and plant productivity. Pp. 27–52. *In* W. W. Weisser and E. Siemann (eds). *Insects and Ecosystem Function*. Springer-Verlag, Berlin Heidelberg.

Hastings, A., J. E. Byers, J. A. Crooks, K. Cuddington, C. G. Jones, J. G. Lambrinos, T. S. Talley, and W. G. Wilson. 2006. Ecosystem engineering in space and time. *Ecology Letters* 10: 153–164.

Haynes, K. F. and M. C. Birch. 1985. The role of other pheromones, allomones and kairomones in the behavioral responses of insects. Pp. 225–255. *In* G. A. Kerkut and L. I. Gilbert (eds). *Comprehensive Insect Physiology, Biochemistry and Pharmacology*. Volume 9. Behavior. Pergamon Press, Oxford.

Heard, S. B. 1994. Pitcher-plant midges and mosquitoes: a processing chain commensalism. *Ecology* 75: 1647–1660.

Hennig, W. 1950. *Grundzüge einer Theroie der phylogenetischen Systematik.* Deutscher Zentralverlag, Berlin. 370 pp.

Hennig, W. 1965. Phylogenetic systematics. *Annual Review of Entomology* 10: 97–116.

Hennig, W. 1966. *Phylogenetic Systematics.* University of Illinois Press, Urbana. 263 pp.

Hennig, W. 1969. *Die Stammesgeschichte der Insekten.* Kramer, Frankfurt am Main, Germany. 436 pp.

Hennig, W. 1981. *Insect Phylogeny.* Academic Press, New York. 536 pp.

Henry, S. M. (ed). 1967. *Symbiosis. Volume II. Associations of Invertebrates, Birds, Ruminants, and Other Biota.* Academic Press, New York. 443 pp.

Herdt, R. W. 1987. Equity considerations in setting priorities for Third World rice biotechnology research. *Development: Seeds of Change* 4: 19–24.

Hewitt, G. M. 1988. Hybrid zones – natural laboratories for evolutionary studies. *Trends in Ecology and Evolution* 3: 158–167.

Hewitt, G. M. 1990. Divergence and speciation as viewed from an insect hybrid zone. *Canadian Journal of Zoology* 68: 1701–1715.

Hill, J. K., C. D. Thomas, R. Fox, M. G. Telfer, S. G. Willis, J. Asher, and B. Huntley. 2002. Responses of butterflies to twentieth century climate warming: implications for future ranges. *Proceedings of the Royal Society. B. Biological Sciences* 269: 2163–2171.

Hodkinson, I. D. and D. Casson. 1991. A lesser predilection for bugs: Hemiptera (Insecta) diversity in tropical rain forests. *Biological Journal of the Linnean Society* 43: 101–109.

Horvitz, H. R., H. M. Ellis, and P. W. Sternberg. 1982. Programmed cell death in nematode development. *Neuroscience Commentaries* 1 (2): 56–65.

Howarth, F. G. 1991. Environmental impacts of classical biological control. *Annual Review of Entomology* 36: 485–509.

Huettel, M. D. and G. L. Bush. 1972. The genetics of host selection and its bearing on sympatric speciation in *Procecidochares* (Diptera: Tephritidae). *Entomologia Experimentalis et Applicata* 15: 465–480.

Huffaker, C. B. (ed) 1971. *Biological Control.* Plenum Publishing, New York. 511 pp.

Hunkapiller, T., H. Huang, L. Hood, and J. H. Campbell. 1982. The impact of modern genetics on evolutionary theory. Pp. 164–189. *In* R. Milkman (ed). *Perspectives on Evolution.* Sinauer Associates, Sunderland, Massachusetts.

Hunter, M. D. 2002. Landscape structure, habitat fragmentation, and the ecology of insects. *Agricultural and Forest Entomology* 4: 159–166.

Hutchins, R. E. 1966. *Insects.* Prentice-Hall, Englewood Cliffs, New Jersey. 324 pp.

Hutchinson, G. E. 1953. The concept of pattern in ecology. *Proceedings of the Academy of Natural Sciences of Philadelphia* 105: 1–12.

Jacobson, M. 1972. *Insect Sex Pheromones.* Academic Press. New York. 382 pp.

Janning, W. J. 1978. Gynandromorph fate maps in *Drosophila.* Pp. 1–28. *In* W. Gehring (ed). *Genetic Mosaics and Cell Differentiation.* Springer-Verlag, Berlin.

Janzen, D. H. 1966. Coevolution of mutualism between ants and acacias in Central America. *Evolution* 20: 249–275.

Janzen, D. H. 1979. How to be a fig. *Annual Review of Ecology and Systematics* 10: 13–51.

Janzen, D. H. 1987. Insect diversity in a Costa Rican dry forest: why keep it, and how? *Biological Journal of the Linnean Society* 30: 343–356.

Jeffries, M. J. and J. H. Lawton. 1984. Enemy free space and the structure of ecological communities. *Biological Journal of the Linnean Society* 23: 269–286.

Johansen, C. A. 1977. Pesticides and pollinators. *Annual Review of Entomology* 22: 177–192.

Johnson, C. G. 1969. *Migration and Dispersal of Insects by Flight.* Methuen and Co., London. 763 pp.

Johnson, C. G. and J. Bowden. 1973. Problems related to the transoceanic transport of insects especially between the Amazon and Congo area. *Tropical Forest Ecosystems in Africa and South America: A Comparative Review.* Smithsonian Institution, Washington, DC. Pp. 207–222.

Jones, C. G., J. H. Lawton, and M. Shachak. 1994. Organisms as ecosystem engineers. *Oikos* 69: 373–386.

Jones, C. G., R. S. Ostfeld, M. P. Richard, E. M. Schauber, and J. O. Wolff. 1998. Chain reactions linking acorns to gypsy moth outbreaks and Lyme disease risk. *Science* 279: 1023–1026.

Jürgens, G. 1987. Segmental organization of the tail region in the embryo of *Drosophila melanogaster*: a blastoderm fate map of the cuticle structures of the larval tail region. *Roux's Archives of Developmental Biology* 196: 141–157.

Kaiser, K., J. W. Sentry, and D. J. Finnegan. 1995. Eukaryotic transposable elements as tools to study gene structure and function. Pp. 69–100. *In* D. J. Sherratt (ed). *Mobile Genetic Elements.* IRC Press at Oxford University Press, Oxford.

Kalmus, H. 1948. *Simple Experiments with Insects.* William Heinemann, London. 132 pp.

Karlson, P. and A. Butenandt. 1959. Pheromones (ectohormones) in insects. *Annual Review of Entomology* 4: 49–58.

Karlson, P. and M. Lüscher. 1959. "Pheromones": a new term for a class of biologically active substances. *Nature* 183: 55–56.

Kato, M. and T. Inoue. 1994. Origin of insect pollination. *Nature* 368: 195.

Kause, E. 1960. Preformed ooplasmic reaction systems in insect eggs. Pp. 302–327. *In* S. Ranzi (ed). *Symposium on Germ Cells and Development.* Institut Internationale d'Embriologie Fondazione A. Baselle, Milan.

Kayser, H. 1985. Pigments. Pp. 367–415. *In* G. A. Kerkut and L. I. Gilbert (eds). *Comprehensive Insect Physiology, Biochemistry and Pharmacology.* Volume 10. Biochemistry. Pergamon Press, Oxford.

Kearns, C. A. and D. W. Inouye. 1997. Pollinators, flowering plants, and conservation biology. Much remains to be learned about pollinators and plants. *BioScience* 47: 297–307.

Kearns, C. A., D. W. Inouye, and N. M. Waser. 1998. Endangered mutualism: the conservation of plant–pollinator interactions. *Annual Review of Ecology and Systematics* 29: 83–112.

Keilin, D. 1925. On cytochrome, a respiratory pigment common to animals, yeast and higher plants. *Proceedings of the Royal Society of London, Series B* 98: 312–339.

Kettlewell, H. B. D. 1959. Darwin's missing evidence. *Scientific American* 200 (3): 48–53.

Kettlewell, H. B. D. 1961. The phenomenon of industrial melanism in Lepidoptera. *Annual Review of Entomology* 6: 245–262.

Kettlewell, H. B. D. 1973. *The Evolution of Melanism: The Study of a Recurring Necessity with Special Reference to Industrial Melanism in the Lepidoptera.* Clarendon Press, London. 423 pp.

Kevan, P. G., E. A. Clark, and V. G. Thomas. 1990. Pollination: a crucial ecological and mutualistic link in Agroforestry and sustainable Agriculture. *Proceedings of the Entomological Society of Ontario* 121: 43–48.

Key, K. H. L. 1968. The concept of stasipatric speciation. *Systematic Zoology* 17: 14–22.

Klein, A.-M. , B. E. Vaissière, J. H. Cane, I. Steffan-Dewenter, S. A. Cunningham, C. Kremen, and T. Tscharntke. 2007. Importance of pollinators in changing landscapes for world crops. *Proceedings of the Royal Society. B. Biological Sciences* 274: 303–313.

Konishi, M. and Y. Ito. 1973. Early entomology in East Asia. Pp. 1–20. *In* R. F. Smith, T. E. Mittler, and C. N. Smith (eds). *History of Entomology*. Annual Reviews, Palo Alto, California.

Koul, O. and G. S. Dhaliwal (eds). 2003. *Predators and Parasitoids.* Taylor and Francis, London. 191 pp.

Kremen, C., N. M. Williams, and R. W. Thorp. 2002. Crop pollination from native bees at risk from agricultural intensification. *Proceedings of the National Academy of Sciences USA* 99: 16812–16816.

Kristensen, N. P. 1981. Phylogeny of insect orders. *Annual Review of Entomology* 26: 133–157.

Lavelle, P. 1997. Faunal activities and soil processes: adaptive strategies that determine ecosystem function. *Advances in Ecological Research.* 27: 93–132.

Lavelle, P. 2002. Functional domains in soils. *Ecological Research* 17: 441–450.

Leach, J. G. 1940. *Insect Transmission of Plant Diseases.* McGraw-Hill Book Co., New York. 615 pp.

Lee, R. E., Jr. and D. L. Denlinger (eds). 1991. *Insects at Low Temperature.* Chapman and Hall, New York. 513 pp.

Levin, M. D. 1983. Value of bee pollination in US agriculture. *Bulletin of the Entomological Society of America* 29: 50–51.

Liebhold, A., J. Elkinton, D. Williams, and R.-M. Muzika. 2000. What cause outbreaks of the gypsy moth in North America? *Population Ecology* 42: 257–266.

Lobo, J. M. 2000. Species diversity and composition of dung beetle (Coleoptera: Scarabaeoidea) assemblages in North America. *Canadian Entomologist* 132: 307–321.

Lockshin, R. A. 1985. Programmed cell death. Pp. 301–317. *In* G. A. Kerkut and L. I. Gilbert (eds). *Comprehensive Insect Physiology, Biochemistry and Pharmacology.* Volume 2. Postembryonic Development. Pergamon Press, Oxford.

Losey, J. E. and M. Vaughan. 2006. The economic value of ecological services provided by insects. *BioScience* 56: 311–323.

Louda, S. M., D. Kendall, J. Connor, and D. Simberloff. 1997. Ecological effects of an insect introduced for the biological control of weeds. *Science* 277: 1088–1090.

Louda, S. M., R. W. Pemberton, M. T. Johnson, and P. A. Follett. 2003. Nontarget effects – the Achilles' heel of biological control? Retrospective analysis to reduce risk associated with biocontrol introductions. *Annual Review of Entomology* 48: 365–396.

Louda, S. M. and P. Stiling. 2004. The double-edged sword of biological control in conservation and restoration. *Conservation Biology* 18: 50–53.

Lounibos, L. P. 2002. Invasions by insect vectors of human disease. *Annual Review of Entomology* 47: 233–266.

Lubbock, J. 1873. *On the Origin and Metamorphosis of Insects.* Macmillan and Co., New York. 108 pp.

Lyonet, P. 1762. *Traité Anatomique de la Chenille Quironge le Bois de Saule.* Gosse and Pinet, La Haye. 616 pp.

Maddrell, S. H. P. and B. D. C. Gardner. 1976. Excretion of alkaloids by Malpighian tubules of insects. *Journal of Experimental Biology* 64: 267–281.

Maddrell, S. H. P. and J. E. Phillips. 1975. Active transport of sulphate ions by Malpighian tubules of the larvae of the mosquito *Aedes campestris. Journal of Experimental Biology* 62: 367–378.

Maddrell, S. H. P. and J. E. Phillips. 1978. Induction of sulphate transport and hormonal control of fluid secretion by Malpighian tubules of larvae of the mosquito *Aedes taeniorhynchus. Journal of Experimental Biology* 72: 181–202.

Majer, J. D. 1987. The conservation and study of invertebrates in remnants of native vegetation. Pp. 333–335. *In* D. A. Saunders, G. W. Arnold, A. A. Burbridge, and A. J. M. Hopkins (eds). *Nature Conservation: The Role of Remnants of Native Vegetation.* Surrey Beatty and Sons, Sydney.

Malcolm, S. B. 1986. Aposematism in a soft-bodied insect: a case for kin selection. *Behavioral Ecology and Sociobiology* 18: 387–393.

Malcolm, S. B. 1990. Mimicry: status of a classical evolutionary paradigm. *Trends in Ecology and Evolution* 5: 57–62.

Mallet, J. 2006. What does *Drosophila* genetics tell us about speciation? *Trends in Ecology and Evolution* 21: 386–393.

Marshall, E. 2000. A renewed assault on an old and deadly foe. *Science* 290: 428–430.

Martin, K., A. Norris, and M. Drever. 2006. Effects of bark beetle outbreaks on avian biodiversity in the British Columbia interior: Implications for critical habitat management. *British Columbia Journal of Ecosystems and Management* 7 (3): 10–24.

Martin, M. M. 1970. The biochemical basis of the fungus-attine ant symbiosis. *Science* 169: 16–20.

Marx, J. 1992. Homeobox genes go evolutionary. *Science* 255: 399–401.

May, R. M. 1988. How many species are there on earth? *Science* 241: 1441–1449.

May, R. M. 1990. How many species? *Philosophical Transactions of the Royal Society. B. Biological Sciences* 330: 293–304.

Mayer, M. S. and J. R. McLaughlin (eds). 1990. *Handbook of Insect Pheromones and Sex Attractants*. CRC Press, Boca Raton, Florida. 992 pp.

Mayr, E. 1942. *Systematics and the Origin of Species*. Columbia University Press, New York. 334 pp.

McCafferty, W. P. 1981. *Aquatic Entomology: the Fisherman's and Ecologists' Illustrated Guide to Insects and Their Relatives*. Science Books International, Boston, Massachusetts. 448 pp.

McGill, W. B. and J. R. Spence. 1985. Soil fauna and soil structure: feedback between size and architecture. *Quaestiones Entomologicae* 21: 645–654.

McKelvey, J. J., Jr. 1973. *Man Against Tsetse. Struggle for Africa*. Cornell University Press, Ithaca, New York. 306 pp.

McNeill, W. H. 1976. *Plagues and People*. Anchor/Doubleday, New York. 355 pp.

Meredith, J., L. Moore, and G. G. E. Scudder. 1984. The excretion of ouabain by the Malpighian tubules of *Oncopeltus fasciatus*. *American Journal of Physiology* 246: R705–R715.

Michener, C. D. 2007. The professional development of an entomologist. *Annual Review of Entomology* 52: 1–15.

Miller, J. C. 1993. Insect natural history, multi-species interactions and biodiversity in ecosystems. *Biodiversity and Conservation* 2: 233–241.

Miller, L. H. and B. Greenwood. 2002. Malaria – a shadow over Africa. *Science* 298: 121–122.

Minchella, D. J. and M. E. Scott. 1991. Parasitism: a cryptic determinant of animal community structure. *Trends in Ecology and Evolution* 68: 250–254.

Mishler, B. D. and M. J. Donoghue. 1982. Species concepts: a case for pluralism. *Systematic Zoology* 31: 491–503.

Moore, B. P., G. E. Woodroffe, and A. R. Sanderson. 1956. Polymorphism and parthenogenesis in a ptinid beetle. *Nature* 177: 847–848.

Moore, J. C. and D. E. Walter. 1988. Arthropod regulation of micro- and mesobiota in below-ground detrital food webs. *Annual Review of Entomology* 33: 419–439.

Moore, L. V. and G. G. E. Scudder. 1985. Ouabain-resistant Na, K-ATPase and cardenolides tolerance in the large milkweed bug, *Oncopeltus fasciatus*. *Journal of Insect Physiology* 32: 27–33.

Morris, R. F. 1951. The importance of insect control in a forest management program. *Canadian Entomologist* 83: 176–181.

Mott, D. G., L. D. Nairn, and J. A. Cook. 1957. Radial growth in forest trees and effects of insect defoliation. *Forest Science* 3: 286–304.

Mueller, U. G., S. A. Rehner, and T. R. Schultz. 1998. The evolution of agriculture in ants. *Science* 281: 2034–2038.

Mullins, D. E. 1985. Chemistry and physiology of the hemolymph. Pp. 355–451. *In* G. A. Kerkut and L. I. Gilbert (eds). *Comprehensive Insect Physiology, Biochemistry and Pharmacology*. Volume 3. Integument, Respiration and Circulation. Pergamon Press, Oxford.

Mumcuoglu, K. Y., A. Ingber, L. Bilead, J. Stessman, R. Friedmann, H. Schulman, H. Bichucher, I. Ioff-Uspensky, J. Miller, R. Galun, and I. Raz. 1999. Maggot therapy for the treatment of intractable wounds. *International Journal of Dermatology* 38: 623–627.

Murphy, D. D., K. E. Freas, and S. B. Weiss. 1990. An environment-metapopulation approach to population viability analysis for a threatened invertebrate. *Conservation Biology* 4: 41–51.

Nachtigall, W. 1974. *Insects in Flight: a Glimpse Behind the Scenes in Biophysical Research*. McGraw Hill, New York. 150 pp.

Nachtigall, W. 1989. Mechanics and aerodynamics of flight. Pp. 1–29. *In* G. J. Goldsworthy and C. H. Wheeler (eds). *Insect Flight*. CRC Press, Boca Raton, Florida.

Nagano, C. D. and W. H. Sakai. 1989. The monarch butterfly. Pp. 367–385. *In* W. J. Chandler, L. Labate, and C. Wille (eds). *Audubon Wildlife Report 1989/1990*. Academic Press, San Diego, California.

Naiman, R. J. 1988. Animal influences on ecosystem dynamics. *BioScience* 38: 750–752.

Nardi, F., G. Spinsanti, J. L. Boore, A. Carapelli, R. Dallai, and F. Frati. 2003. Hexapod origins: monophyletic or paraphyletic? *Science* 299: 1887–1889.

Nault, L. R. 1994. Transmission biology, vector specificity and evolution of planthopper-transmitted plant viruses. Pp. 429–448. *In* R. F. Denno and T. J. Perfect (eds). *Planthoppers: Their Ecology and Management*. Chapman and Hall, New York.

Naylor, R. L. and P. R. Ehrlich. 1997. Natural pest control services and agriculture. Pp. 151–174. *In* G. C. Daily (ed). *Nature's Services. Societal Dependence on Natural Ecosystems*. Island Press, Washington, DC.

Nei, M., T. Maruyama, and R. Chakraborty. 1975. The bottleneck effect and genetic variability in populations. *Evolution* 29: 1–10.

Neville, A. C. 1975. *Biology of the Arthropod Cuticle*. Springer-Verlag, Berlin. 448 pp.

New, T. R. 1988. *Associations between Insects and Plants*. New South Wales University Press, Kensington, NSW. 113 pp.

New, T. R. 1995. *An Introduction to Invertebrate Conservation Biology*. Oxford University Press, Oxford. 194 pp.

Nielsen, E. S. and L. A. Mound. 2000. Global diversity of insects: the problems of estimating numbers. Pp. 213–222. *In* P. H. Raven and T. Williams (eds). *Nature and Human Society: the Quest for a Sustainable World*. National Academy Press, Washington, DC.

Noyce, K. V., P. B. Kannowski, and M. R. Riggs. 1997. Black bears as ant-eaters: seasonal associations between bear myrmecophagy and ant ecology in north-central Minnesota. *Canadian Journal of Zoology* 75: 1671–1686.

Oberlander, H. 1985. The imaginal discs. Pp. 151–182. *In* G. A. Kerkut and L. I. Gilbert (eds). *Comprehensive Insect Physiology, Biochemistry and Pharmacology*. Volume 2. Postembryonic Development. Pergamon Press, Oxford.

Odum, E. P., C. D. Connell, and L. B. Davenport. 1962. Population energy flow of three primary consumer components of old-field ecosystems. *Ecology* 43: 88–96.

Orians, G. H. 1966. Food of nestling yellow-headed blackbirds, Cariboo Parklands, British Columbia. *Condor* 68: 321–337.

Ouweneel, W. J. 1975. Developmental genetics of homoeosis. *Advances in Genetics* 18: 179–248.

Park, T. 1948. Experimental studies of interspecific competition. I. Competition between populations of the flour beetles. *Tribolium confusum* Duval and *Tribolium castaneum* Herbst. *Ecological Monographs* 18: 265–308.

Park, T. 1954. Experimental studies of interspecific competition. II. Temperature, humidity, and competition in two species of *Tribolium*. *Physiological Zoology* 27: 177–238.

Patel, N. H. 1994. Developmental evolution: insights from studies of insect segmentation. *Science* 266: 581–590.

Pearson, D. E. and R. M. Callaway. 2006. Biological control agents elevate hantavirus by subsidizing deer mouse populations. *Ecology Letters* 9: 443–450.

Pellmyr, O. 1992. Evolution of insect pollination and Angiosperm diversification. *Trends in Ecology and Evolution* 7: 46–49.

Pennisi, E. 1991. Robots go buggy. Engineers eye biology for better robot designs. *Science News* 140: 361–363.

Petersen, A. 1956. *Fishing with Natural Insects*. Spuhr and Glenn, Columbus, Ohio. 176 pp.

Petersen, H. and M. Luxton. 1982. A comparative analysis of soil fauna populations and their role in decomposition processes. *Oikos* 39: 287–388.

Peterson, I. 1995. Butterfly blue: packaging a butterfly's iridescent sheen. *Science News* 148: 296–297.

Peterson, R. K. D. 1995. Insects, disease, and military history: the Napoleonic campaigns and historical perception. *American Entomologist* 41: 147–160.

Pfadt, R. E. (ed). 1962. *Fundamentals of Applied Entomology*. Macmillan Co., New York. 668 pp.

Phelan, P. L. 1992. Evolution of sex pheromones and the role of asymmetric tracking. Pp. 265–314. *In* B. D. Roitberg and M. B. Isman (eds). *Insect Chemical Ecology: an Evolutionary Approach*. Chapman and Hall, New York.

Philogene, B. J. R. 2006. Insect's strategies for survival. *Global Ecology and Biogeography* 15: 110–111.

Pimentel, D. (ed). 2002. *Biological Invasions: Economic and Environmental Costs of Alien Plant, Animal, and Microbe Species*. CRC Press, Boca Raton, Florida. 369 pp.

Polis, G. A., M. E. Power, and G. R. Huxel (eds). 2004. *Food Webs at the Landscape Level*. University of Chicago Press, Chicago, Illinois. 548 pp.

Powell, J. A. 1992. Interrelationships of yuccas and yucca moths. *Trends in Ecology and Evolution* 7: 10–15.

Price, P. W. 1981. Semiochemicals in evolutionary time. Pp. 251–279. *In* D. A. Nordlund, R. L. Jones, and W. J. Lewis (eds). *Semiochemicals: Their Role in Pest Control*. Wiley, New York.

Price, P. W. 1984. *Insect Ecology*. Second Edition. John Wiley and Sons, New York. 607 pp.

Proctor, M., P. Yeo, and A. Lack. 1996. *The Natural History of Pollination*. Timber Press, Portland, Oregon. 479 pp.

Punnett, R. C. 1915. *Mimicry in Butterflies*. University Press, Cambridge. 188 pp.

Pyle, R., M. Bentzien, and P. Opler. 1981. Insect conservation. *Annual Review of Entomology* 26: 233–258.

Qin, W. and V. K. Walker. 2006. *Tenebrio molitor* antifreeze protein gene identification and regulation. *Gene* 367: 142–149.

Rafaeli-Bernstein, A. and W. Mordue. 1978. The transport of the cardiac glycoside ouabain by the Malpighian tubules of *Zonocerus variegates*. *Physiological Entomology* 3: 59–63.

Raff, M. C. 1994. Cell death genes: *Drosophila* enters the field. *Science* 264: 668–669.

Raloff, J. 1996. Growers bee-moan shortage of pollinators. *Science News* 149: 406–407.

Ramsay, J. A. 1955. The excretion of sodium, potassium and water by the Malpighian tubules of the stick insect *Dixippus morosus* (Orthoptera, Phasmidae). *Journal of Experimental Biology* 32: 200–216.

Reitz, S. R. and J. T. Trumble. 2002. Competitive displacement among insects and arachnids. *Annual Review of Entomology* 47: 435–465.

Remington, C. L. 1968. The population genetics of insect introduction. *Annual Review of Entomology* 13: 415–426.

Resh, V. H. and T. T. Cardé. 2003. *Encyclopedia of Insects*. Academic Press, San Diego, California. 266 pp.

Rice, W. R. 1987. Speciation via habitat specialization. *Evolutionary Ecology* 1: 301–314.

Rice, W. R. and G. W. Salt. 1988. Speciation via disruptive selection on habitat preference: experimental evidence. *American Naturalist* 131: 911–917.

Richardson, J. S. 1993. Limits to productivity in streams: evidence from studies of macroinvertebrates. Pp. 9–15. *In* R. J. Gibson and R. E. Cutting (eds). *Production of Juvenile Atlantic Salmon, Salmo salar, in Natural Waters*. Canada Special Publication of Fisheries and Aquatic Sciences 118. National Research Council of Canada and Department of Fisheries and Oceans, Ottawa.

Ricketts, T. H. 2001. The matrix matters: effective isolation in fragmented landscapes. *American Naturalist* 158: 87–99.

Ring, R. A. 1981. The physiology and biochemistry of cold tolerance in arctic insects. *Journal of Thermal Biology* 6: 219–229.

Ritcher, P. O. 1958. Biology of Scarabaeidae. *Annual Review of Entomology* 3: 311–334.

Ritland, D. B. and L. P. Brower. 1991. The viceroy butterfly is not a Batesian mimic. *Nature* 350: 497–498.

Roberts, D. B. 2006. *Drosophila melanogaster*: the model organism. *Entomologia Experimentalis et Applicata* 121: 93–103.

Roitberg, B. D. and M. Mangel 1988. On the evolutionary ecology of marking pheromones. *Evolutionary Ecology* 2: 289–315.

Ross, H. H. 1973. Evolution and phylogeny. Pp. 171–184. *In* R. F. Smith, T. E. Mittler, and C. N. Smith (eds). *History of Entomology*. Annual Reviews, Palo Alto, California.

Royce, L. A. and P. A. Rossignol. 1990. Honey bee mortality due to tracheal mite parasitism. *Parasitology* 100: 147–151.

Rutowski, R. L. 1981. The function of pheromones. *Journal of Chemical Ecology* 7: 481–484.

Samways, M. J. 1993. Insects in biodiversity conservation: some perspectives and directives. *Biodiversity and Conservation* 2: 258–282.

Samways, M. J. 1994. *Insect Conservation Biology*. Chapman and Hall, London. 358 pp.

Samways, M. J. 1997. Classical biological control and biodiversity conservation: what risks are we prepared to accept? *Biodiversity and Conservation* 6: 1309–1316.

Samways, M. J. 2005. *Insect Diversity Conservation*. Cambridge University Press, New York. 342 pp.

Samways, M. J. 2007a. Insect conservation: a synthetic management approach. *Annual Review of Entomology* 52: 465–487.

Samways, M. J. 2007b. Implementing ecological networks for conserving insects and other biodiversity. Pp. 127–143. *In* A. Stewart, O. Lewis, and T. R. New (eds). *Insect Conservation Biology*. CABI, Wallingford, United Kingdom.

Sander, K., H. O. Gutzeit, and H. Jäckle. 1985. Insect embryogenesis: morphology, physiology, genetical and molecular aspects. Pp. 319–385. *In* G. A. Kerkut and L. I. Gilbert (eds). *Comprehensive Insect Physiology, Biochemistry and Pharmacology*. Volume 1. Embryogenesis and Reproduction. Pergamon Press, Oxford.

Sanderson, A. R. 1960. The cytology of a diploid bisexual spider beetle, ''*Ptinus clavipes*'' Panzer and its triploid gynogenetic form ''mobilis'' Moore. *Proceedings of the Royal Society of Edinburgh. B. Biological Sciences* 67: 333–350.

Sato, S. and A. F. G. Dixon. 2004. Effect of intraguild predation on the survival and development of three species of aphidophagous ladybirds: consequences for invasive species. *Agricultural and Forest Entomology* 6: 21–24.

Schubiger, G. and S. M. Newman, Jr. 1982. Determination in *Drosophila* embryos. *American Zoologist* 22: 47–55.

Schwerdtfeger, F. 1941. Uber die Ursachen des Massenwechsels der Insekten. *Zeitschrift für Angewandte Entomologie* 28: 254–30.

Scott, M. P. and A. J. Weiner. 1984. Structural relationships among genes that control development: sequence homology between the antennapedia, ultrabithorax, and fushi tarazu loci of *Drosophila*. *Proceedings of the National Academy of Sciences USA* 81: 415–419.

Scudder, G. G. E. 1974. Species concepts and speciation. *Canadian Journal of Zoology* 52: 1121–1134.

Scudder, G. G. E. 1976. Are insects perfect? *Bulletin of the Entomological Society of Canada* 8: 2–6.

Scudder, G. G. E. 1983. A review of factors governing the distribution of two closely related corixids in the saline lakes of British Columbia. *Hydrobiologia* 105: 143–154.

Seidel, F. 1924. Die Geschlechtsorgane in der embryonalen Entwicklung von *Pyrrhocoris apterous*. *Zeitschrift für Morphologie und Okologie der Tiere* 1: 429–506.

Shearn, A. 1985. Analysis of transdetermination. Pp. 183–199. *In* G. A. Kerkut and L. I. Gilbert (eds). *Comprehensive Insect Physiology, Biochemistry and Pharmacology*. Volume 2. Postembryonic Development. Pergamon Press, Oxford.

Shepherd, V. E. and C. A. Chapman. 1998. Dung beetles as secondary seed dispersers: impact on seed predation and germination. *Journal of Tropical Ecology* 14: 199–215.

Sherratt, D. J. (ed). 1995. *Mobile Genetic Elements*. IRC Press at Oxford University Press, Oxford. 179 pp.

Shurin, J. B., D. S. Gruner, and H. Hillebrand. 2005. All wet or dried up? Real differences between aquatic and terrestrial food webs. *Proceedings of the Royal Society. B. Biological Sciences* 273: 1–9.

Simon, A., K. Sofka, G. Wiszniewsky, G. Blaser, U. Bode, and G. Fleischhack. 2006. Wound care with antibacterial honey (Medihoney) in pediatric haematology-oncology. *Supportive Care in Cancer* 14: 91–97.

Simpson, G. G. 1953. *The Major Features of Evolution*. Columbia University Press, New York. 434 pp.

Smith, B. P. 1988. Host–parasite interaction and impact of larval water mites on insects. *Annual Review of Entomology* 33: 487–507.

Smith, D. S. 1965. The flight muscles of insects. *Scientific American* 212 (6): 77–88.

Smith, K. G. V. (ed). 1987. Charles Darwin's entomological notes, with an introduction and comments. *Bulletin of the British Museum (Natural History) (Historical Series)* 14: 1–143.

Smol, J. P. 1992. Paleolimnology: an important tool for effective ecosystem management. *Journal of Aquatic Ecosystem Health* 1: 49–58.

Smol, J. P. and J. R. Glew. 1992. Paleolimnology. *Encyclopedia of Earth System Science* 3: 551–564.

Somps, C. and M. Luttges 1985. Dragonfly flight: novel uses of unsteady separate flows. *Science* 228: 1326–1329.

Southwick, E. E. and L. Southwick, Jr. 1992. Estimating the economic value of honey bees (Hymenoptera: Apidae) as agricultural pollinators in the Untied States. *Journal of Economic Entomology* 85: 621–633.

Staines, C. L., Jr., M. J. Rothchild, and R. B. Trumble. 1990. A survey of the Coccinellidae (Coleoptera) associated with nursery stock in Maryland. *Proceedings of the Entomological Society of Washington* 92: 310–313.

Sternberg, L. da S. L., M. C. Pinzon, M. Z. Moreira, P. Moutinho, E. I. Rojas and E. A. Herre. 2007. Plants use macronutrients accumulated in leaf-cutting ant nests. *Proceedings of the Royal Society. B. Biological Sciences* 274: 315–321.

Steffan-Dewenter, I., S. G. Potts, and L. Packer. 2005. Pollinator diversity and crop pollination services are at risk. *Trends in Ecology and Evolution* 20: 651–652.

Stinner, R. E., C. S. Barfield, J. L. Stimac, and L. Dohse. 1983. Dispersal and movement of insect pests. *Annual Review of Entomology* 28: 319–335.

Stork, N. E. 1988. Insect diversity: facts, fiction and speculation. *Biological Journal of the Linnean Society* 35: 321–337.

Stork, N. E. 1993. How many species are there? *Biodiversity and Conservation* 2: 215–232.

Stork, N. E. and K. G. Gaston. 1990. Counting species one by one. *New Scientist* 1729: 43–47.

Strong, D. R., Jr., J. H. Lawton, and T. R. E. Southwood. 1984. *Insects on Plants: Community Patterns and Mechanisms.* Blackwell Scientific, Oxford. 313 pp.

Sturtevant, A. H. 1965. *A History of Genetics.* Harper and Row, New York. 165 pp.

Styrsky, J. D. and M. D. Eubanks. 2007. Ecological consequences of interactions between ants and honeydew-producing insects. *Proceedings of the Royal Society. B. Biological Sciences* 274: 151–164.

Suttle, K. B. and M. S. Hoddle. 2006. Engineering enemy-free space: an invasive pest that kills its predators. *Biological Invasions* 8: 639–649.

Swaine, J. M. 1933. The relation of insect activities to forest development as exemplified in the forests of eastern North America. *Forestry Chronicle* 9 (4): 5–32.

Tamarin, R. H. (ed). 1978. *Population Regulation.* Dowden, Hutchinson and Ross, Stroudsburg, Pennsylvania. 391 pp.

Tamaki, Y. 1985. Sex pheromones. Pp. 145–191. *In* G. A. Kerkut and L. I. Gilbert (eds). *Comprehensive Insect Physiology and Pharmacology.* Volume 9. Behavior. Pergamon Press, Oxford.

Tauber, C. A. and M. J. Tauber. 1981. Insect seasonal cycles: genetics and evolution. *Annual Review of Ecology and Systematics* 12: 281–308.

Taylor, S. W. and A. L. Carroll. 2004. Disturbance, forest age, and mountain pine beetle outbreak dynamics. Pp. 41–51. *In* T. L. Shore, J. E. Brooks, and J. E. Stone (eds). Mountain Pine Beetle Symposium: Challenges and Solutions. October 30–31, 2003. Information Report BC-X-399. Kelowna, British Columbia Natural Resources Canada, Canadian Forest Service, Pacific Forestry Centre, Victoria, British Columbia.

Templeton, A. R. 1981. Mechanisms of speciation – a population genetic approach. *Annual Review of Ecology and Systematics* 12: 23–48.

Thoday, J. M. 1972. Disruptive selection. *Proceedings of the Royal Society of London. Series B. Biological Sciences* 182: 109–143.

Thomas, C. D., E. J. Bodsworth, R. J. Wilson, A. D. Simmons, Z. G. Davies, M. Musche, and L. Conradt. 2001. Ecological and evolutionary processes at expanding range margins. *Nature* 411: 577–581.

Thompson, R. M. 2006. Embedding food webs in ecological landscapes. *Global Ecology and Biogeography* 15: 109.

Throckmorton, L. H. 1977. *Drosophila* systematics and biochemical evolution. *Annual Review of Ecology and Systematics* 8: 235–254.

Townsend, G. F. and E. Crane. 1973. History of agriculture. Pp. 387–406. *In* R. F. Smith, T. E. Mittler, and C. N. Smith (eds). *History of Entomology.* Annual Reviews, Palo Alto, California.

Turnock, W. J., I. L. Wise, and F. O. Matheson. 2003. Abundance of some native coccinellines (Coleoptera: Coccinellidae) before and after the appearance of *Coccinella septempunctata. Canadian Entomologist* 135: 391–404.

Tyshenko, M. G., D. Doucet, P. L. Davies, and V. K. Walker. 1997. The antifreeze potential of the spruce budworm thermal hysteresis protein. *Nature Biotechnology* 15: 887–890.

Tyshenko, M. G. and V. K. Walker. 2004. Hyperactive spruce budworm antifreeze protein expression in transgenic *Drosophila* does not confer cold shock tolerance. *Cryobiology* 49: 28–36.

Valladares, G., A. Salvo, and L. Cagnolo. 2006. Habitat fragmentation effects on trophic processes in insect-plant food webs. *Conservation Biology* 20: 212–217.

Van Driesche, R. G. and T. S. Bellows (eds). 1996. *Biological Control.* Chapman and Hall, New York. 539 pp.

Vansell, G. H. and W. H. Griggs. 1952. Honey bees as agents of pollination. Pp. 88–107. *In Insects. The Yearbook of Agriculture 1952.* United States Department of Agriculture, Washington, DC.

Varley, G. C., G. R. Gradwell, and M. P. Hassell. 1973. *Insect Population Ecology: An Analytical Approach.* Blackwell, Oxford. 212 pp.

Vogel, S. and C. Westerkamp. 1991. Pollination: an integrating factor in biocenoses. Pp. 159–170. *In* A. Seitz and V. Loeschcke (eds). *Species Conservation: A Population-Biological Approach.* Birkhäuser Verlag, Basel.

Wahlberg, N., T. Klemetti, V. Selonen, and I. Hanski. 2002. Metapopulation structure and movements in five species of checkerspot butterflies. *Oecologia* 130: 33–43.

Waldbauer, G. 1996. *Insects Through the Seasons.* Harvard University Press, Cambridge, Massachusetts. 289 pp.

Waldbauer, G. 1998. *The Birder's Bug Book.* Harvard University Press, Cambridge, Massachusetts. 290 pp.

Waldbauer, G. 2003. *What Good Are Bugs? Insects in the Web of Life.* Harvard University Press, Cambridge, Massachusetts. 366 pp.

Wardle, D. A. 2002. *Communities and Ecosystems: Linking the Above- and Below Ground Components.* Princeton University Press, Princeton, New Jersey. 392 pp.

Watanabe, M. E. 1994. Pollination worries rise as honey bees decline. *Science* 265: 1170.

Waterhouse, D. F. 1974. The biological control of dung. *Scientific American* 230 (4): 100–109.

Weber, N. A. 1966. Fungus-growing ants. *Science* 153: 587–604.

Weis-Fogh, T. 1960. A rubberlike protein in insect cuticle. *Journal of Experimental Biology* 37: 889–907.

Weismann, A. 1864. Die nachembryonale Entwicklung der musciden nach Beobachtungen an *Musca vomitoria* und *Sarcophaga cararia. Zeitschrift fur Wissenschaftliche Zoologie* 14: 187–326.

Weiss, R. 1989. New dancer in the hive. *Science News* 136: 282–283.

Weisser, W. W. and E. Siemann. 2004. The various effects of insects on ecosystem functioning. Pp. 3–24. *In* W. W. Weisser and E. Siemann (eds). *Insects and Ecosystem Function.* Springer-Verlag, Berlin Heidelberg.

Wheeler, Q. D. 1990. Insect diversity and cladistic constraints. *Annals of the Entomological Society of America* 83: 1031–1047.

Wheeler, W. C., P. Cartwright, and C. Y. Hayashi. 1993. Arthropod phylogeny: a combined approach. *Cladistics* 9: 1–39.

Wheeler, W. C., M. Whiting, Q. D. Wheeler, and J. M. Carpenter. 2001. The phylogeny of the extant hexapod orders. *Cladistics* 17: 113–169.

White, D., Jr., K. C. Kendall, and H. D. Picton. 1998. Grizzly bear feeding activity at alpine army cutworm moth aggregation sites in northwest Montana. *Canadian Journal of Zoology* 76: 221–227.

White, K., M. E. Grether, J. M. Abrams, L. Young, K. Farrell, and H. Steller. 1994. Genetic control of programmed cell death in *Drosophila. Science* 264: 677–683.

White, M. J. D. 1973. *Animal Cytology and Evolution.* Third Edition. Cambridge University Press Cambridge. 961 pp.

Whittaker, R. H. and P. P. Feeny. 1971. Allelochemics: chemical interaction between species. *Science* 171: 757–770.

Wiebes, J. T. 1979. Co-evolution of figs and their pollinators. *Annual Review of Ecology and Systematics* 10: 1–12.

Wiegert, R. G. and F. L. Evans. 1967. Investigations of secondary productivity in grasslands. Pp. 499–518. *In* K. Petrusewicz (ed). *Secondary Productivity in Terrestrial Ecosystems.* Institute of Ecology, Polish Academy of Science, Warsaw.

Wigglesworth, V. B. 1964. *The Life of Insects.* Weidenfeld and Nicholson, London. 360 pp.

Wigglesworth, V. B. 1976. *Insects and the Life of Man. Collected Essays in Pure Science and Applied Biology.* Chapman and Hall, London. 217 pp.

Wigglesworth, V. B. 1985. Historical perspective. Pp. 1–24. *In* G. A. Kerkut and L. I. Gilbert (eds). *Comprehensive Insect Physiology, Biochemistry and Pharmacology.* Volume 7. Endocrinology I. Pergamon Press, Oxford.

Wiley, E. O. 1981. *Phylogenetics: The Theory and Practice of Phylogenetic Systematics.* John Wiley and Sons, New York. 439 pp.

Williams, D. W. and A. M. Liebhold. 1997. Latitudinal shifts in spruce budworm (Lepidoptera: Tortricidae) outbreaks and spruce–fir forest distributions with climate change. *Acta Phytopathologica et Entomologica Hungarica* 32: 205–215.

Williams, D. W. and A. M. Liebhold, 2002. Climate change and the outbreak ranges of two North American bark beetles. *Agricultural and Forest Entomology* 4: 87–99.

Wilson, E. O. 1971. *The Insect Societies.* Belknap Press, Cambridge, Massachusetts. 548 pp.

Wilson, E. O. 1987. The little things that run the world (the importance and conservation of invertebrates). *Conservation Biology* 1: 344–346.

Wilson, E. O. 1992. *The Diversity of Life.* W. W. Norton, New York. 424 pp.

Wilson, R. J., Z. G. Davies, and C. D. Thomas. 2007. Insects and climate change: processes, patterns and implications for conservation. Pp. 245–279. *In* A. J. A. Stewart, T. R. New, and O. T. Lewis (eds). *Insect Conservation Biology.* CAB International, Wallingford, United Kingdom.

Winston, M. L. 1992. Semiochemicals and insect sociality. Pp. 315–333. *In* B. D. Roitberg and M. D. Isman (eds). *Insect Chemical Ecology: An Evolutionary Approach.* Chapman and Hall, London.

Wootton, R. 2000. From insects to microvehicles. *Nature* 403: 144–145.

Wootton, R. J. and C. P. Ellington. 1991. Biomechanics and the origin of insect flight. Pp. 99–112. *In* J. M. V. Rayner and R. J. Wootton (eds). *Biomechanics in Evolution.* Cambridge University Press, Cambridge.

Wootton, R. J., J. Kukalova-Peck, D. J. S. Newman, and J. Muzon. 1998. Smart engineering in the Mid-Carboniferous: how well could Paleozoic dragonflies fly? *Science* 282: 749–751.

Zinsser, H. 1934. *Rats, Lice and History.* Routledge, London. 301 pp.

Part I

Insect Biodiversity: Regional Examples

The Nearctic Region, occupying an area of nearly 20 million km^2 (Fig. 3.1), includes many different environments inhabited by insects, ranging from arctic tundra to lush forests. Published and unpublished information and extrapolations and corroborations by specialists suggest that about 144,000 species of insects occur in the region, but only about 65% of the species have been formally described in the scientific literature. Only about 6% of the species are estimated to have been described in the immature stages. The state of knowledge varies widely among different groups; families that contain many species of economic importance are better known. High Arctic regions are least rich; southern areas (especially northern Mexico, Arizona, Florida, and Texas) contain the greatest percentage of the fauna. The Nearctic Region

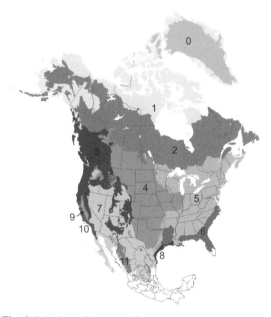

Fig. 3.1 Ecological biomes of the Nearctic Region, adapted from data available from the World Wildlife Fund at www.worldwildlife.org/science/ecoregions/item1847.html, as described by Olson et al. (2001). 0 – Permanent Ice; 1 – Tundra; 2 – Boreal Forest; 3 – Moist Temperate Forest; 4 – Temperate Grasslands; 5 – Temperate Broad-leaf and Mixed Forest; 6 – Temperate Coniferous Forest (Southeast); 7 – Xeric Shrublands and Deserts; 8 – Subtropical Grasslands (Gulf Coastal Grasslands); 9 – Temperate Grasslands (Central California); 10 – Mediterranean (California Chaparral and Woodlands); 11 – Warm Temperate Evergreen (Oak–Pine) Forest.

has lower insect biodiversity relative to other biogeographical realms and is most similar faunistically to the Palearctic Region, with many overlapping taxa at the lower taxonomic levels (genera and species). The fauna overlaps significantly with that of the Neotropical Region at the generic level. Basic taxonomic research on Nearctic insects is more advanced than for most other regions, but is still in the discovery and development stages for the vast majority of insect taxa. Most Nearctic insect species are known only from a basic description and antiquated distributional records or are completely unknown to science. Only a small percentage of Nearctic insect taxa have been given a modern and thorough taxonomic treatment to describe and diagnose them properly, map out their distributions through space and time, and assess their evolutionary relationships with other taxa.

This chapter examines the biodiversity of insects in the Nearctic Region, based on extant species, and the state of knowledge about them. Basic information on the numbers of species in this region results from educated bookkeeping exercises; these estimated numbers depend on fewer assumptions than are necessary to make similar estimates in tropical regions of much greater biodiversity, so that considerable efforts are not made here to refine the estimates in detail. Rather, some attempt is made to synthesize and interpret the general patterns that emerge.

The Nearctic Region comprises the northern part of the New World. The core of the fauna lives in North America, but Nearctic faunal elements occupy the mountains of Mexico and Central America, at least as far south as the montane pine–oak forests of Honduras and northern Nicaragua (Halffter 1974, 1987, 2003). The nucleus of northern, more-or-less cold-adapted forms in North America is supplemented by many forms of subtropical affinity, especially in the southern USA. Additional species occur in the Mexican highlands, and the inclusion of these taxa would increase the numbers of species reported from the 'Nearctic Region'. However, detailed faunal data are adequate only for North America north of Mexico. Therefore, most of the information presented below comes from this somewhat smaller area.

The Nearctic fauna is closely linked with that of the adjacent Palearctic fauna, from which it has received many components, especially in the north. The amount of overlap varies by taxon; a recent study of Holarctic spiders found that of 13,800 species, about 1% occurred naturally in both the Palearctic and Nearctic

Regions (Marusik and Koponen 2005). On the other hand, no known species of scarab beetles have a natural distribution across the two realms (Smith 2003, Löbl and Smetana 2006). In the South, the Nearctic Region has received elements from the adjacent Neotropical fauna, but it also has contributed significantly to the Neotropical fauna.

North America occupies an area of about 19 million km^2. The continent is so large that conditions change greatly from south to north and from west to east, spanning about 58 degrees of latitude and more than 105 degrees of longitude. Temperatures fall toward the north, and summer temperatures are higher and winter temperatures lower in the middle of the continent. Consequently, North America includes environments inhabited by insects, ranging in mean July temperature from below 5°C (Canadian High Arctic) to 33°C (Phoenix, Arizona) and in mean January temperature from below −30°C (High Arctic) to about 20°C (southern Florida). Regions vary in mean annual rainfall from less than 100 or 200 mm (polar deserts; southwestern deserts, e.g., Reno, Nevada) to 2,000 – 3,000 mm on the northwest coast (e.g., Yakutat, Alaska: 3,348 mm) and northern Cordillera (e.g., Prince Rupert: 2,399 mm) (Bryson and Hare 1974). Correspondingly, habitats range from hot desert and cold arctic fellfield to lush coastal temperate rainforest. The many different vegetation formations in this area have been classified or categorized in several different ways (e.g., Taktajan 1986).

The current Nearctic fauna also is influenced by history (Matthews 1979). Northern North America was glaciated in Pleistocene times, and the faunas of large areas in the northern part of the continent were exterminated by ice sheets 1–2 km thick, which receded only 10,000–15,000 years ago. More recently, invasive species have become a significant component of the fauna (Lindroth 1957, Turnbull 1979, Pimentel 2002, Foottit et al. 2006), with some effects on native species. Reductions also have occurred, some attributable to long-term habitat changes and some attributable to human activities.

Nearctic biodiversity is low relative to other biogeographic realms. For example, 2,586 Nearctic scarab beetles are known (Smith 2003). This total is similar to that for Europe (2,318), but is approximately four times less than the total for the entire Palearctic Region (Löbl and Smetana 2006). The greater species richness in the Palearctic Region is due mainly to the much larger land area with a greater variety of ecozones. Nevertheless, a tremendous amount of insect biodiversity exists in the Nearctic Region, including endemic families such as the Pleocomidae and Diphyllostomatidae (Coleoptera), with low overlap, at the species and genus levels, with insects in other realms.

INFLUENCE OF INSECT BIODIVERSITY ON SOCIETY IN THE NEARCTIC REGION

The importance of entomology and the study of insect biodiversity emerged in North America during the 1800s. One of the main driving forces behind this emergence was the unprecedented expansion of agriculture in the eastern half of the continent. Many native and invasive pest insects were destroying crops, which led to great thirst for knowledge about these species. Naturalists and farmers realized that knowledge of natural history, life cycles, and biodiversity of insects was crucial to controlling pests. These challenges are still faced today, with globalization and climate change bringing new pest species into the Nearctic with ever-increasing frequency. A major problem faced by early entomologists in North America was that all of the entomological expertise, collections, and libraries were in Europe and practically inaccessible to those in the New World. Early pioneers in North American entomology, such as Thomas Say (1787–1834), began publishing and establishing collections in North America (Stroud 1992). Entomological research published in North America began to proliferate in later decades as natural history societies and collections sprang up across the USA and Canada (Spencer 1964, Barnes 1985). In the twentieth century, major institutions such as the Smithsonian Institution, California Academy of Sciences, Canadian National Collection of Insects, American Museum of Natural History, Harvard University Museum of Comparative Zoology, and others came to prominence for their leading-edge collections and research programs.

Even though North America hosts a large proportion of the most eminent and active insect systematics programs and collections, major challenges still exist to describing the insect biodiversity of the Nearctic Region. A sobering number of Nearctic insect species are still unknown to science – an estimated 50,000, or 35% of Nearctic insect species are undescribed (Table 3.1). The underfunding of taxonomy and biodiversity research has been well documented and has affected the ability of the scientific community to get the job done – even

Table 3.1 Census of Nearctic insects. Many estimates are based on tenuous evidence and provide indications only; however, estimates are generally conservative. Numbers for Coleoptera are from Arnett and Thomas (2000) and Arnett et al. (2002); those for Diptera are from Thompson (2008). Other numbers are based on Kosztarab and Schaefer (1990a), with adjustments based on additional research and expert comment (see text for further explanation).

Order	Number of Species Known	Estimated Number of Species Unknown	Estimated Total Number of Species	Number of Species with Immatures Described	% of Total Known Adults	% of Total Known Immatures
Microcoryphia	35	23	58	7	60	12
Thysanura	30	20	50	6	60	12
Ephemeroptera	555	59	614	461	90	75
Odonata	415	2	417	355	>99	85
Plecoptera	578	0	578	232	100	40
Blattodea (Dictyoptera)	66	10	76	8	87	11
Isoptera (Dictyoptera)	41	3	44	0	93	0
Mantodea (Dictyoptera)	20	2	22	0	91	0
Grylloblattodea	13	0	13	2	100	15
Dermaptera	23	0	23	0	100	0
Orthoptera	1804	513	2317	360	78	16
Phasmatodea	31	5	36	0	86	0
Embiidina	13	1	14	0	93	0
Zoraptera	3	5	8	0	37	0
Hemiptera	10,804	5105	15,909	1470	68	9
Thysanoptera	700	233	933	35	75	4
Psocoptera	257	110	367	10	70	4
Anoplura (Phthiraptera)	76	68	144	35	53	24
Mallophaga (Phthiraptera)	700	655	1355	1	52	<1
Strepsiptera	109	20	129	6	84	5
Megaloptera	43	1	44	30	98	68
Raphidioptera	21	0	21	2	100	10
Neuroptera	310	85	395	64	78	16
Coleoptera	25,160	2516	27,676	3000	91	11
Mecoptera	75	18	93	13	81	14
Siphonaptera	258	20	278	55	93	20
Diptera	21,356	19,562	40,918	833	52	2
Trichoptera	1340	200	1540	370	87	24
Lepidoptera	11,300	2700	14,000	1113	81	8
Hymenoptera	17,429	18,571	36,000	529	48	1
TOTAL INSECTA	93,565	50,507	144,072	8997	65	6

in the Nearctic Region, which has arguably been the hotspot for taxonomy and biodiversity research since World War II. Other factors have compounded efforts to describe the insect fauna of the Nearctic Region. Many Nearctic species were described in the Victorian Era, with type material deposited in European museums.

Modern taxonomy often is inhibited by the difficult task of tracking down old type material. Poor record keeping, lost or destroyed collections, poor specimen labeling, specimen exchange without documentation, theft, and chronic underfunding of natural history collections are all impediments.

The limited knowledge of the taxonomy of many groups causes problems even with basic characterizations and estimates of biodiversity. Species diversity of many poorly known groups is far greater than numbers found in the scientific literature because large proportions of the fauna are unknown to science. Newton et al. (2000), for example, noted that the number of species of Staphylinidae beetles in North America increased by more than 700 species between 1963 and 2000, an increase of about 17% in 37 years. Likewise, Grissell (1999) found that new species of Nearctic Hymenoptera were being described at a rate of at least 1,500 per decade toward the last half of the twentieth century. This rate peaked in the 1960s when more than 2,500 new species of Hymenoptera were described from the Nearctic. Limited taxonomic knowledge also has led to the opposite problem – overestimations of species diversity have occurred in groups where careless or sloppy taxonomic work has resulted in species being described multiple times. Perhaps the most stunning examples of this were due to Thomas Lincoln Casey (1831–1896). Subsequent workers discovered that more than 90% of the 'species' of carabid beetles Casey described from North America actually were previously described (Lindroth 1969a, b). Casey described *Tomarus gibbosus* (De Geer), a common and widespread species of scarab beetle, 23 times (Saylor 1946)! The limited knowledge of the taxonomy of many groups can cause both overestimations and underestimations in species diversity, further undermining our abilities to have even basic levels of understanding of biodiversity for the most diverse groups of organisms. Because taxonomy and biodiversity research are the foundation of other biological sciences, the paucity of basic knowledge of Nearctic insect biodiversity has a profound effect on other endeavors.

INSECT CONSERVATION

Conservation of natural habitats and protection of species at risk are strong desires of people across North America. However, the use of insects in conservation studies and the recognition of insect species at risk have barely begun. Bossart and Carlton (2002) and Spector (2006) presented summaries indicating that insects are better than any other group of organisms as ecological indicators or in monitoring biotic changes in the environment. However, they pointed out that only a tiny fraction of insect diversity from only a few select taxa have been used for this purpose (notably dragonflies,

butterflies, and selected groups of moths and beetles). Population studies of British Lepidoptera have demonstrated that insects are effective at detecting biodiversity declines (Thomas 2005, Conrad et al. 2006). Insect conservation studies, however, have a long way to go before they are comparable with vertebrate and plant conservation research.

More than 1,000 insect species in the USA are species of greatest concern among those designated as 'at risk' (Bossart and Carlton 2002). This list consists of insect species of highest priority for conservation, including 160 species that are legally protected by state statutes in the USA. Some endangered insect species in the Nearctic have been studied extensively, such as the American burying beetle (Bedick et al. 1999, Sikes and Raithel 2002). However, the poor state of scientific knowledge about most Nearctic insects indicates that the vast majority of at-risk insect species have not been detected.

SPECIES DIVERSITY AND THE STATE OF KNOWLEDGE

Assembling the Data

Several previous attempts to inventory the numbers of species in the North American fauna have been made (e.g., Danks 1979, Kosztarab and Schaefer 1990a, Poole and Gentili 1996a, 1996b, 1996c, 1997). Although these efforts are of varying quality and accuracy, they provide estimates that can be examined and compared in detail to evaluate their reliability and the reliability of other estimates. Information on biodiversity requires knowledge of the number of species reported in each group, and an estimate of what remains to be discovered. Depending on the group, different procedures might be required to obtain this information, and figures that are exactly comparable with other groups might not be possible to obtain. All estimates are based on the knowledge accumulated by individual specialists.

Efforts to enumerate taxa in small rather than large blocks, typically at or below the level of family, produce more accurate totals. For example, an estimate of the Canadian insect fauna derived from careful family-by-family analysis (Danks 1979) was about half that obtained from earlier general judgments made chiefly at the ordinal level (Downes 1974). The increased precision resulted because numbers had to be derived in a more detailed way and because in large orders, each specialist is familiar only with certain individual families. Nevertheless, data for families that lack a specialist will be less precise, so that the total numbers of species known and estimated are always approximate.

Four components are involved in developing such estimates. Published information, albeit incomplete, is available for most groups. In the best-known groups, a recent catalog might list all known species and clarify the status of available names. In other groups, comparison between an older catalog and available recent generic revisions shows how the number of described taxa has increased. Unpublished information is available to most specialists in the form of distinct species recognized in collections but known to be undescribed, and from the specialist's own or others' opinions, as yet unpublished.

Published and unpublished information allows the number of named species that occur in a region to be tabulated. Extrapolation from the tabulations then provides an estimate of the total number of species expected. A basic method of extrapolation is to apply across-the-group increases from older names revealed by selected recent revisions. Provided the sample taxa cover a range of ecological and taxonomic types, and so represent the diversity of the group, reliable estimates can be expected (Danks and Foottit 1989, Danks 1993a). In some groups, estimates can be based on the number of host species available (e.g., Kim et al. 1990 for ectoparasites). Extrapolations also can be based on analysis of the collecting effort rather than of the taxa. Thus, the biodiversity of areas from which adequate collections have not been made can be estimated from preliminary information and reflected in global estimates. For example, Schaefer (1990) noted that relatively little collecting effort for Heteroptera has been made in New Mexico. The diversity of small, obscure ground-dwelling Hymenoptera, such as proctotrupids, is large relative to the numbers known because these forms are seldom collected unless appropriate ground-level traps are used (Masner and Huggert 1989).

More complex extrapolations are required when knowledge is markedly deficient. They can be based on assumptions about trophic status, host specificity of herbivores, and information on plant diversity (e.g., estimates of tropical biodiversity by Erwin 1982, 1991 and Stork 1988). Estimates made by extrapolation sometimes can be corroborated through parallel evidence. For example, the ratio of insect species between Britain, a temperate area where nearly all the species are known, and Canada, another temperate area, is between 1:2 and 1:3 in well-known groups (J. R. Vockeroth cited by McAlpine 1979). Similar ratios should be obtained in groups where the Canadian fauna is less well known. Deriving estimates, however, is an inexact process. Schaefer (1990) pointed out that one cannot quantify the unknown, but only guess at it intelligently. In estimating Nearctic insect diversity, knowledge is deficient in many taxa, information is available in different formats for different groups (e.g., recent catalogs versus estimates made from scattered collections), specialists are not available for some groups, and different people make different judgments in developing their estimates. Some estimates for larger groups are given to the nearest ten species, or to the nearest 5% or 10%, because they are known to be approximate; however, an estimate such as 86, which has been derived from the belief that 30% of species are still unknown in a group with 200 known species, thereby gains an apparent but potentially misleading precision. Estimates tend to be conservative in groups for which large numbers of species are expected to be unknown (Danks 1979) because individuals are more comfortable with estimates such as one species known: three species unknown than with estimates such as 1:10. In a few instances, some individuals specializing in different but related families might use different concepts of what degree of structural difference in the specimens they study should be formalized into species. Finally, the most diverse groups are likely to produce the least reliable estimates, disproportionately affecting the reliability of any global estimates. For example, estimates for the inadequately known and diverse parasitic Hymenoptera are necessarily approximate. Totals for the Hymenoptera are dominated by this group of many thousands of species. In turn, how these estimates are made greatly influences the total for Insecta. The family Cicadellidae is the largest of the Auchenorryncha (Kosztarab et al. 1990). Consequently, a 10% difference in the estimated fraction of unknown species in this family (as appears between the text and table in this source) changes the estimated number of unknown species by 500.

The difficulties of enumerating the numbers of immature stages and different sexes 'known' are even greater because the quality of descriptions varies so widely, and much of the information is scattered in a variety of nontaxonomic papers. In many groups, therefore, such information is likely to be inaccurate or unavailable. Given these limitations, the estimates that follow can be characterized as credible, but inexact. More effort, however, should be spent actually studying the species in particular groups than in trying to refine such estimates of their numbers.

Synopsis of Biodiversity

The numbers of species of insects estimated to occur in the Nearctic Region are shown in Table 3.1. Orders that are combined by some authors (e.g., Mallophaga and Anoplura as Phthiraptera) have been kept separate in the tables to assist the reader in viewing the data, and not to make any statement with respect to the higher classification of the arthropods. The data are drawn chiefly from the volume edited by Kosztarab and Schaefer (1990b), with many additions and corrections. An effort has been made to complete or extrapolate the tabulations, even approximately, because gaps in the data produce misleading totals. For example, three groups, the first with 100 known species and an estimated 10 unknown species, the second with 50 known species and an estimated 3 unknown species, and the third with 200 species and no estimate available for unknown species, should not be totaled as 350 known species and 13 estimated unknown species. Therefore, data missing from earlier tabulations, or incorrectly reported as 0, have been obtained, and specific figures have been substituted for a range of estimates. The additional data were acquired by asking specialists for further information and by extrapolating from estimates available for parts of the Nearctic fauna, notably for Canada in the work of Danks (1979) and in certain monographs. Information in Table 3.1, thus, relies on estimates made by different people, partly using different criteria, so that the figures are approximate, even for adults. For most groups, information on the immatures is even more approximate than for the adults, because if some account of the larva of a species is available, it may or may not have been judged as 'described', whether or not this allows it to be distinguished from the larvae of other known species in the genus. In summary, the reliability of the data depends on how well the groups are known. The state of knowledge varies widely, as discussed in the following section, confirming that the estimates are approximate.

State of Knowledge

Requirements for the study of Nearctic arthropods depend on the state of knowledge about the taxa, as well as on the number of included species. Differences in knowledge about the groups in Table 3.1 stem from the classification, morphology, ecology, and economic importance of the species included in each group. Thus, for adults, diverse groups in which taxonomic work is difficult because many similar or closely related species have to be distinguished are least known. For example, for the Ichneumonidae, only 42% of the 8,000 species are estimated to be known, and for the Chalcidoidea, 44% of 5,000 species are estimated known (Masner 1990). Groups with few species generally are better known: Dermaptera, 100% of 23 species; Phasmatodea, 86% of 36 species (Table 3.1). Groups of large, colorful insects are well known. Nearly all the 765 species of butterflies (Covell 1990) and 417 species of Odonata are known (Table 3.1). Some families of beetles also are well known (Arnett and Thomas 2000, Arnett et al. 2002). Conversely, groups containing chiefly small, dull-colored species are less well studied, such as several families of small beetles (Arnett et al. 2002). Species that are easy to find and capture are better known, such as some families of large moths (Munroe 1979), whereas inconspicuous species in habitats that are more difficult to sample, such as the soil, tend to be less well known, as in many groups of beetles, especially Staphylinidae and Curculionidae. Groups containing many species of economic importance, especially those of medical–veterinary significance, tend to be well known; for example, 94% of the 176 estimated species of the Culicidae are known (Thompson 1990).

Real taxonomic knowledge might exist to different degrees, even when many species names are available. Additional difficulties stem from studies focusing on regional or Nearctic faunas that have tended to generate synonyms among the many Holarctic species (cf. Benson 1962, Larson and Nilsson 1985). Thompson (1990) recognized six levels of taxonomic knowledge: (1) species descriptions only, (2) unreliable keys to few species, (3) reliable keys to some species, (4) good keys to most species, (5) complete revisions, and finally (6) monographs that also include immature stages. When groups first become reasonably well known and most species in them have been discovered, the number of species listed tends to decrease as duplicate names are synonymized by additional taxonomic study. In the Nearctic Region, only 2 of the 121 families of Diptera, representing only 190 of the estimated total of nearly 42,000 species in the order, were judged by Thompson (1990) to have reached the final level of taxonomic knowledge.

Taxonomic information also is required to identify stages other than those on which descriptions and identifications primarily are based. Knowledge of different sexes varies chiefly according to the biology of the group – some species are parthenogenetic and some have sexes with different habits – and on the taxonomic procedures used, which in many insect groups emphasizes the genitalia of males. Conversely, classification in some groups of large and colorful species is not based on a full range of characters, such as genitalia. The longevity of adults and the typical broad similarity of the sexes in the Coleoptera enhance the knowledge of both sexes; both sexes are known in many of the described species. Arrhenotoky and unusual methods of sex determination and unusual sex ratios occur frequently in Hymenoptera, and both sexes are known in only one-third of the described species (Kosztarab and Schaefer 1990b, Table 3.1), although the order is inadequately known for other reasons.

Taxonomic analysis of immature stages requires detailed sampling, rearing, and other specialized study. Overall, only 6% of Nearctic insects are known in the immature stages, but as with the adults, the state of knowledge varies widely from group to group. In several groups, no larvae are known (Table 3.1). Only in a few well-studied groups with relatively few species (e.g., Odonata and Megaloptera) are most of the larvae known. Even groups well known in the adult stage can be little known taxonomically as larvae. For example, in the Siphonaptera, 93% of adults are known but only 20% of larvae, and in the Mantodea, 91% of adults versus 0% of larvae are known. For the Coleoptera, which are well known as a whole (90%), only 11% of the larvae are estimated to have been described. Identification of larvae usually has been based on the final instar. This instar is most often encountered because it is the largest and usually lasts the longest, thereby eating the most. Earlier instars, except for the first one in many groups, tend to be similar to the final instar. All of the earlier instars, therefore, seldom require separate descriptions. In some exopterygotes, too, the developing characters of the adults can be used to identify some of the species (e.g., Nickle et al. 1987).

In summary, taxonomic knowledge of adults and larvae of most Nearctic terrestrial insects is incomplete. Knowledge about the species is also deficient in the ecological arena. Although many general ecological elements can be predicted from taxonomic affinities (e.g., presence of phytophagy and overwintering stage), detailed information has been collected for relatively few species. Data provided by Kosztarab and Schaefer (1990a) suggest that some biological information is available for 1–20% of species in most groups. In the four largest orders, biological information covers only some 5–10% of the species. The biology of the species is well known in only a few groups, chiefly selected families of economic importance. Typically, the biology of many species has not been studied even in groups with good taxonomic knowledge of both adults and larvae. Even simple records of habitat association are deficient.

The assessment of knowledge of Nearctic insect biodiversity, from a taxonomic viewpoint, is confirmed from other perspectives. Deficiencies in collecting are suggested by the fact that only some 22% of the estimated 11,000 species of aquatic insects reported from Canada are known from marshes, which would be expected to contain many widely distributed aquatic species (Danks and Rosenberg 1987). Deficiencies in taxonomic knowledge are confirmed by the observation that less than half of the species of insects collected from two species of bracket fungi in one small area could be named to genus or even family (Pielou and Verma 1968, Matthewman and Pielou 1971).

VARIATIONS IN BIODIVERSITY

Nearctic insect diversity is considerable, but the scale of the task of collecting and understanding the species depends not only on their raw numbers, but also on how the biodiversity varies from place to place. The real requirements of personnel and resources for characterizing the fauna depend on the differences among regions, the complexity of communities, the divergence among populations, and other measures of variation in diversity in North America. Some broad spatial patterns in Nearctic biodiversity are indicated briefly in this section.

Regional Variation

The diversity and composition of the fauna in the Nearctic Region change in particular ways from north to south and from east to west. The changes are complex and depend on ecogeoclimatic features and history; nonetheless, several major trends can be determined. From south to north within the region, a general reduction occurs in the number of species. For example, more than 93,000 insect species have been named from North America (Table 3.1), about 30,000 from Canada (Danks 1979), about 12,000 from the North American boreal zone (the northern coniferous forest) (Danks and Foottit 1989), and only about 1,650 from the North American Arctic, north of the treeline (Danks 1981, 1990). The High Arctic, the Queen Elizabeth Islands of the Canadian Arctic Archipelago, has only about 242 reported insect species (Danks 1981, 1990).

Although the latitudinal ranges of many species coincide with major boundaries within the continent, such as the treeline, communities of species do not simply follow major vegetation formations. Rather, at least in northern North America, individual species appear to react with some degree of independence from one another to changes of climate, growing season, vegetation, and other factors, so that they drop out gradually from south to north even within a single vegetation zone such as the boreal (e.g., Danks and Foottit 1989, Danks 1993a). Moreover, the latitudinal trends are disrupted because habitat changes are not uniform everywhere. For example, substantial numbers of species in the arctic or subarctic live at higher elevations on southern mountains, where ecological conditions are similar (e.g., Danks 1981).

The composition, as well as the diversity of faunas, also changes with latitude. For example (Danks 1988b, 1993a), the Coleoptera are well represented at lower latitudes (about one-third of the insect fauna of North America), but they comprise only 3% of the High Arctic fauna. Conversely, Diptera are best represented at higher latitudes, comprising 61% of the High Arctic insect fauna (especially Chironomidae and Muscidae of the genus *Spilogona*), compared with only 19% of the North American fauna. Several taxa of southern or subtropical affinity, including some families of the Coleoptera (e.g., Hydroscaphidae), Lepidoptera (e.g., Dalceridae and Manidiidae), and even orders such as the Zoraptera, occur only in the southernmost parts of the continent.

From east to west, the trend is toward increasing biodiversity. In general, compared with eastern North America, western parts of the continent tend to be milder for their latitude, and the complexity of habitats is greater because of Cordilleran topography and other ecological and historical factors. Thus, species of the order Grylloblattodea occur only in the western mountains. Most species of tenebrionids and other arid-adapted forms are confined to the western deserts.

Many groups, at the specific or generic level, are more strongly represented in the West. Regional variations in the North American fauna, therefore, reflect the integration of south–north, east–west, and habitat differences. These variations can be documented through the occurrence of species in a particular province, state, or zone.

In Canada, data have been assembled for hundreds of species in a sample of higher taxa of various taxonomic and ecological types, presumed to be representative of Canadian insect diversity (Danks 1993a). These data show that biodiversity is greatest in the west, where 61% of the Canadian fauna in the sample groups occurs in western coastal and Cordilleran zones, notably in British Columbia, where 57% of the Canadian fauna occurs. Moreover, 37% of Canadian species in the sample groups are confined to the western half of the country (Manitoba westward), compared with only 28% confined to the eastern half (Manitoba eastward). Considerable diversity occurs in the eastern half, however, in the deciduous plus transitional (mixed deciduous-coniferous) forest (57% of the Canadian fauna), which has its greatest area in Ontario. These figures suggest a moderate degree of regional differentiation, but the more striking conclusion is that, in these northern life zones, much of the fauna is widely distributed: 34% of Canadian species occur in both the east and west, and more than half of the country's species occur in a single large province such as Ontario or British Columbia.

A preliminary analysis of more than 3,000 species in a number of sample groups from North America as a whole (adding state-by-state information from USA) shows that most of the species are southern and that more of them have a restricted distribution in the south than do the species of the Canadian fauna alone (Danks 1994, unpublished data). Biodiversity is greatest in the southwest (Arizona and Texas) and also is high in California and the southeast (Florida). The prevalence of species in southern and western habitats means that nearly three-quarters of the species occur in the southernmost part of the continent (Arizona to Oklahoma, Louisiana to Georgia and Florida) and that more of the fauna occurs only in its western half than in the eastern half alone.

Regional variations in the Nearctic fauna also can be documented through patterns in the ranges of individual species. Such a treatment is beyond the scope of this chapter, but sample information is available in the works of Danks (1981), Downes (1981), Lafontaine

(1982), Ball (1985), Downes and Kavanaugh (1988), and Danks and Foottit (1989).

Habitats

Biodiversity also varies from place to place on a smaller scale, from terrestrial to aquatic habitats, from one habitat to another, and even from one microhabitat to another. These small-scale differences are especially significant because they make true assessments of biodiversity difficult. Specialized collecting techniques and expert knowledge of any given group are required to obtain samples of the full range of species at one place. Although some species are generalists, occurring in many places (e.g., most species from shallow, lentic habitats are widely distributed; Danks and Rosenberg 1987), many other species are specialists restricted to particular conditions. Significant fractions of the faunas of discrete and more or less isolated habitats such as springs (Danks and Williams 1991) are confined to those habitats. Most monophagous herbivores can be captured only on their specific host plants. If they feed on specific seasonal plant parts, the herbivores can be captured only during a limited time period.

Habitat differences have a longer-term temporal component. For example, disturbance that increases the local heterogeneity of habitats contributes to local biodiversity. Thus, fire and spring flooding are important in maintaining the variety of habitats in the boreal zone (Danks and Foottit 1989). The total biodiversity in a given region, therefore, is difficult to sample effectively because systems are dynamic.

CONCLUSIONS AND NEEDS

The Nearctic fauna includes more than 144,000 species of terrestrial insects (Table 3.1). Little more than half of these species have been described (some inadequately), few are known in the larval stage, and only a few hundred people are available to study their systematics. Support for staff and for the maintenance of collections and of reference materials is small relative to the diversity. These deficiencies vary with the group and region, however, and knowledge of the Nearctic fauna has many positive features. Considerable excellent literature, albeit incomplete, exists, and millions of specimens are preserved for study in collections.

Two points are worth particular emphasis. First, real needs related to biodiversity depend not just on the raw numbers of species in the fauna, but also on variations in occurrence, distribution, and ecology across the many different areas, habitats, and subhabitats of the continent (e.g., Danks 1994). These variations compound the difficulties of assessing an already enormous diversity. Second, although basic taxonomic knowledge of each species is the major building block for subsequent information (and hence tabulating the numbers of species known and unknown is especially informative), the full value of that knowledge comes from its use in providing a logical framework for communicating information, and for synthesis, interpretation, and application of information in an ecological context (Danks 1988a, Danks and Ball 1993, Goldstein 2004). In particular, systematics work is widely relevant to understanding ecosystems through the taxonomic and ecological database it supports. Biological knowledge grows by the steady, long-term accumulation of information about species, through systematics research. It is accessible only through species names, which allow identification, and through classifications, which allow synthesis.

Major tasks to ensure adequate future knowledge of the Nearctic arthropod fauna likewise fall into two classes. The first seeks more systematic knowledge. For example, additional collecting is required, especially in relatively unexplored areas and in regions of particular interest, such as the arctic (Danks 1992), and areas with unusually high biodiversity (Masner 1990). Refinement of collecting techniques for this purpose overcomes the complexities of habitat and insect behavior. Basic faunistic work is lacking for most insects groups even in the most heavily populated areas of the Nearctic. Sikes (2003), for example, recently reported a staggering 656 new state species records and 13 new state family records for Coleoptera in Rhode Island. Similarly, more than 100 new provincial records for Coleoptera from the Maritime Provinces in Canada recently have been reported (Majka 2006, 2007, Majka and Jackman 2006, Majka and McCorquodale 2006, and more than 40 additional publications). Both of the above faunistic projects were small-scale projects run essentially by one individual. Great discoveries and progress could be made by well-funded teams of researchers conducting faunistic studies on Nearctic insects. Moreover, collections that house the specimens collected have to be developed, more fully supported, and given the means to take advantage of modern techniques; many specimens already in collections need to be curated more fully to make them available for study.

The second major task for the future would seek to establish a wider appreciation and use of systematics outside the systematics community. Systematics work has to be coordinated to a greater degree than hitherto with work in other disciplines so that, for example, ecosystem studies include both ecological and taxonomic specialists in the design, as well as in the execution of projects. Long-term studies are required, including the establishment of meaningful baselines by careful taxonomic analysis, a requirement that has not usually been evident in environmental appraisals. They allow us to interpret and respond to environmental degradation and climate change.

All biologists require an appreciation of diversity and evolution and a foundation in whole organism biology to understand their discipline. Maintaining the systematics faculty required to train students in biology favors their proper education, maintains the health of systematics, and provides opportunities for other systematists. The importance of systematics work, therefore, will be recognized properly only by insisting on its broader implications and applications. The study of biodiversity in the Nearctic Region, just as in other parts of the world, is necessary to permit wise use and protection of the earth's resources (Danks and Ball 1993). Acceptance of this long-term view would help buffer the essential long-term work of systematics from short-term political interference and would provide the resources required to understand the biodiversity of insects.

ACKNOWLEDGMENTS

We thank the many specialists whose expert knowledge made the estimates of Nearctic diversity possible, including the authors of chapters in Danks (1979) and Kosztarab and Schaefer (1990a), as well as the following colleagues who provided further information: R. S. Anderson, G. E. Ball, Y. Bousquet, M. L. Buffington, A. T. Finnamore, G. A. P. Gibson, J. H. Huber, D. J. Lafontaine, S. A. Marshall, F. W. Stehr, F. C. Thompson, and G. B. Wiggins. Drs. Anderson and Ball provided very helpful comments on an earlier version of the manuscript. This publication was supported, in part, by an NSF/BS&I grant (DEB–0342189).

REFERENCES

Arnett, R. H. and M. C. Thomas (eds). 2000. *American Beetles.* Volume 1. Archostemata, Myxophaga, Adephaga, Polyphaga: Staphyliniformia. CRC Press, Boca Raton, Florida. xv + 443 pp.

Arnett, R. H., M. C. Thomas, P. E. Skelley, and J. H. Frank (eds). 2002. *American Beetles.* Volume 2. Polyphaga: Scarabaeoidea through Curculionoidea. CRC Press, Boca Raton, Florida. xv + 861 pp.

Ball, G. E. (ed). 1985. *Taxonomy, Phylogeny and Zoogeography of Beetles and Ants: A Volume Dedicated to the Memory of Philip Jackson Darlington, Jr. (1904–1983).* W. Junk, Dordrecht. 514 pp.

Barnes, J. K. 1985. Insects in a new nation: a cultural context for the emergence of American entomology. *Bulletin of the Entomological Society of America* 35: 21–30.

Bedick, J. C., B. C. Ratcliffe, W. W. Hoback, and L. G. Higley. 1999. Distribution, ecology, and population dynamics of the American burying beetle [*Nicrophorus americanus* Olivier (Coleoptera, Silphidae)] in south-central Nebraska, USA. *Journal of Insect Conservation* 3: 171–181.

Benson, R. B. 1962. Holarctic sawflies (Hymenoptera: Symphyta). *Bulletin of the British Museum (Natural History): Entomology* 12: 381–409.

Bossart, J. L. and C. E. Carlton. 2002. Insect conservation in America. *American Entomologist* 48: 82–92.

Bryson, R. A. and F. K. Hare. 1974. *World Survey of Climatology.* Volume 11. Climates of North America. Elsevier, New York. x + 420 pp.

Conrad, K. F., M. S. Warren, R. Fox, M. S. Parsons, and I. P. Woiwod. 2006. Rapid declines of common, widespread British moths provide evidence of an insect biodiversity crisis. *Biological Conservation* 132: 279–291.

Covell, C. V. 1990. The status of our knowledge of the North American Lepidoptera. Pp. 211–230. *In* M. Kosztarab and C. W. Schaefer (eds). *Systematics of the North American Insects and Arachnids: Status and Needs.* Blacksburg, Virginia Agricultural Experiment Station Information Series 90-1.

Danks, H. V. (ed). 1979. Canada and its insect fauna. *Memoirs of the Entomological Society of Canada* 108: 1–573.

Danks, H. V. 1981. *Arctic Arthropods: A Review of Systematics and Ecology with Particular Reference to the North American Fauna.* Entomological Society of Canada, Ottawa. viii + 608.

Danks, H. V. 1988a. Systematics in support of entomology. *Annual Review of Entomology* 33: 271–296.

Danks, H. V. 1988b. *Insects of Canada. A Synopsis Prepared for Delegates to the XVIIIth International Congress of Entomology (Vancouver, 1988).* Biological Survey of Canada, Ottawa.

Danks, H. V. 1990. Arctic insects: instructive diversity. Pp. 444–470. *In* C. R. Harington (ed). *Canada's Missing Dimension: Science and History in the Canadian Arctic Islands.* Volume 2. Canadian Museum of Nature, Ottawa.

Danks, H. V. 1992. Arctic insects as indicators of environmental change. *Arctic* 45: 159–166.

Danks, H. V. 1993a. Patterns of diversity in the Canadian insect fauna. *In* G. E. Ball and H. V. Danks (eds). Systematics and entomology: diversity, distribution, adaptation and application. *Memoirs of the Entomological Society of Canada* 165: 51–74.

Danks, H. V. 1994. Regional diversity of insects in North America. *American Entomologist* 40: 50–55.

Danks, H. V and G. E. Ball. 1993. Systematics and entomology: some major themes. *In* G. E. Ball and H. V. Danks (eds). Systematics and entomology: diversity, distribution, adaptation and application. *Memoirs of the Entomological Society of Canada* 165: 257–272.

Danks, H. V. and R. G. Foottit. 1989. Insects of the boreal zone of Canada. *Canadian Entomologist* 121: 626–690.

Danks, H. V. and D. M. Rosenberg. 1987. Aquatic insects of peatlands and marshes in Canada: Synthesis of information and identification of needs for research. *In* D. M. Rosenberg and H. V. Danks (eds). Aquatic insects of peatlands and marshes in Canada. *Memoirs of the Entomological Society of Canada* 140: 163–174.

Danks, H. V. and D. D. Williams. 1991. Arthropods of springs, with particular reference to Canada: synthesis and needs for research. *In* D. D. Williams and H. V. Danks (eds). Arthropods of springs with particular reference to Canada. *Memoirs of the Entomological Society of Canada* 155: 203–217.

[Downes, J. A.]. 1974. *A Biological Survey of the Insects of Canada*. Entomological Society of Canada, Ottawa.

Downes, J. A. (ed). 1981. Temporal and spatial changes in the Canadian insect fauna. *Canadian Entomologist* 112: 1089–1238.

Downes, J. A. and D. H. Kavanaugh (eds). 1988. Origins of the North American insect fauna. *Memoirs of the Entomological Society of Canada* 144: 1–168.

Erwin, T. L. 1982. Tropical forests: their richness in Coleoptera and other arthropod species. *Coleopterists Bulletin* 36: 74–75.

Erwin, T. L. 1991. How many species are there?: Revisited. *Conservation Biology* 5: 330–333.

Foottit, R. G., S. E. Halbert, G. L. Miller, E. Maw, and L. M. Russell. 2006. Adventive aphids (Hemiptera: Aphididae) of America north of Mexico. *Proceedings of the Entomological Society of Washington* 108: 583–610.

Grissell, E. E. 1999. Hymenopteran biodiversity: some alien notions. *American Entomologist* 45: 235–244.

Goldstein, P. Z. 2004. Systematic collection data in North American invertebrate conservation and monitoring programmes. *Journal of Applied Ecology* 41: 175–180.

Halffter, G. 1974. Elements anciens de l'entomofaune neotropicale: ses implications biogéographiques. *Quaestiones Entomologicae* 10: 223–262.

Halffter, G. 1987. Biogeography of the montane entomofauna of Mexico and Central America. *Annual Review of Entomology* 32: 95–114.

Halffter, G. 2003. Biogeografía de la entomofauna de montaña de México y América Central. Pp. 87–97. *In* J. J. Morrone and J. Liorente Bousquets (eds). *Una Perspectiva Latinoamericana de la Biogeografía*. Universidad Nacional Autónoma de México, Mexico City.

Kim, K. C., K. C. Emerson, and R. Traub. 1990. Diversity of parasitic insects: Anoplura, Mallophaga, and Siphonaptera. Pp. 91–103. *In* M. Kosztarab and C. W. Schaefer (eds). *Systematics of the North American Insects and Arachnids: Status and Needs*. Virginia Agricultural Experiment Station Information Series 90-1.

Kosztarab, M., L. B. O'Brien, M. B. Stoetzel, L. L. Dietz, and P. H. Freytag. 1990. Problems and needs in the study of Homoptera in North America. Pp. 109–145. *In* M. Kosztarab and C. W. Schaefer (eds). *Systematics of the North American Insects and Arachnids: Status and Needs*. Virginia Agricultural Experiment Station Information Series 90-1.

Kosztarab, M. and C. W. Schaefer (eds). 1990a. *Systematics of the North American Insects and Arachnids: Status and Needs*. Virginia Agricultural Experiment Station Information Series 90-1.

Kosztarab, M. and C. W. Schaefer. 1990b. Conclusions. Pp. 241–247. *In* M. Kosztarab and C. W. Schaefer (eds). *Systematics of the North American Insects and Arachnids: Status and Needs*. Virginia Agricultural Experiment Station Information Series 90-1.

Lafontaine, J. D. 1982. Biogeography of the genus *Euxoa* (Lepidoptera: Noctuidae) in North America. *Canadian Entomologist* 114: 1–53.

Larson, D. J. and A. N. Nilsson. 1985. The Holarctic species of *Agabus* (*sensu lato*) Leach (Coleoptera: Dytiscidae). *Canadian Entomologist* 117: 119–130.

Lindroth, C. H. 1957. *The Faunal Connections between Europe and North America*. Wiley, New York. 344 pp.

Lindroth, C. H. 1969a. The ground-beetles (Carabidae excl. Cicindelinae) of Canada and Alaska. *Part 6. Opuscula Entomologica Supplementum* 34: 945–1192.

Lindroth, C. H. 1969b. The ground-beetles (Carabidae exc. Cicindelinae) of Canada and Alaska. *Part 1. Opuscula Entomologica Supplementum* 34: I–XLVIII.

Löbl, I. and A. Smetana (eds). 2006. *Catalogue of Palaearctic Coleoptera*. Volume 3. Scarabaeoidea – Scirtoidea – Dascilloidea – Buprestoidea – Byrrhoidea. Apollo Books, Stenstrup.

Majka, C. G. 2006. The checkered beetles (Coleoptera: Cleridae) of the Maritime Provinces of Canada. *Zootaxa* 1385: 13–46.

Majka, C. G. 2007. The Erotylidae and Endomychidae (Coleoptera: Cucujoidea) of the Maritime Provinces of Canada: new records, zoogeography, and observations on beetle–fungi relationships and forest health. *Zootaxa* 1546: 39–50.

Majka, C. G. and J. A. Jackman. 2006. The Mordellidae (Coleoptera) of the Maritime provinces of Canada. *Canadian Entomologist* 138: 292–304.

Majka, C. G. and D. B. McCorquodale. 2006. The Coccinelidae (Coleoptera) of the Maritime provinces of Canada: new records, biogeographical notes, and conservation concerns. *Zootaxa* 1154: 49–68.

Marusik, Y. M. and S. Koponen. 2005. A survey of spiders (Araneae) with Holarctic distribution. *Journal of Arachnology* 33: 300–305.

Masner, L. 1990. Status report on taxonomy of Hymenoptera in North America. Pp. 231–240. *In* M. Kosztarab and C. W. Schaefer (eds). *Systematics of the North American Insects and Arachnids: Status and Needs.* Virginia Agricultural Experiment Station Information Series 90-1.

Masner, L. and L. Huggert. 1989. World review and keys to genera of the subfamily Inostemmatinae with reassignment of the taxa to the Platygastrinae and Sceliotrachelinae (Hymenoptera: Platygastridae). *Memoirs of the Entomological Society of Canada* 147: 1–214.

Matthewman, W. G. and D. P. Pielou. 1971. Arthropods inhabiting the sporophores of *Fomes fomentarius* (Polyporaceae) in Gatineau Park, Quebec. *Canadian Entomologist* 103: 775–847.

Matthews, J. V. 1979. Tertiary and quaternary environments: historical background for an analysis of the Canadian insect fauna. *In* H. V. Danks (ed). Canada and its insect fauna. *Memoirs of the Entomological Society of Canada* 108: 31–86.

McAlpine, J. F. 1979. Diptera. *In* H. V. Danks (ed). Canada and its insect fauna. *Memoirs of the Entomological Society of Canada* 108: 389–395.

Munroe, E. 1979. Lepidoptera. *In* H. V. Danks (ed). Canada and its insect fauna. *Memoirs of the Entomological Society of Canada* 108: 427–481.

Newton, A. F., M. K. Thayer, J. S. Ashe, and D. S. Chandler. 2000. Staphylinidae Latreille, 1802. *In* R. H. Arnett and M. C. Thomas (eds). *American Beetles.* Volume 1. Archostemata, Myxophaga, Adephaga, Polyphaga: Staphyliniformia. CRC Press, Boca Raton, Florida.

Nickle, D. A., M. A. Brusven, and T. J. Walker. 1987. Order Orthoptera. Pp. 147–170. *In* F. W. Stehr (ed). *Immature Insects, Volume 1.* Kendall/Hunt Publishing Company, Iowa.

Olson, D. M., E. Dinerstein, E. D. Wikramanayake, N. D. Burgess, G. V. N. Powell, E. C. Underwood, J. A. D'amico, I. Itoua, H. E. Strand, J. C. Morrison, C. J. Loucks, T. F. Allnutt, T. H. Ricketts, Y. Kura, J. F. Lamoreux, W. W. Wettengel, P. Hedao, and K. R. Kassem. 2001. Terrestrial ecoregions of the world: a new map of life on earth. *BioScience* 51: 933–938.

Pielou, D. P. and A. N. Verma. 1968. The arthropod fauna associated with the birch bracket fungus, *Polyporus betulinus*, in eastern Canada. *Canadian Entomologist* 100: 1179–1199.

Pimentel, D. (ed). 2002. *Biological Invasions: Economic and Environmental Costs of Alien Plant, Animal, and Microbe Species.* CRC Press, Boca Raton, Florida. 384 pp.

Poole, R. W. and P. Gentili. 1996a. *Nomina Insecta Nearctica. A Checklist of the Insects of North America.* Volume 1. Coleoptera, Strepsiptera. Entomological Information Services, Rockville, Maryland.

Poole, R. W. and P. Gentili. 1996b. *Nomina Insecta Nearctica. A Checklist of the Insects of North America.* Volume 2. Hymenoptera, Mecoptera, Megaloptera, Neuroptera, Raphidioptera, Trichoptera. Entomological Information Services, Rockville, Maryland.

Poole, R. W. and P. Gentili. 1996c. *Nomina Insecta Nearctica. A Checklist of the Insects of North America.* Volume 3. Diptera, Lepidoptera, Siphonaptera. Entomological Information Services, Rockville, Maryland.

Poole, R. W. and P. Gentili. 1997. *Nomina Insecta Nearctica: A Checklist of the Insects of North America.* Volume 4. Non-Holometabolous Orders. Entomological Information Services, Rockville, Maryland.

Saylor, L. W. 1946. Synoptic revision of the United States scarab beetles of the subfamily Dynastinae, no. 3: tribe Oryctini (part). *Journal of the Washington Academy of Sciences* 36: 41–45.

Schaefer, C. W. 1990. The Hemiptera of North America: what we do and do not know. Pp. 105–118. *In* M. Kosztarab and C. W. Schaefer (eds). *Systematics of the North American Insects and Arachnids: Status and Needs.* Blacksburg, Virginia Agricultural Experiment Station Information Series 90-1.

Sikes, D. S. 2003. The beetle fauna of the state of Rhode Island, USA (Coleoptera): 656 new state records. *Zootaxa* 340: 1–38.

Sikes, D. S. and C. J. Raithel. 2002. A review of hypotheses of decline of the endangered American burying beetle. *Journal of Insect Conservation* 6: 103–113.

Smith, A. B. T. 2003. Checklist of the Scarabaeoidea of the Nearctic Realm, Version 3. Available from: http://www.unl.edu/museum/research/entomology/nearctic.htm [Accessed 28 January 2008].

Spector, S. 2006. Scarabaeine dung beetles (Coleoptera: Scarabaeidae: Scarabaeinae): an invertebrate focal taxon for biodiversity research and conservation. *Coleopterists Society Monographs Patricia Vaurie Series* 5: 71–83.

Spencer, G. J. 1964. A century of entomology in Canada. *Canadian Entomologist* 96: 33–59.

Stork, N. E. 1988. Insect diversity: facts, fiction and speculation. *Biological Journal of the Linnean Society* 35: 321–337.

Stroud, P. T. 1992. *Thomas Say: New World Naturalist.* University of Pennsylvania Press, Philadelphia, Pennsylvania. xv + 340 pp.

Taktajan, A. 1986. *Floristic Regions of the World.* University of California Press, Los Angeles, California. 522 pp.

Thomas, J. A. 2005. Monitoring changes in the abundance and distribution of insects using butterflies and other indicator groups. *Philosophical Transactions of the Royal Society B, Biological Sciences* 360: 339–357.

Thompson, F. C. 1990. Biosystematic information – Dipterists ride the third wave. Pp. 179–201. *In* M. Kosztarab and

C. W. Schaefer (eds). *Systematics of the North American Insects and Arachnids: Status and Needs.* Virginia Agricultural Experiment Station Information Series 90-1.

Thompson, F. C. 2008. The Biosystematic Database of World Diptera, Version 8.5. Available from: http://www.sel.barc.usda.gov/diptera/biosys.htm [Accessed 28 January 2008].

Turnbull, A. L. 1979. Recent changes to the insect fauna of Canada. *In* H. V. Danks (ed). Canada and its insect fauna. *Memoirs of the Entomological Society of Canada* 108: 180–194.

Chapter 4

AMAZONIAN RAINFORESTS AND THEIR RICHNESS OF COLEOPTERA, A DOMINANT LIFE FORM IN THE *CRITICAL ZONE* OF THE NEOTROPICS

Terry L. Erwin and Christy J. Geraci

Department of Entomology, National Museum of Natural History, Smithsonian Institution, Washington DC

Patterns are the grist of the natural historian, phylogenist, and biogeographer; patterns are found in the synthesized data points of the ecologist and ethologist; patterns allow reasoning hominids to understand their surroundings, their past, and for the more clever among them, their future. Patterns are all about us, in our daily lives, in world events, and in the life forces of the biosphere. As biologists, we are lucky in that we can spend our lives observing these biospheric life forces, and through them catch glimpses of past events, which shaped the world in which we live today.

—*Erwin 1985*

Insect Biodiversity: Science and Society, 1st edition. Edited by R. Foottit and P. Adler
© 2009 Blackwell Publishing, ISBN 978-1-4051-5142-9

Near-term habitability of our planet now teeters between 'business as usual' and mitigation of a plethora of human-caused *Critical Zone* injuries, some on a global scale (e.g., climate warming and waste accumulation), but most on regional and local scales (e.g., species extinction, deforestation, conversion of wetlands and watersheds, desertification, and continental shelf freshwater pollution). The *Critical Zone* (Admundson et al. 2007) 'is the portion of the biosphere that lies at the interface of the lithosphere, atmosphere, and hydrosphere, and it encompasses soils and terrestrial ecosystems'. Now, more than ever, as humankind enters its third century of massive conversion of the pre-human planet, an understanding of the functional relationships among the abiotic and biotic players in the *Critical Zone* is paramount. To understand and manage the *Critical Zone*, we must find and interpret patterns. According to Admundson et al. (2007), the First Order Biota, nano-organisms aligned with their counterpart abiotic elements, are key life forms stabilizing the lithosphere and its processes in close cooperation with fundamental soil elements. Second Order Biota – insects and their kin – are particularly important because of their role as an interface between the macrobiota and nano-organisms. Our chapter, particularly, focuses on a quarter of the species in this Second Order Biota, that is, the Coleoptera, or beetles, about which we know little in the Neotropical Realm compared with other biogeographical regions. Yet, they and other terrestrial arthropods are the transformers of macroscopic life forms (e.g., plants and vertebrates, the Third Order Biota) to nutrients useful to nano life

forms living at the near-edges of the lithosphere. Perhaps the beetles and their brethren are the 'middle men' in human 'business as usual' parlance; at least we can think of them in this way here. Rainforests and their kin, the dominant suite of biotopes of Amazonian South America, are incredibly complex architecturally and chemically. Their canopies provide a multitude of fractal universes wherein dwell a magnificent richness of arthropod species, especially those belonging to the order Coleoptera, the dominant insect life form on the planet (more than 25% of described species are beetles). Evidence of the evolutionary history of this Amazon area is visibly obvious (for those inclined to observe and discern species): here sits the richest center of biodiversity on the planet. The Amazon, and by extension, the Neotropics, is also the least known area biologically.

Attempting to write about neotropical beetle biodiversity creates fears similar to those described by Douglas Adams (1980) in his *Total Perspective Vortex*. The *Vortex* was designed to give sentient beings a harrowing 'sense of proportion' regarding their place in the Universe. Entomologists working in Amazonia understand this humbling feeling, as the dense forests and sheer abundance of arthropods together form a *Total Perspective **Biodiversity** Vortex*. The Amazon Basin (Fig. 4.1) contains the largest contiguous tropical rainforest on earth. Each hectare of currently unconverted rainforest houses up to 3×10^{10} individuals of terrestrial arthropods representing some 100,000+ species (Erwin 2004) at any point in time. Its usable natural resources are important

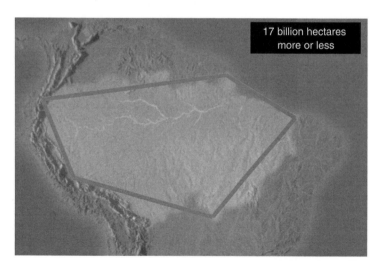

17 billion hectares
more or less

Fig. 4.1 Extent of the Amazon Basin.

to the peoples and economies of nine countries. Its millions of species, their interactions, and the delicate balance in which they live are important to the peoples of the world. The enormity of this ecosystem-level vortex complicates efforts to characterize and conserve neotropical arthropods. This is not to say that we know nothing about insect diversity in the tropics, because major taxonomic works on a number of groups have been published (Table 4.1). Nevertheless, one can go into any museum collection and find that a large proportion of Amazonian specimens are not identified to a formally described and published species name, but rather carry a morphospecies code or merely 'sp.' on their labels.

Recent studies on tree communities and insect faunas in Old World and temperate forests have provided methods, data, and hypotheses that could be applied to studying biodiversity patterns in Amazonian forest communities (Table 4.2). These studies have focused largely on phytophagous insect groups because their life histories are linked to those of their host trees. The question of how the degree of host specificity and niche partitioning relate to insect speciation rates is central to modeling biodiversity of insects in the Neotropics. There are insect groups whose life histories and biodiversity patterns are neither presently related, nor have they ever been tightly linked to, those of the tree-canopy community. The characterization of host trees, therefore, is not useful for predicting their true biodiversity in the Neotropics. But many insect species are associated with tree canopies in some way (e.g., herbivores at any life stage, pollinators, predators on host-specific herbivores, fungivores on canopy specialists, parasites, and parasitoids on species that are host-tree specialists), and this association holds true for beetles.

Kitching (2006) noted, 'Over the past few decades, we have built up an impressive body of data that identifies biodiversity patterns in tropical rainforests. There remains much to be done, but there is enough information that we can look for general mechanistic explanations of the patterns'. The more we know about the relationships between insects and their hosts in different tropical ecosystems, the more accurately we can extrapolate survey data across spatial scales and predict species richness and densities in the Amazon Basin. Kitching's (2006) perspective addressed a concentration of recent studies on the link between insect and host-tree diversity; however, none of these studies included the Amazonian lowland insect fauna (Table 4.2). Estimates of local species diversity of Coleoptera were made for a New Guinea lowland rainforest (Novotny et al. 2004, 2006), but not using canopy-fogging techniques. This chapter addresses the diversity-generating mechanisms (Kitching 2006) that we predict are driving beetle patterns in the Amazonian lowland rainforests. To illustrate this point, we summarized our knowledge of tree and beetle diversity (species richness and abundance) in only two hectares of Ecuadorian forest canopy, arguably now the best-known two hectares in the entire Basin. Our ultimate goal is to cast light on how Amazonian insect biodiversity can be characterized better by future collecting efforts and field experiments that connect mechanisms and patterns.

Yearly, 14 to 16 million hectares of tropical forests are converted to other land uses, mostly agricultural, and to degraded latersols rampant with erosional processes leading to desertification. The principal agents of deforestation – those individuals and companies who are cutting down the forests – include slash-and-burn farmers, commercial farmers, ranchers, loggers, firewood collectors, infrastructure developers, and

Table 4.1 Selected major taxonomic monographs and books published since 1960 for neotropical insect groups.

Group	Reference
Arachnida and Myriapoda	Goloboff 1995, Adis 2002, Lourenco 2002
Coleoptera	Howden 1976, Martins 1998, 1999, 2003, Spangler and Santiago 1987
Collembola	Heckman 2001
Diptera	Mathis 1980, Guimarães 1997, Borkent and Spinelli 2007, Coscarón and Coscarón Arias 2007
Ephemeroptera	Heckman 2002, Dominguez et al. 2006
Hemiptera	Kormilev 1975, Godoy et al. 2006
Hymenoptera	Porter 1967, Kempf 1972
Lepidoptera	Gates Clarke 1971, D'Abrera 1981, Heppner 1984, Pinas Rubio and Manzano Pesantez 1997, Pinas et al. 1997, Bollion and Onore 2001, Cuevara et al. 2002
Odonata	Heckman 2006
Plecoptera	Heckman 2003
Thysanoptera	Mound and Marullo 1996
Trichoptera	Flint et al. 1999

Table 4.2 Recent studies on the relationships between insect diversity, host specialization, and tropical forest communities.

Reference	Focus	Group Examined	Location
Kitching et al. 2003	Canopy-insect assemblages and relatedness in relation to host-tree phylogeny	Phytophagous Coleoptera	Australia
Novotny et al. 2004	Local species richness of leaf-chewing insects in 1 hectare of lowland rainforest	Lepidoptera, Coleoptera, orthopteroids	New Guinea
Pokon et al. 2005	Species assemblages of beetles and host trees	Chrysomelidae (Coleoptera)	New Guinea
Novotny et al. 2006	Diversification of herbivorous insects	Phytophagous insects	New Guinea
Nyman et al. 2006	Evolution of larval host specificity	Nematine sawflies (Hymenoptera)	Holarctic
Odegaard 2006	Species diversity, host specificity, and turnover comparison between two lowland forests	Phytophagous Coleoptera	Panama
Weiblen et al. 2006	Insect-community structure versus host-plant phylogeny	Lepidoptera, Coleoptera, orthopteroids	New Guinea

government initiatives (http://www.rcfa-cfan.org). Monitoring and mitigating this impact is of paramount importance to indigenous peoples who depend on the rainforest for their sustenance and culture, to global human generations that will follow those now living, and to the plethora of dependent biota, such as beetles. Armed with more data, we can make stronger arguments and smarter plans for the Amazon forest's long-term conservation (Salo and Pyhälä 2007, Schulman et al. 2007, Kremen et al. 2008). We hope that the following shows both how little we actually know of what lives in the entire 17 billion hectares of the forested Amazon Basin, and what we can infer about diversity-generating mechanisms from what data we do have.

THE CLIMATIC SETTING AND *CRITICAL ZONE* ESTABLISHMENT

To put data from our Ecuador forest plots in context, we needed to begin with the *Critical Zone* features (Valencia et al. 2004) that form the bases of tree occurrence, then progress through our data on a logical trajectory. One cannot escape 'rock meets biota', if one wants to understand the big picture of how arthropod populations are assembled. Amazonian lowland rainforest biotopes are sandwiched between, and continuous with, the Neotropical (subtropics), Nearctic, and Neaustral Realms to the north or south, respectively, the highlands of the Andes and the high

Tepuis to the west and north, the Mata Atlântica (southern Atlantic forest) in the southeast, and the Atlantic Ocean due east (Fig. 4.1). Below an altitude of about 700 m, the climate is equatorial without frost, and in most places, with a distinctive wet and dry season. The Antarctic circumglobal weather generator sometimes pumps cold air north along the Andes as far as middle Perú, lowering the temperature to 8°C during these cold spells. Some Amazonian areas, in the dry season, peak at around 40°C and even higher where desertification has taken hold, particularly in Brazil, due to ranching practices that have destroyed forests and their ameliorating effects on temperature.

The Andes have undergone orogeny for the last 40 million years, with all Amazonian rivers flowing eastward for the last 12 million years. The general regional landscape for terrestrial arthropods has remained mostly unchanged for that time, with minor fluctuations of wet and dry epochs, expanding and contracting forest cover alternating with savannah (Haffer 1982, de Fátima Rossettia et al. 2005, Campbell et al. 2006), and a mosaic pattern of soil deposition from Andean erosional processes (Erwin and Adis 1982). Until sometime in the middle of the twentieth century, unpolluted air measures were obtained near Manaus, in the middle of the Amazon Basin, indicating that a healthy atmosphere existed there. This is no longer true; no towns or cities in Amazonia have controls for auto-exhaust pollution strongly in place. Transport and local temporary storage (e.g., in oxbow lakes) of alluvium from the Andes to the Atlantic Ocean across 4 000 km of the Amazon Basin, besides being one of the greatest wonders of the world, provide a vast multitude of lithosphere diversity of historical soil types. The hydrosphere includes Varzea waters flowing from the Andes, carrying substantial amounts of alluvial particles, acid waters seeping slowly from the Igapó forests of rivers such as the Rio Negro running from Venezuela's southern Orinoco watershed south to Manaus, and clear-water rivers such as the Tapajos running across stone and pebble beds from the crystalline arcs underlying parts of the upper Amazon watershed (*cf.* Erwin and Adis 1982). The *Critical Zone* of the Amazon Basin (atmosphere + lithosphere + hydrosphere) is unmatched in complexity, and we found that this also is so for the arthropod life in our samples. That complexity ultimately depends on *Critical Zone* health for its sustainability, and in the long term, its evolution.

CHARACTERIZATION OF TYPICAL LOWLAND RAINFOREST COMPOSITION IN THE WESTERN BASIN

Insects live in all strata of the rainforest from deep in the soil (e.g., cicada nymphs tunneling along deep tree roots) to the tops of the trees (e.g., leaf-chewing and sucking insects on leaves receiving direct insolation). Data on insects presented in this chapter principally come from two hectares of rich hardwood rainforest at an altitude of about 250 m in the eastern part of Amazonian Ecuador (Fig. 4.3). The two study plots, named Piraña (or Onkone Gare, OG) and Tiputini Biodiversity Station (TBS), are 21 km apart as the macaw flies and lie almost on the equator (Lucky et al. 2002, Erwin et al. 2005). Local and regional climate, soils, and topography surrounding our plots are described by Valencia et al. (2004). We examined tree biodiversity data from these plots because primary consumer insects depend on trees and evolve based on their chemistry and physical attributes. Subsequently, the predators and parasites depend on the evolutionary histories of the primary consumers.

In a 2001 study, Pitman et al. attempted to gain perspective on the assembly rules of Amazonian rainforest tree communities. Previously, when forest ecologists looked at the Amazon Basin they saw complexity, not pattern. After examining

Fig. 4.2 Canopy-fogging technique, showing collecting sheet and fogger.

tree-distribution data more closely, however, Pitman et al. (2001) found that the common trees formed predictable oligarchic hierarchies that were present across the complex landscape. In both their Ecuador and Perú plots, they found dozens of common species and hundreds of rare species. The common tree species oligarchy comprised only 15% of the species inventoried; the other 85% were rare species that the authors posited are strict habitat specialists. The common species appeared in high densities locally and in high frequencies over large tracts of forest, making their occurrence at least somewhat predictable. Even the rarer species usually belonged to what they considered a predictable set of families and genera. The authors argued that

> '...tropical tree communities are not qualitatively different from their temperate counterparts, where a few common species concentrated in a few higher taxa can dominate immense areas of forest'.

The oligarchy hypothesis is not universally accepted (Tuomisto et al. 2003), but it was supported by the

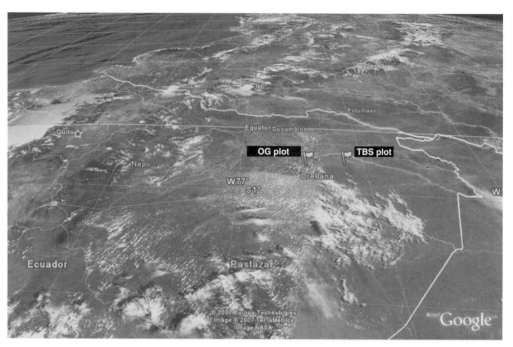

Fig. 4.3 Map generated using GoogleEarth, showing locations of the two forest plots under study in eastern Ecuador (OG = Piraña, TBS = Tiputini Biodiversity Station).

results of a broad-scale study by Macia and Svenning (2005). Entomologists studying groups whose life histories are intimately related to those of trees might ask whether Amazonian insect communities are organized into oligarchies of common taxa, and how we detect oligarchic patterns in the canopy. Pitman's group used a density threshold of one individual per hectare to define common trees, but this method is not straightforwardly applicable to insects that can have varying abundances across time and space. Yet, faced with an overwhelming richness of species yet to be described, perhaps canopy insect systematists should prioritize describing, mapping, and monitoring the commonest lineages. Two major challenges limiting the application of biomonitoring strategies to Amazonian rainforests is the sheer number of species to be identified, and, as a result, the lack of taxonomic resolution in survey data. Reaching an appropriate level of taxonomic resolution is critical to understanding ecological patterns and detecting disturbances (e.g., road building) in the form of changes in any insect community (Lenat and Resh 2001).

We begin to address insect diversity patterns by summarizing the tree community for our two Ecuador plots. Within the ten transect zones of each plot in Yasuni National Park, all trees with diameter at breast height >10 cm were identified to the lowest possible taxonomic unit. At Piraña, 669 trees represent 250 species and 51 families (data from N. Pitman, personal communication). The most common tree families (i.e., represented by more than 19 trees each) include (in order of abundance) Fabaceae (87 individuals), Arecaceae (60), Bombacaceae (57), Moraceae (46), Lecythidaceae (43), Burseraceae (36), Cecropiaceae (24), Myristacaceae (22), and Euphorbiaceae (19). The commonest tree is *Matisia malacocalyx s. lat.* (Bombacaceae), which is represented by 42 individuals. The 623 trees at Tiputini represent 252 species and 48 families (data from N. Pitman, personal communication). The most common tree families here include Fabaceae (82), Moraceae (57), Arecaceae (41), Bombacaceae (27), Cecropiaceae (34), and Myristacaceae (34), Lecythidaceae (30), Euphorbiaceae (27), Sapotaceae (27), Lauraceae (26), Meliaceae (20), and Burseraceae (19). The commonest species is the palm *Iriartea deltoidea*, which is represented by 20 individuals. This species of palm is the commonest canopy tree species in the Western Amazon Basin (Pitman et al. 2001).

The difference in number of individual trees at Piraña (669) and Tiputini (623) is accounted for by a blow-down along two transects between the time of setting up the plot and its inventory by Pitman's team. Some 30 trees were felled in this blow-down and were not included on the identification list; otherwise, the plots are nearly identical in number of stems and amount of standing volume. Even so, both of our study sites represent accurately the expected tree composition of the *terra firme* (upland) forests in the equatorial Amazon Basin. The Piraña and Tiputini sites contain 80.5% and 82.9%, respectively, of the total 42 common species that Pitman et al. (2001) identified across the Western Amazon Basin. Every one of the tree species on the list of Pitman et al. (2001) was found in at least one of the Yasuni plots. However, although the most common tree species of Western Amazon *terra firme* forests are represented in our plots, only a lesser part of the total regional tree species is represented. Four species (*Celtis schippii, Warscewiczia coccinea, Guarea pterorhachis*, and *Pouroumaminor*) were not present in the Piraña and TBS plots. When we examined the data at the genus level, we found that three each were missing from Piraña (*Jacaratia, Lindackeria*, and *Celtis*) and from TBS (*Helicostylis, Iryanthera*, and *Celtis*). In some cases, a species Pitman et al. (2001) noted as common across Western Amazonia was not as common in Yasuni as were other species of the same genus (or it was absent entirely).

Here, we compared merely the 42 common tree species among plots, not the hundreds of rare species that are also present. Pitman's team studied 22 other 1-ha^2 forest plots across the Yasuni region, from which they identified a total of 1176 tree species. The two plots we studied have about 250 species each; thus, each plot represents 21.42% (Tiputini) and 21.26% (Piraña) of regional tree species richness. However, because each of our plots has a different composition, together they represent about 34% of tree species richness in the Yasuni region. Tree composition at the two sites is similar at the family level (Complementarity Index, CI = 0.26); Fabaceae, Arecaceae, Moraceae, Bombacaceae, Lecythidaceae, Burseraceae, Cecropiaceae, Myristacaceae, and Euphorbiaceae are the commonest families at both sites. However, at the generic level, the difference is more pronounced (CI = 0.52) and at the species level, dramatically so (CI = 0.73).

SAMPLING ARTHROPOD BIODIVERSITY IN AMAZONIAN FORESTS

Adis and Schubart (1984) developed numerous techniques for sampling soil-dwelling arthropods in the Manaus area of the Central Amazon Basin, including those terrestrial species whose members climb trees during flood stage and return down when the soil dries. We studied canopy-dwelling beetles in the Western Amazon Basin in Perú and Ecuador. Our sampling by insecticidal fogging (Fig. 4.2) was designed to gather specimens to determine species turnover of one component of species living in the canopy/understory fractal universes, a measure of at least two-thirds of the biodiversity at this site (*cf.* Erwin 1982). To determine the insect biodiversity of the forested sites, we established two study plots in 1994 (Piraña/OG) and in 1997 (Tiputini/TBS) (Fig. 4.3). Each was a 100 m × 1000 m area that consisted of ten 10 m × 100 m transects spaced at 100 m from each other along the plot (e.g. Fig. 4.4). Ten collection stations, each 3 m × 3 m, were located randomly within each 10 m × 100 m zone arrayed on both sides of a centerline. Fogging stations at each site were numbered 1–100 at the beginning of the study and were then, each one, used repeatedly for nine sampling events across three years, three seasons per year (dry, wet, and wet-to-dry transitional). Canopy/understory sampling spanned eight years from January 1994 through July 2002 (three consecutive years at the Piraña site, and parts of five years at the Tiputini site to account for trans-annual variation). Intermittent funding, as well as sporadic political unrest in Ecuador extended the fieldwork at Tiputini by two years to obtain seasonal samples equal to what we took at the Piraña site.

To account for seasonal variation, sampling took place in January/February (the dry season), June/July (the wet season), and October (a transitional period between the wet and dry seasons). The fogging techniques have been described previously (Erwin 1983a, 1983b). The only modifications made for this study, as opposed to those previously undertaken, were the use of 3 m × 3 m sampling sheets suspended by nylon string tied to trees or stakes and arranged 1 m above the ground at each station, and fogging undertaken from ground level (Figs. 4.5, 4.6). Canopy foggings were made from just above each sheet as a column (1 m to 'n' m high) into the canopy at 0345–0500 hours. Height of fog ascendancy was measured for each fogging event. Canopy strata at our study sites

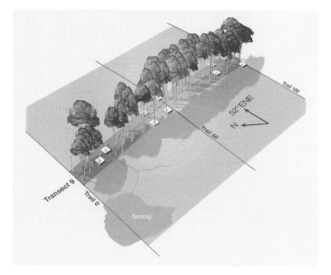

Fig. 4.4 Diagrammatic view: oblique forest cross transect, 100 m × 10 m.

vary from one to three levels up to 37 m, with some super-emergent trees rising above 40 m. Our sampling target was a column of forest including the understory above 1 m and as much of the canopy strata as possible at each station, depending on local microclimate each morning. In total, the area sampled within each of the two plots (Piraña and Tiputini) was 9250 m², a mere 1.1% of the entire plot area.

The primary purpose of the Yasuni area sampling was to monitor road-building activities of Maxus Oil Company and their effect on arthropods, as part of a team effort designed to mitigate damage to the primary forests of the region. A by-product of that study provided abundant specimens (Table 4.3) and data that have enabled us to ask basic science-driven questions. One such question dealt with species turnover between the Piraña and Tiputini sites. Samples collected from each fogging event were sorted and 14 target taxa were extracted, including the Coleoptera. Subsequently, a portion of the beetles from both sites was re-sorted to acquire the 17 family-group taxa for the present project, which is ongoing. We then sorted at least 600 samples (300 or more from Piraña and 300 from Tiputini from like seasons), of family-group taxa to the level of morphospecies (Erwin et al. 2005). Two of these family-group taxa had been removed previously from all samples for other studies (Carabidae, *cf.* Lucky et al. 2002, and the Cleridae for taxonomic treatment by our colleague J. Mawdsley). Although this project

is far from finished, we report preliminary results on a few coleopteran families. Our data demonstrate why it is so difficult to describe arthropod biodiversity of the Amazon Basin, indeed, of the entire Neotropical Realm.

Fig. 4.5 Diagrammatic view: sampling column.

Fig. 4.6 Sheet (3 m × 3 m) used to catch falling arthropods.

RICHNESS OF VARIOUS LINEAGES AND GUILDS

Some arthropod taxa across the Amazon Basin, particularly lineages with few species, have been documented in major taxonomic works (Table 4.1). Even so, their locality records are sparse. No taxa that are reasonably diverse or hyperdiverse are well known with regard even to which species are present, to say nothing about their distributions. Our study relies on morphospecies estimates and presently focuses only on beetles. If 25% of the 1.7 million species described are beetles (with an estimated 20% synonymy, Hammond 1992), then of the 100,000+ estimated arthropod species per hectare (Erwin et al. 2004), 25,000+ per hectare are beetle species. Of that, 80% are estimated to be new to science (according to several participating taxonomists working with the fogged specimens). Given these estimates, tabulation of morphospecies is the best we can do to achieve a preliminary understanding of the actual biodiversity of this miniscule part of the Amazon Basin. This estimate of morphospecies 'biorichness' is a beginning. Biodiversity estimates require more than lists of taxonomic units or numbered barcode clusters; these are hypotheses of species identity that must be confirmed by taxonomists. True characterization of biodiversity patterns includes estimates of abundance, evenness of spatial and temporal distributions, and other factors used in micro- and macroecology (Price 2003). Biodiversity, as we broadly consider it here, is a complex concept that implies knowledge of evolutionary and ecological relationships among species, and the timing of diversification events (e.g., McKenna and Farrell 2006).

GENERAL PATTERNS

The canopy sampling regime of 100 forest columns at Piraña resulted in a capture of some 1.7 million counted specimens belonging to 14 target taxa selected specifically for the Maxus Oil Company monitoring project. The total sampling regime acquired, based on randomly counted samples of all other arthropod taxa, an estimated 3.8 million total specimens, 49% of which were ants. Beetles in the 900 Piraña samples accounted for 262,751 counted adult individuals, nearly 19% of the total abundance of fogging captures (Table 4.3, Fig. 4.7). This count was second in abundance to Diptera at 33%, although this dipteran abundance was because of one family, the Mycetophilidae, whose populations typically burst out in the rainy season when fungi are available. The Tiputini samples likely will be a reflection of the Piraña samples, once counts are completed. We detected no significant seasonal differences for taxon abundance at the order level in our target groups across three trans-annual fogging events at Piraña (ANOVA: $F = 0.39$, $P > 0.05$); thus, our sampling regime by fogging was not affecting faunal balance.

To obtain a preliminary estimate of beetle richness in our plots, we chose 300 samples from Piraña, plus

Table 4.3 Arthropod abundances in the Piraña (Onkone Gare) 1994–1996 canopy samples: OR = Orthoptera; DS = dry season; WS = wet season; TS = transitional season.

Taxon		1994 January DS	1994 June/July WS	1994 October TS	1995 February DS	1995 July WS	1995 October TS	1996 February DS	1996 July WS	1996 October TS	Grand total
Araneida		12,769	15,549	15,357	17,951	24,790	20,119	18,534	24,239	21,864	171,172
Blattaria	Nymphs	5146	6524	6342	7763	8009	7029	5973	8476	8219	63,481
	Adults	644	594	866	1153	867	781	1310	804	1051	8070
OR:Acrididae	Nymphs	423	726	467	507	558	559	523	539	473	4775
	Adults	379	182	153	301	290	202	345	188	220	2260
OR:Tettigoniidae	Nymphs	2020	2136	2402	2108	3085	2934	1908	2405	2559	21,557
	Adults	136	152	240	427	332	258	315	214	496	2570
OR:Gryllidae	Nymphs	2096	2263	3358	3011	3276	3963	3302	3410	3594	28,273
	Adults	214	426	467	553	557	612	595	574	670	4668
Mantodea	Nymphs	114	93	176	185	157	143	119	171	156	1314
	Adults	10	14	19	34	27	25	19	19	29	196
Phasmatodea	Nymphs	123	165	144	123	135	94	88	107	132	1111
	Adults	23	27	42	47	35	32	36	25	30	297
Psocoptera	Nymphs	1921	2309	3443	11,811	5354	12,586	9679	8764	12,355	68,222
	Adults	1591	763	5156	8801	1907	5039	5002	3050	4582	35,891
Hemiptera	Nymphs	1588	1319	2227	3004	3508	4550	2340	2925	4027	25,488
	Adults	2260	2217	4232	3941	4779	5886	4671	5104	5656	38,746
Homoptera	Nymphs	12,164	7455	8200	10,883	10,212	13,030	10,443	9926	11,893	94,206
	Adults	10,324	10,178	14,470	15,794	18,669	25,533	21,004	13,602	23,286	152,860
Coleoptera	Larvae	626	1040	1159	703	1672	2049	1379	2035	1896	12,559
	Adults	20,063	19,724	26,405	21,859	29,597	35,245	32,887	28,967	48,004	262,751
Lepidoptera	Larvae	2205	2499	5445	2981	4440	7477	3711	5209	7747	41,714
	Adults	728	1126	1716	1623	2177	2276	2081	1780	1993	15,500
Hymenoptera	Adults	10,080	10,528	14,764	16,507	31,563	67,524	29,325	22,050	42,484	244,825
Diptera	Adults	13,610	31,038	29,649	20,832	89,139	92,557	53,866	55,947	76,310	462,948
TOTALS		**101,257**	**119,047**	**146,899**	**152,902**	**245,135**	**310,503**	**209,455**	**200,530**	**279,726**	**1,765,454**

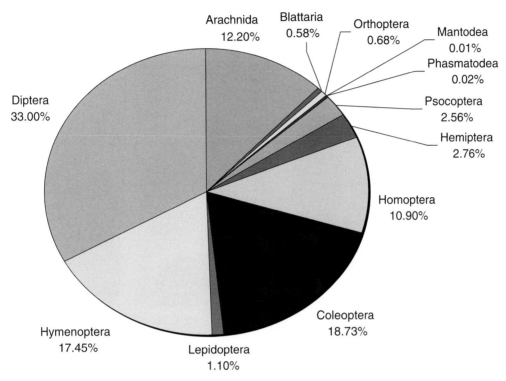

Percentages of total abundance for Piraña (OG) target taxa 1994–1996

Arachnida 12.20%
Blattaria 0.58%
Orthoptera 0.68%
Mantodea 0.01%
Phasmatodea 0.02%
Psocoptera 2.56%
Hemiptera 2.76%
Homoptera 10.90%
Coleoptera 18.73%
Diptera 33.00%
Hymenoptera 17.45%
Lepidoptera 1.10%

Fig. 4.7 Adult arthropod target-taxa abundances for the Piraña (Onkone Gare) 1994–1996 canopy samples.

another 300 like samples at the control site, Tiputini. We found that 15,126 specimens of 14 family-group beetle taxa represented 2010 morphospecies. A selection of these families is illustrated in (Fig. 4.8). Based on species accumulation curve estimates (Erwin et al. 2005), a consistent set of patterns emerged in the data for the family-group beetle taxa thus far investigated. In these taxa, even with the rigorous sampling regime, more samples than 600 are needed to know the universe of canopy/understory species in the local area, and certainly more than 300 samples at one site, even for the less diverse of families. The only exception among these family-group taxa were the otidocephaline Curculionidae (Fig. 4.8i); combining 300 samples from both sites for a total of 600 proved adequate to nearly reach an asymptote for the species accumulation curve for the area. The Carabidae (Fig. 4.8c) at

Piraña were not adequately sampled with 900 samples; however, the 300 samples of carabids that have been processed thus far from Tiputini provided enough additional species for a species accumulation curve to reach an asymptote for the region. When these two family-group taxa reaching near asymptote were analyzed for complementarity between the two sites, the values were CI = 0.29 for the otidocephaline Curculionidae and CI = 0.69 for the Carabidae.

For the rest of the families examined, at least 300 or more samples will need to be added to the database for the species accumulation curve to reach an asymptote. Thus, observed and predicted numbers of morphospecies (Fig. 4.8) likely will increase with additional collecting at Tiputini. In addition, some taxa already have produced more morphospecies to date than were predicted from previous accumulation curve estimates

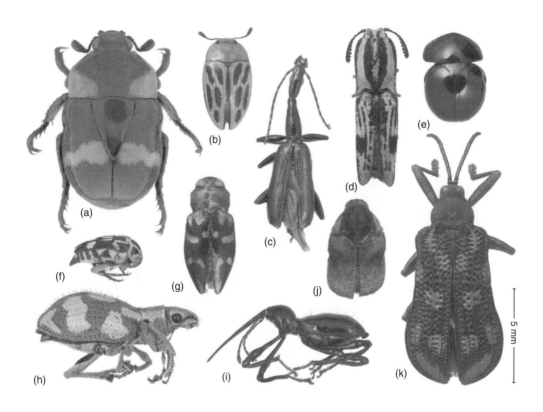

| | species richness | | adult feeding guilds |
	empirical	predicted	
a) Scarabaeidae	58	133	herbivores, leaf feeders
b) Erotylidae	214	168	fungivores
c) Carabidae	458	477	predators
d) Elateridae	150	138	detritivores
e) Ceratocanthidae	11	36	termitophiles
f) Mordellidae	388	361	pollen feeders
g) Buprestidae	209	234	pollen feeders
h) Entiminae (Curculionidae)	41	45	herbivores, leaf feeders
i) Otidocephalini (Curculionidae)	35	37	herbivores
j) Cryptocephalinae (Chrysomelidae)	43	227	herbivores, leaf feeders
k) Hispinae (Chrysomelidae)	227	316	herbivores, leaf feeders
l) Alticinae (Chrysomelidae) not figured	400	n/a	herbivores, leaf feeders
m) Cleridae, not figured	122	139	predators
	total = 2315		

Fig. 4.8 Coleoptera taxa examined in Yasuni from 1994 to 2006. Images were taken with the EntoVision extended focus photography system (Alticinae and Cleridae not figured). For each taxon, the number of observed morphospecies (as of 2008) is listed, as well as predicted numbers of species based on accumulation curves and ICE calculations (Erwin et al. 2005). (*See color plate.*)

(Fig. 4.8, Erwin et al. 2005), while richness for other taxa remains lower than estimated. This discrepancy reflects the varying degree of taxonomic treatment by specialists in each group. The tumbling flower beetles of the family Mordellidae (Fig. 4.8f) is an example of a group that is not particularly well known from the Neotropics but has high morphospecies richness in the canopy. This situation demonstrates the importance of samples taken for applied conservation purposes as a significant new source of material for basic systematics research, and highlights the need for proper vouchering and databasing of individual specimens in ecological studies.

MORPHOSPECIES RICHNESS TO BIODIVERSITY

Estimating morphospecies richness in an ecosystem such as the Amazon is a necessary first step toward understanding its biodiversity patterns. As our taxonomic resolution increases and the technology to map patterns of richness and abundance data at the specimen level becomes more sophisticated, our ability to characterize assembly rules for tropical forests will strengthen. For example, we can use distribution patterns to hypothesize which species are tourists and which are true canopy dwellers. Figure 4.9 illustrates this point for the family Mordellidae, which we use as a case study. Worldwide there are about 1500 total described species, with 205 known from North America (Jackman and Lu 2002). Compared to other phytophagous beetle clades, little is known about the family Mordellidae in the Neotropics. North American mordellids are broadly classified as pollen feeders (especially of Apiaceae and Asteraceae) as adults, whereas the larvae are endophagous feeders of woody and herbaceous stems, decaying wood, and fungi (Jackman and Lu 2002). Two subfamilies and five tribes currently are described worldwide, but their phylogenetic relationships are not understood. In temperate zones, larvae in the tribe Mordellini bore into woody host plants, whereas larvae in the tribe Mordellistini use herbaceous hosts (Jackman and Lu 2002). A third tribe, the Conaliini, is found in the Neotropics, but the life histories of adults and larvae are not well characterized.

The morphospecies distribution curve for Mordellidae (Fig. 4.9a) shows a pattern of few abundant taxa, with a long tail representing doubletons and singletons. To understand the mechanism driving this pattern we should know how the abundant and rare species are related to each other, and what life history characteristics they share (either via similar ecology or shared recent ancestry). The utility of this approach has been shown for comparing physiological traits in aquatic insects (Buchwalter et al. 2008). Preliminary examination of the Mordellidae morphospecies collected from Piraña (OG) from 1994 to 1996 suggests that an oligarchy of dominant species exists (Fig. 4.9b–d): six morphospecies were found in numbers greater than 50 in the 900 total samples, plus a large number of rare species. When we put the six abundant morphospecies numbers into phylogenetic context, we found that they fit into only three genera in two tribes. The tribe Conaliini also is known from Amazonia but conaliine morphospecies were rare. Little is known about the life histories of adult and larval Mordellidae in the Amazonian canopy, so we cannot say what processes are driving this pattern other than to suggest that some of the rare taxa probably are understory species that were captured as tourists. We do know, however, that the Mordellini and Mordellistenini were dominant in the canopy samples from the Piraña plot from 1994 to 1996. Morphospecies designations need to be confirmed and valid species names applied before accurate statistics can be calculated, but genus-level keys are available only for the North American fauna. Work is underway to describe the putative common oligarchy morphospecies and, in the process, update the genus-level classifications and keys.

Histogram curves of other representative beetle groups (Fig. 4.9a) show different lineages with varying patterns of species distributions, as we would expect. These patterns might reflect different life histories and ecosystem requirements, or they could be artifacts of varying taxonomic scrutiny. The general pattern we see, however, among those families that have been identified to morphospecies, is that many rare species are in the canopy samples. Collections in the understory ecosystem in these plots will reveal whether the rare canopy taxa are abundant elsewhere, and whether the Mordellidae fauna is stratified vertically. We know little in general about what the overlap is between ground and canopy faunas, and this relationship can differ even among beetle subfamilies. For example, the beetle family Carabidae has a distinct canopy fauna (i.e., arboreal species) and distinct ground fauna (i.e., terrestrial species) (Erwin 1991). Of the nearly 500 species recorded thus far from the canopy of our two plots, only a dozen or

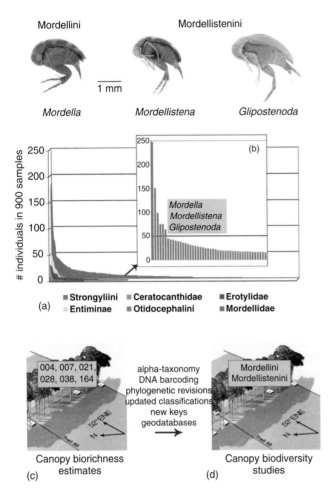

Fig. 4.9 Histograms (a) showing morphospecies distribution curves for six Coleoptera groups from the 900 canopy samples collected from 1994 to 1996 from the Piraña (OG) plot (Otidocephalinae and Entiminae = Curculionidae; Strongyliini = Tenebrionidae). The shape of the curve for the Mordellidae (b) suggests oligarchic dominance by six morphospecies in three genera. These morphospecies, represented by numerical codes (c), were present in abundances of greater than 50 individuals in 900 samples. Further research and incorporation of new technologies will increase taxonomic resolution and put morphospecies into phylogenetic context (d). Images were taken with the EntoVision extended-focus photography system.

so are terrestrial true species, the so-called tourists. Amazonian biodiversity cannot be generalized until we know how much overlap occurs at both large and small spatial scales. Does the roughly 25% overlap in tree species between Piraña and Tiputini predict a certain amount of overlap between the insect communities? How does the degree of overlap differ among insect lineages?

The above case study demonstrates that for many neotropical insect groups, phylogenetic relationships, classifications, and keys at the genus level need to be revised before we can reliably assign valid names or associate males and females to confirm that morphospecies designations are accurate. A team of taxonomists is already doing this research with the specimens from these two hectare plots. Their contribution to

increasing taxonomic resolution will allow us to compare more accurately pre- and post-road building canopy richness between the control (TBS) and impact (OG) plots. Thus, advancing taxonomic knowledge will enable us to test whether canopy samples taken 12 years after road building show the same relative species abundances over time, and if the distributions of species and genera among transects have changed.

BEETLES: LIFE ATTRIBUTES HAVE LED TO CONTEMPORARY HYPERDIVERSITY

True beetles appear for the first time in the Triassic (ca. 240 mybp). A marked radiation occurred in the Jurassic (210 to 145 mybp), with the appearance

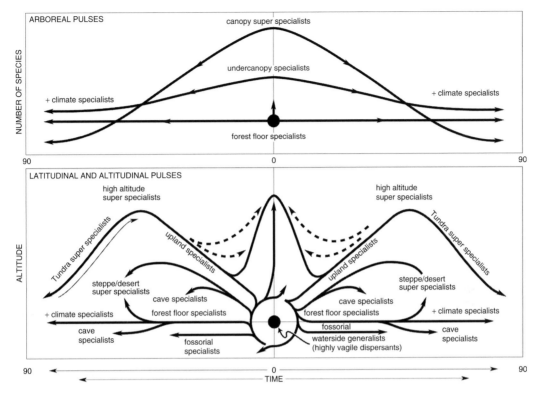

Fig. 4.10 Taxon pulse model (Erwin 1979, 1985).

of several families still in existence. Then, in the Cretaceous (145 to 65 mybp) an incredible radiation occurred in parallel with the spread of angiosperm plants. Across a time span of 100 million years, beetles invaded all sorts of biotopes and habitats, including deep soil, caves, freshwater, and forest canopies (Fig. 4.10). Table 4.4 depicts the guilds represented in neotropical rain forests (Lawrence and Britton 1994). To comprehend the true nature of neotropical beetle diversity, a selection of the more hyperdiverse guilds will need to be explored in depth across the commonest of Amazonian biotopes and habitats in the Basin and its surroundings. New evidence and methodology/technology (McKenna and Farrell 2006) seem to support the taxon pulse model (Erwin 1979, 1985) (Fig. 4.10) in demonstrating downtimes in the evolution of radiating beetles alternating with surge times. If this evolutionary process is connected to common regional *Critical Zone* events, and thus synchronous across family-group taxa, patterns can be discovered, processes discerned, hypotheses tested,

and predictions made. If no connection exists to such common events, and the evolutionary process is asynchronous across family-group taxa, chaos theory must be brought to bear for explanatory insights.

SUMMARY AND GUIDE TO FUTURE RESEARCH, OR 'TAKING A SMALL STEP INTO THE VORTEX'

A faunistic and floristic inventory can be an important and necessary tool for planning and creating protected areas. Such inventories establish the basic information for monitoring community dynamics in space and time. Our study provides methodological information that could assist in the design and creation of protocols for entomofauna inventories in the Amazonian *terra firme* forests and elsewhere. We have demonstrated that it is possible to adequately sample insect family-group taxa for a local area in a relatively short time, inexpensively, even in an area of incredible biodiversity. A series of

Table 4.4 Beetle guilds.

Microhabitat	Basic Food Sources	Guilds
Living vegetative surfaces	Leaves, flowers, cones, seeds, pods	Herbivores, predators, fungivores, pollinivores
Bark and wood surfaces	Wood, detritus	Borers, predators, fungivores, scavengers
Herbaceous plant tissues	Stems, gall and leaf interiors, roots	Borers, herbivores, predators, fungivores
Leaf litter, ground debris	Dead vegetation, rotting wood, decaying flowers and fruits	Predators, fungivores, scavengers, detritivores, seed-predators
Animal nests	Dung, carrion, nest construction materials	Predators, fungivores, detritivores, scavengers
Caves	Carrion, algae	Predators, detritivores, scavengers
Soil	Worms, nematodes, bacteria, etc.	Predators, detritivores, scavengers
Fungi	Fruiting bodies, mycelia	Fungivores, predators
Carrion	Carrion	Predators, detritivores, scavengers
Dung	Pass-throughs	Predators, fungivores, scavengers, detritivores, seed-predators
Littoral zone	Algae, debris, carrion, herbaceous plant tissues	Predators, detritivores, scavengers
Lentic zones	Aquatic plants, detritus	Predators, detritivores, scavengers

fogging plots sampled rigorously across a region at various floristically determined locales, as in the case of Piraña and Tiputini, might better describe the apparent mosaic distribution of canopy and understory (e.g., in plants by Tuomisto et al. 2003, in birds by Terborgh et al. 1990, and Blake 2007). To complement the fogging regime and sample for all terrestrial arthropod species and other organisms, larger and more diversely sampled plots will be necessary. Geodatabasing and DNA barcoding will help streamline future canopy-sampling efforts.

The results of canopy sampling from 1994 to 1996 hint at a variable degree of beetle-species turnover across feeding guilds and across a forest mosaic within short distances. The distribution of beetle species in the Western Amazon Basin likely is arrayed in a mosaic or discontinuous pattern rather than being evenly distributed across the local landscape. How that pattern is shaped by the distribution of trees or whether it constitutes an oligarchy remains to be tested. Pitman and colleagues (2001) predicted that the tree oligarchy pattern was driven by varying degrees of habitat specialization, particularly to soil type. With our present knowledge, we cannot say whether subtle forest, soil, or climatic differences, or perhaps something historical, accounts for the patterns we observed in species turnover or morphospecies richness. However,

we predict that a central ecological factor driving beetle-diversity patterns in Amazonian rainforest canopies is the level of tree (either as food or habitat) host specificity of adult and immature stages. If we accept the oligarchy hypothesis that the same suite of tree species and genera are found consistently throughout Amazonia at densities of at least one individual per hectare, then we predict that beetle lineages associated with those trees also are found in oligarchic hierarchies of taxa, probably at the level of genus or tribe. The taxonomic data further suggest that, over evolutionary time, beetle radiation patterns have been shaped by taxon pulses.

Biodiversity research lies at the crossroads of taxonomy, ecology, and evolutionary biology. We need to combine experimental research on ecological and evolutionary mechanisms with taxonomic survey data to make more precise estimates of insect richness in neotropical forests. Lists of morphospecies compiled either via cursory observation of morphological characters or DNA-sequence variations are merely estimates of biorichness unless they are put into evolutionary context, that is, given names and classified. Biodiversity research requires phylogenetic perspective. Testing the oligarchy hypothesis, or any hypothesis of insect community structure, is not possible without more alpha-taxonomic work on neotropical phytophagous insects. Proper use of molecular tools and geospatial databasing of museum-vouchered specimens, along with sound alpha-taxonomy will arm the next generation of biodiversity scientists to monitor the effects of disturbances on Amazonian rainforests.

ACKNOWLEDGMENTS

Foggers: The following (in order of number of times of participation) were those who treaded into the forest at 0300 hrs to set up collection sheets and clear paths for operations, a truly dedicated bunch. **First team** (also includes substantial specimen sorting): Pablo E. Araujo, Ma. Cleopatra Pimienta, Sandra Enríquez, Fabian Bersosa, Ruben Carranco, María Teresa Lasso, Vladimir Carvajal, Ana Maria Ortega, Paulina Rosero, Andrea Lucky, Sarah Weigel, Valeria Granda. **Sometimes or one-time field assistants** (in alphabetical order): Mila Coca Alba, Gillian Bowser, Franklin, Paulo Guerra, Henry, Peter Hibbs, Amber Jonker, Pella Larsson, Keeta DeStefano Lewis, Jennifer Lucky, Ana Mariscal, Marinez Marques, Mayer, Raul F. Medina, Wendy Moore, Karen Ober, Monica O'Chaney, Kristina Pfannes, Mike G. Pogue, Wendy Porras, Theresia Radtke, Jennifer Rogan, Leah Russin, Mercedes Salgado, Linda Sims, Dawn Southard, George L. Venable, Joe Wagner Jr., Winare. **Museum team** (in alphabetical order): The following individuals provided all those important things in the Museum that allowed production: Ma. Cleopatra Pimienta, Valeria Aschero, Gary H. Hevel, Jonathan Mawdsley, Mike G. Pogue, Linda Sims, Warren Steiner, George L. Venable, Carol Youmans. **Technical assistance**: Grace Servat for assistance with the Sigma Plot program for graphing our results; Valeria Aschero for developing the initital GIS system for Mordellidae; Rob Colwell for assistance with the EstimateS program; Karie Darrow for digital image preparation; Charles Bellamy, Mary Liz Jameson, Paul Johnson, Jonathan Mawdsley, Brett Ratcliff, Paul Skelly, and Warren Steiner for generic determinations of some beetles used in our study. **Grants**: The following provided the important thing in the National Museum of Natural History that allowed everything, funding: the Programs NLRP (Richard Vari, P.I.), BSI (George Zug, P.I.), the Department of Entomology, and the Casey Fund (Entomology). Field support from Ecuambiente Consulting Group S.A. in Quito, Ecuador, allowed participation of several Ecuadorian students at Piraña Station, as well as logistics.

REFERENCES

Adams, D. 1980. *The Restaurant at the End of the Universe.* Random House, New York. 208 pp.

Adis, J. (ed) 2002. *Amazonian Arachnida and Myriapoda: Identification Keys to All Classes, Orders, Families, Some Genera, and Lists of Known Terrestrial Species.* Pensoft Publishers, Sofia, Bulgaria. 590 pp.

Adis, J. and H. O. R. Schubart. 1984. Ecological research on arthropods in central Amazonian forest ecosystems with recommendations for study procedures. Pp. 111–144. *In* J. H. Cooley and F. B. Golley (eds). *Trends in Ecological Research for the 1980s.* NATO Conference Series. Series I: Ecology. Plenum Press, New York.

Admundson, R., D. D. Richter, G. S. Humphreys, E. G. Jobbágy, and J. Gaillardet. 2007. Coupling between biota and earth materials in the critical zone. *Elements* 3: 327–332.

Blake, J. G. 2007. Neotropical forest bird communities: a comparison of species richness and composition at local and regional scales. *Condor* 109: 237–255.

Bollion, M. and G. Onore. 2001. *Mariposas del Ecuador. Volume 10a.* Centro de Biodiversidad y Ambiente, Ecuador. 171 pp.

Borkent, A. and G. R. Spinelli. 2007. *Neotropical Ceratopogonidae (Diptera: Insecta).* Pensoft Publishers, Sofia, Bulgaria. 198 pp.

Buchwalter, D. B., D. J. Cain, C. A. Martin, L. Xie, S. N. Luoma, and T. Garland. 2008. Aquatic insect ecophysiological traits reveal phylogenetically based differences in dissolved cadmium susceptibility. *Proceedings of the National Academy of Sciences USA* 105: 8321–8326.

Campbell, K. E., C. D. Frailey, and L. Romero-Pittman. 2006. The Pan-Amazonian Ucayali Peneplain, late Neogene sedimentation in Amazonia, and the birth of the modern Amazon River system. *Palaeogeography Palaeoclimatology Palaeoecology* 239: 166–219.

Coscarón, S. and C. L. Coscarón Arias. 2007. *Neotropical Simuliidae (Diptera: Insecta)* Pensoft, Sofia, Bulgaria. 685 pp.

Cuevara, D., A. Iorio, and P. Backhuys. 2002. *Mariposas del Ecuador (Continental y Galapagos). Volume 17A: Sphingidae.* Backhuys Publishers BV, Leiden, The Netherlands. 234 pp.

D'Abrera, B. 1981. *Butterflies of the Neotropical Region.* Lansdowne Editions in association with E.W. Classey, East Melbourne, Australia. 172 pp.

de Fátima Rossettia, D., P. Mann de Toledob, and A. M. Góes. 2005. New geological framework for Western Amazonia (Brazil) and implications for biogeography and evolution. *Quaternary Research* 63: 78–89.

Dominguez, E., C. Molineri, M. L. Pescador, M. D. Hubbard, and C. Nieto. 2006. *Ephemeroptera of South America.* Pensoft Publishers, Sofia, Bulgaria. 646 pp.

Erwin, T. L. 1979. Thoughts on the evolutionary history of ground beetles: hypotheses generated from comparative faunal analyses of lowland forest sites in temperate and tropical regions. Pp. 539–592. *In* T. L. Erwin, G. E. Ball, D. R. Whitehead, and A. L. Halpern (eds). *Carabid Beetles:*

Their Evolution, Natural History, and Classification. Proceedings of the First International Symposium of Carabidology. Dr. W. Junk Publishers, The Hague.

Erwin, T. L. 1982. Tropical forests: their richness in Coleoptera and other arthropod species. *Coleopterists Bulletin* 36: 74–75.

Erwin, T. L. 1983a. Beetles and other arthropods of the tropical forest canopies at Manaus, Brasil, sampled with insecticidal fogging techniques. Pp. 59–75. *In* S. L. Sutton, T. C. Whitmore, and A. C. Chadwick (eds). *Tropical Rain Forests: Ecology and Management.* Blackwell Scientific Publications, Oxford, United Kingdom.

Erwin, T. L. 1983b. Tropical forest canopies, the last biotic frontier. *Bulletin of the Entomological Society of America* 29: 14–19.

Erwin, T. L. 1985. The taxon pulse: a general pattern of lineage radiation and extinction among carabid beetles. Pp. 437–472. *In* G. E. Ball (ed). *Taxonomy, Phylogeny, and Zoogeography of Beetles and Ants: A Volume Dedicated to the Memory of Philip Jackson Darlington Jr. 1904–1983.* Dr. W. Junk Publishers, The Hague.

Erwin, T. L. 1991. Natural history of the carabid beetles at the BIOLAT Rio Manu Biological Station, Pakitza, Perú. *Revista Peruana de Entomologia* 33: 1–85.

Erwin, T. L. 2004. The biodiversity question: how many species of terrestrial arthropods are there? Pp. 258–269. *In* M. Lowman and B. Brinker (eds). *Forest Canopies.* Academic Press London, United Kingdom.

Erwin, T. L. and J. Adis. 1982. Amazon inundation forests: their role as short-term refuges and generators of species richness and taxon pulses. Pp. 358–371. *In* G. Prance (ed). *Biological Diversification in the Tropics.* Columbia University Press, New York.

Erwin, T. L., M. C. Pimienta, O. E. Murillo, and V. Aschero. 2005. Mapping patterns beta-diversity for beetles across the Western Amazon basin: a preliminary case for improving inventory methods and conservation strategies. *Proceedings of the California Academy of Sciences, Series 4* 56: 72–85.

Flint, O. S., Jr., R. W. Holzenthal, and S. C. Harris. 1999. *Catalog of the Neotropical Caddisflies (Trichoptera).* Ohio Biological Survey, Columbus, Ohio. 239 pp.

Gates Clarke, J. F. 1971. Neotropical Microlepidoptera. XIX. Notes on and new species of Oecophoridae (Lepidoptera). *Smithsonian Contributions to Zoology* 95: 1–39.

Godoy, C., X. Miranda, and K. Nishida. 2006. *Membrácidos de la América Tropical/Treehoppers of Tropical America.* INBio: The Costa Rican Biodiversity Institute, Costa Rica. 352 pp.

Goloboff, P. A. 1995. A revision of the South American spiders of the family Nemesiidae (Aranea, Mygalomorphae). Part 1: Species from Peru, Chile, Argentina and Uruguay. *Bulletin of the American Museum of Natural History* 224: 1–189.

Guimarães, J. H. 1997. *Systematic Database of Diptera of the Americas South of the United States.* Editora Pleiade/FAPESP, São Paulo. 286 pp.

Haffer, J. 1982. General aspects of the refuge theory. Pp. 6–24. *In* G. Prance (ed). *Biological Diversification in the Tropics.* Columbia University Press, New York.

Hammond, P. 1992. Species inventory. Pp. 17–39. *In* B. Groombridge (ed). *Global Biodiversity: Status of the Earth's Living Resources.* Chapman and Hall, London.

Heckman, C. W. 2001. *Collembola: Illustrated Keys to Known Families, Genera, and Species in South America.* Kluwer Academic Publishers, Boston, Massachusetts. 418 pp.

Heckman, C. W. 2002. *Ephemeroptera: Illustrated Keys to Known Families, Genera, and Species in South America.* Kluwer Academic Publishers, Boston, Massachusetts. 428 pp.

Heckman, C. W. 2003. *Plecoptera: Illustrated Keys to Known Families, Genera, and Species in South America.* Kluwer Academic Publishers, Boston, Massachusetts. 338 pp.

Heckman, C. W. 2006. *Odonata – Anisoptera. Illustrated Keys to Known Families, Genera, and Species in South America.* Springer, Dordrecht, The Netherlands. 733.

Heppner, J. B. 1984. *Atlas of Neotropical Lepidoptera.* W. Junk, Boston, Massachusetts.

Howden, A. T. 1976. *Pandeletius* of Venezuela and Colombia (Curculionidae: Brachyderinae: Tanymechini). *Memoirs of the American Entomological Institute* 24: 1–316.

Kempf, W. W. 1972. *Catálago das Formigas da Região Neotropical (Hym. Formicidae).* Vozes, Rio de Janeiro, Brasil. 344 pp.

Kitching, R. L. 2006. Crafting the pieces of the diversity jigsaw puzzle. *Science* 313: 1055–1057.

Kitching, R. L., K. Hurley, and L. Thalib. 2003. Tree relatedness and the similarity of canopy insect assemblages: pushing the limits? Pp. 329–340. *In* Y. Basset, V. Novotny, S. E. Miller, and R. L. Kitching (eds). *Arthropods of Tropical Forests: Spatio-Temporal Dynamics and Resource Use in the Canopy.* Cambridge University Press, Cambridge, United Kingdom.

Kormilev, N. A. 1975. Neotropical Aradidae in the collections of the California Academy of Sciences, San Francisco (Hemiptera: Heteroptera). *Occasional Papers of the California Academy of Sciences* 122: 1–28.

Kremen, C., A. Cameron, A. Moilanen, S. J. Phillips, C. D. Thomas, H. Beentje, J. Dransfield, B. L. Fisher, F. Glaw, T. C. Good, G. J. Harper, R. J. Hijmans, D. C. Lees, E. Louis, R. A. Nussbaum, C. J. Raxworthy, A. Razafimpahanana, G. E. Schatz, M. Vences, D. R. Vieites, P. C. Wright, and M. L. Zjhra. 2008. Aligning conservation priorities across taxa in Madagascar with high-resolution planning tools. *Science* 320: 222–226.

Lawrence, J. F. and E. B. Britton. 1994. *Australian Beetles.* Melbourne University Press, Australia. 184 pp.

Lenat, D. R. and V. H. Resh. 2001. Taxonomy and stream ecology – the benefits of genus- and species-level identifications. *Journal of the North American Benthological Society* 20: 287–298.

Lourenco, W. R. 2002. *Scorpions of Brazil.* Les Editions de l'If. 308 pp.

Lucky, A., T. L. Erwin, and J. D. Witman. 2002. Temporal and spatial diversity and distribution of arboreal Carabidae

(Coleoptera) in a western Amazonian rain forest. *Biotropica* 34: 376–386.

Macia, M. J. and J. C. Svenning. 2005. Oligarchic dominance in western Amazonian plant communities. *Journal of Tropical Ecology* 21: 613–626.

Martins, U. R. 1998. *Cerambycidae Sul-Americanos (Coleoptera): Taxonomia. Volumes 1 and 2*. Sociedade Brasileira de Entomologia, Brasil. 217 and 195 pp.

Martins, U. R. 1999. *Cerambycidae Sul-Americanos (Coleoptera): Taxonomia Volume 3*. Pensoft Publishers, Sofia, Bulgaria. 418 pp.

Martins, U. R. 2003. *Cerambycidae Sul-Americanos (Coleoptera): Taxonomia Volume 6*. Pensoft Publishers, Sofia, Bulgaria. 232 pp.

Mathis, W. N. 1980. Studies of Ephydrinae (Diptera: Ephydridae). III: Revisions of some Neotropical genera and species. *Smithsonian Contributions to Zoology* 303: 1–50.

McKenna, D. D. and B. D. Farrell. 2006. Tropical forests are both evolutionary cradles and museums of leaf beetle diversity. *Proceedings of the National Academy of Sciences USA* 103: 10947–10951.

Mound, L. A. and R. Marullo. 1996. *The Thrips of Central and South America: An Introduction (Insecta: Thysanoptera)*. Associated Publishers, Gainesville, Florida. 487 pp.

Novotny, V., Y. Basset, S. E. Miller, R. L. Kitching, M. Laidlaw, P. Drozd, and L. Cizek. 2004. Local species richness of leaf-chewing insects feeding on woody plants from one hectare of a lowland rainforest. *Conservation Biology* 18: 227–237.

Novotny, V., P. Drozd, S. E. Miller, M. Kulfan, M. Janda, Y. Basset, and G. D. Weiblen. 2006. Why are there so many species of herbivorous insects in tropical rainforests? *Science* 313: 1115–1118.

Nyman, T., B. D. Farrell, A. G. Zinovjev, and V. Vikberg. 2006. Larval habits, host-plant associations, and speciation in nematine sawflies (Hymenoptera: Tenthredinidae). *Evolution* 60: 1622–1637.

Odegaard, F. 2006. Host specificity, alpha- and beta-diversity of phytophagous beetles in two tropical forests in Panama. *Biodiversity and Conservation* 15: 83–105.

Pinas, F., I. Manzano, and S. Rab-Green. 1997. *Mariposas del Ecuador. Volume 20. Arctiidae: Arctiinae y Pericopinae*. Centro de Biodiversidad y Ambiente, Ecuador. 215 pp.

Pinas Rubio, F. and I. Manzano Pesantez. 1997. *Mariposas del Ecuador. Volume 1: Generos*. Centro de Biodiversidad y Ambiente, Ecuador. 115 pp.

Pitman, N. C. A., J. W. Terborgh, M. R. Silman, P. Nunez, D. A. Neill, C. E. Ceron, W. A. Palacios, and M. Aulestia. 2001. Dominance and distribution of tree species in upper Amazonian terra firme forests. *Ecology* 82: 2101–2117.

Pokon, R., V. Novotny, and G. A. Samuelson. 2005. Host specialization and species richness of root-feeding chrysomelid larvae (Chrysomelidae, Coleoptera) in a New Guinea rain forest. *Journal of Tropical Ecology* 21: 595–604.

Porter, C. C. 1967. A revision of the South American species of *Trachysphyrus* (Hymenoptera: Ichneumonidae). *Memoirs of the American Entomological Institute* 10: 1–368.

Price, P. W. 2003. *Macroevolutionary Theory on Macroecological Patterns*. Cambridge University Press, Cambridge, United Kingdom. 302 pp.

Salo, M. and A. Pyhala. 2007. Exploring the gap between conservation science and protected area establishment in the Allpahuayo-Mishana National Reserve (Peruvian Amazonia). *Environmental Conservation* 34: 23–32.

Schulman, L., K. Ruokolainen, L. Junikka, I. E. Saaksjarvi, M. Salo, S. K. Juvonen, J. Salo, and M. Higgins. 2007. Amazonian biodiversity and protected areas: do they meet? *Biodiversity and Conservation* 16: 3011–3051.

Spangler, P. J. and S. Santiago. 1987. A revision of the Neotropical aquatic beetle genera *Disersus, Pseudodisersus*, and *Potamophilops* (Coleoptera: Elmidae). *Smithsonian Contributions to Zoology* 446: 1–40.

Terborgh, J., S. K. Robinson, T. A. Parker, C. A. Munn, and N. Pierpont. 1990. Structure and organization of an Amazonian forest bird community. *Ecological Monographs* 60: 213–238.

Tuomisto, H., K. Ruokolainen, and M. Yli-Halla. 2003. Dispersal, environment, and floristic variation of western Amazonian forests. *Science* 299: 241–244.

Valencia, R., R. B. Foster, G. Villa, R. Condit, J. C. Svenning, C. Hernandez, K. Romoleroux, E. Losos, E. Magard, and H. Balslev. 2004. Tree species distributions and local habitat variation in the Amazon: large forest plot in eastern Ecuador. *Journal of Ecology* 92: 214–229.

Weiblen, G. D., C. O. Webb, V. Novotny, Y. Basset, and S. E. Miller. 2006. Phylogenetic dispersion of host use in a tropical insect herbivore community. *Ecology* 87: S62–S75.

Chapter 5

INSECT BIODIVERSITY IN THE AFROTROPICAL REGION

Clarke H. Scholtz and Mervyn W. Mansell

Department of Zoology and Entomology, University of Pretoria, Pretoria 0002, South Africa

Insect Biodiversity: Science and Society, 1st edition. Edited by R. Foottit and P. Adler
© 2009 Blackwell Publishing, ISBN 978-1-4051-5142-9

The Afrotropical Region is renowned for its charismatic megafauna and diverse botanical wealth, but these aspects are overshadowed by the sheer diversity of insects that occupy the continent of Africa and its associated islands. The entomofauna of this zoological realm presents enormous challenges to scientists because of the profusion of insect taxa that have evolved in the multitude of ecological systems of the region and because of historical and sociopolitical issues that afflict the continent. Insects and other invertebrates drive the ecosystems on which the megafauna and flora and, ultimately, humans depend. Vertebrate animals depend on ecosystems for survival – terrestrial ecosystems depend on insects for their establishment and maintenance. Insects provide numerous primary environmental services, from recycling of nutrients to pollination, besides their fundamental contribution to food resources of many vertebrate animals. Insects, consequently, should be at the core of any country's commitment to the International Convention on Biodiversity.

Africa's rich biodiversity, however, is not matched by its economic wealth, as it has the dubious distinction of being home to most of the world's poorest nations. A discussion of insect biodiversity science, consequently, should be seen in this context. The few studies undertaken so far on African insect biodiversity (Miller and Rogo 2001) have arrived at similar conclusions: species numbers are vast, but not accurately known; insect habitats are fast declining; few African countries house insect collections or employ taxonomists (Table 5.1); and if collections of African insects exist, they are housed in foreign museums, either those of the ex-colonial powers, mostly in Europe, or in those based on more recent collections such as some in the USA (Table 5.2). No collections have been returned to the countries from which the insects were collected, but some foreign countries, such as Belgium, that have African collections are preparing to repatriate the information. We believe that before poor countries start to recognize the importance of insect biodiversity studies, they need to be convinced of the material benefit of such research.

Insects have direct, easily understandable value, as well as less tangible value, in poor countries. Examples of the former are insects harvested for food (e.g., mopane caterpillars in large parts of southern Africa, termite alates and locusts over much of the continent, and Ephemeroptera in the Rift Valley Lake systems) and those that produce useful products (e.g., honey and silk

Table 5.1 Major entomological collections in the Afrotropical Region (based on Miller and Rogo 2001).

Country	City	Institution
Angola	Dundo	Museu do Dundo
Kenya	Nairobi	National Museums of Kenya
Mozambique	Maputo	Museum Nacionale de Mozambique
Namibia	Windhoek	State Museum of Namibia
Senegal	Dakar	Institut Fondamental d'Afrique Noire
South Africa	Cape Town	Iziko Museums
South Africa	Grahamstown	Albany Museum
South Africa	Pietemaritzburg	Natal Museum
South Africa	Pretoria	Plant Protection Research Institute
South Africa	Pretoria	Transvaal Museum
Uganda	Kampala	Kawanda Research Station
Zimbabwe	Bulawayo	Natural History Museum

from native silk worms). Examples of the latter include pollinators, pest-control agents, and biodegraders in the form of dung beetles, termites, and flies. Furthermore, insects are critical elements of the food chain for fish, the most important source of animal protein for humans in large parts of Africa. In contrast, some insects have negative effects as vectors of disease agents and as agricultural pests, and these taxa also require further study and monitoring.

Miller and Rogo (2001) reviewed some of the aspects of insect biodiversity science in an excellent synthesis of African and Afrotropical insect diversity. They set the scene by highlighting the insect richness of the region (more than 100,000 described species) and its potential for sustainable use if comprehensively understood and documented. They also provided a detailed background to current and historical resources, including a list of institutions in Africa, Europe, and the USA that hold the collections of Afrotropical insects. They emphasized the difficulty of locating specimens and literature references, and provided a preliminary list of resources where such information could be obtained. Most importantly, they highlighted a series of challenges that face insect biodiversity science in Africa, which they had identified in an earlier study (Miller et al. 2000).

Table 5.2 Some major entomological collections of Afrotropical insects outside Africa (based on Miller and Rogo 2001).

Country	City	Institution
Austria	Vienna	Naturhistorisches Museum Wien
Belgium	Tervuren	Royal Museum of Central Africa
England	London	The Natural History Museum
England	Oxford	Hope Museum
France	Paris	Museum National d'Histoire Naturelle
Germany	Berlin	Museum für Naturkunde der Humboldt Universität
Germany	Munich	Staatssammlungen der Bayerischen Staates
Hungary	Budapest	National Museum
Italy	Florence	Centro di Studio per la Faunistica ed Ecologia Tropicali
Sweden	Stockholm	Naturhistoriska Riksmuseet
USA	Chicago	Field Museum
USA	New York	American Museum of Natural History
USA	Pittsburgh	Carnegie Museum
USA	San Francisco	California Academy of Science
USA	Washington	Smithsonian Institution

Miller et al. (2000) proposed a biodiversity plan for Africa, with three main components: (1) an information-management program to organize and make available the large volume of data that already exists but is not generally accessible, (2) a series of field projects to evaluate the use of insects as indicator organisms and to quantify their roles in ecosystem processes, and (3) training and participatory technology transfer. To these, we add two more components: (4) to emphasize Africa as the center of biodiversity of many higher insect taxa, with examples of sentinel groups, and (5) the importance of identifying insects with potential to become major, Africa-wide pests, and training appropriate taxonomists to identify them. We will develop these points further.

The purpose of the current chapter is not to repeat the information provided by Miller and collaborators, but to try to address some of the questions they raised and challenges they identified, and to add new insights into African biodiversity science by means of insect groups with which we are familiar.

WHAT DO WE KNOW ABOUT AFROTROPICAL INSECTS?

All insect orders, except the Grylloblattodea, are present in Africa. One of these orders, the recently described Mantophasmatodea, occurs only in Africa. Despite this richness, few taxa have been collected extensively and even fewer have been studied comprehensively; those that have been studied have mostly been studied outside of Africa. South Africa is the one exception, with reasonable collections and a history of taxonomic research. Table 5.1 provides a list of African museums with reasonably well-maintained and well-curated collections. This situation, however, reflects the poor state of affairs – only 8 countries, of more than 50, have museums housing insect collections. It is a barometer of the state of insect taxonomic exploration in Africa.

AN INFORMATION-MANAGEMENT PROGRAM

The most immediate challenge to understanding and applying Africa's insect biodiversity lies in documenting new data, including the description of species, and in the collation of taxonomic and biological data in existing historical resources and their coordination with preserved specimens in various institutions scattered throughout the world. This major issue was raised by Miller and Rogo (2001, p. 200) when they posed the question 'But how can these resources be unlocked to make information readily available for use?' An attainable solution exists to this challenge.

Vast amounts of African specimen-related data are incarcerated in the biological collections of the world, but the information has been difficult to access because of logistical and physical constraints. These constraints include the disparate nature of collections and the time and expense involved in visiting different museums in many parts of the globe. The challenge lies in the development of a system to assemble and collate data from many sources, with the simultaneous incorporation of burgeoning new information. Until now, conventional filing systems, such as card catalogs, were unable to accommodate the vast amount of potential data or its collation across multiple platforms – from specimen data, to taxonomy, historical literature references, geographical distributions, phenology, host plants, molecular data, and graphics, to name but a few relevant parameters. Electronic computerization did not initially resolve these issues because complex and expensive programming put these resources beyond the reach of most biologists, and personal computers were initially compromised by software limitations.

The advent of Relational Databases, on the basis of mathematical concepts (Codd 1970), has now placed powerful resources within reach of all biologists. Relational database management systems (RDBMS), such as Microsoft Access and My SQL, enable biologists to design systems to record and collate vast amounts of taxonomic and other data within highly flexible and adaptable applications that are universally standard and easily understood. Relational databases, consequently, have become the cornerstones of sound curatorial practice, information exchange, and dissemination in many museums and collections, and considerable advances have been made in this field. The advantages of virtual, electronically collated collection data are obvious, as numerous products can be generated that advance the relevance of museum collections and justify their continued accrual and maintenance. Products that can be generated from well-designed and integrated relational databases include faunal inventories, distributional data for conservation applications and environmental impact assessments, Geographical Information Systems (GIS) modeling (e.g., climate change, conservation area selection, biodiversity hotspots, endemicity), host plant associations and disease vectors for application in agriculture, and development of Web-based products, as well as printed catalogs and capacity building (e.g., student training). Added advantages include large amounts of integrated data now becoming instantly available and electronic databases, which are also virtual collections of specimens that provide a backup insurance against disasters.

The establishment of the Global Biodiversity Information Facility (GBIF), Species 2000, and other international initiatives have brought databasing into sharp focus because the electronic recording of specimen and taxonomic data are fundamental to the objectives of these international programs. Database activities in the African context are sporadic or nonexistent and many institutions are hesitant to embark on programs because of the perceived enormity of the task or reluctance to provide public access to their data. A general misconception exists that the specimen data

in collections are a marketable resource that belongs strictly to a particular institution. This attitude is a major impediment to documentation of the African insect fauna because such data remain beyond the mainstream flow of biodiversity information and are essentially irrelevant and increasingly difficult to justify in terms of maintenance expenses. This parochial attitude is further negated by current accessibility to substantial funding for the electronic recording of museum specimens from GBIF and the South African Node of GBIF, the South African Biodiversity Information Facility (SABIF).

The South African National Collections of Insects and Arachnids of the Agricultural Research Council (ARC) is the leading institution in Africa, with regard to progress in the databasing of invertebrate collections. The most extensive dataset, although not strictly of insects, is the African Arachnid Database (AFRAD) that now includes more than 6000 taxonomic records, which can be viewed on the ARC website. This initiative has also attracted substantial funding and is a collaborative program with the newly established South African National Biodiversity Institute (SANBI). Substantial progress also has been made in the electronic recording of the insect collections, with comprehensive datasets available for African Tephritidae (fruit flies), Isoptera (termites), and Neuroptera (lacewings), and steadily increasing electronic documentation of other groups, including Scarabaeoidea (Coleoptera), Sternorrhyncha (Hemiptera), Thysanoptera, and Apoidea and Chalcidoidea (Hymenoptera).

The Neuroptera dataset can be used as an example of the manner in which data can be recorded, collated, and disseminated for application in fields including biological control and pollination biology (agriculture), conservation biology, and biogeography, as well as fundamental scientific research. This database comprises several discrete databases: Specimens, Taxonomy, Localities, Institutions, Persons, and Bibliographies that are interlinked through key fields. This arrangement facilitates simultaneous queries across the multiple platforms that can be designed to answer specific questions. The database also accommodates hyperlinks to graphics and to PDF files of associated literature, ensuring that all historical and current resources pertaining to specimens (in all museums), taxonomy, and literature, as well as associated persons and graphics, can be accommodated in one reference source. This information can be translated directly into searchable web-based products, as well as electronic and paper catalogs. The first product is a Catalogue of Lacewings (Neuroptera) of the Afrotropical Region, which is accessible through the GBIF portal. This was a GBIF-funded initiative and the results will also be published in hard-copy through SANBI.

Another of the constraints that a database program is designed to address is the compilation of catalogs of additional insect groups. Such catalogs form the basis of research on all biological organisms. At present, only two comprehensive catalogs are available for entire orders of Afrotropical insects. The first, Catalogue of Diptera of the Afrotropical Region (Crosskey 1980), is now 27 years out of date and not available in electronic format. The second, Catalogue of Lacewings (Neuroptera) of the Afrotropical Region, is currently available only in electronic format. The advantages of electronic catalogs are that they accommodate continual updates, so the data remain current. A forerunner to this catalog can be viewed on the Rhodes University website (www.ru.ac.za/academic/departments/zooento/martin/neuroptera.html). The advent of the GBIF and SABIF has provided further incentives and facilities to produce such catalogs.

Compilation of catalogs requires a distillation of all published information on a particular taxon and a realistic, current assessment of the status and extent of knowledge of a particular group, often with remarkable consequences. In the compilation of the Afrotropical lacewing catalog, the number of valid species of Neuroptera described from the Afrotropics has decreased considerably. For example, in the Chrysopidae, one of the largest families of Neuroptera, 286 species have been described, but only 183 of these names are valid, reducing the number of described, valid species by 103 (36%, or more than one-third). Sixty-seven of these synonyms were created by a single author, the Spanish cleric Longinos Navás (1858–1938). The legacy of Navás and several other early authors of Afrotropical lacewings left a trail of taxonomic confusion that continues to impede progress toward an accurate assessment of the Neuroptera of this region. The number of taxonomic names lost to synonymy probably will be offset by new discoveries, but before most of these descriptions are possible, all existing names have to be checked thoroughly against type specimens, many of which have been destroyed, lost, or deposited in obscure collections. These constraints are intractable to real progress that affects many other insect groups as

well. On the positive side, the advent of electronic web documentation of lacewing taxa and literature, particularly by John D. Oswald of Texas A & M University (http://insects.tamu.edu/research/neuropterida/neur_bibliography/bibhome.html) has greatly facilitated research on lacewings by concentrating all literature and taxonomic names into convenient reference sources. This project has been a major contribution to research on Afrotropical lacewings, further emphasizing the need to encourage sound taxonomic procedures and ongoing electronic documentation and dissemination. The prevailing situation for the lacewings substantiates claims by Scholtz and Chown (1995) that total insect numbers for southern Africa (but extrapolated to the whole of Africa) are unlikely to change significantly over what is currently known because synonymy more or less compensates for new species described.

THE ROLE OF INSECTS IN ECOSYSTEM PROCESSES AND AS INDICATORS OF ENVIRONMENTAL QUALITY – DUNG BEETLES AS A CASE STUDY

The principal importance of dung beetles lies in their maintenance of pasture health by burying dung, which has the effect of removing surface wastes, recycling the constituent nutrients, and reducing exposure of livestock to internal parasites. Negative environmental effects that result from a lack of dung beetles were seen in Australia before the introduction of dung beetles adapted to cattle dung. These effects included the loss of grass cover due to the persistence of unburied dung pads, the growth of unpalatable grass around these pads, the leaching of nutrients in surface rainwater runoff, and the buildup of large populations of dung-breeding flies (Waterhouse 1974).

Much of the traditional farming in most of Africa is of a subsistence nature, based on small areas of planted crops per household and on livestock including camels, cattle, donkeys, goats, horses, pigs, and sheep that graze on communal land by day and are penned at night. Human sewage is often deposited near villages where it is removed by dogs, pigs, and dung beetles. Cattle are conservatively estimated to number about 100 million in Africa, while small stock such as sheep and goats probably number three times more. Considerably smaller numbers of beasts of burden are found, mainly in regions with a long tradition of association between them and their owners. All of these animals, however, produce dung. Camel and cattle dung is collected, usually during the dry season, from the field or from the pens where the animals spend the night. It is dried and used as fuel for fires in regions where trees have been destroyed and, in rehydrated form, as building material. In some areas, dung from accumulations in livestock pens is used as fertilizer in crop fields. Although the total quantities of dung produced or the amount used for fuel, mortar, or fertilizer is impossible to estimate, a safe estimate is that the dung produced by day while the animals are out grazing contributes about half the total.

Because we have a reasonable estimate of cattle and small stock numbers in Africa, and the amount of dung they produce, is known, we can calculate the amount of dung available to dung beetles for feeding and breeding, activities that greatly enhance soil nitrogen, soil porosity, and water penetration (Losey and Vaughan 2006). Cattle produce, on average, about 9000 kg of dung each year. If we assume that half of this dung is collected and used for fuel, mortar, and fertilizer, about 4500 kg per animal remains on the ground. If this amount is multiplied by 100 million cattle, we are presented with the staggering amount of 450 million metric tons of dung deposited on the continent per year. Sheep and goats each produce about 900 kg of dung per year, and if the same algorithm based on a number of about 30 million small stock units is applied for calculating the amount returned to the soil, the total amount is about 135 million metric tons. Add to this the dung produced by humans themselves, as well as their pigs, dogs, and beasts of burden, and the total is immense.

Indicators of environmental quality in agroecosystems have received considerable attention in recent years (Riley 2001). Although an acceptable framework for their application is still lacking because of various degrees of incompatibility among different systems, the use of indicators remains desirable as a basis for managing agroecosystems. Here we describe and review the role of the scarabaeine (Scarabaeidae: Scarabaeinae) dung beetles as agricultural indicators.

One of the uses of dung beetles as indicators is to define biogeographical regions and faunal distribution centers to facilitate and validate comparisons among species assemblages in natural and intensively farmed systems. This has been done in some detail in South Africa (Davis 1993; Davis et al. 1999, 2002a, 2004;

Davis and Scholtz 2004). A second role is to characterize pasturage from natural to heterogeneous, to completely transformed, in terms of dung beetles (Davis et al. 2004).

In a review of terrestrial insects as bio-indicators, McGeoch (1998) recommended a nine-step protocol for assessing the value of the insect group. These nine steps can be reduced to three steps, assessing (1) the category (biodiversity, ecological, or environmental) and scale (regional or local) of indication, (2) the specific objectives of the application, and (3) the data collection and rigorous statistical testing of the group to ensure that it fulfills the study objectives and permits well-supported decision making.

Scarabaeine dung beetles are an obvious choice as indicators because they are an integral part of livestock pasture ecology in the warmer, moister climatic regions where they are centered. These regions are situated within the 45° latitudinal limits in areas receiving more than 250 mm of annual rainfall and subject to mean annual temperatures greater than 15°C. The rainfall limit for dung beetles is also roughly the limit for the growth of pasturage suitable for cattle, although small stock survives in drier areas.

Dung beetles as indicators of regional biodiversity

Biogeographical differences among regions can be defined by comparing data on species richness, taxonomic composition, and species abundance structures of local assemblages (Davis 1997, Davis et al. 1999, 2002a, 2004). Such differences are usually related to climatic factors, although edaphic and vegetational characteristics also vary at a regional scale, for example, in rainforests, winter-rainfall shrublands, and deep desert sands. Whereas vegetation largely depends on climate, edaphic factors can be independent of climate. Analysis of such data permits one to determine natural regional boundaries, define the distribution centers of groups of species, and identify centers of endemism.

A global comparison of 46 widely separated local faunas provides insight into some subcontinental patterns (Davis et al. 2002b). Parsimony analysis of endemism, using these data, suggests a subcontinental centering according to ecoclimatic zones, primarily forest, woody savanna, highland grass, and arid or winter-rainfall climate.

Dung beetles as indicators of habitat transformation

Dung beetles are useful indicators of effects related to local transformation from natural habitat to farmland. The consequences are primarily related to modification of natural vegetation (Estrada and Coates-Estrada 2002, Halffter and Arellano 2002) and the loss of indigenous mammals, particularly large monogastric species that void large fibrous droppings. To conserve local biodiversity in a heterogeneous environment, sufficiently large fragments must remain to support specialists, to maximize species richness, and to maintain pasture health. However, the reality is a heterogeneous system comprising reserves, naturally farmed patches, and intensively farmed areas in various proportions. Relative naturalness can be categorized by surveying differences in dung-beetle assemblages between reserves and natural to disturbed farm habitats.

AFRICA-WIDE PESTS AND TRAINING APPROPRIATE TAXONOMISTS – FRUIT FLIES AS A CASE STUDY

Insect pests account for about 50% of crop losses in Africa, from plant establishment through plant growth, crop maturation, and storage. Fruit flies are some of the world's most devastating crop pests, causing millions of dollars in production loss each year. In Africa, many species attack fruits, vegetables, and native hosts. These pests include at least 12 species that are endemic to Africa, including the world's worst fruit pest, *Ceratitis capitata*, the Mediterranean fruit fly or medfly.

Fruit flies not only constrain fruit export from Africa, but also cause extensive crop losses that severely affect sustainable rural livelihoods and food security in many impoverished regions of Africa. Four species of Asian fruit flies that recently have invaded Africa are of particular concern. *Bactrocera invadens* and *B. cucurbitae* are now widespread in Africa and are devastating pests that attack a wide variety of fruit and vegetable crops including mango, citrus, guava, and tomato, as well as a large number of indigenous hosts. Of the other two invasive species, *Bactrocera zonata* (peach fly) is currently known only from Egypt, while *B. latifrons* (solanum fly) was recently detected in Tanzania. The African Fruit Fly Initiative (AFFI) was established at the International Centre of Insect Physiology and Ecology

(ICIPE), Nairobi, Kenya, to address the above concerns (Ekesi and Billah 2006). The program was designed to cover a number of aspects, including surveys and monitoring, development of control methods, training of scientists and agricultural personnel in the identification and management of fruit flies, and information and support for small-scale farmers. During the initial phases, traps and lures were provided to several countries, especially the East African community of Kenya, Tanzania, Uganda, and Zanzibar. This initial survey was responsible for the first detection of the highly invasive *B. invadens* in Africa (Lux et al. 2003). A number of training courses also have been provided by the AFFI, and a comprehensive manual entitled 'A Field Guide to the Management of Economically Important Tephritid Fruit Flies in Africa' has emanated from the program.

Few surveys to detect and monitor fruit flies are otherwise in progress in Africa, and few professionally competent taxonomists are available to provide identifications of flies detected in these surveys. With one exception, Africa depends on specialists in Britain and Europe for taxonomic expertise to underpin fruit fly research on the continent. Training of tephritid taxonomists beyond the parataxonomic level is, consequently, an urgent priority for fruit fly research and management in Africa.

Apart from the countries surveyed by the AFFI, only, Benin, Namibia, and South Africa conduct regular monitoring programs. Most attention now is focused on the invasive species, in particular *B. invadens*. This species is destined to have a major effect on fruit and vegetable production in Africa and ultimately on sustained agricultural production on the continent. Ideally, intensive continent-wide programs should be launched to monitor the spread of this and other pest species and to inform agricultural authorities of its presence and threat to their countries. Once again, the AFFI has been proactive in this regard, with a major proposal for funding having been submitted to the Food and Agriculture Organization (FAO) to deal with the threat posed by the invasive species.

Invasive species of concern in Africa

Bactrocera invadens, the invader fly, recently was described by Drew et al. (2005). It is native to Sri Lanka, but recently invaded Africa and has spread rapidly across the continent and the Comores Islands.

Since its first detection in Kenya in 1993 (Lux et al. 2003), *B. invadens* has spread to 28 countries in Africa and infests at least 40 host plants. *Bactrocera invadens* belongs to a complex of superficially similar Oriental species. Three members of this 'dorsalis' complex are serious to extremely devastating pests, while other members of the complex are fairly benign. Distinguishing the species requires the services of an experienced taxonomist. *Bactrocera invadens* attacks a wide range of hosts, including mango, guava, citrus, papaya, and the wild hosts *Strychnos* and especially marula (*Sclerocarya birrea*). It also infests citrus and cashew nuts. The potentially wide host range also makes it an extremely dangerous invasive species, further exacerbated by its aggressive behavior, whereby indigenous species are displaced and multiple ovipositions can occur in one fruit. Males are attracted to methyl eugenol.

Bactrocera cucurbitae, the melon fly, is native to Asia, but has been introduced into Africa and is currently known from Egypt, Kenya, Senegal, and Tanzania, as well as Mauritius and Réunion. *Bactrocera cucurbitae* is a serious pest of the Cucurbitaceae, but has been reared from more than 125 species of plants, including many noncucurbits such as tomato, beans, quince, jackfruit, avocado, fig, granadilla, mango, papaya, peach, and tree tomato. The potentially wide host range makes this invasive species extremely dangerous. Males are attracted to cue lure.

Bactrocera (*Bactrocera*) *zonata* (Saunders), the peach fruit fly, has extended its range from Asia and is currently known from Egypt and Mauritius in the Afrotropical Region. It attacks many crops including peach, guava, apple, bitter gourd, date palm, papaya, pomegranate, quince, mango, tropical almond, citrus, and watermelon. The wide host range and the danger of the species spreading southward from Egypt make it a serious potentially invasive species in Africa. Males are attracted to methyl eugenol.

Bactrocera (*Bactrocera*) *latifrons* (Hendel), the solanum fruit fly, only recently has been detected in Tanzania, on egg plant, and is currently known only from Kenya and Tanzania. Three of the species, *B. invadens*, *B. zonata*, and *B. latifrons*, are morphologically similar and easily confused. An adventive population is known from Hawaii, and the species is also known from China, India, Laos, Malaysia, Pakistan, Sri Lanka, Taiwan, and Thailand. It is mainly a pest of solanaceous crops, but the known host list includes *Capsicum annuum* (chilli), *Solanum incanum*, *S. nigrum* (black nightshade), *S. torvum* (terongan), *S. melongena* (egg plant), *S. sisymbriifolium*, *Lycopersicon pimpinellifolium* (currant tomato), *L. esculentum* (tomato), *Baccaurea motleyana* (rambai), and *Malus domestica* (apple). A new lure, latilure, has been developed as an attractant of this fly.

African indigenous fruit flies of economic importance

Ceratitis (*Ceratitis*) *capitata* (Wiedemann), the Mediterranean fruit fly or medfly, is native to Africa and has become one of the world's most notorious crop pests. It has spread into many parts of the world, and adventive populations occur in all continental regions except tropical Asia and North America, where it has been eradicated several times. Millions of dollars are spent annually in preventing establishment in North America. It is also present on most Indian Ocean islands and even Hawaii. The species has an extensive host range (White and Elson-Harris 1992) and is particularly injurious to tropical and subtropical fruits. Males are readily attracted to trimedlure and terpinyl acetate.

Ceratitis (*Pterandrus*) *rosa* Karsch, the Natal fruit fly, is a major pest species that can be as devastating as the medfly in areas where it occurs. It is native to southern and eastern Africa, extending as far north as southern Kenya. Adventive populations exist in Mauritius and Réunion. It is also a major pest of tropical and subtropical fruits. White and Elson-Harris (1992) provided an extensive host list. Males are attracted to trimedlure and terpinyl acetate.

Ceratitis (*Pterandrus*) *fasciventris* Bezzi previously has been confused with *C. rosa*, and many country records for the Natal fruit fly are actually of this species. It replaces *C. rosa* in western Kenya, across to western Africa, and has a similar host range. Males are attracted to trimedlure and terpinyl acetate.

Ceratitis (*Pterandrus*) *anonae* Graham is a western African species with an extensive host range (White and Elson-Harris 1992). No effective lure is available for this species.

Ceratitis (*Ceratalaspis*) *cosyra* (Walker), the marula fruit fly, is widespread in eastern Africa, as is its native host, marula (*Sclerocarya birrea*). The marula fruit fly has a more limited host range than do other *Ceratitis* species, but it is, nonetheless, a serious pest of mango (*Mangifera indica*). It is attracted to terpinyl acetate but not trimedlure.

The above *Ceratitis* species have a vast collective range of cultivated and wild hosts. They are all serious pests of mango and other tropical and subtropical fruits. Mango is a sentinel crop in Africa, both in terms of export potential and sustainable livelihoods, and many rural families depend on this fruit as an essential component of their diet. These *Ceratitis* pests, now exacerbated by the invasive species, are a constraint to agricultural development and food security in Africa.

A number of African *Dacus* species, including *Dacus (Dacus) bivittatus* (Bigot) (pumpkin fly), *Dacus (Didacus) ciliatus* Loew (lesser pumpkin fly), *Dacus (Didacus) vertebratus* Bezzi (jointed pumpkin fly), and *Dacus (Didacus) frontalis* Becker, infest a wide range of the Cucurbitaceae. They are more limited in their host ranges than are the *Ceratitis* species, but nonetheless have a similar effect on rural agriculture and food security. Curcubitaceae are widely cultivated and are an important dietary component in many African countries, and are used extensively as container gourds in many rural households.

Bactrocera (Daculus) oleae (Gmelin), the olive fly, is a unique pest because of its stenophagous host range, which is limited to olives (*Olea*). It is widely distributed in Africa and has spread throughout the olive zone of the Mediterranean. Besides its impact on table olives, presence of its larvae in fruit also can reduce the quality of the oil if the fruits are stored for a long period (White and Elson-Harris 1992).

Trirhithrum nigerrimum (Bezzi) and *Trirhithrum coffeae* Bezzi, the coffee flies, are also widespread in Africa and are major pests of coffee, constraining the cultivation and export of coffee from the continent. They consequently impede the generation of export revenue in several coffee-producing African countries.

SENTINEL GROUPS

The Afrotropical Region is the second most species-rich zoogeographical region in the world (Cowling and Hilton-Taylor 1994). Despite this, the level of focused biodiversity exploration for insects has been almost minimal for most countries, and few invertebrate groups are well known across the entire region. Only the two above-mentioned catalogs encompassing complete orders of Afrotropical insects are available. The Catalogue of Diptera of the Afrotropical Region (Crosskey 1980) was not based on dedicated surveys for Diptera throughout the region, but rather on serendipitous collection records and publications. The Catalogue of Lacewings (Neuroptera) of the Afrotropical Region suffers from the same deficiencies as the Diptera Catalogue, with regard to most African countries, although a dedicated long-term survey for lacewings has been carried out in the southern African subregion as part of the Southern African Lacewing Monitoring Programme (Mansell 2002). Further comprehensive collecting was done during the Belgian colonial period in the now Democratic Republic of Congo, and these extensive collections are housed in the unique Royal Museum for Central Africa in Tervuren, Belgium, an invaluable resource for the study of Afrotropical lacewings and other African insects. Sporadic contributions were made during the colonial era of many African countries, but all materials from these collections are deposited in museums in Europe and can be accessed only at great expense by scientists from Africa. Besides Tervuren, museums that hold significant collections of African lacewings include The Natural History Museum, London; Museum National d'Historie, Paris; Naturhistorisches Museum, Wien; Humboldt University Museum, Berlin; Copenhagen Museum; Lund University Museum; and Barcelona Museum. Crucial collections were destroyed with the Hamburg Museum during World War II. Several of these collections, which house critical type material, were also inaccessible to South African scientists for several decades, for political reasons, which severely hampered progress in the documentation of Afrotropical insects.

Neuroptera

The order Neuroptera is comparatively well known in the Afrotropical Region, especially in the southern African subregion, and can be used as an example to highlight further key issues in the exploration of Afrotropical insects. Thirteen of the 18 families of Neuroptera occur in the Afrotropics, making this the richest zoogeographical region in the world for lacewings. The southern African subregion is especially rich, and all 13 families are represented in the area south of the Zambezi and Cunene Rivers. The lacewing fauna of this area, and South Africa in particular, manifests distinct geographical trends. An eastern fauna comprises taxa that occur in central and eastern Africa and extend their geographical range along the eastern tropical corridor into South Africa. The western fauna occupies the Cape Provinces of South Africa,

Namibia, and western Botswana, with almost 90% of the species being endemic to this area. Two groups of lacewings, the family Nemopteridae and the tribe Palparini (Myrmeleontidae), have undergone extensive speciation in southern Africa, while the related order Megaloptera is represented by a unique relict fauna that extends along the eastern mountain ranges from the Cedarberg, South Africa, to Zimbabwe. The Megaloptera are unknown from other countries of Afrotropical Africa. The Cape Provinces and Namibia harbor 62 of the world's known fauna of 142 species of Nemopteridae, and 47 of these are Cape endemics. The number of endemic species awaiting description will extend South Africa's total to more than half of the world fauna. Extensive, dedicated surveys have facilitated these assessments, which are aimed at evaluating the conservation status of southern Africa's lacewings. These surveys have revealed the precarious existence of this unique lacewing fauna that is severely threatened by burgeoning habitat destruction through urbanization and agriculture. The massive radiation of the Nemopteridae in southern Africa, and the Western and Northern Cape Provinces especially, has undoubtedly been driven by parallel evolution with the vast Cape Flora. All adult nemopterids are obligate pollen feeders and evidence is increasing that many of the endemic taxa are pollinators of specific plants. The relation between these pollinators and plants raises serious conservation concerns because many of the plant communities are equally threatened, highlighting the value of extensive surveys, elucidation of the biology of the species, and conservation assessments.

The tribe Palparini includes the largest and most spectacular of all antlions, with wingspans up to 17 cm. All are characterized by beautifully patterned wings, the patterns having evolved as camouflage to protect these large insects from predation. Although most species are nocturnal, their wing patterns provide protection during diurnal resting among plants (Mansell and Erasmus 2002). Currently, 100 valid species of Palparini are distributed throughout the Afrotropical, Oriental, and southern Palearctic Regions, but are absent, without vestige, from the extensive Australasian Region and the Western Hemisphere, indicating a post-Gondwana radiation. Of the 100 species, 45 occur in southern Africa, with 35 being endemic to the area. The unique vegetation and biomes of southern Africa are thought to have influenced the extensive radiation of the Palparini in the southern subcontinent.

Dung beetles (Coleoptera: Scarabaeidae: Scarabaeinae)

The Afrotropical Region is home to roughly half of the world's genera and species, and most of the tribes are centered in, or restricted to, the region. This distribution pattern can be attributed to evolutionary history and to the suitability of two principal ecological factors that influence tribal, generic, and species distribution patterns (Davis and Scholtz 2001, Davis et al. 2002b). These factors comprise suitable climate and the number of dung types. At species and generic levels, a strong correlation exists between the richness of dung-beetle taxa and the area of suitable climate. However, at the tribal level, taxon richness and taxon composition are strongly correlated with both climatic area and the number of dung types. Dung-type diversity, on the other hand, varies according to the evolutionary history of mammals (Davis and Scholtz 2001, Davis et al. 2002b).

The global dung-beetle fauna is divided into 12 tribes (Cambefort 1991), one of which comprises 3 subtribes. Two tribes (Canthonini and Dichotomiini) are widespread, with principal generic richness in large southerly regions (Afrotropical, Neotropical, Madagascar, Australia + New Guinea), representing a Gondwanaland pattern. Three tribes (Phanaeini, Eucraniini, and Eurysternini) are restricted to the Americas, whereas the remaining tribes and the three subtribes of the Oniticellini are either restricted to Africa – Eurasia (Gymnopleurini, Onitini, and Scarabaeini) or are centered in Africa (Onthophagini, Coprini, Sisyphini, and the three subtribes of the Oniticellini: Oniticellina, Drepanocerina, and Helictopleurina – the latter endemic to Madagascar). The basal 'tunneling' tribe and the basal 'rolling' tribe are generically and specifically dominant in the Western Hemisphere, although the basal relict taxa of both are restricted to Africa (Philips et al. 2004). Besides the three endemic American tribes, all other tribes are numerically dominant in Africa or Madagascar (Helictopleurina).

The scarabaeine dung beetles, consequently, are essentially a tropical group with Gondwanaland ancestry. The tribes with a southern, primarily tropical, Gondwanaland position (Dichotomiini and Canthonini) occupy a basal position in a published phylogeny (Philips et al. 2004). In Africa, they are reduced to minority status, with many, although not all genera, centered in southern forests or cool southern climates. The tropics have been described as museums of diversity (Gaston and Blackburn 1996), owing to the persistence of old taxa. Persistence and dominance of Gondwanaland groups is greatest where present spatial factors (large areas of tropical climate) and trophic factors (low numbers of dung types) show the least differences to those in the early Cenozoic, which were warm, moist, and dominated by forest (Axelrod and Raven 1978), with low cross-latitudinal thermal stratification (Parrish 1987). This observation is consistent with studies indicating that higher biological taxa remain constrained by adaptations to environmental conditions that occurred historically in their centers of diversification (Farrell et al. 1992), leading to ecophysiological limitations on their current spatial distribution patterns (Lobo 2000).

Dung-beetle taxa with older ancestry have been superseded largely in numerical dominance by younger tribes (Davis et al. 2002b), where present climatic conditions and trophic factors are more dissimilar to earlier conditions. From the mid-Cenozoic, cooler, drier climate has resulted in the contraction of tropical forest in Africa (Axelrod and Raven 1978) and the increase in savannas and grasslands. A concomitant increase occurred in the number of dung types as a result of the diversification of coarse-dung-producing mammals (Perissodactyla at 71 mya and some Artiodactyla and Proboscidea at 60 mya; Penny et al. 1999), which have achieved large body size and void large droppings. The later addition of the highly diverse ruminant Bovini in the late Miocene (10 mya), which produce a wide spectrum of large fine-fibered dung pads and small dung types such as pellets lead to increased diversification of younger dung-beetle groups.

CONCLUSIONS

The Afrotropical Region is endowed with exceptional biological diversity because of its high numbers of biomes including deserts, forests, savannas, lake systems, the eastern mountain arc, and the tremendously rich Cape region. The region is also home to Africa's richest and the world's third most species-rich country, South Africa (Cowling and Hilton-Taylor 1994) where the high diversity is largely a result of the rich Cape Floristic Kingdom, the world's hottest plant diversity hot-spot (Cowling and Hilton-Taylor 1994), with about 52% of southern Africa's plant endemics. This value amounts to about 3.5% of the world's flora on only about 0.2% of the earth's surface, a ratio

considerably higher than in even tropical rainforests. The Cape region is home not only to high insect species endemicity, as expected from the close association between plants and insects, but also to higher-level endemicity, such as the recently described order Mantophasmatodea, as well as numerous endemic families, subfamilies, tribes, and genera (Scholtz and Holm 1985).

The African savanna biome, which comprises about half of the continent's surface area, carries the earth's greatest diversity of ungulates, more than in any other biome or continent. This exceptional diversity is directly linked to the high spatial heterogeneity of African savanna ecosystems (du Toit and Cumming 1999). Although general studies of African savanna insects are lacking, their relative diversity and ecological importance probably are at least equal to those of the ungulates for some of the same reasons but also because some insect groups, such as dung beetles, are considered so diverse in Africa because of the high ungulate diversity.

Although forests occupy a sizeable portion of tropical Africa (about 10%), and comprehensive collections of insects are housed in the two large Belgian museums (Institut Royal des Sciences Naturelles and Royal Museum for Central Africa, Tervuren), as a legacy of the colonial occupation of the Belgian Congo (now Democratic Republic of Congo), few studies appear to have considered the African tropics in the same detail as, for instance, the New World Tropics. The neglect of the African tropics is probably largely a result of the decayed infrastructure and civil strife in the 'Congo' that has made this huge area, in one of Africa's largest countries and one that encompasses most of the biome, largely inaccessible to researchers.

The eastern arc mountains that form the backbone of Africa and run in a series of mountain chains from the Cape to Ethiopia have been identified as areas of high endemicity, particularly of relict groups of invertebrates. The area is thought to have served as a corridor for temperate species to spread between the Palearctic Region and southern Africa during several favorable periods in the past. Intervening periods of unfavorable conditions would have restricted adaptable fauna to isolated patches of suitable habitat (Lawrence 1953, Stuckenburg 1962, Griswold 1991). Such abundance provides intimidating challenges to documenters and, a host of additional impediments prevent the execution of the matters raised.

The refrain heard about insect biodiversity study throughout Africa, which is where the bulk of the

Old World's biodiversity occurs, remains more or less unchanged despite the noble sentiments expressed at international biodiversity meetings held every few years. Many African countries were signatories, amid much fanfare, to the Rio Convention on biodiversity. Although some of these countries formulated 'biodiversity policies' as a result of committing to do so at Rio, we are unaware of any that have actually attempted to survey their insect faunas. The official reasons given for inaction remain the same – lack of resources and a shortage of skilled and committed personnel. However, the one fact to which none will admit is the lack of political will by most governments to actually implement their commitments.

African insect habitats are being destroyed at an unprecedented rate. Conservation generally is applied, if at all, only to tourist-attracting large mammals, and political instability or virtual collapse of road infrastructure discourages collecting from large swaths of some of the most biologically rich areas on the continent. Inane laws administered by often corrupt or incompetent bureaucrats discourage serious biodiversity studies by foreigners, who most often have the interest and the means to undertake the studies. The perception often held is that foreigners are intent on plundering the commercial biological potential of the fauna and flora. Rather than allowing this, collecting is discouraged by an often-hostile bureaucracy that would rather learn nothing of the fauna than allow representative specimens to be collected.

Even if sufficient resources and human capital were available to survey the insects, various, increasingly serious factors mitigate against gaining an idea of the historical biodiversity because of human practices. One of the most serious of these practices, which leads to total land transformation, is the practice of charcoal production, which already has led to huge tracts of land in Angola, Mozambique, and Zambia, and possibly other countries, being transformed from dense savanna woodland, one of the biome jewels of the African continent (du Toit and Cumming 1999), to open scrub and grassland. The effects on native insects are unknown but are likely to be severe.

South Africa appears to be the only African country with a serious commitment to biodiversity study and is the only African country with a legal conduit such as this for managing its biodiversity since it recently (2004) implemented a Biodiversity Act providing for the management and conservation of its biodiversity through biodiversity planning and

monitoring, protection of threatened ecosystems and species, control and management of alien and invasive species, regulation of bioprospecting, fair and equitable benefit sharing, and regulation of permits. The Act allowed for the establishment of the SANBI whose mandate is to respond appropriately to biodiversity-related global policy and national priorities, make systematic contributions to the development of national biodiversity priorities, and demonstrate the value of conserving diversity and the relevance of biodiversity to the improvement of the quality of life of all South Africans

Despite the noble wording of the act, major impediments still exist to scientists being able to study South African insect diversity. Some of these impediments are in the form of bureaucracy, complicating the process of issuing permits by the various provinces to which control of biodiversity has devolved. The inefficiency of most of the provincial permitting departments is compounded by draconian penalties for illegally collecting specimens. A recent example of this was the overnight imprisonment of a foreign researcher in the Western Cape Province for collecting a spider without the necessary permit. An admission-of-guilt plea in court and a substantial fine were the immediate outcome, but a permanent criminal record was the more serious and long-term result. The reason for the fiasco was that elephants, rhinoceroses, and invertebrates are covered by the same biodiversity laws, so collecting an insect is legally equivalent to poaching a rhino.

However, as a result of concerns expressed by the biodiversity community in South Africa about the impediments to biodiversity study, a series of meetings and workshops involving representatives of SANBI, provincial permitting authorities, tertiary academic institutions, museums, and amateur insect collectors is currently underway to address the issues. The biodiversity scientists have the interest and the will to succeed, but the question remains whether the bureaucracy can be overcome. Only time will tell.

REFERENCES

Axelrod, D. I. and P. H. Raven. 1978. Late Cretaceous and Tertiary vegetation of Africa. Pp. 77–130. *In* M. J. A. Werger and A. C. van. Bruggen (eds). *Biogeography and Ecology of Southern Africa*. Dr W Junk, The Hague.

Cambefort, Y. 1991. Biogeography and evolution. Pp. 51–67. *In* I. Haski and Y. Cambefort (eds). *Dung Beetle Ecology*. Princeton University Press, Princeton, New Jersey.

Codd, E. F. 1970. A relational model of data for large shared data banks. *Communications of the ACM* 13: 377–387.

Cowling, R. M. and C. Hilton-Taylor. 1994. Patterns of plant diversity and endemism in southern Africa: an overview. Pp. 31–52. *In* B. J. Huntley (ed). *Botanical Diversity in Southern Africa. Strelitzia 1.* National Botanical Institute, Pretoria.

Crosskey, R. (ed). 1980. *Catalogue of the Diptera of the Afrotropical Region.* 1–1437. British Museum (Natural History), London.

Davis, A. L. V. 1993. Biogeographical groups in a southern African, winter rainfall, dung beetle assemblage (Coleoptera: Scarabaeidae) – consequences of climatic history and habitat fragmentation. *African Journal of Ecology* 31: 306–327.

Davis, A. L. V. 1997. Climatic and biogeographical associations of southern African dung beetles (Coleoptera: Scarabaeidae). *African Journal of Ecology* 35: 10–38.

Davis, A. L. V., C. H. Scholtz, and S. L. Chown. 1999. Species turnover, community boundaries and biogeographical composition of dung beetle assemblages across an altitudinal gradient in South Africa. *Journal of Biogeography* 26: 1039–1055.

Davis, A. L. V., R. J. van Aarde, C. H. Scholtz, and J. H. Delport. 2002a. Increasing representation of localised dung beetles across a chronosequence of regenerating vegetation and natural dune forest in South Africa. *Global Ecology and Biogeography* 11: 191–209.

Davis, A. L. V., C. H. Scholtz, and T. K. Philips. 2002b. Historical biogeography of scarabaeine dung beetles. *Journal of Biogeography* 29: 1217–1256.

Davis, A. L. V. and C. H. Scholtz. 2001. Historical versus ecological factors influencing global patterns of scarabaeine dung beetle diversity. *Diversity and Distributions* 7: 161–174.

Davis, A. L. V. and C. H. Scholtz. 2004. Local and regional species ranges of a dung beetle assemblage from the semi arid Karoo/Kalahari margins, South Africa. *Journal of Arid Environments* 57: 61–85.

Davis, A. L. V., C. H. Scholtz, P. W. Dooley, N. Bham, and U. Kryger. 2004. Scarabaeine dung beetles as indicators of biodiversity, habitat transformation and pest control chemicals in agro-ecosystems. *South African Journal of Science* 100: 415–424.

Drew, R. A. I., K. Tsuruta, and I. M. White. 2005. A new species of pest fruit fly (Diptera: Tephritidae) from Sri Lanka and Africa. *African Entomology* 13: 149–154.

du Toit, J. T. and Cumming, D. H. M. 1999. Functional significance of ungulate diversity in African savannas and the ecological implications of the spread of pastoralism. *Biodiversity and Conservation* 8: 1643–1661.

Ekesi, S. and M. K. Billah (eds). 2006. *A Field Guide to the Management of Economically Important Tephritid Fruit Flies in Africa.* The International Centre of Insect Physiology and Ecology (ICIPE), Nairobi, Kenya.

Estrada, A. and R. Coates-Estrada. 2002. Dung beetles in continuous forest, forest fragments and in an agricultural mosaic habitat island at Los Tuxtlas, Mexico. *Biodiversity and Conservation* 11: 1903–1918.

Farrell, B. D., C. Mitter, and D. J. Futuyma. 1992. Diversification at the insect-plant interface, insights from phylogenetics. *Bioscience* 42: 34–42.

Gaston, K. J. and T. M. Blackburn. 1996. The tropics as a museum of biological diversity: an analysis of the New World avifauna. *Proceedings of the Royal Society, London, Series B:* 263: 63–68.

Griswold, C. E. 1991. Cladistic biogeography of Afromontane spiders. *Australian Systematic Botany* 41: 73–89.

Halffter, G. and L. Arellano. 2002. Response of dung beetle diversity to human-induced changes in a tropical landscape. *Biotropica* 34: 144–154.

Lawrence, R. F. 1953. *The Biology of the Cryptic Fauna of Forests – with Special Reference to the Indigenous Forests of South Africa.* A. A. Balkema, Cape Town. 408 pp.

Lobo, J. M. 2000. Species composition and diversity of dung beetles (Coleoptera: Scarabaeoidea) assemblages in North America. *Canadian Entomologist* 132: 307–321.

Losey, J. E. and M. Vaughan. 2006. The economic value of ecological services provided by insects. *Bioscience* 56 (4): 311–323.

Lux, S. A., R. S. Copeland, I. M. White, A. Manrakhan, and M. K. Billah. 2003. A new invasive fruit fly species from the *Bactrocera dorsalis* (Hendel) group detected in East Africa. *Insect Science and its Application* 23: 355–360.

McGeoch, M. A. 1998. The selection, testing and application of terrestrial insects as bioindicators. *Biological Reviews* 73: 181–201.

Mansell, M. W. 2002. Monitoring lacewings (Insecta: Neuroptera) in southern Africa. *Acta Zoologica Academiae Scientiarum Hungaricae* 48 (Supplement 2): 165–173.

Mansell, M. W. and B. F. N. Erasmus. 2002. Southern African biomes and the evolution of Palparini (Insecta: Neuroptera: Myrmeleontidae). *Acta Zoologica Academiae Scientiarum Hungaricae* 48 (Supplement 2): 175–184.

Miller, S. E., B. Gemmill, L. Rogo, M. Allen, and H. R. Herren. 2000. Biodiversity of terrestrial invertebrates in tropical Africa: assessing the needs and plan of action. Pp. 204–212. *In* P. H. Raven and T. Williams (eds). *Nature and Human Society: the Quest for a Sustainable World. Proceedings of the 1997 Forum on Biodiversity.* National Academy Press, Washington, DC. 625 pp.

Miller, S. E. and L. M. Rogo. 2001. Challenges and opportunities in understanding and utilization of African insect diversity. *Cimbebasia* 17: 197–218.

Parrish, J. T. 1987. Global palaeogeography and palaeoclimate of the late Cretaceous and the early Tertiary. Pp. 51–73. *In* E. M. Friis, W. G. Chaloner, and P. R. Crane (eds). *The Origins of Angiosperms and Their Biological Consequences.* Cambridge University Press, New York.

Penny, D., M. Hasegawa, P. J. Waddell, and M. D. Hendy. 1999. Mammalian evolution: timing and implications from using the log-determinant transform for proteins of differing amino acid composition. *Systematic Biology* 48: 76–93.

Philips, T. K., E. Pretorius, and C. H. Scholtz. 2004. A phylogenetic analysis of dung beetles (Scarabaeidae: Scarabaeinae): unrolling an evolutionary history. *Invertebrate Systematics* 18: 53–88.

Riley, J. 2001. Indicator quality for assessment of impact of multidisciplinary systems. *Agriculture, Ecosystems and Environment* 87: 121–128.

Scholtz, C. H. and S. L. Chown. 1995. Insects in southern Africa: how many species are there? *South African Journal of Science* 91: 124–126.

Scholtz, C. H. and E. Holm. 1985. *Insects of Southern Africa.* Butterworths, Durban. 502 pp.

Stuckenburg, B. R. 1962. The distribution of the montane palaeogenic element in the South African invertebrate fauna. *Annals of the Cape Provincial Museums* 2: 190–205.

Waterhouse, D. F. 1974. The biological control of dung. *American Scientist* 230: 101–108.

White, I. M. and M. M. Elson-Harris. 1992. *Fruit Flies of Economic Significance: Their Identification and Bionomics.* CAB International, Wallingford. 600 pp.

Chapter 6

BIODIVERSITY OF AUSTRALASIAN INSECTS

Peter S. Cranston

Department of Entomology, University of California at Davis, Davis, California 95616 USA

Insect Biodiversity: Science and Society, 1st edition. Edited by R. Foottit and P. Adler

M uch Australasian biodiversity, lacking the decimating effects of the predominantly northern hemisphere Pleistocene glaciations, remains intact – from a much deeper geological history. To the alert and traveled northern hemisphere scientist, the biodiversity to the south of the equator is much more extravagant than to the north. For example, a higher taxonomic diversity of ants is recorded for a modestly revegetated hill outside the Australian National Insect Collection Laboratories in Canberra than in all of the UK. Furthermore, Australasian biodiversity displays biogeographic patterns uniting the southern continents, as recognized by early explorer–scientists such as coleopterist Wilhelm F. Erichson (1842) and polymath Hooker (1853), both of whose placements of the fauna and flora of 'Van Diemen's Land' in a global context were remarkably prescient. Southern hemisphere-based entomologists such as Mackerras (1925), Tillyard (1926), and Harrison (1928) recognized austral biodiversity relationships prior even to knowing of Wegener's (1915, 1924) theory of continental drift. The lesson is that a detailed knowledge of the geology and biodiversity of the southern lands, with a comparative perspective, allowed intellectually untrammeled minds to understand the evolution of the biota, despite contrary geological orthodoxy. We might consider if similar constraints are hindering our deeper understanding of the diversification of austral insect biodiversity (McCarthy 2005).

AUSTRALASIA – THE LOCALE

Australasia, for the purposes of this overview, comprises the series of oceanic islands, from massive to minute, in the southwestern Pacific (Fig. 6.1). It corresponds with the geological Australian section of the Indo-Australian Plate, and thus is flanked to the west by the Indian Ocean and to the south by the Antarctic/Southern Ocean. The island continent of Australia (including Tasmania) is the largest landmass, with the globally largest 'official' island of New Guinea to its north, and New Zealand to the southeast. The Australian islands of Lord Howe and Norfolk lie successively to the east, with New Caledonia northeastward, then Vanuatu (New Hebrides) and further eastward, Fiji in the eastern Melanesian Region of the Central Pacific, almost astride the boundary between the Australian and Pacific plates. Tonga lies southeast on its own microplate (terrane).

The biotically defined northwestern boundary to Australasia often is treated as Wallace's Line, named for Alfred Wallace who recognized a faunal discontinuity between the 'Australasian' islands of Celebes (Sulawesi) and Lombok (to the east) and Indo-Malaysian islands including Borneo, Java, and Bali (to the west). The Indonesian islands on the Australian side, termed Wallacea, are separated by deep water from both Indo-Malaysia and the Australia-New Guinean continental shelf. Wallace's noted discontinuities are less evident among insects, or at least the patterns are

Fig. 6.1 The Australasian Region.

so complex, especially concerning Sulawesi, that a simplistic interpretation is not possible (Vane-Wright 1991, New 2002) – an issue also interpretable as 'too many lines...' (Simpson 1977). This region is perhaps best understood as transitional – having an Australian/New Guinean faunal resemblance that is variably attenuated with distance westward, and overlies a rich and highly endemic biota, often with continental Asian affinities that attenuate eastward. In this area, only Christmas Island (10°30′S 105°40′E), an Indian Ocean offshore Territory of Australia, will be considered in this chapter.

Although New Zealand lies close to and astride the eastern margin of the Pacific Plate, with the Alpine fault plate boundary running southwest–northeast across its South Island, here all of New Zealand and its offshore, including subantarctic, islands are considered. These latter represent remnant emergent islands of the Campbell Plateau, lying between Antarctica and South Island New Zealand, and carry biotas exhibiting both endemic and widespread elements (Michaux and Leschen 2005).

SOME HIGHLIGHTS OF AUSTRALASIAN INSECT BIODIVERSITY

Insects 'down under' contain iconic exemplars. The bush fly *Musca vetustissima* provokes the famous Aussie 'wave' and promotes the tourist's souvenir cork-lined hat. The attentions of one of the most abundant and diverse ant faunas in the world authenticate the famous 'Aussie barbecue'. The landscape of northern Australia is characterized by termite mounds of diverse shapes and sizes, among the most striking of which are the flattened and regularly north–south orientated structures produced by the magnetic termites (*Amitermes* spp.). Public attention was captured by the rediscovery and recovery of the presumed extinct Lord Howe Island stick insect (*Dryococelus australis*), dubbed the 'land lobster' on account of its size.

Selections from Australian insect biodiversity are harvested by Aboriginal Australians. The Arrente people use *yarumpa* (honey ants) and *udnirringitta* (witchety grubs), which have now entered the nonindigenous food chain in restaurants catering to wealthy tourists and political entertainment. Adult bogong moths, *Agrotis infusa* (Noctuidae), that gather in their millions in aestivating sites, in narrow caves, and in crevices on mountain summits in southeastern

Australia, once formed another important Aboriginal food. The nocturnal illumination of a new Parliament building in the nation's capital acted as a giant light-trap for migrating bogongs in such numbers that the insects entered the politicians' consciousness and inspired the design for a Canberra-based meeting of the Australian Entomological Society (Fig. 6.2). Australia is not unique in its ambivalent attitude to insects. New Zealanders take warped pride in the voraciousness of their South Island coastal *namu* or sandflies (*Austrosimulium* spp., Simuliidae) and justifiable pride in the endemic radiations of their *wetās* (Anostostomatidae and Rhaphidophoridae). These latter – outsized, flightless orthopterans – include the Little Barrier Island giant wetā (*Deinacrida heteracantha*) or *wetāpunga* (Maori for 'god of ugly things'), one of the heaviest insects, weighing up to 70 g (Trewick and Morgan-Richards 2004). About 1 million tourists each year visit Waitomo Caves (Doorne 1999) to share an extravagant and potentially educational entomological experience – the bioluminescence displayed by glowworms, larvae of the dipteran *Arachnocampa luminosa* (Keroplatidae). For conservation of native biodiversity, Aotearoa (the Maori name for New Zealand) has one of the highest profiles of any country, following recognition of the devastation caused to the unique native invertebrates and birds after the arrival of humans, both Polynesians and European colonists, with their attendant rats and peridomestic animals. Invertebrates feature in many species conservation plans, which depend substantially

Fig. 6.2 Logo for the Annual Australian Entomological Society meeting, incorporating bogong moths and National Parliament Building Canberra.

on elimination of vertebrate vermin from offshore islands.

Central to many of the southern biogeographical studies has been the island of Nouvelle Calédonie (New Caledonia). The French have a long history of biodiversity studies in their erstwhile colony, with entomologists prominent among researchers. Early studies, notably of the flora, suggested the existence of an ancient, relictual fauna that survived from the Cretaceous Gondwanan megacontinent. Although that thesis stimulated global scientific interest, the scenario is not quite as once believed, at least for insects. Nonetheless, this does not negate the evidently high levels of endemism, and multiplicity of evolutionary origins of the biota.

To the north, New Guinea, the second largest island on the planet, is separated from the Australian continent only by the relatively shallow and recent Torres Strait. This geologically (and ethnically) diverse island might have been a lesser target for pioneering exploratory entomologists, compared to the rainforests of Southeast Asia and the Amazon. However, early insect collectors discovered many gaudy, and perhaps saleable specimens such as the largest known butterfly in the world, the Queen Alexandra's birdwing (*Ornithoptera alexandrae*); the world's largest moth, the Hercules moth (*Coscinocera hercules*); and giant stick insects (*Eurycnema goliath* and species of *Acrophylla*). Biologist–explorers of New Guinea, including Wallace (1860) and D'Albertis (1880), collected and recorded extraordinary insect diversity for a nineteenth-century European audience eager for natural history tales and specimens. Papua New Guinea (PNG) (eastern New Guinea) ranks twelfth globally in terms of endemism of large butterflies (Papilionidae, Pieridae, Nymphalidae), with 56 of 303 species endemic. Economic benefits can flow from a perhaps sustainable trade in such live lepidopterans for the butterfly houses of the affluent world. More recently, the island has been the site of some laborious and intensive studies of plant-feeding (phytophagous) insects in an attempt to establish how selective these insects are in their use of host plants such as the diverse figs (*Ficus*).

In keeping with the predictions of island biogeographic theory, Pacific Islands, including New Zealand, have disharmonic biotas, typically with major taxa erratically absent, some groups showing evidence of endemic radiations, and generally high levels of endemism. These islands, and especially Hawai'i, are the theaters in which the processes of species formation

and species extinction can be studied, for mixed in with the natives are substantial alien introductions that threaten all aspects of biodiversity retention.

The Lord Howe Island stick insect

The Lord Howe Island Group is located in the Western Pacific Ocean, some 700 km northeast of Sydney and comprises the main island of Lord Howe, as well as Admiralty Islands, Mutton Bird Islands, Ball's Pyramid, and many reefs. Geologically, the main island is the eroded remnant of a large shield volcano that erupted intermittently from the sea floor in the Late Miocene (some 6.5–7 mya). What remains are the exposed peaks of a 65-km long and 24-km wide volcanic seamount that rises from ocean depths of nearly 2 km. The seamount is near the southern end of a chain of such seamounts, most of which are submerged; the chain extends for more than 1000 km. The Group was inscribed on the World Heritage List in 1982 as an outstanding example of an oceanic island of volcanic origin, with a unique biota and important and significant natural habitats for *in situ* conservation of biological diversity, including those containing species of plants and animals of outstanding universal significance from the point of view of science and conservation. Although no insects were specified in the nomination, subsequently the island has attained some fame as the site of the rediscovery of a large, flightless, stick insect (Fig. 6.3). The Lord Howe Island stick insect, *D. australis* (Montrouzier), was known locally as the 'land lobster' or 'tree-lobster'

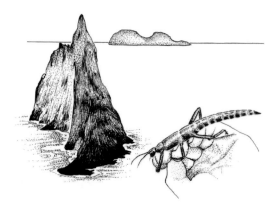

Fig. 6.3 The Lord Howe Island Stick insect against its island home.

on account of its shape and size (up to 12 cm long). This phasmatid was known for its ability to run on the ground. It was common and easily observed sheltering in banyan trees (*Ficus macrophylla*) on the island early last century. Its demise appears to have been connected with the release of black rats on the island when a supply ship ran aground in 1918, triggering extinction of five flightless birds; the land lobster was believed to have become extinct by the 1930s, and was entered as such in the IUCN (International Union for Conservation of Nature and Natural Resources) Red List. Subsequent surveys failed to find any further evidence, until a rock climber on the offshore Ball's Pyramid photographed an adult female; its rare presence was confirmed in the late 1960s. Ball's Pyramid is aptly named, being a pyramidal structure rising 500 m above sea level from a base of only 1100 × 400 m at sea level. The idea that such a basaltic column, lacking both large woody vegetation and refugial hollows, could support a population of a large phasmatid was greeted with skepticism. Numerous attempts to rediscover the species failed until in 2001 a biological survey revealed three specimens associated with the endemic shrubby paperbark (*Melaleuca howeana*). The following year, 24 individuals were observed associated with the same, small patch of paperbark, believed to be the only suitable habitat on the island. Retreats were clefts in the rock–root interface where seepages occurred, in contrast to the tree holes used on the mainland prior to extinction. The rediscoverers (Priddel et al. 2003) pointed out that the Ball's Pyramid locality is too small to maintain a population against any future environmental change and made a plea for a captive-breeding program, concurrent with elimination of rats from Lord Howe, to allow reintroduction. Action plans prepared by the New South Wales National Parks and Wildlife Service (NSW NP and WS) followed the Commonwealth Government listing of the species as critically endangered (from extinct). Two pairs of the insects were captured and transported to the mainland. One pair went to Melbourne Zoo, an institution with an enviable record of breeding invertebrates, and the second was housed by one of Australia's leading stick insect experts. Rearing has been successful using garden-grown *M. howeana* as the food stock, with banyan in reserve (Zoos Victoria 2006). The necessary elimination of rats from Lord Howe Island, required for any reintroduction to succeed, remains under consideration for feasibility and cost.

Australasian birdwing conservation

The world's largest butterfly, the Queen Alexandra's birdwing (*O. alexandrae*) of PNG, is a regional success story. This spectacular species, whose caterpillars feed only on *Aristolochia diehlsiana* vines, is limited to a small area of lowland rainforest in northern PNG. Under PNG law, this birdwing species has been protected since 1966, and international commercial trade was banned by endangered listing on Appendix 1 of the Convention on International Trade in Endangered Species of Wild Fauna and Flora (CITES). Dead specimens in good condition command a high price, which can be more than US$2000. In 1978, the PNG governmental Insect Farming and Trading Agency (IFTA), in Bulolo, Morobe Province, was established to control conservation and exploitation and act as a clearinghouse for trade in Queen Alexandra's birdwings and other valuable butterflies. Local cultivators, numbering some 450 village farmers associated with IFTA, ranch their butterflies. Farmers plant appropriate host vines, often on land already cleared for vegetable gardens at the forest edge, thereby providing food plants for a chosen local species of butterfly. Wild adult butterflies emerge from the forest to feed and lay their eggs; hatched larvae feed on the vines until pupation when they are collected and protected in hatching cages. According to species, the purpose for which they are being raised, and conservation legislation, butterflies can be exported live, as pupae, or dead as high-quality collector specimens. IFTA, a nonprofit organization, has sold some $400,000 worth of PNG insects yearly to collectors, scientists, and artists around the world, generating an income for a society that struggles for cash. Local people recognize the importance of maintaining intact forests as the source of the parental wild-flying butterflies of their ranched stock. In this system, the Queen Alexandra's birdwing butterfly has acted as a flagship species for conservation in PNG, and this success story attracts external funding for surveys and reserve establishment. In addition, conserving PNG forests for this and related birdwings undoubtedly results in conservation of much biodiversity under the umbrella effect. The UK Darwin Initiative in 2005 funded a 3 year project 'Socio-economics of insect farming in PNG' to assess and train in sustainable insect collecting. Early reviews suggest that burdensome licensing, permits, fees, and paperwork threaten the viability of legal trade and encourage a growing illegal trade especially in abundant species that ought to need no CITES listing.

New Guinean insect conservation efforts need a commercial incentive to provide impoverished people with some recompense for protecting natural environments. Commerce need not be the sole motivation; however, the aesthetic appeal of native birdwing butterflies flying wild in local neighborhoods, combined with local education programs in schools and communities, has saved the subtropical Australian Richmond birdwing butterfly, *Troides* or *O. richmondia*. Larvae of Richmond birdwings eat *Pararistolochia* or *Aristolochia* vines, choosing from three native species to complete their development (Sands et al. 1997). However, much coastal rainforest habitat supporting native vines has been lost, and the alien South American *A. elegans* (Dutchman's pipe), introduced as an ornamental plant and escaped from gardens, has been luring females to lay eggs on it as a prospective host. This oviposition mistake is deadly because this plant is toxic to young caterpillars. The answer to this conservation problem has been an education program to encourage the removal of Dutchman's pipe from native vegetation, from sale in nurseries, and from gardens and yards. Replacement with native *Pararistolochia* was encouraged after a massive effort to propagate the vines. Community action throughout the native range of the Richmond birdwing appears to have reversed its decline, without any requirement to designate land as a reserve (Sands et al. 1997).

DROWNING BY NUMBERS? HOW MANY INSECT SPECIES ARE IN AUSTRALASIA?

Australia

For the total Australian insect diversity, Taylor (1983) estimated about 110,000 species. Even then, this figure clearly was an underestimate. After questioning of practicing taxonomists, Monteith (1990) determined a figure of about 20,000–40,000 beetle species, although what proportion of the total biota this constituted was unknown. Suggesting that the Australian dipteran fauna of some 10,000–12,000 species and the Australian insect fauna as a whole both comprise about 5% of their respective world totals, Colless and McAlpine (1991) extrapolated to estimate global diversity in the low- to mid-range of existing estimates. Stork (1993) then suggested the existence of some 400,000 Australian insect

species, of which three-quarters were undescribed. In extrapolating from comparatively well-known geographically constrained regions to generate global biodiversity estimates, Gaston and Hudson (1994) included the Australian insect biota in their analyses as 'well known'. They argued that the more modest Australian species richness estimates supported their own estimated low value for total global species richness. Most recently, a combination of expert opinion with extrapolation from the rate of species discovery by recent revisionary taxonomists, led Yeates et al. (2003) to a figure of near 205,000 insect species for Australia. Within this estimate, numbers of described and estimated species range from only 10 undescribed species among the estimated 330 Odonata, to some 60,000 undescribed Coleoptera, with only 22,000 described. The major (megadiverse) orders include estimates of 12,000 Hemiptera; 20,000 Lepidoptera; 30,000 Diptera; and 40,000 Hymenoptera. Coleoptera comprise about 40–50% of the total numbers of insects. The authors acknowledge shortcomings, including some related to revisionary taxonomists' choices of study taxa – some selected their taxa due to high levels of perceived novelty (upwardly biasing extrapolates), whereas others, reviewing already well-known taxa (e.g., butterflies), contributed little in the way of species discovery (biasing downward). These figures, nonetheless, seem to reflect a reality in which Australia, with some 6% of the earth's land area, supports about 5–6% of the global insect species diversity of some 4 million species.

New Zealand (Aotearoa), Chatham Islands, and Subantarctic Islands

New Zealand, comprising two main islands with an area much smaller than Australia (0.27 versus 7.6 million square kilometers), supports a commensurately lower insect biodiversity. A preliminary assessment (for Species 2000 New Zealand) provides an estimate of some 20,000 species, among them slightly more than 10,000 described species, and nearly 9000 species awaiting description. Kuschel's (1990) claim of 40,000 species, based on extrapolation from a rather exaggerated Coleoptera diversity, seems too high, especially because Leschen et al. (2003) assess the New Zealand Coleoptera at approximately 10,000 species.

Insects on New Zealand's offshore islands have been subject to continuing study, especially the beetles.

Thus, addition of 131 species of Coleoptera to the known fauna of Chatham Islands (*ca.* 40°S) increased the total to 286 (Emberson 1998). Smaller, perhaps younger, and increasingly southerly (subantarctic) islands, emergent from the Campbell Plateau, have more modest insect biodiversity, at least as far as numbers alone are concerned. Of 150 insect species reported from the Antipodes Islands (*ca.* 49°S) (Marris 2000), Coleoptera comprise only 25 species in 13 families. In comparison, the Snares Islands (48°S) have a reported 25 species in 14 families; the Bounty Islands (*ca.* 47°S), 9 species in 7 families; the Auckland Islands (*ca.* 50°S), 57 species in 17 families; and Campbell Island (*ca.* 52°S), 40 species in 15 families. The most austral island in the area, the Australian Macquarie Island (54.30°S), has eight species in only two families (Williams 1982, Greenslade 1990, Young 1995, Klimaszewski and Watt 1997, Marris 2000).

These data suggest that the island faunas are species poor and, under these circumstances, the use of standard multipliers from the known Coleoptera fauna to the total Insecta fauna might be misleading. This scenario is especially so on oceanic islands with unbalanced faunas and different histories and distances from sources. The question of endemism and biogeographic and phylogenetic relationships of the insect faunas of New Zealand will be considered later.

New Guinea

From ratios of the New Guinea (NG) fauna to the world fauna for well-known groups, rates of description of new taxa, and the size of the Australian insect fauna, Miller et al. (1995) calculated a possible 300,000 species of insects for New Guinea. They qualified this figure as potentially overestimated or underestimated by 100,000 species. More recently, biodiversity surveys assessing richness of phytophagous (plant-feeding) insects have delivered further insights. Insect herbivores sampled from the foliage of 15 species of figs (*Ficus*, Moraceae) in rainforest and coastal habitats in the Madang area, Papua New Guinea, revealed 349 species of leaf-chewing insects (13,193 individuals) and 430 species of sap-sucking insects (44,900 individuals) (Basset and Novotny 1999). Despite a high sampling intensity, the species-accumulation curve did not reach an asymptote. Evidence from studies such as this, concerning host specificity of phytophagous insects in New Guinea,

led Novotny et al. (2002) to challenge previous large multipliers based on high levels of monophagy. Based on an estimate that New Guinean Coleoptera and Lepidoptera constitute 5% of the global biodiversity of each order, a proportion derived from Sekhran and Miller (1996), new regional and global biodiversity estimates were derived from host-specificity ratios. For butterflies, for which figures of 959 described species in New Guinea (Parsons 1999) and 15,000–20,000 species globally are likely to be accurate, Novotny et al. (2002) calculated an estimated 1179 species in New Guinea and 23,500 worldwide.

New Caledonia and the West Pacific

New Caledonia (NC) lies on the Tropic of Capricorn, between 19°S and 23°S, 1200 km due east of Capricornia, Australia, and 1500 km northwest of New Zealand. The island is a recognized biodiversity hot spot. Some 4000 insect species had been cataloged at the time of Chazeau's (1993) review of the terrestrial fauna of the island, but an estimate of 8000–20,000 insect species is realistic. Endemism is variably high – the more than 70 native species of butterflies and 300 species of moths exhibit endemism of only 38% (Chazeau 1993). According to the Department of Entomology at The Swedish Museum of Natural History (2003), endemism of Trichoptera is especially high – for example, all but 2 of 111 described species are endemic.

For the other land areas of the region – Vanuatu, the Solomons, and Fiji – estimates of insect species diversity can be little more than 'guestimates', given the lack of modern survey. Robinson (1975) calculated the total number of insect species inhabiting the Fiji group of islands as in excess of 3500. This figure is to be tested, because insect biodiversity of the Fijian islands is being inventoried in detail under the National Science Foundation–Fiji Terrestrial Arthropod Survey (Evenhuis and Bickel 2005).

AUSTRALASIAN INSECT BIODIVERSITY – OVERVIEW AND SPECIAL ELEMENTS

Australia

Australia, in keeping with its size as a continent occupying some 5–6% of the earth's landmass, exhibits all major elements of insect biodiversity. Among the few departures from proportionality are the Isoptera (termites), with perhaps more than 10% of the global diversity; this might be so also for Phasmatodea (stick and leaf insects). The three absent ordinal level taxa, Grylloblattodea, Mantophasmatodea, and Zoraptera, are geographically restricted minor orders of low species diversity, and are not present anywhere in Australasia. Strepsiptera and Embiidina, although represented in Australia, are absent from New Zealand. At the level of family, distinctive patterns emerge with some local endemics and many more restricted to the southern hemisphere ('Gondwanan').

Typical diversity patterns are shown by the aquatic orders. In Odonata, the Hemiphlebiidae, comprising a single charismatic species sometimes given the higher rank of Hemiphleboidea, is restricted to a few southern Australian pools (Trueman et al. 1992). The family Petaluridae, represented by Jurassic fossils, has five extant Australian species, including *Petalura ingentissima*, with a wingspan of up to 160 mm, making it the largest of living dragonflies. The small odonate families Hypolestidae, Diphlebiidae, and Cordulephyidae are endemic to Australia, whereas others are widespread, of northern origin, or patchily present in other austral continents ('Gondwanan').

Early recognition of the austral relationships among the Ephemeroptera by Edmunds (1972) has been substantiated and reinforced, even as family concepts change. The family Teloganodidae (which rather unusually includes South Africa and Madagascar in its Gondwanan pattern) and many clades, including Ameletopsidae, Nesameletidae, Oniscigastridae, and Rallidentidae, suggest austral radiation, as do the remarkably diverse Australasian Leptophlebiidae (Christidis 2006). Three of four Australasian families of Plecoptera (Eustheniidae, Gripopterygidae, and Austroperlidae) belong to the monophyletic Gondwanan suborder Antarctoperlaria (McLellan 2006). In the Trichoptera, the Hydrobiosidae are predominantly austral Plectrotarsidae; Antipodoecidae are endemic to Australia; and Chathamiidae, Oeconesidae, and Conoesucidae are restricted to Australia and New Zealand; Kokiriidae to South America, Australia, New Zealand, and New Caledonia; Tasimiidae to Australia and South America; and Helicophidae to New Caledonia, New Zealand, Australia, Chile, and Southwest Argentina (Wiggins 2005). Although these aquatic orders and their predominantly running-water habitats are quite well studied, only recently has a novel aquatic diversity been uncovered – that of diving beetles (Dytiscidae) in underground aquifers beneath arid Australia (e.g., Humphreys 2001, Cooper et al. 2002, Balke et al. 2004). This unexpectedly diverse system originated perhaps when aquatic species from temporary habitats, and from several different lineages, evaded drought by entering calcretes (the hyporheic zone in limestone areas), essentially driven underground by historical aridification of the continent in the Late Miocene/Early Pliocene some 5 mya (Leys et al. 2003). Many true flies (Diptera) with immature stages associated with minor aquatic habitats such as seeps or waterfalls, including the Blephariceridae (Zwick 1977, Gibson and Courtney 2007), Chironomidae (e.g., Cranston et al. 2002), Empididae (Sinclair 2003), and Thaumaleidae (Austin et al. 2004), have diversified from Gondwanan ancestors. The Nannochoristidae, the only Mecoptera with an aquatic larva, likewise are present in streams of New Zealand, southeastern Australia, and Chile. The high-ranking hemipteran suborder Coleorrhyncha comprises only the Gondwanan family Peloriidae (mossbugs), which live among sphagnum and liverworts, especially in forests of southern beeches (*Nothofagus*).

Patterns of familial diversity, endemism, and regional distribution, such as those shown by these aquatic insects, are mirrored within the more terrestrial orders (Cranston and Naumann 1991, Austin et al. 2004). Some strong diversifications of insects are associated with Australia's major plant radiations, for example the Mimosaceae, including *Acacia*; Myrtaceae especially *Eucalyptus*; and to a lesser extent the Casuarinaceae. Proteaceae seem, for the main part, to have avoided insect associations. Majer et al. (1997) estimated that there could be between 15,000 and 20,000 species of phytophagous insects on *Eucalyptus* in Australia, including members of Hemiptera (especially psylloids and coccoids), Coleoptera, Diptera, and Lepidoptera (Austin et al. 2004), but there is a conspicuous paucity of Thysanoptera (Mound 2004). The Chrysomelinae (leaf beetles) demonstrate a radiation associated with eucalypts of some 750 species, but are virtually lacking on Proteaceae (C. A. M. Reid, personal communication). Notable Australian plant-associated radiations include several independent originations of gall-inducing Coccoidea (scale insects) (Cook and Gullan 2004, Gullan et al. 2005); the phlaeothripid Acacia thrips (Crespi et al. 2004); the ecologically significant, lerp-forming spondyliaspine Psylloidea (Hollis 2004, Taylor 2006);

the cecidomyiid gall midges (Kolesik et al. 2005) and near-endemic Fergusoninidae (Diptera) (Scheffer et al. 2004, Taylor 2004); gall-inducing Chalcidoidea (Hymenoptera) (La Salle 2004); and the several thousand (perhaps 5000) species of oecophorine moths (Oecophoridae), whose larvae consume mainly fallen myrtaceous leaves (Common 1994). Other disproportionately represented insect groups include the Phasmatodea, cicadas, and pselaphine Staphylinidae. Among the social insects, the short-tongued bees (Colletinae and Halictinae) and eumenine vespid wasps are strongly represented, as are the termites, especially in arid and northern Australia. Above all, the Formicidae – the ants – rule. With a described 1275 species and subspecies (Shattuck and Barnett 2001) and perhaps as many as 4000–5000 endemic species, ants dominate ecologically in all but tropical rainforest. This ant diversity and biomass, particularly in more arid Australia, is evident to any modestly observant citizen.

Relictuality, which refers to the appearance of isolated high-ranking taxa for which phylogenetic or fossil evidence exists of past wider distribution and diversity, is a notable feature of Australasian insects. Thus, the species-rich bulldog ants (subfamily Myrmeciinae), now restricted to Australia, indubitably include fossil taxa from Argentina and the Baltic (Ward and Brady 2003). Mastotermitidae, the once-diverse sister group to the remaining termites, is reduced now to *Mastotermes darwiniensis*, a pest in northern Australia. The biting midge family Austroconopidae, which is abundant and diverse in Lebanese and other Cretaceous-age ambers, is represented now by a single species that feeds on early morning golfers in suburban Perth, Western Australia. The cicada family Tettigarctidae (the sister group to all other cicadas) is known from only two extant Australian species, although several Mesozoic northern hemisphere fossil genera are described. Many other examples suggest that the southern hemisphere, perhaps Australia especially, has served as a long-term refuge from the effects of extinctions in the northern hemisphere (Cranston and Gullan 2005, Grimaldi and Engel 2005).

Australia's offshore islands support some enigmatic biodiversity, perhaps none more so than Lord Howe Island. In a report provided by the Australian Museum to NSW NP and WS in 2003, some 1800 terrestrial and freshwater invertebrate species were reported for Lord Howe Island. Additional survey has brought the Coleoptera fauna to more than 500 species, with minimally 23 species of at least 10 mm in length. Typically, for offshore islands, about half the species are flightless and many of these are probably extinct on the main island, whereas offshore islands and rocks still harbor some species. A series of observations from first surveys in 1916 (by Arthur Lea, 2 years before rats arrived), three surveys in the 1970s, and more recent ones show that extinction for many species is probable but not definite, because species such as a large flightless scarab and blaberid cockroach have been rediscovered in sites where rats have been present for many decades (C. A. M. Reid, personal communication). The Lord Howe Island wood-feeding cockroach (*Panesthia lata*) and stick insect (*D. australis*) are two of only ten insect species listed as endangered under the New South Wales Threatened Species Conservation Act.

The more remote, and far more degraded Norfolk Island only developed its fauna when regional volcanic activity ceased, probably within the last 2.3 million years (Holloway 1977). The fauna comprises endemics and a mix of derivatives with relationships to proximate areas; of the 98 species of Macrolepidoptera, 22 species and subspecies are endemic. Among the endemics, most have affinities with Australia and New Caledonia, only two with New Zealand (Holloway 1977). The single Norfolk Island endemic species of the otherwise New Zealand cicada genus *Kikihia*, *K. convicta*, apparently derives from a recent long-distance dispersal from New Zealand (Arensburger et al. 2004b). Half of the 20 species of Orthoptera on Norfolk Island have associations with Australia, the remainder with New Caledonian or Southeast Asia; three native Blattodea are known, with five introduced species especially on the degraded Phillip Island (Rentz 1988).

New Zealand

One of the strongest contrasts between insect biodiversity in New Zealand and Australia is the modest diversity and ecological insignificance of New Zealand ants – only 7 genera and 11 species of native ants (Brown 1958). That this is increasingly supplemented by invasive tramp ants, in common with so many Pacific Islands, does not alter what is, by any definition, an impoverished ant fauna, simply an attenuated subset of Australia's ant fauna, with no high-level endemism. Biogeographic explanations include the ecological – the generally more humid and cool climate is unfavorable – and the historic – ants failed to recolonize after major past extinction associated with submersion of the islands.

A second strong difference is the high rate of aptery (winglessness) among New Zealand insects, especially those living on offshore and subantarctic islands. According to Larivière and Larochelle (2004), about 25% of the New Zealand Heteroptera fauna is flightless, with rates of 65–70% in the Aradidae and Rhyparochromidae. All endemic Blattodea (cockroaches) are flightless, as are many Coleoptera and Orthoptera.

An absence in New Zealand of the major Australian plant radiations means a lack of the associated diversifications that are seen in Australia. The four native species of *Nothofagus*, southern beeches, which often form natural monoculture forests, have a greater insect diversity than might be expected (Hutcheson et al. 1999). *Nothofagus* might host numerous sap-sucking insects, especially certain beech scales (Coccoidea) whose exudates provide an important food resource in temperate forest ecosystems (Moller and Tilley 1989, Sessions 2001). Minor insect radiations are associated with the podocarps, including some yponomeutid moth genera shared with Tasmania (Dugdale 1996a, McQuillan 2003) and, as in Australia, many moth species develop in forest leaf litter (Dugdale 1996b). In contrast to the rather low forest insect diversity, more open habitats support endemic cicadas, orthopterans (both acridid grasshoppers and the wetās), and cockroaches. These have diversified, notably on the South Island, apparently associated with the Alpine orogeny-induced climate changes and the development of more open upland and offshore habitats, including tussock grasslands and scree fields (Buckley et al. 2001, Trewick 2001, Arensburger et al. 2004b, Chinn and Gemmell 2004).

Although New Zealand's aquatic insects belong largely to families, and often genera, that are present in Australia or New Caledonia, Chile, and Patagonia, they display high species endemism (90–99% according to Collier 1993, Harding 2003). The mayfly family Siphlaenigmatidae is endemic to New Zealand. The presence of glaciers on the South Island, presumably originating with Pliocene uplift, seems to have triggered little associated endemism, excepting perhaps a monotypic endemic genus of midge, *Zelandochlus*, the ice worm found on and in ice caves on the Fox Glacier (Boothroyd and Cranston 1999).

New Caledonia, New Guinea, and Melanesia

New Caledonia is renowned for an extraordinary botanical diversity, with both relicts and radiations of several plant groups, including *Araucaria* and *Agathis* and other rare and endemic gymnosperms. To test if phytophagous insect diversity might track this plant diversity, Holloway (1993) assessed the diversity and relationships of New Caledonian Lepidoptera, especially the Macrolepidoptera. Levels of endemism were modest, relationships tended to be directly with proximate Australian, New Guinean, and, less often, New Zealand taxa, and species richness was lower than expected for the area (Holloway 1993). Although exhibiting a radiation, New Caledonia's *Rhytidoponera* ants also appear to be derived recently from Australian relatives (Ward 1984, Lattke 2003). Entomological contributors to the treatment by Najt and Grandcolas (2002) surveyed insect taxa with some diversity, but failed to find any substantial or ancient radiations, and in general Holloway's impressions have been confirmed by studies of other groups. Thus, Murienne et al. (2005) found the diversification of ten New Caledonian species of the cockroach genus *Angustonicus* (Blattidae: subfamily Tryonicinae) dated to not older than 2 million years. New Caledonia is implicated in the evolution of West Pacific cicadas in the clade *Kikihia*, *Maoricicada*, and *Rhodopsalta* distributed among New Zealand, New Caledonia, and eastern Australia. Molecular phylogenetic studies by Arensburger et al. (2004a) indicated the group originated (1) from a New Caledonian ancestor, (2) gave rise to one or several New Caledonian clades, or (3) arose from an Australian ancestor that colonized both New Zealand and New Caledonia. Whichever is correct, the divergence of New Zealand genera from the Australian and New Caledonian genera evidently took place in the last 11–12 million years (Arensburger et al. 2004a). Proposed adaptive radiations in New Caledonia, such as the micropterigid Lepidoptera (Gibbs 1983) that show older, vicariant relationships between the island and New Zealand, evidently need application of molecular techniques to assess implied datings.

The numbers of phytophagous insects of New Guinea show that this geologically complex island harbors an expected high diversity such as that associated with tropical regions. Among these phytophages, the leaf-rolling weevil family Attelabidae, notably some species groups in the genus *Euops*, are associated with *Nothofagus* (Riedel 2001a), although with Myrtaceae in

Australia (Riedel 2001b). Exceptional diversity is found in the Phasmatodea – with more than 200 species of phasmids in 58 different genera, comprising more than 6% of the world's described fauna (van Herwaarden 1998), and the Odonata with more than 400 species, comprising about 10% of the global fauna (Kalkman 2006). Evolutionary biologists have uncovered a diversity of flies (Diptera) with 'antlers' in New Guinea and northern Australia, among which the tephritid genus *Phytalmia* is exceptionally endowed (McAlpine and Schneider 1978).

The Fijian insect fauna is poorly known (Fiji Department of the Environment 1997), but in keeping with its oceanic location, it evidently has a fraction of the megadiversity of New Guinea. Robinson (1975) surveyed the Lepidoptera and estimated 600 species of Microlepidoptera and 400 of Macrolepidoptera. In the absence of an explicit phylogeny, it is unclear if the estimation of seven endemic genera, with modest radiation (although high for a Pacific Island), and several other intrinsic radiations is correct. An estimated 88 ant species from Fiji (Wilson and Hunt 1967) is an underestimate, with some 180 species likely including 30 exotics (E. Sarnat, personal communication). Of 33 species of Odonata recorded from Fiji, 22 (67%) were endemic (Tillyard 1924). Macrolepidoptera, cicadas, dolichopodid flies, and some beetle families show relationships with New Guinea, but several unexpected relationships link Fiji with the New World (Evenhuis and Bickel 2005). The Fijian cicada fauna includes 14 (93%) endemic species, including 1 endemic genus, *Fijipsalta* (Duffels 1988). The cicadas, among the best studied of the regional insects, include the tribe Cosmopsaltriina (Cicadidae) found in Fiji, Vanuatu, and Tonga, and forming the sister group of the Southeast Asian tribe Dundubiina, having no close relatives in Australia, New Caledonia, or New Zealand (de Boer and Duffels 1996, Duffels and Turner 2002).

Vanuatu seems to have many single species representatives of genera that are otherwise widely distributed in the Asia-Pacific Region, although some modest Miocene-dated radiations in platynine Carabidae have been identified (Liebherr 2005). This finding of a platynine radiation is supported by Hamilton's (1981) study of the endemic aphrophorine Cercopidae (Hemiptera), but no other such radiations have been documented. Although 105 of the 364 recorded Vanuatuan species of Macrolepidoptera are

endemic, Robinson (1975) believed that none had its closest relatives in this Archipelago.

THREATENING PROCESSES TO AUSTRALASIAN INSECT BIODIVERSITY

The processes threatening insect biodiversity vary somewhat across and within Australasia, but there are several unifying causes. These are habitat loss, introduced animals including invasive ants, potential effects of climate change, and a minor possibility of overexploitation.

Land clearance and alteration

Clearance of native vegetation, including deforestation, remains the single most significant threat to all terrestrial biodiversity. Worldwide, only four countries exceed the ongoing rate of clearance of native vegetation in Australia (Williams et al. 2001). Rates of deforestation in New Zealand now are low, an inevitable consequence, given an estimated 80% loss of forest cover since human settlement. Forest loss in New Guinea, the Solomons, and Fiji continues, driven by demand from developed countries and weak central controls over logging permits and exports, and revenue collection (FAO 2000). In New Guinea, and throughout Melanesia (the Solomons, Bismarck Archipelago, and Vanuatu), Polhemus et al. (2004) have documented freshwater biotas threatened by alteration of aquatic environments by logging, as well as mining and rapid human population growth. Effects of these activities on all regional freshwater insects include loss of shading by riparian vegetation, leading to elevated water temperatures, increased sedimentation and nutrient inputs, and more variable flows, including increased susceptibility to drought and flood. These modifications all lead to loss of native biodiversity, increased abundance and biomass of tolerant, sometimes alien, species, and growth of algae and often nonnative macrophytes. Replacement of native woody debris in streams with alien woody debris, such as from pines in Australian afforestation programs, has detrimental biodiversity effects, even as a riparian structure is retained (McKie and Cranston 1998). New Zealand's deforestation appears to have caused diminished taxonomic richness, range restriction, and likely extinction of certain stream insects, as assessed at the nearly denuded, but endemism-rich Banks Peninsula (Harding 2003).

Conversion of native forests and pastures to impoverished agroecosystems dominated by alien annual grasses results in the loss of native insect biodiversity. In this context, the day-flying castniid sun moths (*Synemon plana*) have been invoked as flagship or umbrella insects (e.g., New 1997, Douglas 2004) for Australia's ever-diminishing native temperate grasslands, which are among its most threatened ecosystems (Specht 1981). In subtropical and tropical Australian grasslands and savanna, threats to biodiversity include overgrazing by stock of native vegetation, especially in drought, and, controversially, land management by fire in simulations of putative natural fire regimes (Latz 1995, Andersen et al. 2005).

Introduced animals

The sad history of the effects on native biodiversity of introduced animals, such as cats, goats, mongooses, rabbits, rats, and many others, especially on islands, is well known (e.g., Atkinson 1989). However, much documentation (and inference) relate to extinction or threats to native vertebrates, perhaps especially to ground-nesting birds. Only more recently has predator pressure on invertebrates (whether charismatic or not) reached the attention of conservationists and a wider public. Experimental exclosure, or removal, of vertebrate pests, shows that not only birds and lizards, but also large, flightless insects such as New Zealand's wetās and regional stick insects, including the Lord Howe stick (on the vertiginous Balls Pyramid), can survive, even thrive, in the absence of four-footed vermin. That feral animals have caused extinction is hard to verify, but can be surmised by evidence such as the loss of the 15-cm wingspan Buller's moth (*Aoraia mairi*) seen last by Sir Walter Buller in the New Zealand's Ruahine Ranges 120 years ago (while he hunted huia birds) (Meads 1990). Examination of fossil coleopteran remains provides evidence that some large species such as the fern weevil *Tymbopiptus valeas* (Kuschel 1987) and a large ulodid beetle (Leschen and Rhode 2002) are now extinct and that large weevils were much more widespread in New Zealand in the past (Kuschel and Worthy 1996). Such evidence, combined with the fact that many similarly large flightless insects are now restricted largely to difficult-to-access, predator-free offshore islands (Worthy 1997, Michaux and Leschen

2005) and Australia's Lord Howe Island (Priddel et al. 2003), supports the view that vertebrate predators have been, and remain, major threats to large and edible insects. In places where eradication of alien vertebrates has been successful, as on an increasing number of New Zealand's offshore islands, the results have been spectacular, especially for native birds. A curious and unexpected effect includes recovery of geckos, whose dependence on scale-insect honeydew had been masked by the presence of alien rats at this food source (Towns 2002).

Although the effects of introduced vertebrates are visible and well-documented disasters for biodiversity, a more insidious, but as serious a threat comes from the ever-expanding distribution of invasive insects. Just how numerous are the introduced insects can be gauged by data from New Zealand, the country with the greatest awareness and best monitoring systems for these threats. Thus, at the millenium, exotic insects included 229 beetles among nearly 1000 species sampled in and around suburban Auckland, 66 thrips in a total fauna of 119 species, and some 100 species of aphids (Emberson 2000). Emberson (2000) argued that, by extrapolation, there must be more than 2500 adventives, comprising about 13% of the New Zealand insect fauna. The point Emberson (2000) was making – that only 2.5% of these adventives were introduced for biological control purposes, with less than 1% comprising host-specific, carefully screened insect species released and established for biological control of weeds – is important, but the numbers of exotic insects are quite staggering and are still increasing. The rain of Australian insects that descends on New Zealand after appropriate meteorological conditions in the Tasman (Close et al. 1978, Fox 1978) evidently leads to some establishment, including of phytophages on forestry plantation *Eucalyptus* (Withers 2001).

Alien ants, none of which are native to Australasia, are among the worst of invasive animals. They create an insidious problem worldwide that threatens insect biodiversity, especially on Pacific Islands where existing disturbance has predisposed habitat to invasion (McGlynn 1999, Le Breton et al. 2005). On New Caledonia, the lethal effect of *Wasmannia auropunctata* on the native forest-ant fauna was observed in invaded plots, where this New World invasive tramp ant represented more than 92% of all pitfall-trapped ants (Le Breton et al. 2003). Eight months after the invasion front arrived, of the 23 prior resident native ant morphospecies (in 14 genera), only 4 cryptic species survived. Of particular concern in this study was that the invaded habitat was dense, pristine rainforest on ultramafic soils, suspected previously of being more resistant to invasion. Such ecosystems, which support high biodiversity and narrow range endemics, undoubtedly are massively disrupted when the native ant community is decimated – with ecosystem-wide repercussions likely to occur (Le Breton et al. 2003). Such observations can be repeated throughout the Australasian Region as an invasive ant fauna, including *Pheidole megacephala* (the big-headed ant), *Solenopsis invicta* (the red imported fire ant), *Anoplolepis gracilipes* (the yellow crazy ant), and *Linepithema humile* (the Argentine ant), threatens to spread. The economic and environmental costs associated with such invasions are well recognized, as when *S. invicta* was discovered in suburban and industrial Brisbane, Australia, in 2001 (Vanderwoude et al. 2004). The species had probably been present undetected for several years already, stemming from two separate breaches of quarantine (Henshaw et al. 2005). With predictions of rapid spread across the continent (Scanlan and Vanderwoude 2006), a massive plan to eradicate the species was quickly put in place as the state (Queensland) and federal governments recognized and listed the presence of the red imported fire ant as a key threatening process to biodiversity. Through intensive baiting with methoprene and piriproxyfen, coupled with a massive public awareness campaign, a 99% reduction had been attained by 2004. Costs for a 6-year program were Aus$175 million, with a cost-benefit analysis showing potential costs over 30 years of noneradication of Aus$8.9 billion.

Other introduced insects that threaten native biodiversity include honey bees (*Apis melifera*), the large earth bumblebee (*Bombus terrestris*), European wasps (*Vespula* spp.), and the Asian and Australian paper wasps (*Polistes chinensis* and *P. humilis*). Honey bees take over hollows in trees for nesting and compete with native animals, including native bees, for floral resources (Goulson 2003), leading to their presence and activities being recognized as a key threatening process in NSW NP and WS, Australia (NSW NP and WS 2004). Recognition of *B. terrestris* as threatening stems from its role as a pollinator of many environmental weeds and a potential disruptor of native plant

pollination, based on experiences in Tasmania and New Zealand (NSW NP and WS 2004).

In New Zealand, wasps pose a major problem: *P. humilis* became abundant in Northland in the 1880s and remains in the north. German wasps (*Vespula germanica*) arrived in the 1940s and spread to the South Island 10 years later; *V. vulgaris* became established in the 1970s, and both are now widespread. The most recent arrival, the Asian paper wasp (*P. chinensis*), was found first near Auckland in 1979, and is rapidly extending its range southward. Problems with effects of wasps on native biodiversity are particularly significant in the *Nothofagus* beech forests where abundant honeydew, produced by endemic coccoids (Coelostomidiinae), provides an abundant source of carbohydrate. This resource, on which a community of honeydew-feeding native birds once thrived, has been hijacked by *Vespula*, particularly *V. vulgaris*. Extraordinary wasp densities of 10,000 workers ha^{-1} and peak biomass of 3.8 kg ha^{-1} can develop by the late summer (Beggs 2001). Wasp demand for protein is nearly insatiable, such that for many invertebrate prey items, such as caterpillars, individual survivorship probability is near zero (Beggs and Rees 1999). Wasp densities need to be reduced by an estimated 80–90% to conserve vulnerable native species, but even mass baiting with the effective fipronil and the introduction of an ichneumonid wasp parasitoid is unlikely to sustain such high levels of control (Beggs 2001). Asian paper wasps in warm and humid northern New Zealand can develop densities of more than 6300 wasps and 200 nests ha^{-1} and consume 1 kg ha^{-1} of invertebrate biomass per season, giving rise to fears of severe effects on native ecosystems (Clapperton 1999). Control of this species is even more difficult than for *Vespula* because nests are difficult to find due to location in dense bush and infrequent traffic of wasp residents, and numbers of wasps are reduced only minimally by baiting and trapping (Toft and Harris 2004).

Climate change

The history of the globe involves cyclical climate change, some of which has been induced by volcanism and solar cycles, interspersed with episodic bolide impacts. The planet is recovering still from the effects of the Pleistocene glaciations of much of the mid-high latitudes of the northern hemisphere. In Australia, the Holocene period has seen reduced temperatures and increased aridity, although any linkage to northern climate variations remains unverified (Turney et al. 2006). Nonetheless, Australian Holocene climate changes affected vegetation (with increased xeric conditions producing rainforest contraction and assisting spread of sclerophylous, *Eucalyptus*-dominated vegetation). The insect biota was affected too, as distributions shifted in latitude and elevation with changing temperatures (Porch and Elias 2000 (Coleoptera), Dimitriadis and Cranston 2001 (Chironomidae)). In New Zealand, the same organisms also show historical changes (Marra et al. 2004, 2006, Woodward and Shulmeister 2007), although the causes and synchronicity might be dissimilar to those in Australia (Alloway et al. 2007), as they relate mostly to extensive volcanism in New Zealand. In both Australia and New Zealand, El Niño-Southern Oscillation (ENSO) events have been important in the past and remain influential. Despite this background variability, the Australian government (a notable nonsignatory to the Kyoto Convention at November 2007) accepted that alteration to Australia's climate already occurs over and above natural variability (DEH 2004). Changes such as long-term spatial and temporal variation in rainfall and temperature patterns are expected to influence Australia's biological diversity (DEH 2004). A widely accepted scenario of 3°C warming by 2050–2100, for example, compared with a 1990 baseline (IPCC 2001, Hughes 2003), means that species with latitudinal ranges of less than about 300 km or with an elevation range less than about 300 m will dissociate totally from their present-day temperature envelopes (Westoby and Burgman 2006). In other words, their present-day distributions could not be maintained. Federal acceptance of the scientific consensus concerning such expected effects on species (and ecosystems) under future climate scenarios came from a growing list of changes consistent with predictions. Inevitably, perhaps because of their ectothermy, some of these changes involve insects.

As in the northern hemisphere, the extensive databases for butterfly locations and flight dates have provided a foundation against which changes can be assessed. Using bioclimatic modeling for 77 species of Australian butterflies, Beaumont and Hughes (2002) showed that, although few species had narrow climatic ranges (<3°C), a majority of species were predicted to suffer range losses, some substantial. Under extreme scenarios, narrow-range endemics would lose their total ranges; especially vulnerable would be those

species that depended on mutualisms such as ant attendance and food-plant specificity. Even potential beneficiaries, such as a pair of Western Australian species whose modeled suitable habitat was predicted to expand to include South Australia, still would have to cross 1000 km of an inhospitable Nullabor to make the necessary range extension. Evidently, specialists are in trouble under climate change, as predicted for Acacia-feeding beetles (Andrews and Hughes 2004) and high-elevation dipterans (Wilson et al. 2007). Species unable to track their changing climate envelopes, due to low dispersivity, including aptery (Yeates et al. 2002), small population sizes, or loss of their host or habitat (e.g., insufficient altitude) will go extinct. Adding to the problem is the fragmentation of the landscape, in which a potentially suitable environmental envelope might be entirely unavailable through anthropogenic conversion. In New Zealand, we can imagine that glacier-associated insects will lose their habitat. Current glacial retreat is one of the clearest indicators of climate change affecting the New Zealand environment. For example, the Franz Joseph glacier has retreated 1500 m since scientific observations began in 1860. The Franz Joseph, Fox, and other glaciers are in an overall pattern of retreat, despite some fluctuations and short-term advances, which is expected to continue and threaten cold-stenothermic insects.

As a perhaps terminal insult, we can reasonably assume that climate change increases the risk to biodiversity from invasive species, in that broad-ranging taxa are potentially invasive and least affected by climate change.

AUSTRALASIAN BIODIVERSITY CONSERVATION

In comparison to the northern hemisphere, insect biodiversity conservation was somewhat late to attain recognition in Australia. Key's (1978) report to the Australian National Parks and Wildlife Service on the conservation status of the nation's insects pioneered in drawing attention to issues of land degradation and its effects on the endemic insect fauna. New's *Insect Conservation: an Australian Perspective* (1984) was influential in furthering awareness, and he has continued to play a leadership role in promoting insect conservation globally (e.g., New et al. 1995, New 1997). Penny Greenslade has been an indefatigable advocate for the inclusion of insects in conservation planning, especially concerning listing of species, in collating invertebrate data on Australia's offshore responsibilities, and in stimulating a conservation ethos in the national entomological society. The review by Greenslade and New (1991), which remains an excellent, far-sighted overview of Australian insect conservation, drew the attention of one of the largest meetings of insect conservationists ever held at that time, the 15th Symposium of the Royal Entomological Society, London, 1989 (Collins and Thomas 1991). Because such meetings indicate the depth of interest in the field, the holding of Invertebrate Biodiversity and Conservation meetings in Australia since the early 1990s is indicative of a wide and growing regional interest. The proceedings of each meeting (Ingram et al. 1994, Yen and New 1997, Ponder and Lunney 1999, Austin et al. 2003) demonstrate 'progressively greater awareness and concern for a great variety of taxonomic groups at scientific, legislative, economic and social levels, and approaches to practical appraisal and management' (New and Sands 2004: p. 258).

Butterflies always have been best known in terms of their taxonomy (with some 650 or so Australian species and subspecies), distributions, life histories, and threats to their existence (New 1990, 1999, Dunn et al. 1994, Sands 1999) – and this remains so today with a range of conservation action and recovery plans in place (e.g., Sands et al. 1997, O'Dwyer and Attiwill 2000, Sands and New 2002). Models of such processes can be seen in the endangered species listing and development of action plans for the Bathurst copper (*Paralucia spinifera*) and the ongoing management plans for the Eltham copper (*P. pyrodiscus lucida*) in Victoria (New and Sands 2004).

In New Zealand, too, assessment of risk to insects often starts with Lepidoptera. Thus, using a suite of evaluation criteria, Patrick and Dugdale (2000) assessed the conservation status for New Zealand's 1685 species of Lepidoptera. Category A (highest priority) threatened species numbered 42 species, with 29 in urgent need of conservation action. Category B (second priority) numbered 42 species, and Category C (third priority) 20 species; another 102 species were regarded as being at risk. Previously, only two lepidopteran species, *Asaphodes stinaria* and *Xanthorhoe bulbulata*, had been listed in Category A. Most of the endangered species are phytophagous as larvae; the more numerous detritivores appear less threatened. Most at-risk Lepidoptera are members of natural shrub–grassland communities,

and are concentrated geographically in Canterbury and Otago where such habitats are more prevalent.

Despite the butterfly-centric bias in insect conservation, an increasing range of taxa are recognized as being of conservation concern, able to arouse a wider concern for environmental conservation and to act as flagships for endangered ecosystems. In New Zealand, some rarer species of wetās have attained such status (e.g., Sherley and Hayes 1993, McGuinness 2001). Concern for the phylogenetically distinct odonate *Hemiphlebia mirabilis* led to assessment of the conservation of certain Australian wetlands (New 1993); other aquatic insects, including large charismatic odonates and restricted-range Plecoptera and torrent midges (Blephariceridae), all featured early in Australian listings for conservation concern.

As is widely recognized, legislative actions concerning conservation, although sometimes bureaucratic and always time consuming, are essential for public education, demonstration of institutional commitment, and putting in place enforcible and long-term protection. This is the interface between conservation science and its application. At the international level, Australasian nations all were early signatories of the Convention on Biological Diversity (CBD), in mid-1992, with ratifications made before the end of 1993. In keeping with the obligations associated with the convention, each country has produced plans for conservation. Australia's *National Strategy for the Conservation of Australia's Biological Diversity* (DEST 1996) is typical in integrating biodiversity conservation with natural resource management, sustainability, identification, and management of threatening processes, combined with advocacy of enhanced research, education, and documentation. The issue of how one can conserve biodiversity on a continental scale without understanding what is present was addressed through a strengthened role for the Australian Biological Resources Study (ABRS), the major funding source for biodiversity inventory work, although a decade later any enhanced role remains too modest for the taxonomic task in hand (Yeates et al. 2003).

Australia's biodiversity strategy led directly or indirectly to some novel approaches concerning invertebrates in biodiversity conservation that are of global relevance. These approaches include efforts to integrate phylogeny into conservation measures (e.g., Faith 2002, exemplified by Faith et al. 2004), the development of modeling tools for conservation-site selection (e.g., Ferrier et al. 1999), and a veritable minor industry of environmental indicators (e.g., ants: Andersen 1990, Andersen et al. 2002; aquatic insects: Bunn 1995, Marchant et al. 1997, Simpson and Norris 2000). Development of a national strategy, in turn, stimulated development of rapid biodiversity assessment protocols (e.g., Cranston and Hillman 1992, Beattie et al. 1993, Oliver and Beattie 1995) and encouraged exploration of the concept of indicator species of biodiversity (Trueman and Cranston 1997, Andersen 1999, Kitching et al. 2000); McGeoch (1998) gave an informative review of some of these issues. Certain lessons from these studies were assessed critically for New Zealand by Ward and Larivière (2004) in relation to their country's biodiversity inventory needs.

New Zealand's biodiversity strategy (DoC 2000) seeks to halt the decline in New Zealand's indigenous biodiversity, noting especially the loss of around 80% of native forest cover and 90% of wetlands. Programs include those to incorporate Maori knowledge and enhance their understanding and involvement in biodiversity conservation. The strategy gives a high profile to the biodiversity threats posed by invasive species; the country is a world leader in the field of invasive species research as a result of its experience with a host of plant and animal pests and its concern with quarantine (Biosecurity) and the maintenance of a healthy environment. With this background, New Zealand was an obvious host for the IUCN Invasive Species Specialist Group (ISSG) of the World Conservation Union. ISSG offers practical advice and aid to smaller biodiverse island nations to eliminate aliens (such as cane toads, cats, and rats), notably through the collaborative Pacific Invasives Initiative.

CONCLUSION

Australasian biodiversity is substantial and perhaps surprisingly well studied. Taylor's (1976) 'taxonomic impediment' hindering incorporation of species data into inventory and conservation planning remains a problem, but this has not deterred entomologists and conservation biologists from making substantial contributions to an understanding of regional insect biodiversity and its conservation. The resources needed for full description of the Australian insect biota, as assessed by Yeates et al. (2003), are quite modest by comparison with big science projects in other areas. Provision of these resources throughout Australasia

would allow the identification of more of the biota, and stimulate study of more of the 'other 99%'. When insect biodiversity coincides with biosecurity issues, as with potential introductions of pests, major funding can be found. Some of this funding would be well spent if allocated to the big pictures of phylogeny, ecology, and conservation. Biodiversity practitioners need to remain aware of the importance of gaining support from the essentially sympathetic, but highly urban populations of Australia and New Zealand. Major inroads have been made with flagship insects, such as the Richmond birdwing, Bathurst copper, and Eltham copper, as schools have become engaged with host planting, alien removals, and population censusing. Furthermore, the affluent Australasian nations need to maintain and expand their regional roles in biodiversity conservation, passing on expertise in areas such as management of invasive species, which seems likely to remain a major threat to regional biodiversity over the next decades.

REFERENCES

Alloway, B. V., D. J. Lowe, D. J. A. Barrell, R. M. Newnham, P. C. Almond, P. C. Augustinus, N. Bertler, L. Carter, N. J. Litchfield, M. S. McGlone, J. Shulmeister, M. J. Vandergoes, P. W. Williams, and NZ-INTIMATE members. 2007. Towards a climate event stratigraphy for New Zealand over the past 30,000 years (NZ-INTIMATE project). *Journal of Quaternary Science* 22: 9–35.

Andersen, A. N. 1990. The use of ant communities to evaluate change in Australian terrestrial ecosystems: a review and a recipe. *Proceedings of the Ecological Society of Australia* 16: 347–357.

Andersen, A. N. 1999. My bioindicator or yours? Making the selection. *Journal of Insect Conservation* 3: 1–4.

Andersen, A. N., B. D. Hoffmann, W. J. Müller, and A. D. Griffiths. 2002. Using ants as bioindicators in land management: simplifying assessment of ant community responses. *Journal of Applied Ecology* 39: 8–17.

Andersen, A. N., G. D. Cook, L. K. Corbett, M. M. Douglas, R. W. Eager, J. Russell-Smith, S. A. Setterfield, R. J. Williams, and J. C. Z. Woinarski. 2005. Fire frequency and biodiversity conservation in Australian tropical savannas: implications from the Kapalga fire experiment. *Austral Ecology* 30: 155–167.

Andrews, N. R. and L. Hughes. 2004. Species diversity and structure of phytophagous beetle assemblages along a latitudinal gradient: predicting the potential impacts of climate change. *Ecological Entomology* 29: 527–542.

Arensburger, P., T. R. Buckley, C. Simon, M. Moulds, and K. E. Holsinger. 2004a. Biogeography and phylogeny of the New Zealand cicada genera (Hemiptera: Cicadidae) based on nuclear and mitochondrial DNA data. *Journal of Biogeography* 31: 557–569.

Arensburger, P., C. Simon, and K. Holsinger. 2004b. Evolution and phylogeny of the New Zealand cicada genus *Kikihia* Dugdale (Homoptera: Auchenorrhyncha: Cicadidae) with special reference to the origin of the Kermadec and Norfolk Islands' species. *Journal of Biogeography* 31: 1769–1783.

Atkinson, I. A. E. 1989. Introduced animals and extinctions. Pp. 54–75. *In* D. C. Western and M. C. Pearl (eds). *Conservation for the Twenty-first Century*. Oxford University Press, New York.

Austin, A. D., D. A. Mackay, and S. J. B. Cooper (eds). 2003. *Invertebrate Biodiversity and Conservation*. Monograph Series 7. Records of the South Australian Museum, Monograph Series 7, Adelaide.

Austin, A. D., D. K. Yeates, G. Cassis, M. J. Fletcher, J. La Salle, J. F. Lawrence, P. B. McQuillan, L. A. Mound, D. Bickel, P. J. Gullan, D. F. Hales, and G. S. Taylor. 2004. Insects 'down under' – diversity, endemism and evolution of the Australian insect fauna: examples from select orders. *Australian Journal of Entomology* 43: 216–234.

Balke, M., C. H. S. Watts, S. J. B. Cooper, W. F. Humphreys, and A. P. Vogler. 2004. A highly modified stygobiont diving beetle of the genus *Copelatus* (Coleoptera, Dytiscidae): taxonomy and cladistic analysis based on mitochondrial DNA sequences. *Systematic Entomology* 29: 59–67.

Basset, Y. and V. Novotny. 1999. Species richness of insect herbivore communities on *Ficus* in Papua New Guinea. *Biological Journal of the Linnean Society* 67: 477–499.

Beattie, A. J., J. D. Majer, and I. Oliver. 1993. Rapid biodiversity assessment: a review. Pp. 4–14. *In* A. J. Beattie (ed). *Rapid Biodiversity Assessment*. Macquarie University, Sydney.

Beaumont, L. J. and L. Hughes. 2002. Potential changes in the distribution of latitudinally restricted Australian butterfly species in response to climate change. *Global Change Biology* 8: 594–971.

Beggs, J. 2001. The ecological consequences of social wasps (*Vespula* spp.) invading an ecosystem that has an abundant carbohydrate resource. *Biological Conservation* 11: 24–37.

Beggs, J. R. and J. S. Rees. 1999. Restructuring of Lepidoptera communities by introduced *Vespula* wasps in a New Zealand beech forest. *Oecologia* 119: 565–571.

Boothroyd, I. and P. S. Cranston. 1999. The 'ice worm' – the immature stages, phylogeny and biology of the glacier midge *Zelandochlus* (Diptera: Chironomidae). *Aquatic Insects* 21: 303–316.

Brown, W. L. 1958. A review of the ants of New Zealand. *Acta Hymenopterologica* 1: 1–50.

Buckley, T. R., C. Simon, and G. K. Chambers. 2001. Phylogeography of the New Zealand cicada *Maoricicada campbelli* based on mitochondrial DNA sequences: ancient clades associated with cenozoic environmental change. *Evolution* 55: 1395–1407.

Bunn, S. E. 1995. Biological monitoring of water quality in Australia: workshop summary and future directions. *Australian Journal of Ecology* 20: 220–227.

Chinn, W. G. and N. J. Gemmell. 2004. Adaptive radiation within New Zealand endemic species of the cockroach genus *Celatoblatta* Johns (Blattidae): a response to Plio-Pleistocene mountain building and climate change. *Molecular Ecology* 13: 1507–1518.

Chazeau, J. 1993. Research on New Caledonian terrestrial fauna: achievements and prospects. *Biodiversity Letters* 1: 123–129.

Christidis, F. 2006. Phylogenetic relationships of the Australian Leptophlebiidae (Ephemeroptera). *Invertebrate Systematics* 19: 531–539.

Clapperton, B. K. 1999. Abundance of wasps and prey consumption of paper wasps (Hymenoptera, Vespidae: Polistinae) in Northland, New Zealand. *New Zealand Journal of Ecology* 23: 11–19.

Close, R. C., N. T. Moar, A. I. Tomlinson, and A. D. Lowe. 1978. Aerial dispersal of biological material from Australia to New Zealand. *International Journal of Biometeorology* 22: 1–19.

Colless, D. H. and D. K. McAlpine. 1991. Diptera. Pp. 717–786. *In* I. D. Naumann (ed). *The Insects of Australia*, Volume 2, Second Edition. Melbourne University Press, Melbourne.

Collier, K. J. 1993. Review of the status, distribution, and conservation of freshwater invertebrates in New Zealand. *New Zealand Journal of Marine and Freshwater Research* 27: 339–356.

Collins, N. M. and J. A. Thomas (eds). 1991. *The Conservation of Insects and their Habitats*. Academic Press, London. 450 pp.

Common, I. F. B. 1994. *Oecophorine Genera of Australia I: The Wingia Group (Lepidoptera: Oecophoridae)*. Monographs on Australian Lepidoptera 3. CSIRO Publishing, Melbourne.

Cook, L. G. and P. J. Gullan. 2004. The gall-inducing habit has evolved multiple times among the eriococcid scale insects (Sternorrhyncha: Coccoidea: Eriococcidae). *Biological Journal of the Linnean Society* 83: 441–452.

Cooper, S. J. B., S. Hinze, R. Leys, C. H. S. Watts, and W. F. Humphreys. 2002. Islands under the desert: molecular systematics and evolutionary origins of stygobiont water beetles (Coleoptera: Dytiscidae) from central Western Australia. *Invertebrate Systematics* 16: 589–598.

Cranston, P. S. and I. D. Naumann. 1991. Biogeography. Pp. 180–197. *In* I. D. Naumann (ed). *Insects of Australia*, Second Edition. Melbourne University Press, Melbourne.

Cranston, P. S. and T. Hillman. 1992. Rapid assessment of biodiversity using 'biological diversity technicians'. *Australian Biologist* 5: 144–154.

Cranston, P. S., D. H. D. Edward, and L. G. Cook. 2002. New status, distribution records and phylogeny for Australian mandibulate Chironomidae (Diptera). *Australian Journal of Entomology* 41: 357–366.

Cranston, P. S. and P. J. Gullan. 2005. Time flies?. *Evolution* 59: 2492–2494.

Crespi, B. J., D. C. Morris, and L. A. Mound. 2004. *Evolution of Ecological and Behavioural Diversity: Australian Acacia Thrips as Model Organisms.* Australian Biological Resources Study and Australian National Insect Collection, CSIRO, Canberra. 321 pp.

D'Albertis, L. M. 1880. New Guinea: what I did and what I saw. Sampson Low, London. [English Translation].

de Boer, A. J. and J. P. Duffels. 1996. Historical biogeography of the cicadas of Wallacea, New Guinea and the West Pacific: a geotectonic explanation. *Palaeogeography, Palaeoclimatology, Palaeoecology* 124: 153–177.

Department of the Environment and Heritage (DEH). 2004. National Biodiversity and Climate Change Action Plan 2004–2007. http://www.environment.gov.au/biodiversity/publications%20/nbccap/index.html [Accessed 13 November 2007].

Department of the Environment, Sport and Territories (DEST). 1996. National Strategy for the Conservation of Australia's Biological Diversity. Department of the Environment and Heritage, Commonwealth of Australia. http://www.deh.gov.au/biodiversity/publications/strategy/ [Accessed 1 July 2006].

Department of Conservation/Ministry for the Environment (DoC/ME). 2000. The New Zealand Biodiversity Strategy. Department of Conservation and the Ministry for the Environment, Wellington, New Zealand. http://www.biodiversity.govt.nz/picture/doing/nzbs/contents.html [Accessed 1 July 2006].

Dimitriadis, S. and P. S. Cranston. 2001. An Australian Holocene climate reconstruction using Chironomidae from a tropical volcanic maar lake. *Palaeogeography, Palaeoclimatology, Palaeoecology* 176: 109–131.

Doorne, S. 1999. *Visitor Experience at the Waitomo Glowworm Cave.* Science for Conservation 95. Department of Conservation, Wellington. 43 pp.

Douglas, F. 2004. A dedicated reserve for conservation of two species of *Synemon* (Lepidoptera: Castniidae) in Australia. *Journal of Insect Conservation* 8: 221–228.

Duffels, J. P. 1988. *The Cicadas of Fiji, Samoa and Tonga Islands, Their Taxonomy and Biogeography.* Entomonograph 10. E.J. Brill/Scandinavian Science Press, Vinderup. 108 pp.

Duffels, J. P. and H. Turner. 2002. Cladistic analysis and biogeography of the cicadas of the Indo-Pacific subtribe Cosmopsaltriina (Hemiptera: Cicadoidea: Cicadidae). *Systematic Entomology* 27: 235–261.

Dugdale, J. S. 1996a. *Chrysorthenches* new genus, conifer-associated plutellid moths (Yponomeutoidea, Lepidoptera) in New Zealand and Australia. *New Zealand Journal of Zoology* 23: 33–59.

Dugdale, J. S. 1996b. Natural history and identification of litter-feeding Lepidoptera larvae (Insecta) in beech forests, Orongorongo Valley, New Zealand, with especial reference to the diet of mice (*Mus musculus*). *Journal of the Royal Society of New Zealand* 26: 251–274.

Dunn, K. L., R. L. Kitching, and E. M. Dexter. 1994. The conservation status of Australian butterflies. Report to Australian National Parks and Wildlife Service, Canberra.

Edmunds, G. F., Jr. 1972. Biogeography and evolution of Ephemeroptera. *Annual Review of Entomology* 17: 21–42.

Emberson, R. M. 1998. The beetle (Coleoptera) fauna of the Chatham Islands. *New Zealand Entomologist* 21: 25–64.

Emberson, R. W. 2000. Endemic biodiversity, natural enemies, and the future of biocontrol. Pp. 875–880. *In* N. R. Spencer (ed.) *Proceedings of the X International Symposium on Biological Control of Weeds, 4–14 July 1999.* Montana State University, Bozeman, Montana.

Erichson, W. F. 1842. Beitrag zur Fauna von Vandiemensland mit besonderer Rücksicht auf die Geographische Verbreitung der Insekten. *Wiegmann's Archiv* 8: 83–287.

Evenhuis, N. L. and D. J. Bickel. 2005. The NSF-Fiji terrestrial arthropod survey: overview. Fiji Arthropods I. *Bishop Museum Occasional Papers* 82: 3–25.

Food and Agriculture Organisation (FAO). 2002. Global Forest Resources Assessment 2000. FAO Forestry Paper 140. 512 pp. http://www.fao.org/DOCREP/004/Y1997E/Y1997E00.HTM [Accessed 1 July 2006].

Faith, D. P. 2002. Quantifying biodiversity: a phylogenetic perspective. *Conservation Biology* 16: 248–252.

Faith, D. P., C. A. M. Reid, and J. Hunter. 2004. Integrating phylogenetic diversity, complementarity, and endemism. *Conservation Biology* 18: 255–261.

Ferrier, S., M. R. Gray, G. A. Cassis, and L. Wilkie. 1999. Spatial turnover in species composition of ground-dwelling arthropods, vertebrates and vascular plants in north-east New South Wales: implications for selection of forest reserves. Pp. 68–76. *In* W. Ponder and D. Lunney (eds). *The Other 99%: The Conservation and Biodiversity of Invertebrates.* Transactions of the Royal Society of New South Wales, Mosman.

Fiji Department of Environment. 1997. Convention on Biological Diversity – 1997 National Report to the Conference of Parties by the Republic of Fiji. Fiji Ministry of Local Government, Housing and the Environment, Suva. 29 pp.

Fox, K. J. 1978. The transoceanic migration of Lepidoptera to New Zealand – a history and a hypothesis on colonisation. *New Zealand Entomologist* 6: 368–380.

Gaston, K. J. and E. Hudson. 1994. Regional patterns of diversity and estimates of global insect species richness. *Biodiversity and Conservation* 3: 493–500.

Gibson, J. F. and G. W. Courtney. 2007. Revision of the net-winged midge genus *Horaia* Tonnoir and its phylogenetic relationship to other genera within the tribe Apistomyiini (Diptera: Blephariceridae). *Systematic Entomology* 32: 276–304.

Gibbs, G. W. 1983. Evolution of Micropterigidae (Lepidoptera) in the SW Pacific. *GeoJournal* 7: 505–510.

Goulson, D. 2003. Effects of introduced bees on native ecosystems. *Annual Review of Ecology, Evolution and Systematics* 34: 1–26.

Greenslade, P. 1990. Notes on the biogeography of the free-living terrestrial invertebrate fauna of Macquarie Island with an annotated checklist. *Papers and Proceedings of the Royal Society of Tasmania* 124: 35–50.

Greenslade, P. and T. R. New. 1991. Australia: conservation of a continental insect fauna. Pp. 11–34. *In* N. M. Collins and J. A. Thomas (eds). *The Conservation of Insects and Their Habitats*. Academic Press, London.

Grimaldi, D. and M. S. Engel. 2005. *Evolution of the Insects*. Cambridge University Press, Cambridge. 755.

Gullan, P. J., D. R. Miller, and L. G. Cook. 2005. Gall-inducing scale insects (Hemiptera: Sternorrhyncha: Coccoidea). Pp. 159–229. *In* A. Raman, C. W. Schaefer, and T. M. Withers (eds). *Biology, Ecology, and Evolution of Gall-Inducing Arthropods*. Science Publishers, Plymouth, United Kingdom.

Hamilton, K. G. A. 1981. Aphrophorinae of the Fiji, New Hebrides and Banks Islands (Rhynchota: Homoptera: Cercopidae). *Pacific Insects* 23: 465–477.

Harding, J. S. 2003. Historic deforestation and the fate of endemic invertebrate species in streams. *New Zealand Journal of Marine and Freshwater Research* 37: 333–345.

Harrison, L. 1928. The composition and origins of the Australian fauna with special reference to Wegener's hypothesis. *Report of the Australian and New Zealand Association for the Advancement of Science* 18: 332–396.

Henshaw, M. T., N. Kunzmann, C. Vanderwoude, M. Sanetra, and R. H. Crozier. 2005. Population genetics and history of the introduced fire ant, *Solenopsis invicta* Buren (Hymenoptera: Formicidae), in Australia. *Australian Journal of Entomology* 44: 37–44.

Hollis, D. 2004. *Australian Psylloidea: Jumping Plantlice and Lerp Insects*. Australian Biological Resources Study, Canberra. 16, 216 pp.

Holloway, J. D. 1977. *Lepidoptera of Norfolk Island, Their Biogeography and Ecology*. Series Entomologica 13. Dr. W. Junk Publishers, The Hague. 292 pp.

Holloway, J. D. 1993. Lepidoptera in New Caledonia: diversity and endemism in a plant-feeding insect group. *Biodiversity Letters* 1: 92–101.

Hooker, J. D. 1853. *The Botany of the Antarctic Voyage of H.M. Discovery Ships Erebus and Terror in the Years 1839–1843. II. Flora Novae-Zelandiae*. Lovell-Reeve, London. Pp. 1–39. [Introductory essay]

Hughes, L. 2003. Climate change and Australia: trends, projections and impacts. *Austral Ecology* 28: 423–443.

Humphreys, W. F. 2001. Groundwater calcrete aquifers in the Australian arid zone: the context to an unfolding plethora of stygal biodiversity. *Records of the Western Australian Museum Supplement* 64: 63–83.

Hutcheson, J., P. Walsh, and D. Given. 1999. Potential value of indicator species for conservation and management of New Zealand terrestrial communities. *Science for Conservation* 109: 1–90.

Ingram, G. J., R. J. Raven, and P. J. F. Davie (eds). 1994. Invertebrate biodiversity and conservation. *Memoirs of the Queensland Museum* 36: 1–239.

IPCC. 2001. *Climate Change 2001: The Scientific Basis*. Contribution of Working Group I to the Third Assessment Report of the Intergovernmental Panel on Climate Change. J. T. Houghton, Y. Ding, D. J. Griggs, M. Noquer, P. J. vander. Linden, X. Dai, K. Maskell, and C. A. Johnson (eds). Cambridge University Press, Cambridge. 881 pp.

Kalkman, V. J. 2006. The Odonata (Dragonflies and Damselflies) of Papua. Papua Insects Foundation. http://www.papua-insects.nl/insect%20orders/Odonata/Odonata%20suborders.htm [Accessed July 2006].

Key, K. H. L. 1978. *Conservation Status of Australia's Insect Fauna*, Occasional Paper No.1. Australian National Parks and Wildlife Service, Canberra.

Kitching, R. L., A. G. Orr, L. Thalib, H. Mitchell, M. S. Hopkins, and A. W. Graham, 2000. Moth assemblages as indicators of environmental quality in remnants of upland Australian rain forest. *Journal of Applied Ecology* 37: 284–297.

Klimaszewski, J. and J. C. Watt. 1997. *Coleoptera: Family-group Review and Keys to Identification*. Fauna of New Zealand 3. Manaaki Whenua Press, Lincoln. 198 pp.

Kolesik, P., R. J. Adair, and G. Eick. 2005. Nine new species of *Dasineura* (Diptera: Cecidomyiidae) from flowers of Australian *Acacia* (Mimosaceae). *Systematic Entomology* 30: 454–479.

Kuschel, G. 1987. The subfamily Molytinae (Coleoptera: Curculionidae): general notes and descriptions of new taxa from New Zealand and Chile. *New Zealand Entomologist* 9: 11–29.

Kuschel, G. 1990. Beetles in a suburban environment: a New Zealand case study: the identity and status of Coleoptera in the natural and modified habitats of Lynfield, Auckland (1974–1989). DSIR Plant Protection Report 3. New Zealand Department of Scientific and Industrial Research, Auckland.

Kuschel, G. and T. H. Worthy. 1996. Past distribution of large weevils (Coleoptera: Curculionidae) in the South Island, New Zealand, based on Holocene fossil remains. *New Zealand Entomologist* 19: 15–22.

Larivière, M. C. and A. Larochelle. 2004. *Heteroptera (Insecta: Hemiptera)*. Fauna of New Zealand 50. Manaaki Whenua Press, Lincoln. 330 pp.

Lattke, J. E. 2003. Biogeographic analysis of the ant genus *Gnamptogenys* Roger in South-East Asia-Australasia (Hymenoptera: Formicidae: Ponerinae). *Journal of Natural History* 37: 1879–1897.

Latz, P. K. 1995. *Bushfires and Bushtucker: Aboriginal Plant Use in Central Australia*. IAD Press, Alice Springs. 400 pp.

La Salle, J. 2004. Biology of gall inducers and evolution of gall induction in Chalcidoidea (Hymenoptera: Eulophidae, Eurytomidae, Pteromalidae, Tanaostigmatidae, Torymidae). Pp. 503–533. *In* A. Raman, C. W. Schaeffer, and

T. M. Withers (eds). *Biology, Ecology, and Evolution of Gall-Inducing Arthropods*, Science Publishers, Enfield.

Le Breton, J., J. Chazeau, and H. Jourdan. 2003. Immediate impacts of invasion by *Wasmannia auropunctata* (Hymenoptera: Formicidae) on native litter ant fauna in a New Caledonian rainforest. *Austral Ecology* 28: 204–209.

Le Breton, J., H. Jourdan, J. Chazeau, O. Jérôme, and A. Dejean. 2005. Niche opportunity and ant invasion: the case of *Wasmannia auropunctata* in a New Caledonian rain forest. *Journal of Tropical Ecology* 21: 93–98.

Leschen, R. A. B. and B. E. Rhode. 2002. A new genus and species of large extinct Ulodidae (Coleoptera) from New Zealand. *New Zealand Entomologist* 25: 57–64.

Leschen, R. A. B., J. F. Lawrence, G. Kuschel, S. Thorpe, and Q. Wang. 2003. Coleoptera genera of New Zealand. *New Zealand Entomologist* 26: 15–28.

Leys, R., C. H. S. Watts, S. J. B. Cooper, and W. F. Humphreys. 2003. Evolution of subterranean diving beetles (Coleoptera: Dytiscidae: Hydroporini, Bidessini) in the arid zone of Australia. *Evolution* 57: 2819–2834.

Liebherr, J. K. 2005. Platynini (Coleoptera: Carabidae) of Vanuatu: Miocene diversification on the Melanesian Arc. *Invertebrate Systematics* 19: 263–295.

Mackerras, I. M. 1925. The Nemestrinidae of the Australian Region. *Proceedings of the Linnean Society of New South Wales* 50: 489–501.

Majer, J. D., H. F. Recher, A. B. Wellington, J. C. Z. Woinarski, and A. L. Yen. 1997. Invertebrates of eucalypt formations. Pp. 278–302. *In* J. E. Williams and J. C. Z. Woinarski (eds). *Eucalypt Ecology. Individuals to Ecosystems.* Cambridge University Press, Cambridge.

Marchant, R., A. Hirst, R. H. Norris, R. Butcher, L. Metzeling, and D. Tiller. 1997. Classification and prediction of macroinvertebrate assemblages from running waters in Victoria, Australia. *Journal of the North American Benthological Society* 16: 664–681.

Marra, M. J., E. G. C. Smith, J. Shulmeister, and R. Leschen. 2004. Late Quaternary climate change in the Awatere Valley, South Island, New Zealand using a sine model with a maximum likelihood envelope on fossil beetle data. *Quaternary Science Reviews* 23: 1637–1650.

Marra, M. J., J. Shulmeister, and E. G. C. Smith. 2006. Reconstructing temperature during the Last Glacial Maximum from Lyndon Stream, South Island, New Zealand using beetle fossils and maximum likelihood envelopes. *Quaternary Science Reviews* 25: 1841–1849.

Marris, J. M. W. 2000. The beetle (Coleoptera) fauna of the Antipodes Islands, with comments on the impact of mice; and an annotated checklist of the insect and arachnid fauna. *Journal of the Royal Society of New Zealand* 30: 169–195.

McAlpine, D. K. and M. A. Schneider. 1978. A systematic study of *Phytalmia* (Diptera, Tephritidae) with description of a new genus. *Systematic Entomology* 3: 159–175.

McCarthy, D. 2005. Biogeography and scientific revolutions. *The Systematist* 5: 3–12.

McGeoch, M. A. 1998. The selection, testing and application of terrestrial insects as bioindicators. *Biological Review* 73: 181–201.

McGlynn, T. P. 1999. The worldwide transfer of ants: geographical distribution and ecological invasions. *Journal of Biogeography* 26: 535–548.

McGuinness, C. A. 2001. *The Conservation Requirements of New Zealand's Nationally Threatened Invertebrates.* Biodiversity Recovery Unit, Department of Conservation, Wellington, New Zealand. 658 pp.

McKie, B. and P. S. Cranston. 1998. Keystone coleopterans? Colonisation by wood-feeding elmids of experimentally-immersed woods in south-east Australia. *Marine and Freshwater Research* 49: 79–88.

McLellan, I. 2006. Endemism and biogeography of New Zealand Plecoptera (Insecta). *Illiesia* 2: 15–23.

McQuillan, P. B. 2003. The giant Tasmanian 'pandani' moth *Proditrix nielseni*, sp. nov. (Lepidoptera: Yponomeutoidea: Plutellidae *s.l.*). *Invertebrate Systematics* 17: 59–66.

Meads, M. J. 1990. *Forgotten Fauna: the Rare, Endangered and Protected Invertebrates of New Zealand.* DSIR Publishing, Wellington. 94 pp.

Michaux, B. and R. A. B. Leschen. 2005. East meets west: biogeology of the Campbell Plateau. *Biological Journal of the Linnean Society* 86: 95–115.

Miller, S., E. Hyslop, G. Kula, and I. Burrows. 1995. Status of Biodiversity in Papua New Guinea. Pp. 67–95. *In* N. Sekhran and S. Miller (eds). *Papua New Guinea Country Study on Biological Diversity.* Department of Environment and Conservation, report to United Nations Environment Program, Waigani. http://www.geocities.com/RainForest/9468/papua_ng.htm [Accessed 20 June 2006].

Moller, H. and Tilley, J. A. V. 1989. Beech honeydew: Seasonal variation and use by wasps, honey bees, and other insects. *New Zealand Journal of Zoology* 16: 289–302.

Monteith, G. 1990. Rainforest insects: biodiversity, bioguesstimation, or just hand-waving?. *Myrmecia* 26: 93–95.

Mound, L. A. 2004. Australian Thysanoptera – biological diversity and a diversity of studies. *Australian Journal of Entomology* 43: 248–257.

Murienne, J., P. Grandcolas, M. D. Piulachs, X. Bellés, C. D'Haese, F. Legendre, R. Pellens, and E. Guilbert. 2005. Evolution on a shaky piece of Gondwana: is local endemism recent in New Caledonia?. *Cladistics* 21: 2–7.

Najt, J. and P. Grandcolas (eds). 2002. Zoologia Neocalédonia 5. Systématique et Endémism en Nouvelle-Calédonie. *Mémoires du Muséum national d'Histoire naturelle* 187: 1–282.

New, T. R. 1984. *Insect Conservation: An Australian Perspective.* Junk, Dordrecht. 204 pp.

New, T. R. 1990. Conservation of butterflies in Australia. *Journal of Research on the Lepidoptera* 29: 237–253.

New, T. R. 1993. *Hemiphlebia mirabilis* Selys: recovery from habitat destruction at Wilson's Promontory, Victoria, Australia, and implications for conservation management (Zygoptera: Hemiphlebiidae). *Odonatologica* 22: 495–502.

New, T. R. 1997. Are Lepidoptera an effective 'umbrella group' for biodiversity conservation?. *Journal of Insect Conservation* 1: 5–12.

New, T. R. 1999. The evolution and characteristics of the Australian butterfly fauna. Pp. 33–52. *In* R. L. Kitching, E. Scheermeyer, R. E. Jones, and N. E. Pierce (eds). *Biology of Australian Butterflies*. CSIRO, Melbourne.

New, T. R. 2002. Neuroptera of Wallacea: a transitional fauna between major geographical regions. *Acta Zoologica Academiae Scientiarum Hungaricae* 48: (Supplement 2), 217–227.

New, T. R., R. M. Pyle, J. A. Thomas, C. D. Thomas, and P. C. Hammond. 1995. Butterfly conservation management. *Annual Review of Entomology* 40: 57–83.

New, T. R. and D. P. A. Sands. 2004. Management of threatened insect species in Australia, with particular reference to butterflies. *Australian Journal of Entomology* 43: 258–270.

NSW NP and WS (New South Wales National Parks and Wildlife Service). 2004. Feral bees. http://www.nationalparks.nsw.gov.au/npws.nsf/Content/feral_bees [Accessed July 2006].

Novotny, V., Y. Basset, S. E. Miller, G. D. Weiblen, B. Bremer, L. Cizek, and P. Drozd. 2002. Low host specificity of herbivorous insects in a tropical forest. *Nature* 216: 841–844.

O'Dwyer, C. and P. M. Attiwill. 2000. Restoration of a native grassland as habitat for the Golden Sun Moth *Synemon plana* Walker (Lepidoptera; Castniidae) at Mount Piper, Australia. *Restoration Ecology* 8: 170–174.

Oliver, I. and A. J. Beattie. 1995. Invertebrate morphospecies as surrogates for species: a case study. *Conservation Biology* 10: 99–109.

Parsons, M. 1999. *The Butterflies of Papua New Guinea*. Academic Press, London. 928 pp.

Patrick, B. H. and J. S. Dugdale. 2000. *Conservation Status of the New Zealand Lepidoptera*. Science for Conservation 136. Department of Conservation, Wellington. 33 pp.

Polhemus, D. A., R. A. Englund, and G. R. Allen. 2004. Freshwater biotas of New Guinea and nearby islands: analysis of endemism, richness, and threats. Bishop Museum Technical Report 31. Final Report Prepared for: Conservation International, Washington, DC.

Ponder, W. and D. Lunney (eds). 1999. *The Other 99%: The Conservation and Biodiversity of Invertebrates*. Royal Zoological Society of New South Wales, Mosman, Australia. 462 pp.

Porch, N. and S. Elias. 2000. Quaternary beetles: a review and issues for Australian studies. *Australian Journal of Entomology* 39: 1–9.

Priddel, D., N. Carlile, M. Humphrey, S. Fellenberg, and D. Hiscox. 2003. Rediscovery of the 'extinct' Lord Howe Island stick-insect (*Dryococelus australis* (Montrouzier))

(Phasmatodea) and recommendations for its conservation. *Biodiversity and Conservation* 12: 1391–1403.

Rentz, D. C. F. 1988. The orthopteroid insects of Norfolk Island, with descriptions and records of some related species from Lord Howe Island. *Invertebrate Taxonomy* 2: 1013–1077.

Riedel, A. 2001a. The *pygmaeus*-group of *Euops* Schoenherr (Coleoptera, Curculionoidea, Attelabidae), weevils associated with *Nothofagus* in New Guinea. *Journal of Natural History* 35: 1173–1237.

Riedel, A. 2001b. Revision of the *Euops quadrifasciculatus*-group (Coleoptera: Curculionoidea: Attelabidae) from the Australian region, with a discussion of shifts between *Nothofagus* and *Eucalyptus* host plants. *Invertebrate Taxonomy* 15: 551–587.

Robinson, G. S. 1975. *The Macrolepidoptera of Fiji and Rotuma: A Taxonomic Study*. E. W. Classey, Faringdon. 374 pp.

Sands, D. P. A. 1999. Conservation status of Lepidoptera: assessment, threatening processes and recovery actions. Pp. 382–387. *In* W. Ponder and D. Lunney (eds). *The Other 99%: The Conservation and Biodiversity of Invertebrates*. Transactions of the Royal Zoological Society of New South Wales, Mosman, Australia.

Sands, D. P. A. and T. R. New. 2002. *The Action Plan for Australian Butterflies*. Environment Australia, Canberra. 378 pp.

Sands, D. P. A., S. E. Scott, and R. Moffatt. 1997. The threatened Richmond birdwing butterfly (*Ornithoptera richmondia* (Gray)): a community conservation project. *Memoirs of the Museum of Victoria* 56: 449–453.

Scanlan, J. C. and C. Vanderwoude. 2006. Modelling the potential spread of *Solenopsis invicta* Buren (Hymenoptera: Formicidae) (red imported fire ant) in Australia. *Australian Journal of Entomology* 45: 1–9.

Scheffer, S. J., R. M. Girlin-Davis, G. S. Taylor, K. A. Davies, M. Purcell, M. L. Lewis, J. Goolsby, and T. D. Center. 2004. Phylogenetic relationships, species limits, and host specificity of gall-forming *Fergusonina* flies (Diptera: Fergusoninidae) feeding on *Melaleuca* (Myrtaceae). *Annals of the Entomological Society of America* 97: 1216–1221.

Sekhran, N. and S. E. Miller (eds). 1996. *Papua New Guinea Country Study on Biological Diversity*. Papua New Guinea Department of Environment and Conservation, Waigani. 124 pp.

Sessions, L. 2001. New Zealand Sweet Stakes. *Natural History* 110: 64–71.

Shattuck, S. O. and N. J. Barnett. 2001. Australian Ants Online. http://www.ento.csiro.au/science/ants/ants_in_australia.htm [Accessed 11 July 2006].

Sherley, G. H. and L. M. Hayes. 1993. The conservation of a giant weta (*Deinacrida* n. sp. Orthoptera: Stenopelmatidae) at Mahoenui, King Country: habitat use, and other aspects of its ecology. *New Zealand Entomologist* 16: 55–68.

Simpson, G. G. 1977. Too many lines: the limits of the Oriental and Australian zoogeographic regions. *Proceedings of the American Philosophical Society* 121: 107–120.

Simpson, J. C. and R. H. Norris. 2000. Biological assessment of river quality: development of AUSRIVAS models and outputs. Pp. 125–142. In J. F. Wright, D. W. Sutcliffe, and M. T. Furse (eds). *Assessing the Biological Quality of Fresh Waters: RIVPACS and Other Techniques*. Freshwater Biological Association, Ambleside.

Sinclair, B. J. 2003. Taxonomy, phylogeny and zoogeography of the subfamily Ceratomerinae of Australia (Diptera: Empidoidea). *Records of the Australian Museum* 55: 1–44.

Specht, R. L. 1981. Conservation of vegetation types. Pp. 393–410. In R. H. Groves (ed). *Australian Vegetation*. Cambridge University Press, Cambridge.

Stork, N. E. 1993. How many species are there?. *Biodiversity and Conservation* 2: 215–232.

Taylor, G. 2006. Book review of Australian Psylloidea: jumping plantlice and lerp insects. *Systematic Entomology* 31: 199–200.

Taylor, G. S. 2004. Revision of *Fergusonina* Malloch gall flies (Diptera: Fergusoninidae) from *Melaleuca* (Myrtaceae). *Invertebrate Systematics* 18: 251–290.

Taylor, R. W. 1976. *Submission to The Australian Senate Standing Committee on Social Environment Inquiry into the Impact on the Australian Environment of the Current Woodchip Industry Programme*. Australian Senate Official Hansard, Canberra. Pp. 3724–3731.

Taylor, R. W. 1983. Descriptive taxonomy: past, present and future. Pp. 93–134. In E. Highley and R. W. Taylor (eds). *Australian Systematic Entomology: A Bicentenary Perspective*. Commonwealth Scientific and Industrial Research Organisation, Melbourne.

The Swedish Museum of Natural History. 2003. Checklist of New Caledonian Trichoptera. http://www2.nrm .se/en/caledonia/trichopteranc.html [Accessed 20 June 2006].

Tillyard, R. J. 1924. The dragonflies (Order Odonata) of Fiji, with special reference to a collection made by Dr. H. W. Simmonds, F. E. S., on the island of Viti Levu. *Transactions of the Entomological Society of London* 71(1923): 305–345.

Tillyard, R. J. 1926. *The Insects of Australia and New Zealand*. Angus and Robertson, Sydney.

Toft, R. J. and R. J. Harris. 2004. Can trapping control Asian paper wasp (*P. chinensis antennalis*) populations?. *New Zealand Journal of Ecology* 28: 279–282.

Towns, D. R. 2002. Interactions between geckos, honeydew scale insects and host plants revealed on islands in northern New Zealand, following eradication of introduced rats and rabbits. Pp. 329–335. In C. R. Veitch and M. N. Clout (eds). *Turning the Tide: Eradication of Invasive Species*. IUCN SSC Invasive Species Specialist Group, IUCN, Gland, Switzerland and Cambridge, United Kingdom.

Trewick, S. A. 2001. Scree weta phylogeography: surviving glaciation and implications for Pleistocene biogeography in New Zealand. *New Zealand Journal of Zoology* 28: 291–298.

Trewick, S. A. and M. Morgan-Richards. 2004. Phylogenetics of New Zealand's tree, giant and tusked weta (Orthoptera: Anostostomatidae): evidence from mitochondrial DNA. *Journal of Orthoptera Research* 13: 185–196.

Trueman, J. W. H. and P. S. Cranston. 1997. Prospects for rapid assessment of terrestrial invertebrate biodiversity. *Memoirs of the Museum of Victoria* 56: 349–354.

Trueman, J. W. H., G. A. Hoye, J. H. Hawking, J. A. L. Watson, and T. R. New. 1992. *Hemiphlebia mirabilis* Selys: new localities in Australia and perspectives on conservation (Zygoptera: Hemiphlebiidae). *Odonatologica* 21: 367–374.

Turney, C. S. M., S. Haberle, D. Fink, A. P. Kershaw, M. Barbetti, M. T. T. Barrows, M. Black, T. J. Cohen, T. Corrège, P. P. Hesse, Q. Hua, R. Johnston, V. Morgan, P. Moss, G. Nanson, T. van Ommen, S. Rule, N. J. Williams, J.-X. Zhao, D'Costa, D. Feng, Y.-X., M. Gagan, S. Mooney, and Q. Xia. 2006. Integration of ice-core, marine and terrestrial records for the Australian Last Glacial Maximum and Termination: a contribution from the OZ INTIMATE group. *Journal of Quaternary Science* 21: 751–761.

van Herwaarden, H. C. M. 1998. A guide to the genera of stick- and leaf-insects (Insecta: Phasmida) of New Guinea and the surrounding islands. *Science in New Guinea* 24: 55–117.

Vanderwoude, C., M. Elson-Harris, J. R. Hargreaves, E. Harris, and K. P. Plowman. 2004. An overview of the red imported fire ants (*Solenopsis invicta*) eradication plan for Australia. *Records of the South Australian Museum Series* 7: 11–16.

Vane-Wright, R. I. 1991. Transcending the Wallace line: do the western edges of the Australian Region and the Australian Plate coincide?. *Australian Systematic Botany* 4: 183–197.

Wallace, A. R. 1860. Notes of a voyage to New Guinea. *Journal of the Royal Geographical Society* 30: 127–177.

Ward, D. F. and M. C. Larivière. 2004. Terrestrial invertebrate surveys and rapid biodiversity assessment in New Zealand: lessons from Australia. *New Zealand Journal of Ecology* 28: 151–159.

Ward, P. S. 1984. A revision of the ant genus *Rhytidoponera* in New Caledonia. *Australian Journal of Zoology* 32: 131–175.

Ward, P. S. and S. G. Brady. 2003. Phylogeny and biogeography of the ant subfamily Myrmeciinae (Hymenoptera: Formicidae). *Invertebrate Systematics* 17: 361–386.

Wegener, A. 1915. *Die Entstehung der Kontinente und Ozeane*. Friedrich Vieweg und Sohn, Braunschweig.

Wegener, A. 1924. *The Origin of Continents and Oceans*. Methuen and Co., London.

Westoby, M. and M. Burgman. 2006. Climate change as a threatening process. *Austral Ecology* 31: 549–550.

Wiggins, G. B. 2005. *Caddisflies, the Underwater Architects*. University of Toronto Press, Toronto, Ontario. 292 pp.

Williams, G. R. 1982. Species-area and similar relationships of insects and vascular plants on the southern outlying islands of New Zealand. *New Zealand Journal of Ecology* 5: 86–96.

Williams, J., C. Read, A. Norton, S. Dovers, M. Burgman, W. Proctor, and H. Anderson. 2001. Biodiversity, Australia State of the Environment Report 2001 (Theme Report). CSIRO Publishing on behalf of the Department of the Environment and Heritage, Canberra.

Wilson, E. O. and G. L. Hunt. 1967. Ant fauna of Futuna and Wallis Islands, stepping stones to Polynesia. *Pacific Insects* 9: 563–584.

Wilson, R. D., J. W. H. Trueman, S. E. Williams, and D. K. Yeates. 2007. Altitudinally restricted communities of schizophoran flies in Queensland's Wet Tropics: vulnerability to climate change. *Biodiversity Conservation* 16: 3163–3177.

Withers, T. M. 2001. Colonization of eucalypts in New Zealand by Australian insects. *Austral Ecology* 26: 467–476.

Woodward, C. and J. Shulmeister. 2007. Chironomid-based reconstructions of summer air temperature from lake deposits in Lyndon Stream, New Zealand spanning the MIS 3/2 transition. *Quaternary Science Reviews* 26: 142–154.

Worthy, T. H. 1997. The quaternary fossil fauna of South Canterbury, South Island, New Zealand. *Journal of the Royal Society of New Zealand* 27: 67–162.

Yeates, D. K., P. Bouchard, and G. B. Monteith. 2002. Patterns and levels of endemism in the Australian wet tropics rainforest: evidence from flightless insects. *Invertebrate Systematics* 16: 605–619.

Yeates, D. K., M. S. Harvey, and A. D. Austin. 2003. New estimates for terrestrial arthropod species-richness in Australia. *Records of the South Australian Museum, Monograph Series* 7: 231–241.

Yen, A. L. and T. R. New (eds). 1997. Proceedings of the conference on invertebrate biodiversity and conservation. *Memoirs of Museum Victoria* 56: 261–675.

Young, E. C. 1995. Conservation values, research and New Zealand's responsibilities for the Southern Ocean islands and Antarctica. *Pacific Conservation Biology* 2: 99–112.

ZoosVictoria. 2006. The Lord Howe Island Stick-insect *Dryococelus australis*. http://www.zoo.org.au/education/factsheets/inv-lhisi.pdf [Accessed July 2006].

Zwick, P. 1977. Australian Blephariceridae (Diptera). *Australian Journal of Zoology Supplementary Series* 46: 1–121.

Chapter 7

INSECT BIODIVERSITY IN THE PALEARCTIC REGION

Alexander S. Konstantinov[1], Boris A. Korotyaev[2], and Mark G. Volkovitsh[2]

[1]Systematic Entomology Laboratory, PSI, Agricultural Research Service, U. S. Department of Agriculture, Smithsonian Institution, Washington, DC

[2]Zoological Institute, Russian Academy of Sciences, Universitetskaya nab. 1, St. Petersburg, Russia

Insect Biodiversity: Science and Society, 1st edition. Edited by R. Foottit and P. Adler
© 2009 Blackwell Publishing, ISBN 978-1-4051-5142-9

For more than 5 years, small dragonflies of apparently one species have occupied the tips of dry branches of a boxelder tree (*Acer negundo* L.) opposite B. A. Korotyaev's window on the first floor of his apartment building in Krasnodar, Russia. They appear in late July after the swifts migrate from the city. Although the traffic in the street is heavy, the dragonflies hunt all day in the polluted air. Liberally paraphrasing Kant's famous statement, two phenomena are most striking for biologists: the tremendous diversity of living organisms and the obvious, though often unapparent, organization of this diversity. To adequately describe insect biodiversity and its organization in the Palearctic Region is a major challenge.

The Palearctic Region is the world's largest biogeographic region and the best known with respect to its overall insect diversity. It also has the longest history of faunistic and biodiversity studies. Nevertheless, reliable estimates of the number of species are available for only some insect orders (Table 7.1). Existing Palearctic (e.g., Aukema and Rieger 1995–2006, Löbl and Smetana 2003–2007) and world catalogs (e.g., Hansen 1998, Heppner 1998, Woodley 2001) improve our understanding of insect biodiversity, but catalogs are unavailable for many groups, and the needed data can be provided only by experts. Some taxa have been treated or otherwise revised recently, while others have remained untouched for the last century or so, meaning that certain data are absent for some groups. In addition, some parts of the Palearctic, such as western and northern Europe, are better known than others, such as North Africa and China. However, exploration of the mountains of China in recent decades has resulted in discoveries of the species-rich alpine faunas of large coleopteran families such as the Carabidae (e.g., Zamotajlov and Sciaky 1996, Kataev and Liang 2005) and Tenebrionidae (e.g., Medvedev 2005a, 2006).

The great variety of views on subdivision of the Palearctic presents another difficulty in describing the region's insect diversity. We follow the main biogeographical units proposed by Emeljanov (1974), which are compatible with many of the existing schemes and are finding increasing support (Volkovitsh and Alexeev 1988, Krivokhatsky and Emeljanov 2000). We concentrate on general features and patterns of the entire insect fauna of the Palearctic Region, providing examples of taxa that are better known to us, mostly Coleoptera, or most typical of entities in various biogeographical units.

The Palearctic biota is an immense and irreplaceable source of organisms of the highest economic value for the entire world, particularly for the temperate regions, and includes cultivated plants and useful animals, their natural enemies (weeds, predators, and pathogens), and enemies of these enemies (the biological control agents). It is, however, experiencing increasing anthropogenic pressures, resulting in the disappearance of many species of plants and animals, as well as entire communities and landscapes.

GEOGRAPHIC POSITION, CLIMATE, AND ZONALITY

The Palearctic Region (Fig. 7.1a) occupies cold, temperate, and subtropical regions of Eurasia and Africa north of the Sahara Desert, together with islands of the Arctic, Atlantic, and Pacific oceans – Azores, Canaries, Iceland, British Isles, and Cape Verde in the Atlantic, and Komandorski, Kurile, and Japan in the Pacific Ocean. The southern border of the Palearctic in Asia lies along the southern border of subtropical forests, leaving the southern part of China, Taiwan, and the Ryukyu Islands of Japan in the Oriental Region. The Himalayas are mostly attributed to the Palearctic, but distributional patterns of some (mostly phytophagous) insect groups suggest that the southern Himalayas could be considered Oriental. The southern border of the Palearctic in East Asia is nearly impossible to define, particularly in China, because the Chinese fauna is poorly known and the distributional patterns of various groups provide conflicting results. The southern border of the Palearctic generally is viewed as a wide band at the southern limits of subtropical forests between the Yantzu and Huanche rivers, where interchanges of the Palearctic and Oriental faunas occur.

The Palearctic, together with the Nearctic, forms a larger zoogeographic division, the Holarctic, which includes all nontropical areas of the northern hemisphere (Kryzhanovsky 1965, 2002, Takhtajan 1978, Lopatin 1989). Overall, the insect faunas of the Palearctic and Nearctic are more similar in the North and more unique in their southern parts (Kryzhanovsky 2002), where Afrotropical, Oriental, and Neotropical elements contribute significantly to the biodiversity. A few high-ranked groups of insects are endemic to the Holarctic: the only endemic order, the Grylloblattida (western North America, Japan, Russian Far East; with one species in the

Table 7.1 Biodiversity of major insect groups in the Palearctic Region (for some insect orders, only selected fields could be filled; families of Diptera are not included).

Order/ Suborder	Family (Superfamily)	Number of Species	Number of Genera	Endemic Family-Group Taxa and their Ranges	Source
Thysanura		22 (USSR)			A. L. Lobanov (personal communication)
Ephemeroptera		300 (USSR)			A. L. Lobanov (personal communication)
Odonata	Amphipterygidae	4	2	.	Steinmann 1997
Odonata	Calopterygidae	72			Steinmann 1997
Odonata	Euphaeidae	20			Steinmann 1997
Odonata	Chlorolestidae	9			Steinmann 1997
Odonata	Lestidae	19			Steinmann 1997
Odonata	Megapodagrionidae	3			Steinmann 1997
Odonata	Pseudolestidae	6	3		Steinmann 1997
Odonata	Coenagrionidae	78			Steinmann 1997
Odonata	Platycnemididae	17			Steinmann 1997
Odonata	Epiophlebiidae	2	1	Epiophlebiidae, Japan and Himalayas	Belyshev and Kharitonov 1981
Odonata	Petaluridae	1	1		Belyshev and Kharitonov 1981
Odonata	Gomphidae	38			Belyshev and Kharitonov 1981
Odonata	Cordulegasteridae	11			Belyshev and Kharitonov 1981
Odonata	Aeshnidae	27			Belyshev and Kharitonov 1981
Odonata	Corduliidae	24			Belyshev and Kharitonov 1981
Odonata	Libellulidae	96			Belyshev and Kharitonov 1981
Odonata	Macrodiplactidae	2			Belyshev and Kharitonov 1981
Blattaria		100 (USSR)	30		L. N. Anisyutkin (personal communication)
Mantodea		35 (USSR)			A. L. Lobanov (personal communication)
Isoptera		7 (USSR)			A. L. Lobanov (personal communication)
Grylloblattodea		3 (USSR)			A. L. Lobanov (personal communication)
Phasmatodea		8 (USSR)			A. L. Lobanov (personal communication)
Orthoptera		802 (USSR)			A. L. Lobanov (personal communication)
Orthoptera/ Ensifera		266 (USSR)			A. L. Lobanov (personal communication)
Orthoptera/ Caelifera		536 (USSR)			A. L. Lobanov (personal communication)

(continued)

Table 7.1 *(continued)*.

Order/ Suborder	Family (Superfamily)	Number of Species	Number of Genera	Endemic Family-Group Taxa and their Ranges	Source
Plecoptera		595	40	1	Zhiltzova 2003
Plecoptera	Scopuridae	5	1	Scopuridae	Zhiltzova 2003
Plecoptera	Taeniopterygidae	58	11		Zhiltzova 2003
Plecoptera	Nemouridae	281	8		Zhiltzova 2003
Plecoptera	Capniidae	100	13		Zhiltzova 2003
Plecoptera	Leuctridae	146	7		Zhiltzova 2003
Dermaptera		80	25		L. N. Anisyutkin (personal communication)
Embioptera		2 (USSR)			A. L. Lobanov (personal communication)
Zoraptera		2 (Tibet)			N. V. Golub and V. G. Kuznetsova (personal communication)
Psocoptera		352 (not including China)			N. V. Golub and V. G. Kuznetsova (personal communication)
Mallophaga		400 (USSR)			A. L. Lobanov (personal communication)
Anoplura		60 (USSR)			A. L. Lobanov (personal communication)
Thysanoptera		500 (USSR)			A. L. Lobanov (personal communication)
Sternorrhyncha/ Psyllina		520 (USSR)			A. L. Lobanov (personal communication)
Sternorrhyncha/ Aphidina		853 (Russia)			A. L. Lobanov (personal communication), A. V. Stekolshchikov (personal communication)
Sternorrhyncha/ Coccina	Aleyrodidae	49 (USSR)			A. L. Lobanov (personal communication)
Sternorrhyncha/ Coccina	Coccidae	~600 (USSR)			A. L. Lobanov (personal communication)
Auchenorrhyncha		>4000	718	Dorysarthriini, Ranissini, Colobocini. Almanini, Bocrini, Adenissini, Ommatidiotini, Durgulini, Adelungiini, Aphrodini, Stegelytrini, Grypotini, Fieberiellini	Nast 1972, A. F. Emeljanov (personal communication)
Hemiptera		8413	1489		

Table 7.1 *(continued).*

Order/ Suborder	Family (Superfamily)	Number of Species	Number of Genera	Endemic Family-Group Taxa and their Ranges	Source
Hemiptera	Aenictopecheidae	1	1		Aukema and Rieger 1995
Hemiptera	Enicocephalidae	15	6		Aukema and Rieger 1995, Schuh and Slater 1995
Hemiptera	Ceratocombidae	11	1		Aukema and Rieger 1995
Hemiptera	Dipsocoridae	14	1		Aukema and Rieger 1995, Schuh and Slater 1995
Hemiptera	Schizopteridae	9	6		Aukema and Rieger 1995, Schuh and Slater 1995
Hemiptera	Nepidae	21	5		Aukema and Rieger 1995
Hemiptera	Belostomatidae	14	5		Aukema and Rieger 1995
Hemiptera	Gelastocoridae	4	1		Aukema and Rieger 1995
Hemiptera	Ochteridae	3	1		Aukema and Rieger 1995
Hemiptera	Corixidae	143	15		Aukema and Rieger 1995
Hemiptera	Naucoridae	9	7		Aukema and Rieger 1995
Hemiptera	Aphelocheiridae	18	1		Aukema and Rieger 1995
Hemiptera	Notonectidae	50	4		Aukema and Rieger 1995
Hemiptera	Pleidae	6	2		Aukema and Rieger 1995
Hemiptera	Helotrephidae	6	5		Aukema and Rieger 1995
Hemiptera	Mesoveliidae	7	2		Aukema and Rieger 1995
Hemiptera	Hebridae	24	4		Aukema and Rieger 1995
Hemiptera	Hydrometridae	14	1		Aukema and Rieger 1995
Hemiptera	Hermatobatidae	2	1		Aukema and Rieger 1995
Hemiptera	Veliidae	64	11		Aukema and Rieger 1995
Hemiptera	Gerridae	99	20		Aukema and Rieger 1995
Hemiptera	Aepophilidae	1	1	Aepophilidae	Aukema and Rieger 1995
Hemiptera	Saldidae	99	14		Aukema and Rieger 1995
Hemiptera	Leptopodidae	11	4		Aukema and Rieger 1995
Hemiptera	Omaniidae	2	2		Aukema and Rieger 1995
Hemiptera	Joppeicidae	1	1		Aukema and Rieger 1996
Hemiptera	Tingidae	473	61		Aukema and Rieger 1996
Hemiptera	Microphysidae	27	3		Aukema and Rieger 1996
Hemiptera	Nabidae	112	10		Aukema and Rieger 1996
Hemiptera	Anthocoridae	181	28		Aukema and Rieger 1996
Hemiptera	Cimicidae	15	5		Aukema and Rieger 1996
Hemiptera	Polyctenidae	3	2		Aukema and Rieger 1996
Hemiptera	Pachynomidae	2	1		Aukema and Rieger 1996
Hemiptera	Reduviidae	808	145		Aukema and Rieger 1996
Hemiptera	Miridae	2808	397		Aukema and Rieger 1999
Hemiptera	Aradidae	204	28		Aukema and Rieger 2001
Hemiptera	Lygaeidae	1001	225		Aukema and Rieger 2001
Hemiptera	Piesmatidae	19	2		Aukema and Rieger 2001
Hemiptera	Malcidae	25	2		Aukema and Rieger 2001
Hemiptera	Berytidae	54	13		Aukema and Rieger 2001
Hemiptera	Colobathristidae	7	2		Aukema and Rieger 2001
Hemiptera	Largidae	8	3		Aukema and Rieger 2001
Hemiptera	Pyrrhocoridae	43	13		Aukema and Rieger 2001
Hemiptera	Rhopalidae	69	14		Aukema and Rieger 2006

(continued)

Table 7.1 *(continued)*.

Order/ Suborder	Family (Superfamily)	Number of Species	Number of Genera	Endemic Family-Group Taxa and their Ranges	Source
Hemiptera	Stenocephalidae	18	1		Aukema and Rieger 2006
Hemiptera	Alydidae	69	26		Aukema and Rieger 2006
Hemiptera	Coreidae	306	84		Aukema and Rieger 2006
Hemiptera	Urostylididae	131	8		Aukema and Rieger 2006
Hemiptera	Thaumastellidae	1	1		Aukema and Rieger 2006
Hemiptera	Parastrachiidae	2	1		Aukema and Rieger 2006, Schuh and Slater 1995
Hemiptera	Cydnidae	167	37		Aukema and Rieger 2006
Hemiptera	Thyreocoridae	4	1		Aukema and Rieger 2006, Schuh and Slater, 1995
Hemiptera	Plataspidae	104	10		Aukema and Rieger, 2006
Hemiptera	Acanthosomatidae	107	9		Aukema and Rieger 2006
Hemiptera	Tessaratomidae	30	12		Aukema and Rieger 2006
Hemiptera	Dinidoridae	19	4		Aukema and Rieger 2006
Hemiptera	Pentatomidae	841	219		Aukema and Rieger 2006
Hemiptera	Incertae sedis	107			Aukema and Rieger 2006
Coleoptera		100,000			I. Löbl (personal communication), A. L. Lobanov (personal communication)
Coleoptera	Crowsoniellidae	1	1	Italy	Löbl and Smetana 2003
Coleoptera	Micromalthidae	1	1		Löbl and Smetana 2003
Coleoptera	Cupedidae	7	1		Löbl and Smetana 2003
Coleoptera	Jurodidae (= Sikhotealiniidae)	1	1	Russian Far East	Löbl and Smetana 2003
Coleoptera	Torridincolidae	1	1		Löbl and Smetana 2003
Coleoptera	Hydroscaphidae	9	1		Löbl and Smetana 2003
Coleoptera	Sphaeriusidae	8	1		Löbl and Smetana 2003
Coleoptera	Gyrinidae	102	7		Löbl and Smetana 2003
Coleoptera	Haliplidae	66	3		Löbl and Smetana 2003
Coleoptera	Trachypachidae	1	1		Löbl and Smetana 2003
Coleoptera	Noteridae	30	5		Löbl and Smetana 2003
Coleoptera	Amphizoidae	2	1		Löbl and Smetana 2003
Coleoptera	Hygrobiidae	2	1		Löbl and Smetana 2003
Coleoptera	Dytiscidae	901	67		Löbl and Smetana 2003
Coleoptera	Rhysodidae	24	5		Löbl and Smetana 2003
Coleoptera	Carabidae	11,333	579		Löbl and Smetana 2003
Coleoptera	Helophoridae	150	1		Löbl and Smetana 2004
Coleoptera	Epimetopidae	4	1		Löbl and Smetana 2004
Coleoptera	Georissidae	19	1		Löbl and Smetana 2004
Coleoptera	Hydrochidae	28	1		Löbl and Smetana 2004
Coleoptera	Spercheidae	5	1		Löbl and Smetana 2004
Coleoptera	Hydrophilidae	559	54		Löbl and Smetana 2004
Coleoptera	Sphaeritidae	3	1		Löbl and Smetana 2004
Coleoptera	Synteliidae	3	1		Löbl and Smetana 2004
Coleoptera	Histeridae	847	107		Löbl and Smetana 2004
Coleoptera	Hydraenidae	785	8		Löbl and Smetana 2004

Table 7.1 *(continued).*

Order/ Suborder	Family (Superfamily)	Number of Species	Number of Genera	Endemic Family-Group Taxa and their Ranges	Source
Coleoptera	Ptiliidae	188	21		Löbl and Smetana 2004
Coleoptera	Agyrtidae	45	8		Löbl and Smetana 2004
Coleoptera	Leiodidae	2172	245		Löbl and Smetana 2004
Coleoptera	Scydmaenidae	1005	26		Löbl and Smetana 2004
Coleoptera	Silphidae	117	15		Löbl and Smetana 2004
Coleoptera	Staphylinidae	15,307	903		Löbl and Smetana 2004
Coleoptera	Lucanidae	332	34		Löbl and Smetana 2006
Coleoptera	Passalidae	25	8		Löbl and Smetana 2006
Coleoptera	Trogidae	79	3		Löbl and Smetana 2006
Coleoptera	Glaresidae	24	1		Löbl and Smetana 2006
Coleoptera	Bolboceratidae	52	13		Löbl and Smetana 2006
Coleoptera	Geotrupidae	299	13		Löbl and Smetana 2006
Coleoptera	Ochodaeidae	26	2		Löbl and Smetana 2006
Coleoptera	Ceratocanthidae	4	2		Löbl and Smetana 2006
Coleoptera	Hybosoridae	16	7		Löbl and Smetana 2006
Coleoptera	Glaphyridae	158	5		Löbl and Smetana 2006
Coleoptera	Scarabaeidae	5787	404		Löbl and Smetana 2006
Coleoptera	Decliniidae	2	1	Russian Far East	Löbl and Smetana 2006
Coleoptera	Eucinetidae	18	7		Löbl and Smetana 2006
Coleoptera	Clambidae	43	3		Löbl and Smetana 2006
Coleoptera	Scirtidae	254	8		Löbl and Smetana 2006
Coleoptera	Dascillidae	40	8		Löbl and Smetana 2006
Coleoptera	Rhipiceridae	15	3		Löbl and Smetana 2006
Coleoptera	Buprestidae	2430	99	Paratassini, Kisanthobiini	Löbl and Smetana 2006, Bellamy 2003
Coleoptera	Byrrhidae	321	22		Löbl and Smetana 2006
Coleoptera	Elmidae	235	35		Löbl and Smetana 2006
Coleoptera	Dryopidae	64	11		Löbl and Smetana 2006
Coleoptera	Limnichidae	66	13		Löbl and Smetana 2006
Coleoptera	Heteroceridae	94	3		Löbl and Smetana 2006
Coleoptera	Psephenidae	92	16		Löbl and Smetana 2006
Coleoptera	Ptilodactylidae	47	9		Löbl and Smetana 2006
Coleoptera	Chelonariidae	11	2		Löbl and Smetana 2006
Coleoptera	Eulichadidae	11	1		Löbl and Smetana 2006
Coleoptera	Callirhipidae	16	3		Löbl and Smetana 2006
Coleoptera	Artematopidae	8	2		Löbl and Smetana 2007
Coleoptera	Cerophytidae	2	1		Löbl and Smetana 2007
Coleoptera	Eucnemidae	132	41		Löbl and Smetana 2007
Coleoptera	Throscidae	48	2		Löbl and Smetana 2007
Coleoptera	Elateridae	3661	233		Löbl and Smetana 2007
Coleoptera	Plastoceridae	1	1		Löbl and Smetana 2007
Coleoptera	Drilidae	44	6		Löbl and Smetana 2007
Coleoptera	Omalisidae	12	2	Europe	Löbl and Smetana 2007
Coleoptera	Lycidae	403	47		Löbl and Smetana 2007
Coleoptera	Phengodidae	27	5		Löbl and Smetana 2007
Coleoptera	Lampyridae	263	19		Löbl and Smetana 2007
Coleoptera	Omethidae	3	2		Löbl and Smetana 2007

(continued)

Table 7.1 *(continued).*

Order/ Suborder	Family (Superfamily)	Number of Species	Number of Genera	Endemic Family-Group Taxa and their Ranges	Source
Coleoptera	Cantharidae	1981	62		Löbl and Smetana 2007
Coleoptera	Derodontidae	14	3		Löbl and Smetana 2007
Coleoptera	Nosodendridae	12	1		Löbl and Smetana 2007
Coleoptera	Dermestidae	520	25		Löbl and Smetana 2007
Coleoptera	Endecatomidae	2	1		Löbl and Smetana 2007
Coleoptera	Bostrichidae	151	46		Löbl and Smetana 2007
Coleoptera	Ptinidae	717	99		Löbl and Smetana 2007
Coleoptera	Jacobsoniidae	3	2		Löbl and Smetana 2007
Coleoptera	Lymexylidae	15	8		Löbl and Smetana 2007
Coleoptera	Phloiophilidae	1	1	Europe	Löbl and Smetana 2007
Coleoptera	Trogossitidae	45	16		Löbl and Smetana 2007
Coleoptera	Thanerocleridae	7	3		Löbl and Smetana 2007
Coleoptera	Cleridae	311	62		Löbl and Smetana 2007
Coleoptera	Acanthocnemidae	1	1		Löbl and Smetana 2007
Coleoptera	Prionoceridae	41	3		Löbl and Smetana 2007
Coleoptera	Melyridae	59	9		Löbl and Smetana 2007
Coleoptera	Dasytidae	764	51		Löbl and Smetana 2007
Coleoptera	Malachiidae	1146	81		Löbl and Smetana 2007
Coleoptera	Sphindidae	14	3		Löbl and Smetana 2007
Coleoptera	Kateretidae	52	8		Löbl and Smetana 2007
Coleoptera	Nitidulidae	687	56		Löbl and Smetana 2007
Coleoptera	Monotomidae	72	11		Löbl and Smetana 2007
Coleoptera	Helotidae	55	1		Löbl and Smetana 2007
Coleoptera	Phloeostichidae	1	1		Löbl and Smetana 2007
Coleoptera	Silvanidae	94	21		Löbl and Smetana 2007
Coleoptera	Passandridae	16	3		Löbl and Smetana 2007
Coleoptera	Cucujidae	16	2		Löbl and Smetana 2007
Coleoptera	Laemophloeidae	82	17		Löbl and Smetana 2007
Coleoptera	Phalacridae	144	16		Löbl and Smetana 2007
Coleoptera	Cryptophagidae	336	30		Löbl and Smetana 2007
Coleoptera	Erotylidae	420	66		Löbl and Smetana 2007
Coleoptera	Byturidae	8	3		Löbl and Smetana 2007
Coleoptera	Biphyllidae	18	2		Löbl and Smetana 2007
Coleoptera	Bothrideridae	132	19		Löbl and Smetana 2007
Coleoptera	Cerylonidae	53	17		Löbl and Smetana 2007
Coleoptera	Alexiidae	42	1	West Palearctic	Löbl and Smetana 2007
Coleoptera	Discolomatidae	37	3		Löbl and Smetana 2007
Coleoptera	Endomychidae	282	56		Löbl and Smetana 2007
Coleoptera	Coccinellidae	1208	112	Lithophilinae (Mediterranean to Nepal and northern China)	Löbl and Smetana 2007
Coleoptera	Corylophidae	76	12		Löbl and Smetana 2007
Coleoptera	Latridiidae	268	18		Löbl and Smetana 2007
Coleoptera	Chrysomelidae	3500 (>2500 in USSR)			A. K. estimate, A. L. Lobanov (personal communication)

Table 7.1 (*continued*).

Order/ Suborder	Family (Superfamily)	Number of Species	Number of Genera	Endemic Family-Group Taxa and their Ranges	Source
Coleoptera	Chrysomelidae, subfamily Bruchinae	>120 (Russia)	15	Rhaebini (Israel to northern-central China); Kytorhinini (subendemic to central and eastern Palearctic, with one species in northwestern North America)	Egorov 1996a
Coleoptera	Cerambycidae	4500 (880 in USSR)			A. L. Lobanov (personal communication)
Coleoptera	Nemonychidae	7	3	Nemonychinae; Europe, NW Africa, Turkey (Anatolia), Armenia, Azerbaijan, Kazakhstan, Turkmenistan	Dieckmann 1974, Alonso-Zarazaga and Lyal 1999
Coleoptera	Urodontidae	>50	2		Alonso-Zarazaga and Lyal 1999
Coleoptera	Anthribidae	118 (West Palearctic and Russian Far East); 170 (Japan)	68		Frieser 1981, Egorov 1996b, Alonso-Zarazaga and Lyal 1999
Coleoptera	Oxycorynidae	1	1		Alonso-Zarazaga and Lyal 1999
Coleoptera	Rhynchitidae	86 (Russia)	38		Legalov 2006
Coleoptera	Attelabidae	29 (Russia)	19		Legalov 2006
Coleoptera	Dryophthoridae	7 (Russia); (37 in Japan) (+9 – western and central Palearctic)	?29		Morimoto 1978, Alonso-Zarazaga and Lyal 1999, B. A. Korotyaev unpublished data
Coleoptera	Brachyceridae	48	3		Alonso-Zarazaga and Lyal 1999, 2002; Arzanov 2005
Coleoptera	Cryptolaryngidae	1	1		Alonso-Zarazaga and Lyal 1999
Coleoptera	Brentidae	31 (27 in Japan, 2 in Russian Far East)	?20 (16) in Japan		Morimoto 1976, Alonso-Zarazaga and Lyal 1999

(*continued*)

Table 7.1 (*continued*).

Order/ Suborder	Family (Superfamily)	Number of Species	Number of Genera	Endemic Family-Group Taxa and their Ranges	Source
Coleoptera	Apionidae	540	57	Ceratapiini; Exapiini (southwestern Palearctic); Metapiini (south of Western and Central Palearctic)	Wanat 1994, Alonso-Zarazaga 1990, Alonso-Zarazaga and Lyal 1999, Friedman and Freidberg 2007
Coleoptera	Nanophyidae	90	14	Corimaliini (subendemic to south of Western and Central Palearctic; also in Namibia and probably in India)	Alonso-Zarazaga 1989, Alonso-Zarazaga and Lyal 1999
Coleoptera	Raymondionymidae	>50	9		Alonso-Zarazaga and Lyal 1999
Coleoptera	Erirhinidae	96	34	Himasthlophallini (south of Russian Far East)	Alonso-Zarazaga and Lyal 1999, 2002
Coleoptera	Curculiondae excluding Scolytinae	?11,000		numerous examples	B. A. K. tentative estimate
Coleoptera	Curculionidae, subfamily Scolytinae	750	74		Wood and Bright 1992
Coleoptera	Platypodidae	35	3		Wood and Bright 1992
Neuroptera	Ascalaphidae	40			V. A. Krivokhatsky (personal communication)
Neuroptera	Berothidae	15			V. A. Krivokhatsky (personal communication)
Neuroptera	Chrysopidae	176			V. A. Krivokhatsky (personal communication)
Neuroptera	Coniopterygidae	50			V. A. Krivokhatsky (personal communication)
Neuroptera	Dilaridae	7			V. A. Krivokhatsky (personal communication)
Neuroptera	Hemerobiidae	80			V. A. Krivokhatsky (personal communication)
Neuroptera	Mantispidae	10			V. A. Krivokhatsky (personal communication)

Table 7.1 *(continued).*

Order/ Suborder	Family (Superfamily)	Number of Species	Number of Genera	Endemic Family-Group Taxa and their Ranges	Source
Neuroptera	Myrmeleontidae	391		Pseudimarini (1 sp.), Gepini (31 spp.), Isoleontini (13 spp.)	V. A. Krivokhatsky (personal communication)
Neuroptera	Nemopteridae	10			V. A. Krivokhatsky (personal communication)
Neuroptera	Crocidae	20			V. A. Krivokhatsky (personal communication)
Neuroptera	Neurorthidae	5			V. A. Krivokhatsky (personal communication)
Neuroptera	Osmylidae	15			V. A. Krivokhatsky (personal communication)
Neuroptera	Sisyridae	5			V. A. Krivokhatsky (personal communication)
Raphidioptera	Inocelliidae	10			V. A. Krivokhatsky (personal communication)
Raphidioptera	Raphidiidae	100			V. A. Krivokhatsky (personal communication)
Megaloptera	Corydalidae	10			V. A. Krivokhatsky (personal communication)
Megaloptera	Sialidae	28			V. A. Krivokhatsky (personal communication)
Hymenoptera/ Symphyta		1384 (USSR)		Megalodontidae (45 spp.), Blasticotomidae (3–4 spp.)	A. L. Lobanov (personal communication)
Hymenoptera	Ichneumonidae	8712			Yu and Horstmann 1997
Hymenoptera	Braconidae	4500	250	Telengainae, Middle Asian deserts (Turkmenistan)	S. A. Belokobylsky (personal communication)
Hymenoptera	Aphidiidae	73 (Russian Far East)	24 (Russian Far East)		Davidian 2007
Hymenoptera	Cynipidae	>500	40		O. V. Kovalev (personal communication)

(continued)

Table 7.1 *(continued).*

Order/ Suborder	Family (Superfamily)	Number of Species	Number of Genera	Endemic Family-Group Taxa and their Ranges	Source
Hymenoptera	Ibaliidae	7	2		O. V. Kovalev (personal communication)
Hymenoptera	Liopteridae	5	2		O. V. Kovalev (personal communication)
Hymenoptera	Figitidae	?150	15		O. V. Kovalev (personal communication)
Hymenoptera	Emarginidae	1	1		O. V. Kovalev (personal communication)
Hymenoptera	Charipidae	>100	4		O. V. Kovalev (personal communication)
Hymenoptera	Eucoilidae	>350	30		O. V. Kovalev (personal communication)
Hymenoptera	Eurytomidae/ Eurytominae	265	5		Zerova 1995
Hymenoptera	Eurytomidae/ Eudecatominae	9	1		Zerova 1995
Hymenoptera	Eurytomidae/ Rileyinae	2	2		Zerova 1995
Hymenoptera	Eurytomidae/ Buresiinae	1	1		Zerova 1995
Hymenoptera	Eurytomidae/ Harmolitinae	100	5		Zerova 1995
Hymenoptera	Torymidae/ Megastigminae	54			Grissell 1999
Hymenoptera	Torymidae/ Toryminae	334	30		Grissell 1995
Hymenoptera	Encyrtidae	1260	163		Trjapitzin 1989
Hymenoptera	Eupelmidae	130	16		Sharkov 1995
Hymenoptera	Eulophidae	318 (Russian Far East)	50 (Russian Far East)		Storozheva et al. 1995
Hymenoptera	Aphelinidae	60 (Russian Far East)	18 (Russian Far East)		Jasnosh 1995
Hymenoptera	Trichogrammatidae	150			Fursov 2007
Hymenoptera	Chrysididae	1236			Kimsey and Bohart 1990
Hymenoptera	Scoliidae	33 (USSR)	3		Lelej 1995a
Hymenoptera	Mutillidae	523	54		Lelej 2002
Hymenoptera	(Apoidea)	3840	112		Yu. A. Pesenko (personal communication)
Hymenoptera	Andrenidae		12		Yu. A. Pesenko (personal communication)
Hymenoptera	Halictidae		26		Yu. A. Pesenko (personal communication)
Hymenoptera	Halictidae/ Nomioidinae	35	1		Pesenko 1983
Hymenoptera	Melittidae		5	Protomelittini	Yu. A. Pesenko (personal communication)
Hymenoptera	Megachilidae		36	Pararhophini	Yu. A. Pesenko (personal communication)

Table 7.1 *(continued).*

Order/ Suborder	Family (Superfamily)	Number of Species	Number of Genera	Endemic Family-Group Taxa and their Ranges	Source
Hymenoptera	Apidae			Ancylini	Yu. A. Pesenko (personal communication)
Hymenoptera	Formicidae	1200	80		Radchenko 1999
Hymenoptera	Pompilidae	285 (USSR)	47		Lelej 1995b
Hymenoptera	Sphecidae	1000 (USSR)	93		Nemkov et al. 1995
Hymenoptera	Vespidae	77 (Russian Far East)	17		Kurzenko 1995
Mecoptera		>42 (USSR)			A. L. Lobanov (personal communication)
Trichoptera		2530	215	Phrygano-psychidae (3 spp.); Thremmatidae (3 spp.; S Europe)	M. L. Chamorro-Lacayo (personal communication), Morse 2008
Lepidoptera		25,000		Catapterigidae (1 gen., 1 sp.) (Crimea) Crinopterygidae (1 gen., 1 sp.) (Hesperian Region) Axiidae (2 genn., 6 spp.) (Hesperian) Endromidae (Transpalearctic) Mirinidae (1 gen., 2 spp.) (Stenopean: Manchuria) Somabrachyidae (3 spp.; N Africa, Syria)	Heppner 1998, S. Yu. Sinev (personal communication)
Lepidoptera	(Micropterigoidea)	78			Heppner 1998
Lepidoptera	(Eriocranioidea)	12			Heppner 1998
Lepidoptera	(Hepialoidea)	37			Heppner 1998
Lepidoptera	(Nepticuloidea)	493			Heppner 1998
Lepidoptera	(Incurvarioidea)	207			Heppner 1998
Lepidoptera	(Tineoidea)	1307			Heppner 1998
Lepidoptera	(Gelechioidea)	3843			Heppner 1998
Lepidoptera	(Copromorphoidea)	64			Heppner 1998
Lepidoptera	(Yponomeutoidea)	485			Heppner 1998
Lepidoptera	(Immoidea)	1			Heppner 1998
Lepidoptera	(Pyraloidea)	2936			Heppner 1998
Lepidoptera	(Pterophoroidea)	315			Heppner 1998
Lepidoptera	(Sesioidea)	382			Heppner 1998
Lepidoptera	(Zygaenoidea)	162			Heppner 1998
Lepidoptera	(Cossoidea)	210			Heppner 1998

(continued)

Table 7.1 *(continued).*

Order/ Suborder	Family (Superfamily)	Number of Species	Number of Genera	Endemic Family-Group Taxa and their Ranges	Source
Lepidoptera	(Tortricoidea)	1606			Heppner 1998
Lepidoptera	(Calluduloidea)	2			Heppner 1998
Lepidoptera	(Uranioidea)	18			Heppner 1998
Lepidoptera	(Geometroidea)	3545			Heppner 1998
Lepidoptera	(Papilionoidea)	1896			Heppner 1998
Lepidoptera	(Drepanoidea)	25			Heppner 1998
Lepidoptera	(Bombycoidea)	291			Heppner 1998
Lepidoptera	(Sphingoidea)	75			Heppner 1998
Lepidoptera	(Noctuoidea)	4475			Heppner 1998
Siphonaptera		900			S. G. Medvedev 1998
Diptera		40,291			Nartshuk 2003, Thompson 2006
TOTAL		**193,057**			

Altai Mountains and one in the Sayan Mountains, southern Siberia); and the beetle families Amphizoidae (mountain streams of North America, and north-western and eastern China) and Sphaeritidae (three species in the taiga of Eurasia, mountains of Sichuan, and northwestern North America). The range of the order Raphidioptera is limited almost entirely to the Holarctic. Lindroth (1957) lists land and freshwater animal species common to Europe and North America. Despite its similarity to the Nearctic, however, the Palearctic is traditionally regarded as a separate region (Sclater 1858, Darlington 1963, Emeljanov 1974, Vtorov and Drozdov 1978, Krivokhatsky and Emeljanov 2000).

Fig. 7.1 (See figure on the following page.)

(a) Main divisions of the Palearctic (after Emeljanov 1974, simplified).
　　I. Arctic (Circumpolar Tundra) Region.
　　II. Taiga (Euro-Siberian) Region.
　　III. European (nemoral) Region.
　　IV. Stenopean (nemoral) Region.
　　V. Hesperian (evergreen forest) Region.
　　　　Va. Macaronesian Subregion.
　　　　Vb. Mediterranean Subregion.
　　VI. Orthrian (evergreen forest) Region.
　　VII. Scythian (Steppe) Region.
　　　　VIIa. West Scythian Subregion.
　　　　VIIb. East Scythian Subregion.
　　VIII. Sethian (Desert) Region.
　　　　VIIIa. Saharo-Arabian Subregion.
　　　　VIIIb. Irano-Turanian Subregion.
　　　　VIIIc. Central Asian Subregion.
(b) NE Russia, Magadan Province, forest-tundra with flowering *Ledum decumbens* (Ericaceae) and a single bush of *Pinus pumila* on the right (Photo D. I. Berman).
(c) NE Russia, Magadan Province, northern taiga (Photo D. I. Berman).
(d) Russia, Smolensk District, Ugra River near Skotinino Village, mixed forest (Photo A. Konstantinov).
(e) Russia, Caucasus (Photo M. Volkovitsh).

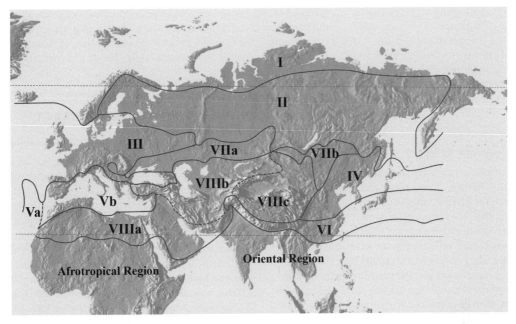

(a)

(b)

(c)

(d)

(e)

Insect biodiversity in the Palearctic is influenced by diverse climatic and other geographic conditions that exhibit a well-developed zonality. Temperature gradients are mainly responsible for the Arctic, Boreal, and Subtropical latitudinal belts. The following main zones are distributed from north to south in the western parts of the Palearctic: tundra, taiga, mixed and broadleaf (nemoral) forests, dry sclerophyll Mediterranean-type forests, wet subtropical forests, steppe, and deserts. Most mountain ranges demonstrate successive series of climatic altitudinal belts similar to lowland zonality. Atmospheric circulation and variations in precipitation yield the Atlantic, Continental, and Pacific longitudinal sectoral groups (Emeljanov 1974). The Atlantic sectoral group is characterized by two types of climate: Mediterranean with maximum precipitation in the winter, and boreal with maximum precipitation in the summer. The Pacific sectoral group has a monsoon climate with maximum precipitation during the summer. The boundary between the Atlantic and Pacific sectoral groups, though rather conventional, is usually drawn along the Yenisei River, Tien Shan mountains, and west of the Indus River in the south. The easternmost sectors of the Atlantic group and westernmost sectors of the Pacific group along this boundary where the oceanic influence is drastically weakened are characterized by lower precipitation and greater temperature fluctuations between summer and winter (continental and supercontinental climate); these sectors can be grouped as the Continental sectoral group. Continental climate is responsible for the taiga spreading southward and the steppe northward, squeezing out nemoral and subtropical zones in the Continental sectors (Emeljanov 1974). Continentality is one of the most important factors in Palearctic faunal differentiation, splitting subtropical and nemoral zones into two isolated fragments with rather different, though, in part, closely related insect faunas. In the easternmost Pacific sectors, high humidity is responsible for the lack of semidesert and desert areas and the large meridional extension of mixed coniferous and broadleaf forests that are impoverished northward because of elimination of broadleaf elements and that gradually change to subtropical forests southward.

GENERAL FEATURES OF INSECT BIODIVERSITY

More than 200,000 species of insects are known in the Palearctic Region – about one-fifth the total number of insect species in the world (Table 7.1). About half of all insects in the Palearctic are beetles. This estimate is far from final; hundreds of new species are described from the Palearctic every year. In recent years, parasitic Hymenoptera have been described most intensively, with 686 new species of Braconidae described from the Far East and neighboring areas in two volumes of the 'Key to Insects of the Russian Far East' by S. A. Belokobylsky and V. I. Tobias (1998, 2000), which include 2593 species. In the largest family of parasitic flies, the Tachinidae, more than 680 species are included in the key to the Far-Eastern fauna (Richter 2004), with 39 species and 9 genera described during the preparation of the key.

Most Palearctic species do not occur outside the region. In the Diptera (Fig. 7.2c), 37,123 species – 92% of the total 40,291 species – are Palearctic endemics (Thompson 2006). The percentage of endemic genera, and especially family-group taxa, is much less. Only four families of Diptera are endemic to the region (Nartshuk 1992): Eurygnathomyiidae, Phaeomyiidae, Risidae, and Stackelbergomyiidae. Of these, only the Risidae are generally accepted as a valid family; the others usually are treated as subfamilies or tribes (Sabrosky 1999).

One of the most obvious features of insect biodiversity in the Palearctic is its sharp increase from

---

Fig. 7.2 (See figure on the following page.)

  A. *Calosoma sycophanta* (L.) (Coleoptera: Carabidae) (Turkey) (Photo A. Konstantinov).
  B. *Nemoptera sinuata* Olivier (Neuroptera: Nemopteridae) (Turkey) (Photo M. Volkovitsh).
  C. *Eristalis tenax* (L.) (Diptera: Syrphidae) (Turkey) (Photo A. Konstantinov).
  D. *Cryptocephalus duplicatus* Suffrian (Coleoptera: Chrysomelidae) (Turkey) (Photo A. Konstantinov).
  E. *Poecilimon* sp. (Orthoptera: Tettigoniidae) (Turkey) (Photo M. Volkovitsh).
  F. *Capnodis carbonaria* (Klug) (Coleoptera: Buprestidae) (Turkey) (Photo M. Volkovitsh).
  G. *Cyphosoma euphraticum* (Laporte et Gory) (Coleoptera: Buprestidae) (southern Russia) (Photo M. Volkovitsh). (*See color plate*).

north to south, corresponding to the most fundamental pattern of life on earth (Willig et al. 2003). For example, the Orthoptera (Figs. 7.2e and 7.3e), Blattaria, Dermaptera, Mantodea, and Phasmatodea are represented by 72 species in the forest, 171 species in the steppe, and 221 species in the desert zone of the former USSR (Iablokov-Khnzorian 1961).

The insect fauna of the Palearctic is slightly depauperate at the order, family, and genus levels (compared to most other regions) but it has high species richness; that is, the number of higher taxa with only a single species in the Palearctic is relatively low and the relative number of species per higher taxon is large. In this regard, it is similar to island faunas characterized by a small number of introductions that were followed by extensive species-level radiations (Magnacca and Danforth 2006). The Palearctic and Oriental regions have about the same number of species of flea beetles (Chrysomelidae: Alticini) (about 3000, although the Oriental fauna is much less known), but differ sharply in generic diversity, with about 60 genera in the Palearctic and 220 in the Oriental. Most flea-beetle species richness in the Palearctic is concentrated in a few large, nearly cosmopolitan genera (e.g., *Aphthona*; see Konstantinov 1998).

Historically, the Palearctic fauna is considered to have been derived from the ancient fauna of Laurasia, dramatically changed by the aridization of the Tertiary, but primarily by the Quaternary glaciation (Lopatin 1989), which includes the largest global glaciations of the upper Pliocene and Pleistocene. Also important were fluctuations of sea level, which led not only to changes in coast lines, but also to formation of a variety of land bridges between continents and various islands, and the Alpine orogenesis during which the largest mountain systems in Europe and Asia appeared (Kryzhanovsky 2002). These climatic and geomorphological changes might explain the appearance of a Tibetan scarab, *Aphodius holdereri* Reitter, in England (Coope 1973); disjunct distributions of *Helophorus lapponicus* Thompson (Coleoptera: Helophoridae) between its main range (Scandinavia to eastern Siberia) and relict populations in mountainous areas of Spain, Transcaucasia, and Israel/Lebanon (Angus 1983); and current restriction of the water beetle *Ochthebius figueroi* Garrido et al. (Coleoptera: Hydraenidae), known from Pleistocene deposits in England, to a small mountain area in northern Spain (Angus 1993).

Insect biodiversity has a particular pattern in space and time. The distribution of specialized herbivores,

closely associated with specific plants and plant communities, often reveals a more distinct pattern than the distribution of insects with other food specializations. Changes in the Curculionoidea (Coleoptera) fauna along a 160-km transect that crosses six types of desert plant communities in the Transaltai Gobi Desert in Mongolia illustrate this pattern (Table 7.2). Two types of plant communities exist at the extremes of the profile – a northern steppefied desert and a southern extra-arid desert. The numbers of weevil species at the extremes do not differ sharply (17 in the north and 11 in the south), but only one species, *Conorhynchus conirostris* (Gebler), occurs in all types of deserts. All 'northern' species (with ranges situated mostly north of the investigation site) gradually disappear southward, being substituted by 'southern' species in accordance with vegetational changes. The distribution of (mostly predatory) carabids and nonspecialized phyto- and detritophagous tenebrionid beetles mostly depends on climate, chemical and mechanical properties of soil, and vegetation density; it follows vegetational changes less closely, but exhibits similar patterns.

Different natural zones are characterized by the dominance of some higher taxa, particularly in the Arctic. The majority of the Arctic chrysomelid fauna is composed of 25 species of a single subfamily, the Chrysomelinae (with 12 species belonging to the genus *Chrysolina*; see Chernov et al. 1994), although a few boreal species of Cryptocephalinae and Galerucinae contribute to the Hypoarctic leaf-beetle fauna (Medvedev and Korotyaev 1980). The taxonomic pattern of the biogeographical regions has a historical background, but the zonal peculiarities of the faunas are largely due to specific requirements of the taxa. In the holometabolous insects, these requirements are related primarily to the environmental conditions appropriate for larval development. An example is illustrated by the chrysomelid fauna of the Transaltai Gobi Desert in Mongolia. Ten species of Chrysomelidae, out of the fourteen species found in the six types of desert plant communities (Korotyaev et al. 2005), belong to the tribes Cryptocephalini, Clytrini, and Cassidini. The case-bearing (Cryptocephalini) and sheltered (Clytrini) larvae apparently are better adapted to the xeric environment than are the ectophytic larvae of the Chrysomelinae and Galerucini or soil-inhabiting larvae of the Eumolpinae and some Alticini.

Another obvious feature of insect biodiversity in the Palearctic is its organization in time. Nearly

Fig. 7.3

  A. *Julodis variolaris* (Pallas) (Coleoptera: Buprestidae) (Kazakhstan) (Photo M. Volkovitsh).

  B. *Julodella abeillei* (Théry) (Coleoptera: Buprestidae) (Turkey) (Photo M. Volkovitsh).

  C. *Mallosia armeniaca* Pic (Coleoptera: Cerambycidae) (Turkey) (Photo M. Volkovitsh).

  D. *Trigonoscelis schrencki* Gebler (Coleoptera: Tenebrionidae) (Kazakhstan) (Photo M. Volkovitsh).

  E. *Saga pedo* Pallas (Orthoptera: Tettigoniidae) (Kazakhstan) (Photo M. Volkovitsh).

  F. *Piazomias* sp. (Coleoptera: Curculionidae) (Kazakhstan) (Photo M. Volkovitsh). (*See color plate*).

Table 7.2 Distribution of Curculionoidea across six types of desert plant communities in Transaltai Gobi, Mongolia. 6–1 = desert plant communities (from north to south): 6, *Anabasis brevifolia* steppefied desert; 5, *Reaumuria soongorica* + *Sympegma regelii* desert; 4, *Haloxylon ammodendron* desert; 3, *Reaumuria soongorica* + *Nitraria sphaerocarpa* desert; 2, extra-arid *Iljinia regelii* desert; 1, extra-arid *Ephedra przewalskii* + *Haloxylon ammodendron* (in dry temporary waterbeds) desert.

Steppe Species (+)	6	5	4	3	2	1	Desert Species (*)
	*		*				*Pseudorchestes furcipubens* (Reitter)
	*	*			*		*Deracanthus faldermanni* Faldermann
Cionus zonovi Korotyaev	+	*	*	*	*	*	*D. hololeucus* Faldermann
Eremochorus inflatus (Petri)	+	*	*			*	*Mongolocleonus gobiensis* (Voss)
Macrotarrhus kiritshenkoi Zaslavsky	+		*	*	*	*	*Pycnodactylus oryx* (Reitter)
Philernus gracilitarsis (Reitter)	+		*			*	*Stephanocleonus helenae* (Ter-Minassian)
Pseudorchestes sp.	+		*		*	*	*Elasmobaris alboguttata* (Brisout)
Stephanocleonus paradoxus (Fåhraeus)	+		*		*	*	*Perapion ?myochroum* (Schilsky)
Fremuthiella vossi (Ter-Minassian)	+	+	*		*	*	*P. centrasiaticum* (Bajtenov)
Stephanocleonus potanini Faust	+	+	*		*	*	*Anthypurinus kaszabi* (Bajtenov)
S. inopinatus (Ter-Minassian)	+	+	*				*Lixus incanescens* Boheman
S. excisus Reitter	+	+	*				*Sibinia* sp. pr. *beckeri* Desbrochers
S. persimilis Faust		+	*				*Cosmobaris scolopacea* (Germar)
Gronops semenovi Faust	+		+			*	*Oxyonyx kaszabi* Bajtenov
Eremochorus mongolicus (Motschulsky)	+	+	+			*	*Platygasteronyx humeridens* (Voss)
Conorhynchus conirostris (Gebler)	+	+	+	+	+	+*	*P. macrosquamosus* Korotyaev
Extrazonal consortium of *Reaumuria soongorica* (□)							
Corimalia reaumuriae (Zherichin)	□	□		□			
Coniatus zaslavskii Korotyaev		□		□			
C. minutus Korotyaev	□	□		□			

all regions of the Palearctic are subjected to strong seasonal changes in temperature and precipitation. Long periods of fall and winter are characterized by minimum insect activity. Some of the most typical winter insects are representatives of the Holarctic family Boreidae, the snow scorpionflies. These small mecopterans of the genus *Boreus* (Fig. 7.4g) appear in the fall and winter and often hop and walk on the snow.

On the Russian Plain, maximum insect biodiversity of herbivores can be observed in late May and in June. The taxonomic aspect of any local fauna also changes with time. A number of species are active only in early spring, particularly those in southern, more xeric regions of the Palearctic. Good examples include many flightless black *Longitarsus* (Chrysomelidae) species that are associated with ephemeral plants,

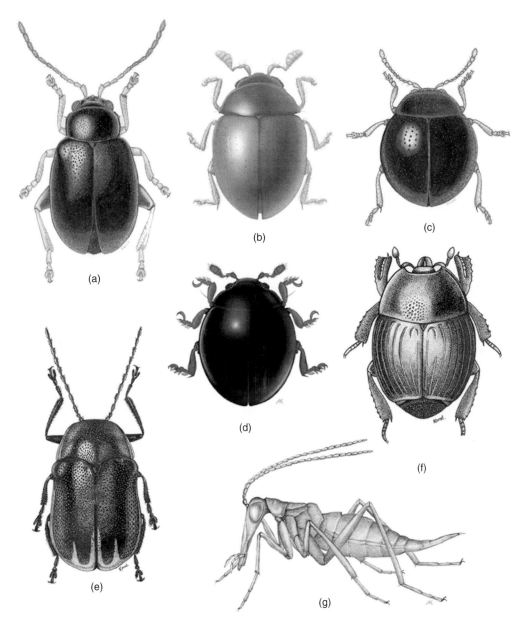

Fig. 7.4

A. *Aphthona nonstriata* Goeze (Coleoptera: Chrysomelidae).
B. *Clavicornaltica dali* Konstantinov and Duckett (Coleoptera: Chrysomelidae).
C. *Mniophila muscorum* Koch (Coleoptera: Chrysomelidae).
D. *Kiskeya baorucae* Konstantinov and Chamorro-Lacayo (Coleoptera: Chrysomelidae).
E. *Cryptocephalus ochroloma* Gebler (Coleoptera: Chrysomelidae).
F. *Margarinotus* (*Kurilister*) *kurbatovi* Tishechkin (Coleoptera: Histeridae).
G. *Boreus hyemalis* (L.) (Mecoptera: Boreidae).

and in their flightless features, are similar to alpine members of the genus (Konstantinov 2005). The maximum biodiversity of adult weevils in the steppe of the Northwestern Caucasus precedes the maximum local air temperature and precipitation, which might mean that the phenology of holometabolous herbivores is adjusted to the maximum supply of warmth and water for the feeding (larval) stage. Late maturation of the hemimetabolous insects (orthopterans, bugs, and leafhoppers) fits this speculation: the development of their larvae and nymphs proceeds in the warmest part of summer.

Several of the largest Palearctic genera of families of most species that are rich in woodlands of other biogeographic regions are confined to open landscapes in the Palearctic Region. The best illustration is the largest genus of the mostly xylophagous family Cerambycidae, the endemic *Dorcadion*, with all 300-plus species occurring in meso- to xerophilic grasslands. No genus of dendrobiont cerambycids has comparable diversity in the Palearctic. The leaf-mining tribe Rhamphini of the Curculioninae, with about 100 Palearctic species in some 10 genera on trees and bushes, has its largest subendemic genus in the region (with one species in Namibia). This subendemic genus, *Pseudorchestes*, has more than 35 described and dozens of yet undescribed species that develop on herbaceous plants and semishrubs of the Asteraceae. Among the predatory beetles, the lady-beetle genus *Tetrabrachys* (= *Lithophilus*) of the endemic subfamily Lithophilinae, with 51 species (Iablokoff-Khnzorian 1974), is the largest in the region. The vast majority of its species are confined to xeric areas of the southern Palearctic, mostly in the western half. The second largest genus in the region, *Hyperaspis*, with 45 species, also has most of its species distributed in the dry open landscapes. The great diversification of the open-landscape Hyperinae and Lixinae, compared with just a few typically woodland species of these large subfamilies of the Curculionidae, is also characteristic, as is the absence of dendrophilous Baridinae in the western and central Palearctic; the baridine fauna associated with herbs, however, is fairly species rich. The same tendency is also obvious in the Alticini (Chrysomelidae: Galerucinae), Cassidinae (Chrysomelidae), and the subfamily Ceutorhynchinae, tribes Cionini and Tychiini of the Curculioninae, and Sitonini of the Entiminae (Curculionidae). The faunal similarity with open landscapes of the Nearctic counterpart is rather low. The Nearctic has greater representation of the predominantly tropical weevil families Anthribidae and Dryophthoridae and the subfamily Conoderinae of the Curculionidae, whereas the Hyperinae, Lixinae, Cyclominae, Rhythirinini, *Tychius*, and many taxa of the Entiminae that dominate grassland communities in the Palearctic are absent or subordinate in the Nearctic.

NOTES ON BIODIVERSITY OF SOME INSECT GROUPS IN THE PALEARCTIC

Most of the large orders have wide representation in the Palearctic. More than 100,000 species of beetles, for example, occur in the region. The other most species-rich orders are Diptera, Hymenoptera, and Lepidoptera. Higher taxa with different types of food specialization gain richness in the Palearctic, including its northern parts. Parasitic wasps, for example, are represented by a greater number of species than is the entire order Coleoptera in the British Isles (LaSalle and Gauld 1991, cited by Sugonyaev and Voinovich 2006). The less-specialized predatory, mycophagous, and phytophagous fungus gnats (Diptera: Mycetophilidae) and highly specialized phytophagous sawflies (Hymenoptera: Tenthredinidae) have enormous faunas, with 645 species in Karelia (Polevoi 2007) and about 800 species in Finland (A. G. Zinovjev, personal communication), respectively. Many higher taxa of noxious blood-sucking insects, such as the orders Siphonaptera and Phthiraptera and the dipteran families Culicidae, Ceratopogonidae, Simuliidae, Tabanidae, Gasterophilidae, Hypodermatidae, Oestridae, and Hippoboscidae, are widely represented in the Palearctic. Of the blood-sucking Diptera that transmit dangerous disease agents, only the Glossinidae with their infamous tse-tse are absent, and the Psychodidae are restricted to the southernmost regions of the Palearctic. The Palearctic Siphonaptera fauna, with 900 species, is the largest in the world, comprising 40% of the world species and genera, although only one monotypic subfamily, with 19 species, is endemic (Medvedev 1998). Scale insects (Sternorrhyncha: Coccina), most conspicuous in the tropical forests, are less abundant and diversified in the Palearctic but are present in all climatic zones, including the Arctic. Entomophagous wasps of the superfamily Chalcidoidea, which are associated closely with scale insects, have developed a special strategy of host use in the high latitudes that differs from the strategy used in the tropics (Sugonyaev and Voinovich 2006).

A few orders, largely contributing to the faunas of the adjacent tropical regions, and all belonging to the Orthopteroidea, are poorly represented in the Palearctic: Blattaria, Dermaptera, Isoptera, Mantodea, and Phasmatodea. Most higher taxa of Palearctic Orthoptera are widely distributed in the tropics, but a few subfamilies are mainly Palearctic (Deracanthinae, Glyphonotinae, Onconotinae, Pamphaginae, and Thrinchinae). Among tribes, the Chrysochraontini, Conophymatini, Drymadusini, Gampsocleidini, and Odonturini are endemic. Almost all the taxa in these groups have their centers of diversity and endemism in the southern Palearctic (Sergeev 1993).

For many large family-group taxa, the Palearctic contains a relatively small percentage of the world fauna. For example, the beetle family Buprestidae (Fig. 7.2f, g) comprises 2430 species in 99 genera in the Palearctic – approximately 17% of the species and 20.2% of the genera in the world (Bellamy 2003, Löbl and Smetana 2006). The large group of phytophagous flea beetles (Chrysomelidae, Alticini), with about 11,000 species and 600 genera worldwide, is represented in the Palearctic by about 2400 species and 64 genera. Yet only a few genera of the Alticini are endemic to the Palearctic; most of them are distributed in the mountains of southern Europe, the Caucasus, and the Mediterranean (Konstantinov and Vandenberg 1996). Many Oriental and some Afrotropical genera are represented by only a few species in the Palearctic at the eastern or southern borders of the region.

The subfamily Ceutorhynchinae of the Curculionidae – similar to the Alticini in many ecological features and range of body size – had 1316 described species as of 2003, and includes Palearctic representatives of 12 of 14 presently distinguished tribes (Colonnelli 2004), with only two Paleotropical tribes (Lioxyonychini and Hypohypurini) absent from the region. About half of the species of this worldwide subfamily and 102 of the total 167 genera occur in the Palearctic; 79 genera and 2 tribes are Palearctic endemics.

For many boreal and temperate groups, the Palearctic has the most species-rich fauna, compared with other zoogeographical regions. The Cecidomyiidae (Diptera), for example, have 3057 species in the Palearctic, compared with only 533 species in

the Neotropics (Gagne 2007). This trend is also true for aphids, which are mostly Holarctic; their complicated life cycle is possibly an adaptation to a temperate climate. Ichneumonidae (Hymenoptera) have a similar distribution of diversity. They are richer in the northern hemisphere, particularly in the Palearctic. The tribe Exenterini, for example, is distributed almost entirely in the Holarctic, with all genera represented in the Palearctic (Kasparyan 1990). Many species have wide transpalearctic ranges that are almost entirely confined to forest regions. Most (Kasparyan 1990) are parasites of various Tenthredinidae (Hymenoptera: Symphyta), which are also rich in species in the Palearctic. Taeger et al. (2006) counted 1386 species of Symphyta in Europe, 220 species in Norway, and 8 in Novaya Zemlya. For China, the estimate is 2600 species and 350 genera (Wei et al. 2006).

Aquatic and amphibiotic insect groups are also rich in the Palearctic. About one-third of the world's blood-sucking Simuliidae, all with aquatic larvae, are distributed in the Palearctic (Adler and Crosskey 2008). Chironomidae, with their predominantly aquatic larvae, also are species rich in the Palearctic, dominating the Arctic aquatic complexes and including the northernmost dipterans (also southernmost in the Antarctic: Nartshuk 2003). For Russia, the following species numbers are available for the largest insect orders (partly including the fauna of neighboring countries): Ephemeroptera – about 300 species (N. Ju. Kluge personal communication), Odonata – 148 species (Kharitonov 1997), Plecoptera –225 species (Zhiltzova 2003), Megaloptera –15 species (Vshivkova 2001), Trichoptera – 652 species (Ivanov 2007), Neuroptera –11 species (Krivokhatsky 2001), Lepidoptera – 8 species (Lvovsky 2001), Coleoptera – about 700 species (Kirejtshuk 2001), and Hymenoptera – 24 species (Kozlov 2001). Plecoptera, with preference for cold water, are quite unique in the Palearctic (Zhiltzova 2003). In particular, the Euholognatha include a large number of taxa endemic to the Palearctic, including the Scopuridae, with a single genus of five species. The Taeniopterygidae and Capniidae contain large numbers of genera endemic to the Palearctic (6 of 13 worldwide and 7 of 17, respectively). Only one family, the Notonemouridae, with 69 species, is absent from the Palearctic, being distributed in the Neotropics, Australia, and Africa south of the Sahara (Zhiltzova 2003). The ranges of many Palearctic species are relatively small, such as that of *Capnia kolymensis* Zhiltzova from the Kolyma River and several species of *Nemoura* from Iturup Island. The distributions of many species associated with cold water are restricted to mountain systems (Alps, Carpathians, Caucasus, Tien Shan). Their endemism at the species level reaches 60% (Zhiltzova 2003). Endemism of other aquatic insect groups, such as Trichoptera, is estimated at 36% (Zhiltzova 2003).

The Palearctic is sometimes among a few places on Earth where a rare group of flies is distributed. The family Canthyloscelidae, for example, has a single genus (*Hyperoscelis*) with three species in the Palearctic and two other genera with four species in southern South America and four species in New Zealand (Nartshuk 1992). Among the widely distributed families of flies, the following are known from the Palearctic but are absent in the Nearctic: Camillidae, Cryptochetidae, Megamerinidae, and Xenasteiidae of the Acalyptratae, and Eugeniidae and Villeneuviellidae of the Calyptratae. Almost all are widely distributed in the Oriental or Afrotropical Regions. The Nearctic has seven families of flies not known from the Palearctic. The recently established family Xenasteiidae in the Palearctic occurs in the Mediterranean and on the islands of the Indian and Pacific oceans. The Apioceridae are known from all regions except the Palearctic (Nartshuk 1992).

Many endemic Palearctic higher taxa have relatives in either the Nearctic or the temperate areas of the Southern Hemisphere. But the weevil subfamily Orobitidinae, in addition to the oligotypic Palearctic genus *Orobitis* associated with *Viola* plants, includes the genus *Parorobitis* (Fig. 7.5c), with a few species (hosts unknown) in tropical South America (Korotyaev et al. 2000) where Violaceae are represented by more than 300 species (Smith et al. 2004).

Small insect families contribute considerably to the insect biodiversity of the Palearctic Region. *Panorpa* species, the scorpionflies (Mecoptera), although neither speciose nor abundant, are common in mixed and nemoral European forests, and the oligotypic Sialidae (Megaloptera) comprise a conspicuous component of the European riparian landscape. A relatively small beetle family, the Trogossitidae, includes the worldwide synanthropic *Tenebrioides mauritanicus* L., which is injurious to stored products. *Ostoma ferrugineum* (L.) and a few species of the genera *Peltis* and *Thymalus* frequently are found under bark in all types of Palearctic forests except the northernmost

taiga. Larvae of one of the two European representatives of the beetle family Byturidae in some years destroy a considerable part of the raspberry harvest. The European *Dascillus cervinus* (L.) and the Caucasian *D. elongatus* Faldermann of the small family Dascillidae are abundant under the forest canopy for short periods of their adult lives. Their Eastern Palearctic ally, *Macropogon pubescens* Motschulsky, is one of the few common beetles around bushes of *Pinus pumila* Regel in the hills of the middle Kolyma basin (Northeast Asia), which has a markedly impoverished insect fauna. *Mycterus curculionoides* (F.) of the small family Mycteridae often dominates assemblages of medium-sized beetles in the mid-summer dry grasslands of the Western Palearctic.

Two myco-detritophagous families of small beetles, Cryptophagidae and Latridiidae, include several hundred brown beetles, all rather uniform in appearance, occurring almost everywhere except the northern tundra. They are especially abundant in wet riparian litter. These families, as well as the predominantly mesohygrophilous and rather uniform Helodidae (= Cyphonidae), illustrate a tendency that the least conspicuous taxa often have the greatest species richness. Many staphylinid beetles fit this trend, including those that live in the open (some *Stenus*) and the cryptobionts (e.g., small Aleocharinae). The largest genus of the Alticini in the Palearctic, *Longitarsus*, with 221 species in the region (29 species only on Mount Hermon in Israel; Chikatunov and Pavliček 2005), is not particularly diverse morphologically. One of the largest genera of weevils in the Palearctic, *Ceutorhynchus*, with about 300 species, is also morphologically less diverse, compared with the showy Oriental *Mecysmoderes* from the same subfamily.

BIODIVERSITY OF INSECT HERBIVORES

Among the herbivores, the greatest species richness is achieved by insects with rhizophagous, soil-inhabiting larvae. These groups include the weevil subfamily Entiminae (Fig. 7.3f) with about 14,000 known species (3500 in the Palearctic), the tribe Alticini of the Chrysomelidae with about 11,000 species worldwide and ca 3000 species in the Palearctic, and the noctuid moths with 27,000 species (5500 in the Palearctic). Of the Entiminae, the largest genus is the Palearctic

Otiorhynchus s.l., with about 1000 species; all are wingless and many have restricted ranges in the mountains of southern Europe, Anatolia, and Middle and Central Asia[1]. The high number of species probably relates little to trophic specializations because many species are apparently polyphagous, which is probably true most of the subfamily Entiminae. Winglessness likely facilitates geographic isolation. *Otiorhynchus* also includes a great number of parthenogenetic forms that are genetically isolated, with their own ranges and ecological associations. Many other mostly wingless genera of the Entiminae include parthenogenetic forms, which probably enables rapid range extensions. The vast territory of the Russian Plain with its short post-glaciation history has faunas of the genera *Otiorhynchus* (18 species) and *Trachyphloeus* (9 species) that consist exclusively of parthenogenetic forms whose known bisexual ancestors, when present, have narrow ranges in the neighboring or rather distant mountain systems. Most of the parthenogenetic forms occur in the temperate and boreal forests and in the steppe zone, whereas no endemic parthenogenetic forms are known from the tundra, and only a few live in the desert zone (Korotyaev 1992). The relict bisexual forms often are localized in the mountains, whereas their parthenogenetic derivatives are in the plains.

The wingless species of *Otiorhynchus* contradict speculation that a limited number of species can be produced within a single genus, presuming that only a certain number of combinations of morphological characters co-occur to give rise to viable forms, whereas many other combinations, being maladaptive, are

[1] Middle Asia is a climatic/natural region distinct from Central Asia (Korotyaev et al. 2005, Medvedev 2005b). It includes the Asian republics of the former USSR and neighboring parts of Afghanistan and Iran. The region is characterized by warm winters and maximum rainfall in spring and autumn. Central Asia is a climatic region that includes Mongolia and a large area of northwestern China. It is characterized by an extreme continental climate with harsh winters and maximum rainfall in late summer. The term Middle Asia is used in the Russian literature, but in the English-language literature Middle Asia is incorporated in Central Asia.

selected against. This idea might apply to planktonic forms (Zarenkov 1976) but is wrong for terrestrial beetles. *Otiorhynchus* in the broad sense has a great number of species that differ only in the proportions of their antennal funicle, sexually dimorphic characters, vestiture, and coloration. Considering that every new taxon potentially could produce a further set of descendants with innumerable combinations of old and new characters, one can hardly imagine reasonable limitations for the diversity of species that could evolve in *Otiorhynchus* and other apterous beetles. Parthenogenesis increases this diversity, but here a limitation does exist: parthenogenetic forms of weevils do not have more than six haploid chromosome sets; hexaploids are terminal products of parthenogenesis.

The diversity of a taxon of specialized herbivores is often proportional to the number of species in the host-plant taxon (although this statement is more hypothetical than verified), but sometimes it is not. Even when it is, many species of the higher host taxon have no insect consortia of their own, whereas a single or a few species harbor many herbivores. This situation is true for *Ephedra* in Mongolia: 10 species of the weevil tribe Oxyonychini (Coleoptera: Curculionidae) (Fig. 7.5d) known from this country (Korotyaev 1982 and unpublished data) live on *Ephedra sinica* Stapf and *E. przewalskii* Stapf, whereas no species of Oxyonychini is apparently associated with 6 (Grubov 1982) other Mongolian species of *Ephedra*. Similarly, 14 species of Oxyonychini (half the species with known hosts) are associated with *Ephedra major* Host (= *E. procera* Fisch. and Mey.) in the Western Palearctic (Colonnelli 2004), which is probably the host plant with one of the greatest numbers of weevils specialized on a single species in the Palearctic Region. Often, only a few, or no, herbivores can be found in localities or regions with a wide variety of potential hosts, while relatively poor habitats or countries can harbor greater insect biodiversity. For example, in Ul'yanovsk Province in the middle Volga area of Russia, 9 species of the weevil genus *Tychius* are found on 9 species of *Astragalus* (Fabaceae) (Isaev 2001), whereas in Mongolia, with

Fig. 7.5 (See figure on the following page.)

A. *Carabus lopatini* Morawitz (Coleoptera: Carabidae).
B. *Cimberis attelaboides* (F.) (Coleoptera: Nemonychidae).
C. *Parorobitis gibbus* Korotyaev, O'Brien and Konstantinov (Coleoptera: Curculionidae).
D. *Theodorinus* sp. (Coleoptera: Curculionidae).

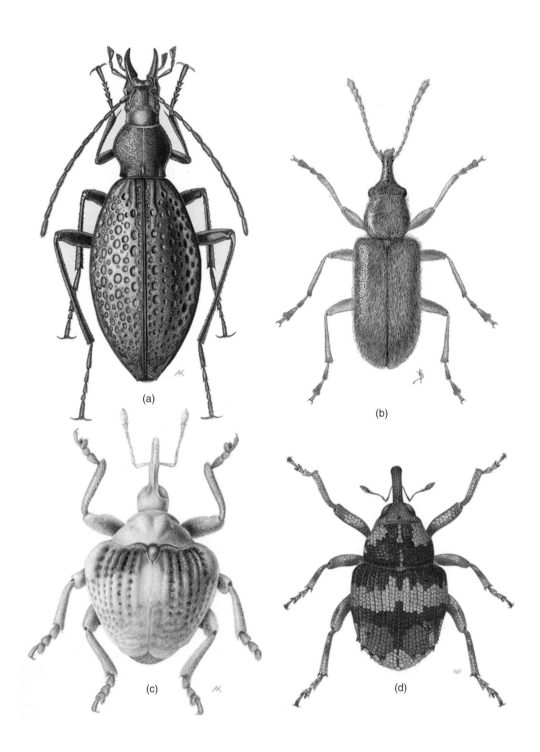

(a)

(b)

(c)

(d)

68 species (Sanchir 1982) of *Astragalus*, only 4 or 5 species of *Tychius* are known to be associated with this host genus.

The number of specialized and occasional arthropod feeders on a particular plant species can be quite high. For example, 175 species are reported on *Lepidium draba* L. (Brassicaceae), with the majority being insects (Cripps et al. 2006).

For the entire superfamily Curculionoidea, excluding scolytines, the plant species to beetle species ratio is 6:1 in the Caucasus and Mongolia. In continental Northeast Asia, in the northern taiga and tundra zones of Magadan Province and Chukchi Autonomous District, this ratio is about 12:1 (13:1 in the impoverished biota of the Kamchatka Peninsula and the Koryak Plateau north of it), and about 30:1 in the Arctic Wrangel Island (B. A. Korotyaev unpublished data). In local steppe areas, the ratio is about 2.5:1 (Korotyaev 2000 – isolated steppe site; Korotyaev, unpublished data – Taman' Peninsula).

The diversity of beetle herbivores is unevenly distributed among higher plant taxa. Grasses and sedges that dominate the vegetation over vast territories in the Palearctic usually possess poor insect consortia, consisting mostly of planthoppers (Auchenorrhyncha), although the genus *Carex* has the greatest number of insect herbivores among herbaceous plants (75; Emeljanov 1967). Most Palearctic insects with chewing mouthparts, such as weevils and leaf beetles, avoid monocots. Among the Palearctic Chrysomelidae, a few genera of flea beetles (e.g., *Chaetocnema* and *Psylliodes*) include a relatively large number of species that feed on monocots. Most other chrysomelids associated with monocots in the Palearctic belong to a few primitive subfamilies, such as the predominantly temperate, aquatic Donaciinae with 62 species in the former USSR (Lopatin et al. 2004). A few Palearctic species of the large, mostly tropical tribe Hispini are associated with grasses. In Belarus, 64 of 351 species of leaf beetles feed on Poaceae and Cyperaceae (Lopatin and Nesterova 2005). Two small genera of stem-mining cerambycids (*Theophilea* and *Calamobius*) and the two largest Palearctic genera with rhizophagous larvae (*Dorcadion* and *Eodorcadion*) are associated with Poaceae. Among weevils, most of the monocot feeders belong to specific family- and genus-group taxa with predominantly extratropical distributions (family Erirhinidae; subfamily Bagoinae, tribe Mononychini, and genera *Prisistus* and *Oprohinus* of the Ceutorhynchinae; and genus *Limnobaris* of the

Baridinae, Curculionidae) and tropical distributions (families Brachyceridae and Dryophthoridae; *Apsis albolineatus* (F.) of the Curculionidae: Myorhinini). Buprestids of the small but morphologically specialized genera *Cylindromorphus* and *Paracylindromorphus* are associated with grasses, sedges, and reed (*Phragmites australis* (Cav.) Trin. ex Steudel). Several species of *Aphanisticus* develop on *Juncus* (Juncaceae), and larvae of the highly specialized genus *Cyphosoma* (Chrysochroinae: Dicercini) (Fig. 7.2g) develop in tubers of *Bolboschoenus* (Ascherson) Palla (Cyperaceae). Orthopterans – 'hexapod horses' with powerful mandibles – are the only insects with chewing mouthparts that are abundant on grasses, although not all of them feed on monocots.

The majority of Palearctic phytophagous insects are associated with seed plants (Spermatopsida). Conifers (Pinophyta) have less diverse insect assemblages than do angiosperms (Magnoliophyta). For example, of 351 species of leaf beetles (Chrysomelidae) in Belarus, only 4 feed on conifers: the monophagous *Cryptocephalus pini* L. (Cryptocephalinae), oligophagous *Calomicrus pinicola* Duftschmidt (Galerucinae), and polyphagous *Cryptocephalus quadripustulatus* Gyllenhal (Cryptocephalinae) and *Luperus longicornis* F. (Galerucinae) (Lopatin and Nesterova 2005). Even at the northern border of taiga in Northeast Asia, only 8 of 130 species of Curculionoidea (excluding Scolytinae) are associated with conifers. Yet, a vast number of wood-borers are associated with conifers, primarily beetles of the families Anobiidae (including Ptininae), Bostrichidae, Buprestidae, Cerambycidae, and Scolytinae of the Curculionidae, all of which possess many species, even in northern taiga, and regularly cause damage to forests and plantations. Lepidoptera developing on foliage also include serious forest pests, and a few Symphyta (Hymenoptera) attack both strobiles and wood. In total, 202 species of strictly oligophagous insects are associated with 5 genera of conifers in the former USSR, according to Emeljanov (1967): *Pinus* (82 species), *Picea* (50), *Larix* (25), *Juniperus* (23), and *Abies* (22). *Ephedra* (Ephedraceae) is unique in having an entire fauna of the endemic weevil tribe Oxyonychini, with 20 genera and about 60 species. Most of the few xylophagous buprestids and longhorn beetles that develop on *Ephedra* also are specific to this genus. Some planthoppers and apparently less mobile Sternorrhyncha also are specialized on this plant. Small predatory coccinellids of the genus *Pharoscymnus* in southern Mongolia occur only on *Ephedra* (B. A. Korotyaev unpublished data), and *Pharoscymnus auricomus* Savoiskaya is associated mainly with these plants in sand deserts of Middle Asia (Savoiskaya 1984). Leaf beetles and weevils do not commonly feed on both conifers and angiosperms, but some large and widely distributed weevil genera (e.g., *Anthonomus, Hylobius,* and *Cossonus*) include species that develop on either conifers or angiosperms.

Ferns and mosses have relatively small numbers of phytophagous insects in the Palearctic. Among leaf beetles, fern feeding is known only in the far south of the region, in the Himalayas where many species of the genus *Manobia* (Alticini) use a variety of ferns. In the New World, fern-feeding flea beetles are assigned to a Caribbean genus (*Normaltica*), with one species in the Dominican Republic and one in Puerto Rico and the genus *Leptophysa*, about 18 species of which are distributed in Central and South America and in the Caribbean. *Leptophysa* species are remarkably similar to those of *Manobia*. Among the Buprestidae, larvae of the mainly Oriental genus *Endelus* (Agrilinae) feed on ferns. Some European *Otiorhynchus* weevils feed on ferns. Mosses are less populated by flea beetles in the northern Palearctic than in the south. The only known moss-living flea beetle in the northern Palearctic is *Mniophila muscorum* Koch (Fig. 7.4c). In the Himalayas, the mountains of Yunnan, and far-ther south in Asia, most of the moss-living flea beetles belong to the same genera as the leaf-litter flea beetles (e.g., *Benedictus, Clavicornaltica* (Fig. 7.4b), *Paraminota,* and *Paraminotella*) (Konstantinov and Duckett 2005), except for *Ivalia* and *Phaelota*, which live in mosses in southern India, but are not found in leaf litter or mosses in the Palearctic (Duckett et al. 2006, Konstantinov and Chamorro-Lacayo 2006). In the New World, the only moss-living flea beetles belong to the genus *Kiskeya* (Fig. 7.4d), with two species known from two neigh-boring mountain systems in the Dominican part of the Caribbean island Hispaniola (Konstantinov and Chamorro-Lacayo 2006). All moss-feeding flea beetles share a similar habitus. They are among the small-est flea beetles and have round bodies, relatively robust appendages, and somewhat clavate antennae. The Hol-arctic flea-beetle genus *Hippuriphila* is unique in its host choice. All four species of the genus feed on *Equisetum* (Equisetopsida) (Konstantinov and Vandenberg 1996). No other leaf beetle in the Palearctic feeds on plants of this taxon, but a few species of *Bagous* weevils and all four species of the Holarctic genus *Grypus* of erirhinid weevils also are associated with *Equisetum*.

It is unknown why some plant groups have diverse phytophagous assemblages while others do not. Some plant taxa that are diverse and significant in plant communities have a low diversity of insect consumers. Two plant genera have similar representation in the Mongolian vegetation but differ sharply in the diversity of their weevil consortia. Thirty species of *Allium* are found in Mongolia. Many of them dominate plant communities and are major sources of food for grazing animals, but only a single weevil of the specialized ceutorhynchine genus *Oprohinus* is associated with them. The wormwood genus *Artemisia* has 65 species in Mongolia (Leonova 1982) and dominates many types of plant communities. These plants have diverse phytophagous assemblages of more than 100 species of Auchenorrhyncha and Coleoptera (Buprestidae, Chrysomelidae, and Curculionidae). Some plants that dominate certain communities have species-rich phytophagous consortia, such as oaks (*Quercus*) among trees, *Artemisia* among herbs and semishrubs, and sedges (*Carex*) among herbs (Emeljanov 1967). Many highly specialized herbivores are associated with plants that have a high level of mechanical and chemical defenses, including many *Euphorbia*-eating *Aphthona* (Chrysomelidae) (Fig. 7.4a), *Perotis cuprata* (Klug) (Buprestidae), and *Oberea* (Cerambycidae); *Aphthona nonstriata* Goeze and *Mononychus punctumalbum* (Herbst) (Curculionidae) on *Iris pseudacorus* L.; *Nastus* and *Lixus* species (Curculionidae) on *Heracleum* species; and a few Buprestidae, *Lema decempunctata* (Gebler), several Alticini in the genera *Psylliodes* and *Epitrix*, two species in two genera of Cassidini (Chrysomelidae), and *Neoplatygaster venustus* (Faust) (Curculionidae: Ceutorhynchinae) on various Solanaceae. Some chrysomelids and weevils specialize almost exclusively on plants that contain toxic secondary compounds, combining unrelated taxa in their host ranges. For example, *Psylliodes* species feed on Brassicaceae, Cannabaceae, Cardueae, and Poaceae, as well as on Solanaceae; and of the four Far-Eastern Russian *Lema* species (Coleoptera: Chrysomelidae), two develop on monocots, one on Solanaceae, and one on Cardueae (Medvedev 1992). An analysis of host specialization of 321 species of flea beetles (Chrysomelidae) in the European part of the former USSR and the Caucasus shows that plant families most popular as hosts for flea beetles are among the most speciose in the flora. The most species-rich families of plants on the Russian Plain are the Asteraceae, Poaceae, Rosaceae,

Fabaceae, Cyperaceae, Lamiaceae, Scrophulariaceae, and Brassicaceae (Alexeev and Gubanov 1980). The plant families most popular for flea beetles are the Brassicaceae, Lamiaceae, Asteraceae, Boraginaceae, Scrophulariaceae, Euphorbiaceae, and Poaceae; for the buprestid tribe Acmaeoderini, they are the Fabaceae, Fagaceae, Anacardiaceae, Apiaceae, Rosaceae, Asteraceae, and Moraceae (Volkovitsh and Lobanov 1997); and for the weevil subfamily Ceutorhynchinae (in the entire Palearctic), they are the Brassicaceae, Ephedraceae, Lamiaceae, Boraginaceae, Asteraceae, Fagaceae, and Liliaceae.

Among plant families, the Brassicaceae are common and numerous in the Palearctic and Nearctic, but their weevil consortia in the two regions are quite different. Nearly half of the Palearctic fauna of the weevil subfamily Baridinae – about 60 species in 6 endemic genera – is associated with crucifers, whereas crucifer-eating Baridinae are lacking in the Nearctic. Another weevil subfamily, the Ceutorhynchinae, is much less species rich in the Nearctic than in the Palearctic, with about 80 species of a single genus Ceutorhynchus, as opposed to more than 220 species of Ceutorhynchus and 6 species of 3 other small genera in the Palearctic. This subfamily includes members of at least 11 Holarctic species groups, totaling about 25–30% of the Nearctic fauna. The Brassicaceae are unique in having at least 450 species of specialized beetles in the Palearctic (Cerambycidae, Chrysomelidae, Urodontidae, and Curculionidae: Baridinae, Ceutorhynchinae, and Lixinae) that feed on its many species. About 60 species of flea beetles feed on the Brassicaceae in France (Doguet 1994).

The plant family Asteraceae has a diverse fauna of phytophagous beetles. About 65 species of flea beetles, for example, feed on various asteraceans in France (Doguet 1994), and 67 species of the leaf beetles feed on Asteraceae in Belarus (Lopatin and Nesterova 2005). Asteraceae-feeding Baridinae (Curculionidae) are present in both the Palearctic and Nearctic, but they belong to different genera or even tribes, as do the Baridinae living on hygrophilous monocots.

Oak consortia probably are richest in the Palearctic (Emeljanov 1967). Some weevil genera (e.g., Curculio) of these consortia are equally represented in the Nearctic, but some, such as several genera of the tribe Rhamphini (Curculioninae) and genus Coeliodes (Ceutorhynchinae), are lacking in the Nearctic.

Some insects demonstrate major differences in their host choice in the Palearctic and other biogeographical regions. Flea beetles in the Palearctic, for example, feed mostly on herbs and grasses, with just a few species on woody plants. In the Oriental Region, flea beetles feed mostly on bushes and trees. Thus, in the Palearctic, they occur mostly in open spaces such as meadows and swamps, whereas in the Oriental Region, they are rich in species in forest communities.

BOUNDARIES AND INSECT BIODIVERSITY

The boundaries between faunal complexes of any rank (e.g., biogeographical realms or assemblages of insects in two adjacent localities) and the distributional limits for particular species typically run across physical gradients. Although the literature on this matter is voluminous (Darlington 1963), we add several new Palearctic examples.

The boundary between the western and eastern parts of the Palearctic for most groups of plants and animals is situated between the Altai Mountains and Lake Baikal; for some taxa, it coincides approximately with the Yenisei River. Two typical Arctic Chrysolina species (Chrysomelidae), Ch. cavigera Sahlberg and Ch. subsulcata Mannerheim, have not been found west of the Yenisei River (Chernov et al. 1994). At the source of the Yenisei, at Kyzyl City in Tuva, where the width of the river barely exceeds 100 m, the fauna differs considerably between the left (western) bank and the right (eastern; in Tuva, northern) bank. Chrysolina jakovlevi Weise occurs mostly north of Tuva in southern Krasnoyarsk Territory and Khakasia. In Tuva, it inhabits only stony slopes in a narrow strip of desert steppe on the right bank of the Yenisei (Medvedev and Korotyaev 1976). Chrysolina tuvensis L. Medvedev is known only from the desert steppe on the right bank near Kyzyl, and Ch. sajanica Jacobson occurs in the dry steppe on the southernmost West Sayans piedmont. The wingless Chrysolina urjanchaica Jacobson, an endemic of the right-bank steppe and the interfluvial area of the Ka-Khem and Bii-Khem rivers (Korotyaev 2001b), is substituted on the left bank by a similar and closely related species, Ch. convexicollis Jacobson, distributed throughout the rest of Tuva and in adjacent northwestern Mongolia. A wingless weevil, Eremochorus zaslavskii Korotyaev, endemic to the right bank, substitutes there for the widely distributed E. sinuatocollis (Faust). These examples demonstrate the isolating role that a relatively narrow river can play. Rivers and their valleys also play a role as distribution pathways. Land insects with both relatively limited and large ranges tend to penetrate farther north or south along valleys of major rivers in the Palearctic.

Another boundary exists along the Ural Mountains and lower course of the Volga River. Isaev (1994) found several species of beetles with eastern distributions reaching the right bank of the Volga in Ul'yanovsk Province, including a steppe weevil, *Ceutorhynchus potanini* Korotyaev, distributed in Siberia from the Western Sayan Mountains to central Yakutia, as well as in Mongolia.

Mountain systems represent obvious borders between faunas. The Carpathians limit distribution to the east for a number of beetle species including carabids (*Carabus auronitens* F. and *C. variolosus* F.) and a chafer (*Hoplia praticola* Duftschmidt). A number of southeastern European species do not reach farther west than the Carpathians (Arnoldi 1958).

Climate might explain why some distributional limits do not coincide with obvious physical boundaries. The southern boundaries of the ranges of several Euro-Siberian weevils, such as *Phyllobius thalassinus* Gyllenhal (Entiminae), *Ceutorhynchus pervicax* Weise, *C. cochleariae* Gyllenhal, and *Trichosirocalus barnevillei* (Grenier) (Ceutorhynchinae), run along the southern slope of the West Sayan Range, but not in the northern foothills of this range, presently impassable for these weevils. The distribution of these species to the south is apparently not limited by the high mountain ranges, but rather by the aridity. *Trichosirocalus barnevillei* is present in the Yenisei flood land below and south of the forest margin, which is a typical distribution for woodland species outside forest massifs.

Two common Euro-Siberian weevils have the southern margins of their ranges on northern slopes of different mountain chains in Tuva. *Hemitrichapion reflexum* (Gyllenhal) is found only on the southernmost ridge of the West Sayan, the Uyukskii Range, whereas *Brachysomus echinatus* (Bonsdorff) is found only on the northern slope of the Eastern Tannu-Ola Range. One is a steppe species, the other is a forest species. Their distributions in Tuva fit the rule formulated by the well-known plant geographer O. E. Agakhaniantz (1987): migrants occupy slopes facing the direction of the country from which they came. Yet the preceding examples of species with the southern boundaries of their ranges

running along the southern slope of West Sayan do not follow this rule.

LOCAL BIODIVERSITY

High local biodiversity is illustrated by five closely related endemic species of the weevil genus *Ptochus* in a small area of Daghestan (Caucasus) (Ismailova 2006). Three species are located along the river, about 30 km apart, and two on the opposite sides of the mountains facing the river. All species are flightless, similar externally, and feed on the same species of *Artemisia*. A similar pattern of local biodiversity is known for some Middle Asian *Prosodes* (Coleoptera: Tenebrionidae). The high level of biodiversity often is maintained by reproductive isolation associated with complicated structure of the genitalia (Medvedev 1990).

Local biodiversity of some insect groups is increased by separation of periods of insect activity, either daily or seasonal. A well-known example is the scarabaeid genus *Chioneosoma* in southern Kazakhstan. Six sympatric species of the genus, in addition to having different habitat associations, have sequential flight periods, with most adults flying in different plant layers. Of the two species that co-occur in the Muyunkum Sands and exhibit no clear differences in flight habits, one flies before sunset and the other half an hour after the flight of the first species ceases (Nikolajev 1988). These differences enhance reproductive isolation of species with similar larval habits. Seasonal differentiation of phytophagous insect assemblages follows changes in vegetational aspects: ephemeral plants possess specific insect consortia often formed by particular insect taxa, occasionally using a specific habit such as gall inducement (Kaplin 1981).

Supposedly nonspecialized feeders also can exhibit high species diversity, such as predatory carabids, coprophagous scarabs, and detritophagous tenebrionids, of which many include genera with more than 500 species (e.g., *Carabus*, Fig. 7.5a). Local biodiversity of parasitoids can be higher than that of herbivores. For example, about 80 species of ichneumonids were collected on one excursion to the Kamchatka Peninsula (D. R. Kasparyan personal communication), whereas the entire weevil fauna of the Peninsula comprises about 50 species (Korotyaev 1976). Yet in farther southern regions, phytophagous insects are locally rich. On the plains of the Northwestern Caucasus, for example, one excursion in May yielded 39 species of the weevil subfamily Ceutorhynchinae. The entire super-family Curculionoidea is represented by 551 species in Berlin (Winkelmann 1991). Wanat (1999) reported 480 species of Curculionoidea (excluding Platypodidae and Scolytinae) in the Białowieża Primeval Forest (eastern Poland), probably the largest broadleaf forest in Europe.

INSECT BIODIVERSITY AND HABITATS

Plant communities (habitats) in the Palearctic are classified as zonal, azonal, or extrazonal (Chernov 1975, Walter and Breckle 1985). Zonal communities are situated on 'the flat elevated areas with deep soil which are neither too porous to water, like sand, nor retain too much water, like clay. ... There must be no influence from ground water' (Walter and Breckle 1985). The particular regional climate has its full effect on such areas, called euclimatopes or 'plakor' in Russian. Azonal vegetation appears when the groundwater table is so high that the whole area is covered with bogs or when vegetation is on sand or alluvial soil. Extrazonal vegetation is zonal vegetation outside its climatic area, for example, steppe meadows in forest zones.

The distribution of insects among these types of communities has some regularity. Zonal communities are not particularly rich in the northern Palearctic, but increase in richness southward (Chernov 1966). They usually are formed of species whose ranges are associated with a particular zone. Ranges tend to be relatively large in northern zonal communities and small in southern communities. Most southern zones (e.g., steppe) have some endemics (e.g., the chrysomelid *Aphthona sarmatica* Ogloblin), whereas azonal communities in the North have a larger number of species with wider ranges.

Insects with larger ranges tend to occur in nonspecific habitats, either intra- or extrazonal. In Mongolia, for example, leaf beetles with transpalearctic ranges commonly occur in ruderal and agricultural habitats, whereas species with smaller, Central Asian ranges occur in deserts or saline habitats specific to Mongolia (Medvedev 1982), where many widespread species also occur in high populations (Korotyaev et al. 1983). Ranges of many taiga insects also can be quite large, covering the entire taiga.

The azonal riparian landscape has a high proportion of insect biodiversity in most of the natural zones, including the tundra (Chernov 1966), taiga (Ivliev

et al. 1968), northern part of the Stenopean forests (Egorov et al. 1996), steppes (Medvedev and Korotyaev 1976), and deserts (Korotyaev et al. 1983), with the proportion of the riparian species in the total fauna increasing northward within the taiga zone.

Azonal, mainly riparian, communities also include species endemic to certain zones. The weevil genus *Dorytomus*, for example, is one of the largest genera of riparian beetle complexes, with 63 species in the Palearctic. Most of its members occur in floodland habitats. In the oceanic sectors, they are distributed across several zones, so that in the Russian Far East, the fauna of the taiga zone is largely the same as that in the nemoral Stenopean forests. Yet in the desert and lower mountain zones of Middle Asia, most *Dorytomus* species are endemic; for example, all four species in southern Tajikistan are endemic to Middle Asia (Nasreddinov 1975). The entire riparian landscape in the desert zone, with its specific type of forests called 'tugai', has characteristic insect complexes that include water beetles and bugs, many amphibionts, xylobionts, and herbivorous insects. Leaf-beetle communities of tugai are highly specific (Lopatin 1977). Most genera endemic to Middle Asia and southern Kazakhstan occur there (e.g., *Atomyria*, *Jaxartiolus*, and *Parnops*). Lopatin (1977) suggested that the tugai leaf-beetle fauna is closely related to the Mediterranean fauna, differing from it in having a number of northern elements (e.g., *Donacia*, *Gastrophysa*, and *Phaedon*), which migrated to Middle Asia around the Pleistocene.

Many insect communities are hidden in the substrate and less conspicuous than the large and beautiful butterflies, dragonflies, acridids, and bees, or the nasty horse flies and mosquitoes. They, nonetheless, play diverse roles, such as preying on injurious and beneficial animals and decomposing dead organisms. Beetles, with their hard bodies, dominate most types of these hidden assemblages except those where rapid larval development in semi-liquid substrates (e.g., dung and carrion) gives an advantage to flies, although there, too, many beetles hunt fly larvae. Dozens of insect species constitute different kinds of coprobiont, necrobiont, and various dendrobiont (mostly under-bark) communities. Several large beetle families are specialized for these sorts of habitats. Most Silphidae (carrion beetles) consume large and medium-sized carrion; Catopinae (Leiodidae) feed on small dead bodies; Trogidae, probably *Necrobia* (Cleridae), and *Dermestes* (Dermestidae) feed on dry skin remnants of vertebrates. Several beetle families are specialized fungivores, including most of the Ciidae, Erotylidae, Endomychidae, many Tenebrionidae (e.g., *Bolitophagus reticulatus* L.), the subfamily Dorcatominae of the Ptinidae, and some Nitidulidae (e.g., *Cyllodes ater* (Herbst) and *Pocadius* species). Beetle consortia of a particular tree fungus in the southern taiga of the Urals and Western Siberia comprise dozens of species. For example, 54 species in 16 families occur on *Daedaleopsis confragosa* (Bolton: Fr.) Schrot (Basidiomycetes) (Krasutskii 2007). In the nemoral zone, especially in the Far East, fungal beetle consortia are more diversified and include representatives of exotic Oriental genera. On Kunashir Island, large erotylids of several species are visible for 20 m in thin forest, where they sit on the stroma of large tree fungi. Leiodidae are far more speciose and abundant in the mild and humid oceanic climate, especially in autumn. *Agathidium laevigatum* Erichson is present even on Bering Island (Lafer 1989).

Diversified insect communities exist under bark and in forest leaf litter. These communitites are dominated by beetles, but several families of flies also contribute to the overall diversity, as do many parasitic Hymenoptera and several families of bugs, the strongly flattened Aradidae being the most characteristic. A considerable part of these communities consists of scolytines, their predators and parasites, and consumers of the fungi and debris in their tunnels. One of the largest genera of beetles in these complexes is *Epuraea* (Nitidulidae), with more than 100 species in the former USSR (Kirejtshuk 1992).

Coprophagous assemblages are particularly diverse, depending on season, landscape, and type and age of the excrement. These assemblages include many well-known scarabs such as *Copris, Scarabaeus*, and *Sisyphus*, as well as countless species of the largest beetle genera *Aphodius* and *Onthophagus*. Up to 13 species of *Aphodius*, plus two species of *Onthophagus*, can be found in just a few neighboring deposits of dung in Leningrad Province, northwestern Russia (O. N. Kabakov, personal communication). In addition to the Scarabaeidae, some Hydrophilidae are common in dung. Numerous coprophagous insects are hunted by fly larvae and predatory beetles of the families Histeridae (Fig. 7.4f) and Staphylinidae.

Animal nests also possess variably specific insect communities where many flies, beetles, bugs, fleas, and lice predominate. Zhantiev (1976) gives extensive data on the occurrence of dermestids in insect and mammal nests, as well as in tree hollows. Nests and colonies of social insects host species-rich communities;

200 species of beetles, for example, occur in bumblebee nests. Myrmecophilous insects include many species with characteristic structural adaptations to life in ant nests, including bare, glabrous, uniformly brown bodies with tufts of setae for tactile or chemical communication with the hosts. The largest genus is probably *Thorictus* of the Dermestidae, which, because of its aberrant habitus, was long kept in a separate family. The termitophilous *Eremoxenus chan* Semenov-Tian-Shansky of Middle Asia and the myrmecophilous *Amorphocephalus coronatus* Germar of the Mediterranean region are the few Palearctic members of a group of commensal Brentidae.

Even in urban environments, insects form species-rich communities. Occasionally, some predatory and saprophagous species irritate humans. In the Academy Archive in St. Petersburg in the mid-1990s, increased activity of larval dermestids (*Attagenus smirnovi* Zhantiev) caused anxiety among the staff. Inspection of the volumes of the Academy Sessions protocols with M. V. Lomonosov's autographs revealed an invasion by *Lasioderma* (Ptinidae), whose remains were then eaten by larvae of the harmless *Attagenus smirnovi*. During the reconstruction of the Stroganov Palace at Nevski Prospect in St. Petersburg in the late 1980s, a dozen beetle species were found in the 40-cm thick larch beams in the attic, including dermestids and wood-borers. Forty-two species of dermestids were listed by Zhantiev (1976) as harmful to stored products in the former USSR. Although some introduced species can be harmful, the majority of the most destructive species of dermestids in every natural zone belong to the local fauna (Zhantiev 1976).

The urban fauna of phytophagous insects is also rich and increasing; 799 species of insects, for example, are known from the city of Moscow (Russia). This fauna is dominated by exposed and partly hidden phyllophagous forms (328 species), followed by exposed and partly hidden Rhynchota (Auchenorrhyncha, Sternorrhyncha, and Heteroptera; 143 species), leaf miners (124), gall inducers (122), and wood-borers (110, mainly scolytines) (Belov 2007).

A specialized insect assemblage, including beetles and flies of three families, occurs along the seashore. The most widely distributed coleopterous group in this assemblage is the tenebrionid tribe Phaleriini, which lives on sandy beaches. A characteristic feature of these fast, small, and medium-sized beetles is their sand-colored integument, with dorsomedial infuscation. Small hydrophilids of the genus *Cercyon* and various Staphylinidae and Histeridae occur under decaying algae. On the Pacific shore, several similar species of the endemic genus *Lyrosoma* of Silphidae are abundant at the shoreline, especially on the Kurile and Komandorski islands. A member of the weevil sub-family Molytinae, *Sthereus ptinoides* (Germar), with a boreal amphipacific distribution, lives on driftwood. In southern Japan, the peculiar weevil genus *Otibazo*, with uncertain affinities, is confined to seashores. The characteristic plant genus *Cakile* (Brassicaceae) occupies the outermost sandy strip of beach where the salt spray reaches. It has a complex of two weevil species in the northern Atlantic basin: *Ceutorhynchus cakilis* Hansen in Europe and *C. hamiltoni* Dietz in North America. These two weevils belong to different species groups, illustrating a tendency of the Ceutorhynchinae (and other weevil taxa) to exploit marginal habitats with pioneer plant communities. In Europe, several nonspecialized crucifer feeders of the Curculionidae occur on *Cakile maritima* Scopoli and *C. euxina* Pobed. A tribe of supralittoral weevils, the Aphelini, is distributed along the Pacific Coast from northern California to the temperate Far East, as well as in Australia.

In Japan and the southernmost portion of the Russian Far East, *Isonycholips gotoi* Chûjo and Voss, another representative of the Aphelini, lives among grasses on dry coastal sand dunes. An additional coastal insect assemblage exists here, including large weevils of the tribe Tanymecini and *Craspedonotus tibialis* Schaum of the fossorial ground-beetle tribe Broscini. In the Atlantic sector, the genus *Onycholips* (two species in northwestern Africa), the fossorial broad-nosed weevil *Philopedon plagiatus* (Schaller) in the tribe Cneorhinini, and several species of the tribe Brachyderini live on sand dunes along the Baltic coast. *Philopedon plagiatus* is parthenogenetic and apparently introduced to inland seashores. Carabids of the genera *Broscus* and *Scarites* and tiger beetles (*Cicindela*) are common on sandy beaches. Coastal rocks in the northern Pacific have a highly specific beetle, *Aegialites stejnegeri* Linell, of the monotypic subfamily Aegialitinae (Salpingidae), which lives in cracks and seabird nests (Nikitsky 1992).

INSECT BIODIVERSITY AND PALEARCTIC MOUNTAINS

The Palearctic has a number of mountain systems, including the Alps, Carpathians, Caucasus, Pamirs, Tien Shan, Urals (dividing the continents of Europe and Asia), Altai, Sayan, Tibet, and northern Himalaya. The majority of Palearctic Mountains belonging to the Alp–Himalayan mountain 'belt' are relatively young, except the Urals. The mountain systems are situated in different parts of the Palearctic, with different climatic and other geographic conditions. Their patterns of insect biodiversity, however, share a number of features.

Insect biodiversity in the Palearctic mountains changes with climate along altitudinal gradients (Walter and Breckle 1985). Generally, from the piedmont to the mountain tops, the following altitudinal belts are recognized: colline–montane (lower and upper), alpine (lower and upper), and nival (Walter and Breckle 1985). Although these altitudinal belts do not correspond precisely to the zones from south to north, insect biodiversity usually decreases from the bottom to the top of the mountains, as it does from south to north in the Palearctic. The degree of endemism, however, increases from bottom to top. An important reason for high biodiversity of mountain-insect faunas is that many groups in the plains find refugia in corresponding mountain belts when climatic conditions change and zones fluctuate. For example, many species widely distributed in the Arctic and Boreal regions occur in the upper mountain belts in southern Europe; this kind of distribution is called arcto-alpine or boreo-montane, depending on the specific features of the range. During the glacial age, these insects moved southward following the glaciers, and found refugia in the high mountains when the climate became warmer and the ice shield retreated. Many mesophilous groups find appropriate environments in mountains during aridization in adjacent plains, accounting for the considerable number of so-called paleoendemics usually represented by one or a few species in remote mountain systems. Examples include carabid beetles of the genus *Broscosoma* scattered along the entire Alp–Himalayan mountain belt (Kryzhanovsky 2002), or the monotypical cerambycid genus *Morimonella* recently described from the Caucasus. The capacity of the mountain systems to accumulate great numbers of endemic species is facilitated by the broad diversity of habitats plus the apparently effective distributional barriers. Wingless species are among the most speciose mountain taxa; they are usually classified as neoendemics. For example, the Caucasian fauna of the largest weevil genus, *Otiorhynchus s.l.*, is provisionally estimated at 250 species (Savitskii and Davidian 2007); 50 small, blind, wingless, forest litter and, partly, endogean species of the carabid tribe Trechini are recorded from the Caucasus, mostly from its western part (Belousov 1998). On the other hand, meridionally oriented mountain chains allow the possibility of deep penetration of some typically Palearctic insect groups into the Oriental Region and vice versa.

Alpine insect communities of Middle Asia are relatively poor, but highly endemic. For example, of approximately 800 leaf beetles in Middle Asia, 175 occur in the alpine belt, of which 150 are endemics (Lopatin 1996). Among them are *Oreomela*, with more than 80 species in the alpine regions of the Tien Shan, Himalayas, Altai, and southwestern ridges of China between 2400 and 4300 m, and *Xenomela*, all 11 species of which are known from the Tien Shan between 1300 and 3000 m (Lopatin and Nesterova 2004). This unusually high level of endemism might be explained by the specific alpine environment and high degree of isolation. Alpine biodiversity further increases because many species with larger ranges are represented by specialized morphs in the mountains that might not have species status, but differ genetically. However, insect groups that are most abundant and species rich elsewhere are not always species rich in the alpine belt. For example, the most speciose leaf-beetle groups in the Palearctic are the Galerucinae *sensu lato*, followed by the Chrysomelinae and Cryptocephalinae (Fig. 7.4e). In the alpine belt of Middle Asia and northwestern China, the most speciose are the Chrysomelinae followed by the Galerucinae, Eumolpinae, and Cryptocephalinae (Lopatin 1996).

In Yakutia, thousands of male hover flies (Syrphidae) sometimes hover over mountain peaks, while females feed on flowers at the bottoms and on the slopes of the mountains. In the late afternoon, with the flow of warm air, females fly to the mountaintops where copulation occurs. A similar kind of behavior ('hilltopping') is known for butterflies and flies in the families Sarcophagidae, Tabanidae, and Tachinidae (Barkalov and Nielsen 2007).

Mountain forest belts usually have higher species biodiversity than does the alpine belt. Mid-altitude forests in the Caucasus and Transcaucasia include a number of endemics, many of which are flightless. Examples include *Psylliodes valida* Weise and *Aphthona testaceicornis* Weise (Chrysomelidae) in the forests of the Northwestern Caucasus and *Altica breviuscula* Weise (Chrysomelidae) in the narrow strip of mountain forest

in Talysh (Azerbaijan). Specialized herbivores (e.g., weevils) are most species rich in the piedmont forest belt in Abkhazia (Caucasus) (Zarkua 1977).

The high-altitude fauna of the Palearctic is also unique, compared with the neighboring Oriental Region. Palearctic montane flea-beetle communities (in the Caucasus, Crimea, and Middle and Central Asia) consist of species from rich, cosmopolitan genera such as *Longitarsus* and *Psylliodes*. In the Oriental Region (southern Himalayas, Western Ghats), however, species occurring at high altitudes belong to genera with limited ranges, and many are confined to the Himalayas or do not occur outside the Oriental Region.

A characteristic feature of the Palearctic highland fauna are weevils of the subfamily Ceutorhynchinae at the upper margin of insect distribution. Several species of *Ceutorhynchus*, *Neophytobius*, *Scleropterus*, and others reach highest altitudes in the upper mountain zone in the Altai, Sayan Mountains, Tien Shan, Sredinnyi Kamchatka Range, Tibet, and Himalayas. Another characteristic feature of the Palearctic alpine fauna of weevils is the concentration of bisexual forms of species that reproduce parthenogenetically at lower elevations (Korotyaev 1992).

The alpine belt in the largest part of the Continental sector lacks the multicolored, tallgrass meadows typical of the European mountains, or if they are present, they are poorly developed. Typical of the Sayan Mountains and ranges to the south and southwest are mountain tundra, shortgrass subalpine meadows with large orange–yellow flowers of *Trollius asiaticus* L. (Ranunculaceae), alpine steppe, and an endemic alpine type of vegetation – dense, mesophilous, meadow-like grasslands dominated by *Cobresia* species (Cyperaceae), from which no specific weevils or leaf beetles are known. Edelweisses (*Leontopodium*, Asteraceae) are common in the West Sayan and other mountains of Tuva, descending in places to the mid-forest belt and occurring in pastures. Intense collecting in Tuva on edelweisses by B. A. Korotyaev in 1969–1972 revealed no herbivores, although Bajtenov (1977) described the weevil *Pseudostyphlus leontopodi* from Altai. The alpine weevil fauna of this mountain country is poor but characteristic, formed largely of representatives of several genera of Apionidae, *Notaris* and *Tournotaris* of the Erirhinidae, *Lepyrus*, several genera of Hyperinae, *Dactylotus globosus* (Gebler), a few *Sitona* species, and the Ceutorhynchinae (Curculionidae). Endemic to this area are many wingless Carabidae, including

Carabus, Nebria, and *Trechus* in the highlands; the montane subgenus *Aeneobyrrhus* of *Byrrhus,* with five species, and one of the three species of the Holarctic genus *Byrrhobolus* (Byrrhidae); and several oligotypic genera and considerable species of Chrysomelidae. The predominantly montane butterfly genus *Parnassius* (Papilionidae) is represented by several species. The multizonal *Aporia crataegi* L. (Pieridae) often occurs in great numbers. In the tangle of mountain ridges of West Altai, at the boundary of Kazakhstan and Russia, a heterogeneous butterfly fauna with 176 species exists, including several pairs of closely related Western and Eastern Palearctic species, as well as representatives of the Middle Asian fauna (Lukhtanov et al. 2007).

Wide variation in altitude is characteristic of the steppes of Mongolia, adjacent Tuva, and the Russian Altai. Steppes occupy plains and bottoms of variably sized depressions, but also south-facing mountain slopes in the forest zone and the highest mountain reaches, usually with southern exposure, adjacent to and alternating with other subalpine and alpine habitats, such as isolated *Larix* stands, subalpine meadows, and mountain tundra. The latter type of steppe, with its characteristic flora, is referred to by botanists as 'alpine steppe'. Insects of alpine steppe include a few characteristic species, along with several common mountain-steppe insects. Among beetles, this kind of distribution is found in *Stephanocleonus* (Curculionidae: Lixinae), *Crosita,* and *Chrysolina* (Chrysomelidae: Chrysomelinae).

Mountain steppe consists of several altitudinal types that are less clearly separated than the alpine steppe from the rest of the steppes. In the most elevated southwestern part of Tuva, adjacent to the Altai, a few subendemic species of weevils and chrysomelids, mostly of the same genera that constitute the bulk of the alpine steppe fauna, occur only in the highest areas of the steppe, usually on stony slopes, but not in the alpine steppe.

A diverse and characteristic fauna is associated with mountain areas that have climates intermediate between that of steppe and desert, and vegetation with shrubs (mostly *Caragana* and *Atraphaxis*) and perennial wormwoods (Medvedev 1990). One of the largest weevil genera subendemic to Central Asia, *Alatavia* (Entiminae) with 13 species, is distributed there, with only three species in central and eastern Mongolia occurring in the plains, which in this area are about 1000 m or more above sea level.

Many insects in the mountains developed common features viewed as adaptations to high-altitude life. A high percentage of alpine insects are flightless (e.g., more than 80% of chrysomelid beetles in the genus *Oreomela*). Mountain beetles, such as *Psylliodes valida* Weise (Chrysomelidae) and *Geotrupes inermis* Ménétriés (Geotrupidae), have a more swollen prothorax than do their lowland relatives. Many alpine beetles have a high level of melanization of the integument, similar to that found in Arctic insects. Their elytra, often covered with ridges, are convex, forming a subelytral cavity that functions as a temperature and humidity buffer for the poorly sclerotized abdominal tergites. Most alpine insects live close to the ground or under rocks and plants, even if their closest relatives occupy other habitats (Lopatin 1971). Numerous alpine leaf beetles are viviparous or oviviviparous (*Oreomela* species; Lopatin 1996), as are some Arctic leaf beetles. Because of the short warm season, the life cycle of some insects exceeds 2 years, and overwintering can occur at various life stages (Lopatin 1996).

INSECT BIODIVERSITY IN MAJOR BIOGEOGRAPHICAL DIVISIONS OF THE PALEARCTIC

Various ideas exist on the subdivision of the Palearctic, beginning in 1876 with that of Wallace (Semenov-Tian-Shansky 1936, Lopatin 1989). We follow the subdivision proposed by Emeljanov (1974), which is based on climatic and other physical geographic conditions most closely reflected by vegetational cover and which largely follows subdivisions proposed by botanists (Lavrenko 1950). Emeljanov (1974) recognized the following eight regions: Circumpolar, Euro-Siberian taiga (boreal), European and Stenopean nemoral, Hesperian (Mediterranean and Macaronesian) and Orthrian evergreen forest (subtropical), Scythian steppe, and Sethian (Saharo-Gobian) desert regions.

Arctic (Circumpolar Tundra) Region

The northernmost part of the Palearctic Region is the Arctic (Fig. 7.1a: I, b). It includes the coldest regions of the Palearctic and its southern border corresponds approximately with the 12°C isotherm of the warmest month (July) (Chernov 2002). In European Russia, the southern border runs along the southern border of the tundra and forest–tundra (Lopatin 1989). Here, we consider the Arctic in the broad sense, including the Hypoarctic (= Subarctic) as its southern subzone.

Altogether, about 3 300 species of insects live in the Arctic (including the American Arctic), representing 0.6% of total insect diversity. Relative species richness of insects in the Arctic is about 15%, compared with 50% for the World (Chernov 2002). Of 25 insect orders, 16 occur in the Arctic. Of the 1800 genera of carabid beetles, 22 occur in the Arctic (Chernov et al. 2000, 2001). Beetles constitute 13% of the Arctic insect fauna, and flies 60% (about 90% in more northern regions) (Chernov 1995). The most common beetles in the Arctic are the Carabidae, Staphylinidae, and Chrysomelidae. The Curculionidae are less common but are represented by a relatively large number of species. Of the remaining large families of Coleoptera, in addition to the entirely aquatic Dytiscidae, the Coccinellidae and Elateridae are most conspicuous in the southern part of the tundra, but they are represented by only a few species. The biodiversity of phytophagous insects decreases more strongly than that of predatory insects from south to north. For example, even an outdated list of Carabidae of Magadan Province and Chukchi Autonomous District (Budarin 1985) includes 161 species, whereas an updated list of Apionidae, Erirhinidae, and Curculionidae of the same area has slightly more than 100 names (B. A. Korotyaev, unpublished data). In the Northwestern Caucasus, with its wide range of mountain belts, the Carabidae (576 species; Zamotajlov 1992) are slightly less species rich than the fauna of listed weevil families (B. A. Korotyaev, unpublished data). Predatory beetles constitute about 70% of the total beetle fauna in the Arctic, whereas in temperate areas they make up about 25% (Chernov et al. 1994).

A characteristic feature of the Asian Arctic is the wide distribution of steppe-plant communities of several types, including the cryophytic steppe and the tundra-steppe, which are especially developed in Northeast Asia. Many insects occur in these Arctic steppes. Some are endemic, many are distributed outside the tundra zone in the southern Siberian and Mongolian steppes and in the mountains of Eastern Middle Asia, and some also occur in dry areas of North America (Berman et al. 2002).

Insect communities of Wrangel Island, north of the Chukchi Peninsula, described in detail by Khruleva (1987), are dominated by Diptera (76 species),

Coleoptera (67), Lepidoptera (54), and Hymenoptera (at least 46). Aquatic insects are well represented by the Ephemeroptera, Plecoptera, and Trichoptera, and the aquatic Diptera are dominated by the Tipulidae and Chironomidae. Among other groups, relatively large forms with well-developed flight predominate. The Lepidoptera include two species of Pieridae (*Colias*), two of Lycaenidae (two genera), seven of Nymphalidae (all *Boloria*), five of Satyridae (three *Erebia* and two *Oeneis*), two of Geometridae, two of Lymantriidae, ten of Noctuidae (six genera), seven of Arctiidae (six genera), and one of Pterophoridae. Of the Hymenoptera, the entire superfamily Apoidea is represented by three species of *Bombus* only (Khruleva 1987, O. A. Khruleva, personal communication).

The absence in the Arctic of a few taxa that are abundant and diverse from the southern boundary of the Palearctic to the taiga zone is noteworthy. These taxa include the Orthoptera, beetles of the family Tenebrionidae, ants (represented by a single species in the riparian shrub tundra; Chernov 1966), and predatory Hymenoptera (e.g., Sphecoidea and Vespoidea).

Forest regions

Forests occupy most of the Palearctic. They are classified into five regions: the taiga, stretching across Eurasia in the boreal climatic zone; the nemoral European and Stenopean regions in the temperate zone; and the subtropical Hesperian and Orthrian regions. The contiguous regions of the different climatic zones are more similar to each other, in many respects, than to their zonal counterparts. Along the same lines, the affinities of the insect faunas of the Hesperian and Orthrian regions are overshadowed by the similarities they have with their respective neighboring regions. As a result, biogeographic attribution of some large areas often becomes problematic. Anatolia, for example, is classified by Emeljanov (1974) as the zone intermediate between the Mediterranean Subregion of the Hesperian Region and the Saharo-Gobian Desert Region, but it is often considered an area with predominantly steppe-type vegetation that is attributed to the boreal biota by Kryzhanovsky (2002). The Anatolian insect fauna includes some taxa typical of each of the three regions but lacks others no less characteristic of them. It also has many endemic species and genera. Anatolia apparently has accumulated natural complexes typical of several climatic zones in a much narrower range of altitudes, compared with the Himalayas where all vegetational types from tropical forests to boreal deserts and nival communities exist.

Insect biodiversity in the forests of the Palearctic is great, compared with that of the Arctic, due mostly to a relatively warm, humid climate that allows a variety of plant communities to thrive. Numerous plant species provide environments for many phytophagous insects, even those not directly associated with woody plants. The number of species of flea beetles, for example, increases from a few in the Arctic to 83 in the taiga, 131 in mixed forests, and 163 in broadleaf forests of the Russian Plain (Konstantinov 1991). Iablokov-Khnzorian (1961) gives a figure of 580 species of beetles in 35 families, feeding on various Pinaceae in the Palearctic. Most species belong to the Scolytinae (Curculionidae) (177), Cerambycidae (108; Fig. 7.3c), other Curculionidae (76), Buprestidae (59), and Anobiidae (37). A good portion of the species (121) is known from Europe and the Mediterranean (84).

The Palearctic fauna of weevils in the genus *Magdalis*, with 57 species associated exclusively with woody plants (Barrios 1986), is distributed across all forest zones. Only three (the largest) of the ten Palearctic subgenera of *Magdalis* are represented in taiga and none of them are endemic to it, whereas seven subgenera occur only in the temperate and subtropical zones. The northern boundary of the range of *Magdalis* is formed by species of the nominotypical subgenus that develop on *Pinus silvestris* L. in Europe and on *P. pumila* Regel in Northeast Asia where this characteristic shrub forms the northern forest margin. The southern boundary also is formed by species of *Magdalis s. str.*, most of which are associated with *Pinus* species (13 of 16 species with known hosts), whereas one species is associated with *Picea*, *Abies*, and *Cedrus*, and two nonspecific feeders develop on *Larix*. No specialized *Magdalis* (or scolytines; M. Yu. Mandelshtam, personal communication) are known from *Pinus sibirica* Du Tour (Barrios 1986), the most valuable Siberian conifer. In the taiga, only *Betula*, *Sorbus*, and *Prunus* among deciduous trees are used by *Magdalis*, whereas *Quercus*, *Ulmus*, *Populus*, and several Rosaceae are hosts of all southern species except those of the nominotypical subgenus.

Of the variety of deciduous trees, the Salicaceae have the most diversified coleopterous fauna in the northern forests (Ivliev et al. 1968, Korotyaev 1976, Medvedev and Korotyaev 1980), whereas *Quercus* has the greatest number of specialized phytophagous insects in the entire Palearctic (Emeljanov 1967).

Dorytomus is the most species-rich genus of weevils in the south of the Russian Far East (the northernmost part of the Stenopean Region). Most weevils, including phyllo- and carpophagous and wood-boring species, differentiate *Salix* from *Populus*, and many distinguish *Populus tremula* L. from other *Populus* species, but some polyphagous species, such as *Saperda populnea* (L.), *Lamia textor* (F.) (Cerambycidae), and *Cryptorhynchus lapathi* (L.) (Curculionidae: Cryptorhynchinae), feed on *Salix* and *Populus*. Among the willows, narrow-leaved riparian species (*Salix viminalis* L. and similar species in Europe, and *S. udensis* Trautv. and C. A. Mey. and *S. schwerinii* E. Wolf in the Far East) possess the largest consortia of weevils and chrysomelids. The broad-leaved *S. caprea* L. in Europe has a considerable number of specialized feeders. The family Salicaceae includes a Far-Eastern monotypic genus (*Chosenia*), in addition to *Populus* and *Salix*. Little is known of the phytophagous consortium of this tall tree in the south of its range in Korea, Japan, northeastern China, and the Russian Far East. In Northeast Russia (Magadan Province and Kamchatka Territory), two species of the genus *Dorytomus* (probably developing in catkins) and a leaf-miner *Rhamphus choseniae* Korotyaev (all Curculionidae) are monophagous on *Chosenia arbutifolia* (Pall.) A. Skvorts. All three are distributed southward at least to Primorskii Territory in Russia. No specialized chrysomelids or cerambycids are known on *Chosenia*; several species from poplars and willows are recorded as occasional feeders.

Highly diversified riparian assemblages of phytophagous insects associated with the Salicaceae are endemic to the Holarctic Region. A large gravel area along the Ma River in the low mountains of Thanh Hoa Province in North Vietnam was investigated by B. A. Korotyaev in 1988. This area is dominated by bushes of *Drypetes salicifolia* Gagnep., which look similar to willows but belong to the family Euphorbiaceae, widely represented in the tropics. The bushes harbor no coleopteran genus typical of the Palearctic riparian shrub.

Temperate and subtropical forest faunas of many insect groups are more diversified in the Far East than in the Western Palearctic, which is separated from the woodland Afrotropical biota by the wide desert zone. Coleoptera are represented in the Stenopean Region by many families unknown from the European Region, such as the Cupedidae of the Archostemata; the aquatic Aspidytidae recently described (Ribera et al. 2002) from South Africa and found in China (Shaanxi: *Aspidytes wrasei* Balke, Ribera, and Beutel); and the Helotidae, Inopeplidae, Ischaliidae, Monommidae, Othniidae, Pilipalpidae, Synteliidae, Trictenotomidae, and Zopheridae of the Polyphaga (Lehr 1992). Many subfamilies and tribes of the largest phytophagous families can be added to this list, such as the Megalopodinae and Chlamysini of the Chrysomelidae, and more than 20 tribes of the Curculionoidea.

Many large weevil taxa with predominantly tropical (Oriental or Paleotropical) distributions gradually decrease in representation in the eastern forest faunas through the subtropical Orthrian and temperate nemoral Stenopean regions, and some do not reach the boreal taiga region. This trend is demonstrated at the family level by the Anthribidae, Brentidae, and Dryophthoridae (Table 7.1) and by the Paleotropical tribe Mecysolobini of the Curculionidae (Molytinae), with far more than 150 species (more than 40 in Vietnam), including 7 in Japan (Morimoto 1962), 4 in Korea (all shared with Japan and Russia), and 2 in the south of the Russian Far East. At the generic level, the northern impoverishment of the fauna is best exemplified by the second largest genus of the weevil subfamily Ceutorhynchinae, *Mecysmoderes*, with more than 100 species in East and Southeast Asia (more than 40 in Vietnam alone), more than 10 in Japan, only 2 in South Korea, and 1 in the Sakhalin and Kunashir islands of the Russian Far East.

On the contrary, large boreal and temperate genera decrease in species numbers southward and reach tropical countries, if at all, as single representatives in the northernmost mountain systems; examples include *Carabus* and *Chrysolina* in Vietnam. The Holarctic weevil genus *Dorytomus* includes 19 species in boreal Northeast Asia, 28 species in the southern (Stenopean) part of the Russian Far East, 10 species in Japan, and 7 in Korea, mostly in the central and south-central Stenopean part of the peninsula (Hong et al. 2001).

Taiga

The taiga (Fig. 7.1a: II, c) is the largest ecozone in the world, stretching from the Atlantic to the Pacific Ocean in the Palearctic. Together with its Nearctic counterpart, it occupies 13% of the earth's landmass (Schultz 1995). Despite its large size, it is relatively uniform floristically. Sochava (1953) recognized three main types of taiga in the former USSR: dark coniferous forests dominated mostly by spruces (*Picea* spp.) and firs (*Abies* spp.); pine forests with *Pinus silvestris*; and larch forests formed in different areas by three species of *Larix*. One of the characteristic features of the taiga is the small number of lianas. *Clematis* (Ranunculaceae) is the only common representative, with an oligotypic, specialized genus of flea beetles (*Argopus*) feeding on it.

Many insects in the taiga are associated with the most common conifers and occur in abundance, sometimes becoming serious pests. Among the lepidopteran pests are *Lymantria monacha* (L.) (Lymantriidae), *Dendrolimus pini* L. (Lasiocampidae), and *Dioryctria abietella* Denis and Schiffermüller (Pyralidae). Common pests also include many scolytines (e.g., *Ips*, *Pityogenes*, and *Polygraphus*). The Scolytinae have a fairly large number of species in the taiga, mostly associated with conifers. In northern Europe, both *Pinus silvestris* and *Picea abies* (L.) Karsten have multispecies assemblages of scolytines (up to 15 species on a single tree; M. Yu. Mandelshtam, personal communication), and in Siberia and the Far East, *Larix* species are heavily attacked, although by few species. Birches, constituting a considerable part of the northern forests, harbor a few species, and alder still fewer. Xylophagous Cerambycidae are not species rich in the taiga but include several species that damage conifers and rather regularly exhibit outbreaks.

Coleoptera constitute a considerable part of the taiga insect fauna, and are dominated by the Staphylinidae, Carabidae, and Curculionidae. Ermakov (2003) listed 592 beetle species in 64 families collected in the northern Urals in 1998–2001, from the plains to the highest point at 1492 m.

Characteristically, many taiga insects have enormous ranges stretching across the Palearctic and have a tendency to become invasive if introduced to the Nearctic. For example, the core of the buprestid fauna (Coleoptera) of the Euro-Siberian taiga is formed mainly of transpalearctic species, such as *Dicerca furcata* (Thunberg) (Chrysochroinae), *Buprestis rustica* (L.), and *Anthaxia quadripunctata* (L.). Two of the most common buprestids are associated with the coniferous genus *Larix*: *Buprestis strigosa* Gebler and *Phaenops guttulatus* (Gebler).

Some high-level forest taxa constituting a considerable part of the nemoral and subtropical faunas are absent from the taiga. Singing cicadas (Auchenorrhyncha: Cicadidae) are one of the most noticeable examples during the day, and crickets

Fig. 7.6
  (A) Russia, Astrakhan Province, near Lake Baskunchak, steppe (Photo M. Volkovitsh).
  (B) Armenia, Vedi Desert (Photo A. Konstantinov).
  (C) Russia, Sakhalin Island, mixed forest (Photo A. Konstantinov).
  (D) Italy, Sicily, mountain nemoral forest with *Fagus* (Photo M. Volkovitsh).
  (E) Nepal, Lantang District, mountain forest (3200 m) (Photo A. Konstantinov).
  (F) Bhutan, Shemgang District (2900 m) (Photo A. Konstantinov).

(Orthoptera: Grylloidea), glow-worms, and fireflies (Coleoptera: Lampyridae) at night. Among the weevils, no Brentidae or Platypodidae are present. Among the Chrysomeloidea, the Chlamysini, Lamprosominae, and Megalopodinae are not represented in the taiga.

Nemoral European and Stenopean Forests

These forests account for 10% of the earth's landmass (Schultz 1995) (Figs. 7.1a: III, IV, d: 7.6c). The Palearctic nemoral zone, with aborigine vegetation dominated by coniferous–broadleaf (or mixed) and broadleaf forests, consists of two currently isolated parts, the European and Stenopean regions (Emeljanov 1974). The fragmentation of the nemoral zone followed an increase of continentality that probably occurred in the Pliocene (Sinitsyn 1965) when the northern tundra and taiga zones expanded southward and the steppe and desert zones drifted northward. An occurrence of numerous closely related vicarious species of buprestids, carabids, cerambycids, histerids, and other insects in both European and Stenopean regions supports this hypothesis (Volkovitsh and Alexeev 1988, Kryzhanovsky 2002). For example, a number of pairs of closely related vicarious species of Buprestidae are in the European and Stenopean nemoral regions, including *Lamprodila rutilans* (F.) – *L. amurensis* (Obenberger), *Eurythyrea aurata* (Pallas) – *E. eoa* Semenov, and *Chrysobothris affinis* (F.) – *Ch. pulchripes* Fairmaire.

In European forests, the most species-rich weevil taxa are associated with the woodland landscape – herbage both under the canopy and, especially, in the openings, forest litter and decaying plant tissues – rather than with the trees themselves. Representative genera include *Acalles* and its allies (Cryptorhynchinae), *Dichotrachelus* (Cyclominae), *Otiorhynchus* (Entiminae), *Plinthus*, and several smaller genera of Plinthini (Molytinae), none of them being represented in the Stenopean Region. On the contrary, major taxa of the Curculionidae with larval development in the leaves, fruits, or wood of woody plants, have poorer representation in the European forests, compared with the Stenopean forests. Examples include the leaf-mining tribe Rhamphini of the Curculioninae (7 genera with 51 species in Japan; Morimoto 1984; and 5 genera with 27 species in Europe; Lohse 1983), the genus *Curculio* in the broad sense (13 species in the south of the Russian Far East, 50 in Japan (Egorov et al. 1996), and 11 in Central (nemoral) Europe). Of the main forest-inhabiting higher taxa, only the tribes Anthonomini of the Curculioninae, Magdalini of the Mesoptiliinae, and Pissodini of the Molytinae are almost equally represented in the European and Stenopean regions. The Western and Eastern Palearctic members of the largest weevil subfamily, Entiminae, with soil-inhabiting larvae, are represented almost exclusively by endemic genera (and subgenera in a few trans- or amphipalearctic genera), except for a small number of widely distributed species. Many other insects associated with trees also are

more species rich in the Stenopean Region than in the European Region. For example, the nitidulid genera *Epuraea*, mostly occurring under bark, and *Meligethes*, associated exclusively with herbaceous woodland plants, are represented in the fauna of the southern Russian Far East by 47 and 20 species, respectively, out of approximately 100 in the former USSR (Kirejtshuk 1992).

Among ancient faunal elements worthy of mention as endemic to the Stenopean Region is the beetle *Sikhotealinia zhiltzovae* Lafer, originally (Lafer 1996) placed in the separate monotypic family Sikhotealini- idae, but later (Kirejtshuk 2000) transferred to the Mesozoic family Jurodidae Ponomarenko (suborder Archostemata), which was thought to be extinct. Other examples include *Declinia relicta* Nikitsky et al. (1993) a species of the recently described coleopteran family Decliniidae; a relict longhorn beetle, *Callipogon* (*Eoxenus*) *relictus* Semenov, with Neotropical affinities; the aquatic beetle *Aspidytes wrasei*; and representatives of the coleopteran family Synteliidae (Histeroidea) known only in this part of the Palearctic.

Hesperian and Orthrian evergreen forests

These forests (Figs. 7.1a: V, VI; 7.6d, e, f) originated from a single region of sclerophyll vegetation that existed at the edge of Arcto-Tertiary and Tropical–Tertiary forests until the end of the Neogene along the Tethys Ocean (Axelrod 1975a). Aridization changed the dominant plant communities from ever- green laurophyllous forests to sclerophyll forests and chaparral-macchia. Insect relicts of the laurophyllous forests currently occur in some semi-arid areas of the Palearctic such as the Canary Islands, Himalayas, and southern China. As a result of Pleistocene glaciation, boreal elements are also present in the Hesperian and Orthrian evergreen forests, mostly at high elevations in various mountain systems.

The Macaronesian Subregion of the Hesperian Region includes islands of Madeira, the Canaries, Azores, and Cape Verde, with typical oceanic climates. The insect fauna of the islands is less rich than the continental fauna. It contains many endemics and includes Mediterranean and Afrotropical elements. For example, the fauna of the Canary Islands, the largest of the region, contains 230 species of ground beetles, of which 140 are endemics; of 83 genera, 14 are endemic to the Canaries and 2 are shared with Madeira. The buprestid fauna of the Canaries includes 20 species in 9 genera, of which 12 species are endemics. The Canary fauna of buprestids is less diverse compared with the 189 species and 29 genera in Spain, and 224 species and 31 genera in Morocco.

The Mediterranean insect fauna is species rich and highly endemic. Oosterbroek (1994) suggested that the Mediterranean region in a broad sense (including Anatolia, Armenia, and Hirkanian) together with the Far East is the most species rich in the Palearctic. The insect fauna of the Mediterranean proper constitutes about 75% of the Western Palearctic fauna (Balletto and Casale 1991). For example, for the flea-beetle genus *Longitarsus*, with about 500 valid species in the world and 221 in the Palearctic, 158 occur in the Mediterranean (in the narrow sense) and about 10% of them are endemics. Of 461 species of Neuroptera in the Mediterranean (in the broad sense), 230 are endemics; of 677 species of Rhopalocera (Lepidoptera), 416 are endemics; and of 498 species of Tipulidae (Diptera), 361 are endemics (Oosterbroek 1994). The most species-rich areas of the Mediterranean are the Balkans and Asia Minor (Oosterbroek 1994). Widespread temperate European and Palearctic species contribute significantly to the overall biodiversity of the region. The Mediterranean biota, however, has been impoverished by human influences over a long period of time, which have completely transformed the region (Mooney 1988).

The Orthrian Region includes the Himalayas (lower and middle altitudinal belts of the southern slope of the Central and Eastern Himalayas up to 2000–3000 m, belonging to the Oriental Region or to territories transitional between the Palearctic and Oriental Regions; Emeljanov 1974) and parts of China, Japan, and southernmost Korea. Because Pleistocene glaciation did not cover the eastern part of its territory, the region contains a large number of floristic and faunistic relicts (Axelrod 1975b). The insect fauna of the Orthrian Region is rich and has many Oriental elements. For example, the leaf-beetle fauna of the small Himalayan country of Nepal contains 797 species, which is nearly half the number of species in the entire fauna of the former USSR (Medvedev and Sprecher-Uebersax 1999) and more than twice that in Mongolia (Medvedev 1982). The most primitive dragonfly family, Epiophlebiidae, occurs in the Orthrian Region (Himalayas and Japan) (Kryzhanovsky 2002).

Faunal connections between the Mediterranean Subregion of the Hesperian Region and the Orthrian Region are of interest. Species of a few buprestid genera are split almost evenly between these zoogeographical entities. *Polyctesis* has two species in the Mediterranean Subregion and two in the Orthrian Region. *Ptosima* has one species in the Mediterranean Subregion and two in the Orthrian Region. Eastern Mediterranean and Irano-Turanian faunal elements are common in the West Orthrian Subregion, but disappear in the eastern portion of the region. Species of the predominantly Western Palearctic and Afrotropical buprestid genera *Julodis* (Fig. 7.3a) and *Julodella* (Fig. 7.3b) occur in the West Orthrian Subregion. The West Orthrian Subregion is also the western limit for Oriental genera such as *Microacmaeodera*, and the northern limit of distribution for the Paleotropical genera *Sternocera* and *Coroebina*. From west to east in the Orthrian Region, the influence of the Oriental fauna increases significantly. In Yunnan, for example, about half of the buprestid fauna is Oriental. An influence of the Holarctic and Stenopean groups increases in the East Orthrian Subregion, as illustrated by a few buprestid genera and subgenera: *Nipponobuprestis* (six species in southern China and Japan) and *Sapaia* (Yunnan and North Vietnam) (Volkovitsh and Alexeev 1988, Kuban et al. 2006).

Scythian (Steppe) Region

The steppe (Figs. 7.1a: VII; 7.6a) stretches from the Hungarian lowland to eastern Siberia, Mongolia, and northern China, with the southern border along the Black Sea, Crimean and Caucasian mountains, and the deserts of Kazakhstan and Middle Asia. The steppe is characterized by the following climatic conditions: well-developed to extreme continentality (Mongolian steppes differ by more than 100°C between the winter minimum and summer maximum), limited and uneven precipitation, and strong winds. Genuine steppe landscape nearly lacks forests, with only some gallery forests situated along the rivers, small isolated forests in depressions (ovragi, balki), and forest belts in the mountains.

The steppe landscape (the steppe proper, prairies, and pampas), occupying only 8% of the land, provides 80% of the cereals and meat and other cattle products (Mordkovich et al. 1997); 66% of the steppes are located in Eurasia. Due to its fertile soils and favorable climate, the steppe has been severely transformed by agriculture. Some remnants of steppe occur in southern Ukraine and the south of European Russia, but in the northern Caucasus, steppe communities are almost entirely gone, except for some fragments on slopes unsuitable for agriculture (Korotyaev 2000); the same is true for the nearly completely cultivated northern Kazakhstan steppes ('Tselina' = virgin soil). Large steppe areas still exist in Tuva and Mongolia.

The steppe is usually subdivided into the West Scythian and East Scythian subregions (Emeljanov 1974), with the boundary in the area between the Altai Mountains and Yenisei River; subregions are further subdivided into provinces and subprovinces. The boundary between the Western and Eastern Palearctic steppe faunas is quite sharp. In the close territories of southeastern West Siberia and in Tuva, the weevil faunas are nearly equal, consisting of 320 species (Krivets 1999) and 311 species, respectively (B. A. Korotyaev, unpublished data), including 1 and at least 35 species, respectively, of *Stephanocleonus* (Coleoptera: Curculionidae).

Strong and frequent winds might be largely responsible for insects with well-developed flight skills or flightless forms having an advantage in the steppe. Flightlessness is particularly common for ground-dwelling beetles. Among them, the cerambycid genus *Dorcadion*, with many species, is characteristic of the western and central parts of the Steppe Region, whereas the eastern part is populated by its close relative *Eodorcadion*. Species of *Dorcadion* are active in the spring, and those of *Eodorcadion* in the middle of the summer, according to the maximum precipitation in the respective regions. Many steppe carabids are also flightless, including the most conspicuous, viz., *Callisthenes* and some *Calosoma* (Fig. 7.2a). Flightless insects are especially common in dry variants of the steppe with sparse vegetation, for example, in Tuva and Mongolia, where the predominance of medium-sized and large wingless orthopterans and beetles is impressive. Some large weevil genera, including almost exclusively fully winged species, are represented in the Palearctic steppe by wingless species. Examples include a wingless species of *Pseudorchestes*, which is a minute leaf miner on semidesert wormwoods (*Artemisia pauciflora* Web. ex Stechm.) in Kazakhstan and Mongolia, and the entire subgenus *Anthonomidius* of the worldwide *Anthonomus*, with four species on *Potentilla*.

A characteristic feature of steppe vegetation is the prevalence of the underground biomass over the above-ground biomass, leading Paczoski (1917) to call the steppe 'the forest upside down'. The proportion of weevils with soil-inhabiting larvae in the Ciscaucasian steppe (20%), nonetheless, is less than that in the six types of the Transaltai Gobi desert communities (40–57%) and at the southern boundary of the adjacent mountain steppes (54%), and equals that in the desert solonchaks and oases (Korotyaev 2000).

A variety of beetles, from carnivores to phytophages, is rich in the steppe. Medvedev (1950) gave an excellent review of the steppe fauna, which is especially detailed for the western steppes. He considered that 5300 species of beetles occur in the steppes from Moldova to Transbaikalia; however, only half of them occur in the true steppe landscape. Weevils (Curculionidae in a broad sense, excepting Scolytinae) are the most species-rich group of beetles in the steppes of the former USSR (763 species), followed by ground beetles (Carabidae, 752 species), rove beetles (Staphylinidae, 657), leaf beetles (Chrysomelidae, 500), and scarabs (Scarabaeoidea, 288) (Medvedev 1950, cited by Iablokov-Khnzorian 1961).

Many groups of saprophagous (copro-, detrito-, and proper saprophagous) beetles are well represented in the steppe. The largest beetle genera, *Aphodius* and *Onthophagus*, dominate the dung-beetle assemblages throughout the steppe zone. Some species of *Aphodius* are associated with burrows of typical steppe rodents, the susliks (ground squirrels) and marmots. Regional faunas of *Aphodius* include as many as 150 species in Kazakhstan and Middle Asia (Nikolajev 1988) and about 50 mostly steppe species in Mongolia (Puntsagdulam 1994). The Mongolian steppe fauna of coprophagous scarabaeids is more diversified (59, 32, and 35 species in the forest steppe, genuine steppe, and desert steppe, respectively) than the desert fauna (seven species) (Puntsagdulam 1994).

The most conspicuous detrito- and phytodetritophagous group of beetles in the steppe is the family Tenebrionidae, of which the most common is *Opatrum sabulosum* (L.). A few species of the genus *Pedinus* also are typical of the Eastern European steppe (*P. femoralis* (L.) being the most common and widely distributed), but missing from the steppes of Kazakhstan, Siberia, and Mongolia, whose fauna is dominated by the genera *Anatolica*, *Penthicus*, *Melanesthes*, *Scythis*, and *Blaps* with 42, 25, 13, 12, and 10 species, respectively, in Mongolia and Tuva (Medvedev 1990).

These genera, except *Blaps*, are poorly represented in the European steppes where no genus of Tenebrionidae has gained a particular diversity. For example, on the Taman' Peninsula, with its predominantly steppe landscape, each of the 12 genera of tenebrionid beetles is represented by only one species (B. A. Korotyaev, unpublished data).

Among pollinators, several medium-sized, hairy, and brightly colored scarabaeoid beetles of the family Glaphyridae, with thin, short elytra (*Amphicoma* and *Glaphyrus*) – apparently mimicking bumblebees – are specific to the western steppe but lacking in the Siberian and Mongolian steppe. Two related families, the Malachiidae and Dasytidae, are common on flowers throughout the steppe zone. Alleculids are conspicuous only in the western steppes.

Steppe herbivores are numerous and specific. Two orders, the Lepidoptera and Coleoptera, are apparently most species rich in the steppe. In the Karadagh Nature Reserve in the Crimean steppe, 1516 species of Lepidoptera have been found (Budashkin 1991); 796 species are recorded from the largely dry-woodland Abrau Peninsula in the Black Sea, and 300 species from the almost forestless Taman' Peninsula (Shchurov 2004). In the steppes of the Northwestern Caucasus near Novoaleksandrovsk, 169 species of weevils dominate the herbivorous insect assemblage, with the second largest group, the Chrysomelidae, barely reaching half the weevil total, and other beetle families (Buprestidae and Cerambycidae) being represented by 6 or 7 species. The Auchenorrhyncha (planthoppers and leafhoppers) and Orthoptera have no more than 20 and 6 species, respectively, in that area (Korotyaev 2000, 2001a, unpublished data).

Most steppe-specific buprestids are representatives of *Sphenoptera* (subgenera *Chilostetha* and *Sphenoptera s. str.*), which are mainly root borers; *Agrilus* (subgenus *Xeragrilus*), associated mainly with *Artemisia*; and endemic Palearctic *Cylindromorphus* that feed on Cyperaceae and Poaceae. Leaf-mining species of the genus *Trachys* are common on Lamiaceae plants in the western steppes, whereas representatives of another leaf-mining genus, *Habroloma*, are associated mainly with *Erodium* (Geraniaceae).

The tribe Dorcadionini, with soil-inhabiting larvae, is the most characteristic group of cerambycid beetles in the steppe. The three cerambycid genera with the greatest numbers of species in the Caucasus – *Dorcadion* (43), *Phytoecia* (40), and *Agapanthia* (16) – comprise more than a quarter of the entire fauna (343 species) of this extensively wooded mountain system. They are associated with herbaceous vegetation, and most (31 of 43 species of *Dorcadion* recorded from the Caucasus; Danilevskii and Miroshnikov 1985) live in the steppes.

Leaf beetles have a number of groups specific to the steppe. No fewer than 50 species of several subgenera of *Chrysolina* occur in the steppe and some large subgenera (e.g., *Pezocrozita*) are subendemic. All species of *Chrysolina* are associated with herbs and semishrubs (mostly *Artemisia* species), and are especially conspicuous and diversified in the steppes of southern Siberia and Mongolia. Many of the central Palearctic steppe *Chrysolina* are wingless, but two tribes of actively flying leaf beetles, the Clytrini and Cryptocephalini (Fig. 7.2d), with no flightless species in the Palearctic, codominate, with *Chrysolina*, the steppe chrysomelid assemblages. In the Cryptocephalini, *Cryptocephalus* is apparently the most species-rich genus of leaf beetles in the steppe and includes many common species on *Artemisia*, *Atraphaxis*, and *Caragana*, the most characteristic steppe semishrubs and shrubs.

The steppe flea-beetle (Alticini) fauna is the most species rich among the flatland flea-beetle faunas of the European part of the former USSR, with 198 species (Konstantinov 1991). This fauna also has the lowest percentage of species with transpalearctic ranges (16.2%) (Konstantinov 1991).

In the Bruchinae, the monotypic tribe Kytorhinini includes about a dozen species associated mostly with *Caragana* (Fabaceae) and distributed from southeastern European Russia to Nepal and the Far East. The genera *Bruchidius*, *Bruchus*, and *Spermophagus* are represented in the steppe by a few species, including some endemics. A Nearctic species, *Acanthoscelides pallidipennis* (Motschulsky), has been introduced together with its American host, *Amorpha fruticosa* L. (Fabaceae), and is now the second most common (after *Spermophagus sericeus* Geoffroy) bruchine in the North Caucasian steppe, having reached China in the east.

The largest phytophagous group, the Curculionoidea, is species rich in the steppes. Of the 11 families represented in the steppe, only the Apionidae and Curculionidae are represented by more than 10 species, but most of the families, except the Nanophyidae and Dryophthoridae, include at least one endemic or subendemic steppe species. The Nemonychidae (Fig. 7.5b) and Brachyceridae occur only in the southeastern European steppes. The Anthribidae and Rhynchitidae are represented by one endemic species each in the

Continental sector on *Caragana* (Fabaceae) and *Spiraea* (Rosaceae), respectively. The Urodontidae include 10 species of the Palearctic genus *Bruchela* in the steppe, of which *B. orientalis* (Strejček) is trans-zonal, *B. exigua* Motschulsky is endemic to extreme south-eastern Europe, and *B. kaszabi* (Strejček) is endemic to southern Mongolia; all are associated with xerophilic Brassicaceae. The Erirhinidae include the riparian endemic trans-zonal *Lepidonotaris petax* (Sahlberg) and East Scythian *Notaris dauricus* Faust. The Attelabidae are represented in eastern Mongolia and Transbaikalia by a few Stenopean species, mostly on *Ulmus pumila* L.

The Apionidae are common throughout the entire steppe zone from its southernmost parts to the cold (= cryophytic) steppes and tundra-steppe of the Siberian Arctic zone. They comprise 10% of the entire Curculionoidea fauna (340 species) of the steppe on Taman' Peninsula (Korotyaev 2004 and unpublished data), and 12% of the 169 species in the steppe of Stavropol Territory in the Northwestern Caucasus (Korotyaev 2000, 2001b, unpublished data). In Siberia and Mongolia, with their poorer steppe flora and extreme continental climate, the Apionidae constitute a smaller part of the steppe-weevil fauna (about 3% in the steppe zone of Tuva, with somewhat over 220 species).

The Curculionidae is the most species-rich family in the entire steppe zone, although only a few estimates are available. Of the two investigated steppe areas in the plains of the Northwestern Caucasus, one is the Taman' Peninsula (Korotyaev 2004), with a variety of habitats. The other steppe area is a fragment of the steppe with a territory of about 2 hectares, completely isolated from the closest sites of native vegetation by fields and orchards (Korotyaev 2000). Both lists of the Curculionoidea are dominated by the Ceutorhynchinae, Entiminae, and Curculioninae, followed by the Apionidae. Aside from the steppes of China, where no specific studies have been conducted, the rest of the Eurasian steppe-weevil fauna is characterized by dominance of four subfamilies, the Ceutorhynchinae, Entiminae, Curculioninae, and Lixinae. The first three of these subfamilies are most species rich in Europe and Anatolia, while the Lixinae overwhelmingly dominate in Mongolia and adjacent parts of eastern Siberia.

The xylophagous Curculionoidea are poorly represented in the steppe. Scolytines are represented mostly by thamnobionts and dendrobionts that develop on steppe bushes such as *Prunus spinosa* L. One scolytid genus, *Thamnurgus*, is herbivorous. One of its species,

Th. russicus Alexeev, is endemic to the meadow-steppe subzone of European Russia, developing on *Delphinium cuneatum* Stev. (Ranunculaceae; Alexeev 1957), and another species, *Th. caucasicus* Reitter, is common on *Carduus* (Asteraceae) in the North Caucasian steppes.

Sethian (Desert) Region

Palearctic deserts form a great belt stretching from Northern Africa to Northwestern China and Western India (Figs. 7.1a: VIII; 7.6b). Depending on the climatic conditions, soil composition, and vegetation, three large subregions usually are distinguished in Palearctic deserts: Saharo-Arabian, Irano-Turanian, and Central Asian (Lavrenko 1950, Emeljanov 1974, Kryzhanovsky 2002). Geographic position and climate partly determine the similarities and distinctions among the insect faunas of all subregions. The Saharo-Arabian Subregion includes a great number of taxa of Afrotropical (Ethiopian) origin. The Irano-Turanian Subregion shares many taxa with the Mediterranean Subregion, particularly with the East Mediterranean Province of the Hesperian Region, while its North Turanian Province has many features of the Central Asian Subregion.

The flatland desert insect fauna is xerophilic and relatively poor, other than some desert-specific groups adapted to xeric conditions. Of approximately 800 species of leaf beetles in Central and Middle Asia, 233 species occur in deserts (Lopatin 1999). Among them, the Cryptocephalinae are the most numerous, followed by the Alticini and Eumolpinae.

Some insect groups are poorly represented or absent in flatland deserts. These groups include grasshoppers (Orthoptera: Ensifera), earwigs (Dermaptera), aphids (Aphidina), some brown lacewings (Hemerobiidae), ground beetles (Carabidae, excluding specialized groups), Megaloptera, Raphidioptera, and Mecoptera. Other insect groups are abundant in the deserts, including some groups of Blattaria, Mantodea, Isoptera, Orthoptera (particularly some groups of locusts: Catantopinae, Oedipodinae, and Pamphaginae), Rhynchota (particularly Psyllina, Auchenorrhyncha, and Heteroptera), and Neuroptera (Ascalaphidae, Mantispidae, Myrmeleontidae, and Nemopteridae (Fig. 7.2b)). The desert fauna has a large percentage of endemic taxa (Kryzhanovsky 1965). In the Orthoptera, the family Acrididae contains a number of endemic tribes (Dericorythini, Diexini, Egnatiini, Iranellini, and

Table 7.3 Distribution of Tenebrionidae across six types of desert plant communities in Transaltai Gobi, Mongolia. 6–1 = desert plant communities (from north to south): 6, *Anabasis brevifolia* steppefied desert; 5, *Reaumuria soongorica* + *Sympegma regelii* desert; 4, *Haloxylon ammodendron* desert; 3, *Reaumuria soongorica* + *Nitraria sphaerocarpa* desert; 2, extra-arid *Iljinia regelii* desert; 1, extra-arid *Ephedra przewalskii* + *Haloxylon ammodendron* (in dry, temporary waterbeds) desert.

Steppe Species (+)	6	5	4	3	2	1	Desert Species (*)
						*	*Dilamus mongolicus* Kaszab
						*	*Penthicus lenczyi* Kaszab
					*		*Psammoestes dilatatus* Reitter
Blaps miliaria Fischer de Waldheim	+		*				*Blaps kiritshenkoi* Semenov and Bogačev
Melanesthes heydeni Cziki	+	+	*	*			*Melanesthes czikii* Kaszab
Platyope mongolica Faldermann	+	+	*	*	*		*Sternoplax mongolica* Reitter
Monatrum prescotti Faldermann	+	+	+*	*	*		*Anatolica mucronata* Reitter
Blaps femoralis medusula Kaszab	+	+	+*	*	*	*	*A. polita borealis* Kaszab
Eustenomacidius mongolicus (Kaszab)	+	+	+*	*	*	*	
Anatolica cechiniae Bogdanov-Kat'kov	+	+	+*	*	*	*	*Blaps kashgarensis gobiensis* Frivaldski
A. sternalis gobiensis Kaszab	+	+	+*	*	*	*	*Trigonoscelis sublaevigata granicollis* Kaszab
Eumilada punctifera amaroides Reichardt	+	+*	+*	*	*	*	*Cyphosthete mongolica* Kaszab
Epitrichia intermedia Kaszab	+*	+	+*	*			*Anatolica amoenula* Reitter
Microdera kraatzi Reitter	+*	+*	+*	+*	+	+*	*Anemia dentipes* Ballion
	*	*	*	*		*	*Cyphogenia intermedia* A. Bogačev
							Pterocoma reitteri Frivaldski

Uvaroviini), mostly in the subfamily Catantopinae, which is common in both the Sahara and Gobi Subregions (Sergeev 1993). Endemics constitute almost 70% of the leaf beetles (Lopatin 1999). Insect biodiversity sharply increases in the mountains, with altitude belts inhabited by mesophilous groups, particularly in the Irano-Turanian Subregion (Kryzhanovsky 2002).

The borders between steppes and deserts in the Palearctic generally are not sharp and usually are represented by an intermediate subzone of semideserts now typically attributed to the steppe zone as the desert steppe. Faunal changes along climatic and vegetational gradients are well-organized phenomena, which is illustrated with examples of three family-group taxa of the Coleoptera in the Transaltai Gobi. Gradual changes were studied in species assemblages of the superfamily Curculionoidea (Table 7.2), Chrysomelidae (except Bruchinae), and Tenebrionidae (Table 7.3) across six types of zonal plant communities, from steppefied deserts (sites 6 and 5) in the north of the 160-km long soil and vegetation profile to extra-arid deserts with only one (site 2) or two species of plants (site 1). The general pattern is a subsequent substitution of species with the greatest portions of their ranges in the mountain steppe or desert steppe north of site 6 by species distributed in true deserts.

The Tenebrionidae differ from the herbivorous taxa in the broader overlapping of the steppe and desert complexes along the profile, such that the number of tenebrionid species is greater than the number of weevils at most sites, although the total number of weevils in the profile is 1.3 times that of tenebrionids. Also distinguishing the Tenebrionidae from the herbivorous beetles is their much larger portion of zonal desert species in the total Transaltai Gobi fauna (see Table 7.3): 59% for the tenebrionids, compared with 37% for the Curculionoidea and 31% for the Chrysomelidae (excluding Bruchinae). A drop in biodiversity occurs at site 6 close to the Dzhinst Mountains, with developed steppe communities, relative to the more distant site 5, and the highest diversity is found at the sandy *Haloxylon* desert (site 4).

Two special faunistic surveys of Coleoptera have been conducted in different parts of the Palearctic deserts. The first was by a French Scientific Mission (Peyerimhoff 1931) along an 800-km route in the Central Sahara, including the Hoggar Plateau, in February–May 1928, and the second was a part of the investigation on insects in the main plant communities of the Transaltai Gobi in July–early October 1981 and in the summer of 1982 by the Joint Soviet–Mongolian Biological Expedition (Korotyaev et al. 1983). Although the periods and organization of the collecting differed significantly, the results (Table 7.4) are essentially similar. Differences in the family sets from the two surveys are of two kinds. The absence of several aquatic families in the Transaltai Gobi is due to the small number and small size of natural water bodies, whereas a rather large river and several freshwater streams were investigated in the Sahara. The other important difference is the presence of the two predominantly tropical families in the Sahara, the Bostrichidae and Brentidae, which can be explained by the location of the area at the southern border of the Palearctic. The presence of a few families (e.g., Alleculidae, Cantharidae, Leiodidae, Nitidulidae, Oedemeridae, and Pselaphidae) with one or two species in only one of the two faunas is probably accidental, resulting mainly from the difference in collecting periods.

Several features are common to the Saharan and Gobian coleopteran faunas and characteristic of the desert Palearctic fauna in general. These are the leading positions of the Curculionidae and Tenebrionidae and the relatively wide representation of other large families such as the Buprestidae, Carabidae, Chrysomelidae, Scarabaeidae, and Staphylinidae, but not the Cerambycidae. Also noteworthy is the absence of the Silphidae from the desert faunas, in sharp contrast to the Mongolian steppes where large black beetles of *Nicrophorus argutor* B. Jakovlev are common around dense colonies of pikas and susliks. The Scolytinae also are absent, although a desert species, *Thamnurgus pegani*

Eggers, occurs in Middle Asia on *Peganum harmala* L. (Peganaceae) in Turkmenistan. Relatives of this species are known from Euphorbiaceae in tropical Africa and the Mediterranean.

To survive in extreme conditions of deserts, insects have developed a number of specialized behavioral, ecological, morphological, and physiological adaptations. A specific sand-desert morphobiological form has arisen in several beetle families (e.g. Dermestidae; Zhantiev 1976). The set of hyperthrophied adaptive characters can mask the affinities to such an extent that a desert dermestid, *Thylodrias contractus* Motschulsky, has been described repeatedly in several families (R. D. Zhantiev, personal communication). Adaptive rearrangements of various body structures are manifested by desert Braconidae (Hymenoptera). In addition to depigmentation of the integument and enlargement of the eyes, in connection with nocturnal activity, the braconid wasps exhibit a smoothening of the body sculpture for reflection of light and heat, shortening of the wings with basal shifting of the cells associated with strong winds in open landscapes, and elongation of the labiomaxillary complex for feeding on flowers of desert plants (Tobias 1968).

ACKNOWLEDGMENTS

This work would not have been possible without help and advice from our friends and colleagues – experts on various insect groups – especially those at the Zoological Institute in St. Petersburg (Russia): L. N. Anisyutkin (Blattaria, Dermaptera), S. A. Belokobylsky (Hymenoptera: Braconidae), A. F. Emeljanov (Auchenorrhyncha, zoogeography), N. V. Golub (Zoraptera, Psocoptera), A. V. Gorokhov (Orthoptera), D. R. Kasparyan (Hymenoptera: Ichneumonidae), the late I. M. Kerzhner (Heteroptera) A. G. Kirejtshuk (Coleoptera), O. V. Kovalev (Hymenoptera: Cynipoidea), V. A. Krivokhatsky (Neuropteroidea), V. G. Kuznetsova (Zoraptera, Psocoptera), A. L. Lobanov (Coleoptera, database), M. Yu. Mandelshtam (Scolytinae), A. Yu. Matov (Lepidoptera, Noctuidae), G. S. Medvedev (Coleoptera: Tenebrionidae), E. P. Nartshuk (Diptera), the late Yu. A. Pesenko (Hymenoptera: Apoidea), V. A. Richter (Diptera: Tachinidae), S. Yu. Sinev (Lepidoptera), A. V. Stekolshchikov (Sternorrhyncha: Aphidina), and from many other institutions: M. L. Chamorro-Lacayo (Trichoptera, Department of Entomology, University

Table 7.4 Number of species of Coleoptera in the Transaltai Gobi and Central Sahara.

Family	Transaltai Gobi	Central Sahara
Carabidae	43	36
Haliplidae	–	1
Dytiscidae	4	19
Gyrinidae	–	2
Georissidae	–	1
Hydrophilidae (including Helophoridae)	5	13 (including Helophoridae and Hydraenidae)
Hydraenidae	3	?
Histeridae	15	11
Leiodidae	1	–
Staphylinidae	30	43
Pselaphidae	–	2
Scarabaeidae *sensu lato*	15 (*Aphodius* – 6, *Onthophagus* – 1)	30
Dryopidae	–	4
Heteroceridae	–	2
Buprestidae	16 (*Anthaxia* – 1, *Acmaeoderella* – 1, *Sphenoptera* – 7, *Agrilus* – 3, *Paracylindromorphus* – 1)	20 (*Anthaxia* – 4, *Acmaeodera* – 4, *Sphenoptera* – 2, *Agrilus* – 1, *Cylindromorphus* – 1)
Elateridae	9 (*Aeoloides* – 1, *Aeloderma* – 1, *Zorochrus* – 1, *Agriotes* – 1, *Cardiophorus* – 2)	8 (*Drasterius* – 2, *Zorochrus* – 1, *Agriotes* – 1, *Cardiophorus* – 1)
Cantharidae	1	–
Dermestidae	12 (*Dermestes* – 4, *Attagenus* – 6–7, *Anthrenus* – 2)	11 (*Dermestes* – 1, *Attagenus* – 5, *Anthrenus* – 1)
Bostrichidae	–	8
Anobiidae	2 (*Xyletinus*)	1 (*Theca*)
Stylopidae	1	–
Cleridae	3 (*Emmepus arundinis* Motsch., *Necrobia rufipes* DeG., *Opetiopalpus sabulosus* Motsch.)	4 (*Emmepus* sp., *Necrobia rufipes*)
Dasytidae	–	5
Melyridae	3	5
Nitidulidae	–	1
Cybocephalidae	2	3
Phalacridae	4	2
Cucujidae	3 (*Airaphilus* – 1)	2 (*Airaphilus* – 1)
Helodidae	–	1
Cryptophagidae	3 (*Cryptophagus* – 2)	1 (*Cryptophagus*)
Coccinellidae	27 (*Hyperaspis* – 1, *Coccinella* – 5, *Scymnus* s. l. – 6, *Pharoscymnus* – 2)	11 (*Hyperaspis* – 1, *Coccinella* – 1, *Scymnus* s. l. – 4, *Pharoscymnus* – 2)
Mordellidae	4 (*Mordellistena* – 3, *Pentaria* – 1)	3 (*Mordellistena* – 1, *Pentaria* – 2)
Rhipiphoridae	1 (*Macrosiagon medvedevi* Iablokov-Khnzorian)	1 (*Macrosiagon*)
Oedemeridae	1 (*Homomorpha cruciata* Sem.)	–
Anthicidae	15 (*Steropes latifrons* Sumakov, *Notoxus* – 2, *Anthicus* s. l. – 11, *Formicomus* sp. – 1)	19 (*Notoxus* – 2, *Anthicus* s. l. – 14)
Meloidae	3	23
Alleculidae	–	1
Tenebrionidae	42	70

(continued)

Table 7.4 *(continued).*

Family	Transaltai Gobi	Central Sahara
Scarptiidae	1 (*Scraptia* sp.)	1 (*Scraptia straminea* Peyer.)
Cerambycidae	4 (*Chlorophorus ubsanurensis* Tsherep., *Ch. obliteratus* Ganglb., *Asias mongolicus* Ganglb., *Eodorcadion kozlovi* Suv.)	6
Chrysomelidae	51 [incl. Bruchinae (6): *Rhaebus* – 1, *Spermophagus* – 2, *Bruchidius* – 3]	23 [incl. Bruchinae (2): *Caryoborus* – 1, *Bruchidius* – 1]
Urodontidae	1	5
Brentidae	–	1
Apionidae	15	13
Curculionidae	73	39

of Minnesota, Minneapolis, MN, USA), O. N. Kabakov (St. Petersburg, Russia), N. Ju. Kluge (Ephemeroptera, Department of Entomology, St. Petersburg University, Russia), A. S. Lelej (Hymenoptera: Mutillidae, Institute of Biology and Soil Science, Far Eastern Branch, Russian Academy of Sciences, Vladivostok, Russia), S. W. Lingafelter, M. Pogue, J. Prena, F. C. Thompson, N. E. Woodley (Systematic Entomology Laboratory, USDA, Washington DC, USA), W. Steiner (Department of Entomology, Smithsonian Institution, Washington DC, USA), I. Löbl (Genéve, Switzerland), I. K. Lopatin (Department of Zoology, Byelorussian State University, Minsk, Belarus), R. D. Zhantiev (Department of Entomology, Moscow State University, Moscow, Russia), and A. G. Zinovjev (Hymenoptera: Tenthredinidae, Boston, MA, USA).

We thank V. I. Dorofeyev (Komarov Botanical Institute, Russian Academy of Sciences, St. Petersburg) for identification of some plants referred to in the text; D. I. Berman and Yu. M. Marusik (Institute of the Biological Problems of the North, Far Eastern Branch, Russian Academy of Sciences, Magadan), O. A. Khruleva (Institute of the Ecology and Evolution, Russian Academy of Sciences, Moscow), and S. A. Kuzmina (Paleontological Institute, Russian Academy of Sciences, Moscow) for useful advice and the many-year supply of interesting material collected in their studies in the North; D. I. Berman for excellent photographs of the northern landscapes; O. Merkl (Hungarian Natural History Museum, Budapest) for help at the Budapest Museum and consultations and excursions with B. A. Korotyaev to several unique landscapes in Hungary; L. Gültekin (Atatürk University, Erzurum, Turkey) for organizing collecting trips to Turkey and the fruitful long-term collaboration;

N. A. Florenskaya (St. Petersburg, Russia) for the habitus illustration of *Theodorinus* sp.; E. Roberts (Systematic Entomology Laboratory, USDA, Washington DC, USA) for the habitus illustration of *Clavicornaltica dali*; and H. Bradford (Systematic Entomology Laboratory, USDA, Washington DC, USA, archives) for the habitus illustration of *Cimberis attelaboides*. The work was performed with the use of the collection of the Zoological Institute, Russian Academy of Sciences (UFC ZIN no. 2–2.20), contract No. 02.452.11.7031 with Rosnauka (2006-RI-26.0/001/070).

We greatly appreciate the advice and constructive suggestions of our friends and colleagues who read this manuscript at various stages of completion: V. Grebennikov (Entomology, Canadian Food Inspection Agency, Ottawa), V. Gusarov (Department of Zoology, Natural History Museum, University of Oslo, Oslo, Norway), S. W. Lingafelter, R. Ochoa, A. Norrbom (Systematic Entomology Laboratory, Washington DC, USA), and A. K. Tishechkin (Department of Entomology, Louisiana State University, Baton Rouge, Louisiana, USA). We thank A. S. Lelej for help with the literature. We are particularly thankful to P. H. Adler and R. Footitt for editing this manuscript and their numerous and valuable suggestions.

This study was supported in part by the Collaborative Linkage Grant No. 981318 of the NATO Life Science and Technology Program. The work of MGV and BAK was partly supported also by the Russian Foundation for Basic Research, Grant nos. 07-04-00482-a, 07-04-10146-k, and 04-04-81026-Bel2004a. BAK's collecting in the steppes of the Northwestern Caucasus in 2007, which provided additional observations and material for this chapter, was supported by a grant from the Systematics Research Fund (London).

REFERENCES

Adler, P. H. and R. W. Crosskey. 2008. World black-flies (Diptera: Simuliidae): a fully revised edition of the taxonomic and geographical inventory. http://entweb.clemson.edu/biomia/pdfs/blackflyinventory.pdf [Accessed 14 May 2007].

Agakhaniantz, O. E. 1987. *One Pamir Year*. Mysl, Moscow. 190 pp. [In Russian].

Alexeev, A. V. 1957. On finding of a scolytine, *Thamnurgus* Eichh. in Kursk oblast'. *Uchenyye zapiski, Orekhovo-Zuevskii pedagogicheskii institut* 5: 159–163. [In Russian].

Alexeev, Y. E. and Gubanov, I. A. 1980. *Flora of the Environs of Puschino-na-Oke*. Izdatel'stvo MGU, Moscow. 103 pp. [In Russian].

Alonso-Zarazaga, M. A. 1989. Revision of the supraspecific taxa in the Palaearctic Apionidae Schoenherr, 1823. I. Introduction and subfamily Nanophyinae Seidlitz, 1891 (Coleoptera, Curculionoidea). *Fragmenta Entomologica, Roma* 21: 205–262.

Alonso-Zarazaga, M. A. 1990. Revision of the supraspecific taxa in the Palaearctic Apionidae Schoenherr, 1823 (Coleoptera, Curculionoidea). 2. Subfamily Apioninae Schoenherr, 1823: Introduction, keys and descriptions. *Graellsia* 46: 19–156.

Alonso-Zarazaga, M. A. and C. H. C. Lyal. 1999. *A World Catalogue of Families and Genera of Curculionoidea (Insecta: Coleoptera)*. Entomopraxis, Barcelona. 315 pp.

Alonso-Zarazaga, M. A. and C. H. C. Lyal. 2002. Addenda et corrigenda to 'A World Catalogue of Families and Genera of Curculionoidea (Insecta: Coleoptera)'. *Zootaxa* 63: 1–37.

Angus, R. B. 1983. Evolutionary stability since the Pleistocene illustrated by reproductive compatibility between Swedish and Spanish *Helophorus lapponicus* Thomson (Coleoptera, Hydrophilidae). *Biological Journal of the Linnean Society* 19: 17–25.

Angus, R. B. 1993. Spanish "endemic" *Ochthebius* as a British Pleistocene fossil. *Latissimus* 2: 24–25.

Arnoldi, L. V. 1958. General review of the insect fauna of the Carpathians. Pp. 30–37. *In* E. N. Pavlovsky (ed). *Animal World of the USSR*. Volume 5. Izdatel'stvo Akademii Nauk SSSR, Moscow, Leningrad. 655 pp. [In Russian].

Arzanov, Yu. G. 2005. Review of the weevil genus *Brachycerus* Olivier (Coleoptera: Brachyceridae) from European Russia, Caucasus and neighboring countries. *Kavkazskii entomologicheskii byulleten'* 1: 65–80. [In Russian].

Aukema, B. and C. Rieger. 1995. *Catalogue of the Heteroptera of the Palaearctic Region. Volume 1. Enicocephalomorpha, Dipsocoromorpha, Nepomorpha, Gerromorpha and Leptopodomorpha*. Netherlands Entomological Society, Amsterdam. 222 pp.

Aukema, B. and C. Rieger. 1996. *Catalogue of the Heteroptera of the Palaearctic Region. Volume 2. Cimicomorpha I*. Netherlands Entomological Society, Amsterdam. 361 pp.

Aukema, B. and C. Rieger. 1999. *Catalogue of the Heteroptera of the Palaearctic Region. Volume 3. Cimicomorpha II*. Netherlands Entomological Society, Amsterdam. 577 pp.

Aukema, B. and C. Rieger. 2001. *Catalogue of the Heteroptera of the Palaearctic Region. Volume 4. Pentatomomorpha I*. Netherlands Entomological Society, Amsterdam. 436 pp.

Aukema, B. and C. Rieger. 2006. *Catalogue of the Heteroptera of the Palaearctic Region. Volume 5. Pentatomomorpha II*. Netherlands Entomological Society, Amsterdam. 550 pp.

Axelrod, D. I. 1975a. Evolution and biogeography of Madrean-Tethyan sclerophyll vegetation. *Annals of the Missouri Botanical Garden* 62: 280–334.

Axelrod, D. I. 1975b. Plate tectonics and problems of angiosperm history. Pp. 72–85. *In* Memoirs du Museum National d'Histoire Naturelle. Nouvelle Serie. Serie A, Zoologie, tome LXXXVIII. XVIIe Congress International de Zoologie, Monaco – 25–30 September 1972. Theme 31. Biogeographie et liaisons intercontinentales au cors du Mesozoique. Paris, Edition du Museum.

Bajtenov, M. S. 1977. Neue Rüsselkäfer-Arten aus der UdSSR (Coleoptera: Curculionidae). *Reichenbachia* 16: 225–228.

Balletto, E. and Casale, A. 1991. Mediterranean insect conservation. Pp. 121–142. *In* N. M. Collins and J. A. Thomas (eds). *The Conservation of Insects and Their Habitats*. Academic Press, London.

Barkalov, A. V. and T. R. Nielsen. 2007. *Platycheirus* species (Diptera, Syrphidae) from Yakutia, Eastern Siberia, with description of two new species. *Volucella* 8: 87–94.

Barrios, H. E. 1986. Review of the Weevil Subfamily Magdalinae (Coleoptera, Curculionidae) of the USSR and Neighboring Countries. Referate of the PhD Dissertation. Zoologicheskii Institut Akademii Nauk SSSR, Nauka, Leningrad Branch. Leningrad. 22 pp. [In Russian].

Bellamy, C. L. 2003. An illustrated summary of the higher classification of the superfamily Buprestoidea (Coleoptera). *Folia Heyrovskyana, Supplementum* 10: 1–197.

Belokobylsky, S. A. and V. I. Tobias. 1998. Order Hymenoptera. Suborder Apocrita. Superfam. Ichneumonoidea. Fam. Braconidae. Pp. 1–657. *In* P. A. Lehr (ed). *Key to Insects of the Russian Far East*. Volume 4. Part 3. Dal'nauka, Vladivostok. 707 pp. [In Russian].

Belokobylsky, S. A. and V. I. Tobias. 2000. Order Hymenoptera. Suborder Apocrita. Superfam. Ichneumonoidea. Fam. Braconidae. Pp. 1–572. *In* A. S. Lelej (ed). *Key to Insects of the Russian Far East*. Volume 4. Part 4. Dal'nauka, Vladivostok. 650 pp. [In Russian].

Belousov, I. A. 1998. *Le complexe générique de Nannotrechus Winkler du Caucase et du la Crimée (Coleoptera, Carabidae, Trechini)*. Series Faunistica, 8. Pensoft, Sofia-Moscow-St. Petersburg. 256 pp.

Belov, D. A. 2007. Species composition and structure of the urban arthropod complexes (Arthropoda) (on example of Moscow). P. 23. *In* A. S. Zamotajlov (ed). *Contribution of Entomology to the Agroindustrial Complex, Forestry and Medicine*. Abstracts. XIII Congress of Russian Entomological

Society, Krasnodar, 9–15 September 2007. Krasnodar. 239 pp. [In Russian].

Belyshev, B. F. and A. Yu. Kharitonov. 1981. *Geography of Dragonflies (Odonata) of Boreal Faunistic Realm.* Nauka, Sibirskoe Otdelenie, Novosibirsk. 275 pp. [In Russian].

Berman, D. I., A. V. Alfimov, and B. A. Korotyaev. 2002. Xerophilic arthropods in the tundra–steppe of the Utyosiki locality (Chukchi Peninsula). *Zoologicheskii Zhurnal* 81: 444–450. [In Russian].

Budarin, A. M. 1985. *Carabids of Magadan Province. Species List.* (A preprint). Akademiya Nauk SSSR, Dal'nevostochnyi Nauchnyi Tsentr. Institut Biologicheskikh Problem Severa, Vladivostok. 21 pp. [In Russian].

Budashkin, Yu. I. 1991. Butterflies and Moths (Lepidoptera) of the Karadagh Nature Reserve; Ecological–Faunistic and Zoogeographical Analysis. Referate of the PhD Dissertation. Zoologicheskii Institut Akademii nauk SSSR, Leningrad. 22 pp. [In Russian].

Chernov, Yu. I. 1966. A brief essay of the animal world of the tundra zone in the USSR. Pp. 52–91. *In* Yu. A. Isakov (ed). *Zonal Characteristics of the Land Animal World.* Nauka, Moscow. 127 pp. [In Russian].

Chernov, Yu. I. 1975. *Natural Zonality and Land Animal World.* Mysl, Moscow. 221 pp. [In Russian].

Chernov, Yu. I. 1995. The order Diptera (Insecta, Diptera) in the Arctic fauna. *Zoologicheskii Zhurnal* 74: 68–83. [In Russian].

Chernov, Yu. I. 2002. Arctic biota: taxonomic diversity. *Zoologicheskii Zhurnal* 81: 1411–1431. [In Russian].

Chernov, Yu. I., K. V. Makarov, and P. K. Eremin. 2000. The family Carabidae in the Arctic fauna. Communication 1. *Zoologicheskii Zhurnal* 79: 1409–1420. [In Russian].

Chernov, Yu. I., K. V. Makarov, and P. K. Eremin. 2001. The family Carabidae in the Arctic fauna. Communication 2. *Zoologicheskii Zhurnal* 80: 285–293. [In Russian].

Chernov, Yu. I., L. N. Medvedev, and O. A. Khruleva. 1994. Leaf beetles (Coleoptera, Chrysomelidae) in the Arctic. *Entomologicheskoe Obozrenie* 73: 152–167. [In Russian].

Chikatunov, V. and T. Pavlíček. 2005. Leaf beetles (Coleoptera: Chrysomelidae) of the West and Southwest facing slopes in the Israeli part of the Hermon Mountains. Pp. 17–42. *In* A. S. Konstantinov, A. K. Tishechkin, and L. Penev (eds). *Contributions to Systematics and Biology of Beetles: Papers Celebrating the 80th Birthday of Igor Konstantinovich Lopatin.* Pensoft, Sofia-Moscow. 388 pp.

Colonnelli, E. 2004. *Catalogue of Ceutorhynchinae of the World with a Key to Genera.* Argania, Barcelona. 124 pp.

Coope, G. R. 1973. Tibetan species of dung beetle from Late Pleistocene deposits in England. *Nature* 245: 335–336.

Cripps, M. G., H. L. Hinz, J. L. McKenney, B. L. Harmon, F. W. Merickel, and M. Schwarzländer. 2006. Comparative survey of the phytophagous arthropod fauna associated with *Lepidium draba* in Europe and the western United States, and the potential for biological weed control. *Biocontrol Science and Technology* 16: 1006–1030.

Danilevskii, M. L. and A. I. Miroshnikov. 1985. *Longhorned Beetles of the Caucasus (Coleoptera, Cerambycidae). A Key.* Krasnodar Kubanskii Ordena Trudovogo Krasnogo Znameni Selskohozyaistvennyi Institut, Krasnodar. 419 pp. [In Russian].

Darlington, P. J. 1963. *Zoogeography: The Geographical Distribution of Animals.* John Wiley and Sons, New York. 657 pp.

Davidian, E. M. 2007. Family Aphidiidae – Aphidiid wasps. Pp. 192–254. *In* A. S. Lelej (ed). *Key to Insects of the Russian Far East.* Volume 4, Neuropteroidea, Scorpionflies, Hymenopterans. Part 5. Dal'nauka, Vladivostok. 1052 pp. [In Russian].

Dieckmann, L. 1974. Beiträge zur Insektenfauna der DDR: Coleoptera – Curculionidae (Rhinomacerinae, Rhynchitinae, Attelabinae, Apoderinae). *Beiträge zur Entomologie* 24: 5–54.

Doguet, S. 1994. *Coléoptères Chrysomelidae.* Volume 2 Alticinae. Faune de France, France et Regions Limitrophes. 80. Fédération Française des Sociétés de Sciences Naturelles, Paris. 694 pp.

Duckett, C. N., K. D. Prathapan, and A. S. Konstantinov. 2006. Notes on identity, new synonymy and larva of *Ivalia* Jacoby (Coleoptera: Chrysomelidae) with description of a new species. *Zootaxa* 1363: 49–68.

Egorov, A. B. 1996a. Family Bruchidae – Seed beetles. Pp. 140–166. *In* P. A. Lehr (ed). *Key to Insects of the Russian Far East*, Volume 3, Coleoptera. Part 3. Dal'nauka, Vladivostok. 555 pp. [In Russian].

Egorov, A. B. 1996b. Family Anthribidae – Anthribids. Pp. 166–199. *In* P. A. Lehr (ed). *Key to Insects of the Russian Far East*, Volume 3, Coleoptera. Part 3. Dal'nauka, Vladivostok. 555 pp. [In Russian].

Egorov, A. B., V. V. Zherikhin, and B. A. Korotyaev. 1996. 112b. Family Curculionidae – Weevils. Pp. 431–517. *In* P. A. Lehr (ed). *Key to Insects of the Russian Far East*, Volume 3, Coleoptera. Part 3. (Supplement). Dal'nauka, Vladivostok. 555 pp. [In Russian].

Emeljanov, A. F. 1967. Some characteristics of the distribution of oligophagous insects over host plants range. Pp. 28–65. *In* E. P. Nartshuk (ed). *Chteniya pamyati N. A. Kholodkovskogo. 1 aprelya 1966.* Zoologicheskii Institut, Leningrad. Volume 19. 85 pp. [In Russian].

Emeljanov, A. F. 1974. Proposals on the classification and nomenclature of ranges. *Entomologicheskoe Obozrenie* 53: 497–522. [In Russian].

Ermakov, A. I. 2003. Coleopterous fauna (Insecta, Coleoptera) of the "Denezhkin Kamen" Nature Reserve. *Trudy Gosudarstvennogo Zapovednika "Denezhkin Kamen".* Akademkniga, Yekaterinburg 2: 79–93. [In Russian].

Friedman, A. L. L. and A. Freidberg. 2007. The Apionidae of Israel and the Sinai Peninsula (Coleoptera; Curculionoidea). *Israel Journal of Entomology* 37: 55–180.

Frieser, R. 1981. Die Anthribiden der Westpaläarktis einschliesslich der Arten der UdSSR (Coleoptera, Anthribidae).

Mitteilungen der Münchener Entomologischen Gesellschaft 71: 33–107.

Fursov, V. N. 2007. Family Trichogrammatidae – Trichogrammatid wasps. Pp. 963–989. *In* A. S. Lelej (ed). *Key to Insects of the Russian Far East*. Volume 4, Neuropteroidea, Scorpionflies, Hymenopterans. Part 5. Dal'nauka, Vladivostok. 1052 pp. [In Russian].

Gagne, R. 2007. Species numbers of Cecidomyiidae (Diptera) by zoogeographical regions. *Proceedings of the Entomological Society of Washington* 109: 499.

Grissell, E. E. 1995. Toryminae (Hymenoptera: Chalcidoidea: Torymidae): a redefinition, generic classification, and annotated world catalog of species. *Memoirs on Entomology, International* 2: 1–470.

Grissell, E. E. 1999. An annotated catalog of World Megastigminae (Hymenoptera: Chalcidoidea: Torymidae). *Contributions of the American Entomological Institute* 43: 1–92.

Grubov, V. I. 1982. *Key to Vascular Plants of Mongolia (with an Atlas)*. Nauka, Leningradskoe Otdelenie, Leningrad. 441 pp. [In Russian].

Hansen, M. 1998. Hydraenidae (Coleoptera). *World Catalogue of Insects*. Volume 1. Apollo Books, Stenstrup. 168 pp.

Heppner, J. B. 1998. Classification of Lepidoptera. *Part 1. Introduction. Holarctic Lepidoptera*. 5, Supplement. Gainesville, FL. 148 pp.

Hong, K.-J., A. B. Egorov, and B. A. Korotyaev. 2001. *Illustrated Catalogue of Curculionidae in Korea (Coleoptera)*. Insects of Korea. Ser. 5. Korea Research Institute of Bioscience and Biotechnology & Center for Insect Systematics, Junghaeng-Sa, Seoul. 337 pp.

Iablokov-Khnzorian, S. M. 1961. *An Attempt of Reconstruction of the Genesis of the Beetle Fauna of Armenia*. Izdatel'stvo Akademii Nauk Armyanskoi SSR, Erevan. 264 pp. [In Russian].

Iablokoff-Khnzorian, S. M. 1974. Monographie der Gattung *Lithophilus* Froelich (Col. Coccinellidae). *Entomologische Arbeiten aus dem Museum G. Frey* 25: 149–243.

Isaev, A. Yu. 1994. Ecological-faunistic review of weevils (Coleoptera: Apionidae, Rhynchophoridae, Curculionidae) of Ul'yanovsk Province. *Priroda Ul'yanovskoi Oblasti* 4: 77.

Isaev, A. Yu. 2001. Trophic associations of weevils of the genus *Tychius* Germ. (Coleoptera, Curculionidae) with *Astragalus* species in the forest-steppe of the middle Volga area. *Entomologicheskoe Obozrenie* 80: 819–822. [In Russian].

Ismailova, M. Sh. 2006. A review of the weevil genus *Ptochus* Schoenh. (Coleoptera, Curculionidae) of the fauna of Daghestan. *Entomologicheskoe Obozrenie* 85: 602–617. [In Russian].

Ivanov, V. D. 2007. Caddisfly (Trichoptera) fauna of Russia. P. 132. *In* A. S. Zamotajlov (ed). *Problems and Perspectives of General Entomology*. Abstracts of the reports in the XIII Congress of Russian Entomological Society, Krasnodar, 9–15 September 2007. Krasnodar. 420 pp. [In Russian].

Ivliev, L. A., D. G. Kononov, and L. N. Medvedev. 1968. Fauna of leaf beetles of Magadan Province and northern Khabarovsk Territory. Pp. 62–87. *In* A. I. Kurentsov and Z. A. Konovalova (eds). *Fauna i Ekologiya Nasekomyh Dal'nego Vostoka*. Biologo-pochvennyi Institut Dal'nevostochnogo Filiala Sibirskogo Otdeleniya Akademii Nauk SSSR, Vladivostok. [In Russian].

Jasnosh, V. A. 1995. Family Aphelinidae – Aphelinid wasps. Pp. 506–551. *In* P. A. Lehr (ed). *Key to Insects of the Russian Far East*. Volume 4, Neuropteroidea, Scorpionflies, Hymenopterans. Part 2. Nauka, St. Petersburg. 597 pp. [In Russian].

Kaplin, V. G. 1981. *Arthropod Complexes in Plant Tissues in Sand Deserts*. Ylym, Ashkhabad, Turkmenia, USSR. 376 pp. [In Russian].

Kasparyan, D. R. 1990. Ichneumonidae, subfamily Tryphoninae: tribe Exenterini; subfamily Adelognathinae. *Fauna of the USSR: Hymenopterous Insects*. Volume 3, part 2. Nauka, Leningrad. 340 pp. [In Russian].

Kataev, B. M. and H.-B. Liang. 2005. New and interesting records of ground beetles of the tribe Harpalini from China (Coleoptera: Carabidae). *Zoosystematica Rossica* 13: 209–212.

Kharitonov, A. Yu. 1997. Order Dragonflies (Odonata). Pp. 222–246. *In* E. P. Nartshuk, D. V. Tumanov, and S. Ya. Tsalolikhin (eds). *Key to Freshwater Invertebrates of Russia and Adjacent Territories*. Volume 3. Aranea and hemimetabolous insects. Zoologicheskii Institut Rossiiskoi Akademii Nauk, St. Petersburg. 439 pp. [In Russian].

Khruleva, O. A. 1987. Invertebrates. Pp. 6–35. *In* V. E. Sokolov (ed). *Flora and Fauna of Reserves of the USSR. Fauna zapovednika "Ostrov Vrangelya"*. Nauka, Moscow. [In Russian].

Kimsey, L. S. and R. M. Bohart. 1990. *The Chrysidid Wasps of the World*. Oxford University Press, New York. 652 pp.

Kirejtshuk, A. G. 1992. 59, 61. Fam. Nitidulidae – Nitidulids. Pp. 114–209. *In* P. A. Lehr (ed). *Key to Insects of the Far East of the USSR*. Volume 3, Beetles. Part 2. Nauka, Leningrad. 704 pp. [In Russian].

Kirejtshuk, A. G. 2000. *Sikhotealinia zhiltzovae* Lafer, 1996 – recent representative of the Jurassic coleopterous fauna (Coleoptera, Archostemata, Jurodidae). *Trudy Zoologicheskogo Instituta Rossiiskoi Akademii Nauk* (1999) 281: 21–26. [In Russian].

Kirejtshuk, A. G. 2001. Coleoptera (coleopterans, or beetles). Pp. 79–367. *In* S. Ya. Tsalolikhin (ed). *Key to Freshwater Invertebrates of Russia and Adjacent Lands*. Volume 5. Nauka, St. Petersburg. 836 pp. [In Russian].

Konstantinov, A. S. 1991. Zonal structure of the flea beetle fauna (Coleoptera, Chrysomelidae, Alticinae) of the European part of the USSR and the Caucasus. Pp. 148–168. *In* I. K. Lopatin and E. I. Khotko (eds). *Fauna i Ekologiya Zhukov Belarusi*. Navuka i Tekhnika, Minsk. 263 pp. [In Russian].

Konstantinov, A. S. 1998. Revision of the Palearctic species of *Aphthona* Chevrolat and cladistic classification of the Aphthonini (Coleoptera: Chrysomelidae: Alticinae). *Memoirs on Entomology, International* 11: 1–429.

Konstantinov, A. S. 2005. New species of Middle Asian *Longitarsus* Latreille with discussion of their subgeneric placement (Coleoptera: Chrysomelidae). *Zootaxa* 1056: 19–42.

Konstantinov, A. S. and M. L. Chamorro-Lacayo. 2006. A new genus of moss-inhabiting flea beetles (Coleoptera: Chrysomelidae) from the Dominican Republic. *Coleopterists Bulletin* 60: 275–290.

Konstantinov, A. S. and C. N. Duckett. 2005. New species of *Clavicornaltica* Scherer (Coleoptera: Chrysomelidae) from continental Asia. *Zootaxa* 1037: 49–64.

Konstantinov, A. S. and N. J. Vandenberg. 1996. Handbook of Palearctic flea beetles (Coleoptera: Chrysomelidae: Alticinae). *Contributions on Entomology, International* 1: 236–439.

Korotyaev, B. A. 1976. Review of weevils (Coleoptera, Curculionidae) of the Kamchatka Peninsula. *Trudy Zoologicheskogo Instituta Akademii Nauk SSSR* 62: 43–52. [In Russian].

Korotyaev, B. A. 1982. Review of weevils of the subtribe Oxyonycina Hoffm. (Coleoptera, Curculionidae), living on *Ephedra*, of the fauna of the USSR and Mongolia. *Trudy Zoologicheskogo Instituta Akademii Nauk SSSR* 110: 45–81. [In Russian].

Korotyaev, B. A. 1992. Role of parthenogenesis in the origin of modern weevil fauna of the Palaearctic. Pp. 404–405. *In* L. Zombori and L. Peregovits (eds). *Proceedings of the Fourth European Congress of Entomology and the XIII. Internationale Symposium für die Entomofaunistik Mitteleuropas*, 1–6 September, 1991, Gödöllö, Hungary. Volume 1. Hungarian Natural History Museum, Budapest. 418 pp.

Korotyaev, B. A. 2000. On an unusually high diversity of rhynchophorous beetles (Coleoptera, Curculionoidea) in steppe communities of the Northern Caucasus. *Zoologicheskii Zhurnal* 79: 242–246. [In Russian].

Korotyaev, B. A. 2001a. Note on the English translation of the paper "On an unusually high diversity of rhynchophorous beetles (Coleoptera, Curculionoidea) in steppe communities of the Northern Caucasus". *Entomological Review* 80 (2000): 1026.

Korotyaev, B. A. 2001b. Urjanchai leaf beetle *Chrysolina urjanchaica* (Jacobson, 1925). P. 140. *In* V. I. Danilov-Danilian (ed). *The Red Data Book of Russian Federation. (Animals).* Moscow. 862 pp. [In Russian].

Korotyaev, B. A. 2004. Rhynchophorous beetles (Coleoptera, Curculionoidea) of the Taman Peninsula. Pp. 41–44. *In* Yu. V. Lokhman (ed). *Ekologicheskiye problemy Tamanskogo poluostrova.* Kubanskii Gosudarstvennyi Universitet, Krasnodar. 218 pp. [In Russian].

Korotyaev, B. A., A. S. Konstantinov, S. W. Lingafelter, M. Yu. Mandelshtam, and M. G. Volkovitsh. 2005. Gall-inducing Coleoptera. Pp. 239–271. *In* A. Raman, C. W. Schaefer, and T. M. Withers (eds). *Biology, Ecology, and Evolution of Gall-Inducing Arthropods.* Volume 1. Science Publishers, Enfield, New Hampshire. 429 pp.

Korotyaev, B. A., A. S. Konstantinov, and C. W. O'Brien. 2000. A new genus of the Orobitidinae and discussion of its relationships (Coleoptera: Curculionidae). *Proceedings of the Entomological Society of Washington* 102: 929–956.

Korotyaev, B. A., A. L. Lvovsky, and A. G. Kirejtshuk. 1983. Insect fauna. Pp. 65–71. *In* V. E. Sokolov, O. Shakhdarsuren, E. M. Lavrenko, P. D. Gunin, V. V. Krynitsky, and I. T. Fedorova. (eds). *A Complex Characteristic of Desert Ecosystems of Transaltai Gobi (Using an Example of the Desert Field Station and "Bolshoi Gobiisky Zapovednik" Nature Reserve).* Puschino Institut Evolutsionnoi Morfologii I Ekologii Zhivotnykh Imeni A. N. Severtsova Akademii Nauk SSSR. 114 pp. [In Russian].

Kozlov, M. A. 2001. Hymenoptera (hymenopterans). Pp. 381–386. *In* S. Ya. Tsalolikhin (ed). *Key to Freshwater Invertebrates of Russia and Adjacent Lands.* Volume 5. Nauka, St. Petersburg. 836 pp. [In Russian].

Krasutskii, B. V. 2007. Coleoptera associated with *Daedaleopsis confragosa* (Bolton: Fr.) Schrot (Basidiomycetes, Aphyllophorales) in the forests of the Urals and Trans-Urals. *Entomologicheskoe Obozrenie* 86: 289–305. [In Russian].

Krivets, S. A. 1999. Ecological–Faunistic Review of Weevils (Coleoptera: Apionidae, Dryophthoridae et Curculionidae) of Southeastern West Siberia. Referate of the Ph. D. Dissertation. Tomskii Gosudarstvennyi Universitet i Tomskii Filial Instituta Lesa Imeni V. N. Sukacheva Sibirskogo Otdeleleniya Rossiskoi Adademii Nauk, Tomsk, 28 pp. [In Russian].

Krivokhatsky, V. A. 2001. Neuroptera. Pp. 369–371. *In* S. Ya. Tsalolikhin (ed). *Key to Freshwater Invertebrates of Russia and Adjacent Lands.* Volume 5. Nauka, St. Petersburg. 836 pp. [In Russian].

Krivokhatsky, V. A. and A. F. Emeljanov. 2000. Usage of the general zoogeographic subdivisions for particular zoogeographic researches exemplified by the Palearctic fauna of antlions (Neuroptera, Myrmeleontidae). *Entomologicheskoe Obozrenie* 79: 557–578. [In Russian].

Kryzhanovsky, O. L. 1965. *The Composition and Origin of Land Fauna of Middle Asia.* Nauka, Moscow-Leningrad. 419 pp. [In Russian].

Kryzhanovsky, O. L. 2002. *Composition and Distribution of the Insect Faunas of the World.* Tovarischestvo Nauchnyh Izdanii KMK, Moscow. 237 pp. [In Russian].

Kuban, V., S. Bílý, E. Jendek, M. Yu. Kalashian, and M. G. Volkovitsh. 2006. Superfamily Buprestoidea, Family Buprestidae. Pp. 40–60, 325–421, 457–670. *In* I. Löbl and A. Smetana (eds). *Catalogue of Palaearctic Coleoptera.* Volume 3. Apollo Books, Stenstrup. 690 pp.

Kurzenko, N. V. 1995. Family Vespidae – Wasps. Pp. 264–324. *In* P. A. Lehr (ed). *Key to Insects of the Russian Far East.* Volume 4, Neuropteroidea, Scorpionflies, Hymenopterans. Part 1. Nauka, St. Petersburg. 606 pp. [In Russian].

Lafer, G. Sh. 1989. 18. Fam. Leiodidae (Anisotomidae). Pp. 318–329. *In* P. A. Lehr (ed). *Key to the Insects of Far East of the USSR*. Volume 3, Beetles. Part 1. Nauka, Leningrad. 572 pp. [In Russian].

Lafer, G. Sh. 1996. Family Sikhotealiniidae. Pp. 390–396. *In* P. A. Lehr (ed). *Key to Insects of the Russian Far East*. Volume 3, Beetles. Part 3. Dal'nauka, Vladivostok. 555 pp. [In Russian].

LaSalle, J. and I. D. Gauld. 1991. Parasitic Hymenoptera and the biodiversity crisis. *Redia* 71: 315–334.

Lavrenko, E. M. 1950. Main features of phytogeographical division of the USSR and adjacent countries. *Problems of Botany* 1: 530–548. [In Russian].

Legalov, A. A. 2006. Annotated list of the leaf-rolling weevils (Coleoptera: Rhynchitidae, Attelabidae) of the Russian fauna. *Trudy Russkogo Entomologicheskogo Obschestva* 77: 200–210. [In Russian].

Lehr, P. A. (ed). 1992. *Key to Insects of the Far East of the USSR*. Volume 3, Beetles. Part 2. Nauka, St. Petersburg. 704 pp. [In Russian].

Lelej, A. S. 1995a. Family Scoliidae – Scoliids. Pp. 193–196. *In* P. A. Lehr (ed). *Key to Insects of the Russian Far East*. Volume 4, Neuropteroidea, Scorpionflies, Hymenopterans. Part 1. Nauka, St. Petersburg. 606 pp. [In Russian].

Lelej, A. S. 1995b. Family Pompilidae – Road wasps. Pp. 211–264. *In* P. A. Lehr (ed). *Key to Insects of the Russian Far East*. Volume 4, Neuropteroidea, Scorpionflies, Hymenopterans. Part 1. Nauka, St. Petersburg. 606 pp. [In Russian].

Lelej, A. S. 2002. *A Catalogue of Mutillid Wasps (Hymenoptera, Mutillidae) of the Palearctic Region*. Dal'nauka, Vladivostok. 171 pp. [In Russian].

Leonova, T. G. 1982. *Artemisia* L. – Sharilzh. Pp. 245–253. *In* V. I. Grubov (ed). *Key to Vascular Plants of Mongolia (with an Atlas)*. Nauka, Leningradskoe Otdelenie, Leningrad. 441 pp. [In Russian].

Lindroth, C. H. 1957. *The Faunal Connections Between Europe and North America*. John Wiley and Sons, New York. 344 pp.

Löbl, I. and A. Smetana (eds). 2003. *Catalogue of Palaearctic Coleoptera*. Volume 1. Archostemata – Myxophaga –Adephaga. Apollo Books, Stenstrup. 819 pp.

Löbl, I. and A. Smetana (eds). 2004. *Catalogue of Palaearctic Coleoptera*. Volume 2. Hydrophiloidea–Staphylinoidea. Apollo Books, Stenstrup. 942 pp.

Löbl, I. and A. Smetana (eds). 2006. *Catalogue of Palaearctic Coleoptera*. Volume 3. Scarabaeoidea, Scirtoidea, Dascilloidea, Buprestoidea and Byrrhoidea. Apollo Books, Stenstrup. 690 pp.

Löbl, I. and A. Smetana (eds). 2007. *Catalogue of Palaearctic Coleoptera*. Volume 4. Elateroidea, Derodontoidea, Bostrichoidea, Lymexyloidea, Cleroidea, and Cucujoidea. Apollo Books, Stenstrup. 935 pp.

Lohse, G. A. 1983. 32. Unterfamilie: Rhynchaeninae. Pp. 283–294. *In* H. Freude, K. W. Harde, and G. A. Lohse (eds). *Die Käfer Mitteleuropas*. Volume 11. Goecke and Evers, Krefeld. 344 pp.

Lopatin, I. K. 1971. High altitude insect fauna of Middle Asia, its ecological characteristics and origin. P. 168. *In* E. L. Gurjeva and O. L. Kryzhanovsky (eds). *Proceedings of XIII International Congress of Entomology*. Moscow, 2–9 August, 1968. Volume 1. Nauka, Leningrad. 581 pp.

Lopatin, I. K. 1977. *Leaf Beetles of Middle Asia and Kazakhstan*. Nauka, Leningrad. 268 pp. [In Russian].

Lopatin, I. K. 1989. *Zoogeography*. Vysheishaya Shkola, Minsk. 316 pp. [In Russian].

Lopatin, I. K. 1996. High altitude fauna of the Chrysomelidae of Central Asia: Biology and biogeography. Pp. 3–12. *In* P. Jolivet and M. Cox (eds). *Chrysomelidae Biology*. Volume 3. General Studies. SPB Academic Publishing, Amsterdam, The Netherlands. 365 pp.

Lopatin, I. K. 1999. Biology and biogeography of desert leaf beetles of Central Asia. Pp. 159–168. *In* M. L. Cox (ed). *Advances in Chrysomelidae Biology 1*. Backhuys Publishers, Leiden, The Netherlands. 671 pp.

Lopatin, I. K., O. R. Aleksandrovich, and A. S. Konstantinov. 2004. *Check List of Leaf-Beetles (Coleoptera, Chrysomelidae) of Eastern Europe and Northern Asia*. Mantis, Olsztyn. 343 pp.

Lopatin, I. K. and O. L. Nesterova. 2004. Biology and ecology of mountainous genera *Oreomela* Jacobson, *Xenomela* Weise and *Crosita* Motschulsky (Coleoptera, Chrysomelidae, Chrysomelinae). Pp. 415–421. *In* P. Jolivet, J. A. Santiago-Blay, and M. Schmitt (eds). *New Developments in the Biology of Chrysomelidae*. SPB Academic Publishing, The Hague, The Netherlands. 803 pp.

Lopatin, I. K. and O. L. Nesterova. 2005. *Insects of Belarus: Leaf Beetles (Coleoptera, Chrysomelidae)*. UP Tekhnoprint, Minsk. 293. [In Russian].

Lukhtanov, V. A., M. S. Vishnevskaya, A. V. Volynkin, and A. V. Yakovlev. 2007. Butterflies (Lepidoptera, Rhopalocera) of the West Altai. *Entomologicheskoe Obozrenie* 86: 337–359. [In Russian].

Lvovsky, A. L. 2001. Lepidoptera (lepidopterans, or butterflies). Pp. 73–77. *In* S. Ya. Tsalolikhin (ed). *Key to Freshwater Invertebrates of Russia and Adjacent Lands*. Volume 5. Nauka, St. Petersburg. 836 pp. [In Russian].

Magnacca, K. N. and B. N. Danforth. 2006. Evolution and biogeography of native Hawaiian *Hylaeus* bees (Hymenoptera: Colletidae). *Cladistics* 22: 393–411.

Medvedev, G. S. 1990. *Keys to the Darkling Beetles of Mongolia*. Trudy Zoologicheskogo Instituta Akademii Nauk SSSR. Volume 220. Leningrad. 253 pp. [In Russian].

Medvedev, G. S. 2005a. New species of the tenebrionid genus *Asidoblaps* Fairm. (Coleoptera, Tenebrionidae) from Chinese provinces Gansu and Sichuan. *Entomologicheskoe Obozrenie* 84: 531–568. [In Russian].

Medvedev, G. S. 2005b. On connections of the sand desert faunas of Tenebrionidae (Coleoptera) of Middle Asia, Iran and Afghanistan. Pp. 299–314. *In* A. S. Konstantinov,

A. K. Tishechkin, and L. Penev (eds). *Contributions to Systematics and Biology of Beetles.* Papers Celebrating the 80th Birthday of Igor Konstantinovich Lopatin. Pensoft, Sofia-Moscow. 388 pp.

Medvedev, G. S. 2006. New data on the China fauna of the tenebrionid genus *Agnaptoria* Rtt. (Coleoptera, Tenebrionidae). *Entomologicheskoe Obozrenie* 85: 176–205. [In Russian].

Medvedev, L. N. 1982. *Leaf Beetles of MPR.* Nauka, Moscow. 302 pp. [In Russian].

Medvedev, L. N. 1992. Fam. Chrysomelidae – Leaf Beetles. Pp. 533–602. *In* P. A. Lehr (ed). *Key to Insects of the Far East of the USSR.* Volume 3, Beetles. Part 2. Nauka, St. Petersburg. 704 pp. [In Russian].

Medvedev, L. N. and B. A. Korotyaev. 1976. On the fauna of chrysomelid beetles (Coleoptera, Chrysomelidae) of the Tuva Autonomous Republic and of the north-western Mongolia, II. *Nasekomye Mongolii* 4: 241–244. [In Russian].

Medvedev, L. N. and B. A. Korotyaev. 1980. Notices on the fauna of chrysomelid beetles (Coleoptera, Chrysomelidae) of Arctic Asia and Kamchatka. Pp. 77–95. *In* G. S. Medvedev and E. G. Matis (eds). *Issledovaniya po Entomofaune Severo-Vostoka SSSR.* Vladivostok. 170 pp. [In Russian].

Medvedev, L. N. and E. Sprecher-Uebersax. 1999. Katalog der Chrysomelidae von Nepal. *Entomologica Basilensia* 21: 261–354.

Medvedev, S. G. 1998. Fauna and host-parasite associations of fleas in the Palearctic (Siphonaptera). *Entomologicheskoe Obozrenie* 77: 295–314. [In Russian].

Medvedev, S. I. 1950. Beetles – Coleoptera. Pp. 294–347. *In* E. N. Pavlovsky and B. S. Vinogradov (eds). *Animal World of the USSR.* Volume 3. Izdatel'stvo Akademii Nauk SSSR. Moscow, Leningrad. 672 pp. [In Russian].

Mooney, H. A. 1988. Lessons from Mediterranean-climate regions. Pp. 157–165. *In* E. O. Wilson (ed). *Biodiversity.* National Academic Press, Washington, DC. 551 pp.

Mordkovich, V. G., A. M. Ghilarov, A. A. Tishkov, and S. A. Balandin. 1997. *The Fate of the Steppes.* Mangazeya, Novosibirsk. 208 pp. [In Russian].

Morimoto, K. 1962. Provisional check list of the family Curculionidae of Japan. II. *Science Bulletin of the Faculty of Agriculture, Kyushu University* 19: 341–368. [In Japanese].

Morimoto, K. 1976. On the Japanese species of the family Brentidae (Coleoptera). *Kontyû* 44: 267–282.

Morimoto, K. 1978. Check-list of the family Rhynchophoridae (Coleoptera) of Japan, with descriptions of a new genus and five new species. *Esakia* 12: 103–118.

Morimoto, K. 1984. The family Curculionidae of Japan. IV. Subfamily Rhynchaeninae. *Esakia* 22: 5–76.

Morse, J. C. 2008. The Trichoptera World Check List. http://entweb.clemson.edu/database/trichopt/[Accessed 11 July 2008].

Nartshuk, E. P. 1992. Analysis of the distribution of fly families (Diptera). *Entomologicheskoe Obozrenie* 71: 464–477. [In Russian].

Nartshuk, E. P. 2003. *Key to Families of Diptera (Insecta) of the Fauna of Russia and Adjacent Countries.* Trudy Zoologicheskogo Instituta Rossiiskoi Akademii Nauk. Volume 294. St. Petersburg. 249 pp. [In Russian].

Nasreddinov, Kh. A. 1975. A brief review of weevils (Coleoptera, Curculionidae) of Southern Tadzhikistan. *Entomologicheskoe Obozrenie* 54: 541–554. [In Russian].

Nast, J. 1972. *Palearctic Auchenorrhyncha (Homoptera). An Annotated Checklist.* Polish Academy of Sciences, Institute of Zoology. Polish Scientific Publishers, Warszawa. 550 pp.

Nemkov, P. G., V. L. Kazenas, E. R. Budris, and A. V. Antropov. 1995. Family Sphecidae – Digger wasps. Pp. 368–480. *In* P. A. Lehr (ed). *Key to Insects of the Russian Far East.* Volume 4, Neuropteroidea, Scorpionflies, Hymenopterans. Part 1. Nauka, St. Petersburg. 606 pp. [In Russian].

Nikitsky, N. B. 1992. 89. Fam. Salpingidae – Salpingids. Pp. 482–493. *In* P. A. Lehr (ed). *Key to Insects of the Russian Far East.* Volume 3, Beetles. Part 2. Nauka, Leningrad. 704 pp. [In Russian].

Nikitsky, N. B., J. F. Lawrence, A. G. Kirejtshuk, and V. G. Grachev. 1993. A new beetle family, Decliniidae fam. n., from the Russian Far East and its taxonomic relationships (Coleoptera, Polyphaga). *Russian Entomological Journal* 2: 3–10.

Nikolajev, G. V. 1988. *Lamellicorn Beetles of Kazakhstan and Middle Asia.* Nauka, Kazakhskaya SSR, Alma-Ata. 231 pp. [In Russian].

Oosterbroek, P. 1994. Biodiversity of the Mediterranean Region. Pp. 289–307. *In* P. I. Forey, C. J. Humphries, and R. I. Vane-Wright (eds). *Systematics and Conservation Evaluation.* Systematic Association Special Volume 50. Clarendon Press, Oxford. 466 pp.

Paczoski, J. 1917. *Description of the Vegetation of Kherson Gubernia. 2. Steppes.* S. N. Ol'khovikov and S. A. Khodushin steam typo-lithography, Kherson. 366 pp. [In Russian].

Pesenko, Yu. A. 1983. *Bees, Halictidae, Subfamily Halictinae, Tribe Nomioidini (Fauna of the Palearctic).* Fauna of the USSR: Hymenopterous Insects. Volume 17. Issue 1. Nauka, Leningrad. 198 pp. [In Russian].

Peyerimhoff, P. 1931. Mission Scientifique du Hoggar envoyee de Fevrier a Mai 1928 par M. Pierre Bordes Gouverneur General de l'Algerie. *Mémoires de la Société Histoire Naturelle l'Afrique du Nord* 2: 1–172.

Polevoi, A. V. 2007. Dipterans (Diptera) of the Karelia fauna. http://www.zin.ru/projects/zinsecta/rus/indikar.asp. [Accessed on 22 November 2007].

Puntsagdulam, Zh. 1994. Lamellicorn beetles (Coleoptera, Scarabaeidae) of Mongolia. Referate of the PhD Dissertation, Zoologicheskii Institut Rossiiskoi Akademii Nauk. St. Petersburg. 30 pp. [In Russian].

Radchenko, A. G. 1999. Ants (Hymenoptera, Formicidae) of the Palearctic (Evolution, Systematics, Faunogenesis). Referate of the DSc Dissertation, Zoologicheskii Institut Imeni I. Shmal'gausena Natsional'noi Akademii Nauk Ukrainy. Kiev. 47 pp. [In Russian].

Ribera, I., R. G. Beutel, M. Balke, and A. P. Vogler. 2002. Discovery of Aspidytidae, a new family of aquatic Coleoptera. *Proceedings of the Royal Society. B. Biological Sciences* 269: 2351–2356.

Richter, V. A. 2004. Fam. Tachinidae – Tachinids. Pp. 148–398. *In* V. S. Sidorenko (ed). *Key to Insects of the Russian Far East.* Volume 6. Part 3, Diptera and Aphaniptera. Dal'nauka, Vladivostok. 658 pp. [In Russian].

Sabrosky, C. W. 1999. *Family-Group Names in Diptera. MYIA. Volume 10.* Backhuys Publishers, Leiden. 576 pp.

Sanchir, Ch. 1982. *Astragalus* L. – Khunchir. Pp. 156–163. *In* V. I. Grubov (ed). *Key to Vascular Plants of Mongolia (with an Atlas).* Nauka, Leningradskoe otdelenie, Leningrad. 441 pp. [In Russian].

Savitskii, V. Yu. and G. E. Davidian. 2007. New data on the taxonomy, distribution, and ecology of the weevil genus *Otiorhynchus* Germar (Coleoptera, Curculionidae) in the Caucasus. *Entomologicheskoe Obozrenie* 86: 185–205. [In Russian].

Savoiskaya, G. I. 1984. Coccinellid Beetles (Coleoptera, Coccinellidae) of the USSR Fauna (Systematics, Biology, Economic Importance). Referate of the DSc Dissertation. Zoologicheskii Institut Akademii Nauk SSSR, Leningrad. 43 pp. [In Russian].

Schuh, R. T. and J. A. Slater. 1995. *True Bugs of the World (Hemiptera: Heteroptera): Classification and Natural History.* Cornell University Press, Ithaca, New York. 336 pp.

Schultz, J. 1995. *The Ecozones of the World: The Ecological Divisions of the Geosphere.* Springer-Verlag, Berlin. 449 pp.

Sclater, P. L. 1858. On the geographical distribution of the class Aves. *Journal of the Linnean Society of London, Zoology* 2: 130–145.

Semenov-Tian-Shansky, A. 1936. *Limits and Zoogeographic Divisions of the Palearctic Region for Land Animals Based on the Distribution of Coleopterous Insects.* Academy of Sciences Publishing House, Moscow-Leningrad. 1–16 pp. [In Russian].

Sergeev, M. G. 1993. The general distribution of Orthoptera in the main zoogeographical regions of North and Central Asia. *Acta Zoologica Cracoviensia* 36: 53–76.

Sharkov, A. V. 1995. Family Eupelmidae – Eupelmid wasps. Pp. 170–177. *In* P. A. Lehr (ed). *Key to Insects of the Russian Far East.* Volume 4. Neuropteroidea, Scorpionflies, Hymenopterans. Part 2. Nauka, St. Petersburg. 597 pp. [In Russian].

Shchurov, V. I. 2004. The butterfly and moth fauna (Insects, Lepidoptera) of the Taman' Peninsula. Pp. 53–68. *In* Yu. V. Lokhman (ed). *Ekologicheskiye Problemy Tamanskogo Poluostrova.* Kubanskii Gosudarstvennyi Universitet, Krasnodar. 218 pp. [In Russian].

Sinitsyn, V. M. 1965. *Ancient Climates of Eurasia. Part 1. Paleocene and Neocene.* Leningrad. 167 pp. [In Russian].

Smith, N., S. A. Mori, A. Henderson, D. Wm. Stevenson, and S. V. Heald. 2004. *Flowering Plants of the Neotropics.* Princeton University Press, Princeton, New Jersey. 594 pp.

Sochava, V. B. 1953. Vegetation of the forest zone. Pp. 7–61. *In* E. N. Pavlovsky (ed). *Animal World of the USSR.* Volume 4. Izdatel'stvo Akademii Nauk SSSR. Moscow, Leningrad. 737 pp. [In Russian].

Steinmann, H. 1997. World Catalogue of Odonata. Volume 1. Zygoptera. *The Animal Kingdom.* A Compilation and Characterization of the Recent Animal Groups. 110. Walter de Gruyter, Berlin. 500 pp.

Storozheva, N. A., V. V. Kostjukov, and Z. A. Efremova. 1995. Family Eulophidae – Eulophid wasps. Pp. 291–505. *In* P. A. Lehr (ed). *Key to Insects of the Russian Far East.* Volume 4. Neuropteroidea, Scorpionflies, Hymenopterans. Part 2. Nauka, St. Petersburg. 597 pp. [In Russian].

Sugonyaev, E. S. and N. D. Voinovich. 2006. *Adaptations of Chalcidoid Wasps (Hymenoptera, Chalcidoidea) for Parasitization Soft Scales (Hemiptera, Sternorrhyncha, Coccidae) under Different Latitude Conditions.* Tovarischestvo Nauchnykh Izdanii KMK, Moscow. 263 pp. [In Russian].

Taeger, A., S. M. Blank, and A. D. Liston. 2006. European sawflies (Hymenoptera: Symphyta) – a species checklist for the countries. Pp. 399–504. *In* S. M. Blank, S. Schmidt, and A. Taeger (eds). *Recent Sawfly Research: Synthesis and Prospects.* Goecke and Evers, Keltern. 701 pp.

Takhtajan, A. L. 1978. *Floristic Regions of the World.* Nauka, Leningrad. 246 pp. [In Russian].

Thompson, F. C. (ed). 2006. Statistics for Palearctic Region. Biosystematic Database of World Diptera. http://www.diptera.org/biosys.htm [Accessed 23 May 2007].

Tobias, V. I. 1968. System, Phylogeny, and Evolution of the Family Braconidae (Hymenoptera). Referate of the DSc Dissertation. Zoologicheskii Institut Akademii Nauk SSSR, Leningrad. 28 pp. [In Russian].

Trjapitzin, V. A. 1989. *Parasitic Hymenoptera of the fam. Encyrtidae of the Palearctic.* Nauka, Leningrad. 487 pp. [In Russian].

Volkovitsh, M. G. and A. V. Alexeev. 1988. Comparative characteristics of the fauna of buprestids (Coleoptera, Buprestidae) of Northern Eurasia. Pp. 42–58. *In* V. V. Zlobin (ed). *Svyazi Entomofaun Severnoi Evropy i Sibiri.* Zoologicheskii Institut Akademii Nauk SSSR, Leningrad. [In Russian].

Volkovitsh, M. G. and A. L. Lobanov. 1997. Data bank of host associations of the buprestid tribe Acmaeoderini (Coleoptera, Buprestidae) of the Palearctic. Pp. 166–187. *In* S. D. Stepanyants, A. L. Lobanov, and M. B. Dianov (eds). *Bazy Dannykh i Komp'yuternaya Grafika v Zoologicheskikh Issledovaniyakh.* Trudy Zoologicheskogo Instituta Rossiiskoi Akademii Nauk, Volume 269. Zoologicheskii Institut Akademii Nauk SSSR, St. Petersburg. 207 pp. [In Russian and English].

Vshivkova, T. S. 2001. Megaloptera. Pp. 373–380. *In* S. Ya. Tsalolikhin (ed). *Key to Freshwater Invertebrates of Russia and Adjacent Lands.* Volume 5. Nauka, St. Petersburg. 836 pp. [In Russian].

Vtorov, P. P. and N. N. Drozdov. 1978. *Biogeography.* Prosveshcheniye, Moscow. 271 pp. [In Russian].

Wallace, A. R. 1876. *The Geographical Distribution of Animals. 2 Volumes.* Macmillan, London. 607 pp.

Walter, H. and S.-W. Breckle. 1985. *Ecological Systems of the Geobiosphere. Volume 1. Ecological Principles in Global Perspective.* Springer-Verlag, Berlin. 242 pp.

Wanat, M. 1994. Systematics and phylogeny of the tribe Ceratapiini (Coleoptera: Curculionoidea: Apionidae). *Genus (International Journal of Invertebrate Taxonomy, Supplement),* Wroclaw. 406 pp.

Wanat, M. 1999. Ryjkowce (Coleoptera: Curculionoidea bez Scolytidae i Platypodidae) Puszczy Bialowieskiej – Charakteristika Fauny. *Parki Narodowe i Rezerwaty Przyrody* 18 (3): 25–47.

Wei, M., H. Nie, and A. Taeger. 2006. Sawflies (Hymenoptera: Symphyta) of China – checklist and review of research. Pp. 503–574. *In* S. M. Blank, S. Schmidt, and A. Taeger (eds). *Recent Sawfly Research: Synthesis and Prospects.* Goecke and Evers, Keltern. 701 pp.

Willig, M. R., D. M. Kaufman, and R. D. Stevens. 2003. Latitudinal gradients of biodiversity: pattern, process, scale, and synthesis. *Annual Review of Ecology, Evolution and Systematics* 34: 273–309.

Winkelmann, H. 1991. Liste der Rüsselkäfer (Col.: Curculionidae) von Berlin mit Angaben zur Gefahrdungssituation ("Rote Liste"). Pp. 319–357. *In* A. Auhagen, R. Platen, and H. Sukopp (eds). *Rote Listen der gefahrdeten Pflanzen und Tiere in Berlin.* Landschaftsentwicklung und Umweltforschung. Special issue 6.

Wood, S. L. and D. E. Bright, Jr.1992. A catalog of Scolytidae and Platypodidae (Coleoptera), Part 2: Taxonomic index. *Great Basin Naturalist Memoirs* 13: 1–833.

Woodley, N. E. 2001. A world catalog of the Stratiomyidae (Insecta: Diptera). *Myia* 11 (8): 1–475.

Yu, D. S. and K. Horstmann. 1997. A catalogue of world Ichneumonidae. Part 1: Subfamilies Acaenitinae to Ophioninae. *Memoirs of the American Entomological Institute* 58: 1–763.

Zamotajlov, A. S. 1992. *Fauna of Ground Beetles (Coleoptera, Carabidae) of the Northwestern Caucasus: A Manual for Students.* Kubanskii Gosudarstvennyi Sel'skokhozyaistvennyi Universitet, Krasnodar. 76 pp. [In Russian].

Zamotajlov, A. S. and R. Sciaky. 1996. Contribution to the knowledge of Patrobinae (Coleoptera, Carabidae) from south-east Asia. *Coleoptera* 20: 1–63.

Zarenkov, N. A. 1976. *Lectures on the Theory of the Systematics.* Izdatel'stvo Moskovskogo Universiteta, Moscow. 140 pp. [In Russian].

Zarkua, Z. D. 1977. Weevils (Coleoptera: Attelabidae and Curculionidae) of Abkhazia. Referate of the PhD Dissertation. Baku. Institut Zoologii Azerbaidzhanskoi Akademii Nauk Baku, 26 pp. [In Russian].

Zerova, M. D. 1995. *Parasitic Hymenoptera – Eurytominae and Eudecatominae of the Palearctic.* Navukova Dumka, Kiev. 456 pp. [In Russian].

Zhantiev, R. D. 1976. *Dermestids of the USSR Fauna.* Izdatel'stvo Moskovskogo Universiteta, Moscow. 181 pp. [In Russian].

Zhiltzova, L. A. 2003. Insecta Plecoptera. Volume 1. Issue 1. *Plecoptera, Group Euholognatha. Fauna of Russia and Neighboring Countries.* New Series N 145. Nauka, St. Petersburg. 537 pp. [In Russian].

Part II

Insect Biodiversity: Taxon Examples

Chapter 8

BIODIVERSITY OF AQUATIC INSECTS

John C. Morse

Department of Entomology, Soils and Plant Sciences, Clemson University, Clemson, South Carolina 29634-0315, USA

Insect Biodiversity: Science and Society, 1st edition. Edited by R. Foottit and P. Adler
© 2009 Blackwell Publishing, ISBN 978-1-4051-5142-9

W ater is critical for life on earth. Living organisms are composed mostly of water and require water for most of their metabolic functions. Terrestrial species have structures and methods for acquiring water periodically, or they live in habitats that are adequately moist to meet their needs. Many organisms, however, live fully submerged in aqueous habitats, so that water for them is unlimited, as long as the water does not evaporate or disappear for other reasons.

Three-quarters of the earth's surface is covered with oceans that are rarely or only marginally inhabited by insects. Several reasons have been proposed to explain why insects are so uncommon in oceans, such as low calcium concentration in seawater, competitive exclusion by Crustacea, or a need to evolve highly sophisticated osmoregulatory and respiration mechanisms simultaneously (Cheng 1976). Apart from oceans, water occurring on and under the surface of the ground is habitat for a highly diverse insect fauna. Unlike ocean water, this water occurring above the level of the oceans is fresh, with usually lower concentrations of salts and other solutes. Below the surface of the ground, water flows slowly in interstitial spaces of sediments and is especially abundant below the water table, often in aquifers, sometimes dissolving rocks such as limestone and creating subterranean cracks and caverns. Above the surface of the ground, habitable water occurs in a variety of lotic (flowing water) habitats including seeps, springs, creeks, and rivers; and many kinds of lentic (standing water) habitats such as pools, ponds, lakes, and the many relatively small quantities of water in puddles, aerial epiphytes, and artificial containers (e.g., footprints, birdbaths, exposed cisterns, open metal cans, old tires).

Water is of particular concern not only because of its critical importance for life, including human life, but also because it is the most obvious habitat into which the effects of terrestrial activities are concentrated. Precipitation carries to streams and groundwater most of the effects of activities occurring on otherwise 'dry land', including deposition of sewage, fertilizers, carcasses, and other nutrients; release of toxic manufactured products and byproducts; mobilization of sediments that result from exposure of soil to precipitation and runoff; and removal of trees and other riparian vegetation.

Because most aquatic insects remain below the surface of the water, they are rarely seen, so that the high diversity of insects that inhabit freshwater ecosystems for at least part of their lives is poorly known by most people. Most of these insect species, especially in tropical parts of the world, are still unknown even to scientists.

Along with other small invertebrates, aquatic insects are an indispensable part of the food web and of nutrient cycling in freshwater ecosystems (or 'spiraling' in streams, e.g., Newbold et al. 1982, 1983), and are an essential component in the diet of fish and amphibians and of many birds and mammals. For this reason, their imitations are attractive to the game of fly-fishing enthusiasts (e.g., McCafferty 1981). Because of the species-specific variation in tolerances that insects have for a wide range of environmental circumstances, insects also are widely used as indicators of the level of pollution in the waters that we drink and use for recreation and many practical purposes (Barbour et al. 1999).

Eggs, larvae, and pupae of some species of wasps (Hymenoptera) and moths (Lepidoptera), and especially of beetles (Coleoptera) and flies (Diptera), occur in freshwater ecosystems, and adults of some of the beetles also live there. Because these groups are discussed in more detail in other chapters, their aquatic species will be mentioned only briefly here. Some groups of insects have only a few species that live in the water or that live closely associated with water, such as springtails (Collembola), grasshoppers (Orthoptera), earwigs (Dermaptera), lice (Phthiraptera), and scorpionflies (Mecoptera). Because they contribute little to the themes of this chapter, they also will be mentioned only briefly. The focus of this chapter, therefore, will be the mayflies (Ephemeroptera), dragonflies and damselflies (Odonata), stoneflies (Plecoptera), bugs (Hemiptera-Heteroptera), hellgrammites (Megaloptera), spongillaflies (Neuroptera-Sisyridae, Nevrorthidae), and caddisflies (Trichoptera).

OVERVIEW OF TAXA

Springtails (Collembola)

The English common name for these small insects refers to their ability to jump by rapid release of a posterior leaping organ (furca or furcula) that is ordinarily folded beneath the abdomen and held in place by a ventral clasp (retinaculum or tenaculum). Some species are common skaters on the surface of temperate and subpolar waters or sprawlers on the shores of these waters. They feed on bits of decaying plants and animals and microflora (Waltz and McCafferty 1979, Christiansen and Snider 1996).

Mayflies (Ephemeroptera)

Mayflies occur globally in a wide variety of lotic and lentic habitats. The life cycle of a mayfly includes egg, larva, subimago (winged immature stage), and imago (male or female adult) stages. Eggs and larvae typically occur in natural lotic and lentic habitats; subimagoes and imagoes are aerial and terrestrial, usually flying or resting in vegetation near the habitat of the eggs and larvae. One life cycle usually is completed per year, but some species have more than one generation each year or require 2 years. Up to 9000 eggs (Fremling 1960) can be laid by a single female on the water surface by touching it or resting on it, below the water surface (a few *Baetis* species), or from the air above the water (a few species). Eggs usually require a few weeks to develop, but can enter diapause for 3–11 months. Larvae usually eat bits of organic matter collected or scraped from the substrate or filtered from suspension in the water; a few rare species are predators (Edmunds and Waltz 1996). A larva undergoes 9–45 molts before swimming to the water surface and emerging as a subimago (Unzicker and Carlson 1982). Usually the subimago flies into the nearby vegetation where it molts to an imago in a few hours. Adults mate usually while flying, typically after a female flies through a male swarm over or near water. They live for a few minutes to a few days and do not feed.

Dragonflies and damselflies (Odonata)

Dragonflies and damselflies are widely distributed and abundant in almost all permanent fresh and brackish water. They are particularly abundant in warmer waters such as those in lowlands of tropical and subtropical regions. A few species are semiaquatic in bog moss, damp leaves, and seepages, and a few tropical species inhabit water in bromeliads or tree holes. The Odonata have egg, larval, and adult stages, with eggs and larvae usually submerged in lentic or lotic habitats and aerial adults flying or resting near the water. Eggs can be endophytic, laid in living or dead plant tissue above or below the water through the use of an ovipositor, or they can be exophytic, laid in sand or silt in shallow areas of a stream or released above the water surface (Westfall and Tennessen 1996). They hatch in 5–30 days (Huggins and Brigham 1982), sometimes with a winter diapause (Westfall and Tennessen 1996). The predatory larvae pass through about 11 or 12 instars and complete a life cycle in as little as 1 month to as long as 5 years, depending on the species and the availability of invertebrate prey (Westfall and Tennessen 1996). Larvae of different species capture prey either by stalking them on the stems of aquatic plants or on the sediment surface, or by hiding in fine sediments and ambushing them. Adult emergence occurs in about half an hour, after a larva crawls out of the water onto some stable terrestrial substrate, often at night. The adult remains close to the water (e.g., Cordulegastridae) or flies away from water for one or more weeks, later returning to its species-specific habitat to feed, mate, and lay eggs (e.g., Aeshnidae). Adults are predatory, feeding on other flying insects captured in flight. Males often guard territory into which females are permitted or attracted. Mating typically occurs in tandem, with a male using the end of its abdomen to grasp the female behind the head while she receives the sperm from special copulatory organs on the male's second or third abdominal sternum, the male having transferred the spermatophore there earlier from the end of its abdomen. A significant monograph on odonate biology was provided by Corbet (1999).

Stoneflies (Plecoptera)

Stoneflies are distributed globally, mostly in cold mountain streams, but with about one-third of all species occurring in the tropics (Zwick 2003). These insects have egg, larval, and adult life history stages, with aquatic eggs and larvae and terrestrial adults. Eggs are laid above water or on the water surface. They sink to the bottom and attach to the substrate by a sticky gelatinous covering or by chorionic anchoring devices. The eggs hatch usually in 3–4 weeks, but can undergo diapause for 3–6 months in intermittent habitats, especially in the families Leuctridae, Nemouridae, Perlidae, and Perlodidae. Larvae are generally univoltine, with one generation each year, but they are sometimes semivoltine, with one generation every 2 or 3 years. Growth rates can vary in different taxa, especially in those that undergo egg or larval diapause in response to a variety of environmental conditions (Stewart and Stark 2002). The number of larval instars is variable in many stonefly species, ranging from 10–22+ instars (Sephton and Hynes 1982, Butler 1984). Larvae sprawl on the substrate or cling to rocks, feeding mostly as shredders of dead vegetation or predators of small arthropods (Stewart and Harper 1996).

Emergence of adult stoneflies is similar to that of adult Odonata – larvae crawl out of the water usually at night, attach to a rock or log, and exit the last larval exuviae in about 5–10 minutes (Stewart and Stark 2002). Adults live 1–4 weeks and some species feed on epiphytic algae or young leaves or buds (Stewart and Harper 1996). Communication for mating is often facilitated by tapping ('drumming') on substrates with the end of the abdomen, the vibrations being detected in the potential mate's subgenual organs (Stewart and Maketon 1991). Mating is accomplished while standing and facing the same direction, with the male standing on top of the female (Stewart and Harper 1996).

Grasshoppers and crickets (Orthoptera)

This familiar group of insects is mostly terrestrial, feeding usually on living plant tissue. A few species feed on plants beside or emerging from streams and ponds (Cantrall and Brusven 1996). One species that feeds on water hyacinth is being bred for possible release as a means to control this weed in the wild (Franceschini et al. 2005, Lhano et al. 2005, Adis et al. 2007).

Earwigs (Dermaptera)

One species of this order, *Anisolabis maritima* (Bonelli), is a predator of small animals beneath litter and driftwood on marine beaches (Langston 1974, Langston and Powell 1975).

Lice (Phthiraptera)

A few species of lice are ectoparasites of birds and mammals that live in freshwater and marine habitats. Some species of Menoponidae, for example, feed on bits of feathers, skin, and blood of geese and ducks (Price 1987); and Echinophthiriidae suck blood of otters, sea lions, seals, and walruses (Kim 1987).

Bugs (Hemiptera)

Most of the cicadas, hoppers, scale insects, and their relatives (suborders Sternorrhyncha and Auchenorrhyncha) in this order live on plants that have little or no association with surface water. However, several families of bugs in the suborder Heteroptera are intimately associated with freshwater habitats. Hemiptera probably evolved an aquatic lifestyle at least five times (Carver et al. 1991): (1) A few species of auchenorrhynchous leafhoppers (Cicadelloidea) feed on the emergent portions of freshwater vascular plants. Among the Hemiptera-Heteroptera (2) the Dipsocoroidea are shore-dwelling burrowers under stones, (3) the Gerromorpha (or 'Amphibicorisae' or 'Amphibicorizae') are all water-surface-dwelling bugs, (4) the Leptopodomorpha are shore-dwelling climbers and burrowers, and (5) the Nepomorpha (or 'Hydrocorisae' or 'Hydrocorizae') mostly swim beneath the water surface. Most of the species diversity for aquatic bugs is in the Gerromorpha and the Nepomorpha. A variable number of eggs is usually laid beneath or at the surface of the water in plants or in sand, fastened to the substrate at one end with a sticky substance or a slender stalk; females of species of the giant water bug genera *Abedus* and *Belostoma* (Belostomatidae) lay eggs on the backs of males, where they are aerated and protected from predation (Yonke 1991). Larvae resemble adults in general body form, although lacking wings and reproductive structures, and live usually in the same aquatic habitats. Most larvae grow through five instars, usually with one generation each year, although some are multivoltine (Sanderson 1982). Surface-dwelling and semiaquatic families are usually most common in late summer, whereas Nepomorpha families are most frequent in late fall. Respiration for Nepomorpha usually is by means of a transported air bubble that is replenished from time to time; a few species of Naucoridae and Aphelocheiridae respire with a plastron (permanent air film); all others are air breathers, including Nepidae and some Belostomatidae that access the air with respiratory siphons or straps (Sanderson 1982, Polhemus 1996). Most Heteroptera reduce their susceptibility to predation with noxious excretions from abdominal (larvae) or thoracic (adults) scent glands, or use these substances to groom themselves, reducing microbial growth on their bodies (Kovac and Maschwitz 1989). All Hemiptera larvae and adults have sucking mouthparts; these are organized into a three- or four-segmented beak for most Heteroptera (Corixidae are an exception) and are used primarily to feed on fluids of plants or animals. The forewings of adult Heteroptera are leathery basally and membranous apically, but many species are apterous as adults and many other species have both macropterous and brachypterous forms. The habitats of the aquatic Heteroptera are highly variable among the families and genera, but most often they involve emergent vegetation in shallow water. Other habitats include the shores of surface freshwaters (many taxa), marine intertidal zones (some Saldidae), the surface of the open sea (genus *Halobates* in Gerridae), ocean beaches, estuaries, brackish and alkaline lentic habitats, high mountain lakes, hot springs, roadside ditches, temporary pools, bogs, swamps, and running water of all types (Polhemus 1996).

Wasps (Hymenoptera)

Eleven families of the Hymenoptera include species that are considered aquatic, parasitizing primarily the eggs or larvae of other aquatic insects (Bennett 2008). The Agriotypinae are external parasitoids of Trichoptera pupae (Elliott 1982). Spider wasps of the genus *Anoplius* (Pompilidae) capture and paralyze submerged spiders of the genus *Dolomedes*, carrying them one at a time to a nest in the bank, and laying an egg externally so that the resulting wasp larva can feed on the spider (Evans and Yoshimoto 1962).

Hellgrammites and alderflies (Megaloptera)

The larvae of the Megaloptera are aquatic, whereas the eggs, pupae, and adults are terrestrial. Several hundreds of the tiny cylindrical eggs are laid in rows or layers in rounded or quadrangular masses on substrates above the water (Evans and Neunzig 1996). Hatchlings drop to and through the water surface and swim to appropriate habitat (Brigham 1982a). Larvae live usually in rivers, permanent and temporary streams, spring seeps, ponds, or swamps, or in lakes near wave-washed shores, occurring in leaf litter, among coarse rubble, or in soft organic sediment. They are predators on a wide variety of macroinvertebrates and require 1–5 years to complete development, with larvae passing through up to 12 instars (Evans and Neunzig 1996). They eventually crawl out of the water to fashion unlined pupation chambers in soil and litter near the larval habitat, under rocks, sometimes in soft, rotting shoreline logs or stumps or in dry stream beds (Brigham 1982a, Neunzig and Baker 1991). Adult mating and egg-laying activity can be diurnal or nocturnal; apparently adults do not feed (Evans and Neunzig 1996).

Nerve-winged insects (Neuroptera)

Most species of Neuroptera (or Planipennia) are terrestrial. However, the larvae of Nevrorthidae and Sisyridae are aquatic. Larvae of at least the European *Nevrorthus* species of Nevrorthidae are eurythermic, living in creeks and small rivers with temperatures ranging from 2°–23.8°C; unlike Sisyridae, they pupate under water and apparently complete one life cycle each year (Zwick 1967, Malicky 1984).

Larvae of the Sisyridae feed on freshwater sponges with stylet-like mouthparts that are a coaptation of their mandibles and maxillae; eggs, pupae, and adults are terrestrial. Larval development is accomplished in three instars; *Climacia areolaris* (Hagen) produces as many as five generations each year (White 1976). Usually, a cluster of a few tiny eggs are deposited in crevices or depressions on objects overhanging the water, such as the undersides of leaves in forks of leaf veins, and covered by a tent of three or four layers of white, silken threads. Hatchlings drop through the water surface and swim in search of sponges where they settle for the remainder of their development (Pennak 1978). They pierce the sponge cells and suck their contents without imbibing the dangerous silicate spicules, sometimes cyclically feeding for a few minutes then wandering for the next several minutes. Larvae camouflage themselves with pieces of sponge impaled on the sharp spines protruding from their bodies, and respire with gills folded beneath the abdomen. They eventually swim or crawl to shore and then walk as much as 15 m inland to pupate in a double-layered silken cocoon in the shelter of a bark or rock crevice. After a few days, they emerge as adults, mate, and lay eggs for about 2 weeks, apparently sustained by pollen and nectar (Brigham 1982b).

Scorpionflies (Mecoptera)

The Nannochoristidae are the only scorpionflies with species having aquatic larvae, occurring only in the Australasian and Neotropical regions. The long, slender, wireworm-like larvae live in mud, and prey on larvae of midges (Diptera: Chironomidae) (Pilgrim 1972).

Beetles (Coleoptera)

The beetles evidently evolved into freshwater habitats several times, with freshwater species occurring globally in a wide range of lentic and lotic habitats or at the margins of these habitats; some beetles occur in marine intertidal situations (White and Brigham 1996). These independent evolutionary events in their various stages of transition to aquatic habitats have resulted in a variety of living strategies for the eggs, larvae, pupae, and adults. Truly aquatic forms have eggs, larvae, and adults, all generally aquatic and pupae generally terrestrial (e.g., Elmidae). Some truly aquatic taxa have only eggs and larvae aquatic, with pupae and adults terrestrial (e.g., Psephenidae). In others, only the adults are aquatic (e.g., Dryopidae, Hydraenidae). Feeding habits are variable, with larvae or adults of the same species sometimes differing in feeding habits; predation, herbivory (either vascular plants or algae scraped from rocks), and detritivory are common (White and Brigham 1996). Respiration by adults is generally by air bubble or plastron (permanent air film), whereas respiration by larvae is generally transcuticular (with or without gills) or involves piercing plant tissues (Eriksen et al. 1996). Eggs are laid in various situations and hatch in 1–2 weeks. Development of larvae is usually accomplished with three to eight instars in 6–8 months (White and Brigham 1996). Pupae usually develop in 2–3 weeks (White and Brigham 1996), although some overwinter (e.g., Chrysomelidae: *Donacia*) (P. Zwick, personal communication). Most beetles are univoltine in temperate regions but can have more than one generation per year in tropical or subtropical climates.

Caddisflies (Trichoptera)

Essentially all caddisflies are aquatic, with eggs, larvae, and pupae occurring in a wide variety of freshwater habitats and adults flying and resting in nearby terrestrial habitats. Most species are oviparous, with tens to hundreds of eggs per distinctively shaped eggmass covered with polysaccharide spumalin and deposited on substrates suspended above water, at the water's edge, or submerged under water after deposition by swimming or crawling females (Unzicker et al. 1982). Larvae typically develop through five instars, completing growth in 2 months to 2 years, with most life cycles univoltine (Wiggins 1996). First-instar diapause has been reported for species in temporary pools (Wiggins 1973) and fifth-instar aestivation for species in temporary streams (Wiggins 1996).

Larval habits, habitats, and feeding strategies are strongly correlated with their architectural behavior, facilitated by use of labial silk (Mackay and Wiggins 1979). Five general architectural strategies for case-making or retreat-making are recognized (Wiggins 2004): (1) Rhyacophiloidea are free-living, building a dome-like shelter in the last instar for pupation, which occurs in an interior cocoon. (2) Glossosomatidae build a portable dome-like shelter precociously in all larval instars, transporting it much like a turtle wears its shell; the transverse ventral strap is removed and the resulting dome cemented with silk to a stone for pupation, which occurs in an interior cocoon. (3) Hydroptilidae make, in only the last instar, a usually portable purse-like depressed or compressed case with silk and often various mineral or plant inclusions; the case is attached to the substrate and the ends are sealed for pupation, which occurs in an interior cocoon. (4) Species of the suborder Integripalpia make a portable case that is essentially tubular, composed of silk and mineral or plant material; the case is usually attached to the substrate, and porous silken sieve plates seal the ends for pupation without a cocoon. (5) Species of the suborder Annulipalpia make a stationary retreat of silk that often has a filtering function to remove suspended nutrient particles from the water; a separate dome-like shelter is constructed usually of mineral material for pupation, without a cocoon.

Larval habits include swimming in the water column or burrowing in the substrate or clinging, sprawling, or climbing on substrate surfaces (Wiggins 1996). Larvae of various species feed as shredding and piercing herbivores, shredding and collecting detritivores, gougers of wood, filterers of food particles from the water column, grazers on substrate biofilm, predators, and parasitoids (Wells 1992, Wiggins 1996).

After 2–3 weeks as a pupa, a mature pupa typically uses large mandibles to cut through the shelter and any cocoon and then swims to the water surface or crawls out of the water (Unzicker et al. 1982) where it emerges in less than a minute as an adult and flies into the sheltering riparian vegetation.

Adults live for a few days to 3 months (Unzicker et al. 1982), subsisting on nectar and water obtained with imbibing mouthparts (Crichton 1957). They fly most readily in early evening after sunset, but some are nocturnal and others are diurnal (Unzicker et al. 1982). Mating is facilitated by pheromones (Wood and Resh 1984), by substrate vibrations (Ivanov and Rupprecht 1993), and by swarming dances (Solem 1984, Ivanov 1993). Sperm transfer is accomplished with male and female end-to-end, facing in opposite directions while standing on shoreside substrate (Ivanov and Rupprecht 1993).

Moths (Lepidoptera)

Most species of Lepidoptera are terrestrial herbivores. Many species feed on the emergent parts of plants that grow in the water. Most species of the crambid subfamily Nymphulinae are truly aquatic, with eggs, larvae, and pupae fully submerged in water. Caterpillars of the nymphuline tribe Nymphulini are herbivores of many freshwater plant species (Brigham and Herlong 1982), whereas those of the tribe Argyractini live under silk tents on rocks, and scrape biofilm from the rock surfaces (Lange 1996).

Flies (Diptera)

All Diptera are aquatic in the broadest sense, requiring a humid environment for development. Eggs, larvae, and pupae of most families of Nematocera and several families of Brachycera develop only when submerged in wet environments, flies having evolved into these habitats at least eight times independently (McAlpine and Wood 1989). These insects live in almost every type of aquatic habitat, including coastal marine and brackish waters and brine pools; shallow and deep lakes; ponds; geyser pools up to 49°C; natural seeps of crude petroleum; stagnant or temporary pools and puddles; water in bromeliads, pitcher plants, and artificial containers; and slow to fast flowing streams and rivers (Courtney et al. 1996a).

Much of human history has been affected by the role of flies in disease transmission, both in determining the outcomes of war and in retarding economic development. Many of the carriers of these disease agents develop in freshwater habitats, especially lakes, ponds, streams, and artificial containers with small amounts of water, such as discarded open cans and old tires. At the same time, in aquatic ecosystems, flies often play a key role in food webs, consuming large amounts of detritus and serving as the primary food source for other freshwater organisms (Courtney et al. 1996b). In freshwater habitats, larvae are planktonic in the water column, burrowers in the substrate, or clingers or sprawlers on substrates (Courtney et al. 1996a). Their feeding strategies are as diverse as those of caddisflies,

including shredding living and dead plant tissue, gouging wood, filtering suspended food particles from the water column, collecting bits of food from the substrate, grazing on biofilms, and preying on or parasitizing other macroinvertebrates (Courtney et al. 1996a).

SPECIES NUMBERS

Table 8.1 provides not only the number of species currently known for the various orders of aquatic insects, but also for some groups, the number that has been estimated actually to exist, including yet-unknown species. In some cases, not more than about 20% of the species are presently known to science, with a substantial majority yet to be discovered.

Geographically, the tropical and subtropical regions include most of the yet-unknown species of aquatic insects. For example, Schmid (1984) estimated that 50,000 species of caddisflies occur in the world (of which only 25% are presently described), with 40,000 of those species occurring in the tropics and subtropics of southern Asia. The density of caddisfly species in the Oriental Region is about twice that of the next most densely speciose region (Neotropical) (Morse 1997), yet species discovery continues in the Oriental Region (e.g., Yang and Morse 2000), without any sign of waning.

SOCIETAL BENEFITS AND RISKS

The benefits and risks of aquatic insect biodiversity cannot be appreciated unless that biodiversity is known. Many species of aquatic insects remain unknown to science, and many have been collected and recognized as yet undescribed. Species accumulation curves suggest that we have not yet approached an asymptote, so that many additional yet-unknown species have not been collected, especially in tropical and subtropical waterways (Morse 2002).

In biomonitoring programs to assess water quality, one of the major impediments is a lack of ability to identify larvae at the species level. Larvae are most commonly collected in stream assessments, but the taxonomy of most groups began with the adults, often with just one gender. For this reason, larvae are poorly known, with not more than about 50% of the species recognizable as larvae in regions where most species are otherwise known to science, and often less than 5%

Table 8.1 Major orders (and Diptera families) of aquatic insects, with estimates of the known number of species and, for selected taxa, the predicted number of species.

Taxa	Number of Known Species	Predicted Total Species
Ephemeroptera	3046[1]	4000[2]
Odonata	5680[3]	7000[3]
Plecoptera	3497[4]	7000[4]
Aquatic Orthoptera	188[5]	
Aquatic Heteroptera	4656[6]	5770[6]
Megaloptera	328[7]	370–400[7]
Aquatic Neuroptera	73[7]	
Aquatic Mecoptera	8[8]	
Aquatic Coleoptera	12,604[9]	18,000[9]
Aquatic Hymenoptera	150[10]	
Trichoptera	12,868[11]	50,000[12]
Aquatic Lepidoptera	740[13]	
Diptera:	46,259–46,269	114,165–120,348
Tipuloidea	15,178[14]	
Ptychopteridae	69[15]	
Thaumaleidae	174[16]	350–520[17]
Blephariceridae	322[15]	500[18]
Deuterophlebiidae	14[19]	20[18]
Nymphomyiidae	7[19]	10[18]
Dixidae	180[20]	500[41]
Scatopsidae	5[21]	
Tanyderidae	41[20]	
Ceratopogonidae	5598[22]	15,000[22]
Chironomidae	4147[23]	10,000[24]–15,000[25]
Chaoboridae	50[26]	
Corethrellidae	97[27]	
Simuliidae	2015[28]	3050[28]
Culicidae	3492[29]	
Axymyiidae	5[18]	10[18]
Athericidae	90[30]–100[31]	150[30,31]
Oreoleptidae	1[18]	
Pelecorhynchidae	46[18]	
Tabanidae	4300[32]	
Aquatic Stratiomyidae	889[33]	1200[30]
Aquatic Psychodidae	1988[20]	19,880[20]
Empididae	671[34]	2000[17]
Dolichopodidae	3182[35]	31,820[35]
Ephydridae	1251[36]	2500–3000[37]
Canacidae	286[37]	450–500[37]
Coelopidae	29[37]	
Heleomyzidae	12[37]	
Heterocheilidae	2[37]	
Aquatic Dryomyzidae	3[37]	
Aquatic Muscidae	599[38]	1200–1500[39]
Aquatic Phoridae	17[40]	51[41]–160[42]

Table 8.1 *(continued)*.

Taxa	Number of Known Species	Predicted Total Species
Aquatic Sciomyzidae	156[43]	306[44]–360[45]
Lonchopteridae	2[46]	
Aquatic Syrphidae	1341[47]	1841[48]
TOTAL	**90,097–90,107**	**207,464–213,677[49]**

[1] Barber-James et al. (2008).
[2] M. Sartori (personal communication).
[3] Kalkman et al. (2008).
[4] Fochetti and de Figueroa (2008).
[5] Amedegnato and Devriese (2008).
[6] Polhemus and Polhemus (2008).
[7] Cover and Resh (2008).
[8] Ferrington (2008a).
[9] Jäch and Balke (2008).
[10] Bennett (2008).
[11] Morse (2008).
[12] Schmid (1984).
[13] Mey and Speidel (2008).
[14] de Jong et al. (2008).
[15] Zwick in Wagner et al. (2008).
[16] Sinclair and Wagner in Wagner et al. (2008).
[17] B. J. Sinclair (personal communication).
[18] G. W. Courtney (personal communication).
[19] Courtney and Wagner in Wagner et al. (2008).
[20] Wagner in Wagner et al. (2008).
[21] Haenni in Wagner et al. (2008).
[22] Borkent in Wagner et al. (2008).
[23] Ferrington (2008b).
[24] P. Cranston (personal communication).
[25] O. Saether (personal communication).
[26] Wagner and Goddeeris in Wagner et al. (2008).
[27] Borkent and Wagner in Wagner et al. (2008).
[28] Adler and Crosskey (2008), Currie and Adler (2008).
[29] Rueda (2008).
[30] Woodley (2001).
[31] D. Webb (personal communication).
[32] J. F. Burger (in P. H. Adler, this volume).
[33] Rozkosny and Woodley in Wagner et al. (2008).
[34] Sinclair in Wagner et al. (2008).
[35] R. L. Hurley (personal communication).
[36] Zatwarnicki in Wagner et al. (2008).
[37] W. Mathis (personal communication).
[38] Pont in Wagner et al. (2008).
[39] A. Pont (personal communication).
[40] Disney (2004).
[41] R. H. L. Disney (personal communication).
[42] B. V. Brown (personal communication).
[43] Rozkosny and Knutson in Wagner et al. (2008).
[44] L. Knutson (personal communication).
[45] R. Rozkosny (personal communication).
[46] Bartak in Wagner et al. (2008).
[47] Rotheray in Wagner et al. (2008).
[48] G. E. Rotheray (personal communication).
[49] Conservative estimate. Where a predicted total estimate for a taxon is not available, the number of known species only is included.

of the species recognizable as larvae in regions where species are less well known (Morse 2002).

When trying to find herbivores that can control invasive aquatic weeds, the countries from which the weeds originally came often must be explored. In many such searches, herbivores are found that could serve this purpose, but they have not yet been described and named. Description and naming is usually the first step toward discovering the bionomics of a species, its potential hosts, and environmental requirements – data that are necessary before deciding whether to import the species into a nonnative country.

Similarly, medically important species, especially in tropical regions, often are poorly known to science. To learn the biology of these vectors of disease agents sufficient to devise appropriate control measures, their diversity and identities must be discovered.

At least three additional reasons for discovering biodiversity in aquatic insects pertain to widespread human values, including (1) our natural affiliation with the diversity of living things, (2) our moral obligations to protect species, and (3) our innate curiosity about our amazingly diverse natural world (Kellert 1996). With respect to the first of these values, Wilson's (1984) Pulitzer-Prize-winning treatise on 'Biophilia' argued that we have a natural affiliation with the diversity of living things, reflecting our evolutionary origins surrounded by them. Many people feel an intense reverence and sense of peace when immersed in a natural setting with high species diversity and abundance

(Fuller et al. 2007). Many people also recognize a moral obligation generally derived from their religious traditions to be good stewards of biological diversity. Such people are convinced that 'all creatures great and small' are innately deserving of protection from harm or extinction (Hamilton and Takeuchi 1991, Berry 2000, Lodge and Hamlin 2006). From both of these perspectives, discovery of biodiversity is implicit because appreciation and stewardship of biodiversity require knowledge of it. The third of these values lies in the need to satisfy our inborn curiosity about the world around us, including the species inhabiting it, with all of their wonderful and amazing shapes and colors and behaviors and functions. A drive to learn as much as possible about the other species with which we share our planet is as fundamental to who we are as humans as any of our other innate characteristics.

A high diversity of aquatic insect species is of value to people for a variety of other reasons, but four are particularly important, including the role of insects in food webs, in biomonitoring, in fishing, and in control of noxious weeds. A few species of aquatic insects pose some risks, as well.

Societal benefits of aquatic insect diversity in food webs

In food webs, aquatic insects capture, use, and make available to other freshwater organisms nutrients that otherwise would be unavailable. In general, they do this by processing nutrients from coarse particulate organic matter (CPOM) and from fine particulate organic matter (FPOM) (Cummins and Merritt 1996).

CPOM occurs in freshwater ecosystems in the form of both living and dead vascular plants and larger animals. Dead plant material is especially abundant, but cannot be consumed by many animals because it is either too large or mostly indigestible. The nutrients from detritus often come indirectly from the CPOM, having been consumed first by communities of bacteria and fungi; in other words, much of this material must first be 'microbially conditioned' by colonizing fungi and bacteria before it can be assimilated (Cummins and Klug 1979). Shredding herbivores, shredding detritivores, predators, and parasites fragment CPOM and release smaller organic particles as feeding scraps and feces; their own bodies assimilate nutrients from CPOM and become available to many predators and parasites.

FPOM occurs in freshwater ecosystems as tiny particles of both living and dead plant and animal material or as flocculated dissolved organic matter (DOM). FPOM is abundant, but cannot be consumed by many animals because it is too small for efficient consumption, and cannot be used by plants because its nutrients are not in solution. Among the aquatic insects, many collecting gatherers and suspension filterers proficiently consume these materials (Wallace and Merritt 1980, Merritt and Wallace 1981) and then become prey for freshwater and nearby terrestrial predators.

Aquatic insects are, thus, important links in food webs to assure that nutrients are passed to other members of the community (Fig. 8.1). Higher diversity and abundance of aquatic insects help these nutrients pass more efficiently to other animals, thereby retaining the nutrients in the ecosystem longer, helping to assure the health of the overall ecosystem (Newbold et al. 1982, 1983).

A high abundance (or density) of aquatic insects helps assure the processing of large amounts of nutrients. A high diversity (or taxa richness) of aquatic insects helps assure diversity of resources (e.g., nutrients, habitats) and effective use of all available resources in both space and time.

Societal benefits of aquatic insect diversity in biomonitoring

In many parts of the world, the quantity of water is adequate to sustain human life, but potable water is scarce. Pollution is often so severe that human health is jeopardized.

Pollution has been defined as changes in the physical, chemical, or biological characteristics of water, air, or soil that can affect the health or activities of living organisms (Miller 1988). Historically, attempts have been made to monitor water pollution by chemical analyses. A water sample is taken at a particular point in time and analyzed for one or more suspected chemical pollutants. Based on the principle in Miller's (1988) definition that pollution is best recognized in living organisms, a more modern technique for monitoring water quality is toxicity testing, whereby surrogate organisms are subjected to different concentrations of test effluents (Thompson et al. 2005). However, scientists in many countries, especially more developed countries, have been using communities of freshwater macroinvertebrates, including

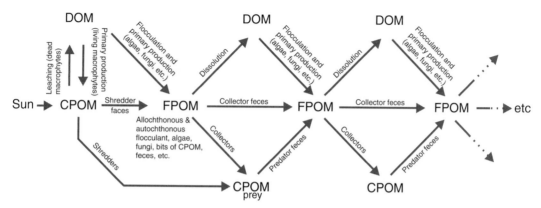

Fig. 8.1 The nutrient spiraling concept provides an understanding of the way that organic nutrients from living or dead plants or animals cascade through coarse particulate organic matter (CPOM, larger than $10 \mu^3$), dissolved organic matter (DOM), and fine particulate organic matter (FPOM, smaller than $10 \mu^3$) by the action of physical, chemical, and biological agents as this matter passes downstream (Newbold et al. 1982, 1983).

insects, as an alternative method to monitor the water biologically (bioassessment or biomonitoring). Unlike chemical analyses, these animals 'sample' the water continuously and 'test' for all biologically relevant substances. Furthermore, these animals respond more meaningfully to the combined (synergistic) effects of these substances. As a consequence, bioassessment reflects the ecological integrity of freshwater ecosystems and is less expensive than chemical testing and toxicity testing. The status of living organisms also is of direct interest to the public as a measure of a pollution-free environment. Biomonitoring is the only practical means for evaluating freshwater ecosystems, where criteria for specific ambient effects do not exist (e.g., where thresholds for ecosystem or human health have not been established) (Barbour et al. 1999).

Biomonitoring protocols have been established in several countries, using not only macroinvertebrates, but also fish and algae. For example protocols have been developed for Australia (Australian River Assessment System 2005), Canada (Rosenberg et al. 2005), the European Union (European Union Water Framework Directive 2000), New Zealand (Stark et al. 2001), UK (RIVPACS 2005), and USA (Barbour et al. 1999).

Biomonitoring efforts that use aquatic insects and other macroinvertebrates have a number of benefits. Macroinvertebrates are good indicators of localized conditions because, unlike fish, they do not migrate appreciably. They integrate effects of short-term environmental variations more effectively than do algae

because they usually have much longer life cycles. Degraded conditions can be detected by experienced biologists, with cursory examination because of ease of identification to family or lower taxonomic levels, unlike for algae. There are usually many species in a given waterway, each with its own ecological requirements, so that collectively the species exhibit a wide range of trophic levels (showing cumulative effects) and pollution tolerances, unlike fish. Sampling is relatively easy and requires few people and inexpensive gear, unlike for fish. In comparison with fish, insects are more abundant and diverse in most streams and lakes, permitting computation of statistically reliable results. In addition, sampling of macroinvertebrates has minimal detrimental effects on the resident biota. They are a primary food source for fish, including recreationally and commercially important species, and most governmental water-quality agencies that undertake biomonitoring routinely collect macroinvertebrates (Barbour et al. 1999). Several developing countries are drafting protocols for using insects in freshwater biomonitoring (e.g., Morse et al. 2007).

The advantages of biomonitoring with insects and other living organisms can be realized only if the natural fauna of the region is known. Where the species diversity of freshwater insects is poorly known, biomonitoring is possible only at relatively crude taxonomic levels, such as order or family. Because different species have different tolerances to pollution, the more inclusive ordinal and familial taxa characterize water quality much too broadly for refined biomonitoring purposes. In other words, before biomonitoring can be most effective, background knowledge of the insect biodiversity at the genus and species levels must be obtained from the wide variety of freshwater habitats of the region.

In many parts of the world, only a small portion of the freshwater insect species is known to science. For example, at least 75% of the caddisfly species of southern China are unknown (Yang and Morse 2000), and despite many years of taxonomic effort in Thailand and surrounding countries, large numbers of freshwater insect species continue to be discovered and described there each year (e.g., Malicky and Prommi 2006).

Government regulatory agencies in most developed countries now have well-established protocols for monitoring the quality of surface water by sampling and identifying their aquatic insects. In many of these countries, private companies with approved laboratories contract their sampling and identification services with these agencies or with private industries and municipalities regulated by the government agencies. Less well-developed countries are at various stages of enacting and enforcing laws and regulations standardizing this technology, staffing and equipping government laboratories implementing it, and encouraging private consulting firms to assist in using it. Common impediments in these countries for using insect biodiversity in freshwater biomonitoring include limited scientific knowledge of the fauna, few or no training opportunities, limited equipment and literature, and poor understanding of the benefits (Morse et al. 2007).

Traditionally, scientists from developed countries have staged expeditions to gain better knowledge of the insect faunas in poorly explored countries, with minimum collaboration in the source countries. These scientists have taken the captured specimens back to their own well-equipped museums and published the results by themselves in their own national journals. More recently, increasing concerns about protecting national biodiversity from commercialization by foreign entrepreneurs has increased the difficulty for foreign scientists to capture and describe the fauna of many countries, including many where water pollution is a serious problem. Consequently, recent biodiversity research has been much more collaborative, involving indigenous scientists in all its phases, and helping to train a new generation of indigenous taxonomists to continue the process of describing the native faunas (Morse 2002).

Societal benefits of aquatic insect diversity in fishing

Because insects are so important in the diets of many fish species, including species that are commonly consumed by humans for food, aquatic insect biodiversity is of considerable interest to society. People who fish with natural or artificial baits have long had particular interest in them. Records of human interest in aquatic insects, with respect to fish food, date to at least 200 CE (Aelianus 1611). An especially well-known book on the subject is Izaac Walton's *The Compleat Angler* (1653), 'one of the three most published books in English literature (the other two are *The Bible* and *The Complete Works of Shakespeare*). *The Compleat Angler* has run to more than 300 editions' (Herd 2006).

Anglers for centuries have attempted to imitate the form and color of various aquatic insects on hooks (or angles) in the hope of tricking fish to swallow them and become snagged. Mayflies, caddisflies, stoneflies, and midges have been groups whose species are most commonly imitated. Larvae, pupae, and adults are imitated and presented to the fish in ways intended to replicate the behavior of those forms as they grow on the bottom substrate, drift in the current, emerge from the water surface (often in synchronized mass emergences, or 'hatches'), or return to the water as egg-laying females or spent adults. A high diversity of these insects in a particular stream, each with its own specific emergence time, assures that food is available to the fish through much of the year and through different times of day. This aquatic insect species diversity and the diversity of behaviors they exhibit have contributed to the sporting challenge of 'matching the hatch' (Caucci and Nastasi 1975).

Societal benefits of aquatic insect diversity in control of noxious weeds

Several species of noxious, invasive weeds have become problems in parts of the world where they outcompete native species, clog otherwise navigable waters and water-intake structures, and exclude food-fish species. Herbicides often are employed to control these weeds, but some success also has resulted from the introduction of insect herbivores (Buckingham 1994). For example, in the USA, alligatorweed (*Alternanthera philoxeroides*), an invasive species from South America, has been controlled successfully by three imported herbivores: alligatorweed stem borer, *Arcola malloi* (Lepidoptera: Crambidae); alligatorweed flea beetle, *Agasicles hygrophila* (Coleoptera: Chrysomelidae); and alligatorweed thrips, *Amynothrips andersoni* (Thysanoptera: Phlaeothripidae) (Center for Aquatic and Invasive Plants 2007a). Another example is the successful control of common water hyacinth, *Eichhornia crassipes*, an invasive species from Brazil, by two imported species of weevil (mottled water hyacinth weevil, *Neochetina eichhorniae*; and chevroned water hyacinth weevil, *N. bruchi* (Coleoptera: Curculionidae)) and one species of an imported moth (Argentine water hyacinth moth, *Sameodes albiguttalis* (Lepidoptera: Crambidae)) (Center for Aquatic and Invasive Plants 2007b).

Societal risks of aquatic insects

Several species of aquatic insects are a nuisance to people, annoying them by biting, landing, or simply flying about. Adults also foul automobile windshields and wet paint (Hickin 1967) and, when present in large numbers, can make highway bridge surfaces dangerously slick (Edmunds et al. 1976). Caddisfly larvae with filter nets can clog water-intake structures of hydropower plants (Ueno 1952). Adults of various aquatic Diptera (especially mosquitoes, black flies, and sand flies) transmit some of the most serious disease agents to humans and their domesticated animals (e.g., Greenberg 1971, 1973). Hairs of adult caddisflies or other aquatic insects can cause allergic reactions (Henson 1966).

Among the aquatic insects, agricultural pests on rice include the rice water weevil (*Lissorhoptrus oryzophilus*, Coleoptera: Curculionidae; Saito et al. 2005), the white stem borer and yellow stem borer (*Scirpophaga innotata* and *S. incertula*, respectively, Lepidoptera: Crambidae; Claridge 2006), and some species of the caddisfly families Leptoceridae (Moretti 1942, Wiggins 2004) and Limnephilidae (Hickin 1967). Also, various species of Lepidoptera (e.g., *Plutella xylostella*, diamondback moth), Coleoptera (e.g., *Phaedon* spp., watercress leaf beetle), and Trichoptera (e.g., *Limnephilus lunatus*, *Drusus annulatus*) are pests of commercial watercress (*Nasturtium nasturtium-aquaticum*) (Gower 1965,

1967, Nakahara et al. 1986, Tanaka and Takahara 1989).

BIODIVERSITY CONCERNS FOR AQUATIC INSECTS

Four general concerns are especially deserving of immediate and long-term attention, including (1) threats to freshwater species as a result of pollution and habitat alteration, (2) the need for discovery of species diversity and description of freshwater larvae especially in countries with severe water-pollution problems, (3) the need for refinement of species definitions and associated biological information among species that transmit disease agents of humans and their animals, and (4) a trend for young people today to know less about species, including species of aquatic insects, and to be less inclined to follow careers intent on discovering them and adding to our knowledge about them and their ecological requirements.

Threats to freshwater species of insects

In the USA, 43% of the species of stoneflies and 18% of dragonflies and damselflies are considered 'species at risk', being either vulnerable (27% and 10%, respectively), imperiled (12% and 5%), critically imperiled (2% and 2%), or possibly/presumed extinct (2% and 0.4%) (Master et al. 2000). Among all groups of organisms considered in the inventory of Master et al. (2000), the five groups with the highest percentage of species at risk (freshwater mussels, crayfishes, stoneflies, freshwater fishes, and amphibians), all rely on freshwater habitats. Between the two groups of aquatic insects that were considered, the stoneflies generally inhabit cold, clear, headwater streams, whereas the Odonata more often live in warmer waters, suggesting that headwater streams are at greater risk in the USA. Morse et al. (1993, 1998) also emphasized the vulnerability of species occurring in headwater streams of the southern Appalachian Mountains.

Although extinctions and evolution of new species have probably always occurred since life began on Earth, 'estimates of the current rate of extinctions are conservatively estimated to be 100–1000 times greater than background levels, and evolution of new species is unlikely to see a similar increase in rate to offset these losses' (Master et al. 2000). Freshwater ecosystems are especially vulnerable. Barber-James et al. (2008) considered deforestation the principal threat to diversity of mayflies and other aquatic insects in the tropics (e.g., Madagascar, Borneo), and pollution, reshaping of banks with disconnection from floodplain, and habitat fragmentation the main threats in temperate regions. Habitat degradation and loss (especially from water developments such as dams and other impoundments) and pollution (including siltation and nutrient inputs, e.g., from agriculture) are major threats to species in the US waterways, along with livestock grazing, road construction and maintenance, logging, and mining (Richter et al. 1997, Wilcove et al. 2000).

Need for biodiversity discovery and description of aquatic insects

The discovery process for aquatic insects should be dramatically accelerated because of the unprecedented rate that species are being lost, along with the potential benefits that they could provide. For those same reasons, description of all life history stages for these insects is needed, particularly of the larval stage, which usually extends over the longest portion of an insect's life and is the stage in which the insect feeds and most intimately interacts with other species and abiotic factors of its environment. The need for ability to diagnose larval aquatic insects is especially pressing in countries with the most severe water-pollution problems, countries where knowledge about the freshwater insect fauna usually is poorest.

Most aquatic insect species were described originally from an adult form, usually the male, so that the application of a name to a larva or other life history form requires a method to associate that yet-unnamed life history form confidently with the previously described and named form. In the past, this association has been accomplished with greatest confidence by a laborious and often unsuccessful rearing process. Modern molecular techniques now have been added to the methods available to accomplish these associations (Zhou 2007, Zhou et al. 2007).

Need to refine definitions of species of aquatic insects

Traditionally, species of aquatic insects have been diagnosed by conspicuous morphological differences visible with a light microscope. Scanning electron microscopy, three-dimensional imagery through computer synthesis of staged digital photographs, confocal imagery, and other modern techniques have enhanced traditional diagnostic methods. Increasingly, however, we are becoming aware that genetically independent populations that look similar are, in fact, distinct species. These cryptic species often have different life history characteristics and behaviors, including different abilities to transmit disease agents among humans and domesticated animals. Therefore, various cytogenetic (e.g., Adler et al. 2004) and molecular (e.g., Zhou 2007, Zhou et al. 2007) techniques have been developed to help distinguish these cryptic species. Many of today's so-called 'species' probably will be found to be complexes of multiple, biologically different species.

Need for new generation of aquatic entomologists

Historically, biodiversity discovery was accomplished by men, seldom women, who were independently wealthy or who had other careers that allowed them to pursue their passion. Teaching and research employment opportunities in entomology emerged widely in Europe in the mid-1800s and in the USA near the beginning of the 1900s (Comstock and Comstock 1895), including many employment opportunities to discover species of insects. Within the past two decades, those employment opportunities became diluted with other responsibilities or were eliminated, so that employment and publishing opportunities for organismal biologists who recognize and study living species is in decline (Agnarsson and Kuntner 2007). Compounding the problem is that as older professional taxonomists retire and die, their expertise in various groups is being lost and not passed to a succeeding generation, so that some groups now have no experts, other groups will soon be without experts, and for more and more groups, no mentors are available to teach younger biodiversity scholars.

Concurrently, there is a trend for young people today to spend less time in the woods and streams of a former rural countryside and more time indoors learning about the natural world indirectly through electronic media (Louv 2005). As a result, they seem less inclined to follow careers in biodiversity discovery. Biological teaching and research in secondary schools and institutions of higher learning is more commonly focused on molecular principles and methods, with less emphasis placed on the recognition and study of living organisms in nature. A result of these trends is fewer professional entomologists capable of recognizing species of aquatic insects or interested in discovering their biological characteristics. As we have observed, this capacity for discovery of species is declining at the very time in history when there is greatest need for biodiversity discovery, before many more species become extinct. A significant change in direction toward increased emphasis on biodiversity education and research for today's youth and toward improved employment opportunities in biodiversity discovery is urgently needed.

REFERENCES

Adis, J., E. Bustorf, M. G. Lhano, C. Amedegnato, and A. L. Nunes. 2007. Distribution of *Cornops* grasshoppers (Leptisminae: Acrididae: Orthoptera) in Latin America and the Caribbean Islands. *Studies on Neotropical Fauna and Environment* 42: 11–24.

Adler, P. H. and R. W. Crosskey. 2008. World Blackflies (Diptera: Simuliidae): A Fully Revised Edition of the Taxonomic and Geographical Inventory. http://entweb.clemson.edu/biomia/pdfs/blackflyinventory.pdf [Accessed 31 March 2008].

Adler, P. H., D. C. Currie, and D. M. Wood. 2004. *The Black Flies (Simuliidae) of North America*. Cornell University Press, Ithaca, NY. xv + 941 pp., 24 color plates.

Aelianus, C. 1611. De Animalium Natura Libri XVII. (originally about 200 CE) Apud. I. Tornaesium, Lyons. pp. 4 + 1018 + 94.

Agnarsson, I. and M. Kuntner. 2007. Taxonomy in a changing world: seeking solutions for a science in crisis. *Systematic Biology* 56: 531–539.

Amedegnato, C. and H. Devriese. 2008. Global diversity of true and pygmy grasshoppers (Acridomorpha, Orthoptera) in freshwater. Pp. 535–543. *In* E. V. Balian, C. Lévêque, H. Segers, and K. Martens (guest eds). *Freshwater Animal Diversity Assessment*. Hydrobiologia 595.

Australian River Assessment System (AUSRIVAS). 2005. AUSRIVAS Bioassessment: Macroinvertebrates. http://ausrivas.canberra.edu.au/Bioassessment/Macroinvertebrates/ [Accessed 2 September 2006].

Barber-James, H. M., J.-L. Gattolliat, M. Sartori, and M. D. Hubbard. 2008. Global diversity of mayflies (Ephemeroptera; Insecta) in freshwater. Pp. 339–350. *In* E. V. Balian, C. Lévêque, H. Segers, and K. Martens (guest eds). *Freshwater Animal Diversity Assessment*. Hydrobiologia 595.

Barbour, M. T., J. Gerritsen, B. D. Snyder, and J. B. Stribling. 1999. *Rapid Bioassessment Protocols for Use in Streams and Wadeable Rivers: Periphyton, Benthic Macroinvertebrates, and Fish*, Second Edition. EPA 841-B-99-002. United States Environmental Protection Agency, Office of Water, Washington, DC.

Bennett, A. M. R. 2008. Global diversity of hymenopterans (Hymenoptera; Insecta) in freshwater. Pp. 529–534. *In* E. V. Balian, C. Lévêque, H. Segers, and K. Martens (guest eds). *Freshwater Animal Diversity Assessment*. Hydrobiologia 595.

Berry, R. J. 2000. *The Care of Creation: Focusing Concern and Action*. Inter-Varsity Press, Leicester. 213 pp.

Brigham, A. R. and D. D. Herlong. 1982. Aquatic and semi-aquatic Lepidoptera. Pp. 12.1–12.36. *In* A. R. Brigham, W. U. Brigham, and A. Gnilka (eds). *Aquatic Insects and Oligochaetes of North and South Carolina*. Midwest Aquatic Enterprises, Mahomet, IL.

Brigham, W. U. 1982a. Megaloptera. Pp. 7.1–7.12. *In* A. R. Brigham, W. U. Brigham, and A. Gnilka (eds). *Aquatic Insects and Oligochaetes of North and South Carolina*. Midwest Aquatic Enterprises, Mahomet, IL.

Brigham, W. U. 1982b. Aquatic Neuroptera Pp. 8.1–8.4. *In* A. R. Brigham, W. U. Brigham, and A. Gnilka (eds). *Aquatic Insects and Oligochaetes of North and South Carolina*. Midwest Aquatic Enterprises, Mahomet, IL.

Buckingham, G. R. 1994. Biological control of aquatic weeds. Pp. 413–479. *In* D. Rosen, F. D. Bennett, and J. L. Capinera (eds). *Pest Management in the Subtropics: Biological Control – a Florida Perspective*. Intercept Ltd., Andover.

Butler, M. G. 1984. Life histories of aquatic insects. Pp. 24–55. *In* V. H. Resh and D. M. Rosenberg (eds). *The Ecology of Aquatic Insects*. Praeger, NY.

Cantrall, I. J. and M. A. Brusven. 1996. Semiaquatic Orthoptera. Pp. 212–216. *In* R. W. Merritt and K. W. Cummins (eds). *An Introduction to the Aquatic Insects of North America*, Third Edition. Kendall/Hunt Publishing Company, Dubuque, IA.

Carver, M., G. F. Gross, and T. E. Woodward. 1991. Hemiptera. Pp. 429–509. *In* I. D. Naumann (ed). *The Insects of Australia: A Textbook for Students and Research Workers*, Volume 1. Cornell University Press, Ithaca, NY. 542 pp.

Caucci, A. and R. Nastasi. 1975. *Hatches*. Comparahatch, NY. 320 pp.

Center for Aquatic and Invasive Plants. 2007a. Biological Control Insects for Aquatic and Wetland Weeds, Alligator Weed. Aquatic and Wetland Plant Information Retrieval System, Institute of Food and Agricultural Sciences, University of Florida. http://plants.ifas.ufl.edu/alligat.html [Accessed 22 September 2007].

Center for Aquatic and Invasive Plants. 2007b. Biological Control Insects for Aquatic and Wetland Weeds, Water hyacinth. Aquatic and Wetland Plant Information Retrieval System, Institute of Food and Agricultural Sciences,

University of Florida. http://plants.ifas.ufl.edu/hyacin.html [Accessed 22 September 2007].

Cheng, L. (ed). 1976. *Marine Insects*. North-Holland Publishing Company, NY. 581 pp.

Christiansen, K. A. and R. J. Snider. 1996. Aquatic Collembola. Pp. 113–125. *In* R. W. Merritt and K. W. Cummins (eds). *An Introduction to the Aquatic Insects of North America*, Third Edition.Kendall/Hunt Publishing Company, Dubuque, IA.

Claridge, M. F. 2006. Integrated Strategies for the Biological Control of Major Insect Pests of Rice in South East Asia. Tropical and Subtropical Agriculture, Third STD Programme (1992–1995). Summary Reports of European Commission Supported STD-3 projects (1992–1995), Published by CTA 1998. http://www.agricta.org/pubs/std/vol1/pages/pdf/018.pdf [Accessed 26 November 2006].

Comstock, J. H. and A. B. Comstock. 1895. *A Manual for the Study of Insects*. The Comstock Publishing Company, Ithaca, NY. 677 pp.

Corbet, P. S. 1999. *Dragonflies: Behavior and Ecology of Odonata*. Cornell University Press, Ithaca, NY. 829 pp.

Courtney, G. W., R. W. Merritt, K. W. Cummins, B. A. Foote, and D. W. Webb. 1996a. Table 22B, Summary of ecological and distributional data for larval aquatic Diptera. Pp. 537–548 *In* R. W. Merritt and K. W. Cummins (eds). *An Introduction to the Aquatic Insects of North America*, Third Edition. Kendall/Hunt Publishing Company, Dubuque, IA.

Courtney, G. W., H. J. Teskey, R. W. Merritt, and B. A. Foote. 1996b. Aquatic Diptera. Part 1. Larvae of aquatic Diptera. Pp. 484–514. *In* R. W. Merritt and K. W. Cummins (eds). *An Introduction to the Aquatic Insects of North America*, Third Edition. Kendall/Hunt Publishing Company, Dubuque, IA.

Cover, M. R. and V. H. Resh. 2008. Global diversity of dobsonflies, fishflies, and alderflies (Megaloptera; Insecta) and spongillaflies, nevrorthids, and osmylids (Neuroptera; Insecta) in freshwater. Pp. 409–417. *In* E. V. Balian, C. Lévêque, H. Segers, and K. Martens (guest eds). *Freshwater Animal Diversity Assessment*. Hydrobiologia 595.

Crichton, M. I. 1957. The structure and function of the mouth parts of adult caddis flies (Trichoptera). *Philosophical Transactions of the Royal Society of London B* 241: 45–91.

Cummins, K. W. and M. J. Klug. 1979. Feeding ecology of stream invertebrates. *Annual Review of Ecology and Systematics* 10: 147–172.

Cummins, K. W. and R. W. Merritt. 1996. Ecology and distribution of aquatic insects. Pp. 74–86. *In* R. W. Merritt and K. W. Cummins (eds). *An Introduction to the Aquatic Insects of North America*, Third Edition. Kendall/Hunt Publishing Company, Dubuque, IA.

Currie, D. C. and P. H. Adler. 2008. Global diversity of black flies (Diptera: Simuliidae) in freshwater. Pp. 469–475. *In* E. V. Balian, C. Lévêque, H. Segers, and K. Martens (guest eds). *Freshwater Animal Diversity Assessment*. Hydrobiologia 595.

de Jong, H., P. Oosterbroek, J. Gelhaus, H. Reusch, and C. Young. 2008. Global diversity of craneflies (Insecta, Diptera: Tipulidea or Tipulidae *sensu lato*) in freshwater. Pp. 457–467. *In* E. V. Balian, C. Lévêque, H. Segers, and K. Martens (guest eds). *Freshwater Animal Diversity Assessment*. Hydrobiologia 595.

Disney, R. H. L. 2004. Insecta: Diptera, Phoridae. Pp. 818–825. *In* C. M. Yule and H. S. Yong (eds). *Freshwater Invertebrates of the Malaysian Region*. Academy of Sciences of Malaysia, Kuala Lumpur.

Edmunds, G. F., Jr., S. L. Jensen, and L. Berner. 1976. *The Mayflies of North and Central America*. University of Minnesota Press, Minneapolis, MN. 330 pp.

Edmunds, G. F., Jr. and R. D. Waltz. 1996. Ephemeroptera. Pp. 126–163. *In* R. W. Merritt and K. W. Cummins (eds). *An Introduction to the Aquatic Insects of North America*, Third Edition. Kendall/Hunt Publishing Company, Dubuque, IA.

Elliott, J. M. 1982. The life cycle and spatial distribution of the aquatic parasitoid *Agriotypus armatus* (Hymenoptera: Agriotypidae) and its caddis host *Silo pallipes* (Trichoptera: Goeridae). *Journal of Animal Ecology* 51: 923–942.

Eriksen, C. H., V. H. Resh, and G. A. Lamberti. 1996. Aquatic insect respiration. Pp. 29–39. *In* R. W. Merritt and K. W. Cummins (eds). *An Introduction to the Aquatic Insects of North America*, Third Edition. Kendall/Hunt Publishing Company, Dubuque, IA.

European Union Water Framework Directive. 2000. WFD, Directive 2000/60/EC. *Official Journal of European Communities* 23: 1–72. http://europa.eu.int/comm/environment/water/waterframework/index_en.html [Accessed 2 September 2006].

Evans, E. D. and H. H. Neunzig. 1996. Megaloptera and aquatic Neuroptera. Pp. 298–308. *In* R. W. Merritt and K. W. Cummins (eds). *An Introduction to the Aquatic Insects of North America*, Third Edition. Kendall/Hunt Publishing Company, Dubuque, IA.

Evans, H. E. and C. M. Yoshimoto. 1962. The ecology and nesting behavior of the Pompilidae (Hymenoptera) of the northeastern United States. *Miscellaneous Publications of the Entomological Society of America* 3: 67–119.

Ferrington, L. C. 2008a. Global diversity of scorpionflies and hangingflies (Mecoptera) in freshwater. Pp. 443–445. *In* E. V. Balian, C. Lévêque, H. Segers, and K. Martens (guest eds). *Freshwater Animal Diversity Assessment*. Hydrobiologia 595.

Ferrington, L. C. 2008b. Global diversity of non-biting midges (Chironomidae; Insecta-Diptera) in freshwater. Pp. 447–455. *In* E. V. Balian, C. Lévêque, H. Segers, and K. Martens (guest eds). *Freshwater Animal Diversity Assessment*. Hydrobiologia 595.

Fochetti, R. and J. M. T. de Figueroa. 2008. Global diversity of stoneflies (Plecoptera; Insecta) in freshwater. Pp. 365–377. *In* E. V. Balian, C. Lévêque, H. Segers, and K. Martens (guest eds). *Freshwater Animal Diversity Assessment*. Hydrobiologia 595.

Franceschini, M. C., S. Capello, M. G. Lhano, J. Adis, and M. L. Wysiecki. 2005. Morfometría de los estadios ninfales de *Cornops aquaticum* Bruner, 1906 (Acrididae: Leptysminae) en Argentina. *Amazoniana* 18(3/4): 373–386.

Fremling, C. R. 1960. Biology of a large mayfly, *Hexagenia bilineata* (Say), of the Upper Mississippi River. *Agricultural and Home Economics Experiment Station, Iowa State University, Research Bulletin* 482: 842–852.

Fuller, R. A., K. N. Irvine, P. Devine-Wright, P. H. Warren, and K. J. Gaston. 2007. Psychological benefits of greenspace increase with biodiversity. *Biology Letters* 3: 390–394.

Gower, A. M. 1965. The life cycle of *Drusus annulatus* Steph. (Trich., Limnephilidae) in watercress beds. *Entomologist's Monthly Magazine* 101: 133–141.

Gower, A. M. 1967. A study of *Limnephilus lunatus* (Trichoptera: Limnephilidae) with reference to its life cycle in watercress beds. *Transactions of the Royal Entomological Society of London* 119: 283–302.

Greenberg, G. B. 1971. *Flies and Disease, Volume 1. Ecology, Classification and Biotic Associations.* Princeton University Press, Princeton, NJ. xii + 856 pp.

Greenberg, G. B. 1973. Flies and Disease. *In Biology and Disease Transmission,* Volume 2. Princeton University Press, Princeton, NJ. 447 pp.

Hamilton, L. S. and H. F. Takeuchi (eds). 1991. *Ethics, Religion, and Biodiversity: Relations between Conservation and Cultural Values.* Pacific Science Congress, Honolulu, Hawaii. 218 pp.

Henson, E. B. 1966. Aquatic insects as inhalant allergens: a review of American literature. *Ohio Journal of Science* 66: 529–532.

Herd, A. N. 2006. A Flyfishing History. http://www .flyfishinghistory.com/walton.htm [Accessed 23 November 2006].

Hickin, N. E. 1967. *Caddis Larvae: Larvae of the British Trichoptera.* Hutchinson and Company, London. 476 pp.

Huggins, D. G. and W. U. Brigham. 1982. Odonata. Pp. 4.1–4.100. *In* A. R. Brigham, W. U. Brigham, and A. Gnilka (eds). *Aquatic Insects and Oligochaetes of North and South Carolina.* Midwest Aquatic Enterprises, Mahomet, IL.

Ivanov, V. D. 1993. Principles of sexual communication in caddisflies (Insecta, Trichoptera). Pp. 609–626. *In* K. Wiese, F. G. Gribakin, A. V. Poppv, and G. Renninger (eds). *Sensory Systems of Arthropods (Advances in Life Sciences).* Birkhäuser-Verlag, Basel, Switzerland.

Ivanov, V. D. and R. Rupprecht. 1993. Substrate vibration for communication in adult *Agapetus fuscipes* (Trichoptera: Glossosomatidae). Pp. 273–278. *In* C. Otto (ed). *Proceedings of the 7th International Symposium on Trichoptera, Umeå, Sweden, 3–8 August 1992.* Backhuys Publishers, Leiden, The Netherlands. 312 pp.

Jäch, M. A. and M. Balke. 2008. Global diversity of water beetles (Coleoptera) in freshwater. Pp. 419–442. *In* E. V. Balian, C. Lévêque, H. Segers, and K. Martens (guest eds). *Freshwater Animal Diversity Assessment.* Hydrobiologia 595.

Kalkman, V. J., V. Crausnitzer, K.-D. B. Dijkstra, A. G. Orr, D. R. Paulson, and J. van Tol. 2008. Global diversity of dragonflies (Odonata) in freshwater. Pp. 351–363. *In* E. V. Balian, C. Lévêque, H. Segers, and K. Martens (guest eds). *Freshwater Animal Diversity Assessment.* Hydrobiologia 595.

Kellert, S. R. 1996. *The Value of Life.* Island Press, Washington, DC.

Kim, K. C. 1987. Order Anoplura. Pp. 224–245. *In* F. W. Stehr (ed). *Immature Insects.* Kendall/Hunt Publishing Company, Dubuque, IA.

Kovac, D. and U. Maschwitz. 1989. Secretion – grooming in the water bug *Plea minutissima*: a chemical defense against microorganisms interfering with the hydrofuge properties of the respiratory region. *Ecological Entomology* 14: 403–411.

Lange, W. H. 1996. Aquatic and semiaquatic Lepidoptera. Pp. 387–398. *In* R. W. Merritt and K. W. Cummins (eds). *An Introduction to the Aquatic Insects of North America,* Third Edition. Kendall/Hunt Publishing Company, Dubuque, IA.

Langston, R. L. 1974. The maritime earwig in California (Dermaptera: Carcinophoridae). *Pan-Pacific Entomologist* 50: 28–34.

Langston, R. L. and J. A. Powell. 1975. The earwigs of California (Order Dermaptera). *Bulletin of the California Insect Survey* 20: 1–25.

Lhano, M. G., J. Adis, M. I. Marques, and L. D. Battirola. 2005. *Conops aquaticum* (Orthoptera, Acrididae, Leptysminae): aceitação de plantas alimentares por ninfas vivendo em *Eichornia azurea* (Pontederiacae) no Pantanal Norte, Brasil. *Amazoniana* 18(3/4): 397–404.

Lodge, D. M. and C. Hamlin. 2006. *Religion and the New Ecology: Environmental Responsibility in a World in Flux.* University of Notre Dame, Notre Dame, IN. 325 pp.

Louv, R. 2005. *Last Child in the Woods: Saving Our Children from Nature-Deficit Disorder.* Algonquin Books of Chapel Hill, NC. 336 pp.

Mackay, R. J. and G. B. Wiggins. 1979. Ecological diversity in Trichoptera. *Annual Review of Entomology* 24: 185–208.

Malicky, H. 1984. Ein Beitrag zur Autoekologie und Bionomie der aquatischen Netzflüglergattung *Neurorthus* (Insecta, Neuroptera, Neurorthidae). *Archiv für Hydrobiologie* 101: 231–246.

Malicky, H. and T. Prommi. 2006. Beschreibung einiger Köcherfliegen aus Süd-Thailand (Trichoptera) (Arbeit Nr. 42 über thailändische Köcherfliegen). *Linzer biologische Beiträge* 38(2): 1591–1608.

Master, L. L., B. A. Stein, L. S. Kutner, and G. A. Hammerson. 2000. Vanishing assets: conservation status of U.S. species. Pp. 93–118. *In* B. A. Stein, L. S. Kutner, and J. S. Adams (eds). *Precious Heritage: the Status of Biodiversity in the United States.* Oxford University Press, Oxford.

McAlpine, J. F. and D. M. Wood (eds). 1989. *Manual of Nearctic Diptera,* Volume 3. Monograph Number 32. Canadian Government Publishing Centre, Canada. Research Branch: Agriculture Canada, Pp. 1333–1581.

McCafferty, W. P. 1981. *Aquatic Entomology: The Fisherman's and Ecologists' Illustrated Guide to Insects and Their Relatives.* Science Books International, Boston, MA. 448 pp.

Merritt, R. W. and J. B. Wallace. 1981. Filter-feeding insects. *Scientific American* 244: 132–144.

Mey, W. and W. Speidel. 2008. Global diversity of butterflies (Lepidotera) in freshwater. Pp. 521–528. *In* E. V. Balian, C. Lévêque, H. Segers, and K. Martens (guest eds). *Freshwater Animal Diversity Assessment.* Hydrobiologia 595.

Miller, G. T., Jr. 1988. *Living in the Environment: An Introduction to Environmental Science.* Fifth Edition. Wadsworth, Belmont, CA.

Moretti, G. 1942. Studii sui tricotteri: XV. Comportamento del *Triaenodes bicolor* Curt. (Trichoptera–Leptoceridae). *Bollettino de Zoologia Agraria Bachicoltura Milano* 11: 89–131.

Morse, J. C. 1997. Checklist of World Trichoptera. Pages 339–342. *In* R. W. Holzenthal and O. S. Flint, Jr. (eds). *Proceedings of the 8th International Symposium on Trichoptera.* Ohio Biological Survey, Columbus, Ohio. 496 pp.

Morse, J. C. 2002. Following a dream. Pp. 3–7. *In* Y. J. Bae (ed). *The 21st Century and Aquatic Entomology in East Asia. Proceedings of the 1st Symposium of the Aquatic Entomological Societies of East Asia.* Korean Society for Aquatic Entomology, Seoul.

Morse, J. C. (ed). 2008. Trichoptera World Checklist. http://entweb.clemson.edu/database/trichopt/index.htm [Accessed 18 June 2008].

Morse, J. C., Y. J. Bae, G. Munkhjargal, N. Sangpradub, K. Tanida, T. S. Vshivkova, B. Wang, L. Yang, and C. M. Yule. 2007. Freshwater biomonitoring with macroinvertebrates in East Asia. *Frontiers in Ecology and the Environment* 5(1): 33–42.

Morse, J. C., B. P. Stark, and W. P. McCafferty. 1993. Southern Appalachian streams at risk: implications for mayflies, stoneflies, caddisflies, and other aquatic biota. *Aquatic Conservation: Marine and Freshwater Ecosystems* 3: 293–303.

Morse, J. C., B. P. Stark, W. P. McCafferty, and K. J. Tennessen. 1998. ('1997'). Southern Appalachian and other southeastern streams at risk: implications for mayflies, dragonflies and damselflies, stoneflies, and caddisflies. Pp. 17–42. *In* G. W. Benz and D. E. Collins (eds). *Aquatic Fauna in Peril: The Southeastern Perspective,* Special Publication 1. Southeast Aquatic Research Institute, Lenz Design and Communications, Decatur, Georgia. 554 pp.

Nakahara, L. M., J. J. McHugh, Jr., C. K. Otsuka, G. Y. Funasaki, and P. Y. Lai. 1986. Integrated control of diamondback moth and other insect pests using an overhead sprinkler system, an insecticide, and biological control agents, on a watercress farm in Hawaii. Pp. 403–404. *In* N. S. Talekar and T. C. Griggs (eds). *Diamondback Moth Management: Proceedings of the First International Workshop.* Asian Vegetable Research and Development Center, Shanhua, Taiwan.

Neunzig, H. H. and J. R. Baker. 1991. Order Megaloptera. Pp. 112–122. *In* F. W. Stehr (ed). *Immature Insects,* Volume 2. Kendall/Hunt Publishing Company, Dubuque, IA.

Newbold, J. D., J. W. Elwood, R. V. O'Neill, and W. Van Winkle. 1983. Resource spiraling: an operational paradigm for analyzing lotic ecosystems. Pp. 3–27. *In* T. D. Fontaine and S. M. Bartell (eds). *Dynamics of Lotic Ecosystems.* Ann Arbor Science Publishers, Ann Arbor, MI.

Newbold, J. D., R. V. O'Neill, J. W. Elwood, and W. VanWinkle. 1982. Nutrient spiraling in streams: implications for nutrient limitation and invertebrate activity. *American Naturalist* 120: 628–652.

Pennak, R. W. 1978. *Fresh-water Invertebrates of the United States,* Second Edition. John Wiley & Sons, New York. 803 pp.

Pilgrim, R. L. C. 1972. The aquatic larva and pupa of *Choristella philpotti* Tillyard, 1917 (Mecoptera: Nannochoristidae). *Pacific Insects* 14: 151–168.

Polhemus, J. T. 1996. Aquatic and semiaquatic Hemiptera. Pp. 267–297. *In* R. W. Merritt and K. W. Cummins (eds). *An Introduction to the Aquatic Insects of North America,* Third Edition. Kendall/Hunt Publishing Company, Dubuque, IA.

Polhemus, J. T. and D. A. Polhemus. 2008. Global diversity of true bugs (Heteroptera; Insecta) in freshwater. Pp. 379–391. *In* E. V. Balian, C. Lévêque, H. Segers, and K. Martens (guest eds). *Freshwater Animal Diversity Assessment.* Hydrobiologia 595.

Price, R. D. 1987. Order Mallophaga. Pp. 215–223. *In* F. W. Stehr (ed). *Immature Insects.* Kendall/Hunt Publishing Company, Dubuque, IA. 754 pp.

Richter, B. D., D. P. Braun, M. A. Mendelson, and L. L. Master. 1997. Threats to imperiled freshwater fauna. *Conservation Biology* 11: 1081–1093.

RIVPACS. 2005. http://dorset.ceh.ac.uk/River_Ecology/River_Communities/Rivpacs_2003/rivpacs_introduction.htm [Accessed 2 September 2006].

Rosenberg, D. M., I. J. Davies, D. G. Cobb, and A. P. Wiens. 2005. *Protocols for Measuring Biodiversity: Benthic Macroinvertebrates in Fresh Waters.* Department of Fisheries and Oceans, Winnipeg, Canada. www.eman-rese.ca/eman/ecotools/protocols/freshwater/benthics/benthic_fresh_e.pdf [Accessed 2 September 2006].

Rueda, L. M. 2008. Global diversity of mosquitoes (Insecta: Diptera: Culicidae) in freshwater. Pp. 477–487. *In* E. V. Balian, C. Lévêque, H. Segers, and K. Martens (guest eds). *Freshwater Animal Diversity Assessment.* Hydrobiologia 595.

Saito, T., H. Kazuo, and M. O. Way. 2005. The rice water weevil, *Lissorhoptrus oryzophilus* Kuschel (Coleoptera: Curculionidae). *Applied Entomology and Zoology* 40: 31–39.

Sanderson, M. W. 1982. Aquatic and semiaquatic Heteroptera. Pp. 6.1–6.94. *In* A. R. Brigham, W. U. Brigham, and A. Gnilka (eds). *Aquatic Insects and Oligochaetes of North and South Carolina.* Midwest Aquatic Enterprises, Mahomet, IL.

Schmid, F. 1984. Essai d'evaluation de la faune mondiale des Trichoptères. P. 337. In J. C. Morse (ed). *Proceedings of the 4th International Symposium on Trichoptera, Clemson, South Carolina, 11–16 July 1983*. Series Entomologica 30. Dr. W. Junk Publishers, The Hague, The Netherlands.

Sephton, D. H. and H. B. N. Hynes. 1982. The numbers of nymphal instars of several Australian Plecoptera. *Aquatic Insects* 4: 153–166.

Solem, J. O. 1984. Adult behaviour of North European caddisflies. Pp. 375–382. In J. C. Morse (ed). *Proceedings of the 4th International Symposium on Trichoptera, Clemson, South Carolina, 11–16 July 1983*. Series Entomologica 30. Dr. W. Junk Publishers, The Hague, The Netherlands.

Stark, J. D., I. K. G. Boothroyd, J. S. Harding, J. R. Maxted, and M. R. Sarsbrook. 2001. Protocols for Sampling Macroinvertebrates in Wadeable Streams. Auckland, New Zealand: New Zealand Ministry for the Environment. http://freshwater.rsnz.org/ProtocolsManual2.pdf [Accessed 3 September 2006].

Stewart, K. W. and P. P. Harper. 1996. Plecoptera. Pp. 217–266. In R. W. Merritt and K. W. Cummins (eds). *An Introduction to the Aquatic Insects of North America*, Third Edition. Kendall/Hunt Publishing Company, Dubuque, IA.

Stewart, K. W. and M. Maketon. 1991. Structures used by Nearctic stoneflies (Plecoptera) for drumming, and their relationship to behavioral pattern diversity. *Aquatic Insects* 13: 33–53.

Stewart, K. W. and B. P. Stark. 2002. *Nymphs of North American Stonefly Genera (Plecoptera)*, Second Edition. The Caddis Press, Columbus, OH. 510 pp.

Tanaka, H. and T. Takahara. 1989. Studies on ecology and control of diamondback moth, *Plutella xylostella* L., on watercress (1): distribution patterns of egg and larval stages and estimation of their densities. *Bulletin of the Osaka Agricultural and Forestry Research Center* 25: 9–14 [In Japanese with English summary].

Thompson, K. C., K. Wadhia, and A. P. Loibner. 2005. *Environmental Toxicity Testing*. CRC Press, Boca Raton, FL. 388 pp.

Ueno, M. 1952. Caddis fly larvae interfering with the flow in the water way tunnels of a hydraulic power plant. *Kontyu* 19: 73–80.

Unzicker, J. D. and P. H. Carlson. 1982. Ephemeroptera. Pp. 3.1–3.97. In A. R. Brigham, W. U. Brigham, and A. Gnilka (eds). *Aquatic Insects and Oligochaetes of North and South Carolina*. Midwest Aquatic Enterprises, Mahomet, IL.

Unzicker, J. D., V. H. Resh, and J. C. Morse. 1982. Trichoptera. Pp. 9.1–9.138. In A. R. Brigham, W. U. Brigham, and A. Gnilka (eds). *Aquatic Insects and Oligochaetes of North and South Carolina*. Midwest Aquatic Enterprises, Mahomet, IL.

Wagner, R. (coordinating author), M. Bartak, A. Borkent, G. Courtney, B. Goddeeris, J. P. Haenni, L. Knutson, A. Pont, G. E. Rotheray, R. Rozkosny, B. Sinclair, N. Woodley, T. Zatwarnicki, and P. Zwick. 2008. Global diversity of dipteran families (Insecta Diptera) in freshwater (excluding Simulidae, Culicidae, Chironomidae, Tipulidae and Tabanidae). Pp. 489–519. In E. V. Balian, C. Lévêque, H. Segers, and K. Martens (guest eds). *Freshwater Animal Diversity Assessment*. Hydrobiologia 595.

Wallace, J. B. and R. W. Merritt. 1980. Filter-feeding ecology of aquatic insects. *Annual Review of Entomology* 25: 103–132.

Walton, I. 1653. *The Compleat Angler or the Contemplative Man's Recreation. Being a Discourse of Fish and Fishing, not Unworthy the Perusal of Most Anglers*. Maxey, London. 246 pp. [Charles Cotton was co-author for the fifth and subsequent editions].

Waltz, R. D. and W. P. McCafferty. 1979. Freshwater springtails (Hexapoda: Collembola) of North America. *Purdue University Agricultural Experiment Station Research Bulletin* 960: 32.

Wells, A. 1992. The first parasitic Trichoptera. *Ecological Entomology* 17: 299–302.

Westfall, M. J., Jr. and K. J. Tennessen. 1996. Odonata. Pp. 164–211. In R. W. Merritt and K. W. Cummins (eds). *An Introduction to the Aquatic Insects of North America*, Third Edition. Kendall/Hunt Publishing Company, Dubuque, IA.

White, D. S. 1976. *Climacia areolaris* (Neuroptera: Sisyridae) in Lake Texoma, Texas and Oklahoma. *Entomological News* 87: 287–291.

White, D. S. and W. U. Brigham. 1996. Aquatic Coleoptera. Pp. 399–473. In R. W. Merritt and K. W. Cummins (eds). *An Introduction to the Aquatic Insects of North America*, Third Edition. Kendall/Hunt Publishing Company, Dubuque, IA.

Wiggins, G. B. 1973. A contribution to the biology of caddisflies (Trichoptera) in temporary pools. *Life Sciences Contributions of the Royal Ontario Museum* 88: 1–28.

Wiggins, G. B. 1996. *Larvae of the North American Caddisfly Genera (Trichoptera)*, Second Edition. University of Toronto Press, Toronto. 457 pp.

Wiggins, G. B. 2004. *Caddisflies – The Underwater Architects*. University of Toronto Press, Toronto, Ontario. 292 pp.

Wilcove, D. S., D. Rothstein, J. Dubow, A. Phillips, and E. Losos. 2000. Leading threats to biodiversity: what's imperiling U. S. species. Pp. 239–254. In B. A. Stein, L. S. Kutner, and J. S. Adams (eds). *Precious Heritage: The Status of Biodiversity in the United States*. Oxford University Press, Oxford. 399 pp.

Wilson, E. O. 1984. *Biophilia*. Harvard University Press, MA. 157 pp.

Wood, J. R. and V. H. Resh. 1984. Demonstration of sex pheromones in caddisflies (Trichoptera). *Journal of Chemical Ecology* 10: 171–175.

Woodley, N. E. 2001. *A World Catalog of the Stratiomyidae (Insecta: Diptera)*. Myia 11: 1–473. Backhuys Publishers, Leiden.

Yang, L. and J. C. Morse. 2000. Leptoceridae (Trichoptera) of the People's Republic of China. *Memoirs of the American Entomological Institute* 64: viii + 309.

Yonke, T. R. 1991. Order Hemiptera. Pp. 22–65. *In* F. W. Stehr (ed). *Immature Insects*, Volume 2. Kendall/Hunt Publishing Company, Dubuque, IA. 975 pp.

Zhou, X. 2007. The larvae of Chinese Hydropsychidae (Insecta: Trichoptera): delimiting species boundaries using morphology and DNA sequences. Ph.D. dissertation, Rutgers University, New Brunswick, NJ.

Zhou, X., K. M. Kjer, and J. C. Morse. 2007. Associating larvae and adults of Chinese Hydropsychidae caddisflies (Insecta: Trichoptera) using DNA sequences. *Journal of the North American Benthological Society* 26: 719–742.

Zwick, P. 1967. Beschreibung der aquatischen Larve von *Neurorthus fallax* (Rambur) und Errichtung der neuen Planipennierfamilie Neurorthidae fam. nov. *Gewässer und Abwässer* 44/45: 65–86.

Zwick, P. 2003. 8. Ordnung Plecoptera, Steinfliegen, Frühlingsfliegen, Uferfliegen. *In* A. Kaestner (ed), *Lehrbuch der Speziellen Zoologie*, 2. Aufl 1(5): 144–154. Spektrum Akademischer Verlag, Heidelberg, Berlin.

Chapter 9

BIODIVERSITY OF DIPTERA

Gregory W. Courtney[1], Thomas Pape[2], Jeffrey H. Skevington[3], and Bradley J. Sinclair[4]

[1] Department of Entomology, 432 Science II, Iowa State University, Ames, Iowa 50011 USA

[2] Natural History Museum of Denmark, Zoological Museum, Universitetsparken 15, DK – 2100 Copenhagen Denmark

[3] Agriculture and Agri-Food Canada, Canadian National Collection of Insects, Arachnids and Nematodes, K.W. Neatby Building, 960 Carling Avenue, Ottawa, Ontario K1A 0C6 Canada

[4] Entomology – Ontario Plant Laboratories, Canadian Food Inspection Agency, K.W. Neatby Building, 960 Carling Avenue, Ottawa, Ontario K1A 0C6 Canada

Insect Biodiversity: Science and Society, 1st edition. Edited by R. Foottit and P. Adler
© 2009 Blackwell Publishing, ISBN 978-1-4051-5142-9

The Diptera, commonly called true flies or two-winged flies, are a familiar group of insects that includes, among many others, black flies, fruit flies, horse flies, house flies, midges, and mosquitoes. The Diptera are among the most diverse insect orders, with estimates of described richness ranging from 120,000 to 150,000 species (Colless and McAlpine 1991, Schumann 1992, Brown 2001, Merritt et al. 2003). Our world tally of more than 152,000 described species (Table 9.1) is based primarily on figures extracted from the 'BioSystematic Database of World Diptera' (Evenhuis et al. 2007).

The Diptera are diverse not only in species richness, but also in their structure (Fig. 9.1), habitat exploitation, life habits, and interactions with humankind (Hennig 1973, McAlpine et al. 1981, 1987, Papp and Darvas 2000, Brown 2001, Skevington and Dang 2002, Pape 2009). The Diptera have successfully colonized all continents, including Antarctica, and practically every habitat except the open sea and inside glaciers. Larval Diptera are legless (Figs. 9.2 and 9.3F-J) and found in a variety of terrestrial and aquatic habitats (Teskey 1976, Ferrar 1987, Hövemeyer 2000, Courtney and Merritt 2008). Larvae of most species can be considered aquatic in the broadest sense because, for survival, they require a moist to wet environment within the tissues of living plants, amid decaying organic materials, as parasites or parasitoids of other animals, or in association with bodies of water. Most larvae are free-living and swim, crawl, or tunnel actively in water (e.g., Chaoboridae, Chironomidae, Culicidae, and Simuliidae), sediments (e.g., Ceratopogonidae, Psychodidae, Tabanidae, and Tipulidae), wood (e.g., Axymyiidae, some Syrphidae and Tipulidae), fruit (e.g., Chloropidae and Tephritidae), or decaying organic material (e.g., Ephydridae, Muscidae, Sarcophagidae, and Sphaeroceridae). Other larvae inhabit the tissues of living organisms (e.g., Acroceridae, Oestridae, Pipunculidae, and Tachinidae). Still others (e.g., larvae of the superfamily Hippoboscoidea) are retained and nourished in the female abdomen until deposited and ready to quickly pupariate. Most of the feeding and accumulation of biomass occurs in the larval stage, and adult Diptera mostly take only what they need to supply their flight muscles with energy. Among those flies that feed extensively, their diets consist of nectar or honeydew (e.g., Blephariceridae and Bombyliidae), pollen (e.g., Nemestrinidae and Syrphidae), vertebrate blood (e.g., Culicidae and Glossinidae), insect hemolymph (e.g., some Ceratopogonidae), and

other organic materials that are liquified or can be dissolved or suspended in saliva or regurgitated fluid (e.g., Calliphoridae, Micropezidae, and Muscidae). The adults of some groups are predaceous (e.g., Asilidae, Empididae, and some Scathophagidae), whereas those of a few Diptera (e.g., Deuterophlebiidae and Oestridae) lack mouthparts completely, do not feed, and live for only a brief time.

As holometabolous insects that undergo complete metamorphosis, the Diptera have a life cycle that includes a series of distinct stages or instars. A typical life cycle consists of a brief egg stage (usually a few days or weeks, but sometimes much longer), three or four larval instars (usually three in Brachycera, four in lower Diptera, more in Simuliidae, Tabanidae, Thaumaleidae, and a few others), a pupal stage of varying length, and an adult stage lasting from less than 2 h (Deuterophlebiidae) to several weeks or even months. The eggs of Diptera are laid singly, in small clusters, or in loose or compact masses, and they can be attached to rocks, vegetation, or other substrata, or deposited on or in the food source. Oviposition sites are usually in or near the larval habitat, which ensures that eggs are placed in a location suitable for larval development, with a notable exception being the human bot fly, *Dermatobia hominis*, which glues its eggs to zoophilous dipterans (e.g., calyptrate flies and Culicidae), thereby ensuring a carrier-mediated infection (Guimarães and Papavero 1999). In some groups, eggs are incubated and hatch during (e.g., Sarcophagidae) or immediately after deposition (e.g., many Tachinidae), or the female is truly viviparous when the larvae are nourished and grow while still inside the female (e.g., Hippoboscoidea, and mesembrinelline Calliphoridae) (Ferrar 1987, Meier et al. 1999). For a given species, all larval instars usually occur in the same habitat. In general, the duration of the first instar is shortest, whereas that of the last instar is much longer, often several weeks or even months. Although most Diptera exhibit sexual reproduction, parthenogenesis occurs in some groups, and reproduction by immature stages (paedogenesis) has been recorded in some gall midges (Cecidomyiidae).

Among the most unusual life histories is that of the Nymphomyiidae. Adults have a larviform appearance, lack mouthparts (Fig. 9.3B), and possess wings that are deciduous, elongate, and fringed with long microtrichia (Courtney 1994). Most species are associated with small headwater streams where larvae, pupae, and copulating adults occur on rocks covered with aquatic mosses. Although few details about mating behavior are available, observations of Appalachian species suggest that adults locate a mate soon after emergence, couple, descend into the water in copula, shed their wings, and crawl to an oviposition site. The female then lays a rosette of eggs around the coupled adults, which die in copula (Courtney 1994).

Another remarkable life history is that of *Fergusonina turneri* (Fergusoninidae), which in an obligate mutualism with the nematode *Fergusobia quinquenerviae* is gall building on the myrtacean plant *Melaleuca quinquenervia* (Taylor 2004). Galls are initiated in buds and young leaves by juvenile nematodes, which are injected by ovipositing female flies, along with their own eggs. When the fly eggs hatch, the larvae form individual cavities in the galls, and nematodes move into these and coexist with the fly larva. The nematodes pass through at least one parthenogenetic generation, and fertilized female nematodes of a later sexual generation invade the late third-instar fly larva. Nematode eggs are deposited in the larval hemolymph and, after hatching, the juvenile nematodes migrate to the fly ovaries. When the adult female fly hatches, it will continue the cycle by depositing new nematodes along with its own eggs.

In the Phoridae, the peculiarly swollen, physogastric females of species in the subfamily Termitoxeniinae, which are all associated with fungus-growing termites, show a post-metamorphic growth in both head and hind legs, which is unique for an adult, nonmolting insect (Disney and Kistner 1995). These termite inquilines were described by Wasmann (1910: 38) as 'a store-house of anomalies, whether we consider them from the point of view of morphologists, anatomists, evolutionists, or biologists. They are exceptions to the laws of entomology'.

Some predaceous species of flies have evolved odd larval lifestyles. Adults of *Oedoparena glauca* (Dryomyzidae) oviposit on closed barnacles during low tide. The eggs hatch during subsequent low-tide periods and larvae enter the barnacles as they open, with the incoming tide. During high tide, larvae feed inside the tissues of the submerged barnacles, and in subsequent low-tide periods they search for new prey (Burger et al. 1980). The larvae of the Vermileonidae are commonly called worm lions because they construct pitfall traps similar to those of ant lions (family Myrmeleontidae) of the order Neuroptera. The worm lion waits buried in the bottom of the pit for an insect prey to tumble in, pounces, sucks out its body juices, and then tosses the victim's corpse from the pit (Wheeler 1930, Teskey 1981). A

number of species in the Keroplatidae (*Orfelia fultoni*, *Arachnocampa* spp., and *Keroplatus* spp.) are bioluminescent and emit a blue-green light as larvae. These glowworms construct mucous tubes from which they hang snares with droplets of oxalic acid that capture and kill prey attracted by their bioluminescence. The larvae are voracious predators and feed on many types of arthropods attracted to their glow (Baker 2002). In some instances, these glowworms congregate in large numbers and form impressive displays. For example, a superb concentration of *Arachnocampa luminosa* in the Waitoma Caves in New Zealand attracts more than 300,000 visitors per year (Baker 2002).

A common mating behavior among the lower Diptera (Chaoboridae, Chironomidae, and others) is the formation of dense and sometimes enormous swarms (Vockeroth 2002). The swarms generally are composed of males, and when females enter the swarm, coupling quickly takes place. Males often exhibit adaptations that enable mate detection. For example, the eyes of males of most species engaged in swarming are enlarged and contiguous above (presumably to aid in spotting females from below) and the antennae have numerous, long, hairlike setae that allow them to detect a female's wing beats. In the Brachycera, premating behavior includes posturing and displays in courtship that can become complex performances with combinations of kneeling, jumping, and flapping (e.g., Struwe 2005). In dance flies (Empididae), which often have mating swarms, the male presents the female with an edible lure or an inedible substitute to initiate mating (Cumming 1994).

Many Diptera congregate at landmarks for the purpose of mating. Landmarks can range from a rock to a tuft of grass, a road, a stream course, a canyon, a bog, an emergent tree (taller than the others), or a hilltop. The difference between simple landmarks and hilltops is that simple landmarks typically support only a single species. However, emergent trees in rainforests are likely immensely important for landmark mating species, although few data are available. Hilltops are significant landmarks because they support many species, often hundreds or, in rare cases, even thousands (Skevington 2008). Hilltops range from massive, rocky mountaintops more than 4000 m high to small hummocks in flat country. The height above the surrounding land must not be too intimidating to exclude many species, while the hilltop must be distinctive and visible at large distances. Some 33 families of Diptera are known to hilltop (Skevington 2008).

OVERVIEW OF TAXA

Lower Diptera

Some of the most common and easily recognized flies (Figs. 9.1A-E, 9.3C), including black flies (Simuliidae), crane flies (Tipuloidea), fungus gnats (e.g., Mycetophilidae and Sciaridae), and mosquitoes (Culicidae), belong to the lower Diptera (also known as the 'Nematocera'). The group contains approximately 40 families and more than 52,000 species worldwide (Evenhuis et al. 2007). Although the Diptera and several subordinate taxa (e.g., Brachycera, Eremoneura, Cyclorrhapha, and Schizophora) are considered monophyletic, the lower Diptera generally are considered a paraphyletic or grade-level grouping (Hennig 1973, Wood and Borkent 1989, Oosterbroek and Courtney 1995, Yeates and Wiegmann 1999, Yeates et al. 2007). Despite this position, a review is useful of some of the features shared by members of this phyletic grade of Diptera. For the most part, adults of lower Diptera are characterized as slender, delicate, long-legged flies with long, multisegmented antennae (e.g., Culicidae, Tanyderidae, and Tipulidae); however, the group also includes some stout-bodied flies with relatively short antennae (e.g., Axymyiidae, Scatopsidae, and Simuliidae). Larvae of most lower Diptera have a well-developed, sclerotized head capsule (Figs. 9.2A-F) (Courtney et al. 2000).

Although a few lineages in the lower Diptera (e.g., Bibionomorpha) occur primarily in terrestrial or semiterrestrial habitats, the vast majority of lower Diptera have larvae and pupae that are aquatic or semi-aquatic (Foote 1987, Brown 2001, Merritt et al. 2003). Aquatic habitats include a wide range of lentic (standing water) and lotic (flowing water) situations (Courtney and Merritt 2008, Courtney et al. 2008). Lakes, cold and hot springs, temporary pools, stagnant waters of ground pools, phytotelmata (tree holes and other plant cavities), and artificial containers (e.g., buckets and tires) are among the many lentic habitats colonized by larvae. The Culicoidea (e.g., Chaoboridae, Culicidae, and Dixidae) are especially well represented in lentic habitats. These families include proficient swimmers that can travel to considerable depths; yet their larvae generally remain near the water surface because of the dependence on atmospheric respiration. The culicid genera *Coquillettidia* and *Mansonia* are unusual because their larvae use their specialized respiratory siphons to obtain oxygen from submerged or floating vegetation (Wood et al.

1979, Clements 1992). Free-swimming larvae of common midges (Chironomidae) and biting midges (Ceratopogonidae) do not depend on atmospheric respiration and can colonize larger and deeper bodies of water. Some chironomids survive at great depths, with one species, *Sergentia koschowi*, known to occur as deep as 1360 m in Lake Baikal (Linevich 1971). Lotic habitats of lower Diptera range from slow, silty rivers to torrential streams to groundwater zones (Courtney and Merritt 2008). The larvae of the net-winged midges (Figs. 9.2F, 9.3E) (Blephariceridae), mountain midges (Fig. 9.2D) (Deuterophlebiidae), and black flies (Simuliidae) are among the most specialized inhabitants of flowing waters; all lack spiracles (exchanging oxygen directly through their cuticle) and have structural modifications that permit survival on current-exposed substrates. Blepharicerid larvae, which frequently occur in current velocities exceeding 2 m/sec, show perhaps the greatest morphological specialization, including ventral suctorial discs used to adhere to smooth rocks (Zwick 1977; Hogue 1981; Courtney 2000a, 2000b). Similar habitats and comparably unusual attachment devices (prolegs with apical rows of hooks) are typical of larval Deuterophlebiidae (Courtney 1990, 1991) and Simuliidae (Crosskey 1990, Adler et al. 2004). Other specialized lotic habitats include seepages on cliff faces and waterfall splash zones, where larval Thaumaleidae and many Chironomidae, Psychodidae, Simuliidae, and Tipulidae can be common (Vaillant 1956, 1961, Sinclair and Marshall 1987, Sinclair 1988, 1989, 2000, Craig and Currie 1999), and saturated wood along stream margins, where larvae of Axymyiidae and certain Tipuloidea (e.g., *Lipsothrix*) reside (Dudley and Anderson 1987, Wood 1981). Groundwater zones are another important but largely unstudied habitat for larval Diptera, particularly for the Chironomidae (McElravy and Resh 1991, Ward 1992, 1994). Finally, larvae of a few lower Diptera (e.g., some members of the Ceratopogonidae, Chironomidae, Culicidae, and Tipulidae) can be abundant in marine and brackish-water environments, including intertidal pools, seaweed beds, lagoons, and estuarine marshes (Hashimoto 1976, Linley 1976, O'Meara 1976, Robles and Cubit 1981, Pritchard 1983, Colbo 1996, Cranston and Dimitriadis 2005, Dimitriadis and Cranston 2007).

The taxonomic and ecological diversity of the lower Diptera is reflected in the wide range of larval feeding habits, which encompass nearly every trophic group.

Many groups consume live plants (e.g., Cecidomyiidae and some Tipuloidea) or decomposing plant fragments or fungi (e.g., Mycetophilidae, Sciaridae, and many Tipuloidea). Others feed on decaying, fine organic matter and associated microorganisms (e.g., many Chironomidae). The larvae of some aquatic families (e.g., Blephariceridae and Thaumaleidae) use specialized mouthparts to graze on the thin film of algae and organic matter on rocks and other substrates (Courtney 2000a, 2000b; Alverson et al. 2001). Many families contain a few predaceous species, whereas the larvae of some groups (e.g., Ceratopogonidae) feed primarily or exclusively on other animals (McAlpine et al. 1981, Hövemeyer 2000). Nearly all of these trophic groups are represented in the diverse family Chironomidae (nearly 7000 species) and superfamily Tipuloidea (more than 15,000 species). Their trophic diversity and numerical abundance make the lower Diptera an important component in aquatic and terrestrial ecosystems, both as primary consumers and as a food resource for other invertebrates, fish, amphibians, reptiles, birds, and mammals. The Chironomidae, which in aquatic ecosystems are often the most abundant organisms in both numbers and biomass, can be especially important in ecosystem functioning (Armitage et al. 1995). The trophic importance of aquatic Diptera extends also to aquaculture programs in which nearly every life stage can be an important component of fish diets.

Brachycera

Lower Brachycera

As in the lower Diptera, the lower Brachycera are paraphyletic, but remain a convenient grade for discussion. This group is also widely known as the 'Orthorrhapha' (referring to the T-shaped opening of the pupal exuviae) and comprises mostly predaceous larvae (except Stratiomyomorpha) and parasitoids of spiders and other insect orders. Adult lower Brachycera are blood feeders, predators, or flower visitors. The lower Brachycera contain some of the largest and most colorful flies, including bee flies, horse flies, mydas flies, and robber flies. This grade includes some 24,000 species comprising 20 families assigned to three infraorders (Stratiomyomorpha, Tabanomorpha, and Xylophagomorpha) and several superfamilies (Asiloidea and Nemestrinoidea) (Yeates et al. 2007).

The Pantophthalmidae are enormous flies (up to 5.5 cm in length), with larvae that dig galleries in dead or living trees and likely feeding on the fermenting sap in the tunnels (Val 1992, D. M. Wood, personal communication). Both the Xylomyidae and Stratiomyidae (Fig. 9.1N) are unique among the lower Brachycera in regard to their scavenging and filter-feeding habits (Rozkošný 1997) and by pupating in the final-instar larval exuviae (comparable to the cyclorrhaphan puparium). The Stratiomyidae larvae can be assigned generally to two groups: terrestrial and aquatic. Terrestrial larvae live in decaying leaves and other plant material, upper layers of soil, manure, under loose bark of decaying trees, and in ant nests. Aquatic larvae (Fig. 9.2L) can be found in saturated moss, littoral zones of ponds, lakes, and marshes, hygropetric situations in spring streams, phytotelmata, roadcuts or similar seepages, saline habitats, and even hot thermal springs (Rozkošný 1997, Sinclair and Marshall 1987, Sinclair 1989).

Feeding on vertebrate blood by female flies has evolved at least two or three times in the lower Brachycera, but is restricted to the Tabanomorpha (Athericidae, Rhagionidae *sensu lato*, and Tabanidae) (Wiegmann et al. 2000, Grimaldi and Engel 2005). The Tabanidae (deer flies and horse flies) are well known to campers and swimmers during the early summer months in northern latitudes due to the voracious blood-sucking behavior of most species. Many species of the subfamily Pangoniinae are characterized by their long mouthparts (known as long-tongues), often stretching longer than their than body length. These groups generally are believed to be nectar feeders (Goldblatt and Manning 2000), but several species also have been observed feeding on warm-blooded (humans – *Philoliche*; Morita 2007) and cold-blooded (caimans – *Fidena*; B.A. Huber, personal communication) vertebrates. Tabanid larvae mostly inhabit swampy biotopes, where they prey on insect larvae. They even are known to feed opportunistically on toads (Jackman et al. 1983).

Adults of many of the remaining families of lower Brachycera are fast-flying flower visitors. *Moegistorhynchus longirostris* (Nemestrinidae) from southern Africa possesses a proboscis nearly five times its body length and is an important pollinator of tubular flowers (Goldblatt and Manning 2000). Bee flies (Bombyliidae) occur worldwide and reach their greatest diversity in Mediterranean climates (Yeates 1994). The female abdomen of several bee fly subfamilies is modified to form an invaginated

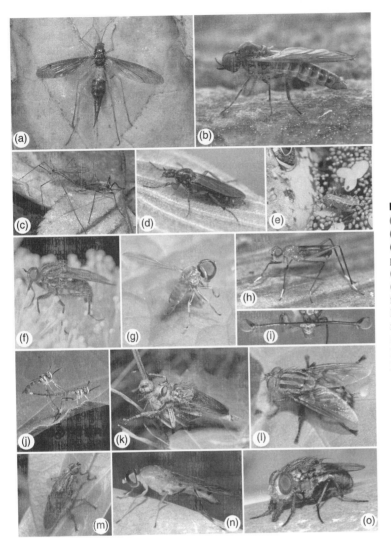

Fig. 9.1 Adult Diptera. (a) Tipulidae (*Tanysipetra*) habitus, dorsal view. (b) Axymyiidae (*Axymyia*), lateral view. (c) Limoniidae (*Prionolabis*) mating pair, oblique-dorsal view. (d) Bibionidae (*Bibio*) habitus, oblique-lateral view. (e) Culicidae (*Culex*) feeding on ranid frog. (f) Empididae (*Empis*) habitus, lateral view. (g) Pipunculidae taking flight, oblique-lateral view. (h) Micropezidae (*Grallipeza*) habitus, lateral view. (i) Diopsidae (*Teleopsis*) head, frontal view. (j) Conopidae (*Stylogaster*) mating pair, lateral view. (k) Asilidae (*Proctacanthus*) feeding on dragonfly, oblique-dorsal view. (l) Sarcophagidae (*Sarcophaga*) habitus, dorsal view. (m) Scathophagidae (*Scathophaga*) habitus, oblique-lateral view. (n) Stratiomyidae habitus, lateral view. (o) Calliphoridae (*Hemipyrella*) habitus, frontolateral view. (*See color plate*). (Images by E. Bernard [a], G. Courtney [b, c, h, i, m], S. Marshall [e, f, g, j, k], M. Rice [d] and I. Sivec [l, n, o].)

sand chamber, which is first filled when they alight on open surfaces (Yeates 1994, Greathead and Evenhuis 1997). The egg is laid in the sand chamber and coated with soil particles before being ejected by the hovering female onto oviposition sites. The larvae of the Bombyliidae are mostly parasitoids of holometabolous insects (e.g., acridoid egg pods, solitary bees, and wasps) (Greathead and Evenhuis 1997). The larvae of small-headed flies (Acroceridae) are internal parasitoids of true spiders. First-stage larvae actively seek out hosts, capable of looping along a single web strand (Nartshuk 1997). In contrast to adults of most lower Brachycera, those of the Asilidae (Fig. 9.1K) are strictly predaceous on insects (Hull 1962). They focus on large prey from a wide variety of insect orders, sometimes taking prey more than twice their size (e.g., dragonflies; Platt and Harrison 1995).

Empidoidea

The dance flies, balloon flies, and other predaceous flies (Fig. 9.1F) that traditionally have been placed in the

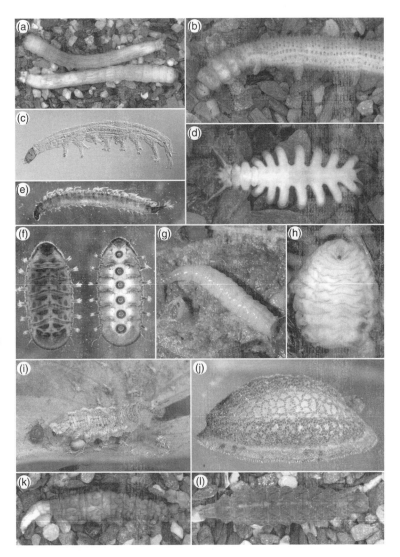

Fig. 9.2 Larval Diptera.
(a) Tipulidae (*Epiphragma*) habitus,
dorsal (top) and ventral (bottom)
views. (b) Ptychopteridae
(*Bittacomorpha*) head, thorax and
abdominal segments I–III, lateral
view. (c) Nymphomyiidae
(*Nymphomyia*) habitus lateral view.
(d) Deuterophlebiidae (*Deuterophlebia*)
habitus, dorsal view. (e) Psychodidae
(*Pericoma*) habitus, lateral view.
(f) Blephariceridae (*Horaia*) habitus,
dorsal (left) and ventral (right) views.
(g) Calliphoridae (*Lucilia*) habitus,
dorsal view. (h) Tephritidae (*Eurosta*)
habitus, ventral view. (i) Syrphidae
(*Syrphus*) feeding on aphids, dorsal
view. (j) Syrphidae (*Microdon*) on
glass, lateral view. (k) Sciomyzidae
(*Tetanocera*) habitus, lateral view.
(l) Stratiomyidae (*Caloparyphus*)
habitus, dorsal view. (*See color plate*).
(Images by G. Courtney [a–f, h, k, j]
and S. Marshall [g, i, l].)

family Empididae are now classified in four families
of Empidoidea, along with the long-legged flies of the
family Dolichopodidae. With approximately 12,000
described species and many more undescribed species,
the Empidoidea are one of the largest superfamilies
of Diptera and the most diverse lineage of predaceous
flies (Sinclair and Cumming 2006). The vast majority
are predators as adults, with the few exceptions being
obligate flower-feeding groups that consume pollen as
their only protein source. They are found in a variety of forested and open habitats where they breed in

moist soils, decaying wood, and dung, and occur in
aquatic habitats. All known larvae appear to be predators on invertebrates. The common name 'dance flies'
is derived from the behavior of members of the large
subfamily Empidinae in which adult males transfer
nuptial gifts to the female during courtship and mating
(Cumming 1994). The Dolichopodidae are common
metallic-colored flies, often observed sitting on leaves
and mud flats. Many possess elaborate leg ornamen-
tations that are used in courtship displays (Sivinski
1997).

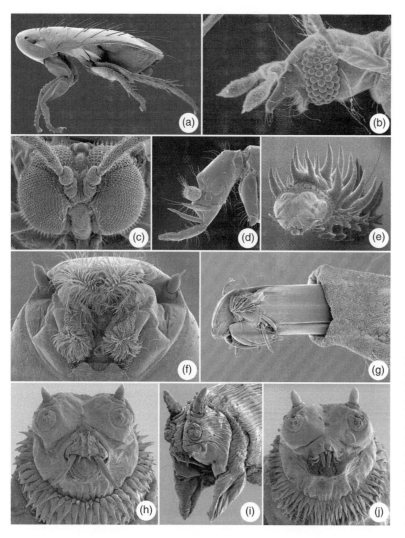

Fig. 9.3 Scanning electron micrographs of Diptera. (a) Phoridae (*Thaumatoxena*) adult habitus, lateral view. (b) Nymphomyiidae (*Nymphomyia*) adult head, lateral view. (c) Blephariceridae (*Blepharicera*) adult head, frontal view. (d) Phoridae (*Termitophilomya*) adult head, lateral view. (e) Blephariceridae (*Agathon*) larva habitus, oblique-frontal view. (f) Ptychopteridae (*Bittacomorpha*) larval mouthparts, ventral view. (g) Athericidae (*Atherix*) larval head, lateral view. (h) Calliphoridae (*Onesia*) larval head, ventral view. (i) Sarcophagidae (*Metopia*) larval head, oblique-ventral view. (j) Calliphoridae (*Bellardia*) larval head, ventral view. (Images by G. Courtney [b, c, e, f, g] and T. Pape [a, d, h–j].)

Lower Cyclorrhapha

This group of taxa has been considered a monophyletic group (Aschiza) by some workers (McAlpine 1989, Disney 1994a). As with the lower Diptera, strong evidence now suggests that the lower Cyclorrhapha are a paraphyletic assemblage (Cumming et al. 1995, Sinclair and Cumming 2006, Moulton and Wiegmann 2007). The seven lower cyclorrhaphan families are discussed below.

The Platypezidae include 250 species, commonly found individually hovering in deep shade or in swarms of dancing males in forest openings. Males of

Microsania ('smoke flies') form epigamic swarms in smoke from forest and campfires. Eggs of platypezids are laid between the gills or in the pores of fungi on which the larvae feed. The Ironomyiidae occur in both dry sclerophyll and rainforest habitats and have been found hilltopping (Skevington 2008). The family contains one described and two undescribed extant species (all in the genus *Ironomyia*), all from Australia (D. K. McAlpine, personal communication). Fifteen fossil species from the Holarctic Region have been described from five additional genera from the Upper Jurassic and Cretaceous periods (McAlpine 1973, Zhang 1987, Mostovski 1995,

Grimaldi and Cumming 1999). The Lonchopteridae are distinctive, pointed-winged, strongly bristled, often yellowish-brown flies, most commonly found in moist environments, along streams or ponds, in bogs, in deciduous forests, or even in alpine meadows or hot springs (Nielsen et al. 1954, Smith 1969). Some species are found in hot, dry meadows (Andersson 1970), while others occur in rocky tidal zones near the coast (Dahl 1960). Most species are bisexual but at least one species, *Lonchoptera bifurcata*, is parthenogenetic in most areas of its nearly cosmopolitan range (Stalker 1956). It is the only species of the family that is found in the neotropics and Australia. Lonchopterids have a predominantly Old World distribution, with 46 of the approximately 60 described species occurring in the Palearctic and Oriental Regions. Larvae are apparently saprophagous, microphagous, or mycetophagous, but more study is needed to confirm these feeding habits.

The Phoridae, or scuttle flies, are one of the most diverse fly families and have been proposed as the most biologically diverse family of insects on Earth (Disney 1994b). Approximately 4000 described species have been described but more than 30,000 species are estimated to exist (Evenhuis et al. 2007, Brown 2008). More than half of the described phorid species belong to the huge genus *Megaselia*. They are found in almost all terrestrial habitats, with the exclusion of exceptionally cold and dry environments. Larvae have extremely diverse tastes, with many saprophagous species, fungivores, and herbivores (including leaf miners, root feeders, and bud, seed, and fruit feeders). The majority of phorid larvae are likely predators, parasitoids, or parasites, and several species occur in highly specialized habitats such as pitcher plants and even intertidal areas (Disney 1998).

The Pipunculidae, or big-headed flies, comprise 1400 described species. Most pipunculids that have been studied are endoparasitoids of several families of Homoptera (Auchenorrhyncha). The only known exceptions are species of the genus *Nephrocerus*, which attack adult crane flies (Tipuloidea) (Koenig and Young 2007). Adults use all terrestrial habitats, but diversity and numbers are greatest in forest openings and along forest edges. Pipunculids also are known for their hilltopping behavior (Skevington 2000, 2001).

Adult flower flies (Syrphidae), also known as hover flies, vary considerably in size and appearance, ranging from 4 to 25 mm long and from small black flies to large wasp or bee mimics. They include about 6000 described species and are among the most abundant and conspicuous flies. Their visibility is partly related to their ability to hover motionless and partly to their frequent flower visitation. They are among the most significant pollinators in the Diptera and should be assessed in comparison to the bees (Ssymank et al. 2008). Some species such as *Episyrphus balteatus* are strong fliers and are migratory. Many larvae are entomophagous, feeding on ant brood, aphids and other soft-bodied Sternorrhyncha, and social wasp larvae (Thompson and Rotheray 1998). Saprophagous syrphid larvae exploit wet or moist conditions and are typically associated with fermenting tree sap, rot holes and fallen wood, decaying vegetation in water or wet compost, and dung. Some larvae are mycophages–phytophages, feeding in pockets of decay in live plants.

Non-Calyptratae Muscomorpha

The acalyptrates are likely another grade, or group of convenience. Treated as the sister group of the calyptrates for years (Hennig 1971, 1973; McAlpine 1989), some evidence suggests that this assemblage is paraphyletic (Griffiths 1972). The early radiation of the more than 80 families of flies in this group appears to have been explosive, which may have obscured the evolutionary history of the group. Griffiths (1972), Hennig (1973), and McAlpine (1989) are the only researchers to have proposed a phylogeny (or at least a phylogenetically based classification) for the entire group. Despite their efforts, evidence indicates that even some of the superfamilies are not monophyletic. This radiation resulted in a remarkable diversity of flies and many of the groups are of considerable importance to society. More than 50% of the acalyptrate species are contained in just six large families: Agromyzidae, Chloropidae, Drosophilidae, Ephydridae, Lauxaniidae, and Tephritidae. Some of the ecological diversity shown by these families and others is discussed below.

Most of the Conopidae are parasitoids of bees and aculeate Hymenoptera. One lineage, the Stylogastrinae, comprises parasitoids of orthopteroid insects. Many of the Neotropical species follow foraging army ant raids, attacking the orthopteroids that flee from the advancing army. All known Psilidae are phytophagous and some are well-known pests (e.g., the carrot rust fly, *Chamaepsila hennigi*; Peacock and Norton 1990, as *Psila rosae*) (Szwejda and Wrzodak 2007). Some Diopsidae are agricultural pests on grasses, but some also are well-known research subjects because of their variously developed eyestalks. Many species are sexually dimorphic for the length of the eyestalks, with males having much longer eyestalks than females. This dimorphism is believed to have evolved because of the mating advantages that they bestow on these males. Females show a strong preference for males with longer eyestalks, and males compete with each other to control lekking aggregation sites by a ritualized contest that involves facing each other and comparing their relative eyespans, often with the front legs spread out to add emphasis. A few other families of acalyptrate flies have members that have developed similar types of apparent runaway sexual selection. For example, some Drosophilidae, Platy-stomatidae, Richardiidae, and Tephritidae have stalked eyes, or large, sometimes antler-like spines extending from their genae. Of these families, the Platystomatidae and Tephritidae are important phytophagous groups. Tephritids in particular include many pest species that are becoming problems because they are spread via global trade (e.g., Oriental fruit fly, *Bactrocera orientalis*, and Mediterranean fruit fly, *Ceratitis capitata*). The Drosophilidae are most famous because of *Drosophila melanogaster*, the subject of considerable genetics research. Drosophilids also are renowned for their expansive radiation in Hawaii, where approximately 1000 species (more than 500 described) radiated from a single colonizer (Kaneshiro 1997, O'Grady et al. 2003).

The Pyrgotidae are bizarre-looking flies, most commonly seen at lights at night. These flies are specialized internal parasitoids of scarab beetles and likely have a pronounced effect in controlling some populations of pest scarabaeids (Steyskal 1987). The Piophilidae are small flies that tend to specialize on carrion. One species, the cheese skipper *Piophila casei*, is a serious pest in the food industry, with larvae found in cured meats, smoked fish, cheeses, and decaying animals. Some species, such as *Protopiophila litigata*, form impressive mating aggregations on discarded cervid antlers (Bon-duriansky and Brooks 1998). The Clusiidae are another of a handful of acalyptrate families known to engage in this type of lekking behavior. Males establish dominance at a lekking site by defending territories from other males on logs or branches to attract females and mate (Lonsdale and Marshall 2004).

Larvae of all Agromyzidae feed on living plant tissues, forming mines that are species specific. Most are either monophagous or oligophagous, and although best known as leaf miners, they attack all parts of plants (Dempewolf 2004).

Sciomyzid larvae are all predators or parasitoids of freshwater or terrestrial molluscs (Berg and Knutson 1978). The Chamaemyiidae are free-living predators of adelgids, aphids, coccids, and scales and have been used in biological control programs (Gaimari and Turner 1996, Vail et al. 2001). The Sphaeroceridae are diverse and associated with all types of organic decay including dung, carrion, fungi, supralittoral seaweed, compost, mammal nests, conifer duff, cave debris, and deposits of dead vegetation (Marshall and Richards 1987). The Ephydridae, or shore flies, are important food for wildlife along both freshwater and saltwater pools. In some wetlands, such as Mono Lake in California, millions of birds are supported almost entirely by ephydrids (Jehl 1986, Rubega and Inouye 1994). Larvae of the Chloropidae have varied food habits (Ferrar 1987). Many are phytophagous, damaging cereals and other grasses. These species include the frit fly (*Oscinella frit*), the wheat stem maggot (*Meromyza americana*), and the gout fly of wheat and barley in Europe (*Chlorops pumilionis*). Others are saprophagous, fungivorous, and even predaceous. *Thaumatomyia glabra* is an important predator on the sugarbeet root aphid *Pemphigus populivenae*. One of the most unusual habits among Chloropidae is that of the species of the Australian genus *Batrachomyia*, whose larvae live under the skin on the back of frogs (Sabrosky 1987).

Calyptratae

The calyptrate flies are generally rather robust, most are strong fliers, and many are in the size range of the common housefly. The group contains some 22,000 species, which are arranged in 10–15 families, depending on the classification (Evenhuis et al. 2007). A large number of species breed in living or decaying plant or fungal material (especially the muscoid families Anthomyiidae, Fanniidae, Muscidae, and Scathophagidae). The biology of species in the large family Muscidae is particularly varied, with habitats including vertebrate dung and carrion; organic debris in nests, burrows, and dens of mammals, birds, and insects; rot holes and decaying wood; sap runs; living plants; fungi; and in or on the edges of ponds and streams (Skidmore 1985, Ferrar 1987). Evolution can take surprising routes, as in the Scathophagidae, whereby a few lineages have evolved from the ancestral plant-feeding life habit into breeding in dung or rotting seaweed and even as predators on caddisfly egg masses or small invertebrates (Kutty et al. 2007).

Large mammals are a rich source of food for many calyptrates, which may suck their blood, imbibe their sweat, eat their dung, or even be true endoparasites. The tse-tse (Glossinidae) and the ectoparasitic louse and bat flies (Hippoboscidae, including the Nycteribiinae and Streblinae) have excelled as highly specialized blood feeders and, like the few other calyptrate bloodsuckers, both males and females take a blood meal. Hosts are mainly mammals and birds but a few feed on large reptiles. The reproductive biology of hippoboscoid calyptrates is remarkable in that the eggs hatch one at a time in the female oviduct, and the larva is nourished by a secretion from the female accessory glands until it is fully fed and near pupariation (Ross 1961, Hill 1963, Marshall 1970, Potts 1973). Obligate parasites of mammals are found particularly in the family Oestridae, with larvae taking up their final position either subdermally (Cuterebrinae, Hypodermatinae, and lower Gasterophilinae), in the gastrointestinal tract (higher Gasterophilinae), or in the nasopharyngeal cavities (Oestrinae) (Zumpt 1965).

Numerous calyptrates are associated with either vertebrate or invertebrate carrion, and some species infest wounds or body orifices of living vertebrates as larvae. Social insects are hosts of many calyptrates. Several blow flies (Calliphoridae) are associated with ants and termites as either kleptoparasites or, more rarely, parasitoids (Ferrar 1987, Sze et al. 2008). Solitary, aculeate Hymenoptera can have their nests usurped by kleptoparasitizing flesh flies (Sarcophagidae) of the subfamily Miltogramminae (Spofford and Kurczewski 1990, Pape 1996). Snails and earthworms are heavily exploited by both flesh flies and blow flies (Keilin 1919, Ferrar 1976, Guimarães 1977, Downes 1986), and pterygote insects are hosts to the exclusively parasitoid species of the large family Tachinidae (Stireman et al. 2006).

SOCIETAL IMPORTANCE

As expected for a ubiquitous group with diverse habits and habitats, the Diptera are of considerable economic importance. Pestiferous groups can have significant effects on agriculture, animal and human health, and forestry. Other groups can be a general nuisance when present in large numbers or because of allergic reactions to detached body setae. Despite these negative effects, flies play a valuable role as scavengers, parasitoids and predators of other insects, pollinators, food for predators, bioindicators of water quality, and tools for scientific research.

Diptera as plant pests (agriculture, silviculture, and floriculture)

A large number of fruit flies (Tephritidae) are capable of causing considerable economic damage to fruits and vegetables, making these flies perhaps the most important dipteran family to agriculture (e.g., Dowell and Wange 1986, McPheron and Steck 1996, Norrbom 2004). The genera *Anastrepha, Bactrocera, Ceratitis, Dacus,* and *Rhagoletis* contain most of the pest species. Economic impact includes direct losses from decreasing yield, increasing costs for control and fruit treatment, and shrinking export markets due to local regulations. Quarantine laws designed to reduce the spread of fruit fly species can severely restrict global commerce of many commercial fruits. Evidence of their economic effects is illustrated by the millions of US dollars spent annually to prevent the Mediterranean fruit fly from entering California (Jackson and Lee 1985, Dowell and Wange 1986, Aluja and Norrbom 1999).

The Agromyzidae, well known for the plant-mining habits of their larvae, also contain a number of important plant pests, including the chrysanthemum leafminer (*Liriomyza trifolii*), the serpentine leafminer (*L. brassicae*), and the vegetable leafminer (*L. sativae*) (Spencer 1973, 1990). The pea leafminer *L. huidobrensis* is a highly polyphagous species that can damage a wide range of field and greenhouse crops, including alfalfa, artichoke, beans, beets, carrots, celery, lettuce, melons, onions, peas, potato, pumpkin, spinach, tomatoes, and several crucifers and ornamentals. Control of established populations can be especially problematic (Steck 2005).

Among lower Diptera, the gall midges (Cecidomyiidae) are perhaps the best-known agricultural pests. A widespread and common group containing mostly plant-feeding species, gall midges are suspected to include several thousand new and undescribed species; however, studies of the tropical fauna are still in their infancy (e.g., Gagné 1994). Because of the feeding habits of their larvae, gall midges include many serious plant pests, especially species that attack cereal crops and conifers. As with many fruit flies and other plant pests, their effects include not only direct damage, but also economic losses related to quarantine issues (Pollard 2000, Gagné et al. 2000).

A few Diptera can be important floricultural pests. Larvae of the black fungus gnats (Sciaridae) *Bradysia coprophila* and *Bradysia impatiens* feed on roots and algae in the upper soil surface, which can cause considerable damage in propagation areas and seedling flats. Larval feeding also can cause wilting and facilitate entry of plant pathogens. Furthermore, adult flies are known to disseminate soil-inhabiting pathogens on their bodies and in their feces (Parrella 2004). The ephydrid *Scatella stagnalis* can be a pest in ornamental nurseries and greenhouses, where adult flies spread fungal spores. Fecal spots left on leaves by resting adults also can cause cosmetic damage to plants (Parrella 2004).

Several Phoridae and Sciaridae and the moth fly *Psychoda phalaenoides* can be important pests of commercial mushroom gardens (Hussey 1960, Rinker and Snetsinger 1984, Somchoudhury et al. 1988, Scheepmaker et al. 1997, Menzel and Mohrig 1999). Although larval feeding can cause moderate damage, the major effect is through the transmission of fungal diseases (White 1981).

The Bombyliidae are usually considered beneficial insects because some species parasitize and prey on cutworms, beetle grubs, and grasshopper egg pods; however, *Heterostylum robustum* kills up to 90% of the larvae of the alkali bee, an important alfalfa pollinator in northwestern USA (Bohart et al. 1960, Bohart 1972). Other pests in apiaries include the European miltogrammine flesh fly *Senotainia tricuspis*, which infects as many as 90% of the adult bees in a hive (Santini 1995a, 1995b, Palmeri et al. 2003). Honeybees are sometimes attacked by species of the phorid *Melaloncha* (Ramírez 1984, Brown 2004, Gonzalez and Brown 2004), and native colonies of meliponine bees can be affected by the phorids *Pseudohypocera kerteszi* and *Megaselia scalaris* (Robinson 1981; Reyes 1983; Hernández and Gutiérrez 2001; Robroek et al. 2003a, 2003b). *Megaselia scalaris* also can cause considerable damage to live arthropod cultures in laboratories, insect zoos, and butterfly houses (Disney 1994b; G. W. Courtney, personal observations).

Medical and veterinary importance

Disease transmission

The mouthparts of many adult Diptera have effective piercing stylets, enabling these flies to 'bite' and suck blood. Major families with piercing and sucking mouthparts include the bat flies, biting midges, black flies, horse and deer flies, louse and bat flies, mosquitoes, phlebotomine sand flies (Psychodidae), tse-tse, and a few muscid flies. Because of their blood-feeding habits, these flies are natural carriers of pathogens and play a major role in the transmission of bacteria, fungi, nematodes, protozoans, viruses, and other parasites. The affinities of some dipterans to carrion and excrement might enhance their capacity to transmit disease agents, and for this reason alone Diptera can be considered the most economically important insect order. Diptera-borne diseases affect humans, as well as livestock worldwide, and the resulting costs are enormous.

Mosquitoes are perhaps the best-known and most-studied blood-feeding dipterans, due largely to their medical and veterinary importance. Of the approximately 3600 known species of mosquitoes, fewer than 150 are pests or vectors of pathogens that cause disease in humans and domesticated animals (Harbach 2007). However, these species, which are largely confined to the genera *Aedes* (traditional, broad sense), *Anopheles*, and *Culex*, are the indirect cause of more morbidity and mortality among humans than any other group of organisms. Mosquitoes are vectors of a number of agents that cause debilitating diseases, including malaria, yellow fever, filariasis, dengue, dog heartworm, the encephalitides, and related viral diseases. For malaria alone, the effects are staggering: 300–500 million people are infected annually; 1.0–1.5 million people die every year (World Health Organization 2007); an African child dies from malaria every 30 sec (WHO 2007); 35 million future life-years are lost because of premature mortality and disability (World Bank 1993); and an annual cost of nearly US$2 billion is incurred in tropical Africa (MicrobiologyBytes 2007). Even in contemporary North America, the effect of mosquito-borne diseases remains acute. A recent (2002–2003) outbreak of West Nile virus in Louisiana came at a price of approximately US$20 million, with slightly more than half the costs related to the illness (e.g., direct medical costs and productivity losses from illness and death) and the remaining costs related to public health responses (e.g., mosquito control, surveillance, and abatement) (KPLC 2004).

Black flies transmit the filarial parasites *Dirofilaria* (in bears) and *Mansonella* and *Onchocerca* (in humans), as well as the protozoans *Leucocytozoon* (in birds) (Crosskey

1990). Phlebotomine sand flies are vectors of the agents that cause leishmaniasis (sand fly fever) and also can transmit filarial parasites (*Icosella neglecta*) to the edible European green frog (Desportes 1942). Biting midges are capable of transmitting at least 66 viral pathogens and a wide range of microorganisms (Borkent 2004), including those responsible for livestock diseases that cause blue-tongue in sheep and cattle, African horse sickness, bovine ephemeral fever, and eastern equine encephalitis (Parsonson 1993, Mellor and Boorman 1995, Wall and Shearer 1997). The Tabanidae serve as vectors of African eye-worm or loa loa that causes loiasis, the bacterium that causes tularaemia, and the Old World trypanosome *Trypanosoma evansi* (Oldroyd 1973).

Calyptrate flies (e.g., house, stable, and blow flies) harbor more than 100 species of pathogenic microorganisms (Greenberg 1971, 1973, Förster et al. 2007, Sawabe et al. 2006). Many species of *Musca* are carriers of bovine and equine filariases, as well as bacteria and viruses. For example, *Musca autumnalis* can transmit *Parafilaria bovicola*, and eyeworms of the genus *Thelazia*, and *Hydrotaea irritans* serves as a mechanical vector of *Corynebacterium pyogenes*, a cause of mastitis in cattle (Neville 1985, Krafsur and Moon 1997). *Musca sorbens* transmits eye diseases such as trachoma and conjunctivitis (Emerson et al. 2000). The horn fly *Haematobia irritans* is a well-established biting cattle pest throughout many tropical and temperate areas of the Northern Hemisphere, while its close relative, the buffalo fly *Haematobia exigua* is particularly important to cattle and dairy industries of Australia.

As important vectors of blood-borne diseases, dipterans have shaped human culture (Harrison 1978). This role is evident especially in Africa, where trypanosome-infected tse-tse can be a serious constraint on livestock practices, and epidemics of sleeping sickness have had profound socioeconomic implications (e.g., Lyons 1992, Hide 1999). In the New World, the introduction of yellow fever has had a comparable impact (Crosby 2006). Between 1904 and 1914, the need to control yellow fever and malaria was a major contributor to the completion of the Panama Canal (Powers and Cope 2000). During military campaigns, diseases such as malaria can account for more casualties than the fighting (Bruce-Chwatt 1988), which explains the interest in vector-borne disease research at military-affiliated institutions (e.g., Walter Reed Army Institute of Research, Walter

Reed Biosystematics Unit). The original transmission of the HIV virus from chimpanzees to humans is suspected to have been caused by stable flies or similar blood-sucking flies (Eigen et al. 2002), adding yet another example of the vast influence of Diptera on human societies. Mosquitoes have even shaped human evolution through their disease-carrying capabilities, with the most notable example from Africa, where the sickle-cell anemia gene became prevalent due to its partial protection against the malaria parasite (Pagnier et al. 1984, Barnes 2005).

Myiasis

The families Calliphoridae (blow flies), Oestridae (bot flies), and Sarcophagidae (flesh flies) are the major producers of myiasis, a term referring to the development of dipteran larvae in a living vertebrate body. Bot flies are involved in dermal, enteric, and nasopharyngeal myiasis of animals and sometimes humans. The larvae of cattle grubs migrate through the host's body and eventually reach the upper back where they cut a small opening in the hide and remain there until ready to pupate. Economic losses result from reduction in milk production, weight loss, and damage to hides (Scholl 1993). In the Northern Hemisphere, *Hypoderma bovis* and *H. lineatum* (Hypodermatinae) are the major pests, whereas in the New World tropics, *Dermatobia hominis* (Cuterebrinae) is the prevalent cattle warble fly (Guimarães et al. 1983). Production losses can be significant, with annual losses in Brazil estimated at US$200–260 million (Grisi et al. 2002). Most bot flies either never attack humans, or do so only accidentally (Zumpt 1965). However, *Dermatobia hominis*, known also as the human bot fly or tórsalo, develops readily in humans. Infections are painful but generally benign, even when the larva is allowed to develop to maturity (Dunn 1930). In rare cases, infections can be lethal (Rossi and Zucoloto 1973, Noutis and Millikan 1994). Damage caused by nasal bot flies (Oestrinae) in camels, goats, and sheep, and stomach bot flies (Gasterophilinae) in donkeys and horses varies from violent reactions (i.e., 'gadding' behavior) caused by the ovipositing flies, to irritation by larvae when burrowing into oral tissues and subsequent interference with digestion. These attacks can reduce growth rates and are particularly harmful to younger individuals (Zumpt 1965).

Myiasis-producing blow flies and flesh flies usually are attracted to the wounds and sores of humans and domestic animals, where larvae feed on necrotic tissue and accidentally can be ingested or invade wounds, causing severe discomfort and subsequent secondary infections. Certain calliphorids can cause severe primary myiasis, particularly *Cochliomyia hominivorax* (the primary or New World screwworm) and *Lucilia cuprina* and *Chrysomya bezziana* in the Old World tropics (Hall and Wall 1995). In recent years, other species of screwworms in the genus *Chrysomya* have been introduced accidentally from the Old World into South America and have spread north into Central America, two even reaching North America (Baumgartner 1993, Tomberlin et al. 2001). Among the flesh flies, species of *Wohlfahrtia* cause myiasis in commercially raised mink (Eschle and DeFoliart 1965), livestock (Hall 1997, Valentin et al. 1997, Farkas and Kepes 2001, Farkas et al. 2001) and, rarely, humans (Hall and Wall 1995, Delir et al. 1999, Iori et al. 1999).

Human urogenital myiasis occasionally is caused by larvae of the Psychodidae and Phoridae (Disney and Kurahashi 1978, Abul-Hab and Salman 1999). Species of the African bot fly genus *Gedoelstia* may larviposit in the eyes of cattle, goats, and sheep, with the encephalitic form often fatal (Zumpt 1965). Cases of human ophthalmomyiasis have been caused by *Gedoelstia* spp., *Hypoderma* spp., and *Oestrus ovis* (Bisley 1972, Masoodi and Hosseini 2004, Lagacé-Wiens et al. 2008).

Invasive alien Diptera

The introduction of alien species can have devastating effects on the native fauna. The effects have been documented for the avifauna of Hawaii, with the introduction of avian malaria and a suitable vector, *Culex quinquefasciatus* (Van Riper et al. 2002). In contrast, the avifauna of the Galápagos Islands is largely intact, but the establishment of this avian vector and potential diseases poses a great threat (Whiteman et al. 2005). The introduction of the nestling parasite *Philornis downsi* to the Galápagos Islands also represents a serious threat to the endemic passerine fauna (Fessl et al. 2006). Numerous dipterans have become invasive around the world, especially members of the Agromyzidae, Anthomyiidae, Calliphoridae, Culicidae, Drosophilidae, Muscidae, Phoridae, Psilidae, Sarcophagidae, and Tephritidae, but our attention to the phenomenon is strongly skewed toward pests and disease vectors. The most detailed studies of the effect of invasive dipterans on local species have thus been for the blow flies (Wells 1991) and mosquitoes (Juliano and Lounibos 2005).

Diptera as a general nuisance

In addition to their significance in myiasis and the transmission of disease agents, flies can be a general nuisance and interfere with human activities (e.g., Cook et al. 1999, Howard 2001). The nuisance problem includes harassment by mosquitoes, gnats, and other flies, and the occasional presence of Diptera in true nuisance numbers (e.g., Westwood 1852). Several modern-day examples of the latter exist. Following European settlement of Australia, the Australian bush fly, *Musca vetustissima*, bred vigorously in the cow dung that accumulated in the absence of native ruminant-adapted dung beetles. South African dung beetles imported during the 1960s–1980s have ameliorated the problem (Ridsdill-Smith 1981, Matthiessen et al. 1984, Ridsdill-Smith et al. 1987). Another example, from Europe, pertains to the chloropid *Thaumatomyia notata*. This fly, in attempting to find suitable overwintering sites, enters apartments by the millions (Nartshuk 2000, Kotrba 2004, Nartshuk and Pakalniškis 2004). Swarms of this species have even been mistaken as smoke, prompting calls to the fire brigade (Kiesenwetter 1857, Letzner 1873). Species from several other families (e.g., Phoridae and Sphaeroceridae) proliferate indoors such that eradication measures are needed (Fredeen and Taylor 1964; Disney 1991, 1994b; Cleworth et al. 1996). Certain Psychodidae can be a nuisance as both larvae and adults, the former when mass occurrences in trickling filters of sewage treatment facilities diminish flow and filtering efficiency, and adults when setae from the wings and body are inhaled, causing a disease similar to bronchial asthma (Gold et al. 1985).

Adults of the aquatic families Chaoboridae and Chironomidae sometimes constitute a nuisance by their sheer numbers emerging from ponds and lakes. When encountering swarms of these flies, avoiding inhaling them or keeping them out of one's eyes can be difficult. Emerging chironomid adults can be a serious nuisance of lakefront settlements and cities (Ali 1991), and massive swarms can cause traffic problems (Lindegaard and Jónasson 1979). At some East African lakes, swarms of emerging *Chaoborus edulis* are so dense as to pose a risk of suffocation should one get trapped within them, particularly if swarms also contain the allergenic chironomid midge *Cladotanytarsus lewisi* (Armitage et al. 1995). Despite these negative effects, local people capture the swarming midges and convert them into round cakes that are dried in the sun for later consumption (Oldroyd 1964, Eibl and Copeland 2005).

Even beneficial species of flies, such as *Sarcophaga aldrichi* ('the friendly fly'), build up to such large numbers that people complain vehemently about their presence. This species is an important parasitoid of the forest tent caterpillar (*Malacosoma disstria*), completely controlling their massive outbreaks (USDA, Forest Service 1985).

Biting midges, black flies, deer flies and horse flies, and mosquitoes, apart from their capacity as disease vectors, are infamous for their sometimes incessant harassment of humans and other animals. In some parts of the world, considerable resources are devoted to reducing the numbers of mosquitoes and black flies along major rivers (e.g., Skovmand 2004) and in cities (e.g., Callaway 2007). Mosquitoes, as well as biting midges, deer flies, and horse flies, also can be abundant along beaches and in salt marshes and mangrove swamps. The latter are especially well-known breeding sites for species of *Culicoides*, abundances of which can manifest in biting rates of several hundred per minute (Borkent 2004).

In many urban areas, the major nuisance fly is the common housefly *Musca domestica*, which often occurs in large numbers around humans or human activities. In addition to the discomfort caused by direct contact with large numbers of flies buzzing around food, garbage, and other items, house flies can be mechanical vectors of various microbial pathogens (Nayduch et al. 2002, Sanchez-Arroyo 2007).

Other dipterans that annoy and interfere with human comfort include certain members of the Chloropidae, especially the genus *Hippelates*, commonly known as eye gnats. Larvae of these flies inhabit the soil, but adults can be a nuisance because they are attracted to sweat, tears, and other secretions around the eyes or exposed skin. Though more of an annoyance than a health risk, eye gnats can serve as vectors of the agents of anaplasmosis, bacterial conjunctivitis, and bovine mastitis (Lindsay and Scudder 1956, Mulla 1965, Tondella et al. 1994).

Diptera in biological control

A large number of Diptera can be beneficial, especially the many predaceous or parasitoid groups that help regulate insect pest populations. Many species are native components of the ecosystems in which they are found, but others are introduced to control native or exotic pests. The gall midge *Feltiella acarisuga* is a widespread and effective predator of spider mites (Tetranychidae) (Gagné 1995). It can successfully control populations of *Tetranychus urticae* in various crops (Opit et al. 1997) and is a potentially useful agent for integrated pest management of spider mites in greenhouses (Gillespie et al. 1998). The aphid predatory midge *Aphidoletes aphidimyza* is another gall midge that has shown promise as an effective predator of aphids. It is an important component of biological control programs for greenhouse crops and is now widely available commercially (Hoffmann and Frodsham 1993). The cottony cushion scale killer *Cryptochetum iceryae* (Cryptochetidae), along with the vedalia beetle *Rodolia cardinalis*, was introduced from Australia to California to successfully control the cottony cushion scale *Icerya purchasi* (DeBach and Rosen 1991, Waterhouse and Sands 2001). Florida citrus growers subsequently introduced *C. iceryae* for the same purpose, and the species is now widespread in warmer parts of the New World (Pitkin 1989). Another group used in the control of coccids and aphids is the family Chamaemyiidae. *Leucopis tapiae*, for example, was introduced into Hawaii to control the Eurasian pine adelgid (Greathead 1995, Vail et al. 2001).

Snail-killing flies (Sciomyzidae) control populations of the intermediate hosts of trematodes causing bilharzia (Berg and Knutson 1978, Maharaj et al. 1992) and multivoltine species have the potential to control pest helicid snails in Australian pastures (Coupland and Baker 1995). Some studies (Graham et al. 2003, Porter et al. 2004) suggest that ant-decapitating species of Phoridae can suppress introduced populations of fire ants. The Pipunculidae are of interest as potential control agents for rice and sugarcane leafhopper pests (Greathead 1983). Species of the muscid genus *Coenosia*, whose larvae and adults are predaceous, are being tested for biological control of agromyzids, ephydrids, sciarids, and white flies in greenhouses (Kühne 2000).

The exclusively parasitic Tachinidae are used extensively in biological control programs, especially against pestiferous Lepidoptera. Success stories include the introduction of *Bessa remota* to control coconut moths in Fiji (DeBach and Rosen 1991) and the use of *Billaea claripalpis*, *Lixophaga diatraeae*, and *Lydella minense* to control sugarcane stem borers (*Diatraea* spp.) in the Neotropical Region (Bennett 1969, Cock 1985, DeBach and Rosen 1991). Other introductions have achieved limited success, such as use of the Palearctic tachinid *Cyzenis albicans* to control the winter moth *Operophtera brumata* in Canadian deciduous forests (Horgan et al. 1999). Although most tachinids attack a narrow spectrum of hosts, a few species (e.g., *Compsilura concinnata*) have been reared from hundreds of different host species. Under the latter circumstances, a potential negative side effect of biological control is the harm to nontarget hosts. For example, *C. concinnata* was introduced to control gypsy moths in New England. The tachinid had a modes effect on the target host but is thought to have led to declining populations of local silk moths (Boettner et al. 2000).

A few Diptera, particularly fruit flies (Tephritidae), have been used for biological control of weeds. Primary targets have been knapweeds and thistles (*Carduus, Centaurea*, and *Cirsium*) and various other genera with noxious species (e.g., *Ageratina, Lantana*, and *Senecio*) (Bess and Haramoto 1972, White and Clement 1987, Harris 1989, White and Elson-Harris 1992, Turner 1996). Other flies used for the control of certain weeds include a few Agromyzidae (Spencer 1973), Syrphidae, and Cecidomyiidae (Gagné et al. 2004).

Pollination

Diptera are major contributors to the maintenance of plant diversity through their participation in many pollination systems and networks (Ssymank et al. 2008). Diptera probably were among the first angiosperm pollinators, and flies might have been influential in spurring the early angiosperm diversification (Labandeira 1998, Endress 2001). Flies visit flowers to obtain nectar for energy and pollen for protein; flowers also provide species-specific rendezvous sites for mating and a beneficial microclimate (Kearns 2002, Kevan 2002). Diptera are among the most common insects that visit flowers (Free 1993), and in Belgium more than 700 plant species in 94 families were visited by flower flies alone (De Buck 1990). Diptera have lower flower-visiting consistency, are generally much less hairy than the aculeate Hymenoptera, and most lack specialized structures for pollen transport. Despite the latter, pollen clings to flies with furry body vestiture and some flies have foretarsal modifications allowing them to gather and eat pollen (e.g., Holloway 1976, Neff et al. 2003). But even generalist flower visitors contribute significantly to plant reproductive success (Kearns 2001, Kevan 2002). Diptera pollinate a significant number of important crops, including apples, cacao, carrots, cashew, cassava, cauliflower, leek, mango, mustard, onions, strawberries, and tea (Heath 1982, Hansen 1983, Clement et al. 2007, Mitra and Banerjee 2007). The highly specialized flowers of cacao are pollinated exclusively by small midges (Ceratopogonidae), particularly of the genus *Forcipomyia* (Young 1986, 1994), and an increasing number of flowering plants are being discovered that depend entirely on dipteran pollinators. Examples include the seed-for-seed mutualism where species of the anthomyiid genus *Chiastocheta* pollinate the closed flowers of *Trollius europaeus* (Pellmyr 1989), and the gall-midge pollination of *Artocarpus*, which is a mutualism involving also a parasitic fungus (Sakai et al. 2000). A significant number of flowers have specialized in being pollinated by carrion flies, including the world's largest flower *Rafflesia arnoldii* (Beaman et al. 1988). Diptera are particularly important pollinators of flat to bowl-shaped flowers in habitats and under conditions where bees are less active. Many flies have adapted well to moist and cool habitats, such as cloud forests and arctic and alpine environments, which have a large proportion of dipteran pollinators (Kevan 1972; Elberling and Olesen 1999; Kearns 1992, 2001). Small Diptera might be the most important pollinators in the forest understory, particularly for shrubs with numerous small, inconspicuous, and dioecious flowers (Larson et al. 2001, Borkent and Harder 2007).

Other ecological services (scavengers and decomposers)

In most terrestrial and freshwater ecosystems, Diptera are more species rich and have a higher biomass than do other insect decomposers (McLean 2000). Representatives of many families, including the Calliphoridae, Coleopidae, Muscidae, Mycetophilidae, Phoridae, Psychodidae, Sarcophagidae, Sciaridae, Sepsidae, Sphaeroceridae, Stratiomyidae, Syrphidae, and Tipuloidea are important decomposers and recyclers of decaying organic matter. The importance of Diptera in recycling dung has been well studied (Laurence 1953, 1954, 1955; Papp 1976, 1985; Papp and Garzó 1985; Skidmore 1991; O'Hara et al. 1999).

The black soldier fly *Hermetia illucens* (Stratiomyidae), a pantropical species that breeds in decaying fruit and other decomposing organic material (James

1935), exemplifies the decomposing capacity and ecological significance of flies. Chicken manure colonized by *H. illucens* often leads to reduced amounts of manure (Sheppard et al. 1994), results in fewer houseflies (Sheppard 1983, Axtell and Arends 1990), and can even provide a food resource (i.e., prepupae) for fish and swine (Newton et al. 1977, Bondari and Sheppard 1981). *Hermetia illucens* also has been implicated as a potentially beneficial species to the citrus industry, where the destruction of orange-peel waste can be a costly endeavor. Pape (2009) recounted a compelling example of this potential use in Costa Rica, where waste from a local orange-growing company dumped hundreds of tons of orange peel on strongly degraded bushland during the dry season. At the onset of the rains, populations of *H. illucens* boomed and the waste was completely decomposed by the larvae in nine months. As an added value, a new indigenous dry forest was resprouting from this 'Biodiversity Processing Ground'.

As effective decomposers of organic material, many Diptera have a high pest potential through their ability to locate and infect stored human food. Provisions particularly prone to become infested are household meats and meat products, which may become 'blown' with blow fly eggs. Cheese and ham also are favored habitats for the cheese skipper *Piophila casei*. An important way of preserving fish practically throughout the world is to dry the meat under the sun, and a number of flies breed in such cured fish, especially blow flies of the genera *Calliphora*, *Chrysomya*, and *Lucilia*. These flies can be a serious problem in many tropical and subtropical societies, causing losses of up to 30% (Haines and Rees 1989, Esser 1991, Wall et al. 2001).

Diptera of forensic, medicolegal, and medical importance

Flies are usually the first insects to arrive at vertebrate carrion, which make especially the Calliphoridae (and species of Fanniidae, Muscidae, Phoridae, Piophilidae, Sarcophagidae, and Stratiomyidae) potential forensic indicators in cases involving dead bodies. A forensic entomologist estimates the time elapsed since a blow fly larva, found on a corpse, hatched from its egg by backtracking the development time, that is, by measuring the number of degree days required to complete development and subtracting this from the known total required for complete larval development (Higley and Haskell 2001), or through phenological information (Staerkeby 2001).

The affinity of blow fly maggots for decaying flesh makes some species ideal for cleaning certain wounds, particularly bedsores and age- and diabetes-related gangrene involving reduced circulation. Even severe burns and extensive abrasions, where small islands of dead tissue scattered over larger areas make physical removal complicated, can be treated in this manner. The use of maggot therapy for treating wounds of humans and livestock is an old discovery (Grantham-Hill 1933, Leclercq 1990, Sherman et al. 2000), but, since it has been improved through sterile breeding of larvae and controlled application under specially developed bandages, the technique has been taken up by many clinics. The maggots provide a dual effect by eating the dead tissue and secreting antiseptic saliva. Even the mechanical stimulus from the active larvae exerts a micromassage promoting the circulation of lymphatic fluids in the recovering tissues (Sherman 2001, 2002, 2003).

Diptera as research tools

Physiology and genetics

The muscle tissues of vertebrates and insects might be only remotely homologous (Mounier et al. 1992), but insect indirect flight muscles bear some functional and physiological resemblances to human heart musculature (Chan and Dickinson 1996, Maughan et al. 1998). Knowledge of functional properties and organization of proteins in *Drosophila* wing-muscle myofibrils (e.g., Vigoreaux 2001) and the expression of heterologous human cytoplasmic actin in *Drosophila* flight muscles carry significant potential for increased understanding of human muscular disorders (Brault et al. 1999). Asynchronous flight muscles, whereby a single nerve impulse causes a muscle fiber to contract multiple times, is the key to the extreme mechanical and physiological efficiency behind the high-frequency wing beat necessary for sustained flight in many insects. Peak performance is found in some Ceratopogonidae in which Sotavalta (1947, 1953) measured frequencies surpassing 1000 Hertz in *Forcipomyia* sp., and by experimentally reducing wing length, more than doubled this frequency. This can be accomplished only when muscles are able to contract in an oscillatory manner, requiring that they are attached to an appropriate mechanically resonant load, which in a fly would be its thorax and wings. Insect flight muscles, therefore, offer insights into muscular operational design, contractile costs, and energy-saving mechanisms (Conley and Lindstedt 2002, Syme and Josephson 2002), which will have implications for human health as well as for biotechnological advances.

Drosophila melanogaster was introduced as a laboratory animal for the geneticist about a century ago (Castle 1906). This introduction turned out to be extremely fruitful, and *D. melanogaster* is now, for many people, the icon of genetic research. Numerous studies using this model species have brought tremendous insight into gene expression, gene regulatory mechanisms, and, more recently, genomics (Ashburner and Bergman 2005). Revealing the *D. melanogaster* genome, which was the second animal genome to be fully sequenced, has been remarkably rewarding. Almost 75% of the candidate human disease genes can be matched by homologues in *Drosophila* (Reiter et al. 2001), and today FlyBase (http://flybase.bio .indiana.edu/), the *Drosophila* community database, is providing one of the highest-quality annotated genome sequences for any organism. The recognition of gene homologues carries a large potential for improved treatments of human disorders ranging from type-II diabetes to alcoholism (e.g., Campbell et al. 1997, Fortini et al. 2000, Brogiolo et al. 2001, Leevers 2001, Morozova et al. 2006). At a higher genetic level, the discovery of the Hox genes in *D. melanogaster* in the early 1980s paved the way for entirely new insights into how organisms regulate the identity of particular segments and body regions by controlling the patterning along the embryonic head-to-tail axis (e.g., Carroll 1995, Akam 1998, Lewis 1998). The short generation time of *D. melanogaster*, which makes it so suitable for genetic studies, also makes it suitable for studies on age-specific and lifetime behavior patterns involved in aging (Carey et al. 2006), as well as on the genetics and physiology of age-related memory impairment (Horiuchi and Saitoe 2005).

Technology

Insect flight uses thin, flexible plates (wings) reinforced by a system of ridges (the veins) that allow for semi-automated deformation, which optimizes aerodynamic forces. Insect wings typically produce two–three times more lift than can be accounted for by conventional aerodynamics, and they produce a high amount of lift while keeping drag at a minimum. These features make insects attractive models for microplane design (Dwortzan 1997, Ellington 1999, Wooton 2000, Bar-Cohen 2005). Dipteran flight has been fine-tuned through millions of years of evolution, and some of the most diligent insect flyers are found among the Diptera, whereby certain species of bee flies, flower flies, pipunculids, and rhiniine blow flies show a range of aerial acrobatics unsurpassed by any other flying animal. Such maneuverability built into so-called micro-air-vehicles (MAVs), apart from obvious military interests, would be of use, for example, in aerial surveillance operations and reconnaissance in confined spaces, for example, by rescue squads dealing with partly collapsed or burning buildings.

Female mosquitoes have excelled for millions of years in their ability to take a blood meal almost without being felt by the victim. As they cut their way through tough vertebrate skin, the apically microserrated mandibular and maxillary stylets provide less friction and require lower insertion forces, compared with conventional human-made syringe needles. Micro-engineers and biomedical engineers, therefore, have turned to the mosquito proboscis in an attempt to develop ultra-narrow syringe needles for minimally invasive (pain-free), micro-electromechanical drug delivery and sampling of body fluids for microdialysis in continued medical monitoring (Cohen 2002, Gattiker et al. 2005).

Diptera in conservation

Bioindicators

Biomonitoring is the use of living organisms or their responses to evaluate environmental quality (Rosenberg and Resh 1993, Resh et al. 1996, Barbour et al. 1999, Moulton et al. 2000, Bonada et al. 2006, Rosenberg et al. 2007) and involves three general areas of investigation: (1) surveys before and after an impact to determine the effects of that impact (e.g., Thomson et al. 2005), (2) regular sampling or toxicity testing to measure compliance with legally mandated environmental quality standards (e.g., Yoder and Rankin 1998, Maret et al. 2003), and (3) large-scale surveys to establish reference conditions or evaluate biological impairment across geographical landscapes and under different land-management practices (e.g., Klemm et al. 2003, Black and Munn 2004).

Although these studies encompass a range of taxa and assessment metrics, many will include in their analyses some evaluation of Diptera diversity and abundance. The inclusion of Diptera has been especially true of water-quality and bioassessment studies to classify the degree of pollution or other impacts in a water body. The mouthparts of many aquatic larvae continuously filter detritus and microorganisms, and the associated habitat and microhabitat specialization means that species associations can be informative about water quality (e.g., Sæther 1979). The Chironomidae are perhaps the most widely used dipterans for these purposes. Larvae of the midge genus *Chironomus* are commonly referred to as 'blood worms' because of hemoglobin in their blood, a trait that permits survival in poorly oxygenated aquatic habitats. These larvae and those of other Chironomidae (e.g., Wiederholm 1980, Raddum and Saether 1981) and other aquatic Diptera (e.g., the moth flies *Psychoda* and rat-tailed maggots *Eristalis*) are often used as indicators of polluted water or water low in oxygen (e.g., Lenat 1993b, Barbour et al. 1999, Courtney et al. 2008). Furthermore, evaluation of morphological deformities in the larval head capsules (e.g., changes in sclerite shape) of dipterans, especially chironomids, has been used extensively to assess environmental stress (Wiederholm 1984; Warwick and Tisdale 1988; Warwick 1989, 1991; Diggins and Stewart 1993; Lenat 1993a). Although the general perception is that most aquatic Diptera are tolerant of environmental impacts, some groups (e.g., Blephariceridae and Deuterophlebiidae) have received tolerance values indicating extreme sensitivity to environmental perturbations (Lenat 1993b, Barbour et al. 1999, Courtney et al. 2008). Despite these general patterns, most families have species that exhibit a range of tolerances from pristine to impaired conditions. The Chironomidae, Culicidae, and Tipuloidea are noteworthy in this respect. Finally, some Diptera (especially Chironomidae) are used commonly in acute and chronic laboratory toxicity studies to compare toxicants and the factors affecting toxicity, and ultimately to predict the environmental effects of the toxicant (Michailova et al. 1998, Karouna-Renier and Zehr 1999).

Diptera are rich in species with specific microhabitat or breeding-site requirements, providing them with a high potential for habitat-quality assessment and conservation planning (Rotheray et al. 2001). Abundance patterns of stiletto flies (Therevidae) can be an indicator of habitat heterogeneity and successional stage in dry areas (Holston 2005). Haslett (1988) used flower flies as bioindicators of environmental stress on ski slopes in Austria, and Sommagio (1999) suggested that their strength would be in the evaluation of landscape diversity. Many saproxylic and fungivorous flies have an association with old-growth forests, which implies a considerable potential as indicators of woodland quality that might help in designing and implementing management strategies such as forest-cutting regimes and tree-species composition (Speight 1986; Økland 1994, 1996, 2000; Good and Speight 1996; Fast and Wheeler 2004; Økland et al. 2004, 2008). Special measures have been taken to conserve saproxylic insects in England (Rotheray and Mac-Gowan 2000). Increasing forest cover and changes in their management in the Netherlands since the 1950s have meant that the saproxylic Syrphidae in general are on the increase (Reemer 2005). A similar situation has been indicated for Germany (Ssymank and Doczkal 1998). The ecologically diverse Dolichopodidae are showing promise as indicators of site value over a range of nonforest habitats (Pollet 1992, 2001; Pollet and Grootaert 1996), and endemic Hawaiian Dolichopodidae, together with selected Canacidae, Chironomidae, and Ephydridae, are potential indicators of valuable aquatic habitats with high native diversity (Englund et al. 2007).

Vanishing species

The world biota is under ever-increasing pressure from humankind. Only four dipterans are on the IUCN Red List of Threatened Species (IUCN 2007), containing species that are either vulnerable, endangered, or critically endangered, and therefore facing a higher risk of global extinction, but this list is probably grimly misleading, as still more species of Diptera are finding their way onto regional red lists (e.g., Falk 1991, Stark 1996, Binot et al. 1998, Pollet 2000). Geographically restricted populations are often particularly vulnerable, and numerous species of Diptera most probably have already disappeared due to the arrival of invasive species on oceanic islands or through major habitat destruction. An example is the single flesh fly endemic for Bermuda, *Microcerella bermuda*, which has not been collected for the last 100 years and was not recovered in the most recent inventory (Woodley and Hilburn 1994). Other Diptera are so rare that they face imminent extinction: the peculiar *Mormotomyia hirsuta*, sole representative of the family Mormotomyiidae, has been found in only a single bat roost in a large, cave-like rock crevice in Kenya (Oldroyd 1964, Pont 1980). The three species of rhino stomach bot flies (*Gyrostigma* spp.) have been experiencing increasing difficulties maintaining healthy populations, with gradually declining host stocks. The situation already could be critical for *Gyrostigma sumatrensis*, which is still known only from larvae expelled from captive Sumatran rhinos in a few European zoos, all before 1950. The African *G. conjungens*, known from the black rhino, has not been captured since 1961. A successful conservation program for the white rhino, however, appears to have had positive effects on *G. rhinocerontis* (Barraclough 2006). The Delhi Sands flower-loving fly, *Rhaphiomidas terminatus abdominalis* (Mydidae), is the first fly to be listed as endangered by the US Endangered Species Act. It is endemic to the Delhi Sands formation, a small area of ancient inland dunes in southern California, where the adults are nectar feeders and the drastic loss of habitat has led to its perceived decline and endangered status (Rogers and Mattoni 1993).

What may be the first dipteran to be eradicated by humans is the European bone skipper *Thyreophora cynophila*. When described by Panzer (1794), this fly was rather common and was often observed in Austria, France, and Germany. A beautiful, redheaded fly, it could be observed walking on big cadavers such as dead dogs, horses, and mules in the early spring (Robineau-Desvoidy 1830). Suddenly, 50 years after its discovery, it disappeared and has never been collected again. Its disappearance might be due to changes in livestock management and improved carrion disposal, following the Industrial Revolution in Europe, but the underlying scenario probably is the reduction of the megafauna, including the near absence of large predators to leave large carcasses with partly crushed long bones, thereby limiting access to the medullar canal and bone marrow, the favored breeding site for *T. cynophila*. The Quaternary megafauna extinctions, which might have had a human component, most probably had fatal consequences for those Diptera that we assume depended on these large animals or their excreta. The stomach bot fly (*Cobboldia russanovi*) of the woolly mammoth disappeared with its host (Grunin 1973), and D. K. McAlpine (2007) envisions a much larger fauna of Australian wombat flies at the time of the large marsupials some 100,000 years ago. A few other species have been declared globally extinct: four species of *Emperoptera*

and one species of *Campsicnemus* (Dolichopodidae), and a single species of *Drosophila* (Drosophilidae), all from Hawaii (Hawaii Biological Survey 2002, IUCN 2007), and the volutine Stoneyian tabanid (*Stonemyia velutina*) from California (IUCN 2007).

Diptera as part of our cultural legacy

Just as much as man might have realized 'that his destiny is coupled to coexistence with a complex biota that also contains Diptera' (Pape 2009), flies are an integral part of our cultural past. Thousands of children have been fascinated by the fairy tale about the brave little tailor, who got 'seven in one blow', and who has not been laughing at jokes where a customer in a restaurant complains that 'there is a fly in my soup!' Some might even have read 'The Fly' by George Langelaan (1957), featuring a human–housefly hybrid, or seen one or more of the several films based on this short story, or any of the sequels to these. Flies were one of the biblical plagues (Exodus 8:21), but fly symbolism spans the entire range from William Golding's (1954) somber novel 'Lord of the Flies' to William Blake's (1794) lyric poem 'The Fly':

Little Fly
Thy summers play,
My thoughtless hand
Has brush'd away.

Am not I
A fly like thee?
Or art not thou
A man like me?

For I dance
And drink and sing;
Till some blind hand
Shall brush my wing.

If thought is life
And strength and breath;
And the want
Of thought is death;

Then am I
A happy fly,
If I live,
Or if I die.

Table 9.1 Families of Diptera and numbers of described species in the world. Family classification and species richness based on Evenhuis et al. (2007).

Suborder	Infraorder	Other Category	Superfamily (or equivalent)	Family	Described Species
'LOWER DIPTERA'	Ptychopteromorpha			Ptychopteridae	74
'LOWER DIPTERA'	Ptychopteromorpha			Tanyderidae	38
'LOWER DIPTERA'	Culicomorpha		Chironomoidea	Ceratopogonidae	5621
'LOWER DIPTERA'	Culicomorpha		Chironomoidea	Chironomidae	6951
'LOWER DIPTERA'	Culicomorpha		Chironomoidea	Simuliidae	2080
'LOWER DIPTERA'	Culicomorpha		Chironomoidea	Thaumaleidae	173
'LOWER DIPTERA'	Culicomorpha		Culicoidea	Chaoboridae	55
'LOWER DIPTERA'	Culicomorpha		Culicoidea	Corethrellidae	97
'LOWER DIPTERA'	Culicomorpha		Culicoidea	Culicidae	3616
'LOWER DIPTERA'	Culicomorpha		Culicoidea	Dixidae	185
'LOWER DIPTERA'	Blephariceromorpha		Blephariceroidea	Blephariceridae	322
'LOWER DIPTERA'	Blephariceromorpha		Blephariceroidea	Deuterophlebiidae	14
'LOWER DIPTERA'	Blephariceromorpha		Nymphomyioidea	Nymphomyiidae	7
'LOWER DIPTERA'	Bibionomorpha		Axymyioidea	Axymyiidae	6
'LOWER DIPTERA'	Bibionomorpha		Bibionoidea	Bibionidae	754
'LOWER DIPTERA'	Bibionomorpha		Bibionoidea	Hesperinidae	6
'LOWER DIPTERA'	Bibionomorpha		Bibionoidea	Pachyneuridae	5
'LOWER DIPTERA'	Bibionomorpha		Sciaroidea	Bolitophilidae	59
'LOWER DIPTERA'	Bibionomorpha		Sciaroidea	Cecidomyiidae	6051
'LOWER DIPTERA'	Bibionomorpha		Sciaroidea	Diadocidiidae	19
'LOWER DIPTERA'	Bibionomorpha		Sciaroidea	Ditomyiidae	93
'LOWER DIPTERA'	Bibionomorpha		Sciaroidea	Keroplatidae	907
'LOWER DIPTERA'	Bibionomorpha		Sciaroidea	Lygistorrhinidae	30
'LOWER DIPTERA'	Bibionomorpha		Sciaroidea	Mycetophilidae	4105
'LOWER DIPTERA'	Bibionomorpha		Sciaroidea	Rangomaramidae	39
'LOWER DIPTERA'	Bibionomorpha		Sciaroidea	Sciaridae	2224
'LOWER DIPTERA'	Psychodomorpha			Anisopodidae	158
'LOWER DIPTERA'	Psychodomorpha			Perissommatidae	5
'LOWER DIPTERA'	Psychodomorpha			Psychodidae	2886
'LOWER DIPTERA'	Psychodomorpha		Scatopsoidea	Canthyloscelidae	16
'LOWER DIPTERA'	Psychodomorpha		Scatopsoidea	Scatopsidae	323
'LOWER DIPTERA'	Psychodomorpha		Scatopsoidea	Valeseguyidae	3
'LOWER DIPTERA'	Tipulomorpha		Tipuloidea	Cylindrotomidae	67
'LOWER DIPTERA'	Tipulomorpha		Tipuloidea	Limoniidae	10,334
'LOWER DIPTERA'	Tipulomorpha		Tipuloidea	Pediciidae	494

(continued)

Table 9.1 *(continued)*.

Suborder	Infraorder	Other Category	Superfamily (or equivalent)	Family	Described Species
'LOWER DIPTERA'	Tipulomorpha		Tipuloidea	Tipulidae	4325
'LOWER DIPTERA'	Tipulomorpha			Trichoceridae	160
				Subtotal ('Lower Diptera')	**52,302**
BRACHYCERA	Stratiomyiomorpha			Pantophthalmidae	20
BRACHYCERA	Stratiomyiomorpha			Stratiomyidae	2666
BRACHYCERA	Stratiomyiomorpha			Xylomyidae	134
BRACHYCERA	Tabanomorpha			Athericidae	122
BRACHYCERA	Tabanomorpha			Austroleptidae	8
BRACHYCERA	Tabanomorpha			Oreoleptidae	1
BRACHYCERA	Tabanomorpha			Rhagionidae	707
BRACHYCERA	Tabanomorpha			Spaniidae	43
BRACHYCERA	Tabanomorpha			Tabanidae	4387
BRACHYCERA	Xylophagomorpha			Xylophagidae	136
BRACHYCERA	Vermileonomorpha			Vermileonidae	59
BRACHYCERA	Muscomorpha	'lower Brachycera'		Nemestrinidae	275
BRACHYCERA	Muscomorpha	'lower Brachycera'	Asiloidea	Acroceridae	394
BRACHYCERA	Muscomorpha	'lower Brachycera'	Asiloidea	Apioceridae	169
BRACHYCERA	Muscomorpha	'lower Brachycera'	Asiloidea	Apsilocephalidae	3
BRACHYCERA	Muscomorpha	'lower Brachycera'	Asiloidea	Apystomyiidae	1
BRACHYCERA	Muscomorpha	'lower Brachycera'	Asiloidea	Asilidae	7413
BRACHYCERA	Muscomorpha	'lower Brachycera'	Asiloidea	Bombyliidae	5030
BRACHYCERA	Muscomorpha	'lower Brachycera'	Asiloidea	Evocoidae	1
BRACHYCERA	Muscomorpha	'lower Brachycera'	Asiloidea	Hilarimorphidae	32
BRACHYCERA	Muscomorpha	'lower Brachycera'	Asiloidea	Mythicomyiidae	346
BRACHYCERA	Muscomorpha	'lower Brachycera'	Asiloidea	Mydidae	463
BRACHYCERA	Muscomorpha	'lower Brachycera'	Asiloidea	Scenopinidae	414
BRACHYCERA	Muscomorpha	'lower Brachycera'	Asiloidea	Therevidae	1125
BRACHYCERA	Muscomorpha	'lower Brachycera'	Asiloidea	Atelestidae	10
BRACHYCERA	Muscomorpha		Empidoidea	Brachystomatidae	145
BRACHYCERA	Muscomorpha		Empidoidea	Dolichopodidae	7118
BRACHYCERA	Muscomorpha		Empidoidea	Empididae	2935
BRACHYCERA	Muscomorpha		Empidoidea	*Homalocnemus* group	7
BRACHYCERA	Muscomorpha		Empidoidea	Hybotidae	1882
BRACHYCERA	Muscomorpha		Empidoidea	*Iteaphila* group	27
BRACHYCERA	Muscomorpha		Empidoidea	*Oreogeton* group	12
BRACHYCERA	Muscomorpha	'lower Cyclorrhapha'	Empidoidea	Ironomyiidae	1

BRACHYCERA	Muscomorpha				
BRACHYCERA	Muscomorpha	'lower Cyclorrhapha'		Lonchopteridae	58
BRACHYCERA	Muscomorpha	'lower Cyclorrhapha'		Opetidae	5
BRACHYCERA	Muscomorpha	'lower Cyclorrhapha'		Phoridae	4022
BRACHYCERA	Muscomorpha	'lower Cyclorrhapha'		Pipunculidae	1381
BRACHYCERA	Muscomorpha	'lower Cyclorrhapha'		Platypezidae	252
BRACHYCERA	Muscomorpha	'lower Cyclorrhapha'		Syrphidae	5935
BRACHYCERA	Muscomorpha	Nerioidea	Schizophora: 'Acalyptrates'	Cypselosomatidae	34
BRACHYCERA	Muscomorpha	Nerioidea	Schizophora: 'Acalyptrates'	Megamerinidae	15
BRACHYCERA	Muscomorpha	Nerioidea	Schizophora: 'Acalyptrates'	Micropezidae	578
BRACHYCERA	Muscomorpha	Nerioidea	Schizophora: 'Acalyptrates'	Neriidae	111
BRACHYCERA	Muscomorpha	Diopsoidea	Schizophora: 'Acalyptrates'	Diopsidae	183
BRACHYCERA	Muscomorpha	Diopsoidea	Schizophora: 'Acalyptrates'	Gobryidae	5
BRACHYCERA	Muscomorpha	Diopsoidea	Schizophora: 'Acalyptrates'	Nothybidae	8
BRACHYCERA	Muscomorpha	Diopsoidea	Schizophora: 'Acalyptrates'	Psilidae	321
BRACHYCERA	Muscomorpha	Diopsoidea	Schizophora: 'Acalyptrates'	Somatiidae	7
BRACHYCERA	Muscomorpha	Diopsoidea	Schizophora: 'Acalyptrates'	Strongylophthalmyiidae	47
BRACHYCERA	Muscomorpha	Diopsoidea	Schizophora: 'Acalyptrates'	Syringogastridae	10
BRACHYCERA	Muscomorpha	Diopsoidea	Schizophora: 'Acalyptrates'	Tanypezidae	21
BRACHYCERA	Muscomorpha	Conopoidea	Schizophora: 'Acalyptrates'	Conopidae	783
BRACHYCERA	Muscomorpha	Tephritoidea	Schizophora: 'Acalyptrates'	Ctenostylidae	10
BRACHYCERA	Muscomorpha	Tephritoidea	Schizophora: 'Acalyptrates'	Lonchaeidae	480
BRACHYCERA	Muscomorpha	Tephritoidea	Schizophora: 'Acalyptrates'	Pallopteridae	66
BRACHYCERA	Muscomorpha	Tephritoidea	Schizophora: 'Acalyptrates'	Piophilidae	82
BRACHYCERA	Muscomorpha	Tephritoidea	Schizophora: 'Acalyptrates'	Platystomatidae	1162
BRACHYCERA	Muscomorpha	Tephritoidea	Schizophora: 'Acalyptrates'	Pyrgotidae	351
BRACHYCERA	Muscomorpha	Tephritoidea	Schizophora: 'Acalyptrates'	Richardiidae	174
BRACHYCERA	Muscomorpha	Tephritoidea	Schizophora: 'Acalyptrates'	Tachiniscidae	3
BRACHYCERA	Muscomorpha	Tephritoidea	Schizophora: 'Acalyptrates'	Tephritidae	4621
BRACHYCERA	Muscomorpha	Tephritoidea	Schizophora: 'Acalyptrates'	Ulidiidae	672
BRACHYCERA	Muscomorpha	Lauxanioidea	Schizophora: 'Acalyptrates'	Celyphidae	116
BRACHYCERA	Muscomorpha	Lauxanioidea	Schizophora: 'Acalyptrates'	Chamaemyiidae	349
BRACHYCERA	Muscomorpha	Lauxanioidea	Schizophora: 'Acalyptrates'	Eurychoromyiidae	1
BRACHYCERA	Muscomorpha	Lauxanioidea	Schizophora: 'Acalyptrates'	Lauxaniidae	1893
BRACHYCERA	Muscomorpha	Sciomyzoidea	Schizophora: 'Acalyptrates'	Coelopidae	35
BRACHYCERA	Muscomorpha	Sciomyzoidea	Schizophora: 'Acalyptrates'	Dryomyzidae	25
BRACHYCERA	Muscomorpha	Sciomyzoidea	Schizophora: 'Acalyptrates'	Helcomyzidae	12
BRACHYCERA	Muscomorpha	Sciomyzoidea	Schizophora: 'Acalyptrates'	Helosciomyzidae	23
BRACHYCERA	Muscomorpha	Sciomyzoidea	Schizophora: 'Acalyptrates'	Heterocheilidae	2
BRACHYCERA	Muscomorpha	Sciomyzoidea	Schizophora: 'Acalyptrates'	Huttonidae	8

(continued)

Table 9.1 (*continued*).

Suborder	Infraorder	Other Category	Superfamily (or equivalent)	Family	Described Species
BRACHYCERA	Muscomorpha	Schizophora: 'Acalyptrates'	Sciomyzoidea	Natalimyzidae	1
BRACHYCERA	Muscomorpha	Schizophora: 'Acalyptrates'	Sciomyzoidea	Phaeomyiidae	3
BRACHYCERA	Muscomorpha	Schizophora: 'Acalyptrates'	Sciomyzoidea	Rhopalomeridae	33
BRACHYCERA	Muscomorpha	Schizophora: 'Acalyptrates'	Sciomyzoidea	Sciomyzidae	604
BRACHYCERA	Muscomorpha	Schizophora: 'Acalyptrates'	Sciomyzoidea	Sepsidae	375
BRACHYCERA	Muscomorpha	Schizophora: 'Acalyptrates'	Opomyzoidea	Agromyzidae	3013
BRACHYCERA	Muscomorpha	Schizophora: 'Acalyptrates'	Opomyzoidea	Anthomyzidae	94
BRACHYCERA	Muscomorpha	Schizophora: 'Acalyptrates'	Opomyzoidea	Asteiidae	132
BRACHYCERA	Muscomorpha	Schizophora: 'Acalyptrates'	Opomyzoidea	Aulacigastridae	18
BRACHYCERA	Muscomorpha	Schizophora: 'Acalyptrates'	Opomyzoidea	Clusiidae	349
BRACHYCERA	Muscomorpha	Schizophora: 'Acalyptrates'	Opomyzoidea	Fergusoninidae	29
BRACHYCERA	Muscomorpha	Schizophora: 'Acalyptrates'	Opomyzoidea	Marginidae	3
BRACHYCERA	Muscomorpha	Schizophora: 'Acalyptrates'	Opomyzoidea	Neminidae	14
BRACHYCERA	Muscomorpha	Schizophora: 'Acalyptrates'	Opomyzoidea	Neurochaetidae	20
BRACHYCERA	Muscomorpha	Schizophora: 'Acalyptrates'	Opomyzoidea	Odiniidae	62
BRACHYCERA	Muscomorpha	Schizophora: 'Acalyptrates'	Opomyzoidea	Opomyzidae	61
BRACHYCERA	Muscomorpha	Schizophora: 'Acalyptrates'	Opomyzoidea	Periscelididae	84
BRACHYCERA	Muscomorpha	Schizophora: 'Acalyptrates'	Opomyzoidea	Teratomyzidae	8
BRACHYCERA	Muscomorpha	Schizophora: 'Acalyptrates'	Opomyzoidea	Xenasteiidae	13
BRACHYCERA	Muscomorpha	Schizophora: 'Acalyptrates'	Carnoidea	Acartophthalmidae	4
BRACHYCERA	Muscomorpha	Schizophora: 'Acalyptrates'	Carnoidea	Australimyzidae	9
BRACHYCERA	Muscomorpha	Schizophora: 'Acalyptrates'	Carnoidea	Braulidae	7
BRACHYCERA	Muscomorpha	Schizophora: 'Acalyptrates'	Carnoidea	Canacidae	119
BRACHYCERA	Muscomorpha	Schizophora: 'Acalyptrates'	Carnoidea	Carnidae	90
BRACHYCERA	Muscomorpha	Schizophora: 'Acalyptrates'	Carnoidea	Chloropidae	2863

BRACHYCERA	Muscomorpha	Schizophora: 'Acalyptrates'	Carnoidea	Cryptochetidae	33
BRACHYCERA	Muscomorpha	Schizophora: 'Acalyptrates'	Carnoidea	Inbiomyiidae	10
BRACHYCERA	Muscomorpha	Schizophora: 'Acalyptrates'	Carnoidea	Milichiidae	276
BRACHYCERA	Muscomorpha	Schizophora: 'Acalyptrates'	Carnoidea	Tethinidae	193
BRACHYCERA	Muscomorpha	Schizophora: 'Acalyptrates'	Sphaeroceroidea	Chyromyidae	106
BRACHYCERA	Muscomorpha	Schizophora: 'Acalyptrates'	Sphaeroceroidea	Heleomyzidae	717
BRACHYCERA	Muscomorpha	Schizophora: 'Acalyptrates'	Sphaeroceroidea	Mormotomyiidae	1
BRACHYCERA	Muscomorpha	Schizophora: 'Acalyptrates'	Sphaeroceroidea	Nannodastiidae	5
BRACHYCERA	Muscomorpha	Schizophora: 'Acalyptrates'	Sphaeroceroidea	Sphaeroceridae	1580
BRACHYCERA	Muscomorpha	Schizophora: 'Acalyptrates'	Ephydroidea	Camillidae	40
BRACHYCERA	Muscomorpha	Schizophora: 'Acalyptrates'	Ephydroidea	Curtonotidae	61
BRACHYCERA	Muscomorpha	Schizophora: 'Acalyptrates'	Ephydroidea	Diastatidae	48
BRACHYCERA	Muscomorpha	Schizophora: 'Acalyptrates'	Ephydroidea	Drosophilidae	3925
BRACHYCERA	Muscomorpha	Schizophora: 'Acalyptrates'	Ephydroidea	Ephydridae	1977
BRACHYCERA	Muscomorpha	Schizophora: 'Acalyptrates'	Hippoboscoidea	Glossinidae	23
BRACHYCERA	Muscomorpha	Schizophora: Calyptratae	Hippoboscoidea	Hippoboscidae	786
BRACHYCERA	Muscomorpha	Schizophora: Calyptratae	Muscoidea	Anthomyiidae	1896
BRACHYCERA	Muscomorpha	Schizophora: Calyptratae	Muscoidea	Fanniidae	319
BRACHYCERA	Muscomorpha	Schizophora: Calyptratae	Muscoidea	Muscidae	5153
BRACHYCERA	Muscomorpha	Schizophora: Calyptratae	Muscoidea	Scathophagidae	392
BRACHYCERA	Muscomorpha	Schizophora: Calyptratae	Oestroidea	Calliphoridae	1524
BRACHYCERA	Muscomorpha	Schizophora: Calyptratae	Oestroidea	Rhiniidae	363
BRACHYCERA	Muscomorpha	Schizophora: Calyptratae	Oestroidea	Mystacinobiidae	1
BRACHYCERA	Muscomorpha	Schizophora: Calyptratae	Oestroidea	Oestridae	150
BRACHYCERA	Muscomorpha	Schizophora: Calyptratae	Oestroidea	Rhinophoridae	147
BRACHYCERA	Muscomorpha	Schizophora: Calyptratae	Oestroidea	Sarcophagidae	2632
BRACHYCERA	Muscomorpha	Schizophora: Calyptratae	Oestroidea	Tachinidae	9629
				Subtotal (Brachycera)	**99,942**
				TOTAL	**152,244**

REFERENCES

Abul-Hab, J. K. and Y. G. Salman. 1999. Human urinogenital myasis by *Psychoda*. *Saudi Medical Journal* 20: 635–636.

Adler, P. H., D. C. Currie, and D. M. Wood. 2004. *The Black Flies (Simuliidae) of North America*. Cornell University Press, Ithaca, New York. xv + 941 pp.

Akam, M. 1998. Hox genes, homeosis and the evolution of segment identity: no need for hopeless monsters. *International Journal of Developmental Biology* 42: 445–451.

Ali, A. 1991. Perspectives on management of pestiferous Chironomidae, an emerging global problem. *Journal of the American Mosquito Control Association* 7: 260–281.

Aluja, M. and A. L. Norrbom. 1999. Preface. Pp. i–iii *In* M. Aluja and A. L. Norrbom (eds). *Fruit Flies (Tephritidae): Phylogeny and Evolution of Behavior*. CRC Press, Boca Raton, Florida. xvi + 944 pp.

Alverson, A. J., G. W. Courtney, and M. R. Luttenton. 2001. Niche overlap of sympatric *Blepharicera* (Diptera: Blephariceridae) larvae from the southern Appalachian Mountains. *Journal of the North American Benthological Society* 20: 564–581.

Andersson, H. 1970. Notes on north European *Lonchoptera* (Dipt., Lonchopteridae) with lectotype designations. *Entomologisk Tidskrift* 91: 42–45.

Armitage, P. D., P. S. Cranston, and L. C. V. Pinder (eds). 1995. *The Chironomidae: Biology and Ecology of Non-Biting Midges*. Chapman and Hall, London. xii + 572 pp.

Ashburner, M. and C. M. Bergman. 2005. *Drosophila melanogaster*: a case study of a model genomic sequence and its consequences. *Genome Research* 15: 1661–1667.

Axtell, R. C. and J. J. Arends. 1990. Ecology and management of arthropod pests of poultry. *Annual Review of Entomology* 35: 101–126.

Baker, C. H. 2002. Dipteran glow-worms: Marvellous maggots weave magic for tourists. *Biodiversity* 3 (4): 23.

Barbour, M. T., J. Gerritsen, B. D. Snyder, and J. B. Stribling. 1999. *Rapid Bioassessment Protocols for Use in Streams and Wadable Rivers: Periphyton, Benthic Macroinvertebrates and Fish*. United States Environmental Protection Agency, Washington, DC. 202 pp.

Bar-Cohen, Y. 2005. Biomimetics: mimicking and inspired-by biology. Proceedings of the EAPAD Conference, SPIE Smart Structures and Materials Symposium, Paper #5759-02. http://ndeaa.jpl.nasa.gov/nasa-nde/lommas/papers/SPIE-05-Biomimetics.pdf [Accessed 15 November 2007].

Barnes, E. 2005. *Diseases and Human Evolution*. University of New Mexico Press, Albuquerque. xii + 484 pp.

Barraclough, D. 2006. Bushels of bots: Africa's largest fly is getting a reprieve from extinction. *Natural History* [2006], June: 18–21.

Baumgartner, D. L. 1993. Review of *Chrysomya rufifacies* (Diptera: Calliphoridae). *Journal of Medical Entomology* 30: 338–352.

Beaman, R. S., P. J. Decker, and J. H. Beaman. 1988. Pollination of *Rafflesia* (Rafflesiaceae). *American Journal of Botany* 75: 1148–1162.

Bennett, F. D. 1969. Tachinid flies as biological control agents for sugar cane moth borers. Pp. 117–148. *In* J. R. Williams, J. R. Metcalfe, R. W. Mungomery, and R. Mathes (eds). *Pests of Sugar Cane*. Elsevier, Amsterdam. xiii + 568 pp.

Berg, C. O. and Knutson, L. 1978. Biology and systematics of the Sciomyzidae. *Annual Review of Entomology* 23: 239–258.

Bess, H. A. and F. H. Haramoto. 1972. Biological control of pamakani *Eupatorium adenophorum*, in Hawaii by a tephritid gall fly, *Procecidochares utilis*. 3. Status of the weed, fly and parasites of the fly in 1961–71 versus 1950–57. *Proceedings of the Hawaiian Entomological Society* 21: 165–178.

Binot, M., R. Bless, P. Boye, H. Grüttke, and P. Prescher (eds). 1998. Rote Liste gefährdeter Tiere Deutschlands. *Schriftenreie für Landschaftspflege und Naturschutz* 55: 1–434.

Bisley, G. G. 1972. A case of intraocular myiasis in man due to the first stage larva of the oestrid fly *Gedoelstia* spp. *East African Medical Journal* 49: 768–771.

Black, R. W. and M. D. Munn. 2004. Using macroinvertebrates to identify biota–land cover optima at multiple scales in the Pacific Northwest, USA. *Journal of the North American Benthological Society* 23: 340–362.

Blake, W. 1794. Songs of Experience. London, 17 pls. [Published by the author. Here from D. V. Erdman (ed). 1988. Electronic edition of the complete poetry and prose of William Blake. Newly revised edition. http://virtual.park.uga.edu/~wblake/eE.html [Accessed 1 March 2008].

Boettner, G. H., J. S. Elkinton, and C. J. Boettner. 2000. Effects of a biological control introduction on three nontarget native species of saturniid moths. *Conservation Biology* 14: 1798–1806.

Bohart, G. E. 1972. Management of wild bees for pollination of crops. *Annual Review of Entomology* 17: 287–312.

Bohart, G. E., W. P. Stephen, and R. K. Eppley. 1960. The biology of *Heterostylum robustum*, a parasite of the alkali bee. *Annals of the Entomological Society of America* 53: 425–435.

Bonada, N., N. Prat, V. H. Resh, and B. Statzner. 2006. Developments in aquatic insect biomonitoring: a comparative analysis of different approaches. *Annual Review of Entomology* 51: 495–523.

Bondari, K. and D. C. Sheppard. 1981. Soldier fly larvae as feed in commercial fish production. *Aquaculture* 24: 103–109.

Bonduriansky, R. and R. J. Brooks. 1998. Male antler flies (*Protopiophila litigata*; Diptera: Piophilidae) are more selective than females in mate choice. *Canadian Journal of Zoology* 76: 1277–1285.

Borkent, A. 2004. Ceratopogonidae. Pp. 113–126. *In* W. C. Marquardt (ed). *Biology of Disease Vectors*. Second edition. Elsevier Academic Press, San Diego. xxiii + 785 pp.

Borkent, C. J. and L. D. Harder. 2007. Flies (Diptera) as pollinators of two dioecious plants: behaviour and implications for plant mating. *Canadian Entomologist* 139: 235–246.

Brault, V., M. C. Reedy, U. Sauder, R. A. Kammerer, U. Aebi, and C. Schoenenberger. 1999. Substitution of flight muscle-specific actin by human (beta)-cytoplasmic actin in the indirect flight muscle of *Drosophila*. *Journal of Cell Science* 112: 3627–3639.

Brogiolo, W., H. Stocker, T. Ikeya, F. Rintelen, R. Fernandez, and E. Hafen. 2001. An evolutionarily conserved function of the *Drosophila* insulin receptor and insulin-like peptides in growth control. *Current Biology* 11: 213–221.

Brown, B. V. 2001. Flies, gnats, and mosquitoes. Pp. 815–826. *In* S. A. Levin (ed). *Encyclopedia of Biodiversity*. Academic Press, London.

Brown, B. V. 2004. Revision of the subgenus *Udamochiras* of *Melaloncha* bee-killing flies (Diptera: Phoridae: Metopininae). *Zoological Journal of the Linnean Society* 140: 1–42.

Brown, B. V. 2008. Phorid Net. http://www.phorid.net/ [Accessed 25 February 2008].

Bruce-Chwatt, L. J. 1988. History of malaria from prehistory to eradication. Pp. 1–59. *In* W. H. Wernsdorfer and I. McGregor (eds). *Malaria. Principles and Practice of Malariology*. Volume 1. Churchill Livingstone, Edinburgh. 912 pp.

Burger, J. F., J. R. Anderson, and M. F. Knudsen. 1980. The habits and life history of *Oedoparena glauca* (Diptera: Dryomyzidae), a predator of barnacles. *Proceedings of the Entomological Society of Washington* 82: 360–377.

Callaway, E. 2007. Dengue fever climbs the social ladder. *Nature* 448: 734–735.

Campbell, H. D., S. Fountain, I. G. Young, C. Claudianos, J. D. Hoheisel, K. S. Chen, and J. R. Lupski 1997. Genomic structure, evolution, and expression of human FLII, a gelsolin and leucine-rich-repeat family member: overlap with LLGL. *Genomics* 42: 46–54.

Carey, J. R., N. Papadopoulos, N. Kouloussis, B. Katsoyannos, H.-G. Müller, J.-L. Wang, and Y.-K. Tseng. 2006. Age-specific and lifetime behavior patterns in *Drosophila melanogaster* and the Mediterranean fruit fly, *Ceratitis capitata*. *Experimental Gerontology* 41: 93–97.

Carroll, S. B. 1995. Homeotic genes and evolution of arthropods and chordates. *Nature* 376: 479–485.

Castle, W. E. 1906. Inbreeding, cross-breeding and sterility in *Drosophila*. *Science* 23: 153.

Chan, W. P. and M. H. Dickinson 1996. *In vivo* length oscillations of indirect flight muscles in the fruit fly *Drosophila virilis*. *Journal of Experimental Biology* 199: 2767–2774.

Clement, S. L., B. C. Hellier, L. R. Elberson, R. T. Staska, and M. A. Evans. 2007. Flies (Diptera: Muscidae: Calliphoridae) are efficient pollinators of *Allium ampeloprasum* L. (Alliaceae) in field cages. *Journal of Economic Entomology* 100: 131–135.

Clements, A. N. 1992. *The Biology of Mosquitoes*. Volume 1. Development, Nutrition, and Reproduction. Chapman and Hall, London. 509 pp.

Cleworth, M. A., R. M. Coombs, and D. H. Davies 1996. The control of *Psychoda alternata* (Psychodidae) in sewage biological filters by application of the insect growth-regulator pyriproxyfen. *Water Research* 30: 654–662.

Cock, M. J. W. (ed). 1985. A review of biological control of pests in the Commonwealth Caribbean and Bermuda up to 1982. *Commonwealth Institute of Biological Control, Technical Communication* 9: 1–218.

Cohen, D. 2002. This won't hurt a bit. *New Scientist* 174 (2337): 21. Also available as 'Painless needle copies mosquito's stinger' at: http://www.newscientist.com/news/news.jsp?id=ns99992121 and http://www.newscientist.com/article.ns?id=dn2121 [Accessed 14 February 2008].

Colbo, M. H. 1996. Chironomidae from marine coastal environments near St Johns, Newfoundland, Canada. *Hydrobiologia* 318: 117–122.

Colless, D. H. and D. K. McAlpine. 1991. Diptera (flies). Pp. 717–786. *In* Commonwealth Scientific and Industrial Research Organisation (Division of Entomology). *The Insects of Australia*. Second Edition. Volume 2. Cornell University Press, Ithaca, NY. vi + 543–1137 pp.

Conley, K. E. and S. L. Lindstedt. 2002. Energy-saving mechanisms in muscle: the minimization strategy. *Journal of Experimental Biology* 205: 2175–2181.

Cook, D. F., I. R. Dadour, and N. J. Keals. 1999. Stable fly, house fly (Diptera: Muscidae), and other nuisance fly development in poultry litter associated with horticultural crop production. *Journal of Economic Entomology* 92: 1352–1357.

Coupland, J. and G. Baker. 1995. The potential of several species of terrestrial Sciomyzidae as biological control agents of pest helicid snails in Australia. *Crop Protection* 14: 573–576.

Courtney, G. W. 1990. Cuticular morphology of larval mountain midges (Diptera: Deuterophlebiidae): implications for the phylogenetic relationships of Nematocera. *Canadian Journal of Zoology* 68: 556–578.

Courtney, G. W. 1991. Life history patterns of Nearctic mountain midges (Diptera: Deuterophlebiidae). *Journal of the North American Benthological Society* 10: 177–197.

Courtney, G. W. 1994. Biosystematics of the Nymphomyiidae (Insecta: Diptera): life history, morphology and phylogenetic relationships. *Smithsonian Contributions to Zoology* 550: 41.

Courtney, G. W. 2000a. Revision of the net-winged midges of the genus *Blepharicera* Macquart (Diptera: Blephariceridae) of eastern North America. *Memoirs of the Entomological Society of Washington* 23: 1–99.

Courtney, G. W. 2000b. A.1. Family Blephariceridae. Pp. 7–30. *In* L. Papp and B. Darvas (eds). *Contributions to a Manual of Palaearctic Diptera*. Appendix. Science Herald, Budapest. 604 pp.

Courtney, G. W. and R. W. Merritt. 2008. Aquatic Diptera: Part one: larvae of aquatic Diptera. Pp. 687–722. *In* R. W. Merritt, K. W. Cummins, and M. B. Berg (eds). *An Introduction to the Aquatic Insects of North America*. Fourth Edition. Kendall/Hunt Publishing, Dubuque, Iowa. 1158 pp.

Courtney, G. W., R. W. Merritt, K. W. Cummins, M. B. Berg, D. W. Webb, and B. A. Foote. 2008. Ecological and distributional data for larval aquatic Diptera. Pp. 747–771. *In* R. W. Merritt, K. W. Cummins, and M. B. Berg (eds). *An Introduction to the Aquatic Insects of North America.* Fourth Edition. Kendall/Hunt Publishing, Dubuque, Iowa. 1158 pp.

Courtney, G. W., B. J. Sinclair, and R. Meier. 2000. 1.4. Morphology and terminology of Diptera larvae. Pp. 85–161. *In* L. Papp and B. Darvas (eds). *Contributions to a Manual of Palaearctic Diptera. Volume I, General and Applied Dipterology.* Science Herald, Budapest. 978 pp.

Craig, D. A. and D. C. Currie. 1999. Phylogeny of the central-western Pacific subgenus *Inseliellum* (Diptera: Simuliidae). *Canadian Journal of Zoology* 77: 610–623.

Cranston, P. S. and S. Dimitriadis. 2005. *Semiocladius*: taxonomy and ecology of an estuarine midge (Diptera, Chironomidae, Orthocladiinae). *Australian Journal of Entomology* 44: 252–256.

Crosby, M. C. 2006. *The American Plague: The Untold Story of Yellow Fever, the Epidemic that Shaped Our History.* Berkley Books, New York. viii + 308 pp.

Crosskey, R. W. 1990. *The Natural History of Blackflies.* John Wiley and Sons, Chichester. ix + 711 pp.

Cumming, J. M. 1994. Sexual selection and the evolution of dance fly mating systems (Diptera: Empididae; Empidinae). *Canadian Entomologist* 126: 907–920.

Cumming, J. M., B. J. Sinclair, and D. M. Wood. 1995. Homology and phylogenetic implications of male genitalia in Diptera – Eremoneura. *Entomologica Scandinavica* 26: 121–151.

Dahl, C. 1960. Musidoridae (Lonchopteridae) Diptera, Cyclorrhapha from the Azores. *Boletim do Museu Municipal do Funchal* 13: 96.

De Buck, N. 1990. Bloembezoek en bestuivingsecologie van Zweefvliegen (Diptera, Syrphidae) in het bijzonder voor Belgie. *Studiendocumenten Royal Belgian Institute of Natural Sciences* 60: 1–167.

DeBach, P. and D. Rosen. 1991. *Biological Control by Natural Enemies.* Second Edition. Cambridge University Press, Cambridge. 440 pp.

Delir, S., F. Handjani, M. Emad, and S. Ardehali. 1999. Vulvar myiasis due to *Wohlfahrtia magnifica*. *Clinical and Experimental Dermatology* 24: 279–280.

Dempewolf, M. 2004. *Arthropods of Economic Importance – Agromyzidae of the World.* CD-ROM series, ETI/ZMA, Amsterdam. Also available at http://nlbif.eti.uva.nl/bis/agromyzidae.php

Desportes, C. 1942. *Forcipomyia velow* Winn. et *Sycorax silacea* Curtis, vecteurs d'*Icosiella neglecta* (Diesing), filaire commune de la grenouille verte. *Annales de Parasitologie Humaine et Comparée* 19: 53–68.

Diggins, T. P. and K. M. Stewart. 1993. Deformities of aquatic larval midges (Chironomidae: Diptera) in the sediments of the Buffalo River, New York. *Journal of Great Lakes Research* 19: 648–659.

Dimitriadis, S. and P. S. Cranston 2007. From the mountains to the sea: assemblage structure and dynamics in Chironomidae (Insecta: Diptera) in the Clyde River estuarine gradient, New South Wales, south-eastern Australia. *Australian Journal of Entomology* 46: 188–197.

Disney, R. H. L. 1991. The aquatic Phoridae (Diptera). *Entomologica Scandinavica* 22: 171–191.

Disney, R. H. L. 1994a. Continuing the debate relating to the phylogenetic reconstruction of the Phoridae (Diptera). *Giornale Italiano di Entomologia* 7: 103–117.

Disney, R. H. L. 1994b. *Scuttle Flies: the Phoridae.* Chapman and Hall, London. xii + 467 pp.

Disney, R. H. L. 1998. Family Phoridae. Pp. 51–79. *In* L. Papp and B. Darvas (eds). *Contributions to a Manual of Palaearctic Diptera. Volume 3. Higher Brachycera.* Science Herald, Budapest. 880 pp.

Disney, R. H. L. and D. H. Kistner. 1995. Revision of the Afrotropical Termitoxeniinae (Diptera: Phoridae). *Sociobiology* 26: 117–225.

Disney, R. H. L. and H. Kurahashi. 1978. A case of urogenital myiasis caused by a species of *Megaselia* (Diptera: Phoridae). *Journal of Medical Entomology* 14: 717.

Dowell, R. V. and L. K. Wange. 1986. Process analysis and failure avoidance in fruit fly programs. Pp. 43–65. *In* M. Mangel, J. R. Carey, and R. E. Plant (eds). *Pest Control: Operations and Systems Analysis in Fruit Fly Management.* NATO Advanced Science Institutes Series G. Ecological Sciences Springer-Verlag, New York 11 xii + 465 pp.

Downes, W. L., Jr. 1986. The Nearctic *Melanomya* and relatives (Diptera: Calliphoridae), a problem in calyptrate classification. *Bulletin of the New York State Museum* 460: v + 1–35.

Dudley, T. L. and N. H. Anderson. 1987. The biology and life cycles of *Lipsothrix* spp. (Diptera: Tipulidae) inhabiting wood in western Oregon streams. *Freshwater Biology* 17: 437–451.

Dunn, L. H. 1930. Rearing the larvae of *Dermatobia hominis* Linn. in man. *Psyche* 37: 327–342.

Dwortzan, M. 1997. It's a Fly! It's a Bug! It's a Microplane! http://www2.technologyreview.com/articles/97/10/reporter21097.asp [Accessed 1 April 2005].

Eibl, J. and R. Copeland. 2005. [Explanation of a phantom phenomenon.] http://www.sel.barc.usda.gov/Diptera/youngent/clouds_explanation.htm [Accessed 8 February 2008.]

Eigen, M., W. J. Kloft, and G. Brandner. 2002. Transferability of HIV by arthropods supports the hypothesis about transmission of the virus from apes to man. *Naturwissenschaften* 89: 185–186.

Elberling, H. and J. M. Olesen. 1999. The structure of a high latitude plant-flower visitor system: the dominance of flies. *Ecography* 22: 314–323.

Ellington, C. P. 1999. The novel aerodynamics of insect flight: applications to micro-air vehicles. *Journal of Experimental Biology* 202: 3439–3448.

Emerson, P. M., R. L. Bailey, and O. S. Mahdi. 2000. Transmission ecology of the fly *Musca sorbens*, a putative vector of trachoma. *Transactions of the Royal Society of Tropical Medicine and Hygiene* 94: 28–32.

Endress, P. K. 2001. The flowers in extant basal angiosperms and inferences on ancestral flowers. *International Journal of Plant Sciences* 162: 1111–1140.

Englund, R. A., M. G. Wright, and D. A. Polhemus. 2007. Aquatic insect taxa as indicators of aquatic species richness, habitat disturbance, and invasive species impacts in Hawaiian streams. *In* N. L. Evenhuis and J. M. Fitzsimons (eds). Biology of Hawaiian streams and estuaries. *Bishop Museum Bulletin in Cultural and Environmental Studies* 3: 207–232.

Eschle, J. L. and G. R. DeFoliart. 1965. Control of mink myiasis caused by the larvae of *Wohlfahrtia vigil. Journal of Economic Entomology* 58: 529–531.

Esser, J. R. 1991. Biology of *Chrysomya megacephala* (Diptera: Calliphoridae) and reduction of losses to the salted-dried fish industry in south-east Asia. *Bulletin of Entomological Research* 81: 33–41.

Evenhuis, N. L., T. Pape, A. C. Pont, and F. C. Thompson (eds). 2007. BioSystematics Database of World Diptera, Version 9.5. http://www.diptera.org/biosys.htm [Accessed 20 January 2008].

Falk, P. 1991. A review of the scarce and threatened flies of Great Britain (part I). *Research and Survey in Nature Conservation* 39: 1–194.

Fast, E. and T. A. Wheeler. 2004. Faunal inventory of Brachycera (Diptera) in an old growth forest at Mont Saint-Hilarie, Quebec. *Fabreries* 29: 1–15.

Farkas, R. and G. Kepes. 2001. Traumatic myiasis of horses caused by *Wohlfahrtia magnifica. Acta Veterinaria Hungarica* 49: 311–318.

Farkas, R., Z. Szanto, and M. R. J. Hall. 2001. Traumatic myiasis of geese in Hungary. *Veterinary Parasitology* 95: 45–52.

Ferrar, P. 1976. Macrolarviparous reproduction in Ameniinae (Diptera: Calliphoridae). *Systematic Entomology* 1: 107–116.

Ferrar, P. 1987. A guide to the breeding habits and immature stages of Diptera Cyclorrhapha. *Entomonograph* 8: 1–907. [2 volumes.]

Fessl, B., B. J. Sinclair, and S. Kleindorfer. 2006. The life-cycle of *Philornis downsi* (Diptera: Muscidae) parasitizing Darwin's finches and its impacts on nestling survival. *Parasitology* 133: 739–747.

Foote, B. A. (coordinator). 1987. Order Diptera. Pp. 690–915. *In* F. W. Stehr (ed). *Immature Insects. Volume 2.* Kendall/Hunt Publishing, Dubuque, Iowa. 975 pp.

Förster, M., S. Klimpel, H. Mehlhorn, K. Sievert, S. Messler, and K. Pfeffer. 2007. Pilot study on synanthropic flies (e.g. *Musca, Sarcophaga, Calliphora, Fannia, Lucilia, Stomoxys*) as vectors of pathogenic microorganisms. *Parasitological Research* 101: 243–246.

Fortini, M. E., M. P. Skupski, M. S. Boguski, and I. K. Hariharan. 2000. A survey of human disease gene counterparts in the *Drosophila* genome. *Journal of Cell Biology* 150: F23–F29.

Fredeen, F. J. H. and M. E. Taylor. 1964. Borborids (Diptera: Sphaeroceridae) infesting sewage disposal tanks, with notes on the life cycle, behaviour and control of *Leptocera caenosa* (Rond.). *Canadian Entomologist* 96: 801–808.

Free, J. B. 1993. *Insect Pollination of Crops.* Second Edition. Academic Press, London. xii + 684 pp.

Gaimari, S. D. and W. J. Turner. 1996. Methods for rearing aphidophagous *Leucopis* spp. (Diptera: Chamaemyiidae). *Journal of Kansas Entomological Society* 69: 363–369.

Gagné, R. J. 1994. *The Gall Midges of the Neotropical Region.* Cornell University Press, Ithaca, New York. xv + 352 pp.

Gagné, R. J. 1995. Revision of tetranychid (Acarina) mite predators of the genus *Feltiella* (Diptera: Cecidomyiidae). *Annals of the Entomological Society of America* 88: 16–30.

Gagné, R. J., H. Blanco-Metzler, and J. Etienne. 2000. A new Neotropical species of *Clinodiplosis* (Diptera: Cecidomyiidae), an important new pest of cultivated peppers (*Capsicum* spp.: Solanaceae). *Proceedings of the Entomological Society of Washington* 102: 831–837.

Gagné, R. J., A. Sosa, and H. Cordo. 2004. A new Neotropical species of Clinodiplosis (Diptera: Cecidomyiidae) injurious to alligatorweed, Alternanthera philoxeroides (Amaranthaceae). *Proceedings of the Entomological Society of Washington* 106: 305–311.

Gattiker, G., K. V. I. S. Kaler, and M. P. Mintchev. 2005. Electronic mosquito: designing a semi-invasive microsystem for blood sampling, analysis and drug delivery applications. *Microsystem Technologies* 12: 44–51.

Gillespie, D. R., B. Roitberg, E. Basalyga, M. Johnstone, G. Opit, J. Rodgers, and N. Sawyer. 1998. Biology and application of *Feltiella acarisuga* (Vallot) (Diptera: Cecidomyiidae) for biological control of twospotted spider mites on greenhouse vegetable crops. Pacific Agri-Food Research Centre (Agassiz). Technical Report 145. Agriculture and Agri-Food Canada.

Gold, B. L., K. P. Mathews, and H. A. Burge. 1985. Occupational asthma caused by sewer flies. *American Review of Respiratory Diseases* 131: 949–952.

Goldblatt, P. and J. C. Manning. 2000. The long-proboscid fly pollination system in southern Africa. *Annals of the Missouri Botanical Garden* 87: 146–170.

Golding, W. 1954. *Lord of the Flies.* Faber and Faber Ltd., London. 248 pp.

Gonzalez, L. and B. V. Brown. 2004. New species and records of *Melaloncha* (Udamochiras) bee-killing flies (Diptera: Phoridae). *Zootaxa* 730: 1–14.

Good, J. A. and M. C. D. Speight. 1996. *Saproxylic Invertebrates and Their Conservation Throughout Europe. Convention on the Conservation of European Wildlife and Natural Habitats.* Council of Europe, Strasbourg. 58 pp.

Graham, L. C., S. D. Porter, R. M. Pereira, H. D. Dorough, and A. T. Kelley. 2003. Field releases of the decapitating fly *Pseudacteon curvatus* (Diptera: Phoridae) for control of imported fire ants (Hymenoptera: Formicidae) in Alabama, Florida, and Tennessee. *Florida Entomologist* 86: 334–339.

Grantham-Hill, C. 1933. Preliminary note on the treatment of infected wounds with the larva of *Wohlfartia* [*sic*] *nuba*. *Transactions of the Royal Society of Tropical Medicine and Hygiene* 27: 93–98.

Greathead, D. J. 1983. Natural enemies of *Nilaparvata lugens* and other leaf- and planthoppers in the tropical agroecosystems and their impact on pest population(s). Pp. 371–383. *In* J. W. Knight, N. C. Pant, T. S. Robertson, and M. R. Wilson (eds). *First International Workshop on Leafhoppers and Planthoppers of Economic Importance*. Commonwealth Institute of Entomology, London. 500 pp.

Greathead, D. J. 1995. The *Leucopis* spp. (Diptera: Chamaemyiidae) introduced for biological control of *Pineus* sp. (Homoptera: Adelgidae) in Hawaii: implications for control of *Pineus boerneri* in Africa. *The Entomologist* 114: 83–90.

Greathead, D. J. and N. L. Evenhuis. 1997. Family Bombyliidae. Pp. 487–512. *In* L. Papp and B. Darvas (eds). *Contributions to a Manual of Palaearctic Diptera. Volume 2. Nematocera and Lower Brachycera*. Science Herald. Budapest. 592 pp.

Greenberg, B. 1971. *Flies and Diseases*. Volume 1. Ecology Classification and Biotic Associations. Princeton University Press, Princeton. 856 pp.

Greenberg, B. 1973. *Flies and Disease*. Volume II. Biology and Disease Transmission. Princeton University Press, Princeton. 447 pp.

Griffiths, G. C. D. 1972. *The Phylogenetic Classification of Diptera Cyclorrhapha with Special Reference to the Structure of the Male Postabdomen*. Series Entomologica 8. Dr. W. Junk, The Hague. 340 pp.

Grimaldi, D. and J. M. Cumming. 1999. Brachyceran Diptera in Cretaceous ambers and Mesozoic diversification of the Eremoneura. *Bulletin of the American Museum of Natural History* 239: 1–124.

Grimaldi, D. and M. S. Engel. 2005. *Evolution of the Insects*. Cambridge University Press, Cambridge. 755 pp.

Grisi, L., C. L. Massard, G. E. Moya-Borja, and J. B. Pereira. 2002. Impacto econômico das principais ectoparasitoses em bovinos no Brasil. *Hora Veterinária* 21: 8–10.

Grunin, K. J. 1973. The first discovery of larvae of the mammoth bot-fly *Cobboldia* (*Mamontia*, subgen. n.) *russanovi* sp.n. (Diptera, Gasterophilidae). *Éntomologicheskoe Obozrenie* 52: 228–233. [In Russian with English subtitle; English translation: Entomological Review, Washington 52: 165–169].

Guimarães, J. H. 1977. A systematic revision of the Mesembrinellidae, stat. nov. (Diptera, Cyclorrhapha). *Arquivos de Zoologia, São Paulo* 29: 1–109.

Guimarães, J. H. and N. Papavero. 1999. *Myiasis in Man and Animals in the Neotropical Region; Bibliographic Database*. Flêiade/FAPESP, São Paulo, Brazil. 308 pp.

Guimarães, J. H., N. Papavero, and A. Prado. 1983. As míiases na região Neotropical (identificação, biologia, bibliografia). *Revista Brasileira de Zoologia* 1: 239–416.

Haines, C. P. and D. P. Rees. 1989. A field guide to the types of insects and mites infesting cured fish. FAO Fisheries Technical Paper 303: 1–33. [Also available at http://www.fao.org/documents/ [Accessed 3 June 2005].

Hall, M. J. R. 1997. Traumatic myiasis of sheep in Europe: a review. *Parasitologia* 39: 409–413.

Hall, M. J. R. and R. Wall. 1995. Myiasis of humans and domestic animals. *Advances in Parasitology* 35: 257–334.

Hansen, M. 1983. *Yuca* (Yuca, Cassava). Pp. 114–117. *In* D. Janzen (ed). *Costa Rican Natural History*. University of Chicago Press, Chicago. xi + 816 pp.

Harbach, R. E. 2007. The Culicidae (Diptera): a review of taxonomy, classification and phylogeny. *Zootaxa* 1668: 591–638.

Harris, P. 1989. The use of Tephritidae for the biological control of weeds. *Biocontrol News and Information* 10: 7–16.

Harrison, G. 1978. *Mosquitoes, Malaria and Man: a History of the Hostilities Since 1880*. First Edition. John Murray, London. viii + 314 pp.

Hashimoto, H. 1976. Non-biting midges of marine habitats (Diptera: Chironomidae). Pp. 377–414. *In* L. Cheng (ed). *Marine Insects*. North-Holland Publishing Co., Amsterdam, The Netherlands. 581 pp.

Haslett, J. R. 1988. Qualitätsbeurteilung alpiner Habitate: Schwebfliegen (Diptera: Syrphidae) als Bioindikatoren für Auswirkungen des intensiven Skibetriebes auf alpinen Wiesen in Österreich. *Zoologischer Anzeiger* 220: 179–184.

Hawaii Biological Survey. 2002. Hawaii's Extinct Species – Insects. http://hbs.bishopmuseum.org/endangered/ext-insects.html [Accessed 4 March 2008].

Heath, A. C. G. 1982. Beneficial aspects of blowflies (Diptera: Calliphoridae). *New Zealand Entomologist* 7: 343–348.

Hennig, W. 1971. Neue untersuchungen ueber die familien der Diptera Schizophora (Diptera: Cyclorrhapha). Stuttgarter Beitraege zur Naturkunde, *Serie A (Biologie)* 226: 1–76.

Hennig, W. 1973. Diptera (Zweiflügler). *Handbuch der Zoologie (Berlin)* 4: 1–200.

Hernández, F. O. and A. A. Gutiérrez. 2001. Avoiding *Pseudohypocera* attacks (Diptera: Phoridae) during the artificial propagation of *Melipona beecheii* colonies (Hymenoptera: Apidae: Meliponini). *Folia Entomologica Mexicana* 40: 373–379.

Hide, G. 1999. History of sleeping sickness in East Africa. *Clinical Microbiology Reviews* 12: 112–125.

Higley, L. G. and N. H. Haskell. 2001. Insect development and forensic entomology. Pp. 287–302. *In* J. H. Byrd and J. L. Castner (eds). *Forensic Entomology*. CRC Press, Boca Raton, Florida. 418 pp.

Hill, D. S. 1963. The life history of the British species of *Ornithomya* (Diptera: Hippoboscidae). *Transactions of the Royal Entomological Society of London* 115: 391–407.

Hoffmann, M. P., and A. C. Frodsham. 1993. *Natural Enemies of Vegetable Insect Pests*. Cooperative Extension, Cornell University, Ithaca, New York. 63 pp.

Hogue, C. L. 1981. Blephariceridae. Pp. 191–197. *In* J. F. McAlpine, B. V. Petersen, G. E. Shewell, H. J. Teskey, J. R. Vockeroth, and D. M. Wood (coordinators). *Manual of Nearctic Diptera*, Volume 1. Research Branch, Agriculture

Canada Monograph 27. Supply & Services Canada, Hull, Quebec.

Holloway, B. A. 1976. Pollen-feeding in hover-flies (Diptera: Syrphidae). *New Zealand Journal of Zoology* 3: 339–350.

Holston, K. C. 2005. Evidence for community structure and habitat partitioning by stiletto flies (Diptera: Therevidae) at the Guadalupe-Nipomo Dune System, California. *Journal of Insect Science* 5 (42): 1–17.

Horgan, F. G., J. H. Myers, and R. Van Meel. 1999. *Cyzenis albicans* (Diptera: Tachinidae) does not prevent the outbreak of winter moth (Lepidoptera: Geometridae) in birch stands and blueberry plots on the lower mainland of British Columbia. *Environmental Entomology* 28: 96–107.

Horiuchi, J. and M. Saitoe. 2005. Can flies shed light on our own age-related memory impairment. *Ageing Research Reviews* 4: 83–101.

Hövemeyer, K. 2000. 1.11. Ecology of Diptera. Pp. 437–489. *In* L. Papp and B. Darvas (eds). *Contributions to a Manual of Palaearctic Diptera. Volume 1. General and Applied Dipterology.* Science Herald, Budapest. 978 pp.

Howard, J. 2001. Nuisance flies around a landfill: patterns of abundance and distribution. *Waste Management and Research* 19: 308–313.

Hull, F. M. 1962. Robber flies of the World. The genera of the family Asilidae. *Bulletin of the United States National Museum* 224: 1–907.

Hussey, N. W. 1960. Biology of mushroom phorids. *Mushroom Science* 4: 260–269 [1959].

Iori, A., B. Zechini, L. Cordier, E. Luongo, G. Pontuale, and S. Persichino. 1999. A case of myiasis in man due to *Wohlfahrtia magnifica* (Schiner) recorded near Rome. *Parasitologia* 41: 583–585.

IUCN. 2007. 2007 IUCN Red List of Threatened Species. http://www.iucnredlist.org/ [Accessed 4 March 2008].

Jackman, R., S. Nowicki, D. J. Aneshansley, and T. Eisner. 1983. Predatory capture of toads by fly larvae. *Science* 222: 515–516.

Jackson, D. S. and B. G. Lee. 1985. Medfly in California 1980–1982. *Bulletin of the Entomological Society of America* 1985: 29–37.

James, M. T. 1935. The genus *Hermetia* in the United States (Diptera: Stratiomyidae). *Bulletin of Brooklyn Entomological Society* 30: 165–170.

Jehl, J. R., Jr. 1986. Biology of red-necked phalaropes (*Phalaropus lobatus*) at the western edge of the Great Basin in fall. *Great Basin Naturalist* 46: 185–197.

Juliano, S. A. and L. P. Lounibos. 2005. Ecology of invasive mosquitoes: effects on resident species and on human health. *Ecology Letters* 8: 558–574.

Kaneshiro, K. Y. 1997. R. C. L. Perkins' legacy to evolutionary research on Hawaiian Drosophilidae (Diptera). *Pacific Science* 51: 450–461.

Karouna-Renier, N. K. and J. P. Zehr. 1999. Ecological implications of molecular biomarkers: assaying sub-lethal stress in the midge *Chironomus tentansusing* heat shock protein 70 (HSP-70) expression. *Hydrobiologia* 401: 255–264.

Kearns, C. A. 1992. Anthophilous fly distribution across an elevation gradient. *American Midland Naturalist* 127: 172–182.

Kearns, C. A. 2001. North American dipteran pollinators: assessing their value and conservation status. *Conservation Ecology* 5 (1): 5. http://www.consecol.org/vol5/iss1/art5/ [Accessed 15 November 2007].

Kearns, C. A. 2002. Flies and flowers: an enduring partnership. *Wings (The Xerces Society)* 25 (2): 3–8.

Keilin, D. 1919. On the life history and larval anatomy of *Melinda cognata* Meigen, parasitic in the snail *Helicella (Heliomanes) virgata* da Costa, with an account of the other Diptera living upon molluscs. *Parasitology* 11: 430–454.

Kevan, P. G. 1972. Insect pollination of high arctic flowers. *Journal of Ecology* 60: 831–847.

Kevan, P. G. 2002. Flowers, pollination, and the associated diversity of flies. *Biodiversity* 3 (4): 16–18.

Kiesenwetter, E. A. 1857. *Chlorops nasuta* Meig. in grossen Schwärmen beobachtet. *Berliner Entomologische Zeitschrift* 1: 171.

Klemm, D. J., K. A. Blocksom, F. A. Fulk, A. T. Herlihy, R. M. Hughes, P. R. Kaufman, D. V. Peck, J. L. Stoddard, W. T. Theony, M. B. Griffith, and W. S. Davis 2003. Development and evaluation of a Macroinvertebrate Biotic Integrity Index (MBII) for regionally assessing Mid-Atlantic Highlands streams. *Environmental Management* 31: 656–669.

Koenig, D. P. and C. W. Young. 2007. First observation of parasitic relations between big-headed flies of the genus *Nephrocerus* (Diptera: Pipunculidae) and crane flies of the genus *Tipula* (Diptera: Tupulidae: Tipulinae), with larval and puparial descriptions of *Nephrocerus atrapilus* Skevington. *Proceedings of the Entomological Society of Washington* 109: 52–65.

Kotrba, M. 2004. Fliegenschwärme im südbayerischen Seengebiet (Diptera: Chloropidae). *Nachrichtenblatt der Bayerischen Entomologen* 53: 32.

KPLC, Lakes Charles, Louisiana. 2004. West Nile virus cost LA $20 million. http://www.kplctv.com/global/story.asp?s=2358859andClientType=Printable [Accessed 20 February 2008].

Krafsur, E. S. and R. D. Moon. 1997. Bionomics of the face fly, *Musca autumnalis*. *Annual Review of Entomology* 42: 503–523.

Kühne, S. 2000. Räuberische Fliegen der Gattung *Coenosia* Meigen, 1826 (Diptera: Muscidae) und die Möglichkeit ihres Einsatzes bei der biologischen Schädlingsbekämpfung. *Studia Dipterologica, Supplement* 9: 1–78.

Kutty, S., M. V. Bernasconi, F. Sifner, and R. Meier. 2007. Sensitivity analysis, molecular systematics, and natural history evolution of Scathophagidae (Diptera: Cyclorrhapha: Calyptratae). *Cladistics* 23: 64–83.

Labandeira, C. C. 1998. How old is the flower and the fly. *Science* 280: 85–88.

Lagacé-Wiens, P. R. S., R. Dookeran, S. Skinner, R. Leicht, D. D. Colwell, and T. D. Galloway. 2008. Human ophthalmomyiasis interna caused by *Hypoderma tarandi*, Northern Canada. *Emerging Infectious Diseases* 14: 64–66.

Langelaan, G. 1957. The Fly. *Playboy Magazine* 42: 16–18.

Larson, B. M. H., P. G. Kevan, and D. W. Inouye. 2001. Flies and flowers: taxonomic diversity of anthophiles and pollinators. *Canadian Entomologist* 133: 439–465.

Laurence, B. R. 1953. Some Diptera bred from cow dung. *Entomologist's Monthly Magazine* 89: 281–283.

Laurence, B. R. 1954. The larval inhabitants of cow pats. *Journal of Animal Ecology* 23: 234–260.

Laurence, B. R. 1955. Flies associated with cow dung. *Entomologist's Record and Journal of Variation* 67: 123–126.

Leclercq, M. 1990. Utilisation de larves de diptères – maggot therapy – en médicine: historique et actualité. *Bulletin et Annales de la Société royale Belge d'Entomologique* 126: 41–50.

Lenat, D. R. 1993a. Using mentum deformities of *Chironomus* larvae to evaluate the effects of toxicity and organic loading in streams. *Journal of the North American Benthological Society* 12: 265–269.

Lenat, D. R. 1993b. A biotic index for the southeastern United States: derivation and list of tolerance values, with criteria for assigning water-quality ratings. *Journal of the North American Benthological Society* 12: 279–290.

Letzner, K. W. 1873. Schwärme der *Chlorops ornata* Meigen. *Jahresbericht der Schlesischen Gesellschaft für Vaterländische Cultur* 50: 193–199.

Leevers, S. J. 2001. Growth control: invertebrate insulin surprises! *Current Biology* 11: 209–212.

Lewis, E. B. 1998. The bithorax complex: the first fifty years. *International Journal of Developmental Biology* 42: 403–415.

Lindegaard, C. and P. M. Jónasson. 1979. Abundance, population dynamics and production of zoobenthos in Lake Mývatn, Iceland. *Oikos* 32: 202–227.

Lindsay, D. R. and H. I. Scudder. 1956. Nonbiting flies and disease. *Annual Review of Entomology* 1: 323–346.

Linevich, A. A. 1971. The Chironomidae of Lake Baikal. *Limnologica (Berlin)* 8: 51–52.

Linley, J. R. 1976. Biting midges of mangrove swamps and salt-marshes (Diptera: Ceratopogonidae). Pp. 335–376. *In* L. Cheng (ed). *Marine Insects.* North-Holland Publishing Co., Amsterdam, The Netherlands. 581 pp.

Lonsdale, O. and S. Marshall. 2004. Clusiidae. Version 10, December 2004. http://tolweb.org/Clusiidae/10628/2004.12.10 In The Tree of Life Web Project: http://tolweb.org/ [Accessed on 7 March 2008].

Lyons, M. 1992. *The Colonial Disease: a Social History of Sleeping Sickness in Northern Zaire, 1900–1940.* Cambridge University Press, New York. 335 pp.

Maharaj, R., C. C. Appleton, and R. M. Miller. 1992. Snail predation by larvae of *Sepedon scapularis* Adams (Diptera: Sciomyzidae), a potential biocontrol agent of snail intermediate hosts of schistosomiasis in South Africa. *Medical and Veterinary Entomology* 6: 183–187.

Maret, T. R., D. J. Cain, D. E. MacCoy, and T. M. Short. 2003. Response of Benthic invertebrate assemblages to metal exposure and bioaccumulation associated with hard-rock mining in northwestern streams, USA. *Journal of the North American Benthological Society* 22: 598–620.

Marshall, A. G. 1970. The life cycle of *Basilia hispida* Theodor 1967 (Diptera: Nycteribiidae) in Malaysia. *Parasitology* 61: 1–18.

Marshall, S. A. and O. W. Richards. 1987. Sphaeroceridae. Pp. 993–1006. *In* J. F. McAlpine, B. V. Peterson, G. E. Shewell, H. J. Teskey, J. R. Vockeroth, and D. M. Wood (coordinators). *Manual of Nearctic Diptera.* Volume 2. Research Branch, Supply & Services Canada, Hull, Quebec, Agricultural Canada Monograph 28.

Masoodi, M. and K. Hosseini. 2004. External ophthalmomyiasis caused by sheep botfly (*Oestrus ovis*) larva: a report of 8 cases. *Archives of Iranian Medicine* 7: 136–139.

Matthiessen, J., L. Hayles, and M. Palmer. 1984. An assessment of some methods for the bioassay of changes in cattle dung as insect food, using the bush fly, *Musca vetustissima*, Walker (Diptera: Muscidae). *Bulletin of Entomological Research* 74: 463–467.

Maughan, D., J. Moore, J. Vigoreaux, B. Barnes, and L. A. Mulieri. 1998. Work production and work absorption in muscle strips from vertebrate cardiac and insect flight muscle fibers. *Advances in Experimental and Medical Biology* 453: 471–480.

McAlpine, D. K. 2007. Review of the Borboroidini or wombat flies (Diptera: Heteromyzidae), with reconsideration of the status of families Heleomyzidae and Sphaeroceridae, and descriptions of femoral gland baskets. *Records of the Australian Museum* 59: 143–219.

McAlpine, J. F. 1973. A fossil ironomyiid fly from Canadian amber (Diptera: Ironomyiidae). *Canadian Entomologist* 105: 105–111.

McAlpine, J. F. 1989. Phylogeny and classification of the Muscomorpha. Pp. 1397–1518. *In* J. F. McAlpine and D. M. Wood (coordinators). *Manual of Nearctic Diptera.* Volume 3. Research Branch, Supply & Services Canada, Hull, Quebec, Agricultural Canada Monograph 32.

McAlpine, J. F., B. V. Peterson, G. E. Shewell, H. J. Teskey, J. R. Vockeroth, and D. M. Wood (coordinators). 1981. *Manual of Nearctic Diptera.* Volume 1. Research Branch, Supply & Services Canada, Hull, Quebec, Agriculture Canada Monograph 27. 674 pp.

McAlpine, J. F., B. V. Peterson, G. E. Shewell, H. J. Teskey, J. R. Vockeroth, and D. M. Wood (coordinators). 1987. *Manual of Nearctic Diptera.* Volume 2. Research Branch, Supply & Services Canada, Hull, Quebec, Agriculture Canada Monograph 28. 675–1332 pp.

McElravy, E. P. and V. H. Resh. 1991. Distribution and seasonal occurrence of the hyporheic fauna in a northern California stream. *Hydrobiologia* 220: 233–246.

McLean, I. F. G. 2000. Beneficial Diptera and their role in decomposition. Pp. 491–517. *In* L. Papp and B. Darvas (eds). *Contributions to a Manual of Palaearctic Diptera. Volume 1. General and Applied Dipterology.* Science Herald, Budapest. 978 pp.

McPheron, B. A. and G. J. Steck (eds). 1996. *Fruit Fly Pests: A World Assessment of Their Biology and Management.* St. Lucie Press, Delray Beach. 586 pp.

Meier, R., M. Kotrba, and P. Ferrar. 1999. Ovoviviparity and viviparity in the Diptera. *Biological Reviews* 74: 199–258.

Mellor, P. S. and J. Boorman. 1995. The transmission and geographical spread of African horse sickness and blue-tongue viruses. *Annals of Tropical Medicine and Parasitology* 89: 1–15.

Menzel, F. and W. Mohrig. 1999. Revision der paläarktischen Trauermücken (Diptera: Sciaridae). *Studia Dipterologica, Supplement* 6: 1–761.

Merritt, R. W., G. W. Courtney, and J. B. Keiper. 2003. Diptera. Pp. 324–340. *In* V. H. Resh and R. T. Cardé (eds). *Encyclopedia of Insects.* Academic Press, London.

Michailova, P., N. Petrova, G. Sella, L. Ramella, and S. Bovero. 1998. Structural-functional rearrangements in chromosome G in *Chironomus riparius* (Diptera, Chironomidae) collected from a heavy metal-polluted area near Turin, Italy. *Environmental Pollution* 103: 127–134.

MicrobiologyBytes. 2007. Malaria. http://www .microbiologybytes.com/introduction/Malaria.html [Last updated 26 April 2007, accessed 10 February 2008].

Mitra, B. and D. Banerjee 2007. Fly pollinators: assessing their value in biodiversity conservation and food security in India. *Records of the Zoological Survey India* 107 (Part 1): 33–48.

Morita, S. I. 2007. Pangonid.net. http://pangonid.net/pages/ philoimages/pages/image1.html [Accessed 3 March 2008].

Morozova, T. V., R. R. H. Anholt, and T. F. C. Mackay. 2006. Transcriptional response to alcohol exposure in *Drosophila melanogaster. Genome Biology* 7 (10): R95.

Mostovski, M. B. 1995. New taxa of ironomyiid flies (Diptera, Phoromorpha, Ironomyiidae) from Cretaceous deposits of Siberia and Mongolia. *Paleontologicheski Zhurnal* 4: 86–103.

Moulton, J. K. and B. M. Wiegmann. 2007. The phylogenetic relationships of flies in the superfamily Empidoidea (Insecta: Diptera). *Molecular Phylogenetics and Evolution* 43: 701–713.

Moulton, S. R., II, J. L. Carter, S. A. Grotheer, T. F. Cuffney, and T. M. Short. 2000. *Methods of analysis by the U.S. Geological Survey National Water Quality Laboratory – processing, taxonomy, and quality control of benthic macroinvertebrates. U.S. Geological Survey Open-File Report 00–212.* United States Geological Survey, Reston, Virginia.

Mounier, N., M. Gouy, D. Mouchiroud, and J. C. Prudhomme. 1992. Insect muscle actins differ distinctly from invertebrate and vertebrate cytoplasmic actins. *Journal of Molecular Evolution* 34: 406–415.

Mulla, M. S. 1965. Biology and control of *Hippelates* eye gnats. *Proceedings and Papers of the Annual Conference of the California Mosquito Control Association* 33: 26–28.

Nartshuk, E. and S. Pakalniškis. 2004. Contribution to the knowledge of the family Chloropidae (Diptera, Muscomorpha) of Lithuania. *Acta Zoologica Lituanica* 14 (2): 56–66.

Nartshuk, E. P. 1997. Family Acroceridae. Pp. 469–485. *In* L. Papp and B. Darvas (eds). *Contributions to a Manual of Palaearctic Diptera. Volume 2. Nematocera and Lower Brachycera.* Science Herald, Budapest. 592 pp.

Nartshuk, E. P. 2000. Periodicity of outbreaks of the predatory fly *Thaumatomyia notata* Mg. (Diptera, Chloropidae) and its possible reasons. *Entomologicheskoe Obozrenie* 79: 771–781. [In Russian; English translation in *Entomological Review* 80: 911–918.]

Nayduch, D., G. P. Noblet, and F. J. Stutzenberger 2002. Vector potential of houseflies for the bacterium *Aeromonas caviae. Medical and Veterinary Entomology* 16: 193–198.

Neff, J. L., B. B. Simpson, N. L. Evenhuis, and G. Dieringer. 2003. Character analysis of adaptations for tarsal pollen collection in the Bombyliidae (Insecta: Diptera): the benefits of putting your foot in your mouth. *Zootaxa* 157: 1–14.

Neville, E. M. 1985. The epidemiology of *Parafilaria bovicola* in the Transvaal Bushveld of South Africa. *Onderstepoort Journal of Veterinary Research* 52: 261–267.

Newton, G. L., C. V. Booram, R. W. Barker, and O. M. Hale. 1977. Dried *Hermetia illucens* larvae meal as a supplement for swine. *Journal of Animal Science* 44: 395–400.

Nielsen, P., O. Ringdahl, and S. L. Tuxen. 1954. 48a. Diptera 1 (exclusive of Ceratopogonidae and Chironomidae). *Zoology Iceland* 3: 1–189.

Norrbom, A. L. 2004. Fruit Fly (Diptera: Tephritidae) Taxonomy Pages. http://www.sel.barc.usda.gov/diptera/ tephriti/tephriti.htm [Accessed 10 February 2008].

Noutis, C. and L. E. Millikan. 1994. Myiasis. *Dermatologic Clinics* 12: 729–736.

O'Grady, P., J. Bonacum, R. DeSalle, and F. Do Val. 2003. The placement of *Engiscaptomyza, Grimshawomyia,* and *Titanochaeta,* three clades of endemic Hawaiian Drosophilidae (Diptera). *Zootaxa* 159: 1–16.

O'Hara, J. E., K. D. Floate, and B. E. Cooper. 1999. The Sarcophagidae (Diptera) of cattle feedlots in southern Alberta. *Journal of the Kansas Entomological Society* 72: 167–176.

Økland, B. 1994. Mycetophilidae (Diptera), an insect group vulnerable to forest practices? A comparison of clearcut, managed and semi-natural spruce forests in southern Norway. *Biodiversity and Conservation* 3: 68–85.

Økland, B. 1996. Unlogged forests: Important sites for preserving the diversity of mycetophilids (Diptera, Sciaroidea). *Biological Conservation* 76: 297–310.

Økland, B. 2000. Management effects on the decomposer fauna of Diptera in spruce forests. *Studia Dipterologica* 7: 213–223.

Økland, B., F. Götmark, B. Nordén, N. Franc, O. Kurina, and A. Polevoi 2004. Regional diversity of mycetophilids (Diptera: Sciaroidea) in Scandinavian oak-dominated forests. *Biological Conservation* 121: 9–20.

Økland, B., F. Götmark, and B. Nordén. 2008. Oak woodland restoration: testing the effects on biodiversity of mycetophilids in southern Sweden. Biodiversity and Conservation, 17 (11): 2599–2616.

Oldroyd, H. 1964. *The Natural History of Flies.* Weidenfield and Nicholson, London. xiv + 324 pp.

Oldroyd, H. 1973. Tabanidae (horse-flies, clegs, deer-flies, etc.). Pp. 195–202. In K. G. V. Smith (ed). *Insects and Other Arthropods of Medical Importance*. British Museum (Natural History), London. xiv + 561 pp.

O'Meara, G. F. 1976. Saltmarsh mosquitoes (Diptera: Culicidae). Pp. 303–333. In L. Cheng (ed). *Marine Insects*. North-Holland Publishing, Amsterdam, The Netherlands. 581 pp.

Oosterbroek, P. and G. W. Courtney. 1995. Phylogeny of the Nematocerous families of Diptera (Insecta). *Zoological Journal of the Linnean Society* 115: 267–311.

Opit, G. P., B. Roitberg, and D. R. Gillespie. 1997. The functional response and prey preference of *Feltiella acarisuga* (Vallot) (Diptera: Cecidomyiidae) for two of its prey: male and female twospotted spider mites, *Tetranychus urticae* Koch (Acari: Tetranychiidae). *Canadian Entomologist* 129: 221–227.

Pagnier, J., J. G. Mears, O. Dunda-Belkhodja, K. E. Schaefer-Rego, C. Beldjord, R. L. Nagel, and D. Labie. 1984. Evidence for the multicentric origin of the sickle cell hemoglobin gene in Africa. *Proceedings of the National Academy of Sciences USA* 81: 1771–1773.

Palmeri, V., S. Longo, and A. E. Carolei. 2003. Osservazioni su "Apimiasi" causate da *Senotainia tricuspis* in Calabria. *Apicoltore Moderno* 88: 183–189.

Panzer, G. W. F. 1794. *Fauna insectorvm germanicae initia oder Deutschlands Insecten. Heft 24*. Felsecker, Nürnberg. 24 pp., 24 pls.

Pape, T. 1996. Catalogue of the Sarcophagidae of the world (Insecta: Diptera). *Memoirs of Entomology International* 8: 1–558.

Pape, T. 2009. 4. Economic importance of Diptera. *In* B. V. Brown, A. Borkent, J. M. Cumming, D. M. Wood, and M. Zumbado (eds). *Manual of Central American Diptera. Volume 1*. NRC Research Press, Ottawa.

Papp, L. 1976. Ecological and zoogeographical data on flies developing in excrement droppings (Diptera). *Acta zoologica Hungarica* 22: 119–138.

Papp, L. 1985. Flies (Diptera) developing in sheep droppings in Hungary. *Acta zoologica Hungarica* 31: 367–379.

Papp, L., and B. Darvas (eds). 2000. *Contributions to a Manual of Palaearctic Diptera. Volume 1. General and Applied Dipterology*. Science Herald, Budapest. 978 pp.

Papp, L. and P. Garzó. 1985. Flies (Diptera) of pasturing cattle: some new data and new aspects. *Folia entomologica Hungarica* 46: 153–168.

Parrella, M. P. 2004. UC IPM pest management guidelines: floriculture and ornamental nurseries. UC ANR Publication 3392. http://www.ipm.ucdavis.edu/PMG/r280301111.html [Accessed 15 October 2004].

Parsonson, I. M. 1993. Bluetongue virus infection of cattle. Pp. 120–125. *In Proceedings of the 97th Annual Meeting of the USAHA*. Cummings Corporation and Carter Printing Company, Richmond. 577 pp.

Peacock, L. and G. A. Norton. 1990. A critical analysis of organic vegetable crop protection in the U.K. *Agriculture, Ecosystems and Environment* 31: 187–197.

Pellmyr, O. 1989. The cost of mutualism: interactions between *Trollius europaeus* and its pollinating parasites. *Oecologia* 78: 53–59.

Pitkin, B. R. 1989. Family Cryptochetidae. P. 666. *In* N. L. Evenhuis (ed). *Catalog of the Diptera of the Australasian and Oceanian Regions*. Bishop Museum Special Publication 86. Bishop Museum Press, Honolulu, and E.J. Brill, Leiden. 1155 pp.

Platt, A. P. and S. J. Harrison. 1995. Robber fly and trout predation on adult dragonflies (Anisoptera: Aeshnidae) and first records of *Aeshna umbrosa* from Wyoming. *Entomological News* 106: 229–235.

Pollard, G. 2000. FAO Assistance to Jamaica for improving hot pepper production, in part through development of an IPM programme for gall midges and broad mite. *Caribbean IPM Knowledge Network, Current Awareness Bulletin* 2: 1–6.

Pollet, M. 1992. Impact of enviromental variables on the occurrence of dolichopodid flies in marshland habitats in Belgium (Diptera: Dolichopodidae). *Journal of Natural History* 26: 621–636.

Pollet, M. 2000. A documented red list of the dolichopodid flies (Diptera: Dolichopodidae) of Flanders. *Communications of the Institute of Nature Conservation* 8: 1–190. [In Dutch, with English summary.]

Pollet, M. 2001. Dolichopodid biodiversity and site quality assessment of reed marshes and grasslands in Belgium (Diptera: Dolichopodidae). *Journal of Insect Conservation* 5: 99–116.

Pollet, M. and P. Grootaert 1996. An estimation of the natural value of dune habitats using Empidoidea (Diptera). *Biodiversity and Conservation* 5: 859–880.

Pont, A. C. 1980. Family Mormotomyiidae. P. 713. *In* R. W. Crosskey (ed). *Catalogue of the Diptera of the Afrotropical Region*. British Museum (Natural History), London. 1437 pp.

Porter, S. D., L. A. Nogueirade Sá, and L. W. Morrison. 2004. Establishment and dispersal of the fire ant decapitating fly *Pseudacteon tricuspis* in north Florida. *Biological Control* 29: 179–188.

Potts, W. H. 1973. Glossinidae (tse-tse flies). Pp. 209–249. *In* K. G. V. Smith (ed). *Insects and Other Arthropods of Medical Importance*. British Museum (Natural History), London. xiv + 561 pp.

Powers, N. R. and S. E. Cope. 2000. William Crawford Gorgas: the great sanitarian. *Wing Beats* 11: 4–28.

Pritchard, G. 1983. Biology of Tipulidae. *Annual Review of Entomology* 28: 1–22.

Raddum, G. G. and O. A. Saether. 1981. Chironomid communities in Norwegian lakes with different degrees of acidification. *Verhandlungen der Internationalen Vereinigung für Theoretische und Angewandte Limnologie.* 21: 399–405.

Ramírez, W. 1984. Biología del género *Melaloncha* (Phoridae), moscas parasitoides de la abeja doméstica (*Apis mellifera* L.) en Costa Rica. *Revista de Biologia Tropical* 32: 25–28.

Reemer, M. 2005. Saproxylic hoverflies benefit by modern forest management (Diptera: Syrphidae). *Journal of Insect Conservation* 9: 49–59.

Reiter, L. T., L. Potocki, S. Chien, M. Gribskov, and E. Bier. 2001. A systematic analysis of human disease-associated gene sequences in *Drosophila melanogaster*. *Genome Research* 11: 1114–1125.

Resh, V. H., M. J. Myers, and M. J. Hannaford. 1996. Use of biotic indices, habitat assessments, and benthic macroinvertebrates in evaluating environmental quality. Pp. 647–667. *In* F. R. Hauer and G. A. Lamberti (eds). *Methods in Stream Ecology*. Academic Press, San Diego, California. 674 pp.

Reyes, F. 1983. A new record of *Pseudohypocera kerteszi*, a pest of honey bees in Mexico. *American Bee Journal* 119–120.

Ridsdill-Smith, T. J. 1981. Some effects of three species of dung beetles (Coleoptera: Scarabaeidae) in south-western Australia on the survival of the bush fly, *Musca vetustissima* Walker (Diptera: Muscidae), in dung pads. *Bulletin of Entomological Research* 71: 425–433.

Ridsdill-Smith, T. J., L. Hayles, and M. Palmer. 1987. Mortality of eggs and larvae of the bush fly, *Musca vetustissima* Walker (Diptera: Muscidae) caused by scarabaeine dung beetles (Coleoptera: Scarabaeidae) in favourable cattle dung. *Bulletin of Entomological Research* 77: 731–736.

Rinker, D. L. and R. J. Snetsinger. 1984. Damage threshold to a commercial mushroom by a mushroom-infesting phorid (Diptera: Phoridae). *Journal of Economic Entomology* 77: 449–453.

Robineau-Desvoidy, J. B. 1830. Essai sur les myodaires. *Mémoires présentés par divers Savans a l' Académie Royal des Sciences de l' Institut de France* 2 (2): 1–813.

Robinson, G. E. 1981. *Pseudohypocera kerteszi* (Enderlein) (Diptera: Phoridae), a pest of the honey bee. *Florida Entomologist* 64: 456–457.

Robles, C. D. and J. Cubit. 1981. Influence of biotic factors in an upper intertidal community: dipteran larvae grazing on algae. *Ecology* 62: 1536–1547.

Robroek, B. J. M., H. de Jong, H. G. Arce, and M. J. Sommeijer. 2003a. The development of *Pseudohypocera kerteszi* (Diptera, Phoridae), a kleptoparasite in nests of stingless bees (Hymenoptera, Apidae) in Central America. *Proceedings of the Section Experimental and Applied Entomology, Nederlandse Entomologische Vereniging, Amsterdam* 12: 71–74.

Robroek, B. J. M., H. de Jong, and M. J. Sommeijer. 2003b. The behaviour of *Pseudohypocera kerteszi* (Diptera, Phoridae) in hives of stingless bees (Hymenoptera, Apidae) in Central America. *Proceedings of the Section Experimental and Applied Entomology, Nederlandse Entomologische Vereniging, Amsterdam* 12: 65–70.

Rogers, R. and M. Mattoni. 1993. Observations on the natural history and conservation biology of the giant flower loving flies, *Rhaphiomidas* (Diptera: Apioceridae). *Dipterological Research* 4 (1–2): 21–34.

Rosenberg, D. M. and V. H. Resh (eds). 1993. *Freshwater Biomonitoring and Benthic Macroinvertebrates*. Chapman and Hall, New York. 488 pp.

Rosenberg, D. M., V. H. Resh, and R. S. King. 2007. Use of aquatic insects in biomonitoring. Pp. 123–137. *In* R. W. Merritt, K. W. Cummins, and M. B. Berg (eds). *An Introduction to the Aquatic Insects of North America*. Fourth Edition. Kendall/Hunt Publishing, Dubuque, Iowa. 1158 pp.

Ross, A. 1961. Biological studies on bat ectoparasites of the genus *Trichobius* (Diptera: Streblidae) in North America, north of Mexico. *Wasmann Journal of Biology* 19: 229–246.

Rossi, M. A. and S. Zucoloto. 1973. Fatal cerebral myiasis caused by the tropical warble fly, *Dermatobia hominis*. *American Journal of Tropical Medicine and Hygiene* 22: 267–269.

Rotheray, G. E. and I. MacGowan. 2000. Status and breeding sites of three presumed endangered Scottish saproxylic syrphids (Diptera, Syrphidae). *Journal of Insect Conservation* 4: 215–223.

Rotheray, G. E., G. Hancock, S. Hewitt, D. Horsfield, I. MacGowan, D. Robertson, and K. Watt. 2001. The biodiversity and conservation of saproxylic Diptera in Scotland. *Journal of Insect Conservation* 5: 77–85.

Rozkošný, R. 1997. Family Stratiomyidae. Pp. 387–411. *In* L. Papp and B. Darvas (eds). *Contributions to a Manual of Palaearctic Diptera. Volume 2. Nematocera and Lower Brachycera*. Science Herald, Budapest. 592 pp.

Rubega, M. and C. Inouye. 1994. Prey switching in Red-necked Phalaropes (*Phalaropus lobatus*): Feeding limitations, the functional response and water management at Mono Lake, California, USA. *Biological Conservation* 70: 205–210.

Sabrosky, C. W. 1987. Chloropidae. Pp. 1049–1067. *In* J. F. McAlpine, B. V. Peterson, G. E. Shewell, H. J. Teskey, J. R. Vockeroth, and D. M. Wood (coordinators). *Manual of Nearctic Diptera*. Volume 2. Research Branch, Supply & Services Canada, Hull, Quebec, Agricultural Canada Monograph 28.

Sæther, O. A. 1979. Chironomid communities as water quality indicators. *Holarctic Ecology* 2: 65–74.

Sakai, S., M. Kato, and H. Nagamasu. 2000. *Artocarpus* (Moraceae) – gall midge pollination mutualism mediated by a male-flower parasitic fungus. *American Journal of Botany* 87: 440–445.

Sanchez-Arroyo, H. 2007. House fly. University of Florida, Publication Number: EENY-48. http://creatures.ifas.ufl.edu/urban/flies/house_fly.htm [Accessed 10 February 2008.]

Santini, L. 1995a. *Senotainia tricuspis* (Meigen) (Diptera, Sarcophagidae) in littoral Tuscany apiaries. *L'Ape nostra Amica* (1995): 4–10.

Santini, L. 1995b. On the "Apimyiasis" caused by *Senotainia tricuspis* (Meigen) (Diptera, Sarcophagidae) in Central Italy. *Apicoltore Moderno* 86: 179–183.

Sawabe, K., K. Hoshino, H. Isawa, T. Sasaki, T. Hayashi, Y. Tsuda, H. Kurahashi, K. Tanabayashi, A. Hotta, T. Saito, A. Yamada, and M. Kobayashi. 2006. Detection and isolation of highly pathogenic H5N1 avian influenza A viruses from blow flies collected in the vicinity of an infected poultry farm in Kyoto, Japan, 2004. *American Journal of Tropical Medicine and Hygiene* 75: 327–332.

Scheepmaker, J. W. A., F. P. Geels, P. H. Smits, and L. J. L. D. Van Griensven. 1997. Location of immature stages of the mushroom insect pest *Megaselia halterata* in mushroom-growing medium. *Entomologia Experimentalis et Applicata* 83: 323–327.

Scholl, P. J. 1993. Biology and control of cattle grubs. *Annual Review of Entomology* 39: 53–70.

Schumann, H. 1992. Systematische Gliederung der Ordnung Diptera mit besonderer Berücksichtigung der in Deutschland vorkommenden Familien. *Deutsche Entomologische Zeitschrift* 1–3: 103–116.

Sheppard, D. C. 1983. House fly and lesser house fly control utilizing the black soldier fly in manure management systems for caged laying hens. *Environmental Entomology* 12: 1439–1242.

Sheppard, D. C., G. L. Newton, and S. A. Thompson. 1994. A value added manure management system using the black soldier fly. *Bioresource Technology* 50: 275–279.

Sherman, R. A. 2001. Maggot therapy for foot and leg wounds. *International Journal of Lower Extremity Wounds* 1: 135–142.

Sherman, R. A. 2002. Maggot vs conservative debridement therapy for the treatment of pressure ulcers. *Wound Repair and Regeneration* 10: 208–214.

Sherman, R. A. 2003. Cohort study of maggot therapy for treating diabetic foot ulcers. *Diabetes Care* 26: 446–451.

Sherman, R. A., M. J. R. Hall, and S. Thomas. 2000. Medicinal maggots: an ancient remedy for some contemporary afflictions. *Annual Review of Entomology* 45: 55–81.

Sinclair, B. J. 1988. The madicolous Tipulidae (Diptera) of eastern North America, with descriptions of the biology and immature stages of *Dactylolabis montana* (O.S.) and *D. hudsonica* Alexander (Diptera: Tipulidae). *Canadian Entomologist* 120: 569–573.

Sinclair, B. J. 1989. The biology of *Euparyphus* Gerstaecker and *Caloparyphus* James occurring in madicolous habitats of eastern North America, with descriptions of adult and immature stages (Diptera: Stratiomyiidae). *Canadian Journal of Zoology* 67: 33–41.

Sinclair, B. J. 2000. Immature stages of Australian *Austrothaumalea* Tonnoir and *Niphta* Theischinger (Diptera: Thaumaleidae). *Australian Journal of Entomology* 39: 171–176.

Sinclair, B. J. and J. M. Cumming. 2006. The morphology, higher-level phylogeny and classification of the Empidoidea (Diptera). *Zootaxa* 1180: 1–172.

Sinclair, B. J. and S. A. Marshall. 1987. The madicolous fauna of southern Ontario. *Proceedings of the Entomological Society of Ontario* 117: 9–14.

Sivinski, J. 1997. Ornaments in the Diptera. *Florida Entomologist* 80: 142–164.

Skevington, J. H. 2000. Pipunculidae (Diptera) systematics: spotlight on the diverse tribe Eudorylini in Australia. Ph. D. thesis. University of Queensland, Brisbane. 446 pp.

Skevington, J. H. 2001. Revision of Australian *Clistoabdominalis* (Diptera: Pipunculidae). *Invertebrate Taxonomy* 15 (5): 695–761.

Skevington, J. H. 2008. Hilltopping. Pp. 1799–1807. *In* J. L. Capinera (ed) *Encyclopedia of Entomology*. 4. Second Edition. Springer.

Skevington, J. H. and P. T. Dang (eds). 2002. Exploring the diversity of flies (Diptera). *Biodiversity* 3: 3–27.

Skidmore, P. 1985. *The Biology of the Muscidae of the World*. Junk, Dordrecht. Series Entomologica 29. xiv + 550 pp.

Skidmore, P. 1991. Insects of the British cow-dung community. Field Studies Council. *Occasional Publication* 21: 1–160.

Skovmand, O. 2004. Bti: control of mosquitoes and blackflies. *Biocontrol Files* (2004, December issue): 7.

Smith, K. G. V. 1969. Diptera. Family Lonchopteridae. *Handbooks for the Identification of British Insects* 10 (2ai): 1–9.

Somchoudhury, A. K., T. K. Das, P. K. Sarkar, and A. B. Mukherjee. 1988. Life history of three dipteran flies infesting edible mushroom *Pleurotus sajor-caju*. *Environment and Ecology* 6: 463–464.

Sommagio, D. 1999. Syrphidae: can they be used as environmental bioindicators? *Agriculture, Ecosystems and Environment* 74: 343–356.

Sotavalta, O. 1947. The flight-tone (wing-stroke frequency) of insects. *Acta Entomologica Fennica* 4: 1–117.

Sotavalta, O. 1953. Recordings of high wing-stroke and thoracic vibration frequency in some midges. *Biological Bulletin of Woods Hole* 104: 439–444.

Speight, M. C. D. 1986. Attitudes to insects and insect conservation. Pp. 369–385. Proceedings of the 3rd European Congress of Entomology.

Spencer, K. A. 1973. Agromyzidae (Diptera) of economic importance. Series Entomologica 9. xi + 418 pp.

Spencer, K. A. 1990. *Host Specialization in the World Agromyzidae (Diptera)*. Kluwer Academic Publishers, Dordrecht. xii + 444 pp.

Spofford, M. G. and F. E. Kurczewski. 1990. Comparative larvipositional behaviours and cleptoparasitic frequencies of Nearctic species of Miltogrammini (Diptera: Sarcophagidae). *Journal of Natural History* 24: 731–755.

Ssymank, A. and D. Doczkal. 1998. Rote Liste der Schwebfliegen (Diptera: Syrphidae). Pp. 65–72. *In* M. Binot, R. Bless, P. Boye, H. Gruttke, and P. Pretscher (eds). *Rote Liste gefährdeter Tiere Deutschlands*. Schriftenreihe für Landschaftspflege und Naturschutz 55.

Ssymank, A., C. A. Kearns, T. Pape, and F. C. Thompson. 2008. Pollinating flies (Diptera): a major contribution to plant diversity and agricultural production. *Biodiversity* 9: 86–89.

Staerkeby, M. 2001. Dead larvae of *Cynomya mortuorum* (L.) (Diptera, Calliphoridae) as indicators of the post-mortem interval – a case history from Norway. *Forensic Science International* 120: 77–78.

Stalker, H. D. 1956. On the evolution of parthenogenesis in *Lonchoptera* (Diptera). *Evolution* 10: 345–359.

Stark, A. 1996. Besonderheiten der Dipterenfauna Sachsen-Anhalts-eine Herausforderung für den Natur- und Umweltschutz. *Berichte des Landesamtes für Umweltschutz Sachsen-Anhalt, Halle* 1996 (21): 100–108.

Steck, G. 2005. Pea leaf miner, *Liriomyza huidobrensis* (Diptera: Agromyzidae). http://www.doacs.state.fl.us/pi/enpp/ento/peamin.html [Last updated March 2006, accessed 13 April 2007].

Steyskal, G. C. 1987. Pyrgotidae. Pp. 813–816. *In* J. F. McAlpine, B. V. Peterson, G. E. Shewell, H. J. Teskey, J. R. Vockeroth, and D. M. Wood (coordinators). *Manual of Nearctic Diptera*. Volume 2. Research Branch, Supply & Services Canada, Hull, Quebec, Agricultural Canada Monograph 28.

Stireman, J. O., J. E. O'Hara, and D. M. Wood. 2006. Tachinidae: evolution, behavior, and ecology. *Annual Review of Entomology* 51: 525–555.

Struwe, I. 2005. Rhythmic dancing as courtship behaviour in *Limnophora riparia* (Fallén) (Diptera: Muscidae). *Studia dipterologica* 11: 597–600.

Syme, D. A. and R. K. Josephson. 2002. How to build fast muscles: synchronous and asynchronous designs. *Integrative and Comparative Biology* 42: 762–770.

Sze, W. T., T. Pape, and D. K. O'Toole. 2008. The first blow fly parasitoid takes a head start in its termite host (Diptera: Calliphoridae, Bengaliinae; Isoptera: Macrotermitidae). *Systematics and Biodiversity* 6: 25–30.

Szwejda, J. and R. Wrzodak. 2007. Phytophagous entomofauna occurring on carrot and plant protection methods. *Vegetable Crops Research Bulletin* 67: 95–102.

Taylor, G. S. 2004. Revision of *Fergusonina* Malloch gall flies (Diptera: Fergusoninidae) from *Melaleuca* (Myrtaceae). *Invertebrate Systematics* 18: 251–290.

Teskey, H. J. 1976. Diptera larvae associated with trees in North America. *Memoirs of the Entomological Society of Canada* 100: 1–53.

Teskey, H. J. 1981. Vermileonidae. Pp. 529–532. *In* J. F. McAlpine, B. V. Peterson, G. E. Shewell, H. J. Teskey, J. R. Vockeroth, and D. M. Wood (coordinators). *Manual of Nearctic Diptera*. Volume 1. Research Branch, Supply & Services Canada, Hull, Quebec, Agriculture Canada Monograph 27.

Thompson, F. C. and G. E. Rotheray. 1998. Family Syrphidae. Pp. 81–139. *In* L. Papp and B. Darvas (eds). *Contributions to a Manual of Palaearctic Diptera. Volume 3. General and Applied Dipterology*. Science Herald, Budapest. 880 pp.

Thomson, J. R., D. D. Hart, D. F. Charles, T. L. Nightengale, and D. M. Winter. 2005. Effects of removal of a small dam on downstream macroinvertebrate and algal assemblages in a Pennsylvania stream. *Journal of the North American Benthological Society* 24: 192–207.

Tomberlin, J. K., W. K. Reeves, and D. C. Sheppard. 2001. First record of *Chrysomya megacephala* (Diptera: Calliphoridae) in Georiga [sic], U.S.A. *Florida Entomologist* 84: 300–301.

Tondella, M. L., C. H. Paganelli, I. M. Bortolotto, O. A. Takano, K. Irino, M. C. Brandileone, B. Mezzacapa Neto, V. S. Vieira, and B. A. Perkins. 1994. Isolation of *Haemophilus aegyptius* associated with Brazilian purpuric fever, of Chloropidae (Diptera) of the genera *Hippelates* and *Liohippelates*. *Revista do Instituto de Medicina Tropical de São Paulo* 36: 105–109.

Turner, C. E. 1996. Tephritidae in the biological control of weeds. Pp. 157–164. *In* B. A. McPheron and G. J. Steck (eds). *Fruit Fly Pests: A World Assessment of Their Biology and Management*. St. Lucie Press, Delray Beach. xxii + 586 pp.

USDA, Forest Service 1985. Insects of Eastern Forests. *Miscellaneous Publications* 1426: 1–608.

Vail, P. V., J. R. Coulson, W. C. Kauffmann, and M. E. Dix. 2001. History of biological control programs in the United States Department of Agriculture. *American Entomologist* 47: 24–49.

Vaillant, F. 1956. Recherches sur la faune madicole (hygropétrique s.l.) de France, de Corse et d' Afrique de Nord. *Mémoirs du Museum d'Histoire Naturelle Paris* 11: 1–258.

Vaillant, F. 1961. Fluctuations d'une population madicole au cours d'une année. *Verhandlungen der Internationalen Vereinigung fur Theoretische und Angewandte Limnologie* 14: 513–516.

Val, F. C. 1992. Pantophthalmidae of Central America and Panama (Diptera). Pp. 600–610. *In* D. Quintero and A. Aiello (eds). *Insects of Panama and Mesoamerica: Selected Studies*. Oxford University Press, Oxford. 692 pp.

Valentin, A., M. P. Baumann, E. Schein, and S. Bajanbileg. 1997. Genital myiasis (Wohlfahrtiosis) in camel herds of Mongolia. *Veterinary Parasitology* 73: 335–346.

Van Riper C., III, S. G. Van Riper, and W. R. Hansen. 2002. Epizootiology and effect of avian pox on Hawaiian forest birds. *Auk* 119: 949–942.

Vigoreaux, J. O. 2001. Genetics of the *Drosophila* flight muscle myofibril: a window into the biology of complex systems. *Bioessays* 23: 1047–1063.

Vockeroth, J. R. 2002. Introducing the ubiquitous Diptera. *Biodiversity* 3 (4): 3–5.

Wall, R. and D. Shearer. 1997. *Veterinary Entomology*. Chapman and Hall, London. 439 pp.

Wall, R., J. J. Howard, and J. Bindu. 2001. The seasonal abundance of blowflies infesting drying fish in south-west India. *Journal of Applied Ecology* 38: 339–348.

Ward, J. V. 1992. *Aquatic Insect Ecology. 1. Biology and Habitat*. John Wiley and Sons, Inc., New York. 438 pp.

Ward, J. V. 1994. Ecology of alpine streams. *Freshwater Biology* 32: 277–294.

Warwick, W. F. 1989. Morphological deformities in larvae of *Procladius* Skuse (Diptera: Chironomidae) and their biomonitoring potential. *Canadian Journal of Fisheries and Aquatic Science* 46: 1255–1270.

Warwick, W. F. 1991. Indexing deformities in ligulae and antennae of *Procladius* larvae (Diptera: Chironomidae): application to contaminant-stressed environments. *Canadian Journal of Fisheries and Aquatic Science* 48: 1151–1166.

Warwick, W. F. and N. A. Tisdale. 1988. Morphological deformities in *Chironomus, Cryptochironomus* and *Procladius* larvae (Diptera: Chironomidae) from two differentially stressed sites in Tobin Lake, Saskatchewan. *Canadian Journal of Fisheries and Aquatic Science* 45: 1123–1144.

Wasmann, E. 1910. *Modern Biology and the Theory of Evolution*. Herder, St. Louis, MO. 539 pp., 8 pls. [Translated by A. M. Buchanan from E. Wasmann. 1906. *Die moderne Biologie und die Entwicklungstheorie*, Third Edition. Freiburg. 528 pp.].

Waterhouse, D. F. and D. P. A. Sands. 2001. Classical biological control of arthropods in Australia. *ACIAR Monograph Series* 77: 1–560.

Wells, J. D. 1991. *Chrysomya megacephala* (Diptera: Calliphoridae) has reached the continental United States: review of its biology, pest status, and spread around the world. *Journal of Medical Entomology* 28: 471–473.

Westwood, J. O. 1852. Observations on the destructive species of dipterous insects known in Africa under the names of tsetse, zimb and tsaltsalya, and on their supposed connection with the fourth plague of Egypt. *Proceedings of the Zoological Society of London* 18: 258–270.

Wheeler, W. M. 1930. *Demons of the Dust*. W.W. Norton and Company, New York. xviii + 378 pp.

White, I. M. and S. L. Clement. 1987. Systematic notes on *Urophora* (Diptera, Tephritidae) species associated with *Centaurea solstitialis* (Asteraceae, Cardueae) and other palaearctic weeds adventive in North America. *Proceedings of the Entomological Society of Washington* 89: 571–580.

White, P. F. 1981. Spread of the mushroom disease *Verticillium fungicola* by *Megaselia halterata* (Diptera: Phoridae). *Protection Ecology* 3: 17–24.

White, I. M. and M. M. Elson-Harris. 1992. *Fruit Flies of Economic Significance: Their Identification and Bionomics*. CAB International, Wallingford. xii + 601 pp.

Whiteman, N. K., S. J. Goodman, B. J. Sinclair, T. Walsh, A. A. Cunningham, L. D. Kramer, and P. G. Palmer. 2005. Establishment of the avian disease vector *Culex quinquefasciatus* Say 1823 (Diptera: Culicidae) on the Galápagos Islands, Ecuador. *Ibis* 147: 844–847.

Wiederholm, T. 1980. Chironomids as indicators of water quality in Swedish lakes. *Acta Universitatis Carolinae-Biologica* 1978: 275–283.

Wiederholm, T. 1984. Incidence of deformed chironomid larvae (Diptera: Chironomidae) in Swedish lakes. *Hydrobiologia* 109: 243–249.

Wiegmann, B. M., S.-C. Tsaur, D. W. Webb, D. K. Yeates, and B. K. Cassel. 2000. Monophyly and relationship of the Tabanomorpha (Diptera: Brachycera) based on 28S ribosomal gene sequences. *Annals of the Entomological Society of America* 93: 1031–1038.

Wood, D. M. 1981. Axymyiidae. Pp. 209–212. In J. F. McAlpine, B. V. Petersen, G. E. Shewell, H. J. Teskey, J. R. Vockeroth, and D. M. Wood (coordinators). *Manual of Nearctic Diptera*. Volume 1. Research Branch, Supply & Services Canada, Hull, Quebec, Agriculture Canada Monograph 27.

Wood, D. M. and A. Borkent. 1989. Phylogeny and classification of the Nematocera. Pp. 1333–1370. In J. F. McAlpine and D. M. Wood (eds). *Manual of Nearctic Diptera* Volume 3. Research Branch, Supply & Services Canada, Hull, Quebec, Agricultural Canada Monograph 32.

Wood, D. M., P. T. Dang, and R. A. Ellis. 1979. *The Mosquitoes of Canada. Diptera: Culicidae. The Insects and Arachnids of Canada, Part 6*. Agriculture Canada, Hull, Quebec. 390 pp.

Woodley, N. E. and D. J. Hilburn. 1994. The Diptera of Bermuda. *Contributions of the American Entomological Institute* 28: 1–64.

Wooton, R. 2000. From insects to microvehicles. *Nature* 403: 144–145.

World Bank. 1993. *World Development Report 1993 – Investing in Health*. Oxford University Press, Oxford. 344 pp.

World Health Organization. 2007. Malaria fact sheet. http://www.who.int/mediacentre/factsheets/fs094/en/index.html [Accessed 10 February 2008].

Yeates, D. K. 1994. Cladistics and classification of the Bombyliidae (Diptera: Asiloidea). *Bulletin of the American Museum of Natural History* 219: 1–191.

Yeates, D. K. and B. M. Wiegmann. 1999. Congruence and controversy: toward a higher-level classification of Diptera. *Annual Review of Entomology* 44: 397–428.

Yeates, D. K., B. M. Wiegmann, G. W. Courtney, R. Meier, C. Lambkin, and T. Pape. 2007. Phylogeny and systematics of Diptera: two decades of progress and prospects. *Zootaxa* 1668: 565–590.

Yoder, C. O. and E. T. Rankin. 1998. The role of biological indicators in a state water quality management process. *Environmental Monitoring and Assessment* 51: 61–88.

Young, A. M. 1986. Cocoa pollination. *Cocoa Growers Bulletin* 37: 5.

Young, A. M. 1994. *The Chocolate Tree, a Natural History of Cacao*. Smithsonian Institution Press, Washington, London. 200 pp.

Zhang, J. F. 1987. Four new genera of Platypezidae. *Acta Palaeontologica Sinica* 26: 595–603.

Zumpt, F. 1965. *Myiasis in Man and Animals in the Old World*. Butterworth, London. xv + 267 pp.

Zwick, P. 1977. Australian Blephariceridae (Diptera). *Australian Journal of Zoology, Supplementary Series* 46: 1–121.

Chapter 10

BIODIVERSITY OF HETEROPTERA

Thomas J. Henry

Systematic Entomology Laboratory, Plant Science Institute, Agriculture Research Service, United States Department of Agriculture, c/o National Museum of Natural History, Smithsonian Institution, Washington, DC 20013–7012

Insect Biodiversity: Science and Society, 1st edition. Edited by R. Foottit and P. Adler
© 2009 Blackwell Publishing, ISBN 978-1-4051-5142-9

The Heteroptera, or true bugs, currently considered a suborder of the Hemiptera, represent the largest and most diverse group of hemimetabolous insects. Much attention has been devoted to the classification of the Hemiptera in recent years, beginning with the realization that the order Homoptera is paraphyletic, based on both morphological and molecular evidence. As a consequence, a larger, more encompassing order Hemiptera, with the four suborders Auchenorrhyncha, Sternorrhyncha, Coleorrhyncha, and Heteroptera (Wheeler et al. 1993), is currently recognized. Additional studies done more recently now suggest that Auchenorrhyncha is not monophyletic and should be separated into the suborders Fulgoromorpha and Cicadomorpha (Bourgoin and Campbell 2002, Brambila and Hodges 2004). Recent use of the confusing term Prosorrhyncha (e.g., Sorensen et al. 1995, Maw et al. 2000, Schaefer 2003), which includes the Heteroptera and the Coleorrhyncha, has not been widely accepted.

Nevertheless, the suborder Heteroptera is considered a monophyletic group generally defined by the wings lying flat over the body, with the forewings partially sclerotized and partially membranous, a piercing-sucking labium that arises anteriorly on the head, four- to five-segmented antennae, a large well-developed scutellum, paired metathoracic scent glands in adults, and dorsal abdominal scent glands in nymphs, among other specializations (Slater 1982, Schuh and Slater 1995). The Heteroptera are separated into seven infraorders (Štys and Kerzhner 1975), two of which are primarily aquatic (Gerromorpha and Nepomorpha), one semiaquatic (Leptopodomorpha), and the remaining four terrestrial (Enicocephalomorpha, Dipsocoromorpha, Cimicomorpha, and Pentatomomorpha).

With the availability of a number of comprehensive world catalogs, recent estimates of the number of described true bugs have taken on a new level of accuracy. We now have world catalogs for nearly all the major families (e.g., Aradidae, Lygaeoidea, Miridae, Reduviidae, Tingidae). In addition, the appearance of large regional catalogs for North America (Henry and Froeschner 1988), Australia (Cassis and Gross 1995, 2002), and the Palearctic (Aukema and Rieger 1995–2006) has added significant documentation of the world fauna. Catalogs of the last two major groups, the Coreidae (M. Webb, personal communication) and Pentatomidae (D. A. Rider, personal communication), are now well underway. Comprehensive reviews, such as the 'True Bugs of the World' (Schuh and Slater 1995), have provided ready access to additional information on Heteroptera. A recent study on global diversity of the water bugs (Polhemus and Polhemus 2008) raised the aquatic bug numbers considerably. Given these resources, the most recent estimate of the number of described Heteroptera ranges from 38,000 (Schuh and Slater 1995, Schaefer 2003) to 39,300 (Cassis and Gross 1995, 2002). Based on the documentation presented in this chapter and my search of the literature, using Zoological Record (since 1995 for the Enicocephalomorpha to the Cimicomorpha and since 2002 for the Pentatomomorpha), the number of described true bugs is now more than 42,300 (Table 10.1), an increase of more than 3000 species.

The question of how many Heteroptera actually occur on the planet remains conjecture. Speculation on the number of insects present in the world has taken on enormous new proportions in recent years. Figures from 650,000 to 1 million described species given in most textbooks, with bold estimates of 2.5 to 10 million species (e.g., Metcalf 1940, Sabrosky 1952), led most to believe that a minimum of half or more of the taxa had been described. A paper by Erwin (1982) with an estimate of up to 30 million insects globally, based on canopy fogging in Panama, realigned most thinking on insect biodiversity in the world, and since then numerous other papers have appeared offering additional hypotheses, some of which predict 80 million or more species will be discovered (e.g., Stork 1988, Erwin 2004). As more and more areas are explored, particularly some of the designated 'hotspots' (e.g., Mittermeier et al. 2004) revealing huge numbers of new Miridae from Australian (Cassis et al. 2006, G. Cassis and R. T. Schuh, personal communication) and Neotropical forest canopies involving numerous families of Heteroptera taken by T. L. Erwin (T. J. Henry, personal observations), we can expect tremendous increases in the number of species.

In this chapter, I present an overview of the seven infraorders of Heteroptera and all 89 of the currently recognized families. For each family, I provide a brief diagnosis, selected information on their habits and economic importance, references to the key literature, and the known numbers of species. Table 10.1 provides the numbers of genera and species for the Australian, Nearctic, and Palearctic regions compared with the world. In my concluding remarks, I discuss the importance of understanding heteropteran biodiversity from a phylogenetic and economic viewpoint, including their role in conservation biology and global warming.

Table 10.1 Summary of the known number of heteropteran genera and species by family and infraorder for the Australian[1], Nearctic[2], and Palearctic[3] regions and the world[4].

Taxon	Australian Genus	Australian Species	Nearctic Genus	Nearctic Species	Palearctic Genus	Palearctic Species	World Genus	World Species
Enicocephalomorpha:								
Aenictopecheidae	2	2	1	1	1	1	10	20
Enicocephalidae	3	5	4	9	6	15	55	405
Total	**5**	**7**	**5**	**10**	**7**	**16**	**65**	**425**
Dipsocoromorpha:								
Ceratocombidae	1	1	1	4	1	11	8	52
Stemmocryptidae	0	0	0	0	0	0	1	1
Dipsocoridae	1	4	1	2	1	14	5	51
Hypsipterygidae	0	0	0	0	0	0	1	4
Schizopteridae	13	61	4	4	6	9	44	229
Total	**15**	**66**	**6**	**10**	**8**	**34**	**59**	**337**
Gerromorpha:								
Gerridae (Gerroidea)	20	113	8	47	20	99	67	751
Hebridae (Hebroidea)	3	8	3	15	4	24	9	221
Hermatobatidae (Gerroidea)	1	2	0	0	1	2	1	9
Hydrometridae (Hydrometroidea)	1	15	1	6	1	14	7	126
Macroveliidae (Hebroidea)	0	0	2	2	0	0	3	3
Mesoveliidae (Mesovelioidea)	4	13	1	3	2	7	12	46
Paraphrynoveliidae (Hebroidea)	0	0	0	0	0	0	1	2
Veliidae (Gerroidea)	20	176	4	31	11	64	61	962
Total	**49**	**327**	**19**	**104**	**39**	**210**	**161**	**2120**
Nepomorpha:								
Aphelocheiridae (Naucoroidea)	1	6	0	0	1	18	1	78
Belostomatidae (Nepoidea)	2	5	3	17	5	14	9	160
Corixidae (Corixoidea)	5	46	18	136	15	143	35	607
Gelastocoridae (Ochteroidea)	1	47	2	7	1	4	3	111
Helotrephidae (Notonectoidea)	1	12	0	0	0	0	21	180
Naucoridae (Naucoroidea)	8	36	4	29	7	9	37	391
Nepidae (Nepoidea)	5	23	3	13	5	21	15	268
Notonectidae (Notonectoidea)	6	92	3	35	4	50	11	400
Ochteridae (Ochteroidea)	2	29	1	6	1	3	3	68
Pleidae (Notonectoidea)	1	6	2	5	2	6	3	38
Potamocoridae (Naucoroidea)	0	0	0	0	0	0	2	8
Total	**32**	**302**	**36**	**248**	**41**	**268**	**140**	**2309**
Leptopodomorpha:								
Aepophilidae (Saldoidea)	0	0	0	0	1	1	1	1
Leptopodidae (Leptopodoidea)	1	4	1	1	4	11	32	39
Omaniidae (Leptopodoidea)	0	1	0	0	2	2	2	6
Saldidae (Saldoidea)	6	23	11	70	14	99	29	335
Total	**7**	**28**	**12**	**71**	**20**	**113**	**64**	**381**
Cimicomorpha:								
Anthocoridae (Cimicoidea)	12	23	20	73	26	170	71	445
Lyctocoridae (Cimicoidea)	1	1	1	8	1	10	1	27
Lasiochilidae (Cimicoidea)	3	5	2	8	1	1	10	62
Cimicidae (Cimicoidea)	1	1	8	15	5	15	24	110
Curaliidae (Velocipedoidea)	0	0	1	1	0	0	1	1
Joppeicidae (Joppeicoidea)	0	0	0	0	1	1	1	1
Medocostidae (Naboidea)	0	0	0	0	0	0	1	1

(continued)

Table 10.1 *(continued)*.

Taxon	Australian Genus	Species	Nearctic Genus	Species	Palearctic Genus	Species	World Genus	Species
Microphysidae (Microphysoidea)	0	0	4	4	3	27	5	25
Miridae (Miroidea)	91	186	223	1930	397	2808	1300	10,400
Nabidae (Naboidea)	7	22	10	34	10	112	31	386
Pachynomidae (Reduvioidea)	0	0	0	0	1	2	4	15
Plokiophilidae (Cimicoidea)	0	0	0	0	0	0	5	13
Polyctenidae (Cimicoidea)	2	2	1	2	2	3	5	32
Reduviidae (Reduvioidea)	100	226	49	184	145	808	981	6878
Thaumastocoridae (Miroidea)	3	11	1	1	0	0	6	19
Tingidae (Miroidea)	56	147	22	154	61	473	260	2124
Velocipedidae (Velocipedoidea)	0	0	0	0	0	0	1	25
Total	**276**	**624**	**342**	**2414**	**653**	**4430**	**2707**	**20,564**
Pentatomomorpha:								
Acanthosomatidae (Pentatomoidea)	17	45	2	4	9	107	46	184
Alydidae (Coreoidea)	7	16	11	30	26	69	45	254
Aphylidae (Pentatomoidea)	2	3	0	0	0	0	2	3
Aradidae (Aradoidea)	38	143	10	123	28	204	233	1931
Artheneidae (Lygaeoidea)	1	2	2	2	4	16	8	20
Berytidae (Lygaeoidea)	6	7	7	12	13	54	36	172
Blissidae (Lygaeoidea)	9	15	3	28	8	55	50	435
Canopidae (Pentatomoidea)	0	0	0	0	0	0	1	8
Colobathristidae (Lygaeoidea)	1	1	0	0	2	7	23	84
Coreidae (Coreoidea)	43	83	33	88	84	306	267	1884
Cymidae (Lygaeoidea)	4	10	2	10	2	17	9	54
Cryptorhamphidae (Lygaeoidea)	2	4	0	0	0	0	2	4
Cydnidae (Pentatomoidea)	21	83	17	84	38	171	120	765
Cyrtocoridae (Pentatomoidea)	0	0	0	0	0	0	4	11
Dinidoridae (Pentatomoidea)	4	6	0	0	4	19	16	65
Geocoridae (Lygaeoidea)			2	28	7	75	25	274
Hyocephalidae (Coreoidea)	2	3	0	0	0	0	2	3
Heterogastridae (Lygaeoidea)	3	5	1	2	11	24	24	100
Henicocoridae (Lygaeoidea)	1	1	0	0	0	0	1	1
Idiostolidae (Idiostoloidea)	2	3	0	0	0	0	3	4
Largidae (Pyrrhocoroidea)	2	4	3	12	3	8	13	106
Lestoniidae (Pentatomoidea)	1	2	0	0	0	0	1	2
Lygaeidae (*sensu stricto*) (Lygaeoidea)	22	81	13	70	29	59	102	968
Malcidae (Lygaeoidea)	0	0	0	0	2	25	3	29
Megarididae (Pentatomoidea)	0	0	0	0	0	0	1	16
Ninidae (Lygaeoidea)	2	2	1	1	2	5	5	13
Oxycarenidae (Lygaeoidea)	1	4	2	10	19	63	23	147
Pachygronthidae (Lygaeoidea)	6	10	3	7	6	17	13	78
Parastrachiidae (Pentatomoidea)	0	0	0	0	0	0	1	2
Pentatomidae (Pentatomoidea)	134	363	60	222	219	841	900	4700
Phloeidae (Pentatomoidea)	0	0	0	0	0	0	2	3
Piesmatidae (Lygaeoidea)	1	4	1	7	2	19	6	44
Plataspidae (Pentatomoidea)	2	20	0	0	10	104	59	560
Pyrrhocoridae (Pyrrhocoroidea)	3	11	2	10	13	43	33	340
Rhopalidae (Coreoidea)	2	6	10	39	14	69	21	209
Rhyparochromidae (Lygaeoidea)	75	185	54	163	135	564	372	1850
Scutelleridae (Pentatomoidea)	10	22	15	34	38	158	81	450
Stenocephalidae (Coreoidea)	1	1	0	0	1	18	1	16
Termitaphidae (Aradoidea)	1	1	0	0	0	0	2	9

Table 10.1 *(continued)*.

Taxon	Australian		Nearctic		Palearctic		World	
	Genus	Species	Genus	Species	Genus	Species	Genus	Species
Tessaratomidae (Pentatomoidea)	12	18	1	1	12	30	55	240
Thaumastellidae (Pentatomoidea)	0	0	0	0	1	1	1	3
Urostylididae (Pentatomoidea)	0	0	0	0	8	131	11	170
Total	**438**	**1164**	**255**	**987**	**750**	**3279**	**2623**	**16,211**
Grand Total	**822**	**2518**	**675**	**3844**	**1518**	**8350**	**5819**	**42,347**

[1]Based on Cassis and Gross (1995, 2002).
[2]Based on chapters in Aukema and Rieger (1995–2006).
[3]Based on chapters in Henry and Froeschner (1988).
[4]Based on a summary of all sources cited in this chapter.

The text is arranged phylogenetically by infraorder from the most plesiomorphic (i.e., Enicocephalomorpha) to the most derived (i.e., the sister infraorders Cimicomorpha and Pentatomomorpha). Within each infraorder, superfamilies and families are alphabetical. Table 10.1 is arranged phylogenetically by infraorder and alphabetically by family, with the superfamily noted in parentheses for each.

OVERVIEW OF THE HETEROPTERA

The suborder Heteroptera has been separated into seven infraorders and at least 24 superfamilies (Schuh and Slater 1995, Henry 1997a). The following overview is based on currently accepted classifications of the Heteroptera by Štys and Kerzhner (1975), Schuh (1979), Štys (1985), Schuh (1986), Schuh and Štys (1991), W. Wheeler et al. (1993), Schuh and Slater (1995), Henry (1997a), and Polhemus and Polhemus (2008).

EUHETEROPTERA

Infraorder Enicocephalomorpha

Only two families are included in this infraorder. These unusual insects are often called unique-headed bugs because of the peculiar bilobed heads.

The Aenictopecheidae possess a protruding, often inflatable phallus and movable parameres, whereas in the Enicocephalidae (Fig. 10.1) the intromittent organ is reduced and noninflatable and the parameres are immobile, among other characters. These small, cryptic bugs range in size from 2 to more than 15 mm and are thought to prey on other

arthropods. One African species of Enicocephalidae feeds on larvae of the ant *Rhoptromyrex transversiodis* Mayr (Bergroth 1915). Enicocephalids are well known for their swarming behavior, sometimes forming clouds of many thousands of individuals (Usinger 1945; Henry, personal observations in Pennsylvania, USA, and Mexico). About 10 genera and 20 species of Aenictopecheidae and 55 genera and 405 species of Enicocephalidae are known (Štys 1995a, 2002; Zoological Record 2003–2007). Wygodzinsky and Schmidt (1991) monographed the New World fauna, Štys (2002) provided a key to the enicocephalomorphan genera of the world, and Kerzhner (1995a) cataloged the Palearctic fauna.

Infraorder Dipsocoromorpha

This infraorder comprises only five families, all of which contain small, cryptically colored species, ranging in size from only 0.5 mm to about 4.0 mm. Most are found in rotting wood, water margins, or in the ground litter layer and similar conditions of tropical forest canopies. These tiny predatory insects are most abundant in the tropics. As Štys (1995b) noted for the Ceratocombidae, 'the number of undescribed species is enormous'.

Ceratocombids (Fig. 10.2) are characterized by long, slender antennae; two- or three-segmented tarsi on all legs; bristlelike setae on the antennae, head, and tibiae; and a short, distinct fracture at the middle of the costa on the hemelytra. Štys (1983) revised the group and provided keys to genera and higher groups. About eight genera and 51 species are known, most of which are tropical (Cassis and Gross 1995, Zoological Record 1996–2007). In the Nearctic only two genera and four species are known (Henry 1988b); in the

Palearctic only one genus and 11 species have been recorded (Kerzhner 1995b).

The Dipsocoridae (Fig. 10.3) usually possess three-segmented tarsi, long slender antennae with bristlelike setae, and a cuneus-like structure on the apical third of each hemelytron. They inhabit moist litter and rocky habitats along streams, ponds, lakes, and swampy areas, where they prey on coexisting arthropods. Emsley (1969) and Štys (1970) provided the most recent treatments. Two genera and 51 species are known (Cassis and Gross 1995, Zoological Record 1996–2007). Only one genus and two species occur in North America (Henry 1988c) and 2 genera and 12 species in the Palearctic Region (Kerzhner 1995b).

The Schizopteridae (Fig. 10.4) represent the largest family of Dipsocoromorpha with more than 42 genera and about 229 species (Cassis and Gross 1995, Zoological Record 1996–2007). Slater (1982) considered the world fauna largely undescribed and estimated that more than 1200 species exist. These bugs range in size from about 0.8 mm to nearly 2.0 mm and are characterized by their often convex, beetlelike forewings, subequally long first and second antennal segments, ventrally enlarged propleura, and hindlegs modified for jumping (Henry 1988h). Key works for this family are McAtee and Malloch's (1925) revision and Emsley's (1969) monograph, including a catalog of the world fauna.

The remaining dipsocoromorphan families are the Hypsipterygidae and Stemmocryptidae. The former contains only one genus of four species, one from Thailand and two from Africa (Štys 1995c, Redei 2007); the latter, based on a monotypic genus from New Guinea (Štys 1983), until recently, was the latest heteropteran family to be described.

NEOHETEROPTERA

Infraorder Gerromorpha

These are the aquatic true bugs, previously included in the Amphibicorisae (Dufour 1833, Leston et al. 1954). All are surface-inhabiting taxa found in a wide array of habitats from the smallest, often temporary, bodies of fresh water to the open oceans of the world. Eight families placed among four superfamilies are represented by more than 2000 species (Polhemus and Polhemus 2008).

Good general references on Gerromorpha include those by Brooks and Kelton (1967), Menke (1979), Andersen (1982), Schuh and Slater (1995), and Polhemus and Polhemus 2007, 2008), and catalogs by Henry and Froeschner (1988), Aukema and Rieger (1995), and Cassis and Gross (1995).

Gerroidea

The Gerridae (Fig. 10.5), frequently called water striders or pond skaters, largely inhabit open water surfaces, including ponds, lakes, rivers, temporary pools, and even the open oceans (Smith 1988a). They are characterized by the elongate to globular, pile-covered bodies and long legs, usually with subapical claws. Wing dimorphism is common. Water striders range in size from about 1.6 mm to more than 36 mm for *Gigantometra gigas* (China) (Andersen 1982). About 67 genera and 751 species are known (Polhemus and Polhemus 2008).

The family Hermatobatidae, sometimes called seabugs, contains only one genus and nine species (Polhemus and Polhemus 2008). They are small elongate-oval marine bugs, ranging from 2.7 to 4 mm long, with relatively large eyes and strong claws for clinging to rocks. They are found along intertidal zones and are mostly predaceous (Cheng 1977).

The Veliidae (Fig. 10.6), commonly known as riffle bugs, small water striders, or broad-shouldered bugs, are characterized by the oval to more elongate body covered with water-repellent pile, relatively short legs, and multiple types of surface locomotion (Smith 1988c). They are found in both freshwater and saltwater habitats and may be found from mud flats and wet rocks to rapidly flowing streams and rivers (D. Polhemus 1997). Veliids represent the largest gerromorphan family, with 58 genera and 962 species (Polhemus and Polhemus 2008).

Hebroidea

The Hebridae (Fig. 10.7), or velvet water bugs, one of three families in the superfamily Hebroidea, are recognized by their apical claws, well-developed scutellum, and a rostral groove on the underside of the head. Hebrids are found in semiaquatic habitats along shorelines or on floating vegetation. They walk either on the water surface or along submerged plants protected

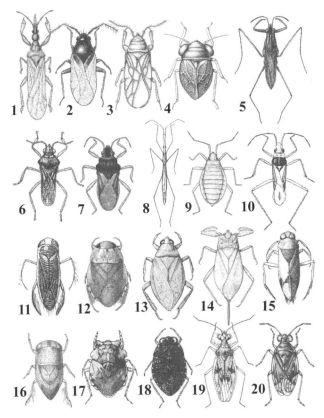

Figs. 10.1–10.20 Enicocephalomorpha, Dipsocoromorpha, Gerromorpha, Nepomorpha, and Letopodomorpha.
10.1, Enicocephalomorpha: *Systelloderes biceps* (Say) [Enicocephalidae]. 10.2–10.4, Dipsocoromorpha: 10.2, *Ceratocombus vagans* McAtee and Malloch [Ceratocombidae]; 10.3, *Cryptostemma uhleri* McAtee and Malloch [Dipsocoridae]; 10.4, *Glyptocombus saltator* Heidemann [Schizopteridae]. 10.5–10.10, Gerromorpha: 10.5, *Gerris marginatus* Say [Gerridae: Gerroidea]; 10.6 *Microvelia ashlocki* Polhemus [Veliidae: Gerroidea]; 10.7, *Hebrus concinnus* Uhler [Hebridae: Hebroidea]; 10.8, *Hydrometra martini* Kirkaldy [Hydrometridae: Hydrometroidea]; 10.9, *Darwinivelia fosteri* Anderson and Polhemus [Mesoveliidae: Mesoveloidea]; 10.10, *Mesovelia mulsanti* White [Mesoveliidae: Mesoveloidea]. 10.11–10.15, Nepomorpha: 10.11, *Sigara hubbelli* (Hungerford) [Corixidae: Corixoidea]; 10.12, *Pelocoris carolinensis* Torre-Bueno [Naucoridae: Naucoroidea]; 10.13, *Belostoma flumineum* Say [Belostomatidae: Nepoidea]; 10.14, *Nepa apiculata* Uhler [Nepinae: Nepidae: Nepoidea]; 10.15, *Notonecta undulata* Say [Notonectidae: Notonectoidea]; 10.16, *Neoplea striola* (Fieber) [Pleidae: Notonectoidea]; 10.17, *Gelastocoris oculatus* (Fabricius) [Gelastocoridae: Ochteroidea]; 10.18, *Ochterus americanus* (Uhler) [Ochteridae: Ochteroidea]. 10.19, 10.20, Leptopodomorpha: 10.19, *Patapius spinosus* (Rossi) [Leptopodidae: Leptopodoidea]; 10.20, *Saldula galapagosana* Polhemus [Saldidae: Saldoidea]. (Figs. 10.1–10.3 after Froeschner 1944; 10.4, Henry 1988h; 10.5, 10.8, 10.11–10.18, Froeschner 1962; 10.6, 10.9, 10.20, Froeschner 1985; 10.7, 10.10, Froeschner 1949; 10.19, Froeschner and Peña 1985).

by their velvety hydrofuge pile (Polhemus and Polhemus 1988a). Nine genera and 221 species are known worldwide (Polhemus and Polhemus 2008).

Macroveliids are a small group of New World semi-aquatic bugs represented by only three genera and three species (Polhemus and Polhemus 2008). They are distinguished from true water striders by having apical claws. Found on surface vegetation, they prey or scavenge on other arthropods. The Paraphrynoveliidae are represented by only two southern African species, ranging in length from 1.7 to 2.4 mm (Andersen 1978, 1982). They resemble wingless hebrids and inhabit wet debris and water-soaked mosses along shorelines.

Hydrometroidea

The Hydrometridae (Fig. 10.8), often referred to as water treaders or water measurers, are the only family in the Hydrometroidea. These slender insects range in length from about 2.7 mm to nearly 22 mm and have long, slender legs and an elongate head. They are found primarily on emergent or floating vegetation, but some are found in more terrestrial habitats or on moist rock faces above streams or pools (Smith 1988b). About seven genera and 126 species are known worldwide (Polhemus and Polhemus 2008).

Mesovelioidae

Mesoveliids (Figs. 10.9, 10.10), often called pond tread-ers or pondweed bugs, are the sole representatives of the superfamily Mesoveloidea. These predatory bugs are small to medium sized, ranging from about 1.2 to 4.2 mm. They occur in diverse habitats, including open ponds and lakes to leaf litter on forest floors, water-soaked mosses, and seeping rock faces. Mesoveliidae are considered the sister group of all other Gerromor-pha (Andersen 1982). Andersen and Polhemus (1980) provided a world checklist and Andersen (1982) gave a detailed morphological and phylogenetic review. Twelve genera and 46 species are known (Polhemus and Polhemus 2008).

PANHETEROPTERA

Infraorder Nepomorpha

The water bugs belonging to this infraorder include the taxa previously placed in the Hydrocorisae (Dufour 1933, Leston et al. 1954). They are characterized by short antennae that are concealed in part or entirely by the eyes. All are predatory, except for some Corixidae. Most can inflict painful bites. Except for the riparian Gelastocoridae and Ochteridae, all are aquatic and have legs modified for swimming (Schuh and Slater 1995).

Corixoidea

Corixids (Fig. 10.11), or water boatmen, ranging in length from 1.5 to 16 mm, are distinguished by their elongate-oval form, short unsegmented labium, one-segmented front tarsi, and short antennae. They swim dorsally using oarlike hind legs (Polhemus et al. 1988a). Most species inhabit fresh water but others have adapted to high concentrations of salt (Scudder 1976), where they forage for algae, protozoa, and metazoa, as well as other prey such as mosquito larvae and brine shrimp (Polhemus et al. 1988a). Thirty-five genera and 607 species are known (Polhemus and Polhemus 2008). Hungerford's (1948) revision of the New World fauna is the single most important reference of this family; Jansson (1986) treated the European corixids.

Naucoroidea

The Aphelocheiridae are a relatively small family of naucorid-like bugs, represented by only one genus containing approximately 78 species (Polhemus and Polhemus 2008). Members of this family represent one of the few insect groups that live their entire lives under-water, including mating, through the use of a plastron respiration system (Polhemus and Polhemus 1988a). The Potamocoridae, like the aphelocheirids, are a small group, previously considered a subfamily of the Nau-coridae (Štys and Jansson 1988). Two genera and eight species are known. Aphelocheirids are found primar-ily in the Old World tropics, whereas potamocorids are known only from the Neotropics. The creeping water bugs, or Naucoridae (Fig. 10.12), represent the largest family of the Naucoroidea, with approximately 37 genera and 391 species found in all zoogeographic regions (Polhemus and Polhemus 2008). They are ovate, strongly dorsoventrally flattened, 5 to 20 mm long, and have raptorial forelegs for capturing prey. They occur in both flowing and still water where they live among submerged plants. Like many nepomor-phans, these bugs can inflict a painful bite if carelessly handled (Polhemus and Polhemus 1988b).

Nepoidea

The Belostomatidae (Fig. 10.13), or giant water bugs, are the largest of the true bugs, ranging in length from 9 mm to more than 110 mm. Ovate and dorsoventrally flattened, they possess powerful raptorial forelegs, and breathe through straplike appendages at the tip of the abdomen. All are voracious predators that can attack prey many times their size, subduing their victims with a powerful hydrolytic enzyme (J. Polhemus et al. 1988b). Eggs of most belostomatids are laid on emergent vegetation and other objects, but species of *Abedus* and *Belostoma* deposit their eggs on the backs of males (Smith 1976). The family classification was established by Laucke and Menke (1961). Nine genera and about 160 species are known (Polhemus and Polhemus 2008).

The Nepidae (Fig. 10.14), or waterscorpions, are a small group of dull brown, slender leaflike or sticklike insects, with raptorial forelegs and a long posterior respiratory siphon. Ranging in size from about 15 to 45 mm, nepids are found in fast- or slow-moving water, but prefer the latter where they wait submerged below the surface on floating plants and other debris for prey (D. Polhemus 1988). Fifteen genera and 268 species are known in the world (Polhemus and Polhemus 2008).

Notonectoidea

The Notonectidae (Fig. 10.15), or back swimmers, are a widespread group of elongate, medium-sized bugs ranging from 5 to 15 mm long. They are ventrally flattened and dorsally convex and have oarlike hind legs for swimming upside down. Backswimmers readily fly and often invade swimming pools where they can become a nuisance with the potential of inflicting painful bites to unwary swimmers (Polhemus and Polhemus 1988b). My only painful heteropteran bite – one that became extremely tender and throbbed for several days – came from naively holding a notonectid close-handed, while general collecting (about age 12) in submergent vegetation. Eleven genera and 400 species are known (Polhemus and Polhemus 2008).

The Pleidae (Fig. 10.16), or pygmy backswimmers, previously placed with the backswimmers until Esaki and China (1928) gave support for family status, are a small group of only three genera and 38 species. Like notonectids, these tiny predaceous insects range from 2 to 3 mm long and swim on their backs. They are strongly convex, possess coleopteriform or helmet-like hemelytra, and have slender hind legs, often set with long hairs (D. Polhemus 1988, Schuh and Slater 1995). The closely related Helotrephidae also were placed in the Notonectidae until given family rank by Esaki and China (1927), based on several unique characters, including the fused head and pronotum. About

21 genera and 180 species are known (Polhemus and Polhemus 2008).

Ochteroidea

Gelastocoridae (Fig. 10.17) are frequently called toad bugs because of their stout, oval bodies, usually roughened or warty upper surface, and ability to hop. They measure 6.0 to 10 mm long and occur along damp open areas near streams, ponds, lakes, and muddy ditches (Cassis and Gross 1995). Members of the genus *Nerthra* have been found in rotting logs and leaf litter far from water. The group occurs primarily in the Southern Hemisphere and is absent from most of the Palearctic. Todd (1955) revised the family. Three genera and 111 species are known (J. Polhemus 1995a, Polhemus and Polhemus 2008).

Ochteridae (Fig. 10.18), known as velvety shore bugs, are compact oval bugs, with nonraptorial forelegs and a velvety or soft-textured dorsum, often with scattered pale spots, frosted bluish areas, and golden setae (Polhemus and Polhemus 1988c, J. Polhemus 1995b). They live in damp areas along ponds, lakes, and streams where they feed on small arthropods. Three genera and 68 species are known (Polhemus and Polhemus 2008).

Infraorder Leptopodomorpha

This infraorder comprises four families separated into the superfamilies Leptopodoidea and Saldoidea, as defined by Schuh and Polhemus (1980).

Leptopodoidea

The Leptopodidae (Fig. 10.19) are generally elongate-oval, fast-moving predatory bugs measuring more than 2 mm long, usually with prominent or subpedunculate eyes. Many are found along streams and other wet areas, but some, such as the Palearctic *Patapius spinosus* (Rossi) introduced into Chile (Froeschner and Peña 1985) and USA (Sissom and Ray 2005), can be found far from water in semiarid conditions. Ten genera and 39 species are known (Polhemus and Polhemus 2008).

The Omaniidae have been called intertidal dwarf bugs. They are small, measuring only 1.15–1.59 mm long, with large prominent eyes, coleopteriform hemelytra, and adhesive pads on the hind coxae. Only two genera and six species are known (Cobben 1970, Schuh and Slater 1995, Polhemus and Polhemus 2008).

Saldoidea

The family Aepophilidae is represented by only one genus and one species, *Aepophilus bonnairei* Signoret, often called the marine bug. These tiny predatory intertidal insects measure only about 2 mm, live among rocks, and survive below high tide with the aid of tiny hairs forming a plastron that traps a layer of air (Schuh and Slater 1995).

The Saldidae (Fig. 10.20), or shore bugs, represent the largest family of Lepopodomorpha with 29 genera and 335 species (Schuh et al. 1987, Polhemus and Polhemus 2008). They are fast-moving, medium-sized, oval to elongate bugs, with long antennae and four to five closed cells in the hemelytral membrane. These primarily predatory bugs are found mainly in littoral habitats along stream, lake, and marine shorelines, but can be found in numerous other situations, including wet vertical rock faces or in dry areas far from water (J. Polhemus 1988).

Infraorder Cimicomorpha

This infraorder is separated into seven superfamilies and 17 families, including the two largest heteropteran families, the Miridae and Reduviidae (Schuh and Štys 1991), and the most recently described heteropteran family, the Curaliidae (Schuh et al. 2008).

Cimicoidea

This superfamily contained only four families before Schuh and Štys (1991) proposed elevating the subfamilies of Anthocoridae to family status based on their hypothesis that shared traumatic insemination – where the male pierces the female body cavity during insemination – has evolved only once and, thus, supporting the monophyletic grouping with the 'nontraumatic' Lasiochilidae as sister to Anthocoridae, Cimicidae, Lyctocoridae, Plokiophilidae, and Polyctenidae. This classification

has not gained general acceptance, however, and was not followed in the recent Australian (Cassis and Gross 1995) and Palearctic (Péricart 1996) catalogs. More recently, Tatarnic et al. (2006) presented another example of traumatic insemination in the Miridae. Herein, I follow Schuh and Štys (1991), with the realization that this classification is still in a state of flux. Eighty-two genera and 534 species of Anthocoridae (*sensu lato*) are known (Cassis and Gross 1995, Zoological Record 1996–2007).

Anthocoridae (Fig. 10.21), frequently called flower bugs or minute pirate bugs, are predatory, ranging in size from about 1.5 mm to 4.5 mm or more. Flower bugs are characterized by the fusiform antennal segments III and IV, the straight or forward-curving ostiolar canal, and the left paramere often grooved to accept the vesica. These bugs exhibit traumatic insemination (Péricart 1972, Schuh and Stys 1991). Many anthocorids are of great agricultural importance in biological control programs, especially the widely occurring genus *Orius* Wolff (Ryerson and Stone 1979, Lattin 1999, Hernández and Stonedahl 1999). Schuh and Slater (1995) provided a key to the tribes and Lattin (1999) reviewed the bionomics.

The Cimicidae (Fig. 10.22), often called bed bugs, are broadly oval, dorsoventrally flattened, wingless ectoparasites of birds and mammals. Cimicids range in size from about 4 mm to 12 mm. Of the 24 genera and 110 known species (Cassis and Gross 1995, Zoological Record 1996–2007), only two, the tropical bed bug *Cimex hemipterus* Fabricius and the human bed bug *C. lectularius* Linnaeus, are permanently associated with humans (Wheeler 1982). Most, however, are associated with birds and bats. All species exhibit traumatic insemination. Usinger (1966) monographed the family, and Ryckman et al. (1981) provided a world checklist and bibliography.

Lasiochilidae are a small family containing 10 genera and about 62 species, with most included in the genus *Lasiochilus*. They are about 3.0 to 4.0 mm long and are characterized by a posteriorly curved metathoracic scent-gland channel, a sickle-shaped left paramere, and nontraumatic insemination (Schuh and Slater 1995). These bugs are considered predators of various insect larvae under bark and in bracket fungi (Kelton 1978).

The family Lyctocoridae (Fig. 10.23) contains 27 species, all of which are placed in the genus *Lyctocoris*. Members of this predatory family resemble Anthocoridae and Lasiochidae, from which they were

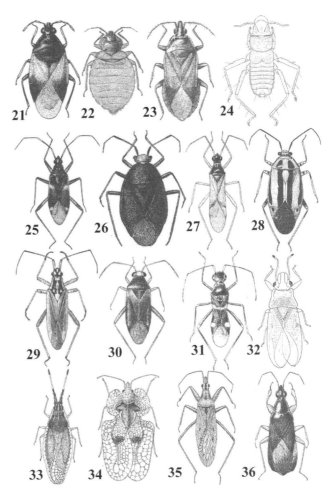

Figs. 10.21–10.36 Cimicomorpha: 10.21, *Orius insidiosus* (Say) [Anthocoridae: Cimicoidea; 10.22, *Cimex lectularius* Linnaeus [Cimicidae: Cimicoidea]; 10.23, *Lyctocoris campestris* (Fabricius) [Lyctocoridae: Cimicoidea]; 10.24, *Hesperoctenes eumops* Ferris and Usinger [Polyctenidae: Cimicoidea]; 10.25, *Fulvius imbecilis* (Say) [Cylapinae: Miridae: Miroidea]; 10.26, *Bothynotus modestus* Wirtner [Deraeocorinae: Miridae: Miroidea]; 10.27, *Dicyphus agilis* (Uher) [Bryocorinae: Miridae: Miroidea]; 10.28, *Poecilocapsus lineatus* (Fabricius) [Mirinae: Miridae: Miroidea]; 10.29, *Leptopterna dolabrata* (Linnaeus) [Mirinae: Miridae: Miroidea]; 10.30, *Ceratocapsus modestus* (Uhler) [Orthotylinae: Miridae: Miroidea]; 10.31, *Cyrtopeltocoris illini* Knight [Phylinae: Miridae: Miroidea]; 10.32, *Xylastodorus luteolus* Barber [Thaumastocoridae: Miroidea]; 10.33, *Atheas mimeticus* Heidemann [Tingidae: Miroidea]; 10.34, *Corythuca ciliata* (Say) [Tingidae: Miroidea]; 10.35, *Nabis americoferus* Carayon [Nabidae: Naboidea]; 10.36, *Pagasa fusca* (Stein) [Nabidae: Naboidea]. (10.21–10.23, 10.25–10.31 after Froeschner 1949; 10.24, Froeschner 1988e; 10.32 redrawn after Barber 1920; 10.33–10.36, Froeschner 1944).

recently separated (Schuh and Štys 1991). A modified vesica, rather than the left paramere, is used during traumatic insemination (Schuh and Slater 1995). This primarily north temperate group of eight species includes one adventive species in the Nearctic (Henry 1988a) and 20, in the Palearctic (Péricart 1996).

The family Plokiophilidae, often called web lovers, includes small bugs about 1.2 to 3.0 mm long. They comprise about 13 species separated into five genera (Štys 1991, Carpintero and Dellapé 2005, Schuh 2006) placed in two subfamilies. Mating involves traumatic insemination. The Embiophilinae, containing only one genus and three species, are associated with the Embioptera, whereas all Plokiophilinae are associated with spiders, where they feed on insects entrapped in their host's webs (Eberhard et al. 1993).

The Polyctenidae (Fig. 10.24), or the bat bugs, are a small group of wingless, viviparous ectoparasites comprising five genera and 32 species. Bat bugs range in size from 3 to 5 mm and are dorsoventrally flattened, lack compound eyes and ocelli, possess setal combs or ctenidia, and have traumatic insemination (Slater 1982, Cassis and Gross 1995, Schuh and Slater 1995). Polyctenids never leave their bat hosts and transfer to other individuals only by direct contact (Marshall 1982). Ferris and Usinger (1939), Maa (1964), and Ueshima (1972) provided key revisions of the family and Ryckman and Sjogren (1980) cataloged the world fauna.

Joppeicoidea

Only one family is included in this superfamily. The family Joppeicidae is represented by one species, *Joppeicus paradoxus* Puton, found only in Israel, Egypt, and Sudan. These tiny anthocorid-like bugs, about 3.0 mm long, possess several unique wing, abdominal, and genital characters that make it difficult to place them phylogenetically (Davis and Usinger 1970). They have been found under stones and other objects, in caves, under bark, and with bat bugs (Cimicoidea: Polyctenidae) (Štys 1971).

Microphysoidea

The Microphysidae are a small group of predatory bugs, only about 1.5 to 3.0 mm long, that resemble certain species of Anthocoridae. Five genera and 25 species are known (Schuh and Slater 1995, Zoological Record 1996–2007). The Palearctic species are all strongly sexually dimorphic, with the females usually brachypterous, whereas the only two endemic Nearctic species, *Chinaola quercicola* Blatchley and *Mallochiola gagates* (McAtee and Malloch), are similar in both sexes. Microphysids are often found on the bark of trees (Péricart 1972) or are associated with lichens and mosses growing on the trunks and larger branches of trees and shrubs (Henry 1988e). *Chinaola quercicola* is associated with foliose lichens growing on *Juniperus virginiana* (Cupressaeae) in South Carolina and Virginia (Wheeler 1992).

Miroidea

The Miroidea contains three families: Miridae, Thaumastocoridae, and Tingidae. The Thaumastocoridae were first placed in the Cimicomorpha by Drake and Slater (1957) and later were considered the sister group of the Tingidae by Schuh and Štys (1991), who included both in the Mirioidea with the Miridae.

The Miridae, or plant bugs, represent the largest family in the Heteroptera, with more than 1300 genera and 10,040 species, or about 25% of the true bugs. They are currently separated into eight subfamilies (Schuh 1995, Cassis and Gross 1995, Cassis et al. 2006). Since the appearance of the Carvalho catalog (1957–1960), considerable attention has been given to the family. Henry and Wheeler's (1988) catalog of the North American fauna and Cassis and Gross's (1995) catalog of the Australian fauna were followed by Schuh's (1995) world catalog, now available online at http://research.amnh.org/pbi/catalog/. More recently, Kerzhner and Josifov (1999) cataloged the Palearctic fauna with great detail devoted to distribution and key literature.

Carvalho's (1955) now largely outdated keys to subfamilies, tribes, and genera are a remarkable attempt to identify the Miridae on a world basis. Since then, changes in the classification and the emergence of numerous workers have made it difficult to update these keys on such a large scale. Schuh and Slater (1995) provided the most recent key to subfamilies, largely following the classification supported by Schuh (1974, 1976). The most recent works on a regional basis are found in the catalogs cited above.

Mirids range in size from about 1.5 mm in certain species of Bryocorinae, Orthotylinae, and Phylinae, especially brachypterous forms, to 15 mm or more in some of the Neotropical species of Restheniini (Mirinae). Many plant bugs are brightly colored red, orange, and yellow, often with spots and stripes, whereas the majority are less spectacular gray, brown, or black, blending in remarkably well with their surroundings (Henry and Wheeler 1988, Wheeler and Henry 2004). Numerous species, such as the cotton fleahopper, *Pseudatomoscelis seriatus* (Reuter), and lygus bugs, *Lygus* spp., in North America, cocoa capsids, *Distantiella theobroma* (Distant) and *Sahlbergella singularis* Haglund, in Africa, and tea bugs, *Helopeltis* spp., in Asia and India, are major agricultural pests (Wheeler 2000a, 2001), whereas many others are predatory and often beneficial in agroecosystems (Wheeler 2000b, 2001).

The habits and biology of the Miridae have been documented in numerous, widely scattered publications. Certain classical treatments, such as those by Kullenberg (1944), Southwood and Leston (1959), Putshkov (1966), and Ehanno (1983–1987), represent important sources of biological information. These and numerous other references have been examined by Wheeler (2001), who synthesized data on biology, hosts, and habits, bringing the knowledge of the family to a new level. The subfamily Isometopinae, often called jumping tree bugs, contains mostly small bugs about 2 to 3 mm, but one giant from Sumatra, *Gigantometopus rossi* Schwartz and Schuh (1990), which is nearly 7.0 mm. About 175 species are known (Herczek 1993, Schuh 1995, Akingbohungbe 1996). They are characterized by their often holoptic eyes, unusually shaped, often anteriorly flattened head, saltatorial hind legs, and the possession of ocelli, a character unique and considered plesiomorphic in the Miridae. Isometopines, once thought to be associated with mosses and lichens, are now known to be specialized predators of scale insects (Hemiptera: Coccoidea) (Wheeler and Henry 1978). Henry (1980) gave a key to the New World genera and Ghauri and Ghauri (1983), a key to world genera. More recently, Herczek (1993) analyzed relationships and Akingbohungbe (1996) revised the African, European, and Middle Eastern faunas.

The Cylapinae (Fig. 10.25) are a small subfamily of primarily tropical bugs characterized by long, slender, apically toothed claws and an unusually long labium extending well onto the abdomen. This subfamily is separated into three tribes (Carvalho 1957, Kerzhner and Josifov 1999), though the Bothriomirini and Fuliviini have been treated as synonyms of Cylapini (Schuh 1995). The New World genus *Cylapus*, quick-moving bugs with unusual stalked eyes, are often associated with fungi on rotting logs and have been long thought to be predatory. Wheeler and Wheeler (1994), however, observed that the gut contents of *Cylapus tenuis* Say from New York and *Cylapus* sp. from Peru contained pyrenomycete spores, seemingly demonstrating that at least some members of this genus are mycophagous. Members of the fulviine genus *Fulvius* Stål, on the other hand, are often found under loose bark and are probably predators of coexisting arthropods (e.g., Kelton 1985, Wheeler 2001). The recently described *Rhyparochromomiris femoratus* Henry and Paula, remarkable in being the only mirid with swollen rhyparochromid-like forefemora, strongly resembles certain myrmecomorphic herdoniine Mirinae (Henry and Paula 2004).

The subfamily Deraeocorinae (Fig. 10.26), the fifth largest subfamily with more than 100 genera separated into six tribes, is recognized by the distinct pronotal collar, usually punctate dorsum, and claws with setiform parempodia and cleft or toothed bases (Schuh and Slater 1995). Members of the genus *Deraeocoris* Kirschbaum are well known for their predatory habits and may be useful in biological control programs. For example, Wheeler et al. (1975) showed that *D. nebulosus* (Uhler) in North America is an effective predator of numerous ornamental pests, including lace bugs, white flies, and mites. The Termatophylini, recently revised by Cassis (1995), are unusual in resembling certain species of Anthocoridae. Most members of the tribe are specialized thrips predators, although a few have been associated with lepidopteran larvae (Cassis 1995). Wheeler (2001) documented the feeding behavior of three New World thrips specialists in the genus *Termatophylidea*. Ferreira (2000, 2001) revised and provided a key to the 17 genera of the New World tribe Clivinematini, many of which prey on ensign scales (Ortheziidae) (Wheeler 2001). Members of the Old World hyaliodine genus *Stethoconus* are tingid specialists, including *S. japonicus* Schumacher, recently reported in North America, which feeds on the azalea lace bug *Stephanitis pyrioides* (Scott) (Henry et al. 1986).

The Bryocorinae (Fig. 10.27), representing the fourth largest subfamily, are a mixed group comprising three tribes, with about 200 genera (Schuh and Slater 1995). The tribe Bryocorini contains only five genera, all of which are restricted to ferns. The largest and most widespread genus, *Monalocoris*, contains 15 species and occurs in all zoogeographic regions. The tribe Eccritotarsini is largely a New World group recognized by large, disc-shaped pulvilli; the largest genus, *Eccritotarsus*, which undoubtedly is not monophyletic, contains about 90 species, most of which are distinguished by the male genitalia and distinct dorsal color pattern of many species (Carvalho and Schaffner 1986, 1987). Eccritotarsines produce characteristic chlorosis or leaf spotting on their hosts, making many of them potentially serious ornamental and crop pests. In the USA, *Halticotoma valida* Townsend severely discolors the foliage of certain species of ornamental *Yucca* spp., often killing entire plants (Wheeler 1976a). Members of the genus *Tenthecoris* are well-known pests of orchids (Hsaio and Sailer 1947), and *Pycnoderes quadrimaculatus* (Guérin-Méneville) causes serious injury to beans and other garden crops (Wehrle 1935, Wheeler 2001). Within the Dicyphini, subtribe Odoniellina, *Distantiella theobroma* (Distant) and *Sahlbergella singularis* Haglund are among the most devastating cocoa pests (Entwistle 1972, Wheeler 2001), whereas in the subtribe Dicyphina, *Macrolophus melanotoma* (Costa) is being used effectively to control whiteflies in European greenhouses (Schelt et al. 1996, Wheeler 2001). Many members of this subtribe live on glandular-hairy plants where they prey on insects entrapped on the viscid surfaces of stems, leaves, and flower clusters (Henry 2000d, Wheeler 2001).

The Mirinae (Figs. 10.28, 10.29) represent the largest subfamily, comprising more than 300 genera and six tribes. Members of this group are characterized by the distinct pronotal collar and widely divergent,

subfamilies of this primarily tropical group are known (Carayon and Villiers 1968).

The Reduviidae (Figs. 10.37–10.41), or assassin bugs, represent the second largest heteropteran family. About 981 genera and more than 6878 species (Froeschner and Kormilev 1989, Maldonado 1990, Cassis and Gross 1995, Putshkov and Putshkov 1996, Zoological Record 1996–2007) are known, with most occurring in the tropics. These predatory bugs range in size from a few millimeters to more than 40 mm for species of the New World harpactorine genus *Arilus* or the heavy-bodied African species of the ectrichodiine genus *Centraspis*. Most reduviids can inflict painful bites if handled carelessly, with the exception of the Triatominae that are capable of painlessly taking blood meals from their vertebrate hosts, including humans, because of the anesthetizing action of the bug's saliva (Lent and Wygodzinsky 1979). The suprageneric classification has been in serious need of study. Putshkov and Putshkov (1985–1989) listed 21 subfamilies, Maldonado (1990) recognized 25, Cassis and Gross (1995) documented 26, and, most recently, Putshkov and Putshkov (1996) listed 23. The Phymatinae, long treated as a separate family (Froeschner and Kormilev 1989, Maldonado 1990), are now considered a subfamily within the Reduviidae (Cassis and Gross 1995, Schuh and Slater 1995, Putshkov and Putshkov 1996). More recently, Weirauch (2008) gave strong support for 21 subfamilies based on the first cladistic analysis of the family using 162 morphological characters.

Of the major subfamilies, the Ectrichodiinae contains some of the largest species. They are usually stout bodied, bright red and black, and sexually dimorphic with females often brachypterous. Approximately 300 species occur circumtropically. All are thought to be obligate milliped predators (Louis 1974). Cook (1977) treated the Asian species and Dougherty (1980), the Neotropical ones.

The Emesinae, or thread-legged bugs, are generally delicate, slender bugs, with long, thread-thin legs. They range from about 3 mm in some species of *Empicoris* to some of the quite long species of *Berlandiana* and *Emesaya*, measuring more than 36 mm. Emesines are mostly nocturnal predators, with some known to live in spider webs (e.g., Hickman 1969), whereas several others feed on mosquitoes and sand flies (Hribar and Henry 2007). Wygodzinsky's (1966) profusely illustrated monograph is the most important source of information on emesine biology and systematics.

The Harpactorinae (Figs. 10.37, 10.38) represent the largest and one of the most diverse reduviid subfamilies with more than 288 genera and more than 2000 species worldwide. They range from the stout, oval-bodied Apiomerini, with modified forelegs, to the large wheel bug, *Arilus cristatus* (Say), named for the large crest-shaped wheel on the pronotum. Certain genera, such as the genus *Graptocleptes*, form a mimicry complex with certain species of wasps. Harpactorines are general predators with many specializing on insect larvae. Barber (1920) and Moul (1945) documented more than 25 different prey for the wheel bug; preserved prey in the USNM collection include a grasshopper, *Melanoplus* sp. (Orthoptera: Acrididae); the Carolina mantid, *Stagmomantis carolina* (Johannson) (Mantodea: Mantidae); a walkingstick, *Diapheromera femorata* (Say) (Phasmatodea: Heteronemiidae); a paper wasp, *Polistes* sp. (Hymenoptera: Vespidae); and honeybees, *Apis mellifera* Linnaeus (Hymenoptera: Apidae).

The Phymatinae (Fig. 10.40), or ambush bugs, until relatively recently were given their own family status, but recent acceptance of evidence for subfamily status within the Reduviidae (Carayon et al. 1958) is now followed by most workers (e.g., Putshkov and Putshkov 1996). Ambush bugs are stout bodied, often tuberculate, and well-camouflaged diurnal predators, frequently lying in wait within flowers or flower clusters to grab prey with their strongly modified, raptorial forelegs. About 26 genera and 281 species are known (Froeschner and Kormilev 1989). The most recent taxonomic treatments are by Kormilev (1962) and Maa and Lin (1956); Froeschner and Kormilev (1989) cataloged the world fauna and gave keys to species for all genera, except *Lophoscutus* and *Phymata*.

The Reduviinae, with about 138 genera and nearly 1000 species, are a cosmopolitan group recognized in part by the presence of ocelli and the absence of a discal cell on hemelytra (Maldonado 1990, Cassis and Gross 1995). Weirauch (2008) has shown that this subfamily is polyphyletic, indicating significant changes will be forthcoming. Most of the taxa now included in the subfamily are generalist predators, but certain species are associated with stored products or humans, such as the masked hunter, *Reduvius personatus* (Linnaeus), that occasionally bites people and has acquired the popular name 'kissing bug'.

The Salyavatinae are a small subfamily of medium-sized, mostly Old World assassin bugs, with only 15 genera and 99 species (Maldonado 1990). Little

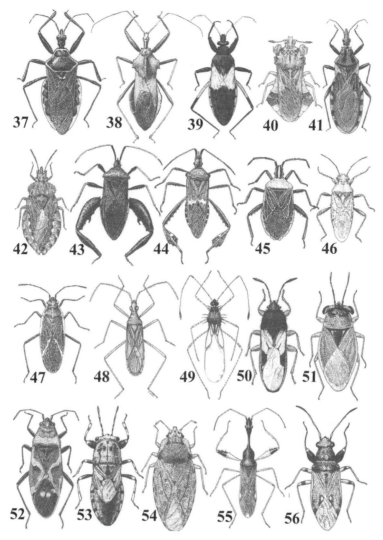

Figs. 10.37–10.56 Cimicomorpha and Pentatomomorpha. 10.37–10.41. Cimicomorpha. 10.37, *Apiomerus crassipes* (Fabricius) [Harpactorinae: Reduviidae: Reduvioidea]; 10.38, *Arilus cristatus* (Say) [Harpactorinae: Reduviidae: Reduvioidea]; 10.39, *Microtomus purcis* (Drury) [Micotominae: Reduviidae: Reduvioidea]; 10.40, *Phymata pennsylvanica* Handlirsch [Phymatinae: Reduviidae: Reduvioidea]; 10.41, *Triatoma sanguisuga* (Leconte) [Triatominae: Reduviidae: Reduvioidea]. 10.42–10.56, Pentatomomorpha. 10.42, *Aradus acutus* Say [Aradidae: Aradoidea]; 10.43, *Acanthocephala femorata* (Fabricius) [Coreinae: Coreidae: Coreoidea]; 10.44, *Leptoglossus phyllopus* (Linnaeus) [Coreinae: Coreidae: Coreoidea]; 10.45, *Chelinidea vittiger* Uhler [Coreinae: Coreidae: Coreoidea]; 10.46, *Arhyssus lateralis* (Say) [Rhopalinae: Rhopalidae: Coreoidea]; 10.47, *Boisea trivittata* (Say) [Serinethinae: Rhopalidae: Coreoidea]; 10.48, *Dicranocephalus insularis* (Dallas) [Stenocephalidae: Coreoidea]; 10.49, *Pronotacantha annulata* Uhler [Berytidae: Lygaeoidea]; 10.50, *Blissus leucopterus* (Say) [Blissidae: Lygaeoidea]; 10.51, *Geocoris punctipes* (Say) [Geocoridae: Lygaeoidea]; 10.52, *Lygaeus kalmii* Stål [Lygaeidae: Lygaeoidea]; 10.53, *Phlegyas abbreviatus* (Uhler) [Pachygronthidae: Lygaeoidea]; 10.54, *Parapiesma cinereum* (Say) [Piesmatidae: Lygaeoidea]; 10.55, *Myodocha serripes* Olivier [Myodochinae: Rhyparochromidae: Lygaeoidea]; 10.56, *Pseudopachybrachius basalis* (Dallas) [Myodochinae: Rhyparochromidae: Lygaeoidea]. (10.37–10.41 after Froeschner 1944; 10.42–10.47, 10.50–10.56, Froeschner 1942; 48, Froeschner 1985; 10.49, Froeschner and Henry 1988.)

information is available on these unusual, often spined bugs, except for the New World *Salyavata variegata* Amyot and Serville, popularly known as the 'fishing bug' (McMahon 2005) because of the way it captures its primary prey, species of *Nasutitermes* termites. The well-camouflaged immatures use previously fed-upon termite carcasses to bait and capture new termite prey from their nests (McMahon 1983).

The Stenopodainae are mostly tropical bugs with about 114 genera and more than 720 species (Maldonado 1990). They are dull colored, with a large closed cell on the hemelytra. Most apparently they are ground dwelling and not often collected, except in Malaise traps or at lights (personal observations).

The Triatominae (Fig. 10.41), or conenoses, though not a large group (about 116 species), are of great medical importance because they are vectors of the trypanosome, *Trypanosoma cruzi*, the causal agent of Chagas' disease (American trypanosomiasis), an often fatal disease of humans in the Neotropics. Another triatomine, *Rhodnius prolixus* Stål, is a well-known test animal in insect physiology studies (Wigglesworth 1972). Lent and Wygodzinsky (1979) monographed the subfamily and gave detailed information about Chagas' disease vectors.

Velocipedoidea

Two families are currently placed in the Velocipedoidea, the Curaliidae and the Velocipedidae. The Curaliidae, represented by only one genus and one species, is the most recently recognized family within the Heteroptera (Schuh et al. 2008). *Curalium cronini* Schuh, Weirauch, and Henry, measuring only 1.75 mm long, is characterized by four or five closed cells on the wing membrane, placing it as the sister group of the Velocipedidae, the only other cimicomorphan family with such cells. This tiny, deep red, anthocorid like, probable predator, coined the 'ruby bug' for its bright color, is known only from lights in Florida and Louisiana.

Velocipedidae are a small, poorly known group of somewhat broadly oval species, measuring about 10 to 15 mm long. Nothing is known of their feeding habits, but their mouthparts are similar to those found in some aquatic families (Kerzhner 1981), suggesting that they are predatory. Four genera and 25 species are known from northeastern India to New Guinea (Schuh and Slater 1995, Doesberg 2004).

Infraorder Pentatomomorpha

Henry (1997a), in a phylogenetic analysis of the Pentatomomorpha, recognized six superfamilies. The monophyly of the Pentatomomorpha, including Aradoidea, is supported by at least six synapomorphies that include the presence of lamellate pulvilli, abdominal trichobothria (lost in the Aradoidea), an apically bulbous spermatheca, similar accessory salivary glands, an embryonic egg burster, and lack of a true operculum. Based on studies of the pregenital abdomen and lack of abdominal trichobothria, however, Sweet (1996) gave the Aradoidea infraorder status. Nevertheless, Schuh (1996) and Henry (1997a) did not follow Sweet's proposal based on the strong character support for placement of the Aradoidea in the Pentatomomorpha. Cassis and Gross (2002) followed the latter conclusion, stating an alternative hypothesis is not well supported. Henry (1997a) provided a key to the pentatomomorphan superfamilies and families composing the newly revised Lygaeoidea.

Aradoidea

The Aradidae (Fig. 10.42), frequently called flat bugs, are generally dull brown to black, strongly dorsoventrally flattened, oblong-oval bugs, measuring from 3 mm to more than 11 mm (Schuh and Slater 1995). Most have a strongly granulate body surface, often with tubercles and deep ridges or punctures, making them remarkably cryptic on and under bark of living and dead trees. Many are wingless or brachypterous. Aradids have unusually long maxillary stylets that coil within the head and possess typical pentatomomorphan characteristics, except for the absence of abdominal trichobothria, which are hypothesized as a character loss (Henry 1997a). Most species have been associated with various kinds of fungi (Usinger 1936, Kormilev and Froeschner 1987, Froeschner 1988a). *Aradus cimamomeus* Panzer, however, feeds on living tissue, causing serious injury to species of *Larix* and *Pinus* (Pinaceae) in Europe (Strawinsky 1925, Helioevaara 1984). About 233 genera and 1931 species of Aradidae are separated into eight subfamilies (Kormilev and Froeschner 1987, Cassis and Gross 2002, Zoological Record 2003–2007).

The family Termitaphidae comprises small, 2–3 mm, scalelike bugs that, like the Aradidae, have long stylets held coiled inside the head cavity. All species are known only from termite nests. Two genera and nine species are known, with most occurring in the Neotropics (Usinger 1942). Myers (1924) and Usinger (1942) reviewed and provided keys to the world fauna, and Myers (1932) and Usinger (1942) gave notes on the biology and habits of several species.

Coreoidea

Schaefer (1965) included the families Alydidae, Coreidae, and Rhopalidae within the superfamily, and Schaefer (1981) and Henry (1997a) supported the addition of the Hyalocephalidae and Stenocephalidae. Overall, the group is defined by the platelike ovipositor, the shortened buccula, and loss of the Y chromosome (Henry 1997a).

The Alydidae, or broad-headed bugs, are a relatively small group comprising about 45 genera and more than 254 species separated into two (Cassis and Gross 1995, Schuh and Slater 1995, Zoological Record 1996–2007) or, more recently, three subfamilies (Schaefer 1999a). Most are slender bugs with rather broad heads. Many, such as members of the genus *Alydus*, are strongly myrmecomorphic, especially in the nymphal stages. Schaefer and Mitchell (1983), in reviewing feeding habits, concluded that most Alydinae specialize on leguminous plants (Fabaceae), whereas most Leptocorisinae are grass (Poaceae) feeders. Wheeler and Henry's (1984) observations on the North American *Esperanza texana* Barber further supported Schaefer and Mitchell's (1983) speculation that micrelytrines specialize on grasses. Schaffner (1964) reviewed many of the alydine genera; Schaefer (e.g., 1996) and others (e.g., Schaefer and Ahmad 2007) have studied various other genera; Ahmad (1965) revised the Leptocorisinae of the world; and Schaefer (1999a) treated the higher classification.

The Coreidae (Figs. 10.43–10.45), often called leaf-footed bugs, pod bugs, or squash bugs based on relatively small groups of genera, are one of only two major families (the other being Pentatomidae) lacking a recent world catalog. The family contains about 267 genera and more than 1884 species worldwide (Schuh and Slater 1995, Zoological Record 1996–2007). Most recent authors have recognized three subfamilies (e.g., Packauskas 1994, Brailovsky and Cassis 1999, Cassis and Gross 2002), but the tribal classification is utterly confusing, ranging from 10 (Cassis and Gross 2002) to 30 tribes (Schuh and

Slater 1995). Packauskas (1994) provided a key to the subfamilies and tribes of the Western Hemisphere; no similar treatment, however, is available for the Eastern Hemisphere. Of the three subfamilies, the Coreinae is by far the largest and contains most of the economically important species. In North America, *Leptoglossus phyllopus* (Linnaeus) is a serious pest of various crops (Mead 1971, Bolkan et al. 1984, Mitchell 2006). Species of the genus *Chelinidea* often become pests of ornamental and rangeland cacti (*Opuntia* spp.: Cactaceae), but several species, including *C. vittiger* Uhler, have been introduced into Australia to help control invasive prickly pear cacti (Dodd 1940). Species of the South Pacific genus *Amblypelta* can become severe pests of cacao, cassava, coconut, guava, and papaya (Mitchell 2000). The pseudophloeine tur pod bug *Clavigralla gibbosa* Spinola frequently causes serious injury or even total loss to various legumes, including pigeon pea, *Cajanus cajan*, in India (Dolling 1978). Schaefer and Mitchell (1983) reviewed the known host plants and Mitchell (2000) reviewed the economically important members of the family, as well as those used in biological control. Froeschner (1988b) cataloged the Nearctic fauna and Dolling (2006), the Palearctic.

The Rhopalidae (Figs. 10.46, 10.47), often called scentless plant bugs, are a small group ranging in length from about 4.0 to more than 15 mm. The common name is based on the greatly reduced ostiolar scent-gland opening. The family comprises about 21 genera and 209 species, separated into two subfamilies (Cassis and Gross 2002, Zoological Record 2003–2007). The Rhopalinae, containing six tribes, are mostly dull brown, punctate, pubescent bugs resembling certain small Coreidae or Lygaeidae (Orsillinae). The Serinethinae are usually brightly colored red and black, mostly glabrous bugs, frequently confused with lygaeine Lygaeidae. Most rhopalids are of little economic importance, but certain Serinethinae can become nuisance pests. In North America, the common boxelder bug, *Boisea trivittata* (Say), feeds on the seeds of boxelder, *Acer negundo*, and other trees of the maple family (Aceraceae). In the fall when the bugs seek overwintering shelter, they often invade homes in enormous numbers, leaving spots on furniture and other objects and giving off an offensive odor when crushed (Wheeler 1982, Henry 1988g). *Niesthrea louisianica* Sailer has been used in the biocontrol of velvet leaf, *Abutilon theophrasti* (Malvaceae) (Spencer 1988). Chopra (1967) studied the higher classification and provided keys to the subfamilies, tribes, and genera, and Göllner-Scheiding (1983) cataloged the family for the world, including reference to several important generic revisions.

The Hyocephalidae, a small family restricted to Australia, is represented by only two genera and three species (Brailovsky 2002, Cassis and Gross 2002). Hyocephalids are relatively large (up to 15 mm long) bugs thought to feed on fallen seeds. They are characterized by the elongate, reddish-brown to black body and the possession of a glandular organ termed an external strainer or pore-bearing plate found laterally on abdominal sternum III (Štys 1964). They have their closest affinity to the Stenocephalidae (Schaefer 1981, Henry 1997a).

The family Stenocephalidae (Fig. 10.48) is represented by only one (*Dicranocephalus*) or two (*Psotilnus*) genera. The number of species has been somewhat controversial as well. Lansbury (1965, 1966) recognized two genera and 36 species, whereas Moulet (1995a, 1995b) considered only one genus and 16 valid species. These bugs are restricted to the Old World, with only one known from Australia. One species described from the Galapagos Islands, *D. insularis* Dallas, is now thought to be an Old World immigrant that was subsequently described as *D. bianchii* (Jakovlev 1902) based on specimens from its native range in northern Africa and southern Asia (Moulet 1995b, Henry and Wilson 2004). Henry and Wilson (2004) inadvertently used the junior synonymic name *D. bianchii* rather than *D. insularis*, the older name. Stenocephalids are elongate, subparallel bugs up to 15 mm long. They possess both coreoid (e.g., numerous hemelytral veins, four-lobed salivary gland) and lygaeoid (e.g., laciniate ovipositor, XY chromosome) characters (Henry 1997a).

Idiostoloidea

The Henicocoridae are represented by only one genus and one species, *Henicocoris monteithi* Woodward, restricted to southern Australia (Cassis and Gross 2002). Woodward (1968a) described the subfamily Henicocorinae for *H. monteithi* within the Lygaeidae *sensu lato*, and later, Henry (1997a) gave the group family status and formally established its sister-group relationship with the Idiostolidae.

The Idiostolidae are a small family of rhyparochromid like bugs comprising three genera and four species, one known only from Argentina and Chile and three from Australia (Woodward 1968b). Idiostolids, first included as a subfamily of the Lygaeidae *sensu lato* (Scudder 1962a), were given family status by Štys (1964). Apomorphic for these bugs are the numerous but short abdominal trichobothria and the absence of a spermatheca (Scudder 1962a, Henry 1997a). Idiostolids are associated with mosses in *Nothofagus* forests and most likely are phytophagous.

Lygaeoidea

The family classification within the Lygaeoidea has fluctuated considerably in recent years. Leston (1958) considered the Lygaeidae paraphyletic and suggested that it should be separated into at least five families. Štys (1967b) and Schaefer (1975), likewise, speculated on the paraphyly of the Lygaeidae and various relationships, particularly within the malcid line that included the Berytidae, Colobathristidae, Cyminae, and Malcidae. Henry (1997a) concluded that the Lygaeidae were polyphyletic and, as a consequence, transferred the Henicocorinae as a family to the Idiostoloidea, and gave family status to 10 subfamilies, forming a more broadly defined Lygaeoidea comprising 15 families. Slater (1964) and Slater and O'Donnell (1995) cataloged the Lygaeidae (*sensu lato*) for the world.

The Artheneidae are a mostly Palearctic group containing 8 genera and 20 species, previously separated into 4 subfamilies (Cassis and Gross 2002). Kerzhner (1997) gave evidence that the only New World representative of the family, *Polychisme poecilus* Spinola (Polychisminae), a position established by Slater and Brailovsky (1986), belongs in the Lygaeidae (*sensu stricto*) as a tribe of the Ischnorhynchinae. Artheneids are small, strongly punctate, oval bugs, with widely explanate, lateral pronotal carinae. Although naturally occurring only in the Old World, two immigrant species, *Chilacis typhae* (Perrin) (Wheeler and Fetter 1987) and *Holcocranum saturejae* (Kolenati) (Hoffman and Slater 1995) are now well established in North America (Wheeler 2002).

The Berytidae (Fig. 10.49), commonly referred to as stilt bugs because of the long, slender legs of most species, comprise 36 genera and 172 species, separated into 3 subfamilies and 6 tribes (Henry and Froeschner 1998, 2000; Henry 2000a, 2007; Kment and Henry 2008). Stilt bugs range in size from 2.3 mm for certain species of the New World genus *Pronotacantha* to more than 16 mm for *Plyapomus longus* Štusak, known only from St. Helena. Most members of the Gampsocorinae and Metacanthinae have long, slender bodies, with even longer, thread-thin legs and antennae (Henry 1997b, 1997c); all species of *Hoplinus* are armed with stout spines on the head, pronotum, and, frequently, the hemelytra (Henry 2000a). Wheeler and Schaefer (1982) provided a world review of host plants, noting that most live on glandular-hairy or viscid plants, and Henry (2000b) reviewed their economic importance, highlighting both phytophagous and zoophagous feeding habits. Morkel (2006) documented kleptoparasitism by four European species, including the obligate associate *Metacanthus annulosus* (Fieber), in the funnel-webs of the spider *Agelena orientalis* Koch (Araneae: Agelenidae). Henry (1997b) provided a phylogenetic analysis and key to the genera of the world, Henry (1997c) monographed the family for the Western Hemisphere, and Henry and Froeschner (1998) provided a world catalog.

The Blissidae (Fig. 10.50), or chinch bugs, comprise about 50 genera and 435 species (Slater and O'Donnell 1995, Cassis and Gross 2002). All are restricted to feeding on monocots, especially Poaceae, and less commonly on Cyperaceae and Restionaceae (Slater 1976). Blissids are broadly oval to elongate, often flattened, and range in size from less than 3.0 mm to more than 15 mm. The family contains a number of serious crop and turf pests. The common chinch bug, *Blissus leucopterus* (Say), perhaps the most important species in the Lygaeoidea, is a major pest of turf grasses, corn, and cereal crops in North America (Sweet 2000a). Slater (1976) reviewed host plants and Sweet (2000a) provided a detailed overview of their economic importance. Slater (1979) monographed the group as a subfamily and Henry (1997a) elevated Blissinae to family status.

The Colobathristidae are a small group of tropical bugs ranging in length from about 6.0 mm to more than 20 mm. They comprise about 23 genera and 84 species separated into two subfamilies found mostly in the Oriental and Neotropical Regions (Schuh and Slater 1995). Colobathristids, characterized by their slender, elongate bodies and relatively long legs, possess characters appearing in part coreoid and lygaeoid. Henry (1997a) reaffirmed their position in the Lygaeoidea as the sister group of Berytidae. All species feed on grasses, and some are pests of sugarcane in Indonesia and Australia (Sweet 2000a). Horváth (1904) monographed the family and Carvalho and Costa (1989) provided a key to the Neotropical genera.

The Cryptorhamphidae, represented by only two genera and four species restricted to the Australian Region, are small yellowish-brown, punctate bugs, with dorsal spiracles (Henry 1997a, Cassis and Gross 2002). Previously placed in the Cyminae, the cryptorhamphids were placed by Hamid (1971) in their own subfamily, which Henry (1997a) elevated to family.

The Cymidae are small, yellowish-brown bugs, comprising about nine genera and 54 species that are most common in the Eastern Hemisphere (Cassis and Gross 2002). Hamid (1975) monographed the group as a subfamily and Henry (1997a) gave Cyminae family status. Cymids feed primarily on monocots, especially Cyperaceae (Hamid 1975, Péricart 1998).

The Geocoridae (Fig. 10.51), or big-eyed bugs, are a widespread group, comprising 25 genera and about 274 species, separated into three subfamilies (Henry 1997a, Cassis and Gross 2002, Zoological Record 2003–2007). They are characterized by their often enlarged kidney-shaped eyes, broad heads, and the posteriorly curved abdominal sutures between segments 4 and 5 and 5 and 6 (Readio and Sweet 1982, Henry 1997a). Most members of the subfamily Geocorinae are predatory and frequently used in biocontrol programs (Sweet 2000b). Many Pamphantinae are strongly myrmecomorphic, such as the genera *Cattarus* and *Cephalocattarus* (Slater and Henry 1999).

The Heterogastridae are primarily an Old World family, comprising 24 genera and 100 species (Cassis and Gross 2002, Zoological Record 2003–2007), with only two occurring in the Nearctic (Ashlock and Slater 1988). Henry (1997a) hypothesized a sister relationship with the Pachygronthinae based on the deeply inserted ovipositor and noninflatable vesica. Péricart (1998) associated several species with the plant families Lamiaceae and Urticaceae. Scudder (1962b) provided a key to the world genera.

The family Lygaeidae (*sensu stricto*) (Fig. 10.52) comprises about 102 genera and 968 species (Slater and O'Donnell 1995, Zoological Record 1996–2007), separated into three subfamilies: the Ischnorhynchinae, Lygaeinae, and Orsillinae (Henry 1997a). Lygaeids are best recognized by the impressed line across the calli, the Y-shaped pattern on the scutellum, and the dorsal position of abdominal spiracles II through VII. The Ishnorhynchinae comprise about 75 species of dull brown to reddish-brown bugs. The most widespread north temperate genus *Kleidocerys* contains the well-known and occasional nuisance pest, the birch catkin bug, *K. resedae* (Panzer) (Wheeler 1976b).

The Orsillinae, represented by about 250 species, contains the widespread pest genus *Nysius*, collectively often called false chinch bugs (Barber 1947, Péricart 1998). The Lygaeinae, with more than 500 species, are among the most recognized and well-studied Lygaeoidea. Many, such as the species of *Lygaeus*, *Oncopeltus*, and *Spilostethus*, are aposematically colored. Slater (1992) treated the New World fauna and Péricart (1998), the western Palearctic.

The Malcidae are dull, punctate bugs about 3.0 to 4.0 mm long, with waxy setae, stylate eyes, and spined or tuberculate immatures (Štys 1967b, Sweet and Schaefer 1985). Three genera and about 29 species are separated into two subfamilies, the Chauliopinae and Malcinae. The biology and hosts are largely unknown, but several species of *Chauliops* are pests of beans (Fabaceae) in Asia and India (Sweet 2000a).

The Ninidae are a small family represented by five genera and 13 species (Cassis and Gross 2002). They measure about 3.0 to 4.0 mm long and are characterized by their broad head, stylate eyes, often translucent or hyaline hemelytra, and the bifid apex of the scutellum (Henry 1997a). Previously included as a tribe of the Cyminae, the group was given family status by Henry (1997a). Scudder (1957) provided a world revision.

The Oxycarenidae comprise 23 genera and 147 species of primarily Eastern Hemisphere bugs (Cassis and Gross 2002). They are characterized by the punctate, porrect head, hyaline often explanate hemelytra, truncate female abdomen, and transverse comb of setae on the male abdomen (Henry 1997a). *Oxycarenus*, by far the largest genus, contains several economically important species (Sweet 2000a), including *O. hyalinipennis* (Costa) introduced into the New World tropics (Slater and Baranowski 1994). Samy (1969) revised the African fauna and Henry (1997a) gave the group family status.

The family Pachygronthidae (Fig. 10.53), containing 13 genera and 78 species, is separated into two subfamilies, the Pachygronthinae and Teracriinae (Cassis and Gross 2002). The pachygronthines are distinguished by the strongly incrassate, spined profemora and the unusually long first antennal segments. Teracriines have shorter antennae and are held together in the family based only on the ventral position of the abdominal spiracles (Henry 1997a). Members of the family feed primarily on monocots (Cassis and Gross 2002). Slater (1955) provided a world revision.

The Piesmatidae (Fig. 10.54), often called ash-gray plant bugs, were considered related to the Tingidae (Cimicomorpha) until Drake and Davis (1958) clarified their position within the Pentatomomorpha; Henry (1997a) gave further support for their placement in the Lygaeoidea. Six genera and about 44 species are separated into two subfamilies, the Piesmatinae and Psamminae (Henry 1997a, Cassis and Gross 2002). Drake and Davis (1958) revised the world genera and Heiss and Péricart (1983) treated the Palearctic fauna. Slater and Sweet (1965) discussed relationships within the Psamminae, and Henry (1997a) transferred the subfamily into the Piesmatidae based on numerous shared characters, including dorsal 'areoles', two-segmented tarsi, presence of trichobothrial pads, and the loss of certain abdominal trichobothria.

The Rhyparochromidae (Figs. 10.55, 10.56) represent the largest lygaeoid family with 372 genera and more than 1850 species, separated into two subfamilies, the Plinthisinae and Rhyparochrominae, and 14 tribes (Slater and O'Donnell 1995, Cassis and Gross 2002, Zoological Record 2003–2007). Cassis and Gross (2002) provided a table giving the number of genera and species for each tribe. Rhyparochromids are recognized by the incomplete abdominal suture between segments 4 and 5 (except Plinthisinae) and the presence of a trichbothrium near each eye (Henry 1997a). The five largest tribes – the Rhyparochromini (370 mostly Old World species), Myodochini (322 primarily Western Hemisphere species), Drymini (280 primarily North American and Eastern Hemisphere species), Ozophorini (175 largely New World and Oriental species), and Lethaeini (153 worldwide species) – represent more than 70% of the family. Nearly all Rhyparochrominae, with the exception of the haematophagous Cleradini, are seed feeders, thus, the common name seed bugs. Most taxa have enlarged forefemora for grasping seeds. Although most rhyparochromids are not considered major pests, several immigrant species can become serious nuisance pests in western North America, when they invade homes and other structures during outbreak populations (Henry and Adamski 1998, Henry 2004).

Pentatomoidea

This superfamily comprises 16 families. The last world pentatomoid catalog by Kirkaldy (1909) is long outdated. The more recent regional catalogs include those of Henry and Froeschner (1988) for the Nearctic, with chapters by Froeschner; Cassis and Gross (2002), for Australia; and Aukema and Rieger (2006) for the Palearctic, with chapters by J. Davidová-Vilímová, U. Göllner-Scheiding, J. A. Lis, and D. A. Rider. D. A. Rider (personal communication) is working toward a world pentatomid catalog and maintains a website (Rider 2008) containing a list of genera, a comprehensive bibliography, and a wealth of other information pertaining to the Pentatomoidea.

The Acanthosomatidae (Fig. 10.57) are a small family comprising about 46 genera and 184 species, separated into three subfamilies (Cassis and Gross 2002; Froeschner 1997, 2000, Zoological Record 2003–2007). Acanthosomatids are characterized by a combination of two-segmented tarsi, hidden spiracles on the second abdominal segment, paired trichobothria on abdominal segments III–VII, a large mesosternal keel, an anteriorly directed spine on abdominal segment III, and a large exposed eighth abdominal segment in males (Kumar 1974). Most species are found in north temperate regions or at higher elevations in the subtropics (Thomas 1991). Females possess paired abdominal structures known as Pendergrast's organs and exhibit strong maternal brooding behavior (Tallamy and Schaefer 1997). Kumar (1974) provided a world revision with keys to subfamilies, tribes, and genera; Thomas (1991) revised the North American fauna.

The Aphylidae, consisting of only two genera and three species, are restricted to the Australian Region. They are 4–5 mm long and characterized by a strongly convex, oval shape, large 'scutellerid-like' scutellum, and long labium extending beyond the metacoxae. Although Gross (1975) and Rider (personal communication in Cassis and Gross 2002) considered aphylids aberrant pentatomids, Schuh and Slater (1995) and Cassis and Gross (2002) maintained their family status. Štys and Davidová-Vilímová (2001) provided the most recent family review, including the description of a new genus and one new species.

The Canopidae are small, round, convex pentatomoids represented by only one genus and eight species in the New World tropics (Schuh and Slater 1995). McAtee and Malloch (1928) revised the group and McDonald (1979) studied the male and female genitalia of several species. McHugh (1994) associated nymphs and adults of two species, *Canopus burmeisteri* McAtee and Malloch and *C. fabricii* McAtee and Malloch, with polypore fungi on rotting logs.

The Cydnidae (Figs. 10.58, 10.59), or burrowing bugs, are a worldwide group of about 120 genera and 765 species, separated into five (Froeschner 1960) to seven subfamilies (Lis 2006), including the Corimelaeninae and Thyreocorinae (Schuh and Slater 1995, Cassis and Gross 2002, Zoological Record 2003–2007; but see Lis 2006). Burrowing bugs range in size from about 2 to 20 mm and are characterized by their round to oval shape, brown to black coloration, and often spiny legs modified for digging in soil. Many cydnids feed on the roots of their hosts, but members of the subfamily Sehirinae feed on plants like most Pentatomidae (Froeschner 1988c). Most species are of little economic importance, but occasionally they become abundant and cause serious injury to crops (Froeschner 1988c, Lis et al. 2000). Froeschner (1960) monographed the Cydnidae of the Western Hemisphere and J. A. Lis (1999) cataloged the Old World fauna and since (e.g., Lis 2000, 2001) has revised numerous genera. Lis et al. (2000) reviewed the economically important species.

The Cyrtocoridae are a small New World group of pentatomoids 6–10 mm long. They are characterized by the oval, strongly convex form, tuberculate scutellum, explanate connexival segments, and scalelike setae (Packauskas and Schaefer 1998). Formerly treated as a subfamily of the Pentatomidae (e.g., Schuh and Slater 1995), the group was accorded family status by Packauskas and Schaefer (1998), who presented keys to the four genera and eleven species. At least one species feeds on leguminous plants (Fabaceae) (Schaefer et al. 2000).

The Dinidoridae, primarily Oriental and Afrotropical, comprise only 16 genera and 65 species, separated into two subfamilies (Rolston et al. 1996). Most species are thought to be phytophagous, with several becoming pests of certain cucurbits (Schaefer et al. 2000). They are 9–27 mm in length and characterized by large, stout bodies, a short scutellum, and keeled heads (Schuh and Slater 1995). Durai (1987) revised the family and Rolston et al. (1996) cataloged the world fauna.

The Australian family Lestoniidae comprises one genus and only two species (McDonald 1969). These round convex bugs, resembling certain tortoise beetles (Chrysomelidae) and measuring from about 3.5 mm to nearly 6.0 mm long, apparently are related to Plataspidae (China and Miller 1959, Schaefer 1993). *Lestonia haustorifera* China feeds on species of Australian

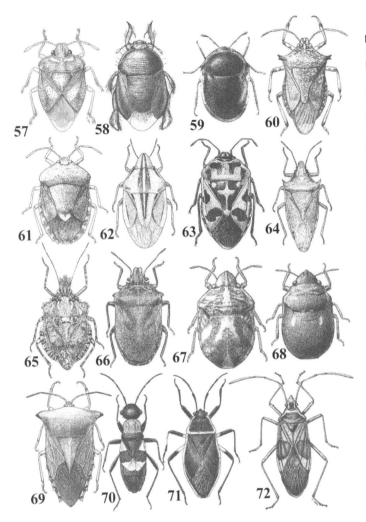

Figs. 10.57–10.72 Pentatomomorpha.
10.57, *Rolstonus rolstoni* Froeschner
[Acanthosomatidae: Pentatomoidea];
10.58, *Scaptocoris castaneus* Perty
[Cephalocteinae: Cydnidae: Pentatomoidea];
10.59, *Corimelaena pulicaria* (Germar)
[Thyreocorinae: Cydnidae: Pentatomoidea];
10.60, *Alcaeorrhynchus grandis* (Dallas)
[Asopinae: Pentatomidae: Pentatomoidea];
10.61, *Edessa florida* Barber [Edessinae:
Pentatomidae: Pentatomoidea]; 10.62, *Aelia
americana* Dallas [Pentatominae:
Pentatomidae: Pentatomoidea];
10.63, *Murgantia histrionica* (Hahn)
[Pentatominae: Pentatomidae:
Pentatomoidea], 10.64, *Oebalus pugnax*
(Fabricius) [Pentatominae: Pentatomidae:
Pentatomoidea]; 10.65, *Parabrochymena
arborea* (Say) [Pentatominae: Pentatomidae:
Pentatomoidea]; 10.66, *Amaurochrous
cinctipes* (Say) [Podopinae: Pentatomidae:
Pentatomoidea]; 10.67, *Camirus porosus*
(Germar) [Scutelleridae: Pentatomoidea];
10.68, *Sphyrocoris obliquus* (Germar)
[Scutelleridae: Pentatomoidea];
10.69, *Piezosternum subulatum* (Thunberg)
[Tessaratomidae: Pentatomoidea]; 10.70,
Arhaphe carolina Herrich-Schaeffer [Largidae:
Pyrrhocoroidea]; 10.71, *Largus succinctus*
(Linnaeus) [Largidae: Pyrrhocoroidea];
10.72, *Dysdercus lunulatus* Uhler
[Pyrrhocoridae: Pyrrhocoroidea].
(Fig. 10.57 after Froeschner 1997;
10.58–10.69, Froeschner 1941; 10.70,
10.71, Froeschner 1944; 10.72, Froeschner
1985).

cypress, *Callitris* spp. (Cupressaceae) (McDonald 1970, Cassis and Gross 2002).

The Megarididae are a small Neotropical family of round, strongly convex bugs with a large globose scutellum covering the wings and abdomen. They are represented by only one genus, *Megaris*, and 16 species (McAtee and Malloch 1928), all of which are about 5.0 mm or less in length. Previously considered a subfamily of the Pentatomidae, megarids have been given family status (McDonald 1979), but their relationship to other families remains unclear (Schuh and Slater 1995).

The Parastrachiidae, containing only one genus and two Oriental and eastern Palearctic species, until recently, were placed as a tribe of the Pentatomidae or a subfamily of the Cydnidae (Schaefer et al. 1988). Sweet and Schaefer (2002), based on wing morphology, scent-gland structure, and male and female genitalia, elevated the group to family status. Parastrachiids are bright red and black bugs, 15–18 mm long. They exhibit subsocial behavior (Tallamy and Schaefer 1997) and maternal egg guarding (Miyamoto 1956).

The family Pentatomidae (Figs. 10.60–10.66), often called stink bugs, is the largest heteropteran group that lacks a modern world catalog. Stink bugs are recognized by the oval body, 4- or 5-segmented antenna, entire hemelytra with 5–12 veins, and 2- or 3-segmented tarsi. Approximately 900 genera and 4700 species are separated into eight or nine subfamilies (Cassis and Gross 2002, Rider 2008). The predatory subfamily Asopinae (Fig. 10.60), with approximately 63 genera and more than 350 species, contains some of the most important biocontrol agents used in agricultural ecosystems, including members of the New World genus *Podisus* (McPherson 1982, De Clercq 2002). Thomas (1992) revised the New World fauna and gave a synopsis of the Old World genera (Thomas 1994). The largest subfamily, Pentatominae (Figs. 10.61–10.65), is represented by more than 404 genera and 2771 species (Schuh and Slater 1995). Among its members are a large number of economically important species, including the Palearctic *Aelia furcula* Fieber, a pest of small grains; the Australian spined citrus bug, *Biprorulus bibax* Breddin; the widespread 'southern green stink bug', *Nezara viridula* (Linnaeus); and New World species of the genus *Euschistus*, which attack various agricultural crops (McPherson 1982, McPherson and McPherson 2000, Panizzi et al. 2000). The Nearctic fauna has been cataloged by Froeschner (1988d), the Australian by Cassis and Gross (2002), and the Palearctic by Rider (2006a).

The Phloeidae are large (20 to 30 mm), mottled brown, flattened bugs with strongly expanded lateral plates around the body, and a color and shape that make them well camouflaged against the lichen-covered bark of their host trees (Lent and Jurberg 1965, 1966). Only two genera and three species are known. The strongly foliate body margins, the three-segmented antenna, and the unique male genitalia distinguish them from all other pentatomoids (Lent and Jurberg 1965, Schuh and Slater 1995). Phloeids exhibit strong maternal egg care and, upon hatching, early instars attach to the undersurface of the female for protection. When disturbed, these bugs emit a stream of clear liquid from the anal opening (Hussey 1934).

The Plataspidae are an Eastern Hemisphere group, most abundant in the Oriental Region. They measure 2–20 mm long and are characterized by their round, strongly convex bodies and large scutellum that covers the abdomen and hemelytra. Fifty-nine genera and about 560 species are known (Jessop 1983, Davidová-Vilímová 2006). All plataspids are phytophagous, with most specializing on leguminous plants (Fabaceae) (Schaefer 1988). Most records of economic damage to various crops pertain to species of the largest genus *Comptosoma*, with 280 species (Schaefer et al. 2000). Jessop (1983) reviewed and provided keys to the genera of the

Libyaspis group and provided a checklist of the species; Davidová-Vilímová (2006) cataloged the Palearctic fauna.

The Scutelleridae (Figs. 10.67, 10.68), or shield bugs, are represented by 81 genera and about 450 species, separated into six subfamilies (Schuh and Slater 1995, Göllner-Scheiding 2006). Shield bugs, measuring from about 5.0 to 20.0 mm long, are distinguished by the oval, often convex shape and large scutellum that covers the abdomen and most of the hemelytra. Most scutellids are dull mottled brown or gray, but many members of the Scutellerinae are brightly colored and often iridescent (Javahery et al. 2000), making them some of the most spectacularly colored of all Heteroptera (Schuh and Slater 1995). Cassis and Vanags (2006) revised the 13 genera and 25 species of Australian jewel bugs. By far the most economically important species is the Sunn pest, *Eurygaster integriceps* Puton, which often devastates wheat and other small grain crops in Asia and the Middle East (Javahery et al. 2000).

The Tessaratomidae (Fig. 10.69) comprise 55 genera and 240 species, separated into three subfamilies (Rolston et al. 1994, Rider 2006b). Most occur in the Old World, with only three species in the Neotropics. Tessaratomids are large robust bugs, some more than 40 mm long. In addition to their large size, they are characterized by the proportionately small head, short labium, and large sternal plate between the middle and hind coxae. The bronze orange bug, *Musgraveia sulciventris* Stål, is the most serious pest in the family, causing economic injury to citrus in Australia. Rolston et al. (1994) cataloged the family. Sinclair (2000) revised the genera of Oncomerinae. Cassis and Gross (2002) cataloged the Australian fauna and Rider (2006b) the Palearctic.

The Thaumastellidae are small, enigmatic bugs, 3.5 mm or less in length, found only in the Afrotropical Region and the Middle East (Schuh and Slater 1995). Only one genus and three species are known. Though they are clearly related to the Cydnidae (Dolling 1981), Jacobs (1989) maintained them as a family. Thaumastellids occur under rocks, where they apparently feed on fallen seeds (Jacobs 1989).

The Urostylididae, comprising 11 genera and about 170 species, are found mostly in southern and eastern Asia (Rider 2006c). They are elongate-oval bugs, with proportionately small heads and long antennae, measuring about 3.5 to 14.0 mm in length. (Schuh and Slater 1995). Rider (2006c) summarized the taxonomic literature and cataloged the Palearctic fauna; Schaefer et al. (2000) reviewed the economic species.

Pyrrhocoroidea

Henry (1997a) reviewed the classification and considered the Pyrrhocoroidea, containing only two families, the sister group of the Coreoidea.

The Largidae (Figs. 10.70, 10.71) are a small, worldwide group, containing 13 genera and about 106 species, separated into two subfamilies, the Larginae and Physopeltinae (Schaefer 2000, Cassis and Gross 2002, Zoological Record 2003–2007). Largids are medium to large (more than 50 mm), often brightly colored bugs, characterized by the lack of ocelli, four-segmented labium and antenna, seven to eight veins on the membrane arising from two closed basal cells, and sometimes fused abdominal segments (Henry 1988d, 1997a; Schuh and Slater 1995). All are phytophagous (Schaefer and Ahmad 2000). The New World subfamily Larginae is separated into two tribes, the Araphini (Fig. 10.70) and Largini (Fig. 10.71) (Schaefer 2000). The former comprises largely the genus *Largus* and the latter, several strongly myrmecomorphic genera, including *Araphe* and *Pararaphe* (Henry 1988d). The Physopeltinae, restricted to the Old World, are large, mostly red, aposematically colored bugs. Hussey (1929) cataloged the world fauna. More recently, Henry (1988d) cataloged the Nearctic, Kerzhner (2001) the Palearctic, and Cassis and Gross (2002) the Australian Region.

The Pyrrhocoridae (Fig. 10.72), or cotton stainers, are represented by about 33 genera and 340 species (Kerzhner 2001, Zoological Record 2003–2007). Pyrrhocorids are medium-sized to large bugs, from about 8.0 to more than 30.0 mm long, and usually aposematically colored red, yellow, and white. They are distinguished by the lack of ocelli, a reduced metathoracic scent-gland opening, two closed cells at the base of the wing membrane, and a platelike ovipositor (Henry 1988f, 1997a, Schuh and Slater 1995). *Dysdercus*, by far the largest genus, occurs nearly worldwide and contains several important cotton pests that cause damage directly by feeding on cotton bolls, and indirectly by introducing bacteria and fungi that cause the bolls to rot (Whitefield 1933, Schaefer and Ahmad 2000). Freeman (1947) treated many of the Old World species and Doesberg (1968) revised the New World fauna. Several members of the genera *Antilochus* and *Raxa* are facultatively predaceous, sometimes becoming obligate predators (Schaefer 1999b).

THE IMPORTANCE OF HETEROPTERAN BIODIVERSITY

Although it is well documented that insects are the most diverse group of organisms, comprising more than half the described species, we are far from knowing the exact number coexisting with us on earth. With Hemiptera considered the fifth largest order, and the suborder Heteroptera containing nearly half of the estimated 90,000 species (Cassis et al. 2006), this diverse group, exhibiting both phytophagous and zoophagous feeding habits (Schuh and Slater 1995), affects nearly every aspect of our environment. The pursuit of a better understanding of true bug diversity, therefore, is paramount. Taxonomy and systematics, as hypothesis-driven disciplines (Wheeler 2004), form the foundation for all biological research through the predictive value they impart from taxon-based studies. For example, newly discovered predatory species related to known beneficial taxa can be expected to have similar biological control potential. If the 10,000 species described during the past 250 years (1758 to present) represent only half the number of Miridae that eventually will be discovered (Henry and Wheeler 1988), an enormous amount of important biological, biogeographic, and host information useful to other disciplines remains to be discovered. Considering the previous rate of only 40 species described per year, the concept of an 'industrialized' effort to combine the global resources of multiple scientists and students, modern technology, and extensive fieldwork in hot-spot areas made possible only through large-scale funding (Cassis et al. 2006) takes on new merit.

That billions of dollars worth of losses to crops are caused by Heteroptera each year reflects, perhaps, the single most important reason to study this diverse suborder. Throughout this chapter, I have given examples of the many true bug attributes, including information on their morphology, distribution, numbers, host associations, and economic importance. The Heteroptera fall primarily into two broad feeding regimes, plant feeders and predators (Schaefer and Panizzi 2000, A. Wheeler 2000a, 2000b, 2001), with many intermediate variations. As a consequence of globalization, greater numbers of true bugs are being transported beyond their native ranges through international commerce, creating new pest situations in foreign lands (Wheeler and Hoebeke 2008), many times involving taxa previously unknown to science (e.g., Dolling 1972). Recent treatments of various families, such as the Coreidae (Mitchell 2000), Lygaeoidea (Sweet 2000a), Miridae (Wheeler 2000a, 2001), Pentatomidae (McPherson and McPherson 2000), and other families presented in Schaefer and Panizzi (2000), have provided much-needed synthesis of widely scattered information important to better understanding certain pest groups.

The effect of global warming on heteropteran distribution also is being reflected through various studies. In eastern North America, the rhopalid, *Jadera haematoloma* (Herrich-Schaeffer), has shown rapid northward dispersal in recent years (Hoffman and Steiner 2005). In Japan, rising temperatures have allowed the poleward range expansion of several rice and fruit-feeding Pentatomidae (Kiritani 2007). Musolin (2007) documented the northern movement of the southern green stink bug, *Nezara viridula* (Linnaeus), in central Japan since the 1960s. The incidence of vector-borne diseases can be affected as well. For example, rising temperatures can expand distributions, accelerate life cycle times, and increase population densities of certain species, as seen in some species of *Triatoma* (Reduviidae) that are vectors of *Trypanosoma cruzi*, the causal agent of Chagas's disease (Curto de Casas and Carcavallo 1995).

Although in most studies the phytophagous bugs are emphasized as agricultural pests, many heteropterans, including all of the Enicocephalomorpha, Dipsocoromorpha, Gerromorpha, Nepomorpha, Leptopodomorpha, and most Cimicomorpha (excepting certain Miridae and all Tingidae), are exclusively or, in large part, predatory and are usually considered beneficial in agricultural situations. From a biological control viewpoint, the families Anthocoridae (*sensu lato*) (Lattin 2000), Geocoridae (Sweet 2000b), Miridae (Wheeler 2000b, 2001), Nabidae (Braman 2000), asopine Pentatomidae (De Clercq 2002), and Reduviidae (Ambrose 2000) contain the most important predatory species. Within these groups are also several external parasitic lineages that feed on vertebrate blood, including the Cimicidae, Polyctenidae, triatomine Reduviidae, and the rhyparochromid tribe Cleradini. Many aquatic bugs are well known for their

value in mosquito control or as a food source for fish and other organisms (Menke 1979).

The Heteroptera are also important in conservation biology. The aquatic and semiaquatic bugs (Gerromorpha, Nepomorpha, and Leptopodomorpha) are well known for their role as water-quality indicators (Jansson 1987). Wheeler (2001) noted that certain Miridae might be of interest to conservation biologists as rare or unique species needing preservation, or as indicators of vitality or changes in ecosystems. For example, the type locality of the poorly known phlox-feeding plant bug, *Polymerus wheeleri* Henry, found in only 12 of 79 eastern US shale barren habitats, has essentially been destroyed by construction (Wheeler 1995, Henry, personal observations). Other rare species, such as the North Amerian *Corixidea major* McAtee and Malloch (Schizopteridae) known from only the holotype and two recent specimens taken in pine barren habitats (Hoffman et al. 2005), serve as valuable indicators of areas needing protection. Aukema (1994) concluded that the preservation and restoration of unique habitats should be made to protect populations of rare and endangered species in the Netherlands.

The overall influence of the Heteroptera as part of the fifth largest insect order is significant. Their roles as plant feeders, bloodsucking parasites, invertebrate predators, or water-quality indicators, make them unquestionably important organisms in our environment. More recent studies addressing the impact of global warming and the influence of Heteroptera in conservation biology reinforce the need for additional study. If even the lowest estimates of the number of Heteroptera prove accurate (Cassis et al. 2006), much challenging work remains to better understand this taxonomically complex and economically important group of fascinating insects.

ACKNOWLEDGMENTS

I dedicate this chapter to the memory of the late Dr. Richard C. Froeschner (Henry 2003), formerly of the Department of Entomology, National Museum of Natural History (USNM), Smithsonian Institution, Washington, DC, for his many contributions to heteropterology, including numerous important publications and his passionate work on the USNM Heteroptera collection; and his talented wife, the late Elsie Herbold Froeschner (Nicholson 2006), for the beautiful and technically accurate scientific illustrations she produced throughout her career, including most of the ones used in this chapter.

I thank Robert G. Foottit (Canadian National Collection of Insects, Agriculture and Agri-Food Canada, Ottawa, Ontario) and Peter H. Adler (Clemson University, Clemson, SC) for their invitation to write this chapter and to be a part of this important contribution covering insect biodiversity in regard to science and society. I am grateful to the Department of Entomology (USNM) for the permission to use illustrations originally published by Richard C. Froeschner as cited under each figure caption (Figs. 10.4, 10.6, 10.9, 10.19, 10.20, 10.24, 10.48, 10.57, and 10.72 were illustrated by Young Sohn, USNM; all others are by Elsie Herbold Froeschner, except Fig. 10.32 as noted). I also thank Michael G. Pogue (Systematic Entomology Laboratory [SEL], ARS, USDA, c/o National Museum Natural History, Washington, DC), Sonja Scheffer (SEL), Michael D. Schwartz (c/o Canadian National Collection of Insects, Agriculture and Agri-Food Canada, Ottawa, Ontario), and Alfred G. Wheeler, Jr. (Clemson University, Clemson, SC) for their careful and constructive reviews of the manuscript.

REFERENCES

Ahmad, I. 1965. The Leptocorisinae (Heteroptera: Alydidae) of the world. *Bulletin on the British Museum (Natural History), Entomology Supplement* 5: 1–156.

Akingbohungbe, A. E. 1996. *The Isometopinae (Heteroptera: Miridae) of Africa, Europe, and the Middle East.* Delar Tertiary Publishers, Ibadan, Nigeria. 170 pp.

Ambrose, D. P. 2000. Assassin bugs (Reduviidae excluding Triatominae). Pp. 695–711. *In* C. W. Schaefer and A. R. Panizzi (eds). *Heteroptera of Economic Importance.* CRC Press, Boca Raton, Florida. 828 pp.

Andersen, N. M. 1978. A new family of semiaquatic bugs for *Paraphrynovelia* Poisson with a cladistic analysis of relationships (Insecta, Hemiptera, Gerromorpha). *Steenstrupia* 4: 211–225.

Andersen, N. M. 1982. *The Semiaquatic Bugs (Hemiptera, Gerromorpha). Phylogeny, Adaptations, Biogeography and Classification.* Entomonograph Volume 3. Scandinavian Science Press, Klampenborg, Denmark. 455 pp.

Andersen, N. M. and J. T. Polhemus. 1980. Four new genera of Mesoveliidae (Hemiptera, Gerromorpha) and the phylogeny and classification of the family. *Entomologica Scandanavica* 11: 369–392.

Ashlock, P. D. and A. Slater. 1988. Family Lygaeidae Schilling, 1829 (=Infericornes Amyot and Serville, 1843;

Myodochidae Kirkaldy, 1899; Geocoridae Kirkaldy, 1902). The seed bugs and chinch bugs. Pp. 167–245. *In* T. J. Henry and R. C. Froeschner (eds). *Catalog of the Heteroptera, or True Bugs, of Canada and the Continental United States.* E. J. Brill, Leiden and New York. 958 pp.

Aukema, B. 1994. Rare terrestrial Heteroptera and nature development. *Entomologische Berichten, Amsterdam* 54: 94–102. [In Dutch, with English abstract].

Aukema, B. and C. Rieger (eds). 1995–2006. *Catalogue of the Heteroptera of the Paleaearctic Region.* Volume 1: Enicocephalomorpha, Dipsocoromorpha, Nepomorpha, Gerromorpha and Leptopodomorpha, 222 pp; volume 2: Cimicomorpha I, 360 pp.; volume 3: Cimicomorpha II, 576 pp.; volume 4: Pentatomomorpha I, 346 pp.; volume 5: Pentatomomorpha II, 550 pp. Netherlands Entomological Society, Amsterdam. [See specific family citations under respective author names.]

Baranowski, R. M. 1958. Notes on the biology of the royal palm bug, *Xylastodoris luteolus* Barber (Hemiptera, Thaumastocoridae). *Annals of the Entomological Society of America* 51: 547–551.

Barber, G. W. 1920. Notes on the oviposition and food of the wheel-bug (*Arilus cristatus* Linn.) (Hemip. Heter). *Entomological News* 31: 107.

Barber, H. G. 1920. A new member of the family Thaumastocoridae. *Bulletin of the Brooklyn Entomological Society* 15: 98–105.

Barber, H. G. 1947. Revision of the genus *Nysius* in the United States and Canada (Hemiptera Heteroptera: Lygaeidae). *Journal of the Washington Academy of Sciences.* 37: 354–366.

Bergroth, E. 1915. Ein neuer Ameisengast aus Südafrika (Hem. Heteropt.). *Wiener Entomologische Zeitung* 34: 291–292.

Bockwinkel, G. 1990. Food resource utilization and population growth of the grassbug *Notostira elongata* (Heteroptera: Miridae: Stenodemini). *Entomologia Generalis* 15: 51–60.

Bolkan, H. A., J. M. Ogawa, R. E. Rice, R. M. Bostock, and J. C. Crane. 1984. Leaffooted bug (Hemiptera: Coreidae) and epicarp lesion of pistachio fruits. *Journal of Economic Entomology* 77: 1163–1165.

Bourgoin, T. and B. C. Campbell. 2002. Inferring a phylogeny for Hemiptera: falling into the 'autapomorphic trap.' *Denisia* 4: 67–82.

Brailovsky, H. 2002. A new species of *Maevius* Stål from Australia and some notes on the family Hyocephalidae (Hemiptera: Heteroptera). *Proceedings of the Entomological Society of Washington* 104: 41–50.

Brailovsky, H. and G. Cassis. 1999. Revision of the tribe Agriopocorini (Hemiptera: Coreidae: Coreinae). *Canadian Entomologist* 131: 293–321.

Braimah, S. A., L. A. Kelton, and R. K. Stewart. 1982. The predaceous and phytophagous plant bugs (Heteroptera: Miridae) found on apple trees in Quebec. *Naturaliste Canadian* 109: 153–180.

Braman, S. K. 2000. Damsel bugs (Nabidae). Pp. 639–656. *In* C. W. Schaefer and A. R. Panizzi (eds). *Heteroptera of Economic Importance*. CRC Press, Boca Raton, Florida. 828 pp.

Brambila, J. and G. S. Hodges. 2004. Bugs (Hemiptera). Pp. 354–371. *In* J. Capinera (ed). *Encyclopedia of Entomology*. Kluwer Academic Publications, Oxford.

Brooks, A. R. and L. A. Kelton. 1967. Aquatic and semiaquatic Heteroptera of Alberta, Saskatchewan, and Manitoba (Hemiptera). *Memoirs of the Entomological Society of Canada* 51: 1–92.

Carayon, J. and A. Villiers. 1968. Étude sure les Hémiptères Pachynomidae. *Annales de la Société Entomologique de France (N. S.)* 4 (3): 703–739.

Carayon, J., R. L. Usinger, and P. Wygodzinsky. 1958. Notes on the higher classification of the Reduviidae, with the description of a new tribe of the Phymatinae (Hemiptera-Heteroptera). *Revue de Zoologie et de Botanique Africaines* 57: 256–281.

Carpintero, D. L. and P. M. Dellapé. 2005. A new species and first record of *Embiophila* (Heteroptera: Plokiophilidae) from Nicaragua. *Studies on Neotropical and Environment* 40: 65–68.

Carpintero, D. L. and P. M. Dellapé. 2006. A new species of *Thaumastocoris* Kirkaldy from Argentina (Heteroptera: Thaumastocoridae: Thaumastocorinae). *Zootaxa* 1228: 61–68.

Carvalho, J. C. M. 1955. Keys to the genera of Miridae of the world (Hemiptera). *Boletim do Museu Paraense Emilio Goeldi* 11 (2): 1–151.

Carvalho, J. C. M. 1957–1960. Catalogue of the Miridae of the world. Arquivos do Museu Nacional, Rio de Janeiro. Part I, Cylapinae, Deraeocorinae, Bryocorinae. Volume 44 No. 1. 1–158 pp. (1957); Part II, Phylinae. Volume 45 No. 2. 1–216 pp. (1958); Part III, Orthotylinae. Volume 47 No. 3. 1–161pp. (1958); Part IV, Mirinae. Volume 48No. 4 1–384 pp. (1959); Part V, Bibliography and Index Volume 51 No. 5. 1–194 pp. (1960).

Carvalho, J. C. M. (in collaboration with G. F. Gross). 1979. The tribe Hyalopeplini of the world (Hemiptera: Miridae). *Records of the South Australian Museum* 17: 429–531.

Carvalho, J. C. M. and L. A. A. Costa. 1989. Chara para identificação do gêneros neotroópicos da família Colobathristidae (Hemiptera). *Revista de Brasileira Biologia* 49: 271–277.

Carvalho, J. C. M. and J. C. Schaffner. 1986. Neotropical Miridae, CCLXXV: Additional species of the genus *Eccritotarsus* Stål (Hemiptera) – Part I. *Anais da Academia Brasileira de Ciências* 58: 303–342.

Carvalho, J. C. M. and J. C. Schaffner. 1987. Neotropical Miridae, CCLXXV: Additional species of the genus *Eccritotarsus* Stål (Hemiptera) – Part II. *Anais da Academia Brasileira de Ciências* 58: 473–488.

Cassis, G. 1995. A reclassification and phylogeny of the Termatophylini (Heteroptera: Miridae: Deraeocorinae), with a taxonomic revision of the Australian species, and a review of the tribal classification of the Deraeocorinae. *Proceedings of the Entomological Society of Washington* 97: 258–330.

Cassis, G. and G. F. Gross. 1995. Hemiptera: Heteroptera (Coleorrhyncha to Cimicomorpha). Pp. 506. *In* W. W. K. Houston and G. V. Maynard (eds).*Zoological Catalogue of Australia*. Volume 27.3A. CSIRO, Melbourne, Australia.

Cassis, G. and G. F. Gross. 2002. Hemiptera: Heteroptera (Pentatomomorpha). Pp. 737. *In* W. W. K. Houston and G. V. Maynard (eds). *Zoological Catalogue of Australia*. Volume 27.3B. CSIRO, Melbourne, Australia.

Cassis, G. and L. Vanags. 2006. Jewel bugs of Australia (Insecta: Heteroptera: Scutelleridae). *Denisia* 50: 275–398.

Cassis, G., M. A. Wall, and R. T. Schuh. 2006. Insect biodiversity and industrialising the taxonomic process: the plant bug case study (Insecta: Heteroptera: Miridae). Pp. 193–212. *In* T. R. Hodkinson and J. A. N. Parnell (eds).*Reconstructing the Tree of Life. Taxonomy and Systematics of Species Rich Taxa*. CRC Press, Boca Raton, Florida.

Cheng, L. 1977. The elusive sea bug *Hermatobates* (Heteroptera). *Pan-Pacific Entomologist* 53: 87–97.

China, W. E. and N. C. E. Miller. 1959. Check-list and keys to the families and subfamilies of Hemiptera-Heteroptera. *Bulletin of the British Museum (Natural History) Entomology* 8: 1–45.

Chopra, N. P. 1967. The higher classification of the family Rhopalidae (Hemiptera). *Transactions of the Royal Entomological Society of London* 119: 363–399.

Cobben, R. H. 1968. *Evolutionary Trends in Heteroptera. Part I. Eggs, Architecture of the Shell, Gross Embryology, and Eclosion*. Centre for Agricultural Publishing and Documentation, Wageningen, Netherlands. 475 pp.

Cobben, R. H. 1970. Morphology and taxonomy of intertidal dwarf-bugs (Heteroptera: Omaniidae fam. nov.). *Tijdschrift Voor Entomologie* 113: 61–90.

Cobben, R. H. 1978. *Evolutionary Trends in Heteroptera. Part 2. Mouthpart-Structures and Feeding Strategies*. Meddlingen Landbouwhogeschool 78-5. H. Veeman, Wageningen, Netherlands. 407 pp.

Cook, M. L. 1977. A key to the genera of Asian Ectrichodinae (Hemiptera: Reduviidae) together with a check-list of genera and species. *Oriental Insects* 11: 63–88.

Curto de Casas, S. I. and R. U. Carcavallo. 1995. Climate change and vector-borne diseases distribution. *Social Science and Medicine* 40: 1437–1440.

Dallas, W. S. 1851–1852. *List of the Specimens of Hemipterous Insects in the Collection of the British Museum*. Taylor and Francis Incorporated, London. 1851. Volume 1. 1–368 pp., plates 1–11; 1852. Volume 2. 369–592 pp., plates 12–15.

Davidová-Vilímová, J. 2006. Family Plataspidae Dallas, 1851. Pp. 150–165. *In* B. Aukema and C. Rieger (eds). *Catalogue of the Heteroptera of the Palaearctic Region*. Volume 5. Pentatomomorpha II. Netherlands Entomological Society, Amsterdam. 550 pp.

Davis, N. T. and R. L. Usinger. 1970. The biology and relationship of the Joppeicidae (Heteroptera). *Annals of the Entomological Society of America* 63: 577–586.

De Clercq, P. 2002. Predaceous stinkbugs (Pentatomidae: Asopinae). Pp. 737–789. *In* C. W. Schaefer and A. R. Panizzi (eds). *Heteroptera of economic importance*. CRC, Boca Raton, Florida. 828 pp.

Dodd, A. P. 1940. *The Biological Campaign Against Prickly-Pear*. A. H. Tucker, Government Printer, Brisbane, Australia. 177 pp.

Doesberg, P. Hvan, Jr. 1968. A revision of the New World species of *Dysdercus* Guérin-Méneville (Heteroptera, Pyrrhocoridae). *Zoologische Verhandelingen* 97: 1–215.

Doesberg, P. Hvan, Jr. 2004. A taxonomic revision of the family Velocipedidae Bergroth, 1891 (Insecta: Heteroptera). *Zoologische Verhandelingen* 347: 1–110.

Dolling, W. R. 1972. A new species of *Dicyphus* Fieber (Hem., Miridae) from southern England. *Entomologist's Monthly Magazine* 107: 244–245.

Dolling, W. R. 1978. A revision of the Oriental pod bugs of the tribe Clavigrallini (Hemiptera: Coreidae). *Bulletin of the British Museum of Natural History* 36: 281–321.

Dolling, W. R. 1981. A rationalized classification of the burrower bugs (Cydnidae). *Systematic Entomology* 6: 61–76.

Dolling, W. R. 2006. Family Coreidae Leach, 1815. Pp. 43–101. *In* B. Aukema and C. Rieger (eds). *Catalogue of the Heteroptera of the Palaearctic Region*. Volume 5. Pentatomomorpha II. Netherlands Entomological Society, Amsterdam. 550 pp.

Dougherty, V. M. 1980. A systematic revision of the New World Ectrichodiinae (Hemiptera: Reduviidae). Ph.D. Dissertation, University of Connecticut, Storrs, Connecticut.

Drake, C. J. and N. T. Davis. 1958. The morphology and systematics of the Piesmatidae (Hemiptera), with keys to world genera and America species. *Annals of the Entomological Society of America* 51: 567–581.

Drake, C. J. and F. A. Ruhoff. 1965. *Lacebugs of the World: A Catalog (Hemiptera: Tingidae)*. United States National Museum Bulletin 213: 1–634.

Drake, C. J. and J. A. Slater. 1957. The phylogeny and systematics of the family Thaumastocoridae (Hemiptera: Heteroptera). *Annals of the Entomological Society of America* 50: 353–370.

Dufour, L. 1833. Recherches anatomiques et physiologiques sure lest Hémiptères, accompagnées de considérations relatives à l'histoire naturelle, et à la classification de ces insectes. *Mémoires présentés par divers savants à l'Academie des Sciences de l'Institut de France* 4: 123–432.

Durai, P. S. S. 1987. A revision of the Dinidoridae of the world (Heteroptera: Pentatomidae). *Oriental Insects* 21: 163–360.

Eberhard, W. G., N. I. Platnick, and R. T. Schuh. 1993. Natural history and systematics of arthropod symbionts (Araneae; Hemiptera; Diptera) inhabiting webs of the spider *Tengella radiata* (Araneae, Tengellidae). *American Museum Novitates* 3065: 1–17.

Ehanno, B. 1983–1987. Le hétéroptères mirides de France. Tome I. Les secteurs biogeographiques. Inventaire Faune Flore. Volume 25. 1–603 pp. (1983); Tome I bis: Les secteurs biogeographiques (suite). Volume 39. 1–96c pp. (1987a); Tome II-A: Inventaire et syntheses ecologiques. Volume 40. 1–647 pp. (1987b); Tome II-B: Inventaire biogeographique et atlas Volume 42: 649–1075 pp. (1987c). Secretariat de la Faune et de la Flore, Paris.

Emsley, M. G. 1969. The Schizopteridae (Hemiptera: Heteroptera) with the description of new species from Trinidad. *Memoirs of the American Entomological Society* 25: 1–154.

Entwistle, P. F. 1972. *Pests of Cocoa*. Longman, London. 779 pp.

Erwin, T. L. 1982. Tropical forests: their richness in Coleoptera and other arthropod species. *Coleopterists Bulletin* 36: 74–75.

Erwin, T. L. 2004. The biodiversity question: how many species of terrestrial arthropods are there? Pp. 259–269. *In* M. D. Lowman and H. B. Rinker (eds). *Forest Canopies*. Second Edition. Elsevier Academic Press, Burlington.

Esaki, T. and W. E. China. 1927. A new family of aquatic Heteroptera. *Transactions of the Entomological Society of London* 1927: 279–295.

Esaki, T. and W. E. China. 1928. A monograph of the Helotrephidae, subfamily Helotrephinae (Hem. Heteroptera). *EOS, Revista Española de Entomologiá* 4: 129–172.

Ferreira, P. S. F. 2000. A taxonomic review of the tribe Clivinematini, with a key to world genera (Heteroptera). *Studies on Neotropical Fauna and Environment* 35: 38–43.

Ferreira, P. S. 2001. Diagnoses and description of the world genera of the tribe Clivinematini (Heteroptera: Miridae). *Studies on Neotropical Fauna and Environment* 36: 227–240.

Ferris, G. F. and R. L. Usinger. 1939. The family Polyctenidae (Hemiptera: Heteroptera). *Microentomology* 4: 1–50.

Freeman, P. 1947. A revision of the genus *Dysdercus* Boisduval (Hemiptera: Pyrrhocoridae), excluding the American species. *Transactions of the Royal Entomological Society of London* 98: 373–424.

Froeschner, R. C. 1941. Contributions to a synopsis of the Hemiptera of Missouri, Pt. I. Scutelleridae, Podopidae, Pentatomidae, Cydnidae, Thyreocoridae. *American Midland Naturalist* 26: 122–146.

Froeshcner, R. C. 1942. Contributions to a synopsis of the Hemiptera of Missouri, Pt. II. Coreidae, Aradidae, Neididae. *American Midland Naturalist* 26: 122–146.

Froeschner, R. C. 1944. Contributions to a synopsis of the Hemiptera of Missouri, Pt. III. Lygaeidae, Pyrrhocoridae, Piesmidae, Tingidae, Enicocephalidae, Phymatidae, Ploiariidae, Reduviidae, Nabidae. *American Midland Naturalist* 31: 638–683.

Froeschner, R. C. 1949. Contributions to a synopsis of the Hemiptera of Missouri, Pt. IV. Hebridae, Mesoveliidae,

Cimicidae, Anthocoridae, Cryptostemmatidae, Isometopidae, Miridae. *American Midland Naturalist* 42: 123–188.

Froeschner, R. C. 1960. Cydnidae of the Western Hemisphere. *Proceedings of the United States National Museum* 111: 337–680.

Froeschner, R. C. 1962. Contributions to a synopsis of the Hemiptera of Missouri, Pt. V. Hydrometridae, Gerridaee, Veliidae, Saldidae, Ochteridae, Gelastocoridae, Naucoridae, Belostomatidae, Nepidae, Notonectidae, Pleidae, Corixidae. *American Midland Naturalist* 67: 208–240.

Froeschner, R. C. 1985. Synopsis of the Heteroptera or true bugs of the Galápagos Islands. *Smithsonian Contributions to Zoology* 407: 1–84.

Froeschner, R. C. 1988a. Family Aradidae Spinola, 1837 (= Dysodiidae Reuter, 1912; Meziridae Oshanin, 1908. Pp. 29–46. *In* T. J. Henry and R. C. Froeschner (eds). *Catalog of the Heteroptera, or True Bugs, of Canada and the Continental United States.* E. J. Brill, Leiden and New York. 958 pp.

Froeschner, R. C. 1988b. Family Coreidae Leach, 1815. The coreid bugs. Pp. 69–92. *In* T. J. Henry and R. C. Froeschner (eds). *Catalog of the Heteroptera, or True Bugs, of Canada and the Continental United States.* E. J. Brill, Leiden and New York. 958 pp.

Froeschner, R. C. 1988c. Family Cydnidae Billberg, 1820. The burrowing bugs. Pp. 119–129. *In* T. J. Henry and R. C. Froeschner (eds). *Catalog of the Heteroptera, or True Bugs, of Canada and the Continental United States.* E. J. Brill, Leiden and New York. 958 pp.

Froeschner, R. C. 1988d. Family Pentatomidae Leach, 1815. The stink bugs. Pp. 544–597. *In* T. J. Henry and R. C. Froeschner (eds). *Catalog of the Heteroptera, or True Bugs, of Canada and the Continental United States.* E. J. Brill, Leiden and New York. 958 pp.

Froeschner, R. C. 1988e. Family Polyctenidae Westwood, 1874. The bat bugs. Pp. 611–612. *In* T. J. Henry and R. C. Froeschner (eds). *Catalog of the Heteroptera, or True Bugs, of Canada and the Continental United States.* E. J. Brill, Leiden and New York. 958 pp.

Froeschner, R. C. 1996. Lace bug genera of the world, I: introduction, subfamily Cantacaderinae (Heteroptera: Tingidae). *Smithsonian Contributions to Zoology* 574: 1–43.

Froeschner, R. C. 1997. *Rolstonus rolstoni,* new genus and new species of Acanthosomatidae from Argentina (Heteroptera: Pentatomidae: Ditomotarsini). *Journal of the New York Entomological Society* 103: 360–363 (1995).

Froeschner, R. C. 2000. Revision of the South American genus *Hellica* Stål (Heteroptera: Acanthosomatidae). *Journal of the New York Entomological Society* 107: 164–170 [1999].

Froeschner, R. C. and T. J. Henry. 1988. Family Berytidae Fieber, 1851 (= Neididae Kirkaldy, 1902; Berytinidae Southwood and Leston, 1959). Pp. 56–60. *In* T. J. Henry and R. C. Froeschner (eds). *Catalog of the Heteroptera, or True Bugs, of Canada and the Continental United States.* E. J. Brill, Leiden and New York. 958 pp.

Froeschner, R. C. and N. A. Kormilev. 1989. Phymatidae or ambush bugs of the world: a synonymic list with keys to species, except *Lophoscutus* and *Phymata* (Hemiptera). *Entomography* 6: 1–76.

Froeschner, R. C. and L. Peña. 1985. First South American record for the circum-Mediterranean *Patapius spinosus* (Rossi) (Heteroptera: Leptopodidae). *Revista Chilena de Entomologia* 12: 223.

Ghauri, M. S. K. and F. Y. K. Ghauri. 1983. A new genus and new species of Isometopidae from North India, with key to world genera (Heteroptera). *Reichenbachia* 21 (3): 19–25.

Göllner-Scheiding, U. 1983. General-Katalog der Familie Rhopalidae (Heteroptera). *Mitteilungen aus dem Zoologischen Museum in Berline* 59: 37–189.

Göllner-Scheiding, U. 2006. Family Scutelleridae Leach, 1815 – shield bugs. Pp. 190–227. *In* B. Aukema and C. Rieger (eds). *Catalogue of the Heteroptera of the Palaearctic Region.* Volume 5. Pentatomomorpha II. Netherlands Entomological Society, Amsterdam. 550 pp.

Gross, G. F. 1975. *Plant-feeding and Other Bugs (Hemiptera) of South Australia. Heteroptera – Part I.* A. B. James, Adelaide. 250 pp.

Hamid, A. 1971. A revision of Cryptorhamphinae (Heteroptera: Lygaeidae) including the description of two new species from Australia. *Journal of the Australian Entomological Society* 10: 163–174.

Hamid, A. 1975. A systematic revision of the Cyminae (Heteroptera: Lygaeidae) of the world with a discussion of the morphology, biology, phylogeny and zoogeography. *Entomological Society of Nigeria, Occasional Publication* No. 14: 1–179.

Harley, K. L. S. and R. C. Kassulke. 1971. Tingidae for biological control of *Lantana camara* [Verbenaceae]. *Entomophaga* 16: 389–410.

Harris, H. M. 1928. A monographic study of the hemipterous family Nabidae as it occurs in North America. *Entomologica Americana (n.s.)* 9: 1–98.

Heiss, E. 1995. Die amerikansische Platanennetzwanze *Corythucha ciliata* – eine Adventivart im Vormarsch auf Europa (Heteroptera, Tingidae). *Stapfia* 37: 143–148.

Heiss, E. and J. Péricart. 1983. Revision of Palaearctic Piesmatidae (Heteroptera). *Mitteilungen der Münchner Entomologischen Gesellschaft* 73: 61–171.

Helioevaara, K. 1984. Alary polymorphism and flight activity of *Aradus cinnamomeus* (Heteroptera, Aradidae). *Annales Entomologica Fenica* 50: 69–75.

Henry, T. J. 1980. Review of *Lidopus* Gibson and *Wetmorea* McAtee and Malloch, descriptions of three new genera and two new species, and key to New World genera (Hemiptera: Miridae: Isometopidae). *Proceedings of the Entomological Society of Washington* 82: 178–194.

Henry, T. J. 1984. Revision of the spider-commensal plant bug genus *Ranzovius* Distant (Heteroptera: Miridae). *Proceedings of the Entomological Society of Washington* 86: 53–67.

Henry, T. J. 1988a. Family Anthocoridae Fieber, 1837. The minute pirate bugs. Pp. 12–28. *In* T. J. Henry and R. C. Froeschner (eds). *Catalog of the Heteroptera, or True Bugs, of Canada and the Continental United States.* E. J. Brill, Leiden and New York. 958 pp.

Henry, T. J. 1988b. Family Ceratocombidae Fieber, 1861. The ceratocombids. Pp. 61–63. *In* T. J. Henry and R. C. Froeschner (eds). *Catalog of the Heteroptera, or True Bugs, of Canada and the Continental United States.* E. J. Brill, Leiden and New York. 958 pp.

Henry, T. J. 1988c. Family Dipsocoridae Dohrn, 1859 (= Cryptostemmatidae McAtee and Malloch, 1925). The dipsocorids. Pp. 130–131. *In* T. J. Henry and R. C. Froeschner (eds). *Catalog of the Heteroptera, or True Bugs, of Canada and the Continental United States.* E. J. Brill, Leiden and New York. 958 pp.

Henry, T. J. 1988d. Family Largidae Amyot and Serville, 1843. The largid bugs. Pp. 159–165. *In* T. J. Henry and R. C. Froeschner (eds). *Catalog of the Heteroptera, or True Bugs, of Canada and the Continental United States.* E. J. Brill, Leiden and New York. 958 pp.

Henry, T. J. 1988e. Family Microphysidae Dohrn, 1859. The microphysids. Pp. 249–250. *In* T. J. Henry and R. C. Froeschner (eds). *Catalog of the Heteroptera, or True Bugs, of Canada and the Continental United States.* E. J. Brill, Leiden and New York. 958 pp.

Henry, T. J. 1988f. Family Pyrrhocoridae Fieber, 1860. The cotton stainers. Pp. 613–615. *In* T. J. Henry and R. C. Froeschner (eds). *Catalog of the Heteroptera, or True Bugs, of Canada and the Continental United States.* E. J. Brill, Leiden and New York. 958 pp.

Henry, T. J. 1988g. Family Rhopalidae Amyot and Serville, 1843 (= Corizidae [Costa 1853]). The scentless plant bugs. Pp. 652–664. *In* T. J. Henry and R. C. Froeschner (eds). *Catalog of the Heteroptera, or True Bugs, of Canada and the Continental United States.* E. J. Brill, Leiden and New York. 958 pp.

Henry, T. J. 1988h. Family Schizopteridae Reuter, 1891. The schizopterids. Pp. 682–683. *In* T. J. Henry and R. C. Froeschner (eds). *Catalog of the Heteroptera, or True Bugs, of Canada and the Continental United States.* E. J. Brill, Leiden and New York. 958 pp.

Henry, T. J. 1994. Revision of the myrmecomorphic plant bug genus *Schaffneria* Knight (Heteroptera: Miridae: Orthotylinae). *Proceedings of the Entomological Society of Washington* 96: 701–712.

Henry, T. J. 1997a. Phylogenetic analysis of family groups within the infraorder Pentatomomorpha (Hemiptera: Heteroptera), with emphasis on the Lygaeoidea. *Annals of the Entomological Society of America* 90: 275–301.

Henry, T. J. 1997b. Cladistic analysis and revision of the stilt bug genera of the world (Heteroptera: Berytidae). *Contributions of the American Entomological Institute* 30 (1): 1–100.

Henry, T. J. 1997c. Monograph of the stilt bugs, or Berytidae (Heteroptera), of the Western Hemisphere. *Memoirs of the Entomological Society of Washington* 19: 1–149.

Henry, T. J. 1999. The spider-commensal plant bug genus *Ranzovius* (Heteroptera: Miridae: Phylinae) revisited: three new species and a revised key, with the description of a new sister genus and phylogenetic analysis. *Acta Societatis Zoologicae Bohemicae* 63: 93–115.

Henry, T. J. 2000a. Review of the stilt bug genus *Hoplinus* with the description of a new species and notes on other Hoplinini (Heteroptera: Berytidae: Gampsocorinae). *Journal of the New York Entomological Society* 110: 182–191.

Henry, T. J. 2000b. Stilt bugs (Berytidae). Pp. 725–735. *In* C. W. Schaefer and A. R. Panizzi (eds). *Heteroptera of Economic Importance.* CRC Press, Boca Raton, Florida. 828 pp.

Henry, T. J. 2000c. Phylogeny of the New World plant bug tribe Ceratocapsini (Heteroptera: Miridae: Orthotylinae).XXI – International Congress of Entomology, Brazil, August 20–26, 2000. Abstract Volume II. 917 pp.

Henry, T. J. 2000d. The predatory Miridae: A glimpse at the other plant bugs. *Wings: Essays on Invertebrate Conservation* 23 (1): 17–20.

Henry, T. J. 2001. Revision of the orthotyline plant bug genus *Hyalochloria*, with a key and descriptions of four new species (Hemiptera: Heteroptera: Miridae). *Journal of the New York Entomological Society* 109: 235–262.

Henry, T. J. 2003. Richard C. Froeschner (1916–2002): biographical sketch, described taxa, and publications. *Proceedings of the Entomological Society of Washington* 105: 1075–1086.

Henry, T. J. 2004. *Raglius alboacuminatus* (Goeze) and *Rhyparochromus vulgaris* (Schilling) (Lygaeoidea: Rhyparochromidae): two Palearctic bugs newly discovered in North America. *Proceedings of the Entomological Society of Washington* 106: 513–522.

Henry, T. J. 2007. A newly discovered Brazilian species of the stilt bug genus *Jalysus* (Hemiptera: Heteroptera: Berytidae) associated with myrmecophytic plants. *Proceedings of the Entomological Society of Washington* 109: 324–330.

Henry, T. J. and D. Adamski. 1998. *Rhparochromus saturnius* (Rossi) (Heteroptera: Lygaeoidea: Rhyparochromidae), a Palearctic seed bug newly discovered in North America. *Journal of the New York Entomological Society* 106: 132–140.

Henry, T. J. and R. C. Froeschner (eds). 1988. *Catalog of the Heteroptera, or True Bugs, of Canada and the Continental United States,* E. J. Brill, Leiden and New York. 958pp. [See specific family citations under respective author names].

Henry, T. J. and R. C. Froeschner. 1998. Catalog of the stilt bugs, or Berytidae, of the world (Insecta: Hemiptera: Heteroptera). *Contributions of the American Entomological Institute* 30 (4): 1–72.

Henry, T. J. and R. C. Froeschner. 2000. Corrections and additions to the "Catalog of the stilt bugs, or Berytidae, of the world (Insecta: Hemiptera: Heteroptera)." *Proceedings of the Entomological Society of Washington* 102: 1003–1009.

Henry, T. J. and A. S. Paula. 2004. *Rhyparochromomiris femoratus*, a remarkable new genus and species of Cylapinae (Hemiptera: Heteroptera: Miridae) from Ecuador. *Journal of the New York Entomological Society* 112: 176–182.

Henry, T. J. and A. G. Wheeler, Jr. 1988. Family Miridae Hahn, 1833 (= Capsidae Burmeister, 1835). The plant bugs. Pp. 251–507. *In* T. J. Henry and R. C. Froeschner (eds). *Catalog of the Heteroptera, or True Bugs, of Canada and the Continental United States*. E. J. Brill, Leiden and New York. 958 pp.

Henry, T. J. and A. G. Wheeler, Jr. 2007. Plant bugs. Pp. 83–91. *In* W. O. Lamp, R. C. Berberet, L. G. Higley, and C. R. Baird (eds). *Handbook of Forage and Rangeland Insects*. Entomological Society of America, Lanham, Maryland. 180 pp.

Henry, T. J. and M. R. Wilson. 2004. First records of eleven true bugs (Hemiptera: Heteroptera) from the Galápagos Islands, with miscellaneous notes and corrections to published reports. *Journal of the New York Entomological Society* 112: 75–86.

Henry, T. J., J. W. Neal, and K. M. Gott, Jr. 1986. *Stethoconus japonicus* (Heteroptera: Miridae): a predator of *Stephanitis* lace bugs newly discovered in the United States, promising in the biocontrol of azalea lace bugs (Heteroptera: Tingidae). *Proceedings of the Entomological Society of Washington* 88: 722–730.

Herczek, A. 1993. *Systematic Position of Isometopinae Fieb. (Miridae, Heteroptera) and Their Intrarelationships*. Uniwersytet šlaski, Katowice. 86 pp.

Hernández, L. M. and G. M. Stonedahl. 1999. A review of the economically important species of the genus *Orius* (Heteroptera: Anthocoridae) in East Africa. *Journal of Natural History* 33: 543–568.

Hickman, V. V. 1969. The biology of two emesine bugs (Hemiptera: Reduviidae) on nests or webs of spiders. *Journal of the Entomological Society of Australia (N.S.W.)* 6: 3–18.

Hoffman, R. L. and J. A. Slater. 1995. *Holcocranum saturejae*, a Palearctic cattail bug established in eastern United States and tropical Africa (Heteroptera: Lygaeidae: Artheneinae). *Banisteria* 5: 12–15.

Hoffman, R. L. and W. E. Steiner. 2005. *Jadera haematoloma*, another insect on its way north (Heteroptera: Rhopalidae). *Banisteria* 26: 7–10.

Hoffman, R. L., S. M. Roble, and T. J. Henry. 2005. The occurrence in Florida and Virginia of *Corixidea major*, an exceptionally rare North American bug (Heteroptera: Schizopteridae). *Banisteria* 26: 18–19.

Horváth, G. 1904. Monographia Colobathrastinarum. *Annales Musei Natiionalis Hungarici* 2: 117–172.

Hribar, L. J. and T. J. Henry. 2007. *Empicoris subparallellus* (Hemiptera: Heteroptera: Reduviidae), a predatory bug new to the fauna of Florida. *Florida Entomologist* 90: 738–741.

Hsaio, T. Y. and R. I. Sailer. 1947. The orchid bugs of the genus *Tenthecoris* Scott (Hemiptera: Miridae). *Journal of the Washington Academy of Science* 37: 64–72.

Hungerford, H. B. 1948. The Corixidae of the Western Hemisphere (Hemiptera). *Kansas University Science Bulletin* 32: 1–827.

Hussey, R. F. 1929. Pyrrhocoridae. Fascicle III. Pp. 1–144. *In* G. Horváth and H. M. Parshley (eds). *General Catalogue of the Hemiptera*. Smith College, Northampton, Massachusetts.

Hussey, R. F. 1934. Observations on *Pachycoris torridus* (Scop.), with remarks on parental care in other Hemiptera. *Bulletin of the Brooklyn Entomological Society* 29: 133–145.

Jacobs, D. H. 1989. A new species of *Thaumastella* with notes on the morphology, biology and distribution of the two southern African species (Heteroptera: Thaumastellidae). *Journal of the Entomological Society of South Africa* 52 (2): 301–316.

Jakovlev, V. E. 1902. New species of the genus *Stenocephalus* Latr. (Hemiptera-Heteroptera, Coreidae). *Horae Societatis Entomologicae Rossicae* 35: 202–209.

Jansson, A. 1986. The Corixidae (Heteroptera) of Europe and some adjacent regions. *Acta Entomologica Fennica* 47: 1–94.

Jansson, A. 1987. Micronectinae (Heteroptera, Corixidae) as indicators of water quality in Lake Vesijärvi, southern Finland, during the period 1976–1986. *Biological Research Report of the University of Jyväskylä* 10: 119–128.

Javahery, M., C. W. Schaefer, and J. D. Lattin. 2000. Shield bugs (Scutelleridae). Pp. 475–503. *In* C. W. Schaefer and A. R. Panizzi (eds). *Heteroptera of Economic Importance*. CRC, Boca Raton, Florida. 828 pp.

Jessop, L. 1983. A review of the genera of Plataspidae (Hemiptera) related to *Libyaspis*, with a revision of *Cantharodes*. *Journal of Natural History* 176: 31–62.

Kelton, L. A. 1978. *The Anthocoridae of Canada and Alaska. Heteroptera: Anthocoridae. The Insects and Arachnids of Canada. Part 4*. Publication 1639. Biosytematics Research Institute, Canada Department of Agriculture, Ottawa, Ontario. 101 pp.

Kelton, L. A. 1985. Species of the genus *Fulvius* Stål found in Canada (Heteroptera: Miridae: Cylapinae). *Canadian Entomologist* 117: 1071–1073.

Kerzhner, I. M. 1981. Fauna of the USSR. Heteroptera of the family Nabidae. *Academy of Sciences USSR, Zoological Institute, Leningrad. (N.S.) No. 127*. 13: 1–326. [In Russian].

Kerzhner, I. M. 1989. On the taxonomy and habits of the genus *Medocostes* (Heteroptera: Nabidae). *Zoologichesky Zhurnal* 68: 150–151.

Kerzhner, I. M. 1995a. Infraorder Enicocephalomorpha. Pp. 1–5. *In* B. Aukema and C. Rieger (eds). *Catalog of the Heteroptera of the Palaearctic Region. Vol. 1. Enicocephalomorpha, Dipsocoromorpha, Nepomorpha, Gerromorpha and Leptopodomorpha*. Netherlands Entomological Society, Amsterdam. 222 pp.

Kerzhner, I. M. 1995b. Infraorder Dipsocoromorpha. Pp. 6–12. *In* B. Aukema and C. Rieger (eds). *Catalog of the Heteroptera of the Palaearctic Region. Vol. 1. Enico-cephalomorpha, Dipsocoromorpha, Nepomorpha, Gerromorpha and Leptopodomorpha.* Netherlands Entomological Society, Amsterdam. 222 pp.

Kerzhner, I. M. 1997. East Palaearctic species of the genus *Artheneis* (Heteroptera: Lygaeidae). *Zoosystematica Rossica* 6: 213–222.

Kerzhner, I. M. 2001. Superfamily Pyrrhocoroidea Amyot and Serville, 1843. Pp. 245–258. *In* B. Aukema and C. Rieger (eds). *Catalog of the Heteroptera of the Palaearctic Region. Vol. 4. Pentatomomorpha I.* Netherlands Entomological Society, Amsterdam. 346 pp.

Kerzhner, I. M. and M. Josifov. 1999. Miridae Hahn, 1833. Pp. 1–577. *In* B. Aukema and C. Rieger (eds). *Catalog of the Heteroptera of the Palaearctic Region. Vol. 3. Cimicomorpha II.* Netherlands Entomological Society, Amsterdam. 577 pp.

Kiritani, K. 2007. The impact of global warming and land-use change on the pest status of rice and fruit bugs (Heteroptera) in Japan. *Global Change Biology* 13: 1586–1595.

Kirkaldy, G. W. 1909. *Catalog of the Hemiptera (Heteroptera) with Biological and Anatomical References, Lists of Foodplants and Parasites, etc. Prefaced by a Discussion on Nomenclature and an Analytical Table of Families. Volume I: Cimicidae.* Felix L. Dames, Berlin. 392 pp.

Kment, P. and T. J. Henry. 2008. Two cases of homonymy in the family Berytidae (Heteroptera). *Proceedings of the Entomological Society of Washington* 110: 811–813.

Kormilev, N. A. 1955. A new myrmecophil family of Hemiptera from the delta of Rio Paraná, Argentina. *Revista Ecuatoriana de Entomologia y Parasitologia* 2 (3–4): 465–477.

Kormilev, N. A. 1962. Revision of the Phymatinae (Hemiptera: Phymatidae). *The Philippine Journal of Science* 89: 287–486.

Kormilev, N. A. and R. C. Froeschner. 1987. Flat bugs of the world: a synonymic list (Heteroptera: Aradidae). *Entomography* 5: 1–246.

Kullenberg, B. 1944. Studien über die Biologie der Capsiden. *Zoologiska Bidrag Uppsula* 23: 1–522.

Kumar, R. 1974. A revision of the world Acanthosomatidae (Heteroptera: Pentatomoidea): keys to and description of subfamilies, tribes and genera, with designation of types. *Australian Journal of Zoology Supplement Series* 34: 1–60.

Lansbury, I. 1965. A revision of the Stenocephalidae Dallas 1852 (Hemiptera-Heteroptera). *Entomologist's Monthly Magazine* 101: 52–92.

Lansbury, I. 1966. A revision of the Stenocephalidae Dallas 1852 (Hemiptera-Heteroptera). *Entomologist's Monthly Magazine* 101: 145–150.

Lattin, J. D. 1989. Bionomics of the Nabidae. *Annual Review of Entomology* 34: 383–440.

Lattin, J. D. 1999. Bionomics of the Anthocoridae. *Annual Review of Entomology* 44: 207–231.

Lattin, J. D. 2000. Minute pirate bugs (Anthocoridae). Pp. 607–637. *In* C. W. Schaefer and A. R. Panizzi (eds). *Heteroptera of Economic Importance.* CRC Press, Boca Raton, Florida. 828 pp.

Laucke, D. R. and A. S. Menke. 1961. The higher classification of the Belostomatidae (Hemiptera). *Annals of the Entomological Society of America* 54: 644–657.

Lent, H. and J. Jurberg. 1965. Contribuição ao conhecimento dos Phloeidae Dallas, 1851, com um estudo sôbre genitália (Hemiptera, Pentatomoidea). *Revista de Brasileira Biologia* 25 (2): 123–144.

Lent, H. and J. Jurberg. 1966. Os estádios larvares de "Phloeophana longirostris" (Spinola 1837) (Hemiptera, Pentatomoidea). *Revista de Brasileira Biologia* 26: 1–4.

Lent, H. and P. Wygodzinsky. 1979. Revision of the Triatominae (Hemiptera, Reduviidae), and their significance as vectors of Chagas' disease. *Bulletin of the American Museum of Natural History* 163: 123–520.

Leston, D. 1958. Chromosome number and the systematics of Pentatomomorpha (Hemiptera). *Proceeding of the Tenth International Congress of Entomology* 2: 911–918.

Leston, D., J. G. Pendergrast, and T. R. E. Southwood. 1954. Classification of the terrestrial Heteroptera (Geocorisae). *Nature* 174: 91.

Lis, B. 1999. Phylogeny and classification of Cantacaderini [= Cantacaderidae stat. nov.] (Hemiptera: Tingoidea). *Annales Zoologici* 49 (3): 157–196.

Lis, J. A. 1999. Burrower bugs of the Old World – a catalogue (Hemiptera: Heteroptera: Cydnidae). *Genus (Wroclaw)* 10 (2): 165–249.

Lis, J. A. 2000. A revision of the burrower-bug genus *Macroscytus* Fieber, 1860 (Hemiptera: Heteroptera: Cydnidae). *Genus (Wroclaw)* 11 (3): 359–509.

Lis, J. A. 2001. A review of the genus *Byrsinus* Fieber, 1860 (Hemiptera: Heteroptera: Cydnidae) in Australia, with a key to the Australo-Pacific species. *Annales Zoologici* 51 (2): 197–203.

Lis, J. A. 2006. Family Thyreocoridae Amyot and Serville, 1843 - negro bugs. Pp. 148–149. *In* B. Aukema and C. Rieger (eds). *Catalogue of the Heteroptera of the Palaearctic Region. Volume 5. Pentatomomorpha II.* Amsterdam. 550 pp.

Lis, J. A., M. Becker, and C. W. Schaefer. 2000. Burrower bugs (Cydnidae). Pp. 405–419. *In* C. W. Schaefer and A. R. Panizzi (eds). *Heteroptera of Economic Importance.* CRC Press, Boca Raton, Florida. 828 pp.

Louis, D. 1974. Biology of Reduviidae of cocoa farms in Ghana. *American Midland Naturalist* 91: 68–89.

Maa, T. C. 1964. A review of the Old World Polyctenidae. *Pacific Insects* 6: 493–516.

Maa, T. C. and K. S. Lin. 1956. A synopsis of the Old World Phymatidae (Hem.). *Quarterly Journal of the Taiwan Museum* 9: 109–154.

Maldonado Capriles, J. 1990. *Systematic Catalog of the Reduviidae of the World.* University of Puerto Rico, Mayaguez. 694 pp.

Marshall, A. G. 1982. The ecology of the bat ectoparasite *Eoctenes spasmae* (Hemiptera: Polyctenidae) in Malaysia. *Biotropica* 14: 50–55.

Maw, H. E. L., R. G. Foottit, K. G. A. Hamilton, and G. G. E. Scudder. 2000. *Checklist of the Hemiptera of Canada and Alaska.* NRC Research Press, Ottawa, Ontario, Canada. 220 pp.

McAtee, W. L. and J. R. Malloch. 1925. Revision of bugs of the family Cryptostemmatidae in the collection of the United States National Museum. *Proceedings of the United States National Museum* 67 (13): 1–42.

McAtee, W. L. and J. R. Malloch. 1928. Synopsis of pentatomoid bugs of the subfamilies Megaridinae and Canopinae. *Proceedings of the United States National Museum* 72 (25): 1–21.

McDonald, F. J. D. 1969. A new species of Lestoniidae (Hemiptera). *Pacific Insects* 11: 187–190.

McDonald, F. J. D. 1970. The morphology of *Lestonia haustorifera* China (Heteroptera: Lestoniidae). *Journal of Natural History* 4: 413–417.

McDonald, F. J. D. 1979. A new species of *Megaris* and the status of the Megarididae McAtee and Malloch and Canopidae Amyot and Serville (Hemiptera: Pentatomoidea). *Journal of the New York Entomological Society* 87: 42–54.

McHugh, J. V. 1994. On the natural history of Canopidae (Heteroptera: Pentatomidae). *Journal of the New York Entomological Society* 102: 112–114.

McIver, J. D. and G. M. Stonedahl. 1987. Biology of the myrmecomorphic plant bug *Coquillettia insignis* Uhler (Heteroptera: Miridae: Phylinae). *Journal of the New York Entomological Society* 95: 258–277.

McMahon, E. A. 1983. Adaptations, feeding preferences, and biometrics of a termite-baiting assassin (Hemiptera: Reduviidae). *Annals of the Entomological Society of America* 76: 483–486.

McMahon, E. A. 2005. Adventures documenting an assassin bug that 'fishes' for termites. *American Entomologist* 51: 202–207.

McPherson, J. E. 1982. *The Pentatomoidea (Hemiptera) of Northeastern North America with Emphasis on the Fauna of Illinois.* Southern Illinois University Press, Carbondale, Illinois. 240 pp.

McPherson, J. E. and R. M. McPherson. 2000. *Stink Bugs of Economic Importance in America North of Mexico.* CRC Press, Boca Raton, Florida. 253 pp.

Mead, F. W. 1971. *Leaffooted bug, Leptoglossus phyllopus (Linnaeus) (Hemiptera: Coreidae).* Florida Department of Agriculture and Consumer Services, Division of Plant Industry, Entomology Circular No. 107. 2 pp.

Menke, A. S. (ed). 1979. The semiaquatic and aquatic Hemiptera of California (Heteroptera: Hemiptera). *Bulletin of the California Insect Survey* 21: 1–166.

Metcalf, Z. P. 1940. How many insects are there in the world?. *Entomological News* 51: 219–222.

Miller, L. T. and W. T. Nagamine. 2005. First records of *Corythucha gossypii* (Hemiptera: Tingidae) in Hawaii, including notes on host plants. *Proceedings of the Hawaiian Entomological Society* 37: 85–88.

Mitchell, P. L. 2000. Leaf-footed bugs (Coreidae). Pp. 337–403. *In* C. W. Schaefer and A. R. Panizzi (eds). *Heteroptera of Economic Importance.* CRC Press, Boca Raton, Florida. 828 pp.

Mitchell, P. L. 2006. Polyphagy in true bugs: A case study of *Leptoglossus phyllopus* (L.) (Hemiptera, Heteroptera, Coreidae). Pp. 1117–1134. *In* W. Rabitsch (ed). *Hug the Bug – for the Love of True Bugs. Festschrift Zum 70.* Geburtstag von Ernst Heiss. Denisia 19: 1–1184.

Mittermeier, R. A., A. Russell, P. Robles Gil, M. Hoffmann, J. Pilgrim, T. Brooks, C. Goettsch Mittermeier, J. Lamoreux, and G. Da Fonseca. 2004. *Hotspots Revisited: Earth's Biologically Richest and Most Endangered Terrestrial Ecoregions.* CEMEX, Mexico City. 392 pp.

Miyamoto, S. 1956. Discovery of *Parastrachia japonensis* eggs. *Kôntyû* 24: 232.

Morkel, C. 2006. On kleptoparasitic stilt bugs (Insecta, Heteroptera: Berytidae) in spider funnel-webs (Arachnidae, Araneae: Agelenidae), with notes on their origin. *Mainzer Naturwissenschaftliches Archiv* 31: 129–143.

Moul, E. T. 1945. Notes on *Arilus critatus* (Linnaeus) in York County, Pennsylvania and on its prey (Heteroptera: Reduviidae). *Entomological News* 56: 57–59.

Moulet, P. 1995a. Hémiptères Coreoidea (Coreidae, Rhopalidae, Alydidae), Pyrrhocoridae, Stenocephalidae, Euro-Méditerranéens. *Faune de France* 81: 1–336.

Moulet, P. 1995b. Synonymies nouvelles dans la famille des Stenocephalidae Latreille, 1825 (Heteroptera, Stenocephalidae). *Nouvelle Revue Entomologie* 11: 353–364 [1994].

Musolin, D. L. 2007. Insects in a warmer world: ecological, physiological and life-history responses of true bugs (Heteroptera) to climate change. *Global Change Biology* 13: 1565–1585.

Myers, J. G. 1924. On the systematic position of the family Termitaphididaae with a description of a new genus and species from Panama. *Psyche* 31: 259–278.

Myers, J. G. 1925. Biological notes on *Arachnocoris albomaculatus* Scott (Hemiptera: Nabidae). *Journal of the New York Entomological Society* 33: 136–145.

Myers, J. G. 1932. Observations on the family Termitaphididae with the description of a new species from Jamaica. *Annals and Magazine of Natural History* 9 (10): 366–372.

Nicholson, T. 2006. Elsie Froesschner, founding member of Guild, passes away. *Guild of Natural Science Illustrators Newsletter* 38: 14.

Packauskas, R. J. 1994. Key to the subfamilies and tribes of the New World Coreidae (Hemiptera), with a checklist of published keys to genera and species. *Proceedings of the Entomological Society of Washington* 96: 44–53.

Packauskas, R. J. and C. W. Schaefer. 1998. Revision of the Cyrtocoridae (Hemiptera: Pentatomoidea). *Annals of the Entomological Society of America* 91: 363–386.

Panizzi, A. R., J. E. McPherson, D. G. James, M. Javahery, and R. M. McPherson. 2000. Stink bugs (Pentatomidae). Pp. 421–474. *In* C. W. Schaefer and A. R. Panizzi (eds). *Heteroptera of Economic Importance.* CRC, Boca Raton, Florida. 828 pp.

Péricart, J. 1972. Hémiptères. Anthocoridae, Cimicidae et Microphyidae de L'Ouest-Palearctique. Faune de L'Europe et du Bassin Mediterraneén. *Masson et Cie, Paris* 7: 1–402.

Péricart, J. 1987. Hémiptères Nabidae d'Europe occidentale et du Maghreb. *Faune de France* 71: 1–169.

Péricart, J. 1996. Family Anthocoridae Fieber, 1836 – flower bugs, minute pirate bugs. Pp. 108–140. *In* B. Aukema and C. Rieger (eds). *Catalogue of the Heteroptera of the Palaearctic Region.* Volume 2. Cimicomorpha I. Netherlands Entomological Society, Amsterdam. 361 pp.

Péricart, J. 1998. Hémiptères Lygaeidae Euro-Méditerranéens. Faune de France. 84 A–C. I, Généralités Systématique: Première Partie, 468 pp.; II, Systématique: Seconde Partie, 453 pp.; III, Systématique: Troisième Partie, 487 pp.

Polhemus, D. A. 1988. Family Pleidae Fieber, 1851. The pygmy backswimmers. Pp. 608–610. *In* T. J. Henry and R. C. Froeschner (eds). *Catalog of the Heteroptera, or True Bugs, of Canada and the Continental United States.* E. J. Brill, Leiden and New York. 958 pp.

Polhemus, D. A. 1997. *Systematics of the Genus Rhagovelia Mayr (Heteroptera: Veliidae) in the Western Hemisphere (Exclusive of the angustipes complex).* Thomas Say Publications in Entomology: Monographs. Entomological Society of America, Lanham, Maryland. 386 pp.

Polhemus, D. A. and J. T. Polhemus. 1988a. The Aphelocheirinae of tropical Asia (Heteroptera: Naucoridae). *Raffles Bulletin of Zoology* 36 (2): 167–300.

Polhemus, D. A. and J. T. Polhemus. 1988b. Family Naucoridae Leach, 1815. The creeping water bugs. Pp. 521–527. *In* T. J. Henry and R. C. Froeschner (eds). *Catalog of the Heteroptera, or True Bugs, of Canada and the Continental United States.* E. J. Brill, Leiden and New York. 958 pp.

Polhemus, D. A. and J. T. Polhemus. 1988c. Family Ochteridae Kirkaldy, 1906. The velvety short bugs. Pp. 541–543. *In* T. J. Henry and R. C. Froeschner (eds). *Catalog of the Heteroptera, or True Bugs, of Canada and the Continental United States.* E. J. Brill, Leiden and New York. 958 pp.

Polhemus, J. T. 1988. Family Saldidae Amyot and Serville, 1843. The shore bugs. Pp. 665–681. *In* T. J. Henry and R. C. Froeschner (eds). *Catalog of the Heteroptera, or True Bugs, of Canada and the Continental United States.* E. J. Brill, Leiden and New York. 958 pp.

Polhemus, J. T. 1995a. Family Gelastocoridae Kirkaldy, 1897 – toad bugs. Pp. 23–25. *In* B. Aukema and C. Rieger (eds). *Catalog of the Heteroptera of the Palaearctic Region.* Volume 1. Enicocephalomorpha, Dipsocoromorpha, Nepomorpha, Gerromorpha and Leptopodomorpha. Netherlands Entomological Society, Amsterdam. 222 pp.

Polhemus, J. T. 1995b. Family Ochteridae Kirkaldy, 1906 – velvety shore bugs. Pp. 25–26. *In* B. Aukema and C. Rieger (eds). *Catalog of the Heteroptera of the Palaearctic Region.* Volume 1. Enicocephalomorpha, Dipsocoromorpha, Nepomorpha, Gerromorpha and Leptopodomorpha. Netherlands Entomological Society, Amsterdam. 222 pp.

Polhemus, J. T. and D. A. Polhemus. 1988a. Family Hebridae Amyot and Serville, 1843. The velvet water bugs. Pp. 152–155. *In* T. J. Henry and R. C. Froeschner (eds). *Catalog of the Heteroptera, or True Bugs, of Canada and the Continental United States.* E. J. Brill, Leiden and New York. 958 pp.

Polhemus, J. T. and D. A. Polhemus. 1988b. Family Notonectidae Latreille, 1802. The backswimmers. Pp. 533–540. *In* T. J. Henry and R. C. Froeschner (eds). *Catalog of the Heteroptera, or True Bugs, of Canada and the Continental United States.* E. J. Brill, Leiden and New York. 958 pp.

Polhemus, J. T. and D. A. Polhemus. 2007. Global trends in the description of aquatic and semiaquatic Heteroptera species, 1758–2004. *Tijdschrift voor Entomologie* 150: 271–288.

Polhemus, J. T. and D. A. Polhemus. 2008. Global diversity of true bugs (Heteroptera; Insecta) in freshwater. *Hydrobiologia* 595: 379–391.

Polhemus, J. T., R. C. Froeschner, and D. A. Polhemus. 1988a. Family Corixidae Leach, 1815. The water boatmen. Pp. 93–118. *In* T. J. Henry and R. C. Froeschner (eds). *Catalog of the Heteroptera, or True Bugs, of Canada and the Continental United States.* E. J. Brill, Leiden and New York. 958 pp.

Polhemus, J. T., D. A. Polhemus, and T. J. Henry. 1988b. Family Belostomatidae Leach, 1815. The giant water bugs or electric light bugs. Pp. 47–55. *In* T. J. Henry and R. C. Froeschner (eds). *Catalog of the Heteroptera, or True Bugs, of Canada and the Continental United States.* E. J. Brill, Leiden and New York. 958 pp.

Polhemus, J. T., A. Jansson, and E. Kanyukova. 1995. Infraorder Nepomorpha – water bugs. Pp. 13. *In* B. Aukema and C. Rieger (eds). *Catalog of the Heteroptera of the Palaearctic Region.* Volume 1. Enicocephalomorpha, Dipsocoromorpha, Nepomorpha, Gerromorpha and Leptopodomorpha. Netherlands Entomological Society, Amsterdam. 222 pp.

Putshkov, V. G. 1966. *The Main Bugs – Plant Bugs – as Pests of Agricultural Crops.* Naukova Dumka, Kiev. 171 pp. [In Ukranian].

Putshkov, V. G. and P. V. Putshkov. 1985–1989. A Catalogue of the Assassin Bugs (Heteroptera, Reduviidae) of the World. Genera (1985). 1–138 pp.; I. Ectrichodiinae (1986): 1–75 [VINITI, No. 4852-B86]; II. Reduviinae, Peiratinae, Phimophorinae, Physoderinae, Saicinae, Salyavatinae, Sphaeridopinae (1987): 1–212 [VINITI, No.

4698-B87]; IV. Bactrodinae, Centrocneminae, Cetherinae, Chryxinae, Elasmodeminae, Manangocorinae, Hammacerainae, Holoptilinae, Stenopodinae, Tribelocephalinae, Triatominae, Vesciinae (1988a): 1–145 [VINITI, No. 287-B88]; III. Harpactorinae (1988b): 1–264 [VINITI, No. 286-B89]. Typescript deposited in VINITI, Moscow and Kiev. [Microfiche copies also deposited in other institutions, including USNM, Washington, DC].

Putshkov, V. G. and P. V. Putshkov. 1996. Family Reduviidae Latreille, 1807 – assassin-bugs. Pp. 148–265. *In* B. Aukema and C. Rieger (eds). *Catalogue of the Heteroptera of the Palaearctic Region*. Volume 2. Cimicomorpha I. Netherlands Entomological Society, Amsterdam. 361 pp.

Readio, J. and M. H. Sweet. 1982. A review of the Geocorinae of the United States east of the 100th Meridian (Hemiptera: Lygaeidae). *Miscellaneous Publication of the Entomological Society of America* 12: 1–91.

Redei, D. 2007. A new species of the family Hypsipterygidae from Vietnam, with notes on the hypsipterygid fore wing venation (Heteroptera: Dipsocoromorpha). *Deutsch Entomologische Zeitschrift* 54: 43–50.

Rider, D. A. 2006a. Family Pentatomidae Leach, 1815. Pp. 233–402. *In* B. Aukema and C. Rieger (eds). *Catalogue of the Heteroptera of the Palaearctic Region*. Volume 5. Pentatomomorpha II. Netherlands Entomological Society, Amsterdam. 550 pp.

Rider, D. A. 2006b. Family Tessaratomidae Stål, 1865. Pp. 182–189. *In* B. Aukema and C. Rieger (eds). *Catalogue of the Heteroptera of the Palaearctic Region*. Volume 5. Pentatomomorpha II. Netherlands Entomological Society, Amsterdam. 550 pp.

Rider, D. A. 2006c. Family Urostylidae Dallas, 1851. Pp. 102–116. *In* B. Aukema and C. Rieger (eds). *Catalogue of the Heteroptera of the Palaearctic Region*. Volume 5. Pentatomomorpha II. Netherlands Entomological Society, Amsterdam. 550 pp.

Rider, D. A. 2008. Pentatomoidea Home Page. North Dakota State University, Fargo. http://www.ndsu.nodak.edu/ndsu/rider/Pentatomoidea/index.htm [Accessed January 2008].

Rolston, L. H., R. L. Aalbu, M. J. Murray, and D. A. Rider. 1994. A catalog of the Tessaratomidae of the world. *Papua New Guinea Journal of Agriculture, Forestry and Fisheries* 36 (2): 36–108.

Rolston, L. H., D. A. Rider, M. J. Murray, and R. L. Aalbu. 1996. Catalog of the Dinidoridae of the world. *Papua New Guinea Journal of Agriculture, Forestry and Fisheries* 39 (1): 22–101.

Ryckman, R. E. and R. D. Sjogren. 1980. A catalogue of the Polyctenidae. *Bulletin of the Society of Vector Ecologists* 5: 1–22.

Ryckman, R. E., D. G. Bentley, and E. F. Archbold. 1981. The Cimicidae of the Americas and oceanic islands, a checklist and bibliography. *Bulletin of the Society of Vector Ecologists* 6: 93–142.

Ryerson, S. A. and J. D. Stone. 1979. A selected bibliography of two species of *Orius*: the minute pirate bugs, *Orius tristicolor*, and *Orius insidiosus* (Heteroptera: Anthocoridae). *Bulletin of the Entomological Society of America* 25: 131–135.

Sabrosky, C. W. 1952. How many insects are there? Insects. Pp. 1–7. *In* A. Stefferud (ed). *The Yearbook of Agriculture 1952*. United States Department of Agriculture, Washington, D. C.

Samy, O. 1969. A revision of the African species of *Oxycarenus* (Hemiptera: Lygaeidae). *Transactions of the Royal Entomological Society of London* 121: 79–165.

Schaefer, C. W. 1965. The morphology and higher classification of the Coreoidea (Hemiptera-Heteroptera): Part III. The families Rhopalidae, Alydidae, and Coreidae. *Miscellaneous Publications of the Entomological Society of America* 5: 1–76.

Schaefer, C. W. 1975. Heteropteran trichobothria (Hemiptera: Heteroptera). *International Journal of Insect Morphology and Embryology* 4: 193–264.

Schaefer, C. W. 1981. The morphology and relationships of the Stenocephalidae and Hyocephalidae (Hemiptera: Heteroptera: Coreoidea). *Annals of the Entomological Society of America* 74: 83–95.

Schaefer, C. W. 1988. The food plants of some "primitive" Pentatomoidea (Hemiptera: Heteroptera). *Phytophaga* 2: 19–45.

Schaefer, C. W. 1993. Notes on the morphology and family relationships of Lestoniidae (Hemiptera: Heteroptera). *Proceedings of the Entomological Society of Washington* 95: 453–456.

Schaefer, C. W. 1996. A new species of *Cydamus*, with a key to the species of the genus (Hemiptera: Alydidae). *Annals of the Entomological Society of America* 89: 37–40.

Schaefer, C. W. 1999a. The higher classification of the Alydinae (Hemiptera: Heteroptera). *Proceedings of the Entomological Society of Washington* 101: 94–98.

Schaefer, C. W. 1999b. Review of *Raxa* (Hemiptera: Pyrrhocoridae). *Annals of the Entomological Society of America* 92: 14–19.

Schaefer, C. W. 2000. Systematic notes on Larginae (Hemiptera: Largidae). *Journal of the New York Entomological Society* 108: 130–145.

Schaefer, C. W. 2003. Prosorrhyncha (Heteroptera and Coleorrhyncha). Pp. 947–965. *In* V. H. Resh and R. T. Cardé (eds). *Encyclopedia of Insects*. Academic Press, Amsterdam. 1266 pp.

Schaefer, C. W. and I. Ahmad. 2000. Cotton stainers and their relatives (Pyrrhocoroidea: Pyrrhocoridae and Largidae). Pp. 271–307. *In* C. W. Schaefer and A. R. Panizzi (eds). *Heteroptera of economic importance*. CRC, Boca Raton, Florida. 828 pp.

Schaefer, C. W. and I. Ahmad. 2007. A revision of *Burtinus* (Hemiptera: Alydidae). *Annals of the Entomological Society of America* 100: 830–838.

Schaefer, C. W. and P. L. Mitchell. 1983. Food plants of Coreoidea (Hemiptera: Heteroptera). *Annals of the Entomological Society of America* 76: 591–615.

Schaefer, C. W. and A. R. Panizzi (eds). 2000. *Heteroptera of Economic Importance*. CRC, Boca Raton, Florida. 828 pp. [See specific family citations under respective author names].

Schaefer, C. W. W. R. Dolling, and S. Tachikawa. 1988. The shieldbug genus *Parastrachia* and its position within the Pentatomoidea (Insecta: Hemiptera). *Zoological Journal of the Linnean Society* 93: 283–311.

Schaefer, C. W., A. R. Panizzi, and D. G. James. 2000. Several small pentatomoid families (Cyrtocoridae, Dinidoridae, Eurostylidae, Plataspidae, and Tessaratomidae). Pp. 505–512. In C. W. Schaefer and A. R. Panizzi (eds). *Heteroptera of economic importance*. CRC, Boca Raton, Florida. 828 pp.

Schaffner, J. C. 1964. A taxonomic revision of certain genera of the tribe Alydini (Heteroptera, Coreidae). Ph.D. dissertation, Iowa State University, Ames, Iowa.

Schelt, J. , J. Klapwijk, M. Letard, and C. Aucouturier.1996. The use of *Macrolophus caliginosus* as a whitefly predator in protected crops. Pp. 515–521. In D. Gerling and R. T. Mayer(eds). *Bemisia: Taxonomy, Biology, Damage, Control and Management*. Intercept, Adover, UK.

Schuh, R. T. 1974. The Orthotylinae and Phylinae (Hemiptera: Miridae) of South Africa with a phylogenetic analysis of the ant-mimetic tribes of the two subfamilies for the world. *Entomologica Americana* 47: 1–332.

Schuh, R. T. 1976. Pretarsal structure in the Miridae (Hemiptera) with a cladistic analysis of relationships within the family. *American Museum Novitates* 2601: 1–39.

Schuh, R. T. 1979. [Review of] Evolutionary trends in Heteroptera. Part II. Mouthpart-structures and feeding strategies. *Systematic Zoology* 28: 653–656.

Schuh, R. T. 1984. Revision of the Phylinae (Hemiptera, Miridae) of the Indo-Pacific. *Bulletin of the American Museum of Natural History* 177: 1–462.

Schuh, R. T. 1986. The influence of cladistics on heteropteran classification. *Annual Review of Entomology* 31: 67–93.

Schuh, R. T. 1995. *Plant Bugs of the World (Insecta: Heteroptera: Miridae). Systematic Catalog, Distributions, Host List, and Bibliography*. New York Entomological Society, New York. 1329 pp.

Schuh, R. T. 1996. Book review. P. 244. In C. W. Schaefer (ed).*Studies on Hemipteran Phylogeny. Thomas Say Publications in Entomology Proceedings*. Entomological Society of America, Lanham, Maryland.

Schuh, R. T. 2006. *Heissophila macrotheleae*, a new genus and new species of Plokiophilidae from Thailand (Hemiptera, Heteroptera), with comments on the family diagnosis. *Denisia* 19: 637–645.

Schuh, R. T. and J. D. Lattin. 1980. *Myrmecophyes oregonensis*, a new species of Halticini (Hemiptera, Miridae) from the western United States. *American Museum Novitates* 2697: 1–11.

Schuh, R. T. and J. T. Polhemus. 1980. Analysis of taxonomic congruence among morphological, ecological, and biogeographic data sets for the Leptopodomorpha (Hemiptera). *Systematic Zoology* 29: 1–25.

Schuh, R. T. and M. D. Schwartz. 1988. Revision of the New World Pilophorini (Heteroptera: Miridae: Phylinae). *Bulletin of the American Museum of Natural History* 187: 101–201.

Schuh, R. T. and J. A. Slater. 1995. *True Bugs of the World (Hemiptera: Heteroptera). Classification and Natural History*. Cornell University Press, Ithaca, New York. 336 pp.

Schuh, R. T. and P. Štys. 1991. Phylogenetic analysis of cimicomorphan family relationships (Heteroptera). *Journal of the New York Entomological Society* 99: 298–350.

Schuh, R. T., G. Cassis, and E. Guilbert. 2006. Description of the first recent macropterous species of Vianaidinae (Heteroptera: Tingidae) with comments on the phylogenetic relationships of the family within the Cimicomorpha. *Journal of the New York Entomological Society* 114: 38–53.

Schuh, R. T., B. Galil, and J. T. Polhemus. 1987. Catalog and bibliography of Leptopodomorpha (Heteroptera). *Bulletin of the American Museum of Natural History* 185 (3): 243–406.

Schuh, R. T., C. Weirauch, T. J. Henry, and S. E. Halbert. 2008. Curaliidae, a new family of Heteroptera (Insecta: Hemiptera) from the Eastern United States. *Annals of the Entomological Society of America* 101: 1–10.

Schwartz, M. D. and R. G. Foottit. 1998. Revision of the Nearctic species of the genus *Lygus* Hahn, with a review of the Palaearctic species (Heteroptera: Miridae). *Memoirs on Entomology, International* 10: 1–428.

Schwartz, M. D. and R. T. Schuh. 1990. The world's largest isometopine, *Gigantometopus rossi*, new genus and new species (Heteroptera: Miridae). *Journal of the New York Entomological Society* 98: 9–13.

Scudder, G. G. E. 1957. A revision of Ninini (Hemiptera-Heteroptera, Lygaeidae) including the description of a new species from Angola. *Publicações Culturais da Companhia de Diamantes de Angola* 34: 91–108.

Scudder, G. G. E. 1962a. Results of the Royal Society Expedition to southern Chile. 1958–59: Lygaeidae (Hemiptera), with the description of a new subfamily. *Canadian Entomologist* 94: 1064–1075.

Scudder, G. G. E. 1962b. New Heterogastrinae (Hemiptera) with a key to the genera of the world. *Opuscula Entomologica* 27: 117–127.

Scudder, G. G. E. 1976. Water-boatmen of saline waters (Hemiptera: Corixidae). Pp. 263–289. In L. Cheng (ed). *Marine Insects*. North Holland/American Elsevier, Amsterdam and New York. 581 pp.

Sinclair, B. P. 2000. A generic revision of the Oncomerinae (Heteroptera: Pentatomoidea: Tessaratomidae). *Memoirs of the Queensland Museum* 46: 307–329.

Sissom, W. D. and J. D. Ray. 2005. *Patapius spinosus* (Rossi) (Hemiptera: Leptopodidae) in the Texas Panhandle, U.S.A. *Entomological News* 116: 117–119.

Slater, A. 1992. A genus level revision of the Western Hemisphere Lygaeinae (Heteroptera: Lygaeidae) with keys to species. *University of Kansas Science Bulletin* 55 (1): 1–56.

Slater, J. A. 1955. A revision of the subfamily Pachygronthinae of the world (Hemiptera: Lygaeidae). *Philippine Journal of Science* 84: 1–160.

Slater, J. A. 1964. *A Catalogue of the Lygaeidae of the World. 2 Volumes.* University of Connecticut, Storrs. 1668 pp.

Slater, J. A. 1976. Monocots and chinch bugs: a study of host plant relationships in the lygaeid subfamily Blissinae (Hemiptera: Lygaeidae). *Biotropica* 8: 143–165.

Slater, J. A. 1979. The systematics, phylogeny, and zoogeography of the Blissinae of the world (Hemiptera, Lygaeidae). *Bulletin of the American Museum of Natural History* 165: 1–180.

Slater, J. A. 1982. Hemiptera. Pp. 417–447. *In*: S. Parker (ed). *Synopsis and Classification of Living Organisms.* Volume 2. McGraw-Hill, New York.

Slater, J. A. and R. M. Baranowski. 1994. The occurrence of *Oxycarenus hyalinipennis* (Costa) (Hemiptera: Lygaeidae) in the West Indies and new Lygaeidae records for the Turks and Caicos islands of Providenciales and North Caicos. *Florida Entomologist* 77: 495–497.

Slater, J. A. and H. Brailovsky. 1986. The first occurrence of the subfamily Artheneinae in the Western Hemisphere with the description of a new tribe (Hemiptera: Lygaeidae). *Journal of the New York Entomological Society* 94: 409–415.

Slater, J. A. and T. J. Henry. 1999. Notes on and descriptions of new Pamphantinae, including four new species of *Cattarus* and a remarkable new myrmecomorphic genus and species (Heteroptera: Lygaeoidea: Geocoridae). *Journal of the New York Entomological Society* 107: 304–330.

Slater, J. A. and J. E. O'Donnell. 1995. *A Catalogue of the Lygaeidae of the World (1960–1994).* New York Entomological Society, New York. 410 pp.

Slater, J. A. and M. H. Sweet. 1965. The systematic position of the Psamminae (Heteroptera: Lygaeidae). *Proceedings of the Entomological Society of Washington* 67: 255–262.

Smith, C. L. 1988a. Family Gerridae Leach, 1815. The water striders. Pp. 140–151. *In* T. J. Henry and R. C. Froeschner (eds). *Catalog of the Heteroptera, or True Bugs, of Canada and the Continental United States.* E. J. Brill, Leiden and New York. 958 pp.

Smith, C. L. 1988b. Family Hydrometridae Billberg, 1820. The marsh treaders. Pp. 156–158. *In* T. J. Henry and R. C. Froeschner (eds). *Catalog of the Heteroptera, or True Bugs, of Canada and the Continental United States.* E. J. Brill, Leiden and New York. 958 pp.

Smith, C. L. 1988c. Family Veliidae Amyot and Serville, 1843. The small water striders. Pp. 734–742. *In* T. J. Henry and R. C. Froeschner (eds). *Catalog of the Heteroptera, or True Bugs, of Canada and the Continental United States.* E. J. Brill, Leiden and New York. 958 pp.

Smith, R. L. 1976. Male brooding behavior of the water bug *Abedus herberti* (Hemiptera: Belostomadtidae). *Annals of the Entomological Society of America* 69: 740–747.

Southwood, T. R. E. and D. Leston. 1959. *Land and water bugs of the British Isles.* Frederick Warne and Co., London. 436 pp.

Sorensen, J. T., B. C. Campbell, and J. D. Steffen-Campbell. 1995. Non-monophyly of Auchenorrhycha ("Homoptera"), based on 18s rDNA phylogeny: eco-evolutionary and cladistic implications within pre-heteropterodea Hemiptera (s.l.) and a proposal for new monophyletic suborders. *Pan-Pacific Entomologist* 71: 31–60.

Spencer, N. R. 1988. Inundative biological control of velvetleaf, *Abutilon theophrasti* (Malvaceae) with *Niesthrea louisianica* (Hem.: Rhopalidae). *Entomophaga* 33: 421–429.

Stonedahl, G. M. 1988. Revision of the mirine genus *Phytocoris* Fallén (Heteroptera: Miridae) for western North America. *Bulletin of the American Museum of Natural History* 188: 1–257.

Stork, N. E. 1988. Insect diversity: facts, fiction and speculation. *Biological Journal of the Linnean Society* 35: 321–337.

Strawinsky, K. 1925. Historja naturalna Korowca sosnowego *Aradus cinnamomeus* Pnz. (Hemiptera-Heteroptera). *Roczniki Nauk Rolniczych I Lesnych* 13: 644–693.

Štys, P. 1964. The morphology and relationship of the family Hyocephalidae (Heteroptera). *Acta Zoologica Academiae Scientiarum Hungaricae* 10: 229–262.

Štys, P. 1967a. Medocostidae – a new family of cimicomorphan Heteroptera based on a new genus and two new species from tropical Africa. I. Descriptive part. *Acta Entomologica Bohemoslovaca* 64: 439–465.

Štys, P. 1967b. Monograph of Malcinae, with reconsideration of morphology and phylogeny of related groups (Heteroptera, Malcidae). *Acta Entomologica Musei Nationalis Pragae* 37: 351–516.

Štys, P. 1970. On the morphology and classification of the family Dipsocoridae s. lat., with particular reference to the genus *Hypsipteryx* Drake (Heteroptera). *Acta Entomologica Bohemoslovaca* 67: 21–46.

Štys, P. 1971. Distribution and habitats of Joppeicidae (Heteroptera). *Acta Faunistica Entomologica Musei Nationalis Pragae* 14 (170): 199–207.

Štys, P. 1983. A new family of Heteroptera with dipsocoromorphan affinities from Papua New Guinea. *Acta Entomologica Bohemoslovaca* 80: 256–292.

Štys, P. 1985 (1984). The present state of beta-taxonomy in Heteroptera. *Práce Slovenskej Entomologickej Spoločnosti* 4: 205–235. [In Czech, with English abstract].

Štys, P. 1991. The first species of Plokiophilidae from Madagascar (Heteroptera, Cimicomorpha). *Acta Entomologica Bohemoslovaca* 88: 425–430.

Štys, P. 1995a. Enicocephalomorpha [including Aenictopecheidae and Enicodephalidae]. Pp. 68–73. *In* R. T. Schuh and J. A. Slater. *True Bugs of the World.* Cornell University Press, Ithaca, New York. 336 pp.

Štys, P. 1995b. Ceratocombidae. Pp. 75–78. *In* R. T. Schuh and J. A. Slater. *True Bugs of the World.* Cornell University Press, Ithaca, New York. 336 pp.

Štys, P. 1995c. Hypsipterygidae. Pp. 80. *In* R. T. Schuh and J. A. Slater. *True Bugs of the World.* Cornell University Press, Ithaca, New York. 336 pp.

Štys, P. 2002. Key to the genus-group taxa of the extant Enicocephalomorpha of the world, their list, and taxonomic changes (Heteroptera). *Acta Universitatis Carolinae Biologica* 45 (2001): 339–368.

Štys, P. and J. Davidová-Vilímová. 2001. A new genus and species of the Aphylidae (Heteroptera: Pentatomoidea) from western Australia, and its unique architecture of the abdomen. *Acta Societas Zoologicae Bohemicae* 65: 105–126.

Štys, P. and A. Jansson. 1988. Check-list of recent family-group names of Nepomorpha (Heteroptera) of the world. *Acta Entomologica Fennica* 50: 1–44.

Štys, P. and I. M. Kerzhner, 1975. The rank and nomenclature of higher taxa in recent Heteroptera. *Acta Entomologica Bohemoslovaca* 72: 65–79.

Sweet, M. H. 1996. Comparative external anatomy of the pregenital abdomen of the Hemiptera. Pp. 119–158. *In* C. W. Schaefer (ed). *Studies on Hemipteran Phylogeny. Thomas Say Publications in Entomology: Proceedings.* Entomological Society of America, Lanham, Maryland. 244 pp.

Sweet, M. H. 2000a. Seed and chinch bugs (Lygaeoidea). Pp. 143–264. *In* C. W. Schaefer and A. R. Panizzi (eds). *2000. Heteroptera of Economic Importance.* CRC Press, Boca Raton, Florida. 828 pp.

Sweet, M. H. 2000b. Economic importance of predation by big-eyed bugs (Geocoridae). Pp. 713–724. *In* C. W. Schaefer and A. R. Panizzi (eds). *2000. Heteroptera of Economic Importance.* CRC Press, Boca Raton, Florida. 828 pp.

Sweet, M. H. and C. W. Schaefer. 1985. Systematics and ecology of *Chauliops fallax* Scott. *Annals of the Entomological Society of America* 78: 526–536.

Sweet, M. H. and C. W. Schaefer. 2002. Parastrachiinae (Hemiptera: Cydnidae) raised to family level. *Annals of the Entomological Society of America* 95: 441–448.

Tallamy, D. W. and C. W. Schaefer. 1997. Maternal care in the Hemiptera: ancestry, alternatives, and current adaptive value. Pp. 94–115. *In* J. C. Choe and B. J. Crespi (eds). *The Evolution of Social Behavior in Insects and Arachnids.* Cambridge University Press, New York. 541 pp.

Tatarnic, N. J., G. Cassis, and D. F. Hochuli. 2006. Traumatic insemination in the plant bug genus *Coridromius* Signoret (Heteroptera: Miridae). *Biology Letters* 2: 58–61.

Thomas, D. B. 1991. The Acanthosomatidae (Heteroptera) of North America. *Pan-Pacific Entomologist* 67: 159–170.

Thomas, D. B. 1992. *Taxonomic Synopsis of the Asopine Pentatomidae (Heteroptera) of the Western Hemisphere.* Thomas Say Foundation Monograph. Volume 16. Entomological Society of America, Lanham, Maryland. 156 pp.

Thomas, D. B. 1994. Taxonomic synopsis of the Old World asopine genera (Pentatomidae: Heteroptera). *Insecta Mundi* 8 (3–4): 145–212.

Todd, E. L. 1955. A taxonomic revision of the family Gelastocoridae (Hemiptera). *University of Kansas Science Bulletin* 37: 277–475.

Tokihiro, G., K. Tanaka, and K. Kondo. 2003. Occurrence of the sycamore lace bug, *Corythucha ciliata* (Say) (Heteroptera: Tingidae) in Japan. *Research Bulletin of the Plant Protection Service, Japan* 39: 85–87.

Ueshima, N. 1972. New World Polyctenidae (Hemiptera), with special reference to Venezuelan species. *Brigham Young University Science Bulletin, Biological Series* 17 (1): 13–21.

Usinger, R. L. 1936. Studies in the American Aradidae with descriptions of new species (Hemiptera). *Annals of the Entomological Society of America* 29: 490–516.

Usinger, R. L. 1942. Revision of the Termitaphididae (Hemiptera). *Pan-Pacific Entomologist* 18: 155–159.

Usinger, R. L. 1945. Classification of the Enicocephalidae (Hemiptera: Reduvioidea). *Annals of the Entomological Society of America* 38: 321–342.

Usinger, R. L. 1966. *Monograph of Cimicidae (Hemiptera-Heteroptera).* Thomas Say Foundation. Volume 7. Entomological Society of America, College Park, Maryland. 585 pp.

Wehrle, L. P. 1935. Notes on *Pycnoderes quadrimaculatus* Guerin (Hemiptera, Miridae) in the vicinity of Tucson, Arizona. *Bulletin of the Brooklyn Entomological Society* 30: 27.

Weirauch, C. 2008. Cladistic analysis of Reduviidae (Heteroptera: Cimicomorpha) based on morphological characters. *Systematic Entomology* 33: 229–274.

Wheeler, A. G., Jr. 1976a. Yucca plant bug, *Halticotoma valida*: authorship, distribution, host plants, and notes on biology. *Florida Entomologist* 59: 71–75.

Wheeler, A. G., Jr. 1976b. Life history of *Kleidocerys resedae* on European white birch and ericaceous shrubs. *Annals of the Entomological Society of America* 69: 459–463.

Wheeler, A. G., Jr. 1982. Bed bugs and other bugs. Pp. 319–351. *In* A. Mallis. *Handbook of Pest Control.* Sixth Edition. Franzak and Foster Co., Cleveland, Ohio. 1101 pp.

Wheeler, A. G., Jr. 1991. Plant bugs of *Quercus ilicifolia*: Myriads of mirids (Heteroptera) in pitch pine-scrub oak barrens. *Journal of the New York Entomological Society* 99: 405–440.

Wheeler, A. G., Jr. 1992. *Chinaola quercicola* rediscovered in several specialized plant communities in the southeastern United States (Heteroptera: Microphysidae). *Proceedings of the Entomological Society of Washington* 94: 249–252.

Wheeler, A. G., Jr. 1995. Plant bugs (Heteroptera: Miridae) of *Phlox subulata* and other narrow-leaved phloxes in eastern United States. *Proceedings of the Entomological Society of Washington* 97: 435–451.

Wheeler, A. G., Jr. 2000a. Plant bugs (Miridae) as plant pests. Pp. 37–83. *In* C. W. Schaefer and A. R. Panizzi (eds). *Heteroptera of Economic Importance.* CRC, Boca Raton, Florida. 828 pp.

Wheeler, A. G., Jr. 2000b. Predacious plant bugs (Miridae). Pp. 657–693. *In* C. W. Schaefer and A. R. Panizzi (eds). *Heteroptera of Economic Importance.* CRC, Boca Raton, Florida. 828 pp.

Wheeler, A. G., Jr. 2001. *Biology of the Plant Bugs (Hemiptera: Miridae): Pests, Predators, Opportunists.* Cornell University Press, Ithaca, New York. 507 pp.

Wheeler, A. G., Jr. 2002. *Chilacis typhae* (Perrin) and *Holcocranum saturejae* (Kolenati) (Hemiptera: Lygaeidae: Artheneidae): updated North American distributions of two Palearctic cattail bugs. *Proceedings of the Entomological Society of Washington* 104: 24–32.

Wheeler, A. G., Jr. and J. E. Fetter. 1987. *Chilacis typhae* (Heteroptera: Lygaeidae) and the subfamily Artheneinae new to North America. *Proceedings of the Entomological Society of Washington* 89: 244–249.

Wheeler, A. G., Jr. and T. J. Henry. 1978. Isometopinae (Hemiptera: Miridae) in Pennsylvania: biology and descriptions of fifth instars, with observations of predation on obscure scale. *Annals of the Entomological Society of America* 71: 607–614.

Wheeler, A. G., Jr. and T. J. Henry. 1980. Seasonal history and host plants of the ant mimic *Barberiella formicoides* Poppius, with description of the fifth-instar (Hemiptera: Miridae). *Proceedings of the Entomological Society of Washington* 82: 269–275.

Wheeler, A. G., Jr. and T. J. Henry. 1984. Host plants, distribution, and description of fifth-instar nymphs of two little-known Heteroptera, *Arhyssus hirtus* (Rhopalidae) and *Esperanza texana* (Alydidae). *Florida Entomologist* 67: 521–529.

Wheeler, A. G., Jr. and T. J. Henry. 1985. *Trigonotylus coelestialium* (Heteroptera: Miridae), a pest of small grains: seasonal history, host plants, damage, and descriptions of adult and nymphal stages. *Proceedings of the Entomological Society of Washington* 87: 699–713.

Wheeler, A. G., Jr. and T. J. Henry. 1992. *A Synthesis of the Holarctic Miridae (Heteroptera): Distribution, Biology, and Origin, with Emphasis on North America.* Thomas Say Foundation Monograph. Volume 15. Entomological Society of America, Lanham, Maryland. 282 pp.

Wheeler, A. G., Jr. and T. J. Henry. 2004. Plant bugs (Hemiptera: Miridae). Pp. 1737–1742. *In* J. Capinera (ed). *Encyclopedia of Entomology.* Kluwer Academic Publications, Oxford.

Wheeler, A. G., Jr. and E. R. Hoebeke. 2009. Adventive (non-native) insects: Importance to science and society. *In* R. G. Foottit and P. H. Adler (eds). *Insect Biodiversity: Science and Society.* Wiley-Blackwell Publishing, Oxford.

Wheeler, A. G., Jr. and J. L. Herring. 1979. A potential insect pest of azaleas. *Quarterly Bulletin of the American Rhododendron Society* 33: 12–14, 34.

Wheeler, A. G., Jr. and J. P. McCaffrey. 1984. *Ranzovius contubernalis*: seasonal history, habits, and description of fifth instar, with speculation on the origin of spider commensalism in the genus *Ranzovius* (Hemiptera: Miridae). *Proceedings of the Entomological Society of Washington* 86: 68–81.

Wheeler, A. G., Jr. and C. W. Schaefer. 1982. Review of stilt bug (Hemiptera: Berytidae) host plants. *Annals of the Entomological Society of America* 75: 498–506.

Wheeler, A. G., Jr., B. R. Stinner, and T. J. Henry. 1975. Biology and nymphal stages of *Deraeocoris nebulosus* (Hemiptera: Miridae), a predator of arthropod pests on ornamentals. *Annals of the Entomological Society of America* 68: 1063–1068.

Wheeler, Q. D. 2004. Taxonomic triage and the poverty of phylogeny. *Philosophical Transactions of the Royal Society. B. Biological Sciences* 359: 571–583.

Wheeler, Q. D. and A. G. Wheeler, Jr. 1994. Mycophagous Miridae? Associations of Cylapinae (Heteroptera) with pyrenomycete fungi (Euascomycetes: Xylariaceae). *Journal of the New York Entomological Society* 102: 114–117.

Wheeler, W. C., R. T. Schuh, and R. Bang. 1993. Cladistic relationships among higher groups of Heteroptera: congruence between morphological and molecular data sets. *Entomologica Scandinavica* 24: 121–137.

Whitefield, F. G. S. 1933. The bionomics and control of *Dysdercus* (Hemiptera) in the Sudan. *Bulletin of Entomological Research* 24: 301–313.

Wigglesworth, V. B. 1972. *The Principles of Insect Physiology.* Seventh Edition. Chapman and Hall, London. 827 pp.

Woodward, T. E. 1968a. A new subfamily of Lygaeidae (Hemiptera-Heteroptera) from Australia. *Proceedings of the Royal Entomological Society (B)* 37: 125–132.

Woodward, T. E. 1968b. The Australian Idiostolidae (Hemiptera: Heteroptera). *Transactions of the Royal Entomological Society of London* 120: 253–261.

Wygodzinsky, P. W. 1966. A monograph of the Emesinae (Reduviidae, Hemiptera). *Bulletin of the American Museum of Natural History* 133: 1–614.

Wygodzinsky, P. W. and K. Schmidt. 1991. Revision of the New World Enicocephalomorpha (Heteroptera). *Bulletin of the American Museum of Natural History* 200: 1–265.

Chapter 11

BIODIVERSITY OF COLEOPTERA

Patrice Bouchard[1], Vasily V. Grebennikov[2], Andrew B. T. Smith[3], and Hume Douglas[2]

[1]Agriculture and Agri-Food Canada, Canadian National Collection of Insects, Arachnids and Nematodes, 960 Carling Avenue, Ottawa, Ontario, K1A 0C6, Canada
[2]Canadian Food Inspection Agency, Canadian National Collection of Insects, Arachnids and Nematodes, 960 Carling Avenue, Ottawa, Ontario, K1A 0C6, Canada
[3]Canadian Museum of Nature, P. O. Box 3443, Station D, Ottawa, Ontario, K1P 6P4, Canada

Insect Biodiversity: Science and Society, 1st edition. Edited by R. Foottit and P. Adler
© 2009 Blackwell Publishing, ISBN 978-1-4051-5142-9

More described species of insects than any other life form exist on this planet, and beetles represent the greatest proportion of described insects. Based on numbers of described species, beetles are the most diverse group of organism on Earth. The estimated number of described species of beetles is between 300,000 and 450,000 (Nielsen and Mound 1999). Our tally of more than 357,000 described species (Table 11.1) is based primarily on figures given in *American Beetles* (Arnett and Thomas 2001, Arnett et al. 2002) and *Immature Insects* (Stehr 1991). Of the described species, many are known only from a single locality or even from a single specimen (Stork 1999a, Grove and Stork 2000). Beetles are so diverse, and most species so poorly known, that an estimate of how many species really exist remains difficult.

Erwin (1982) first proposed an estimate of the total number of beetle species on the planet, based on field data rather than on catalog numbers. The technique used for his original estimate, possibly as many as 12,000,000 species, was criticized, and revised estimates of 850,000–4,000,000 species were proposed (Hammond 1995, Stork 1999b, Nielsen and Mound 1999). Some 70–95% of all beetle species, depending on the estimate, remain undescribed (Grove and Stork 2000).

The beetle fauna is not known equally well in all parts of the world. For example, Yeates et al. (2003) estimated that the known beetle diversity of Australia includes 23,000 species in 3265 genera and 121 families. This estimate of species is slightly lower than that reported for North America, a land mass of similar size: 25,160 species in 3526 genera and 129 families (Marske and Ivie 2003). While Marske and Ivie (2003) predicted that there could be as many as 28,000 species in North America, including currently undescribed species, a realistic estimation of the true diversity of the little-studied Australian beetle fauna could be 80,000–100,000 (Yeates et al. 2003).

Studies of beetle communities in restricted areas such as oceanic islands (Peck 2005), large administrative units (Peck and Thomas 1998, Sikes 2004, Peck 2005, Carlton and Bayless 2007), and specific habitats (Hammond 1990, Chandler and Peck 1992, Carlton and Robison 1998, Anderson and Ashe 2000) can provide important data on biodiversity at finer scales. Comprehensive species lists from well-defined areas or habitat types are useful not only because they give a snapshot of current ecosystem health and function, but also because they can be compared to lists

generated at other periods to monitor changes over time (Howden and Howden 2001). Analysis of changes in species composition allows us to better understand the effects of humans on natural and managed ecosystems and provides powerful arguments to support land use and conservation decisions.

To explain the great diversity of beetles is difficult. One of the most important factors was proposed to be the development of the forewings into sclerotized elytra (Lawrence and Britton 1994). In most beetles, the elytra cover the membranous flight wings and the abdomen. In this way, the elytra are thought to protect beetles against environmental stresses and predation (Hammond 1979). In addition, the close historical association of some of the most diverse groups of beetles, such as weevils (Curculionoidea), longhorn beetles (Cerambycidae), and leaf beetles (Chrysomelidae), with flowering plants during their own period of diversification is thought to be one of the main reasons explaining their great evolutionary success (Farrell 1998, Barraclough et al. 1998; Fig. 11.1a).

Patterns of beetle diversity can illustrate factors that have led to the success of the group as a whole. Based on estimates for all 165 families (Table 11.1), more than 358,000 species of beetles have been described and are considered valid. Most species (62%) are in six megadiverse families (Fig. 11.2), each with at least 20,000 described species: Curculionidae, Staphylinidae, Chysomelidae, Carabidae, Scarabaeidae, and Cerambycidae. The smaller families of Coleoptera account for 22% of the total species in the group, and include 127 families with 1–999 described species and 29 families with 1000–6000 described species. So, the success of beetles as a whole is driven not only by several extremely diverse lineages, but also by a high number of moderately successful lineages. The patterns seen today indicate that beetles went through a massive adaptive radiation early in their evolutionary history, with many of the resulting lineages flourishing through hundreds of millions of years to the present. The adaptive radiation of angiosperms helped drive the diversification of beetles, as four of the six megadiverse families of beetles are primarily angiosperm feeders (Curculionidae, Chysomelidae, Scarabaeidae, and Cerambycidae). However, even without the phytophagous groups, lineages of predators, scavengers, and fungivores are tremendously successful.

Beetles occur in most terrestrial and freshwater habitats (Lawrence and Britton 1994), and a few occupy marine environments (Doyen 1976, Moore

Table 11.1 Families of Coleoptera with estimated number of described species in the world for each family.

Suborder	Series	Family	Number of Species
Archostemata		Cupedidae	30^1
Archostemata		Ommatidae	4^2
Archostemata		Crowsoniellidae	1^2
Archostemata		Micromalthidae	1^1
Archostemata		Jurodidae	1^3
Myxophaga		Lepiceridae	2^2
Myxophaga		Sphaeriusidae	19^1
Myxophaga		Hydroscaphidae	13^1
Myxophaga		Torridincolidae	27^2
Adephaga		Aspidytidae	2^4
Adephaga		Meruidae	1^5
Adephaga		Rhysodidae	170^1
Adephaga		Carabidae	$30,000^2$
Adephaga		Gyrinidae	700^1
Adephaga		Haliplidae	200^1
Adephaga		Trachypachidae	6^1
Adephaga		Noteridae	230^1
Adephaga		Amphizoidae	6^1
Adephaga		Hygrobiidae	5^2
Adephaga		Dytiscidae	4000^1
Polyphaga	Staphyliniformia	Hydrophilidae	2803^1
Polyphaga	Staphyliniformia	Sphaeritidae	4^1
Polyphaga	Staphyliniformia	Synteliidae	5^2
Polyphaga	Staphyliniformia	Histeridae	3900^1
Polyphaga	Staphyliniformia	Hydraenidae	1200^1
Polyphaga	Staphyliniformia	Ptiliidae	550^1
Polyphaga	Staphyliniformia	Agyrtidae	61^1
Polyphaga	Staphyliniformia	Leiodidae	3000^1
Polyphaga	Staphyliniformia	Scydmaenidae	4500^1
Polyphaga	Staphyliniformia	Silphidae	175^1
Polyphaga	Staphyliniformia	Staphylinidae	$47,744^{13}$
Polyphaga	Scarabaeiformia	Lucanidae	800^1
Polyphaga	Scarabaeiformia	Diphyllostomatidae	3^1
Polyphaga	Scarabaeiformia	Passalidae	500^1
Polyphaga	Scarabaeiformia	Glaresidae	50^1
Polyphaga	Scarabaeiformia	Trogidae	300^1
Polyphaga	Scarabaeiformia	Pleocomidae	26^1
Polyphaga	Scarabaeiformia	Geotrupidae	300^1
Polyphaga	Scarabaeiformia	Belohinidae	1^1
Polyphaga	Scarabaeiformia	Ochodaeidae	80^1
Polyphaga	Scarabaeiformia	Hybosoridae	365^1
Polyphaga	Scarabaeiformia	Glaphyridae	68^1
Polyphaga	Scarabaeiformia	Scarabaeidae	$27,800^1$
Polyphaga	Elateriformia	Decliniidae	1^6
Polyphaga	Elateriformia	Eucinetidae	37^1
Polyphaga	Elateriformia	Clambidae	70^1
Polyphaga	Elateriformia	Scirtidae	600^1
Polyphaga	Elateriformia	Dascillidae	80^1
Polyphaga	Elateriformia	Rhipiceridae	57^1
Polyphaga	Elateriformia	Schizopodidae	7^1

(continued)

Table 11.1 *(continued).*

Suborder	Series	Family	Number of Species
Polyphaga	Elateriformia	Buprestidae	14,600[1]
Polyphaga	Elateriformia	Byrrhidae	290[1]
Polyphaga	Elateriformia	Elmidae	1100[2]
Polyphaga	Elateriformia	Dryopidae	230[2]
Polyphaga	Elateriformia	Lutrochidae	12[1]
Polyphaga	Elateriformia	Limnichidae	225[1]
Polyphaga	Elateriformia	Heteroceridae	300[1]
Polyphaga	Elateriformia	Psephenidae	27[2]
Polyphaga	Elateriformia	Cneoglossidae	7[2]
Polyphaga	Elateriformia	Ptilodactylidae	500[1]
Polyphaga	Elateriformia	Chelonariidae	300[1]
Polyphaga	Elateriformia	Eulichadidae	21[1]
Polyphaga	Elateriformia	Callirhipidae	16[1]
Polyphaga	Elateriformia	Artematopodidae	60[1]
Polyphaga	Elateriformia	Brachypsectridae	3[2]
Polyphaga	Elateriformia	Cerophytidae	10[2]
Polyphaga	Elateriformia	Eucnemidae	1300[2]
Polyphaga	Elateriformia	Throscidae	152[1]
Polyphaga	Elateriformia	Elateridae	10,000[1]
Polyphaga	Elateriformia	Plastoceridae	2[2]
Polyphaga	Elateriformia	Drilidae	80[2]
Polyphaga	Elateriformia	Omalisidae	10[2]
Polyphaga	Elateriformia	Lycidae	3500[1]
Polyphaga	Elateriformia	Telegeusidae	8[1]
Polyphaga	Elateriformia	Phengodidae	250[1]
Polyphaga	Elateriformia	Lampyridae	2000[1]
Polyphaga	Elateriformia	Omethidae	32[1]
Polyphaga	Elateriformia	Cantharidae	5083[1]
Polyphaga	Elateriformia	Podabrocephalidae	1[7]
Polyphaga	Elateriformia	Rhinorhipidae	1[8]
Polyphaga	Bostrichiformia	Jacobsoniidae	18[1]
Polyphaga	Bostrichiformia	Derodontidae	22[1]
Polyphaga	Bostrichiformia	Nosodendridae	62[1]
Polyphaga	Bostrichiformia	Dermestidae	700[1]
Polyphaga	Bostrichiformia	Endecatomidae	4[1]
Polyphaga	Bostrichiformia	Bostrichidae	550[1]
Polyphaga	Bostrichiformia	Anobiidae	2200[1]
Polyphaga	Cucujiformia	Lymexylidae	50[1]
Polyphaga	Cucujiformia	Phloiophilidae	2[2]
Polyphaga	Cucujiformia	Trogossitidae	600[1]
Polyphaga	Cucujiformia	Chaetosomatidae	9[2]
Polyphaga	Cucujiformia	Cleridae	3366[1]
Polyphaga	Cucujiformia	Acanthocnemidae	1[2]
Polyphaga	Cucujiformia	Phycosecidae	4[2]
Polyphaga	Cucujiformia	Prionoceridae	50[9]
Polyphaga	Cucujiformia	Melyridae	6000[1]
Polyphaga	Cucujiformia	Protocucujidae	3[2]
Polyphaga	Cucujiformia	Sphindidae	61[1]
Polyphaga	Cucujiformia	Kateretidae	11[1]
Polyphaga	Cucujiformia	Nitidulidae	2800[1]
Polyphaga	Cucujiformia	Smicripidae	6[1]
Polyphaga	Cucujiformia	Monotomidae	220[1]

Table 11.1 *(continued).*

Suborder	Series	Family	Number of Species
Polyphaga	Cucujiformia	Boganiidae	11[10]
Polyphaga	Cucujiformia	Helotidae	100[2]
Polyphaga	Cucujiformia	Phloeostichidae	8[2]
Polyphaga	Cucujiformia	Silvanidae	470[1]
Polyphaga	Cucujiformia	Passandridae	105[1]
Polyphaga	Cucujiformia	Cucujidae	42[1]
Polyphaga	Cucujiformia	Agapythidae	1[11]
Polyphaga	Cucujiformia	Cyclaxyridae	3[11]
Polyphaga	Cucujiformia	Myraboliidae	7[11]
Polyphaga	Cucujiformia	Priasilphidae	11[11]
Polyphaga	Cucujiformia	Tasmosalpingidae	2[11]
Polyphaga	Cucujiformia	Laemophloeidae	400[1]
Polyphaga	Cucujiformia	Propalticidae	25[2]
Polyphaga	Cucujiformia	Phalacridae	504[1]
Polyphaga	Cucujiformia	Hobartiidae	2[2]
Polyphaga	Cucujiformia	Cavognathidae	5[2]
Polyphaga	Cucujiformia	Cryptophagidae	600[1]
Polyphaga	Cucujiformia	Lamingtoniidae	1[1]
Polyphaga	Cucujiformia	Languriidae	1037[1]
Polyphaga	Cucujiformia	Erotylidae	2500[1]
Polyphaga	Cucujiformia	Byturidae	16[1]
Polyphaga	Cucujiformia	Biphyllidae	200[1]
Polyphaga	Cucujiformia	Bothrideridae	300[1]
Polyphaga	Cucujiformia	Cerylonidae	300[1]
Polyphaga	Cucujiformia	Alexiidae	50[10]
Polyphaga	Cucujiformia	Discolomatidae	400[2]
Polyphaga	Cucujiformia	Endomychidae	1300[1]
Polyphaga	Cucujiformia	Coccinellidae	6000[1]
Polyphaga	Cucujiformia	Corylophidae	284[1]
Polyphaga	Cucujiformia	Latridiidae	1050[1]
Polyphaga	Cucujiformia	Mycetophagidae	200[1]
Polyphaga	Cucujiformia	Archeocrypticidae	50[1]
Polyphaga	Cucujiformia	Pterogeniidae	8[2]
Polyphaga	Cucujiformia	Ciidae	550[1]
Polyphaga	Cucujiformia	Tetratomidae	155[1]
Polyphaga	Cucujiformia	Melandryidae	430[1]
Polyphaga	Cucujiformia	Mordellidae	1500[1]
Polyphaga	Cucujiformia	Ripiphoridae	500[1]
Polyphaga	Cucujiformia	Colydiidae	1000[2]
Polyphaga	Cucujiformia	Zopheridae	425[1,2]
Polyphaga	Cucujiformia	Perimylopidae	8[2]
Polyphaga	Cucujiformia	Chalcodryidae	6[2]
Polyphaga	Cucujiformia	Trachelostenidae	2[6]
Polyphaga	Cucujiformia	Tenebrionidae	19,000[1]
Polyphaga	Cucujiformia	Prostomidae	20[1]
Polyphaga	Cucujiformia	Synchroidae	8[1]
Polyphaga	Cucujiformia	Oedemeridae	1500[1]
Polyphaga	Cucujiformia	Stenotrachelidae	20[1]
Polyphaga	Cucujiformia	Meloidae	2500[1]
Polyphaga	Cucujiformia	Mycteridae	160[1]
Polyphaga	Cucujiformia	Boridae	4[1]

(continued)

Table 11.1 *(continued)*.

Suborder	Series	Family	Number of Species
Polyphaga	Cucujiformia	Trictenotomidae	12[2]
Polyphaga	Cucujiformia	Pythidae	50[2]
Polyphaga	Cucujiformia	Pyrochroidae	200[1]
Polyphaga	Cucujiformia	Salpingidae	300[1]
Polyphaga	Cucujiformia	Anthicidae	3000[1]
Polyphaga	Cucujiformia	Aderidae	1000[1]
Polyphaga	Cucujiformia	Scraptiidae	400[1]
Polyphaga	Cucujiformia	Cerambycidae	20,000[1]
Polyphaga	Cucujiformia	Megalopodidae	81[1]
Polyphaga	Cucujiformia	Orsodacnidae	19[1]
Polyphaga	Cucujiformia	Chrysomelidae	36,350[1]
Polyphaga	Cucujiformia	Nemonychidae	65[2]
Polyphaga	Cucujiformia	Anthribidae	4000[1]
Polyphaga	Cucujiformia	Belidae	150[2]
Polyphaga	Cucujiformia	Attelabidae	1914[1]
Polyphaga	Cucujiformia	Brentidae	1200[2]
Polyphaga	Cucujiformia	Caridae	5[12]
Polyphaga	Cucujiformia	Ithyceridae	1[1]
Polyphaga	Cucujiformia	Curculionidae	60,000[1]
		TOTAL	359,891

Family classification is based on Lawrence and Newton (1995), with species numbers from the following sources: [1]Arnett and Thomas (2001), Arnett et al. (2002); [4]Balke et al. (2003), Lawrence [10](1982), [8](1988), [2](1991); Lawrence et al. [6](1995), [7](1999); [9]Lawrence (personal communication); [11]Leschen et al. (2005); [3]Kirejtshuk (1999); [12]Kuschel (1995); [5]Spangler and Steiner (2005); [13]Thayer (2005).

and Legner 1976). Their body length ranges from 0.4 mm (Sörensson 1997) to more than 17 cm. Larvae of some of the larger beetles can weigh more than 140 g and are the heaviest insects known (Acorn 2006). The most common life cycle type in beetles is holometaboly, whereby individuals emerge from eggs as larvae, develop through several instars, pupate, and eventually emerge as adults. Sexual reproduction is predominant, although parthenogenesis (i.e., production of viable, unfertilized eggs) also occurs. More specialized or unusual life cycles, which include the occurrence of active and inactive larval instars in parasitoid species, are also known in Coleoptera (Lawrence and Britton 1994).

Perhaps most people might not recognize that we live in the age of beetles (Evans and Bellamy 1996). Beetles are important parts of most natural terrestrial and freshwater ecosystems, have important effects on agriculture and forestry, and are useful model organisms for many types of science. A better understanding of beetle biodiversity will enhance our knowledge of the world and provide many practical applications.

Coverage of all aspects of beetle biodiversity that affect science and society is not possible in a single chapter. We, therefore, have chosen to cover each major taxonomic group of beetles and provide some examples to illustrate important aspects of beetle biodiversity.

OVERVIEW OF TAXA

Suborders Archostemata and Myxophaga

The suborder Archostemata includes about 40 species of small to medium-sized beetles in five families (Ommatidae, Crowsoniellidae, Micromalthidae, Cupedidae, Jurodidae) (Hörnschemeyer 2005). Most larvae develop in fungus-infested wood, and the mouthparts of adults suggest that most species feed on plant pollen or sap, although the biologies of most species are unknown. Myxophagans are small beetles (usually shorter than 2.5 mm) that feed on algae or blue green algae in freshwater and riparian habitats (Jäch 1998, Beutel 2005). The four myxophagan families include

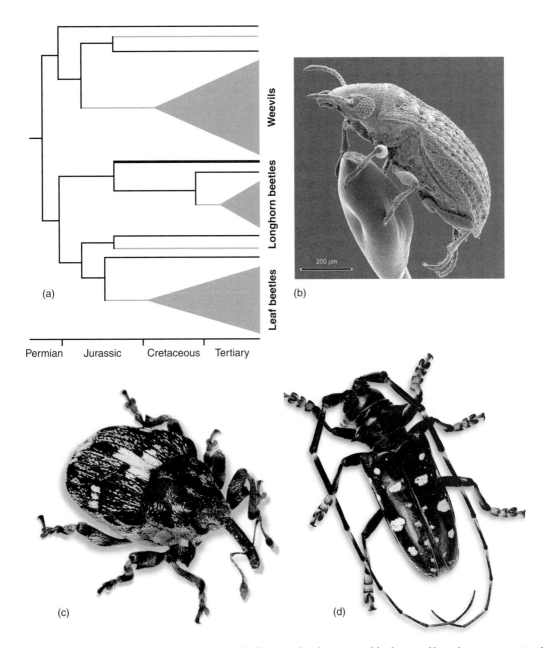

Fig. 11.1 (a) Phylogeny of the Phytophaga, showing the diversity of each group (width of terminal branches are proportional to species richness), host relationships (gray line = lineage feeding on conifer and/or cycads, black branch = lineage feeding on flowering plants, thicker branch = nonherbivorous lineage), and period of radiation (modified from Farrell 1998 and Barraclough et al. 1998). (b) Scanning electron micrograph of *Meru phyllisae*, Spangler and Steiner (2005), the only representative of the recently discovered family Meruidae (Source: W. Steiner; copyright: Government of the United States of America). (c) *Mogulones crucifer* (Pallas), a weevil willingly introduced to North America as a biological control agent of the weed *Cynoglossum officinale* (Linnaeus) (Source: H. Goulet; copyright: Government of Canada). (d) *Anoplophora glabripennis* (Motschulsky), a long-horned beetle unwillingly introduced to North America and Europe from Asia, which feeds on several deciduous tree species (Source: K. Bolte; copyright: Government of Canada).

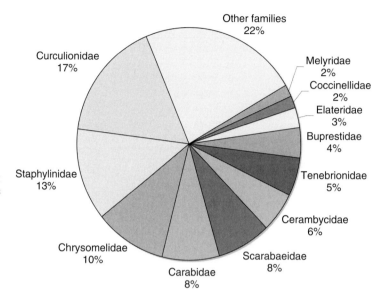

Fig. 11.2 Coleoptera biodiversity, using the estimated number of described species (from Table 11.1). The 11 largest beetle families (each with 6000+ described species) are shown. The remaining 154 families (each with fewer than 6000 described species) are lumped into 'other families' but together represent a significant proportion of beetle diversity.

approximately 100 species worldwide. Larval stages of one family (Lepiceridae) are unknown.

For a robust understanding of Coleoptera classification and phylogenetics, an understanding of the morphology and evolution of the suborders Archostemata and Myxophaga is essential (Beutel and Haas 2000). Direct effects of these two suborders on humans are minor. Examples of these include one species, *Micromalthus debilis*, which was probably transported throughout the world via the timber trade (Philips 2001) and members of the archostematan family Cupedidae that live in structural timber (Nebois 1968, 1984).

Suborder Adephaga

This suborder includes 11 families and more than 40,000 species worldwide (Ball and Bousquet 2001, Beutel and Ribera 2005). Most species are predators in both the larval and adult stages, although several species of Carabidae are granivores or parasitoids (Ball and Bousquet 2001). Three terrestrial families (Carabidae, Rhysodidae, and Trachypachidae) and eight families are associated with freshwater habitats (Gyrinidae, Haliplidae, Noteridae, Amphizoidae, Hygrobiidae, Dytiscidae, Aspidytidae, and Meruidae). Aspidytidae and Meruidae (Fig. 11.1b) were described from recently discovered beetles (Ribera et al. 2002, Spangler and Steiner 2005, respectively), highlighting

the need for continued exploration of our planet's microhabitats (Beutel et al. 2006).

The most diverse family of adephagan beetles associated with freshwater habitats is the Dytiscidae, with approximately 4000 species worldwide. These beetles occur in a variety of microhabitats from large bodies of water to springs and temporary water pools (Roughley and Larson 2001). Specialized groups have evolved to occupy groundwater in aquifers several meters below the surface (Balke et al. 2004, Watts and Humphreys 2006), whereas others have lost their ability to swim and are restricted to deep, wet leaf litter in montane forests of Australia, New Caledonia, and the Himalayas (Brancucci 1979, Brancucci and Monteith 1996). Because of their predatory habits, dytiscids are thought to play an important role in the control of larval mosquitoes that, as adults, often are infected with a variety of disease agents transmittable to humans (Dubitskii et al. 1975, Lopez et al. 1997, Bellini et al. 2000, Lundkvist et al. 2003). An unusual traditional practice using adephagan water beetles recently was reported from East Africa. In this region, young girls collect local water beetles (Gyrinidae and Dytiscidae) and hold them to their nipples to stimulate a biting reflex from these beetles. The biting behavior of the beetles and the simultaneous release of their defensive secretions are thought to stimulate breast growth in these girls (Kutalek and Kassa 2005).

The Carabidae, commonly referred to as 'ground beetles', represent the most diverse family in this suborder, with an estimated 40,000 known species (Erwin 1991). Ground beetles are by far the most important adephagans in terms of anthropogenic interactions. A good understanding of species level taxonomy, especially in northern temperate regions (Lindroth 1961–1969, 1974, 1985–1986), has enabled scientists to use species in this diverse group as tools to address many ecological and evolutionary questions. Studies of ground beetles have enhanced our understanding of the ecological effects of agricultural practices (Varchola and Dunn 1999, Freuler et al. 2001, Duan et al. 2006), forestry practices (Niemela et al. 1993, Magura et al. 2003), habitat fragmentation (Magagula 2003), fire management of natural habitats (Larsen and Work 2003), pollution of natural ecosystems (Freitag 1979), and many more human activities. Many carabid species occurring in agroecosystems are recognized as beneficial natural enemies of agricultural pests (Lövei and Sunderland 1996). Ground beetles are also important for research on habitat conservation (Bouchard et al. 2006) and biogeography (Darlington 1943, Marshall and Liebherr 2000).

Defensive secretions produced by adephagans are usually composed of quinones and hydrocarbons in prothoracic and abdominal glands. Secretions of the extensively studied bombardier beetles have revealed that fluids are usually ejected at temperatures near 60°C. Although the primary function of these secretions is defense against predators, they also are thought to have important antimicrobial and antifungal properties (Roth and Eisner 1962, Blum 1981). Noterids are of little economic importance, although they might be important nutrient recyclers in tropical systems (Young 1967). Life cycles are not known for any of the Nearctic noterid species (Roughley 2001).

Suborder Polyphaga

Series Staphyliniformia

This series includes the superfamilies Hydrophiloidea (including Histeroidea by some authors) and Staphylinoidea. Phylogenetic evidence is mounting that the superfamily Scarabaeoidea also should be included in this series (Hansen 1997, Beutel and Komarek 2004, Korte et al. 2004, Beutel and Leschen 2005b, Caterino et al. 2005, Hughes et al. 2006). However, the exact relationship of the Scarabaeoidea to the Staphyliniformia and the monophyly of the series Staphyliniformia, as presently defined, have

yet to be determined. We have decided to treat the Scarabaeiformia as a separate series to follow current classification schemes. However, additional molecular phylogenetics research inevitably will provide strong evidence that the series Scarabaeiformia should be combined with the series Staphyliniformia. The total number of described species of Staphyliniformia is about 90,000, or roughly a quarter of all beetles and considerably more than all Vertebrata combined.

The superfamily Hydrophiliodea unites four families: Hydrophilidae *sensu lato* (including Helophorinae, Epimetopinae, Georissinae, Hydrochinae, Spercheinae, and Hydrophilinae), Histeridae, Sphaeritidae, and Synteliidae (Archangelsky et al. 2005). The Sphaeritidae and Synteliidae include fewer than 20 species each. These families are associated with decaying organic matter and each includes a single genus distributed in both North America and Eurasia. The Histeridae (330 genera and 4000 species) and Hydrophilidae (170 genera and 2800 species) are two large, cosmopolitan families that are most diverse in the tropics (Beutel and Leschen 2005a).

Hansen's (1991) work on hydrophiloid beetles, which includes taxonomic treatment down to the genus level and a rigorous phylogenetic analysis, is richly illustrated and still the best introduction to the family. These beetles are primarily aquatic, although the Sphaeridiinae are secondarily terrestrial and found in dung and leaf litter. Adults are poor swimmers and normally crawl underwater, feeding mainly on plant material. Larger species are predaceous on similar-sized organisms. Eggs are normally stored in variably shaped silk cases, which have a 'mast' and can float; females of the Epimetopinae carry egg cases attached to their abdomens. Larvae of the Hydrophilidae are normally aquatic, go through three instars, and are predators; pupation, however, takes place in the soil. Davis (1994) considered South African dung-inhabiting Hydrophilidae as biocontrol agents against dung-breeding flies in Australia.

Unlike the Hydrophilidae, the Histeridae are strictly terrestrial predators in the adult stage, and their larvae have only two instars. Fly maggots in dung, carcasses, or rotten vegetation are the main diet of most histerid species. Adults of some species catch flies by hiding in crevasses in dung and seizing them with their mouthparts (Kovarik and Caterino 2005). Many Hetaeriinae and Chlamydopsinae histerids are myrmecophilous or termitophilous, with unusual morphology and strong host dependency. The chapter on Histeridae by Kovarik and Caterino (2005) is a good introduction to the systematics, distribution, and biology of the family. The Histeridae were used repeatedly as control agents against house flies (*Musca domestica*) (Kaufman et al. 2000) and other flies with dung-inhabiting larvae (Davis 1994). *Teretrius nigrescens*, a histerid beetle, was released as a biocontrol agent against the bostrichid beetle *Prostephanus truncatus* to protect maize stores in equatorial Africa, although success was only moderate (Holst and Meikle 2003).

The superfamily Staphylinoidea includes seven families. Beutel and Leschen (2005b) hypothesized that the Hydraenidae and Ptiliidae form a sister group of the Agyrtidae plus Leiodidae, while these four families are together a sister group of the Staphylinidae and Silphidae plus Scydmaenidae. Six of these families are worldwide in distribution, while the Agyrtidae are known only from northern temperate areas, with one genus in New Zealand (Newton 1997). The family Hydraenidae comprises some 1300 described detritus-feeding species in 40 genera; however, the true species richness is probably higher (Jäch et al. 2005). Most hydraenid species occur in freshwater habitats; some are terrestrial and either riparian or associated with wet forest leaf litter. Their sister group, the Ptiliidae (featherwing beetles) are strictly terrestrial and mycophagous and include more than 550 described species in about 70 genera (Hall 2005). The Ptiliidae include the smallest known beetles, with most species shorter than 1.0 mm in body length, and a documented minimum length of about 0.4 mm (Sörensson 1997). Other remarkable features of some Ptiliidae include polymorphism in wing and eye development, association with social insects, obligate parthenogenesis, large egg size (approximately 50% of female body volume), and giant sperm (subequal in length to that of the adult). The Agyrtidae represent a small family with about 60 species in 8 genera, feeding on decayed organic matter (Newton 1997). Their sister group, the Leiodidae, is a larger family with about 3460 species in about 340 genera (Newton 2005). This family is variable in food and habitat requirements, although the majority of species inhabit forested areas and feed on decaying organic matter including slime molds. Leiodids exhibit the only known case of true parasitism by beetles, as shown by *Platypsyllus castori* on beavers (Peck 2006). Some species of the cave-dwelling genera *Speonomus* and *Leptodirus* exhibit a unique evolutionary oddity in that their larval development is shortened to a single, nonfeeding larval instar (Deleurance-Glaçon 1963). Newton's (1998) work on the Leiodidae is a

good introduction to the diversity of this family. The influence of the Hydraenidae, Ptiliidae, Agyrtidae, and Leiodidae on humans is not obvious, although they are important decomposers of organic matter and occupy a unique and definite niche in natural and agricultural ecosystems.

The second clade of Staphylinoidea beetles consists of three families and comprises more than 50,000 described species. The Silphidae, also known as the large carrion beetles, include fewer than 200 species in 15 genera (Sikes 2005) and are efficient carrion decomposers. The Scydmaenidae mainly prey on mites and have about 82 genera with more than 4600 species, more than half of which belong to the single genus *Euconnus* (O'Keefe 2005). Probably both the Silphidae and Scydmaenidae are derived from within the Staphylinidae (rove beetles); thus, their classification as separate families might cause paraphyly of the latter (Beutel and Leschen 2005b). This problem most likely will be solved by reduction of both groups in taxonomic rank and their incorporation as subfamilies in the already enormous Staphylinidae *sensu lato*, a strategy exemplified by the inclusion of the former Pselaphidae (Newton and Thayer 1995). The chapter by Thayer (2005) is a good introduction to the Staphylinidae. This family includes 47,744 described species and rivals the weevils (Curculionidae) in diversity. Thirty-one subfamilies of rove beetles are currently recognized; their phylogenetic relationships have only recently become a subject of investigation (Newton and Thayer 1995). The Staphylinidae are predominantly mycophagous, saprophagous, or predaceous, although some species are phytophagous. Rove beetles are among the most commonly encountered beetles in nature, particularly in moist terrestrial habitats. As predators, some are potentially useful to humans in controlling phytophagous and nuisance arthropods. Some *Aleochara* species parasitize flies, suggesting biocontrol applications. Most importantly, the Staphylinidae are organic components of healthy ecosystems, both natural and modified by humans.

Series Scarabaeiformia

The series Scarabaeiformia, as currently classified, consists of 12 families (Pleocomidae, Geotrupidae, Belohinidae, Passalidae, Trogidae, Glaresidae, Diphyllostomatidae, Lucanidae, Ochodaeidae, Hybosoridae, Glaphyridae, and Scarabaeidae), 43 subfamilies, and 118 tribes (Smith 2006). Recent estimates of the number of described species in this group range from 31,000 to 35,000 (Jameson and Ratcliffe 2002, Scholtz and Grebennikov 2005). Gaston (1991) gave a conservative estimate that the actual total number of species might be around 50,000 species. Scarab beetles are terrestrial, with their larvae generally inhabiting soil, detritus, or decaying wood. Scarab species can be found in most terrestrial habitats worldwide and are most diverse in tropical forests.

Several phylogenetic analyses for taxa within the series Scarabaeiformia have been published, based on morphological characters (Howden 1982, Browne and Scholtz 1995, 1998, 1999, Grebennikov and Scholtz 2004), and a preliminary phylogenetic analysis based on molecular characters is also available (Smith et al. 2006). The 12 families are well defined, but phylogenetic evidence for the relationships among many of these families was weak or contradictory among studies. The Scarabaeidae is by far the most diverse family, having an estimated 91% of the described species of the Scarabaeiformia (Ratcliffe et al. 2002). Two major clades occur in the family Scarabaeidae: the dung-beetle clade and the phytophagous scarab clade (Smith et al. 2006). The phytophagous scarab clade is highly diverse, containing the majority of scarabaeoid species. Members of this clade primarily feed on angiosperm plant foliage, flowers, nectar, pollen, or fruit during their adult stage; and roots, humus, detritus, or deadwood as larvae. The dung-beetle clade is another highly successful and diverse lineage of scarab beetles. The shift to dung feeding is the second most successful trophic adaptation, following angiosperm plant feeding in the evolution of scarab beetles, when looking at total biodiversity within the group. Among the other major lineages of scarab beetles are groups that are primarily dead-wood feeders (Lucanidae and Passalidae), carcass feeders (Trogidae), fungivores (Geotrupidae), and detritivores (Hybosoridae).

Scarab beetles often have been used as a focal taxon for evolution, biodiversity, and conservation research. Dung beetles, in particular, have been used in ecological and biodiversity studies (Spector 2006). Dung beetles have been used to compare habitats (Nummelin and Hanski 1989, Jankielsohn et al. 2001, Estrada and Coates-Estrada 2002), to study diversity across landscapes and continents (Lobo and Davis 1999, Arellano and Halffter 2003, Lobo et al. 2006), as indicator taxa of ecosystem health (Halffter and Arellano 2002, Davis et al. 2004, Quintero and Roslin 2005, Scheffler 2005), and for many other purposes.

Scarab beetles are an extremely diverse, ubiquitous, and widely distributed group and, as such, they have a large influence on human activities and welfare. Hundreds of species of scarab beetles are nectar feeders and pollinators of a diverse assemblage of plant species. The benefits of dung removal and burial by dung beetles worldwide are well documented from the perspective of pastureland productivity, nutrient recycling, and health (Edwards and Aschenborn 1987, Mittal 1993, Tyndale-Biscoe 1994). In addition, deadwood feeding and detritovorous scarabs play a major role in nutrient cycling; burrowing scarabs play a significant role in seed burial and germination for some plant species (Vulinec 2002, Chapman and Chapman 2003), and scarabs are a major food source for many predators. Many phytophagous species are considered pests of turf-grass, ornamental plants, and agriculture. Numerous invasive species damage plants and ecosystems (Potter and Held 2002, Jackson and Klein 2006) and might outcompete native species. Some scarab species cause human health problems as carriers of diseases (Graczyk et al. 2005).

Series Elateriformia

This group of beetles contains five superfamilies (Scirtoidea, Dascilloidea, Buprestoidea, Byrrhoidea, and Elateroidea) (Lawrence et al. 1995), 32 families (Lawrence and Newton 1995), and at least 40,000 species. Of these, about 30,000 species belong to the three largest families, the Buprestidae, Elateridae, and Cantharidae. Elateriform beetles are found throughout the world and are most diverse in tropical regions. Chapters on Elateriformia in Beutel and Leschen (2005a) are good introductions to the classification and biology of the series and its superfamilies (except Elateroidea, which is not yet available).

With the most recent morphological test of elateriform monophyly, Lawrence et al. (1995) expanded the group to include the Scirtoidea. This finding was contradicted by two molecular studies indicating that the Elateriformia are monophyletic, excluding the Scirtoidea (Caterino et al. 2005, Hughes et al. 2006), or more broadly paraphyletic (Hughes et al. 2006). Beutel (1995) analyzed interfamily relationships of the Elateriformia (*sensu* Lawrence 1987), based on 27 larval characters. Membership of the Byrrhoidea was redefined to reflect a monophyletic group, based on a morphological analysis by Costa et al. (1999). The membership and internal relationships of the Elateroidea were addressed using morphological tools by Lawrence (1987), Calder et al. (1993), Muona (1995), and Branham and Wenzel (2003). Elateroid monophyly was supported in a molecular study by Caterino et al. (2005), based on 7 in-groups and 103 out-groups. Nelson and Bellamy (1991) examined buprestoid relationships, but did not test superfamily monophyly. The monophyly and internal relationships of the Scirtoidea and Dascilloidea have not been examined phylogenetically, although Friedrich and Beutel (2006) have identified two new possible synapomorphies for the Scirtoidea excluding the Decliniidae.

Although the major known economic effects of elateriform beetles are negative (e.g., Buprestidae as forest pests and Elateridae as agricultural pests), many are likely beneficial because, although the larval prey of many predaceous Elateridae, Cantharidae, and Lampyridae are unknown, they probably include pest organisms. Among the beneficial species, only a few elaterids have been introduced for biological control (of Scarabaeidae: Bianchi 1937, Clausen 1978). Similarly, the role of elateroids and buprestoids as pollinators is probably also undervalued.

Elateriform beetles are potentially valuable to evolutionary and ecological research. Their diverse natural history includes seven apparent independent colonizations of aquatic habitats (Lawrence 1987). They have a wider diversity of larval feeding habits than for any other series, which include feeding on deadwood (dry, e.g., Buprestidae; submerged e.g., Elmidae), herbivory (including leaf mining and stem boring, e.g., Buprestidae; moss feeding, e.g., Byrrhidae), fungivory (e.g., Elateridae, Luterek 1966), carnivory (as predators of invertebrates such as termites, Costa et al. 1992; and vertebrates such as sea turtle eggs by Elateridae, Donlan et al. 2004), and ectoparasitism (e.g., of cicadas by Rhipiceridae). Their value to evolutionary and ecological research also includes their role as primary pollinators for at least one plant species (Peter and Johnson 2005), chemical defenses, mimicry, and multiple evolutionary origins of bioluminescence (Branham and Wenzel 2003, Sagegami-Oba et al. 2007).

Series Bostrichiformia

The Bostrichiformia are composed of seven families, each containing relatively few species. The Jacobsoniidae, Derodontidae, Nosodendridae, and Endecatomidae each contain fewer than 100 species worldwide, whereas the Bostrichidae, Dermestidae, and Anobiidae contain between 500 and 2500 species each. Some of the species in the less diverse families are tiny (<1 mm), their biology is essentially unknown,

and they are collected only by using specialized tools such as the Berlese funnel. The global biodiversity of these little-known groups is difficult to assess (Philips et al. 2002). Overall, the species-poor families Jacobsoniidae, Derodontidae, Nosodendridae, and Endecatomidae have diverse feeding habits and habitat preferences, although few affect humans directly. They occur, for example, in fungi, in rotten wood, under bark, in leaf litter, and in bat guano (Lawrence and Hlavac 1979, Leschen 2002, Ivie 2002a). Species in the derodontid genus *Laricobius* are predators, and one species was introduced in North America as a biological control agent of the hemlock woolly adelgid (Hemiptera: Adelgidae) (Lamb et al. 2006).

The larvae of most Bostrichidae are wood borers and several are major pests. The grain borers (*Prostephanus* and *Rhyzopertha*), powder-post beetles (*Lyctus* and *Trogoxylon*), and bamboo powder-post beetles (*Dinoderus*) are especially important economically (Ivie 2002b). One species also damages books (Hoffman 1933). The Dermestidae include primarily scavengers of dried animal and plant material. This family includes some of the most important pests of stored products and museum specimens. In addition, some dermestids are used in forensic entomology.

The family Anobiidae, which is most diverse in tropical and subtropical habitats, includes more than 2000 species worldwide. The larvae of the Anobiidae can be separated into two groups. The first group of species bore into plant materials such as bark, seeds, dry wood, and galls, while the other group feeds on dried animal and plant detritus (Howe 1959, Philips 2002). Some species are found in association with vertebrate nests, whereas others breed in dung. Economically important species include the death watch beetle (*Anobium punctatum*), which damages furniture, woodwork, book bindings, and other cellulose products (Peters and Fitzgerald 1996, Philips 2002). Other species of Anobiidae are pests of stored products including drugs, food commodities, spices, tobacco, and wool (Howe 1959, Bousquet 1990).

Series Cucujiformia

The series Cucujiformia, as defined by Lawrence and Newton (1995), includes six superfamilies: Cleroidea, Lymexyloidea, Cucujoidea, Tenebrionoidea, Chrysomeloidea, and Curculionoidea. The superfamily Lymexyloidea is the smallest, containing the single family Lymexylidae of unknown phylogenetic affinities. More than 50 predominantly tropical species are organized into seven genera (Wheeler 1986). Adults of at least some species transmit spores of ambrosia fungi in specialized membranous cavities on their ovipositors (Wheeler 1986). Adults of some species are longer than 60 mm and have shortened elytra resembling those of rove beetles (Staphylinidae). Larvae of at least some Lymexylidae burrow into wood and cultivate ambrosia fungi believed to be their only larval food source. This larval lifestyle causes some damage to forestry in temperate areas, whereas in the tropics, *Promelittomma insulare* is a pest of coconut palms (Brown 1954).

The superfamily Cleroidea is a group of 10 families and about 10,000 species, although some authors (Majer 1994, Kolibáč 2004) advocate splitting the largest family, the Melyridae, into several smaller families. Adults and larvae of most species are predators. The larvae of some species pursue and attack wood-boring insects inside their tunnels, benefitting the forestry industry. The Acanthocnemidae have a single extant species, *Acanthocnemus nigricans*, native to Australia (Lawrence et al. 1999). The Chaetosomatidae include four genera in New Zealand and Madagascar (Lawrence et al. 1999). Cleridae is the second largest family in this superfamily, with about 3500 species worldwide (Kolibáč 2004). The Melyridae (*sensu lato*, including Dasytidae and Malachiidae *sensu* Majer 1994, 2002) have about 5000 species worldwide (Kolibáč 2004). The Metaxinidae recently were established as a separate family (Kolibáč 2004) to accommodate a single species, *Metaxina ornata*, from New Zealand. The Phloiophilidae include the single European and North African genus *Phloiophilus*; the larvae feed on fungi (Lawrence et al. 1999). The Phycosecidae include a single genus, *Phycosecis*, endemic to Australia and New Zealand (Beutel and Pollock 2000). The Prionoceridae have at least three genera in the Oriental and Afrotropical regions, and are pollen-feeders in the adult stage (Lawrence et al. 1999). The Thanerocleridae, with 29 species in 7 predominantly tropical genera, are known from all continents except Antarctica (Kolibáč 1992, 2004). The Trogossitidae were redefined by Ślipiński (1992) so that the family has fewer than a thousand species, is the third largest in the Cleroidea, and is worldwide in distribution (Kolibáč 2005).

The superfamily Cucujoidea includes about 20,000 species in about 1500 genera (Pakaluk et al. 1994), which are classified in the following 35 families (Leschen et al. 2005): Agapythidae, Alexiidae, Biphyllidae, Boganiidae, Bothrideridae, Byturidae, Cavognathidae, Cerylonidae, Coccinellidae, Corylophidae, Cryptophagidae, Cucujidae, Cyclaxyridae,

Discolomatidae, Endomychidae, Erotylidae, Helotidae, Hobartiidae, Keteretidae, Laemophloeidae, Latridiidae, Lamingtoniidae, Monotomidae, Myraboliidae, Nitidulidae, Passandridae, Phalacridae, Phloeostichidae, Priasilphidae, Propalticidae, Protocucujidae, Silvanidae, Smicripidae, Sphindidae, and Tasmosalpingidae. Some species, particularly of lady beetles (Coccinellidae), prey on phytophagous arthropods, benefitting humans by protecting beneficial plants. Other species are invasive or otherwise detrimental. The small hive beetle, *Aethina tumida* (Nitidulidae), is a pest of honeybee colonies and recently spread to the USA (1998), Canada (2002), and Australia (2002) from southern Africa (Hood 2004). *Harmonia axyridis*, the multicolored Asian lady beetle, was released in North America to control aphids; however, it began preying on nontarget arthropods, ecologically displacing native coccinellids, and becoming a nuisance pest by entering human houses in search of hibernation places (Pervez and Omkar 2006). Most cucujoid species are, however, of little direct relevance to human economy because they are associated with decaying vegetation and fungi. Significant taxonomic progress has been made toward stabilizing the number, as well as taxonomic limits, of the beetle families in this superfamily. The works by Pakaluk et al. (1994) and Leschen et al. (2005) are important and serve as entry points into more detailed taxonomic treatments.

The superfamily Tenebrionoidea is composed of 27 families, 19 of which are minor, with only 5–500 species described in each (Archeocrypticidae, Boridae, Melandryidae, Mycetophagidae, Mycteridae, Perimylopidae, Prostomidae, Pterogeniidae, Pyrochroidae, Pythidae, Ripiphoridae, Salpingidae, Scraptiidae, Stenotrachelidae, Synchroidae, Tetratomidae, Trachelostenidae, Trictenotomidae, and Ulodidae). Many of these beetles usually feed, for example, under bark of dead trees, in fungi, in decaying wood, and in litter. Larvae of the Ripiphoridae have unusual feeding habits for beetles; they are endoparasites of immature stages of other insects (bees, wasps, and cockroaches). Species that have a more direct effect on society are stored-product pests and vectors of disease organisms in the family Mycetophagidae (Bousquet 1990, Hald et al. 1998). One species of *Hemipeplus* in the family Mycteridae is a pest of palms (Lepesme 1947, Zelazny and Pacumbaba 1982).

The families Aderidae, Anthicidae, Ciidae, Meloidae, Mordellidae, Oedemeridae, and Zopheridae are moderately diverse within the Tenebrionoidea, each containing 550–3000 described species. The adults of some are important pollinators of wild and cultivated plants. A large proportion of these beetles develop in dead wood, fungi, and leaf litter and are uncommonly encountered by people. Several species of the Zopheridae (subfamily Colydiinae) are pests of stored-food products (Bousquet 1990), whereas *Nacerdes melanura* (Oedemeridae) is a pest of wooden marine structures and archaeological timber (Pitman et al. 1993, 2003). Beetles of the families Oedemeridae and Meloidae produce antipredator secretions called cantharidins that can harm humans or kill domesticated animals. A few species in the family Anthicidae prey on the eggs and small larvae of pest species and can be useful as biological control agents (McCutcheon 2002).

By far the most diverse tenebrionoid family is the Tenebrionidae, with nearly 20,000 described species worldwide. This family is expected to be much more diverse than current totals would suggest, especially in tropical regions. Recent studies of tropical groups have resulted in large increases in the number of known species (e.g., Bouchard 2002). Distantly related groups of the Tenebrionidae have evolved remarkable adaptations to survive in some of the most extreme habitats in the world. The production of glycerol-like compounds in adults of *Upis ceramboides* has enabled them to survive under bark of trees during prolonged freezing periods to $-50°C$ (Miller 1978, Lundheim and Zachariassen 1993). Conversely, unusual morphological and behavioral adaptations in other tenebrionids enable them to survive in some of the hottest deserts on the planet (Cloudsley-Thompson 1964, McClain et al. 1985, Zachariassen 1991). Detailed studies of these adaptations might yield important applications for humans. Tenebrionids are a major faunal component of most world deserts and play a significant role in the food webs (Crawford et al. 1993). Shifts in the abundance and range of xerophiles, versus moist-forest species, have been used as indicators of climate change (Geisthardt 2003). Several tenebrionid species have been transported across the world for centuries by humans because of their association with stored products (Andres 1931, Chaddick and Leek 1972). These beetles cause billions of dollars in losses to stored products every year and much money is spent to control them. One of the most important pests, *Tribolium castaneum*, is one of the most studied of all insects. Similar to adephagans, many tenebrionids produce antipredator secretions containing a combination of repulsive agents (e.g., quinones, phenols), wetting agents, and spreading agents (e.g., hydrocarbons) or other compounds with unknown purposes (Hurst et al. 1964, Tschinkel 1975). These secretions are thought to provide antibacterial or antifungal properties in addition to their repulsive action (Blum 1981). Some Tenebrionidae, especially those that have lost their ability to fly, are recognized as endangered and in need of habitat conservation to avoid extinction.

The superfamily Chrysomeloidea is a group of seven beetle families (Beutel and Leschen 2005a) with more than 50,000 species arranged in the cerambyciform (Cerambycidae, Vesperidae, Oxypeltidae, Disteniidae) and chrysomeliform (Megalopodidae, Orsodacnidae, Chrysomelidae) lineages. Because they are almost exclusively phytophagous, many of these beetles are important to humans, mainly as plant pests or biological agents against unwanted plants. The families Cerambycidae (longhorn beetles) and Chrysomelidae (leaf beetles) are by far the largest in this superfamily and include worldwide more than 20,000 (Turnbow et al. 2002) and 38,000 (Jolivet 1997) species, respectively. Reid (2000) recently reviewed phylogenetic analyses of the chrysomeliform lineage and provided a key to subfamilies based on adults and larvae. Švácha et al. (1997) provided the most recent phylogenetic hypothesis of the cerambyciform lineage. Although the majority of species in the Cerambycidae are winged and have arboreal habits, specialized adaptations to ground-dwelling habits are known in the flightless genus *Dorcadion* and related genera (Plavilstshikov 1958). The remaining five families combined include fewer than 1000 species. The Disteniidae have fewer than 100 species (Švácha et al. 1997) in all major regions of the world, except Australia, Europe, and New Zealand (Lawrence et al. 1999). The Oxypeltidae include three large and brightly colored South American species in two genera (Lawrence et al. 1999). The Vesperidae have about 60 species in the warmer regions of the world, among which the soil-dwelling larvae of *Philus antennatus* are notorious pests of pine plantations in China (Švácha et al. 1997). The pantropical Megalopodidae include some 350 species, while the Orsodacnidae have fewer than 30 species in 3 genera in the Nearctic, Neotropical, and Palearctic regions (Clark and Riley 2002).

The herbivorous superfamily Curculionoidea contains 7 smaller families with fewer than 4500 described species each (Nemonychidae, Anthribidae, Belidae,

Attelabidae, Caridae, Brentidae, and Ithyceridae), as well as the most diverse family of beetles, the Curculionidae. Most species in this superfamily have a distinctive cylindrical extension of the head that leads to the buccal cavity. This adaptation, the rostrum, is used to prepare cavities in plant tissues for the deposition of eggs. The evolution of the rostrum is thought to be one of the reasons for the success of this group of beetles because it allows species of the Curculionoidea to exploit plant tissues that other groups cannot access (Anderson 1995).

Members of the seven smaller families of the Curculionoidea are recognized as more 'primitive' in an evolutionary context. These beetles generally are associated with plants that also are considered 'primitive' (e.g., conifers, cycads, and ferns) (Farrell 1998). Because of the long-lasting and specialized associations between these primitive weevils and their host plants, they are considered important in their pollination (Kevan and Baker 1983). The anthribid *Araecerus fasciculatus* (coffee-bean weevil) feeds on a variety of dried plant materials and has become a serious pest of stored products such as coffee and cocoa beans (Childers and Woodruff 1980, Bousquet 1990). Some species of the subfamily Apioninae (Brentidae) are specific to a single or a small number of host plants and have been introduced in new areas of the world as biological tools to control the spread of invasive weeds (O'Brien 1995).

The family Curculionidae is one of the most diverse groups of organisms, with more than 60,000 species described and thousands more to be described. Their close association with flowering plants is one of the main factors explaining their great diversity (Fig. 11.1). Curculionids feed on plants of any terrestrial or freshwater habitat and on any plant tissue, from roots to seeds (Anderson 2002). Because of these characteristics, curculionids are important economically. Some species are serious pests of agricultural and forestry plants; others feed on stored-plant tissue. Curculionids have been used to control populations of invasive plants (O'Brien 1995). The subfamilies Scolytinae and Platypodinae (considered separate families until recently) are specialized in exploiting injured, dying or weakenened woody plants (Anderson 2002). They either feed on the phloem of the inner bark of their hosts or feed on symbiotic ambrosia fungi that they cultivate in the xylem of their hosts. Several species in these two subfamilies are major forestry pests.

SOCIETAL BENEFITS AND RISKS

Beetles of economic importance

Negative effects of beetles

Agriculture

Each year, beetles have a major effect on the world's agriculture. Hundreds of species of beetles, including many in the families Curculionidae, Chrysomelidae, Elateridae, and Scarabaeidae, feed on crops and ornamental plants in their larval or adult stages or both (Campbell et al. 1989). At least one species probably attacks each cultivated plant species around the world. One of the best-studied pests is the boll weevil (*Anthonomus grandis grandis*), which reduces cotton production. This species is native to tropical and subtropical America, but has been established in the USA since the late 1800s (Burke et al. 1986). In the last 100 years, yield losses and control costs against this species have been estimated at more than $22 billon (Kaplan 2003). A boll weevil eradication program that started in 1983 (Dickerson et al. 2001), coordinated by the US Department of Agriculture and State agencies, has resulted in the eradication of the weevil from more than 80% of the cotton production area in the USA.

Blister beetles that contaminate animal feed also affect agriculture. These meloid beetles are sometimes abundant in hay fields and possess cantharidin compounds that they use for defense against predators. Farm animals of several species (cows, emus, goats, horses, and sheep) have become ill or died after consuming alfalfa contaminated by dead blister beetles (Capinera et al. 1985). The incidence of this condition, termed cantharidiasis, has increased recently due to the common practice of hay conditioning, which crushes the beetles inside the hay and incorporates them into the bales (Capinera et al. 1985, Blodgett et al. 1992).

Beetles are important vectors of pathogens to crops and livestock (Harris 1981). One species, the striped cucumber beetle (*Acalymma vittatum*), is a specialist herbivore that feeds on plants of the cucumber family Cucurbitaceae. The association of these beetles with the wilt-inducing plant bacterium *Erwinia tracheiphila* has led to major losses in the past (Garcia-Salazar et al. 2000). Management of this bacterial wilt relies heavily on vector management (Ellers-Kirk and Fleischer

2006) through field and laboratory studies of the bionomics of the striped cucumber beetle and the effects of varying agricultural practices. The lesser mealworm beetle (*Alphitobius diaperinus*) is a common species associated with poultry production. It colonizes poultry houses where it feeds on shed feathers, droppings, and spilled poultry meal, as well as on dead, dying, or newly born chicks (Harris 1966, Kumar 1988, Watson et al. 2001). Although they have a minor beneficial effect in consuming chicken mites and housefly maggots, the benefits are more than offset by their role in the transmission of chicken diseases such as *Campylobacter*, *Metarhizum*, and *Salmonella* (Davies and Wray 1995, Alves et al. 2004, Strother et al. 2005).

Hinton (1945) stated that the Coleoptera was the most important order of insects attacking stored products. More than 600 species of beetles are associated with stored products around the world. Although many of these species actually eat the stored products, others feed on the fungi or other animals that inhabit warehouses (Bousquet 1990). Some of the most problematic groups of beetles in stored products are the Laemophloeidae, Dermestidae, and Tenebrionidae. The red flour beetle (*T. castaneum*), one of the most frequently encountered pests of stored products, can feed on a wide variety of products in the adult and the larval stages. Because of its economic importance and the ease of rearing it, the red flour beetle is used extensively as a tool in genetics research.

Museum collections

Museums assemble millions of specimens for research and public displays. Beetles, especially those belonging to the family Dermestidae, are one of the greatest threats to the world's museum collections. Species of *Anthrenus* and *Dermestes* commonly attack and destroy preserved animals such as pinned insects and preserved pelts and skins, as well as cultural artifacts that incorporate hair, wool, fur, leather, feathers, and other animal derivatives (Armes 1988, Campbell et al. 1989). The control of museum beetle pests is becoming increasingly difficult because pesticides and fumigants used in the past can cause health problems for museum workers. Museum-pest research is now focusing largely on preventive conservation (Goldberg 1996).

Forestry

Many species of Coleoptera are highly relevant to humans through their activity as forestry pests. The

assemblage of forest pests varies among countries. Up to 45% of the annual wood volume grown in Sweden was estimated to be lost to only two bark beetles: *Tomicus piniperda* and *Ips typographus* (Eidmann 1992). *Dendroctonus frontalis*, feeding on loblolly pine (*P. taeda*), kills about 40,000 trees with a value exceeding $10 million in the southeastern USA in just 18 months (Strom et al. 2004). For these reasons alone, overall losses might justifiably be termed significant.

Weevils of the subfamilies Scolytinae and Platypodinae (Curculionidae), with 5812 and 1463 species, respectively (Wood and Bright 1992), are the most infamous forestry pests. Phloem-feeding, tree-killing bark beetles constitute most of the diversity of the Scolytinae. Female bark beetles normally burrow in phloem and lay eggs, from which larvae hatch and construct further feeding burrows. Phloem of the dying or freshly killed trees is the typical environment for the bark beetles. Many bark beetles use chemical signals (pheromones) to schedule a synergetic attack by many individuals on a single tree, thus overcoming its resistance and killing it, making it suitable for beetle reproduction (Paine et al. 1997).

One of the most notorious phloem feeders is, however, not a scolytine weevil, but rather the emerald ash borer, *Agrilus planipennis*, a jewel beetle (Buprestidae). This species, native to the North Asia-Pacific Region, recently was accidentally introduced and established in North America. The larvae feed on phloem of ash trees (*Fraxinus* spp.) and cause widespread decline and mortality of the trees. The species was first detected in July 2002 in Detroit (Michigan, USA) and Windsor (Ontario, Canada). The cumulative death toll of ash trees is estimated to be as high as 15 million, many of which are high-value urban trees (Poland and McCullough 2006). The genus *Agrilus* is among the most species-rich genera of all organisms, with 2767 species worldwide (Bellamy 2003), many of which have potential for introduction and forest destruction abroad.

Phloem-feeding scolytine beetles are among the species most easily introduced accidentally to new regions, primarily through dunnage and casewood (crating) used in international trade. Island countries like New Zealand are particularly vulnerable to introductions because of their isolated and mostly endemic native biota. Some 103 species of scolytine weevils were intercepted at New Zealand's borders between 1950 and 2000 (Brockerhoff et al. 2006), including high-risk species such as *D. ponderosae* and *I. typographus*. The mainland Eurasian species *D. micans* was first recorded in Wales in 1982 and since then spread throughout the UK (Gilbert et al. 2003). The North American species *D. valens* was introduced to China in 2001 and currently infests more than 0.5 million hectares of pine forest (Cognato et al. 2005). The pine shoot beetle (*T. piniperda*) is another Palearctic bark beetle recently introduced to North America. Since its discovery in Ohio in 1992, this beetle has become established in Ontario and Quebec and in 13 states in the northcentral and northeastern USA (Morgan et al. 2004). Haack (2006) reviewed the history of bark- and wood-boring beetles introduced to the USA for the last 20 years and concluded that 2 exotic Buprestidae, 5 Cerambycidae, and 18 Scolytinae species have been introduced and are expanding their ranges in North America.

Although most forestry-important Coleoptera species are phloem feeders, many beetles also tunnel through sapwood, weakening or killing the host tree. Ambrosia weevils of the Scolytinae and Platypodinae have evolved a type of agriculture, cultivating ambrosia fungi in their burrows (Farrell et al. 2001). The Asian longhorn beetle *Anoplophora glabripennis* (Fig. 11.1d) emerged recently as the top-profile beetle in the news in North America and, partly, Western Europe. In the early 1990s, this species hitchhiked with international trade across the Pacific Ocean and infested urban forests in the USA and Canada, with the potential of rapidly killing trees of many species. This alarming news prompted government agencies to cut down thousands of trees in New Jersey, New York, and Toronto in an effort to eradicate the species. The genus *Anoplophora* has some 35 other species in Northwest Asia, which remained poorly studied taxonomically until recently revised by Lingafelter and Hoebeke (2002). Their study indicates that other *Anoplophora* species also could become introduced pests overseas.

Forestry in the Southern Hemisphere is based mainly on the hardwood *Eucalyptus* trees, which also have an associated complex of beetle pests. The Christmas beetle *Anoplognathus chloropyrus* (Scarabaeinae), and the leaf beetles *Paropsis atomaria*, *P. charybdis*, and some other taxonomically poorly known Paropsinae (Chrysomelidae) are significant defoliators of commercially grown eucalypts in Queensland, Australia (Johns et al. 2004, Nahrung 2004).

Many highly destructive fungal forest pathogens are transmitted by beetles. Black-stain root disease of conifers in western North America, for example, is caused by the ophiostomatoid fungus *Leptographium wageneri*. Spores of this species were detected in 37%

of bark beetles in California, with individual beetles carrying up to 100,000 spores (Schweigkofler et al. 2005). The mountain pine beetle (*D. ponderosae*) and its mutualistically associated blue-stain fungi (*Ophiostoma* spp.) have benefited from unusually high winter temperatures, which have lowered larval mortality in recent years. These factors created the largest beetle epidemic in Canada's history, primarily killing lodgepole pines (*Pinus contorta* var. *latifolia*) in British Columbia (Kim et al. 2005). This phenomenon of beetle-fungus mutualistic association is not restricted to bark and ambrosia beetles, but has evolved independently in a few other beetle taxa.

An understanding of the biodiversity of tree-associated beetles is relevant to the world forest industry. To date, only a few groups, such as the European and North American bark and ambrosia beetles, are adequately known taxonomically. Most tree-living beetle species beyond these two geographical regions remain inadequately known.

Positive effects of beetles

Biological control of weeds

Plant-feeding beetles, primarily of the families Curculionidae and Chrysomelidae, have been used successfully to control the spread of invasive alien plant species throughout the world. Weeds have been transported accidentally throughout the world for centuries and the trend is increasing with global trade (Mason et al. 2002). Once established in a new area, weeds often spread quickly, primarily because of the absence of insect herbivores associated with them. Weed biological control, or the introduction and manipulation of natural enemies to reduce the spread of invasive weeds, is often used as a pesticide-free control method. Careful studies of the biodiversity and biology of beetles in the weed's country of origin are necessary for safe and efficient control (O'Brien 1995, Rea 1997, Lindgren et al. 2002). Recent success stories include control of houndstongue (*Cynoglossum officinale*), using a weevil (Fig. 11.1c) (De Clerck-Floate et al. 2005), and control of purple loosestrife (*Lythrum salicaria*), using two species of leaf beetles (Lindgren et al. 2002).

Dung removal

A single cow can produce more than 9000 kg of solid waste every year (Fincher 1981). The accumulation of dung from agricultural activities pollutes waterways, reduces pasture quality, and creates microhabitats for the development of flies and other pests. Dung beetles (members of the families Scarabaeidae and Geotrupidae) promote manure decomposition by burying it in the ground for larval food. Dung beetles annually avert economic losses to the USA of US$0.38 billion by accelerating burial of livestock feces (Losey and Vaughan 2006). Dung beetles have been imported into countries, such as Australia, where an increase of livestock dung could not be removed by native species (Hughes et al. 1975). Dung beetles also provide a vital ecosystem service in urban environments where they use dog dung, thereby reducing stream pollution and improving nutrient-poor soils (Wallace 2005).

Pollination

Beetles, along with other insect groups, are critically important in the pollination of cultivated and wild plants. Pollination by beetles is often referred to as cantharophily. The Coleoptera are considered the most primitive pollinators (Kevan and Baker 1983). Beetles have been associated with flowering plants over millions of years, leading in some cases to the evolution of specialized structures in the host-plant flowers and in their beetle pollinators (Barth 1985, Fenster et al. 2004). Thirty-four families of flowering plants contain at least one species that is pollinated primarily by beetles (Bernhardt 2000). In some ecosystems such as cloud forests, more than 45% of palms and herbs rely on beetles for pollination (Seres and Ramírez 1995). A recent decline in native and nonnative pollinators worldwide was attributed to pesticide use, habitat loss, plant biodiversity loss, and poor agricultural practices (Allen-Wardell et al. 1998). If this trend continues, we can expect threats to our food supply and global biodiversity (Allen-Wardell et al. 1998).

Beetles of cultural importance

Beetles have cultural significance to societies around the world. The most ubiquitous of these cultural practices is the use of beetles as a food and the traditions around that use. Beetles are recognized by most indigenous societies as a good source of food that adds protein to the diet. Onore (1997) surveyed human entomophagy in Ecuador and found that at least 30 species of beetles were eaten. Onore's list includes many of the large and common beetles found near settlements in the Ecuadorian highlands. Smith and Paucar (2000) provide a detailed account of the traditions surrounding the eating of the scarab beetle *Platycoelia lutescens*. Based on these accounts, the practice of entomophagy appears widespread and important to Ecuadorian society, especially among the rural people. Ecuadorians are generally familiar with how and when to collect and prepare these beetles for food (Smith and Paucar 2000). In another example, Utsunomiya and Masumoto (1999) found striking similarities in beetle consumption practices in northern Thailand and Ecuador. In northern Thailand, approximately 100 different species of beetles were eaten, with 89% of respondents listing 'taste' as the answer to the question of why insects are eaten. Cultural traditions surrounding beetle consumption are not restricted to Ecuador and Thailand. They also have been observed in many other world cultures. However, these traditions may be in decline because of conflict with taboos of Western culture against eating beetles. For more information, the website *Insects as Food* (www.food-insects.com), by Gene DeFoliart, is a tremendous resource about the use of insects as food.

Perhaps the best-known use of beetles in mythology and religion concerns the sacred scarab of the ancient Egyptians. The symbol of a scarab beetle was used prominently over a period spanning approximately 3000 years, starting with the 1st Dynasty (about 5000 years ago) (Cambefort 1994) and ending with the conquest of the land of the Pharaohs by the Roman Empire. Ancient Egyptians likened the rolling of dung balls by dung beetles to the sun rolling across the sky each day. Because the sun was thought to be reborn each day, scarab beetles became powerful symbols of resurrection and eternal life, which were prominent aspects of ancient Egyptian mythology. Through 3000 years of ancient Egyptian cultural development, an elaborate mythology developed around scarab beetles, including the incorporation of scarab pupation (likened to mummification) and other aspects of their natural history and life cycle into mythological stories. Cambefort (1994) gives a detailed review of all facets of the use of scarab beetles by ancient Egyptians. Kritsky (1991) and Cambefort (1994) reported that the carvings and symbols of Buprestidae and Elateridae also held meaning to the ancient Egyptians. They hypothesized that buprestids symbolized the Egyptian myth of rebirth, perhaps due to the emergence of adult beetles from the trunks of living trees.

In many countries, keeping personal beetle collections and live beetles are popular leisure activities, especially in Europe, Japan, and North America. Beetles are a major component of the multimillion dollar insect trade. This trade is conducted by mail and internet or at annual insect fairs in cities including Los Angeles, Paris, Prague, and Tokyo. Keeping live beetles is popular in Japan, where an industry provides not only live beetles, but also the required food and other paraphernalia for keeping them alive. Evidence for the popularity of this pastime in Japan can be seen on the website www.youtube.com, which features numerous Japanese videos of live beetles.

Beetles of medical and legal importance

Medical entomology

Blistering of human skin (dermatosis) can be caused by beetles of the families Meloidae, Oedemeridae, and Staphylinidae (*Paederus* and *Paederinus*) (Nicholls et al. 1990). Dermatosis occurs when a beetle's hemolymph is released onto the skin after it is crushed or brushed accidentally. The vesicant chemical compound in the body fluids of the Meloidae and Oedemeridae is cantharidin, whereas pederin is the vesicant in beetles of the genera *Paederus* and *Paederinus* (Mackie et al. 1945, Nichols et al. 1990, Piel et al. 2005). These chemicals are generally used by the beetles against predators (Pinto and Bologna 2002).

Pederin belongs to a group of complex compounds found only in these staphylinid genera and a few marine sponges (Piel et al. 2004a). This compound is found only in females of some species in these genera and is thought to be produced by symbiotic bacteria (Kellner 2001). Research on the biosynthesis and mode of action of pederin has increased in recent years since it was found that related compounds have potent antitumor properties (Piel et al. 2004a, b). Pederin inhibits protein biosynthesis in tumor cells, and the isolation of pederin-producing genes in these symbiotic bacteria might lead to the development of anticancer drugs (Piel et al. 2005).

Canthariasis is a term used to describe the infection of human internal organs by beetles. The most common type of canthariasis occurs when people accidentally or voluntarily ingest beetles. Accidental ingestions generally occur when people eat foods contaminated by stored-product pests. Adults, larvae, and cast skins can

be ingested in this manner, resulting in irritation of the digestive system or allergic responses. The inadvertent ingestion of larder beetle larvae (*Dermestes lardarius*, Dermestidae), which are covered by long, narrow, and barbed setae, can cause diarrhea, abdominal pain, and perianal itch (Goddard 2000).

Deliberate ingestion of beetles for medicinal purposes also has also been reported. One of the better-known insect-derived medicinal compounds – in the families Meloidae and Oedemeridae – is cantharidin. The insect referred to as Spanish fly (*Lytta vesicatoria*) is a European meloid beetle. Dried, crushed beetles containing cantharidins have been ingested as a vesicant to treat various ailments and to serve as an aphrodisiac for millennia (Karras et al. 1996). Cantharidin can be toxic to humans, and illness caused by its abuse is significant (Sandorini 2001). Ingestion of live darkling beetles (*Ulomoides dermestoides*) for similar purposes was reported from Southeast Asia (Sandorini 2001). The practice of ingesting beetles for food or medicine is risky because some beetles are intermediate hosts for tapeworms and nematodes that later can infect vertebrates (Mackie et al. 1945, Halffter and Matthews 1966, Lethbridge 1971), possibly including humans (Chu et al. 1977).

One type of canthariasis, called scarabiasis, refers to the short-term infestation of the human gut by adult dung beetles (Scarabaeidae). This condition usually affects preschool children in the tropics. The infestation begins when dung beetles, attracted to lights at night, fly into dwellings with naked, sleeping children. Adult dung beetles, usually shorter than 1 cm, follow the smell of feces and are thought to enter the anus and feed internally (Arrow 1931, Halffter and Matthews 1966). Adult beetles are often seen flying away from newly passed stool (Rajapakse 1981). Other types of canthariasis involve the infrequent urinary, ocular, nasal, and cutaneous infections to humans caused by beetle eggs, larvae, or adults (Mackie et al. 1945).

Several families of elateroid beetles produce light, using a biochemical reaction (Lloyd 1983, Viviani 2002). Light is produced when enzymes called luciferases catalyze the oxidisation of luciferin compounds. Beetles principally use this cold light, called bioluminescence, for sexual communication purposes, although other possible functions, including aposematic signals, attraction of prey, and defense, are possible (Lloyd 1983, Underwood et al. 1997). Studies of firefly (Lampyridae) light production have resulted in the development of several medical applications, based on the use of luciferases and their associated genes. These applications range from studies of monitoring the progress of infections such as HIV (Contag et al. 1997) to visualizing living cells in human embryonic development (Greer and Szalay 2002). Luciferases have played a significant role in the development of more efficient drugs for many diseases (Viviani 2002).

Forensic entomology

The major contribution of insects in criminal cases involving homicide is to estimate the limits of the postmortem interval (time between death and discovery of the body). Flies are a dominant group on human corpses, although several groups of beetles are also associated with dead bodies, either preying on other arthropods developing on carcasses or actually feeding on the body (Smith 1986, Catts and Haskell 1990). The postmortem interval is usually estimated from experimental studies using pigs or other vertebrates, and then applied to human corpses in criminal cases (Tabor et al. 2004, Arnaldos et al. 2005). The period of development of the beetles and the succession of species that colonize carcasses in various stages of decomposition provide data for estimating the postmortem interval (Franc et al. 1989, Moura et al. 1997, Carvalho et al. 2000, Turchetto and Vanin 2004). Succession of beetle species tends to follow a rough pattern from an initial fresh stage of decomposition, through bloated and active decay stages to a final dry stage. The Dermestidae (skin beetles) and Cleridae (bone beetles) are among the most common beetles on corpses and have provided important postmortem information, especially for finds of dry skeletal remains (de Souza and Linhares 1997, Kulshrestha and Satpathy 2001).

Although most experimental postmortem-interval studies use bodies on the ground surface, some research has examined buried or submerged bodies (Smith 1986). A few beetles (e.g., Histeridae, Silphidae, and Staphylinidae) are associated with buried bodies (Payne et al. 1968, VanLaerhoven and Anderson 1999, Bourel et al. 2004). Molecular and toxicological analyses show that necrophagous beetles also could be informative for criminal cases involving badly decomposed bodies (DiZinno et al. 2002), movement of bodies (Benecke 1998), and bodies with toxic substances (e.g., drugs, heavy metals, and poisons) (Bourel et al. 2001, Gagliano-Candela and Aventaggiato 2001).

Beetles as research tools

Beetles are used widely as research tools in biophysics and related disciplines. Because they are the most diverse animal order, beetles possess great potential for bioengineering studies. Geometry and mechanics of elytral opening and closing are studied by aeronautic and astronautic engineers (Frantsevich et al. 2005). Whirligig beetles (Gyrinidae), roughly 1 cm in length, swim on the water surface at $55\,\mathrm{cm\,s^{-1}}$ and are capable of making 12 horizontal rotations per second (Fish and Nicastro 2003), which would be a dream performance for any human-made autonomous device. Aquatic larvae of *Hydrobius fuscipes* (Hydrophilidae) demonstrate how simple side-to-side body movements in swimming can easily be adapted, through the use of tracheal gills, as anchors in a novel kind of limbless skating on the lower surface of the water (Brackenbury 1999). Study of walking mechanics in *Pachnoda marginata* (Scarabaeidae) gave better understanding of friction forces between the tarsal claw systems and walking substrates (Dai et al. 2002). Specialized insect adhesive devices, such as the arolium, euplantulae, pulvilli, and tarsal hairs (Gorb 2001), inspired engineers to develop novel adhesive surfaces. Legs of the leaf beetle *Gratiana spadicea* (Chrysomelidae) attach to leaf surfaces by matching tarsal claw aperture with that of pointed rays of the host-plant trichomes (Medeiros and Moreira 2002). Wide, bilobed tarsi of rove beetle in the genus *Stenus* (Staphylinidae) allow the beetles to run well on solid surfaces or water (Betz 2002).

The high diversity of beetles makes them useful tools for physiological research. Studies of muscle function in arthropods, using the beetle *Cotinus mutabilis* (Scarabaeidae), suggest that asynchronous flight muscles can provide greater power output than synchronous muscles for operation at the high-contraction frequencies of insect flight (Josephson et al. 2001). Some dung-rolling scarab beetles (*Scarabaeus* species) possess a highly evolved sense of polarized light, a sense lacking in humans. The crepuscular beetle *Scarabaeus zambesianus* rolls dung balls away from the dung pile to avoid competition, navigating by polarized skylight sensed by specialized ommatidia of the dorsal rims of its eyes (Dacke et al. 2003). Different biophysical and behavioral aspects of bioluminescence of glowworms (Phengodidae), click beetles (Elateridae), and fireflies (Lampyridae) have fueled research on the evolution of color vision (Stolz 2003, Booth et al. 2004). Recent discoveries of the abilities of tiger and scarab beetles (Cicindelidae and Scarabaeidae) to use paired membranous ears to detect airborne sounds (Forrest et al. 1997) shed light on the evolution of this sense in the Animalia.

Many phytophagous beetles rely on aggregation pheromones to coordinate different aspects of their biology. Bark beetles use pheromones to coordinate an attack on the host tree (Raffa and Berryman 1983). Recent discovery of the first pheromone for the Colorado potato beetle, *Leptinotarsa decemlineata*, is unusual because it is the first male-produced pheromone known in the Chrysomelidae (Dickens et al. 2002). In longhorn beetles (Cerambycidae), males use sex pheromones to attract females or aggregate (Lacey et al. 2004).

Biochemistry and DNA research are areas in which beetles are particularly important as research tools. *Tribolium castaneum*, the red flour beetle (Tenebrionidae), is a common insect pest of stored grains worldwide. This species is similar to the famous fruit fly, *Drosophila melanogaster*, in that it is used widely in genetics and developmental biology research. The red flour beetle is, to date, the only beetle for which a complete genome sequence is published. Its genome consists of about 200 million nucleotides arranged in a haploid set of 10 chromosomes. For comparison, the human genome has about 15 times more nucleotides and a haploid set of 23 chromosomes. *T. castaneum* was the first animal reported to produce inhibitors of prostaglandin synthetase, which were purified from the beetle's defensive secretions (Howard et al. 1986). These substances are used widely in aspirin-like anti-inflammatory drugs. This beetle is also intensively studied as a model for understanding the mechanisms of insect resistance to insecticides.

Beetles, however, are underrepresented in genomic databases. A total of 170,611 beetle nucleotide sequences were found in the National Center for Biotechnology Information (www.ncbi.nln.noh.gov; entry date 25 January 2007), about 60% of which are from *Tribolium*. This figure is much lower than that for Diptera (2,619,203 entries; 1,747,519 from *Drosophila*) and Lepidoptera (582,918) and just slightly higher than for Hymenoptera (159,989). Mitochondrial and RNA-coding genes are sequenced most often (primarily for phylogenetic comparison) for beetles, while sequenced protein-coding genes remain few (Theodorides et al. 2002). Thus, coleopteran genes, particularly protein-coding regions, remain a huge area for fruitful investigation.

Using beetles as research tools sheds light on bizarre aspects of evolution, some of which are unique. For example, *M. debilis*, the sole North American member of the family Micromalthidae, possesses one of the most bizarre life cycles in the Metazoa. This species combines both thelytokous and viviparous larviform diploid females and rare haploid males, which eat and kill their own mothers from inside her body (Pollock and Normark 2002). The second known case of haplodiploid sex determination in beetles is that of some scolytine weevils including *Ozopemon* from Southeast Asia, which is also the only beetle genus having neotenic and strangely modified larviform males (once thought to be histerids) (Jordal et al. 2002). The huge diversity of ecological relationships with other organisms makes beetles ideal research tools for understanding such evolutionary phenomena as sociality (Scott 1998), parasitism (Weber et al. 2006), symbiosis (Kellner 2003), and phoresy (Bologna and Pinto 2001).

Groups of beetles with well-resolved phylogenies are powerful tools in biogeographical and paleogeographical reconstructions. Most commonly, these reconstructions are achieved using molecular data for the species with limited dispersal capacities, such as flightless *Scarabaeus* dung beetles (Scarabaeidae), which show grades of colonization of the Namib Desert in the Miocene (Sole et al. 2005). Phylogenetic trees inferred from mtDNA of darkling beetles of the genus *Nesotes* (Tenebrionidae) provided insight into how the genus colonized the Canary Islands (Rees et al. 2001). The well-resolved phylogeny of endemic Iberian diving beetles (Dytiscidae) indicated that their speciation was induced by repeated fragmentation of populations during glacial and interglacial periods (Ribera and Vogler 2004). Patterns of insect colonization of Pacific Islands were deduced based on the distribution and phylogeny of Colymbetinae diving beetles (Dytiscidae) on New Caledonia and Fiji (Balke et al. 2007). Analysis of fossil and subfossil chitinous remains of beetles in Quaternary sediments (1.8 Ma–present) provides insights into past environments and climates, particularly quantitative estimates of temperatures and precipitation levels (Porch and Elias 2000). All these studies capitalized on the diversity and relative abundance of beetles to focus on more inclusive natural phenomena.

Fireflies have been used extensively for research on bioluminescence (McElroy and DeLuca 1983, Viviani 2002). In addition to the important medical applications that have resulted from such studies, bioluminescence research has yielded several applications

Table 11.2 Coleoptera on the 'Red List of Threatened Species' (IUCN 2006).

Family	Scientific Name	Common Name	Status	Distribution
Anthicidae	Anthicus sacramento	Sacramento beetle	Endangered	United States
Buprestidae	Buprestis splendens	Goldstreifiger	Vulnerable	Albania, Austria, Belarus, Denmark, Finland, Greece, Poland, Russian Federation, Serbia and Montenegro, Spain, Sweden
Carabidae	Carabus olympiae		Vulnerable	France (int), Italy
Carabidae	Cicindela columbica	Columbia River tiger beetle	Vulnerable	United States
Carabidae	Cicindela puritana	Puritan tiger beetle	Endangered	United States
Carabidae	Elaphrus viridis	Delta green ground beetle	Critically endangered	United States
Carabidae	Mecodema punctellum		Extinct	New Zealand (RE)
Cerambycidae	Cerambyx cerdo	Cerambyx longicorn	Vulnerable	Algeria, Armenia, Austria, Azerbaijan, Belarus, Czech Republic, France, Georgia, Germany, Hungary, Islamic Republic of Iran, Republic of Moldova, Morocco, Poland, Spain, Sweden, Switzerland, Tunisia, Turkey, Ukraine, United Kingdom
Cerambycidae	Morimus funereus		Vulnerable	Belgium, Czech Republic, Germany, Hungary, Moldova, Republic of Romania, Serbia and Montenegro, Slovakia, Ukraine
Cerambycidae	Rosalia alpina	Rosalia longicorn	Vulnerable	Algeria, Armenia, Austria, Azerbaijan, Belarus, Bulgaria, Czech Republic, Denmark, France, Georgia, Germany, Greece, Hungary, Islamic Republic of Iran, Israel, Italy, Jordan, Lebanon, Liechtenstein, Morocco, Netherlands, Poland, Portugal, Romania, Russian Federation, Serbia and Montenegro, Spain, Sweden, Switzerland, Syrian Arab Republic, Tunisia, Ukraine
Cerambycidae	Xylotoles costatus	Pitt Island longicorn beetle	Endangered	New Zealand (Chatham Is.)
Cerambycidae	Macrodontia cervicornis		Vulnerable	Brazil, Peru

Family	Species	Common name	Status	Distribution
Cucujidae	Cucujus cinnaberinus		Vulnerable	Austria, Belarus, Czech Republic, Estonia, Finland, Germany, Hungary, Latvia, Lithuania, Norway, Poland, Romania, Russian Federation, Serbia and Montenegro, Slovakia, Sweden
Curculionidae	Dryophthorus distinguendus		Extinct	United States [RE] (Hawaiian Is. [RE])
Curculionidae	Dryotribus mimeticus		Extinct	United States [RE] (Hawaiian Is. [RE])
Curculionidae	Karocolens tuberculatus		Extinct	New Zealand [RE]
Curculionidae	Macrancylus linearis		Extinct	United States [RE] (Hawaiian Is. [RE])
Curculionidae	Oedemasylus laysanensis		Extinct	United States [RE] (Hawaiian Is. [RE])
Curculionidae	Rhyncogonus bryani		Extinct	United States [RE] (Hawaiian Is. [RE])
Curculionidae	Gymnopholus lichenifer	Lichen weevil	Vulnerable	Papua New Guinea
Curculionidae	Pentarthrum blackburni	Blackburn's weevil	Extinct	United States [RE] (Hawaiian Is. [RE])
Curculionidae	Trigonoscuta rossi	Fort Ross weevil	Extinct	United States [RE]
Curculionidae	Trigonoscuta yorbalindae	Yorba Linda Weevil	Extinct	United States [RE]
Dytiscidae	Acilius duvergeri		Vulnerable	Algeria, Italy, Morocco, Portugal, Spain
Dytiscidae	Agabus clypealis		Endangered	Denmark, Germany, Latvia, Poland, Russian Federation, Sweden
Dytiscidae	Agabus discicollis		Endangered	Ethiopia
Dytiscidae	Agabus hozgargantae		Endangered	Spain
Dytiscidae	Deronectes aljibensis		Endangered	Spain
Dytiscidae	Deronectes depressicollis		Vulnerable	Spain
Dytiscidae	Deronectes ferrugineus		Vulnerable	Portugal
Dytiscidae	Dytiscus latissimus		Vulnerable	Austria, Belarus, Belgium [RE], Bosnia and Herzegovina, Croatia [RE?], Czech Republic, Denmark, Finland, France [RE], Germany [RE], Hungary [RE?], Italy, Latvia, Luxembourg [RE], Netherlands [RE], Norway, Poland, Romania [RE?], Russian Federation, Slovakia [RE?], Sweden, Switzerland [RE?], Ukraine

(continued)

Table 11.2 (continued).

Family	Scientific Name	Common Name	Status	Distribution
Dytiscidae	Graphoderus bilineatus		Vulnerable	Austria, Belgium [RE?], Bosnia and Herzegovina, Czech Republic, Denmark, Finland, France, Germany, Hungary, Italy, Latvia, Luxembourg, Netherlands, Norway, Poland, Russian Federation, Serbia and Montenegro, Slovakia, Switzerland, Turkmenistan, Ukraine, United Kingdom [RE]
Dytiscidae	Graptodytes delectus		Endangered	Spain (Canary Is.)
Dytiscidae	Hydrotarsus compunctus		Critically endangered	Spain (Canary Is.)
Dytiscidae	Hydrotarsus pilosus		Endangered	Spain (Canary Is.)
Dytiscidae	Hygrotus artus	Mono Lake diving beetle	Extinct	United States [RE]
Dytiscidae	Megadytes ducalis		Extinct	Brazil [RE]
Dytiscidae	Meladema imbricata		Critically endangered	Spain (Canary Is.)
Dytiscidae	Meladema lanio		Vulnerable	Portugal (Madeira)
Dytiscidae	Rhantus alutaceus		Endangered	New Caledonia
Dytiscidae	Rhantus novacaledoniae		Extinct	New Caledonia [RE]
Dytiscidae	Rhantus orbignyi		Extinct	Argentina [RE], Brazil [RE]
Dytiscidae	Rhantus papuanus		Extinct	Papua New Guinea (RE)
Dytiscidae	Rhithrodytes agnus		Endangered	Portugal
Dytiscidae	Siettitia balsetensis	Perrin's cave beetle	Extinct	France [RE]
Elmidae	Stenelmis gammoni	Gammon's riffle beetle	Vulnerable	United States
Leiodidae	Glacicavicola bathysciodes	Blind cave beetle	Vulnerable	United States
Lucanidae	Colophon barnardi		Endangered	South Africa
Lucanidae	Colophon berrisfordi		Critically endangered	South Africa
Lucanidae	Colophon cameroni		Vulnerable	South Africa
Lucanidae	Colophon cassoni		Critically endangered	South Africa
Lucanidae	Colophon eastmani		Endangered	South Africa
Lucanidae	Colophon haughtoni		Endangered	South Africa

Family	Species	Common name	Conservation status	Distribution
Lucanidae	*Colophon montisatris*		Critically endangered	South Africa
Lucanidae	*Colophon neli*		Vulnerable	South Africa
Lucanidae	*Colophon primosi*		Critically endangered	South Africa
Lucanidae	*Colophon stokoei*		Vulnerable	South Africa
Lucanidae	*Colophon thunbergi*		Endangered	South Africa
Lucanidae	*Colophon westwoodi*		Vulnerable	South Africa
Lucanidae	*Colophon whitei*		Endangered	South Africa
Scarabaeidae	*Aegialia concinna*	Ciervo scarab beetle	Vulnerable	United States
Scarabaeidae	*Aegialia crescenta*	Crescent dune scarab beetle	Vulnerable	United States
Scarabaeidae	*Prodontria lewisi*	Cromwell chafer beetle	Critically endangered	New Zealand
Scarabaeidae	*Osmoderma eremita*	Hermit beetle	Vulnerable	Austria, Belarus, Belgium, Czech Republic, Denmark, Estonia, Finland, France, Germany, Greece, Hungary, Italy, Latvia, Liechtenstein, Lithuania, Republic of Moldova, Netherlands, Norway, Poland, Russian Federation, Serbia and Montenegro, Slovakia, Spain, Sweden, Switzerland, Ukraine
Scarabaeidae	*Pseudocotalpa giulianii*	Giuliani's dune scarab beetle	Vulnerable	United States
Silphidae	*Nicrophorus americanus*	American burying beetle	Critically endangered	United States, Canada [RE]
Tenebrionidae	*Coelus globosus*	Globose dune beetle	Vulnerable	Mexico, United States
Tenebrionidae	*Coelus gracilis*	San Joachin dune beetle	Vulnerable	United States
Tenebrionidae	*Polposipus herculeanus*	Frigate Island giant tenebrionid beetle	Critically endangered	Seychelles

Species are listed alphabetically by family. RE = regionally extinct, int = introduced.

commonly used by humans. For example, luciferases are routinely used as environmental biosensors. Several applications have been developed to enhance the monitoring and detection of pollutants such as agrochemicals, lead, and mercury (Naylor 1999).

Threatened Beetles

Many beetles are especially vulnerable to local and global extinction. These beetles often have low powers of dispersal (flightless), only occur in extreme or specific microhabitats, or occur over small geographical areas. Beetles of oceanic islands, isolated dunes, caves, mountains, and other ecological islands fit into this category. In addition to beetles considered threatened locally or globally (Table 11.2), 69 globally threatened species appear on the International Union for Conservation of Nature and Natural Resources' 'Red List of Threatened Species' either as vulnerable (27), endangered (16), critically endangered (10), or extinct (16). These species belong to 13 families and occur in 60 countries. Human activities, such as habitat destruction and the introduction of invasive alien species, continue to threaten many of the world's natural ecosystems and the myriad of beetle species in them (Spence and Spence 1988, Kamoun 1996, Martikainen and Kouki 2003, Munks et al. 2004, Abellan et al. 2005, Davis and Philips 2005, Bouchard et al. 2006, Talley and Holyoak 2006). We argue that the number of species currently listed as vulnerable, threatened, or extinct represents a gross underestimation of the number that should be targeted for conservation.

CONCLUSIONS

Beetles are a superdiverse group of arthropods that occur in most habitats on this planet. Their influence on science and society is great. Beetles provide essential ecological services and are used as tools in many scientific endeavors, some with large effects on humans. On the other hand, beetles continue to have negative effects on vital industries such as agriculture and forestry. Studies on beetle biodiversity and the conservation of their habitats are necessary to ensure the sustainability of natural ecosystems and critical human activities.

REFERENCES

Abellan, P., D. Sanchez-Fernandez, J. Velasco, and A. Millan. 2005. Conservation of freshwater biodiversity: a comparison of different area selection methods. *Biodiversity and Conservation* 14: 3457–3474.

Acorn, J. 2006. The world's biggest bug is a grub. *American Entomologist* 52: 270–272.

Allen-Wardell, G., P. Bernhardt, R. Bitner, A. Burquez, S. Buchmann, J. Cane, P. Cox, V. Dalton, P. Feinsinger, M. Ingram, D. Inouye, C. Jones, K. Kennedy, P. Kevan, H. Koopowitz, R. Medellin, S. Medellin-Morales, G. Nabhan, B. Pavlik, V. Tepedino, P. Torchio, and S. Walker. 1998. The potential consequences of pollinator declines on the conservation of biodiversity and stability of food crop yields. *Conservation Biology* 12: 8–17.

Alves, L. F. A., V. S. Alves, D. F. Bressan, P. M. O. J. Neves, and S. B. Alves. 2004. Natural occurrence of *Metarhizium anisopliae* (Metsch.) Sorok. on adults of the lesser mealworm (*Alphitobius diaperinus*) (Panzer) (Coleoptera: Tenebrionidae) in poultry houses in Cascavel, PR, Brazil. *Neotropical Entomology* 33: 793–795.

Anderson, R. S. 1995. An evolutionary perspective on diversity in Curculionoidea. *Memoirs of the Entomological Society of Washington* 14: 103–118.

Anderson, R. S. 2002. Curculionidae Latreille 1802. Pp. 722–815. *In* R. H. Arnett, Jr., M. C. Thomas, P. E. Skelley, and J. H. Frank (eds). *American Beetles*, Volume 2. CRC Press, New York.

Anderson, R. S. and J. S. Ashe. 2000. Leaf litter inhabiting beetles as surrogates for establishing priorities for conservation of selected tropical montane cloud forests in Honduras, Central America (Coleoptera; Staphylinidae, Curculionidae). *Biodiversity and Conservation* 9: 617–653.

Andres, A. 1931. Catalogue of the Egyptian Tenebrionidae. *Bulletin de la Société Royale Entomologique d'Egypte* 15: 74–125.

Archangelsky, M., R. G. Beutel, and A. Komarek. 2005. Hydrophiloidea. Introduction, Phylogeny. Pp. 157–183. *In* R. G. Beutel and R. A. B. Leschen (eds). *Handbook of Zoology: A Natural History of the Phyla of the Animal Kingdom*, Volume IV. Arthropoda: Insecta, Part 38. Coleoptera, Beetles, Volume 1. Morphology and Systematics. Walter de Gruyter, Berlin.

Arellano, L. and G. Halffter. 2003. Gamma diversity: derived from and a determinant of alpha diversity and beta diversity. An analysis of three tropical landscapes. *Acta Zoologica Mexicana (Nueva Serie)* 90: 27–76.

Armes, N. J. 1988. The seasonal activity of *Anthrenus sarnicus* and some other beetle pests in the museum environment. *Journal of Stored Products Research* 24: 29–37.

Arnaldos, M. I., M. D. Garcia, E. Romera, J. J. Presa, and A. Luna. 2005. Estimation of postmortem interval in real cases based on experimentally obtained entomological evidence. *Forensic Science International* 149: 57–65.

Arnett, R. H., Jr. and M. C. Thomas (eds). 2001. *American Beetles*, Volume 1. Archostemata, Myxophaga, Adephaga, Polyphaga: Staphyliniformia. CRC Press, New York. 443 pp.

Arnett, R. H., M. C. Thomas, P. E. Skelley, and J. H. Frank. 2002. *American Beetles*, Volume 2. Scarabaeoidea through Curculionoidea. CRC Press, Boca Raton, FL. 861 pp.

Arrow, G. J. 1931. Coleoptera: Lamellicornia Part 3. *In The Fauna of British India, Including Ceylon and Burma*. Taylor and Francis, London. 428 pp.

Balke, M., I. Ribera, and R. G. Beutel. 2003. Aspidytidae: on the discovery of a new beetle family: detailed morphological analysis, description of a second species, and a key to fossil and extant adephagan families (Coleoptera). Pp. 53–66. *In*: M. A. Jäch and L. Ji (eds). *Water Beetles of China*. Volume III. Wien, Zoologisch-Botanische Gesellschaft in Österreich and Wiener Coleopterologenverein. vi + 572 pp.

Balke, M., C. H. S. Watts, S. J. B. Cooper, W. F. Humphreys, and A. P. Vogler. 2004. A highly modified stygobiont diving beetle of the genus *Copelatus* (Coleoptera, Dytiscidae): taxonomy and cladistic analysis based on mitochondrial DNA sequences. *Systematic Entomology* 29: 59–67.

Balke, M., G. Wewalka, Y. Alarie, and I. Ribera. 2007. Molecular phylogeny of Pacific island Colymbetinae: radiation of New Caledonian and Fijian species (Coleoptera, Dytiscidae). *Zoological Scripta* 36: 173–200.

Ball, G. E. and Y. Bousquet. 2001. Carabidae Latreille, 1810. Pp. 32–132. *In* R. H. Arnett, Jr. and M. C. Thomas (eds). *American Beetles*, Volume 1. Archostemata, Myxophaga, Adephaga, Polyphaga: Staphyliniformia. CRC Press, New York.

Barraclough, T. G., M. V. L. Barclay, and A. P. Vogler. 1998. Species richness: does flower power explain beetle–mania?. *Current Biology* 8: 843–845.

Barth, F. B. 1985. *Insects and Flowers: The Biology of a Partnership*. Princeton University Press, Princeton, NJ.

Bellamy, C. L. 2003. An illustrated summary of the higher classification of the superfamily Buprestoidea (Coleoptera). *Folia Heyrovskyana, Supplement* 10: 1–197.

Bellini, R., F. Pederzani, R. Pilani, R. Veronesi, and S. Maini. 2000. *Hydroglyphus pusillus* (Fabricius) (Coleoptera Dytiscidae): its role as a mosquito larvae predator in rice fields. *Bollettino dell'Istituto di Entomologia "Guido Grandi" della Universita degli Studi di Bologna* 54: 155–163.

Benecke, M. 1998. Random amplified polymorphic DNA (RAPD) typing of necrophagous insects (Diptera, Coleoptera) in criminal forensic studies: validation and use in practice. *Forensic Science International* 98: 157–168.

Bernhardt, P. 2000. Convergent evolution and adaptive radiation of beetle-pollinated angiosperms. *Plant Systematics and Evolution* 222: 293–320.

Betz, O. 2002. Performance and adaptive value of tarsal morphology in rove beetles of the genus *Stenus* (Coleoptera, Staphylinidae). *Journal of Experimental Biology* 205: 1097–1113.

Beutel, R. G. 1995. Phylogenetic analysis of Elateriformia (Coleoptera: Polyphaga) based on larval characters. *Journal of Zoological Systematics and Evolutionary Research* 33: 145–171.

Beutel, R. G. 2005. Myxophaga Crowson, 1955. Pp. 43–52. *In* R. G. Beutel and R. A. B. Leschen (eds). *Handbook of Zoology: A Natural History of the Phyla of the Animal Kingdom*, Volume IV. Arthropoda: Insecta, Part 38. Coleoptera, Beetles, Volume 1. Morphology and Systematics. Walter de Gruyter, Berlin.

Beutel, R. G. and F. Haas. 2000. Phylogenetic relationships of the suborders of Coleoptera (Insecta). *Cladistics* 16: 1–39.

Beutel, R. G. and A. Komarek. 2004. Comparative study of thoracic structures of adults of Hydrophiloidea and Histeroidea with phylogenetic implications (Coleoptera, Polyphaga). *Organisms, Diversity and Evolution* 4: 1–34.

Beutel, R. G., and R. A. B. Leschen (eds). 2005a. *Coleoptera, Volume 1: Morphology and Systematics (Archostemata, Adephaga, Myxophaga, Polyphaga partim)*. Walter De Gruyter, Berlin.

Beutel, R. G. and R. A. B. Leschen. 2005b. Phylogenetic analysis of Staphyliniformia (Coleoptera) based on characters of larvae and adults. *Systematic Entomology* 30: 510–548.

Beutel, R. G. and D. A. Pollock. 2000. Larval head morphology of *Phycosecis litoralis* (Pascoe) (Coleoptera: Phycosecidae) with phylogenetic implications. *Invertebrate Taxonomy* 14: 825–835.

Beutel, R. G. and I. Ribera. 2005. Adephaga Schellenberg, 1806. Pp. 53–55. *In* R. G. Beutel and R. A. B. Leschen (eds). *Handbook of Zoology: A Natural History of the Phyla of the Animal Kingdom*, Volume IV. Arthropoda: Insecta, Part 38. Coleoptera, Beetles, Volume 1. Morphology and Systematics. Walter de Gruyter, Berlin.

Beutel, R. G., M. Balke, and W. E. Steiner, Jr. 2006. The systematic position of Meruidae (Coleoptera, Adephaga) based on a cladistic analysis of morphological characters. *Cladistics* 22: 102–131.

Bianchi, F. A. 1937. Notes on a new species of *Pyrophorus* introduced into Hawaii to combat *Anomala orientalis* Waterhouse. *Hawaii Plant Record* 41: 319–333.

Blodgett, S. L., J. E. Carrel, and R. A. Higgins. 1992. Cantharidin contamination of alfalfa hay. *Journal of Medical Entomology* 29: 700–703.

Blum, M. S. 1981. *Chemical Defenses of Arthropods*. Academic Press, New York.

Bologna, M. A. and J. D. Pinto. 2001. Phylogenetic studies of Meloidae (Coleoptera), with emphasis on the evolution of phoresy. *Systematic Entomology* 26: 33–72.

Booth, D., A. J. A. Stewart, and D. Osorio. 2004. Color vision in the glow-worm *Lampyris noctiluca* (L.) (Coleoptera: Lampyridae): evidence for a green-blue chromatic mechanism. *Journal of Experimental Biology* 207: 2373–2378.

Bouchard, P. 2002. Phylogenetic revision of the flightless Australian genus *Apterotheca* Gebien (Coleoptera:

Tenebrionidae: Coelometopinae). *Invertebrate Systematics* 16: 449–554.

Bouchard, P., T. A. Wheeler, and H. Goulet. 2006. Ground beetles (Coleoptera: Carabidae) from alvar habitats in Ontario. *Journal of the Entomological Society of Ontario* 136: 3–23.

Bourel, B., G. Tournel, V. Hédoin, M. L. Goff, and D. Gosset. 2001. Determination of drug levels in two species of necrophagous Coleoptera reared on substrates containing morphine. *Journal of Forensic Sciences* 46: 600–603.

Bourel, B., G. Tournel, V. Hédoin, and D. Gosset. 2004. Entomofauna of buried bodies in northern France. *International Journal of Legal Medicine* 118: 215–220.

Bousquet, Y. 1990. *Beetles Associated with Stored Products in Canada: An Identification Guide.* Canada Department of Agriculture Publication, Ottawa, Ontario.

Brackenbury, J. 1999. Water skating in the larvae of *Dixella aestivalis* (Diptera) and *Hydrobius fuscipes* (Coleoptera). *Journal of Experimental Biology* 202: 845–863.

Brancucci, M. 1979. *Geodessus besucheti* n. gen., n. sp. le premier Dytiscide terrestre (Cole., Dytiscidae, Bidessini). *Entomologica Basiliensia* 4: 213–218.

Brancucci, M. and G. B. Monteith. 1996. A second *Terradessus* species from Australia (Coleoptera, Dytiscidae). *Entomologica Basiliensia* 19: 585–591.

Branham, M. A. and J. W. Wenzel. 2003. The origin of photic behaviour and the evolution of sexual communication in fireflies (Coleoptera: Lampyridae). *Cladistics* 19: 1–22.

Brockerhoff, E. G., J. Bain, M. Kimberley, and M. Knížek. 2006. Interception frequency of exotic bark and ambrosia beetles (Coleoptera: Scolytinae) and relationship with establishment on New Zealand and worldwide. *Canadian Journal of Forest Research* 36: 289–298.

Brown, E. S. 1954. The biology of the coconut pest *Melittomma insulare* (Col., Lymexylonidae), and its control in the Seychelles. *Bulletin of Entomological Research* 45: 1–66.

Browne, D. J. and C. H. Scholtz. 1995. Phylogeny of the families of Scarabaeoidea (Coleoptera) based on characters of the hindwing articulation, hindwing base and wing venation. *Systematic Entomology* 20: 145–173.

Browne, D. J. and C. H. Scholtz. 1998. Evolution of the scarab hindwing articulation and wing base: a contribution toward the phylogeny of the Scarabaeidae (Scarabaeoidea: Coleoptera). *Systematic Entomology* 23: 307–326.

Browne, D. J. and C. H. Scholtz. 1999. A phylogeny of the families of Scarabaeoidea (Coleoptera). *Systematic Entomology* 24: 51–84.

Burke, H. R., W. E. Clark, J. R. Cate, and P. A. Fryxell. 1986. Origin and dispersal of the boll weevil. *Bulletin of the Entomological Society of America* 32: 228–238.

Calder, A. A., J. F. Lawrence, and J. W. H. Trueman. 1993. *Austrelater*, gen. nov. (Coleoptera: Elateridae), with a description of the larva and comments on elaterid relationships. *Invertebrate Taxonomy* 7: 1349–1394.

Cambefort, Y. 1994. *Le Scarabée et les Dieux.* Boubée, Paris.

Campbell, J. M., M. J. Sarazin, and D. B. Lyons. 1989. *Canadian Beetles (Coleoptera) Injurious to Crops, Ornamentals, Stored Products, and Buildings.* Research Branch Agriculture Canada Publication 1826, Ottawa, Ontario.

Capinera, J. L., D. R. Gardener, and F. R. Stermitz. 1985. Cantharidin levels in blister beetles (Coleoptera: Meloidae) associated with alfalfa in Colorado. *Journal of Economic Entomology* 78: 1052–1055.

Carlton, C. E. and V. M. Bayless. 2007. Documenting beetle (Arthropoda: Insecta: Coleoptera) diversity in Great Smoky Mountains National Park; beyond the halfway point. *Southeastern Naturalist* 6 (Special Issue 1): 183–192.

Carlton, C. E. and H. W. Robison. 1998. Diversity of litter-dwelling beetles in deciduous forests of the Ouachita highlands of Arkansas (Insecta: Coleoptera). *Biodiversity and Conservation* 7: 1589–1605.

Carvalho, L. M., P. J. Thyssen, A. X. Linhares, and F. A. Palhares. 2000. A checklist of arthropods associated with pig carrion and human corpses in southeastern Brazil. *Memorias do Instituto Oswaldo Cruz* 95: 135–138.

Caterino, M. S., T. Hunt, and A. P. Vogler. 2005. On the constitution and phylogeny of Staphyliniformia (Insecta: Coleoptera). *Molecular Phylogenetics and Evolution* 34: 655–672.

Catts, E. P. and N. H. Haskell. 1990. *Entomology and Death: A Procedural Guide.* Joyce's Print Shop, Clemson, SC.

Chaddick, P. R. and F. F. Leek. 1972. Further specimens of stored product insects found in ancient Egyptian tombs. *Journal of Stored Products Research* 8: 83–86.

Chandler, D. S. and S. B. Peck. 1992. Seasonality and diversity of Leiodidae (Coleoptera) in an old-growth and 40 year-old forest in New Hampshire. *Environmental Entomology* 21: 1283–1293.

Chapman, C. A. and L. J. Chapman. 2003. Fragmentation and alteration of seed dispersal process: an initial evaluation of dung beetles, seed fate, and seedling diversity. *Biotropica* 35: 382–393.

Childers, C. C. and R. E. Woodruff. 1980. A bibliography of the coffee bean weevil *Araecerus fasciculatus* (Coleoptera: Anthribidae). *Bulletin of the Entomological Society of America* 26: 384–394.

Chu, G. S., J. R. Palmieri, and J. T. Sullivan. 1977. Beetle-eating: a Malaysia folk medical practice and its public health implications. *Tropical and Geographical Medicine* 29: 422–427.

Clark, S. M. and E. G. Riley. 2002. Megalopodidae Latreille 1802 – Orsodacnidae Thomson 1859. Pp. 609–616. *In* R. H. Arnett, M. C. Thomas, P. E. Skelley, and J. H. Frank (eds). *American Beetles,* Volume 2. Scarabaeoidea through Curculionoidea. CRC Press, Boca Raton, FL.

Clausen, C. P. 1978. Scarabaeidae. Pp. 277–292. *In* C. P. Clausen (ed). *Introduced Parasites and Predators of Arthropod Pests and Weeds: A World Review.* United States Department of Agriculture Handbook No. 480. Washington, DC.

Cloudsley-Thompson, J. L. 1964. On the function of the sub-elytral cavity in desert Tenebrionidae (Col.). *Entomologist's Monthly Magazine* 100: 148–151.

Cognato, A. I., J.-H. Sun, M. A. Anducho-Reyes, and D. R. Owen. 2005. Genetic variation and origin of red turpentine beetle (*Dendroctonus valens* LeConte) introduced to the People's Republic of China. *Agricultural and Forest Entomology* 7: 87–94.

Contag, C. H., S. D. Spilman, P. R. Contag, M. Oshiro, B. Eames, P. Dennery, D. K. Stevenson, and D. A. Benaron. 1997. Visualizing gene expression in living mammals using a bioluminescent reporter. *Photochemistry and Photobiology* 66: 523–531.

Costa, C., S. Casari-Chen, and A. Vanin. 1992. On the larvae of Tetralobini (Coleoptera, Elateridae). *Revista Brasiliera de Entomologia* 36: 879–888.

Costa, C., S. A. Vanin, and S. Ide. 1999. Systematics and bionomics of Cneoglossidae with a cladistic analysis of Byrrhoidea Sensu Lawrence and Newton (1995) (Coleoptera, Elateriformia). *Arquivos de Zoologia* 35: 231–300.

Crawford, C. S., W. P. Mackay, and J. G. Cepeda-Pizarro. 1993. Detrivores of the Chilean arid zone (27–32°S) and the Namib Desert: a preliminary comparison. *Revista Chilena de Historia Natural* 66: 283–289.

Dacke, M., P. Nordström, and C. H. Scholtz. 2003. Twilight orientation to polarised light in the crepuscular dung beetle *Scarabaeus zambezianus*. *Journal of Experimental Biology* 206: 1535–1545.

Dai, Z., S. N. Gorb, and U. Schwarz. 2002. Roughness-dependent friction force of the tarsal claw system in the beetle *Pachnoda marginata* (Coleoptera, Scarabeidae). *Journal of Experimental Biology* 205: 2479–2488.

Darlington, P. J., Jr. 1943. Carabidae of mountains and islands: data on the evolution of isolated faunas, and on atrophy of wings. *Ecological Monographs* 13: 37–61.

Davies, R. H. and C. Wray. 1995. Contribution of the lesser mealworm beetle (*Alphitobius diaperinus*) to carriage of *Salmonella enteritidis* in poultry. *Veterinary Record* 137: 407–408.

Davis, A. L. V. 1994. Associations of Afrotropical Coleoptera (Scarabaeidae: Aphodiidae: Staphylinidae: Hydrophilidae: Histeridae) with dung and decaying matter: implications for selection of fly-control agents from Australia. *Journal of Natural History* 28: 383–399.

Davis, A. L. V. and T. K. Philips. 2005. Effect of deforestation on a southwest Ghana dung beetle assemblage (Coleoptera: Scarabaeidae) at the periphery of Ankasa Conservation Area. *Environmental Entomology* 34: 1081–1088.

Davis, A. L. V., C. H. Scholtz, P. W. Dooley, N. Bham, and U. Kryger. 2004. Scarabaeinae dung beetles as indicators of biodiversity, habitat transformation and pest control chemicals in agro-ecosystems. *South African Journal of Science* 100: 415–424.

De Clerck-Floate, R. A., B. M. Wikeem, and R. S. Bourchier. 2005. Early establishment and dispersal of the weevil, *Mogulones cruciger* (Coleoptera: Curculionidae) for biological control of houndstongue (*Cynoglossum officinale*) in British Columbia, Canada. *Biocontrol Science and Technology* 15: 173–190.

de Souza, A. M. and A. X. Linhares. 1997. Diptera and Coleoptera of potential forensic importance in southeastern Brazil: relative abundance and seasonality. *Medical and Veterinary Entomology* 11: 8–12.

Deleurance-Glaçon, S. 1963. Recherches sur les Coléoptères troglobies de la sous-famille des Bathysciinae. *Annales des Sciences Naturelles (Zoologie)* 5: 1–172.

Dickens, J. C., J. E. Oliver, B. Hollister, J. C. Davis, and J. A. Klun. 2002. Breaking a paradigm: male-produced aggregation pheromone for the Colorado potato beetle. *Journal of Experimental Biology* 205: 1925–1933.

Dickerson, W. A., A. L. Brashear, J. T. Brumley, F. L. Carter, W. J. Grefenstette, and F. A. Harris (eds). 2001. *Boll Weevil Eradication Through 1999*. The Cotton Foundation, Memphis, TN.

DiZinno, J. A., W. D. Lord, M. B. Collins-Morton, M. R. Wilson, and M. L. Goff. 2002. Mitochondrial DNA sequencing of beetle larvae (Nitidulidae: Osmatida) recovered from human bone. *Journal of Forensic Sciences* 47: 1337–1339.

Donlan, E. M., J. H. Townsend, and E. A. Golden. 2004. Predation of *Caretta caretta* (Testudines: Cheloniidae) eggs by larvae of *Lanelater sallei* (Coleoptera: Elateridae) on Key Biscayne, Florida. *Caribbean Journal of Science* 40: 415–420.

Doyen, J. T. 1976. Marine beetles (Coleoptera excluding Staphylinidae). Pp. 497–519. *In* L. Cheng (ed). *Marine Insects*. North-Holland Publishing, Amsterdam.

Duan, J. J., M. S. Paradise, J. G. Lundgren, J. T. Bookout, C. J. Jiang, and R. N. Wiedenmann. 2006. Assessing nontarget impacts of Bt corn resistant to rootworms: tier-1 testing with larvae of *Poecilus chalcites* (Coleoptera: Carabidae). *Environmental Entomology* 35: 135–142.

Dubitskii, A. M., R. T. Akhmetbekova, and V. V. Nazarov. 1975. Aquatic beetles (Coleoptera: Dytiscidae) in mosquito control. *Izvestiya Akademii Nauk Kazakhskoi SSR Seriya Biologicheskaya* 13: 47–51.

Edwards, P. B. and H. H. Aschenborn. 1987. Patterns of nesting and dung burial in *Onitis* dung beetles: implications for pasture productivity and fly control. *Journal of Applied Ecology* 24: 837–851.

Eidmann, H. H. 1992. Impact of bark beetle on forests and forestry in Sweden. *Zeitschrift für Angewandte Entomologie* 114: 193–200.

Ellers-Kirk, C. and S. J. Fleischer. 2006. Development and life table of *Acalymma vittatum* (Coleoptera: Chrysomelidae), a vector of *Erwinia tracheiphila* in cucurbits. *Environmental Entomology* 35: 875–880.

Erwin, T. L. 1982. Tropical forests: their richness in Coleoptera and other arthropod species. *Coleopterists Bulletin* 36: 74–75.

Erwin, T. L. 1991. Natural history of the carabid beetles at the BIOLAT Biological Station, Rio Manu, Pakitza, Peru. *Revista Peruana de Entomologia* 33: 1–85.

Estrada, A. and R. Coates-Estrada. 2002. Dung beetles in continuous forest, forest fragments and in an agricultural mosaic habitat island at Los Tuxtlas, Mexico. *Biodiversity and Conservation* 11: 1903–1918.

Evans, A. V. and C. L. Bellamy. 1996. *An Inordinate Fondness for Beetles*. Henry Holt and Co., New York. 208 pp.

Farrell, B. D. 1998. "Inordinate fondness" explained: why are there so many beetles?. *Science* 281: 553–557.

Farrell, B. D., A. S. Sequeira, B. O'Meara, B. B. Normark, J. H. Chung, and B. H. Jordal. 2001. The evolution of agriculture in beetles (Curculionidae: Scolytinae and Platypodinae). *Evolution* 55: 2011–2027.

Fenster, C. B., W. S. Armbruster, P. Wilson, M. R. Dudash, and J. D. Thomson. 2004. Pollination syndromes and floral specialization. *Annual Review of Ecology, Evolution and Systematics* 35: 375–403.

Fincher, G. T. 1981. The potential value of dung beetles in pasture ecosystems. *Journal of the Georgia Entomological Society* 16: 301–316.

Fish, F. E. and A. J. Nicastro. 2003. Aquatic turning performance by the whirligig beetle: constraints on maneuverability by a rigid biological system. *Journal of Experimental Biology* 206: 1649–1656.

Forrest, T. G., M. P. Read, H. E. Farris, and R. R. Hoy. 1997. A tympanal hearing organ in scarab beetles. *Journal of Experimental Biology* 200: 601–606.

Franc, V., P. Hrasko, and M. Snopko. 1989. Use of entomologic findings in forensic medicine and criminology. *Soudni Lekarstvi* 34: 37–45.

Frantsevich, L., Z. Dai, W. Y. Wang, and Y. Zhang. 2005. Geometry of elytra opening and closing in some beetles (Coleoptera, Polyphaga). *Journal of Experimental Biology* 208: 3145–3158.

Freitag, R. 1979. Carabid beetles and pollution. Pp. 485–506. *In* T. L. Erwin, G. E. Ball, D. R. Whitehead, and A. L. Halpern (eds). *Carabid Beetles: Their Evolution, Natural History, and Classification*. Junk, The Hague.

Freuler, J., G. Blandenier, H. Meyer, and P. Pignon. 2001. Epigeal fauna in a vegetable agroecosystem. *Mitteilungen der Schweizerischen Entomologischen Gesellschaft* 74: 17–42.

Friedrich, F. and R. G. Beutel. 2006. The pterothoracic skeletomuscular system of Scirtoidea (Coleoptera: Polyphaga) and its implications for the high-level phylogeny of beetles. *Journal of Zoological Systematics and Evolutionary Research* 44: 290–315.

Gagliano-Candela, R. and L. Aventaggiato. 2001. The detection of toxic substances in entomological specimens. *International Journal of Legal Medicine* 114: 197–203.

Garcia-Salazar, C. G., F. E. Gildow, S. J. Fleischer, D. Cox-Foster, and F. L. Lukezic. 2000. ELISA versus immunolocalization to determine the association of *Erwinia tracheiphila* in *Acalymma vittatum* (F.) (Coleoptera: Chrysomelidae). *Environmental Entomology* 29: 542–550.

Gaston, K. J. 1991. The magnitude of global insect species richness. *Conservation Biology* 5: 283–296.

Geisthardt, M. 2003. Tenebrionidae (Insecta, Coleoptera) as an indicator for climatic changes on the Cape Verde Islands. *Special Bulletin of the Japanese Society of Coleopterology* 6: 331–337.

Gilbert, M., N. Fielding, H. F. Evans, and J.-C. Grégoire. 2003. Spatial pattern of invading *Dendroctonus micans* (Coleoptera: Scolytidae) populations in the United Kingdom. *Canadian Journal of Forest Research* 33: 712–725.

Goddard, J. 2000. *Physician's Guide to Arthropods of Medical Importance*, Third Edition. CRC Press, Boca Raton, FL.

Goldberg, L. 1996. A history of pest control measures in the anthropology collections, National Museum of Natural History, Smithsonian Institution. *Journal of the American Institute for Conservation* 35: 23–43.

Gorb, S. 2001. *Attachment Devices of the Insect Cuticle*. Kluwer Academic Press, Dordrecht, Sweden.

Graczyk, T. K., R. Knight, and L. Tamang. 2005. Mechanical transmission of human protozoan parasites by insects. *Clinical Microbiology Reviews* 18: 128–132.

Grebennikov, V. V. and C. H. Scholtz. 2004. The basal phylogeny of Scarabaeoidea (Insecta: Coleoptera) inferred from larval morphology. *Invertebrate Systematics* 18: 321–348.

Greer, L. and A. A. Szalay. 2002. Imaging of light emission from the expression of luciferases in living cells and organisms: a review. *Luminescence* 17: 43–74.

Grove, S. J. and N. E. Stork. 2000. An inordinate fondness for beetles. *Invertebrate Taxonomy* 14: 733–739.

Haack, R. A. 2006. Exotic bark- and wood-boring Coleoptera in the United States: recent establishments and interceptions. *Canadian Journal of Forest Research* 36: 269–288.

Hald, B., A. Olsen, and M. Madsen. 1998. *Typhaea stercorea* (Coleoptera: Mycetophagidae), a carrier of *Salmonella enterica* serovar *infantis* in a Danish broiler house. *Journal of Economic Entomology* 91: 660–664.

Halffter, G. and L. Arellano. 2002. Response of dung beetle diversity to human-induced changes in a tropical landscape. *Biotropica* 34: 144–154.

Halffter, G. and E. G. Matthews. 1966. The natural history of dung beetles of the subfamily Scarabaeinae (Coleoptera, Scarabaeidae). *Folia Entomologica Mexicana* 12–14: 3–312.

Hall, W. E. 2005. Ptiliidae Erichson, 1845. Pp. 251–261. *In* R. G. Beutel and R. A. B. Leschen (eds). *Handbook of Zoology: A Natural History of the Phyla of the Animal Kingdom*, Volume IV. Arthropoda: Insecta, Part 38. Coleoptera, Beetles, Volume 1. Morphology and Systematics. Walter de Gruyter, Berlin.

Hammond, P. M. 1979. Wing-folding mechanisms of beetles, with special reference to special investigations of adephagan phylogeny (Coleoptera). Pp. 113–180. *In* T. L. Erwin, G. E. Ball, D. R. Whitehead, and A. L. Halpern (eds). *Carabid*

Beetles: Their Evolution, Natural History, and Classification. Junk, The Hague.

Hammond, P. H. 1990. Insect abundance and diversity in the Dumoga-Bone National Park, N. Sulawesi, with special reference to the beetle fauna of lowland rain forest in the Toraut region. Pp. 197–254. *In* W. J. Knight and J. D. Holloway (eds). *Insects and the Rain Forests of South East Asia (Wallacea).* Royal Entomological Society of London, London.

Hammond, P. M. 1995. Described and estimated species numbers: an objective assessment of current knowledge. Pp. 29–71. *In* D. Allspp, D. L. Hawkesworth, and R. R. Colwell (eds). *Microbial Diversity and Ecosystem Function.* CAB International, Wallingford.

Hansen, M. 1991. The hydrophiloid beetles: phylogeny, classification and a revision of the genera (Coleoptera, Hydrophiloidea). *Biologiske Skrifter, Kongelige Danske Videnskabernes Selkab* 40: 1–367.

Hansen, M. 1997. Phylogeny and classifcation of the staphyliniform beetle families. *Biologiske Skrifter, Kongelige Danske Videnskabernes Selkab* 48: 1–339.

Harris, F. 1966. Observations on the lesser mealworm, *Alphitobius diaperinus* (Panz.). *Journal of the Georgia Entomological Society* 1: 17–18.

Harris, K. F. 1981. Arthropod and nematode vectors of plant viruses. *Annual Review of Phytopathology* 19: 391–426.

Hinton, H. E. 1945. *A Monograph of the Beetles Associated with Stored Products.* British Museum, London. 443 pp.

Hoffman, W. A. 1933. *Rhizopertha dominica* F. is a library pest. *Journal of Economic Entomology* 26: 293–294.

Holst, N. and W. G. Meikle. 2003. *Teretrius nigrescens* against larger grain borer *Prostephanus truncatus* in African maize stores: biological control at work?. *Journal of Applied Ecology* 40: 307–319.

Hood, W. M. 2004. The small hive beetle, *Aethina tumida*: a review. *Bee World* 85: 51–59.

Hörnschemeyer, T. 2005. Archostemata Kolbe, 1908. Pp. 29–42. *In* R. G. Beutel and R. A. B. Leschen (eds). *Handbook of Zoology: A Natural History of the Phyla of the Animal Kingdom,* Volume IV. Arthropoda: Insecta, Part 38. Coleoptera, Beetles, Volume 1. Morphology and Systematics. Walter de Gruyter, Berlin.

Howard, R. W., R. A. Jurenka, and G. J. Blomquist. 1986. Prostaglandin synthetase inhibitors in the defensive secretion of the red flour beetle *Tribolium castaneum* (Herbst) (Coleoptera: Tenebrionidae). *Insect Biochemistry* 16: 757–760.

Howden, H. F. 1982. Larval and adult characters of *Frickius* Germain, its relationship to the Geotrupini, and a phylogeny of some major taxa in the Scarabaeoidea (Insecta: Coleoptera). *Canadian Journal of Zoology* 60: 2713–2724.

Howden, H. and A. Howden. 2001. Change through time: a third survey of the Scarabaeinae (Coleoptera: Scarabaeidae) at Welder Wildlife Refuge. *Coleopterists Bulletin* 55: 356–362.

Howe, R. W. 1959. Studies on beetles of the family Ptinidae: XVII. Conclusions and additional remarks. *Bulletin of Entomological Research* 50: 287–326.

Hughes, J., S. J. Longhorn, A. Papadopoulou, K. Theodorides, A. de Riva, M. Mejia-Chang, P. G. Foster, and A. P. Vogler. 2006. Dense taxonomic EST sampling and its applications for molecular systematics of the Coleoptera (beetles). *Molecular Biology and Evolution* 23: 268–278.

Hughes, R. D., P. Ferrar, A. Macqueen, P. Durie, G. T. McKinney, and F. H. W. Morley. 1975. Introduced dung beetles and Australian pasture ecosystems: papers presented at a symposium during the meeting of the Australian and New Zealand Association for the Advancement of Science at Canberra in January 1975. *Journal of Applied Ecology* 12: 819–837.

Hurst, J. J., J. Meinwald, and T. Eisner. 1964. Defense mechanisms of arthropods – XII. Glucose and hydrocarbons in the quinone-containing secretion of *Eleodes longicollis. Annals of the Entomological Society of America* 57: 44–46.

International Union for Conservation of Nature and Natural Resources (IUCN). 2006. IUCN Red List of Threatened Species. www.iucnredlist.org [Accessed 2 March 2007].

Ivie, M. A. 2002a. Nosodendridae Erichson 1846. Pp. 224–227. *In* R. H. Arnett, M. C. Thomas, P. E. Skelley, and J. H. Frank (eds). *American Beetles,* Volume 2. Scarabaeoidea through Curculionoidea. CRC Press, Boca Raton, FL.

Ivie, M. A. 2002b. Bostrichidae Latreille 1802. Pp. 223–244. *In* R. H. Arnett, M. C. Thomas, P. E. Skelley, and J. H. Frank (eds). *American Beetles,* Volume 2. Scarabaeoidea through Curculionoidea. CRC Press, Boca Raton, FL.

Jäch, M. A. 1998. Annotated check list of aquatic and riparian/littoral beetle families of the world (Coleoptera). Pp. 25–42. *In* M. A. Jäch and L. Ji (eds). *Water Beetles of China,* Volume 2. Zoologisch-Botanische Gesellschaft in Österreich and Wiener Coleopterologenverein, Wien.

Jäch, M. A., R. G. Beutel, J. A. Delgado, and J. A. Diaz. 2005. Hydraenidae Mulsant, 1844. Pp. 224–251. *In* R. G. Beutel and R. A. B. Leschen (eds). *Handbook of Zoology: A Natural History of the Phyla of the Animal Kingdom,* Volume IV. Arthropoda: Insecta, Part 38. Coleoptera, Beetles, Volume 1. Morphology and Systematics. Walter de Gruyter, Berlin.

Jackson, T. A. and M. G. Klein. 2006. Scarabs as pests: a continuing problem. *Coleopterists Society Monographs* 5: 102–119.

Jameson, M. L. and B. C. Ratcliffe. 2002. Series Scarabaeiformia Crowson 1960. (= Lamellicornia): Superfamily Scarabaeoidea Latreille 1802: Introduction. Pp. 1–5. *In* R. H. Arnett, M. C. Thomas, P. E. Skelley, and J. H. Frank (eds). *American Beetles,* Volume 2. Scarabaeoidea through Curculionoidea. CRC Press, Boca Raton, FL.

Jankielsohn, A., C. H. Scholtz, and S. V. Louw. 2001. Effect of habitat transformation on dung beetle assemblages: a comparison between a South African nature reserve and neighboring farms. *Environmental Entomology* 30: 474–483.

Johns, C. V., C. Stone, and L. Hughes. 2004. Feeding preferences of the Christmas beetle *Anoplognathus chloropyrus* (Coleoptera: Scarabaeidae) and four paropsine species (Coleoptera: Chrysomelidae) on selected *Eucalyptus grandis* clonal foliage. *Australian Forestry Journal* 67: 184–190.

Jolivet, P. 1997. Biologie des Coléoptères Chrysomélides. Société Nouvelle des Éditions Boubée, Paris.

Jordal, B. H., R. A. Beaver, B. B. Normark, and B. D. Farrell. 2002. Extraordinary sex ratios and the evolution of male neoteny in sib-mating *Ozopemon* beetles. *Biological Journal of the Linnean Society* 75: 353–360.

Josephson, R. K., J. G. Malamud, and D. R. Stokes. 2001. The efficiency of an asynchronous flight muscle from a beetle. *Journal of Experimental Biology* 204: 4125–4139.

Kamoun, S. 1996. Occurence of the threatened *Cicindela senilis frosti* Varas-Arangua in an inland salt marsh in Riverside County, California (Coleoptera, Cicindelidae). *Coleopterists Bulletin* 50: 369–371.

Kaplan, J. K. 2003. We don't cotton to boll weevil 'round here anymore. *Agricultural Research* February 2003, 51: 4–8.

Karras, D. J., S. E. Farrell, R. A. Harrigan, F. M. Henretig, and L. Gealt. 1996. Poisoning from "Spanish fly" (cantharidin). *American Journal of Emergency Medicine* 14: 478–483.

Kaufman, P. E., S. J. Long, D. A. Rutz, and C. S. Glenister. 2000. Prey- and density-mediated dispersal in *Carcinops pumilio* (Coleptera: Histerida), a predator of house fly (Diptera: Muscidae) eggs and larvae. *Journal of Medical Entomology* 37: 929–932.

Kellner, R. L. L. 2001. Suppression of pederin biosynthesis through antibiotic elimination of endosymbionts in *Paederus sabaeus*. *Journal of Insect Physiology* 47: 475–483.

Kellner, R. L. L. 2003. Stadium-specific transmission of endosymbionts needed for pederin biosynthesis in three species of *Paederus* rove beetles. *Entomologia Experimentalis et Applicata* 107: 115–124.

Kevan, P. and H. G. Baker. 1983. Insects as flower visitors and pollinators. *Annual Review of Entomology* 28: 407–453.

Kim, J.-J., E. A. Allen, L. M. Humble, and C. Breuil. 2005. Ophistomatoid and basidiomycetous fungi associated with green, red, and grey lodgepole pines after mountain pine beetle (*Dendroctonus ponderosae*) infestation. *Canadian Journal of Forest Research* 35: 274–284.

Kirejtshuk, A. G. 1999. *Sikhotealinia zhiltzovae* (Lafer, 1996)–recent representative of the Jurassic coleopterous fauna (Coleoptera, Archostemata, Jurodidae). *Proceedings of the Zoological Institute of the Russian Academy of Science* 281: 21–26.

Kolibáč, J. 1992. Revision of Thanerocleridae n. stat. (Coleoptera, Cleridae). *Mitteilungen der Schweizerischen Entomologischen Gesellschaft* 65: 303–340.

Kolibáč, J. 2004. Metaxinidae fam. nov., a new family of Cleroidea (Coleoptera). *Entomologica Basiliensia* 26: 239–268.

Kolibáč, J. 2005. A review of the Trogossitidae. Part 1: Morphology of the genera (Coleoptera, Cleroidea). *Entomologica Basiliensia et Collectionis Frey* 27: 39–159.

Korte, A., I. Ribera, R. G. Beutel, and D. Bernhard. 2004. Interrelationships of Staphyliniform groups inferred from 18S and 28S rDNA sequences, with special emphasis on Hydrophiloidea (Coleoptera, Staphyliniformia). *Journal of Zoological Systematics and Evolutionary Research* 42: 281–288.

Kovarik, P. W. and M. S. Caterino. 2005. Histeridae Gyllenhal, 1808. Pp. 190–222. *In* R. G. Beutel and R. A. B. Leschen (eds.). *Coleoptera, Beetles*, Volume 1. Morphology and Systematics (Archostemata, Adephaga, Myxophaga, Polyphaga partim). Walter de Gruyter, Berlin.

Kritsky, G. 1991. Beetle gods of ancient Egypt. *American Entomologist* 37: 85–89.

Kulshrestha, P. and D. K. Satpathy. 2001. Use of beetles in forensic entomology. *Forensic Science International* 120: 15–17.

Kumar, P. 1988. Flesh eating behaviour of *Alphitobius diaperinus* Panz. (Tenebrionidae; Coleoptera). *Indian Journal of Entomology* 48: 113–115.

Kuschel, G. 1995. A phylogenetic classification of Curculionoidea to families and subfamilies. *Memoirs of the Entomological Society of Washington* 14: 5–33.

Kutalek, R. and A. Kassa. 2005. The use of gyrinids and dytiscids for stimulating breast growth in East Africa. *Journal of Ethnobiology* 25: 115–128.

Lacey, E. S., M. D. Ginzel, J. G. Mallar, and L. M. Hanks. 2004. Male-produced aggregation pheromone of the cerambycid beetle *Neoclytus acuminatus acuminatus*. *Journal of Chemical Ecology* 30: 1493–1507.

Lamb, A. B., S. M. Salom, L. T. Kok, and D. L. Mausel. 2006. Confined field release of *Laricobius nigrinus* (Coleoptera: Derodontidae), a predator of the hemlock woolly adelgid, *Adelges tsugae* (Hemiptera: Adelgidae), in Virginia. *Canadian Journal of Forest Research* 36: 369–375.

Larsen, K. J. and T. W. Work. 2003. Differences in ground beetles (Coleoptera: Carabidae) or original and reconstructed tallgrass prairies in northeastern Iowa, USA, and impact of 3-year spring burn cycles. *Journal of Insect Conservation* 7: 153–166.

Lawrence, J. F. 1982. Coleoptera. Pp. 482–553. *In* S. P. Parker (ed). *Synopsis and Classification of Living Organisms*, Volume 2. McGraw-Hill, New York.

Lawrence, J. F. 1987. Rhinorhipidae, a new beetle family from Australia, with comments on the phylogeny of Elateriformia. *Invertebrate Taxonomy* 2: 1–53.

Lawrence, J. F. 1988. Rhinorhipidae, a new beetle family from Australia, with comments on the phylogeny of the Elateriformia. *Invertebrate Taxonomy* 2: 1–53.

Lawrence, J. F. 1991. Order Coleoptera. *In* F. W. Stehr (ed). *Immature Insects*, Volume 2. Kendall/Hunt Publishing, Dubuque, IA. 975 pp.

Lawrence, J. F. and E. B. Britton. 1994. *Australian Beetles*. Melbourne University Press, Carlton.

Lawrence, J. F. and T. Hlavac. 1979. Review of the Derodontidae (Coleoptera: Polyphaga) with new species from North America and Chile. *Coleopterists Bulletin* 33: 369–414.

Lawrence, J. F. and A. F. Newton, Jr. 1995. Families and subfamilies of Coleoptera (with selected genera, notes, references and data on family-group names). Pp. 779–1006. *In* J. Pakaluk and S. A. Ślipiński (eds). *Biology, Phylogeny, and Classification of Coleoptera: Papers Celebrating the 80th Birthday of Roy A. Crowson*. Muzeum i Instytut Zoologii PAN, Warszawa.

Lawrence, J. F., A. M. Hastings, M. J. Dallwitz, T. A. Paine, and E. J. Zurcher. 1999. *Beetles of the World*, Version 1.0. CSIRO Division of Entomology, Canberra.

Lawrence, J. F., N. B. Nikitsky, and A. G. Kirejtshuk. 1995. Phylogenetic position of Decliniidae (Coleoptera: Scirtoidea) and comments on the classification of Elateriformia (sensu lato). Pp. 375–410. *In* J. Pakaluk and S. A. Ślipiński (eds). *Biology, Phylogeny, and Classification of Coleoptera: Papers Celebrating the 80th Birthday of Roy A. Crowson*. Muzeum i Instytut Zoologii PAN Warszawa.

Lepesme, P. 1947. *Les Insects des Palmiers*. Lecavalier, Paris.

Leschen, R. A. B., J. F. Lawrence, and S. A. Ślipiński. 2005. Classification of basal Cucujoidea (Coleoptera: Polyphaga): cladistic analysis, keys and review of new families. *Invertebrate Systematics* 19: 17–73.

Leschen, R. A. B. 2002. Derodontidae LeConte 1861. Pp. 221–223. *In* R. H. Arnett, M. C. Thomas, P. E. Skelley, and J. H. Frank (eds). *American Beetles*, Volume 2. Scarabaeoidea through Curculionoidea. CRC Press, Boca Raton, FL.

Lethbridge, R. C. 1971. The hatching of *Hymenolepis diminuta* eggs and penetration of the hexacanths in *Tenebrio molitor*. *Parasitology* 62: 445–456.

Lindgren, C. J., J. Corrigan, and R. A. De Clerck-Floate. 2002. *Lythrum salicaria* L., purple loosestrife (Lynthaceae). Pp. 383–390. *In* P. G. Mason and J. T. Huber (eds). *Biological Control Programmes in Canada, 1981–2000*. CABI Publishing, Wallingford.

Lindroth, C. H. 1961–1969. The ground-beetles (Carabidae, excl. Cicindelinae) of Canada and Alaska. *Opuscula Entomologica, Supplementa* 20, 24, 29, 33, 34, 35: 1–1192+1–48.

Lindroth, C. H. 1974. *Coleoptera Family Carabidae*. Royal Entomological Society of London, London.

Lindroth, C. H. 1985–1986. *The Carabidae (Coleoptera) of Fennoscandia and Denmark*. Scandinavian Science Press Ltd., Copenhagen.

Lingafelter, S. W. and E. R. Hoebeke. 2002. *Revision of the Genus Anoplophora (Coleoptera: Cerambycidae)*. Entomological Society of Washington, Washington, DC.

Lloyd, J. E. 1983. Bioluminescence and communication in insects. *Annual Review of Entomology* 28: 131–160.

Lobo, J. M. and A. L. V. Davis. 1999. An intercontinental comparison of dung beetle diversity between two Mediterranean-climatic regions: local versus regional and historical influences. *Diversity and Distributions* 5: 91–103.

Lobo, J. M., J. Hortal, and F. J. Cabrero-Sañudo. 2006. Regional and local influence of grazing activity on the diversity of a semi-arid dung beetle community. *Diversity and Distributions* 12: 111–123.

Lopez, P. D., E. Lugo, S. Valle P. Espinosa, M. M. M. Lopez, M. Delgado, P. Rivera et al. 1997. Aquatic insects as biological control agents of mosquito larvae in Nicaragua. *Revista Nicaraguense de Entomologia* 39: 27–30.

Losey, J. E. and M. Vaughan. 2006. The economic value of ecological services provided by insects. *BioScience* 56: 311–323.

Lövei, G. L. and K. D. Sunderland. 1996. Ecology and behavior of ground beetles (Coleoptera: Carabidae). *Annual Review of Entomology* 41: 231–256.

Lundheim, R. and K. E. Zachariassen. 1993. Water balance of over-wintering beetles in relation to strategies for cold tolerance. *Journal of Comparative Physiology B* 163: 1–4.

Lundkvist, E., J. Landin, M. Jackson, and C. Svensson. 2003. Diving beetles (Dytiscidae) as predators of mosquito larvae (Culicidae) in field experiments and in laboratory tests of prey preference. *Bulletin of Entomological Research* 93: 219–226.

Luterek, D. 1966. Observations on the larvae of some species of click beetles (Col., Elateridae) feeding on mushrooms. *Polskie Pismo Entomologiczne B* 3–4: 341–345.

Mackie, T. T., G. W. I. Hunter, and C. B. Worth. 1945. *A Manual of Tropical Medicine*. W. B. Saunders, Philadelphia, PA.

Magagula, C. N. 2003. Changes in carabid beetle diversity within a fragmented agricultural landscape. *African Journal of Ecology* 41: 23–30.

Magura, T., B. Tothmeresz, and Z. Elek. 2003. Diversity and composition of carabids during a forestry cycle. *Biodiversity and Conservation* 12: 73–85.

Majer, K. 1994. A review of the classification of the Melyridae and related families (Coleoptera, Cleroidea). *Entomologica Basiliensia* 17: 319–390.

Majer, K. 2002. Subfamilial classification of the family Malachiidae (Coleoptera, Cleroidea). *Entomologica Basiliensia* 24: 179–244.

Marshall, C. J. and J. K. Liebherr. 2000. Cladistic biogeography of the Mexican transition zone. *Journal of Biogeography* 27: 203–216.

Marske, K. A. and M. A. Ivie. 2003. Beetle fauna of the United States and Canada. *Coleopterists Bulletin* 57: 495–503.

Martikainen, P. and J. Kouki. 2003. Sampling the rarest: threatened beetles in boreal forest biodiversity inventories. *Biodiversity and Conservation* 12: 1815–1831.

Mason, P. G., J. T. Huber, and S. M. Boyetchko. 2002. Introduction. Pp. 11–14. *In* P. G. Mason and J. T. Huber (eds). *Biological Control Programmes in Canada, 1981–2000*. CABI Publishing, Wallingford.

McClain, E., M. K. Seely, N. F. Hadley, and V. Gray. 1985. Wax blooms in tenebrionid beetles on the Namib Desert: correlations with environment. *Ecology* 66: 112–118.

McCutcheon, G. S. 2002. Consumption of tobacco budworm (Lepidoptera: Noctuidae) by hooded beetle (Coleoptera: Anthicidae) and bigeyed bug (Hemiptera: Lygaeidae). *Journal of Agricultural and Urban Entomology* 19: 55–61.

McElroy, W. D. and M. A. DeLuca. 1983. Firefly and bacterial luminescence: basic science and applications. *Journal of Applied Biochemistry* 5: 197–209.

Medeiros, L. and G. R. P. Moreira. 2002. Moving on hairy surfaces: modifications of *Gratiana spadicea* larval legs to attach on its host plant *Solanum sisymbriifolium*. *Entomologia Experimentalis et Applicata* 102: 295–305.

Miller, L. K. 1978. Physical and chemical changes associated with seasonal alterations in freezing tolerance in the adult northern tenebrionid, *Upis ceramboides*. *Journal of Insect Physiology* 24: 791–796.

Mittal, I. C. 1993. Natural manuring and soil conditioning by dung beetles. *Tropical Ecology* 34: 150–159.

Moore, I. and E. F. Legner. 1976. Intertidal rove beetles (Coleoptera: Staphylinidae). Pp. 521–551. *In* L. Cheng (ed). *Marine Insects*. North-Holland Publishing, Amsterdam.

Morgan, R. E., P. de Groot, and S. M. Smith. 2004. Susceptibility of pine plantations to attack by the pine shoot beetle (*Tomicus piniperda*) in southern Ontario. *Canadian Journal of Forest Research* 34: 2528–2540.

Moura, M. O., C. J. de Carvalho, and E. L. Monteiro-Filho. 1997. A preliminary analysis of insects of medico-legal importance in Curitiba, State of Parana. *Memorias do Instituto Oswaldo Cruz* 92: 269–274.

Munks, S., K. Richards, J. Meggs, M. Wapstra, and R. Corkrey. 2004. Distribution, habitat and conservation of two threatened stag beetles, *Hoplogonus bornemisszai* and *H. vanderschoori* (Coleoptera: Lucanidae) in north-east Tasmania. *Australian Zoologist* 32: 586–596.

Muona, J. 1995. The phylogeny of Elateroidea (Coleoptera), or which tree is best today. *Cladistics* 11: 317–341.

Nahrung, H. F. 2004. Biology of *Chrysophtharta agricola* (Chapuis), a pest of eucalypt plantations in Victoria and Tasmania. *Australian Forestry Journal* 67: 59–66.

Naylor, L. H. 1999. Reporter gene technology: the future looks bright. *Biochemical Pharmacology* 58: 749–757.

Nebois, A. 1968. Larva and pupa of *Cupes varians* Lea, and some observations on its biology (Coleoptera: Cupedidae). *Memoirs of the National Museum of Victoria* 28: 17–19.

Nebois, A. 1984. Reclassification of *Cupes* Fabricius (s. lat.) with descriptions of new genera and species (Cupedidae: Coleoptera). *Systematic Entomology* 9: 443–477.

Nelson, G. H. and C. L. Bellamy. 1991. A revision and phylogenetic re-evaluation of the family Schizopodidae (Coleoptera, Buprestoidea). *Journal of Natural History* 25: 985–1026.

Newton, J. A. F. 1977. Review of Agyrtidae (Coleoptera), with a new genus and species from New Zealand. *Annales Zoologici* 47: 111–156.

Newton, J. A. F. 1998. Phylogenetic problems, current classification and generic catalog of world Leiodidae (including Cholevidae). Pp. 41–178. *In* P. M. Giachino and S. B. Peck (eds). *Phylogeny and Evolution of Subterranean and Endogean Cholevidae (= Leiodidae)*. Proceedings of the 20th International Congress of Entomology, Florence, 1996. Museo Regionale di Scienze Naturali, Torino. 295 pp.

Newton, J. A. F. 2005. Leiodidae Fleming, 1821. Pp. 269–280. *In* R. G. Beutel and R. A. B. Leschen (eds). *Handbook of Zoology: A Natural History of the Phyla of the Animal Kingdom*, Volume IV. Arthropoda: Insecta, Part 38. Coleoptera, Beetles, Volume 1. Morphology and Systematics. Walter de Gruyter, Berlin.

Newton, J. A. F. and M. K. Thayer. 1995. Protopselaphinae new subfamily for *Protopselaphus* new genus from Malaysia, with a phylogenetic analysis and review of the Omaliine Group of Staphylinidae including Pselaphidae (Coleoptera). Pp. 219–320. *In* J. Pakaluk and S. A. Ślipiński (eds). *Biology, Phylogeny and Classification of Coleoptera: Papers Celebrating the 80th Birthday of Roy A. Crowson*. Muzeum I Instytut Zoologii PAN, Warszawa.

Nichols, D. S., T. I. Christmas, and D. E. Greig. 1990. Oedemerid blister beetle dermatosis: a review. *Journal of the American Academy of Dermatology* 22: 815–819.

Nielsen, E. S. and L. A. Mound. 1999. Global diversity of insects: the problems of estimating numbers. Pp. 213–222. *In* P. H. Raven and T. Williams (eds). *Nature and Human Society: The Quest for a Sustainable World*. National Academy Press, Washington, DC.

Niemela, J., J. R. Spence, D. Langor, Y. Haila, and H. Tukia. 1993. Logging and boreal ground-beetle assemblages on two continents: implications for conservation. Pp. 29–50. *In* K. J. Gaston, T. R. New, and M. J. Samways (eds). *Perspectives on Insect Conservation*. Intercept Ltd, Andover.

Nummelin, M. and I. Hanski. 1989. Dung beetles of the Kibale Forest, Uganda; comparison between virgin and managed forests. *Journal of Tropical Ecology* 5: 549–552.

O'Brien, C. W. 1995. Curculionidae, premiere biocontrol agents (Coleoptera: Curculionidae). *Memoir of the Entomological Society of Washington* 14: 119–128.

O'Keefe, S. T. 2005. Scydmaenidae Leach, 1815. Pp. 280–288. *In* R. G. Beutel and R. A. B. Leschen (eds). *Handbook of Zoology: A Natural History of the Phyla of the Animal Kingdom*, Volume IV. Arthropoda: Insecta, Part 38. Coleoptera, Beetles, Volume 1. Morphology and Systematics. Walter de Gruyter, Berlin.

Onore, G. 1997. A brief note on edible insects in Ecuador. *Ecology of Food and Nutrition* 36: 277–285.

Paine, T. D., K. F. Raffa, and T. C. Harrington. 1997. Interactions among scolytid bark beetles, their associated fungi, and live host conifers. *Annual Review of Entomology* 42: 179–206.

Pakaluk, J., S. A. Ślipiński, and J. F. Lawrence. 1994. Current classification and family-group names in Cucujoidea (Coleoptera). *Genus* 5: 223–268.

Payne, J. A., E. W. King, and G. Beinhart. 1968. Arthropod succession and decomposition of buried pigs. *Nature* 219: 1180–1181.

Peck, S. B. 2005. A checklist of the beetles of Cuba with data on distributions and bionomics (Insecta: Coleoptera). *Arthropods of Florida and Neighboring Areas* 18: 1–241.

Peck, S. B. 2006. Distribution and biology of the ectoparasitic beaver beetle *Platypsyllus castoris* Ritsema in North America (Coleoptera: Leiodidae: Platypsyllinae). *Insecta Mundi* 20: 85–94.

Peck, S. B. and M. C. Thomas. 1998. A distributional checklist of the beetles (Coleoptera) of Florida. *Arthropods of Florida and Neighboring Areas* 16: 1–180.

Pervez, A. and Omkar 2006. Ecology and biological control application of multicoloured Asian ladybird, *Harmonia axyridis*: a review. *Biocontrol Science and Technology* 16: 111–128.

Peter, C. I. and S. D. Johnson. 2005. Anther cap retention prevents self-pollination by elaterid beetles in the South African orchid *Eulophia foliosa*. *Annals of Botany* 97: 345–355.

Peters, B. C. and C. J. Fitzgerald. 1996. Anobiid pests of timber in Queensland: a literature review. *Australian Forestry* 59: 130–135.

Philips, T. K. 2001. A record of *Micromalthus debilis* (Coleoptera: Micromalthidae) from Central America and a discussion of its distribution. *Florida Entomologist* 84: 159–160.

Philips, T. K. 2002. Anobiidae Fleming 1821. Pp. 245–260. *In* R. H. Arnett, M. C. Thomas, P. E. Skelley, and J. H. Frank (eds). *American Beetles*, Volume 2. Scarabaeoidea through Curculionoidea. CRC Press, Boca Raton, FL.

Philips, T. K., M. A. Ivie, and J. J. Giersch. 2002. Jacobsoniidae Heller 1926. Pp. 219–220. *In* R. H. Arnett, M. C. Thomas, P. E. Skelley, and J. H. Frank (eds). *American Beetles*, Volume 2. Scarabaeoidea through Curculionoidea. CRC Press, Boca Raton, FL.

Piel, J., I. Höfer, and D. Hui. 2004a. Evidence for a symbiosis island involved in horizontal acquisition of pederin biosynthetic capabilities by the bacterial symbiont of *Paederus rufipes* beetles. *Journal of Bacteriology* 186: 1280–1286.

Piel, J., G. Wen, M. Platzer, and D. Hui. 2004b. Unprecedented diversity of catalytic domains in the first four modules of the putative pederin polyketide synthase. *ChemBioChem* 8: 93–98.

Piel, J., D. Butzke, N. Fusetani, D. Hui, M. Platzer, G. Wen, and S. Matsunaga. 2005. Exploring the chemistry of uncultivated bacterial symbionts: antitumor polyketides of the pederin family. *Journal of Natural Products* 68: 472–479.

Pinto, J. D. and M. A. Bologna. 2002. Meloidae Gyllenhal 1810. Pp. 522–529. *In* R. H. Arnett, M. C. Thomas, P. E. Skelley, and J. H. Frank (eds). *American Beetles*, Volume 2. Scarabaeoidea through Curculionoidea. CRC Press, Boca Raton, FL.

Pitman, A. J., A. M. Jones, and E. B. Gareth Jones. 1993. The wharf-borer *Nacerdes melanura* L., a threat to stored archaeological timbers. *Studies in Conservation* 38: 274–284.

Pitman, A. J., E. B. Gareth, P. Jones, and P. Oevering. 2003. An overview of the biology of the wharf borer beetle (*Nacerdes melanura* L., Oedemeridae) a pest of wood in marine structures. *Biofouling* 19: 239–248.

Plavilstshikov, N. N. 1958. Coleoptera. Cerambycidae, Part 3. Lamiinae, Part 1. Fauna of USSR 23: 592pp. [In Russian].

Poland, T. M. and D. G. McCullough. 2006. Emerald ash borer: invasion of the urban forest and the threat to North America's ash resource. *Journal of Forestry* 104: 118–124.

Pollock, D. A. and B. B. Normark. 2002. The life cycle of *Micromalthus debilis* LeConte (1878) (Coleoptera: Archostemata: Micromalthidae): historical review and evolutionary perspective. *Journal of Zoological Systematics and Evolutionary Research* 40: 105–112.

Porch, N. and S. Elias. 2000. Quaternary beetles: a review and issue for Australian studies. *Australian Journal of Entomology* 39: 1–9.

Potter, D. A. and D. W. Held. 2002. Biology and management of the Japanese beetle. *Annual Review of Entomology* 47: 175–205.

Quintero, I. and T. Roslin. 2005. Rapid recovery of dung beetle communities following habitat fragmentation in central Amazonia. *Ecology* 86: 3303–3311.

Raffa, K. F. and A. A. Berryman. 1983. The role of host plant resistance in the colonization behavior and ecology of bark beetles (Coleoptera: Scolytidae). *Ecological Monographs* 53: 27–49.

Rajapakse, S. 1981. Beetle marasmus. *British Medical Journal* 283: 1316–1317.

Ratcliffe, B. C., M. L. Jameson, and A. B. T. Smith. 2002. Scarabaeidae Latreille 1802. Pp. 39–81. *In* R. H. Arnett, M. C. Thomas, P. E. Skelley, and J. H. Frank (eds). *American Beetles*, Volume 2. Scarabaeoidea through Curculionoidea. CRC Press, Boca Raton, FL.

Rea, N. 1997. Biological control: premises, ecological input and *Mimosa pigra* in the wetlands of Australia's top end. *Wetlands Ecology and Management* 5: 227–242.

Rees, D. J., B. C. Emerson, P. Oromí, and G. M. Hewitt. 2001. Reconciling gene trees with organism history: the mtDNA phylogeography of three *Nesotes* species (Coleoptera: Tenebrionidae) on the western Canary Islands. *Journal of Evolutionary Biology* 14: 139–147.

Reid, C. A. M. 2000. Spilopyrinae Chapuis: a new subfamily in the Chrysomelidae and its systematic placement. *Invertebrate Taxonomy* 14: 837–862.

Ribera, I., R. G. Beutel, M. Balke, and A. Vogler. 2002. Discovery of Aspidytidae, a new family of aquatic Coleoptera. *Proceedings of the Royal Society of London Series B, Biological Sciences* 269: 2351–2356.

Ribera, I. and A. P. Vogler. 2004. Speciation of Iberian diving beetles in Pleistocene refugia (Coleoptera, Dytiscidae). *Molecular Ecology* 13: 179–193.

Roth, L. M. and T. Eisner. 1962. Chemical defenses of arthropods. *Annual Review of Entomology* 7: 107–136.

Roughley, R. E. 2001. Noteridae C. G. Thomson, 1857. Pp. 147–152. *In* R. H. Arnett, Jr. and M. C. Thomas (eds). *American Beetles*, Volume 1. Archostemata, Myxophaga, Adephaga, Polyphaga: Staphyliniformia. CRC Press, New York.

Roughley, R. E. and D. J. Larson. 2001. Dytiscidae Leach, 1815. Pp. 156–186. *In* R. H. Arnett, Jr. and M. C. Thomas (eds). *American Beetles*, Volume 1. Archostemata, Myxophaga, Adephaga, Polyphaga: Staphyliniformia. CRC Press, New York.

Sagegami-Oba, R., Y. Oba, and H. Ôhira. 2007. Phylogenetic relationships of click beetles (Coleoptera: Elateridae) inferred from 28S ribosomal DNA: insights into the evolution of bioluminescence in Elateridae. *Molecular Phylogenetics and Evolution* 42: 410–421.

Sandorini, P. 2001. Aphrodisiacs past and present: a historical review. *Clinical Autonomic Research* 11: 303–307.

Scheffler, P. Y. 2005. Dung beetle (Coleoptera: Scarabaeidae) diversity and community structure across three disturbance regimes in eastern Amazonia. *Journal of Tropical Ecology* 21: 9–19.

Scholtz, C. H. and V. V. Grebennikov. 2005. Scarabaeiformia Crowson, 1960. Pp. 345–366. *In* R. G. Beutel and R. A. B. Leschen (eds). *Coleoptera, Beetles*, Volume 1. Morphology and Systematics (Archostemata, Adephaga, Myxophaga, Polyphaga partim). Walter de Gruyter, Berlin.

Schweigkofler, W., W. J. Otrosina, S. L. Smith, D. R. Cluck, K. Maeda, K. G. Peay, and M. Garbelotto. 2005. Detection and quantification of *Leptographium wageneri*, the cause of black-stain root disease, from bark beetles (Coleoptera: Scolytidae) in Northern California using regular and real-time PCR. *Canadian Journal of Forest Research* 35: 1798–1808.

Scott, M. P. 1998. The ecology and behavior of burying beetles. *Annual Review of Entomology* 43: 595–618.

Seres, A. and N. Ramírez. 1995. Biologia floral y polinizacion de algunas monocotiledoneas de un bosque nublado venezolano. *Annals of the Missouri Botanical Garden* 82: 61–81.

Sikes, D. S. 2004. The beetle fauna of Rhode Island. An annotated checklist. *The Biota of Rhode Island*. Volume 3. Rhode Island Natural History Survey, Kingston. 6+295 pp.

Sikes, D. S. 2005. Silphidae Latreille, 1807. Pp. 288–296. *In* R. G. Beutel and R. G. B. Leschen (eds). *Handbook of Zoology: A Natural History of the Phyla of the Animal Kingdom*, Volume IV. Arthropoda: Insecta, Part 38. Coleoptera, Beetles, Volume 1. Morphology and Systematics. Walter de Gruyter, Berlin.

Ślipiński, S. A. 1992. Larinotinae – a new subfamily of Trogossitidae (Coleoptera), with notes on the constitution of Trogossitidae and related families of Cleroidea. *Revue Suisse de Zoologie* 99: 439–463.

Smith, A. B. T. 2006. A review of the family-group names for the superfamily Scarabaeoidea (Coleoptera) with corrections to nomenclature and a current classification. *Coleopterists Society Monographs* 5: 144–204.

Smith, K. G. V. 1986. *A Manual of Forensic Entomology*. Cornell University Press, Ithaca, NY.

Smith, A. B. T. and A. Paucar. 2000. Taxonomic review of *Platycoelia lutescens* (Scarabaeidae: Rutelinae: Anoplognathini) and a description of its use as food by the people of the Ecuadorian Highlands. *Annals of the Entomological Society of America* 93: 408–414.

Smith, A. B. T., D. C. Hawks, and J. M. Heraty. 2006. An overview of the classification and evolution of the major scarab beetle clades (Coleoptera: Scarabaeoidea) based on preliminary molecular analyses. *Coleopterists Society Monographs* 5: 35–46.

Sole, C. L., C. H. Scholtz, and A. D. S. Bastos. 2005. Phylogeography of the Namib Desert dung beetles *Scarabaeus* (*Pachysoma*) MacLeay (Coleoptera: Scarabaeidae). *Journal of Biogeography* 32: 75–84.

Sörensson, M. 1997. Morphological and taxonomical novelties in the world's smallest beetles, and the first Old World record of Nanosellini (Coleoptera: Ptiliidae). *Systematic Entomology* 22: 257–283.

Spangler, P. J. and W. E. Steiner Jr. 2005. A new aquatic beetle family, Meruidae, from Venezuela (Coleoptera: Adephaga). *Systematic Entomology* 30: 339–357.

Spector, S. 2006. Scarabaeine dung beetles (Coleoptera: Scarabaeidae: Scarabaeinae): an invertebrate focal taxon for biodiversity research and conservation. *Coleopterists Society Monographs* 5: 71–83.

Spence, J. R. and D. H. Spence. 1988. Of ground-beetle and men: introduced species and the synanthropic fauna of western Canada. *Memoirs of the Entomological Society of Canada* 144: 151–168.

Stehr, F. W. 1991. *Immature Insects*, Volume 2. Kendall Hunt Publishing, Dubuque, IA.

Stolz, U., S. Velez, K. V. Wood, M. Wood, and J. L. Feder. 2003. Darwinian natural selection for orange bioluminescent color in a Jamaican click beetle. *Proceedings of the National Academy of Sciences of the United States of America* 100: 14955–14959.

Stork, N. E. 1999a. The magnitude of biodiversity and its decline. Pp. 3–32. *In* J. Cracraft and M. Novacek (eds). *The Living Planet in Crisis: Biodiversity, Science and Policy*. American Museum of Natural History, New York.

Stork, N. E. 1999b. Estimating the number of species on Earth. Pp. 1–7. *In* W. Ponder and D. Lunney (eds). *The Other 99%: The Conservation and Biodiversity of Invertebrates*. Royal Zoological Society of New South Wales, Sydney.

Strom, B. L., S. R. Clarke, and P. J. Shea. 2004. Efficacy of 4-allylanisole-based products for protecting individual loblolly pines from *Dendroctonus frontalis* Zimmermann (Coleoptera: Scolytidae). *Canadian Journal of Forest Research* 34: 659–665.

Strother, K. O., C. D. Steelman, and E. E. Gbur. 2005. Reservoir competence of lesser mealworm (Coleoptera: Tenebrionidae) for *Campylobacter jejuni* (Campylobacterales: Campylobacteraceae). *Journal of Medical Entomology* 42: 42–47.

Švácha, P., J.-J. Wang, and S.-C. Chen. 1997. Larval morphology and biology of *Philus antennatus* and *Heterophilus punctulatus*, and systematic position of the Philinae (Coleoptera: Cerambycidae and Vesperidae). *Annales de la Société Entomologique de France* 33: 323–369.

Tabor, K. L., C. C. Brewster, and R. D. Fell. 2004. Analysis of the successional patterns of insects on carrion in southwestern Virginia. *Journal of Medical Entomology* 41: 785–795.

Talley, T. S. and M. Holyoak. 2006. The effects of dust on the federally threatened valley elderberry longhorn beetle. *Environmental Management* 37: 647–658.

Thayer, M. K. 2005. Staphylinidae Latreille, 1802. Pp. 296–344. *In* R. G. Beutel and R. A. B. Leschen (eds). *Handbook of Zoology: A Natural History of the Phyla of the Animal Kingdom, Volume IV. Arthropoda: Insecta, Part 38. Coleoptera, Beetles, Volume 1. Morphology and Systematics.* Walter de Gruyter, Berlin.

Theodorides, K., A. De Riva, J. Gómez-Zurita, P. G. Foster, and A. P. Vogler. 2002. Comparison of EST libraries from seven beetle species: towards a framework for phylogenomics of the Coleoptera. *Insect Molecular Biology* 11: 467–475.

Tschinkel, W. R. 1975. A comparative study of the chemical defensive system of tenebrionid beetles: chemistry of the secretions. *Journal of Insect Physiology* 21: 753–783.

Turchetto, M. and S. Vanin. 2004. Forensic entomology and globalization. *Parassitologia* 46: 187–190.

Turnbow, R. H. and M. C. Thomas. 2002. Cerambycidae Leach, 1815. Pp. 568–601. *In* R. H. Arnett, M. C. Thomas, P. E. Skelley, and J. H. Frank (eds). *American Beetles*, Volume 2. Scarabaeoidea through Curculionoidea. CRC Press, Boca Raton, FL.

Tyndale-Biscoe, M. 1994. Dung burial by native and introduced dung beetles (Scarabaeidae). *Australian Journal of Agricultural Research* 45: 1799–1808.

Underwood, T. J., D. W. Tallamy, and J. D. Pesek. 1997. Bioluminescence in firefly larvae: a test of the aposematic display hypothesis (Coleoptera: Lampyridae). *Journal of Insect Behavior* 10: 365–370.

Utsunomiya, Y. and K. Masumoto. 1999. Edible beetles (Coleoptera) from northern Thailand. *Elytra* 27: 191–198.

VanLaerhoven, S. L. and G. S. Anderson. 1999. Insect succession on buried carrion in two biogeoclimatic zones of British Columbia. *Journal of Forensic Sciences* 44: 32–43.

Varchola, J. M. and J. P. Dunn. 1999. Changes in ground beetle (Coleoptera: Carabidae) assemblages in farming systems bordered by complex or simple roadside vegetation. *Agriculture Systems and Environment* 73: 41–49.

Viviani, V. R. 2002. The origin, diversity, and structure function relationships of insect luciferases. *Cellular and Molecular Life Sciences* 59: 1833–1850.

Vulinec, K. 2002. Dung beetle communities and seed dispersal in primary forest and disturbed land in Amazonia. *Biotropica* 34: 297–309.

Wallace, M. G. 2005. Observations of urban dung beetles utilizing dog feces (Coleoptera: Scarabaeidae). *Coleopterists Bulletin* 59: 400–401.

Watson, D. W., P. E. Kaufman, D. A. Rutz, and C. S. Glenister. 2001. Impact of the darkling beetle *Alphitobius diaperinus* (Panzer) on establishment of the predaceous beetle *Carcinops pumilio* (Erichson) for *Musca domestica* control in caged-layer poultry houses. *Biological Control* 20: 8–15.

Watts, C. H. S. and W. F. Humphreys. 2006. Twenty-six new Dytiscidae (Coleoptera) of the genera *Limbodessus* Guignot and *Nirripirti* Watts and Humphreys, from underground waters in Australia. *Transactions of the Royal Society of South Australia* 130: 123–185.

Weber, D. C., D. L. Rowley, M. H. Greenstone, and M. M. Athanas. 2006. Prey preference and host suitability of the predatory and parasitoid carabid beetle, *Lebia grandis*, for several species of *Leptinotarsa* beetles. *Journal of Insect Science* 6: 1–14.

Wheeler, Q. D. 1986. Revision of the genera of Lymexylidae (Coleoptera: Cucujiformia). *Bulletin of the American Museum of Natural History* 183: 113–210.

Wood, S. L. and D. E. Bright. 1992. A catalog of Scolytidae and Platypodidae (Coleoptera): part 2: taxonomic index (volumes A and B). *Great Basin Naturalist Memoirs* 13: 1–1553.

Yeates, D. K., M. S. Harvey, and A. D. Austin. 2003. New estimates for terrestrial arthropod species-richness in Australia. *Records of the South Australian Museum, Monograph Series* 7: 231–241.

Young, F. N. 1967. A possible recycling mechanism in tropical forests. *Ecology of Food and Nutrition* 48: 506.

Zachariassen, K. E. 1991. Routes of transpiratory water loss in a dry-habitat tenebrionid beetle. *Journal of Experimental Biology* 157: 425–437.

Zelazny, B. and E. Pacumbaba. 1982. Phytophagous insects associated with cadang-cadang infected and healthy coconut palms in southeastern Luzon, Philippines. *Ecological Entomology* 7: 113–120.

Chapter 12

BIODIVERSITY OF HYMENOPTERA

John T. Huber

Canadian Forestry Service, K. W. Neatby Bldg., 960 Carling Ave.,
Ottawa, Ontario K1A 0C6 Canada

Insect Biodiversity: Science and Society, 1st edition. Edited by R. Foottit and P. Adler
© 2009 Blackwell Publishing, ISBN 978-1-4051-5142-9

The most widely recognized hymenopterans – bees, ants, and wasps or hornets – have long been part of art, ritual, and folklore worldwide (Proverbs 6: 6; Deuteronomy 1: 44; Exodus 23: 28; Chauvin 1968; Hanson and Gauld 1995, 2006; Turillazzi and West-Eberhard 1996). Bees have been recorded since stone-age times, as seen in cave paintings (Gore and Lavies 1976, Valli and Summers 1988), and the first recorded human casualty as a result of a wasp sting, 4500 years ago, was King Menes, Pharaoh of Egypt (Spradberry 1973). Fig wasps (Figs. 12.1 A and B), despite their small size and inconspicuous habits, were known to the ancient Greeks and Egyptians (Kevan and Phillips 2001).

The order Hymenoptera contains far more, and more diverse, species than simply ants, bees, and wasps. Other common names for some hymenopterans include paper wasp, potter wasp, yellow jacket, bumblebee, velvet ant, wood wasp, and horntail. These names have a narrower meaning than 'wasp' or 'ant', but they are still collective names for multiple species in a single genus. Common names are given to few identifiable species of Hymenoptera, usually pests. Most Hymenoptera belong to groups unknown to the general public. These are small (usually less than 5 mm long) parasitic species that go about their lives unnoticed by all but the insect taxonomist, ecologist, biological control researcher, or dedicated naturalist, and are often unnamed, even scientifically. Most also lack English group names, though they might have anglicized group names based on their scientific name (e.g., ichneumon wasp). If they have a common name, it is usually for species that have secondarily reverted to plant feeding or pollinating habits, such as 'gall wasp' and 'fig wasp'.

EVOLUTION AND HIGHER CLASSIFICATION

Hymenoptera first appear in the fossil record in the late Triassic, about 230 million years ago. The primitive lineages were plant feeding; the parasitoid mode of life and stinging Hymenoptera did not appear until about 210 and 155 million years ago, respectively (Grimaldi and Engel 2005). Some groups are extinct, and today several small, archaic, rarely collected hymenopteran families are found mainly or exclusively in the southern hemisphere. The order, nonetheless, diversified tremendously over the past 150–200 million years.

Both extant and extinct Hymenoptera are classified into two broad groups, Symphyta and Apocrita (Ronquist 1999a). The Symphyta (sawflies, wood wasps, and horntails) include the most primitive hymenopterans and comprise about 7% of the Hymenoptera. The group grades into the Apocrita through the Orussidae, the only symphytan family with parasitic larvae (Vilhelmsen 2001, 2003). Most symphytans have a conspicuous, free-living, caterpillar-like larva that feeds on leaves; the relatively few leaf-mining and stem- and wood-boring species have a more grub-like larva. Because certain symphytans occasionally can be pests of trees and shrubs, they have received considerable attention from foresters, horticulturalists, and home gardeners. The Apocrita comprise about 93% of the Hymenoptera and are subdivided into the Aculeata (stinging hymenopterans), which include familiar species such as ants, social wasps, and bees, and the Parasitica, a diverse and abundant group of usually small, inconspicuous species, most of which parasitize insects and spiders. Because they differ fundamentally from true parasites, which spend their entire lives inside or on a host and do not usually kill it, they are better known as parasitoids – insects with free-living adults whose larvae develop singly or gregariously on or in a single host individual, eventually killing it. The larvae of Aculeata are usually grublike and are not free living.

NUMBERS OF SPECIES AND INDIVIDUALS

Hymenoptera have diversified into a variety of morphological forms and ways of life and might be the largest order of insects (Gaston 1993, Austin and Dowton 2000), though currently the order ranks second or third after the Coleoptera and Lepidoptera (Stork 1997). Pagliano and Scaramozzino (1990) counted approximately 125,000 described species and more than 17,000 generic names but, based on species-area relationships, Ulrich (1999a) estimated that the number of species could be one million. Pagliano (2006) enumerated 140,470 valid species, 12,167 valid subspecies, 46,213 species synonyms, and 1712 subspecies synonyms, for a total of 200,562 names. These names are classified in 8359 valid extant genera and 715 valid fossil genera. Recent counts (Table 12.1) total about 144,695 extant species. Most described species are of temperate distribution, where a far higher proportion of the fauna has been described, compared with tropical areas (Gaston 1993). Among the five temperate regions that Gaston compared, the best known fauna is Britain's, where an estimated 82% of the species are named, and the most poorly known is Switzerland's, where only 21% are named. Not all the remaining species in these areas need to be described because they might have been described from nearby countries, but the numbers demonstrate that even in temperate zones, the proportions of named species can be relatively low. Furthermore, Gaston's (1993) figures might be optimistic because taxonomists estimate that in groups such as the small parasitic wasps much less than 10% of the species have been described worldwide.

The number of individuals of various hymenopterans at the family or superfamily level that can be collected at a given locality varies considerably. Two surveys in Europe, one in a forest for all Hymenoptera and another in agroecosystems for Parasitica only, indicate relative numbers. Martínez de Murgía et al. (2001) collected 78,229 specimens representing 12 superfamilies (35 families) in Malaise traps in a forest in Spain. The numbers of specimens in the three major divisions of Hymenoptera were 440 (0.6%) Symphyta, 1270 (1.6%) Aculeata, and 76,519 (97.8%) Parasitica. The last-named group contained the following: Ichneumonoidea 31,110 (39.8%), Proctotrupoidea 19,990 (25.6%), Chalcidoidea 13,097 (16.7%), Ceraphronoidea 5395 (6.9%), Platygastroidea 5287 (6.8%), and Cynipoidea 1640 (2.1%). Two families, the Ichneumonidae and Diapriidae, constituted 57% (43,903 specimens) of the Parasitica. Schmitt (2004) surveyed the Parasitica, using 1-m^2 light boxes on the ground in cereal fields, fallow fields, and variously managed grasslands in Germany. The 23,781 collected specimens represented 6 superfamilies (24 families): Platygastroidea 8986 (37.8%), Chalcidoidea 8815 (37.1%), Ichneumonoidea 2860 (12%), Ceraphronoidea 1966 (8.3%), Proctotrupoidea 1005 (4.2%), and Cynipoidea 149 (0.6%). Results for the Parasitica are not strictly comparable because of different collecting methods, but the numbers indicate tendencies in the relative proportions of the major groups. A third survey treated a tropical area in Sulawesi, Indonesia, and concluded that the diversity of all families of Hymenoptera, except possibly the Symphyta and gall-forming Cynipidae, were almost certainly more species rich in Sulawesi than in Britain (Noyes 1989). The diversity of Parasitica as a whole

appeared to be greater in tropical than in temperate areas, although tropical areas vary considerably.

Aculeates and sawflies generally are collected in smaller numbers than are the Parasitica, but this trend is an artifact due partly to collecting techniques. No passive collecting methods are efficient at collecting social Hymenoptera, especially the soil-dwelling groups such as ants. Individual workers can be collected in traps but, because colonies are patchy, most individuals are missed unless, by chance, a trap is placed next to a colony. Consequently, the number of individuals of social species is greatly underestimated. Ants occur in huge numbers in all regions except the highest latitudes. Bees and social wasps are abundant in the tropics. Hölldobler and Wilson (1994) cited a study by German ecologists near Manaus, Brazil, who estimated that ants, termites (Isoptera), stingless bees, and polybiine wasps accounted for 80% of the insect biomass and one-third of the biomass of all animals in the area. Even when termites are excluded, the vast numbers and correspondingly high total weight of the social Hymenoptera, compared with other animals, are astonishing. In all these surveys for Hymenoptera, the actual numbers are not particularly meaningful without reference to the total number of insects collected, but they provide a reasonable idea of the relative proportions of the major groups in the order, despite the biases introduced by collecting technique. Ulrich (1999b) confirmed previous estimates that the Hymenoptera constitute about 20% of the temperate insect fauna.

MORPHOLOGICAL AND BIOLOGICAL DIVERSITY

Hymenopteran morphology is so diverse that some species might not be recognized as Hymenoptera (Figs. 12.1 and 12.2). The structure of any particular body part (e.g., mandibles) and surface sculpture is remarkably varied, as shown by scanning electron micrographs of one group only, the ants (Bolton 1994, 2000). Color covers the visible spectrum and color combinations and patterns are often remarkable, partly because showy, contrasting colors such as yellow or red and black (Manley and Pitts 2007, Buck et al. 2008) effectively advertise that stinging Hymenoptera are dangerous and best left alone. Hymenoptera range from 0.13 to 75.00 mm in body length (Mymaridae and Pelecinidae, respectively), but if the extended ovipositor and antennae are also

Table 12.1 Numbers of described species of Hymenoptera, listed by superfamily and family.

Superfamily	Family	Number of Valid Described, Extant Species (numbers valid on date given)
Xyeloidea[1]	Xyelidae	52
Tenthredinoidea[1] (7434)	Argidae	1000
	Blasticotomidae	8
	Cimbicidae	75
	Diprionidae	120
	Pergidae	431
	Tenthredinidae	5800
Cephoidea[1]	Cephidae	100
Pamphilioidea[1] (650)	Megalodontesidae	50
	Pamphiliidae	600
Siricoidea[1] (116)	Anaxyelidae	1
	Siricidae	115
Xiphydrioidea[1]	Xiphydriidae	112
Orussoidea[2]	Orussidae	77
Megalyroidea[3]	Megalyridae	50
Ceraphronoidea[4] (603)	Ceraphronidae	302
	Megaspilidae	301
Chalcidoidea[5] (22,740)	Agaonidae (sensu Bouček)	762
	Aphelinidae	1192
	Chalcididae	1469
	Encyrtidae	4058
	Eucharitidae	427
	Eulophidae	4969
	Eupelmidae	931
	Eurytomidae	1453
	Leucospidae	134
	Mymaridae	1437
	Ormyridae	125
	Perilampidae	284
	Pteromalidae	3544
	Rotoitidae	2
	Signiphoridae	78
	Tanaostigmatidae	94
	Torymidae	900
	Trichogrammatidae	881
Cynipoidea (3001)	Austrocynipidae[6]	1
	Cynipidae[7]	1370
	Figitidae[8]	960
	Ibaliidae[6]	20
	Liopteridae[8]	200
Evanioidea (1135)	Aulacidae[1]	178
	Evaniidae[9]	455
	Gasteruptiidae[1]	502

(continued)

Table 12.1 (*continued*).

Superfamily	Family	Number of Valid Described, Extant Species (numbers valid on date given)
Mymarommatoidea[10]	Mymarommatidae	10
Ichneumonoidea[11] (41,938)	Braconidae	17,605
	Ichneumonidae	23,331
Stephanoidea[12]	Stephanidae	326
Proctotrupoidea[4] (2486)	Austroniidae	3
	Diapriidae	2049
	Heloridae	23
	Maamingidae	2
	Monomachidae	27
	Pelecinidae	3
	Peradeniidae	2
	Proctorenyxidae	1
	Proctotrupidae	359
	Roproniidae	20
	Vanhorniidae	5
Platygastroidea[4] (4697)	Platygastridae	1311
	Scelionidae	3386
Trigonaloidea[13]	Trigonalidae	100
Apoidea (25,906)	Ampulicidae[14]	198
	Andrenidae[15]	2330
	Apidae[15]	5130
	Colletidae[15]	2000
	Crabronidae[14]	8636
	Halictidae[15]	3500
	Heterogynaidae[14]	8
	Megachilidae[15]	3170
	Melittidae[15]	182
	Sphecidae[14]	731
	Stenotritidae[15]	21
Chrysidoidea (6516)	Bethylidae[16]	2325
	Chrysididae[17]	2500
	Dryinidae[18]	1598
	Embolemidae[18]	47
	Plumariidae[19]	20
	Sclerogibbidae[20]	20
	Scolebythidae[21]	6
Vespoidea (15,646)	Bradynobaenidae[22]	200
	Formicidae[23]	11,946
	Mutillidae[19]	3500
	Pompilidae[24]	4850
	Rhopalosomatidae[25]	39
	Sapygidae[26]	67
	Scoliidae[27]	560

(*continued*)

Table 12.1 *(continued)*.

Superfamily	Family	Number of Valid Described, Extant Species (numbers valid on date given)
	Sierolomorphidae[28]	10
	Tiphiidae[17]	2000
	Vespidae[29]	4918

Superfamily totals are given (in parentheses) only if more than one included family is listed.
[1] D. R. Smith (personal communication, 22.v.2006).
[2] L. Vilhelmsen (personal communication, 3.i.2007).
[3] S. Shaw (personal communication, 2.i.2007).
[4] N. F. Johnson (personal communication, 13.iv.2006).
[5] J. S. Noyes (personal communication, 5.ix.06).
[6] Ronquist (1999b).
[7] Melika (2006).
[8] M. Buffington (personal communication, 31.viii.2006).
[9] Deans (2005).
[10] Gibson et al. (2007).
[11] Yu et al. (2005).
[12] Aguiar (2004).
[13] Carmean and Kimsey (1998).
[14] Pulawski (2006).
[15] http://stri.discoverlife.org/mp/20q?guide'Apoidea_speciesandflags [Accessed 19.i.2007].
[16] C. Azevedo (personal communication, 23.x.2006).
[17] Kimsey in Hanson and Gauld (2006).
[18] M. Olmi (personal communication, 10.x.2006).
[19] Brothers (personal communication, 19.i.2007).
[20] Olmi (2005), Sclerogibbidae.
[21] Beaver (2002), Scolebythidae.
[22] Hanson and Gauld (2006).
[23] Agosti and Johnson (2005).
[24] Engel and Grimaldi (2006).
[25] A. Guidotti (personal communication, 11.i.2007).
[26] Bennett and Engel (2005).
[27] Osten (2005).
[28] Hanson (2006).
[29] J. Carpenter (personal communication, 20.xi.06).

included, female *Megarhyssa* species (Ichneumonidae) are up to 20 cm long (antennae 2.5 cm, body 4.5 cm, ovipositor 13 cm), making them the world's longest hymenopterans. Titans such as *Pepsis* species (Pompilidae), with a body length up to 62 mm and a wingspan of 110 mm, are among the heaviest species. The nesting structures created by some Aculeata are equally varied, ranging from dainty clay pots to massive, yet intricate paper nests and enormous, complex underground dwellings.

General entomology texts with comprehensive treatments of Hymenoptera are those by Marshall (2006) and Triplehorn and Johnson (2005) for North America and Naumann (1991) for Australia. Books devoted exclusively to the entire order for particular regions are those by Gauld and Bolton (1996) for Britain and Hanson and Gauld (1995, 2006) for Costa Rica. Other important works include a discussion of hymenopteran diversity and importance (LaSalle and Gauld 1993), overviews of Symphyta (Benson 1950), biology of Parasitica (Quicke 1997) and solitary Aculeata (O'Neill 2001), and identification keys and diagnoses (Goulet and Huber 1993 for the world, Fernández and Sharkey 2007 for Latin America).

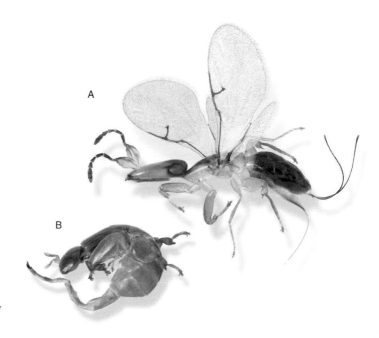

Fig. 12.1 *Pleistodontes addicotti* Wiebes (Agaoninae). A. Female. B. Male. Image by Klaus Bolte.

IMPORTANCE TO HUMANS

Stings and bites, food and other products

The first serious encounter most people have with Hymenoptera is personal and usually memorable. Every year thousands receive the unwanted attention of bees, wasps, or ants. More often than not, more than one individual, and often many, 'attend' to the victim, especially when an entire nest or colony inadvertently has been disturbed. Depending on the species, a single individual may deliver one or several stings, bites or, sometimes both, as with bulldog ants in Australia and

fire ants in the Western Hemisphere. The insect is not, of course, 'angry', though that possibility might be debatable; rather it or, correctly, she – only females have stingers and most do the biting – is defending herself or her nest from what is perceived as a threat to life or colony. Regardless, the memory of the event is usually long-lasting and results in the victim being more cautious when hymenopteran nests are known to be in the vicinity. Most stings and bites, of course, occur precisely because the person does not know and accidentally steps on an individual hymenopteran or inadvertently approaches a nest hanging in a bush, above a door, or under a balcony or verandah. Less

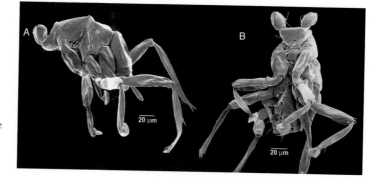

Fig. 12.2 *Dicopomorpha echmepterygis* Mockford, paratype (Mymaridae). A. Entire body, lateral. B. Ventral (with fungal hypha exiting mouth opening). Image by Klaus Bolte.

often, the individual knows a bee or wasp is around and starts flailing wildly at it – precisely the action that increases the chances of being stung (Schmidt and Boyer Hassen 1996). Stings can cause local or systemic, hypersensitive reactions that lead to death (Schmidt 1986, 1992, Goddard 1996, Levick et al. 2000, McGain et al. 2000). A survey of about 2000 physicians in the southern USA in 1988 indicated that 22,755 patients had been treated for reactions to fire ant stings; 2% of them suffered anaphylactic shock (Goddard 1996).

Most females of the Aculeata are capable of stinging in self-defense but the solitary, predatory species use their stingers almost exclusively to deliver venom to paralyze insect or spider prey. Humans, therefore, rarely are stung by solitary wasps, and their venom is not designed to cause pain, though there are notable exceptions in the Bethylidae, Mutillidae, and Pompilidae. Some *Pepsis* species and the bullet ant *Paraponera clavata* deliver the most painful stings known (Schmidt 2004). Schmidt's advice to any victim stung by one is to 'lie down and scream'. The intense pain, however, lasts only a few minutes. The venoms and stings of highly social Hymenoptera are painful because they are designed for colony defense against vertebrate predators (Schmidt 1990), and a painful sting discourages these predators. A few species of Parasitica also can 'sting'. The relatively large, nocturnal, orange species of *Ophion* and *Enicospilus* (Ichneumonidae) can deliver a painful jab with the tip of their short, sharp ovipositor. The importance of bee and wasp stings has resulted in the commercial production of life-saving antivenins. Pharmacological research on venom constituents might yield other kinds of useful medications (Beleboni et al. 2004).

Hymenoptera have biting mouthparts that they use for feeding, capturing prey, building nests, or, when necessary, defense. In tropical countries, stingless bees defend their nests from human intruders by swarming over the body, pinching the skin, and pulling the hair. One species in tropical America also ejects a burning liquid from its mandibles (Wilson 1971). When ants bite a person, they might not let go even when decapitated, and hundreds of bites, occurring simultaneously, can have a considerable effect (Moffett 1986).

By far, the most important food produced by any hymenopteran is honey, harvested from several bee species. Other honey-bee products – bee wax, propolis, and royal jelly – also are important, in the production of cosmetics, for example. The larvae of social bees or wasps and the repletes of certain ant species (Wilson 1971) are eaten by some people. Specialty products such as mead (honey beer) and liquor distilled from honey and chocolate-coated ants are commercially available.

Ecological importance

Hymenoptera occur in all terrestrial, and some aquatic, habitats. They play critical ecological roles, as summarized by Hanson and Gauld (1995, 2006), Gauld et al. (1990), LaSalle and Gauld (1993), and, for ants, by Hölldobler and Wilson (1994). Grissell (2001) wrote eloquently about the importance of insects, including Hymenoptera, in home gardens where many people are likely to observe them. Pollination by bees, fig wasps, and pollen wasps is the most beneficial role, without which many plants, including most crops, would disappear. Parasitism and predation of insects are important activities of many Hymenoptera that prevent numerous insects from becoming crop and forest pests. When food webs are intensively investigated, many Hymenoptera are found to be involved and the webs can be complex (Eveleigh et al. 2007). Many hymenopterans have been used successfully in biological control (J. Heraty, this volume). The role of polydnaviruses, a group of viruses found so far only in some Ichneumonoidea, has implications for biological control because of their immunosuppression action (Beckage 1998). Polydnaviruses are also of considerable interest to basic research in virology (Frederici and Bigot 2003). *Wohlbachia*, a genus of bacteria that is sexually transmitted and inherited through the female line, strongly skews sex ratios in various parasitic wasps (Tagami et al. 2001). Finding ways to manipulate or eliminate the bacteria might affect the biological control potential of their host insects (Floate et al. 2006).

Ants play an immensely important role in soil formation by mixing and aerating it. They also are the predominant scavengers of other organisms their own size, and help recycle dead insects that are carried back to their nests as food. Ants alter the abundance and distribution of flowering plants, especially in arid areas, by transporting seeds and discarding some of them uneaten around nests or by protecting some plant species from plant-feeding insects.

Hymenoptera have nuisance or deleterious roles as well. Some ants feed on seeds and, since ancient times, have been reported as grain pests. Some chalcid wasps develop inside seeds of conifers and other plants, greatly reducing the numbers of viable seeds for plantation forestry. Symphyta and ants feed on leaves or burrow in wood; they are among the most serious defoliators of conifers in temperate regions and broadleaved trees in the tropics, respectively. Some leaf-cutter bees can be nuisance defoliators of garden plants. Ants protect sap-sucking insects from parasitism by other insects, allowing them to increase in numbers sufficiently to cause plant damage. Many ants have become worldwide nuisances in human dwellings and on agricultural land where livestock can be adversely affected. Gall-forming chalcids (LaSalle 2005) and gall wasps (Kjellberg et al. 2005) deform or kill plants, and wood-tunneling bees (carpenter bees) damage wood dwellings.

TAXONOMIC DIVERSITY

Taxonomic and biological information on the main hymenopteran groups, with important recent references to each, are given below. Older taxonomic and biological literature is listed under the appropriate family in Goulet and Huber (1993). Many Internet sites on Hymenoptera provide useful information on particular families.

Symphyta

Adults of the 8000–10,000 species of symphytans in 14 families (6 superfamilies) are relatively well known taxonomically (except some Nematinae) for the Western Hemisphere, but much less so for the Eastern Hemisphere tropics. The larvae of all families need taxonomic work at the species level, particularly because they represent the most conspicuous and damaging stage. Abe and Smith (1991) cataloged the genera of Symphyta, and Schmidt and Smith (2006) cataloged the Pergidae. The Anaxyelidae, Cephidae, Siricidae, and Xiphydriidae have larvae that bore in wood (Blasticotomidae, many Cephidae), ferns, or grass stems. The Orussidae parasitize wood-boring larvae, primarily those of beetles (Vilhelmsen 2003). The Xyelidae include the most primitive living Hymenoptera; the larvae feed on leaves or pollen-producing cones of pines. The remaining families, the Argidae, Cimbicidae, Diprionidae, Megalodontesidae, Pamphiliidae, Pergidae, and Tenthredinidae have leaf-feeding or leaf-mining larvae.

Parasitica

Most species in the 11 superfamilies of Parasitica are parasitoids of insects and spiders. The next five super-families, each with only one included family (except the Evanioidea, with three), are relatively small but diverse in morphology, hosts, and habits. They contain an unusual number of rare, archaic species, many of them associated with wood. Apart from the Evaniidae, they are not important economically but are scientifically important in biology and conservation and for inferring phylogenies. The last six superfamilies include the most commonly encountered species, many of them important in natural and biological control of pest insects.

Stephanoidea

Aguiar (2004) cataloged the family Stephanidae, which consists of rarely collected, solitary ectoparasitoids of wood-boring insect larvae, mainly of beetles (Buprestidae and Cerambycidae).

Megalyroidea

Shaw (1990a) discussed the phylogeny of the Megalyridae. Most species are rarely collected and are associated primarily with ancient tropical forests; the group is most diverse in *Eucalyptus* and *Acacia* habitats of Australia (Shaw 1990b).

Trigonaloidea

Carmean and Kimsey (1998) treated the 15 genera (about 100 species) of Trigonalidae (or Trigonalyidae). Most species are parasitoids of moths or social wasps; some are hyperparasitoids through parasitic flies (Tachinidae).

Mymarommatoidea

The Mymarommatidae contain 11 described, extant species, most less than 0.5-mm long. Members of this enigmatic, worldwide group have a unique wing structure and two-segmented petiole (Gibson et al. 2007). Their biology is unknown.

Evanioidea

Three disparate families are included in this super-family. Jennings and Austin (2004) reviewed the biology and host relationships of the Aulacidae and Gasteruptiidae. Aulacids are parasitoids of wood-wasps (Xiphydriidae) and beetles (Cerambycidae and Buprestidae). Smith (2001) cataloged the family. The Evaniidae, or ensign wasps, so called because their abdomens in profile resemble semaphore flags – they also waggle their abdomens up and down as they walk – are solitary predators inside cockroach oothecae. Deans and Huben (2003) provided keys to genera, and Deans (2005) cataloged the species. The Gasteruptiidae, with six genera in two subfamilies, are parasitoids of bees and pollen wasps. Jennings and Austin (2002) treated the genera and species of Hyptiogastrinae.

Ichneumonoidea

Two families comprise this superfamily, the Braconidae and the Ichneumonidae, and it is by far the largest, with almost 2600 genera. Yu et al. (2005) cataloged the information on this superfamily from almost 29,000 scientific papers.

Ichneumonidae

Gauld et al. (2002) recognized 37 subfamilies of Ichneumonidae. They suggested that the total global-species richness for the Ichneumonidae would exceed 100,000 species. The overwhelming majority feed as larvae on or in the larvae or pupae of other holometabolous insects or on immature or adult spiders. Most species parasitize the larvae and pupae of moths and butterflies. Others parasitize the pre-pupae or pupae of aculeate Hymenoptera or wood-boring beetles. Despite their abundance and widespread occurrence in most habitats, especially in forests rather than in grasslands and agricultural habitats, comparatively few ichneumonids have been used successfully in biological control, though they must be important in natural control. For example, budworms, which include some of the most devastating forest pests in North America, are parasitized by more than 110 species of Ichneumonidae (Bennett 2008). The 16 described species of the Agriotypinae (Bennett 2001) are ectoparasitoids of the prepupae and pupae of caddisflies in fast-running streams. The long respiratory filament of an *Agriotypus* pupa that pokes out of the host case is unique.

The role of polydnaviruses, found so far only in the Ichneumonoidea, has implications for biological control because of the immunosuppression action (Beckage 1998). Polydnaviruses are of considerable interest to virologists because their genome is an integral part of the host genome (Frederici and Bigot 2003).

Braconidae

Though fewer in number of species and generally smaller in body size than the Ichneumonidae, the Braconidae are morphologically more diverse and better studied because a greater proportion of the species are important to agriculture and have been used in biological control. Species of the 35 or so currently recognized subfamilies are 1–30 mm in length. Braconids commonly parasitize the larvae of butterflies and moths, beetles, and flies, but hosts in other orders are known. They can be endo- or ectoparasitoids and solitary or gregarious; some are polyembryonic, a few are hyperparasitoids, and even fewer are phytophagous (Wharton 1993). Wharton et al. (1997) provided keys to the more than 400 Western Hemisphere genera and summarized the biology for each subfamily. Yu et al. (2005) cataloged the species.

Cynipoidea

This superfamily contains five families (Ronquist 1999b). The Austrocynipidae, Ibaliidae, and Liopteridae together contain fewer than 200 relatively large, rarely collected, archaic species that parasitize insect larvae, usually in wood. The single species of Austrocynipidae develops in *Araucaria* cones. Ronquist (1995) revised the genera of Liopteridae, and Nordlander et al. (1996), the species of Ibaliidae. The numerous, poorly known Figitidae are all parasitoids of other insects. The species of Cynipidae (gall wasps and their inquilines) are phytophagous and have been known since ancient times because of their often conspicuous, strangely shaped, or brightly colored galls formed on plants, especially oaks and roses (Csóka et al. 2005). Oak galls were used in Europe during the Middle Ages for tanning, ink production, and sometimes medicine.

Proctotrupoidea

This morphologically and biologically diverse superfamily consists of 11 extant families, but only 2, the Diapriidae and Proctotrupidae, contain most of the species. The remainder (Austroniidae,

Heloridae, Maamingidae, Monomachidae, Pelecinidae, Peradaeniidae, Proctorenyxidae, Roproniidae, and Vanhorniidae) are small, relict families with about 75 species in total. Most species of Diapriidae, the largest family, are parasitoids of flies, but some morphologically bizarre species occur in army ant or termite nests. Masner and García (2002) treated the Western Hemisphere genera of Diapriinae, one of the four subfamilies. The Belytinae parasitize fungus gnats and make up a huge proportion of parasitic Hymenoptera at ground level in forests. The Proctotrupidae, revised by Townes and Townes (1981), are parasitoids of beetle larvae in soil or rotting wood. The Heloridae parasitize larvae of lacewings. Johnson and Musetti (1999) revised the species of Pelecinidae, which parasitize June beetles, and Musetti and Johnson (2004) revised the Western Hemisphere Monomachidae, which parasitize soldier flies. The Maamingidae, from New Zealand, is the latest family of extant Hymenoptera described (Early et al. 2001). Johnson (1992) cataloged the Ceraphronoidea, Platygastroidea (excluding Platygastridae), and Proctotrupoidea.

Platygastroidea

The two included families, the Platygastridae and Scelionidae, contain small species, most of which (except Platygastrinae) are egg parasitoids. Masner (1976) and Masner and Huggert (1989) gave keys to the genera of Scelionidae and part of the Platygastridae, respectively. Austin et al. (2005) reviewed the current state of knowledge for the superfamily. Vlug (1995) cataloged the Platygastridae. The Telenominae (Scelionidae), which include the most important species for biological control of economically important pests, especially moths, represent the greatest taxonomic challenge.

Ceraphronoidea

Dessart and Cancemi (1986) provided keys to the genera for the two extant families, the Megaspilidae and Ceraphronidae. Johnson and Musetti (2004) cataloged the species and briefly summarized their biology. Species of both families parasitize hosts in several insect orders.

Chalcidoidea

This superfamily of mostly small to minute parasitic wasps has a greater range of biological diversity than that of any other superfamily (Plate 6), except perhaps the Vespoidea. Askew (1971), Bouček (1988), Gauld and Bolton (1996), and Grissell and Schauff (1997) gave general accounts of chalcid biology. Whereas most chalcids are parasitoids of insects and spiders, species in six families have reverted to plant feeding in galls or seeds (LaSalle 2005) and some are important pests. The small size of chalcids means they are mostly overlooked and relatively poorly studied despite their abundance in most habitats and their importance in biological control. Only 2 of the 19 currently recognized families have common names. The chalcids, perhaps deservedly, are considered the most difficult group to identify to family; chalcid taxonomists themselves cannot always agree on family limits. Keys to chalcid genera exist for only two continents: North America (Gibson et al. 1997) and Australia (Bouček 1988), excluding four families. Gibson (2005) listed online resources and the major taxonomic literature, including the few family-level world treatments. Noyes (2002) cataloged the world chalcid fauna, summarizing the taxonomic, biological, morphological, and distributional information from more than 39,000 literature references. Aspects of chalcid morphology, biology, and economic importance for only three taxa are given here.

Mymaridae

One of only two family groups of chalcids that have a common name, the Mymaridae or fairyflies, is so called because of their minute size and delicate, long-fringed wings. Though never noticed by most people, they are among the most common chalcids, presumably because almost all parasitize host eggs, the most abundant stage of their host insects. The diminutive, 0.13-mm long males of *Dicopomorpha echmepterygis* (Figs. 12.2 A and B), whose females are typical hymenopterans, are wingless and eyeless, their antennal segments are coalesced into melon-shaped blobs, their mouthparts have apparently vanished leaving only a hole, their tarsi are lost, and the only part of the pretarsus that remains is the pad between the claws, greatly modified into a suction cup, presumably to hang onto females long enough to mate.

Agaonidae

Fig wasps are obligate pollinators of the 700-plus species of figs (*Ficus*), an important component of tropical forests. The mutualistic interactions between the Agaoninae and figs are a textbook example of coevolution (Kjellberg et al. 2005). Males are wingless (Fig. 12.1B) and have some of the most unusual morphology among the Hymenoptera, including slender, snakelike forms that literally swim inside figs and humpbacked, turtlelike forms that fight each other when females become rare. Females are winged (Fig. 12.1A) and disperse to other conspecific fig trees, pollinating them. Thus, fig wasps are responsible for a yearly, multimillion dollar crop of Smyrna figs (Sisson 1970) and the survival of all other fig species. However, the great majority of edible figs come from varieties of *Ficus carica*, the common fig, which do not require pollinators.

Trichogrammatidae

These minute parasitoids of the eggs of other insects are best known because of the genus *Trichogramma*. Certain species are mass-produced by the millions around the world on factitious hosts for biological control against lepidopteran pests of agriculture and forestry. An entire journal was devoted to *Trichogramma*, until it expanded its scope to include other egg parasitoids (Hertz et al. 2004). Though many aspects of *Trichogramma* biology are understood, the species are taxonomically well known only for North America. The rest of the Trichogrammatidae remain poorly known. Pinto (2006) gave keys to genera of the Western Hemisphere.

Aculeata

All Hymenoptera that sting are in the Aculeata, classified in three superfamilies: the Chrysidoidea, Vespoidea, and Apoidea (Brothers 1999). O'Neill (2001) treated the behavior and natural history of the solitary species, and Wilson (1971) treated all the social groups, with an analysis of sociality. Despite their relatively small numbers, the approximately 11,500 truly social species have had a disproportionately important role in human affairs, both economically and scientifically. The social groups consist of all ant species (Formicidae), 1000 bee species (Apidae and Halictidae), about 800 wasp species (Vespidae), and the genus *Microstigmus* (Crabronidae). Attenborough (2005, Chap. 5) provided an entertaining account of the lives of some species.

Chrysidoidea

This superfamily consists of three moderately species-rich families (Bethylidae, Chrysididae, and Dryinidae) and four small, rare ones (Embolemidae, Plumariidae, Sclerogibbidae, and Scolebythidae). Azevedo (1999) provided keys to the species of Scolebythidae, all of which are assumed to parasitize wood-boring beetles. Olmi (1995, 2005) revised and gave keys to the species of the Embolemidae and Sclerogibbidae, respectively. Members of these two families are ectoparasitoids of leafhoppers and related groups, and nymphs and adults of webspinners, respectively. The Plumariidae are restricted to dry areas of southern Africa and southern South America. Males are winged and females are wingless, subterranean, and exceedingly rarely collected.

Chrysididae

These wasps are the peacocks of the hymenopteran world, rivaled in brilliance only by the orchid bees (Euglossinae) and some Chalcidoidea. Most species are iridescent green, blue, or coppery-red, or combinations of these. The largest genus, *Chrysis*, with at least 1000 described species, almost half the family total, is known popularly as cuckoo wasps in North America or gold wasps in Europe where many species have a coppery-red abdomen. Most species are parasitoids or cleptoparasitoids in the nests of bees and other wasps, hence the name 'cuckoo wasp'. Adults are capable of rolling into a ball when the wasp owner of a burrow they enter tries to sting them. Because they have a hard cuticle and can withdraw their antennae and legs tightly against the body, the owner cannot sting them but will grab them by the wings and eject them from their burrow, presumably unharmed. The Amiseginae and Loboscelidiinae parasitize eggs of walking sticks; the Cleptinae parasitize sawfly prepupae. Kimsey and Bohart (1990) revised the genera and listed the species.

Dryinidae

This family contains females that are usually wingless and antlike, but with pincerlike foretarsal claws and with winged males that usually lack pincerlike claws. The species are ectoparasitiods of leafhoppers and their relatives (Olmi 1984).

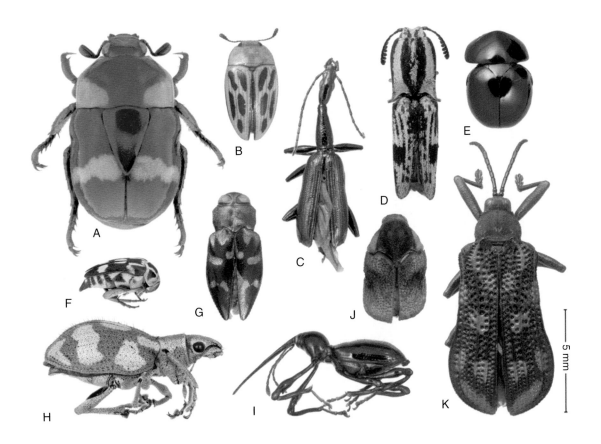

	species richness		adult feeding guilds
	empirical	predicted	
A) Scarabaeidae	58	133	herbivores, leaf feeders
B) Erotylidae	214	168	fungivores
C) Carabidae	458	477	predators
D) Elateridae	150	138	detritivores
E) Ceratocanthidae	11	36	termitophiles
F) Mordellidae	388	361	pollen feeders
G) Buprestidae	209	234	pollen feeders
H) Entiminae (Curculionidae	41	45	herbivores, leaf feeders
I) Otidocephalini (Curculionidae)	35	37	herbivores
J) Cryptocephalinae (Chrysomelidae)	43	227	herbivores, leaf feeders
K) Hispinae (Chrysomelidae)	227	316	herbivores, leaf feeders
L) Alticinae (Chrysomelidae) not figured	400	n/a	herbivores, leaf feeders
M) Cleridae, not figured	122	139	predators
total = 2315			

Plate 1

Coleoptera taxa examined in Yasuni National Park, Ecuador, from 1994 to 2006. Images were taken with the EntoVision extended focus photography system (Alticinae and Cleridae not figured). For each taxon, the number of observed morphospecies (as of 2008) is listed, as well as predicted numbers of species based on accumulation curves and ICE calculations (Erwin et al. 2005).

Plate 2

A. *Calosoma sycophanta* (L.) (Coleoptera: Carabidae) (Turkey) (Photo A. Konstantinov). B. *Nemoptera sinuata* Olivier (Neuroptera: Nemopteridae) (Turkey) (Photo M. Volkovitsh). C. *Eristalis tenax* (L.) (Diptera: Syrphidae) (Turkey) (Photo A. Konstantinov). D. *Cryptocephalus duplicatus* Suffrian (Coleoptera: Chrysomelidae) (Turkey) (Photo A. Konstantinov). E. *Poecilimon* sp. (Orthoptera: Tettigoniidae) (Turkey) (Photo M. Volkovitsh). F. *Capnodis carbonaria* (Klug) (Coleoptera: Buprestidae) (Turkey) (Photo M. Volkovitsh). G. *Cyphosoma euphraticum* (Laporte et Gory) (Coleoptera: Buprestidae) (southern Russia) (Photo M. Volkovitsh).

Plate 3

A. *Julodis variolaris* (Pallas) (Coleoptera: Buprestidae) (Kazakhstan) (Photo M. Volkovitsh). B. *Julodella abeillei* (Théry) (Coleoptera: Buprestidae) (Turkey) (Photo M. Volkovitsh). C. *Mallosia armeniaca* Pic (Coleoptera: Cerambycidae) (Turkey) (Photo M. Volkovitsh). D. *Trigonoscelis schrencki* Gebler (Coleoptera: Tenebrionidae) (Kazakhstan) (Photo M. Volkovitsh). E. *Saga pedo* Pallas (Orthoptera: Tettigoniidae) (Kazakhstan) (Photo M. Volkovitsh). F. *Piazomias* sp. (Coleoptera: Curculionidae) (Kazakhstan) (Photo M. Volkovitsh).

Plate
(capti

Plate 8

Dragonfly (female *Erythemis simplicicollis*) consuming a bee-fly pollinator *Bombylius* sp. (Diptera: Bombyliidae).

Plate 7

A sample of lepidopteran biodiversity. 1, *Oxydia geminata* Maassen (Geometridae). 2, *Ectoedemia virgulae* (Braun) (Nepticulidae). 3, *Lieinix nemesis* Latreille (Pieridae). 4, *Charaxes cithaeron* C. & R. Felder (Nymphalidae). 5, *Ethmia monticola* (Walsingham) (Elachistidae). 6, *Apantesis virgo* (Linnaeus) (Noctuidae). 7, *Gloveria gargamelle* Strecker (Lasiocampidae). 8, *Catacrosis lithosialis* Ragonot (Pyralidae). 9, *Sparganothis reticulatana* (Clemens) (Tortricidae). 10, *Plagodis serinaria* Herrich-Schäffer (Geometridae). 11, *Atteva punctella* Cramer (Yponomeutidae). 12, *Feltia jaculifera* (Guenée) (Noctuidae). 13, *Epimartyria pardella* (Walsingham) (Micropterigidae). 14, *Decantha tistra* Hodges (Oecophoridae). 15, *Oligocentria semirufescens* (Walker) (Notodontidae). 16, *Nemapogon defectella* (Zeller) (Tineidae). 17, *Noctua pronuba* Linnaeus (Noctuidae). 18, *Olethreutes arcuella* (Clerck) (Tortricidae). 19, *Schinia florida* (Guenée) (Noctuidae). 20, *Euclea delphinii* (Boisduval) (Limacodidae). 21, *Diastictis robustior* Munroe (Crambidae). 22, *Dyseriocrania griseocapitella* (Walsingham) (Eriocraniidae). 23, *Stilbosis tesquella* Clemens (Cosmopterigidae). 24, *Xestia dolosa* Franclemont (Noctuidae). 25, *Filatima obidenna* Clarke (Gelechiidae). Prepared by David Adamski.

Bethylidae

This largest of the chrysidoid families consists mostly of inconspicuous, brown or black, tropical species that are common in many habitats, such as forest litter and deserts. They parasitize moths or beetles. Some species attack stored-products pests and can be found in homes. Sexual dimorphism is slight to extreme, and the usually wingless females can be confused with ants. Gordh and Móczár (1990) cataloged the world fauna.

Vespoidea

This superfamily contains five small, relatively rare families (Bradynobaenidae, Rhopalosomatidae, Sapygidae, Scoliidae, and Sierolomorphidae), and five large, almost ubiquitous families (Formicidae, Mutillidae, Pompilidae, Tiphiidae, and Vespidae) that are discussed individually here. About 20,000 mainly tropical species of vespoids have been described, less than half the estimated number.

Mutillidae

Members of this family are known as velvet ants because the usually wingless females are covered with dense, brightly colored hairs and resemble colorful ants (Manley and Pitts 2007). Velvet ants are ectoparasitoids of the enclosed larvae and pupae of other insects, especially Aculeata. Many occur in arid areas, searching on sand for their soil-inhabiting hosts. Because of sexual dimorphism, conspecific, or even congeneric, males and females are not easy to associate unless they are captured during mating. The separate keys to genera of males and females do not necessarily coincide. One sex often is much more rarely collected than the other, so keys are based almost exclusively on males or on females. The powerful sting of females has given them the common name 'cow killer' in North America.

Pompilidae

Known as spider wasps, pompilids are commonly seen running erratically on sand or bare earth, flicking their usually dark-colored wings. Most are predators of spiders, including the largest ones, the tarantulas. Species of *Pepsis*, spectacular wasps known as tarantula hawks, are strongly dimorphic sexually and many species form complex mimicry groups, both Batesian and Muellerian. Their taxonomic popularity has resulted in numerous species being described many times. Vardy (2000, 2002, 2006) synonymized 374 names among the 133 species treated.

Tiphiidae

Taxonomic problems in this cosmopolitan family are similar to those of the Mutillidae because of the strong sexual dimorphism in many species. Separate keys to each sex are often needed, and these are often not congruent, such that the number of genera for males differs from that for females; the names applied to each sex also can differ (Kimsey and Wasbauer 2006). The species are solitary ectoparasitoids on larvae of soil-dwelling beetles, especially scarabs.

Vespidae

After the bees and ants, this family of wasps is the best known because of the social species and their often spectacular paper nests and ability to defend these vigorously by stinging. Three of the six recognized subfamilies – the Vespinae, Polistinae, and Stenogastrinae – contain the social species. Social wasps are central to the study of behavioral evolution because the Polistinae (paper wasps) mark a clear transition between solitary and highly eusocial behavior. Most observations have involved *Polistes*, a cosmopolitan genus of about 200 species (Turilazzi and West-Eberhard 1996). Members of *Polistes* are well suited for experimental and theoretical investigations of social behavior because their relatively small, open colonies and ease of location (often on human constructions) allow for detailed observations (Pickett and Wenzel 2004). The solitary Eumeninae or potter wasps, so called because of the beautiful clay pots constructed by *Eumenes*, are the largest subfamily, with about 3000 species. Spradberry (1973) discussed the natural history of some. Recent generic keys exist only for the Eumeninae of the Western Hemisphere and Vespidae of northeastern North America (Carpenter and Garcete-Barrett 2003, Carpenter 2004, Buck et al. 2008). The Masarinae, or pollen wasps, include about 300 species. They are unusual among the Vespoidea because they all provision their nests with pollen. They occur in Mediterranean and arid climates and are most diverse in southern Africa, which contains more than half the described species (Gess 1996).

Formicidae

Ants rival the bees in being the best-known hymenopterans. All are social and form colonies. Queen ants can live long, the record being 29 years in one captive colony. Bolton (1994, 1995, 2003) provided generic keys, cataloged the species, and reclassified and keyed the subfamilies and tribes, respectively. Hölldobler and Wilson (1990) estimated 20,000 species in total. Two popular accounts of ants (Hölldobler and Dawson 1984, Hölldobler and Wilson 1994) illustrate their diversity. Their effect on the earth's ecosystems and human society is immense (e.g., Majer et al. 2004), especially in the tropics where most species occur. Numerous parallels with various aspects of human society have been made, including the best (agriculture) and worst (slavery, thievery, warfare). Different species can be grouped informally as follows: (1) *harvesters/gatherers* – the seed harvester, *Messor barbatus*, is perhaps the species referred to in Proverbs 6: 6 (Kloets and Kloets 1959), (2) *farmers/herders* – many species collect honeydew from sap-sucking insects and protect them from predators, (3) *mushroom growers* – leaf-cutting ants make fungus gardens in their underground nests, with chemical control of unwanted fungi and garden sanitation resulting in refuse piles outside the nest, (4) *hunters/scavengers* – army and driver ants epitomize the hunters and scavengers, (5) *building engineers* – wood ants and leafcutter ants are the premiere building engineers. Food storage, using living bodies of their own species swollen with a sweet liquid drawn upon during times of scarcity, is unique to ants. Slavery among ants usually involves one species making slaves of another, but a few species make slaves of members of their own species from different colonies.

Apoidea

Bees are best known among the Hymenoptera because of their ubiquity in all types of habitats from arctic tundra to tropical forests and deserts, their relatively large size and sometimes bright colors (e.g., orchid bees) or fuzzy appearance (bumblebees), the painful sting that members of the highly social species deliver to the unwary, their purposeful 'busy-ness', and their beneficial role as pollinators. The number of described bee species is estimated at 17,000. In addition to bees, the Apoidea include a group of solitary wasps formerly classified as the Sphecoidea. Some of the solitary species

are gregarious and nest in large numbers in sandy or bare soil, but generally they are poorly known. Two syntheses (Bohart and Menke 1976, Michener 2000) treat the classification and biology of bees and sphecoid (now apoid) wasps, respectively. Pulawski (2006) cataloged the 'sphecoid' wasps, currently classified as four families (Ampulicidae, Crabronidae, Heterogynaeidae, and Sphecidae). The bees are now classified in seven families (Andrenidae, Apidae, Colletidae, Halictidae, Megachilidae, Melittidae, and Stenotritidae). Michener (1974) gave an account of bee behavior.

The sphecoid wasps are all carnivorous. Females provision their nests with a wide variety of insects, paralyzed and placed in larval cells they construct in the ground, in holes in wood, or in mud constructions of their own making. As predators, they are presumably beneficial, destroying many insects, including potential pests. Bees are basically vegetarian wasps that provision their nests with pollen, except for species of *Trigona*, which use carrion (Michener 2000). About 30% of human food is derived from bee-pollinated plants (O'Toole 1993). Numerous bee species besides honeybees pollinate crops, and bees are the most important group of pollinators in a wide variety of natural habitats. In many areas, populations of wild bees (and other Aculeata) have been reduced by human activity (Kevan and Viana 2004).

The social bees (*Bombus* and *Apis*, and Meliponini, the stingless bees) are the best-known bees. The approximately 250 *Bombus* species, adapted to cool climates, are primitively social and mainly Holarctic; a few occur in mountains in the tropics, and some extend to the southern tip of South America. Some are social parasites in nests of other *Bombus* species. Ruttner (1988) discussed the biogeography and taxonomy of *Apis*, and Engel (1999) clarified their scientific names. Taxonomic problems still exist but the mess of names – about 170, 90 associated with *A. mellifera* alone – for the 6 fossil and 4 extant species has been cleaned up. No hymenopteran has been better studied than the honeybee. From crop pollination to honey and wax production to its use in scientific studies of behavior and genetics and in the folklore of people around the world, no other insect has had such a direct and visible influence on human society. Probably since time immemorial, humans have raided bee colonies for food, both honey and larvae. An African strain of *A. mellifera* is the most aggressive of the various subspecies, probably because early humans and mammal predators exerted a sufficiently strong selective pressure over enough time to produce this characteristic. Such is the attraction of the honeybee that even a blind man, the Swiss entomologist François Huber, carried out 'observations' on them, through the eyes of his manservant, François Burnens, who accurately observed and recorded his own observations and told them to his employer. Huber would then think about these observations and plan further experiments to be carried out by his surrogate eyes. Huber's son offered to write up a second volume of his father's accumulated but unpublished 'observations' because he thought 'they would be of interest to naturalists'. The combined results are a two-volume set of observations and 12 beautifully rendered plates of drawings of bee parts, hives, and nest architecture (Huber 1814).

Apis mellifera was introduced into the Western Hemisphere by European colonists, with tremendous benefits for agriculture. In 1956, the African strain, already fairly aggressive compared with the relatively docile European strains, was introduced into quarantine for breeding experiments. Twenty-six queens escaped in 1957 and their progeny became the Africanized bee, spreading from Brazil north to the southern United States. This nemesis of people throughout Latin America has killed countless animals and many people (Gore and Lavies 1976). In recent decades honeybees have been plagued by a variety of widely spread arthropods and diseases (Morse and Nowogrodzki 1990, Elzen et al. 1999). Colony collapse disorder is a serious apicultural problem of uncertain cause, although a virus in bee colonies imported from Australia to North America has now been implicated. Great loss to apiculturalists and farmers who depend on honeybees to pollinate their crops has resulted. With perfect hindsight, such panzootics in a single host species may be seen as inevitable, the result of overly heavy dependence on one species that has been bred, managed intensively, and spread around the world for generations. The up side is that the great importance of numerous other bee species in maintaining plant diversity in natural ecosystems is now better recognized (Delaplane and Mayer 2000). Ecologists knew this, and bee taxonomists knew that in some crop systems, certain other bees were better pollinators than honeybees. The new pressures on honeybees have made the industry (and perhaps politicians) more aware of the need to study, conserve, and use other bee species. This awareness, in turn, requires new taxonomic and biological work to determine which species are most important as native pollinators of specific crops in different regions (Kevan 1988).

Hymenoptera other than bees also need to be conserved (Shaw and Hochberg 2001). Attempts to do this are underway in Europe (Gauld et al. 1990), but much more could and should be done.

SOCIETAL BENEFITS AND DETRIMENTS OF HYMENOPTERA

The importance of Hymenoptera to applied science and society is varied because of their diverse activities. Most effects are indirect, either beneficial (pollination, parasitism, predation, and scavenging) or deleterious (plant feeding), but a few are direct (stings and food). If Hymenoptera appear better at pollination and parasitism than other insect orders that have members similarly engaged, it is largely because a much higher proportion of the species are involved and they are often host specific and efficient. In contrast, the importance of Hymenoptera to basic science is often because of their almost unique features, the study of which might lead eventually to important applications, though perhaps not so generally applicable to human welfare as, for example, pollination. Their best-known feature is the halpodiploid method of reproduction in which unfertilized eggs yield males and fertilized eggs yield females. It is the basis for their intricate social interactions. Studies on the behavior of honeybees led to a Nobel prize for Karl von Frisch in 1973 and an enormous literature on genetics, sex determination, kin selection, and biochemistry (e.g., royal jelly and pheromones involved in colony organization).

A serious problem in hymenopteran taxonomy is the morphological plasticity of individuals within species, often combined with their small size (most parasitic species) and the huge number of species, especially in the tropics. We know this diversity exists from preliminary taxonomic studies of Hymenoptera in major insect collections, though the number of cryptic species that cannot be identified readily by morphology alone might be badly underestimated. Specialized collecting methods might be needed to collect certain groups or species, such as canopy dwellers that rarely or never come within range of insect traps. Collections alone, therefore, do not necessarily represent the fauna of an area, giving another reason for biodiversity underestimates. The result is an enormous taxonomic impediment in trying to identify and describe hymenopteran biodiversity. Most species are still unnamed and the biology and host associations of relatively few species are adequately known. Given this problem, we cannot estimate their importance to human welfare in any other than general terms. If half the Hymenoptera species became extinct in the next decade, what would be the consequences? We do not know what half is – or any other percentage. We would not be able to predict much of anything accurately and, in any case, few would notice their disappearance, even of large and colorful species such as bumblebees. However, we can confidently surmise that more problems than benefits would result if that biodiversity disappeared. As global trade increases, both through human population growth and greater individual demands and expectations for more products, more pest species such as sawflies, wasps, and ants inevitably will be spread to new places, causing both foreseeable (local reduction or loss of biodiversity) and unforeseeable problems. At the same time, beneficial species will be reduced in numbers or locally, even globally, extirpated, often before we find out what their role in nature was. Their loss eventually would be recognized indirectly by the lack of services they provide.

Habitat contamination or complete destruction affects different guilds of Hymenoptera differently, depending on their biology. The hymenopteran guilds or activities given above can be further subdivided, with varying degrees of host specificity within each, for example, oligolectic versus polylectic pollinators, predators nesting in sand versus rock faces, monophagous parasitoids of woodborers versus polyphagous parasitoids of leafminers. The species in different guilds might be differentially susceptible to extirpation or extinction.

Intuitively, one expects that generalist species with wide geographic distributions or habitat requirements are less likely to be endangered than are specialists with restricted distributions (island distributions especially) and narrow requirements. Few studies (e.g., Ulrich 2001) have examined the aspects of this empirically. Aculeates, particularly pollinators that are host–plant specific, are likely the most vulnerable to extirpation or extinction because they need particular nesting sites as well as nectar from plant species that might, themselves, be threatened with extinction. If either the plants they pollinate or the nesting microhabitats vanish, so too will the pollinators. Aculeates that nest in sandy habitats are at greater risk of extirpation than those nesting in holes in wood (Buck et al. 2008) because pristine dune systems are often threatened by human development.

In Canada, *Bombus affinis* seems to have disappeared, and some other western Canadian *Bombus* species are threatened. The conspicuous, 3-cm long *Teredon cubensis* (Siricidae) from Cuba is known from only four specimens, despite intensive efforts to collect more. Probably many similar examples could be given. Yet, no Hymenoptera are listed on the International Union for Conservation of Nature (IUCN) Red List of Threatened Species, though some might be on regional or country lists.

Even if no new economic problems arise as a result of species losses, humanity would be impoverished simply because Hymenoptera, or at least some of the large, showy species that are a noticeable part of our lives and cultures, would no longer be seen. The obvious solution, if it could be implemented reasonably, is to preserve a greater number and diversity of habitats and minimize the use of synthetic chemicals as much as possible.

CONCLUSIONS

Our present knowledge of Hymenoptera – ants and bees in particular, but also sawflies, and parasitic, predatory, and plant-feeding wasps – is considerable. Enormous gaps, however, remain. Cataloging the described species of Hymenoptera is well ahead of taxonomic knowledge, especially at the species level. The taxonomically and biologically best-known faunas are those of Australia, Europe, and North America, primarily because most Hymenoptera taxonomists live there. The classification of Hymenoptera at the family level is relatively well understood and agreed upon, though considerable debate still occurs about relationships. Species-level taxonomy is moderately good for most Aculeata and Symphyta – at least half the species are described – particularly for the small families, but is abysmal for the Parasitica, particularly the larger superfamilies, for which less than 10% of the fauna is estimated to be described. For most species, biological knowledge is poor or lacking, especially for the Parasitica. Conservation is in its infancy but needs to be carried out more vigorously in the larger context of habitat conservation and sustainable agriculture and forestry. No recent species of Hymenoptera are known for certain to have gone extinct, but some have been extirpated from large areas (Godsoe 2004), and human activity might have reduced the numbers of many species.

The morphological and biological diversity and economic and scientific value of Hymenoptera need more promotion so more people can learn to appreciate and enjoy them, rather than ignore or fear them.

ACKNOWLEDGMENTS

Many colleagues provided me with the numbers of species and recent references for their groups of expertise; space constraints did not allow all references to be included. I thank M. Buffington and D. Smith (National Museum of Natural History, Washington, DC), J. Carpenter (American Museum of Natural History, New York, NY), A. Deans (North Carolina State University, Raleigh, NC), N. Johnson (Ohio State University, Columbus, OH), C. Rasmussen (University of Illinois, Urbana, IL), J. Schmidt (Southwestern Biological Institute, Tucson, AZ), S. Shaw (University of Wyoming, Laramie, WY), and C. Vardy (retired from The Natural History Museum, London) for drawing my attention to the references. My colleagues A. Bennett, H. Goulet, L. Masner, and G. Gibson (Canadian National Collection of Insects, Agriculture and Agri-Food Canada, Ottawa) and two unknown reviewers made useful suggestions and corrections on drafts of this chapter.

REFERENCES

Abe, M. and D. R. Smith. 1991. The genus-group names of Symphyta (Hymenoptera) and their type species. *Esakia* 31: 1–115.

Agosti, D. and N. F. Johnson (eds). 2005. Antbase, World Wide Web electronic publication. Antbase.org, version (05/2005) [Accessed 18 January 2007].

Aguiar, A. P. 2004. World catalogue of the Stephanidae (Hymenoptera: Stephanoidea). *Zootaxa* 753: 1–120.

Askew, R. R. 1971. *Parasitic Insects.* Heinemann Educational Books, London, United Kingdom.

Attenborough, D. 2005. *Life in the Undergrowth.* Princeton University Press, New Jersey.

Austin, A. D. and M. Dowton. 2000. The Hymenoptera: an introduction. Pp. 3–7. *In* A. D. Austin and M. Dowton (eds). *Hymenoptera: Evolution, Biodiversity and Biological Control.* CSIRO Publishing, Collingwood, Victoria, Australia.

Austin, A. D., N. F. Johnson, and M. Dowton. 2005. Systematics, evolution, and biology of scelionid and platygastrid wasps. *Annual Review of Entomology* 50: 553–582.

Azevedo, C. O. 1999. A key to world species of Scolebythidae (Hymenoptera: Chrysidoidea), with description of a new species of *Dominibythus* from Brazil. *Journal of Hymenoptera Research* 8: 1–5.

Beaver, R. A. 2002. A new species of *Ycaploca* (Hym., Scolebythidae) from Fiji. *Entomologist's Monthly Magazine* 138: 139–142.

Beckage, N. E. 1998. Parasitoids and polydnaviruses. *Bioscience* 48: 305–311.

Beleboni, R. de O., A. B. Pizzo, A. C. K. Fontana, G. C. de O. Ruither, J. Coutinho-Netto, and W. F. dos Santos. 2004. Spider and wasp neurotoxins: pharmacological and biochemical aspects. *European Journal of Pharmacology* 493(1–3): 1–17.

Bennett, A. M. R. 2001. Phylogeny of the Agriotypinae (Hymenoptera: Ichneumonidae), with comments on the subfamily relationships of the basal Ichneumonidae. *Systematic Entomology* 26: 329–356.

Bennett, A. M. R. 2008. A review and identification keys to the ichneumonid parasitoids (Hymenoptera: Ichneumonidae) of Nearctic *Choristoneura* species (Lepidoptera: Tortricidae). *Canadian Entomologist* 140: 1–1.

Bennett, D. J. and M. S. Engel. 2005. A primitive wasp in Burmese amber (Hymenoptera: Sapygidae). *Acta Zoologica Cracoviensia* 48B: 1–9.

Benson, R. B. 1950. An introduction to the natural history of British sawflies. *Transactions of the Society for British Entomology* 10: 46–142.

Bohart, R. M. and A. S. Menke. 1976. *Sphecid Wasps of the World: A Generic Revision.* University of California Press, Berkeley, California.

Bolton, B. 1994. *Identification Guide to the Ant Genera of the World.* Harvard University Press, Cambridge, Massachusetts.

Bolton, B. 1995. *A New General Catalogue of the Ants of the World.* Harvard University Press, Cambridge, Massachusetts.

Bolton, B. 2000. The ant tribe Dacetini. *Memoirs of the American Entomological Institute* 65: 1–1028.

Bolton, B. 2003. Synopsis and classification of Formicidae. *Memoirs of the American Entomological Institute* 71: 1–370.

Bouček, Z. 1988. *Australasian Chalcidoidea (Hymenoptera). A Biosystematic Revision of Genera of Fourteen Families, with a Reclassification of Species.* C.A.B. International, Wallingford, United Kingdom.

Brothers, D. J. 1999. Phylogeny and evolution of wasps, ants and bees (Hymenoptera, Chrysidoidea, Vespoidea and Apoidea). *Zoologica Scripta* 28 (1–2): 233–249.

Buck, M., S. A. Marshall, and D. K. B. Cheung. 2008. Identification atlas of the Vespidae (Hymenoptera, Aculeata) of the northeastern Nearctic region. *Canadian Journal of Arthropod Identification* 5: 492.

Carmean, D. and L. Kimsey. 1998. Phylogenetic revision of the parasitoid wasp family Trigonalidae (Hymenoptera). *Systematic Entomology* 23: 35–76.

Carpenter, J. M. 2004. *Ancistroceroides* de Saussure, a potter wasp genus new for the United States, with a key to the genera of Eumeninae of America North of Mexico (Hymenoptera: Vespidae). *Journal of the Kansas Entomological Society* 77: 721–741.

Carpenter, J. M. and B. R. Garcete-Barrett. 2003. A key to the Neotropical genera of Eumeninae (Hymenoptera: Vespidae). *Boletín del Museo Nacional de Historia Natural del Paraguay* 14(1–2): 52–73.

Chauvin, R. 1968. *Traité de Biologie de l'Abeille. V. Histoire, Ethnographie et Folklore.* Masson et Cie, Paris, France.

Csóka, G., G. N. Stone and G. Melika. 2005. Biology, ecology, and evolution of gall-inducing Cynipidae. Pp. 573–642. *In* R. Anantanarayanan, C. W. Schaefer and T. M. Withers (eds). *Biology, Ecology and Evolution of Gall-Inducing Arthropods.* Science Publishers, Enfield, New Hampshire.

Deans, A. R. 2005. Annotated catalog of the world's ensign wasp species (Hymenoptera: Evaniidae). *Contributions of the American Entomological Institute* 34: 1–164.

Deans, A. R. and M. Huben. 2003. Annotated key to the ensign wasp (Hymenoptera: Evaniidae) genera of the world, with descriptions of three new genera. *Proceedings of the Entomological Society of Washington* 105: 859–875.

Delaplane, K. S. and D. F. Mayer. 2000. *Crop Pollination by Bees.* CAB International, Wallingford, United Kingdom.

Dessart, P. and P. Cancemi. 1986. Tableau dichotomique des genres de Ceraphronoidea (Hymenoptera) avec commentaires et nouvelles espèces. *Frustula Entomologica* 7-8: 307–372.

Early, J. W., L. Masner, I. D. Naumann, and A. D. Austin. 2001. Maamingidae, a new family of proctotrupoid wasp (Insecta: Hymenoptera) from New Zealand. *Invertebrate Taxonomy* 15: 341–352.

Elzen, P. J., J. R. Baxter, D. Westervelt, C. Randall, K. S. Delaplane, L. Cutts, and W. T. Wilson. 1999. Field control and biology studies of a new pest species, *Aethina tumida* Murray (Coleoptera: Nitidulidae), attacking European honey bees in the Western Hemisphere. *Apidologie* 30: 361–366.

Engel, M. S. 1999. The taxonomy of recent and fossil honey bees (Hymenoptera: Apidae; *Apis*). *Journal of Hymenoptera Research* 8: 165–196.

Engel, M. S. and D. A. Grimaldi. 2006. The first Cretaceous spider wasp (Hymenoptera: Pompilidae). *Journal of the Kansas Entomological Society* 79: 359–368.

Eveleigh, E. S., K. S. McCann, P. C. McCarthy, S. J. Pollock, C. J. Lucarotti, B. Morin, G. A. McDougall, D. B. Strongman, J. T. Huber, J. Umbanhowar, and L. D. B. Faria. 2007. Fluctuations in density of an outbreak species drive diversity cascades in food webs. *Proceedings of the National Academy of Sciences USA* 104: 16976–16981.

Fernández, F. and M. J. Sharkey (eds). 2007. *Introducción a los Hymenoptera de la región Neotropical.* Sociedad Columbiana de Entomología y, Universidad de Colombia, Bogotá D.C., Colombia 893 pp.

Floate, K. D., G. K. Kyei-Poku, and P. G. Coghlin. 2006. Overview and relevance of *Wolbachia* bacteria in biocontrol research. *Biocontrol Science and Technology* 16: 767–788.

Frederici, B. A. and Y. Bigot. 2003. Origin and evolution of polydnaviruses by symbiogenesis of insect DNA viruses in endoparasitic wasps. *Journal of Insect Physiology* 49: 419–432.

Gaston, K. J. 1993. Spacial patterns in the description and richness of the Hymenoptera. Pp. 277–293. *In* J. LaSalle and I. D. Gauld (eds). *Hymenoptera and Biodiversity.* CAB International. Wallingford, United Kingdom.

Gauld, I., N. M. Collins, and M. G. Fitton. 1990. *The Biological Significance and Conservation of Hymenoptera in Europe. No. 44.* Publishing and Documentation Service, Council of Europe, Strasbourg, Belgium.

Gauld, I., R. Sithole, J. U. Gómez, and C. Godoy. 2002. The Ichneumonidae of Costa Rica, 4. *Memoirs of the American Entomological Institute* 66: 1–768.

Gauld, I. and B. Bolton. 1996. *The Hymenoptera*, Second Edition. Oxford University Press, New York.

Gess, S. K. 1996. *The Pollen Wasps. Ecology and Natural History of the Masarinae.* Harvard University Press, Cambridge, Massachusetts.

Gibson, G. A. P. 2005. About Chalcidoidea (Chalcid wasps). http://canacoll.org/hym/staff/gibson/chalcid.htm

Gibson, G. A. P., J. T. Huber, and J. B. Woolley. 1997. *Annotated Keys to the Genera of Nearctic Chalcidoidea.* NRC Research Press, Ottawa, Canada.

Gibson, G. A. P., J. Read, and J. T. Huber. 2007. Diversity, classification and higher relationships of Mymarommatoidea (Hymenoptera). *Journal of Hymenoptera Research* 16: 51–146.

Goddard, J. 1996. *Physician's Guide to Arthropods of Medical Importance*, Second Edition. CRC Press, Boca Raton, Florida.

Godsoe, W. 2004. Evidence for the extirpation of *Ceropales bipunctata* Say (Hymenoptera: Pompilidae) in Ontario. *Journal of the Entomological Society of Ontario* 134: 135–140.

Gordh, G. and L. Móczár. 1990. A catalog of the world Bethylidae (Hymenoptera: Aculeata). *American Entomological Institute* 46: 1–364.

Gore, R. and B. Lavies. 1976. Those fiery Brazilian bees. *National Geographic* 149(4): 491–501.

Goulet, H. and J. T. Huber (eds). 1993. *Hymenoptera of the World: An Identification Guide to Families.* Research Branch, Agriculture Canada, Ottawa, Canada.

Grimaldi, D. and M. S. Engel. 2005. *Evolution of the Insects.* Cambridge University Press, Cambridge, Massachusetts.

Grissell, E. 2001. *Insects and Gardens.* Timber Press, Portland, Oregon.

Grissell, E. E. and M. E. Schauff. 1997. Chalcidoidea. Pp. 45–116. *In* G. A. P. Gibson, J. T. Huber and J. B. Woolley (eds). *Annotated Keys to the Genera of Nearctic Chalcidoidea (Hymenoptera).* NRC Research Press, Ottawa, Canada.

Hanson, P. E. 2006. Familia Sierolomorphidae. Hymenoptera de la Región Neotropical. *Memoirs of the American Entomological Institute* 77: 1–944.

Hanson, P. E. and I. D. Gauld. 1995. *The Hymenoptera of Costa Rica.* Oxford University Press, Oxford, United Kingdom.

Hanson, P. E. and I. D. Gauld (eds). 2006. Hymenoptera de la región neotropical. *Memoirs of the American Entomological Institute* 77: 1–944.

Hertz, A., O. Zimmermann, and S. A. Hassan. 2004. Egg. *Parasitoid News* 16: 1–56.

Hölldobler, B. and J. D. Dawson. 1984. The wonderfully diverse ways of the ant. *National Geographic* 165(6): 779–813.

Hölldobler, B. and E. O. Wilson. 1990. *The Ants.* Belknap Press of Harvard University Press, Cambridge, Massachusetts.

Hölldobler, B. and E. O. Wilson. 1994. *Journey to the Ants: A Story of Scientific Exploration.* Belknap Press of Harvard University Press, Cambridge, Massachusetts.

Huber, F. 1814. *Nouvelles Observations sur les Abeilles,* Séconde Edition. Tome premier. J. Paschoud, Geneva, Switzerland.

Jennings, J. T. and A. D. Austin. 2002. Systematics and distribution of world hyptiogastrine wasps (Hymenoptera: Gasteruptiidae). *Invertebrate Systematics* 16: 735–811.

Jennings, J. T. and A. D. Austin. 2004. Biology and host relationships of aulacid and gasteruptiid wasps (Hymenoptera: Evanioidea): a review. Pp. 187–215. *In* K. Rajmohana, K. Sudeer, P. Girish Kumar and S. Santhosh (eds). *Perspectives on Biosystematics and Biodiversity.* Prof. T. C. Narendran Commemoration Volume. University of Calicut, SERSA, Calicut, India.

Johnson, N. F. 1992. Catalog of world species of Proctotrupoidea, exclusive of Platygastridae (Hymenoptera). *Memoirs of the American Entomological Institute* 51: 1–825.

Johnson, N. F. and L. Musetti. 1999. Revision of the proctotrupoid genus *Pelecinus* Latreille (Hymenoptera: Pelecinidae). *Journal of Natural History* 33: 1513–1543.

Johnson, N. F. and L. Musetti. 2004. Catalog of systematic literature on the superfamily Ceraphronoidea (Hymenoptera). *Contributions of the American Entomological Institute* 33 (2): 1–149.

Kevan, P. G. 1988. Alternative pollinators for Ontario's crops: prefatory remarks to papers presented at a workshop held at the University of Guelph, 12 April, 1986. *Proceedings of the Entomological Society of Ontario* 118: 109–110.

Kevan, P. G. and T. P. Phillips. 2001. The economic impacts of pollinator declines: an approach to assessing the consequences. *Conservation Ecology* 5(1): 8. [online] URL: http://www.consecol.org/vol15/iss1/art8 16 pp.

Kevan, P. G. and B. F. Viana. 2004. The global decline of pollination services. *Biodiversity* 4: 3–8.

Kimsey, L. S. 2006. Familia Tiphiidae. Hymenoptera de la Región Neotropical. *Memoirs of the American Entomological Institute* 77: 575–583.

Kimsey, L. S. and R. M. Bohart. 1990. *The Chrysidid Wasps of the World.* Oxford University Press, Oxford, United Kingdom.

Kimsey, L. S. and M. S. Wasbauer. 2006. Phylogeny and checklist of the nocturnal tiphiids of the Western Hemisphere (Hymenoptera: Tiphiidae: Brachycistidinae). *Journal of Hymenoptera Research* 15: 9–25.

Kjellberg, F., E. Jousselin, M. Hossaert-McKey and J.-Y. Rasplus. 2005. Biology, ecology, and evolution of fig-pollinating wasps (Chalcidoidea, Agaonidae). Pp. 539–572. *In* R. Anantanarayanan, C. W. Schaefer and T. M. Withers (eds). *Biology, Ecology and Evolution of Gall-Inducing Arthropods.* Science Publishers, Enfield, New Hampshire.

Kloets, A. B. and E. B. Kloets. 1959. *Living Insects of the World.* Hamish Hamilton, London, United Kingdom.

LaSalle, J. 2005. Biology of gall inducers and evolution of gall induction in Chalcidoidea (Hymenoptera: Eulophidae, Eurytomidae, Pteromalidae, Tanaostigmatidae, Torymidae). Pp. 507–537. *In* R. Anantanarayanan, C. W. Schaefer and T. M. Withers (eds). *Biology, Ecology and Evolution of Gall-Inducing Arthropods.* Science Publishers, Enfield, New Hampshire.

LaSalle, J. and I. D. Gauld. 1993. Hymenoptera: their diversity, and their impact on the diversity of other organisms. Pp 1–26. *In* J. LaSalle and I. D. Gauld (eds). *Hymenoptera and Biodiversity.* CAB International. Wallingford, United Kingdom.

Levick, N. R., J. O. Schmidt, J. Harrison, G. S. Smith and K. D. Winkel. 2000. Review of bee and wasp sting injuries in Australia and the USA. Pp. 437–447. *In* A. D. Austin and M. Dowton (eds). *Hymenoptera: Evolution, Biodiversity and Biological Control.* CSIRO Publishing, Collingwood, Victoria, Australia.

Majer, J. D., S. O. Shattuck, A. N. Andersen, and A. J. Beattie. 2004. Australian ant research: fabulous fauna, functional groups, pharmaceuticals, and the Fartherhood. *Australian Journal of Entomology* 43: 235–247.

Manley, D. G. and J. P. Pitts. 2007. Tropical and subtropical velvet ants of the genus *Dasymutilla* Ashmead (Hymenoptera: Mutillidae) with descriptions of 45 new species. *Zootaxa* 1487: 1–128.

Marshall, S. A. 2006. *Insects: Their Natural History and Diversity, with a Photographic Guide to Insects of Eastern North America.* Firefly Books, Richmond Hill, Ontario, Canada.

Masner, L. 1976. Revisionary notes and keys to world genera of Scelionidae (Hymenoptera: Proctotrupoidea). *Memoirs of the Entomological Society of Canada* 97: 1–87.

Masner, L. and J. J. Garcia. 2002. The genera of Diapriinae (Hymenoptera: Diapriidae) in the New World. *Bulletin of the American Museum of Natural History* 268: 1–138.

Masner, L. and L. Huggert. 1989. World review and keys to genera of the subfamily Inostemmatinae with reassignment of the taxa to the Platygastrinae and Sceliotrachelinae (Hymenoptera: Platygastridae). *Memoirs of the Entomological Society of Canada* 147: 1–214.

Martínez de Murgía, L., M. Angeles Vazquez, and J. L. Nieves-Aldrey. 2001. The families of Hymenoptera (Insecta) in an heterogenous acidofilous forest in Artikutza (Navarra, Spain). *Frustula Entomologica (nuova serie)* 24(37): 81–98.

McGain, F., J. Harrison, and K. D. Winkel. 2000. Wasp sting mortality in Australia. *Medical Journal of Australia* 173: 198–200.

Melika, G. 2006. *Gall Wasps of Ukraine. Cynipidae*. Volume 1. Vestnik Zoologii, Kyiv, 300 pp.

Michener, C. D. 1974. *The Social Behavior of the Bees: A Comparative Study*. Belknap Press of Harvard University Press, Cambridge, Massachusetts.

Michener, C. D. 2000. *The Bees of the World*. Johns Hopkins University Press, Baltimore, Maryland.

Moffett, M. W. 1986. Marauders of the jungle floor. *National Geographic* 170(2): 273–286.

Morse, R. A. and R. Nowogrodzki (eds). 1990. *Honey Bee Pests, Predators, and Diseases*, Second Edition. Comstock Publishing, Ithaca, New York.

Musetti, L. and N. F. Johnson. 2004. Revision of the New World species of the genus *Monomachus* Klug (Hymenoptera: Proctotrupoidea, Monomachidae). *Canadian Entomologist* 136: 501–552.

Naumann, I. D. 1991. Hymenoptera (wasps, bees, ants, sawflies). Pp. 916–1000. *In* Division of Entomology, CSIRO. *The Insects of Australia*, Second Edition. Melbourne University Press, Carlton, Victoria, Australia.

Nordlander, G., Z. Liu, and F. Ronquist. 1996. Phylogeny and historical biogeography of the cynipoid wasp family Ibaliidae (Hymenoptera). *Systematic Entomology* 21: 151–166.

Noyes, J. S. 1989. The diversity of Hymenoptera in the tropics with special reference to Parasitica in Sulawesi. *Ecological Entomology* 14: 197–207.

Noyes, J. S. 2002. *Interactive Catalogue of World Chalcidoidea 2001*, Second Edition. Taxapad, Vancouver, Canada (CD-ROM: www.taxapad.com).

Olmi, M. 1984. A revision of the Dryinidae (Hymenoptera). *Memoirs of the American Entomological Institute* 37: 1–1913.

Olmi, M. 1995. A revision of the world Embolemidae (Hymenoptera Chrysidoidea). *Frustula Entomologica (nuova serie)* 18(19): 85–146.

Olmi, M. 2005. A revision of the world Sclerogibbidae (Hymenoptera Chrysidoidea). *Frustula Entomologica (nuova serie)* 26–27(39–40): 46–193.

Osten, T. 2005. Checkliste der Dolchwespen der Welt (Insecta: Hymenoptera, Scoliidae). *Bericht der Naturforschenden Gesellschaft Augsburg* 62: 1–62.

O'Neill, K. M. 2001. *Solitary Wasps: Behavior and Natural History*. Comstock Publishing Associates, Ithaca, New York.

O'Toole, C. 1993. Diversity of native bees and agroecosystems. Pp. 169–196, *In* J. LaSalle and I. D. Gauld (eds). *Hymenoptera and Biodiversity*. CAB International, Wallingford, United Kingdom.

Pagliano, G. 2006. Gli Imenotteri del Piemonte. http://www.storianaturale.org/anp/pagliano.htm [Accessed 2 January 2007].

Pagliano, G. and P. Scaramozzino. 1990. Elenco dei generi di Hymenoptera del Mondo. *Memorie della Società Entomologica Italiana* 68: 1–210.

Pickett, K. M. and J. W. Wenzel. 2004. Phylogenetic analysis of the New World *Polistes* (Hymenoptera: Vespidae: Polistinae) using morphology and molecules. *Journal of the Kansas Entomological Society* 77: 742–760.

Pinto, J. D. 2006. A review of the New World genera of Trichogrammatidae (Hymenoptera). *Journal of Hymenoptera Research* 15: 38–163.

Pulawski, W. J. 2006. http://www.calacademy.org/research/entomology/Entomology_Resources/Hymenoptera/sphecidae/Number_of_Species.htm [Accessed 2 January 2007].

Quicke, D. L. J. 1997. *Parasitic Wasps*. Chapman and Hall, London, United Kingdom.

Ronquist, F. 1995. Phylogeny and classification of the Liopteridae, an archaic group of cynipoid wasps (Hymenoptera). *Entomologica Scandinavica (Supplement)* 46: 1–74.

Ronquist, F. 1999a. Phylogeny of the Hymenoptera (Insecta): the state of the art. *Zoologica Scripta* 28(1–2): 3–11.

Ronquist, F. 1999b. Phylogeny, classification and evolution of the Cynipoidea. *Zoologica Scripta* 28(1–2): 139–164.

Ruttner, F. 1988. *Biogeography and Taxonomy of Honeybees*. Springer-Verlag, Berlin, Germany.

Schmidt, J. O. 1986. Allergy to Hymenoptera venoms. Pp. 509–546. *In* T. Piek (ed). *Venoms of the Hymenoptera*. Academic Press, London, United Kingdom.

Schmidt, J. O. 1990. Hymenopteran venoms: striving toward the ultimate defense against vertebrates. Pp. 387–419. *In* D. L. Evans and J. O. Schmidt (eds). *Insect Defenses: Adaptive Mechanisms and Strategies of Prey and Predators*. State University of New York Press, Albany, New York.

Schmidt, J. O. 1992. Allergy to venomous insects. Pp. 1209–1269. *In* J. M. Graham (ed). *The Hive and the Honey Bee*. Dadant and Sons, Hamilton, Illinois.

Schmidt, J. O. 2004. Venom and the good life in tarantula hawks (Hymenoptera: Pompilidae): how to eat, not be eaten, and live long. *Journal of the Kansas Entomological Society* 77: 402–413.

Schmidt, J. O. and L. V. Boyer Hassen. 1996. When Africanized bees attack: what you and your clients should know. *Veterinary Medicine* 91: 924–927.

Schmidt, S. and D. R. Smith. 2006. An annotated systematic world catalogue of the Pergidae (Hymenoptera). *Contributions of the American Entomological Institute* 34(3): 1–207.

Schmitt, G. 2004. Parasitoid communities (Hymenoptera) in the agricultural landscape: effects of land use types and cultivation methods on structural parameters. PhD thesis. Technische Universität Dresden, Tharandt, Germany. [In German].

Shaw, S. R. 1990a. Phylogeny and biogeography of the parasitoid wasp family Megalyridae (Hymenoptera). *Journal of Biogeography* 17: 569–581.

Shaw, S. R. 1990b. A taxonomic revision of the long-tailed wasps of the genus *Megalyra* Westwood (Hymenoptera: Megalyridae). *Invertebrate Taxonomy* 3: 1005–1052.

Shaw, M. R. and M. E. Hochberg. 2001. The neglect of parasitic Hymenoptera in insect conservation strategies: the

British fauna as a prime example. *Journal of Insect Conservation* 5: 253–263.

Sisson, R. F. 1970. The wasp that plays cupid to a fig. *National Geographic* 138(5): 690–697.

Smith, D. R. 2001. World catalog of the family Aulacidae (Hymenoptera). *Contributions on Entomology, International* 4: 261–291.

Spradberry, J. P. 1973. *Wasps: An Account of the Biology and Natural History of Solitary and Social Wasps.* University of Washington Press, Seattle, Washington.

Stork, N. E. 1997. Measuring global biodiversity and its decline. Pp. 41–68. *In* M. L. Reaka-Kudla, D. E. Wilson and E. O. Wilson (eds). *Biodiversity II.* Joseph Henry Press, Washington, DC.

Tagami, Y., K. Miura, and R. Stouthamer. 2001. How does infection with parthenogenesis-inducing *Wohlbachia* reduce the fitness of *Trichogramma?. Journal of Invertebrate Pathology* 78: 267–271.

Townes, H. and M. Townes. 1981. A revision of the Serphidae (Hymenoptera). *Memoirs of the American Entomological Institute* 32: 1–541.

Triplehorn, C. A. and N. F. Johnson. 2005. *Borror and DeLong's Introduction to the Study of Insects,* Seventh Edition. Thomson Brooks/Cole, Belmont, California.

Turillazzi, S. and M. J. West-Eberhard (eds). 1996. *Natural History and Evolution of Paper-Wasps.* Oxford University Press, Oxford, United Kingdom.

Ulrich, W. 1999a. The number of species of Hymenoptera in Europe and assessment of the total number of Hymenoptera in the world. *Polskie Pismo Entomologiczne* 68: 151–164.

Ulrich, W. 1999b. Regional and local faunas of parasitic Hymenoptera. *Polskie Pismo Entomologiczne* 68: 217–230.

Ulrich, W. 2001. Differences in temporal variability and extinction probabilities between species of guilds of parasitic Hymenoptera. *Polskie Pismo Entomologiczne* 70: 9–30.

Valli, E. and D. Summers. 1988. Honey hunters of Nepal. *National Geographic* 174 (5): 661–671.

Vardy, C. R. 2000. The New World tarantula-hawk wasp genus *Pepsis* Fabricius (Hymenoptera: Pompilidae). Part 1. Introduction and the *P. rubra* species-group. *Zoologische Verhandelingen* 332: 1–86.

Vardy, C. R. 2002. The New World tarantula-hawk wasp genus *Pepsis* Fabricius (Hymenoptera: Pompilidae). Part 2. The *P. grossa-* to *P. deaurata*-groups. *Zoologische Verhandelingen* 338: 1–135.

Vardy, C. R. 2006. The New World tarantula-hawk wasp genus *Pepsis* Fabricius (Hymenoptera: Pompilidae). Part 3. The *P. inclyta-* to *P. auriguttata*-groups. *Zoologische Mededelingen* 79-5: 1–305.

Vilhelmsen, L. 2001. Phylogeny and classification of the extant basal lineages of the Hymenoptera (Insecta). *Zoological Journal of the Linnean Society* 131: 393–442.

Vilhelmsen, L. 2003. Phylogeny and classification of the Orussidae (Insecta: Hymenoptera), a basal parasitic wasp taxon. *Zoological Journal of the Linnean Society* 139: 337–418.

Vlug, H. J. 1995. Catalogue of the Platygastridae (Platygastroidea) of the world. *Hymenopterorum Catalogus (nova editio)* 19: 1–168.

Wharton, R. A. 1993. Bionomics of the Braconidae. *Annual Review of Entomology* 38: 121–143.

Wharton, R. A., P. M. Marsh, and M. J. Sharkey. 1997. Manual of the New World genera of the family Braconidae (Hymenoptera). *International Society of Hymenopterists, Special Publication* 1: 1–439.

Wilson, E. O. 1971. *The Insect Societies.* Belknap Press of Harvard University Press, Cambridge, Massachusetts.

Yu, D. S., K. van Achterberg and Horstmann. 2005. World Ichneumonoidea 2004. *Taxonomy, Biology, Morphology and Distribution. CD/DVD.* Taxapad, Vancouver, Canada. www.taxapad.com.

Chapter 13

BIODIVERSITY OF LEPIDOPTERA

Michael G. Pogue

Systematic Entomology Laboratory, PSI, Agricultural Research Service, U. S. Department of Agriculture, c/o Smithsonian Institution, P.O. Box 37012, NMNH, MRC-168, Washington, DC, 20013-7012, USA

Insect Biodiversity: Science and Society, 1st edition. Edited by R. Foottit and P. Adler
© 2009 Blackwell Publishing, ISBN 978-1-4051-5142-9

Wilson (1997) defines biodiversity as 'everything'. Biodiversity encompasses the genes in a single population or species, all of the species in a local community, and all of the communities that make up the many ecosystems of the world. Biodiversity, as defined in this chapter, equates to species richness of the various superfamilies and families of Lepidoptera. In recent years, the biggest question confronting systematists seems to be, simply, 'how many species are there?' This question seems to have become the essence of biodiversity research. Systematists have been trying to answer this question since Westwood (1833) mentioned estimates by the seventeenth-century naturalist John Ray of 20,000 species and in the early nineteenth century by the Reverend William Kirby who guessed at 400,000–600,000 species. These guesses have continued to the present, ranging from 2.5 million to 80 million species (Erwin 2004) and indicating that no one has a clear idea of how many species exist on earth.

Many authors have estimated global biodiversity, based on a variety of taxa and methods that include taxon ratios, size ratios, catalog extrapolations, canopy samples, counting descriptions, discussions with taxonomists, and herbivore-plant extrapolations (Sabrosky 1952; Erwin 1982; May 1988, 1990, 1992; Stork 1988; Stork and Gaston 1990; Gaston 1991; Hammond 1992; Hodkinson and Casson 1991; Ødergaard 2000; Novotny et al. 2002). Most of these methods were not based on sound scientific sampling until Erwin (1982) fogged the canopy and estimated the number of arthropod species to be 30 million, based on the number of herbivore species per tree species in an acre of tropical forest. No one tested Erwin's hypothesis until Ødergaard (2000) and Novotny et al. (2002) did so in Panama and Papua New Guinea, respectively. Ødergaard (2000) used various studies that documented the different parameters in Erwin's estimate, including the number of tree, liana, and epiphyte species; proportion of phytophagous beetles associated with plant species; number of beetle species in the canopy; number of beetle species; and number of arthropods. Using various extrapolations, an estimate of 4.9 million species of arthropods was calculated. Novotny et al. (2002) reared Coleoptera and Lepidoptera from different tree species and found the number of herbivore species per tree was much less than Erwin's estimate. This procedure resulted in an estimate of 3.7 million (based on Coleoptera) and 5.8 million (based on Lepidoptera) arthropod species.

Erwin (2004) mentioned one of the key elements in estimating biodiversity: whether globally, regionally, or within ecosystems, one must examine species turnover (ß-diversity) between sites. Colwell and Coddington (1994) reported a method to estimate species richness suitable for megadiverse groups, based on sampled richness and relative abundance. More statistically based estimates can be predicted using species richness and complementarity (difference in species composition between sampling areas). The accuracy of the method will be limited to the amount of sampling and the equation(s) used in the estimation of local species richness (Erwin 2004). This method is rigorous in both collecting and documentation effort and probably will be the most accurate.

Estimating global species richness in almost any insect group, especially a large one, is impossible. In Lepidoptera, the number of described species ranges from 146,277 (Heppner 1991) to 174,250 (Kristensen 1998a). Not even a clear understanding exists of the number of described species of Lepidoptera. Most authors of the various families of Lepidoptera in Kristensen's (1998a) treatment simply gave estimates of the numbers of described species. The precise number of described species in Lepidoptera cannot be known without producing a complete catalog for the entire order. In recent years, catalogs and family revisions have been produced in which the numbers of described species are more accurately known (Davis 1975; Davis and Nielsen 1980, 1984; Poole 1989; Miller 1994; Robbins and Opler 1997; Adamski and Brown 2001; Adamski 2005; Brown 2005; Davis and Stonis 2007). The estimate of 155,181 species presented here is based, in part, on the numbers in Kristensen (1998a) as a base, plus the number of described species minus the number of synonyms from Zoological Record 1992–2006. The discrepancy between my estimate and that of Kristensen (1998a) is based on the overestimation of Noctuoidea. Poole's (1989) catalog, plus data from the Zoological Record from 1982 to 2006, yielded approximately 45,890 species versus the 70,000 estimated by Kristensen (1998a).

One method of estimating global lepidopteran species richness is by faunal comparisons between the number of described or estimated species of butterflies and moths (Lafontaine and Wood 1997). This method might rival the statistical methods of Colwell and Coddington (1994) in estimating species richness in a local area or region, but it is inaccurate in estimating global species richness. Global butterfly richness is 17,461 described species, representing about 90% of the estimated butterfly fauna of 19,401 species (Robbins and Opler 1997). In well-collected areas in northern Europe and Ottawa, Ontario, Canada, the butterfly fauna is 4–5% of lepidopteran species (Lafontaine and Wood 1997). In the Nearctic Region, the butterfly fauna is 6.63% of lepidopteran species (Hodges 1983). By taking this procedure a step further, the percentage of the butterfly fauna in the Neotropics is 17.11%, in the Palearctic 8.18%, in the Ethiopian 16.73%, in the Oriental 15.51%, and in the Australasian 6.47% (Heppner 1991). Using the percentage for the Neotropics yields an estimated global species richness of 113,390, and using 4.5% based on northern Europe and Canada yields 431,133 species. Thus, faunal comparisons on a global basis are not accurate. The problem is that one must know the fauna well before this type of estimator can provide accuracy. These estimates show that the Neotropical butterfly fauna is well known and the moth fauna is poorly known. This trend is also true in Africa. We do not have a good picture of the butterfly–moth faunal comparison in the tropics. If 19,401 species of butterflies are in the world, and the approximate known lepidopteran fauna is between 155,181 and 174,250 species, then the percentage of butterflies is 12.5% to 11.1%, respectively. Using the lepidopteran species richness of 155,181 presented here and the 174,250 of Kristensen (1998a), with the estimated butterfly species richness of 19,401, yields a global estimate of 155,208 to 174,312. This estimate is low, based on the numbers of new species being discovered in the tropics. Again, using faunal comparisons on a global scale fails to give an accurate estimate of world lepidopteran species richness.

Lepidoptera offer advantages in biodiversity studies because they are highly diverse (Plate 7), relatively easy to identify, especially the butterflies and larger moths, and amenable to quantitative sampling. They are found in many habitats and niches, making ecological comparisons possible, and they can indicate areas of endemism (Solis 1997).

Systematics is a diverse discipline that includes the discovery, description, and naming of species to document biodiversity, and then uses these species to infer phylogenies that can be used to predict classifications, biogeographic relationships, and character evolution. Life-history studies are also an integral part of systematics.

This chapter emphasizes the most recent species totals for the Lepidoptera, based on the numbers of described taxa at the superfamily and family levels, and the way these totals relate to the present phylogeny of lepidopteran superfamilies and higher taxa. Examples of estimating lepidopteran biodiversity at specific sites and the criteria for establishing and analyzing an inventory of Lepidoptera also are discussed.

PRIMARY NEEDS FOR ENHANCING LEPIDOPTERAN BIODIVERSITY STUDIES

Evidence of large-scale lepidopteran biodiversity can be documented from museum collections. The number of described species comes from the literature, which is based on specimens in museums or sometimes in personal collections. A substantial contribution to systematics is the publication of catalogs that document the names from the literature. By examining these catalogs, biodiversity can be traced through time.

One way to document biodiversity is by collecting specimens and obtaining distributional data from museum specimens. To be most useful to society, locality and other data associated with these specimens should be databased and made publicly available. Of the 1.5 million described species of organisms, detailed distributions are unknown for 86% (Stork 1997), suggesting that if distributional data on specimens in museum collections were documented, all species would have some distributional data at a specific point in time. And the number of species of all described organisms would be known. LINNE (Legacy Infrastructure Network for Natural Environments, www.flmnh.ufl.edu/linne) is an initiative that will promote access to specimens and associated data, using analytical tools (e.g., databases, digital imaging, and internet access) to accelerate taxonomic research and to make available reliable information on biological diversity to science and society. Having access to label data on all specimens of Lepidoptera in natural history museums throughout the world would have a major effect on the ability to track potential invasive species, understand the distribution of agricultural pests, and save time and effort in locating biological control agents. It also would establish a baseline for all future biodiversity research.

The world collections do not have adequate material to document the detailed distributions of most known species. The only regions that approach these detailed distributions are well-collected areas such as Europe, Japan, and North America. Even in North America, disjunct distributions could be the result of inadequate collecting (Lafontaine and Wood 1997). Because of the inadequate collecting in tropical areas, only a fraction of the world's biodiversity is currently preserved in collections.

Another serious impediment to biodiversity research is the lack of a central database containing images of the known lepidopteran type specimens. The type specimens are scattered in museums and private collections throughout the world. Some museums have initiated the task of illustrating their type holdings on the worldwide web. The Museum of Comparative Zoology at Harvard University has begun imaging their entire type collection and presenting the images on a website (mcz-28168.oeb.harvard.edu/mcztypedb.htm). The National Museum of Natural History, Smithsonian Institution, has begun imaging the Geometridae type collection, but it is not yet available on the worldwide web. The problem of type access is not new. Clarke (1955–1970) published an eight-volume set on the Meyrick types in The Natural History Museum, London. Meyrick described more than 14,000 species of Microlepidoptera, of which more than 5000 were illustrated by Clarke. This work had a tremendous influence on future Microlepidoptera systematics. Making type collections accessible would virtually stop the wear and tear on loaning specimens, increase the accuracy of systematics, speed up the descriptive process, and standardize genitalic dissections for better comparisons.

Substantial research on any group of Lepidoptera is needed before a systematist can begin to describe new species. One of the first tasks a researcher must undertake is a study of the types of the species involved, including the holotypes associated with all names, both valid names and synonyms. If images of types were readily available online, a tremendous amount of time and energy could be saved. We do not want to create taxonomic problems like those of the past when types were not routinely examined before describing new species.

Synonyms are often equated with mistakes, though not always. John Ray was a seventeenth-century philosopher and theologian often referred to as the Father of Natural History in Britain. He believed in 'natural theology', the doctrine that the wisdom and power of God could be understood by studying the natural world – His creation (www.ucmp.berkely.edu/history/ray.html). This doctrine was widely accepted

throughout the eighteenth and early nineteenth centuries until Darwin's theory of evolution. Species were believed to be individually created, so that any geographical or local variations often were described in the context of separate species. In the late nineteenth and early twentieth centuries when evolutionary concepts were more widely accepted and larger collections showed the reality of genetic variation, this perceived variation was usually formalized by scientific names, often proposed as varieties and forms. In the middle of the twentieth century, the subspecies concept had replaced the more nebulous terms 'variety' and 'form'. By the end of the twentieth century, the subspecies concept had become less frequently used because of the fluidity of the concept, the lack of consensus on a workable definition, and the overuse of the concept through excessive 'splitting' (D. Lafontaine, personal communication). These concepts were partially to blame for the large amount of synonymy in many groups of Lepidoptera when a formal revision was completed.

The Noctuidae can be analyzed with regard to synonymy rate because they have been cataloged (Poole 1989) and databased. Linnaeus described the first noctuid in 1758. In the eighteenth century, 1376 names were proposed, with a 35.2% synonymy rate. Prominent lepidopteran taxonomists of the time, with their synonymy rates, were Linnaeus (10.9%), Cramer (23.5%), Stoll (20.9%), Denis and Schiffermüller (33.3%), and Fabricius (38%). The nineteenth century saw a significant increase in noctuid descriptions, with 14,339 names proposed and a synonymy rate of 36.9%. Prominent taxonomists of the time, with their synonymy rates included Guenée (27.6%), Herrich-Schäffer (35.0%), Walker (40.2%), Felder and Rogenhofer (44%), and Boisduval (47.9%). In the twentieth century, substantial numbers of noctuids were described, with 20,402 names proposed and a synonymy rate of 23.6%.

A pattern exists for lepidopteran taxonomists active in the first half of the twentieth century. Those working on the North American fauna had high rates of synonymy, such as Smith (47.0%), Barnes and Benjamin (41.9%), and Barnes and McDunnough (23.3%). Schaus and Dyar, working predominantly on the Neotropical fauna, had low rates of synonymy because the fauna was poorly known. Schaus had a 7.6% synonymy rate for 1313 proposed noctuid names. Of the 728 noctuid names proposed by Dyar, 83% were Neotropical and 17% were Nearctic. His Neotropical synonymy rate was 6.9%, but his Nearctic rate was 55.6%. Probably the best taxonomist of the early twentieth century was Hampson, who proposed 4160 names, with a synonymy rate of 13.9%. He described noctuids from all over the world instead of from one or two geographical regions. The synonymy rate dropped significantly for most taxonomists in the last half of the twentieth century. This decrease might have been due to several factors, including better access to and comparisons of types, working in areas with high diversity and a poorly known fauna, and the use of genitalic characters to recognize species. Berio (8.2%) and Viette (5.2%), for example, worked on the African fauna, and Holloway (2%) worked on the Oriental fauna. Species described from the two least-diverse regions, the Nearctic and Palearctic, have the highest percentages of synonyms. In the Palearctic Region, Wiltshire, describing species from the Middle East, had a 32.5% synonymy rate and Ronkay a 51.0% rate. In the Nearctic, Hampson had a synonymy rate of 40.1%, Smith 47.0%, and Dyar 55.6%. Modern Nearctic workers have been more thorough, with lower synonymy rates. Hardwick had a rate of 32.5%, Todd 19.6%, Franclemont 8.9%, and Lafontaine 8.7%; all of these synonymy rates treat subspecies as synonyms.

LEPIDOPTERA BIODIVERSITY

A general knowledge of the higher level classification of the Lepidoptera is necessary to understand lepidopteran biodiversity. The 46 currently recognized superfamilies include 121 families (Carter and Kristensen 1998,

Lafontaine and Fibiger 2006). The classification summarized here is derived from morphological characters, and the phylogenetic trees in Figs. 13.1–13.3 are modified from those of Kristensen and Skalski (1998).

The Microlepidoptera comprise an artificial grouping of moth families that are considered more primitive and are usually the 'smaller moths'. No synapomorphic characters define the Microlepidoptera. The Macrolepidoptera is also an unnatural group, and includes the 'butterfly assemblage' of superfamilies (Papilinoidea, Hesperioidea, Hedyloidea, Calliduloidea, and Axioidea), plus the 'macromoths' (Mimallonoidea, Lasiocampoidea, Bombycoidea, Drepanoidea, Geometroidea, and Noctuoidea) (Kristensen and Skalski 1998). More systematic work is needed to understand the higher level classification of the Lepidoptera.

The data in Table 13.1 are compiled from numerous sources, which are detailed in the discussion of each superfamily and can be referred to when comparing diversity of the various superfamilies and families in the following discussion. These data are based on numbers of species extracted from the work of Kristensen (1998a). To update these numbers, all new species and new synonyms were obtained for all superfamilies from the Zoological Record, Part D, Lepidoptera, for the years 2000–2004, and the net numbers of new species were added to the figures from Kristensen (1998a). For some groups, recent catalogs were available and were used to obtain species numbers.

Another consideration in determining the number of described species of Lepidoptera concerns the number of synonyms discovered each year. Based on data from the

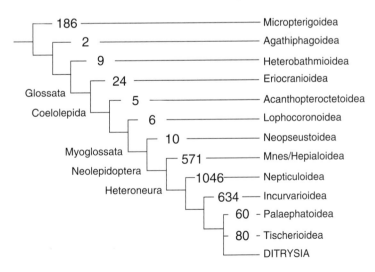

Fig. 13.1 Species richness of superfamilies in the phylogeny of Lower Lepidoptera (Kristensen 1998a). Mnes = Mnesarchaeoidea

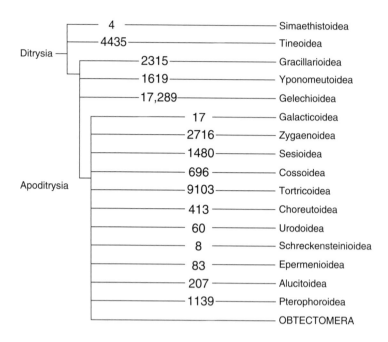

Fig. 13.2 Species richness of superfamilies in the phylogeny of Lower Ditrysia and Apoditrysia (Kristensen 1998a)

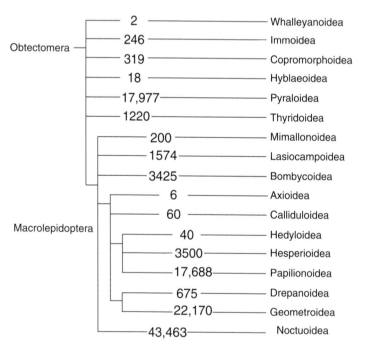

Fig. 13.3 Species richness of superfamilies in the phylogeny of the 'pyraloid grade' and Macrolepidoptera (Kristensen 1998a)

Table 13.1 Numbers of genera and species in superfamilies and families of Lepidoptera.

Superfamily, Family	Number of Genera	Number of Species
MICROPTERIGOIDEA	8	193
Micropterigidae		
AGATHIPHAGOIDEA	1	2
Agathiphagidae		
HETEROBATHMIOIDEA	1	9
Heterobathmiidae		
ERIOCRANIOIDEA	6	25
Eriocraniidae		
ACANTHOPTEROCTETOIDEA	2	5
Acanthopteroctetidae		
LOPHOCORONOIDEA	1	6
Lophocoronidae		
NEOPSEUSTOIDEA	3	10
Neopseustidae		
MNESARCHAEOIDEA	1	8
Mnesarchaeidae		
HEPIALOIDEA	67	625
Anomosetidae	1	1
Neotheoridae	1	1
Prototheoridae	1	12
Palaeosetidae	4	9
Hepialidae	60	602
NEPTICULOIDEA	19	1055
Nepticulidae	11	859
Opostegidae	6	196
INCURVARIOIDEA	74	638
Heliozelidae	12	117
Adelidae	32	278
Prodoxidae	12	89
Cecidosidae	6	7
Incurvariidae	11	139
Crinopterygidae	1	1
PALAEPHATOIDEA	7	60
Palaephatidae		
TISCHERIOIDEA	1	80
Tischeriidae		
SIMAETHISTOIDEA	2	4
Simaethistidae		
TINEOIDEA	373	4388
Tineidae	322	3025
Eriocottidae	5	71
Lypusidae	1	1
Acrolophidae	5	270
Arrhenophanidae	6	10
Psychidae	34	1011
GRACILLARIOIDEA	93	2343
Roeslerstammidae	10	40
Douglasiidae	2	28
Bucculatricidae	4	261
Gracillariidae	77	2014

(continued)

Table 13.1 *(continued)*.

Superfamily, Family	Number of Genera	Number of Species
YPONOMEUTOIDEA	**66**	**1627**
Yponomeutidae	14	597
Ypsolophidae	3	131
Plutellidae	6	53
Acrolepidae	3	95
Glyphipterigidae	19	424
Heliodinidae	14	94
Bedelliidae	1	18
Lyonetiidae	6	215
GELECHIOIDEA	**1428**	**16,581**
Elachistidae	165	3287
Xyloryctidae	86	1255
Chimabachidae	2	6
Glyphidoceridae	1	151
Schistonoeidae	2	2
Oecophoridae	327	3195
Lecithoceridae	90	527
Batrachedridae	6	100
Deoclonidae	3	4
Coleophoridae	47	1444
Autostichidae	67	308
Peleopodidae	6	25
Amphisbatidae	11	65
Cosmopterigidae	106	1642
Gelechiidae	509	4570
GALACTICOIDEA	**3**	**17**
Galacticidae		
ZYGAENOIDEA	**495**	**2757**
Epipyropidae	3	40
Cyclotornidae	1	12
Himantopteridae	3	40
Anomoeotidae	5	40
Megalopygidae	22	260
Somabrachyidae	2	3
Aididae	2	6
Limacodidae	290	1063
Dalceridae	11	88
Lacturidae	10	138
Heterogynidae	1	10
Zygaenidae	145	1057
SESIOIDEA	**152**	**1427**
Brachodidae	6	100
Sesiidae	112	1152
Castniidae	34	175
COSSOIDEA	**124**	**718**
Cossidae	123	712
Dudgeoneidae	1	6
TORTRICOIDEA	**979**	**9416**
Tortricidae		
CHOREUTOIDEA	**12**	**416**
Choreutidae		

(continued)

Table 13.1 *(continued).*

Superfamily, Family	Number of Genera	Number of Species
URODOIDEA	3	60
Urodidae		
SCHRECKENSTEINIOIDEA	2	8
Schreckensteiniidae		
EPERMENIOIDEA	7	111
Epermeniidae		
ALUCITOIDEA	20	208
Tineodidae	11	19
Alucitidae	9	189
PTEROPHOROIDEA	93	1192
Pterophoridae		
WHALLEYANOIDEA	1	2
Whalleyanidae		
IMMOIDEA	6	245
Immidae		
COPROMORPHOIDEA	29	320
Copromorphidae	9	46
Carposinidae	20	274
HYBLAEOIDEA	2	18
Hyblaeidae		
PYRALOIDEA	1972	18,050
Pyralidae	1067	6233
Crambidae	905	11,817
THYRIDOIDEA	92	1223
Thyrididae		
MIMALLONOIDEA	30	200
Mimallonidae		
LASIOCAMPOIDEA	158	1586
Anthelidae	8	74
Lasiocampidae	150	1512
BOMBYCOIDEA	465	3583
Eupterotidae	50	298
Bombycidae	40	351
Endromidae	2	2
Mirinidae	1	2
Saturniidae	165	1535
Carthaeidae	1	1
Lemoniidae	2	22
Brahmaeidae	5	24
Sphingidae	200	1348
AXIOIDEA	2	6
Axiidae		
CALLIDULOIDEA	8	60
Callidulidae		
HEDYLOIDEA	1	40
Hedylidae		
HESPERIOIDEA	546	3500
Hesperiidae0		
PAPILIONOIDEA	1468	13,961
Papilionidae	26	612
Pieridae	74	1049

(continued)

Table 13.1 *(continued)*.

Superfamily, Family	Number of Genera	Number of Species
Lycaenidae	727	6528
Nymphalidae	641	5772
DREPANOIDEA	**129**	**677**
Epicopeiidae	9	25
Drepanidae	120	652
GEOMETROIDEA	**2070**	**21,831**
Sematuridae	6	40
Uraniidae	90	698
Geometridae	1974	21,093
NOCTUOIDEA	**5628**	**45,890**
Oenosandridae	4	8
Doidae	2	6
Notodontidae	736	3546
Micronoctuidae	40	300
Noctuidae	4846	42,030
TOTAL	**16,650**	**155,181**

Zoological Record from 1995 to 2006, 11,819 species were described and 5041 synonyms were proposed, for a net gain of 6778 species (Table 13.2). The average synonymy rate for this period was 42.7%. The percentage of synonyms is unknown for estimates given by Kristensen (1998a).

LEPIDOPTERA CLASSIFICATION

Non-Glossatan Lepidoptera

The Non-Glossatan moths represent a convenient grouping that includes the most ancient clades in the Lepidoptera that have mandibles and lack a coilable proboscis, which is apomorphic for all other Lepidoptera. These ancient clades represent only 0.1% of lepidopteran biodiversity. The three clades share many plesiomorphic traits, including a head with rough vestiture of hairlike scales, mandibles

with strong muscles and well-developed articulations with the head capsule, and five-segmented maxillary palps that are long and folded (Kristensen 1998b).

Micropterigoidea The clade considered the most primitive is the Micropterigoidea, which includes a single family, the Micropterigidae, with 8 genera and 193 described species (Table 13.1). Probable autapomorphies include antennae with ascoid sensilla, the labrum extensively desclerotized, and strong asymmetrical mandibles with an apical incisor cusp on the left mandible (Kristensen 1998b). This family is by far the most diverse of the primitive Lepidoptera. The two major lineages of micropterygids are the genus *Micropterix*, in the Palearctic Region, and the *Sabatinca* group of genera, which is most diverse in the Southern Hemisphere, but includes species in North America, Japan, China, and Taiwan. *Micropterix* consists of 67 described species distributed across the Palearctic from

Table 13.2 Numbers of new species described, synonyms proposed, and percentage synonymy for Lepidoptera, 1995–2006 (from Zoological Record).

	1995	1996	1997	1998	1999	2000	2001	2002	2003	2004	2005	2006	Total
New species	1093	1016	892	1019	864	1296	1181	819	937	908	940	854	11,819
Synonyms	353	487	1244	259	226	320	304	430	500	261	498	159	5041
Net total	740	529	352	760	638	976	877	389	437	647	442	695	6778
% Synonymy	32.3	47.9	139.5	25.4	26.2	24.7	25.7	52.5	53.4	28.7	53.0	18.6	42.7

Europe to Japan. The *Sabatinca* group is most diverse in the Southwest Pacific, especially New Caledonia, with 50-plus species, most of which are undescribed (Gibbs 1983, Kristensen 1998b).

Agathiphagoidea The Agathiphagoidea consist of a single family, the Agathiphagidae. Probable apomorphies include the absence of ocelli and antennae with bifurcate sensilla (Kristensen 1998b). The two species in this group were described from Fiji and Queensland, Australia, and have a distribution around the southwestern Pacific in Vanuatu, the Solomon Islands, and New Caledonia (Scoble 1992).

Heterobothmioidea The Heterobothmioidea include the family Heterobathmiidae and the genus *Heterobathmia*, with at least nine species, although only three have been described (Table 13.1). Probable autapomorphies include forewing veins Sc and R simple and forewing scales on the obverse surface with smooth rounded plates on inter-ridge areas (Kristensen 1998b). These moths occur in southern Argentina (Kristensen and Nielsen 1979).

Glossata

The clade Glossata is characterized primarily by a coiled proboscis and includes 99.9% of lepidopteran species. The homoneurous grade includes superfamilies that have similar wing venation in both the fore- and hindwings and have retained the microtrichiated wing surfaces and distinct jugal lobes on the forewings. All but one of the six superfamilies consist of a single family.

Eriocranioidea The most plesiomorphic superfamily of the Glossata is the Eriocranioidea, with the single family Eriocraniidae containing 6 genera and 25 described species (Table 13.1). This group is defined primarily by a proboscis that lacks intrinsic musculature. The Eriocranioidea are restricted to the Holarctic, with about equal species richness in the Nearctic and Palearctic. In the Palearctic, they are most diverse in Western Europe, but two species are recorded from Japan. In the Nearctic, the Eriocranioidea are restricted to the western North American mountain ranges and eastern Appalachian Highlands (Davis 1978).

Coelolepida

The clade Coelolepida includes the Acanthopteroctetoidea, the Lophocoronoidea, and all other Lepidoptera, and is defined primarily by several characters, including 'normal-type' scales on the wings and the 'normal-type' first thoracic spiracle in the adult (Kristensen 1998c).

Acanthopteroctetoidea The Acanthopteroctetoidea contain a single family, the Acanthopteroctetidae, which was originally placed in the Eriocranioidea (Davis 1978). Probable autapomorphies include an absence of an epiphysis on the fore tibia, labial palps greatly shortened, wings narrow, and vein M1 stalked with Rs in both wing pairs (Kristensen 1998c). This family is most diverse in the western USA, with four described species in *Acanthopteroctetes* and a single species described from Crimea in the genus *Catapterix* (Zagulajev and Sinev 1988) (Table 13.1).

Lophocoronoidea The Lophocoronoidea are endemic to Australia and include a single monotypic family, Lophocoronidae, with six described species (Table 13.1). Probable autapomorphies include mandibles completely devoid of musculature and the absence of an epiphysis on the fore tibia (Kristensen 1998c). These moths occur in arid conditions in southern Australia from Coolgardie in Western Australia along the Great Australian Bight to the Eyre Peninsula in South Australia. In Canberra (Australian Capital Territory), they are associated with mallee and dry sclerophyll forests, and with dry sclerophyll forests northwest of Wollongong in New South Wales (Nielsen and Common 1991).

Myoglossata

The Myoglossata clade includes all of the Glossata-Coelolepida except the Acanthopteroctetoidea and Lophocoronoidea. The clade is probably monophyletic, characterized primarily by the proboscis having intrinsic musculature.

Neopseustoidea. The Neopseustoidea are the most primitive superfamily of the Myoglossata, with the single family Neopseustidae including 3 genera and 10 described species (Table 13.1). Probable autapomorphies include paired facial scales restricted to paired lateral, swollen patches and antennae submoniliform or subserrate and usually longer than the forewing (Kristensen 1998c). The Neopseustoidea have a disjunct distribution. In the Old World, they occur in the Assam region of northeastern India, Myanmar, central China, and Taiwan (Davis 1975). In the New World, they are found in the temperate regions of Chile and Argentina (Davis and Nielsen 1980, 1984).

Neolepidoptera

The Neolepidoptera include all the Myoglossata, except the Neopseustoidea, and are characterized by adecticous, obtect pupae and crochet-bearing larval prolegs on abdominal segments 3–6 and 10.

Neolepidoptera – Exoporia

The Exoporia are homoneurous Neolepidoptera and include the two superfamilies Mnesarchaeoidea and Hepialoidea. This clade is a well-established monophyletic group, with the primary character being the unique configuration of the female genital apparatus. Two genital openings are present, one for copulation and the other for oviposition, as in the Ditrysia, but the internal arrangement is different. The Exoporia have no ductus seminalis, so the spermatozoa deposited in the bursa copulatrix must travel to the spermatheca externally from the copulatory pore via the ovipore along a seminal groove or tract between these two openings (Scoble 1992, Kristensen 1998c).

Mnesarchaeoidea   The Mnesarchaeoidea include a single family, the Mnesarchaeidae, and are endemic to New Zealand, with a singe genus and eight species (Table 13.1). Probable autoapomorphies include narrow, laceolate wings; vein Rs1+2 in both wing pairs represented by a single vein; and wing coupling achieved by a band of long piliform scales on the lower side of the forewing dorsum entangling with a corresponding band of piliform scales on the hindwing costa (Kristensen 1998c).

Hepialoidea   The Hepialoidea represent the most diverse superfamily of all primitive Lepidoptera,

with 5 families, 67 genera, and 625 described species (Nielsen et al. 2000) (Table 13.1). A well-recognized autapomorphy is the strong regression of the proboscis, which is at most as long as the head capsule and can be much shorter or absent, and is never truly coilable (Kristensen 1998c). Four of the five families contain only a few genera and species.

Anomosetidae   This monobasic family occurs in Australian rainforests of southern Queensland and New South Wales (Kristensen 1998c).

Neotheoridae   This monobasic family is known from a single incomplete female specimen from Matto Grosso, Brazil (Kristensen 1998c).

Prototheoridae   This small family has a single genus and nine described species (Table 13.1) from southernmost Africa north to Natal (Kristensen 1998c).

Palaeosetidae   This small family includes four genera and nine described species (Table 13.1). They have a disjunct distribution, with a single species in Colombia, two in Queensland, Australia, one in Assam, India, and four in Taiwan and Thailand (Kristensen 1998c).

Hepialidae   This diverse family, with 60 genera and 602 described species (Table 13.1), has a worldwide distribution. The Neotropical Region has 17 genera and 132 species, 29 of which are endemic to southern South America (Nielsen and Robinson 1983). The Nearctic fauna has 2 genera and 20 species (Davis 1983), and is numerically similar to that of Europe, which has 6 genera and 16 species (de Freina 1996). New Zealand is quite diverse for its size, with 7 genera and 27 species (Dugdale 1994). Australia has a diverse fauna of 10 genera and 119 species (Nielsen 1996a).

Heteroneura

The infraorder Heteroneura comprises the monotrysian heteroneuran clade and the Ditrysia. This monophyletic group is characterized by heteroneurous wing venation, retinaculum–frenulum wing-coupling structures, and an unsclerotized first abdominal sternite (Nielsen and Common 1991). Four superfamilies make up the monotrysian Heteroneura, which is a questionably monophyletic group of distantly related groups believed to be the basal lineage of the Heteroneura. These superfamilies retain the primitive monotrysian female reproductive system, with a single opening for both copulation and oviposition (Davis 1998).

Non-Ditrysian Heteroneura

Nepticuloidea The Nepticuloidea are the smallest moths in the order, with forewing lengths of 1.5–8.3 mm. They are also the most diverse, with 19 genera and 1055 described species (Table 13.1). The Nepticuloidea are characterized by the vertex of the head rough, with erect piliform scales; ocelli and chaetostemmata absent; and antennae with the scape typically expanded to form an eyecap (Davis 1998).

Nepticulidae This family has 11 genera and 859 described species (Table 13.1) and is distributed worldwide. Europe has 210 described species (van Nieukerken 1996a), North America north of Mexico has 82 (Davis 1983), New Zealand has 28, of which 27 are endemic (Donnor and Wilkinson 1989), and Australia has 22, but could exceed 300, with descriptions of new species (Nielsen 1996b).

Opostegidae This family has 6 genera and 196 described species (Table 13.1) and is distributed worldwide (Davis and Stonis 2007). Europe has 6 described species (van Nieukerken 1996b), North America north of Mexico has 7 (Davis 1983), and Australia has 19 described species, but could possibly reach 80 species (Nielsen 1996c). Hundreds of new species undoubtedly are yet to be described in this superfamily and the diversity could exceed 1500 species.

Incurvarioidea The Incurvarioidea represent the second most diverse superfamily of the monotrysian heteroneura, with 6 families, 74 genera, and 638 described species (Table 13.1). Some autapomorphies include an extensible, piercing ovipositor; two or more pairs of retractor apodemes; the enclosure of the female eighth abdominal segment largely inside the greatly enlarged seventh sternum; and sagittate juxta in the male (Davis 1998).

Heliozelidae This moderately species-rich family has 12 genera and 117 described species (Table 13.1) of small (forewing length 1.7–7.0 mm), seldom-collected, diurnal moths. They are found worldwide except in Antarctica and New Zealand, and are poorly known in the Neotropical Region (Davis 1998). The family is diverse in North America, with 31 described species (Davis 1983). Europe has 8 described species (Wojtusiak 1996a) and Australia has 36, with a possible fauna of more than 50 species (Nielsen 1996d).

Adelidae This family is the most diverse in the Incurvarioidea, with 32 genera and 278 described species (Table 13.1). The family is distributed worldwide except in Antarctica and New Zealand (Davis 1998). Europe has 52 described species (Wojtusiak 1996b), North America has 19 (Davis 1983), and Australia has 16 (Nielsen 1996e).

Prodoxidae This family includes 12 genera and 89 described species (Table 13.1) that are almost exclusively restricted to the Holarctic Region, with a single species from southern South America. Europe has 20 species (Wojtusiak 1996c) and North America and Mexico have 29 (Davis 1983, Pellmyr et al. 2006).

Cecidosidae This is a small family with six genera and seven described species (Table 13.1) restricted to southern South America (five species) and South Africa (two species).

Incurvariidae This cosmopolitan family has 11 genera and 139 species (Table 13.1), most of which are undescribed. It is most diverse in Australia, with more than 100 species (Nielsen 1996e). An additional 12 occur in Europe (Wojtusiak 1996d), 22 in North America (Davis 1983), 4 in temperate South America (Nielsen and Davis 1981), and 1 in Africa (Scoble 1980).

Crinopterygidae This monobasic family is found in southern France and Sicily (Davis 1998).

Palaephatoidea The Palaephatoidea are represented by a single family, the Palaephatidae, and are distributed equally between South America and

Australia (Davis 1998). The group is relatively small, with 7 genera and 60 described species (Table 13.1). Autapomorphies include the median ridge of the ovipositor covered with campaniform sensilla and a mesal plate on the abdominal prolegs of the larva (Davis 1998).

Tischerioidea   The Tischerioidea include a single family, the Tischeriidae, with a single genus and 82 described species (Table 13.1). The Tischerioidea are small moths with forewing lengths of 2.7–5.0 mm. The vertex of the head is rough with slender to moderately broad scales directed mostly forward, the frons is smooth, and the ocelli are absent (Davis 1998). The greatest biodiversity is in the USA, with 47 described species. The other species are described from the Ethiopian, Neotropical, and Oriental regions; none has been described from the Australasian Region or Oceania (Davis 1998).

The 13 superfamilies treated above are the plesiomorphic, non-ditrysian Lepidoptera, with 2633 species, making up only 1.7% of the total Lepidoptera. The remaining 98.3%, or 152,576 species, belong to the Ditrysia. The ditrysians are characterized primarily by having separate copulatory and oviposition openings in the female. The superfamily relationships in the Ditrysia are mostly unresolved with the use of morphological characters. Perhaps superfamily relationships will become more resolved when a thorough molecular study is completed.

Lower Ditrysia

The lower Ditrysia (Fig. 13.2) include the more primitive ditrysian superfamilies, but constitute a paraphyletic group in that the higher Ditrysia super-families are derived from within the lower ditrysian clade. The superfamilies include the Simaethistoidea, Tineoidea, Gracillarioidea, Yponomeutoidea, and Gelechoidea. They contain 25,662 species, or about 16.5% of the Lepidoptera. They are mostly small moths that have not received the taxonomic attention accorded the larger Lepidoptera. Because of this lack of taxonomic attention and few systematists working on these groups, the total fauna of the lower Ditrysia could easily double and possibly triple.

Simaethistoidea   The   Simaethistoidea   are   a small superfamily with uncertain affinities and are characterized by the lack of apomorphies diagnostic of other ditrysian superfamilies. They are superficially most similar to some Crambidae-Pyraustinae, but lack a scaled proboscis and tympanic organs. They share various characters with a number of superfamiles including the Tineoidea, Alucitoidea, and Thryridoidea, but lack other important apomorphies to place them in any of these groups. A single family, the Simaethistidae, occurs in China, northern India, and Australia and includes two genera and four described species (Table 13.1) (Dugdale et al. 1998b).

Tineoidea   The Tineoidea are considered the most plesiomorphic of the Ditrysia, and include 373 genera and 4388 described species (Table 13.1). Autapomorphies include the presence of a slender pair of ventral pseudapophyses within A10 of most females and an unusually long ovipositor. Other characters that help distinguish the Tineoidea are the presence of erect scales on the frons, labial palps with lateral bristles, and a haustellum with short, disassociated galeae.

Tineidae   This family is the most species rich in the group, with 322 genera and 3025 described species (Table 13.1). The family has a worldwide distribution.

Eriocottidae   The Eriocottidae are mainly an Old World family with 5 genera and 71 described species (Table 13.1) that range from the southern Palearctic to Australia and from southern Africa to Taiwan.

Lypusidae   This family is monobasic and is represented by a single Palearctic species.

Acrolophidae   This New World family includes 5 genera and 270 described species and is most diverse in the Neotropics.

Arrhenophanidae   This small family has 6 genera and 10 described species (Table 13.1) that are found mostly in the New World, with 2 species in a new genus reported from Taiwan.

Psychidae   The bagworm moths constitute a large group, with 34 genera and 1011 described species

(Table 13.1). They have a worldwide distribution, but more than 85% are restricted to the Old World (Davis and Robinson 1998).

Gracillarioidea The Gracillarioidea are represented by 4 families, 92 genera, and 2343 species (Table 13.1). The moths generally are small, with forewing lengths of 2–10 mm. Some apomorphies include the partially extruded pupa from the cocoon prior to eclosion, and abdominal tergal spines and a smoothly scaled frons in the adults (Davis and Robinson 1998).

Roeslerstammidae This small family has 10 genera and 40 described species (Table 13.1) that are mainly from the Australasian, Oriental, and Palearctic regions. One genus is recorded from Colombia, South America (Moriuti 1978).

Douglasiidae This small family has 2 genera and 28 described species (Table 13.1). Its greatest diversity is in the Palearctic Region, with 20 described species; another 8 species occur in the Nearctic and 1 species is in Australia.

Bucculatricidae This family has 4 genera and 261 described species, with virtually all species in the genus *Bucculatrix*, except for 3 monotypic genera, 2 of which are in Australia (Nielsen 1996f) and 1 in South Africa (Davis and Robinson 1998). The Nearctic is the most diverse region, with 100 species in *Bucculatrix*.

Gracillariidae This family is the most diverse in the group, with 77 genera and 2014 described species (Table 13.1). These tiny moths have a worldwide distribution, and account for more than 85% of the species in the superfamily. The larvae are leaf miners. The Palearctic Region is the most diverse, with about 450 described species. The Nearctic has about 280 described species but, including undescribed species, the fauna could exceed 400 species (D. Davis, personal communication).

Yponomeutoidea The Yponomeutoidea are a moderately species-rich group with 66 genera and 1627 described species (Table 13.1). Controversy exists over the monophyly of the Yponomeutoidea, but the principal autapomorphy is the presence, in males, of posterior expansions of pleuron VIII. These pleural lobes enclose the genitalia and are correlated with the size of the genitalia (Dugdale et al. 1998a).

Yponomeutidae This family is the most diverse in the group, with 14 genera and 597 described species and a worldwide distribution (Table 13.1).

Ypsolophidae This diverse family has 3 genera and 131 described species (Table 13.1) in the Holarctic Region.

Plutellidae This family includes 6 genera and 53 described species worldwide (Table 13.1). Larvae feed mostly on Brassicaceae, living in loose webs and skeletonizing leaves (Dugdale et al. 1998a).

Acrolepidae The Acrolepidae were originally included in the Plutellidae, but because of a lack of good synapomorphies, they are treated as a separate family (Dugdale et al. 1998a). They include 3 genera and 95 described species (Table 13.1) and are distributed worldwide, with more than half of the species in the Palearctic.

Glyphipterigidae This moderate-sized group of diurnal moths includes 19 genera and 424 described species worldwide (Table 13.1).

Heliodinidae This small family has 14 genera and 94 described species, with a worldwide distribution (Table 13.1). A majority (90%) of the species are in the New World (Hsu and Powell 2005).

Bedelliidae The Bedelliidae are represented by the genus *Bedellia*, which has radiated in Hawaii. The family consists of 14 species, 10 of which are new and 13 endemic.

Lyonetiidae This leaf-mining family has 6 genera and 215 described species (Table 13.1). It is cosmopolitan but only one species is represented in the Afrotropical Region (Heppner 1991).

Gelechioidea   The Gelechioidea are the most species rich of the Microlepidoptera, with 1428 genera and 16,581 described species (Table 13.1). A single apomorphy defines the Gelechioidea: the haustellum has overlapping scales on the dorsal surface from the base extending variously to half its length. This character, however, also occurs in the Pyraloidea and Choreutoidea; thus, no single apomorphy defines the Gelechioidea, but the combination of the scaled proboscis, lack of abdominal tympanic organs (present in Pyraloidea), and the absence of naked, minute half-segmented maxillary palps (present in Choreutoidea) help define this superfamily. Other characters present in most Gelechioidea are a four-segmented maxillary palp that is scaled and folded over parallel with the base of the haustellum, labial palps upturned with the third segment long and acute, and the head smooth scaled (Hodges 1998).

This group is under-collected and under-studied, with the potential to increase significantly in number of species. The Gelechioidea are the least well described of the large Lepidoptera superfamilies (Hodges 1998). Hodges (1998) presents the percentage of described species in several families and subfamilies of Nearctic Gelechioidea: 30% of the Gelechiidae are described, 10% of the Scythridinae, 20% of the Coleophorinae, 25% of the Momphinae, 60–70% of the Cosmopterigidae, and 30% of the Elachistidae. For the Neotropics, about 50% of the Stenomatinae and less than 20% of the Gelechiidae might be described. The Oecophorinae of Australia are about 50% described (Hodges 1998). The Afrotropical Region is represented by only 1027 described species of Gelechioidea, mostly in the Gelechiidae and Lecithoceridae, which is a small part of the fauna (Vári and Kroon 1986).

There are 15 families in the Gelechioidea, the most of any superfamily. Hodges (1998) reclassified the Gelechioidea at the family level, using morphological characters and phylogenetic analysis. As a result, many former families became subfamilies or families. I, therefore, list the subfamilies included in each of the families of Gelechioidea to aid interpretation of older literature.

Elachistidae   This large family has 165 genera and 3287 described species (Table 13.1) and is distributed worldwide. Subfamilies included are the Stenomatinae, Ethmiinae, Drepressariinae, Elachistinae,

Agonoxeninae, Hypertrophinae, Deuterogoniinae, and Aeolanthinae. Among the most diverse subfamilies is the Stenomatinae, with more than 1200 species and the center of its diversity in the New World, especially the Neotropics. The Drepressariinae are distributed worldwide, but are absent from many island groups and contain more than 600 species. The Ethmiinae and Elachistinae each have about 250 species. The remaining subfamilies have fewer than 100 species (Hodges 1998).

Xyloryctidae   The Xyloryctidae include 2 subfamilies, 86 genera, and 1255 described species (Table 13.1) and are distributed worldwide. The Xyloryctinae are distributed in Africa south of the Sahara, Indoaustralia, and Polynesia, with more than 500 species. The Scythridinae occur worldwide, with more than 700 species. Many species of Scythridinae are undescribed and, using the above calculations based on the number of described species in the Nearctic, more than 4000 species of Scythridinae might exist in the world (Hodges 1998).

Chimabachidae   The Chimabachidae are a small group with two genera and six described species (Table 13.1) distributed in the Palearctic from Western Europe to Japan, with one species introduced to northwestern North America (Hodges 1998).

Glyphidoceridae   This relatively small family contains the genus *Glyphidocera*, with 151 described species (Table 13.1) and a New World distribution (Hodges 1983, Becker 1984, Adamski and Brown 2001, Adamski 2005). This number could easily double, with many undescribed species, especially in the Neotropics (D. Adamski, personal communication).

Schistonoeidae   The Schistonoeidae include two subfamilies, two genera, and two described species (Table 13.1) and are pantropical in distribution (Hodges 1998).

Oecophoridae   Another large family with 327 genera and 3195 described species distributed worldwide (Table 13.1), the Oecophoridae are the dominant group of Lepidoptera in Australia, with more than 2300 species and at least that many undescribed

(Common 1996a). The Oecophorinae are the dominant subfamily, with 300 genera and 3000 species, including nearly 2000 described species in 250 genera in Australia and numerous undescribed species (Common 1996a). The Stathmopodinae are pantropical and subtropical, with a worldwide distribution. This group is most diverse in Australia, with 104 species in 15 genera, and is widely distributed across Australia, particularly in the Northeast (Common 1996a).

Lecithoceridae This moderate-sized family of 90 genera and 527 described species (Table 13.1) is distributed in the southern Palearctic and Africa, but is most diverse in the Indoaustralasian Region. Australia has 47 described and numerous undescribed species (Common 1996b).

Batrachedridae This relatively small family has 6 genera and 100 described species (Table 13.1) distributed worldwide. The family is divided into 2 subfamilies, the Epimarptinae, with a single species from western India, and the Batrachedrinae, which is distributed worldwide, with 5 genera and more than 100 species.

Deoclonidae This small family has only three genera and four described species (Table 13.1). The subfamily Deocloninae has two genera and three species from California to Argentina, and the subfamily Syringopainae is monotypic and occurs from the Mediterranean region to India (Hodges 1998).

Coleophoridae This family is species rich, currently with 47 genera and 1444 described species (Table 13.1). The Nearctic fauna consists of 145 species, or about 20% of the described coleophorid fauna (Hodges 1998), yielding an estimated Nearctic total of 725 species. Four subfamilies are in the Coleophoridae; all were treated as families at one time. The most species rich of these are the Coleophorinae, with 1050 species in 13 genera. They have a worldwide distribution, but are most diverse in the Holarctic. Another predominantly Holarctic group is the Momphinae, with 6 genera and 60 species; a few species occur in the Neotropics and New Zealand. The Blastobasinae are a moderately species-rich group, with more than 300 described species and

a worldwide distribution, primarily in the New World. Many species remain to be described in the Blastobasinae (D. Adamski, personal communication). The Pterolonchinae are a small group of two genera and eight species in the Mediterranean region and South Africa, with introduced species in the Nearctic.

Autostichidae This moderately species-rich family with 4 subfamilies, 67 genera, and 308 described species (Table 13.1) is distributed in the Palearctic Region, Australasia, Polynesia, and the Nearctic Region. The Holcopogoninae are distributed from the Mediterranean eastward in xeric areas, with 18 species in 7 genera. The Autostichinae contain more than 120 species in 18 genera and are distributed in the eastern Palearctic Region, Australasia, and Polynesia. The Symmocinae have 170 species in 42 genera, and are distributed in the Palearctic from the Mediterranean east to China in xeric areas and in the Nearctic Region (Hodges 1998).

Peleopodidae This small family has 6 genera and 25 described species (Table 13.1) in the New World and Oriental Region (Hodges 1998).

Amphisbatidae Another moderately small family with 11 genera and 65 described species (Table 13.1), the Amphisbatidae have a New World and Palearctic distribution (Hodges 1998).

Cosmopterigidae This species-rich group has 106 genera and 1642 described species (Table 13.1) worldwide. The Chrysopeleiinae are distributed worldwide except in Oceania and contain 22 genera and 270 species. The Cosmopteriginae contain the bulk of the species, with 80 genera and 1350 species throughout the world. The Antequerinae have four genera and eight species in North America and England (Hodges 1998).

Gelechiidae This family is the largest in the clade, with 509 genera, 4570 described species, and a cosmopolitan distribution. The 633 described Nearctic species represent about 30% of the gelechiid fauna (Hodges 1998) and 14.8% of the world fauna. Extrapolating gives an estimated 2110 Nearctic species of Gelechiidae and more than 14,000 species worldwide. Four subfamilies make up the Gelechiidae. The Physoptilinae are distributed in India, Java, Borneo, and Australia, and include a single genus and seven species. The Gelechiinae are the most diverse, with 480 genera, 3400 species, and a worldwide distribution. The Dichomeridinae are also worldwide in distribution, with 4 genera and more than 1000 species. The Pexicopiinae are a pantropical group, also found in the Palearctic and New Zealand, with 21 genera and 110 species (Hodges 1998).

Apoditrysia

All Ditrysian moths, except the Lower Ditrysia, are tentatively classified in the clade Apoditrysia and are considered monophyletic based on the shortened apodemes with enlarged bases on sternum II (Minet 1991). The Obtectomera (Fig. 13.3) include all Apoditrysia with the first four pupal abdominal segments immobile and the dorsal edge of the pulvillus in the adult pretarsus having a dorsal lobe or protrusion (Minet 1991). The immobility of the first four abdominal segments was first mentioned by Mosher (1916), but has considerable homoplasy, occurring in the Epermenioidea and Alucitoidea. However, these superfamilies do not have a modified adult pulvillus, which might be a better synapomorphy for the Obtectomera than is the immobility of the first four pupal abdominal segments (Kristensen and Skalski 1998).

Non-Obtectomeran Apoditrysia

The non-obtectomeran Apoditrysia consist of 11 superfamilies (Fig. 13.2) referred to collectively as the 'tortricid grade'. The most diverse of these superfamilies is the Tortricoidea. The classification of the superfamilies in this grade is not completely understood (Kristensen and Skalski 1998).

Galacticoidea   The Galacticoidea consist of a single family with 3 genera and 17 described species (Table 13.1). Autapomorphies for the group have not been well studied, but both sexes have postabdominal modifications, which include tergum VIII in the male forming a hood over the genitalia, and the ostium bursae in the female being a minute orifice on a curved, sclerotized tube that projects from the membrane behind segment VII. The Galacticidae are distributed in the Old World from North Africa through Asia to Australia, with undescribed species in New Caledonia and the Aftrotropics (Dugdale et al. 1998b).

Zygaenoidea   The Zygaenoidea include 12 families, 495 genera, and 2757 described species (Table 13.1). The monophyly of the group is not clearly defined, but the families Aididae, Dalceridae, Limacodidae, Megalopygidae, and Somabrachyidae make up the 'limacodid group' (Epstein et al. 1998). Several characters support the monophyly of this group in both the immature and adult stages. Synapomorphies include crochets on abdominal segments II–VII of the larva, a sculptured eye flange in the pupa, and dense sensilla trichodea on all female legs and an absence of ocelli (Epstein 1996).

Epipyropidae   Larvae of this family are ectoparasites of homopterans. The family is distributed worldwide, mainly in tropical and subtropical regions, but is most diverse in the Indoaustralian Region. The group is relatively small, with 3 genera and 40 described species (Table 13.1). A single species occurs in the Nearctic Region (Davis 1987).

Cyclotornidae   This small family is endemic to Australia, with a single genus and 12 species (Epstein et al. 1998).

Himantopteridae   Known from the Afrotropical and Oriental regions, the Himantopteridae include 3 genera and 40 described species (Table 13.1) (Epstein et al. 1998).

Anomoeotidae   This family has 5 genera and 40 described species (Table 13.1) with an Afrotropical and Oriental distribution (Epstein et al. 1998).

Megalopygidae   This family is endemic to the New World, with 22 genera and 260 described species (Table 13.1). It is especially diverse in the Neotropics (Epstein et al. 1998).

Somabrachyidae   This small family has two genera and three described species (Table 13.1). It is distributed in the Mediterranean Region and South

Africa; many new taxa are yet undescribed from South Africa (Epstein et al. 1998).

Aididae This small family with two genera and six described species (Table 13.1) originally was placed as a subfamily in the Megalopygidae, but is now accorded family rank after a phylogenetic analysis (Epstein 1996). It is endemic to the Neotropics.

Limacodidae This family is a dominant group in the Zygaenoidea, with 290 genera and 1063 described species (Table 13.1). It is distributed worldwide, reaching its greatest diversity in the tropics (Epstein et al. 1998).

Dalceridae This family, with 11 genera and 88 described species, is almost exclusively endemic to the Neotropics, with one species in Arizona (Miller 1994) (Table 13.1).

Lacturidae The Lacturidae are brightly colored nocturnal moths, with 10 genera and 136 described species (Table 13.1). They are found in tropical or subtropical areas in the Afrotropical, Australian, Neotropical, and Oriental regions (Epstein et al. 1998).

Heterogynidae This small monotypic family has 10 described species (Table 13.1) endemic to the western part of the Mediterranean (Epstein et al. 1998).

Zygaenidae A dominant group with 145 genera and 1057 described species (Table 13.1), the Zygaenidae are distributed worldwide, but are most diverse in subtropical and tropical Asia and in the Palearctic, with many more species yet to be described (Epstein et al. 1998).

Sesioidea The Sesioidea are mostly day-flying moths with prominent ocelli. The group contains 3 families, 152 genera, and 1427 described species (Table 13.1). Probable autapomorphies include an ocular diaphragm most strongly pigmented anteriorly, large patagia that extend ventrad beyond the anteroventral corners of the pronotum, and the posterior tendons of the metafurcal apophyses elongated caudad or dosocaudad (Edwards et al. 1998).

Brachodidae These moths, including 6 genera and 100 described species, are mostly diurnal and occur in all geographic realms except the Nearctic (Table 13.1).

Sesiidae Clear-winged moths are infrequently observed in nature. The larvae bore into the branches, roots, stems, trunks, and, in some cases, seeds of their host plants. Many are pests of orchard trees. The family Sesiidae is the most species rich in the superfamily, with 112 genera and 1152 described species distributed worldwide (Table 13.1).

Castniidae These large day-flying moths resemble butterflies, with their knobbed antennae, although current morphological studies have not shown a close relationship with butterflies. The 34 genera and 175 described species (Table 13.1) are found mainly in tropical, subtropical, and warm temperate climates and have a probable Gondwanan origin, as they are distributed in Australia, the Neotropics, and Southeast Asia. No species are known from New Guinea or the Indonesian Islands east of Sumatra (Edwards et al. 1998).

Cossoidea The Cossoidea are large robust moths whose larvae are woodborers. The superfamily contains 2 families, 124 genera, and 718 described species (Table 13.1). The included families are phenetically similar, but none of the suggested apomorphies have been confirmed. The boring and feeding behavior in woody tissue could be a possible synapomorphy for the Cossoidea (Edwards et al. 1998).

Cossidae This family is the predominant member of the group, with 123 genera and 712 described species (Table 13.1) and a worldwide distribution except in New Zealand.

Dudgeoneidae This small group has a single genus and six species distributed in Africa, Madagascar, India, Southeast Asia, New Guinea, and northern Australia (Edwards et al. 1998).

Tortricoidea The Tortricoidea are the most diverse of the non-obtectomeran Apoditrysia, with 979 genera and 9416 described species (Table 13.1) (Brown 2005). The superfamily contains only one family, the

Drepanoidea The Drepanoidea comprise 2 families, 129 genera, and 677 described species (Table 13.1). Synapomorphies include the complete or nearly complete prespiracular tergosternal sclerites at the base of the adult abdomen and the modified mandibles of the larva, each with a large, flat, lateral area delimited ventrally by a well-defined carina (Minet and Scoble 1998).

Epicopelidae This small group consists of 9 genera and 25 described species (Table 13.1) in temperate and tropical Asia

Drepanidae This moderate-sized family has 120 genera and 652 described species (Table 13.1). They are most diverse in the Oriental Region and absent from the Neotropics (Heppner 1991).

Geometroidea The superfamily Geometroidea is the second most species-rich group of Macrolepidoptera, with 2070 genera and 21,831 described species (Table 13.1). A reliable synapomorphy is the shape of the larval labium, whereby the spinneret is shorter than the prementum along its midline (Minet and Scoble 1998). The superfamily has a worldwide distribution.

Sematuridae The Sematuridae comprise a small group with 6 genera and 40 described species (Table 13.1). They are mainly Neotropical, except for a single species in Arizona and one genus in South Africa (Minet and Scoble 1998).

Uraniidae This group has 90 genera and 698 described species (Table 13.1). The Uraniidae are most diverse in the Old World tropics, especially in the Oriental and Australasian regions (Heppner 1991).

Geometridae The geometrids are the second most species-rich family of Lepidoptera, with 1974 genera and 21,093 species (Table 13.1). They have a worldwide distribution, but are most diverse in the tropics and subtropics.

Noctuoidea The Noctuoidea are the most diverse superfamily of Lepidoptera, with 5 families, 5628 genera, and 45,890 described species (Table 13.1). The monophyly of the noctuoids is characterized by the presence of metathoracic tympanal organs and their associated abdominal structures. The function of these tympanal organs is thought to be the perception of echolocation signals of bats, or reception of mating signals (Kitching and Rawlins 1998).

Oenosandridae This small group has four genera and eight described species (Table 13.1) restricted to Australia (Edwards 1996).

Doidae The Doidae consists of two genera and six described species (Table 13.1) distributed from western North America to northern South America.

Notodontidae This species-rich family has 736 genera and 3546 described species (Table 13.1) worldwide that are most diverse in the tropics, especially the Neotropics.

Micronoctuidae This recently proposed family of Noctuoidea (Fibiger and Lafontaine 2005) has 40 genera and 300 described species, with many more undescribed. The most obvious apomorphy is a two-branched cubital vein in the hindwing. The family is known only from the Old World tropics and subtropics, except for a single European species (Fibiger 1997).

Noctuidae A new classification of the Noctuidae, which includes evidence from morphological and molecular characters, treats the families Arctiidae, Lymantriidae, Nolidae, and Pantheidae of Kitching and Rawlins (1998) as subfamilies in the family Noctuidae (Lafontaine and Fibiger 2006). The monophyly of the Noctuidae is supported by the quadrifid condition of the forewing, whereby the Cu vein appears four-branched. This new concept of the Noctuidae includes 5628 genera and 45,890 described species (Table 13.1). The dominant subfamily of the Noctuidae is the Arctiinae, with more than 11,000 species; the Neotropical fauna is large and includes more than half of the species (Heppner 1991). The Noctuidae have a worldwide distribution, with the greatest diversity in the Neotropics where about a third of the world's fauna is found (Heppner 1991).

LEPIDOPTERA INVENTORIES

Lepidoptera are important in biodiversity studies because they are the major group of plant-feeding insects. Inventories can indicate the ecological health of plant communities, and long-term inventories can show changes in these communities over time. Inventories are, for the most part, not comparable in a meaningful way because of differences in collection methods, size of study areas, duration of collecting, and numerous other factors (Powell 1995).

Species inventories should be designed to maximize species richness at any given site. All sampling methods must be used, including black lights with a trap or sheet, mercury vapor lights, baits, pheromones, and larval rearing. Specimens should be given a unique number and a label that minimally includes the collection locality, latitude and longitude, date, and collector. All information should be databased for ease of analysis.

Although comparisons among inventories can be difficult, they can show general trends such as ecological succession. Brown (2001), for example, showed a trend in the faunal turnover of the Tortricidae over a 100-year period on Plummers Island, Maryland. A comparison of the fauna from two decades (1900–1909 and 1990–1999) showed that species richness of 71 from 1900 to 1909 declined to 59 from 1990 to 1999, a reduction of 17%. The faunal turnover was 54%, with 41 apparent species extinctions and 29 apparent species colonizations. The best explanation for these trends involves plant-community succession. Over this 100-year period, the island went from open juniper grassland to subclimax hickory–maple–oak woodland.

Species richness estimators are statistical algorithms used to predict total species richness, with abundance-based and incidence-based estimators (Colwell and Coddington 1994). Abundance-based estimators use the number of individuals for each species in each sample, whereas incidence estimators use the occurrence (presence/absence) of each species in each sample. These estimators are included in the program EstimateS (Colwell 2005). Both types of estimators calculate species richness based on the number of singletons and doubletons. Singletons are the number of species represented by a single individual for the abundance-based estimators and are the number of localities represented by a single species for the incidence estimators. Doubletons represent how many species or localities are represented by two

individuals per species or two species per locality. The maximum species richness is reached when either of these parameters reaches zero, which, in reality, would be unlikely because singletons and doubletons are represented most often by vagrant or nonresident species that happen to be caught.

Collecting effort also affects species richness estimates. This effort can be measured by the number of specimens collected or by the number of samples. The more specimens or samples collected, the fewer the number of singletons. In George Washington and Monongahela National Forests, West Virginia, for example, 135,271 specimens were collected, including 438 species of Lepidoptera, of which 8% were singletons (Butler et al. 2001). In Great Smoky Mountains National Park, 8930 specimens of 492 species of Noctuidae were collected, of which 10.8% were singletons, and at Plummers Island, Maryland, 2489 specimens of 212 species of Noctuidae were collected, of which 22.6% were singletons. As the percentage of singletons (or doubletons) approaches zero, the species richness estimators become more accurate.

Inventories can be used to estimate species numbers at a site by using the species richness of one taxon to extrapolate the species richness of unrelated taxa. For example, species richness of butterflies at Pakitza, Peru, is 1300 species (Robbins et al. 1996). This number is approximately 6.61% of the butterfly fauna of the world (Robbins and Opler 1997), and within the estimate by Powell (1995) for the percentage of butterflies out of all Lepidoptera in various regional samples. Dividing the butterfly fauna (1300) by 6.61% yields an estimated 19,667 species of moths at Pakitza. A brief inventory of the moths of Pakitza produced 1006 species among 1731 individuals in just 7 collecting days. The estimate of 19,667 species of moths is, therefore, reasonable.

Lepidopteran inventories should be planned with a specific goal, whether to obtain maximum species richness at a site or to document ecological change or environmental health of an area over time. The following issues should be addressed when planning an inventory: (1) sampling strategy must be determined, (2) sampling effort should be quantified (e.g., collecting dates, UV samples, rearing lots), (3) first records of each species should be documented to determine species-discovery rates, (4) voucher specimens should be kept for future study, and (5) specimens should receive an individual number corresponding to all associated data (Powell 1995).

National parks, state parks, and other areas of conservation importance should, as a priority, conduct inventories of all of their biological resources. Few data document how many species of Lepidoptera are in any state, county, or other locality in North America. Several states have published lists of Lepidoptera, including Florida (Kimball 1965), Kentucky (Covell 1999), Maine (Brower 1974, 1983, 1984), and New York (Forbes 1923, 1948, 1954, 1960). Other states with well-known faunas, but unpublished lists, include Connecticut, Maryland (Macrolepidoptera and Pyraloidea), Michigan, Mississippi, Ohio, and Texas. Local inventories include state parks, wildlife refuges, and conservation areas such as those in Nantucket and Martha's Vineyard, Massachusetts (Jones and Kimball 1943); Mount Desert Island, Maine (Proctor 1946); Welder Wildlife Refuge, Sinton, Texas (Blanchard et al. 1985); California (McFarland 1965, J. Powell, unpublished data); and Ohio (Metzler 1988, 1989a, 1989b, 1990, Rings and Metzler 1992, Rings et al. 1987, 1991).

Is our most abused national park our most biologically diverse? Great Smoky Mountains National Park is our most-visited and most-polluted (both air and water) national park. The All Taxa Biological Inventory currently being conducted in Great Smoky Mountains National Park will provide important baseline data for monitoring the park's natural resources far into the future. The Lepidoptera inventory of the park has documented more than 1600 species. A similar Lepidoptera inventory has been conducted in Rocky Mountain National Park since 1990, but has recorded only 1200 species (P. Opler, personal communication). Of all national parks, only two are currently conducting basic biodiversity inventories for Lepidoptera.

Documenting biodiversity is the first step in understanding the world's biodiversity. The most immediate threat to lepidopteran diversity is habitat destruction, especially of tropical forests. At least two-thirds of lepidopteran species in the world are in tropical regions (Heppner 1991). Because the majority of Lepidoptera have a close association with plants, either through larval feeding or adult nectaring, tropical deforestation is a serious threat to overall lepidopteran diversity. Inventories of protected areas such as national parks and preserves will provide not only baseline data for which species are present, but also specimens for revisionary systematics. The number of potential species of Lepidoptera has been estimated at 255,000 (Heppner 1991) to about 500,000 (Gaston 1991). Most of these species have not yet been collected. If the majority of new species are in the tropics, these areas must be sampled before they are destroyed. Should our species estimates be based on the rate of tropical deforestation and lack of collecting in these areas? If so, the numbers we currently have might not increase significantly. Are we extirpating thousands of species before they are discovered?

CONCLUSIONS

Butterflies are the most recognized group of insects in the world. They have influenced art, literature, and religious and mythical traditions. Both butterflies and moths have been represented in art from Egyptian temples, in Chinese amulets, and Aztec ceramics, as well as in drawings, gem carvings, glass, paintings, sculptures, and textiles. Lepidoptera have had a symbolic connection to the soul. In Russian, the word for butterflies and moths means 'little soul', or in Greek, simply 'soul'. Pre-Columbian cultures of Central America respected butterflies and moths in religious and mythical traditions in which they represented souls of the dead, new plant growth, the heat of fire, sunlight, and various other transformations of nature (www.answers.com/topic/lepidoptera-1).

The systematist plays the most important role in the study of lepidopteran biodiversity by providing identification guides (field guides, faunal studies, and journal papers), catalogs, checklists, and distributional data. These guides and lists provide an essential background for ecological studies and conservation, especially when butterflies and Macrolepidoptera are used. Being poorly studied, Microlepidoptera have been used far less in ecological or conservation research (New 2004).

Lepidoptera are targeted for individual conservation management and as an effective tool for detecting environmental changes. Understanding habitat requirements of rare or endangered species can result in the manipulation of areas to increase the chances of their survival. Many of the rare and endangered butterflies and moths have restricted habitat requirements (Fowles et al. 2004, Howe et al. 2004) or the caterpillars feed on a single species of host plant (Byers 1989, Waring 2004). Habitats and rare host plants often are threatened by environmental perturbations by humans, including overgrazing, logging (especially in the tropics), draining of wetlands for development, and suppression of fire in prairie and forest communities.

Lepidoptera have a significant effect on human survival, both pro and con. They play a significant role in plant pollination, biological control of weeds, the human diet in many parts of the tropics, and silk production. Conversely, crop losses, forest damage, and the use of pesticides to control lepidopteran pests amount to billions of dollars annually.

To make significant progress in the study of lepidopteran biodiversity, inventorying and monitoring must continue for ecologically sensitive areas of conservation concern. Government agencies and private foundations must support biodiversity research in areas of conservation, ecology, and systematics. With the threat of major climatic and ecological change, baseline inventories will be essential for determining if and when these changes occur.

REFERENCES

Ackery, P. R., R. de Jong, and R. I. Vane-Wright. 1998. The Butterflies: Hedyloidea, Hesperioidea and Papilonoidea. Pp. 263–300. *In* N. P. Kristensen (ed). *Lepidoptera, Moths and Butterflies. Volume I. Evolution, Systematics, and Biogeography*. Walter de Gruyter, Berlin.

Adamski, D. and R. L. Brown. 2001. *Glyphidocera* Walsingham (Lepidoptera: Gelechioidea: Glyphidoceridae) of Cerro de la Neblina and adjacent areas in Amazonas, Venezuela. *Proceedings of the Entomological Society of Washington* 103: 968–998.

Adamski, D. 2005. Review of *Glyphidocera* Walsingham of Costa Rica (Lepidoptera: Gelechioidea: Glyphidoceridae). *Zootaxa* 858: 1–205.

Becker, V. O. 1984. Blastobasidae. Pp. 41–42. *In* J. B. Heppner (ed). *Atlas of Neotropical Lepidoptera Checklist: Part 1, Micropterigoidea–Immoidea*. Dr. W Junk, The Hague.

Blanchard, A., J. E. Gillaspy, D. F. Hardwick, J. W. Johnson, R. O. Kendall, E. C. Knudson, and J. C. Schaffer. 1985. Checklist of Lepidoptera of the Rob and Bessie Welder Wildlife Refuge, near Sinton, Texas. *Southwestern Entomologist* 10: 195–214.

Brower, A. E. 1974. *A List of the Lepidoptera of Maine. Part 1. The Macrolepidoptera*. Life Sciences and Agricultural Experiment Station, University of Maine, Orono, Maine Technical Bulletin 66. 136 pp.

Brower, A. E. 1983. *A list of the Lepidoptera of Maine. Part 2. The Microlepidoptera. Section 1. Limacodidae through Cossidae*. Maine Department of Conservation, Maine Forest Service, Division of Entomology, Augusta, Maine and the Department of Entomology, Maine Agricultural Experiment Station, University of Maine, Orono, Maine, Technical Bulletin 109. 60 pp.

Brower, A. E. 1984. *A list of the Lepidoptera of Maine. Part 2. The Microlepidoptera. Section 2. Cosmopterigidae through Hepialidae*. Maine Department of Conservation, Maine Forest Service, Division of Entomology, Augusta, Maine and the Department of Entomology, Maine Agricultural Experiment Station, University of Maine, Orono, Maine, Technical Bulletin 114. 70 pp.

Brown, J. W. 2001. Species turnover in the leafrollers (Lepidoptera: Tortricidae) of Plummers Island, Maryland: assessing a century of inventory data. *Proceedings of the Entomological Society of Washington* 103: 673–685.

Brown, J. W. 2005. Tortricidae (Lepidoptera). *World Catalogue of Insects* 5: 1–741.

Butler, L., V. Kondo, and J. Strazanac. 2001. Light trap catches of Lepidoptera in two central Appalachian forests. *Proceedings of the Entomological Society of Washington* 103: 879–902.

Byers, B. A. 1989. Biology and immature stages of *Schinia masoni* (Noctuidae). *Journal of the Lepidopterists' Society* 43: 210–216.

Carter, D. J. and N. P. Kristensen. 1998. Classification and keys to higher taxa. Pp. 27–40. *In* N. P. Kristensen (ed). *Lepidoptera, Moths and Butterflies. Volume 1. Evolution, Systematics, and Biogeography*. Walter de Gruyter, Berlin.

Clarke, J. F. G. 1955–1970. *Catalogue of the Type Specimens of Microlepidoptera in the British Museum (Natural History) Described by Edward Meyrick*. Trustees of the British Museum, London. Volumes I–VIII.

Colwell, R. K. 2005. EstimateS: Statistical estimation of species richness and shared species from samples. Version 7.5. User's Guide and application. http://purl.oclc.org/estimates

Colwell, R. K. and J. A. Coddington 1994. Estimating terrestrial biodiversity through extrapolation. *Philosophical Transactions of the Royal Society of London (Series B)* 345: 101–118.

Common, I. F. B. 1996a. Oecophoridae. Pp. 59–89. *In* E. S. Nielsen, E. D. Edwards, and T. V. Rangsi (eds). *Checklist of the Lepidoptera of Australia*. CSIRO Division of Entomology, Canberra, Australia Monographs on Australian Lepidoptera 4.

Common, I. F. B. 1996b. Lecithoceridae. *In* E. S. Nielsen, E. D. Edwards, and T. V. Rangsi (eds). Pp. 116–117. *Checklist of the Lepidoptera of Australia*. CSIRO Division of Entomology, Canberra, Australia Monographs on Australian Lepidoptera 4.

Covell, C. O. 1999. *Butterflies and Moths (Lepidoptera) of Kentucky: an Annotated Checklist*. Scientific and Technical Series Number 6. Kentucky State Nature Preserves Commission, Frankfort, Kentucky.

Davis, D. R. 1975. Systematics and zoogeography of the family Neopseustidae with the proposal of a new superfamily (Lepidoptera: Neopseustoidea). *Smithsonian Contributions to Zoology* 210: 1–45.

Davis, D. R. 1978. A revision of the North American moths of the superfamily Eriocranioidea with the proposal of a new family, Acanthopteroctetidae (Lepidoptera). *Smithsonian Contributions to Zoology* 251: 1–131.

Davis, D. R. 1983. Exoporia, Hepialoidea, Hepialidae. P. 2. *In* R. W. Hodges, T. Dominick, D. R. Davis, D. C. Ferguson, J. G. Franclemont, E. G. Munroe, and J. A. Powell (eds). *Check List of the Lepidoptera of America North of Mexico.* University Press, Cambridge.

Davis, D. R. 1987. Epipyropidae. Pp. 456–460. *In* F. W. Stehr (ed). *Immature Insects.* Volume 1. Kendall/Hunt, Dubuque, Iowa.

Davis, D. R. 1998. The Monotrysian Heteroneura. Pp. 65–90. *In* N. P. Kristensen (ed). *Lepidoptera, Moths and Butterflies.* Volume 1. Evolution, Systematics, and Biogeography. Walter de Gruyter, Berlin.

Davis, D. R. and E. S. Nielsen. 1980. Descriptions of a new genus and two new species of Neopseustidae from South America, with discussion of phylogeny and biological observations (Lepidoptera: Neopseustoidea). *Steenstrupia* 6: 253–289.

Davis, D. R. and E. S. Nielsen. 1984. The South American neopseustid genus *Apoplania* Davis: a new species, distribution records and notes on adult behaviour (Lepidoptera: Neopseustina). *Entomologica Scandinavica* 15: 497–509.

Davis, D. R. and J. R. Stonis. 2007. Biodiversity and systematics of the New World plant mining moths of the family Opostegidae (Lepidoptera: Nepticuloidea). *Smithsonian Contributions to Zoology* 625: 1–212.

Davis, D. R. and G. S. Robinson. 1998. Tineoidea and Gracillarioidea. Pp. 91–117. *In* N. P. Kristensen (ed). *Lepidoptera, Moths and Butterflies.* Volume 1. Evolution, Systematics, and Biogeography. Walter de Gruyter, Berlin.

de Freina, J. F. 1996. Hepialoidea, Hepialidae. Pp. 20–21. *In* O. Karsholt and J. Razowski (eds). *The Lepidoptera of Europe: A Distributional Checklist.* Apollo Books, Stenstrup, Denmark.

de Jong, R., R. I. Vane-Wright, and P. R. Ackery. 1996. The higher classification of butterflies (Lepidoptera): problems and prospects. *Entomologica Scandinavica* 27: 65–101.

Donnor, H. and C. Wilkinson. 1989. Nepticulidae (Insecta: Lepidoptera). *Fauna of New Zealand* 16: 1–88.

Dugdale, J. S. 1994. Hepialidae (Insecta: Lepidoptera). *Fauna of New Zealand* 30: 1–163.

Dugdale, J. S., N. P. Kristensen, G. S. Robinson, and M. J. Scoble. 1998a. The Yponomeutoidea. Pp. 119–130. *In* N. P. Kristensen (ed). *Lepidoptera, Moths and Butterflies.* Volume 1. Evolution, Systematics, and Biogeography. Walter de Gruyter, Berlin.

Dugdale, J. S., N. P. Kristensen, G. S. Robinson, and M. J. Scoble. 1998b. The smaller microlepidopterous-grade superfamilies. Pp. 216–232. *In* N. P. Kristensen (ed). *Lepidoptera, Moths and Butterflies.* Volume 1. Evolution, Systematics, and Biogeography. Walter de Gruyter, Berlin.

Edwards, E. D. 1996. Oenosandridae. P. 271. *In* E. S. Nielsen, E. D. Edwards, and T. V. Rangsi (eds). *Checklist of the Lepidoptera of Australia.* CSIRO Division of Entomology, Canberra, Australia Monographs on Australian Lepidoptera 4.

Edwards, E. D. and K. D. Fairey. 1996. Anatelidae. Pp. 258–260. *In* E. S. Nielsen, E. D. Edwards, and T. V. Rangsi (eds). *Checklist of the Lepidoptera of Australia.* CSIRO Division of Entomology, Canberra, Australia Monographs on Australian Lepidoptera 4.

Edwards, E. D., P. Gentili, M. Horak, N. P. Kristensen, and E. S. Nielsen. 1998. The Cossoid/Sesioid Assemblage. Pp. 181–197. *In* N. P. Kristensen (ed). *Lepidoptera, Moths and Butterflies.* Volume 1. Evolution, Systematics, and Biogeography. Walter de Gruyter, Berlin.

Epstein, M. E. 1996. Revision and phylogeny of the limacodid-group families, with evoluntionary studies on slug caterpillars (Lepidoptera: Zygaenoidea). *Smithsonian Contributions to Zoology* 582: 1–102.

Epstein, M. E., H. Geertsema, C. M. Naumann, and G. M. Tarman. 1998. The Zygaenoidea. Pp. 159–180. *In* N. P. Kristensen (ed). *Lepidoptera, Moths and Butterflies.* Volume 1. Evolution, Systematics, and Biogeography. Walter de Gruyter, Berlin.

Erwin, T. L. 1982. Tropical forests, their richness in Coleoptera and other arthropod species. *Coleopterists Bulletin* 36: 74–75.

Erwin, T. L. 2004. The biodiversity question: how many species of terrestrial arthropods are there? Pp. 259–269. *In* M. D. Lowman and H. B. Rinker (eds). *Forest Canopies,* Second Edition. Elsevier Academic Press, Burlington.

Fibiger, M. 1997. *Micronoctua karsholti* gen. et sp. n.: an astonishing small nocutid moth (Noctuidae). *Nota Lepidopterologica* 20: 23–30.

Fibiger, M. and J. D. Lafontaine. 2005. A review of the higher classification of the Noctuoidea (Lepidoptera) with special reference to the Holarctic fauna. *Esperiana* 11: 1–205.

Forbes, W. T. M. 1923. The Lepidoptera of New York and neighboring states: primitive forms Microlepidoptera, pyraloids, bombyces. *Cornell University Agricultural Experiment Station Memoir* 68: 1–729.

Forbes, W. T. M. 1948. The Lepidoptera of New York and neighboring states. Part II. Geometridae, Sphingidae, Notodontidae, Lymantriidae. *Cornell University Agricultural Experiment Station Memoir* 274: 1–263.

Forbes, W. T. M. 1954. The Lepidoptera of New York and neighboring states. Part III. Noctuidae. *Cornell University Agricultural Experiment Station Memoir* 329: 1–433.

Forbes, W. T. M. 1960. The Lepidoptera of New York and neighboring states. Part IV. Agaristidae through Nymphalidae, including butterflies. *Cornell University Agricultural Experiment Station Memoir* 371: 1–188.

Fowles, A. P., M. P. Bailey, and A. D. Hale. 2004. Trends in the recovery of a rosy marsh moth *Coenophila subrosea* (Lepidoptera, Noctuidae) population in response to fire and conservation management on a lowland raised mire. *Journal of Insect Conservation* 8: 149–158.

Gaston, K. J. 1991. The magnitude of global insect species richness. *Conservation Biology* 5: 283–296.

Gibbs, G. W. 1983. Evolution of Micropterigidae (Lepidoptera) in the SW Pacific. *GeoJournal* 7: 505–510.

Gielis, C. 2003. *World Catalogue of Insects. Volume 4. Pterophoroidea and Alucitoidea (Lepidoptera)*. Apollo Books, Stenstrup, Denmark.

Hammond, P. 1992. Species inventory. Pp. 17–39. *In* B. Groombridge (ed). *Global Biodiversity: Status of Earth's Living Resources*. Chapman and Hall, London.

Heppner, J. B. 1991. Faunal regions and the diversity of Lepidoptera. *Tropical Lepidoptera* 2: (Supplement 1), 1–85.

Hodges, R. W. 1983. Blastobasidae. Pp. 14–15. *In* R. W. Hodges, T. Dominick, D. R. Davis, D. C. Ferguson, J. G. Franclemont, E. G. Munroe, and J. A. Powell (eds). *Check List of the Lepidoptera of America North of Mexico*. E. W. Classey Ltd., Farringdon, UK.

Hodges, R. W. 1998. The Gelechioidea. Pp. 131–158. *In* N. P. Kristensen (ed). *Lepidoptera, Moths and Butterflies. Volume 1. Evolution, Systematics, and Biogeography*. Walter de Gruyter, Berlin.

Hodkinson, I. D. and D. Casson. 1991. A lesser predilection for bugs, Hemiptera (Insecta) diversity in tropical forests. *Biological Journal of the Linnean Society* 43: 101–109.

Horak, M. 1998. The Tortricoidea. Pp. 199–215. *In* N. P. Kristensen (ed). *Lepidoptera, Moths and Butterflies. Volume 1. Evolution, Systematics, and Biogeography*. Walter de Gruyter, Berlin.

Howe, M. A., D. Hinde, D. Bennett, and S. Palmer. 2004. The conservation of the belted beauty *Lycia zonaria britannica* (Lepidoptera, Geometridae) in the United Kingdom. *Journal of Insect Conservation* 8: 159–166.

Hsu, Y.-F. and J. A. Powell. 2005. Phylogenetic relationships within Heliodinidae and systematics of moths formerly assigned to *Heliodines* Stainton (Lepidoptera: Yponomeutoidea). *University Publications in Entomology* 124: 1–158.

Jones, F. M. and C. P. Kimball. 1943. The Lepidoptera of Nantucket and Martha's Vineyard. *Publications of the Nantucket Maria Mitchell Association* 4: 1–217.

Kimball, C. P. 1965. *Lepidoptera of Florida. Arthropods of Florida and Neighboring Land Areas*. Volume 1. Division of Plant Industry, State of Florida Department of Agriculture, Gainesville, Florida.

Kitching, I. J. and J. E. Rawlins. 1998. The Noctuoidea. Pp. 355–401. *In* N. P. Kristensen (ed). *Lepidoptera, Moths and Butterflies. Volume 1. Evolution, Systematics, and Biogeography*. Walter de Gruyter, Berlin.

Kristensen, N. P. (ed). 1998a. *Lepidoptera, Moths and Butterflies. Volume 1. Evolution, Systematics, and Biogeography*. Walter de Gruyter, Berlin.

Kristensen, N. P. 1998b. The Non-glossatan moths. Pp. 41–49. *In* N. P. Kristensen (ed). *Lepidoptera, Moths and Butterflies. Volume 1. Evolution, Systematics, and Biogeography*. Walter de Gruyter, Berlin.

Kristensen, N. P. 1998c. The homoneurous Glossata. Pp. 51–63. *In* N. P. Kristensen (ed). *Lepidoptera, Moths and Butterflies. Volume 1. Evolution, Systematics, and Biogeography*. Walter de Gruyter, Berlin.

Kristensen, N. P. and E. S. Nielsen. 1979. A new subfamily of moths from South America. A contribution to the morphology and phylogeny of the Micropterigidae, with a generic catalogue of the family (Lepidoptera: Zeugloptera). *Steenstrupia* 5: 69–147.

Kristensen, N. P. and A. W. Skalski. 1998. Phylogeny and Palaeontology. Pp. 7–25. *In* N. P. Kristensen (ed). *Lepidoptera, Moths and Butterflies. Volume 1. Evolution, Systematics, and Biogeography*. Walter de Gruyter, Berlin.

Lafontaine, J. D. and M. Fibiger. 2006. Revised classification of the Noctuidae (Lepidoptera). *Canadian Entomologist* 138: 610–635.

Lafontaine, J. D. and D. M. Wood. 1997. Butterflies and moths (Lepidoptera) of the Yukon. Pp. 723–785. *In* H. V. Danks and J. A. Downes (eds). *Insects of the Yukon*. Biological Survey of Canada (Terrestrial Arthropods), Ottawa, Ontario.

Lemaire, C. and J. Minet. 1998. Bombycoidea and their relatives. Pp. 321–353. *In* N. P. Kristensen (ed). *Lepidoptera, Moths and Butterflies. Volume 1. Evolution, Systematics, and Biogeography*. Walter de Gruyter, Berlin.

May, R. M. 1988. How many species are there on earth?. *Science* 241: 1441–1449.

May, R. M. 1990. How many species?. *Philosophical Transactions of the Royal Society of London, Series B* 330: 293–304.

May, R. M. 1992. How many species inhabit the earth? *Scientific American* 267 (4): 18–24.

McFarland, N. 1965. The moths (Macroheterocera) of a chaparral plant association in the Santa Monica Mountains of southern California. *Journal of Research on the Lepidoptera* 4: 43–74.

Metzler, E. H. 1988. Preliminary annotated checklist of the Lepidoptera of Atwood Lake Park, Ohio. *Ohio Journal of Science* 88: 159–168.

Metzler, E. H. 1989a. Preliminary annotated checklist of the Lepidoptera of Mohican State Forest and Mohican State Park Ashland County, Ohio. *Ohio Journal of Science* 89: 78–88.

Metzler, E. H. 1989b. The Lepidoptera of Cedar Bog II. A check list of moths (in part) of Cedar Bog. Pp. 29–32. *In* R. C. Glotzhober, A. Kochman, and W. T. Schultz (eds). *Cedar Bog Symposium II*. Ohio Historical Society, Columbus, Ohio.

Metzler, E. H. 1990. The Lepidoptera of Fowler Woods State Nature Preserve, Richland County, Ohio. *Great Lakes Entomologist* 23: 43–56.

Miller, S. E. 1994. Systematics of the Neotropical moth family Dalceridae (Lepidoptera). *Bulletin of the Museum of Comparative Zoology* 153: 301–495.

Minet, J. 1991. Tentative reconstruction of the ditrysian phylogeny (Lepidoptera: Glossata). *Entomologica Scandinavica* 22: 69–95.

Minet, J. 1998. Axioidea and Calliduloidea. Pp. 257–261. *In* N. P. Kristensen (ed). *Lepidoptera, Moths and Butterflies. Volume 1. Evolution, Systematics, and Biogeography.* Walter de Gruyter, Berlin.

Minet, J. and M. J. Scoble. 1998. The drepanoid/geometroid assemblage. Pp. 301–320. *In* N. P. Kristensen (ed). *Lepidoptera, Moths and Butterflies. Volume 1. Evolution, Systematics, and Biogeography.* Walter de Gruyter, Berlin.

Moriuti, S. 1978. Amphitheridae (Lepidoptera): four new species from Asia, *Telethera blepharacma* Meyrik new to Japan and Formosa, and *Sphenograptis* Meyrick transferred to family. *Bulletin of the University of Osaka Prefecture, Series B* 30: 1–17.

Mosher, E. 1916. A classification of the Lepidoptera based on characters of the pupa. *Bulletin of the Illinois State Laboratory of Natural History* 12: 17–159. [Reprinted, 1969, Entomological Reprint Specialists, East Lansing, Michigan].

Munroe, E. and M. A. Solis. 1998. The Pyralioidea. Pp. 231–256. *In* N. P. Kristensen (ed). *Lepidoptera, Moths and Butterflies. Volume 1. Evolution, Systematics, and Biogeography.* Walter de Gruyter, Berlin.

New, T. R. 2004. Moths (Insecta: Lepidoptera) and conservation: background and perspective. *Journal of Insect Conservation* 8: 79–94.

Nielsen, E. S. 1996a. Exoporia, Hepialoidea, Hepialidae s. lat. Pp. 24–26. *In* E. S. Nielsen, E. D. Edwards, and T. V. Rangsi (eds). *Checklist of the Lepidoptera of Australia.* CSIRO Division of Entomology, Canberra, Australia Monographs on Australian Lepidoptera 4.

Nielsen, E. S. 1996b. Nepticulidae. P. 27. *In* E. S. Nielsen, E. D. Edwards, and T. V. Rangsi (eds). *Checklist of the Lepidoptera of Australia.* CSIRO Division of Entomology, Canberra, Australia Monographs on Australian Lepidoptera 4.

Nielsen, E. S. 1996c. Opostegidae. P. 28. *In* E. S. Nielsen, E. D. Edwards and T. V. Rangsi (eds). *Checklist of the Lepidoptera of Australia.* CSIRO Division of Entomology, Canberra, Australia Monographs on Australian Lepidoptera 4.

Nielsen, E. S. 1996d. Heliozelidae. P. 29. *In* E. S. Nielsen, E. D. Edwards, and T. V. Rangsi (eds). *Checklist of the Lepidoptera of Australia.* Monographs on Australian Lepidoptera 4.

Nielsen, E. S. 1996e. Incurvariidae. P. 31. *In* E. S. Nielsen, E. D. Edwards, and T. V. Rangsi (eds). *Checklist of the Lepidoptera of Australia.* Monographs on Australian Lepidoptera 4.

Nielsen, E. S. 1996f. Bucculatricidae. P. 45. *In* E. S. Nielsen, E. D. Edwards, and T. V. Rangsi (eds). *Checklist of the Lepidoptera of Australia.* Monographs on Australian Lepidoptera 4.

Nielsen, E. S. and I. F. B. Common. 1991. Lepidoptera (moths and butterflies). Pp. 817–915. *In* Division of Entomology (ed). *Insects of Australia.* Volume II. Commonwealth Scientific and Industrial Research Organization. Melbourne University Press, Carlton, Australia.

Nielsen, E. S. and D. R. Davis. 1981. A revision of the Neotropical Incurvariidae s. str. with the description of two new genera and two new species (Lepidoptera: Incurvarioidea). *Steenstrupia* 7: 25–57.

Nielsen, E. S. and G. S. Robinson. 1983. Ghost moths of southern South America. *Entomonograph* 4: 1–192.

Nielsen, E. S., G. S. Robinson, and D. L. Wagner. 2000. Ghostmoths of the world: a global inventory and bibliography of the Exoporia (Mnesarchaeoidea and Hepialoidea) (Lepidoptera). *Journal of Natural History* 34: 823–878.

Nieukerken, E. J. van. 1996a. Nepticulidae. Pp. 21–27. *In* O. Karsholt and J. Razowski (eds). *The Lepidoptera of Europe: A Distributional Checklist.* Apollo Books, Stenstrup, Denmark.

Nieukerken, E. J. van. 1996b. Opostegidae. P. 27. *In* O. Karsholt and J. Razowski (eds). *The Lepidoptera of Europe: A Distributional Checklist.* Apollo Books, Stenstrup, Denmark.

Novotny, V., Y. Basset, S. E. Miller, G. D. Welbens, B. Bremer, L. Cizek, and P. Drozd. 2002. Low host specificity of herbivorous insects in a tropical forest. *Nature* 416: 841–844.

Ødergaard, F. 2000. How many species of arthropods? Erwin's estimate revised. *Biological Journal of the Linnean Society* 71: 583–597.

Pellmyr, O., M. Balcázar-Lara, D. M. Althoff, K. A. Segraves, and J. Leebens-Mack. 2006. Phylogeny and life history evolution of *Prodoxus* yucca moths (Lepidoptera: Prodoxidae). *Systematic Entomology* 31: 1–20.

Poole, R. W. 1989. *Fascicle 118, Noctuidae. Lepidopterorum Catalogus (New series).* E. J. Brill and Flora and Fauna Publications, Leiden.

Powell, J. A. 1995. Lepidoptera inventories in the continental United States. Pp. 168–170. *In* E. T. LaRoe, G. S. Faris, C. E. Puckett, P. D. Doran, and M. J. Mac (eds). *Our Living Resources.* United States Department of the Interior, Washington, DC.

Proctor, W. 1946. *Biological Survey of the Mount Desert Region. Part VII. The Insect Fauna.* Wistar Institute of Anatomy and Biology, Philadelphia, Pennsylvania.

Rings, R. W., R. M. Ritter, R. W. Hawes, and E. H. Metzler. 1987. A nine-year study of the Lepidoptera of The Wilderness Center, Stark County, Ohio. *Ohio Journal of Science* 87: 55–61.

Rings, R. W., E. H. Metzler, and D. K. Parshall. 1991. A checklist of the Lepidoptera of Fulton County, Ohio with special reference to the moths of Goll Woods State Nature Preserve. *Great Lakes Entomologist* 24: 265–280.

Rings, R. W. and E. H. Metzler. 1992. A check list of the Lepidoptera of Beaver Creek State Park, Columbiana County, Ohio. *Great Lakes Entomologist* 25: 115–131.

Robbins, R. K. and P. A. Opler. 1997. Butterfly diversity and a preliminary comparison with bird and mammal diversity. Pp. 69–82. *In* M. L. Reaka-Kudla, D. E. Wilson, and E. O. Wilson (eds). *Biodiversity II: Understanding and Protecting Our Biological Resources*. Joseph Henry Press, Washington, DC.

Robbins, R. K., G. Lamas, O. H. H. Mielke, and M. A. Casagrande. 1996. Taxonomic composition and ecological structure of the species-rich butterfly community at Pakitza, Parque Nacional del Manu, Peru. Pp. 217–252. *In* D. E. Wilson and A. Sandoval (eds). *Manu, the Biodiversity of Southeastern Peru*. Smithsonian Institution, Washington, DC.

Sabrosky, C. W. 1952. How many insects are there? *In* A. Strefferud (ed). *Yearbook of Agriculture 1952: Insects*. Government Printing Office, Washington, DC.

Scoble M. J. 1980. A new incurvariine leaf-miner from South Africa, with comments on structure, life-history, phylogeny, and the binominal system of nomenclature (Lepidoptera: Incurvariidae). *Journal of the Entomological Society of Southern Africa* 43: 77–88.

Scoble, M. J. 1986. The structure and affinities of the Hedyloidea: a new concept of the butterflies. *Bulletin of the British Museum (Natural History). Entomology Series* 53: 251–286.

Scoble, M. J. 1992. *The Lepidoptera: Form, Function and Diversity*. Oxford University Press, Oxford. 404 pp.

Solis, M. A. 1997. Snout moths: unraveling the taxonomic diversity of a speciose group in the Neotropics. Pp. 231–242. *In* M. L. Reaka-Kudla, D. E. Wilson, and E. O. Wilson (eds). *Biodiversity II: Understanding and Protecting Our Biological Resources*. Joseph Henry Press, Washington, DC.

Stork, N. E. 1988. Insect diversity: facts, fiction, and speculation. *Biological Journal of the Linnean Society* 35: 321–337.

Stork, N. E. 1997. Measuring global biodiversity and its decline. Pp. 41–68. *In* M. L. Reaka-Kudla, D. E. Wilson, and E. O. Wilson (eds). *Biodiversity II: Understanding and Protecting Our Biological Resources*. Joseph Henry Press, Washington, DC.

Stork, N. E. and K. J. Gaston. 1990. Counting species one by one. *New Scientist* 127 (1729): 43–47.

Vári, L. and D. Kroon. 1986. *Southern African Lepidoptera*. Lepidopterists' Society of Southern Africa and the Transvaal Museum, Pretoria. 198 pp.

Waring, P. 2004. Successes in conserving the Barberry Carpet moth *Pareulype berberata* (D. and S.) (Geometridae) in England. *Journal of Insect Conservation* 8: 167–171.

Westwood, J. O. 1833. On the probable number of species of insects in the creation; together with descriptions of several minute Hymenoptera. *Magazine of Natural History and Journal of Zoology, Botany, Mineralogy, Geology, and Meteorology* 6: 116–123.

Wilson, E. O. 1997. Introduction. Pp. 1–3. *In* M. L. Reaka-Kudla, D. E. Wilson, and E. O. Wilson (eds). *Biodiversity II: Understanding and Protecting Our Biological Resources*. Joseph Henry Press, Washington, DC.

Wojtusiak, J. 1996a. Heliozelidae. P. 27. *In* O. Karsholt and J. Razowski (eds). *The Lepidoptera of Europe: A Distributional Checklist*. Apollo Books, Stenstrup, Denmark.

Wojtusiak, J. 1996b. Adelidae. Pp. 28–29. *In* O. Karsholt and J. Razowski (eds). *The Lepidoptera of Europe: A Distributional Checklist*. Apollo Books, Stenstrup, Denmark.

Wojtusiak, J. 1996c. Prodoxidae. Pp. 29–30. *In* O. Karsholt and J. Razowski (eds). *The Lepidoptera of Europe: A Distributional Checklist*. Apollo Books, Stenstrup, Denmark.

Wojtusiak, J. 1996d. Incurvariidae. P. 30. *In* O. Karsholt and J. Razowski (eds). *The Lepidoptera of Europe: A Distributional Checklist*. Apollo Books, Stenstrup, Denmark.

Zagulajev, A. K. and S. Y. Sinev. 1988. Catapterigidae, a new family of lower Lepidoptera (Dacnonypha). *Entomologicheskoe Obozrenie* 67: 593–601. [In Russian; English translation, 1989. *Entomological Review* 68: 35–43].

Part III

Insect Biodiversity: Tools and Approaches

Chapter 14

THE SCIENCE OF INSECT TAXONOMY: PROSPECTS AND NEEDS

Quentin D. Wheeler

International Institute for Species Exploration, Arizona State University,
P.O. Box 876505, Tempe, AZ 85287-6505 USA

'If all mankind were to disappear, the world would regenerate back to the rich state of equilibrium that existed ten thousand years ago. If insects were to vanish, the environment would collapse into chaos'.

—*E. O. Wilson (1985)*

'I do believe that an intimacy with the world of crickets and their kind can be salutary – not for what they are likely to teach us about ourselves but because they remind us, if we will let them, that there are other voices, other rhythms, other strivings and fulfillments than our own'.

—*Howard E. Evans (1968)*

Insect Biodiversity: Science and Society, 1st edition. Edited by R. Foottit and P. Adler
© 2009 Blackwell Publishing, ISBN 978-1-4051-5142-9

Despite unprecedented challenges, the prospects for insect taxonomy have never been brighter. The science of taxonomy is in the midst of transformation, undergoing remarkable theoretical and technological changes. No subdiscipline stands to benefit more from these advances than insect taxonomy, for the simple reasons that most living insect species are unknown to science, most 'known' insect species are poorly and infrequently tested, and there are so many insect species. The translation of Hennig's theories into English (1966) initiated a theoretical revolution, transforming Linnaean classifications and names into testable reflections of phylogenetic history and an improved 'general reference system' (Hennig 1966, Eldredge and Cracraft 1980, Nelson and Platnick 1981, Schoch 1986, Schuh 2000). Hennig's theory was built on traditional strengths of the comparative method of taxonomy, with added or enforced explicitness and testability (Nelson and Platnick 1981). Although the influence of Hennig's *Phylogenetic Systematics* is evident in most biological subdisciplines today, his revolution is far from complete, cut short by a preoccupation with molecular techniques and a focus on global parsimony at the expense of individual character analysis (Rieppel 2004, Nelson 2004, Williams 2004, Wägele 2004, Williams and Forey 2004). A return to unresolved theoretical and practical challenges associated with complex and evolutionarily interesting characters is inevitable and overdue. Taxonomy is as much about character analysis as it is about species descriptions, phylogeny, and classification. Molecular methods have added to the many sources of data routinely synthesized by taxonomists such as morphology, palaeontology, ontogeny, and ethology (Simpson 1961), thus enriching our understanding of 'holomorphology' (Hennig 1966). Molecular data are used to construct cladograms, identify species, and 'fingerprint' individual organisms. Perhaps the greatest contributions of molecular data are yet to come in the form of tests of complex morphological characters by unraveling the 'black box' of developmental biology and bridging genome and phenotype.

Wägele (2004, p. 116) reminds us that 'complexity of characters is the most significant criterion of homology'. Yet our knowledge of complex characters is limited severely. This point is highlighted by the recent discovery of a new organ in the mouse (Terszowski et al. 2006), one of the most intensively anatomically studied species. The literature on few insect species approaches such a detailed level of study. Even if comparably detailed morphological and anatomical studies were completed for over 1 million named insect species, the morphology of 75% of living insect species would remain completely unknown. Morphological characters are testable because they are potentially falsified by contradictory evidence. Although one aspect of contradiction arises from the congruence of characters on a cladogram, 'contradictory evidence is to be derived from a critical discussion of character hypotheses in themselves, not merely from the reciprocal relationships among all characters' (Rieppel 2004, p. 89). The hard work of such thoughtful comparative morphology demands additional attention to complete the Hennigian revolution. This is so because cladistic analysis, like taxonomy generally, is as much about characters as it is about relations among species.

THE WHAT AND WHY OF TAXONOMY

What is taxonomy? That such a question should seem necessary two and a half centuries after Linnaeus is indicative of the myths and misperceptions that surround taxonomy (Wheeler and Valdecasas 2007). Without getting embroiled in formal definitions of taxonomy and systematics (Wheeler 1995a), it is fair to say that an implicit duality in taxonomy contributes to the confusion. Taxonomy is, on one hand, an information science responsible for erecting formal classifications and maintaining names that make possible the storage, retrieval, organization, and communication of billions of facts about millions of species. On the other hand, taxonomy is a hypothesis-driven science concerned with testable predictions about characters, character distributions, species limits, and phylogenetic relationships among species (Wiley 1981, Nelson and Platnick 1981, Schoch 1986, Schuh 2000). Many users of taxonomic information, from agriculturalists to ecologists, simply want the ability to identify species and refer to them by name. Understandably, therefore, such users are frustrated by name changes and the fact that most insect species on earth remain unidentifiable (the basis of the so-called taxonomic impediment). Those name changes, of course, reflect improved hypotheses about species and their relationships that make Linnaean names more, not less, useful (Dominguez and Wheeler 1997). At root, perhaps the greatest cause for misunderstanding arises from the nonexperimental and historical focus of taxonomy (Nelson and Platnick 1981; Wheeler 1995a, 2004). While this does not detract from its standing as a rigorously testable science (Farris 1979, Gaffney 1979, Nelson and Platnick 1981), it does seem to illuminate a misunderstanding among many experimental biologists about what is or is not science.

Anticipated advances in cyber-infrastructure promise a new generation of tools with which to integrate and add efficiency to nearly every aspect of taxonomy (Atkins et al. 2003; Page et al. 2005; Wheeler, 2004, 2008b, 2008c). Seizing this historic development, taxonomy has the opportunity to overcome many of the limitations of the present, including constraints used as excuses for the reduction of funding for studies integrating diverse evidence. As a result, the prospects for successfully exploring the species and character diversity of 'mega-diverse' taxa, such as insects, are better than ever – even in the face of imminent threats of mass species extinctions (Wilson 1985, 1992).

The goals of taxonomy sound deceptively simple: to discover, describe, and critically test species; to describe and interpret the origin and diversification of their characters; to place them in classifications that reflect phylogenetic hierarchy; to make them identifiable; to give them unambiguous names; and to make information about them reliable and accessible. Such goals, however, are not easily attained. In the case of insects, these goals seem particularly difficult. Articulating and testing the numerous hypotheses associated with taxonomy (e.g., characters, species, and monophyletic groups) represents an enormous amount of currently time-consuming work for any taxon. These challenges are magnified by the sheer numbers of insect species. To put this into perspective, Grimaldi and Engel (2005) estimate that about 925,000 species of insects have been described to date (updated to 1,004,898, introduction to this volume) and that those awaiting discovery might number 3 million or more. Their estimate is reasonable but conservative, some authors suggesting an order of magnitude larger number of undescribed living species (e.g., Erwin 1982). Given the tempo of the biodiversity crisis and the fact that existing tools and practices required nearly 250 years to describe about 1 million species of insects, something has to change if entomologists are going to succeed in an exploration of insect biodiversity before a great number of unique species and clades have disappeared.

Reasons to learn our world's species are largely self evident, including the fact that species are the elements that comprise both sustainable ecosystems and evolutionary history (Wheeler 1999). The ability to accurately identify species that are themselves corroborated hypotheses about character distribution is a prerequisite for credible research. Stated bluntly, unless species are known to science and recognizable by scientists, no appreciable progress is possible in basic or applied biology. Prudent steps to assure our security from emerging diseases of humans, plants, and animals; bioterrorism; and the ravages of introduced pestiferous nonnative species similarly depend on reliable taxonomic information.

Reasons to document species and clades soon to be extinct are equally compelling, if somewhat less evident to nontaxonomists. Our ability to interpret newly discovered characters (from morphology to genomics) and to continue to test and refine boundaries of species and the accuracy of phylogenies and classifications depends upon as comprehensive an understanding of species and character diversity as possible. Long after species are extinct in the wild, specimen and tissue collections will continue to permit taxonomists to recognize and fix mistakes as well as improve and refine classifications, names, and hypotheses of characters. As Hennig (1966) rightly observed, the only thing that the vast diversity of life on earth shares in common is evolutionary history. The more comprehensive our collections and complete our knowledge of species, the better our classifications. A relatively complete inventory of species and their distributions also provides invaluable baseline data for detecting increases or decreases in biodiversity, as well as failures and successes in conservation.

We know that insects are among the most spectacularly successful living organisms. We know, too, that without reliable insect taxonomy, the goals of many other fields would be seriously compromised, including those of agriculture, vector biology, ecology, ecosystem science, conservation biology, and genetics, to name only a few. We know that virtually every process and phenomenon in nature can be approached through studies of insects and that, for many purposes, insects provide the best model organisms. Without pursuing the aims of insect taxonomy, no deep understanding of either the evolutionary history of life or ecosystem functions are possible – at least for most terrestrial and freshwater systems. Yet, in spite of the critical importance of advancing insect taxonomy, the field does

Fig. 14.1 *Phobaeticus chani* Bragg, held by Dr. George Beccaloni of London's Natural History Museum, is the longest insect ever recorded (Hennemann and Conle 2008). Photo: Courtesy of Dr. George Beccaloni.

not yet have suitable support for its research agenda, infrastructure, or an adequate workforce.

Beyond the informed guesses of experts such as Grimaldi and Engel (2005), we do not know what or exactly how many species of insects are living today. Although the majority of insect species are less than 10 mm in length, our ignorance is so great that it is not confined to small-sized insects. This point is made impressively by a photograph of Dr. George Beccaloni of the Natural History Museum (London) holding the largest insect ever seen on earth, as measured in total body length (Fig. 14.1). At the time the photograph was taken in 2006, the specimen was of an undescribed species! Imagine the parallel in botany or vertebrate zoology – discovering the equivalent of a redwood tree or blue whale in 2006. We do not yet know enough about insect biodiversity to characterize precisely the magnitude of our ignorance. While many insects awaiting discovery are closely related to known forms, there remain, without doubt, numerous unique subclades to be brought to light. The magnitude of the challenge facing insect taxonomists can be appreciated by comparing the estimated several millions of undescribed insect species to the total number of species in the classes Mammalia (about 5000) and Aves (about 9000). Theodosius Dobzhansky (1973) famously observed that 'Nothing in biology makes sense except in the light of evolution'. With millions of unknown species, there is no hope of casting evolution's light on most of life until many more insect species have been discovered, described, and placed in the context of phylogenetic classifications. Advancing our knowledge of insect biodiversity is a scientific imperative.

Biology is wholly dependent on taxonomy for species identifications, information retrieval, precise words with which to communicate about species and clades, and phylogenetic classifications that predict what characters to expect among species, including those not yet studied. Entomologists can anticipate that any beetle will have elytra, that any true fly will have two developed wings and anteromotorism, and that any species of Holometabola will exhibit complete metamorphosis only because insect classifications reflect phylogeny approximately correctly.

Entomologists of the future, interested in the evolution of some protein, social behavior, physiological tolerance, or other feature of insects, will be dependent on the extent to which taxonomy's agenda is fulfilled in the twenty-first century. The biodiversity crisis threatens much of the evidence of evolutionary history. As species go extinct, so too goes the character evidence of evolutionary history. While it is true that specimens collected today can and will continue to be studied for hundreds of years to come, it is equally true that a growth in taxonomic knowledge will make that inventory process itself far more efficient and successful. Next to comprehensive collections, nothing is more important than nonstop revisionary studies. Without such incessant testing, the reliability of 'known' species gradually decays and both their names and information content slowly become less reliable. Yet the biological community forges ahead in blissful ignorance of the fact that our neglect of descriptive taxonomy is slowly eroding the things that we assume we know and obliterating our chances to explore efficiently the even greater number of species of which we know nothing.

Before Hennig, insect taxonomy was sometimes viewed as little more than a service, providing identifications and names so that other biologists could do and communicate their work. The current emphasis on molecular trees has unwittingly also reduced phylogenetics to a mere service. When cladistic analyses are done as part of taxonomic research, a central goal is the interpretation of characters and character transformations. Today, many 'trees' are presented without enumerated characters because the consumers are other biologists who have some character or process of their own that they are trying to interpret in a historical context. As a mere service, these trees have become divorced from a rigorous and integrative approach to taxonomy and users as well as taxonomists are ultimately less well served. The result is that the general reference system is compromised as the results of phylogenetic analyses are no longer routinely used to improve formal classifications (Wheeler 2004, Franz 2005) or our understanding of morphology (Wheeler 2008d).

Because of the frightening implications of species loss and rapid environmental change, great emphasis is appropriately placed on ecological and conservation biology studies. However, advances in ecology and conservation depend on access to taxonomic information. Unless species are known to exist and can be identified they cannot be understood or conserved. Because of this dependence of environmental sciences on taxonomy, the latter is rightfully seen by many as an environmental science. Conceptually, however, taxonomy is more evolutionary than environmental. Hypotheses about characters, species, and clades are not limited by ecosystem or epoch but instead transcend the geographic, ecological, and temporal constraints on most biological subdisciplines. Thus, taxonomists need access to all relevant museum specimens, without regard to the country or time in which those specimens were collected.

The trend toward increasingly protective laws that limit the collection of specimens is a shortsighted one, deleterious to the advance of taxonomy. Countries that seek to close their borders to species explorers only succeed in assuring a greater level of ignorance about their own faunas. That is because no fauna can be understood properly if studied in isolation. Taxonomic knowledge and museum specimens should be seen and treated as intellectual property of humanity, because no nation can do taxonomy to a high level of excellence without comparative studies of specimens and species found only in other countries. 'Protecting' faunas by prohibiting taxonomic inventories only assures the absence of the reliable taxonomic information needed to make informed and appropriate decisions and policies about the conservation and management of biodiversity.

Such prohibitions on taxonomic collecting are increasingly justified on the basis of ethics: that it is wrong to kill animals great or small. Most entomologists I know respect and deeply care for the insects they study and take no pleasure in killing specimens for the sake of doing so. Rather, properly prepared specimens are essential for the advance of knowledge, and that means collecting and killing specimens for museum collections. Ironically, collecting insects is becoming difficult in many countries while the pest control industry is booming. Anyone morally opposed to killing insects has likely never had head lice or shared an urban apartment with a few thousand cockroaches.

The simple fact is that a single automobile on many stretches of highway kills more insects in a given year than an insect taxonomist is likely to collect for the purpose of scientific study. So-called bug zappers are used in parks where entomologists are denied collecting permits, simultaneously retarding science and indiscriminately killing insects to a questionable end (Nasci et al. 1983). Unless we are prepared to ban automobile travel, concede a lion's share of our crops to pests, and open our homes to numerous insect pests, we shall have a difficult time obtaining the moral high ground needed to oppose insect collecting for science on ethical grounds.

It is not coincidental that Theodore Roosevelt was both an avid outdoorsman and the president who championed the US national park system. Because he knew and enjoyed Nature intimately, he cared enough to conserve it for future generations. Similarly, children encouraged to make insect collections are far more likely to grow up to be voting citizens who appreciate the beauty and importance of insects and who will care whether or not they are conserved and represented in museums. Entomologists face simultaneously the control of the destructive minority of insect species and conservation of the beneficial majority (Samways 2005).

Museums are, first and foremost, taxonomic research resources. Collections are intended to be comprehensive assemblages of species and the variation within and among them. Rare and unusual species are collected preferentially. There is no pretense that collections reflect either local abundance in ecosystems or the shifting gene frequencies of concern to population biologists; both ecology and genetics are approached through field observation and experimentation. Only taxonomy is so heavily tied to collections, as they alone afford the opportunity to compare specimens from many times and places at once. Museums provide crucial historical baseline data for geographic distributions so that we may detect changes in levels of biodiversity. They are a primary data source for host associations and biogeography for most insect taxa. They are essential references for the identification of rare species. They are the best possible documentation of biodiversity in general. Increasingly, we find new uses for museum specimens, from detecting heavy metals in fishes collected a century ago to retrieving DNA from rare taxa to uncovering evidence of past species distributions. With these and many other uses of collections, it is important to keep in mind their primary purpose as the basis of taxonomy.

The most efficient way to do taxonomy is to work in one of a handful of globally comprehensive collections or to temporarily assemble a complete collection of one taxon by borrowing thousands of specimens from many museums. This will remain the case. However, with the advent of remotely operable digital microscopes and high-resolution digital image libraries, we can now envision a future in which type and rare specimens are potentially instantly available for study regardless of where on earth they are. The current process of borrowing such fragile specimens or traveling to many cities to see them takes weeks or months. With cyber-infrastructure, access times for specimens can be compressed to hours or minutes. International consortia of museums can coordinate their collection growth and development in such a way as to be complementary, and assure that collectively their holdings contain as many species and populations as possible, freeing limited time and resources from duplicative efforts. Collection development and invention of new cyber-infrastructure-enabled tools for taxonomy depend on the recognition of the unique needs of taxonomy.

Humanity, one argument goes, has survived to the present day with a fractional knowledge of insect biodiversity. So why is it so important that we document the rest of the world's insect fauna now? Even many biologists question putting resources into naming species that will soon be extinct, arguing that funds be directed instead at saving natural places and deferring descriptive taxonomy until some future date. Environmental changes that threaten the survival of millions of species should be answer enough. It is perhaps worth observing that humanity also managed to survive centuries believing the earth to be flat, the sun to revolve around the earth, and flies to generate spontaneously from rotting meat, while being totally ignorant of atoms or molecules. Does this justify denying funding to physics, astronomy, geography, developmental biology, and molecular genetics simply because experience suggests that we could get by without them? Ignorance may be survivable, but in science it is no virtue. Few other sciences are asked to be content with so much ignorance and fewer still have as much direct relevance to human and environmental welfare. No science faces the prospect of a greater loss of evidence in the decades immediately ahead.

Environments are changing so rapidly and so unpredictably around the world that the idea of conserve-it-first, study-it-later is unrealistic. How can we set conservation priorities, implement conservation plans, or even know when we have a conservation success unless we know what species exist in a given ecosystem from the outset? We might succeed in conserving a few easily recognized and putatively 'key' species, but we cannot know with any certainty just how intact such systems are or are not. Even if unlimited economic resources existed for conservation purposes, it is unlikely that we could conserve anything close to all species. The old, implicit conservation goal of maintaining the status quo is simply untenable (Wheeler, 1995b). Literally saving all species is a vacuous slogan; we face much harder decisions about which and how many species to conserve and how to undertake an ambitious inventory of species. While many branches of biology, including taxonomy, have major roles to play in assuring the continued existence of sustainable tracts of wild places and agroecosystems, taxonomists alone bear the enormous challenge of exploring all species on a rapidly changing planet, including – in some senses, especially – those species soon to become extinct.

Species exploration will move us closer to answers for taxonomy's 'big questions' (Cracraft 2002, Page et al. 2005), as well as countless lesser ones. At the same time, benefits will accrue to other sciences and to humanity as a whole. Among the many reasons to support a major taxonomic study of the world's insect fauna are the following.

Improving Biology's 'General Reference System'
The point of phylogenetic systematic theory was to make classifications and names reflect evolutionary history. As Hennig (1966) correctly reasoned, patterns of descent with modification are the only logical organizing framework for such a general reference system. Put simply, the more species and character states that we know, the better, more complete, more informative, and more corroborated will be our classifications.

Inter-Generational Ethics   Because no future generation will have access to as many species and clades as we do, we have a special responsibility to expand and develop collections. There is no doubt that this is the single most important and noble thing that we can contribute to science and humanity. Because we are aware of the challenge and capable of meeting it, we have an ethical obligation to do so. Our generation is contributing heavily to the decimation of the world's species. While some mitigation is taking place, including a slowing of deforestation in some developing countries, our descendants will face countless challenges due to environmental change. The cost and effort to support taxonomy's goals are not particularly great; the ignorance resulting from inaction will ultimately be far more costly in lost opportunities than any investment in taxonomy and museums.

Fulfilling Our Intellectual Manifest Destiny In completing an inventory of species, we do more than assure that classifications continue to be refined. We fulfill a human intellectual manifest destiny. We have as a species, from our very earliest days, practiced various levels of taxonomy. Initially sorting food from foe and eventually contemplating the mysteries of the origin of species. Through it all, we have been drawn by a deep intellectual curiosity about biodiversity, the patterns of similarities and differences among species, and our place in that diversity. We stand to lose priceless details about the origin and diversification of life and condemn all future generations to wonder what else we might have understood had we bothered to document species diversity with well-chosen museum specimens. Astronomers are rapidly providing answers to similar age-old questions about the heavens. Unlike astronomy, the subjects of taxonomic curiosity have a 'sell-by date' that is rapidly approaching. Anyone who questions the importance of preserving evidence of species, in the form of specimens, should ask himself or herself how valuable it is that we know that dinosaurs ever existed. How impoverished would our thinking be about the origin and diversification of reptiles, birds, and mammals without such fossils? We can assure that curious humans far into the future have a reasonably complete snapshot of the current diversity of species if we inventory, describe, and classify species today. For most of the insect species facing potential extinction, there is effectively no chance that they will leave fossils behind without our helping hand.

Solving Problems Time and again, we return to Nature in search of solutions to problems. Almost without exception, we are rewarded. Insects with their adaptations and attributes are a cornucopia of potential solutions, from biological control agents to sources of chemicals, proteins, behaviors, natural products, physiological processes, and cures for diseases, insects hold incomparable clues. The possibilities for economic returns and human welfare are virtually without limit, yet we have barely scratched the surface. Why? Before accessing possible silks, adhesives, shellacs, and pharmaceuticals from the insects, we must be able to identify them. To explore the properties of millions of species, one must have a roadmap; in taxonomy, that means a predictive classification.

Model Organisms Although insects have served and continue to serve as model organisms – such as *Drosophila melanogaster* in genetics and evolutionary development or *Manduca sexta* in insect endocrinology – biologists have not even begun to exploit the possibilities. Interested in the evolution of social behavior? The greatest nonhuman success stories are found among termites, ants, bees, and wasps. Interested in extreme physiology? Darkling beetles can live up to 20 years in harsh deserts. Want to know more about plasticity in ontogenetic pathways? At various times, populations of *Micromalthus debilis* are bisexual or parthenogenetic, egg laying or live birthing, and sexually mature as adults or larvae (Pollock and Normark 2002). Curious about biogeographic history? Insects can tell us about the spatial history of evolution from scales as fine as individual islands and mountaintops to patterns tied to continental drift. Wonder about rates of evolution? Some fossils reveal insect species unchanged over millions of years, while examples exist of insects that have completed speciation since the Wisconsin glacial maximum about 12,000 years ago.

Aesthetics Even conceding that beauty lies in the eye of the beholder, most people draw great inspiration from the beauty of insects. Few would deny the breathtaking image of a morpho butterfly fluttering in the undergrowth of a tropical forest, a metallic wood-boring beetle alit on a fallen tree, or the lacelike wings of an owlfly by camp light. Who among us is so little a poet as to not be moved to compare the ephemeral nature of our own lives with that of a mayfly whose adult life span is measured in fleeting minutes? As a rule, those insects that people have not appreciated are those they have never seen under the right conditions, sufficiently magnified to reveal their morphological detail, closely enough observed to unmask their remarkable behaviors. Artists, poets, authors, naturalists, hikers, gardeners, and the child in all of us will continue to marvel at the beauty of insect species as long as they survive and, given adequately preserved collections, much, much longer.

Creating the Vocabulary and Syntax of a Language of Biodiversity   Scientific names, the means by which we refer to hypotheses about the kinds of insects, constitute the vocabulary of a sufficiently rich taxonomic language that we are capable of clarity and precision in talking about millions of species. If Linnaean names are the vocabulary of this rich language, then phylogenetic classifications are its syntax, making special sense of these words by virtue of phylogenetic relationships among the biological entities to which they refer. This remarkable system of nomenclature, begun by Linnaeus, works so well that we sometimes take its power and necessity for granted. We should instead celebrate its incredible accomplishments and potential and push for its continued refinement.

Mapping the Biosphere   Imagine exploring the narrow streets of the Latin Quarter in Paris or the vast expanses of the Brazilian Amazon Region without a map. Descriptive taxonomy empowers people to explore and enjoy Nature on their own, up close and personal. Such experiential access to insect species engenders a visceral appreciation for biodiversity that, in turn, builds a basis for supporting conservation efforts, natural history museums, and taxonomy generally.

INSECT TAXONOMY MISSIONS

Systematics Agenda 2000 (Anonymous 1994) restated clearly the tripartite mission of taxonomy: to discover and describe earth's species; to place species in a predictive, phylogenetic classification; and to make species, and all that is known about them, accessible. Discovering and describing species is not, as some have intimated (Janzen 1993), a one-time process. Rather, it is a continuing, arduous process of hypothesis testing. This testing entails both the search for species new to science as well as the repeated critical testing of 'known' species and their characters in light of new specimens, populations, species, and evidence. The falsificationism of Karl Popper (1968; also Gaffney 1979, Nelson and Platnick 1981) seems simplistic for complex experimentalism, but applies far better to elegant scientific hypotheses. Taxonomy uses such elegant hypotheses almost exclusively. All-or-nothing statements about character distributions and clade inclusivity are the norm. Such statements are subject to refutation as new characters and specimens are found.

The second mission, predictive classification, involves the realization of Hennig's (1966) vision of a 'general reference system' as discussed earlier. To make classifications reflect descent with modification, taxonomists undertake cladistic analyses. Although technically possible for any biologist to plug data into phylogenetic software (Godfray 2007), such 'analyses' rarely result in as well-corroborated estimates of cladistic patterns as when such studies are undertaken as objects of interest in and of themselves. Traditionally, the distinction is evident in the extent to which attention is paid to the analysis and interpretation of individual characters. Ontological protestations notwithstanding (de Queiroz 1988), character evidence is what separates testable species from Roswellian extraterrestrial life forms (e.g., Nixon and Wheeler 1990, 1992). Hennig's (1966) admonition that how data are analyzed is more important than their source, and the tradition of including all relevant sources of data (Simpson 1961) applies still and the best taxonomy is integrative (Will et al. 2005).

I interpret the third mission, making what is known of species accessible, in two ways, both resting on the descriptive and classificatory activities in the first two missions. The first involves the application of Linnaean ranks and names in formal classifications. Such nomenclature is absolutely key to clarity and precision in communication and in information search and recovery (Franz 2005). The second involves the use of the best available information systems. In the past, this meant the synthesis and interpretation of taxonomic information in printed publications. It is increasingly coming to mean the use of the Web and electronic means to make information accessible. In addition to the many reasons that Linnaean classifications and names remain our best communication tool after two and one-half centuries (Forey 2002, Nixon et al. 2003), Linnaean names are a superb basis for increasingly sophisticated information search strategies (Patterson et al. 2006).

INSECT TAXONOMY'S GRAND-CHALLENGE QUESTIONS

The 'big questions' in insect taxonomy are the same as those for taxonomy in general (Cracraft 2002, Page et al. 2005). The expansiveness of the taxonomic vision and its grand-challenge questions are not widely appreciated. While taxonomy deals with fine scales, enumerating the elements of ecosystems and making species identifiable at field study sites, it deals, too, with questions on massive scales. Just as ecosystem scientists must account for global carbon budgets, taxonomists must account for the history of billions of characters and millions of species worldwide. Taxonomy is as much a planetary scale science as plate tectonics. Taxa, at least large and diverse ones, seldom recognize geopolitical, ecosystem, or continental borders. To understand a taxon is to transcend such geographic and ecological boundaries and to explore taxa across scales of time from the historic to the geologic. New tools and practices, including those on massively larger scales, are needed to gather, analyze, store, retrieve, and disseminate taxonomic information. Cyber-infrastructure, by opening access to museum specimens through images and distance microscopy, makes it possible to envision a virtual 'species observatory'. With such an 'instrument', taxonomy's big questions can be pursued on every appropriate scale. What are these big questions?

What Is a Species? Insects have played a central role in theories of species and studies of speciation and will continue to provide valuable 'model' organisms in the refinement of our understanding of what species are and how they evolve. The comparatively few arguments for sympatric speciation models have been based largely on insects, and insects are among the archetypal examples of evolutionary radiations, such as Hawaii's drosophilids (Otte and Endler 1989).

Entomologists have attempted to apply most species concepts at one time or another and have proposed several of them. The biological species concept (BSC) was long the most widely cited concept in entomology, but this reflects (in my view) historic inertia rather than convincing theoretical argument in its favor, and I believe has significantly begun to change. The BSC was effectively advocated by architects of the 'new systematics', particularly Ernst Mayr, over a period of six decades (Mayr 1942, 1963, 2000). The concept is based on arguments about interbreeding that introduce process assumptions that today seem unnecessary to recognize species (Nelson and Platnick 1981). As Rosen (1979) observed, interbreeding is a shared ancestral condition that gives little insight into species divergence except through its absence. Botanists never widely accepted the biological species concept and since the publication of Hennig (1966), increasing numbers of zoologists have questioned its general applicability (e.g.,

Wheeler and Platnick 2000). An improved version of this concept was advocated by Hennig (1966; see also Meier and Willmann 2000), but most of the concerns about the concept remain.

The BSC is emblematic of a larger number of concepts, each of which makes some particular assumptions about the underlying processes associated with speciation. While geneticists have convincingly demonstrated a range of causal factors associated with speciation (e.g., Mayr 1963, Otte and Endler 1989), it remains speculative to ascribe a process to a particular case. Recognizing species based on patterns does provide a logical starting point that is consistent with some processes but inconsistent with others (Cracraft 1989). Many of the arguments in the literature about species could be avoided were we to focus initially on patterns that reveal species, and then undertake studies to attempt to understand the processes that might have been responsible for their divergence. A species concept exists that reliably identifies species prior to cladistic analysis and that is consistent with any speciation process.

The Phylogenetic Species Concept (PSC), as articulated by Eldredge and Cracraft (1980), Cracraft (1983), and Nelson and Platnick (1981) and restated by Nixon and Wheeler (1990, 1992) and Wheeler and Platnick (2000), is such a concept. It is applicable to any taxon and any combination of evolutionary processes because it merely recognizes patterns of character distribution that are the outcome of evolutionary history. Two concepts have vied for authority to use the name 'phylogenetic species concept'. One is consistent with phylogenetic theory but not dependent on cladistic analysis (Nelson and Platnick 1981, Cracraft 1983, Nixon and Wheeler 1992, Wheeler and Platnick 2000); the other is more appropriately called the Autapomorphic Species Concept (ASC) in that it claims (unnecessarily) 'monophyly' for species (de Queiroz and Donoghue 1988, Mishler and Theriot 2000). An early version explicitly portrayed species as minimum autapomorphic units (Hill and Crane 1982).

The latter point is of some concern in current efforts to use DNA sequence data to diagnose and recognize species. All species are evolutionarily significant groupings, but not all evolutionarily significant groupings are species. Efforts to recognize every unusually wide gap in similarity (genetic or otherwise) as a species (e.g., Pons et al. 2006) trivializes species into simplistic formulae divorced from an explicit, testable concept such as the PSC. Many such efforts construct branching diagrams using cladistic methods and claim to recognize "monophyletic" groups of individuals as species. Aside from the fact that monophyly was defined by Hennig (1966) as a relationship among species – not populations or individual organisms – two simple examples illustrate why this insistence on 'monophyly' is misguided. First, there is no reason to suppose that ancestral species do not occasionally coexist with one or more of their own daughter species. Such an ancestral species must, by definition, have apomorphies shared also by all of its descendant species. Had autapomorphies evolved in such an ancestral species, it would not be recognizable as an ancestor and would be treated methodologically like any other term. Only when no autapomorphies are known would it be a candidate ancestral species. Second, speciation comes about as a result of extinctions that eliminate ancestral polymorphism (Nixon and Wheeler 1992). Given two alternative morphs in an ancestral species and the extinction of alternate morphs in two allopatric daughter species, in what sense is one of the morphs present in a daughter species more or less apomorphic than the other?

Additional reasons to prefer the PSC relate to applicability and testability. Because it is based on character-distribution patterns, it is compatible with all potential underlying processes, including sexual and asexual reproduction (Wheeler and Platnick 2000). The PSC is testable. Predicted character distributions are tested critically when new specimens or characters are discovered. Because the PSC recognizes the least inclusive or elemental group definable by a unique combination of characters, the PSC is a unit species concept and permits comparisons of taxa shaped by very different causal forces. It is important for evolutionary biology that varied causal factors can be compared. For example, without such a unit concept of species, it is impossible to answer questions such as 'Does asexual evolution result in greater or lesser numbers of descendant species over comparable periods of evolutionary time?' Or, for that matter, can numbers of species in different taxa even be compared? If species of insects mean something evolutionarily different from species of plants or microbes, what possible basis is there to compare the diversity of taxa?

Arguments in favor of pluralism (e.g., Mishler and Donoghue 1982) confuse pattern and process, supposing that different processes demand different species concepts. A simpler solution is to avoid process assumptions in species concepts. This reasoning is familiar to cladists who reconstruct patterns of relative recency of common ancestry by avoiding unnecessary process assumptions (Eldredge and Cracraft 1980, Nelson and Platnick 1981).

What (and How Many) Insect Species Are There?
The appropriate question is 'What species of insects live or have lived on Earth?' Expressing the question instead as 'How many insect species are there?' suggests that the important thing is some absolute number. Because such an absolute number is obviously unknowable, cynical observers may suggest that we might as well simply approximate that number by some indirect measure such as overall gene divergence. While most entomologists would confess a curiosity about that number, they know that the number itself is not all that important. Discovering and describing as many species as possible, on the other hand, is both achievable and meaningful. To succeed in this question we must recognize it as 'big science'. This is a taxonomic question, and a combination of the strengths of existing taxonomy and innovative tools and practices can solve it. We can learn much from the preliminary successes of National Science Foundation's (NSF's) Planetary Biodiversity Inventory (PBI) projects (Knapp 2008, Page 2008) that put together large international teams of taxon specialists and museums to accelerate species discovery and description. The result is the description or redescription of about 5000 species in 5 years.

What Is the Phylogeny of Insects?   With notable exceptions, comparative morphological, paleontological, and molecular studies have confirmed a great deal of the higher classification of insects (Grimaldi and Engel 2005). The relationships among major subdivisions of insects are demonstrably approximately correct, although some enigmatic relationships remain. For example, the position of Strepsiptera relative to other holometabolous orders remains controversial, as does the exact position of the Hymenoptera and the interrelationships among extant Palaeopterous orders. Because three-quarters of living insects are unknown and many exciting fossils continue to be discovered, some surprises no doubt remain. Examples include the Mantophasmatodea (Grimaldi and Engel 2005) and a living species of Lepidotrichidae, a family thought extinct for millions of years (Wygodzinsky 1961). However, it is remarkable that with several thousand new species described each year, the higher classification of insects remains largely intact.

What Are the Histories of Character Transformations in Insects? Among the most intellectually challenging and rewarding aspects of taxonomy is explaining the evolutionary history of complex characters (Cracraft 1981). Insects present intricate and sometimes intuitively improbable characters in astounding abundance, making them incredibly rewarding to study. Even the sexiest questions about character evolution in insects remain controversial. For example, what is the origin of insect wings? Among all animals to have taken to flight, insects are unique in having evolved wings in addition to ancestral ambulatory appendages. The ultimate answer to this question is likely to include clues from paleontology, developmental biology, molecular genetics, and more detailed and inclusive comparative anatomy and morphology. Other major events in insect evolution that demand similarly extensive additional research include tagmatization, endognathy, wing flexion, male and female genitalia, holometaboly, and on and on. Such questions extend downward through all subclades to the species level where incredible 'leaps' sometimes are seen between closely related species. Examples of the latter include a remarkable array of sexual dimorphisms and contrivances by which insects advertise and perceive chemical signals, 'songs', and visual displays. Advances in digital technologies will soon revolutionize how morphological character data are gathered, analyzed, visualized, and communicated.

Where Are Insect Species Distributed? For the vast majority of insect species, what we know of their geographic and ecological distributions are recorded in a few taxonomic publications and databases and on specimen labels. As ecology has appropriately become quantitative and exacting, what used to be called 'natural history' has fallen increasingly to taxonomy. We doubtfully will ever have the resources to complete serious ecological studies of more than a handful of insect species. Taxonomy is, therefore, complementary to ecology, recording broad but shallow insights into insect associations, including simple co-occurrences in ecological and geographic space. Insect geographic distributions matter. Within a single year of the first detection of West Nile virus in the USA, it was obvious that it had the potential to spread rapidly (Rappole et al. 2000). We must establish baseline knowledge of what insect species exist, where, and in what combinations. Otherwise, what hope do we have of efficiently detecting or monitoring pests, vectors, invasive species, instances of bioterrorism, or increases or decreases in biodiversity?

For insects, well-annotated voucher specimens must be deposited in collections. As these collections develop and their data are made retrievable electronically, countless new applications and ways of analyzing those data will emerge. As one example, such data combined with ecological information and remote sensing data can support predictive modeling of invasions of pest species (Roura-Pascual et al. 2006).

How Have Insect Distributions Changed Through Time? Because many insects leave a sparse fossil record, detailed answers to such questions can be difficult to provide even for higher taxa. Among Pleistocene species are subfossils that have provided remarkably detailed pictures of the effect of glacial events on insect populations (e.g., Coope 1979). Chironomid heads from ancient lake sediments have provided another view into changes in insect populations and a novel way to study climate change (Brooks and Birks 2000). What is currently most critical is establishing baseline knowledge of insect distributions so that changes in the future can be detected.

How Can Insect Classifications and Names Be Most Predictive and Informative? Hennig (1966) provided the theoretical answer to this question in his 'Phylogenetic Systematics'. That referential framework is, in its most useful form, a Linnaean classification. Highly informative (Farris 1979), such phylogenetic classifications aid in both information retrieval and prediction (Nelson and Platnick 1981). The challenges ahead are more practical ones related to work on a larger and more international scale, as well as the need to add efficiency to what taxonomists already do well. A renewed emphasis on and support for descriptive taxonomy is immediately and urgently needed.

TRANSFORMING INSECT TAXONOMY

The rate at which human knowledge doubles is accelerating at a blinding speed, yet the rate at which our knowledge of earth's species doubles is slowing. Taxonomy must undergo a fundamental transformation if it is to meet the challenges of the biodiversity crisis. Put in stark terms, we have a matter of decades to discover, describe, and classify three times more insect species than we have described in the past two and one-half centuries. This is literally a race against time. It is a virtual certainty that many species, some the last vestiges of unique branches of the insect tree of life, will soon go extinct (Wilson 1992). Because we get only one chance to document many parts of insect life, we must get it right; we must preserve those aspects of entomological taxonomic practice that are good while accelerating their pace and adding innovations.

Hennig's revolution in taxonomy was cut short (Williams and Forey 2004). Completing it will involve a returned emphasis on character analysis, including morphology (Wheeler 2008d), the relationship between cladograms and Linnaean ranks and names, the adoption of a uniform theory of species, and the integration of all relevant sources of evidence (Will et al. 2005). Hennig's revolution had begun to undo the damage of the New Systematics that had attempted to recast taxonomy in the image of modern genetics (Wheeler 1995a, 2008a). A renewed emphasis on Hennig (1966), combined with cyber tools (Atkins et al. 2003), has the potential to forge a 'New Taxonomy' focused on its own agenda and strengths (Wheeler 2008c).

The concept of cyber-enabled, international, taxon-focused 'knowledge communities' is central to the transformation of taxonomy. Such 'collaboratories' will build on the success of the PBI model (Knapp 2008, Page 2008) and take advantage of the potential of domain-specific cyber-infrastructures for efficient collaboration. All science builds on prior work, of course, but the chain of scholarship is more explicitly evident in taxonomy because of the nomenclatural mandate to consider all relevant names published since 1758. Taxon communities have a strong start and can build upon this centuries-old chain of scholarship, facilitating the acquisition, testing, and growth of knowledge. Where printed libraries have served as the central repository for growth of taxon knowledge for centuries, cyber-enabled taxon 'knowledge banks' will provide for more up-to-date, complete, and reliable information for users.

Challenges associated with building such collaboratories are great. There are technical challenges in fine-tuning cyber-infrastructures to meet the unique needs of taxonomists (particularly in regard to specimen access and character visualization and communication) and users of taxonomic information. There are sociological challenges in team and collaborative work

practices for a discipline with a long tradition of fierce individualism (Knapp 2008). While the challenges are great, there is every reason for optimism. It is entirely realistic to imagine a great taxonomic infrastructure linking the world's taxonomists, research resources, and museums within a decade, utterly transforming the discipline. Foundation stones for this new taxonomy infrastructure are being set in place today.

INSECT TAXONOMY: NEEDS AND PRIORITIES

Many needs of insect taxonomy are not unlike those of other sciences such as educating a new generation of experts and funding research projects. Other needs, however, differ markedly from those of sister biological sciences. These address the special needs for a taxonomy-specific cyber-infrastructure that networks resources required for broadly comparative studies of specimens, species inventories, collection growth and development, unprecedented international collaboration, and revisionary studies of species by taxon 'knowledge communities'. This approach represents a sea change in thinking for insect taxonomists and suggests a number of challenges (e.g., Wheeler and Valdecasas 2005). Let us examine some of these needs.

Education   Although cyber tools will allow us to work more efficiently and collaboratively, taxonomy will remain highly dependent on a well-educated workforce of taxon experts. We need to overcome the political correctness and fads that skew the workforce away from revisionary taxonomy and toward molecular phylogenetics divorced from descriptive work. Rather than coercing all students to work with molecular data, why not encourage and support students to work at the highest levels of excellence in all aspects of taxonomy: morphology, paleoentomology, ontogeny, molecular, comparative ethology, and so forth? We talk a lot about 'interdisciplinarity' and 'transdisciplinarity' and then expect utter conformity among students. Forming teams of specialists who each bring a unique set of skills, knowledge, and perspectives to insect taxonomy will greatly enrich the field and ultimately make it useful to a much wider range of communities. Given the massive amount of work to be done in exploring, describing, and classifying our planet's insect fauna, more than enough work exists for specialists of every kind.

Taxonomy's agenda cannot be realized without the support of others. It is equally imperative that we educate biologists in general to understand the epistemic basis of taxonomy and to distinguish good taxonomy from expedient substitutes. Further, we need to share the wonders of taxonomy with the public at large. Cyber tools will open many opportunities to do so through both formal and informal (public) education activities. Efforts to open access to taxonomic information, such as the recently announced Encyclopedia of Life initiative (www.eol.org and Wilson 1993) will greatly expand appreciation for taxonomy. Such portals to species information depend on active taxonomists for supply of content.

Amateurs have always played a major role in taxonomy and should be invited, encouraged, and enabled to continue to do so into the future. There shall never be a taxonomic workforce sufficiently large to deal with the insects and their relatives adequately. Knowledgeable amateurs can contribute specimens, data, analyses, and insights into most facets of taxonomy. Cybertaxonomy promises to make many of the research resources formerly reserved for professionals openly available to amateurs. This should make insect taxonomy more productive and rewarding to amateurs and enable them to work toward higher standards of excellence.

Last, but not the least, taxonomic education is essential in developing countries with a disproportionate number of undescribed species. The historic north–south divide, with major collections and libraries in Europe and North America and most species in the tropics and southern hemisphere, is rapidly evaporating. Soon, all the same resources will be open to students and taxon experts around the globe via Web-based sources. Building expertise where the bulk of undescribed species live is only common sense and will fuel a much more diverse and interactive world taxonomic workforce.

Planetary Scale Projects and Virtual Species Observatories Taxonomy must be recognized as a planetary scale science. For practical reasons, both collections and many taxonomic studies have been regional in focus. Such fragments of the total picture of species diversity now can be pieced together in cyberspace to enable taxonomists to virtually 'see' species and clades across multiple scales of space and time. The implications are profound for speed and quality of taxonomic work and for the potential to mobilize comprehensive knowledge of large clades for science. Intensive, internationally coordinated megaprojects are needed to tackle large insect taxa. We shall never gain a comprehensive knowledge of major insect taxa unless we mount truly large-scale scientific efforts to inventory, describe, and classify species; unless the world's museums work together to assure comprehensive collections; and unless we pool our collective expertise to speed species exploration. Taxonomy is asking global questions and, therefore, demands large-scale solutions. In reality, we need the full range of project scopes from the individual researcher to planetary scale teams; each has a part to play, and all deserve our support and encouragement. The full potential of virtual 'species observatories' is not yet appreciated but will transform taxonomy.

Cybertaxonomy Infrastructure Generally, what is needed by insect taxonomists is a specialized (taxonomy-specific) cyber-infrastructure that opens access to all those things that a specialist working on a particular taxon needs to work smarter, faster, and better. Rather than attempting a detailed description of what such a taxonomic cyber-infrastructure might look like (but see Page et al. 2005), let us examine some vignettes that illustrate what taxonomy might be like a decade or less from now.

- Electronic sensors scattered through remote regions of Australia alert a heteropterist that accumulated degree-days indicate that host plants are about to flower and she leads a field team to collect poorly known species and document their hosts.
- A team of collectors in the Congo sit around a campfire after a long day of work. Lek clouds of dance flies were sampled that day. A portable, digital, remotely operable microscope permits the world's authority on the group to examine a specimen from his office in Paris. He concludes that it is an exciting find, and the team is requested to collect frozen specimens the following day to support DNA studies.
- A student in Brazil working on her doctoral thesis needs to see a type specimen in London's Natural History Museum (NHM). She checks the NHM's online library of digital images but finds that an important character cannot be seen. On her command, a robotic arm retrieves the specimen with the precision of a watchmaker and places it under a remotely operable digital microscope.

Moments later, she is manipulating the specimen and taking detailed digital images. Because she is the first specialist to look at the specimen, she is in the best position to document key features. Her images answer her question and are added to the NHM's digital library for future researchers.

- A specimen is intercepted at a US port of entry. A shipment worth tens of millions of dollars is immobilized on the docks, its cargo rapidly approaching spoilage. The specimen is mounted on a similar instrument and a specialist at the Systematic Entomology Laboratory in Washington examines it and determines it is harmless. Millions of dollars of wastage are avoided and a problem is solved in minutes that might have taken a day or more by current standards.
- A couple packs their bags for an ecotour of an Amazonian tributary. They have become especially interested in dragonflies and look forward to seeing as many species as possible. They log onto a Web site that is the portal to the world's odonate knowledge community. As they click the 'field guide' option at the dragonfly Web site, software seamlessly incorporates a species named days earlier in a pictorial identification guide and distribution map.
- An ecologist sampling leaf litter in a rainforest in Malaysia needs to identify obscure and numerous species of staphylinid beetles. Bulk samples are placed in an insect processor. Specimens are mounted and labeled automatically and sorted into morphotypes. Those matching the target search image are sent to an expert for human inspection. Limited human resources are targeted where judgment is needed.
- A school child interested in insects makes a collection. She wanders through a garden to the towering sunflowers where she is amazed at the diversity of wasps and flies. Using a handheld device, she snaps a digital photographic image of a beautiful fly. With the click of a button, that image is transmitted via satellite to a computer that matches it up to three possible species for her GIS position. She compares the images and easily identifies the fly.
- A fish ecologist wants to know which species of mayflies are being fed on by fish. He takes a fragment of an insect and sequences a DNA 'barcode' that is compared against a library of known barcodes representing corroborated species and quickly answers his question.

Web-Based Revisions, Taxon Knowledge Communities, and Taxon Knowledge Banks Taxonomic revisions and monographs have long been the gold standard for taxonomic research. Such comprehensively comparative studies are the most efficient way for taxonomists to test large numbers of species and character hypotheses simultaneously and they are the most authoritative sources of taxonomic information. The new model will involve taxon knowledge banks that exist in cyberspace and that hold the sum total of all we know of a taxon. These banks will be maintained by an international community of taxonomists who severally and jointly test, improve, and expand knowledge. As such, taxon knowledge banks will serve the same functions as monographs, yet be dynamic and up-to-date. From such banks, users will request on-demand 'publications' from checklists to field guides, distribution maps, monographs, phylogenies, and homology tables.

Collection Development and Growth No aspect of taxonomy is more important or deserving of adequate support than the development, growth, and care of insect collections. An estimated 3 billion specimens in natural history collections, perhaps half of which are arthropods, are the most extensive documentation of biological diversity. The most important thing that our generation can bequeath to future generations is a comprehensive collection of species. Collections divorced from active taxonomy are soon little more than relics and the information associated with their collections rapidly becomes less and less reliable (e.g., Meier and Dikow 2004).

INTEGRATIVE INSECT TAXONOMY

The relative merits of molecules versus morphology continue to be debated, but this misses the point. It is not a matter of whether classifications should be based exclusively on DNA (Tautz et al. 2003) or not (Lipscomb et al. 2003, Prendini 2005, Wheeler 2005), but rather how the various sources of evidence about insects can best be integrated into taxonomy (Will et al. 2005). This view is not new. Such integration has long been the aim of taxonomy (e.g., Simpson 1961). The fact that most available evidence in the past happened to be morphological has been misconstrued as a deliberate attempt to make classifications dominantly or wholly morphological. Good taxonomic monographs summarize all that is known of the biology of a taxon (e.g., Lent and Wygodzinski 1979) and 'holomorphology' was central to Hennig's (1966) arguments. Molecular evidence is held by some to be necessary for phylogeny reconstruction. This position mirrors a similar, and similarly wrong, position once held by paleontologists who believed fossils to be the only means to reconstruct phylogeny (Nelson 2004). In reality, all relevant sources of evidence are to be prized and pursued.

ACCELERATING DESCRIPTIVE TAXONOMY

Every step in the process of doing descriptive taxonomy can benefit from a taxonomy-specific cyberinfrastructure. Figure 14.2 is an artist's depiction of a 'taxon space' where taxonomists of the future might work. At the core of this space are research resources, such as collections of specimens, digital instruments, visualization and communication tools, robotics, and collection-based data. Collectively, these resources and underlying cyber-infrastructure (Atkins et al. 2003) constitute the taxonomy cyber-infrastructure. The orbiting 'electrons' in the diagram represent individual taxonomists around the world who are experts on this particular taxon. Some work singly on particular subtaxa, while others align their efforts to make rapid progress in a shared taxon space, much like current PBI projects (Knapp 2008, Page 2008). Results are deposited in an openly accessible taxon knowledge bank (Fig. 14.3) from which users can extract information as needed.

Let us examine steps involved in doing taxonomy. At every junction, impediments to rapid progress have existed. Cyber-enabled taxonomy or cybertaxonomy has the potential to overcome each of these obstacles. The numbers for the paragraphs that follow correspond to numbers between steps in Fig. 14.4.

(1) Inventories to Collections Comparatively little has been written on improved ways to undertake taxonomic collecting as compared to other kinds of collecting, such as ecological sampling. Taxonomists are concerned with species limits and phylogenetic and biogeographic patterns. As such, they effectively require presence–absence data. Where taxonomic and ecological questions coincide, it makes sense to gather appropriate quantitative data. However, unless those data meet specified needs of a particular ecological question, devotion of the resources required for quantitative sampling over long periods of time at one locality does not make sense for taxonomy. Mobilizing existing museum data through efforts such as GBIF help identify gaps in collections. There is no reason that remote sensors, automated mass-collecting traps, and robots cannot vastly speed collecting for certain taxa or that digital transmitting microscopes and communications from the field cannot allow one expert to choreograph several collecting teams at one time.

Processing specimens is often a highly repetitive and labor-intensive activity that should be amenable to mechanization. With computer image recognition, aspects of sorting and pinning can, in principle, be automated. For large-scale inventory efforts, protocols are needed that maximize the number of characters ultimately retrievable. When possible, alternative preservation methods should be designed to capture external morphology, internal anatomy, and tissues for molecular sequencing.

(2) Species Descriptions A high-tech 'cockpit' will place all the instruments, remote access, analysis, visualization and communication tools, data, literature, and colleagues needed immediately in front of the taxonomist of tomorrow. Whether this taxonomist is working alone or as part of a distributed, international team, she will draw from and contribute to the knowledge of a taxon. Many details are to be worked out by the community regarding management of such taxon knowledge banks, priority for access to specimens or instruments, peer review of online contributions, and so forth, yet none are in principle insoluble. Even the peer review process can be expedited through automated notification of reviewers and online response.

(3) Species Tests Species descriptions are not one-time propositions. Instead, species are based on explicitly predicted unique combinations of characters (Wheeler and Platnick 2000). From these hypotheses, the same pattern of characters is predicted among newly found characters or specimens. In the past, such tests were completed through comprehensive revisions or monographs. For many insect taxa, revisions were completed once or twice per century. Such tests now can be conducted effectively as fast as new data become available. This approach will more rapidly and efficiently corroborate or falsify 'known' species as well as detect new ones. What remains critical overall, and especially in steps 2 and 3, is the need for deep knowledge by taxonomists. There is no substitute for years

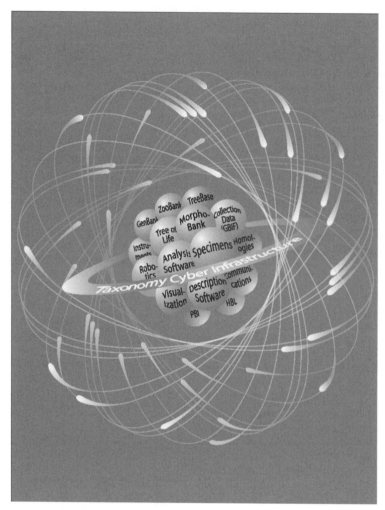

Fig. 14.2 A representation of a taxon 'knowledge community'. Taxonomists are represented as orbits around a 'nucleus' of taxonomy-specific cyber-infrastructure. Taxonomist 'orbits' vary through time. At one point, a given taxonomist is working solo on a taxon; other times variable numbers of taxonomists converge to focus their collective attention on rapid progress in species description or testing, as in the NSF-funded PBI projects. The nucleus of cyber-infrastructure includes natural history collections (specimens), literature, databases, digital instrumentation, robotics, descriptive software, and so forth. Advances in knowledge are immediately stored and made accessible through various electronic means and databases, collectively known as a taxon 'knowledge base'. This is analogous to traditional return of improved and expanded knowledge through publications, with the exception that it is immediately and openly accessible, up-to-date, and can be delivered in a user-defined format. Taxonomists can generate traditional monographs; morphologists make comparisons of homologs in MorphoBank; phylogeneticists access and edit full character matrices; ecotourists generate field guides; field researchers engage a rich toolbox of species identification aides including interactive keys delivered to handheld devices, DNA 'barcodes', and so forth. Graphic by Frances Fawcett, after Wheeler 2008c.

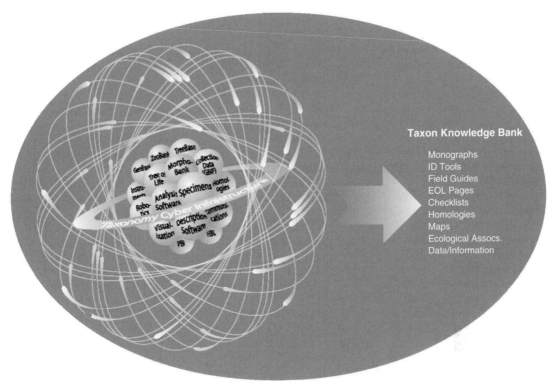

Fig. 14.3 Taxon 'knowledge community' and proposed taxon knowledge bank from which user-specified 'publications' may be generated on demand. Graphic by Frances Fawcett, after Wheeler (2008c).

of practice and research to develop this knowledge; however, cyber-infrastructure will make that learning process vastly more efficient than in the past.

(4) Species Tests to Databases With cyber-infrastructure, the traditional period of time between critical tests of species and the availability of results to users is compressed. Species are corroborated, refuted, and revised as appropriate, and such decisions are instantly available to the user community. The net result is to verify that the best and most current understanding of species and their attributes and distributions are made available to users.

(5) Collection Data Access to existing specimen-associated data is important. It allows taxonomists to find and correct mistakes (Meier and Dikow 2004) and museums to prioritize collection growth. And, of course, it is available to researchers for application to countless questions about biodiversity.

(6) Cladistic Analysis Most users of cladistic information simply want a cladogram with which to interpret some facts or phenomena. As such, most users want rapid access to a corroborated cladogram. Users have access to published cladograms already (e.g., TreeBase: www.treebase.org). Advanced users should be able to generate cladograms on demand based on the latest data and chosen parameters. Efforts are under-way to increase our capacity to deal with ever-larger data matrices (e.g., CIPRES: www.phylo.org).

(7, 8, 9, 10) Phylogenetic Classifications, Names, and Identifications Cladograms are the basis for phylogenetic classifications that are the general reference system (Hennig 1966). Classifi-cations should be dynamic, constantly changing to reflect growth in knowledge. As concepts change, so too do the meanings of names. Approaches are being developed (e.g., Franz 2005) to track *concepts* of species and higher taxa. The language of taxonomy (nomenclature) will become more transparent, more uniformly up-to-date, and more easily interpreted by users as cybertaxonomy progresses.

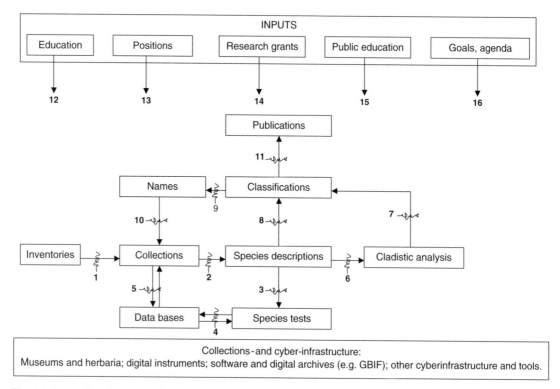

Fig. 14.4 Flowchart showing the bottlenecks in the process of taxonomic research. See text for explanation.

The most fundamental example of applied taxonomy is species identification. In the past, identifications frequently required an expert or synoptic collection. Cybertaxonomy will give users a wide array of identification tools. These will include interactive 'keys' accessible on handheld computers, computerized image recognition systems (e.g., MacLeod 2007, 2008), and DNA barcodes (Little and Stevenson 2007). Access to species is also enhanced by the growing number of electronic catalogs. Registration of names (Polaszek et al. 2005) has the potential to assure that all available names are easily accessed.

(12–16) Inputs The inventory and classification of millions of species of insects will require substantial investments in education, positions, research funding, and collections. It is urgent that funding for collections and descriptive taxonomy be immediately increased. If the taxonomy and collections community speaks with one voice and sets clearly stated goals, such funding is realistic and within reach.

BEWARE SIRENS OF EXPEDIENCY

Excellent taxonomy is hard work that requires special knowledge. Many who do not understand or appreciate the epistemology of taxonomy advocate quick fixes, short cuts, and inferior substitutes. For those who simply want taxonomic information, the appeal of apparent speed and ease are easily understood. Ultimately, however, users of taxonomy are better served by rigorous taxonomic practices.

Molecular data are presumed by some to be independent of other data sources and, to a significant degree, selectively neutral. That said, how data are analyzed is still more important than the kind of data collected (Hennig 1966). Disproportionate funding appears to be driven by a perception that molecular data are superior. That perception appears to be based on modernity, technology, and association with biomedical trends rather than on theoretical arguments. Much is to be learned from the comparative study of molecular, fossil, morphological, and ontogenetic data, and much is to be gained from integrative taxonomy (Will et al. 2005). Further, historical (phylogenetic) information is most faithfully stored in the form of complex characters, many the result of multiple genes and epigenetic factors not simply read from amino acid sequences. Yet to be seen is whether the current practice of using as much DNA as possible (idealized as whole genome sequencing) and trusting overall phenetic similarity to mirror phylogenetic relatedness is ultimately the most successful strategy. It should give us pause that the phenetics program of the 1970s was thoroughly discredited. The fundamental concepts of phylogenetic systematics still include special similarity and homology (Hennig 1966).

Several current proposals would undermine what continues to make taxonomy excellent. Among these are DNA taxonomy and DNA barcoding (Tautz et al. 2003, Hebert and Gregory 2005; but see Lipscomb et al. 2003, Seberg et al. 2003, Prendini 2005, Wheeler 2005); the PhyloCode (de Qeuiroz and Gauthier 1990; but see Forey 2002, Carpenter 2003, Nixon et al. 2003); and Felsenstein's (2001, 2004) preposterous suggestion that formal classifications are no longer needed (but see Franz 2005). These are dangerous distractions at a time when the full attention of taxonomists is needed to confront the biodiversity crisis, create a legacy of specimens and knowledge for posterity, and transform taxonomy into an efficient, responsive, cyber-equipped science for the twenty-first century. It is not an accident that most of these proposals are originated or promoted by individuals who have done little if any descriptive, revisionary, or monographic taxonomy and who presume to solve taxonomic problems with little understanding of the theories, epistemology, history, or practices of good taxonomy.

CONCLUSIONS

The science of insect taxonomy has never had greater prospects for fulfilling its potential. Advances in theory and technology are arming taxonomy with the most powerful tools in its long history and vastly more powerful tools are within reach. The biodiversity crisis has added a sense of urgency to our purpose and sharply focused the great challenges facing us. Insect taxonomists stand at a crossroads and given the right decisions and priorities, can make unprecedented, timely, and lasting contributions.

The time has arrived for insect taxonomists to think big. For the first time in history, we can realistically envisage a generation of tools that enable entomologists to study whole clades efficiently. The fragmented research resources, provincial collections, and individual scholar models of the past can give way to elevation of taxonomy to an appropriate global scale. A combination of cyber-infrastructure (Atkins et al. 2003, Wheeler et al. 2004) and the kind of teamwork exhibited in the NSF's PBIs (Knapp 2008, Page 2008) point to a clear path to success.

REFERENCES

Anonymous. 1994. *Systematics Agenda 2000: Charting the Biosphere. Technical Report.* Society of Systematic Biologists, American Society of Plant Taxonomists, Willi Hennig Society, Association of Systematic Collections. New York.

Atkins, D. E., K. K. Droegemeier, S. I. Feldman, H. Garcia-Molina, M. L. Klein, D. G. Messerschmitt, P. Massina, J. P. Ostriker, and M. H. Wright. 2003. *Revolutionizing Science and Engineering through Cyberinfrastructure. Report of the National Science Foundation* Blue-ribbon Advisory Panel on Cyberinfrastructure. U. S. National Science Foundation, Arlington, Virginia.

Brooks, S. J. and H. J. B. Birks. 2000. Chironomid-inferred late-glacial and early-Holocene mean July air temperatures for Kråkenes Lake, western Norway. *Journal of Paleolimnology* 23: 77–89.

Carpenter, J. M. 2003. Critique of pure folly. *Botanical Review* 69: 79–92.

Coope, G. R. 1979. Late Cenozoic fossil Coleoptera: evolution, biogeography and ecology. *Annual Review of Ecology and Systematics* 10: 247–267.

Cracraft, J. 1981. The use of functional and adaptive criteria in phylogenetic systematics. *American Zoologist* 21: 21–36.

Cracraft, J. 1983. Species concepts and speciation analysis. *Current Ornithology* 1: 159–187.

Cracraft, J. 1989. Speciation and its ontology: the empirical consequences of alternative species concepts for understanding patterns and processes of differentiation. Pp. 28–59. *In* D. Otte and J. A. Endler (eds). *Speciation and Its Consequences.* Sinauer Associates, Sunderland, Massachusetts.

Cracraft, J. 2002. The seven great questions of systematic biology: an essential foundation for conservation and sustainable use of biodiversity. *Annals of the Missouri Botanical Garden* 89: 127–144.

de Queiroz, K. 1988. Systematics and the Darwinian revolution. *Philosophy of Science* 55: 238–259.

de Queiroz, K. and M. Donoghue. 1988. Phylogenetic systematics and the species problem. *Cladistics* 4: 317–338.

de Qeuiroz, K. and J. Gauthier. 1990. Phylogeny as a central principle in taxonomy: phylogenetic definitions of taxon names. *Systematic Zoology* 39: 307–322.

Dobzhansky, T. 1973. Nothing in biology makes sense except in the light of evolution. *American Biology Teacher* 35: 125–129. [reprinted].

Dominguez, E. and Q. D. Wheeler. 1997. Taxonomic stability is ignorance. *Cladistics* 13: 367–372.

Eldredge, N. and J. Cracraft. 1980. *Phylogenetic Patterns and the Evolutionary Process*. Columbia University Press, New York.

Erwin, T. L. 1982. Tropical forests: their richness in Coleoptera and other arthropod species. *Coleopterists Bulletin* 36: 74–75.

Evans, H. E. 1968. *Life on a Little-Known Planet*. E. P. Dutton, New York.

Farris, J. S. 1979. The information content of the phylogenetic system. *Systematic Zoology* 28: 483–519.

Felsenstein, J. 2001. The troubled growth of statistical phylogenetics. *Systematic Biology* 50: 465–467.

Felsenstein, J. 2004. A digression on history and philosophy. Pp. 123–146. *In* J. Felsenstein (ed). *Inferring Phylogenies*. Sinauer Associates, Sunderland, Massachusetts.

Forey, P. L. 2002. Phylocode: pain, no gain. *Taxon* 51: 43–54.

Franz, N. 2005. On the lack of good scientific reasons for the growing phylogeny/classification gap. *Cladistics* 21: 495–500.

Gaffney, E. S. 1979. An introduction to the logic of phylogenetic reconstruction. Pp. 79–111. *In* J. Cracraft and N. Eldredge (eds). *Phylogenetic Analysis and Paleontology*. Columbia University Press, New York.

Godfray, H. C. J. 2007. Linnaeus in the information age. *Nature* 446: 259–260.

Grimaldi, D. A. and M. Engel. 2005. *The Evolution of Insects*. Cambridge University Press, Cambridge.

Hebert, P. D. N. and T. R. Gregory. 2005. The promise of DNA barcoding for taxonomy. *Systematic Biology* 54: 852–859.

Hennemann, F. H. and O. V. Conle. 2008. Revision of Oriental Phasmatodea: The tribe Pharnaciini Günther, 1953, including the description of the world's longest insect, and a survey of the family Phasmatidae Gray, 1835 with keys to the subfamilies and tribes (Phasmatodea: "Anareolatae": Phasmatidae). *Zootaxa* 1906: 1–316.

Hennig, W. 1966. *Phylogenetic Systematics*. University of Illinois Press, Urbana, Illinois.

Hill, C. R. and P. R. Crane. 1982. Evolutionary cladistics and the origin of angiosperms. Pp. 269–361. *In* K. A. Joysey and A. E. Friday (eds). *Problems in Phylogenetic Reconstruction. Systematics Association Special Volume Series*. Academic Press, New York.

Janzen, D. H. 1993. Taxonomy: universal and essential infrastructure for development and management of tropical wildland biodiversity. Pp. 100–113. *In* O. T. Sandlund and P. J. Schei (eds). *Proceedings of the Norway/UNEP Expert Conference on Biodiversity, NINA*. Trondheim, Norway.

Knapp, S. 2008. Taxonomy as a team sport Pp. 33–53. *In* Q. D. Wheeler (ed). *The New Taxonomy. Systematics Association Special Volume Series*. CRC Press, Boca Raton, Florida.

Lent, H. and P. Wygodzinsky. 1979. Revision of the Triatominae (Hemiptera, Reduviidae) and their significance as vectors of Chagas' Disease. *Bulletin of the American Museum of Natural History* 163: 123–520.

Lipscomb, D., N. Platnick, and Q. D. Wheeler. 2003. The intellectual content of taxonomy: a comment on DNA taxonomy. *Trends in Ecology and Evolution* 18: 65–66.

Little, D. P. and D. W. Stevenson. 2007. A comparison of algorithms for identification of specimens using DNA barcodes: examples from gymnosperms. *Cladistics* 23: 1–21.

MacLeod, N. (ed). 2007. *Automated Taxon Identification in Systematics: Theory, Approaches and Applications. Systematics Association Special Volume Series, 74*. CRC Press, Boca Raton, Florida.

MacLeod, N. 2008. Understanding morphology in systematic contexts: three-dimensional specimen ordination and recognition. Pp. 143–209. *In* Q. D. Wheeler (ed). *The New Taxonomy. Systematics Association Special Volume Series*. CRC Press, Boca Raton, Florida.

Mayr, E. 1942. *Systematics and the Origin of Species*. Columbia University Press, New York.

Mayr, E. 1963. *Animal Species and Evolution*. Belknap Press, Cambridge.

Mayr, E. 2000. The biological species concept. Pp. 17–29. *In* Q. D. Wheeler and R. Meier (eds). *Species Concepts and Phylogenetic Theory: A Debate*. Columbia University Press, New York.

Meier, R. and R. Willmann. 2000. The Hennigian Species Concept. Pp. 30–43. *In* Q. D. Wheeler and R. Meier (eds). *Species Concepts and Phylogenetic Theory: A Debate*. Columbia University Press, New York.

Meier, R. and T. Dikow 2004. The significance of specimen databases from taxonomic revisions for estimating and mapping the global species diversity of invertebrates and repatriating reliable and complete specimen data. *Conservation Biology* 18: 478–488.

Mishler, B. D. and M. J. Donoghue. 1982. Species concepts: a case for pluralism. *Systematic Zoology* 31: 491–503.

Mishler, B. D. and E. Theriot. 2000. The Phylogenetic Species Concept (*sensu* Misher and Theriot). Pp. 44–54. *In* Q. D. Wheeler and R. Meier (eds). *Species Concepts and Phylogenetic Theory: A Debate*. Columbia University Press, New York.

Nasci, R. S., C. W. Harris, and C. K. Porter. 1983. Failure of an insect electrocution device to reduce mosquito biting. *Mosquito News* 43: 180–184.

Nelson, G. 2004. Cladistics: its arrested development. Pp. 127–147. *In* D. Williams and P. Forey (eds). *Milestones in Systematics. Systematics Association Special Volume Series, 67*. CRC Press, Boca Raton, Florida.

Nelson, G. and N. I. Platnick. 1981. *Systematics and Biogeography: Cladistics and Vicariance.* Columbia University Press, New York.

Nixon, K. C. and Q. D. Wheeler. 1990. An amplification of the phylogenetic species concept. *Cladistics* 6: 211–223.

Nixon, K. C. and Q. D. Wheeler. 1992. Extinction and the origin of species. Pp. 119–143. *In* M. J. Novacek and Q. D. Wheeler (eds). *Extinction and Phylogeny.* Columbia University Press, New York.

Nixon, K. C., J. M. Carpenter, and D. W. Stevenson. 2003. The phylocode is fatally flawed, and the "Linnaean" system can easily be fixed. *Botanical Review* 69: 111–120.

Otte, D. and J. A. Endler (eds). 1989. *Speciation and Its Consequences.* Sinauer Associates, Sunderland, Massachusetts.

Page, L. M. 2008. Planetary biodiversity inventories as models of the new taxonomy. Pp. 55–62. *In* Q. D. Wheeler (ed). *The New Taxonomy. Systematics Association Special Volume Series.* CRC Press, Boca Raton, Florida.

Page, L. M., H. L. Bart, Jr., R. Beaman, L. Bohs, L. T. Deck, V. A. Funk, D. Lipscomb, M. A. Mares, L. A. Prather, J. Stevenson, Q. D. Wheeler, J. B. Woolley, and D. W. Stevenson. 2005. *LINNE: Legacy Infrastructure Network for Natural Environments.* Illinois Natural History Survey, Urbana.

Patterson, D. J., D. Remsen, W. A. Marino, and C. Norton. 2006. Taxonomic indexing – extending the role of taxonomy. *Systematic Biology* 55: 367–373.

Polaszek, A., D. Agosti, M. Alonso-Zarazaga, G. Beccaloni, P. de Place Bjørn, P. Bouchet, D. J. Brothers, the Earl of Cranbrook, N. Evenhuis, H. C. J. Godfray, N. F. Johnson, F. T. Krell, D. Lipscomb, C. H. C. Lyal, G. M. Mace, S. Mawatari, S. E. Miller, A. Minelli, S. Morris, P. K. L. Ng, D. J. Patterson, R. L. Pyle, N. Robinson, L. Rogo, F. C. Thompson, J. van Tol, Q. D. Wheeler, and E. O. Wilson. 2005. A universal register for animal names. *Nature* 437: 477.

Pollock, D. A. and B. B. Normark. 2002. The life cycle of *Micromalthus debilis* LeConte 1878. (Coleoptera: Archostemata: Micromalthidae): historical review and evolutionary perspective. *Journal of Zoological Systematics and Evolutionary Research* 40: 105–112.

Pons, J., T. Barraclough, J. Gomez-Zurita, A. Cardoso, D. Duran, S. Hazell, S. Kamoun, W. Sumlin, and A. Vogler. 2006. Sequence-based species delimitation for the DNA taxonomy of undescribed insects. *Systematic Biology* 55: 595–609.

Popper, K. 1968. *Logic of Scientific Discovery.* Second English Edition. Harper and Row, New York.

Prendini, L. 2005. Comment on "Identifying spiders through DNA barcodes". *Canadian Journal of Zoology* 83: 498–504.

Rappole, J. H., S. R. Derrickson, and Z. Hubalek. 2000. Migratory birds and spread of West Nile virus in the Western Hemisphere. *Emerging Infectious Diseases* 6: 319–328.

Rieppel, O. 2004. What happens when the language of science threatens to break down in systematics: a Popperian perspective. Pp. 57–100. *In* D. E. Williams and P. L. Forey (eds). *Milestones in Systematics. Systematics Association Special Volume Series, 67.* CRC Press, Boca Raton, Florida.

Rosen, D. E. 1979. Fishes from the uplands and intermontane basins of Guatemala: revisionary studies and comparative geography. *Bulletin of the American Museum of Natural History* 162: 269–375.

Roura-Pascual, N., A. V. Suarez, K. McNyset, C. Gomez, P. Pons, Y. Touyama, A. L. Wild, F. Gascon, and A. T. Peterson. 2006. Niche differentiation and fine-scale projections for Argentine ants based on remotely sensed data. *Ecological Applications* 16: 1832–1841.

Samways, M. J. 2005. *Insect Conservation Biology.* Cambridge University Press, Cambridge.

Schoch, R. M. 1986. *Phylogeny Reconstruction in Paleontology.* Van Nostrand Reinhold, New York.

Schuh, R. T. 2000. *Biological Systematics.* Cornell University Press, Ithaca, New York.

Seberg, O., C. J. Humphries, S. Knapp, D. W. Stevenson, G. Peterson, N. Scharff, and N. M. Anderson. 2003. Shortcuts in systematics? A commentary on DNA-based taxonomy. *Trends in Ecology and Evolution* 18: 63–65.

Simpson, G. G. 1961. *Principles of Animal Taxonomy.* Columbia University Press, New York.

Tautz, D., P. Arctander, A. Minelli, R. H. Thomas, and A. P. Vogler. 2003. A plea for DNA taxonomy. *Trends in Ecology and Evolution* 18: 70–74.

Terszowski, G., S. M. Muller, C. C. Bleul, C. Blum, R. Schirmbeck, J. Reimann, L. du Pasquier, T. Amagai, T. Boehm, and H. R. Rodewald. 2006. Evidence for a functional second thymus in mice. *Science* 312: 284–287.

Wägele, J.-W. 2004. Hennig's phylogenetic systematics brought up to date. Pp. 101–125. *In* D. M. Williams and P. L. Forey (eds). *Milestones in Systematics. Systematics Association Special Volume Series, 67.* CRC Press, Boca Raton, Florida.

Wheeler, Q. D. 1995a. The "Old Systematics": classification and evolution. Pp. 31–62. *In* J. Pakaluk and S. A. Slipinski (eds). *Biology, Phylogeny, and Classification of Coleoptera. Papers Celebrating the 80th Birthday of Roy A. Crowson.* Muzeum i Instytut Zoologii PAN, Warszawa.

Wheeler, Q. D. 1995b. Systematics and biodiversity policies at higher levels. *BioScience* 45 (Supplement): 21–28.

Wheeler, Q. D. 1999. Why the phylogenetic species concept? Elementary. *Journal of Nematology* 31: 134–141.

Wheeler, Q. D. 2004. Taxonomic triage and the poverty of phylogeny. *Philosophical Transactions of the Royal Society. B.* 359: 571–583.

Wheeler, Q. D. 2005. Losing the plot: DNA "barcodes" and taxonomy. *Cladistics* 21: 405–407.

Wheeler, Q. D. 2008a. Introductory: toward the new taxonomy. Pp. 1–17. *In* Q. D. Wheeler (ed). *The New Taxonomy. Systematics Association Special Volume Series.* CRC Press, Boca Raton, Florida.

Wheeler, Q. D. 2008b. Taxonomic shock and awe. Pp. 211–226. *In* Q. D. Wheeler (ed). *The New Taxonomy.*

Systematics Association Special Volume Series. CRC Press, Boca Raton, Florida.

Wheeler, Q. D. (ed). 2008c. *The New Taxonomy. Systematics Association Special Volume Series.* CRC Press, Boca Raton, Florida.

Wheeler, Q. D. 2008d. Undisciplined thinking: morphology and Hennig's unfinished revolution. *Systematic Entomology* 33: 2–7.

Wheeler, Q. D. and N. I. Platnick. 2000. The Phylogenetic Species Concept (*sensu* Wheeler and Platnick). Pp. 55–69. *In* Q. D. Wheeler and R. Meier (eds). *Species Concepts and Phylogenetic Theory: A Debate.* Columbia University Press, New York.

Wheeler, Q. D. and A. G. Valdecasas. 2005. Ten challenges to transform taxonomy. *Graellsia* 61: 151–160.

Wheeler, Q. D. and A. G. Valdecasas. 2007. Taxonomy: myths and misconceptions. *Anales del Jardín Botánico de Madrid* 64: 237–241.

Wheeler, Q. D., P. H. Raven, and E. O. Wilson. 2004. Taxonomy: impediment or expedient? *Science* 303: 285.

Wiley, E. O. 1981. *Phylogenetics.* John Wiley and Sons, New York.

Will, K. W., B. D. Mishler, and Q. D. Wheeler. 2005. The perils of DNA barcoding and the need for integrative taxonomy. *Systematic Biology* 54: 844–851.

Williams, D. M. 2004. Homologues and homology, phenetics and cladistics: 150 years of progress. Pp. 191–224. *In* D. M. Williams and P. L. Forey (eds). *Milestones in Systematics. Systematics Association Special Volume Series, 67.* CRC Press, Boca Raton, Florida.

Williams, D. M. and P. L. Forey (eds). 2004. *Milestones in Systematics. Systematics Association Special Volume, 67.* CRC Press, Boca Raton, Florida.

Wilson, E. O. 1985. The biological diversity crisis: a challenge to science. *Issues in Science and Technology* 2: 20–29.

Wilson, E. O. 1992. *The Diversity of Life.* W. W. Norton, New York.

Wilson, E. O. 1993. The encyclopedia of life. *Trends in Ecology and Evolution* 18: 77–80.

Wygodzinsky, P. 1961. On a surviving representative of the Lepidotrichidae (Thysanura). *Annals of the Entomological Society of America* 54: 621–627.

Chapter 15

INSECT SPECIES – CONCEPTS AND PRACTICE

Michael F. Claridge

School of Biosciences, Cardiff University, Cardiff, Wales, UK

Insect Biodiversity: Science and Society, 1st edition. Edited by R. Foottit and P. Adler

Insect taxonomists face the enormous task of describing, classifying, and providing means for the non-expert to identify the vast numbers of entirely new or little known species that we know to exist, often in extremely endangered environments. The greatly increased interest in biological diversity over the past 20 years or so and the realization of the urgent need to conserve as much as possible of it make these tasks the more urgent. A major problem with this agenda is the ongoing debate and disagreement about species concepts. No topic in evolutionary and systematic biology has been more contentious and controversial than the nature and meaning of species. In the words of the late Ernst Mayr (1982) 'There is probably no other concept in biology that has remained so consistently controversial as the species concept'.

Species are usually the primary units of biodiversity and conservation (Wilson 1992), so it is obviously important that, so far as possible, we agree on their nature. One of the aspects of the species problem that has made it on the one hand so intractable, but on the other so rewarding, is that it is at the same time a very practical problem for all taxonomists and biologists, and also a deeply philosophical and theoretical one, as recently emphasized by Hey (2006). The continuing interest in these problems over the past 10 or 20 years has led to the publication of several relevant books and review articles, notably by Ereshefsky (1992), Paterson (1993), Wilson (1999), and Wheeler and Meier (2000). However, these publications are mostly concerned with the diversity of theoretical and philosophical approaches to the issues, with little reference to real practical problems.

About 10 years ago, together with my colleagues Hassan Dawah and Mike Wilson, I edited and contributed to a multiauthor volume under the title *Species: the Units of Biodiversity* (Claridge et al. 1997a). In this volume, we brought together practicing taxonomists from across the breadth of organismal diversity, including workers on different groups of animals, plants, fungi, and microorganisms, in an attempt to find common ground. The results were interesting, but sadly showed just how difficult it is to have a biologically meaningful and unified species concept across such diverse organisms. However, insects, which though they represent about 60–65% of all living biodiversity (Hammond 1992), are just one subset, one class of one phylum, of all living organisms and thus share many genetic mechanisms and more of a common heritage. Maybe, therefore, we can be more optimistic in attempting to agree on a common species concept for these mega-diverse organisms? Here I shall attempt to review species concepts in a historical context and to look for a unified concept that may, at the same time, be of practical use for insects.

EARLY SPECIES CONCEPTS – LINNAEUS

The term species is an old one and derives from the writings of classical Greek philosophers, most notably Aristotle (Cain 1958). It was natural for scholars and naturalists of the seventeenth and eighteenth centuries to adopt the systems of Aristotelian logic in attempting to classify and make sense of the natural world. Technical terms from this system that were used in attempts to classify living organisms included *definition, genus, differentia*, and *species*. Here, the *genus* referred to the general kind, while *species* referred to the particular kind within the genus, as qualified by the *differentia*.

Carl von Linné, better known to us as Linnaeus and the founder of the binomial system of nomenclature that we still use for living organisms, most clearly documented his principles and practice in producing classifications of both animals and plants (Cain 1958 for a full account). The binomial system itself is a result of the use of the Aristotelian system. Authors previous to Linnaeus, and indeed Linnaeus himself in his early works, had given multinomial names to organisms – the genus being one word, but qualified by a descriptive phrase, the *differentia*, to describe and delineate the species itself. Linnaeus, from the 10th edition of the *Systema Naturae* in 1758 and probably because of the pressures to describe such large quantities of new material, reduced the differentia for animal species to a single word – the specific name – and so invented the binomial that has been used for plants and animals to this day. It has been challenged unsuccessfully at various times in the past and is currently under attack by proponents of the phylocode, but it is likely once again to prevail (Wheeler 2004)!

To Linnaeus, species were simply the lowest category of particular kinds in his classifications, though he also often recognized *varieties* within species. Not only did Linnaeus publish classifications and descriptions of many animals and plants, but he also wrote books in which he detailed his methods and philosophy (e.g., Linnaeus 1737). In a detailed study of these many writings, Ramsbottom (1938) showed that in developing a practical concept of species, Linnaeus recognized three main criteria. Species were (1) distinct and monotypic, (2) immutable and created as such, and (3) true breeding. Criteria 1 and 2 here are to be expected in preevolutionary philosophy; 3 may be a little more surprising. The idea that species had a single norm of morphological variation and were clearly each distinct from all other species within the same genus was widely accepted. Linnaeus was, of course, working in exciting times when European explorers were traveling widely in regions of the world previously unknown to them and bringing back large collections of plants and animals for study. For obvious reasons, these samples consisted of dead and often poorly preserved material. Thus, to describe new species and to classify them, early taxonomists had little recourse but to use morphological characters. The total immutability of species was clearly widely accepted in the eighteenth and early nineteenth centuries, but even Linnaeus later in his life developed some complicated theories of speciation by hybridization (Cain 1993).

Before the enormous influx of largely tropical material into European museums, most early taxonomists, including Linnaeus, were field naturalists themselves and familiar with the organisms on which they worked as living entities. They certainly knew that the species they described from local faunas and floras were biologically distinct and differed in obvious features of their natural history, as well as their morphology. The swamping of museums by large collections from overseas inevitably led to the almost exclusive use of morphological differences both to describe and to recognize species. This obviously remains so today for most groups of animals and particularly for species-rich groups, such as insects. The morphological species or morphospecies has evolved from these early classifications mostly for reasons of convenience. Today, such 'morphospecies' are being developed by the use of molecular characters for recognizing species (Tautz et al. 2003, Blaxter 2004). The morphospecies, however, is not a philosophical concept, but simply a practical category. No doubt Linnaeus himself would have been unhappy that his philosophy should be reduced to such a purely practical matter. Following Linnaeus, taxonomists were forced more and more to use the practical morphospecies for what they largely knew only as dead museum specimens. The amount of difference required to recognize and separate species became inevitably more and more subjective, as illustrated by the well-known quote from Regan (1926): 'A species is a community, or a number of related communities, whose distinctive morphological characters are, in the opinion of a competent systematist, sufficiently definite to entitle it or them to a specific name'.

Because taxonomy and systematics, more and more through force of circumstance, were based only on morphological differences among dead museum specimens, the two quite different traditions of studying the natural world diverged, with what we might call the morphologists on one side and the naturalists on the other. Naturalists, even in the late eighteenth century were well aware that species had some real biological basis in the field. For example, the naturalist Gilbert White (1789) first showed that several morphologically similar species of song birds in Britain of the genus *Phylloscopus*, the warblers, were clearly separated in the field by their distinctive male songs, now known to function as important elements of their specific mate recognition systems (SMRSs). This interest in breeding barriers and species as reproductive communities was an essential element of the naturalist tradition. Later in the nineteenth century English-speaking world, most notable as part of the naturalist tradition were Charles Darwin and Alfred Russell Wallace. It was from this tradition that they independently developed the theory of evolution by natural selection and accumulated overwhelming evidence for descent with modification. After the general acceptance of evolution, species were recognized as the end terms of different lines of descent. The controversies around evolution itself meant that the nature of species was not regarded at that time as a high-priority subject. Darwin himself seems often to have regarded species as more or less arbitrary stages in the process of evolutionary divergence.

BIOLOGICAL SPECIES CONCEPTS

In addition to developing the idea of species as morphologically discrete entities, Linnaeus also had the admittedly more vague idea of species as breeding units that generally breed true (Ramsbottom 1938). However, it was not until the late nineteenth and early twentieth centuries that these ideas were clarified and became central to species thinking. Probably the most important contributor to this way of thinking was Sir Edward Poulton. First in 1904, in his Presidential Address to the Entomological Society of

London, and later in a volume of essays on various aspects of evolution (Poulton 1908), he made the most important advance toward what has since become known as the biological species concept. He emphasized the importance of interbreeding in the field as the most critical species criterion. This criterion was what later Mayr (1942, 1963) termed crossability and contrasted strongly with interfertility or simple ability to hybridize. Poulton was one of the first authors to make this clear differentiation and thus effectively to develop the modern biological species concept, for which generally he receives insufficient credit (Claridge 1960, Mallet 1995).

During the early- and mid-twentieth century the revolution in evolutionary thinking, often known as the Evolutionary Synthesis (Mayr and Provine 1980), was developed by the attempted unification of systematics, genetics, and evolution, exemplified by the publication of major seminal volumes, including *Genetics and the Origin of Species* (Dobzhansky 1937) and *Systematics and the Origin of Species* (Mayr 1942). The so-called biological species concept was central to these ideas and received many definitions, perhaps the most useful and certainly most cited of which is 'species are groups of actually or potentially interbreeding natural populations which are reproductively isolated from other such groups' (Mayr 1942, p. 120). Reproductive isolation in nature, as also for Poulton, was the key factor in identifying and maintaining species. Such reproductive isolation was maintained by what Dobzhansky (1937) termed isolating mechanisms, which were any attributes of species populations that reduced the likelihood of interbreeding between them. Such a category is a wide one, including not only all sorts of postmating genetic incompatibilities, as well as behavioral and ecological differences that act before mating and fusion of gametes, but also totally extrinsic geographical barriers. The latter are clearly not properties of the organisms and are not now generally classified with the intrinsic factors. To Dobzhansky, speciation was the origin of reproductive isolating mechanisms and thus of reproductive isolation. Dobzhansky's system of classification of isolating mechanisms was followed and modified by many authors during the twentieth century, including particularly Mayr (1942, 1963, 1982) and Cain (1954).

A major set of criticisms of the biological species concept has been developed over some years by Hugh Paterson (Paterson 1985, 1993). One of his main concerns is with the concept of species isolating mechanisms and with the implication that they have evolved as adaptations under natural selection to achieve and maintain reproductive isolation. Paterson must be correct, at least for all so-called postmating mechanisms, which logically cannot be due to such adaptation (Mallet 1995). Avise (1994), and recently Coyne and Orr (2004), have suggested the more neutral term, 'isolating barrier' to replace isolating mechanism. This suggestion is useful, but does not deal with Paterson's further criticisms. Maybe there is no need anyway for a term to include such a diversity of phenomena?

Paterson regards species as groups of organisms with common fertilization systems. 'We can, therefore, regard species as that most inclusive population of individual biparental organisms which share a common fertilization system' (Paterson 1985, p. 25). He recognized an important subset of the fertilization system that he termed the specific mate-recognition system (SMRS), 'which is involved in signaling between mating partners and their cells'. Thus, the often complicated reciprocal signals and signaling systems of mating and courtship (well documented by ethologists including Tinbergen 1951, Eibl-Eibesfeldt 1970, and Brown 1975) have the essential function of ensuring specific mate recognition. In such behavior sequences, successive signals release, in turn, successive responses via tuned receptors in the opposite sex. These sequences are usually, but not exclusively, initiated by males. Unless appropriate responses are received at each stage and the signals are recognized as appropriate, exchange will be terminated and ultimate exchange of gametes will not occur. The exchanges of signals between partners may be broken off at any stage. Thus, to Paterson species are defined by their unique SMRSs and the evolution of new species, speciation, is the origin of new SMRSs. He has argued at length over the years that his concept of species is quite distinct from the biological species *sensu* Mayr. He terms the latter the 'isolation concept', because it is defined by reproductive isolation from other species, and terms his own system the 'recognition concept'. Lambert and Spencer (1995) and Vrba (1995) have strongly supported this line of argument, but others have doubted the clear demarcation between isolation and recognition concepts (Claridge 1988, 1995a, Claridge et al. 1997a, Coyne et al. 1988, Mayr in Wheeler and Meier 2000, Coyne and Orr 2004).

Thus, in practice a broadened biological species concept would recognize that different species are characterized by distinct SMRSs that result in the levels of reproductive isolation between sympatric species observed in the field. However, species are only rarely recognized by direct studies of the SMRSs, which are themselves equally rarely understood with certainty, but which must be the final arbiters for determining biological species boundaries (Claridge 1988, 1995a, Claridge et al. 1997a, b). Much more usually, biological species are recognized by markers that are thought to indicate the existence of reproductive isolation. In the past, these were normally morphological markers so that taxonomists hypothesized that the differences they recognized and used to separate morphospecies were also indicators of biological species boundaries. More recently, morphological markers have been supplemented by a wide range of others, including cytological, behavioral, and biochemical ones. In particular, now molecular markers involving characters derived from the amino acid sequences of specific pieces of DNA are more and more being used (Avise 1994). Indeed, it has even been suggested that all species should be diagnosed by such molecular differentiation (e.g., Tautz et al. 2003, Blaxter 2004). However, under the biological species concept, the enormous diversity of markers now available to taxonomists is simply an indicator of levels of reproductive isolation. Levels of gene flow between populations and, therefore, levels of reproductive isolation are now routinely estimated by molecular divergence.

A particular feature of the biological species concept is that reproductive isolation can occur between species populations without any obvious accompanied morphological differentiation. This phenomenon of real biological species existing in nature without any obvious differentiation to the human observer has been recognized since early in the twentieth century and was well discussed by Mayr (1942). Such species are usually known as sibling, or cryptic, species and have been clearly demonstrated in many groups of insects (Thorpe 1940, Mayr 1942, Claridge 1960, 1988, Claridge et al. 1997b).

During the heyday of the biological species in the middle years of the twentieth century, a large body of opinion among museum taxonomists was nevertheless opposed to it (e.g., Sokal and Crovello 1970) and preferred either some overtly morphological species approach or a purely phenetic one. Such views have always been strongly held by botanists on the grounds that interspecific hybridization is so common in plants that reproductive isolation is not a useful criterion

(Gornall 1997, but see Mayr 1992). However, entomologists have also often led the criticism of the biological species.

Two important problems in the use of biological species are acknowledged by all of its proponents. These concern the status of (1) asexual and parthenogenetic forms and (2) geographically or spatially isolated (allopatric) populations:

Agamospecies The biological species in its various manifestations can only be applied to biparental sexually reproducing organisms in which a distinctive SMRS leads to reproductive isolation. Neither asexual nor parthenogenetic organisms have a functional mate-recognition system that leads to the fusion of gametes, so the biological species concept cannot apply to them. These organisms exist as clones that can differ in morphology, biochemistry, cytology, behavior, ecology, and other features (Foottit 1997). Distinctive and diagnosable clones are often described as species, but they cannot be biological species. They are thus practical categories like the morphological species, which have been given the useful term agamospecies by Cain (1954). Many groups of living organisms, including many microorganisms, can only be agamospecies.

Allopatric forms A real problem with applying the biological species concept is that reproductive isolation in the field can only be determined for sympatric populations where alone there are possibilities of testing the effectiveness of SMRSs in the field. Geographical variation and the status of allopatric populations have long been of major interest to both taxonomists and evolutionary biologists. Degrees of observable differentiation between allopatric populations vary from almost nothing to large differences, at least comparable with those observed between distinct sympatric species of the same taxonomic group, but the criterion of gene flow and reproductive isolation in the field cannot be conclusively tested. Experimental crossings of allopatric forms under laboratory and experimental conditions yield results of only limited value. The polytypic nature of biological species has for long been recognized and a series of taxonomic categories, from superspecies to subspecies has been developed to describe such essentially continuous geographical variation (Mayr 1942, Cain 1954). This continuum on the one hand has provided vital data for the development of theories of allopatric, or geographical, speciation (Mayr 1942,

Cain 1954), but on the other has led many also to regard the species as no more than a rather arbitrary stage in the divergence of local populations. For example, Alfred Russell Wallace (1865, p. 12), when confronted with the bewildering range of geographical variation and polymorphism in the swallowtail butterflies of Southeast Asia, in a widely cited quote, said: 'Species are merely those strongly marked races or local forms which, when in contact, do not intermix, and when inhabiting distinct areas are generally believed to... be incapable of producing a fertile hybrid offspring... it will be evident that we have no means whatever of distinguishing so-called "true species" from the several modes of variation... into which they so often pass by an insensible gradation'. Wallace and, of course Darwin also, were impressed with variation within and between natural populations as the basic material for evolution by natural selection. Equally today, the allocation of allopatric populations within the superspecies/subspecies range is largely subjective. Drawing a line through any continuum must be to some extent arbitrary. This problem is undoubtedly a weakness of the biological species, but I would argue equally that it is a weakness of all discrete species concepts.

These obvious complications, together with a frequent desire to eliminate the priority given to one set of organismal characters – the SMRS and reproductive isolation – over all others, have persuaded many systematists to abandon completely the biological species concept in favor of what we termed a general phylogenetic species concept (Claridge et al. 1997a). But before considering phylogenetic concepts in more detail, it is appropriate to discuss the interesting ideas of Mallet (1995) on his 'genotypic cluster criterion' or concept of species. Mallet is a geneticist, who has worked on widely distributed species and populations of tropical butterflies. He is concerned that the biological species is absolute and does not allow for interspecific hybridization and intergrading, which he has recently extensively reviewed (Mallet 2005, 2008). However, because intergradation is the basis of the polytypic biological species espoused by Mayr, Cain, and others, this criticism cannot be a real problem. Even in sympatric interactions, reproductive isolation does not need to be absolute to maintain species integrity. Indeed, the acceptance of the reality of evolution demands that species cannot always be completely reproductively isolated. Intermediates and intergradation must be expected. Thus, all realistic species concepts must allow

for such intergradation and the broadly conceived biological species concept certainly does this. Another interesting contribution, along similar lines to those of Mallet (1995), is the 'genomic integrity species definition' of Sperling (2003). Here species are 'populations that maintain their genomic integrity when they contact each other, even if they occasionally exchange genes'. These various attempts to formulate more inclusive and realistic species concepts appear to me to be quite compatible with and, indeed similar to, the broadly based biological species concept advocated here.

PHYLOGENETIC SPECIES CONCEPTS

A revolution in the philosophy and practice of systematics took place in the English-speaking world after the publication of Hennig's *Phylogenetic Systematics* in translation in 1966. Few systematists today do not use some variant of the cladistic methodologies pioneered by Hennig. Coincident with this widespread acceptance of cladistic methods for determining phylogenies and making robust classifications, came increased published dissatisfaction with, and rejection of, the biological species concept by systematists (e.g., most authors in Wheeler and Meier 2000). Oddly enough, Hennig himself thought of species as reproductive communities, so his species concept was broadly similar to the biological species concept of Mayr (1942). Hennig, of course, was interested primarily in extending species back in time as diagnosable clades, for which the biological species is not suited. In this, he was developing what Simpson (1951) had begun as a broader evolutionary species concept, which has since been taken up by many others (e.g., Cain 1954, Wiley 1978, Mayden 1997, Wiley and Mayden in Wheeler and Meier 2000, and see Hey 2006).

Cladists certainly have not spoken with one voice on the nature of species. Hennig (1966) saw species as that unique level in the taxonomic hierarchy at and above which cladistic methods could be applied to determine phylogenies and below which they could not. Within species, interbreeding relationships dominate and these Hennig differentiated from phylogenetic relationships as tokogenetic ones, a term that has not been widely adopted. Wheeler and Nixon (1990) supported the idea that species are uniquely different from higher level taxa, on the grounds that they do not have resolvable internal phylogenetic structure. On the other hand

Nelson (1989), in stating that 'A species is only a taxon', expressed clearly the view that species simply represent one level in the taxonomic hierarchy and are of no more significance than any others, such as genus, family, or order. In criticism of this view, Wheeler (1999) commented 'That species exist in nature is one aspect of species about which I can agree with Mayr (1963)' and most taxonomists also seem broadly to agree, though Mallet (1995) casts doubt on this assertion. Oddly, modern molecular phylogeographic techniques suggest also that this sharp differentiation in terms of phylogenetic pathways above and below species might not be as absolute as Hennig, Wheeler, and others have suggested. A full introduction to these novel methods is given by Avise (2000).

Many authors have attempted to formulate an expressly phylogenetic species concept. In a valuable volume devoted to a debate about species concepts and phylogenetic theory, proponents of two different phylogenetic concepts presented their arguments and disagreements, not only over the biological species concept and what is termed the Hennigian species concept, but also with each other – Mischler and Theriot on the one hand and Wheeler and Platnick on the other (in Wheeler and Meier 2000). Despite these disagreements, there is some practical consensus, and perhaps the most widely cited definition of the phylogentic concept is that of Cracraft (1983, 1997), who stated that the species is the 'smallest diagnosable cluster of individual organisms within which there is a parental pattern of ancestry and descent'. Some critics have suggested that this definition does not apply to populations, a view fiercely refuted by Cracraft (1997). Nixon and Wheeler (1990) emphasized this when they defined phylogenetic species as 'the smallest aggregation of populations (sexual) or lineages (asexual) diagnosable by a unique combination of character states in comparable individuals'. Thus, it seems to me that the essence of the phylogenetic concept in its various forms involves the recognition of diagnosable clades. The major question then has to be just exactly what is diagnosable? How different do two populations or lineages have to be to be diagnosably and recognizably distinct? These judgements must surely be subjective, particularly because what is distinct to one taxonomist might well not be to another.

Leaving aside the latter difficulty, any of the markers discussed above as useful for delimiting biological species, including molecular and behavioral ones, also can be used to characterize phylogenetic species, though in most groups such characters have tended to be exclusively morphological. Claridge et al. (1997a) concluded that the practical differences between a phylogenetic concept and a broadly biological one were not great, a view which I still hold. To me, the great disadvantage of the phylogenetic concept is the difficulty in agreeing on what precisely is a diagnosable difference. The advantage of the biological concept is that it attempts to identify real, reproductively isolated populations, even though isolation might not always be complete. The phylogenetic concept can be applied to asexual or parthenogenetic lineages, which are effectively agamospecies (Cain 1954). Also for the phylogentic species, the problems of differentiating allopatric populations are not different from sympatric ones. Thus, diagnosably distinct allopatric populations will be regarded as separate species. The result will almost always be that more allopatric populations will be recognized as distinct species than will an application of the polytypic biological concept. For example, the well-known analysis of the birds-of-paradise (Aves: Paradisaeidae) by Cracraft (1992), using his phylogentic concept, established more than twice as many species (90) than had previous applications of the biological species concept to the same data set. However, the judgement will be essentially subjective under both concepts, though various authors, from Mayr (1942) to Sperling (2003), have attempted to provide more objective criteria for assessing the species status of totally allopatric populations. Though there are fundamental differences of philosophy and theory between the biological and phylogenetic concepts, I see little difference generally when applied in practice.

Despite the apparent advantages of the phyologenetic concept in breadth of application, in my view it has one major practical disadvantage compared to biological concepts and that is the improbability that its application will reveal the existence of complexes of sibling species. The philosophy of the phylogenetic species gives no incentive or reason to search for further divisions once diagnosably distinct forms have been established. On the other hand, the emphasis of the biological species on reproductive isolation and specific mate recognition means that sibling species will be revealed by its diligent application. Among insects, and many other organisms, sibling species are now widely known and are often of great biological significance.

SPECIES CONCEPTS AND SPECIATION – A DIGRESSION?

Although not strictly within the remit of this book, theories of speciation have been closely tied to the development of particular species concepts, so that a brief review is appropriate here. Most modern authors will agree that in recognizing and describing species, taxonomists are providing a framework for understanding the diversity of living organisms and their evolutionary relationships. However, the philosophical interactions between different species concepts and particular theories of speciation are long standing and still not fully resolved. A system for describing observed diversity should be independent of the various possible modes by which that diversity might have evolved (but see Bush 1994, 1995, Claridge 1995b).

Probably the most widely accepted mode of animal speciation is that of geographical or allopatric speciation (Mayr 1942, 1963, Cain 1954). The essence of such theories is that an ancestral population is subsequently divided into at least two daughter populations, isolated in space where they diverge and develop genetic isolation prior to any subsequent meeting and sympatry. The most extreme view of this is that the daughter species must have diverged to the extent that they do not interbreed on meeting, that is they have developed completely separate SMRSs, in the terminology of Paterson (1985, 1993), perhaps the strongest current supporter of this view. Contrary to this theory of speciation in complete allopatry, many authors, particularly Wallace (1889) and Dobzhansky (1940), have developed theories of the reinforcement of species isolating mechanisms in sympatry after the partial divergence of incipient species populations in allopatry. This theory is still controversial and has been well reviewed (Coyne and Orr 2004).

Quite distinct from the various theories of allopatric speciation are those of sympatric speciation, whereby no period of allopatry is necessary for two species to diverge from a previous one, normally by powerful disruptive selection. Though not strongly supported in the early years of speciation theory, such ideas have always been advanced by some entomologists and others working with large groups of sympatric specialist feeders, including parasites and herbivores (Walsh 1864, Bush 1975, 1993, 1994). Here, descendant species diverge within the range of the ancestral species and, therefore, all stages of such divergent populations can be expected to exist in the field together. These ideas

have become more and more acceptable in recent years (Coyne and Orr 2004) to the extent that even Ernst Mayr, the strongest opponent of such theories since his 1942 book, in his final work accepted that sympatric speciation is probable at least in some parasites (Mayr 2004). A developing consensus is that there may be a continuum from pure allopatric to pure sympatric speciation, whereby intense natural selection might outweigh the swamping effects of gene flow by hybridization.

Whatever the final consensus on speciation, there is little doubt that the nature of our species concept should not depend on the mode of speciation. Thus, in principle, I agree with most cladists on the particular point that we should describe the patterns of diversity that we see in nature, so far as possible, independent of the theories concerning the evolution of such patterns (Wheeler and Nixon 1990). However, I cannot agree with these authors that 'the responsibility for species concepts lies *solely* with systematists'. Aside from the essential arrogance of such a statement, an evolutionary view of species inevitably must involve at least genetics and evolutionary biology, in addition to systematics. If we accept the generality of evolution and species as the end results of evolutionary divergence, then the species concept itself must be an evolutionary one. Simpson (1951) first attempted to fuse the then biological species, the agamospecies, and the palaeospecies into a unitary, all-embracing evolutionary concept. Cain (1954) developed further and clarified these ideas, as later, particularly following the general acceptance of cladistic methodologies, did Wiley (1978), Mayden (1997, 1999), and Wiley and Mayden (in Wheeler and Meier 2000). Such theories provide a reasonably satisfactory philosophical fusion of the variety of species concepts that account for the diversity of living organisms and their relationships over time, but do not help much in the purely practical recognition and identification of species.

INSECT SPECIES – PRACTICAL PROBLEMS

About 99% of all insects are estimated to be biparental and reproduce sexually, involving the meeting of males and females and the exchange of gametes by copulation in often complicated sequences of courtship behavior. Here, in principle, the criterion of reproductive isolation is applicable and, thus, the biological species should provide a basis for recognizing and establishing species limits for most insect groups.

Species, host races, and biotypes

The enormous numbers of species of insects dominate terrestrial ecosystems. Despite this diversity, insects are remarkably conservative in morphology – to the non-entomologist, they all tend to look the same. The result is that morphological taxonomy has to be based on relatively small differences that are often difficult to appreciate, so that species taxonomy is generally difficult. Despite their morphological conservatism, however, insects are ecologically extremely diverse and usually species specific in their habits. Thus, species often differ most obviously in features of ecology and behavior. About 50% of insects are estimated to be herbivores (Strong et al. 1984) and a high proportion of the remainder are probably parasitoids, largely attacking other insects. Thus, most insects are effectively parasites (Price 1980), either on plants or other insects (parasitoids), and tend to be narrow ecological specialists. The result is that, in well-studied groups, species are most obviously characterized by differences in food exploitation and behavior (Claridge et al. 1997b). Perhaps not surprisingly, the idea of sibling, or cryptic, species has been explored widely, particularly by entomologists. After sibling species have been recognized by appropriate biological studies, markers may then be found that enable identification. In traditional taxonomy, these markers can be very small, but consistent, morphological differences, as for example in the fine structure of genitalia (Claridge et al. 1997b), but today are also likely to be molecular ones (Tautz et al. 2003, Al-Barrak et al. 2004). The latter have the enormous advantage that all stages of the life cycle can be identified, otherwise not usually easy even in well-known organisms.

The specialized relations among insect parasites, including herbivores, and their hosts often make species delimitation and identification difficult. Host-associated populations can show varying degrees of phenotypic differentiation in terms of the markers used by taxonomists, including both morphological and molecular ones. Difficulties in determining the status of such populations have resulted in the widespread use of such quasitaxonomic terms as 'host race', 'biological race', and 'biotype' (e.g., Walsh 1864, Thorpe 1930, Claridge and den Hollander 1983, Diehl and Bush

1984, Claridge 1988, Claridge et al. 1997b, Drès and Mallet 2002) for stages intermediate between completely panmictic populations and distinct biological species. The status of such populations is intimately involved in arguments over the role of sympatric speciation in the evolution of insect parasites (Bush 1994). In my view, these hypotheses should not affect directly discussions on the species status of host-associated populations (Claridge 1995b). Of course, species are the result of splitting lineages, so that at some stage during the process of speciation, diverging populations will not be completely reproductively isolated from each other. Nevertheless, many examples of supposed host races and biotypes, when closely analyzed, have been found to represent separate biological species.

The phenomenon of complexes of sibling species was first recognized by workers on insect vectors of disease organisms of both humans and domestic animals (Lane 1997). The reasons for this are obvious – these insects have been intensively studied because of their importance to human health. Malarial mosquitoes of the genus *Anopheles* were the first to be intensively studied and recognized as complexes of sibling species with differing abilities to transmit various forms of malarial parasites (modern summary by Linton et al. 2003). Experimental breeding and crossing showed that a wide variety of markers might be used to identify genetically distinct populations, including morphology of all life-history stages, chromosome banding patterns, allozyme markers, and particularly now, DNA markers (Linton et al. 2003). An important set of techniques, those of multivariate analysis, have enabled the use of small and variable characters to be used to great effect (Sorensen and Foottit 1992). These techniques are now widely used in the analysis of all sorts of previously intractable insect problems.

A major problem of interpreting the variation of insects on different host plants is that of differentiating between those characters that are the direct result of induced responses to living and feeding on particular hosts and those that reflect real genetic differences between populations. Ideally, differentiation is best achieved by experimental rearing and transfer between hosts (Claridge and Gillham 1992). For example Gillham and Claridge (1994) reported results of multivariate analyses of morphological characters of the common polyphagous, tree-feeding leafhopper, *Alnetoidia alneti* (Dahl.) in Europe. Differences in body size and color of populations from different tree species had earlier suggested that *A. alneti* might consist of a complex of sibling species. We found that populations from different tree species were statistically separable by these techniques. However, the transfer of first-instar nymphs from one host to another resulted in adults similar to those normally found on the host plant to which they were transferred. Thus, there is no evidence that the differences between host-associated populations of this insect represent any significant genetic differentiation, and we concluded that *A. alneti* is a truly polyphagous species.

A further example is provided by the widely distributed and studied Asian brown planthopper, *Nilaparvata lugens* (Ståa°l), a pest of rice, *Oryza* species, in Asia and Australasia. The major strategy for control of this pest has been the use of host-plant resistance developed to great effect mainly at the International Rice Research Institute (IRRI) in the Philippines. Populations evolved in the field rapidly during the 1970s and 1980s in different rice-growing areas of Asia that were able to overcome previously resistant varieties. Some of these virulent populations were reared in laboratory cages and termed biotypes 1, 2, 3, and so forth, depending on their patterns of virulence to particular resistant varieties (IRRI 1979). Working at IRRI, Saxena and Rueda (1982) and Saxena et al. (1983) demonstrated significant morphometric differences among these biotypes and concluded that they were host races and represented intermediate stages in a process of incipient speciation. We confirmed that there were indeed significant differences between these biotype populations, but when they were all reared on one susceptible rice variety under the same conditions, the differences among them disappeared after only one generation (Claridge et al. 1984). These biotypes are simply locally adapted populations with relatively little genetic differentiation among them, as confirmed also by selection and crossing experiments (Claridge and den Hollander 1980, 1982). They certainly do not merit any special taxonomic recognition.

Most controversy has surrounded the subject of possible host races and biotypes, with respect to theories of sympatric speciation. Drès and Mallet (2002) published a full and critical review of insect host races in the context of sympatric speciation. They concluded that at least some supposed examples of host races do represent specialized genetically differentiated populations that still interbreed regularly, but retain their identities by strong disruptive selection and, therefore, may be regarded as true host races. Other populations

represent definite biological species and yet others, such as the biotypes of *N. lugens* on cultivated rices, show little differentiation and are effectively parts of a panmictic species.

Specific mate recognition and sibling species

A helpful method of approaching the problems of interpreting different degrees of host- or habitat-associated variation is to use the biological species concept, with its emphasis on reproductive isolation in the field achieved by distinct SMRS. Mate finding and courtship in insects is usually complex, with a sequential exchange of signals and responses in tuned receptors in the two sexes. Any complete sequence is likely to involve various types of signals and receptors, including chemical, visual, mechanical, and auditory senses. However, few complete SMRS sequences have been described and analyzed for any insect.

Because of the predominance of chemical senses in insects generally, specific chemical signals – pheromones – are usually important. These are well known among the Lepidoptera, but also in the Coleoptera, Diptera, and Hymenoptera of the large orders. Few examples of detailed studies on pheromone systems in groups of related specialist-feeding insects have been made. One of the most instructive is that of the small ermine moths, *Yponomeuta* species. The larvae of these insects feed on the foliage of their host plants, mostly broad-leaved trees, where they often cause extensive defoliation. Thorpe (1929), in a classic study, showed how populations from *Crataegus* and *Malus* hybridized freely in the laboratory, but the same populations showed both feeding and oviposition preferences for the plant from which they had been reared. He concluded that they were biological or host races. More recent study conclusively shows these two forms not to interbreed in the field and to be separate biological species (Menken 1980). Nine species now are recognized in northern Europe (review by Menken et al. 1992). Like many moths, virgin females of these insects 'call' by liberating a specific pheromone. Responsive males are attracted maximally to the pheromone of their own species (Hendrikse 1979, 1986). When in close proximity to a calling female, males produce their own specific sex pheromone, which may elicit mating behavior (Hendrikse et al. 1984). The sequence of exchanges of signals and responses means that interspecific matings are rare. Chemical interactions are thus central to the SMRS of these insects.

Chemical signals are generally difficult to study. Most work has been done on single important pest species in which pheromones may be used to manipulate the behavior of the pest and, thus, can be used in control strategies. Visual and acoustic systems of communication seem to be rarer in insects, but are generally easier for the human observer to study. Acoustic systems have received considerable attention in recent years and are more widespread than previously thought (reviews in Drosopoulos and Claridge 2006). The Auchenorrhyncha is a large species-rich group exclusively of herbivores. All species, so far as known, use acoustic (including vibrational) signals in mate finding and courtship (Claridge 1985). The larger cicadas (Cicadidae) are well known for their often loud male songs that are usually species specific and attractive to virgin females. The much more abundant and generally smaller species of other families, such as the Cicadellidae, Delphacidae, and Membracidae signal between males and females by low-intensity vibrational calls that are transmitted through their substrate, normally the host plant.

One of the best studies on a group of closely related insect herbivores is that of Wood (1993) on the complex of morphologically almost identical treehoppers (Membracidae) in North America, known as *Enchenopa binotata*. Wood started his long-term studies with the hypothesis that the eight or nine host-associated populations were host races. A series of elegant electrophoretic and field experimental studies on these insects demonstrated that *E. binotata* consists of at least nine reproductively isolated biological species. Recent and continuing studies have shown that these insects communicate by substrate-transmitted acoustic signals that are certainly central to specific mate recognition and maintaining reproductive isolation (Hunt 1994, Rodríguez et al. 2004, Cocroft and McNett 2006).

A final example of extreme sibling species is provided by the planthopper, *N. lugens*, and its rice-feeding 'biotypes', which were shown to be populations locally adapted to particular cultivars of rice that incorporate distinctive genes for resistance. However, populations morphologically attributable to *N. lugens* also have been found widely feeding on the wild grass *Leersia hexandra*, a relative of rice, *Oryza*, both of which frequently grow in close proximity. These populations were at first described as 'nonvirulent biotypes' of

N. lugens in the Philippines (Saxena et al. 1983). Rice and *Leersia*-associated populations of *N. lugens* occur regularly in close proximity in the field throughout much of Asia and northern Australia (Claridge et al. 1985b, 1988). Like other planthoppers, these insects exchange substrate-transmitted acoustic signals during mating and courtship. The signals of sympatric males and females differ consistently in call characteristics (Claridge et al. 1985a, b, 1988), which act as barriers to interbreeding and are important parts of the SMRSs. Mate choice and call playback experiments confirm that the call differences are responsible for the lack of detected field hybridization between these forms when in sympatry and, thus, show that they should be regarded as different biological species. In the laboratory, hybridization between the two host-associated species is easily achieved in the absence of a choice of mates, and result in viable F1 and F2 generations, with little indication of hybrid inviability. Preliminary molecular studies on some of these populations also suggest that they are closely related, with little obvious genetic divergence (Jones et al. 1996, Sezer and Butlin 1998). The small, but consistent, differences between calling songs of both males and females of the two closely related species of what must now be regarded as the *N. lugens* complex are the only real differences or markers yet identified, other than host-plant preference. These are, thus, extreme examples of sibling species. Allopatric populations of each of the host-plant associated species also show variability in song characters, such that many do not easily interbreed in the laboratory. Thus, some might also be regarded as further separate species of the complex.

If the real diversity of insects in the field is to be recognized by our system of taxonomy, then the biological species, despite the difficulties outlined here, will incorporate more useful information than will other concepts available to us.

Parthenogenetic insects

About 1% of all insects are parthenogenetic, and an agamospecies concept has to be used in the recognition and description of 'species' (Foottit 1997). Parthenogenetic organisms exist as clones, which contrary to some opinion, can show considerable genetic variation (Loxdale and Lushai 2003). It is essential in such groups to have names for distinctive entities that may be regarded as species. They often differ in important features of behavior, such as feeding preferences and ecology, from other agamospecies (De Bach 1969). From a practical viewpoint, species in these insects should be treated in essentially the same way as biparental species. They may be discriminated by any of the phenotypic markers normally used for biological species, including particularly morphological and molecular markers.

CONCLUSIONS

Most insects are biparental, sexually reproducing organisms so that application of the broad-based biological species concept, as advocated here, will lead to the recognition of more biological diversity than will a purely morphological and molecular approach. Sibling species are important ecological entities about which significant generalizations can be made. In practice, the application of a phylogenetic species concept by taxonomists sensitive to the diversity of markers that are available will often produce similar results to those using the biological species concept. The main problem is that the phylogenetic approach gives no incentive to expose the existence of sibling species within an already diagnosably distinct species. Thus, the extent of the real biodiversity of insects, enormous as it is, may be dramatically underestimated.

Whichever approach is favored by any particular taxonomist, the enormity of the task we face in attempting to document insect diversity is clear. The lack of support for taxonomy over recent years has been based on the misconception that it is in some way not real science. While we shall certainly disagree about the precise species concept, I can only conclude by agreeing wholeheartedly with the recent sentiments of Wheeler (2004) on this: 'Taxonomists synthesise and interpret billions of facts about millions of species, make those species identifiable, provide the vocabulary to talk about them, critically test the evolutionary units of biological diversity, and make accessible and predictable all that we know of life on Earth. It has a rich and proven epistemic basis that makes its hypotheses testable and its results as rigorously scientific as any.'

REFERENCES

Al-Barrak, M., H. D. Loxdale, C. P. Brookes, H. Dawah, D. G. Biron, and O. Alsagair. 2004. Molecular evidence using enzyme and RAPD markers for sympatric evolution in British species of *Tetramesa* (Hymenoptera: Eurytomidae). *Biological Journal of the Linnean Society* 83: 509–525.

Avise, J. C. 1994. *Molecular Markers, Natural History and Evolution.* Chapman and Hall, New York.

Avise, J. C. 2000. *Phylogeography: The History and Formation of Species.* Harvard University Press, Harvard.

Blaxter, M. L. 2004. The promise of a DNA taxonomy. *Philosophical Transactions of the Royal Society Series B, Biological Sciences* 359: 669–679.

Brown, J. L. 1975. *The Evolution of Behavior.* W. W. Norton and Co., New York.

Bush, G. L. 1975. Modes of animal speciation. *Annual Review of Ecology and Systematics* 6: 339–364.

Bush, G. L. 1993. A reaffirmation of Santa Rosalia, or why are there so many kinds of animals? Pp. 229–249. *In* D. R. Lees and D. Edwards (eds) *Evolutionary Patterns and Processes.* Academic Press, London.

Bush, G. L. 1994. Sympatric speciation. *Trends in Ecology and Evolution* 9: 285–288.

Bush, G. L. 1995. Species and speciation. *Trends in Ecology and Evolution* 10: 38.

Cain, A. J. 1954. *Animal Species and Their Evolution.* Hutchinson, London.

Cain, A. J. 1958. Logic and memory in Linnaeus's system of taxonomy. *Proceedings of the Linnean Society of London* 169: 144–163.

Cain, A. J. 1993. Linnaeus's *Ordines naturales. Archives of Natural History* 20: 405–415.

Claridge, M. F. 1960. The biospecies in entomology. *Nature* 188: 1172–1173.

Claridge, M. F. 1985. Acoustic signals in the Homoptera: behaviour, taxonomy and evolution. *Annual Review of Entomology* 30: 297–317.

Claridge, M. F. 1988. Species concepts and speciation in parasites. Pp. 92–111. *In* D. L. Hawkesworth (ed). *Prospects in Systematics.* Clarendon Press, Oxford.

Claridge, M. F. 1995a. Species concepts and speciation in insect herbivores: planthopper case studies. *Bolletino di Zoologia* 62: 53–58.

Claridge, M. F. 1995b. Species and speciation. *Trends in Ecology and Evolution* 10: 38.

Claridge, M. F., H. A. Dawah, and M. R. Wilson. 1997a. Practical approaches to species concepts for living organisms. Pp 1–15. *In* M. F. Claridge, H. A. Dawah, and M. R. Wilson (eds). *Species: The Units of Biodiversity.* Chapman and Hall, London.

Claridge, M. F., H. A. Dawah, and M. R. Wilson. 1997b. Species in insect herbivores and parasitoids – sibling species, host races and biotypes. Pp. 247–272. *In* M. F. Claridge, H. A. Dawah, and M. R. Wilson (eds). *Species: The Units of Biodiversity.* Chapman and Hall London.

Claridge, M. F. and J. den Hollander. 1980. The '"biotypes" of the rice brown planthopper, *Nilaparvata lugens. Entomologia Experimentalis et Applicata* 27: 23–30.

Claridge, M. F. and J. den Hollander. 1982. Virulence to rice cultivars and selection for virulence of the brown planthopper, *Nilaparvata lugens. Entomologia Experimentalis et Applicata* 32: 213–221.

Claridge, M. F. and J. den Hollander. 1983. The biotype concept and its application to pests of agriculture. *Crop Protection* 2: 85–95.

Claridge, M. F., J. den Hollander, and D. Haslam. 1984. The significance of morphometric and fecundity differences between the "biotypes" of the brown planthopper, *Nilaparvata lugens. Entomologia Experimentalis et Applicata* 36: 107–114.

Claridge, M. F., J. den Hollander, and J. C. Morgan. 1985a. Variation in courtship signals and hybridization between geographically definable populations of the rice brown planthopper, *Nilaparvata lugens* (Stål). *Biological Journal of the Linnean Society* 24: 35–49.

Claridge, M. F., J. den Hollander, and J. C. Morgan. 1985b. The status of weed-associated populations of the brown planthopper, *Nilaparvata lugens* (Stål) – host race or biological species? *Zoological Journal of the Linnean Society* 84: 77–90.

Claridge, M. F., J. den Hollander, and J. C. Morgan. 1988. Variation in host plant relations and courtship signals of weed-associated populations of the brown planthopper, *Nilaparvata lugens* (Stål), from Australia and Asia: a test of the recognition species concept. *Biological Journal of the Linnean Society* 35: 79–93.

Claridge, M. F. and M. C. Gillham. 1992. Variation in populations of leafhoppers and planthoppers (Auchenorrhyncha): biotypes and biological species. Pp. 241–259. *In* J. T. Sorensen and R. G. Foottit (eds). *Ordinations in the Study of Morphology, Evolution and Systematics of Insects; Applications and Quantitative Genetic Rationales.* Elsevier, Amsterdam.

Cocroft, R. B. and G. D. McNett. 2006. Vibratory communication in treehoppers (Hemiptera: Membracidae). Pp. 305–317. *In* S. Drosopoulos and M. F. Claridge (eds). *Insect Sounds and Communication: Physiology, Behaviour, Ecology and Evolution.* Taylor and Francis, New York.

Coyne, J. A., H. A. Orr, and D. J. Futuyma. 1988. Do we need a new species concept? *Systematic Zoology* 37: 190–200.

Coyne, J. A. and H. A. Orr. 2004. *Speciation.* Sinauer Associates, Sunderland, MA. 545 pp.

Cracraft, J. 1983. Species concepts and speciation analysis. Pp. 159–187. *In* R. Johnston (ed). *Current Ornithology.* Plenum, New York.

Cracraft, J. 1992. The species of the birds-of-paradise (Paradisaeidae): applying the phylogenetic species concept to complex patterns of diversification. *Cladistics* 8: 1–43.

Cracraft, J. 1997. Species concepts and speciation analysis – an ornithological viewpoint. Pp. 325–339. *In* M. F. Claridge and H. A. Dawah (eds). *Species: The Units of Biodiversity.* Chapman and Hall, London.

De Bach, P. 1969. Uniparental, sibling and semi-species in relation to taxonomy and biological control. *Israel Journal of Entomology* 4: 11–27.

Diehl, S. R. and G. L. Bush. 1984. An evolutionary and applied perspective of insect biotypes. *Annual Review of Entomology* 29: 471–504.

Dobzhansky, T. 1937. *Genetics and the Origin of Species.* Columbia University Press, New York.

Dobzhansky, T. 1940. Speciation as a stage in evolutionary divergence. *American Naturalist* 74: 302–321.

Drès, M. and J. Mallet. 2002. Host races in plant feeding insects and their importance in sympatric speciation. *Philosophical Transactions of the Royal Society Series B, Biological Sciences* 357: 471–492.

Drosopoulos, S. and M. F. Claridge (eds). 2006. *Insect Sounds and Communication: Physiology, Behaviour, Ecology and Evolution.* Taylor and Francis, New York.

Eibl-Eibesfeldt, I. 1970. *Ethology: The Biology of Behavior.* Holt, Rinehart and Winston, New York.

Ereshefsky, M. (ed). 1992. *The Units of Evolution: Essays on the Nature of Species.* MIT Press, New York.

Foottit, R. G. 1997. Recognition of parthenogenetic insect species. Pp. 291–307. *In* M. F. Claridge, H. A. Dawah, and M. R. Wilson (eds). *Species: The Units of Biodiversity.* Chapman and Hall, London.

Gillham, M. C. and M. F. Claridge. 1994. A multivariate approach to host plant associated morphological variation in the polyphagous leafhopper, *Alnetoidia alneti* (Dahlbom). *Biological Journal of the Linnean Society* 53: 127–151.

Gornall, R. J. 1997. Practical aspects of the species concept in plants. Pp. 357–380. *In* M. F. Claridge, H. A. Dawah, and M. R. Wilson (eds). *Species: The Units of Biodiversity.* Chapman and Hall, London.

Hammond, P. M. 1992. Species inventory. Pp. 17–39. *In* B. Groombridge (ed). *Global Biodiversity: Status of the Earth's Living Resources.* Chapman and Hall, London.

Hendrikse, A. 1979. Activity patterns and sex pheromone specificity as isolating mechanisms in eight species of *Yponomeuta* (Lepidoptera : Yponomeutidae). *Entomologia Experimentalis et Applicata* 25: 172–180.

Hendrikse, A. 1986. Intra- and interspecific sex-pheromone communication in the genus *Yponomeuta. Physiological Entomology* 11: 159–169.

Hendrikse, A., C. E. van der Laan, and L. Kerkhof. 1984. The role of male abdominal brushes in the sexual behaviour of small ermine moths (*Yponomeuta* Latr., Lepidoptera). *Mededelingen van de Faculteit Landbouwetenschappen Rijksuniversiteit Gent* 49: 719–729.

Hennig, W. 1966. *Phylogenetic Systematics.* University of Illinois Press, Chicago, IL.

Hey, J. 2006. On the failure of modern species concepts. *Trends in Ecology and Evolution* 21: 447–450.

Hunt, R. E. 1994. Vibrational signals associated with mating behavior in the treehopper, *Enchenopa binotata* Say

(Hemiptera: Homoptera: Membracidae). *Journal of the New York Entomological Society* 102: 266–270.

IRRI. 1979. *Brown Planthopper: Threat to Rice Production in Asia.* International Rice Research Institute, Philippines.

Jones, P., P. Gacesa, and R. Butlin. 1996. Systematics of the brown planthopper and related species using nuclear and mitochondrial DNA. Pp. 133–148. *In* W. O. C. Symondson and J. E. Liddell (eds). *The Ecology of Agricultural Pests: Biochemical Approaches.* Chapman and Hall, London.

Lambert, D. M. and H. G. Spencer (eds). 1995. *Speciation and the Recognition Concept.* Johns Hopkins University Press, Baltimore, MD.

Lane, R. 1997. The species concept in blood-sucking vectors of human diseases. Pp. 273–289. *In* M. F. Claridge, H. A. Dawah, and M. R. Wilson (eds). *Species: The Units of Biodiversity.* Chapman and Hall, London.

Linnaeus, C. 1737. *Critica Botanica.* Ray Society, London. (Translated by A. Hort, 1938).

Linton, Y.-M., L. Smith, G. Koliopoulos, A. Samanidou-Voyadjoglou, A. K. Kounos, and R. Harbach. 2003. Morphological and molecular characterization of *Anopheles (Anopheles) maculipennis* Meigen, type species of the genus and nominotypical member of the Maculipennis Complex. *Systematic Entomology* 28: 39–55.

Loxdale, H. D. and G. Lushai. 2003. Rapid changes in clonal lines: the death of a "sacred cow". *Biological Journal of the Linnean Society* 79: 3–16.

Mallet, J. 1995. A species definition for the modern synthesis. *Trends in Ecology and Evolution* 10: 294–304.

Mallet, J. 2005. Hybridization as an invasion of the genome. *Trends in Ecology and Evolution* 20: 229–237.

Mallet, J. 2008. Hybridization, ecological races and the nature of species: empirical evidence for the ease of speciation. *Philosophical Transactions of the Royal Society(B)* 363: 2971–2986.

Mayden, R. L. 1997. A hierarchy of species concepts: the denouement in the saga of the species problem. Pp. 381–424. *In* M. F. Claridge, H. A. Dawah, and M. R. Wilson (eds). *Species: The Units of Biodiversity.* Chapman and Hall, London.

Mayden, R. L. 1999. Consilience and a hierarchy of species concepts: advances toward closure on the species puzzle. *Journal of Nematology* 31: 95–116.

Mayr, E. 1942. *Systematics and the Origin of Species from the Viewpoint of a Zoologist.* Columbia University Press, New York.

Mayr, E. 1963. *Animal Species and Evolution.* Harvard University Press, Cambridge, MA.

Mayr, E. 1982. *The Growth of Biological Thought.* Harvard University Press, Cambridge, MA.

Mayr, E. 1992. A local flora and the biological species concept. *American Journal of Botany* 79: 222–238.

Mayr, E. 2004. *What Makes Biology Unique? Considerations on the Autonomy of a Scientific Discipline.* Cambridge University Press, Cambridge, MA.

Mayr, E. and W. B. Provine (eds). 1980. *The Evolutionary Synthesis: Perspectives on the Unification of Biology.* Harvard University Press, Cambridge, MA.

Menken, S. B. J. 1980. Inheritance of allozymes in *Yponomeuta.* II. Interspecific crosses within the *padellus*-complex and reproductive isolation. *Proceedings of the Koninklijke Nederlandse Akademic van Weterschapper, Series C* 83: 424–431.

Menken, S. B. J., W. M. Herrebout, and J. T. Wiebes. 1992. Small ermine moths (*Yponomeuta*): their host relations and evolution. *Annual Review of Entomology* 37: 41–66.

Nelson, G. 1989. Cladistics and evolutionary models. *Cladistics* 6: 211–223.

Nixon, K. and Q. D. Wheeler. 1990. An amplification of the phylogenetic species concept. *Cladistics* 6: 211–223.

Paterson, H. E. H. 1985. The recognition concept of species. Pp. 21–29. *In* E. S. Vrba (ed). *Species and Speciation.* Transvaal Museum, Pretoria, South Africa.

Paterson, H. E. H. 1993. *Evolution and the Recognition Concept of Species: Collected Writings of H. E. H. Paterson.* S. F. McEvey (ed). Johns Hopkins University Press, Baltimore, MD.

Poulton, E. B. 1908. What is a species? Pp. 46–94. *In Essays on Evolution 1889–1907,* Clarendon Press, Oxford.

Price, P. W. 1980. *Evolutionary Biology of Parasites.* Princeton, NJ.

Ramsbottom, J. 1938. Linnaeus and the species concept. *Proceedings of the Linnean Society of London* 150: 192–219.

Regan, C. T. 1926. Organic evolution. *Report of the British Association for the Advancement of Science* 1925: 75–86.

Rodríguez, R. L., L. E. Sullivan, and R. B. Cocroft. 2004. Vibrational communication and reproductive isolation in the *Enchenopa binotata* species complex of treehoppers (Hemiptera: Membracidae). *Evolution* 58: 571–578.

Saxena, R. C. and L. M. Rueda. 1982. Morphological variations among three biotypes of the brown planthopper *Nilaparvata lugens* in the Philippines. *Insect Science and its Application* 6: 193–210.

Saxena, R. C., M. V. Velasco, and A. A. Barrion. 1983. Morphological variations between brown planthopper biotypes on *Leersia hexandra* and rice in the Philippines. *International Rice Research Newsletter* 18: 3.

Sezer, M. and R. K. Butlin. 1998. The genetic basis of host plant adaptation in the brown planthopper (*Nilaparvata lugens*). *Heredity* 80: 499–508.

Simpson, G. G. 1951. The species concept. *Evolution* 5: 285–298.

Sokal, R. R. and T. J. Crovello. 1970. The biological species concept: a critical evaluation. *American Naturalist* 104: 127–153.

Sorensen, J. T. and R. G. Foottit (eds). 1992. *Ordination in the Study of Morphology, Evolution and Systematics of Insects. Applications and Quantitative Genetic Rationales.* Elsevier, Amsterdam.

Sperling, F. 2003. Butterfly molecular systematics: from species definitions to higher-level phylogenies. Pp. 431–458. *In* C. L. Boggs, W. B. Watt, and P. R. Ehrlich (eds). *Butterflies,*

Ecology and Evolution Taking Flight. University of Chicago Press, Chicago, IL.

Strong, D. R., J. H. Lawton, and R. Southwood. 1984. *Insects on Plants: Community Patterns and Mechanisms.* Blackwell, Oxford.

Tautz, D., P. Arctander, A. Minelli, R. H. Thomas, and A. P. Vogler. 2003. A plea for DNA taxonomy. *Trends in Ecology and Evolution* 18: 70–74.

Tinbergen, N. 1951. *The Study of Instinct.* Oxford University Press, Oxford.

Thorpe, W. H. 1929. Biological races in *Hyponomeuta padella* L. *Zoological Journal of the Linnean Society* 36: 621–634.

Thorpe, W. H. 1930. Biological races in insects and allied groups. *Biological Reviews* 5: 177–212.

Thorpe, W. H. 1940. Ecology and the future of systematics. Pp. 341–364. *In* J. Huxley (ed). *The New Systematics.* The Oxford University Press, London.

Vrba, E. 1995. Species as habitat specific, complex systems. Pp. 3–44. *In* D. M. Lambert and H. G. Spencer (eds). *Speciation and the Recognition Concept.* Johns Hopkins University Press, Baltimore, MD.

Wallace, A. R. 1865. On the phenomena of variation and geographical distribution as illustrated by the Papilionidae of the Malayan region. *Transactions of the Linnean Society* 25: 1–71.

Wallace, A. R. 1889. *Darwinism: An Exposition of the Theory of Natural Selection.* MacMillan, London.

Walsh, B. J. 1864. On phytophagic varieties and phytophagic species. *Proceedings of the Entomological Society of Philadelphia* 3: 403–430.

Wheeler, Q. 1999. Why the phylogenetic species concept? – elementary. *Journal of Nematology* 31: 134–141.

Wheeler, Q. D. 2004. Taxonomic triage and the poverty of phylogeny. *Philosophical Transaction of the Royal Society Series B, Biological Sciences* 359: 571–583.

Wheeler, Q. D. and R. Meier (eds). 2000. *Species Concepts and Phylogenetic Theory, a Debate.* Columbia University, New York.

Wheeler, Q. D. and K. C. Nixon. 1990. Another way of looking at the species problem: a reply to de Queroz and Donoghue. *Cladistics* 6: 77–81.

White, G. 1789. *The Natural History of Selbourne.* White Cochran, London.

Wiley, E. 1978. The evolutionary species concept reconsidered. *Systematic Zoology* 27: 17–26.

Wilson, E. O. 1992. *The Diversity of Life.* Harvard University Press, Cambridge, MA.

Wilson, R. A. (ed). 1999. *Species, New Interdisciplinary Essays.* MIT Press, Cambridge, Massachusetts.

Wood, T. K. 1993. Speciation of the *Enchenopa binotata* complex (Insecta: Homoptera: Membracidae). Pp. 289–317. *In* D. R. Lees and D. Edwards (eds). *Evolutionary Patterns and Processes.* Academic Press, London.

Chapter 16

MOLECULAR DIMENSIONS OF INSECT TAXONOMY

Felix A. H. Sperling and Amanda D. Roe

Department of Biological Sciences, CW405a Biological Sciences Centre, University of Alberta, Edmonton, Alberta T6G 2E9, Canada

Insect Biodiversity: Science and Society, 1st edition. Edited by R. Foottit and P. Adler
© 2009 Blackwell Publishing, ISBN 978-1-4051-5142-9

At their most fundamental level, molecular methods are simply a subset of the ways that taxonomists gain access to characteristics used to distinguish groups of organisms. The value of any class of characters is ultimately measured by how accurately and conveniently these characters distinguish the units that matter in taxonomy. Diverse genetic and biochemical techniques can be used to supply the identifying features that characterize individuals and delimit groups. Molecular characters have varied strengths and drawbacks, as do any other sources of character information such as morphology. The purpose of this chapter is to briefly review the uses, as well as misuses, of molecular characters in insect taxonomy.

In the study of biodiversity, the units that matter most are species. The staggering abundance of insect species, both beautiful and pestilential, has made them the proving ground for taxonomic information-management systems since the dawn of recorded human history. Most insects also present technical challenges inherent in simply seeing differences between diminutive creatures. Added to that are the conceptual challenges of defining species consistently across a biological lineage that has diversified over the course of 400 million years. Consequently, insect taxonomy is a work in progress and will remain a primary source of new insights in biodiversity informatics for a long time. In this chapter, we focus on the technical and conceptual advances by which molecular methods have incrementally assisted in seeing, recording, and using insect diversity. As illustrative examples, we use the case studies that we are most familiar with, often from our own research, on the principle that it is better to explore and understand the nuances of a few cases than to describe the surface of a broader array of examples.

Molecular methods now dominate many aspects of insect taxonomy, and the field has long passed the point where the contribution of such methods can be reviewed comprehensively in a single book chapter. An overview published in *Annual Review of Entomology* (Caterino et al. 2000) concluded with a call to focus on DNA sequences from only a small number of genes across all of insect systematics, to avoid a 'Tower of Babel' in which different studies do not effectively relate to each other. Mitochondrial protein-coding genes provided the most useful markers at the species level, and the choice of the COI gene by Caterino et al. (2000) was echoed and amplified by Gleeson et al. (2000) on the basis of the demonstrated value of this gene for insect species diagnostics. In vertebrates, DNA-based identification has generally relied on other mitochondrial protein-coding genes, especially cytochrome B (e.g., Parson et al. 2000). Insect species comprise three quarters of animal species and half of all described life, and so it was logical to extend recommendations for standardization beyond the insects to the rest of life, whether as DNA taxonomy (Tautz et al. 2002, Vogler and Monaghan 2006) or DNA barcoding (Floyd et al. 2002, Hebert et al. 2003a), in the explicit hope that the application of these particular molecular methods would relieve the global taxonomic impediment.

We take the view that molecular taxonomy is not, and should not be, a parallel approach to insect taxonomy. Rather, molecular methods are an important part of a more holistic, integrative taxonomy, with such methods providing diverse additional character sets for addressing the classical problems of identifying specimens, discovering and delimiting species, and determining relationships (Dayrat 2005, Will et al. 2005, Valdecasas et al. 2008). Taxonomy is a venerable discipline that remains both vibrant and fundamental to the rest of biology; the use of molecules of all kinds adds depth and new dimensions.

CHALLENGES IN TAXONOMY

Practicing taxonomists encounter an array of logical puzzles whose complexity is matched only by the profound satisfaction of resolving them. Despite endless variety in the details of these puzzles, they can be reduced to variations on four basic themes: (1) determination of the identity of specimens of known species, (2) discovery of new species, (3) delimitation of species boundaries, and (4) phylogenetic reconstruction. Molecular methods are helpful, to varying degrees, in resolving each of these four kinds of problems (the first letters of which comprise a euphonious acronym – D3P).

Determination

The process of identifying specimens as members of previously demarcated species has benefited enormously from the addition of molecular characters, at least in those cases where morphological characters are insufficient. This is a logically simple problem in which the integrity of the boxes (species) is not at issue, but our ability to put a specimen into the correct box is limited by a paucity of useable characters. Molecular methods provide a whole array of new characters that can assist identification. As long as their utility is tested beforehand against established diagnostic characters, any source of information such as a single nucleotide, an enzyme allele, or a hydrocarbon variant can serve to identify different life stages, isolated body parts, or specimens bereft of morphologically diagnostic information.

The oft-ignored essence of effective identification is to survey the natural range of character variation well enough ahead of time, so that the identification of a single specimen has a high probability of being correct, regardless of where the specimen came from (Medina et al. 2007, Muirhead et al. 2008). But such surveys can go on forever, and this begs the question – how well should the range of variation of a character system be known? For cases that really matter, one benchmark is whether the identification would hold up in a court of law. Such a standard is not frivolous because insect identifications are often used to hold up large perishable shipments at ports, to implement expensive eradications of invasive species, or to convict murderers, using forensic evidence. A 95% probability of correct identification will neither convince a judge who is weighing the human cost of a mistake, nor is it likely to deter a sharp corporate lawyer defending a shipping company.

Forensic entomology provided the framework for an early effort to use molecular characters to identify insects in a legal context. The reliability of different fly species as indicators of successive stages of decay (postmortem interval) is reasonably well accepted in court. One drawback is that maggots can be difficult to identify to species; they must be reared to the adult stage and a significant amount of time may elapse before information derived from them can be used to focus an investigation. Such delay can be circumvented by using DNA to identify maggots, an approach first described by Sperling et al. (1994). However, despite extensive exploration of the approach by J. Wells and others (e.g., Wells and Sperling 2000, 2001, Wells et al. 2001a, b), DNA has not yet become the primary method for identification of insects in actual court cases, instead serving only as confirmation of identification by other means. This situation is largely due to the growing realization that standard mitochondrial DNA (mtDNA) sequences often do not form conveniently interpreted clusters that are congruent with species limits determined by other means (Wells et al. 2007, Whitworth et al. 2007, Wells and Stevens 2008). Molecular characters remain highly useful for identifying forensically important species in a great variety of contexts, but it is imperative to first perform a fine-grained survey of population variation across the full geographic range of every fly species that is used.

Verification of character utility is usually performed informally for morphological characters, with explicit examination of numerous specimens and an unconscious assessment of the probability of developmental or evolutionary conversion of one structure into another. A classic 'good taxonomist' is particularly talented in such analyses of shapes, gradually building up a mental model of transformation probabilities that can be applied to correctly identify a particular specimen, with high confidence. The problem is that the mental model is rarely made explicit and is usually difficult to apply across different taxonomic groups. This capability also is difficult to transfer between people; the demise of an experienced taxonomist inevitably means the extinction of a large amount of unique knowledge.

Molecular characters are often advocated as a solution to this problem, because at some levels it can be much easier to train a technician to use molecular methods for diagnostics than to apply morphological methods. What is not so well appreciated is that molecular characters have generally not received as much informal testing as have morphological characters before they are employed in practical situations, both because fewer specimens have been surveyed and because there may be less intuitive understanding of the underlying probabilities of character-state transformation (Rubinoff et al. 2006).

Ermine moths of the genus *Yponomeuta* provide an example of both the utility and limitations of molecular identification in agricultural pest species. Several species of ermine moths cannot be distinguished reliably on the basis of adult morphology, but numerous studies on their ecology, behavior, and genetics in their native range in Europe have shown that they are biologically distinct species that maintain their integrity upon contact (Menken et al. 1993). Several *Yponomeuta* species have now invaded North America, with significant economic consequences. The apple ermine moth *Y. malinellus* was introduced to both British Columbia and Washington by the early 1980s, whereas the cherry ermine moth *Y. padella* was first found in southwestern British Columbia in 1993 (Sperling et al. 1995). The only way to distinguish specimens of these two species with certainty was through pheromone attraction of adults and feeding preferences of the larvae. Neither of these assays was applied in time to determine definitively whether the cherry ermine moth also occurred in Washington State in 1993, and further evaluation had to wait until the next growing season. This situation created a potential economic and political problem, because there was no clear indication that the cherry ermine moth was present in Washington, and as a precautionary step the US Department of Agriculture (USDA) was prepared to stop importation of all nursery trees from Canada that might carry diapausing immatures of the cherry ermine moth. However, it took only a couple of months to develop a DNA-based method for distinguishing the cherry and apple ermine moths, using reared museum specimens with known host associations to calibrate the technique (Sperling et al. 1995). This application of DNA diagnostics showed that the cherry ermine moth had already been collected in Washington, thereby obviating the need for a trade embargo. By the summer of 1994, the protocol for the DNA-based diagnostic test was transferred to the USDA, and later evaluation of the genotype of 800 moths collected in pheromone traps verified the method (Lagasa et al. 1995). This DNA assay, however, might be reliable only in the Pacific Northwest where it was developed and tested. Later sequencing of a few specimens of the same species from part of their native range in The Netherlands has indicated that the genotypes that are species specific in North America are not diagnostic in Europe (Turner et al. 2004).

Discovery

One of the most obvious applications of molecular methods in taxonomy is to provide new sets of characters that show clear discontinuities in assemblages that were previously seen as more or less continuous (Bickford et al. 2007). However, this approach is not commonly employed; the main use of molecular characters lies not in the initial discovery of new species, population units, or evolutionary relationships but in confirming their existence by providing a new set of characters in a hypothesis-testing framework. Because of the greater cost or time that is generally required to assay molecular characters across large numbers of specimens, the existence of most new species is

almost always first suspected on the basis of ecological, behavioral, or morphological variation. Furthermore, assaying molecular characters inevitably necessitates some degree of destruction of specimens, though with continued refinement of techniques that destruction can be minimized (Dean and Ballard 2004, Rowley et al. 2007, Hunter et al. 2008).

For example, the discovery of new species of *Bemisia* whiteflies (Bellows et al. 1994), *Archips* leafrollers (Kruse 2000), and *Phytomyza* leafminers (Scheffer and Hawthorne 2007) was prompted by prior recognition of ecologically distinct biotypes, followed by a search for characters of any kind that would diagnose these previously detected units (Dres and Mallet 2002). The same applies to the discovery of ten cryptic butterfly species in one previously named *Astraptes* species (Hebert et al. 2004, but see Brower 2006), a molecular study that would not have been done without the prior documentation of larval coloration and host associations by coauthor Dan Janzen and associates. Molecular characters as a class are not fundamentally different from any other kinds of characters, although some particular molecular markers such as fast-evolving gene sequences might more likely allow the detection of new species. However, for practical and economic reasons, molecular analyses are generally not performed in the absence of prior information that suggests interesting or important biological discontinuities.

The urgency of biodiversity documentation in disappearing habitats, combined with the frustratingly time-consuming nature of traditional methods for species identification, has created enormous pressure to develop new methods for species discovery, whether molecular or computer based (Godfray 2002). For bacteria and other microorganisms, the shortage of useful characters that are visible by microscope or staining has meant that most new discoveries have relied on DNA sequences for the last decade or more (e.g., Embley and Stackebrandt 1997). However, for rapid processing of large numbers of insect specimens in biodiversity research, molecular methods have not yet caught up to the speed and cost efficiency of visual searches by a student with only a moderate amount of taxonomic training (Cameron et al. 2006).

As molecular processing speeds do catch up to basic visual methods, which will eventually be the case for DNA sequencing and other analyses, there will be a strong temptation to rely on only those molecular markers that are easily assayed. In an extreme form, this sets up a single-marker taxonomic system whereby the new units that are discovered are self-referentially consistent but may have little relevance to genomic diversity (e.g., DNA taxonomy of Monaghan et al. 2006, Pons et al. 2006). We believe that it is better for the stability of communication in taxonomy to leave the discovery of a divergent gene lineage (as is increasingly common for mtDNA) for further evaluation using other characters (e.g., Sperling and Hickey 1994, Nazari and Sperling 2007), rather than drawing immediate and potentially disruptive taxonomic conclusions that are highly vulnerable to falsification (but contrast Franz 2005). As for the identification of previously defined species, the use of molecular methods to discover new species relies on effective integration of molecular information with a number of other kinds of information (Knowles and Carstens 2007, Schlick-Steiner et al. 2007, Wheeler 2008).

Delimitation

Taxonomists commonly delimit several different kinds of units, although species are the fundamental unit in most studies. Any reference to the process of delimiting species raises the question – what is a species? This question seems usually to invite an endless cycle of discussion and disagreement (reviewed recently by Coyne and Orr 2004). Claridge (this volume) has provided an overview for the current volume, allowing us to focus on species delimitation as an exercise in determining the degree of permeability of genetic boundaries between population lineages (Harrison 1998, De Quieroz 2007).

Reduced permeability to gene flow between populations is an essential part of most widely used species concepts. We have found it practical in our own work to explicitly recognize the potential for rare exchange of genes, as well as the fundamentally different process of delimiting species that do not contact each other, through a 'genomic integrity' species definition (Sperling 2003a). This two-part definition relies, first, on inference of the maintenance of genomic integrity of populations when they contact each other, based on a variety of data. Second, for populations that are not in contact and for which evaluation of the maintenance of genomic integrity is not necessarily meaningful, differences between populations are calibrated against the extent of differences between pairs of sister species that do contact each other. Such

allopatric populations are ranked as distinct species when they have levels of character divergence equivalent to the mean for sister-species pairs that are related to the allopatric pairs. This definition is explicitly genetic in its conception and is designed to accommodate molecular data in conjunction with other sources of information. The discrete nature and abundance of most molecular characters are well suited to the kind of quantification that is implied by ranking divergences between populations.

Regardless of the nuances of the particular species definition that may be employed by a working taxonomist, it is important to view each species as a testable hypothesis. Such a hypothesis should be tested against multiple data types, preferably with disparate analytical properties. Here again, molecular data have proven highly useful. For example, Sperling (1987) found that clustering of individual swallowtail butterflies on the basis of morphological character scores and allozyme alleles gave population groupings that had distinct larval host associations and corresponded broadly to previously recognized species. However, neither morphology nor allozymes by themselves delimited species as clearly as they did in conjunction, because there were no 100% frequency differences in individual morphological characters or allozyme loci. Surveys of mtDNA (Sperling and Harrison 1994) not only largely supported the earlier work, but also demonstrated gene leakage of mitochondrial haplotypes into a few populations that nonetheless clearly maintained their overall genomic integrity. An integrated approach to species delimitation remains essential to resolving taxonomic problems in recently diverged species groups (e.g., Rubinoff and Sperling 2004, Cognato and Sun 2007, Roe and Sperling 2007a).

The degree of divergence between populations (e.g., genetic distances calculated as percent similarity or dissimilarity), as well as of taxonomic groups at the level of species and above, provides important information for the delimitation of taxa. However, this approach has serious limitations due to the large amount of overlap in divergences between different taxonomic ranks (Cognato 2006, Nazari et al. 2007). On the other hand, character-based approaches to delimitation of species and other taxa show a great deal of promise for molecular data, in part because they are so conceptually compatible with classical taxonomic practice (DeSalle et al. 2005, Rach et al. 2008). Burns et al. (2007) found that mtDNA phylogenies produced several nonmonophyletic groupings when they were based on distances derived from all available COI sequences, including those that differed slightly in length from the standard barcode size. However, distance-based analysis of a consistent segment of sequence that was invariant in length gave Burns et al. (2007) much better resolution and provided clear synapomorphies for particular nucleotide sites, despite less than 1% divergence between species.

As with the identification and discovery of species, species delimitation based on single genes can be highly prone to error (Funk and Omland 2003, Meier et al. 2006, Monaghan et al. 2006, Elias et al. 2007, Linnen and Farrell 2007, Twewick 2007). Nonetheless, molecular characters have provided enormous benefit to taxonomy when analyzed in the context of other characters and the biology of the whole organism. For many insects, particularly in temperate zones, a good first draft of the species that occur in a region is available based on standard morphological methods. The undescribed remainder is a combination of rare species and species groups with poor morphological distinctions or legitimately messy species boundaries. These difficult-to-delimit species often matter a great deal from an economic viewpoint, and yet such species complexes are also the ones most likely to be resistant to effective characterization using single-gene systems of any kind (Sperling 2003b).

Molecular methods and the genes themselves are wonderfully diverse, and there is effectively no limit to the new kinds of taxonomic information that can be derived from them. For example, early hopes that mtDNA variation would provide a disproportionately effective marker for species limits in Lepidoptera (and other taxa with heterogametic females) have not been borne out, though X-linked genes might instead serve this role (Sperling 1993, Roe and Sperling 2007a).

Phylogeny

Molecular data have had mixed utility in reconstructing phylogenies above the level of species. On the one hand, the availability of large numbers of characters, particularly DNA nucleotides, has allowed the development of sophisticated quantitative methods for phylogeny reconstruction (Felsenstein 2004). On the other hand, different character sets such as DNA from different genes can conflict with each other. Such differences can be due to legitimately conflicting gene phylogenies that result from differential sorting of

gene lineages within species, hybridization leading to gene introgression between species, or selection on maternally inherited symbionts (Ballard and Whitlock 2004, Hurst and Jiggins 2005). Alternatively, conflicts may arise due to sampling errors based on the use of short DNA sequences (e.g., Roe and Sperling 2007b) or unreliable methods for phylogeny reconstruction (Felsenstein 2004). Probably because of the relative ease of obtaining mtDNA sequences compared with nuclear gene sequences, it has only recently become the norm to obtain sequences from multiple, unlinked genes for phylogenetic analysis (e.g., Giribet and Edgecomb 2005, Mallarino et al. 2005, Beltran et al. 2007, Nazari et al. 2007).

Deep phylogenies pose a particular challenge to efforts to reconstruct the evolutionary history of a group, in large part because of the overlay of multiple character substitutions over time (= saturation) and the short time period represented by internodal distances relative to total branch length (Whitfield and Kjer 2008). For this reason, most molecular methods have been consistently applied only near the species level. DNA sequences can provide much deeper information, but the quality of phylogenetic information in mtDNA usually declines rapidly beyond the level of genera (e.g., Nazari et al. 2007). However DNA sequences for some genes have divergence rates that are well suited to retaining traces of their cladistic pattern of nested subsets of mutations. The challenge is to find and consistently sequence the genes that are most informative at a particular taxonomic level, as well as to employ the most appropriate analytical methods. We are currently in a phase of testing numerous genes for deep phylogenetic information (e.g., LepTree Team 2008). One of the key problems is the practical challenge of finding informative genes that do not require RNA extraction and reverse transcriptase PCR, which necessitate reliance on only those specimens with high-quality nucleotide preservation. Wahlberg and Wheat (2008) have made good use of genomics resources to develop a series of intronless genes for use in molecular analysis of recently collected museum specimens, which greatly expands the range of rare taxa that can be sampled effectively.

Distance-based methods for phylogeny reconstruction can be prone to systemic errors caused by factors such as variation in rates of evolutionary change. Character-based methods also can exhibit consistent errors, due to phenomena such as long-branch attraction (Felsenstein 2004). In general, simple distance-based methods, such as neighbor joining, can serve as a quick, but 'dirty', first approximation of relationships; some analyses go no further (e.g., Hebert et al. 2003b). Model-based methods such as maximum likelihood and Bayesian analysis use more of the information inherent in the characters, but can be sensitive to the selection of models (Felsenstein 2004).

One of the advantages of distance methods is that they lend themselves more directly than character-based methods to the use of molecular clocks for estimating divergence dates. However, a key problem with the use of molecular clocks is that rate variation has been widely demonstrated across many taxa, and rate variation between different genes is the norm. For example, mtDNA divergence rates seem to be rapid in *Pissodes* weevils compared with flies and moths, in contrast to allozyme divergence (Langor and Sperling 1997, Boyer et al. 2007). Molecular clocks have thus been widely and justifiably criticized (e.g., de Jong 2007). Their use nonetheless remains attractive in taxonomy (e.g., Avise and Mitchell 2007) because they provide at least some information about relative branching times when fossils are absent or nearly so.

An additional, albeit tenuous, application of distances in taxonomy lies in the quantification of percent DNA-sequence divergences in calibrating the rank of both species level and higher taxa. For example, Hebert et al. (2003a) have used a 3% mtDNA sequence cutoff for recognizing lineage clusters as species, a practice that is poorly supported in insects (Cognato 2006, Hickerson et al. 2006, Meier et al. 2006). Nonetheless, just as neighbor joining can serve as a rough method for estimating phylogenetic relationships, the use of distances as an approximation of taxonomic rank remains attractive because it is simple and rapid. This approach has been used even less above the level of species than at the species level, but the principles remain the same (Nazari et al. 2007). The challenge is to remain cognizant that this practice provides only one weak line of evidence that must be supplemented and tested with ample additional evidence for each group.

SURVEY OF MOLECULAR METHODS

The vast majority of recent studies that have used molecular methods in taxonomy have relied on DNA sequences or sequence length variation (Caterino et al. 2000, Schlötterer 2004, Behura 2006), though it is important to keep in mind that DNA provides only one of many forms of molecular characters. We provide a brief survey of the diversity of molecular methods in current use since 2000, with some commentary on their applications (Table 16.1).

All of the molecular techniques described in Table 16.1 produce data that can be analyzed as either distances or characters. The choice of basic analytical approach generally depends on whether the investigators are geneticists and biodiversity researchers who are in a hurry, in which case they favor distance methods, or whether they are systematists and aiming for high-impact publications, in which case they use character-based methods such as cladistics or explicit models such as maximum likelihood. Molecular techniques that produce only distance data, particularly immunological techniques and DNA-DNA hybridization (Hillis et al. 1996), have fallen by the wayside in insect taxonomy.

The use of biochemical markers also has declined greatly in insect taxonomy, though they may still be employed in groups where they have a long history or where these molecular markers remain relevant for practical reasons (e.g., wasp venoms: Bruschini et al. 2007). Wing-pigment variation has been influential in butterfly taxonomy (e.g. Ford 1944), and analysis of eye pigment diversity remains an active area of investigation; however, assays of these pigments are now performed via the DNA sequences for their genes (Frentiu et al. 2007). Pheromone characterization remains an active field of endeavor in insect control and population surveys, but the relationship between taxonomy and pheromone variation at the level of individuals has been surprisingly little investigated (e.g., Sperling et al. 1996, Cognato et al. 1999).

Of the molecular markers in Table 16.1 that do not rely directly on DNA, both cuticular hydrocarbons and allozymes have the advantage of providing information that is directly relevant to the adaptive phenotype of the whole organism (e.g., Dapporto 2007). Allozymes have the additional advantage of being amenable to classic Mendelian analyses. Both, however, require equipment and training that is distinct from that in standard use in molecular biology. There is still considerable potential to employ allozymes and cuticular hydrocarbons to find out more about the role of genetic variation in the behavior and ecology of individuals among populations and between species (e.g., Jenkins et al. 2000, Dalecky et al. 2007, Foley et al. 2007, Grill et al. 2007, Hay-Roe et al. 2007,

Table 16.1 Survey of techniques currently used to provide molecular characters for insect taxonomy and systematics.

Method	Description	Primary Strength(s)	Primary Weakness(es)	Taxonomic Application	Recent Examples
Cuticular hydrocarbons	Chemical composition of cuticle lipids using gas chromatography/mass spectrometry (GC/MS)	• Linked to species recognition	• Often too labile to be easily compared across taxa • Requires different training and equipment from DNA work	Populations, species	• Dapporto (2007) • Hay-Roe et al. (2007)
Allozymes	Enzyme variants separated by size and charge using gel electrophoresis	• Variation is functional • Can be analyzed as Mendelian markers	• Allele homology is uncertain • Must be analyzed on same gel	Populations, species	• Foley et al. (2007) • Grill et al. (2007)
Karyotypes	Chromosome numbers, banding, and inversions	• Relatively inexpensive	• Requires distinct training and preservation methods	Populations, species	• Martin (2006) • Kandul et al. (2007)
RFLP	Restriction digests of genomic DNA or PCR products to form diagnostic banding on gels	• Inexpensive • Needs only simple equipment and training	• Assays only one or few base pairs at a time • Low amount of information relative to time input	Populations, species	• Naegele et al. (2006) • Beebe et al. (2007) • Li et al. (2007)
(RAPD) DNA	Arbitrarily primed amplification of regions throughout genome to give many short PCR bands	• Inexpensive • Produces numerous markers • No need for prior genetic information	• Poor reproducibility • Needs good quality DNA • Difficult to compare across studies	Populations, species	• Kumar et al. (2001) • Vandewoestijne and Baguette (2002) • Al-Barrak et al. (2004)
AFLP	Restriction digestion of genomic DNA before arbitrarily primed amplification	• Produces numerous markers • No need for prior genetic information	• Medium reproducibility • Needs good quality DNA • Difficult to compare across studies	Populations, species	• Gompert et al. (2006) • Scheffer and Hawthorne (2007) • Mock et al. (2007)
Microsatellites	Short tandem repeats of DNA throughout genome	• Highly variable • Occur throughout genome	• Prone to 'slippage' errors • Difficult to obtain from some insects (e.g., Lepidoptera)	Kinship, populations	• Meglécz et al. 2007 • Zakharov and Hellmann (2008)

Method	Description	Advantages	Disadvantages	Taxonomic level	References
Direct nucleotide sequencing	PCR amplification of genomic DNA and visualization of nucleotide sequence	• Precise resolution of genetic variation and homology • Comparable across taxa	• Moderately expensive • Only surveys a short-targeted fragment	All taxonomic levels	• Elias et al. (2007) • Linnen and Farrell (2007) • Nazari et al. (2007)
SNPs	Single variable nucleotide positions throughout the genome	• Easy to score • Easily automated • Ubiquitous in genome	• Moderately expensive • Limited information per site • Time consuming to identify and develop numerous loci	Genotypes, kinship, Populations	• Morin and McCarthy (2007) • Niehuis et al. (2007) • Wondji et al. (2007)
ESTs	Single-direction read of cDNA sequence cloned randomly from mRNA	• Source for transcribed molecular markers • Can target functional subsets across genome	• Misreads and contaminations due to single read sequencing • Expensive to develop • Computationally intensive	All taxonomic levels	• Papanicolaou et al. (2008) • Wahlberg and Wheat (2008)
Microarray	DNA hybridization to numerous short probes to assay polymorphisms	• Fast assay of large numbers of markers	• Expensive to develop • Computationally intensive	Populations, species	• Turner et al. (2005) • Frey and Pfunder (2006)

Mullen et al. 2007, Pecsenye et al. 2007), but the current trend in molecular taxonomy has largely passed them by. Nonetheless, Avise (1994) aptly notes that if allozyme variation had been discovered only recently it might today be touted as a technique superior to DNA studies for tying organisms more directly to their environment.

A similar case can be made for the use of karyotypes in taxonomy and genetics. Because access to chromosome numbers, banding patterns, and inversions required little more than a good microscope and meticulous attention to detail, these techniques reached a high level of development in the first half of the twentieth century. Recent applications to insect taxonomy continue to use both chromosome counts and bands (e.g., Phasuk et al. 2005, Martin et al. 2006, Brown et al. 2007, Kandul et al. 2007). However, chromosome visualization may well make a comeback via techniques like fluorescent *in situ* hybridization (FISH) for evaluating the relationship between genomic architecture across different phylogenetic lineages (Heckel 2003, Traut et al. 2008).

Restriction fragment polymorphisms provided one of the first ways in which a broad spectrum of evolutionary biologists could access variation directly at the DNA level. Initially, these methods required the use of radioactively labeled probe DNA that was hybridized against DNA from other specimens and species after it was cut with restriction enzymes and the fragments separated by size in large gels. The methods were moderately labor intensive but affordable for the time, and individual restriction sites could be treated as homologous characters if an extra effort was made to convert fragment patterns into restriction site maps. This method has now largely been superseded in taxonomy, but the use of restriction sites to cheaply characterize PCR fragments remains current (Naegele et al. 2006, Beebe et al. 2007, Gariepy et al. 2007, Li et al. 2007).

There was much excitement when randomly amplified polymorphic DNAs (RAPDs) were first described as a method to cheaply produce hundreds of DNA bands by using short (10 bp) primers to arbitrarily produce bands from numerous locations on the genome of an individual (Williams et al. 1990). Analysis of the resulting data was admittedly problematic because heterozygotes and homozygotes could not consistently be distinguished (i.e., they had a 'dominant' mode of inheritance). Multiple nonhomologous bands of the same length were also difficult to distinguish,

requiring hybridization with labeled probes. However, this method promised a cheap, easy, and rapid means to generate large amounts of genetic data, without necessarily needing prior genetic information about the focal species or close relatives. The difficulty of reproducing RAPD bands from the same DNA in different labs or different thermal cyclers soon became clear. The use of this technique is now largely restricted to characterization of host races when there are no comparisons between labs (e.g., Kumar et al. 2001, Vandewoestijne and Baguette 2002, Al-Barrak et al. 2004), or to the generation of bands for later use as sequence-tagged markers (Behura 2006). A variation of this technique uses the rapid evolution of microsatellite loci to generate RAPD-like markers: ISSR-PCR (Inter-simple-sequence repeat-PCR). Inter-simple-sequence repeat markers use primers that anneal to the short repeat regions of microsatellites, and then amplify the region between two closely spaced but reversed microsatellite loci (e.g., Roux et al. 2007).

Amplified fragment length polymorphisms (AFLPs) promised to combine the best features of Restriction fragment length polymorphisms (RFLP) and RAPD techniques (Vos et al. 1995). Genomic DNA was digested with restriction enzymes before some fragments were arbitrarily amplified to produce a fingerprint fragment pattern that provided far more markers than from RFLPs alone. The method produced more reproducible bands than RAPDs, but unfortunately the improvement was not sufficiently robust for suboptimal DNA extractions. AFLPs are generally not used for insect taxonomy above the species level, though they are commonly used to distinguish host races and to provide low-density whole-genome linkage maps for insects where there is no problem obtaining high-quality DNA (Emelianov et al. 2004, Dopman et al. 2005, Gompert 2006, Mock et al. 2007, Scheffer and Hawthorne 2007).

Microsatellites have had a longer and more consistent application in insect population biology, with occasional application to taxonomy. Here, regions of DNA with short, tandem repeats (often two bases) give ample allelic variation that can be consistently reproduced from suboptimally preserved DNA. The fact that this variation can be easily analyzed as Mendelian markers is an added bonus, and a substantial proportion of the research papers and primer notes published in the field of ecological genetics now use microsatellites (Selkoe and Toonen 2006). The amount of effort and cost required to first apply the method to a particular species or species group is large, and there is limited and uneven success in applying microsatellite primers to species other than those for which they have been developed (Barbara et al. 2007, Meglécz et al. 2007). For some groups, such as Lepidoptera, consistent problems with gene conversion of the conserved flanking regions make it difficult to distinguish loci (Zhang 2004). This technique has seen relatively little use in insect taxonomy because microsatellite loci show such rapid mutation rates that it is difficult to be certain of the homologies of allelic variation between species. Nonetheless, microsatellites have proven useful for confirming the morphology-based assignment of a peripheral population of a butterfly species, despite introgression of mtDNA from another species (Zakharov and Hellmann 2008).

Direct nucleotide sequencing has become the dominant molecular method in insect taxonomy, whether at the species level or for reconstructing phylogenies. The method has proven to be sufficiently robust that it is increasingly used to determine the species identity of the meals of predators (King et al. 2008) or even herbivores (Miller et al. 2006). The cost of sequencing a short PCR fragment of 500–1000 bp was moderately high a decade ago, but has now come down to about $3 and is sure to continue to decline. Thus, hundreds or even thousands of specimens can now be sequenced as part of a taxonomic project. This situation has led to a problem where large numbers of sequences have been generated but cannot be effectively compared because they are not from homologous genes; some standardization is the obvious answer (Caterino et al. 2000). Nonetheless, the pendulum has recently swung too far toward standardization by focusing on a single 650-bp region of mitochondrial COI in the Barcode of Life Project (Rubinoff and Holland 2005). The COI gene also has serious limitations as a phylogenetic marker above the level of genera, a fact that is now widely recognized even by barcoding advocates (Hajibabaei et al. 2006a). Use of a larger fragment or the whole 1.4 kb of COI would have reduced some of these problems (Roe and Sperling 2007b).

One positive feature of the Barcode project is the consistent availability of images of voucher specimens (Ratnasingham and Hebert 2007, Floyd et al. this volume), as well as recognition of the need to remain compliant with established Linnean taxonomy (Hebert and Barrett 2005, Hebert and Gregory 2005), rather than establishing a parallel DNA taxonomy (Tautz

et al. 2003, Vogler and Monaghan 2006). The primary strength of DNA barcoding lies in the identification of specimens to known species, as well as the initial detection of possible cryptic species whose existence is then tested by other means, but not as a method for species delimitation or phylogenetic inference. As was already well understood long before the use of mtDNA sequences for biological identifications, which was dubbed 'DNA barcoding', such sequences are most informative when their application includes good characterization of genetic variation across the whole genome, broad population sampling across the whole range of species, and comprehensive sampling of all species across large taxonomic groups. Under these conditions, it is inevitable and even unremarkable that DNA barcoding will become a valued component of the normal operating procedure of most taxonomists. If, however, full integration with traditional taxonomy and sampling at multiple levels does not take place, then the barcoding project will ultimately be unable to deliver on its promise of simplifying our means of assaying biodiversity (Rubinoff 2006, Rubinoff et al. 2006).

Single nucleotide polymorphisms (SNPs) compose a rapidly growing application of DNA technology to produce robust surveys of genome architecture, populations, and sometimes species (Behura 2006, Black and Vantas 2007, Morin and McCarthy 2007). SNP markers are much like RFLPs, except that the particular nucleotide site can be targeted and assayed not just because it happens to be polymorphic but also because it can be related to functional variation in a gene (i.e., the phenotype). Expressed sequence tag (EST) libraries are one common means used to produce a preliminary foundation of sequence from which to develop large numbers of such molecular markers; this involves obtaining single-direction reads of cDNA cloned randomly from the mRNA of an organism (Bouck and Vision 2007). Although various other techniques can be used to develop SNPs (e.g., Garvin and Gharrett 2007), substantial effort usually is required to characterize a sufficient number of SNP markers for genetic mapping. However, to develop and test a small number of SNPs for characterizing genotypes or populations can be quite simple if there is even a limited amount of prior sequence information available for the species of interest. Once such diagnostic tests have been developed, their use is usually limited to the species or populations for which they were developed (but see Shaffer and Thompson 2007). In contrast, diagnostic use of longer DNA sequences is easier to scale up across broader taxonomic ranges. Thus, the growing numbers of EST libraries available for different species compose a valuable resource for developing and testing a diversity of genes for phylogenetic studies (Papanicolaou et al. 2008).

Microarrays are a set of many different DNA sequences that have been bound to different sections of the grid on a plate (Gibson 2002). Sample DNA is then hybridized with the DNA on the plate under precise conditions, lighting up the sections where hybridization occurs. Easily automated, the technique is currently used to assay bacterial diversity or to explore functional questions relating to gene networks in model organisms (e.g., Turner et al. 2005). It has also been suggested as a way to identify a large array of different insect species (Frey and Pfunder 2006). One problem is that microarrays are expensive to develop, even when different genes from the same species compose the array. If the array is made up of the same gene from many different species, sequence similarity would be much greater between sections and identification would be especially sensitive to hybridization conditions. This problem would be especially acute for the most closely related species, even if multiple loci were used. Although the potential for automation and rapid identification are high, the expense of development and problems with sensitivity are likely to mean that direct sequencing of known genes will remain the preferred method of molecular taxonomic identification for some time to come.

CONCLUSIONS AND OUTLOOK FOR THE FUTURE

Diverse molecular methods have been employed in insect taxonomy. The currently dominant method involves sequencing known regions of DNA that are sufficiently conserved to allow use of 'universal' primers, which at the same time have a sufficiently rapid mutation rate that species can be distinguished. Mitochondrial DNA has long been used to identify species and to trace the genealogical histories of populations (Avise 1994). Recent focus on mtDNA-based identification, as DNA barcoding, has raised hopes that molecular methods will lift the perceived burden of biodiversity identification at the species level (Hebert et al. 2003a, Waugh 2007). A variety of authors suggest that as much as one-quarter of species cannot easily be characterized using mtDNA sequences (Funk

and Omland 2003, Elias et al. 2007, Weimers and Fiedler 2007). These problems are likely to be most acute in the identifications for which molecular methods are most needed – closely related species that have limited morphological distinctions (Sperling 2003b). Consequently, both at the species level and above, other character information is clearly needed, including other gene sequences, to test the reality of both species and higher taxonomic groups (e.g., for endangered species: Rubinoff and Sperling 2004, Fallon 2007). Molecular methods can play a vital role in providing characters for such tests, but ultimately the copious number of these characters provides the most hope for the future of taxonomy, not any kind of intrinsic superiority of such characters.

Several practical problems are on the horizon for the future of molecular methods in insect taxonomy. One of the most immediate of these is how to store all the tissue samples and DNA extractions that have been generated in the course of molecular studies (Whitfield 1999, Corthals and DeSalle 2005). These samples constitute an extremely valuable resource that is currently only available as *ad hoc* assemblages in the freezers of the principal investigators. Unlike regular museum specimens, little option is available for browsing among the samples themselves, and repeated freeze-thaw cycles eventually result in degradation of the samples. Yet, it is important to test preliminary taxonomic conclusions, such as those based on single mtDNA gene sequences, against other genes (inevitably nuclear ones). A better standard is gradually emerging for archiving and imaging the voucher morphological specimens from molecular work. It is time that museums also start to invest significantly in the archiving and databasing of legacy DNA samples, as well as in developing consistent policies for the inevitably destructive use of these samples.

A second problem for molecular methods is how to accomplish reasonably comprehensive sampling (Seberg and Petersen 2007). Unlike character systems such as genitalic morphology, which can be investigated in even the oldest museum samples, DNA degradation is a reality that cannot easily be circumvented (Wandeler et al. 2007). It is often possible to PCR-amplify short DNA sequences of less than 200–300 bp where longer amplifications are no longer feasible (Hajibabaei et al. 2006b, Min and Hickey 2007), and some museum specimens have fortuitously been preserved in a way that allows successful

DNA extraction and amplification from specimens that are several decades old (Morin and McCarthy 2007). However, short fragments bring greater risk of problems due to contamination (King et al. 2008). Recent studies have also shown that damage to specimens due to DNA extraction can be minimized by sonication in buffer (Rowley et al. 2007, Hunter et al. 2008). However, by far the largest proportion of the cost of sampling taxa lies in collecting the specimens and their preliminary identification and curation (Cameron et al. 2006). Thus, finishing the job of sampling a taxon will take extraordinary effort for any large taxon (May 2004). In our experience, getting samples of the first 20% of species in a large group is a relatively small task, but each successive increment of 20% involves at least a doubling of the time, energy, and cost. Getting the sampling proportion over 80% for a species-rich taxon will be a major accomplishment, and more than 95% will be an extraordinary challenge. And yet, if we hope to be able to claim more than 95% accuracy in identifications (as might be expected in a legal challenge), it would be logical first to have sampled at least 95% of the species. Furthermore, the proportion of apparently correct identifications might decline as sampling becomes more complete. That is because the gaps between clusters will get smaller as sampling of the existing genetic landscape (or morphospace, for that matter) becomes more comprehensive. Early estimates of identification accuracy are thus sure to be biased by a variety of factors (e.g., Hebert et al. 2003a, Hajibabaei et al. 2006c), because preliminary studies are based on geographically limited sampling that does not fully characterize the true range of variation within and between closely related species (Ekrem et al. 2007).

A third problem for molecular taxonomy will be to get the cost of processing each sample down to that of traditional sight-based identification. Even though the lab cost (excluding labor) of obtaining sequence for a specimen is now in the vicinity of $3.00 or less, and the cost of labor is much reduced with automation, that is still a long way from the pennies that it costs to obtain identifications when there are externally visible morphological characters. Such comparisons can be misleading, because training is more transferable between groups for molecular identifications, but the challenge remains painfully real to anyone doing biodiversity studies that require processing of thousands of samples. New methods of sequencing, such as 454 pyrosequencing or Solexa

SBS sequencing (Hudson 2008) can still greatly reduce the cost of sequencing, but these methods are currently mainly used for whole-genome sequencing projects, not multiple samples from different specimens (though sequencing of environmental DNA slurries is well established in microbiology). Nonetheless, whole-genome sequencing is gradually getting closer to the budget of midsized labs (e.g., Vera et al. 2008), and with incentives like the X-Prize to bring down the speed and cost of sequencing a genome (http://genomics.xprize.org/), single-gene molecular taxonomy might soon be a thing of the past.

Finally, and most exciting, are the challenges that await those people who are taxonomists first and prefer only to use molecular methods as a means to an end. We are entering an era when simple molecular work, integrated with traditional analysis of morphology and biological characters, such as host associations and phenology, will free up energy and creativity for work on the residue of taxonomic problems (perhaps 25% of species) that really need the work. The majority of routine identifications might well be done by relatively untrained technicians once a high-quality database of molecular characters has been achieved. Thus, molecular methods could free us to do better morphological work, as well as to apply new and more complex molecular methods to problems in species determination, discovery, delimitation, and phylogeny. Eventually, we can even develop a field of molecular morphology within taxonomy, devoted to understanding the three-dimensional structures and functions of the proteins into which our DNA sequences translate (e.g., Frentiu et al. 2007, Gaucher et al. 2008). It is sure to be more satisfying than continuing to interpret DNA as long and boring strings of equivalent four-state characters.

ACKNOWLEDGMENTS

This paper has been funded by an NSERC Discovery Grant to Sperling. We thank two anonymous reviewers for their helpful comments.

REFERENCES

Al-Barrak, M., H. D. Loxdale, C. P. Brookes, H. A. Dawah, D. G. Biron, and O. Alsagair. 2004. Molecular evidence using enzyme and RAPD markers for sympatric evolution in British species of *Tetramesa* (Hymenoptera: Eurytomidae). *Biological Journal of the Linnean Society* 83: 509–525.

Avise, J. C. 1994. *Molecular Markers, Natural History and Evolution.* Chapman and Hall NY. 511 pp.

Avise, J. C. and D. Mitchell. 2007. Time to standardize taxonomies. *Systematic Biology* 56: 130–133.

Ballard, J. W. O. and M. C. Whitlock. 2004. The incomplete natural history of mitochondria. *Molecular Ecology* 13: 729–744.

Barbara, T., C. Palma-Silva, G. M. Paggi, F. Bered, M. F. Fray, and C. Lexer. 2007. Cross-species transfer of nuclear microsatellite markers: potential and limitations. *Molecular Ecology* 16: 3759–3767.

Beebe, N. W., P. I. Whelan, A. F. V. D. Hurk, S. A. Ritchie, S. Corcoran, and R. D. Cooper. 2007. A polymerase chain reaction-based diagnostic to identify larvae and eggs of container mosquito species from the Australian region. *Journal of Medical Entomology* 44: 376–380.

Behura, S. 2006. Molecular marker systems in insects: current trends and future avenues. *Molecular Ecology* 15: 3087–3113.

Bellows, T. S., Jr., T. M. Perring, R. J. Gill, and D. H. Headrick. 1994. Description of a species of *Bemisia* (Homoptera: Aleyrodidae). *Annals of the Entomological Society of America* 87: 195–206.

Beltran, M., C. D. Jiggins, A. V. Z. Brower, E. Birmingham, and J. Mallet. 2007. Do pollen feeding, pupal-mating and larval gregariousness have a single origin in *Heliconius* butterflies? Inferences from multilocus DNA sequence data. *Biological Journal of the Linnean Society* 92: 221–239.

Bickford, D., D. J. Lohman, N. S. Sodhi, P. K. L. Ng, R. Meier, K. Winker, K. K. Ingram, and I. Das. 2007. Cyptic species as a window on diversity and conservation. *Trends in Ecology and Evolution* 22: 148–155.

Black, W. C., IV, and J. G. Vantas. 2007. Affordable assays for genotyping single nucleotide polymorphisms in insects. *Insect Molecular Biology* 16: 377–387.

Bouck, A. and T. Vision. 2007. The molecular ecologist's guide to expressed sequence tags. *Molecular Ecology* 16: 907–924.

Boyer, S. L., J. M. Baker, and G. Giribet. 2007. Deep genetic divergences in *Aoraki denticulata* (Arachnida, Opiliones, Cyphophthalmi): a widespread 'mite harvestman' defies DNA taxonomy. *Molecular Ecology* 16: 4999–5016.

Brower, A. V. Z. 2006. Problems with DNA barcodes for species delimitation: 'ten species' of *Astraptes fulgerator* reassessed (Lepidoptera: Hesperiidae). *Systematics and Biodiversity* 4: 127–132.

Brown, K. S., Jr., A. V. L. Freitas, N. Wahlberg, B. V. Schoultz, A. O. Saura, and A. Saura. 2007. Chromosomal evolution in the South American Nymphalidae. *Hereditas* 144: 137–148.

Bruschini, C., R. Cervo, F. R. Dani, and S. Turillazzi. 2007. Can venom volatiles be a taxonomic tool for *Polistes* wasps? *Journal of Zoological Systematics and Evolutionary Research* 45: 202–205.

Burns, J. M., D. H. Janzen, M. Hajibabaei, W. Hallwachs, and P. D. N. Hebert. 2007. DNA barcodes of closely related (but morphologically and ecologically distinct) species of skipper butterflies (Hesperiidae) can differ by only one to three nucleotides. *Journal of the Lepidopterists' Society* 61: 138–153.

Cameron, S., D. Rubinoff, and K. Will. 2006. Who will actually use DNA barcoding and what will it cost? *Systematic Biology* 55: 844–847.

Caterino, M. S., S. Cho, and F. A. H. Sperling. 2000. The current state of insect molecular systematics: a thriving Tower of Babel. *Annual Review of Entomology* 2000: 1–54.

Cognato, A. I. 2006. Standard percent DNA sequence difference for insects does not predict species boundaries. *Journal of Economic Entomology* 99: 1037–1045.

Cognato, A. I., S. J. Seybold, and F. A. H. Sperling. 1999. Incomplete barriers to mitochondrial gene flow between pheromone races of the North American pine engraver, *Ips pini* (Say) (Coleoptera: Scolytidae). *Proceedings of the Royal Society, Series B, Biological Sciences* 266: 1843–1850.

Cognato, A. I. and J. H. Sun. 2007. DNA based cladograms augment the discovery of a new *Ips* species from China (Coleoptera: Curculionidae: Scolytinae). *Cladistics* 23: 1–13.

Corthals, A. and R. DeSalle. 2005. An application of tissue and DNA banking for genomics and conservation: the Ambrose Monell Cryo-collection (AMCC). *Systematic Biology* 54: 819–823.

Coyne, J. A. and H. A. Orr. 2004. *Speciation.* Sinauer Associates,, Sunderland, MA. 545 pp.

Dalecky, A., M. Renucci, A. Tirard, G. Debout, M. Roux, F. Kjellberg, and E. Provost. 2007. Changes in composition of cuticular biochemicals of the facultatively polygynous ant *Petalomyrmex phylax* during range expansion in Cameroon with respect to social, spatial and genetic variation. *Molecular Ecology* 16: 3778–3791.

Dapporto, L. 2007. Cuticular lipid diversification in *Lasiommata megera* and *Lasiommata paramegaera*: the influence of species, sex, and population (Lepidoptera: Nymphalidae). *Biological Journal of the Linnean Society* 91: 703–710.

Dayrat, B. 2005. Towards integrative taxonomy. *Biological Journal of the Linnean Society* 85: 407–415.

Dean, M. D. and J. W. O. Ballard. 2004. Factors affecting mitochondrial DNA quality from museum preserved *Drosophila simulans*. *Entomologia Experimentalis et Applicata* 98: 279–283.

de Jong, R. 2007. Estimating time and space in the evolution of the Lepidoptera. *Tijdschrift voor Entomologie* 150: 319–346.

De Quieroz, K. 2007. Species concepts and species delimitation. *Systematic Biology* 56: 879–886.

DeSalle, R., M. G. Egan, and M. Siddall. 2005. The unholy trinity: taxonomy, species delimitations and DNA barcoding. *Philosophical Transactions of the Royal Society, Series B, Biological Sciences* 360: 1905–1916.

Dopman, E. B., L. Pérez, S. M. Bogdanowicz, and R. G. Harrison. 2005. Consequences of reproductive barriers for genealogical discordance in the European corn borer. *Proceedings of the National Academy of Sciences, USA* 102: 14706–14711.

Dres, M., and J. Mallet. 2002. Host-races in plant-feeding insects and their importance in sympatric speciation. *Philosophical Transactions of the Royal Society, Series B, Biological Sciences* 357: 471–492.

Elias, M., R. I. Hill, K. R. Willmott, K. K. Dasmahapatra, A. V. Z. Brower, J. Mallet, and C. D. Jiggins. 2007. Limited performance of DNA barcoding in a diverse community of tropical butterflies. *Proceedings of the Royal Society, Series B, Biological Sciences* 274: 2881–2889.

Embley, T. M. and E. Stackebrandt. 1997. Species in practice: exploring uncultured prokaryote diversity in natural samples. Pp. 61–82. *In* M. F. Claridge, H. A. Dawah, and M. R. Wilson (eds). *Species: The Units of Biodiversity*. Chapman and Hall, London.

Emelianov, I., F. Marec, and J. Mallet. 2004. Genomic evidence for divergence with gene flow in host races of the larch budmoth. *Proceedings of the Royal Society, Series B, Biological Sciences* 271: 97–105.

Ekrem, T., E. Willassen, and E. Stur. 2007. A comprehensive DNA sequence library is essential for identification with DNA barcodes. *Molecular Phylogenetics and Evolution* 43: 530–542.

Fallon, S. M. 2007. Genetic data and the listing of species under U.S. Endangered Species Act. *Conservation Biology* 21: 1186–1195.

Felsenstein, J. 2004. *Inferring Phylogenies*. Sinauer Associates, Sunderland, MA. 664 pp.

Floyd, R., E. Abebe, A. Papert, and M. Blaxter. 2002. Molecular barcodes for soil nematode identification. *Molecular Ecology* 11: 839–850.

Foley, D. H., J. H. Bryan, and R. C. Wilkerson. 2007. Species-richness of the *Anopheles annulipes complex* (Diptera: Culicidae) revealed by tree and model-based allozyme clustering analyses. *Biological Journal of the Linnean Society* 91: 523–539.

Ford, E. B. 1944. Studies on the chemistry of pigments in the Lepidoptera, with reference to their bearing on systematics. 4. The classification of the Papilionidae. *Transactions of the Royal Entomological Society of London* 94: 201–223.

Franz, N. M. 2005. On the lack of good scientific reasons for the growing phylogeny/classification gap. *Cladistics* 21: 495–500.

Frentiu, F. D., G. D. Bernard, C. I. Cuevas, M. P. Sison-Mangus, K. L. Prudic, and A. D. Briscoe. 2007. Adaptive evolution of color vision as seen through the eyes of butterflies. *Proceedings of the National Academy of Sciences USA* 104 (Supplement 1): 8634–8640.

Frey, J. E., and M. Pfunder. 2006. Molecular techniques for identification of quarantine insects and mites: the potential of microarrays. Pp. 141–163. *In* J. R. Rao, C. C. Fleming, and J. E. Moore (eds). *Molecular Diagnostics: Current Technology and Applications*. Horizon Bioscience, Norfolk.

Funk, D. J. and K. E. Omland. 2003. Species-level paraphyly and polyphyly: frequency, causes, and consequences, with insights from animal mitochondrial DNA. *Annual Review of Ecology, Evolution and Systematics* 34: 397–423.

Gariepy, T. D., U. Kuhlmann, C. Gillott, and M. Erlandson. 2007. Parasitoids, predators and PCR: the use of diagnostic molecular markers in biological control of arthropods. *Journal of Applied Entomology* 131: 225–240.

Garvin, M. R. and A. J. Gharrett. 2007. DEco-TILLING: an inexpensive method for single nucleotide polymorphism discovery that reduces ascertainment bias. *Molecular Ecology Notes* 7: 735–746.

Gaucher, E. A., S. Govindarajan, and O. K. Ganesh. 2008. Palaeotemperature trend for Precambrian life inferred from resurrected proteins. *Nature* 451: 704–8.

Gibson, G. 2002. Microarrays in ecology and evolution. *Molecular Ecology* 11: 17–24,

Giribet, G. and G. D. Edgecomb. 2005. Conflict between datasets and phylogeny of centipedes: an analysis based on seven genes and morphology. *Proceedings of the Royal Society, Series B. Biological Sciences* 273: 531–538.

Gleeson, D., P. Holder, R. Newcomb, R. Howatt, and J. Dugdale. 2000. Molecular phylogenetics of leafrollers: application to DNA diagnostics. *New Zealand Plant Protection* 53: 157–162.

Godfray, H. C. J. 2002. Challenges for taxonomy. *Nature* 417: 17–19.

Gompert, Z., J. A. Fordyce, M. L. Forister, A. M. Shapiro, and C. C. Nice. 2006. Homoploid hybrid speciation in an extreme habitat. *Science* 314: 1923–5.

Grill, A., L. E. L. Raijmann, W. Van Ginkel, E. Gkioka, and S. B. J. Menken. 2007. Genetic differentiation and natural hybridization between the Sardinian endemic *Maniola nurag* and the European *Maniola jurtina*. *Journal of Evolutionary Biology* 20: 1255–70.

Hajibabaei, M., G. A. C. Singer and D. A. Hickey. 2006a. Benchmarking DNA barcodes: an assessment using available primate sequences. *Genome* 49: 851–854.

Hajibabaei, M., M. A. Smith, D. H. Janzen, J. Rodriguez, J. B. Whitfield, and P. D. N. Hebert. 2006b. A minimalist barcode can identify a specimen whose DNA is degraded. *Molecular Ecology Notes* 6: 959–964.

Hajibabaei, M., D. H. Janzen, J. M. Burns, W. Hallwachs, and P. D. N. Hebert. 2006c. DNA barcodes distinguish species of tropical Lepidoptera. *Proceedings of the National Academy of Sciences USA* 103: 968–971.

Harrison, R. G. 1998. Linking evolutionary pattern and process: the relevance of species concepts for the study of speciation. Pp. 19–31. *In* D. J. Howard and S. H. Berlocher (eds). *Endless Forms: Species and Speciation*. Oxford University Press, NY.

Hay-Roe, M. M., G. Lamas, and J. L. Nation. 2007. Pre- and postzygotic isolation and Haldane rule effects in reciprocal crosses of *Danaus erippus* and *Danaus plexippus* (Lepidoptera: Danainae), supported by differentiation of cuticular hydrocarbons, establish their status as separate species. *Biological Journal of the Linnean Society* 91: 445–453.

Hebert, P. D. N. and R. D. H. Barrett. 2005. Reply to the comment by L. Prendini on "Identifying spiders through DNA barcodes." *Canadian Journal of Zoology* 83: 505–506.

Hebert, P. D. N. and T. R. Gregory. 2005. The promise of DNA barcoding for taxonomy. *Systematic Biology* 54: 852–859.

Hebert, P. D. N., A. Cywinska, S. L. Ball, and J. R. deWaard. 2003a. Biological identifications through DNA barcodes. *Proceedings of the Royal Society, Series B, Biological Sciences* 270: 313–322.

Hebert, P. D. N., S. Ratnasingham, and J. deWaard. 2003b. Barcoding animal life: cytochrome c oxidase subunit 1 divergences among closely related species. *Proceedings of the Royal Society, Series B, Biological Sciences* 270: S96–S99.

Hebert, P. D. N., E. H. Penton, J. M. Burns, D. H. Janzen, and W. Hallwachs. 2004. Ten species in one: DNA barcoding reveals cryptic species in the semitropical skipper butterfly *Astraptes fulgerator*. *Proceedings of the National Academy of Sciences USA* 101: 14812–14817.

Heckel, D. G. 2003. Genomics in pure and applied entomology. *Annual Review of Entomology* 48: 235–260.

Hickerson, M. J., C. P. Meyer, and C. Moritz. 2006. DNA barcoding will often fail to discover new animal species over broad parameter space. *Systematic Biology* 55: 729–739.

Hillis, D. M., C. Moritz, and B. K. Mable. 1996. *Molecular Systematics*, Second Edition. Sinauer Associates, Sunderland, MA. 655 pp.

Hudson, M. E. 2008. Sequencing breakthroughs for genomic ecology and evolutionary biology. *Molecular Ecology Resources* 8: 3–17.

Hunter, S. J., T. I. Goodall, K. A. Walsh, R. Owen, and J. C. Day. 2008. Nondestructive DNA extraction from blackflies (Diptera: Simuliidae): retaining voucher specimens for DNA barcoding projects. *Molecular Ecology Resources* 8: 56–61.

Hurst, G. D. D. and F. M. Jiggins. 2005. Problems with mitochondrial DNA as a marker in population, phylogeographic and phylogenetic studies: the effects of inherited symbionts. *Proceedings of the Royal Society, Series B, Biological Sciences* 272: 1525–1534.

Jenkins, T. M., M. I. Haverty, C. J. Basten, L. J. Nelson, M. Page, and B. T. Forschler. 2000. Correlation of mitochondrial haplotypes with cuticular hydrocarbon phenotypes of sympatric *Reticulitermes* species from the southeastern United States. *Journal of Chemical Ecology* 26: 1525–1542.

Kandul, N. P., V. A. Lukhtanov, and N. E. Pierce. 2007. Karyotypic diversity and speciation in *Agrodiaetus* butterflies. *Evolution* 61: 546–559.

King, R. A., D. S. Read, M. Traugott, and W. O. C. Symondson. 2008. Molecular analysis of predation: a review of best practice for DNA-based approaches. *Molecular Ecology* 17: (online early publication doi: 10.1111/j.1365-294X.2007.03613.x)

Knowles, L. L. and B. C. Carstens. 2007. Delimiting species without monophyletic gene trees. *Systematic Biology* 56: 887–895.

Kruse, J. 2000. *Archips goyerana*, n. sp. (Lepidopera: Tortricidae) an important pest of baldcypress (Taxodiaceae) in Louisiana and Mississippi. *Proceedings of the Entomological Society of Washington* 102: 759–764.

Kumar, L. S., A. S. Sawant, and V. S. Gupta. 2001. Genetic variation in Indian populations of *Scirpophaga incertulas* as revealed by RAPD-PCR analysis. *Biochemical Genetics* 39: 43–57.

Lagasa, E., D. Prasher, and E. Bruntjen. 1995. Cherry ermine moth (Lepidoptera: Yponomeutidae) occurrence and survey methods evaluation in Washington State. 1995 Project Report – Washington State Department of Agriculture, Washington, DC.

Langor, D. W. and F. A. H. Sperling. 1997. Mitochondrial DNA sequence divergence in weevils of the *Pissodes strobi* species complex (Coleoptera: Curculionidae). *Insect Molecular Biology* 6: 255–265.

LepTree Team. 2008. Molecular Project Overview|Leptree. net. http://www.leptree.net/molecular. [Accessed 25 February 2008]

Li, S., L. Sun, C. Y. Oseto, and V. R. Ferris. 2007. Phylogenetic analysis and a method for rapid molecular diagnosis of two sunflower seed weevils (Coleoptera: Curculionidae). *Annals of the Entomological Society of America* 100: 649–654.

Linnen, C. R. and B. D. Farrell. 2007. Mitonuclear discordance is caused by rampant mitochondrial introgression in *Neodiprion* (Hymenoptera: Diprionidae) sawflies. *Evolution* 61: 1417–1438.

Mallarino, R., E. Bermingham, K. R. Willmott, A. Whinnett, and C. D. Jiggins. 2005. Molecular systematics of the butterfly genus *Ithomia* (Lepidoptera: Ithomiinae): a composite phylogenetic hypothesis based on seven genes. *Molecular Phylogenetics and Evolution* 34: 625–644.

Martin, J., E. N. Andreeva, I. I. Kiknadze, and W. F. Wülker. 2006. Polytene chromosomes and phylogenetic relationships of *Chironomus atrella* (Diptera: Chironomidae) in North America. *Genome* 49: 1384–1392.

May, R. M. 2004. Tomorrow's taxonomy: collecting new species in the field will remain the rate limiting step. *Philosophical Transactions of the Royal Society, Series B, Biological Sciences* 359: 733–734.

Medina, R. F., P. Barbosa, M. Christman, and A. Battisti. 2007. Number of individuals and molecular markers to use in genetic differentiation studies. *Molecular Ecology Notes* 6: 1010–1013.

Meier, R., S. Kwong, G. Vaidya, and P. K. L. Ng. 2006. DNA barcoding and taxonomy in Diptera: a tale of high intraspecific variability and low identification success. *Systematic Biology* 55: 715–728.

Meglécz, E., S. J. Anderson, D. Bourguet, R. Butcher, A. Caldas, A. Cassel-Lundhagen et al. 2007. Microsatellite flanking region similarities among different loci within insect species. *Insect Molecular Biology* 16: 175–185.

Menken, S. B. J., W. M. Herrebout, and J. T. Wiebes. 1993. Small ermine moths (*Yponomeuta*): their host relations and evolution. *Annual Review of Entomology* 37: 41–66.

Miller, M. A., G. C. Müller, V. D. Kravchenko, M. Junnila, K. K. Vernon, C. D. Matheson, and A. Hausmann. 2006. DNA-based identification of Lepidoptera larvae and plant meals from their gut contents. *Russian Entomological Journal* 15: 427–432.

Min, X. J. and D. A. Hickey. 2007. Assessing the effect of varying sequence length on DNA barcoding of fungi. *Molecular Ecology Notes* 7: 365–373.

Mock, K. E., B. J. Bentz, E. M. O'Neill, J. P. Chong, J. Orwin, and M. E. Pfrender. 2007. Landscape-scale genetic variation in a forest outbreak species, the mountain pine beetle (*Dendroctonus ponderosae*). *Molecular Ecology* 16: 553–568.

Monaghan, M. T., M. Balke, J. Pons, and A. P. Vogler. 2006. Beyond barcodes: complex DNA taxonomy of a South Pacific island radiation. *Proceedings of the Royal Society, Series B, Biological Sciences* 273: 887–893.

Morin, P. A. and M. McCarthy. 2007. Highly accurate SNP genotyping from historical and low-quality samples. *Molecular Ecology Notes* 7: 937–946.

Muirhead, J. R., D. K. Gray, D. W. Kelly, S. M. Ellis, D. D. Heath, and H. J. Macisaac. 2008. Identifying the source of species invasions: sampling intensity vs. genetic diversity. *Molecular Ecology* 17: 1020–1035.

Mullen, S. P., T. C. Mendelson, C. Schal, and K. L. Shaw. 2007. Rapid evolution of cuticular hydrocarbons in a species radiation of acoustically diverse Hawaiian crickets (Gryllidae: Trigonidiinae: *Laupala*). *Evolution* 61: 223–231.

Naegele, M. P., P. I. D. Costa, and J. A. D. Rosa. 2006. Polymorphism of the ITS-2 region of the ribosomal DNA of the Triatominae *Rhodnius domesticus, R. pictipes, R. prolixus,* and *R. stali. Medical and Veterinary Entomology* 20: 353–357.

Nazari, V. and F. A. H. Sperling. 2007. Mitochondrial DNA divergence and phylogeography in western Palaearctic Parnassiinae (Lepidoptera, Papilionidae): how many species are there? *Insect Systematics and Evolution* 38: 121–138.

Nazari, V., E. V. Zakharov, and F. A. H. Sperling. 2007. Phylogeny, historical biogeography, and taxonomic ranking of Parnassiinae (Lepidoptera, Papilioinidae) based on morphology and seven genes. *Molecular Phylogenetics and Evolution* 42: 131–56.

Niehuis, O., A. Judson, J. Werren, W. B. Hunter, P. M. Dang, S. E. Dowd, B. Grillenberger, L. Beukeboom, and J. Gadau. 2007. Species diagnostic EST-derived SNP and STS markers for the wasp genus *Nasonia* Ashmead, 1904 (Hymenoptera: Pteromalidae). *Molecular Ecology Notes* 7: 1033–1036.

Papanicolaou, A., S. Gebauer-Jung, M. L. Blaxter, W. O. McMillan, and C. D. Jiggins. 2008. ButterflyBase: a platform for lepidopteran genomics. *Nucleic Acids Research* 36: D582–D587 (advance access).

Parson, W., K. Pegoraro, H. Niederstätter, M. Föger, and M. Steinlechner. 2000. Species identification by means of the cytochrome b gene. *International Journal of Legal Medicine* 114: 23–28.

Pecsenye, K., J. Bereczki, B. Tihanyi, A. Toth, L. Peregovits, and Z. Varga. 2007. Genetic differentiation among the *Maculinea* species (Lepidoptera: Lycaenidae) in eastern Central Europe. *Biological Journal of the Linnean Society* 91: 11–21.

Phasuk, J., J. Chanpaisaeng, P. H. Adler, and G. W. Courtney. 2005. Chromosomal and morphological taxonomy of larvae of *Simulium* (*Gomphostilbia*) (Diptera: Simuliidae) in Thailand. *Zootaxa* 1052: 49–60.

Pons, J., T. G. Barraclough, J. Gomez-Zurita, A. Cardoso, D. P. Duran, S. Hazell, S. Kamoun, W. D. Sumlin, and A. P. Vogler. 2006. Sequence-based species delimitation for the DNA taxonomy of undescribed insects. *Systematic Biology* 55: 595–609.

Rach, J., R. DeSalle, I. N. Sarker, B. Schierwater, and H. Hadrys. 2008. Character-based DNA barcoding allows discrimination of genera, species and populations in Odonata. *Proceedings of the Royal Society, Series B, Biological Sciences* 275: 237–247.

Ratnasingham, S. and P. D. N. Hebert. 2007. BOLD: the barcoding of life data system (www.barcodinglife.org). *Molecular Ecology Notes* 7: 355–364.

Roe, A. D. and F. A. H. Sperling. 2007a. Population structure and species boundary delimitation of cryptic *Dioryctria* moths: an integrative approach. *Molecular Ecology* 16: 3617–3633.

Roe, A. D. and F. A. H. Sperling. 2007b. Patterns of evolution of mitochondrial cytochrome *c* oxidase I and II DNA and implications for DNA barcoding. *Molecular Phylogenetics and Evolution* 44: 325–345.

Roux, O., M. Gevrey, L. Arvanitakis, C. Gers, D. Bordat, and L. Legal. 2007. ISSR-PCR: tool for discrimination and genetic structure analysis of *Plutella xylostella* populations native to different geographical areas. *Molecular Phylogenetics and Evolution* 43: 240–250.

Rowley, D. L., J. A. Coddington, M. W. Gates, A. L. Norrbom, R. A. Ochoa, N. J. Vandenberg, and M. H. Greenstone. 2007. Vouchering DNA-barcoded specimens: test of a nondestructive extraction protocol for terrestrial arthropods. *Molecular Ecology Notes* 7: 915–924.

Rubinoff, D. 2006. DNA barcoding evolves into the familiar. *Conservation Biology* 20: 1548–1549.

Rubinoff, D. and F. A. H. Sperling. 2004. Mitochondrial DNA sequence, morphology and ecology yield contrasting conservation implications for two threatened buckmoths (*Hemileuca*: Saturniidae). *Biological Conservation* 118: 341–351.

Rubinoff, D., S. Cameron, and K. Will. 2006. A genomic perspective on the shortcomings of mitochondrial DNA for "barcoding" identification. *Journal of Heredity* 97: 581–594.

Rubinoff, D. and B. S. Holland. 2005. Between two extremes: mitochondrial DNA is neither the panacea nor the nemesis of phylogenetic and taxonomic inference. *Systematic Biology* 54: 952–961.

Scheffer, S. J. and D. J. Hawthorne. 2007. Molecular evidence of host-associated genetic divergence in the holly leafminer *Phytomyza glabricola* (Diptera: Agromyzidae): apparent discordance among marker systems. *Molecular Ecology* 16: 2627–2637.

Schlick-Steiner, B. C., B. Seifert, C. Stauffer, E. Christian, R. H. Crozier, and F. M. Steiner. 2007. Without morphology, cryptic species stay in taxonomic crypsis following discovery. *Trends in Ecology and Evolution* 22: 391–392.

Schlötterer, C. 2004. The evolution of molecular markers – just a matter of fashion? *Nature Reviews Genetics* 5: 63–69.

Seberg, O. and G. Petersen. 2007. Assembling the tree of life: magnitude, shortcuts and pitfalls. Pp. 33–46. *In* T. R. Hodkinson and J. A. N. Parnell (eds). *Reconstructing the Tree of Life*. CRC Press, Boca Raton, FL.

Selkoe, K. A. and R. J. Toonen. 2006. Microsatellites for ecologists: a practical guide to using and evaluating microsatellite markers. *Ecology Letters* 9: 615–629.

Shaffer, H. B. and R. C. Thompson. 2007. Delimiting species in recent radiations. *Systematic Biology* 56: 896–906.

Sperling, F. A. H. 1987. Evolution of the *Papilio machaon* species group in western Canada. *Quaestiones Entomologica* 23: 198–315.

Sperling, F. A. H. 1993. Mitochondrial DNA variation and Haldane's rule in the *Papilio glaucus* and *P. troilus* species groups. *Heredity* 70: 227–233.

Sperling, F. A. H. 2003a. Butterfly molecular systematics: from species definitions to higher-level phylogenies. Pp. 431–458. *In* C. Boggs, P. Ehrlich, and W. Watt (eds). *Ecology and Evolution Taking Flight: Butterflies as Model Study Systems*. University of Chicago Press, Chicago, IL.

Sperling, F. 2003b. DNA barcoding: deus ex machina. *Newsletter of the Biological Survey of Canada (Terrestrial Arthropods)* 22: 50–53. (http://www.biology.ualberta.ca/bsc/news22_2/opinionpage.htm)

Sperling, F. A. H., G. S. Anderson, and D. A. Hickey. 1994. A DNA-based approach to identification of insect species used for postmortem interval estimation. *Journal of Forensic Sciences* 39: 418–427.

Sperling, F. A. H. and R. G. Harrison. 1994. Mitochondrial DNA variation within and between species of the *Papilio machaon* species group. *Evolution* 47: 408–422.

Sperling, F. A. H. and D. A. Hickey. 1994. Mitochondrial DNA sequence variation in the spruce budworm species complex (*Choristoneura*: Lepidoptera). *Molecular Biology and Evolution* 11: 656–665.

Sperling, F. A. H., J.-F. Landry, and D. A. Hickey. 1995. DNA-based identification of introduced ermine moth species in North America (Lepidoptera: Yponomeutidae). *Annals of the Entomological Society of America* 88: 155–162.

Sperling, F., R. Byers, and D. Hickey. 1996. Mitochondrial DNA sequence variation among pheromotypes of the dingy cutworm, *Feltia jaculifera* (Gn.) (Lepidoptera: Noctuidae). *Canadian Journal of Zoology* 74: 2109–2117.

Tautz, D., P. Arctander, A. Minnelli, R. H. Thomas, and A. P. Vogler. 2002. DNA points the way ahead in taxonomy. *Nature* 418: 479.

Tautz, D., P. Arctander, A. Minelli, R. H. Thomas, and A. P. Vogler. 2003. A plea for DNA taxonomy. *Trends in Ecology and Evolution* 18: 70–74.

Traut, W., K. Sahara, and F. Marec. 2008. Sex chromosomes and sex determination in Lepidoptera. *Sexual Development* 1: 332–346.

Turner, H., N. Lieshout, W. van Ginkel, and S. B. J. Menken. 2004. Molecular phylogeny of the small ermine moth genus *Yponomeuta* (Lepidoptera, Yponomeutidae). GenBank (http://www.ncbi.nlm.nih.gov/entrez/viewer.fcgi?db=nuccore&id=46405071). Accessed January 15, 2008.

Turner, T. L., M. W. Hahn, and S. V. Nuzhdin. 2005. Genomic islands of speciation in *Anopheles gambiae*. *Public Library of Science Biology* 3: e285.

Twewick, S. A. 2007. DNA barcoding is not enough: mismatch of taxonomy and genealogy in New Zealand grasshoppers (Orthoptera: Acrididae). *Cladistics* 23: 1–15.

Valdecasas, A. G., D. Williams, and Q. D. Wheeler. 2008. 'Integrative taxonomy' then and now: a response to Dayrat (2005). *Biological Journal of the Linnean Society* 93: 211–216.

Vandewoestijne, S. and M. Baguette. 2002. The genetic structure of endangered populations in the cranberry fritillary, *Boloria aquilonaris* (Lepidoptera, Nymphalidae): RAPDs vs allozymes. *Heredity* 89: 439–445.

Vera, J. C., C. W. Wheat, H. W. Fescemyer, M. J. Frilander, D. L. Crawford, I. Hanski, and J. H. Marden. 2008. Rapid transcriptome characterization for a nonmodel organism using 454 pyrosequencing. *Molecular Ecology* 17: 1636–1647.

Vogler, A. P. and M. T. Monaghan. 2006. Recent advances in DNA taxonomy. *Journal of Zoological Systematics and Evolutionary Research* 45: 1–10.

Vos, P., R. Hogers, M. Bleeker M. Reijans, M. Van de Lee, T. Hornes et al. 1995. AFLP: a new technique for DNA fingerprinting. *Nucleic Acids Research* 23: 4407–4414.

Wahlberg, N. and C. W. Wheat. 2008. Genomic outposts serve the phylogenomic pioneers: designing novel nuclear primers for genomic DNA extractions of Lepidoptera. *Systematic Biology* 57: 231–242.

Wandeler, P., P. E. A. Hoeck, and L. F. Keller. 2007. Back to the future: museum specimens in population genetics. *Trends in Ecology and Evolution* 22: 634–642.

Waugh, J. 2007. DNA barcoding in animal species: progress, potential and pitfalls. *BioEssays* 29: 188–197.

Weimers, M. and K. Fiedler. 2007. Does the barcoding gap exist? A case study in blue butterflies (Lepidoptera: Lycaenidae). *Frontiers in Zoology* 4: 8.

Wells, J. D., F. Introna, G. Di Vella, C. P. Campobasso, J. Hayes, and F. A. H. Sperling. 2001a. Human and insect mitochondrial DNA analysis from maggots. *Journal of Forensic Sciences* 46: 685–687.

Wells, J. D., T. Pape, and F. A. H. Sperling. 2001b. DNA-based identification and molecular systematics of forensically important Sarcophagidae (Diptera). *Journal of Forensic Sciences* 46: 1098–1102.

Wells, J. D. and F. A. H. Sperling. 2000. Commentary on: FAH Sperling, GS Anderson and DA Hickey, "A DNA-based approach to the identification of insect species used for postmortem interval estimation." *Journal of Forensic Sciences* 39: 418–427.

Wells, J. D. and F. A. H. Sperling. 2001. DNA-based identification of forensically important Chrysomyinae (Diptera: Calliphoridae). *Forensic Science International* 120: 110–115.

Wells, J. D., R. Wall, and J. R. Stevens. 2007. Phylogenetic analysis of forensically important *Lucilia* flies based on cytochrome oxidase I sequence: a cautionary tale for forensic species determination. *International Journal of Legal Medicine* 121: 229–233.

Wells, J. D. and J. R. Stevens. 2008. Application of DNA-based methods in forensic entomology. *Annual Review of Entomology* 53: 103–120.

Wheeler, Q. D. 2008. Undisciplined thinking: morphology and Hennig's unfinished revolution. *Systematic Entomology* 33: 2–7.

Whitfield, J. B. 1999. Destructive sampling and information management in molecular systematic research: an entomological perspective. Pp. 301–314. *In* D. A. Metsger and S. C. Byers (eds). *Managing the Modern Herbarium: An Interdisciplinary Approach.* Society for Preservation of Natural History Collections and Royal Ontario Museum. Pp. 384.

Whitfield, J. B. and K. M. Kjer. 2008. Ancient rapid radiations of insects: challenges for phylogenetic analysis. *Annual Review of Entomology* 53: 449–472.

Whitworth, T. L., R. D. Dawson, H. Magalon, and E. Baudry. 2007. DNA barcoding cannot reliably identify species of the blowfly genus *Protocalliphora* (Diptera: Calliphoridae). *Proceedings of the Royal Society, Series B, Biological Sciences* 274: 1731–1739.

Will, K. W., B. D. Mishler, and Q. D. Wheeler. 2005. The perils of DNA barcoding and the need for integrative taxonomy. *Systematic Biology* 54: 844–851.

Williams, J. G., A. R. Kubelik, and K. J. Livak. 1990. DNA polymorphisms amplified by arbitrary primers are useful as genetic markers. *Nucleic Acids Research* 18: 6531–6535.

Wondji, C. S., J. Hemingway, and H. Ranson. 2007. Identification and analysis of single nucleotide polymorphisms (SNPs) in the mosquito *Anopheles funestus*, malaria vector. *BMC Genomics* 8: 5.

Zakharov, E. V. and J. J. Hellmann. 2008. Genetic differentiation across a latitudinal gradient in two co-occuring butterfly species: revealing population differences in a context of climate change. *Molecular Ecology* 17: 189–208.

Zhang, D.-X. 2004. Lepidopteran microsatellite DNA: redundant but promising. *Trends in Ecology and Evolution* 19: 506–509.

Chapter 17

DNA BARCODES AND INSECT BIODIVERSITY

Robin M. Floyd, John J. Wilson, and Paul D. N. Hebert

Biodiversity Institute of Ontario and Department of Integrative Biology, University of Guelph, Guelph, Ontario, Canada N1G 2W1

I know this little thing A myriad men will save, O Death, where is thy sting? Thy victory, O Grave?

— *Sir Ronald Ross (1857–1932)*

Insect Biodiversity: Science and Society, 1st edition. Edited by R. Foottit and P. Adler
© 2009 Blackwell Publishing, ISBN 978-1-4051-5142-9

About 3500 species of mosquitoes (Diptera: Culicidae) have been described worldwide. In 1897, Ronald Ross, a Scottish physician working in India, discovered that only members of one mosquito genus, *Anopheles*, carry the *Plasmodium* parasite, the single-celled organism that causes malaria in humans. This revelation reflected painstaking efforts, involving the dissection of stomachs from vast numbers of mosquitoes. It was a key breakthrough that paved the way for Ross to demonstrate the life cycle of the parasite in the laboratory, work rewarded by the 1902 Nobel prize in Medicine. Unusually for a scientist, Ross was also a poet, playwright, and novelist; the preceding verse was written in response to this breakthrough in the understanding of malaria (Carey 1995).

Sadly, Ross's hope that this knowledge would quickly allow malarial control proved too optimistic; the disease still causes more than 1 million deaths per year, mainly in tropical Africa and Asia, despite numerous eradication efforts (Greenwood et al. 2005). This situation continues, in part, because the evolutionary dynamics of both *Plasmodium* and its insect vectors are far more complicated than initially realized. Only a limited number of species in the genus *Anopheles* transmit the agents of the human form of malaria. *Anopheles gambiae* (*sensu stricto*), the most important vector of the *Plasmodium* parasite in humans, belongs to a complex of morphologically indistinguishable sibling species that nevertheless differ markedly in their habitat preferences, behavior, and ability to transmit malarial agents (della Torre et al. 2002, Lehmann et al. 2003). Although these species are likely in the midst of speciation (a process expected to result in morphologically cryptic species complexes), they can be readily discriminated on the basis of their ribosomal DNA sequences (Masendu et al. 2004, della Torre et al. 2005, Guelbeogo et al. 2005). Plans are underway to control populations of *A. gambiae* by introducing transgenes, a strategy that will depend on knowledge of gene flow and population dynamics within and among these sibling species (Cohuet et al. 2005, Tripet et al. 2005).

The message from this story is clear: cryptic biological diversity matters. *Anopheles* serves as a pertinent example of the challenges faced by those concerned with biodiversity. Life exists in an immense number of forms, which are often tiny, difficult to study, and even more difficult to discriminate. Yet, this subtle variation can be crucially important; paraphrasing one article on the subject, what we do not know can hurt us (Besansky et al. 2003).

Insects constitute the most diverse group of animals on the planet, with more than 1 million described species (1,004,898; introduction to this volume) and millions more either awaiting description or simply undiscovered (Grimaldi and Engel 2005). They affect human society in myriad ways, both harmful (e.g., disease vectors, crop pests) and helpful (e.g., pollinators, biological control agents). Research of insects has added immensely to our understanding of evolution, ecology, and the genetic control of development. Yet, a fundamental requirement in gaining useful knowledge about any organismal group is the ability to describe, classify, and subsequently identify its member taxa.

Groups, such as insects, present great challenges to the taxonomic enterprise simply because of their diversity. The identification of species by traditional morphological methods is complex and usually requires specialist knowledge. The recognition, description, and naming of new species is more so; yet, the number of undescribed insect species far outweighs the number of taxonomic specialists (Grissell 1999), whose workforce is in decline (Godfray 2002). New approaches are needed to overcome this 'taxonomic impediment' (Weeks and Gaston 1997, Giangrande 2003). These concerns are not purely academic, but have significant practical implications. Agricultural pests cause immense damage. Total annual crop losses due to insect pests in North America have been estimated at US$7.5 billion and far more in the developing world (Yudelman et al. 1998), making it vital to quickly identify destructive species before invasions become uncontrollable. There is also a basic scientific need to describe biological diversity before the destruction of natural habitats by human activity causes the loss of species on a massive scale. We need a rapid way of assembling species catalogs, so that conservation programs can protect those areas of greatest importance before they are lost (Myers et al. 2000).

SPECIES CONCEPTS AND RECOGNITION

Although species have long been considered the basic 'units of biodiversity' (Claridge et al. 1997), and the only 'real' grouping in the taxonomic hierarchy, the issue of how best to delimit species remains controversial. Mayden (1997) listed 22 species concepts that have appeared in the literature (though some are essentially synonymous) and that employed varied criteria from ecological niches, mate recognition, genetic cohesion, and evolutionary history. These diverse criteria necessarily lead to ambiguity, which can have important implications for studies of biodiversity and conservation, as differing species concepts can produce widely varying estimates of taxon richness (Agapow et al. 2004) Although reproductive isolation is often considered the most important indicator of species status, it is seldom directly tested and fails to address asexual organisms. In practice, most species continue to be recognized by the presence of one or more apparently fixed or nonoverlapping diagnostic differences (Davis and Nixon 1992). For most insect groups, detailed examination of genital morphology has represented the gold standard for species definition for nearly a century, due to the observation of a general phenomenon of rapid and pronounced divergence in the genitalia between species of animals (Eberhard 1985). Actual application of this criterion, however, is hampered by lack of an appropriate methodology to quantify shape variation (Arnqvist 1998) and by questionable homology assessments. All these factors collectively make species identification an extremely specialized and time-consuming science, and even expert taxonomists can have difficulty reaching consensus. Moreover, this reliance on diagnostic characters that are present only in the adult life stage creates a serious constraint on identification, as many specimens lack these characters (Balakrishnan 2005). The life-history stages most commonly intercepted at ports of entry are larvae and pupae (Scheffer et al. 2006), and damage to specimens collected in the field often makes identification difficult or impossible.

Another option exists – species can be diagnosed by the genetic changes that arise between reproductively isolated lineages as a result of genetic drift or selection. The use of DNA sequences to gain information about the taxonomic affinities of an unknown specimen saw its earliest adoption in the least morphologically tractable groups such as viruses and bacteria (Theron and Cloete 2000, Nee 2003). More recently, it has been applied to plants (Chase et al. 2005), to simple metazoan animals such as nematode worms (Floyd et al. 2002), and even to charismatic megafauna such as birds, fish, and mammals (Ward et al. 2005, Clare et al. 2007, Kerr et al. 2007). This approach relies on the use of algorithms enabling DNA-sequence comparison, such as Basic Local Alignment Search Tool (BLAST) (Altschul et al. 1990), in conjunction with DNA databases such as GenBank.

Many authors refer to 'operational taxonomic units' (OTUs) when delimiting taxa by purely phenetic or heuristic means (Sokal and Sneath 1963). OTUs may or may not correspond to species in the strict sense, but can be used in instances where speed and ease of application are of more practical importance than theoretical considerations (and if there are reasons to believe that theory and practicality are not directly in conflict). Taxa diagnosed or delimited by phenetic DNA-sequence divergences can be termed 'molecular operational taxonomic units' (MOTUs – Floyd et al. 2002, Blaxter et al. 2005). This approach has become the standard for environmental surveys of bacteria and other microorganisms, which could be seen as a capitulation to necessity, because these groups are virtually impossible to address in any other way (Hagström et al. 2002, Martinez et al. 2004, Hanage et al. 2005). However, correspondence between MOTUs and species can be examined in a number of ways. One approach, tested with Lepidoptera, is the correlation with previously unassociated morphological or ecological traits, for example, host plants and caterpillar phenotypes in *Astraptes fulgerator* (Hebert et al. 2004). Where morphological or ecological information is unavailable, a common situation in many taxonomic studies, congruence with an appropriate nuclear gene is an objective way to delineate interbreeding groups, and has been investigated in tropical beetles (Monaghan et al. 2005) and tachinid flies (Smith et al. 2006). Seven of the nine methods of delimiting species boundaries recently reviewed by Sites and Marshall (2003) require molecular data, which could imply that molecular markers are becoming increasingly important tools applied by taxonomists, possibly due to objectivity, speed, and increased discriminatory capacity.

DNA BARCODING

In this chapter, we deal with DNA barcoding, the use of short standardized genomic segments as markers for species identification. Just as species differ in morphology, ecology, and behavior, they also differ in their DNA sequences. Hence, at least in principle, a particular gene or gene fragment can be used to identify a given species in much the same way that retail barcodes uniquely identify each consumer product. In practice we would not expect DNA barcoding to work in such a simple manner – real DNA sequences are subject to all the natural complexities of molecular evolution, and can show considerable variation within species (Mallet and Willmott 2003). They are not systematically 'assigned' to entities one by one as retail barcodes are. Nevertheless, if successful, DNA barcoding promises the ability to automate the identification of specimens by determining the sequence of the barcode region, avoiding the complexities inherent in morphological identifications, and prompting advocates to argue for the establishment of a system that ultimately might be applied to all life (Tautz et al. 2003, Blaxter 2004, Savolainen et al. 2005).

The particular genomic region used as a barcode is an important choice. It must be homologous between the organisms compared and have a rate of evolution fast enough to show variation between closely related species, and it also must have sufficient regions of sequence conservation to allow a limited set of PCR primers to amplify the target gene region from broad sections of the tree of life. The resultant sequence information also must generate a robust alignment so that sequences can be compared. In the animal kingdom, attention has been focused on a ~650 base-pair region near the 5′ end of the mitochondrial cytochrome c oxidase subunit I (COI) gene (Hebert et al. 2003a). COI provides an ideal species-identification marker in insects, due to its lack of introns, simple alignment, limited exposure to recombination, and the availability of robust primer sites. Sequence variation in this region generally shows large interspecific, but small intraspecific, divergences, meaning that species frequently form clearly distinguishable clusters on a distance-based or phylogenetic tree. The homogenization of mitochondrial DNA sequences within a species, regardless of population size, is an intriguing phenomenon that has prompted study and speculation as to its evolutionary origin and significance (Bazin et al. 2006). The resulting 'barcoding gap' appears to represent a 'genetic signature' for species (Monaghan et al. 2005). Boundaries signaled by this molecular marker are strongly concordant with species units recognized through past studies of morphological and behavioral characters in a number of specific cases where they have been examined (Hebert et al. 2003a, 2003b).

Important advantages of a sequence-based approach to identification include the digital nature of a DNA sequence, which allows it to be gathered and interpreted objectively. Furthermore, DNA extracts from any life stage of an organism – egg, larva, or adult – or from fragments of dead material will generate a similar identification, whereas traditional identification keys (at least for insects that undergo complete metamorphosis from larval to adult forms) are often dependent on adult features. Additionally, social insects such as ants and termites often exhibit highly divergent caste morphologies that, in some cases, have been diagnosed incorrectly as distinct species – DNA barcoding promises to remove such ambiguities, allowing such forms to be associated (Smith et al. 2005). Sexual dimorphism, too, has long been a source of complications for taxonomists: Janzen et al. (2005) described a case where each sex of the butterfly *Saliana severus* was recorded as a separate species in an inventory, until barcoding revealed that males and females had the same COI sequences, and subsequently led to the recognition of a single, highly sexually dimorphic species.

Extraction and amplification of DNA from insects, including eggs and larvae, presents no technical challenge (Ball and Armstrong 2006). Recent advances in high-throughput DNA sequencing technology (Shendure et al. 2004) and reductions in costs (Hajibabaei et al. 2005) have made the generation of large volumes of DNA data straightforward. Sequences can be produced in the laboratory from a sample within a few hours in a largely automated fashion. While the 'Star Trek' vision of a handheld instant species-identification device (Janzen 2004, Savolainen et al. 2005) remains a speculative (yet attractive) notion for the future, promising advances have been made in reducing the size of the equipment needed to gather barcode data (Blazej et al. 2006). Sequencing is not always a necessary step for rapid identification with DNA, especially when analysis is focused on a small assemblage of closely allied species. Diagnostic restriction-digest enzyme patterns of cytochrome b PCR products were used to distinguish *Bombus ruderatus* and *Bombus hortorum*, two cryptic species of bumblebees, one in decline in the UK and illustrated with the same diagram of male genitalia in an identification key (Ellis et al. 2006). A simple PCR assay has been suggested as a molecular diagnostic tool for the swede midge (*Contarinia nasturtii*), an agricultural pest (Frey et al. 2004).

Full exploitation of DNA barcodes for species identification will be possible only after a comprehensive databank linking organisms and their sequences has been assembled (Savolainen et al. 2005). Databases presently have very uneven sequence coverage among taxa. Intensely studied groups and model organisms (e.g., *Drosophila melanogaster*) have many sequences or even entire genomes available. A few genes have been sequenced for many taxa, but the vast majority of species lack any sequence data (Sanderson et al. 2003). This void raises the possibility that a poor match between a sequence derived from a newly encountered species and an incomplete reference library could result in spurious species diagnoses (Baker et al. 1996). Best BLAST hit, the simplest method of taxonomic assignment, is 'essentially useless' when no relatives have appropriate sequences in GenBank (Tringe and Rubin 2005). Tautz et al. (2003) suggested that an attempt be made to provide a DNA sequence as a component of all future species descriptions; the current barcoding initiatives could go a long way to bridge the gap, at least for major eukaryote groups. Some authors have argued that GenBank is unsuitable for taxonomic purposes due to its failure to include morphological, biogeographical, and ecological information associated with each sequence record (Tautz et al. 2003). However, the concept of 'type sequences' with voucher specimens authenticated by experts on the taxa and with associated taxonomic data is becoming reality. In 2004, the National Centre for Biotechnology Information (NCBI), GenBank's home organization, sealed a partnership with the Consortium for the Barcode of Life, whereby 'barcode standard' DNA sequences with relevant supporting data, including name of the identifier and collection location, can now be archived with the International Nucleotide Sequence Database Collaborative (Hanner 2005, Savolainen et al. 2005), with the keyword 'BARCODE' attached. This approach provides standardization of DNA regions, which previously was lacking and hindering progress in insect molecular systematics (Caterino et al. 2000).

The concept of DNA barcoding has been controversial in the taxonomic community (Moritz and Cicero 2004, Smith 2005). Criticisms of the approach have included questioning whether a single genetic marker has sufficient resolution to discriminate species reliably (Will and Rubinoff 2004, Will et al. 2005); potential problems caused by differing patterns of inheritance between nuclear and mitochondrial genes, which could confound the association between sequence and species (Funk and Omland 2003, Rubinoff 2006); and the feared marginalizing of morphological taxonomy (Lipscomb et al. 2003, Seberg et al. 2003). Other authors have emphasized the benefits of barcoding and DNA-assisted taxonomy in general (Tautz et al. 2002, Blaxter 2004, Hebert and Gregory 2005, Vogler and Monaghan 2006). As with any new concept or methodology in science, barcoding can be judged only by its success in facilitating new research and leading to new and useful knowledge. With the application to insects, and the endeavor to build a systematic database of 'DNA barcodes' linked to data about the species they represent, the barcoding movement has begun to gather real momentum (Hebert et al. 2003a). We, therefore, move to discuss some specific cases in which barcoding has been applied to particular insect groups, and examine how it has advanced our knowledge of biodiversity.

APPLICATIONS OF BARCODING

Lepidoptera

The Lepidoptera are a diverse and charismatic group of insects that have received significant taxonomic and systematic attention. One might think DNA barcoding has little to offer an order for which bright wing patterns and extensive, previous taxonomic attention suggest a group with a well-resolved species taxonomy. However, this perspective would be overly optimistic; approximately 165,000 species of Lepidoptera have been described, representing about 10% of the roughly 1.5 million known animal species (Wilson 2003). Another 150,000 to 1,250,000 species of Lepidoptera are thought to await description. These species do not all reside in hyperdiverse tropical settings; more than 10,000 species in Australia are still undescribed. Lack of taxonomists, problems with the way species are recognized, and extensive morphological convergence mean that most species are undescribed and numbers can only be estimated, particularly in the tropics.

Lepidoptera have now become the model group for barcoding studies since Hebert et al. (2003a) used North American moths to demonstrate the ability of COI to discriminate among specimens of different species. Since then, research with Lepidoptera has demonstrated the potential of molecular diagnostics and practical applications of DNA barcoding. Barcodes have enabled the linking of the varied life stages of the Lepidoptera, as well as the males and females of sexually dimorphic species (Janzen et al. 2005). This advancement is particularly relevant in the identification of pest and invasive species, as many are intercepted as eggs or larvae (Ball and Armstrong 2006).

One of the most significant potential uses of DNA barcoding lies in facilitating biodiversity surveys. Lepidoptera are a model group in ecology and a 'flagship' group for invertebrate conservation. Macromoths and butterflies, in particular, have been used to indicate environmental quality (e.g., habitat degradation), to partition habitat diversity, and as indicators of climate change (Scoble 1992). Their role as model organisms in surveys, however, is limited by lack of taxonomic support. Barcoding could provide a new level of efficiency and comparability to ecological surveys, with DNA barcode records enabling more relevant and meaningful correlation of studies carried out by different experts at different locations and times, rather than the use of arbitrary morphological designators such as Noctuid sp. 01.

Work on the Neotropical skipper *Astraptes fulgerator* provides a prime example of the way in which DNA barcoding can aid species discovery, especially when coupled with morphological and ecological studies. Barcoding of 484 specimens from the Area Conservación Guanacaste (ACG) in Costa Rica revealed that the *A. fulgerator* group comprises a complex of sister species, confirming and extending earlier suspicions gained through studies of adult morphology and larval morphologies. Hebert et al. (2004) hypothesized 'ten species in one' based on COI divergences and association with caterpillar morphology and food plant. Brower (2006) reanalyzed the original DNA dataset under a different framework and also concluded that the *Astraptes fulgerator* sample contained multiple species but was critical of the methods used in the original study. However, other investigators who reanalyzed the same dataset supported the conclusion of 10 taxa (Nielsen and Matz 2006). The ideal framework for the use of barcodes in species delineation requires further research. This example, however, demonstrates the power of the DNA sequences themselves, once submitted to publicly available databases as unambiguous digital data immune from subjective assessment and open to repeated analysis and testing of the species and phylogenetic hypotheses generated.

Barcoding studies on the lepidopteran fauna from one region (the ACG) of Costa Rica are now well underway. Hajibabaei et al. (2006) sequenced more than 4000 individuals from 521 species belonging to 3 families (Hesperiidae, Saturniidae, and Sphingidae) and found that 97.9% of the individuals could be identified to species based on COI divergence patterns. The expanded ACG project now has the goal of barcoding every species of butterfly and moth in the preserve (about 9600 species) within 3 years, representing the first large-scale regional barcoding project. Because parallel initiatives (www.lepbarcoding.org)

seek to barcode all Lepidoptera from two continents (North America, Australia), and all species in two families (Saturniidae and Sphingidae), the Lepidoptera are poised to become the first 'barcode-complete' order of insects. The realization of this goal will not only bring many advances in barcode data collection and analysis, but also provide a newly detailed framework for species delineation and research in molecular evolution.

Diptera

The flies (Diptera) constitute another hyperdiverse insect order, with around 150,000 described species (Grimaldi and Engel 2005, Beutel and Pohl 2006). Among insects, their members have the greatest negative impact on human health and livestock, with groups such as mosquitoes and tse tse acting as vectors of the agents for several major diseases including malaria, sleeping sickness, and filariasis (Yeates and Wiegmann 2005). Even before the establishment of DNA barcoding per se, many molecular diagnostic tools had been applied to the identification of mosquito species, including allozyme electrophoresis (Green et al. 1992), DNA hybridization (Beebe et al. 1996), and restriction fragment length polymorphism (RFLP) (Fanello et al. 2002). Sequencing-based approaches have also been used extensively, albeit mainly focusing on genes other than COI (e.g. Kent et al. 2004, Marrelli et al. 2005, Michel et al. 2005). However, a number of recent studies have shown that the standard COI barcode marker also serves effectively for species-level discrimination in surveys of Canadian (Cywinska et al. 2006) and Indian (Kumar et al. 2007) mosquitoes. Foley et al. (2007) constructed a molecular phylogeny of the Australian *Anopheles annulipes* species complex based on four different loci, both nuclear and mitochondrial (COI, COII, ITS2, and EF-1α). Despite using a shorter fragment of COI (258 bp) than the standard barcode region (658 bp), it was found in this study that 11 of the 17 sibling species (65%) had unique COI sequences, and the authors concluded that 'DNA barcoding holds some promise for diagnosing species within the Annulipes Complex, and perhaps for other anophelines'.

One of the earliest applications of a DNA-based approach to species identification involved fly species important to forensic science. Blow flies (Calliphoridae) and flesh flies (Sarcophagidae) lay eggs on corpses shortly after death. Because each species has a timeframe for development from egg to adult, the particular life stage associated with a corpse can provide key evidence in determining time of death (postmortem interval, or PMI) (Smith 1986, Catts and Haskell 1990). However, because different species of flies have different development rates, accurate species identifications are necessary to make an accurate estimate of the PMI. Because only adults can be placed reliably to species, maggots previously had to be collected and reared to adults, constituting a significant time delay to the process (Nelson et al. 2007). Forensic entomologists were quick to realize the potential of DNA-based methods to distinguish species from any life stage and from dead, preserved material. As a result, extensive literature now exists, detailing how DNA sequences (mainly COI) can accurately discriminate fly species of forensic importance (Sperling et al. 1994, Malgorn and Coquoz 1999, Vincent et al. 2000, Wallman and Donnellan 2001, Wells and Sperling 2001, Wells et al. 2001).

Leafmining flies (family Agromyzidae) are economically important agricultural pests whose periodic population outbreaks are capable of destroying entire crops, particularly potatoes (Shepard et al. 1998). They are also a group for which considerable information on species limits is available (Scheffer and Wiegmann 2000). COI sequences were generated from 258 individuals belonging to three species of invasive leafminers in the Phillipines: *Liriomyza huidobrensis*, *L. trifolii*, and *L. sativae* (Scheffer et al. 2006). As is commonly observed in introduced or invasive populations, fewer mitochondrial haplotypes were found than in the endemic ranges of these species, and those seen were often highly divergent even within a species. This pattern is due to population bottlenecks that tend to occur during introduction (Nei et al. 1975), an effect that is particularly relevant for a marker such as mitochondrial DNA, which is both haploid and maternally inherited. Sequence analysis was able to place all specimens in the correct morphospecies as currently diagnosed. This study also illustrated some of the complexities that the barcoding approach must take into account. Certain mitochondrial sequences in both the *L. trifolii* and *L. sativae* groups were sufficiently divergent that they might suggest new, cryptic species, but no data other than COI-sequence divergence supported this conclusion. Based on existing knowledge of this group, these species are expected to contain highly divergent mitochondrial lineages; therefore, depending on which reference sequences were used, barcoding might overestimate

the number of species present. Future research, however, possibly will reveal that these divergent lineages do represent distinct species. Although some ambiguity might be associated with barcoding in complex cases, these cases normally should be possible to resolve by combining COI data with information from other sources, such as morphology, behavior, or complementary DNA regions.

Insect parasitoids are not only a major component of global biodiversity, but also have significant demographic effects on their host species. Parasitoids also conceal a large diversity of morphologically cryptic species, distinguished by strong host specificity (Godfray 1994). Flies of the family Tachinidae are endoparasitoids of other insects, often lepidopteran larvae. A recent study of the tachinid genus *Belvosia* from northwestern Costa Rica examined their diversity by rearing specimens from wild-caught caterpillars, recording their morphology, and sequencing their COI genes (Smith et al. 2006). DNA sequences were able not only to discriminate 17 known host-specific species of the genus *Belvosia*, but also raised the number of species to 32 by revealing that 3 species, each believed to be host generalists, were complexes of highly host-specific cryptic species. Again, this study illustrates the power of DNA barcoding to reveal unknown diversity in morphologically difficult groups.

Finally, the dipteran family Chironomidae (nonbiting midges) is a species-rich group whose freshwater larval stages are often used in environmental monitoring. However, the connection of larval stages to known species (whose descriptions are mainly based on adult morphology) is a difficult challenge; but DNA barcoding has helped to address this problem in recent studies (Ekrem et al. 2007, Pfenninger et al. 2007). The former paper nevertheless cautions us that in order to use barcoding as a tool to identify unknown individuals by their COI sequence, a comprehensive library of known sequences is necessary for such identifications to be reliable.

Coleoptera

In a famous, though possibly apocryphal, incident, when geneticist J. B. S. Haldane was asked what the study of nature revealed about the mind of God, he answered: 'an inordinate fondness for beetles'. One out of every five animals on the planet is thought to be a beetle. As a consequence, the Coleoptera represent a group where the taxonomic enterprise has been overwhelmed by diversity. Although 350,000 species of beetles have been described (including many economically important pest species), as many as 5–8 million might exist in total. With so many unknown species, major barcoding research with beetles has focused on the use of DNA-based methods in species discovery and delineation.

Monaghan et al. (2005) used the 3′ end of COI and the nuclear gene 28S rRNA to identify clusters of beetles in dung beetles of the genus *Canthon* and water beetles of the family Hydrophilidae. An exact match of nuclear genotypes and mitochondrial clusters suggested that the mtDNA groupings were not misleading due to introgression, and the clusters likely correlated with previously described or undescribed species. The results indicated that COI provides a largely accurate picture of species boundaries in these two beetle groups and provides validation for its use in species discovery.

Barcoding research with water beetles was continued by Monaghan et al. (2006), using the genus *Copelatus* from Fiji. Four DNA markers (three mitochondrial regions: COI, cytochrome b, and 16S rRNA, and the nuclear histone 3 gene) were sequenced for 118 specimens from 20 islands. This effort was seen as a particularly challenging test case for barcoding because many lineages on oceanic islands have undergone rapid 'radiations', resulting in large numbers of recently diverged species with complex gene histories. Beetle taxa were clustered using the concatenated DNA sequences and separately with traditional morphological methods (i.e., male genital morphology). Although the clustering pattern was largely incongruent using the two approaches, the authors concluded that if the morphological approach had been followed with a Linnaean system of naming, it would have formalized, at best, a partial taxonomic resolution, with limited evolutionary understanding of lineage diversification (Monaghan et al. 2006). The morphological approach is time intensive and requires specialized knowledge of character differences associated with species-level classification. Subsequent identification of the 'species', using morphology, thus can be problematic due to ambiguous descriptions and difficulties obtaining the type specimens, the situation encountered with the five formerly described species of *Copelatus* from Fiji. The sequencing approach, combined with phylogenetic analysis, provides an extensive summary of evolutionary history, and once sequences have been submitted to databases (in this case, EMBL), the data are easily accessible and subsequent repeat analyses can be done by anyone. Monaghan et al. (2006) suggest DNA sequences themselves could constitute a system of taxonomic grouping and communication without need of a formal Linnaean classification. The study suggests DNA sequencing could make the task of global species classification achievable when standard morphological methods are inadequate or too time consuming.

Hymenoptera

With about 125,000 described species, the Hymenoptera are the fourth largest insect order after Coleoptera, Lepidoptera, and Diptera (Grimaldi and Engel 2005, Beutel and Pohl 2006). Given the number of cryptic species suspected to exist, its true species richness might even surpass the 'big three' (Grissell 1999).

Ants (Formicidae) constitute the major component of arthropod biomass in many of the world's ecosystems. They are important in nutrient recycling, and their activities within soil create widely varying nutrient microhabitats influencing plant succession, growth, and distribution (Hölldobler and Wilson 1990). In Madagascar, the ant fauna represents a hyperdiverse group, currently estimated to include about 1000 species, of which 96% are thought to be endemic. However, only 25% of this estimated total has been described, presenting a major obstacle to studying their biogeography, conservation status, and roles in ecosystem processes. A recent case study (Smith et al. 2005) examined the question of whether DNA barcoding could act as an effective surrogate for morphological species identifications. A total of 280 specimens from four localities were collected and independently identified to morphospecies and sequenced for COI. The specimens were classified both into MOTUs based on their sequence data and morphospecies based on their morphological traits, allowing the two methods to be directly compared. Additionally, two different sequence-divergence thresholds (2% and 3%) were tested for MOTU assignment.

Although instances of incongruities occurred between the molecular and morphological taxon assignments, strong correlations, nevertheless, existed between the two. A total of 90 morphospecies, 117 3% MOTU and 126 2% MOTU were found. Morphological species designations, therefore, tended

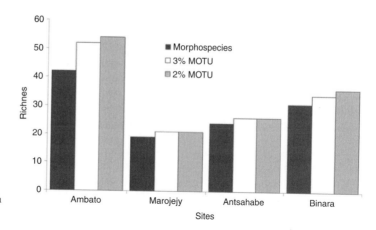

Fig. 17.1 Estimates of taxonomic richness (based on both MOTU and morphospecies) from a survey of ants across four sites in Madagascar (from Smith et al. 2005, used with permission).

to lump specimens that were split by the molecular approach. As in many cases, molecular markers detect cryptic taxa that are difficult or impossible to detect by morphology alone. These cryptic taxa may or may not correspond to true 'species', but they form a starting point for further investigation of their status. After examining the patterns of taxon richness across the four sites, no significant differences occurred between the data shown by MOTUs and morphospecies (Fig. 17.1). Additionally, whether MOTUs were defined on the basis of 2% or 3% divergence altered only the absolute number of taxa delineated, and did not make a significant difference to the overall patterns of diversity observed. This finding is important because it suggests that MOTUs can be used as effective surrogates for traditional species – although they will not necessarily delineate exactly the same taxonomic groupings, they will identify the same general patterns such as the most versus least diverse sites. Studies that delineate taxa by DNA-sequence arrays alone would enable surveys across much larger geographical regions and taxonomic groups than would be possible if the slow and laborious process of morphological identification were required, without losing resolution or information content.

Collembola

Springtails (Collembola) are not true insects but are basal members of the same superclass, Hexapoda (Grimaldi and Engel 2005). Phylogenetically, they are potentially important in unraveling the relationships of higher taxa (Mayhew 2002). They are

among the most diverse and numerically abundant of all soil arthropods (Petersen and Luxton 1982) and have the widest distribution of any hexapod group, occurring throughout the world, including Antarctica. There are about 7000 known species, but many more likely remain undiscovered (Hopkin 1997). In particular, the Arctic regions appear to have a vast uncataloged diversity (Danks 1981). As is typical for such groups, a great diversity combined with difficulty of identification and a global lack of taxonomic specialists creates a severe impediment to understanding their diversity.

Springtails show all the hallmarks of a group for which DNA barcoding could prove highly informative. A recent study (Hogg and Hebert 2004) tested whether COI was able to resolve species differences among a set of Collembola sampled from the Canadian Arctic. In all cases – 19 species in 13 genera – COI sequences were able to discriminate species, with between-species divergences above 8% in all cases and within-species divergences generally below 1%. The single exception to this pattern was that several individuals identified as *Folsomia quadrioculata* showed divergences of up to 13%, likely representing a case of an undescribed and morphologically cryptic sister species, which is a well-known phenomenon among Collembola (Stevens and Hogg 2003).

Ephemeroptera

Mayflies (Ephemeroptera) are an insect order whose larval stages develop in freshwater habitats. They are important in aquatic research, particularly in biomonitoring of water quality: the particular species composition of mayfly and other insect larvae are useful indicators of chemical pollution in rivers (Lenat and Resh 2001). Identifications are often problematic, however, as frequently only larvae are available, and species-level identification keys normally depend on adult features. A DNA-based system allowing identification from any life stage, therefore, would be highly beneficial.

One recent study applied the standard COI barcoding method to a test set of Ephemeroptera specimens (Ball et al. 2005). Sequences were generated from 150 individuals – initially 80 reference specimens that were used to create a profile matching sequences to named species, followed by a further 70 specimens that were used to test whether the correct species assignments could be made on the basis of their COI sequences. All but one of the 70 test specimens were correctly identified, with a mean sequence divergence within species of 1% and mean divergence among congeneric species an order of magnitude greater (18%). The sole exception was an individual identified morphologically as *Maccaffertium modestum*, which showed deep genetic divergence from other *M. modestum* specimens, again suggesting an undescribed sister species.

CONCLUSIONS

Taxonomy is the framework by which we name and classify biological diversity into the groupings used in all areas of biology. As a component of modern systematic science, taxonomy seeks to recognize natural evolutionary groupings – those that are monophyletic. This effort has led to countless revisions as different systematists uncover new data or interpret character distributions in different ways, but despite years of attention, the monophyly of many insect groups remains questionable. Even in intensively studied groups, the evolutionary history is unresolved and the Linnaean hierarchy only adds to the confusion. DNA barcodes represent data points able to be integrated into the traditional Linnaean system (Dayrat 2005), yet at the same time independent from it, for accumulating ecological, geographical, morphological, and other data about organisms. Once submitted to online databases, nucleotide sequences represent a freely available taxonomic resource that allows species recognition to be accomplished in a uniform manner by nonexperts. Barcoding has the potential to become a universal communication tool in a way that complicated and often incomprehensible morphological descriptions cannot be, especially in developing countries where the majority of biodiversity resides (Agosti 2003).

Barcoding need not be restricted to a single gene region. From the point of view of both economics (sequencing is still relatively expensive, though costs are dropping yearly) and simplicity of use, a system based on a single sequencing 'read' per specimen could be established. In some instances, this single marker will fail to discriminate taxa, most often when dealing with recently diverged sister species, which are the most difficult to discriminate in any system (Mallet and Willmott 2003, Hickerson et al. 2006, Meier et al. 2006, Whitworth et al. 2007). In such problematic cases, a single sequence will narrow the options to a small number of closely related taxa, and additional sequence or other data can be added to provide species-level resolution. One of the most important scientific outcomes of large-scale barcoding initiatives will be the production of a library of genomic DNA extracts from archived voucher specimens, which can serve as a basis for numerous future lines of research besides the generation of the initial barcode sequence. Commentators who have criticized barcoding on the basis of its costs (Cameron et al. 2006) have generally ignored such collateral benefits of the research.

Sequence information is easy to obtain, unambiguous, and makes species identification possible by nonspecialists unfamiliar with the intricacies of morphology. MOTU, good species or not, depending on the species concept applied, nevertheless, can be a suitable surrogate for identifying units of diversity in biodiversity studies. This approach enables users to obtain the information much faster than with the traditional morphological taxonomic process, making surveys scalable across much larger taxonomic groupings and wider geographical regions (Smith et al. 2005). Yet, the appeal of barcoding stems not only from speed and operationality; it reflects the increasingly held view that DNA-sequence analysis is as appropriate a mechanism for recognizing and delimiting evolutionary units as morphological comparisons. Although it does not automatically follow from this premise that a barcoding system based on single-gene comparisons will always delimit species-level groups, the studies we have cited offer evidence that it often will be the case.

Support for a large-scale barcoding initiative has grown rapidly; in 2004, the Consortium for the Barcode of Life (CBOL) was established to act as a central organizing body for the barcoding effort (www.barcoding.si.edu). Based at the Smithsonian Institution in Washington, CBOL represents an international collaboration of more than 120 organizations, including many prominent museums. National organizations, such as the Canadian Barcode of Life Network (Dooh and Hebert 2005), have established specimen supply chains and centralized facilities for the generation and analysis of sequence data, including such tools as the Barcode of Life Data Systems (BOLD; www.barcodinglife.org), a central repository for barcode records in conjunction with various analytical tools (Ratnasingham and Hebert 2007). The present total of all insect barcode sequences deposited in BOLD currently stands at 206,434 (including those gathered from GenBank), from 26,262 different species, but most are in the process of final taxonomic validation. As a simple illustration of diversity within insect COI sequences, Fig. 17.2 shows a neighbor-joining tree of sequences for a selection of 4675 validated and published records.

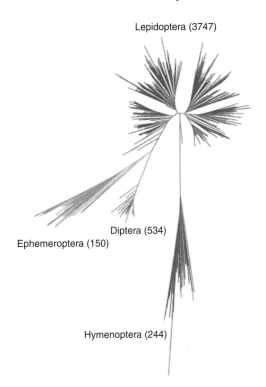

Lepidoptera (3747)

Diptera (534)

Ephemeroptera (150)

Hymenoptera (244)

Fig. 17.2 A neighbor-joining tree, based on Kimura 2-parameter distance, of 4675 cytochrome *c* oxidase I sequences from across the class Insecta.

To describe approximately 1.5 million species, using traditional approaches, has taken taxonomy two centuries. DNA-assisted species discovery has the potential to rapidly accelerate this process, an advantage that cannot be ignored in the light of the current biodiversity crisis affecting our planet (Eldredge 1992). By allowing more rapid detection and monitoring of agricultural pests and disease vectors, the pragmatic significance of barcoding can hardly be over emphasized. Perhaps more importantly in the long term, barcoding promises a blossoming of 'bioliteracy' (Janzen 2004) by shifting the accessibility of taxonomic knowledge from the realm of the specialist into the wider public domain. DNA sequencing technology is still relatively expensive and hence accessible only to well-funded labs, mainly in the developed world. Like any emerging technology, it is expected to become cheaper, faster, and simpler in the future, as has been the case with personal computers, GPS units, and mobile phones. One can envisage a time when a handheld DNA barcoding device allows any curious child to scan some interesting organism and gain immediate access to a library of information – not only the organism's name, but also its biology, ecology, conservation status, and more. Human beings preserve those things we value, and we can only value those things we perceive; for biological diversity to become something valued by all, it must be made visible and understandable in all its complexity – a goal for which barcoding can play a significant role in making reality.

ACKNOWLEDGMENTS

This contribution was supported through funding to the Canadian Barcode of Life Network from Genome Canada (through the Ontario Genomics Institute), NSERC, and other sponsors listed at www.bolnet.ca. We are grateful to Gregory Downs, Robert Dooh, and Sujeevan Ratnasingham for IT/bioinformatics support. We thank Robert Hanner, M. Alex Smith, and two anonymous reviewers for constructive comments on drafts of the manuscript; M.A. Smith also kindly provided the data used in Fig. 17.1.

REFERENCES

Agapow, P. M., O. R. Bininda-Emonds, K. A. Crandall, J. L. Gittleman, G. M. Mace, J. C. Marshall, and A. Purvis.

2004. The impact of species concept on biodiversity studies. *Quarterly Review of Biology* 79: 161–179.

Agosti, D. 2003. Encyclopaedia of life: should species description equal gene sequence? *Trends in Ecology and Evolution* 18: 273–273.

Altschul, S. F., W. Gish, W. Miller, E. W. Myers, and D. J. Lipman. 1990. Basic local alignment search tool. *Journal of Molecular Biology* 215: 403–410.

Arnqvist, G. 1998. Comparative evidence for the evolution of genitalia by sexual selection. *Nature* 393: 784–786.

Baker, C. S., F. Cipriano, and S. R. Palumbi. 1996. Molecular genetic identification of whale and dolphin products from commercial markets in Korea and Japan. *Molecular Ecology* 5: 671–685.

Balakrishnan, R. 2005. Species concepts, species boundaries and species identification: a view from the tropics. *Systematic Biology* 54: 689–693.

Ball, S. L. and K. F. Armstrong. 2006. DNA barcodes for insect pest identification: a test case with tussock moths (Lepidoptera : Lymantriidae). *Canadian Journal of Forest Research* 36: 337–350.

Ball, S. L., P. D. N. Hebert, S. K. Burian, and J. M. Webb. 2005. Biological identifications of mayflies (Ephemeroptera) using DNA barcodes. *Journal of the North American Benthological Society* 24: 508–524.

Bazin, E., S. Glemin, and N. Galtier. 2006. Population size does not influence mitochondrial genetic diversity in animals. *Science* 312: 570–572.

Beebe, N. W., D. H. Foley, R. D. Cooper, J. H. Bryan, and A. Saul. 1996. DNA probes for the *Anopheles punctulatus* complex. *American Journal of Tropical Medicine and Hygiene* 54: 395–398.

Besansky, N. J., D. W. Severson, and M. T. Ferdig. 2003. DNA barcoding of parasites and invertebrate disease vectors: what you don't know can hurt you. *Trends in Parasitology* 19: 545–546.

Beutel, R. G. and H. Pohl. 2006. Endopterygote systematics – where do we stand and what is the goal (Hexapoda, Arthropoda)? *Systematic Entomology* 31: 202–219.

Blaxter, M., J. Mann, T. Chapman, F. Thomas, C. Whitton, R. Floyd, and E. Abebe. 2005. Defining operational taxonomic units using DNA barcode data. *Philosophical Transactions of the Royal Society. B. Biological Sciences* 360: 1935–1943.

Blaxter, M. L. 2004. The promise of a DNA taxonomy. *Philosophical Transactions of the Royal Society. B. Biological Sciences* 359: 669–679.

Blazej, R. G., P. Kumaresan, and R. A. Mathies. 2006. Microfabricated bioprocessor for integrated nanoliter-scale Sanger DNA sequencing. *Proceedings of the National Academy of Sciences USA* 103: 7240–7245.

Brower, A. V. Z. 2006. Problems with DNA barcodes for species delimitation: 'ten species' of *Astraptes fulgerator* reassessed (Lepidoptera: Hesperiidae). *Systematics and Biodiversity* 4: 127–132.

Cameron, S., D. Rubinoff and K. Will. 2006. Who will actually use DNA barcoding and what will it cost? *Systematic Biology* 55: 844–847.

Carey, J. 1995. *The Faber Book of Science.* Faber and Faber, London.

Caterino, M. S., S. Cho, and F. A. H. Sperling. 2000. The current state of insect molecular systematics: A thriving Tower of Babel. *Annual Review of Entomology* 45: 1–54.

Catts, E. P. and N. H. Haskell. 1990. *Entomology and Death: A Procedural Guide.* Joyce's Print Shop, Clemson, SC.

Chase, M. W., N. Salamin, M. Wilkinson, J. M. Dunwell, R. P. Kesanakurthi, N. Haidar, and V. Savolainen. 2005. Land plants and DNA barcodes: short-term and long-term goals. *Philosophical Transactions of the Royal Society. B. Biological Sciences* 360: 1889–1895.

Clare, E. L., B. K. Lim, M. D. Engstrom, J. L. Eger, and P. D. N. Hebert. 2007. DNA barcoding of Neotropical bats: species identification and discovery within Guyana. *Molecular Ecology Notes* 7: 184–190.

Claridge, M. F., H. A. Dawah, and M. R. Wilson (eds). 1997. *Species: The Units of Biodiversity.* Chapman and Hall, London.

Cohuet, A., I. Dia, F. Simard, M. Raymond, F. Rousset, C. Antonio-Nkondjio, P. H. Awono-Ambene, C. S. Wondji, and D. Fontenille. 2005. Gene flow between chromosomal forms of the malaria vector *Anopheles funestus* in Cameroon, Central Africa, and its relevance in malaria fighting. *Genetics* 169: 301–311.

Cywinska, A., F. F. Hunter, and P. D. N. Hebert. 2006. Identifying Canadian mosquito species through DNA barcodes. *Medical and Veterinary Entomology* 20: 413–424.

Danks, H. V. 1981. *Arctic Arthropods : A Review of Systematics and Ecology with Particular Reference to the North American Fauna.* Entomological Society of Canada, Ottawa.

Davis, J. I. and K. C. Nixon. 1992. Populations, genetic variation, and the delimitation of phylogenetic species. *Systematic Biology* 41: 421–435.

Dayrat, B. 2005. Towards integrative taxonomy. *Biological Journal of the Linnean Society* 85: 407–415.

della Torre, A., Z. J. Tu, and V. Petrarca. 2005. On the distribution and genetic differentiation of *Anopheles gambiae* s.s. molecular forms. *Insect Biochemistry and Molecular Biology* 35: 755–769.

della Torre, A., C. Costantini, N. J. Besansky, A. Caccone, V. Petrarca, J. R. Powell, and M. Coluzzi. 2002. Speciation within *Anopheles gambiae* – the glass is half full. *Science* 298: 115–117.

Dooh, R. and P. D. N. Hebert. 2005. The Canadian Barcode of Life Network. http://www.bolnet.ca.

Eberhard, W. G. 1985 *Sexual Selection and Animal Genitalia.* Harvard University Press, Cambridge, Massachusetts.

Eldredge, N. (ed.) 1992. *Systematics, Ecology, and the Biodiversity Crisis.* Columbia University Press, New York.

Ellis, J. S., M. E. Knight, C. Carvell, and D. Goulson. 2006. Cryptic species identification: a simple diagnostic tool for discriminating between two problematic bumblebee species. *Molecular Ecology Notes* 6: 540–542.

Ekrem, T., E. Willassen, and E. Stur. 2007. A comprehensive DNA sequence library is essential for identification with DNA barcodes. *Molecular Phylogenetics and Evolution* 43: 530–542.

Fanello, C., F. Santolamazza, and A. della Torre. 2002. Simultaneous identification of species and molecular forms of the *Anopheles gambiae* complex by PCR-RFLP. *Medical and Veterinary Entomology* 16: 461–464.

Floyd, R., E. Abebe, A. Papert, and M. Blaxter. 2002. Molecular barcodes for soil nematode identification. *Molecular Ecology* 11: 839–850.

Foley, D. H., R. C. Wilkerson, R. D. Cooper, M. E. Volovsek, and J. H. Bryan. 2007. A molecular phylogeny of *Anopheles annulipes* (Diptera: Culicidae) sensu lato: the most species-rich anopheline complex. *Molecular Phylogenetics and Evolution* 43: 283–297.

Frey, J. E., B. Frey, and R. Baur. 2004. Molecular identification of the swede midge (Diptera: Cecidomyiidae). *Canadian Entomologist* 136: 771–780.

Funk, D. J. and K. E. Omland. 2003. Species-level paraphyly and polyphyly: frequency, causes, and consequences, with insights from animal mitochondrial DNA. *Annual Review of Ecology, Evolution, and Systematics* 34: 397–423.

Giangrande, A. 2003. Biodiversity, conservation, and the 'taxonomic impediment'. *Aquatic Conservation: Marine and Freshwater Ecosystems* 13: 451–459.

Godfray, H. C. 2002. Challenges for taxonomy. *Nature* 417: 17–19.

Godfray, H. C. J. 1994. *Parasitoids: Behavioral and Evolutionary Ecology*. Princeton University Press, Princeton, New Jersey.

Green, C. A., L. E. Munstermann, S. G. Tan, S. Panyim, and V. Baimai. 1992. Population genetic evidence for species-A, species-B, species-C and species-D of the *Anopheles dirus* complex in Thailand and enzyme electromorphs for their identification. *Medical and Veterinary Entomology* 6: 29–36.

Greenwood, B. M., K. Bojang, C. J. Whitty, and G. A. Targett. 2005. Malaria. *Lancet* 365: 1487–1498.

Grimaldi, D. A. and M. S. Engel. 2005. *Evolution of the Insects*. Cambridge University Press, Cambridge.

Grissell, E. E. 1999. Hymenopteran biodiversity: some alien notions. *American Entomologist* 45: 235–244.

Guelbeogo, W. M., O. Grushko, D. Boccolini, P. A. Ouedraogo, N. J. Besansky, N. F. Sagnon, and C. Costantini. 2005. Chromosomal evidence of incipient speciation in the Afrotropical malaria mosquito *Anopheles funestus*. *Medical and Veterinary Entomology* 19: 458–469.

Hagström, A., T. Pommier, F. Rohwer, K. Simu, W. Stolte, D. Svensson, and U. L. Zweifel. 2002. Use of 16S ribosomal DNA for delineation of marine bacterioplankton species. *Applied and Environmental Microbiology* 68: 3628–3633.

Hajibabaei, M., D. H. Janzen, J. M. Burns, W. Hallwachs, and P. D. N. Hebert. 2006. DNA barcodes distinguish species of tropical Lepidoptera. *Proceedings of the National Academy of Sciences USA* 103: 968–971.

Hajibabaei, M., J. R. de Waard, N. V. Ivanova, S. Ratnasingham, R. T. Dooh, S. L. Kirk, P. M. Mackie, and P. D. N. Hebert. 2005. Critical factors for assembling a high volume of DNA barcodes. *Philosophical Transactions of the Royal Society. B. Biological Sciences* 360: 1959–1967.

Hanage, W., C. Fraser, and B. Spratt. 2005. Fuzzy species among recombinogenic bacteria. *BMC Biology* 3: 6.

Hanner, R. 2005. Proposed Standards for BARCODE Records in INSDC (BRIs). Consortium for the Barcode of Life. http://www.barcoding.si.edu/PDF/DWG_data_standards-Final.pdf.

Hebert, P. D. N. and T. R. Gregory. 2005. The promise of DNA barcoding for taxonomy. *Systematic Biology* 54: 852–859.

Hebert, P. D. N., E. H. Penton, J. M. Burns, D. H. Janzen, and W. Hallwachs. 2004. Ten species in one: DNA barcoding reveals cryptic species in the neotropical skipper butterfly *Astraptes fulgerator*. *Proceedings of the National Academy of Sciences USA* 101: 14812–14817.

Hebert, P. D. N., A. Cywinska, S. L. Ball, and J. R. deWaard. 2003a. Biological identifications through DNA barcodes. *Proceedings of the Royal Society. B. Biological Sciences* 270: 313–321.

Hebert, P. D. N., S. Ratnasingham, and J. R. deWaard. 2003b. Barcoding animal life: cytochrome *c* oxidase subunit 1 divergences among closely related species. *Proceedings of the Royal Society. B. Biological Sciences* 270 (Supplement): S96–S99.

Hickerson, M. J., C. P. Meyer, and C. Moritz. 2006. DNA barcoding will often fail to discover new animal species over broad parameter space. *Systematic Biology* 55: 729–739.

Hogg, I. D. and P. D. N. Hebert. 2004. Biological identification of springtails (Hexapoda: Collembola) from the Canadian Arctic, using mitochondrial DNA barcodes. *Canadian Journal of Zoology* 82: 749–754.

Hölldobler, B. and E. O. Wilson. 1990. *The Ants*. Belknap Press of Harvard University Press, Cambridge, Massachusetts.

Hopkin, S. P. 1997. *Biology of the Springtails (Insecta, Collembola)*. Oxford University Press, Oxford.

Janzen, D. 2004. Now is the time. *Philosophical Transactions of the Royal Society. B. Biological Sciences* 359: 731–732.

Janzen, D. H., M. Hajibabaei, J. M. Burns, W. Hallwachs, E. Remigio, and P. D. N. Hebert. 2005. Wedding biodiversity inventory of a large and complex Lepidoptera fauna with DNA barcoding. *Philosophical Transactions of the Royal Society. B. Biological Sciences* 360: 1835–1845.

Kent, R. J., A. J. West, and D. E. Norris. 2004. Molecular differentiation of colonized human malaria vectors by 28S ribosomal DNA polymorphisms. *American Journal of Tropical Medicine and Hygiene* 71: 514–517.

Kerr, K. C. R., M. Y. Stoeckle, C. J. Dove, L. A. Weigt, C. M. Francis, and P. D. N. Hebert. 2007. Comprehensive DNA barcode coverage of North American birds. *Molecular Ecology Notes* 7: 535–543.

Kumar, N. P., A. R. Rajavel, R. Natarajan, and P. Jambulingam. 2007. DNA barcodes can distinguish species of Indian mosquitoes (Diptera: Culicidae). *Journal of Medical Entomology* 44: 1–7.

Lehmann, T., M. Licht, N. Elissa, B. T. A. Maega, J. M. Chimumbwa, F. T. Watsenga, C. S. Wondji, F. Simard, and W. A. Hawley. 2003. Population structure of *Anopheles gambiae* in Africa. *Journal of Heredity* 94: 133–147.

Lenat, D. R. and V. H. Resh. 2001. Taxonomy and stream ecology – the benefits of genus- and species-level identifications. *Journal of the North American Benthological Society* 20: 287–298.

Lipscomb, D., N. Platnick, and Q. Wheeler. 2003. The intellectual content of taxonomy: a comment on DNA taxonomy. *Trends in Ecology and Evolution* 18: 65–66.

Malgorn, Y. and R. Coquoz. 1999. DNA typing for identification of some species of Calliphoridae. An interest in forensic entomology. *Forensic Science International* 102: 111–119.

Mallet, J. and K. Willmott. 2003. Taxonomy: renaissance or Tower of Babel? *Trends in Ecology and Evolution* 18: 57–59.

Marrelli, M. T., L. M. Floeter-Winter, R. S. Malafronte, W. P. Tadei, R. Lourenco-de-Oliveira, C. Flores-Mendoza, and O. Marinotti. 2005. Amazonian malaria vector anopheline relationships interpreted from ITS2 rDNA sequences. *Medical and Veterinary Entomology* 19: 208–218.

Martinez, J., L. Martinez, M. Rosenblueth, J. Silva, and E. Martinez-Romero. 2004. How are gene sequence analyses modifying bacterial taxonomy? The case of *Klebsiella*. *International Microbiology* 7: 261–268.

Masendu, H. T., R. H. Hunt, J. Govere, B. D. Brooke, T. S. Awolola, and M. Coetzee. 2004. The sympatric occurrence of two molecular forms of the malaria vector *Anopheles gambiae* Giles sensu stricto in Kanyemba, in the Zambezi Valley, Zimbabwe. *Transactions of the Royal Society of Tropical Medicine and Hygiene* 98: 393–396.

Mayden, R. L. 1997. A hierarchy of species concepts: the denouement in the saga of the species problem. Pp. 381–424. *In* M. F. Claridge, H. A. Dawah, and M. R. Wilson (eds). *Species: The Units of Biodiversity*. Chapman and Hall, London.

Mayhew, P. J. 2002. Shifts in hexapod diversification and what Haldane could have said. *Proceedings of the Royal Society. B. Biological Sciences* 269: 969–974.

Meier, R., K. Shiyang, G. Vaidya, and P. K. Ng. 2006. DNA barcoding and taxonomy in Diptera: a tale of high intraspecific variability and low identification success. *Systematic Biology* 55: 715–728.

Michel, A. P., W. M. Guelbeogo, O. Grushko, B. J. Schemerhorn, M. Kern, M. B. Willard, N. Sagnon, C. Costantini, and N. J. Besansky. 2005. Molecular differentiation between chromosomally defined incipient species of *Anopheles funestus*. *Insect Molecular Biology* 14: 375–387.

Monaghan, M. T., M. Balke, J. Pons, and A. P. Vogler. 2006. Beyond barcodes: complex DNA taxonomy of a South Pacific Island radiation. *Proceedings of the Royal Society. B. Biological Sciences* 273: 887–893.

Monaghan, M. T., M. Balke, T. R. Gregory, and A. P. Vogler. 2005. DNA-based species delineation in tropical beetles using mitochondrial and nuclear markers. *Philosophical Transactions of the Royal Society. B. Biological Sciences* 360: 1925–1933.

Moritz, C. and C. Cicero. 2004 DNA barcoding: promise and pitfalls. *PLoS Biology* 2: e354.

Myers N., R. A. Mittermeier, C. G. Mittermeier, G. A. B. da Fonseca, and J. Kent. 2000. Biodiversity hotspots for conservation priorities. *Nature* 403: 853–858.

Nee, S. 2003. Unveiling prokaryotic diversity. *Trends in Ecology and Evolution* 18: 62–63.

Nei, M., T. Maruyama, and R. Chakraborty. 1975. The bottleneck effect and genetic variability in populations. *Evolution* 29: 1–10.

Nielsen, R. and M. Matz. 2006. Statistical approaches for DNA barcoding. *Systematic Biology* 55: 162–169.

Nelson, L. A., J. F. Wallman, and M. Dowton. 2007. Using COI barcodes to identify forensically and medically important blowflies. *Medical and Veterinary Entomology* 21: 44–52.

Petersen, H. and M. Luxton. 1982. A comparative analysis of soil fauna populations and their role in decomposition processes. *Oikos* 39: 287–388.

Pfenninger, M., C. Nowak, C. Kley, D. Steinke, and B. Streit. 2007. Utility of DNA taxonomy and barcoding for the inference of larval community structure in morphologically cryptic *Chironomus* (Diptera) species. *Molecular Ecology* 16: 1957–1968.

Ratnasingham, S. and P. D. N. Hebert. 2007. The Barcode of Life Data System (www.barcodinglife.org). *Molecular Ecology Notes* 7: 355–364.

Rubinoff, D. 2006. Utility of mitochondrial DNA barcodes in species conservation. *Conservation Biology* 20: 1026–1033.

Sanderson, M. J., A. C. Driskell, R. H. Ree, O. Eulenstein, and S. Langley. 2003. Obtaining maximal concatenated phylogenetic data sets from large sequence databases. *Molecular Biology and Evolution* 20: 1036–1042.

Savolainen, V., R. S. Cowan, A. P. Vogler, G. K. Roderick, and R. Lane. 2005. Towards writing the encyclopedia of life: an introduction to DNA barcoding. *Philosophical Transactions of the Royal Society. B. Biological Sciences* 360: 1805–1811.

Scheffer, S. J., M. L. Lewis, and R. C. Joshi. 2006. DNA barcoding applied to invasive leafminers (Diptera: Agromyzidae) in the Philippines. *Annals of the Entomological Society of America* 99: 204–210.

Scheffer, S. J. and B. M. Wiegmann. 2000. Molecular phylogenetics of the holly leafminers (Diptera: Agromyzidae: *Phytomyza*): species limits, speciation, and dietary specialization. *Molecular Phylogenetics and Evolution* 17: 244–255.

Scoble, M. J. 1992. *Lepidoptera: Form, Function, and Diversity*. Oxford University Press, Oxford.

Seberg, O., C. J. Humphries, S. Knapp, D. W. Stevenson, G. Petersen, N. Scharff, and N. M. Andersen. 2003. Shortcuts in systematics? A commentary on DNA-based taxonomy. *Trends in Ecology and Evolution* 18: 63–65.

Shendure, J., R. D. Mitra, C. Varma, and G. M. Church. 2004. Advanced sequencing technologies: methods and goals. *Nature Reviews Genetics* 5: 335–344.

Shepard, B. M., Samsudin, and A. R. Braun. 1998. Seasonal incidence of *Liriomyza huidobrensis* (Diptera : Agromyzidae) and its parasitoids on vegetables in Indonesia. *International Journal of Pest Management* 44: 43–47.

Sites, J., W. Jack, and J. C. Marshall. 2003. Delimiting species: a Renaissance issue in systematic biology. *Trends in Ecology and Evolution* 18: 462–470.

Smith, K. G. V. 1986. *Manual of Forensic Entomology.* British Museum (Natural History), London.

Smith, M. A., N. E. Woodley, D. H. Janzen, W. Hallwachs, and P. D. N. Hebert. 2006. DNA barcodes reveal cryptic host-specificity within the presumed polyphagous members of a genus of parasitoid flies (Diptera: Tachinidae). *Proceedings of the National Academy of Sciences USA* 103: 3657–3662.

Smith, M. A., B. L. Fisher, and P. D. N. Hebert. 2005. DNA barcoding for effective biodiversity assessment of a hyperdiverse arthropod group: the ants of Madagascar. *Philosophical Transactions of the Royal Society. B. Biological Sciences* 360: 1825–1834.

Smith, V. S. 2005. DNA barcoding: perspectives from a "Partnerships for Enhancing Expertise in Taxonomy" (PEET) debate. *Systematic Biology* 54: 841–844.

Sokal, R. R. and P. H. A. Sneath. 1963. *Principles of Numerical Taxonomy.* W. H. Freeman and Co, San Francisco, California.

Sperling, F. A., G. S. Anderson, and D. A. Hickey. 1994. A DNA-based approach to the identification of insect species used for postmortem interval estimation. *Journal of Forensic Sciences* 39: 418–427.

Stevens, M. I. and I. D. Hogg. 2003. Long-term isolation and recent range expansion from glacial refugia revealed for the endemic springtail *Gomphiocephalus hodgsoni* from Victoria Land, Antarctica. *Molecular Ecology* 12: 2357–2369.

Tautz, D., P. Arctander, A. Minelli, R. H. Thomas, and A. P. Vogler. 2002. DNA points the way ahead in taxonomy. *Nature* 418: 479.

Tautz, D., P. Arctander, A. Minelli, R. H. Thomas, and A. P. Vogler. 2003. A plea for DNA taxonomy. *Trends in Ecology and Evolution* 18: 70–74.

Theron, J. and T. E. Cloete. 2000. Molecular techniques for determining microbial diversity and community structure in natural environments. *Critical Reviews in Microbiology* 26: 37–57.

Tringe, S. G. and E. M. Rubin. 2005. Metagenomics: DNA sequencing of environmental samples. *Nature Reviews Genetics* 6: 805–814.

Tripet, F., G. Dolo, and G. C. Lanzaro. 2005. Multilevel analyses of genetic differentiation in *Anopheles gambiae* s.s. reveal patterns of gene flow important for malaria-fighting mosquito projects. *Genetics* 169: 313–324.

Vincent, S., J. M. Vian, and M. P. Carlotti. 2000. Partial sequencing of the cytochrome oxydase b subunit gene I: a tool for the identification of European species of blow flies for postmortem interval estimation. *Journal of Forensic Sciences* 45: 820–823.

Vogler, A. and M. T. Monaghan. 2006. Recent advances in DNA taxonomy. *Journal of Zoological Systematics and Evolutionary Research,* published online, doi: 10.1111/j.1439–0469.2006.00384.x.

Wallman, J. F. and S. C. Donnellan. 2001. The utility of mitochondrial DNA sequences for the identification of forensically important blowflies (Diptera: Calliphoridae) in southeastern Australia. *Forensic Science International* 120: 60–67.

Ward, R. D., T. S. Zemlak, B. H. Innes, P. R. Last, and P. D. N. Hebert. 2005. DNA barcoding Australia's fish species. *Philosophical Transactions of the Royal Society. B. Biological Sciences* 360: 1847–1857.

Weeks, P. J. D. and K. J. Gaston. 1997. Image analysis, neural networks, and the taxonomic impediment to biodiversity studies. *Biodiversity and Conservation* 6: 263–274.

Wells, J. D., T. Pape, and F. A. Sperling. 2001. DNA-based identification and molecular systematics of forensically important Sarcophagidae (Diptera). *Journal of Forensic Sciences* 46: 1098–1102.

Wells, J. D. and F. A. Sperling. 2001. DNA-based identification of forensically important Chrysomyinae (Diptera: Calliphoridae). *Forensic Science International* 120: 110–115.

Whitworth, T. L., R. D. Dawson, H. Magalon, and E. Baudry. 2007. DNA barcoding cannot reliably identify species of the blowfly genus *Protocalliphora* (Diptera: Calliphoridae). *Proceedings of the Royal Society. B. Biological Sciences* 274: 1731–1739.

Will, K. and D. Rubinoff. 2004. Myth of the molecule: DNA barcodes for species cannot replace morphology for identification and classification. *Cladistics* 20: 47–55.

Will, K. W., B. D. Mishler, and Q. D. Wheeler. 2005. The perils of DNA barcoding and the need for integrative taxonomy. *Systematic Biology* 54: 844–851.

Wilson, E. O. 2003. The encyclopedia of life. *Trends in Ecology and Evolution* 18: 77–80.

Yeates, D. K. and B. M. Wiegmann. 2005. *Evolutionary Biology of Flies.* Columbia University Press, New York.

Yudelman, M., A. Ratta, and D. F. Nygaard. 1998. *Pest Management and Food Production: Looking to the Future.* International Food Policy Research Institute, Washington, DC.

Chapter 18

INSECT BIODIVERSITY INFORMATICS

Norman F. Johnson

Department of Entomology, The Ohio State University, Columbus, OH 43212, USA

Insect Biodiversity: Science and Society, 1st edition. Edited by R. Foottit and P. Adler

The range of research tools available to students of biodiversity has greatly expanded over the past 25 years. Anatomical character data are now routinely accompanied by molecular data. New tools for visualization, such as extended-focus software and confocal microscopes, supplement the traditional light microscope. Phylogenetic analyses are conducted on the desktop, employing computing power and algorithms that were unimagined a short time ago. These tools have been adopted because they enable us to do things faster and better, or make it possible to ask new questions and find better answers to those questions. Information technologies are part of the next wave of tools to be incorporated into systematics, offering both the increased effectiveness at managing data and the ability to tackle new research topics.

Informatics, or information science, is a relatively new term with a variety of definitions. Most of these include aspects of the acquisition, storage, dissemination, and analysis of data, particularly using electronic technologies (e.g., Soberón and Peterson 2004). The term 'biodiversity informatics' nominally restricts this field to the data associated with biodiversity studies, but it includes an array of domains. Primary specimen-occurrence data document the time and place at which an organism was collected or observed, as well as the taxon to which it belongs (Chapman 2005a), and these data can be supplemented by additional information such as collecting method, habitat observations, and the names of the persons making the collection or observation. Taken as a whole, these data document the distribution in time and space of the known species of insects. Character data describing the features of specimens or taxa are widely used in identification procedures and phylogenetic analyses. Imagery can include photographs, drawings, movies, and sounds. Other relevant domains include the literature, published and unpublished, electronic and hard-copy; and analytical products such as models of geographic distributions and phylogenetic hypotheses. Some of the issues associated with these data types have been reviewed (Johnson 2007, Walter and Winterton 2007).

Two organizations are particularly important in helping to keep abreast of the rapid developments in the field of biodiversity informatics. The Global Biodiversity Information Facility, headquartered in Copenhagen, is an international organization whose goal is to facilitate the digitization and dissemination of biodiversity data for the benefit of all (www.gbif.org).

The mission of Biodiversity Information Standards (TDWG) is to develop the standards and protocols that make the sharing of data possible (www.tdwg.org).

Nothing is peculiar to insect biodiversity informatics that sets it apart from any taxon-group informatics. In practice, the size and nature of insects and the size of insect collections imposes some practical constraints. However, the conceptual issues involved are similar across groups. My goal in this chapter is to address the major components included in the definition of informatics – acquisition, storage, dissemination and analysis – drawing examples from studies of insects.

PRIMARY SPECIMEN-OCCURRENCE DATA

The primary tool used for data storage and retrieval is a relational database, and numerous books are available on relational theory (e.g., Date 2004). A number of applications have been developed for the management of primary specimen-occurrence data that provide a software layer between the database-management system and the user. Many of these are listed by Berendsohn (2005), and new applications continue to be developed. The most widely used applications include Specify (www.specifysoftware.org/Specify), KE EMu (www.kesoftware.com/emu/index.html), and Biota (Colwell 1996). These packages sometimes are criticized as being overly complicated, but this assessment overlooks the fairly complex nature of biodiversity data. Effectively handling the taxonomic names associated with specimens requires consideration of issues of synonymy, homonymy, and alternative and changed classifications. Geographic names and the boundaries of the areas they represent change with time and governments. Recording the names of the people that make the collections or determining the identity of the specimens is important to many users. The time of collection may be expressed as a single day, a range of days, or even a specific time of day. Finally, each taxonomic group has its own idiosyncratic data requirements, such as host or microhabitat. Developing a single piece of software to meet such varied needs almost inevitably results in more complexity than that required by any single user.

The choice of a database-management system and a database application to facilitate the process of data entry and retrieval should, first, be based on the needs of the user. As the needs and desires for the system grow over time, a more robust and general solution may be able to accommodate such growth more easily than a relatively simple, but specific application. Migration of data from one system to another can be done, but it is never a pleasant process. A user may find it necessary to create a new custom software application to meet the needs of a project. Modular software design and open-source development may help to accelerate the further development of biodiversity informatics software.

Two different approaches to information management will be encountered in the literature and on the Internet. These may be termed a taxon-based approach versus an event-based approach. As its name implies, a taxon-based database is organized on the basis of the taxon involved, aggregating the information on distribution, phenology, ecology, and other attributes. Two prominent examples of this approach are the Fauna Europaea project (www.fauneur.org) and NatureServe (www.natureserve.org). An event-based database, such as the caterpillars of the Guanacaste Conservation Area (Janzen and Hallwachs 2005), documents individual events, such as a collection or accession of a specimen or rearing of a parasitoid. These data then can be aggregated and the results displayed in the same format as a taxon-based database. An event-based approach also provides the ability to organize the data along many other dimensions. This flexibility and power is achieved at the cost of processing time, programming requirements, and overall database size.

Discussions of specimen databases are usually done in the context of collection management. A database can be an extremely valuable administrative tool, enabling a curator to keep abreast of the status of loans, document the holdings of the collection in detail, and respond to queries for information. However, such a database is also valuable for the practicing biodiversity scientist. It can facilitate the management of incoming loans of specimens, annotation of individual specimen records (e.g., for anomalous characters), correlation of features with geographic and temporal data, production of accurate and consistent documentation of the material examined in a study, mapping and comparison of geographic distributions, and quantitative documentation of the flight period of adults. For the consumer of scientific publications, databased specimen-level information provides an 'audit trail' by which readers can judge whether the conclusions in the paper are supported by the evidence.

Prospective data capture of specimen-occurrence data (i.e., recording the collecting information at the time that specimens are initially processed) is a relatively straightforward process. A typical collecting regimen, such as a Malaise trap sample, will have large numbers of specimens, all of which share the same data on the time, place, method, habitat, and names of collectors. Those people with first-hand knowledge of the collecting event are often available to resolve any ambiguities that arise. Retrospective data capture, recording data from existing collections, perhaps with specimens collected more than a century ago, brings many additional problems. Because of the small size of insect labels, important data elements may be abbreviated or omitted entirely and assumed to be implicit. Dates may be recorded in formats that cause confusion. The meaning of the string '4/9/06' could be April 9, 2006, or September 4, 1906, or some other combination. Handwritten labels are often difficult to decipher. These ambiguities and the time needed to resolve them raise the cost of data capture. In some cases, these issues can be resolved only by those with a detailed understanding of the collectors, the institutions, or the taxa involved. The Entomological Collections Network endorsed the idea that retrospective data capture be incorporated as part of the systematic research process (Thompson 1994). The scientists involved in revisionary studies are often in the best position to interpret the sometimes cryptic information on insect labels.

Georeferencing is the conversion of textual descriptions of places into a coordinate system. A powerful argument can be made that a specimen that is not georeferenced is of practically no use in any biologically interesting sense. Plotting of the georeferenced localities on a map makes egregious data entry errors immediately apparent. In addition, the process of searching for latitude and longitude for named places reveals alternative or incorrect spellings of locality names. Chapman and Wieczorek (2006) have developed a best practices document for georeferencing. A number of sources are available on-line to determine latitude and longitude for named places (Chapman and Wieczorek 2006, Johnson 2007). The coordinates of named places, of course, usually do not represent the precise location where a specimen was collected. Therefore, additional calculations might be necessary to indicate more accurately the position of the collecting locality. A web-based application for this purpose was developed as part of the Mammal Networked Information System (MaNIS) and is accessible at www.manis.org/gc.html. A measure of probable error, typically represented as a radius from a point (Wieczorek et al. 2004), is essential for assessing the 'fitness for use' of the data (Chapman 2005b).

One of the implications of a specimen-based relational database is that the information for each record must be individually identifiable; that is, it needs some sort of unique identifier. This identifier is then linked to the physical specimen, normally by adding a label. This process is a fairly traditional accession practice in many types of collections. In an insect collection having perhaps several million specimens, however, it is definitely not standard procedure. The use of barcode labels has been promoted as a mechanism to facilitate the link between data in the computer and the specimen to which it pertains (e.g., Janzen 1992, Thompson 1994). The Entomological Collections Network (ECN) advocated the use of a Code 49 stacked bar code for insect specimens, primarily because of its compact size. Subsequently, though, two-dimensional formats have been developed that combine small size with a greater capacity for information storage. ECN proposed that the identifier consist of an abbreviation for the institution generating the record (not necessarily the owner of the specimen) and a number, the combination of which should be unique. This information should be both encoded in the bar code and printed in human-readable form. In practice, the bar code itself and its format are not the critical elements. Rather, the uniqueness of the identifier is the critical factor. The advantage of the bar code format is that it allows the identifier to be entered automatically into a computer, thus reducing greatly the possibility of input error. A locally unique identifier, however, can be generated easily, requiring only a minimal amount of bookkeeping to avoid the inadvertent reuse of a number.

Chapman (2005b) provides a review of the issues involved in the quality assurance. Mistakes in data entry are unavoidable. They may be reduced by the use of careful protocols and good user interface design, but they will always occur. Many herbaria are experimenting with imaging entire herbarium sheets, extracting the part of the image containing the labels and annotations, and using optical character recognition to convert the image into text (e.g., the HERBIS project, www.herbis.org). Such technologies have not yet been applied to insect specimens. Drop-down pick-lists are commonly touted as a cure for typing errors. This may reduce the error rate, but substitutes mistakes that look 'good' for spelling errors that may be easier to locate and correct. Errors are simply inevitable,

and a protocol for data review and correction must be established. Guralnik and Neufeld (2005) describe one such methodology used to identify georeferencing errors, and Lampe and Striebing (2005) describe the data capture and quality assurance protocol for a large insect collection.

HTML forms and CGI scripts provide a simple mechanism for linking the information in a database with users on the Internet. The output of primary specimen-occurrence data is usually produced in one of two formats, as text or as maps. Text has traditionally been designed to be human-readable. Mapping applications typically have been customized solutions for each data source, providing views of the data in a range of resolutions. Google Maps now offers a popular programming interface (www.google.com/apis/maps) that can display geographically referenced data on maps at a wide range of resolutions and also incorporate satellite imagery.

Most portals offer access to only a single information resource, and the format of the output is not standardized across these Web sites. Therefore, a user attempting to integrate information from several independent resources must do this on an *ad hoc* basis. To achieve the same result programmatically entails a process called screen-scraping, that is, location and extraction of data from a display designed for a person to see and read. Such a solution is unstable, as any change in the page formatting might well result in an incorrect parsing by the scraping application.

The advantages of being able to integrate data from different Internet resources are numerous. Each data provider is independently responsible for the curation of its own information and can manage issues of access and intellectual property rights as they see fit. The work associated with the digitization and curation of records from hundreds of millions of specimens around the world is shared among a much larger group. As an illustration of the scientific gains to be realized, consider the four maps illustrated in Fig. 18.1. These present the geographic distribution of the four institutions with the numerically greatest holdings of specimens of the parasitic wasp species *Pelecinus polyturator* (Drury) (Hymenoptera: Pelecinidae). No single collection provides an adequate and unbiased representation of the distribution of this species. Every collection, from those of the small regional institutions to the national museums of natural history, contributes vital information to understanding the distribution of biodiversity.

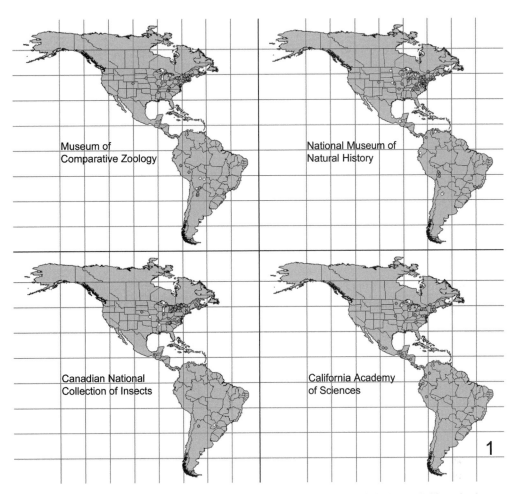

Fig. 18.1 Comparison of data from specimens of *P. polyturator* from four collections with the largest holdings for this species.

The technological limitations to the integration are tractable and are rapidly being resolved. An essential first step is to decide collectively what type of information is to be disseminated and shared. One such specification is the Darwin Core (digir.net/schema/conceptual/darwin/2003/1.0/darwin2.xsd), a simple, flat representation of the important data elements associated with a specimen. The Darwin Core is generally implemented in XML (extended markup language). A small sample of a record might look like as follows:

```
<record>
<darwin:Country>United States</darwin:Country>
<darwin:StateProvince>New York
```

```
    </darwin:StateProvince>
<darwin:InstitutionCode>OSUC
    </darwin:InstitutionCode>
<darwin:CollectionCode>Insects
    </darwin:CollectionCode>
<darwin:CatalogNumber>OSUC 141975
    </darwin:CatalogNumber>
<darwin:ScientificName>Aradophagus fasciatus
    </darwin:ScientificName>
</record>
```

The Darwin Core schema defines a set of data elements for objects in a natural history collection. An alternative schema, much richer in its elements and, as a result, more complicated, is the Access to Biological Collections Data schema (ABCD, www.bgbm.org/TDWG/CODATA/Schema/). These two schemas (and there could be any number more) provide a structure for the answer to a request to databases for information about specimens or observations. The databases, however, do not need to be structured in this same way and typically they are not. The curators of the database match objects in their own schema to, for example, those in the Darwin Core, and then return the response to the data request in Darwin-Core-flavored XML. Thus, the inner structure of the database is hidden from external users.

The Distributed Generic Information Retrieval protocol, DiGIR, was designed to provide a common mechanism to pose queries to a database (digir.net). If a database is capable of delivering data, using the Darwin Core schema, then a request for information about specimens of the genus *Aradophagus* in the USA would be represented in the following XML fragment:

```
<query> <and>
<darwin:Genus>Aradophagus</darwin:Genus>
<darwin:Country>United States</darwin:Country>
</and> </query>
```

The user can specify in the DiGIR request the structure of the desired answer, for example, the latitude and longitude of the collecting locality for the specimens so that a distribution map can be generated. The DiGIR protocol is being superceded by TAPIR (ww3.bgbm.org/protocolwiki/FrontPage), providing more flexibility and the ability to query ABCD data providers, but the fundamental principles remain the same. In an SQL query, the standard interface with a relational database, the user is required to know details about the underlying structure of the database, whereas in a DiGIR query, such complexities are hidden. The combination of Darwin Core and DiGIR, or ABCD and its query protocol BIO-CASE (www.biocase.org) provide a layer of abstraction between the details of individual database schemas and the user. As long as this layer remains relatively stable, the problems of integration of data resources are greatly reduced.

The existence of information resources providing such web services is of limited utility if no one knows they are there, what services are provided, and how to access these services. The UDDI (universal description, discovery, and integration) protocol is designed to meet these needs. GBIF maintains a UDDI registry for databases providing information on primary specimen occurrences (registry.gbif.net/uddi/web).

With increased integration of data sources, the issue of the uniqueness of the identifiers used for objects becomes important. Several databases may store information on the same specimen. If they use different identifiers, redundancy in the data is impossible to detect. Conversely, two databases may use the same identifier to refer to different objects. The ECN strategy of affixing an abbreviation of an institution (such as the collection codens; Evenhuis and Samuelson 2004) to a number might work in the entomological community, as long as it agrees to a standard set of abbreviations. However, this strategy will not ensure uniqueness in a wider biological context. GBIF has recently been grappling with the issue of globally unique identifiers. The preferred technology seems to be life science identifiers or LSIDs (Clark et al. 2004). These identifiers are needed in all aspects of biodiversity informatics to uniquely refer to specimens, taxonomic names (Page 2006), images, and characters.

An LSID is a name identifying an object. The actual location of the object, and information about it, might change over time. An LSID resolving service is required to determine the location of the resource from its name. The format of an LSID is (e.g., Clark et al. 2004) urn:lsid:tigr.org:AT.locusname:AT1G67550. The 'urn:lsid': indicates that this string is a universal resource name, and, more specifically, a life sciences identifier. The next section 'tigr.org': specifies the authority for the LSID; 'AT.locusname' is the authority namespace identifier; and 'AT1G67550' is the identifier for the object within that namespace. A colon followed by an optional revision number may be appended. The namespace and object identifiers are specific to the authority and are likely completely meaningless. The meaning emerges by requesting data and metadata for the object from the authority. The object to which this identifier points is intended to be unchanging so that two workers, accessing the object at different times, can be assured that they have acquired exactly the same object. However, information about the object, its metadata, may change. For this method, the objects in any data resource can be precisely and consistently identified at a global scale.

One of the primary analytical products that emerges from primary specimen-occurrence data is a hypothesis of geographic distribution. The algorithms that are used seek to extrapolate from 'known' places of occurrence to predict the presence of a taxon in unsampled areas. Ecological niche modeling is a complex and developing field, reviewed by Guisan and Zimmermann (2000) and Stockwell (2006). The relevance to natural history collections is discussed by Graham et al. (2004). Common free software packages used are DIVA-GIS (www.diva-gis.org) and DesktopGarp (nhm.ku.edu/desktopgarp), employing different modeling algorithms.

The entomological community lags behind other taxonomic groups in applying these technologies to make data widely available. One of the major constraints faced by insect collections in comparison to other disciplines is the sheer magnitude of the task at hand. In the USA, many land-grant universities have collections, and these quite ordinarily contain more than 1 million specimens. The size of the major national collections around the world is typically one or even two orders of magnitude larger. If the cost to capture the data for a specimen is around US$1, then the total cost of digitizing a collection is overwhelming. Such numbers are daunting, particularly when curatorial and technical staff and budgets are limited.

TAXONOMIC NAMES

The systematics community is the custodian of the 'dictionary' of names of organisms and all of their complexities: homonyms, synonyms, basionyms, replacement names, *nomina nuda*, and others. This dictionary is a critically important resource for the entire biological research community and society at large. All the users rely on the systematics community to produce, document, and manage biological nomenclature so that those names can be used effectively in communicating information about organisms.

On-line resources for information associated with taxonomic names and concepts are developing in a two-tiered approach. At one level, taxonomic specialists harvest and document the names of taxa within their area of specialization. Examples include resources for names in the Collembola (Bellinger et al. 2006), Odonata (Schorr et al. 2005), Orthoptera (Eades et al. 2006), Zoraptera (Hubbard 2004), Embiidina (Ross 1999), Neuropterida (Oswald 2003), Trichoptera (Morse 2008), Siphonaptera (Medvedev et al. 2005), Mecoptera (Penny 1997), Diptera (Thompson 2006), and Hymenoptera (Johnson 2005, Noyes 2005, Taeger and Blank 2006). An even larger number of

databases and lists are available for more restricted geographic areas or taxonomic groups.

At a higher tier, comprehensive data aggregators work to bring together these data into a single data portal. These two tiers are interdependent. The specialists of a taxon have both the interest and the expertise to verify and validate the data. The aggregators manage large amounts of data and provide tools that can be used to investigate names across the entire spectrum of life.

Several data aggregators are currently active, such as ITIS (www.itis.org), Species 2000 (www.sp2000.org), uBio (www.ubio.org), the Electronic Catalog of Names of Known Organisms (www.gbif.org/prog/ecat), Nomina Insecta Nearctica (www.nearctica.com/nomina/main.htm), Nomenclator Zoologicus (www.ubio.org/NomenclatorZoologicus), and Fauna Europaea (www.faunaeur.org).

Polaszek et al. (2005) have proposed that formal registration of any newly proposed names be adopted as an additional criterion for availability under the International Code of Zoological Nomenclature (ICZN 1999). All registered names would be accessible through an on-line resource called ZooBank. This registration then would ensure that the systematics community is aware of all new names that appear in the published literature. Prototype development of ZooBank is currently underway.

The ICZN only provides rules governing the structure and availability of names. It does not regulate the connection between names and the biological entities they are meant to denote. The understanding of what is represented by a taxonomic name can change quite significantly through time and space. Effective communication requires an appreciation of these differences.

The Taxonomic Concept Schema (TCS, tdwg.napier.ac.uk/index.php) is a thoughtful and thorough attempt to codify these ideas and to develop a standard (in XML) for data exchange. A taxonomic concept in the TCS is basically a combination of text (a name) and an 'according to' clause. This clause points to some sort of circumscription of the taxon under consideration. Its implementation in the biodiversity informatics community is still in its earliest stages, but its importance and promise is indicated by its adoption as a formal standard by the Taxonomic Databases Working Group (TDWG).

A fair amount of variation is possible in the data structure used to model taxonomic names. Thompson (2006) and Pyle (2004) have described two robust approaches to the issues involved.

LITERATURE

The published scientific literature contains a tremendous amount of valuable information on biodiversity. But like a collection that is organized taxonomically, the literature can be difficult to mine for data relevant to the question at hand.

The first issue is that much of the literature exists only in hard-copy formats. Having a set of physical objects, such as books and journals, provides insurance against the permanent loss of data or shifts in storage media (as, e.g., away from 5 1/4-inch floppy disks). Printed copies, however, are finite in number, restricting general access to the information. Several parallel efforts are underway by groups to produce electronic copies of publications on their taxonomic specialties, consistent with copyright restrictions. Prominent examples of such resources include the Walter Reed Biosystematics Unit Mosquito Literature (for Culicidae, wrbu.si.edu/MosqLit.html), the William L. Brown Memorial Digital Library (for Formicidae, ripley.si.edu/ent/nmnhtypedb/wlb/), and the Bibliography of the Neuropterida (Oswald 2006). A new venture, the Biodiversity Heritage Library (www.biodiversitylibrary.org) is a collaboration of ten major institutions from the USA and the UK to digitize the biodiversity literature in their holdings. Electronic copies of an extensive number of biodiversity books can be found in AnimalBase (www.animalbase.de) and, for French literature, in Gallica (gallica.bnf.fr).

Many of the electronic versions of published works available in digital libraries are images of the pages in the original documents. The next step in providing access to the information included is to transform these into text via optical character recognition (OCR). The contents may then be canvassed programmatically to find and extract information of interest. Work is now underway to develop XML schemas by which such texts can be 'marked up', that is embedded with tags that designate the meaningful elements in the document (Weitzman and Lyal 2004, Koning et al. 2005).

As in specimen databasing, such markup could be either retrospective or prospective. Marking up the rich heritage of existing literature is an overwhelming task, and will be attempted only with the use of software tools that assist in locating and defining the relevant data elements. One issue with legacy literature is that authors often mix together different kinds of information; for example, a description of character variation may have geographic distribution and phenology information intermixed. This complicates the markup process, particularly with a structure-sensitive mechanism such as XML. Prospective markup, once standards are in place, can be much more rapid, assisted particularly by the understanding on the part of authors of the data elements that are to be tagged.

Many journals are now published only in electronic formats. Taxonomic papers lag somewhat in this trend because of the ICZN requirement that newly proposed names be published in hard-copy. However, the Code does not require that this be the only form in which the papers are produced, and several venues are publishing both paper and electronic versions, e.g., *Zootaxa* and *Fiji Arthropods*. URLs can be embedded in electronic versions that allow authors to take advantage of an exciting array of web services, such as linking to high-resolution images and production of dynamic distribution maps.

CHARACTERS

Handling characters is an old issue, dating to the beginning of the use of computers in phenetic and phylogenetic analyses in the 1950s and 1960s. A matrix-based approach predominates, usually dividing the attributes of organisms into characters and, within each of those, nonoverlapping character states. A number of formats are still in use, depending on the application for which the data are intended. For example, NEXUS (Maddison et al. 1997) is widely used as a data input format for phylogenetic analyses, DELTA for the production of natural-language descriptions and multi-entry keys (Dallwitz 1980, Dallwitz et al. 1993, Dallwitz and Paine 2005). Walter and Winterton (2007) have reviewed one of the major applications of character data, identification tools. The Structure of Descriptive Data XML schema, or SDD, is being developed to provide a general means to exchange data on the attributes of organisms (wiki.tdwg.org/twiki/bin/view/SDD/WebHome). It is a highly structured format capable of handling qualitative, quantitative, and meristic characters of any kind. If successfully implemented, SDD will provide the ability to exchange data between applications without loss of meaning, enhance collaborative efforts, enable data to be mined and used for purposes far beyond those for which they were originally intended, and greatly facilitate the ability to produce descriptions of taxa in multiple languages.

One significant barrier to communication in biodiversity science is the highly specialized and sometimes arcane vocabulary used. Glossaries of entomological terms, such as Torre-Bueno (1989) and Gordh and Headrick (2001) are useful for human readers, and analogous solutions are needed for computer applications. The medical community has developed tools such as the Unified Medical Language System (UMLS), a controlled vocabulary to enable computer systems to ' . . . behave as if they "understand" the language of biomedicine and health' (www.nlm.nih.gov/research/umls/umlsdoc.html). With this resource, an ontology, terms are defined and relationships between them are documented. Examples of such relationships in UMLS are 'ingredient_of', 'contains', and 'tradename_of'. Applications can use these relationships between terms to find literature that is relevant to a subject, but which might not explicitly include the predefined set of search terms. A few such ontologies are available for entomology, including one for the gross anatomy of *Drosophila* (www.flybase.org) and both ticks and mosquitoes (www.anobase.org). The nascent Hymenoptera ontology (ceb.csit.fsu.edu/ronquistlab/ontology) is a collaboration between domain specialists and MorphBank (www.morphbank.net) that seeks not only to define terms and their relationships, but also to document them with images.

The characters of organisms are the data on which phylogenetic hypotheses are based. This domain thus edges into bioinformatics in its traditional sense, as well as into phyloinformatics, concerning phylogenetic inference. These fields are beyond the scope of this chapter. The Cyberinfrastructure for Phylogenetic Research project (CIPRES, www.phylo.org) provides an entry point for phyloinformatics.

ENCYCLOPEDIA OF LIFE, TREE OF LIFE

Two major informatics efforts are underway to organize and disseminate knowledge on the richness of life. The Tree of Life project (tol.org) began in 1995 and is a community-wide collaborative effort to provide information on the diversity, evolutionary history, and characteristics for all groups of organisms, from species to kingdoms. Its organizing principle is the hierarchical pattern of relationships among organisms. The Encyclopedia of Life (www.eol.org) began in 2007 with great fanfare and substantial start-up funding.

Many of its goals substantially overlap with those of the Tree of Life – aggregation of information on all of life, free Internet dissemination of material to users around the world, and a collaborative authorship of pages including both specialists and amateurs.

Recognizing this overlap, the two projects are now collaborating. The Encyclopedia of Life will concentrate on its goal to develop a page on the Internet for every species on the planet. The Tree of Life will focus on the characteristics and relationships of the monophyletic taxa of which these species are members. Both will generate new content, but also will be important aggregators of information from other sources in the published literature and on the Internet. The continued development and implementation of data exchange standards will greatly assist the monumental tasks they have set for themselves and the biodiversity research community.

CONCLUSIONS AND PROSPECTS

The investment in time needed to adopt biodiversity informatics tools is amply compensated by the increased analytical power and greatly enhanced ability to disseminate data and results to the worldwide community. One of the problems with traditional taxonomic monographs is that they quickly become obsolete by the discovery of new distributional data, new characters to discriminate taxa, and new species. Ideally these discoveries are fueled by the monograph itself; thus, a good monograph actually contributes substantially to its own obsolescence. Informatics tools offer the possibility of developing 'living' monographs in which new data can be incorporated as they are found. Traditional constraints of publications such as size, the number of illustrations, and the use of color, are no longer relevant. The possibilities are only beginning to be appreciated, and portend a significant enhancement in how biodiversity studies are conducted, communicated, and appreciated.

ACKNOWLEDGMENTS

Sincere thanks are extended to L. Musetti, D. Agosti, L. Speers, and F. C. Thompson for discussions, ideas, and stimulating interest in this area. This material is based, in part, on work supported by the National Science Foundation under DEB-9521648, DEB-0344034, and DEB-0614764.

REFERENCES

Bellinger, P. F., K. A. Christiansen, and F. Janssens. 2006. Checklist of the Collembola of the World. www.collembola.org [Accessed 20 October 2008].

Berendsohn, W. (ed). 2005. Standards, Information Models, and Data Dictionaries for Biological Collections. www.bgbm.org/TDWG/acc/Referenc.htm [Accessed 20 October 2008].

Chapman, A. D. 2005a. Uses of Primary Species-occurrence Data, Version 1.0. Report for the Global Biodiversity Information Facility, Copenhagen. www.gbif.org/prog/digit/data_quality [Accessed 20 October 2008].

Chapman, A. D. 2005b. *Principles of Data Quality, Version 1.0.* Global Biodiversity Information Facility, Copenhagen. 58 pp.

Chapman, A. D. and J. Wieczorek (eds). 2006. *Guide to Best Practices for Georeferencing.* Global Biodiversity Information Facility, Copenhagen. 80 pp.

Clark, T., S. Martin, and T. Liefeld. 2004. Globally distributed object identification for biological knowledgebases. *Briefings in Bioinformatics* 5: 59–70.

Colwell, R. K. 1996. *Biota: The Biodiversity Database Manager.* Sinauer, Sunderland, MA. 574 pp. viceroy.eeb.uconn.edu/Biota [Accessed 20 October 2008].

Dallwitz, M. J. 1980. A genera system for coding taxonomic descriptions. *Taxon* 29: 41–46.

Dallwitz, M. J. and T. A. Paine. 2005. Definition of the DELTA Format. www.delta-intkey.com/www/standard.pdf [Accessed 20 October 2008].

Dallwitz, M. J., T. A. Paine, and E. J. Zurcher. 1993. User's Guide to the DELTA System: A General System for Processing Taxonomic Descriptions, Fourth Edition. www.delta-intkey.com [Accessed 20 October 2008].

Date, C. J. 2004. *An Introduction to Database Systems,* Eight Edition. Pearson/Addison Wesley, Boston, MA. 983 pp.

Eades, D. C., D. Otte, and P. Naskrecki. 2006. Orthoptera Species File Online, Version 2.6. osf2.orthoptera.org [Accessed 20 October 2008].

Evenhuis, N. L. and G. A. Samuelson. 2004. The Insect and Spider Collections of the World Website. hbs.bishopmuseum.org/codens/codens-r-us.html [Accessed 20 October 2008].

Gordh, G. and D. H. Headrick. 2001. *A Dictionary of Entomology.* CAB International, Wallingford, Oxon. 1042 pp.

Graham, C. H., S. Ferrier, F. Huettman, C. Moritz, and A. T. Peterson. 2004. New developments in museum-based informatics and applications in biodiversity analysis. *Trends in Ecology and Evolution* 19: 497–503.

Guisan, A. and N. E. Zimmermann. 2000. Predictive habitat distribution models in ecology. *Ecological Modelling* 135: 147–186.

Guralnick, R. P. and D. Neufeld. 2005. Challenges building online GIS services to support global biodiversity mapping and analysis: lessons from the Mountain and Plains Database and Informatics project. *Biodiversity Informatics* 2: 57–69.

Hubbard, M. 2004. Catalog of the Order Zoraptera. www.famu.org/zoraptera/catalog.html [Accessed 20 October 2008].

International Commission on Zoological Nomenclature (ICZN). 1999. *International Code of Zoological Nomenclature,* Fourth Edition. International Trust for Zoological Nomenclature, London.

Janzen, D. H. 1992. Information on the bar code system that INBio uses in Costa Rica. *Insect Collection News* 7: 24.

Janzen, D. H. and W. Hallwachs. 2005. Dynamic Database for an Inventory of the Macrocaterpillar Fauna, and Its Food Plants and Parasitoids, of Area de Conservacion Guanacaste (ACG), Northwestern Costa Rica (nn–SRNP–nnnnn voucher codes) janzen.sas.upenn.edu [Accessed 20 October 2008].

Johnson, N. F. (ed). 2005. Hymenoptera Name Server, Version 1.0. www.purl.oclc.org/net/hymenoptera/hns [Accessed 20 October 2008].

Johnson, N. F. 2007. Biodiversity informatics. *Annual Review of Entomology* 52: 421–438.

Koning, D., T. Moritz, D. Agosti, K. Böhm, N. Johnson, B. Morris, S. Spector, E. Nichols, N. Sarkar, and G. Sautter. 2005. NSF Taxonomic Literature Project. research.amnh.org/informatics/taxlit/ [Accessed 20 October 2008].

Lampe, K.-H. and D. Striebing. 2005. How to digitize large insect collections: preliminary results of the DIG project. Pp. 385–393. *In* B. A. Huber, B. J. Sinclair, and K.-H. Lampe (eds). *African Biodiversity: Molecules, Organisms, Ecosystems.* Springer Science/Business Media, New York.

Maddison, D. R., D. L. Swofford, and W. P. Maddison. 1997. NEXUS: an extensible file format for systematic information. *Systematic Biology* 46: 590–621.

Medvedev, S., A. Lobanov, and I. Lyangouzov. 2005. Parhost: World Database of Fleas, Version 2.0. www.zin.ru/Animalia/Siphonaptera/dbas.htm [Accessed 20 October 2008].

Morse, J. C. 2008. Trichoptera World Checklist. entweb.clemson.edu/database/trichopt/index.html [Accessed 20 October 2008].

Noyes, J. S. 2005. Universal Chalcidoidea Database. internt.nhm.ac.uk/jdsml/perth/chalcidoids/ [Accessed 20 October 2008].

Oswald, J. D. 2003. Index to the Neuropterida Species of the World, Version 1.00. insects.tamu.edu/research/neuropterida/neur_sp_index/ins_search.html [Accessed 20 October 2008].

Oswald, J. D. 2006. Bibliography of the Neuropteridae. insects.tamu.edu/research/neuropterida/neur_bibliography/bibhome.html [Accessed 20 October 2008].

Page, R. D. M. 2006. Taxonomic names, metadata, and the semantic web. *Biodiversity Informatics* 3: 1–15.

Penny, N. D. 1997. World Checklist of Extant Mecoptera Species. www.calacademy.org/research/entomology/

Entomology_Resources/mecoptera/index.htm [Accessed 20 October 2008].

Polaszek, A. D., D. Agosti, M. Alonso-Zarazaga, G. Beccaloni, P. de P. Bjørn, P. Bouchet, D. J. Brothers, Earl of Cranbrook, N. Evenhuis, H. C. J. Godfray, N. F. Johnson, F.-T. Krell, D. Lipscomb, C. H. C. Lyal, G. M. Mace, S. Mawatari, S. E. Miller, A. Minelli, S. Morris, P. K. L. Ng, D. J. Patterson, R. L. Pyle, N. Robinson, L. Rogo, J. Taverne, F. C. Thompson, J. van Tol, Q. D. Wheeler, and E. O. Wilson. 2005. A universal register for animal names. *Nature* 437: 477.

Pyle, R. L. 2004. Taxonomer: a relational data model for managing information relevant to taxonomic research. *Phyloinformatics* 1: 1–54.

Ross, E. S. 1999. World List of Extant and Fossil Embiidina (=Embioptera). www.calacademy.org/research/etnomology/Entomology_Resources/embiilist/embiilist.html [Accessed 20 October 2008].

Schorr, M., M. Lindeboom, and D. Paulson. 2005. World Odonata List. http:/www.ups.edu/x6140.xml [Accessed 20 October 2008].

Soberón, J. and A. T. Peterson. 2004. Biodiversity informatics: managing and applying primary biodiversity data. *Philosophical Transactions of the Royal Society, Series B, Biological Sciences* 359: 689–698.

Stockwell, D. 2006. *Niche Modeling: Predictions from Statistical Distributions*. Taylor and Francis, Boca Raton, FL. 224 pp.

Taeger, A. and S. M. Blank. 2006. ECatSym – Electronic World Catalog of Symphyta (Insecta, Hymenoptera). Data Version 2 – Digital Entomological Information, München berg. www.zalf.de/home_zalf/institute/dei/php_e/ecatsym/ecatsym.php [Accessed 11 August 2006].

Thompson, F. C. 1994. Bar codes for specimen data management. *Insect Collection News* 9: 2–4.

Thompson, F. C. (ed). 2006. Biosystematic Database of World Diptera. www.diptera.org/biosys.htm [Accessed 20 October 2008].

Torre-Bueno, J. R. de la. 1989. *The Torre-Bueno Glossary of Entomology*. New York Entomological Society, New York. 840 pp.

Walter, D. E. and S. Winterton. 2007. Keys and the crisis in taxonomy: extinction or reinvention. *Annual Review of Entomology* 52: 193–208.

Weitzman, A. L. and C. H. C. Lyal. 2004. An XML schema for taxonomic literature: taXMLit. www.sil.si.edu/digitalcollections/bca/documentation/taXMLitv1-3Intro.pdf [Accessed 20 October 2008].

Wieczorek, J., Q. Guo, and R. Hijmans. 2004. The point-radius method for georeferencing locality descriptions and calculating associated uncertainty. *International Journal of Geographical Information Science* 18: 745–767.

Chapter 19

PARASITOID BIODIVERSITY AND INSECT PEST MANAGEMENT

John Heraty

Department of Entomology, University of California, Riverside, California 92521, USA

Insect Biodiversity: Science and Society, 1st edition. Edited by R. Foottit and P. Adler
© 2009 Blackwell Publishing, ISBN 978-1-4051-5142-9

The regulation of insects by other insects involves predators, parasites, or parasitoids. Predators consume multiple hosts, and parasites take their nutrition from a host but do not necessarily kill it, whereas parasitoids use and kill a single host individual (Askew 1971). An understanding of the species richness of parasitoids, where they are most diverse, and how they interact are necessary for implementing effective insect pest management (IPM). The diversity of natural enemies is usually lowest or nonexistent for newly introduced pests. In a system out of balance with its natural enemies, pest populations can explode with devastating consequences. Niche specialization, keystone species, potential cascade effects with parasitoid extinction, and the importance of parasitoids are all factors affecting this delicate balance (Greathead 1986, Hawkins et al. 1992, LaSalle and Gauld 1993, Hawkins 1994). By the addition of one or more natural enemies from the country of origin, or the effect of host shifts by native predators or parasitoids, introduced pests may be brought under control (Huffaker 1969, Clausen 1978, Caltagirone 1981, Smith 1993, Van Dreische and Bellows 1996). Biological control programs in agricultural systems over the last century provide key insights into future problems that will be caused by habitat fragmentation, climate change, and loss of primary habitat in all ecosystems. Not only does this information help us to manage pests within urban, forest, or agricultural systems, but it can also provide the impetus for preserving native habitat and conserving parasitoid biodiversity.

Is parasitoid diversity necessary for effective pest management? The number of parasitoid species necessary to provide control remains an important and controversial issue (Myers et al. 1989, Rodríguez and Hawkins 2000, Muller and Brodeur 2002, Cardinale et al. 2003, Bianchi et al. 2006). Suppression of introduced pests can be achieved through the introduction of a single parasitoid. Cassava mealybug was controlled by *Apoanagyrus lopezi* (Encyrtidae) (Neuenschwander 2001) and the Rhodesgrass mealybug by *Neodusmetia sangwani* (Encyrtidae) (Clausen 1978). A combination of predator and parasitoid might be necessary for control. The Comstock mealybug is controlled by a predator, *Rodolia cardinalis* (Coccinellidae), and the parasitoid *Cryptochetum iceryae* (Cryptochetidae) (Huffaker 1969). A complex of several parasitoids might be necessary, as in control of the woolly whitefly by *Amitus spinifrons* (Platygastridae) and *Cales noacki* (Aphelinidae) (DeBach and Rose 1971), and the citrus leafminer

by a diverse array of both introduced and native parasitoids (Peña et al. 1996, 2000). One species might be sufficient, but in general, better control can be achieved with a diverse array of parasitoids and surrounding habitat diversity (Bianchi et al. 2006).

At what level does parasitoid biodiversity and distribution affect our ability to control pest insects? A common assumption in biological control is that pest populations are regulated by one or more insects in their native range. Importance, therefore, is placed on replacing this balance of native control agents on the introduced pest. Simple replacement, however, might not be effective for a number of reasons. Even within a single species, genetic, behavioral, and even associated endosymbiont attributes across the native range can differ significantly, making matches between hosts and effective parasitoids difficult. Morphologically similar (cryptic) species can interfere with each other through ineffective mating, decreasing their ability to lower pest populations (Stouthamer et al. 2000b). Different environmental conditions in the new host range can favor the host over the most effective natural enemy, and in some cases it becomes more important to find different populations or species more suitable to these conditions. Lastly, in large monocultures with little access to secondary habitats, the biodiversity of natural enemies can decline and affect natural insect control (Kruess and Tschartnke 1994, Fisher 1998, Landis et al. 2005, Bianchi et al. 2006). This last factor affects not only the introduced populations of a pest that arrive without their natural enemies, but also can lead to novel pest status for species in their native range as parasitoid reservoirs are depleted. Under these circumstances, conservation of natural enemies within native ranges assumes even greater importance.

How many parasitoids are there? Parasitoid abundance and diversity in most ecosystems remains a mystery, with the proportion of described versus undescribed (estimated) species surprisingly small. Taxonomists remain focused on morphological species recognition, but now molecular methods are discovering numerous morphologically distinct but reproductively isolated cryptic species of parasitoids. The expected decline and extinction of these parasitoids, with habitat fragmentation, agricultural expansion, and climate change will have a dramatic effect on pest abundance and species richness. Without any real idea of the diversity and abundance of parasitoids, we have no idea what we are losing. We must, as a society, develop a better understanding of parasitoids and their diversity, distribution, and behavior if we can ever expect to control the primary trophic levels of insect diversity, especially before these changes become irreversible on a global scale.

WHAT IS A PARASITOID?

Parasitism is defined as the relationship between two organisms, whereby the parasite obtains some or all of its nutrients from the other (Askew 1971). Among insect parasites, hosts generally are confined to the animal kingdom, although in its most extreme form, parasitism has been defined to include specialist herbivorous insects (Price 1980). This latter definition has not been accepted. The parasitoid lifestyle is a specialized subset of parasitism that includes those species that feed on and kill a single arthropod host, although even here the boundaries can be vague and problematic. The act of parasitism (i.e., how hosts are located and eggs deposited) is essentially the same for parasites and parasitoids, and the use of a distinct term such as 'parasitoidism' (Eggleton and Belshaw 1992) is unwarranted. Eggleton and Gaston (1990) applied the term parasitoid only to an organism that 'develops on or in another single ("host") organism, extracts nourishment from it, and kills it as a direct or indirect result of that development'. Eggleton and Belshaw (1992) restricted their definition to exclude (1) facultative relationships, in which an organism feeds on a host that is already in the process of dying but not dying as a direct result of parasitism, (2) species that feed on multiple individuals, such as eggs, in an enclosed space, (3) social insect parasites, in which an organism is responsible for killing a colony, (4) species that castrate but do not kill their host, and (5) herbivores, including seed predators. Phorid flies in the genus *Apocephalus* provide a good example of facultative parasitoids. Adult flies lay their eggs in recently dead or dying ants – they are not the direct cause of death (Morehead et al. 2001). The ancestral habit of related phorids in this case might be either saprophagy or parasitism. Social insect parasites are largely aculeate wasps that specifically target reproductive females in a host colony; these social parasites have invariably developed from a closely related, nonparasitic ancestor. Castration or behavioral modification interfering with reproductive success occurs commonly in strepsipteran parasites. In Strepsiptera, the Mengenillidae may be true parasitoids of Thysanoptera

and kill the host, but the remaining Strepsiptera act as castrators, with the host surviving the parasitic event.

Certain parasitoid groups excluded by the above restrictions have a clear phylogenetic association with a parasitoid ancestor. Species that attack eggs in enclosed sacs or multiple larvae in insect galls or seeds usually have been considered specialized predators, but in many cases, especially in Hymenoptera, they are considered to have developed from a parasitoid ancestor. Seed predators such as *Megastigmus* (Torymidae) and *Bruchophagus* (Eurytomidae) feed individually on seeds of Rubaceae or Fabaceae, respectively. Both genera are descended from ancestors that were insect parasitoids, and in many respects they are simply specialized parasitoids. Cleptoparasitoids are a subset of insect parasitoids that use both the host and its provisioned resources or only its provisioned resources for growth, and have evolved multiple times in Coleoptera, Diptera, and Hymenoptera. Parasitoids are generally terminal evolutionary units not evolving into other lifestyles, except for those groups within Hymenoptera that have subsequently evolved into gall-forming and endophytic phytophages, host-directed cleptoparasites, predators of eggs and larvae, provisioning predators (Aculeata), and provisioning omnivores (Formicidae) (Eggleton and Belshaw 1992). Irrespective of some of the fine nuances of distinguishing a parasitoid from a predator, saprophage, or parasite, the semantics are important. The issues affecting ecological biodiversity, host usage, and phylogenetic radiation can be distinctly different, assuming different levels of taxon inclusiveness (cf. Wiegmann et al. 1993). Among insects, true parasitoids are found only in the Holometabola, with only larvae acting as the parasitic stage.

Parasitoids can be solitary, gregarious (multiple eggs deposited), or polyembryonic (multiple larvae developing from a single egg). All life stages of a host can be attacked, although eggs and immature stages are the most commonly used host stages. Parasitoid larvae can develop on a single host stage (idiobiont) or through multiple host stages (koiniobiont) (Quicke 1997). Generally, most idiobionts are ectoparasitoids (external) and most koiniobionts are endoparasitoids (internal). More rarely, a parasitoid can use a combination of strategies. For example, in hyperparasitic Perilampidae (Hymenoptera), the first-instar larva is endoparasitic in multiple instars of the primary host, then ectoparasitic on the primary parasitoid larva, and finally ectoparasitic on the parasitoid pupa (Smith 1912, Laing and Heraty 1981). Parasitoid lifestyles are further complicated by species that attack previously parasitized hosts (hyperparasitoids), place eggs of different sexes in individuals of the same parasitoid species (autoparasitism), or place eggs in different host species (heteronomy) (Quicke 1997, Hunter and Woolley 2001). Together, these life-cycle strategies are embodied in a tremendously diverse community of parasitoids attacking virtually every possible host niche available (Hawkins et al. 1992, Hawkins 1994).

BIODIVERSITY AND SUCCESS OF INSECT PARASITOIDS

The number of insect species is staggering. Conservative estimates range from 2.4 to 10 million species (Gaston 1991, LaSalle and Gauld 1992, Gaston et al. 1996, Grissell 1999, Noyes 2000). Of these, just over 750,000 species have been described (Wilson 1985) (updated to 1,004,898; introduction to this volume). Eggleton and Belshaw (1992) estimated the total number of insect parasitoids to be 87,000 species, or roughly 10% of all described insects. A review of recent estimates suggests that there could be as many as 700,000 species of parasitoids (Table 19.1). The biology and biodiversity of parasitoids has been thoroughly reviewed by Clausen (1940), Askew (1971), and Quicke (1997), with an excellent summary of their biology and biodiversity in Eggleton and Belshaw (1992).

Parasitoids are distributed among seven different orders of Holometabola (Coleoptera, Lepidoptera, Diptera, Neuroptera, Strepsiptera, Trichoptera, and Hymenoptera), with by far the greatest species biodiversity and numerical abundance in Hymenoptera. Parasitoids are proposed to have developed only once in Hymenoptera, Strepsiptera, Neuroptera, and Trichoptera, but independently at least 10 times in Coleoptera, twice in Lepidoptera, and 21 times in Diptera (Eggleton and Belshaw 1992). Parasitism of a single host is not necessarily a universally successful adaptation. Parasitoids are rare in the Lepidoptera, Neuroptera, and Trichoptera, and most common in the Diptera and Hymenoptera. Parasitic Hymenoptera, and Tachinidae and Phoridae within the Diptera, are species rich and numerically abundant, but most other parasitic groups are not diverse and rarely encountered. To determine if parasitism was a successful

Table 19.1 Described and estimated species of parasitoids.

	Described	Estimated
Neuroptera	50[3]	50[3*]
Trichoptera	10[12]	10[12*]
Lepidoptera	11[3]	11[3*]
Strepsiptera	10[3]	10[3]
Coleoptera	1600[3*]	1600[3*]
Diptera	15,600[3*]	50,600[*]
Tachinidae	9989[13]	20,000[13]
Phoridae	3657[14]	30,000–50,000[14]
Hymenoptera	60,000[3*]	630,000[*]
Ichneumonidae	23,331[11]	60,500[2,5]
Braconidae	17,605[11]	20,000–25,000[2,5,8]
Platygastroidea	4022[2,4]	10,000[5]
Cynipoidea	2827[6]	25,000[1]
Chalcidoidea	22,000[10]	375,000–500,000[7,9,10]
Rough total	77,000	690,000

[1]Nordlander (1984). [2]Gaston (1991). [3]Eggleton and Belshaw (1992). [4]Gaston (1993). [5]Goulet and Huber (1993). [6]Ronquist (1995). [7]Noyes (2000). [8]Dolphin and Quicke (2001). [9]Heraty and Gates (2003). [10]Noyes (2007). [11]Yu et al. (2004). [12]Wells (2005). [13]Stireman (2006). [14]Brown (personal communication) estimates that as many as half of all phorids (30,000–50,000 species) are parasitic. *The numbers of estimated parasitoid species were not estimated by these authors; the value for Diptera and Hymenoptera was increased by the included estimates from other papers.

adaptation, Wiegmann et al. (1993) compared the relative species richness of parasitic and nonparasitic sister groups for 15 lineages, including lice, fleas, and other groups that are not true parasitoids. Some parasitic groups, such as Phoridae, were not considered in their analyses because of uncertain phylogenetic relationships between parasitic and nonparasitic taxa in the family. Only six of the parasitic groups were more diverse than their nonparasitic sister taxa, suggesting that parasitism overall is not a strategy leading consistently to rapid diversification of a group. Even when restricted to 11 'parasitoid' taxa, only 4 (Mantispinae, Sarcophagidae (*Blaesoxipha*), Tachinidae, and Hymenoptera) were more diverse than their nonparasitoid sister taxa. Although the evolution of a parasitoid lifestyle might not be consistently successful, the numerical abundance and diversification, of especially parasitic flies and wasps, is difficult to dispute and worth examining further.

In terms of their overall contribution to parasitoid biodiversity, the Hymenoptera, Phoridae, and

Tachinidae deserve special attention. These three groups contain roughly 74,000 described species, with potentially as many as 670,000 morphologically distinct species estimated (Table 19.1). The Tachinidae and Hymenoptera are by far the most important groups in agroecosystem pest management (Genier 1988, LaSalle 1993). Other parasitoid groups have importance in the control of their insect hosts, but few of these have been used for biological control (*cf.* Huffaker 1969, Clausen 1978, Caltagirone 1981). Some rare cases of agroeconomic importance include the release of the moth parasitoid *Chalcoela* (Pyralidae) against *Polistes* (Vespidae) in the Galapagos Islands, and *C. iceryae* (Cryptochaetidae) against the vedalia beetle (Coccinellidae).

Hymenoptera (Apocrita)

More than 80% of all parasitoid species (about 115,000 species described) belong to the Hymenoptera (Eggleton and Belshaw 1992, Quicke 1997). The order contains 20 superfamilies and 89 families, with its highest biodiversity in the suborder Apocrita (Goulet and Huber 1993). The Hymenoptera comprise two suborders, the paraphyletic Symphyta, which are either herbivores or mycophages, and the Orussidae + Apocrita (Vespina *sensu* Rasnitsyn 1988), with the evolution of true parasitoids as a shared feature of both groups. Within Apocrita, phytophagy, egg-sac predation, provisioning predators (Sphecidae, Pompilidae), and provisioning omnivores (ants, bees) have evolved several times (Eggleton and Belshaw 1992). The vast species richness and numerical abundance of parasitoids are contained in just four superfamilies, the Ichneumonoidea, Cynipoidea, Platygastroidea, and Chalcidoidea, which together contain more than 54,000 described species (Table 19.1). The number of estimated species in each of these groups is vastly out of proportion with the numbers of described species, with as many as 620,000 species estimated (Table 19.1). Values as high as 6 million species have been proposed, using estimates based on the biodiversity of all insect species (LaSalle and Gauld 1993). Chalcidoidea are by far the least understood and potentially most diverse group of parasitoids. The taxonomy of this superfamily has been hindered by their small size (1–2 mm on average) and extremely high biodiversity around the world. Estimates based on comparing tropical and temperate faunas have established a range of 375,000 to 500,000 morphologically distinct species (Noyes 2000, 2007; Heraty and Gates 2003).

Paralleling the biodiversity of hymenopteran parasitoids, the most economically important groups are found in two families of Ichneumonoidea (Ichneumonidae and Braconidae) and four families of Chalcidoidea (Encyrtidae, Aphelinidae, Eulophidae, and Pteromalidae) (Fig. 19.1). In part, their importance for pest management is determined by their tendency to attack economically important pest groups of sessile Hemiptera (aphids, scales, and whiteflies), Lepidoptera, Coleoptera, and phytophagous Hymenoptera. More than 800 species of Chalcidoidea have been used in biological control programs and overall have had the greatest rate of establishment and success (Noyes 2007).

Phoridae

Phoridae are minute flies that can be scavengers, detritivores, facultative 'scavengers' that attack dying or recently dead ants, or true parasitoids (Disney 1994, Morehead et al. 2001, Folgariat et al. 2005). Of the 3,657 described species, 611 are true parasitoids (B. V. Brown personal communication). Most of the described parasitoids belong to the genera *Apocephalus*, *Melaloncha*, and *Myriophora*, which together might contain another 500 undescribed species (B. V. Brown, personal communication), but many of the thousands of undescribed species of the genus *Megaselia* might be parasitoids as well. Phoridae are an immense group and their total biodiversity could reach 30,000–50,000 species (Gaston 1991, B. V. Brown, personal communication). Robinson (1971) proposed that, at least in the genus *Megaselia*, new host records tended to favor a reduction in the overall proportion of true parasitoids in the group; however, Brown suggests that potentially half of the family might be parasitoids – nearly 15,000 species! Phorids are parasitic on a number of insect groups, including bees, beetles, scale insects, millipedes, and adult and pupal coccinellids, but their greatest importance and influence are as parasitoids of adult ants. Phorids attack either dead and dying ants or living adults. In the latter case, their effect is not only direct mortality, but also changes in the behavior of foraging ants to protect the susceptible castes (soldiers), which come at a general cost to the colony by reducing worker foraging rates by as much as 75% (Askew

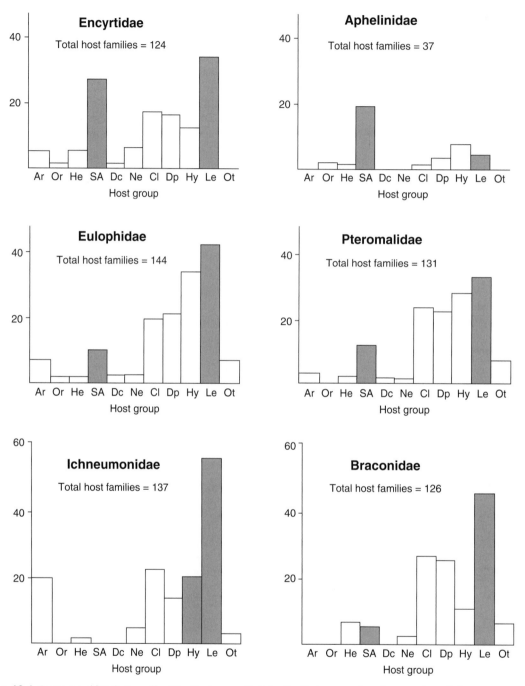

Fig. 19.1 Summary of distribution of host families by parasitoid family. Host groups: Ar = Araneae; Or = Orthoptera; He = Heteroptera; SA = Sternorrhyncha (primarily) and Auchenorryncha; Dc = Dictyoptera; Ne = Neuroptera; Cl = Coleoptera; Dp = Diptera; Hy = Hymenoptera; Le = Lepidoptera; Ot = other groups. Gray bars indicate host groups with the largest numbers of economically important pest groups. Redrawn with permission from Noyes and Hayat (1994).

1971, Feener and Brown 1992). At least two species of *Pseudacteon* have been released against the imported fire ant in the southeastern USA. Both have established and become abundant, but at least for *P. tricuspis*, the flies have not yet been shown to have a significant effect on reducing fire ant populations (Morrison and Porter 2005).

Tachinidae

Tachinidae are predominantly parasitoids of Lepidoptera and other herbivores including Heteroptera, Hymenoptera, and Orthoptera, but they also parasitize a wide range of other arthropods including centipedes, scorpions, and spiders (Stireman et al. 2006). All Tachinidae are endoparasitic, with most species attacking the larval stage and emerging from the pupa; 5–10% attack adult arthropods (Stireman et al. 2006). Recent estimates suggest that 9899 species of Tachinidae have been described in 1530 genera (Stireman et al. 2006). This total potentially represents half of the estimated species richness of Tachinidae, with most species occurring in tropical countries. Most tachinids are large and showy insects, and perhaps as much as 90% of the species in the northern hemisphere are known. However, the large biodiversity of tropical species remains untouched. Of 300 species sampled at one forest site in Costa Rica, 80% are undescribed (Janzen and Hallwachs 2005, Stireman et al. 2006).

Tachinid flies are an important factor in the management of macrolepidopteran pests, and more than 100 species have been used in biological control programs for crop or forest pests with partial or complete success (Greathead 1986, Genier 1988). Many of the agriculturally important tachinid species share the same host species or use alternate hosts in native habitats, which then act as a parasitoid refuge (Marino et al. 2006). Tachinids also have been used as an example of biological control at its worst. *Compsilura concinnata* originally was introduced for control of the gypsy moth in the northeastern USA and, along with another tachinid, *Blepharipa scutellata*, might effect a combined mortality of as much as 50% on the moth. However, *Compsilura* is highly polyphagous and has subsequently been found attacking more than 200 species in dozens of families and 3 orders of native insects, and is having a direct effect on native biodiversity (Boettner et al. 2000).

Other Groups

In the Lepidoptera, *Chacoela* (Pyralidae) are larval or pupal parasitoids of *Polistes* wasps (Hodges et al. 2003), *Sthenauge* (Pyralidae) feed as ectoparasites of saturniid larvae, and Epipyropidae are parasitoids of hemipteran nymphs or lepidopteran larvae (Eggleton and Belshaw 1992). Mantispidae (Neuroptera) are ectoparasitoids of larvae and pupae of Coleoptera, Lepidoptera, or Hymenoptera in soil, or egg sacs of spiders (Eggleton and Belshaw 1992). Trichoptera in the genus *Orthotrichia* (Hydroptilidae) develop as parasitoids of pupae of *Chimarra* (Philopotamidae) (Wells 1992, 2005). In the Strepsiptera, Eggleton and Belshaw (1992) included only the Mengenillidae (about ten species) as true parasitoids. The array of parasitoids in Coleoptera and Diptera is reviewed by Eggleton and Belshaw (1992), but beyond phorids and tachinids, parasitoids in these orders are comparatively uncommon, with only an estimated 1600 species of Coleoptera and 5600 species of Diptera (Table 19.1).

WHERE ARE PARASITOIDS MOST DIVERSE?

A number of factors affect parasitoid biodiversity at the community level and include characteristics of the host and its food plants, ecological complexity, successional stage of the community, plant architecture, and climate (Price 1991, 1994, Hawkins 1994, Marino et al. 2006). In general, agricultural landscapes lack the biodiversity and ecological complexity of natural ecosystems and support a lower overall biodiversity of parasitoids (Benton et al. 2003, Landis et al. 2005, Bianchi et al. 2006). Agricultural intensification, reduction of forests and hedgerows, and continuous planting schemes are all having an effect on native parasitoid and host biodiversity. In temperate agricultural systems, the surrounding vegetation is important in supporting parasitoids that attack pest species. Late successional stage vegetation most commonly supports a higher proportion of generalist species shared with agricultural systems (Landis et al. 2005, Marino et al. 2006). Early successional vegetation, however, might support a higher proportion of oligophagous specialist parasitoids that are also of importance (Marino et al. 2006). The landscape surrounding agricultural monocultures is an essential resource for maintaining parasitoid abundance and biodiversity (Kruess and Tschartke 1994, Landis et al. 2005, Bianchi et al. 2006). Some studies have argued that a simplified agricultural system provides equal or greater control than do complex systems (Rodríguez and Hawkins 2000). However, in a general survey of 15 studies comparing ecologically complex versus simple agricultural landscapes, 74% had higher natural enemy populations and 45% had lower pest pressure (Bianchi et al. 2006).

LEAFMINING PARASITOIDS AND NATIVE LANDSCAPES

Parasitoids of smaller gracillariid leafmining Lepidoptera are a good example of the interaction between native indigenous parasitoids and leafmining pests. In southern Ontario, a survey of 38 species of gracillariids on 14 host plants produced 65 species of parasitoids, of which 63% were eulophid wasps (Fig. 19.2a). Of the 28 parasitoid species reared from apple leafminer, *Phyllonorycter blancardella* (Fig. 19.2b), the overall proportion of species was similar, with most species (78%) overlapping between agricultural and native habitats. A similar composition of parasitoids attacked gracillariid leafminers in southern California (Fig. 19.2c). The invasive citrus leafminer has attracted more than 90 species of indigenous parasitoids worldwide, but no braconid parasites have yet been reported (Fig. 19.2d). In Florida, native parasites of eight genera of Eulophidae parasitized more than 50% of citrus leafminer larvae in citrus in the first few years of its establishment (Peña et al. 1996, 2000). In southern California, several indigenous species have already shifted to citrus leafminer since it arrived in 2000, including some, such as *Closterocerus utahensis* (Eulophidae), which were only rarely encountered prior to the invasion of leafminer (Heraty unpublished). Four genera of Eulophidae (*Cirrospilus, Pnigalio, Sympiesis,* and *Zagrammosoma*) are common in each of the leafminer systems (Fig. 19.2). Species in these genera are generalist idiobiont parasitoids that attack a wide variety of hosts including leafmining Agromyzidae and other parasitic Hymenoptera as hyperparasites. Similar to a study focused on macrolepidopteran pests of agricultural crops (Marino et al. 2006), late successional vegetation (trees and shrubs) was correlated with the greatest species richness and

numerical abundance of generalist parasitoids and is, therefore, a potentially beneficial resource for control of pest populations. In addition to providing a parasitoid refuge, hedgerows and native ecosystems provide a source of flowers, nectar, and other factors that can help enhance both the numbers and species richness of parasitoids (Corbett and Rosenheim 1996, Bostanian et al. 2004, Lavandero et al. 2006). With agricultural intensification, we need to consider planning for diverse landscapes that promote parasitoid biodiversity (Landis et al. 2005, Bianchi et al. 2006).

ARE PARASITOIDS MORE DIVERSE IN TROPICAL VERSUS TEMPERATE CLIMATES?

When we consider parasitoid communities in agricultural landscapes, do we need to modify our expectations based on latitudinal gradients? Should we expect higher biodiversity in equatorial ecosystems? The biodiversity of Tachinidae increases dramatically toward the equator, with a forest site in Costa Rica yielding more than 300 species, of which 80% were undescribed (Stireman et al. 2006). Phoridae are extremely diverse in both temperate and tropical latitudes, although there is a change in composition of taxa and a greater proportion of parasitoid taxa in tropical regions. Small parasitic Hymenoptera and egg parasitoids increase in diversity in tropical regions (Hespenheide 1979, Noyes 1989). Increased biodiversity, for the most part, is correlated with a higher proportion of host taxa. However, the Ichneumonidae either lack the distinct latitudinal gradient (Janzen 1981) or even decline in diversity toward the tropics (Owen and Owen 1974, Gaston et al. 1996). Both the Ichneumonidae and Tachinidae attack similar hosts, but the Tachinidae are less susceptible to the chemical defenses accumulated by the larval hosts and the 'nasty host hypothesis' has been used to explain the ichneumonid anomaly (Gauld et al. 1992, Hawkins et al. 1992, Sime and Brower 1998). A number of issues surround these trends in biodiversity, and in the Ichneumonidae, different subfamilies may have opposite relationships between biodiversity and latitudinal gradients (Gaston et al. 1996). Overall, parasitoids increase in biodiversity toward the tropics. The influence of greater parasitoid biodiversity should transfer to agroecosystems where habitat biodiversity is left intact. In Cacao plantations in Brazil, parasitoid biodiversity can be both abundant and diverse, and

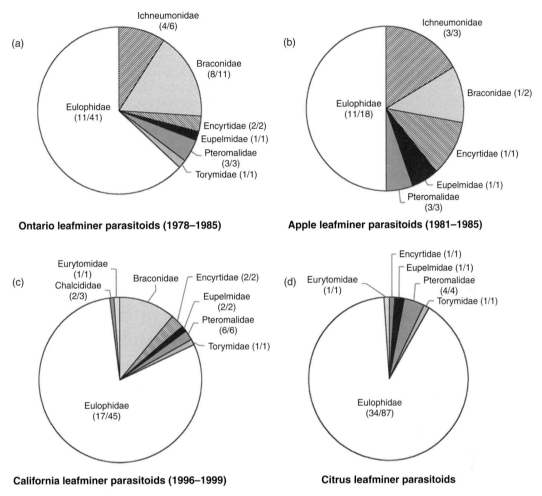

Fig. 19.2 Proportion of higher level hymenopteran parasitoids attacking (a) lepidopteran leafminers of broadleaf plants in southern Ontario (Heraty and Laing unpublished), (b) *Phyllonorycter blancardella* in an unsprayed apple orchard in Ontario (Heraty and Laing unpublished), (c) lepidopteran leafminers of broadleaf plants in California (Braconidae not broken out to genus and species) (Gates et al. 2002), (d) *Phyllocnistis citrella* worldwide (no Ichneumonoidea parasitoids) (Schauff et al. 1998). Figures (a), (b), and (d) are based on number of species sampled, (c) on number of individuals. Numbers of genera and species are in parentheses. Years refer to sampling periods.

similar in complexity to that in native forests (Sperber et al. 2004).

Does increased control in agricultural landscapes correlate with increased parasitoid diversity as we approach the tropics? Hawkins et al. (1992) found no significant correlation between parasitoid richness in agricultural hosts and latitudinal variability, although koinobionts had greater richness in northern North America. However, this finding does not address the control of hosts in the same context. In the leafminer example (Fig. 19.2), parasitoid diversity is higher both for native leafminers in California (USA) and for the citrus leafminer (which has a general tropical distribution) than for apple leafminers and a more northern leafminer landscape in Ontario (Canada). Both insects remain pests in their respective systems, although parasitoids can usually effect control below an economically important threshold in mature orchards.

SYSTEMATICS, PARASITOIDS, AND PEST MANAGEMENT

The importance of valid identification of both pest and parasitoid for successful establishment and cost savings in biological control programs has been discussed numerous times in the literature (Compere 1969, Rosen and DeBach 1973, Heraty 1998, Schauff and LaSalle 1998). Cassava mealybug was a major pest in Africa and threatened the major food source across the continent. Because of an initial misidentification, foreign exploration efforts were misdirected to the wrong host in northern South America, and parasitoids introduced from these efforts did not accept the African host. The correct host species, *Phenacoccus manihoti*, was known to occur only in southern South America. After correct identification by taxonomists, refocused collection efforts resulted in the establishment of a single parasite species, *A. lopezi* (Encyrtidae), and long-term control of the pest (Neuenschwander 2001). Control of Florida red scale on citrus was delayed because of misidentification of the parasite, *Aphytis lingnanensis* (Aphelinidae), attacking California red scale; the correct parasite, *A. holoxanthus*, which occurs in Israel and Hong Kong, was imported to Florida and economically important levels of control were achieved (Clausen 1978). Augmentative control by *Trichogramma* egg parasitoids is often stymied by a lack of proper identification of the correct species collected, reared, or released (Stouthamer et al. 2000b).

Currently, most of our pest management decisions, including choice of biological control agents, are based on a morphological understanding of species. The involvement of taxonomists early in a project can solve some of these problems, but even if expertise is available, the taxon of interest might be poorly surveyed or inherently difficult to identify (Schauff and LaSalle 1998). Beyond identification, systematics is focused on the collection, databasing, and maintenance of parasitoid collections from all habitats (museum based studies), and phylogenetic studies are gaining importance for understanding host relationships and the evolution of behavior and other traits of interest (Gauld 1986, Heraty 1998, 2004). Merging research from museum collections and the literature, two important databases, now incorporate a vast amount of data on nomenclature, geographic distribution, biology, and host relationships for Ichneumonoidea (Yu et al. 2004), and Chalcidoidea (Noyes 2007). The ichneumonoid database includes information on more than 63,000 taxonomic names and 28,000 literature citations. The chalcidoid database has information on more than 35,000 taxonomic names and incorporates more than 40,000 literature citations. These databases provide a good start toward understanding what we know about the biodiversity of parasitoids, host associations, and their potential effects in natural and agricultural ecosystems.

MOLECULES AND PARASITOID BIODIVERSITY

Molecular methods offer a new ability to identify species, albeit not without some of the same caveats facing morphological discrimination. A variety of molecular markers are available for diagnosing all levels of divergence in insect parasitoids (Unruh and Woolley 1999, Heraty 2004, Macdonald and Loxdale 2004). Comparative nucleotide sequences are currently the most common choice for species recognition, identification, and phylogenetic analysis. For both hosts and parasites, a variety of nuclear and mitochondrial ribosomal transcript regions and protein-coding genes have been used to identify and discriminate insect populations. For species level discrimination, these have usually involved comparisons of nuclear 28S rDNA transcript and the associated Internal Transcribed Spacer regions (ITS1 and ITS2) and mitochondrial COI, COII, 16S rDNA, and 12S rDNA. The ribosomal transcript regions (28S, 16S, and 12S) are conserved and considered appropriate for comparisons from the subfamily to the species level, whereas the ITS and cytochrome oxidase regions (COI and COII) are more variable, offering information from the population or species level and above (Heraty 2004). Because of rapid divergence and changes in sequence length, ITS rarely can be aligned for taxa beyond closely related species. All of these gene regions occur in numerous identical copies in the genome, making their extraction and amplification relatively trivial, even for poorly preserved insect samples. Additional single copy nuclear genes being explored among parasitoids include EF-1α (Rokas et al. 2002), phosphoenolpyruvate carboxykinase (PEPCK) and DOPA decarboxylase (DDC), Arginine Kinase, Long-Wave Length Opsin, and wingless (Rokas et al. 2002, Desjardins et al. 2007, Heraty et al. 2007, Burks, personal communication). While these single copy genes are important for phylogenetic purposes, they are more difficult to obtain and will not likely be used as extensively as 28S, ITS, and COI for purposes of species recognition.

Molecular methods will have an increasing influence on research into agricultural pests, identification of pests and parasitoids, and surveys of biodiversity at the pest management level. The use of molecular markers may be essential for the discovery of cryptic species complexes in parasitoids. There is also a drive to develop a standardized method of barcoding in all insects, not only for the recognition of existing species, but also for the discovery of new, even morphologically distinct, species. Important questions can be addressed with these new tools. What species are shared between agricultural and native habitats? How far do parasitoids disperse between habitats? Do cryptic species interfere or complement each other for pest management? Do cryptic species fare better in different climatic zones? Am I releasing the correct species? These are all questions that directly affect our ability to monitor parasitoids and control insect pests in urban, agricultural, and native landscapes. However, these tools are being applied against the tremendous background noise of undescribed taxa and the associated geographic variability. The discovery of cryptic species and identification, using molecular tools, will have a tremendous effect on our ability to use and manipulate insect parasitoids.

CRYPTIC SPECIES

Cryptic species are morphologically similar but reproductively isolated populations that can differ in important physiological, ecological, or behavioral traits. Numerous cryptic species have been recognized over the past few decades through better discrimination of populations, using molecular techniques. In the past, these species have been discovered through differences in behavior, reproductive incompatibility (Heraty et al. 2007), or isozyme or RAPD marker differences (Unruh et al. 1989, Antolin et al. 1996, Heraty 2004). Two cryptic species of phasiine tachinids were discovered by their attraction to different host pheromones (Aldrich and Zary 2002). However, the direct comparison of nucleotide sequences has led to even greater discovery of cryptic species complexes in a variety of parasitoids. COI was useful in discriminating 17 morphospecies of *Belvosia* tachinid flies, as well as discovering an additional 15 cryptic species among

them (Smith et al. 2006). Cryptic species in the *Cotesia melitaearum* complex (Braconidae) associated with different host butterflies were recognized by as little as 1.6% divergence in COI and microsatellites (Kankare et al. 2005a, 2005b). Cryptic species of *Nasonia* (Pteromalidae) were distinguished with ribosomal DNA sequences (ITS2 and 28S-D2) (Campbell et al. 1993). Morphologically indistinct species of *Encarsia* (Aphelinidae) can be separated using mitochondrial (COI) or nuclear (28S or ITS) gene regions (Babcock and Heraty 2000, Babcock et al. 2001, Polaszek et al. 2004, Monti et al. 2005). Monti et al. (2005) used a COI-sequence divergence of 10% to separate morphologically indistinguishable Pakistani and Spanish populations of *Encarsia sophia* that differ in reproductive compatibility and karyotype; similar morphologically distinct species of *Encarsia* differ by as little as 6.5–8.7%. *Trichogramma minutum* and *T. platneri* (Trichogrammatidae) can be distinguished only by a single fixed COI substitution (Stouthamer et al. 1998, 2000a).

Combinations of gene regions might be necessary to differentiate species. Sympatric, cryptic species of *Gonatocerus* (Mymaridae) were recognized using a combination of 28S-D2, COI, COII, ITS1, and ITS2, with all but 28S highly variable within reproductively compatible populations (Triapitsyn et al. 2006). The relationships and identity of six reproductively isolated species in the *Aphelinus varipes* complex (Aphelinidae) could be recognized only through a combination of six gene regions (28S, ITS1, ITS2, COI, COII, and Arginine Kinase) (Fig. 19.3; Heraty et al. 2007). The genetic differences between cryptic species are often minute, but in *Cales noacki* (Chalcidoidea), a parasite of woolly whitefly, cryptic species have been discovered that differ by enough base pairs (3% for 28S and 17% for COI) that they could be considered distinct genera or even subfamilies, compared with other Chalcidoidea.

The effect of the discovery of cryptic species complexes on our estimates of the biodiversity of insect parasitoids is staggering. The above examples of cryptic species complexes involve the chalcidoid families Mymaridae (1424 species), Aphelinidae (1160 species), and Pteromalidae (3506 species) (species numbers from Noyes 2007). Surveys of tropical rainforest canopies have already suggested a much greater biodiversity of morphological species, with samples in Sulawesi yielding as many as 150 morphologically distinct species of mostly undescribed species of *Encarsia*

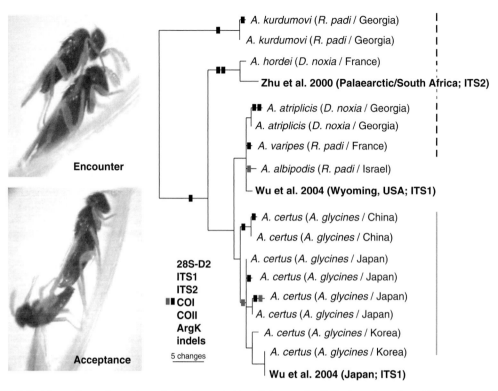

Fig. 19.3 Problems of barcoding, using only COI. Presented is a phylogram for distinct genotypes from different geographic populations of cryptic species in the *Aphelinus varipes* complex, with their source aphid host, geographic locality of collection, and whether they are reproductively compatible (gray line) or reproductively incompatible (dashed line) (Heraty et al. 2007). The phylogeny is based on six genes and insertion/deletion events, and includes three populations (bold) from published partial sequences as indicated. Only COI changes are indicated, with black bars (unique changes) and gray bars (homoplastic changes). COI changes are not correlated across the phylogeny with reproductive incompatibility, which supports a minimum of five species. The three previously published populations would not be correctly associated or discriminated using COI alone. The populations studied by Zhu et al (2000) and from Israel were not tested for compatibility. Populations tested by Wu et al. (2004) and the Georgia (*D. noxia*), China, and Japan populations tested by Heraty et al. (2007) were partially compatible with each other based on hybrid dysgenesis in backcrosses to Japan. In all cases, males will pursue and court heterospecific females, but females reject heterospecific males before mating (Hopper unpublished). (Photographs courtesy of Keith Hopper.)

(Noyes 1989). Even under the most conservative schemes, we might have to double or triple our estimates of extant species in all parasitoid taxa currently recognized by morphological criteria (Table 19.1).

DNA BARCODING AND BIODIVERSITY OF PARASITOIDS

Considerable discussion has occurred over the use of molecular methods to barcode species for easier recognition (Cameron et al. 2006). Sequencing

is becoming cheaper and more reliable, and the taxonomist shortage might be circumvented by comparing sequences to those obtained from known vouchers of identified species (Wheeler et al. 2004). Genetic markers such as 28S-D2, ITS, and COI are most appropriate for the recognition of insect species or populations. *Trichogramma* species are minute and difficult to identify based on features of the genitalia, but identification keys to species can be built, using differences in ITS length and sequence divergence (Ciociola et al. 2001, Pinto et al. 2002). These markers not only are useful for the identification of known species or populations by nonspecialists, but also can be used in studies of larval parasitism. Typically, the dissection or rearing of hosts to assess internal parasites is laborious. COI and ITS markers can be used to discriminate between the host and its parasitoids in both laboratory trials and field-collected hosts (Tilmon et al. 2000, Zhu et al. 2000, Zhu and Williams 2002, Macdonald and Loxdale 2004, Ashfaq et al. 2004, Agusti et al. 2005, Greenstone 2006).

The ease of DNA sequencing and utility of markers for species identification is developing into a movement to barcode numerous individuals of vouchered material for use in the general identification of all species and for the discovery of new species (Hebert et al. 2003, Hajibabaei et al. 2005). Agriculturally important pests or parasitoids are strong candidates for a barcoding initiative. The species involved are generally known and could be sampled for a number of genes across their entire geographic ranges. These markers would be useful for identification of pests at quarantine points, verification of natural enemies for importation, culture and release from quarantine, and post-release tracking of parasitoids. Barcoding can and does work efficiently in some cases, but there are a number of problems. Genetic variation within and between populations makes it difficult to assess the level of species discrimination and association (Meier et al. 2006), and considerable difference in the rates of variation between taxa might make it difficult to choose a single appropriate gene (Heraty 2004). The problem is exacerbated with the inclusion of unknown species. Further, the choice of gene, number of genes chosen, number of taxa sampled over a geographical range, representation of taxa in current databases, and mistakes inherent in existing sequences in databases all cause further problems (Cameron et al. 2006, Hickerson et al. 2006, Meier et al. 2006).

Among parasitoids, genetic divergence within a single gene region may or may not correspond with reproductive compatibility. Two morphologically cryptic species of *Trichogramma* were separated by two fixed-substitution differences in COI, but not by the usually more variable ITS2 region (Stouthamer et al. 2000a, Pinto et al. 2002, Stouthamer, personal communication). Similarly, COI has proved useful for separating cryptic species of *Encarsia* (Monti et al. 2005). Within the *Aphelinus varipes* species complex (Aphelinidae), six gene regions and their indel events were necessary to discriminate and determine relationships

among the different populations (Heraty et al. 2007). Although the most basal and genetically divergent lineages were reproductively isolated from all others, in the more derived lineages, the eastern Palearctic populations were all reproductively isolated from each other, whereas the Asian populations were not (Fig. 19.3). COI changes in the *A. varipes* complex are sparse and scattered and do not have a direct correlation with isolation, and would not be useful in placing populations previously cited in the literature that (1) do not have any COI information available, and (2) could not be separated from other lineages if only COI sequences were available for comparison (Fig. 19.3). COI variation can have no bearing on reproductive isolation. In *Megastigmus transvaalensis*, a seed parasite of various species of *Rhus* and *Schinus* in Africa, 29 haplotypes were found for COI that were distinct enough to provide bootstrap support correlated with the different countries of origin (Scheffer and Grissell 2003). In *Gonatocerus ashmeadi*, a parasitoid of the glassy-winged sharpshooter, reproductively compatible populations were fixed for 28S-D2, but highly variable for COI (9 substitutions) and ITS2 (19 substitutions) (Triapitsyn et al. 2006). In contrast, morphologically distinct and geographically isolated species of *Pnigalio* (Eulophidae) are identical for 28S-D2, but COI readily distinguishes both morphologically distinct and cryptic species (M. Gebiola, personal communication).

Admittedly, these problems of associating molecular divergence of COI with speciation are most problematic for discriminating recently evolved species. However, any barcoding initiative will have to deal with the problems of intraspecific variation, which can be excessive in some species. Molecular sequences can be used to discriminate distinctive or cryptic species, even with a single gene region, but often these findings can be verified only in conjunction with traditional methods of observing morphological divergence, behavioral distinctness, geographic separation, and studies of reproductive compatibility. With nondestructive PCR extraction methods, each specimen in a given research project conceivably might be vouchered by DNA sequences (GenBank deposition) and digital images (MorphBank deposition), with all information freely available through online databases. As these data accumulate, perhaps we can better evaluate the use of barcoding tools and our general interpretation of speciation processes and boundaries.

CAN MOLECULAR MARKERS BE APPLIED TO UNDERSTANDING BIODIVERSITY?

Parasitoids are extremely diverse. At our best guess, there are at least 680,000 morphologically distinct species (Table 19.1). If we begin to revise our estimate to include the vast number of cryptic species that also must be discovered, we must assume that even with a low estimate, there are easily more than 1 million species. Documenting molecular markers for species on a grand scale will be costly, requiring samples of multiple markers, multiple individuals, and multiple populations (Meier et al. 2006). Molecular sequences can be used to discriminate distinctive or cryptic species, even with a single gene region, but often these findings can be verified only in conjunction with traditional methods of observing morphological divergence, behavioral distinctness, geographic separation, and studies of reproductive compatibility.

SUMMARY

More than 680,000 morphologically distinct species of parasitoids are estimated, with the possibility of a vast underlying biodiversity of cryptic species. Most of these species are unknown to science. Molecular tools will be important in the recognition and tracking, especially of these new cryptic species, but as with other methods, they are merely a tool to be used along with more traditional methods of morphological systematics, behavioral and host studies, and community relationships. Parasitoids are more diverse in ecologically complex systems, compared with simple landscapes. Generalist parasitoids may be more common in late-succession communities (trees and shrubs), compared with oligophagous species, which dominate in early-succession communities (ruderals and shrubs). Latitudinal gradients affect parasitoid biodiversity, generally with increasing biodiversity toward the tropics; however, within each group, there are shifts or distinctly different trends among the members of a group, such as in different subfamilies of Ichneumonidae that can be differentiated into different host groups. Biological control has demonstrated that many pests can be controlled by one or more introduced parasitoids. There is still controversy over continuing to increase parasitoid biodiversity against these pests over and above the previously established species, especially if they are already providing control below economic thresholds. However, studies of natural and adjacent agricultural systems suggest that an increased biodiversity of parasitoids is important for controlling pests (Bianchi et al. 2006, Marino et al. 2006).

Agriculture is usually synonymous with monoculture. Ecologically diverse native vegetation is suffering under agricultural intensification and potentially reducing the pool of available parasitoids that are available for control of newly introduced or resident pest species. To support conservation in association with agricultural practices, we need more research on the shared pool of parasitoid biodiversity. Many pest species have come from a different country, arriving without their parasitoids. However, if a general habitat decline is associated with habitat loss, we might see an increasing number of pests developing in their native area of origin. Biological control programs of the past may give us insights into strategies necessary to deal with these emergent pests. In addition, we are entering a new age of transgenics in crops, and their effect on crops and pests and associated effects on parasitoids is far from being understood (White and Andow 2005, Beale et al. 2006, Davidson et al. 2006).

ACKNOWLEDGMENTS

I would like to thank Bob Foottit, Peter Adler, Peter Mason, Jason Mottern, and an anonymous reviewer for comments on the manuscript, and Brian Brown for information on phorid flies. This work was supported, in part, by National Science Foundation grant DEB-010108245.

REFERENCES

Agusti, N., D. Bourguet, T. Spataro, M. Delos, N. Eychenne, L. Folcher, and R. Arditi. 2005. Detection, identification and geographical distribution of European corn borer larval parasitoids using molecular markers. *Molecular Ecology* 14: 3267–3274.

Aldrich, J. R. and A. Zary. 2002. Kairomone strains of *Euclytia flava* (Townsend), a parasitoid of stink bugs. *Journal of Chemical Ecology* 28: 1565–1582.

Antolin, M. F., D. S. Guertin, and J. J. Petersen. 1996. The origin of gregarious *Muscidifurax* (Hymenoptera: Pteromalidae) in North America: an analysis using molecular markers. *Biological Control* 6: 76–82.

Ashfaq, M., L. Braun, D. Hegedus, and M. Erlandson. 2004. Estimating parasitism levels in *Lygus* spp. (Hemiptera: Miridae) field populations using standard and molecular techniques. *Biocontrol Science and Technology* 14: 731–735.

Askew, R. R. 1971. *Parasitic Insects*. Heinemann Educational Books, London.

Babcock, C. S. and J. M. Heraty. 2000. Molecular markers distinguishing *Encarsia formosa* and *Encarsia luteola* (Hymenoptera: Aphelinidae). *Annals of the Entomological Society of America* 93: 738–744.

Babcock, C. S., J. M. Heraty, P. J. De Barro, F. Driver, and S. Schmidt. 2001. Preliminary phylogeny of *Encarsia* Forster (Hymenoptera: Aphelinidae) based on morphology and 28S rDNA. *Molecular Phylogenetics and Evolution* 18: 306–323.

Beale, M. H., M. A. Birkett, T. J. A. Bruce, K. Chamberlain, L. M. Field, A. K. Huttly, J. L. Martin, R. Parker, A. L. Phillips, J. A. Pickett, I. M. Prosser, P. R. Shewry, L. E. Smart, L. J. Wadhams, C. M. Woodcock, and Y. H. Zhang. 2006. Aphid alarm pheromone produced by transgenic plants affects aphid and parasitoid behavior. *Proceedings of the National Academy of Sciences of the United States of America* 103: 10509–10513.

Benton, T. G., J. A. Vickery, and J. D. Wilson. 2003. Farmland biodiversity: is habitat heterogeneity the key? *Trends in Ecology and Evolution* 18: 182–188.

Bianchi, F., C. J. H. Booij, and T. Tscharntke. 2006. Sustainable pest regulation in agricultural landscapes: a review on landscape composition, biodiversity and natural pest control. *Proceedings of the Royal Society. B. Biological Sciences* 273: 1715–1727.

Boettner, G. H., J. S. Elkinton, and C. J. Boettner. 2000. Effects of a biological control introduction on three nontarget native species of saturniid moths. *Conservation Biology* 14: 1798–1806.

Bostanian, N. J., H. Goulet, J. O'Hara, L. Masner, and G. Racette. 2004. Towards insecticide free apple orchards: flowering plants to attract beneficial arthropods. *Biocontrol Science and Technology* 14: 25–37.

Caltagirone, L. E. 1981. Landmark examples in biological control. *Annual Review of Entomology* 26: 213–232.

Cameron, S., D. Rubinoff, and K. Will. 2006. Who will actually use DNA barcoding and what will it cost? *Systematic Biology* 55: 844–847.

Campbell, B. C., J. D. Steffen-Campbell, and J. H. Werren. 1993. Phylogeny of the *Nasonia* species complex (Hymenoptera: Pteromalidae) inferred from an internal transcribed spacer (ITS2) and 28S rDNA sequences. *Insect Molecular Biology* 2: 225–237.

Cardinale, B. J., C. T. Harvey, K. Gross, and A. R. Ives. 2003. Biodiversity and biocontrol: emergent impacts of a multi-enemy assemblage on pest suppression and crop yield in an agroecosystem. *Ecology Letters* 6: 857–865.

Ciociola, A. I., R. B. Querino, R. A. Zucchi, and R. Stouthamer. 2001. Molecular key to seven brazilian species of *Trichogramma* (Hymenoptera: Trichogrammatidae) using sequence of the ITS2 and restriction analysis. *Neotropical Entomology* 30: 259–262.

Clausen, C. P. 1940. *Entomophagous Insects*. McGraw-Hill, London.

Clausen, C. P. 1978. Introduced parasites and predators of arthropod pests and weeds: a world review. *United States Department of Agriculture, Agricultural Handbook* 480: 1–551.

Compere, C. P. 1969. The relation of taxonomy to biological control. *Journal of Economic Entomology* 35: 744–748.

Corbett, A. and J. A. Rosenheim. 1996. Impact of a natural enemy overwintering refuge and its interaction with the surrounding landscape. *Ecological Entomology* 21: 155–164.

Davidson, M. M., R. C. Butler, S. D. Wratten, and A. J. Conner. 2006. Impacts of insect-resistant transgenic potatoes on the survival and fecundity of a parasitoid and an insect predator. *Biological Control* 37: 224–230.

DeBach, P. and M. Rose. 1971. Biological control of woolly whitefly. *California Agriculture* 30: 4–7.

Desjardins, C. A., J. C. Regier, and C. Mitter. 2007. Phylogeny of pteromalid wasps (Hymenoptera: Pteromalidae): initial evidence from four protein-coding nuclear genes. *Molecular Phylogenetics and Evolution* 45: 454–469.

Disney, R. H. L. 1994. *Scuttle Flies – the Phoridae*. Chapman and Hall, London.

Dolphin, K. and D. L. J. Quicke. 2001. Estimating the global species richness of an incompletely described taxon: an example using parasitoid wasps (Hymenoptera: Braconidae). *Biological Journal of the Linnean Society* 73: 279–286.

Eggleton, P. and R. Belshaw. 1992. Insect parasitoids: an evolutionary overview. *Philosophical Transactions of the Royal Society. B. Biological Sciences* 337: 1–20.

Eggleton, P. and K. J. Gaston. 1990. Parasitoid species and assemblages – convenient definitions or misleading compromises. *Oikos* 59: 417–421.

Feener, D. H. and B. V. Brown. 1992. Reduced foraging of *Solenopsis geminata* (Hymenoptera, Formicidae) in the presence of parasitic *Pseudacteon* spp (Diptera, Phoridae). *Annals of the Entomological Society of America* 85: 80–84.

Fisher, B. L. 1998. Insect behavior and ecology in conservation: preserving functional species interactions. *Annals of the Entomological Society of America* 91: 155–158.

Folgarait, P. J., O. Bruzzone, S. D. Porter, M. A. Pesquero, and L. E. Gilbert. 2005. Biogeography and macroecology of phorid flies that attack fire ants in south-eastern Brazil and Argentina. *Journal of Biogeography* 32: 353–367.

Gaston, K. J. 1991. The magnitude of global insect species richness. *Conservation Biology* 5: 283–296.

Gaston, K. J. 1993. Spatial patterns in the description and richness of the Hymenoptera. Pp. 277–294. *In* J. LaSalle and I. D. Gauld (eds). *Hymenoptera and Biodiversity*. CAB International, Wallingford.

Gaston, K. J., I. D. Gauld, and P. Hanson. 1996. The size and composition of the hymenopteran fauna of Costa Rica. *Journal of Biogeography* 23: 105–113.

Gates, M. E., J. M. Heraty, M. E. Schauff, D. L. Wagner, J. B. Whitfield, and D. B. Wahl. 2002. Survey of the parasitic Hymenoptera on leafminers in California. *Proceedings of the Entomological Society of Washington* 11: 213–270.

Gauld, I. D. 1986. Taxonomy, its limitations and its role in understanding parasitoid biology. Pp. 1–21. *In* J. Waage and D. Greathead (eds). *Insect Parasitoids*. Academic Press, London.

Gauld, I. D., K. J. Gaston, and D. H. Janzen. 1992. Plant allelochemicals, tritrophic interactions and the anomalous diversity of tropical parasitoids: the 'nasty' host hypothesis. *Oikos* 65: 353–357.

Genier, S. 1988. Applied biological control with tachinid flies (Diptera, Tachinidae): a review. *Anzeiger für Schädling Pflanzenschutz Umweltschutz* 51: 49–56.

Goulet, H. and J. T. Huber. 1993. *Hymenoptera of the World: An Identification Guide to Families*. Agriculture Canada Research Branch Publication 1894/E., Ottawa.

Greathead, D. J. 1986. Parasitoids in classical biological control. Pp. 289–318. *In* J. K. Waage and D. Greathead (eds). *Insect Parasitoids*. Academic Press, London.

Greenstone, M. H. 2006. Molecular methods for assessing insect parasitism. *Bulletin of Entomological Research* 96: 1–13.

Grissell, E. E. 1999. Hymenopteran biodiversity: some alien notions. *American Entomologist* 45: 235–244.

Hajibabaei, M., J. R. DeWaard, N. V. Ivanova, S. Ratnasingham, R. T. Dooh, S. L. Kirk, P. M. Mackie, and P. D. N. Hebert. 2005. Critical factors for assembling high volume of DNA barcodes. *Philosophical Transactions of the Royal Society. B. Biological Sciences* 360: 1959–1967.

Hawkins, B. A. 1994. *Pattern and Process in Host–Parasitoid Interactions*. Cambridge University Press, United Kingdom.

Hawkins, B. A., M. R. Shaw, and R. R. Askew. 1992. Relations among assemblage size, host specialization, and climatic variability in North-American parasitoid communities. *American Naturalist* 139: 58–79.

Hebert, P. D. N., A. Cywinska, S. L. Ball, and J. R. de Waard. 2003. Biological identification through DNA barcodes. *Proceedings of the Royal Society. B. Biological Sciences* 270: 313–322.

Heraty, J. M. 1998. Systematics: Science or Service? California Conference on Biological Control I, 1998. Pp. 187–190.

Heraty, J. M. 2004. Molecular systematics, Chalcidoidea and biological control. Pp. 39–71. *In* L. E. Ehler, R. Siorza, and T. Mateille (eds). *Genetics, Evolution and Biological Control*. CAB International, London.

Heraty, J. M. and M. E. Gates. 2003. Biodiversity of Chalcidoidea of the El Edén Ecological Reserve, Mexico. Pp. 277–292. *In* A. Gómez-Pompa, M. F. Allen, S. L. Fedick, and J. J. Jiménez-Osornio (eds). *Proceedings of the 21st Symposium in Plant Biology, "Lowland Maya Area: Three Millenia at the Human-Wildland Interface*. Haworth Press, New York.

Heraty, J. M., J. B. Woolley, K. M. Hopper, D. L. Hawks, J.-W. Kim, and M. Buffington. 2007. Phylogenetic relationships of cryptic species in the *Aphelinus varipes* complex (Hymenoptera: Aphelinidae). *Molecular Phylogenetics and Evolution* 45: 480–493.

Hespenheide, H. A. 1979. Are there fewer parasitoids in the tropics? *American Naturalist* 113: 766–769.

Hickerson, M. J., C. P. Meyer, and C. Moritz. 2006. DNA barcoding will often fail to discover new animal species over broad parameter space. *Systematic Biology* 55: 729–739.

Hodges, A. C., G. S. Hodges, and K. E. Espelie. 2003. Parasitoids and parasites of *Polistes metricus* Say (Hymenoptera: Vespidae) in northeast Georgia. *Annals of the Entomological Society of America* 96: 61–64.

Huffaker, C. 1969. *Biological Control*. Plenum Publishing, New York.

Hunter, M. S. and J. B. Woolley. 2001. Evolution and behavioral ecology of heteronomous aphelinid parasitoids. *Annual Review of Entomology* 46: 251–290.

Janzen, D. H. 1981. The peak in North American ichneumonid species richness lies between 38° and 42° N. *Ecology* 62: 532–537.

Janzen, D. H. and W. Hallwachs. 2005. Dynamic database for an inventory of the macrocaterpillar fauna, and its food plants and parasitoids, of the Area de Conservacion Guanacaste (ACG), northwestern Costa Rica. http://janzen.sas.upenn.edu/Wadults/searchpara.lasso

Kankare, M., C. Stefanescu, S. Van Nouhuys, and M. R. Shaw. 2005a. Host specialization by *Cotesia* wasps (Hymenoptera: Braconidae) parasitizing species-rich Melitaeini (Lepidoptera: Nymphalidae) communities in north-eastern Spain. *Biological Journal of the Linnean Society* 86: 45–65.

Kankare, M., S. Van Nouhuys, and I. Hanski. 2005b. Genetic divergence among host-specific cryptic species in *Cotesia melitaearum* aggregate (Hymenoptera: Braconidae), parasitoids of checkerspot butterflies. *Annals of the Entomological Society of America* 98: 382–394.

Kruess, A. and T. Tscharntke. 1994. Habitat fragmentation, species loss, and biological-control. *Science* 264: 1581–1584.

Laing, J. E. and J. M. Heraty. 1981. The parasite complex of the overwintering population of *Epiblema scudderiana* (Lepidoptera: Olethreutidae) in southern Ontario. *Proceedings of the Entomological Society of Ontario* 112: 59–67.

Landis, D. A., F. D. Menalled, A. C. Costamagna, and T. K. Wilkinson. 2005. Manipulating plant resources to enhance beneficial arthropods in agricultural landscapes. *Weed Science* 53: 902–908.

LaSalle, J. 1993. Parasitic Hymenoptera, biological control and the biodiversity crisis. Pp. 197–216. *In* J. LaSalle and I. D. Gauld (eds). *Hymenoptera and Biodiversity*. CAB International, Wallingford.

LaSalle, J. and I. D. Gauld. 1992 [1991]. Parasitic Hymenoptera and the biodiversity crisis. *Redia* 74: 315–334.

LaSalle, J. and I. D. Gauld. 1993. Hymenoptera: their diversity, and their impact on the diversity of other organisms. Pp. 1–26. *In* J. LaSalle and I. D. Gauld (eds). *Hymenoptera and Biodiversity*. CAB International, Wallingford.

Lavandero, B., S. D. Wratten, R. K. Didham, and G. Gurr. 2006. Increasing floral diversity for selective enhancement of biological control agents: A double-edged sward? *Basic and Applied Ecology* 7: 236–243.

Macdonald, C. and H. D. Loxdale. 2004. Molecular markers to study population structure and dynamics in beneficial insects (predators and parasitoids). *International Journal of Pest Management* 50: 215–224.

Marino, P. C., D. A. Landis, and B. A. Hawkins. 2006. Conserving parasitoid assemblages of North American pest Lepidoptera: does biological control by native parasitoids depend on landscape complexity. *Biological Control* 37: 173–185.

Meier, R., K. Shiyang, G. Vaidya, and P. K. Ng. 2006. DNA barcoding and taxonomy in Diptera: a tale of high intraspecific variability and low identification success. *Systematic Biology* 55: 715–728.

Monti, M. M., A. G. Nappo, and M. Giorgini. 2005. Molecular characterization of closely related species in the parasitic genus *Encarsia* (Hymenoptera: Aphelinidae) based on the mitochondrial cytochrome oxidase subunit I gene. *Bulletin of Entomological Research* 95: 401–408.

Morehead, S. A., J. Seger, D. H. Feener, and B. V. Brown. 2001. Evidence for a cryptic species complex in the ant parasitoid *Apocephalus paraponerae* (Diptera: Phoridae). *Evolutionary Ecology Research* 3: 273–284.

Morrison, L. W. and S. D. Porter. 2005. Testing for population-level impacts of introduced *Pseudacteon tricuspis* flies, phorid parasitoids of *Solenopsis invicta* fire ants. *Biological Control* 33: 9–19.

Muller, C. B. and J. Brodeur. 2002. Intraguild predation in biological control and conservation biology. *Biological Control* 25: 216–223.

Myers, J. H., C. Higgins, and E. Kovacs. 1989. How many insect species are necessary for the biological-control of insects. *Environmental Entomology* 18: 541–547.

Neuenschwander, P. 2001. Biological control of the cassava mealybug in Africa: a review. *Biological Control* 21: 214–229.

Nordlander, G. 1984. Vad vet vi on parasitika Cynipoidea. *Entomologisk Tidskrift* 105: 36–40.

Noyes, J. S. 1989. The diversity of Hymenoptera in the tropics with special reference to Parasitica in Sulawesi. *Ecological Entomology* 14: 197–207.

Noyes, J. S. 2000. Encyrtidae of Costa Rica (Hymenoptera: Chalcidoidea), 1. The subfamily Tetracneminae, parasitoids of mealybugs (Homoptera: Pseudococcidae). *Memoirs of the American Entomological Institute* 62: 1–355.

Noyes, J. S. 2007. Universal Chalcidoidea Database. http://www.nhm.ac.uk/research-curation/projects/chalcidoids/

Noyes, J. S., and M. Hayat. 1994. *Oriental Mealybug Parasitoids of the Anagyrini (Hymenoptera: Encyrtidae)*. CAB International, London.

Owen, D. F. and J. Owen. 1974. Species diversity in temperate and tropical Ichneumonidae. *Nature* 249: 583–584.

Peña, J., R. Duncan, and H. Browning. 1996. Seasonal abundance of *Phyllocnistis citrella* (Lepidoptera: Gracillariidae) and its parasitoids in South Florida Citrus. *Environmental Entomology* 25: 698–702.

Peña, J. F., A. Hunsberger, and B. Schaffer. 2000. Citrus leafminer (Lepidoptera: Gracillariidae) density: effect on yield of 'Tahiti' lime. *Journal of Economic Entomology* 93: 374–379.

Pinto, J. D., A. B. Koopmanschap, G. R. Platner, and R. Stouthamer. 2002. The North American *Trichogramma* (Hymenoptera: Trichogrammatidae) parasitizing certain Tortricidae (Lepidoptera) on apple and pear, with ITS2 DNA characterizations and description of a new species. *Biological Control* 23: 134–142.

Polaszek, A., S. Manzari, and D. L. J. Quicke. 2004. Morphological and molecular taxonomic analysis of the *Encarsia meritoria* species-complex (Hymenoptera, Aphelinidae), parasitoids of whiteflies (Hemiptera, Aleyrodidae) of economic importance. *Zoologica Scripta* 33: 403–421.

Price, P. 1980. *Evolutionary Biology of Parasites*. Princeton University Press, Princeton, New Jersey.

Price, P. 1991. Evolutionary theory of host and parasitoid interactions. *Biological Control* 1: 83–93.

Price, P. 1994. Evolution of parasitoid communities. Pp. 473–491. *In Parasitoid Community Ecology*. Oxford University Press, Oxford.

Quicke, D. L. J. 1997. *Parasitic Wasps*. Chapman and Hall, London.

Rasnitsyn, A. P. 1988. An outline of evolution of the hymenopterous insects (Order Vespida). *Oriental Insects* 22: 115–145.

Robinson, W. H. 1971. Old and new biologies of *Megaselia* species (Diptera, Phoridae). *Studia Entomologica* 14: 321–368.

Rodríguez, M. A. and B. A. Hawkins. 2000. Diversity, function and stability in parasitoid communities. *Ecology Letters* 3: 35–40.

Rokas, A., J. A. A. Nylander, F. Ronquist, and G. N. Stone. 2002. A maximum-likelihood analysis of eight phylogenetic markers in gallwasps (Hymenoptera: Cynipidae): implications for insect phylogenetic studies. *Molecular Phylogenetics and Evolution* 22: 206–219.

Ronquist, F. 1995. Phylogeny and early evolution of the Cynipoidea (Hymenoptera). *Systematic Entomology* 20: 309–335.

Rosen, D. and P. DeBach. 1973. Systematics, morphology and biological control. *Entomophaga* 18: 215–222.

Schauff, M. E. and J. LaSalle. 1998. The relevance of systematics to biological control: protecting the investment in research. Pp. 425–436. *In Pest Management – Future Challenges.* Volume 1. Proceedings of the 6th Australian Applied Entomological Conference, Brisbane.

Schauff, M. E., J. LaSalle, and G. A. Wijesekara. 1998. The genera of chalcid parasitoids (Hymenoptera: Chalcidoidea) of citrus leafminer *Phyllocnistis citrella* Stainton (Lepidoptera: Gracillariidae). *Journal of Natural History* 32: 1001–1056.

Scheffer, S. J. and E. E. Grissell. 2003. Tracing the geographical origin of *Megastigmus transvaalensis* (Hymenoptera: Torymidae): an African wasp feeding on a South American plant in North America. *Molecular Ecology* 12: 415–421.

Sime, K. R. and A. V. Z. Brower. 1998. Explaining the latitudinal gradient anomaly in ichneumonid species richness: evidence from butterflies. *Journal of Animal Ecology* 67: 387–399.

Smith, H. S. 1912. Technical results from the gypsy moth parasite laboratory. IV. The chalcidoid genus *Perilampus* and its relation to the problem of parasite introduction. *United States Department of Agricultrue Technical Series* 19: 33–69.

Smith, M. A., N. E. Woodley, D. H. Janzen, W. Hallwachs, and P. D. N. Hebert. 2006. DNA barcodes reveal cryptic host-specificity within the presumed polyphagous members of a genus of parasitoid flies (Diptera: Tachinidae). *Proceedings of the National Academy of Sciences USA* 103: 3657–3662.

Smith, S. M. 1993. Insect parasitoids – a Canadian perspective on their use for biological-control of forest insect pests. *Phytoprotection* 74: 51–67.

Sperber, C. F., K. Nakayama, M. J. Valverde, and F. D. Neves. 2004. Tree species richness and density affect parasitoid diversity in cacao agroforestry. *Basic and Applied Ecology* 5: 241–251.

Stireman, J. O., J. E. O'Hara, and D. M. Wood. 2006. Tachinidae: evolution, behavior, and ecology. *Annual Review of Entomology* 51: 525–555.

Stouthamer, R., Y. Gai, A. B. Koopmanschap, G. R. Platner, and J. D. Pinto. 2000a. ITS-2 sequences do not differ for the closely related species *Trichogramma minutum* and *T. platneri*. *Entomologia Experimentalis et Applicata* 95: 105–111.

Stouthamer, R., J. Hu, F. J. P. M. van Kan, G. R. Platner, and J. D. Pinto. 1998. The utility of internally transcribed spacer 2 DNA sequences of the nuclear ribosomal gene for distinguishing sibling species of *Trichogramma*. *Biocontrol* 43: 421–440.

Stouthamer, R., P. Jochemsen, G. R. Platner, and J. D. Pinto. 2000b. Crossing incompatibility between *Trichogramma minutum* and *T. platneri* (Hymenoptera: Trichogrammatidae): implications for application in biological control. *Environmental Entomology* 29: 832–837.

Tilmon, K. J., B. N. Danforth, W. H. Day, and M. P. Hoffman. 2000. Determining parasitoid species composition in a host population: a molecular approach. *Annals of the Entomological Society of America* 93: 640–647.

Triapitsyn, S. V., D. B. Vickerman, J. M. Heraty, and G. A. Logarzo. 2006. A new species of *Gonatocerus* (Hymenoptera: Mymaridae) parasitic on proconiine sharpshooters (Hemiptera: Cicadellidae) in the New World. *Zootaxa* 1158: 55–67.

Unruh, T. and J. B. Woolley. 1999. Molecular methods in biological control. Pp. 57–85. *In* T. S. Bellows and T. W. Fischer (eds). *Handbook of Biological Control.* Academic Press, San Diego, California.

Unruh, T. R., W. White, D. Gonzalez, and J. B. Woolley. 1989. Genetic relationships among 17 *Aphidius* (Hymenoptera, Aphidiidae) populations, including 6 species. *Annals of the Entomological Society of America* 82: 754–768.

Van Dreische, R. G. and T. S. Bellows, Jr. 1996. *Biological Control.* Chapman and Hall, New York.

Wells, A. 1992. The 1st parasitic Trichoptera. *Ecological Entomology* 17: 299–302.

Wells, A. 2005. Parasitism by hydroptilid caddisflies (Trichoptera) and seven new species of Hydroptilidae from northern Queensland. *Australian Journal of Entomology* 44: 385–391.

Wheeler, Q. D., P. H. Raven, and E. O. Wilson. 2004. Taxonomy: impediment or expedient. *Science* 303: 285.

White, J. A. and D. A. Andow. 2005. Host–parasitoid interactions in a transgenic landscape: spatial proximity effects of host density. *Environmental Entomology* 34: 1493–1500.

Wiegmann, B. M., C. Mitter, and B. Farrell. 1993. Diversification of carnivorous parasitic insects – extraordinary radiation or specialized dead-end. *American Naturalist* 142: 737–754.

Wilson, E. O. 1985. The biological diversity crisis. *Bioscience* 35: 700–706.

Wu, Z. S., K. R. Hopper, R. J. O'Neil, D. J. Voegtlin, D. R. Prokrym, and G. E. Heimpel. 2004. Reproductive compatibility and genetic variation between two strains of *Aphelinus albipodus* (Hymenoptera: Aphelinidae), a parasitoid of the soybean aphid, *Aphis glycines* (Homoptera: Aphididae). *Biological Control* 31: 311–319.

Yu, D. S., K. van Achterberg, and K. Horstmann. 2004. World Ichneumonoidea 2004, Taxonomy, biology, morphology and distribution. http://www.taxapad.com/

Zhu, Y. C., J. D. Burd, N. C. Elliott, and M. H. Greenstone. 2000. Specific ribosomal DNA marker for early polymerase chain reaction detection of *Aphelinus hordei* (Hymenoptera: Aphelinidae) and *Aphidius colemani* (Hymenoptera: Aphidiidae) from *Diuraphis noxia* (Homoptera: Aphididae). *Annals of the Entomological Society of America* 93: 486–491.

Zhu, Y. C. and L. Williams III. 2002. Detecting the egg parasitoid *Anaphes iole* (Hymenoptera: Mymaridae) in tarnished plant bug (Heteroptera: Miridae) eggs by using a molecular approach. *Annals of the Entomological Society of America* 95: 359–365.

Chapter 20

THE TAXONOMY OF CROP PESTS: THE APHIDS

Gary L. Miller[1] and Robert G. Foottit[2]

[1] Systematic Entomology Laboratory, PSI, Agricultural Research Service, U.S. Department of Agriculture, Bldg. 005, BARC-W, Beltsville, MD 20705, USA;
[2] Agriculture and Agri-Food Canada, Canadian National Collection of Insects, Ottawa, ON, K1A 0C6 Canada

'For the most part, the most economically important insect and mite pests are known to science, and their position in our classification system is resolved'.

— Anonymous, USDA, ARS
Research Action Plan, 2004

Insect Biodiversity: Science and Society, 1st edition. Edited by R. Foottit and P. Adler
© 2009 Blackwell Publishing, ISBN 978-1-4051-5142-9

There is a perception that certain insect pests of crops are well-studied biologically and that their taxonomy is in good order. This perception might lead to the impression by those who are unfamiliar with the intricacies of acquiring taxonomic understanding of biological diversity that we know all there is to know. In fact, science is a discipline that continually builds on its previous discoveries and technologies as it advances our knowledge. Much research is needed even in economically important insect groups whose taxonomy might be regarded as advanced. In this chapter, using the Aphidoidea as an example and, in particular, using the early works of North American aphidology as background, we explore various dimensions of taxonomic knowledge in this pest group.

HISTORICAL BACKGROUND

Plants have been cultivated and traded since perhaps 8000 BC (Huxley 1978) and insect pests have long plagued humans and their crops. Ancient civilizations recorded swarms of locusts and other insect pestilence (Harpaz 1973, Konishi and Ito 1973). As humans expanded crop cultivation, associated insect problems soon followed. The European colonists of the New World faced their own set of insect-related problems with the cultivation of both native and introduced crops. For example, tobacco, which is native to the New World, experienced insect damage from hornworms and flea-beetles from the outset of its cultivation (Garner 1946). The introduction of new plants also began early during European colonization. Sugarcane was transported from the Canary Islands to Hispaniola on Columbus's second voyage in 1493 (Deerr 1949). Some of these early introduced plants also had their associated pests, including aphids. The close association of aphids with their hosts meant those insects and their eggs were being transported through commerce as well (Howard 1898). The cabbage aphid, *Brevicoryne brassicae* (Linnaeus), was noted in North America as early as 1791 (Miller et al. 2006). Early entomologists were well aware that commerce and travel were responsible for the transport of some of these pests. In 1856, Asa Fitch speculated that *B. brassicae* was brought along with cabbage plants on shipboard cargo (Miller et al. 2006).

ECONOMIC IMPORTANCE AND EARLY TAXONOMY

Aphids are small, soft-bodied insects mostly ranging between 1.5 and 3.5 mm in length (Blackman and Eastop 2000); they feed on plants with piercing-sucking mouthparts. Besides the mechanical damage they cause by this action, aphids also serve as the largest group of vectors of plant viruses (Eastop 1977, Chan et al. 1991). The damage is further compounded by fouling of the host plant with honeydew. Noted as long ago as in Réaumur's (1737) work, honeydew is excreted from the anus and is high in plant sugars and other compounds. Besides having an influence on predators (Glen 1973) and parasitoids (Faria 2005), it serves as a substrate for the growth of fungal complexes that cause sooty mold (Westcott 1971). In addition to reducing the photosynthetic ability of plants, sooty mold reduces a plant's aesthetic market value (Worf et al. 1995).

More than 250 species of the Aphidoidea (in the families Adelgidae, Phylloxeridae, and Aphididae) feed on agricultural or horticultural crops (Blackman and Eastop 2000). While this figure only represents approximately 5% of the world aphid fauna, the economic consequences of aphid damage are huge. For example, Wellings et al. (1989) estimated for 13 selected crops that aphids contribute about 2% of the total losses attributed to insects and believed that figure was a gross underestimate. Aphids are one of the important vegetable pests (Capinera 2002). Of the 80 groups of vascular plants in the world only 8 lack aphids, and those groups represent only 3% of the plant species (Eastop 1978).

Most aphid references refer to economic impact and some of the very earliest works of the sixteenth and seventeenth centuries have been noted by Blackman and Eastop (2000). Although aphids have extraordinary damaging horticultural effects, some of society's early interest involved their beneficial or positive attributes. Galls of certain aphid species (e.g., *Baizongia pistaciae* [Linnaeus] and *Schlechtendalia chinensis* [Bell]) have been used for centuries in medicine, tanning, and dyeing (Fagan 1918). Woodcuts as early as 1570 illustrate galls of *B. pistaciae* on *Pistacia* (Blackman and Eastop 1994) and by 1596 'Chinese galls' of *Melaphis chinensis* [= *S. chinensis*] on *Rhus javanica* (a sumac) were noted as insect induced (Eastop 1979).

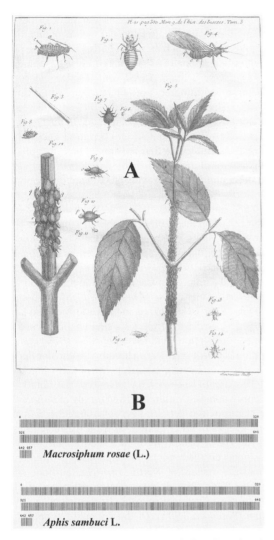

Fig. 20.1 (a) (top). Some of the early aphid work produced by Réaumur (1737) included his woodcuts, which were used for identification. Linnaeus (1758) referenced both of the aphid species in his work. (Figs. 1–4. *Aphis rosae* (= *Macrosiphum rosae*). Figs. 5–15. *Aphis sambuci*.) (b) (bottom). Nearly three centuries later, short DNA sequences from a uniform locality of the genome (barcodes) for *Macrosiphum rosae* and *Aphis sambuci* are being explored

Prior to Linnaeus's (1758) work, many of the papers concerning aphids had little taxonomic value. A notable exception is Réaumur's (1737) 'Mémoires pour servir à l'histore des insectes'. This work included information on aphid life history and biology as well as detailed illustrations. Linnaeus (1758) used Réaumur's (1737) work as a reference in connection with a number of the species he named. The nominal species, *Aphis sambuci* Linnaeus (1758), is illustrated *in habitus* and as several detailed figures on one of Réaumur's (1737) plates (Fig. 20.1a). For aphids (*sensu lato*), Linnaeus (1758) described 1 genus (*Aphis*) and 25 species, as well as the genus *Chermes*, which contained 1 adelgid species.

EARLY APHID STUDIES – A NORTH AMERICAN EXAMPLE

Some of the earliest, if not the earliest, systematic work on North American aphids was that of Rafinesque[1] (1817, 1818), who described 36 species and 4 subgenera. Rafinesque's (1817) interest and intent 'to study all the species of this genus [*Aphis*] found in the United States' was initiated by his observations that they were 'often highly injurious' to their host plants. Other early North American workers often treated the economic importance of aphids. For example, Harris' (1841) report on insects injurious to vegetation includes sections on 'plant-lice'. While little taxonomic information was included, Harris did incorporate observations on life history, biology, predators, host plants, and control. Fitch's work (e.g., Fitch 1851) in the mid-1800s not only included life history information on aphids but also new species descriptions. Between 1851 and 1872, Fitch proposed names for 58 species of Aphidoidea (Barnes 1988). Walsh's (1863) treatment of the Aphididae (*sensu lato*) included his 'Synoptical Table of U.S. Genera', which was essentially a key. By the 1870s–1880s, entomologists such as Thomas (1877), Monell (1879), and Oestlund (1886) were specializing in the study of aphids.

Difficulty in distinguishing aphids was noted as early as Linnaeus (Walsh 1863) and early workers in North America also lamented the lack of study and knowledge of the aphids. Walsh (1863), in Illinois, complained about the need for 'Public Scientific Libraries' that his more fortunate 'Eastern brethren' had. He added that the 'specific distinctions' of the aphids themselves were 'generally evanescent in the dried specimen'. Thomas (1879) reiterated Walsh's comments and believed the reasons for the neglected study of aphids rested on

---

[1] Hottes (1963) proposed to suppress Rafinesques's aphid names and subsequent workers (e.g., Remaudière and Remaudière 1997) have recorded his names as unavailable.

two issues, the difficulty in preserving specimens and the paucity of systematic works, most of which were European. With delays of months to years to procure a reference work (Oestlund 1886), the situation for some was daunting. In 1886, Oestlund still considered the systematics of the aphids 'unsatisfactory' but regarded the lack of literature as the greatest want for the 'frontier naturalist'. Later, Oestlund's (1919) tone changed when his concern focused on the then- recent aphid classification difficulties 'on account of the great number of new genera and species made known'.

Especially noteworthy with Oestlund's (1886) earlier concern about the state of aphid study is his failure to mention the difficulty of preserving specimens. By the 1860s, North Americans, along with their European counterparts, were making progress in preserving pinned insect specimens in cabinets (Sorensen 1995). Instead of being pinned or glued on small boards, aphids were routinely preserved on microscope slides.[2] Changes in the way aphids were being studied and preserved were accompanied by changes and improvements in species descriptions and tools for identification. The earliest North American aphid descriptions (e.g., Rafinesque 1817, 1818; Haldemann 1844) were based almost entirely on coloration, general appearance, and host association. Subsequent workers (e.g., Walsh 1863, Riley 1879, see also Miller et al. 2006) relied on descriptions of general appearance but also routinely included measurements of body length and wing length in their species descriptions. Walsh's (1863) generic descriptions added comparisons to other morphological structures (e.g., relative length of siphunculi in comparison to tarsal length), a practice that was uncommon at the time. By the late

1880s, a new dimension was added to the descriptions of aphid morphology. Relative descriptors of antennal segment length, such as 'about half as long as the preceding' (e.g., Monell 1879) or 'subequal' (e.g., Oestlund 1886), were being replaced with discrete measurements in one-hundredth of a millimeter (e.g., Oestlund 1887). These changes generally corresponded with major advances that were being made in microscopy, especially the development and design of microscope objectives and lenses that maximized both magnification and resolution.

Other changes were taking place in aphid taxonomy in North America. Earliest works reflected Linnaeus's (1758) simple classification (e.g., Haldemann 1844). The list of aphid species referable to Linnaeus's single genus, *Aphis*, was being expanded. The taxonomic works of European aphidologists such as Kaltenbach (1843), Koch (1854), Passerini (1860), and Buckton (1876) influenced the early taxonomic studies of the North American workers (Walsh 1863, Thomas 1877, 1878, 1879). While these works included keys to genera and tribes, species treatments consisted of simple descriptions and lists. Monell's (1879) contribution is worth mentioning because he also developed identification keys for related species of select genera.

The use of species keys advanced the ability to identify aphids. Oestlund's (1887) study of Minnesota aphids provided detailed keys to subfamilies, genera, and species of select genera. In what was a synthesis of the knowledge of North American aphids, he also included detailed species descriptions, an up-to-date literature review of North American authors, and a host plant list. It was also a reflection of the fruits of government-sponsored entomology at that time (Sorensen 1995). Of the 65 publications listed in Oestlund's (1887) aphid bibliography, nearly 75% of the works reflect government-sponsored or government-associated entomology.

A major work published at the beginning of the twentieth century, Hunter's (1901) catalog 'The Aphididae of North America', provided much of the pertinent literature on and taxonomy of North American aphids. Various authors had published lists of described species but Hunter (1901) not only contributed an expanded species list but also included the known systematic and economic literature referable to the species, along with host plant information. Knowledge of North American aphids had grown from 36 species proposed by Rafinesque (1817, 1818) to 166 species identified by Monell (1879), an increase of nearly five

---

[2] Pergande's ledger and card file at the United States Aphidoidea Collection, Beltsville, Maryland, provides an excellent record of the progression of preserved aphid specimens. His earliest ledger entries, while he was in St. Louis, Missouri (1877), record collected aphid specimens as being pinned, mounted on boards, or preserved in alcohol. By 1878 in Washington, DC, he noted aphids as being 'mounted in balsam' and preserved in alcohol. Other aphidologists were also using balsam-mounted specimens, as Pergande noted the receipt of '*Pemphigus aceris* n. sp.' from aphidologist Monell that were 'mounted on slide' in 1878. At the USNM, Pergande was evidently still gluing specimens to pieces of board as late as 1880 but he also 'mounted some on a slide'. In 1903, Pergande recorded that he examined 'the old Fitch collection of Aphides' for *Aphis mali* Fitch, which were 'pinned' and then 'mounted all of them in balsam'. Slide mounting had indeed become a standard way to study and preserve the aphids by the turn of the century.

times, to 325 species identified by Hunter (1901), an increase of more than nine times. The compilation of the North American aphid fauna would continue to grow to 1416 species (Foottit et al. 2006).

RECOGNIZING APHID SPECIES

As the number of recognized aphid species has grown since Linnaeus (1758) (Fig. 20.2), there have been difficulties in recognizing or even defining an aphid species (e.g., Shaposhnikov 1987). The conceptual and operational use of species concepts and definitions in aphid taxonomy throughout the world is complicated by their reproductive biology. Aphids are characterized by cyclical parthenogenesis, but there may also be purely anholocyclic populations not manifesting the sexual phase of the life cycle. Recommendations have been made for the taxonomic treatment of aphid populations that are permanently parthenogenetic;

formal species status could be given to a biologically recognizable anholocyclic group derived from a sexual ancestor (Blackman and Brown 1991, Foottit 1997, Havill and Foottit 2007).

One way to observe trends in aphid systematics is to compare the rates of synonymy and the accumulation of new taxa; as more research is done, more species are described and new synonymies are discovered (Fig. 20.2). From 1758 (when Linnaeus's work was published) until 1840, the number of described valid aphid species (*sensu lato*) and cumulative aphid names was only 109 and the difference between the two parameters remained relatively small. Starting around 1841, shortly before Kaltenbach's (1843) work, the difference between cumulative aphid names and cumulative valid names over time began to increase, albeit gradually, until about the late 1910s. The number of valid species increased nearly eightfold from 129 to 1011 species between 1841 and 1919. Between 1840 and 1949, there were nearly twice as many

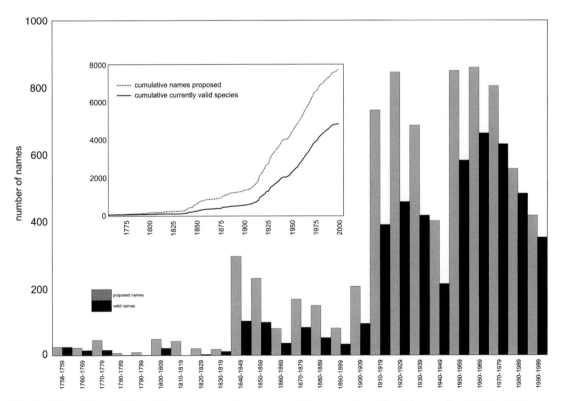

Fig. 20.2 (Inset). Cumulative aphid names proposed versus cumulative currently valid aphid names from 1758 to 2000. (Main). Number of proposed aphid names versus valid aphid names from 1758 to 2000

proposed names as valid names (3891 vs. 2000). During the twentieth century, a steady and dramatic rise in both cumulative and valid names was realized, with the exception of the period of World War II. From 1920 to the present, the number of valid aphid names has increased from 1026 to 4885, an increase of almost five times. Activities slowed in the last several decades of the twentieth century and early twenty-first century showed a slowing of activity. From 1950 to 2000 there were 1.3 times the number of proposed names versus valid names (3495 vs. 2721) (Fig. 20.2 main). This trend might reflect a slowing of activity but it could also reflect a lack of opportunity to reassess previous work. Taxonomic study continues on the Aphidoidea with description of new species and reassessment of old synonymies (Eastop and Blackman 2005). Historically and even recently, the number of aphid species (*sensu lato*) has been estimated (e.g., Oestlund 1886, Kosztarab et al. 1990), but the numbers have always been too low. With close to 5000 valid aphid species currently (e.g., Remaudière and Remaudière 1997), more remains to be done.

Data on the synonymies of aphid species can be further extracted from the most recent world aphid catalogs. Eastop and Hille Ris Lambers's (1976) catalog listed 21 species with 10 or more synonyms (Ilharco and Van Harten 1987). Remaudière and Remaudière's (1997) catalog recorded 28 species with 10 or more synonyms. The five species with the highest number of synonyms, as listed in Remaudière and Remaudière's (1997) catalog, include polyphagous, economically   important   species: *Brachycaudus helichrysi* (Kaltenbach) (47 synonyms), *Aphis gossypii* Glover (42), *Aulacorthum solani* (Kaltenbach) (37), *Aphis fabae* Scopoli (36), and *Myzus persicae* (Sulzer) (32). The large number of synonyms for some species (e.g., *B. helichrysi*) could be explained partly by polyphagy and morphological variation on different hosts (Hunter 1901, Ilharco and Van Harten 1987) or by poor communication between aphid workers (Ilharco and Van Harten 1987). Because aphid morphology is strongly influenced by environmental factors, establishing valid species boundaries and determining synonymies remain problematic (Eastop and Blackman 2005). Other reasons for synonymy could include a lack of funding for follow-up investigations on samples of limited geographic range previously collected or the 'publish or perish' syndrome that might pressure taxonomists into describing new species from small samples (Eastop and Blackman 2005).

THE FOCUS BECOMES FINER

Technological advances in nineteenth-century microscopy were followed by equally significant advances in twentieth-century microbiology and statistical analyses. An early pioneer in aphid genetic work was Nobel Laureate Thomas Hunt Morgan, who is better known for his later work with *Drosophila melanogaster* and establishment of the chromosome theory of heredity. Some of Morgan's (1906) earlier studies included karyological drawings of various phylloxerids. Unfortunately, of the earliest studies on aphid karyology, many are suspect due to uncertainties in proper identification of the respective aphid species (Kuznetsova and Shaposhnikov 1973) and lack of voucher material. Subsequent karyotype studies of aphids have been used at various taxonomic levels. For example, the karyotype is often stable at the generic level, although exceptions do occur (e.g., *Amphorophora*), and in several genera, differences in gross chromosomal morphology are taxonomically important (Blackman 1980a). Importance of the aphid (*sensu lato*) karyotype and chromosomal numbers has been addressed in detailed reviews by several authors (e.g., Kuznetsova and Shaposhnikov 1973; Blackman 1980a, 1980b, 1985; Hales et al., 1997). By the early to mid-1980s, aphid karyotype analysis had moved toward measuring the density of the stained nucleus as a tool for determining DNA content (Blackman 1985). Molecular genetic techniques however were shifting in the late 1980s. DNA restriction fragment length polymorphisms (RFLP) were being applied to aphid taxonomy and systematics (Foottit et al. 1990). The early to mid-1990s saw aphid systematics further benefiting from the use of polymerase chain reaction (PCR) and sequencing techniques (e.g., Sorensen et al. 1995, von Dohlen and Moran 1995). These techniques continue to be refined today (e.g., von Dohlen et al. 2006, Havill et al. 2007, Coeur d'acier et al. 2008).

Morphometrics, the quantitative characterization, analysis, and comparison of biological form (Roth and Mercer 2000), have been used to examine morphological variation in aphids and adelgids and have influenced taxonomic decision making. Several such studies have examined geographic variation and morphological character variation in aphids (e.g., *Pemphigus* spp. by Sokal (1962), Sokal et al. (1980); *Adelges piceae* (Ratzeburg) by Foottit and Mackauer (1980); and *Cinara* spp. by Foottit and Mackauer (1990), Foottit (1992), and Favret and Voegtlin (2004)). Morphometric approaches have been increasingly used to analyze morphological patterns in complexes of pest species combined with analysis of other types of data (e.g., *Myzus* spp. by Blackman (1987), *Rhopalosiphum maidis* (Fitch) by Blackman and Brown (1991), and *Myzus antirrhinii* (Macchiati) by Hales et al. (2000)).

Increasingly, molecular approaches are being used to resolve taxonomic problems throughout the Aphidoidea at all taxonomic levels. The rapid and recent historical development in the use of these techniques includes RFLP (Foottit et al. 1990; Valenzuela et al. 2007), sequencing of nuclear and mitochondrial markers (Havill et al. 2007), microsatellites (Hales et al. 2000), DNA barcoding (Foottit et al. in press; Fig. 20.1b), and other molecular markers (Hales et al. 1997).

The resolution is finer using molecular techniques (Fig.20.1b) but workers are still uncovering problems that require even newer approaches. Evidence suggests that phytophagous insects such as aphids acquire new plant hosts and adapt rapidly to new conditions (Raymond et al. 2001). This results in genetic diversity among aphid populations and even in sibling species making it difficult, if not impossible, to determine this diversity using comparative morphological techniques alone (Eastop and Blackman 2005). A combination of classical approaches and new molecular genetic applications likely will prove necessary to determine the extent of diversity in aphid populations and species complexes (e.g., Lozier et al. 2008). Morphologically indistinguishable species that are differentiated genetically will require a reevaluation of species concepts and the handling of clonal lineages (Foottit 1997). These situations may require a workable nomenclatural system of indexable names for infraspecific taxa (Kim and McPheron 1993).

ADVENTIVE APHID SPECIES

Society depends on agronomic, horticultural, and forest plants for its survival, growth, and development. As phytophagous insects, aphids are intimately tied to their host plants and cause significant economic crop losses through direct feeding damage and transmission of plant viruses. With increased international trade and the consequent increased movement of commodities, the connection between aphids and their hosts has resulted in increased rates of introductions (Foottit et al. 2006). In the absence of natural control measures, some of these aphids have had a major economic impact. In North America alone, in recent years, the establishment of the soy bean aphid, *Aphis glycines* Matsumura, the brown citrus aphid, *Toxoptera citricidus* (Kirkaldy) and the Russian wheat aphid, *Diuraphis noxia* (Kurdjumov), has resulted in millions of dollars in crop losses (Foottit et al. 2006).

Although some notable world treatments address adventive aphids (e.g., Blackman and Eastop 1994, Blackman and Eastop 2000, Blackman and Eastop 2006) as do recent regional taxonomic inventories (e.g., Teulon et al. 2003, Foottit et al. 2006, Mondor et al. 2007), thorough taxonomic analyses of other regional faunas are needed. Where aphid faunas have been developed, the proportion of adventive species is high. For example, percentages of adventive species range from 19% of the North America aphid fauna (Foottit et al. 2006) to 100% of the Hawaiian aphid fauna as adventive (Mondor et al. 2007). Adventive aphids represent an increasing threat in most regions of the world; their detection will require new approaches such as DNA barcoding (Armstrong and Ball 2005).

Historically, questions concerning the taxonomic determination of conspecific aphid species from different biogeographic regions has concerned aphid taxonomists (e.g., Hunter 1901, Foottit et al. 2006). The biogeographic origins of some adventive aphids can be complicated or ill defined. An example of these, among many (Foottit et al. 2006), is that of the woolly apple aphid, *Eriosoma lanigerum* (Hausman). Also known as the American blight, it gained notoriety as an apple pest in Europe where it was considered as originating from America. Although generally considered native to North America (e.g., Smith 1985), its reported origin has long been questioned (Harris 1841, Eastop 1973). The ability to identify the pest aphid species from the source region as well as the region of introduction relies on accurate taxonomic information. We need extensive sampling to encompass the range of aphid variability from different hosts in different regions. To make predictions for possible future introductions and formulate necessary regulations, timely identifications based on sound taxonomic science is critical.

CONCLUSIONS

From examination of the taxonomic history of the Aphidoidea, several conclusions can be drawn. Extensive study at all levels, including faunal studies, revisionary work, and development of a stable classification system for species and genera is needed. Given the complex life cycles of aphids and their parthenogenetic mode of reproduction, taxonomy has to be developed at the infraspecific level.

Society has increasing needs, particularly in understanding processes involving adventive species, managing and protecting crops, and assessing the effects of climate change. Given these needs, it is important to deliver timely and accurate taxonomic information. This delivery can be accomplished most efficiently through an accessible Web-based system.

While aphids might be considered a well-studied pest group, much work remains to be done. Those aphids that garnered Réaumur's (Fig 20.1) attention nearly three centuries ago remain a group in need of study, albeit in finer detail.

REFERENCES

Anonymous. 2004. ARS Crop Protection and Quarantine National Program (304). Action Plan. http://www.ars .usda.gov/research/programs/programs.htm?np_code= 304&docid=355&up [Accessed 22 June 2006].

Armstrong, K. F. and S. L. Ball. 2005. DNA barcodes for biosecurity: Invasive species identification. Philosophical Transactions of the Royal Society B.(Biological Sciences), 360, 813–1823.

Barnes, J. K. 1988. Asa Fitch and the emergence of American entomology, with an entomological bibliography and a catalog of taxonomic names and type specimens. *New York State Museum Bulletin* 461: 1–120.

Blackman, R. L. 1980a. Chromosome numbers in the Aphididae and their taxonomic significance. *Systematic Entomology* 5: 7–25.

Blackman, R. L. 1980b. Chromosomes and parthenogenesis in aphids. Pp. 133–148. *In* R. L. Blackman, G. M. Hewitt, and M. Ashburner (eds). *Insect Cytogenetics.* Symposium of the Royal Entomological Society of London Blackwell Scientific Publications, Oxford, Pp. 10.

Blackman, R. L. 1985. Aphid cytology and genetics. Pp. 171–237. *In* H. Szel giewicz (ed). *Evolution and Biosystematics of Aphids.* Proceedings of the International Aphidological Symposium at Jablonna, 5–11 April 1981. Polska Akademia Nauk Instytut Zoologii. 510 pp.

Blackman, R. L. 1987. Morphological discrimination of a tobacco-feeding form from *Myzus persicae* (Sulzer) (Hemiptera: Aphididae), and a key to New World *Myzus* (*Nectarosiphon*) species. *Bulletin of Entomological Research* 77: 713–730.

Blackman, R. L. and P. A. Brown. 1991. Morphometric variation within and between populations of *Rhopalosiphum maidis* with a discussion of the taxonomic treatment of permanently parthenogenetic aphids (Homoptera: Aphididae). *Entomologia Generalis* 16: 97–113.

Blackman, R. L. and V. F. Eastop. 1994. *Aphids on the World's Trees: An Identification and Information Guide.* CAB International, Wallingford, UK. 1004 pp.

Blackman, R. L. and V. F. Eastop. 2000. *Aphids on the World's Crops: An Identification and Information Guide.* Second Edition. John Wiley & Sons, England. 466 pp.

Blackman, R. L. and V. F. Eastop. 2006. Aphids on the world's herbaceous plants and shrubs. 2 Vols. John Wiley & Sons Ltd., England. 1439 pp.

Buckton, G. B. 1876. Monograph of the British Aphides. *Ray Society, London* 1: 1–193.

Capinera, J. L. 2002. North American vegetable pests: the pattern of invasion. *American Entomologist* 48: 20–39.

Chan, C. K., A. R. Forbes, and D. A. Raworth. 1991. Aphid-transmitted viruses and their vectors of the world. *Agriculture Canada Technical Bulletin* 1991-3E: 1–216.

Coeur d'acier, A., G. Cocuzza, E. Jousselin, V. Cavalieri, and S. Barbagallo. 2008. Molecular phylogeny and systematics in the genus *Brachycaudus* (Homoptera: Aphididae): insights from a combined analysis of nuclear and mitochondrial genes. *Zoologica Scripta* 37: 175–193.

Deerr, N. 1949. *The History of Sugar.* Volume 1. Chapman and Hall, London. 258 pp.

Eastop, V. F. 1973. Aphids and psyllids. Pp. 112–132. *In Viruses and Invertebrates. North-Holland Research Monographs. Frontiers of Biology.* North Holland Publishing, Amsterdam-London. 31.

Eastop, V. F. 1977. Worldwide importance of aphids as virus vectors. Pp. 4–62. *In* K. F. Harris and K. Maramorosh (eds). *Aphids as Virus Vectors.* Academic Press, New York. 559 pp.

Eastop, V. F. 1978. Diversity of the Sternorryncha within major climatic zones. Pp. 71–88. *In* L. A. Mound and N. Waloff (eds). *Diversity of Insect Faunas.* Symposia of the Royal Entomological Society of London Blackwell Scientific Publications, Oxford, Pp. 9.

Eastop, V. F. 1979. Sternorrhyncha as angiosperm taxonomists. *Symbolae Botanicae Upsaliensis* 22(4): 120–134.

Eastop, V. F. and R. L. Blackman. 2005. Some new synonyms in Aphididae (Hemiptera: Sternorrhyncha). *Zootaxa* 1089: 1–36.

Eastop, V. F. and D. Hille Ris Lambers. 1976. *Survey of the World's Aphids.* Dr. W. Junk, Publishers, The Hague. 573 pp.

Fagan, M. M. 1918. The uses of insect galls. *American Naturalist* 52: 155–176.

Faria, C. A. 2005. The nutritional value of aphid honeydew for parasitoids of lepidopteran pests. Doctorate of Natural Sciences dissertation. University of Neuchâtel 124 pp. http://doc.rero.ch/lm.php?url=1000,40,4,20051104114823-UY/1_these_FariaA.pdf [Accessed 6 September 2006].

Favret, C. and D. J. Voegtlin. 2004. Host-based morphometric differentiation in three *Cinara* species (Insecta: Hemiptera: Aphididae) feeding on *Pinus edulis* and *P.* monophylla. *Western North American Naturalist* 64: 364–375.

Fitch, A. 1851. Catalogue with references and descriptions of the insects collected and arranged for the State Cabinet of Natural History. Pp. 43–69. *In* Fourth Annual Report of the Regents of the University, on the condition of the State Cabinet of Natural History, and the historical and antiquarian collection, annexed thereto. Made to the Senate, January 14, 1851. Albany. 146 pp.

Foottit, R. G. 1992. The use of ordination methods to resolve problems of species discrimination in the genus *Cinara* Curtis [Homoptera: Aphidoidea: Lachnidae]. Pp. 192–221. *In* J. T. Sorensen and R. G. Foottit (eds). *Ordination in the Study of Morphology, Evolution and Systematics of Insects. Applications and Quantitative Rationales.* Elsevier, Amsterdam. 418 pp.

Foottit, R. G. 1997. Recognition of parthenogenetic insect species. Pp. 291–307. *In* M. F. Claridge, H. A. Dawah, and M. R. Wilson (eds). *Species: The Units of Biodiversity.* Chapman and Hall, London. 439 pp.

Foottit, R. G. and M. Mackauer. 1980. Morphometric variation between populations of the balsam woolly aphid, *Adelges piceae* (Ratzburg) (Homoptera: Adelgidae), in North America. *Canadian Journal of Zoology* 58: 1494–1503.

Foottit, R. G. and M. Mackauer. 1990. Morphometric variation within and between populations of the pine aphid, *Cinara nigra* (Wilson) (Homoptera: Aphidoidea: Lachnidae), in western North America. *Canadian Journal of Zoology* 68: 1410–1419.

Foottit, R. G., G. Galvis, and L. Bonen. 1990. The application of mitochondrial DNA analysis in aphid systematics. *Acta Phytopaphologica et Entomologica Hungarica* 25(1–4): 211–219.

Foottit, R. G., S. E. Halbert, G. L. Miller, H. E. L. Maw, and L. M. Russell. 2006. Adventive aphids (Hemiptera: Aphididae) of America North of Mexico. *Proceedings of the Entomological Society of Washington* 108: 583–610.

Foottit, R. G., H. E. L. Maw, C. D. von Dohlen, and P. D. N. Hebert. Species identification of aphids (Insecta: Hemiptera: Aphididae) through DNA barcodes. *Molecular Ecology Resources.* 8: 1189–1201.

Garner, W. W. 1946. *The Production of Tobacco.* The Blakiston Company, Philadelphia, Pennsylvania. 516 pp.

Glen, D. M. 1973. The food requirements of *Blepharidopterus angularus* Heteroptera Miridae as a predator of the lime aphid *Eucallipterus tiliae.* Entomologia Experimentalis et Applicata 16(2): 255–267.

Haldemann, S. S. 1844. Descriptions of several species of *Aphis* inhabiting Pennsylvania. *Proceedings of the Boston Society of Natural History* 1: 168–169.

Hales, D. F., J. Tomiuk, K. Wöhrmann, and P. Sunnucks. 1997. Evolutionary and genetic aspects of aphid biology: a review. *European Journal of Entomology* 94: 1–55.

Hales, D., A. C. C. Wilson, J. M. Spence, and R. L. Blackman. 2000. Confirmation that *Myzus antirrhinii* (Macchiati) Hemiptera: Aphididae occurs in Australia, using morphometrics, microsatellite typing and analysis of novel karyotypes by fluorescence *insitu* hybridisation. *Australian Journal of Entomology* 39: 123–129.

Harpaz, I. 1973. Early entomology in the Middle East. Pp. 21–36. *In History of Entomology.* Annual Reviews, Palo Alto, California. 517 pp.

Harris, T. W. 1841. 2. Plant-lice. (Aphididae). Pp. 186–198. *In* A Report on the Insects of Massachusetts Injurious to Vegetation. Folsom, Wells, and Thurston. Cambridge, Massachusetts 459 pp.

Havill, N. P. and R. G. Foottit. 2007. Biology and evolution of Adelgidae. *Annual Review of Entomology* 52: 325–349.

Havill. N. P., R. G. Foottit, and C. D. von Dohlen. 2007. Evolution of host specialization in the Adelgidae (Insecta: Hemiptera) inferred from molecular phylogenetics. *Molecular Phylogenetics and Evolution.* 44: 357–370.

Hottes, F. C. 1963. Aphid names of Rafinesque: proposed suppression under the plenary powers (Insects, Hemiptera, Aphididae). Z. N. (S.) 327, *Bulletin of Zoological Nomenclature* 20: 128–133.

Howard, L. O. 1898. Danger of importing insect pests. Yearbook of the United States Department of Agriculture 1897, pp. 529–52. Government Printing Office, Washington, DC.

Hunter, W. D. 1901. The Aphididae of North America. *Iowa Agriculture College Experiment Station Bulletin* 60: 63–138.

Huxley, A. 1978. *An Illustrated History of Gardening.* Paddington Press, New York. 252 pp.

Ilharco, F. A. and A. Van Harten. 1987. Systematics. Pp. 51–77. *In* A. K. Minks and P. Harrewijn (eds). *Aphids: Their Biology, Natural Enemies and Control.* Volume A. World Crop Pests 2A. Elsevier Science Publishers, Amsterdam, The Netherlands. 450 pp.

Kaltenbach, J. H. 1843. Monographie der Familien der Pflanzenläuse. (Phytophthires.). *Aachen* 1843: 1–223.

Kim, K. C. and B. A. McPheron. 1993. Biology of variation: epilogue. Pp. 453–468. *In* K. C. Kim and B. A. McPheron (eds). *Evolution of Insect Pests: Patterns of Variation.* John Wiley & Sons, New York. 479 pp.

Konishi, M. and Y. Ito. 1973. Early entomology in East Asia. Pp. 1–20. *In History of Entomology.* Annual Reviews Inc. Palo Alto, California. 517 pp.

Koch, C. L. 1854. Die Pflanzenläuse Aphiden, getreu nach dem Leben abgebildet und beschrieben. *Nürnberg.* Hefts I-IV: 1–134.

Kosztarab, M., L. B. O'Brien, M. B. Stoetzel, L. L. Dietz, and P. H. Freytag. 1990. Problems and needs in the study of Homoptera in North America. Pp. 119–145. *In* M. Kosztarab and C. W. Schaefer (eds). *Systematics and the North American Insects and Arachnids: Status and Needs.* Virginia Agricultural Experiment Station Information Series Blacksburg, Virginia 90-1.

Kuznetsova, V. G. and G. K. Shaposhnikov. 1973. The chromosome numbers of the aphids (Homoptera: Aphidinea) of the world fauna. *Entomological Review, Washington* 52: 78–96.

Linnaeus, C. 1758. II. Hemiptera. Systema Naturae per regna tria naturae, secundum clases, ordines, genera, species, cum characteribus, differentiis, synonymis, locis. *Editio decima, reformata* 1: 1–824.

Lozier, J. D., R. G. Foottit, G. L. Miller, N. J. Mills, and G. K. Roderick. 2008. Molecular and morphological evaluation of the aphid genus *Hyalopterus Koch* (Insecta: Hemiptera: Aphididae), with a description of a new species. *Zootaxa* 1688: 1–19.

Miller, G. L., E. C. Kane, J. Eibl, and R. W.Carlson. 2006. Resurrecting Asa Fitch's aphid notes: historical entomology for application today. http/ars.usda.gov/serices/doc.htm?docid=125851 [Accessed 22 June 2006].

Mondor, E. B., M. N. Tremblay, and R. H. Messing. 2007. Morphological and ecological traits promoting aphid colonization of the Hawaiian Islands. *Biological Invasions* 9: 87–100.

Monell, J. 1879. Notes on the Aphididae of the United States, with descriptions of species occurring west of the Mississippi. Part II. Notes on Aphidinae, with descriptions of new species. *Bulletin of the United States Geological and Geographical Survey of the Territories* 5: 18–32.

Morgan, T. H. 1906. The male and female eggs of the phylloxerans of the hickories. *Biological Bulletin* 10: 201–206.

Oestlund, O. W. 1886. List of the Aphididae of Minnesota, with descriptions of some new species. *Annual Report of the Geological Natural History Survey of Minnesota* 14: 17–56.

Oestlund, O. W. 1887. Synopsis of the Aphididae of Minnesota. *Geological and Natural History Survey of Minnesota Bulletin* 4: 1–100.

Oestlund, O. W. 1919. Contributions to knowledge of the tribes and higher groups of the family Aphididae (Homoptera). *Report of the State Entomologist of Minnesota* 17: 46–72.

Passerini, G. 1860. Gli afidi con un prospetto dei generi ed alcune specie nuove Italiane. *Parma* 1860: 1–40.

Rafinesque, C. S. 1817. Specimens of several new American species of the genus *Aphis*. *American Monthly Magazine* 1: 360–361.

Rafinesque, C. S. 1818. Second memoir on the genus *Aphis*, containing the descriptions of 24 new American species. *American Monthly Magazine* 3: 15–18.

Raymond, B., J. B. Searle, and A. E. Douglas. 2001. On the process shaping reproductive isolation in aphids of the *Aphis fabae* (Scop.) complex (Aphididae: Homoptera). *Biological Journal of the Linnaean Society* 74: 205–215.

Réaumur, R. P. 1737. *Mémoires pour servir a l'histore des insectes. Tome troisieme.* de l' Imprimerie Royale Paris. 532 pp. + plates.

Remaudière, G. and M. Remaudière. 1997. *Catalogue des Aphididae du monde (Homoptera Aphidoidea).* INRA Editions, Paris. 473 pp.

Riley, C. V. 1879. Notes on the Aphididae of the United States, with descriptions of species occurring west of the Mississippi. Part I. Notes on the Pemphiginae, with descriptions of new species. *Bulletin of the United States Geological and Geographical Survey of the Territories* 5(1): 1–17.

Roth, V. L. and J. M. Mercer. 2000. Morphometrics in development and evolution. *American Zoologist* 40: 801–810.

Shaposhnikov, G. C. 1987. Organization (structure) of populations and species, and speciation. Pp. 415–432. *In* A. K. Minks and P. Harrewijn (eds). *Aphids. Their Biology, Natural Enemies and Control.* Volume A. Elsevier, Amsterdam. 450 pp.

Smith, C. F. 1985. *Pemphiginae* in North America. Pp. 277–302. *In* H. Szel giewicz (ed). *Evolution and biosystematics of aphids. Proceedings of the International Aphidological Symposium at Jablonna,* 5–11 April, 1981. Polska Akademia Nauk Instytut Zoologii. 510 pp.

Sokal, R. R. 1962. Variation and covariation of characters of alate *Pemphigus populi-transversus* in eastern North America. *Evolution* 16: 227–245.

Sokal, R. R., J. Bird, and B. Riska. 1980. Geographic variation in *Pemphigus populicaulis* (Insecta: Aphididae) in eastern North America. *Biological Journal of the Linnean Society* 14: 163–200.

Sorensen, J. T., B. C. Cambell, R. J. Gill, and J. D. Steffen-Cambell. 1995. Non-monophyly of Auchenorrhyncha ("Homoptera"), based upon 18S rDNA phylogeny: eco-evolutionary and cladistic implications within pre-Heteropterodea Hemiptera (s.l.) and a proposal for new monophyletic suborders. *Pan-Pacific Entomologist* 71: 431–460.

Sorensen, W. C. 1995. *Brethren of the Net: American Entomology, 1840–1880.* University of Alabama Press, Tuscaloosa. 357 pp.

Teulon, D. A. J., V. F. Eastop, and M. A. W. Stufkens. 2003. Aphidoidea – aphids and their kin. *In* D. P. Gordon (ed). *The New Zealand Inventory of Biodiversity: A Species 2000 Symposium Review.* Canterbury University Press, Christchurch, New Zealand.

Thomas, C. 1877. *Notes of the Plant-lice found in the United States.* Transactions of the Illinois State Horticultural Society, 1876. Pp. 137–212.

Thomas, C. 1878. A list of the species of the tribe Aphidini, family Aphidae, found in the United States, which have been heretofore named with descriptions of some new species. *Illinois State Laboratory Natural History Bulletin* 2: 3–16.

Thomas, C. 1879. Noxious and beneficial insects of the state of Illinois. *Report of the State Entomologist (Illinois)* 8: 1–212.

Valenzuela, I., A. A. Hoffman, M. B. Malipatil, P. M. Ridland, and A. R. Weeks. 2007. Identification of aphid species (Hemiptera: Aphididae: Aphidinae) using a rapid polymerase chain reaction restriction fragment length polymorphism method based on the *cytochrome oxidase* subunit I gene. *Australian Journal of Entomology* 46: 305–312.

Von Dohlen, C. D. and N. A. Moran. 1995. Molecular phylogeny of the Homoptera: a paraphyletic taxon. *Journal of Molecular Evolution* 41: 211–223.

Von Dohlen, C. D., C. A. Rowe, and O. E. Heie. 2006. A test of morphological hypotheses for tribal and subtribal relationships of Aphidinae (Insecta: Hemiptera: Aphididae) using DNA sequences. *Molecular Phylogenetics and Evolution* 38: 316–329.

Walsh, B. D. 1863. On the genera of Aphidae found in the United States. *Proceedings of the Entomological Society of Philadelphia* 1: 294–311.

Westcott, C. 1971. Plant disease handbook. Van Nostrand Reinhold Co., New York. 843 pp.

Worf, G. L., M. F. Heimann, and P. J. Pellitteri. 1995. *Sooty mold*. University of Wisconsin Cooperative Extension Publication Madison Wisconsin A2637. 2 pp.

Wellings, P. W., S. A. Ward, A. F. G. Dixon, and R. Rabbinge. 1989. Crop loss assessment. Pp. 49–64. *In* A. K. Minks and P. Harrewijn (eds). *Aphids. Their Biology, Natural Enemies and Control.* Volume C. Elsevier, Amsterdam. 312 pp.

Chapter 21

ADVENTIVE (NON-NATIVE) INSECTS: IMPORTANCE TO SCIENCE AND SOCIETY

Alfred G. Wheeler, Jr.[1] *and E. Richard Hoebeke*[2]

[1]Department of Entomology, Soils & Plant Sciences, Clemson University, Box 340315, 114 Long Hall, Clemson, South Carolina 29634-0315 USA
[2]Department of Entomology, Cornell University, Ithaca, New York 14853-2601 USA

Insect Biodiversity: Science and Society, 1st edition. Edited by R. Foottit and P. Adler
© 2009 Blackwell Publishing, ISBN 978-1-4051-5142-9

Much of invasion biology's conceptual framework rests on Darwinian thought (Williamson 1996, Ludsin and Wolfe 2001). An awareness of immigrant species in North America predates Darwin's work (Inderjit et al. 2005). Before Darwin (1859) published his treatise on the origin of species, entomologists had warned about the establishment of European plant pests in the New World, a concern motivated by a desire to protect agriculture from foreign pests rather than to conserve native biodiversity. George Marsh, however, was aware not only of the presence of immigrant insects in the USA but also other human-induced changes to the environment. His book *Man and Nature* (Marsh 1864) 'revolutionized environmental thought' (Lowenthal 1990) and presaged the disciplines of conservation biology and invasion ecology.

Two classic works inspired interest in adventive species: Elton's (1958) *The Ecology of Invasions by Animals and Plants*, which initiated the science of invasion biology (Parker 2001), and *The Genetics of Colonizing Species* (Baker and Stebbins 1965). The books differ in their emphasis. Elton's book deals mainly with faunal history, population ecology, and conservation. The book Baker and Stebbins edited stresses evolutionary rather than ecological issues and does not address the effects of adventive species on environmental conservation (Davis 2006).

Bates (1956) examined the role of humans as agents in dispersing organisms ranging from microbes to vertebrates. He observed that anthropogenic influences, such as modification of environmental factors and movement of organisms, offer opportunities for experimental studies that could contribute to issues in theoretical ecology and clarify evolutionary mechanisms. Invasion biology assumed prominence during the 1980s (Kolar and Lodge 2001, Davis 2006), receiving impetus from the Scientific Committee on Problems of the Environment of the International Council of Scientific Unions and its early symposia on biological invasions (Macdonald et al. 1986, Mooney and Drake 1986, Drake et al. 1989). Invasion biology now plays a central role in biotic conservation, and invasive species are used as tools for biogeographic, ecological, and evolutionary research (Vitousek et al. 1987, Williamson 1999, Sax et al. 2005, Davis 2006).

An increased mobility of humans and their commodities, coupled with human-induced habitat disturbances, enables plants and animals to breach once insurmountable geographic barriers and become established in distant lands and waters (Soulé 1990, Mack et al. 2000, Mooney and Cleland 2001). Human colonization has increased the geographic scope, frequency, and taxonomic diversity of biotic dispersal (U.S. Congress 1993, Vitousek et al. 1997, Mack et al. 2000). A global estimate of the number of adventive species, including microbes, approaches a half million (Pimentel et al. 2001). The spread of adventive organisms ranks only behind habitat destruction as the greatest threat to biodiversity (Wilson 1992, Wilcove et al. 1998).

Invasion biology, featured in both scientific and popular writings (Simberloff 2004), is fraught with misconceptions and characterized by polemical writing, emotionalism, and controversy. Debate continues over such issues as the patterns and processes affecting the movement and success of invaders, invasibility of mainland areas compared to islands, ecological consequences of invaders, and relative importance of direct compared to indirect effects on ecosystems. Should biological invasions be viewed generally as part of ecological change, and as enriching, rather than impoverishing biodiversity? Whether all invasions should be considered bad and whether a global decline in biodiversity necessarily is bad (Lodge 1993b) depend, in part, on perspective: scientific, or moral and social (Brown and Sax 2004; *cf.* Cassey et al. 2005). The extent to which the effects of invaders are tempered over time and current ecological changes resolved through evolution and succession in the new ecosystems also is uncertain (Daehler and Gordon 1997, Morrison 2002, Strayer et al. 2006).

We cannot treat all facets of a field as diverse and complex as invasion biology or all cultural, ethical, historical, management, philosophical, political, psychological, and socioeconomic aspects of the invasive-species problem. We treat adventive insects that are immigrant (not deliberately introduced) or introduced (deliberately so). Our coverage emphasizes North America. Examples deal mainly with human-assisted movement of insects between countries, even though intracountry changes in range are common among immigrant taxa (e.g., the glassywinged sharpshooter (*Homalodisca vitripennis*) and western corn rootworm (*Diabrotica virgifera*) within the USA). Such range extensions can be as detrimental as those between countries (Simberloff 2000, McKinney 2005). We exclude immigrants that arrive on their own by active flight or passive conveyance on convective air currents (Southwood 1960), on strong winds associated with El Niño events (Roque-Albelo and

Causton 1999), or are able to spread as the result of global climate change (Burckhardt and Mühlethaler 2003, Musolin and Fujisaki 2006, Musolin 2007).

TERMINOLOGY

Invasion ecology's status and public appeal is due partly to its emotive and militaristic language, including the words *alien, exotic,* and *invader* (Colautti and MacIsaac 2004, Larson 2005, Coates 2006). An emphasis on 'headline invaders' (Davis et al. 2001) also has contributed to the discipline's prominence. Elton (1958) did not define the terms *invader* and *invasion,* which permeate the literature on invasive species (Richardson et al. 2000, Rejmánek et al. 2002). Terms relating to the concept of 'not native' are used interchangeably, even though they are not strictly synonymous (Simberloff 1997, Mack et al. 2000, Sax et al. 2005); nonnative species are designated as adventive, alien, exotic, immigrant, or introduced, sometimes in the same paper (e.g., Sailer 1978, Devine 1998, Clout 1999). *Newcomer,* a more neutral term than *invader,* has gained some recent favor (Coates 2006, Acorn 2007). The term *neozoa* is used mainly in the European literature to refer to nonnative animals intentionally or unintentionally introduced since 1492 (Occhipinti-Ambrogi and Galil 2004, Rabitsch and Essl 2006).

Entomologists have not been as involved as botanists and plant ecologists in trying to clarify terminology. Zimmerman (1948) categorized insects not native to Hawaii as either 'immigrant', unintentionally brought in by humans, or 'introduced', for instance, for biological control. Frank and McCoy (1990) similarly reserved *introduced* for species deliberately introduced, and used *immigrant* for hitchhikers and stowaways, as well as species that disperse under their own power. Atkinson and Peck (1994), however, regarded bark beetles that have colonized southern Florida by natural dispersal – immigrant according to Frank and McCoy's (1990) terminology – as native. It can be impossible to determine if the arrival of even clearly adventive species involved deliberate human intervention (Simberloff 1997). Certain predators and parasitoids introduced for biocontrol were already established, but undetected, as immigrants at the time of their release (Frick 1964, Turnbull 1979, 1980).

We distinguish adventive taxa as either immigrant or introduced (Frank and McCoy 1990, 1995b; Frank 2002), and follow Cowie and Robinson (2003) by using *vector* to refer to the vehicle or mechanism that transports a species and *pathway* for the activity or purpose by which a species is introduced (*cf.* Carlton and Ruiz 2005). Table 21.1 gives definitions of these and other key terms used herein.

DISTRIBUTIONAL STATUS: NATIVE OR ADVENTIVE?

Immigrant insects typically are associated with disturbed habitats but can be found in relatively pristine communities and in isolated areas (Wheeler 1999, Klimaszewski et al. 2002, Gaston et al. 2003). Whether a species should be considered native or adventive can be problematic (Claassen 1933, Buckland 1988, Whitehead and Wheeler 1990, Woods and Moriarty 2001). An apparent immigrant of restricted geographic range in its area of invasion poses a conservation dilemma if eradication of the potentially ecologically disruptive species is considered; an effort to resolve distributional status should be made before any attempt is made to eliminate what actually might be a rare precinctive ('endemic') species (Deyrup 2007).

Distributional status is particularly difficult to evaluate in the case of vertebrate ectoparasites, pests of stored products, certain ants, cockroaches, and other cosmopolitan insect groups (Buckland et al. 1995, McGlynn 1999, Kenis 2005). By the late eighteenth century, the honeybee (*Apis mellifera*) had become so common in the USA that it appeared to be native to the New World (Sheppard 1989). The distributional status – native or immigrant – of major North American pests (Webster 1892), such as the Hessian fly (*Mayetiola destructor*) (Riley 1888, Pauly 2002), as well a species officially listed as endangered in the UK (Samways 1994), remains in doubt. Certain insects once thought to be Holarctic likely are immigrant in North America (Turnbull 1979, 1980; Wheeler and Henry 1992). Certain immigrant insects have been thought initially to represent new species (e.g., Thomas et al. 2003) or have been described as new. Thus, a species is not necessarily native to the continent or island from which it was described (Cox and Williams 1981, Green 1984, Gagné 1995). The status of certain insects described from North America can be immigrant, the species being conspecific with

Table 21.1 Some key terms as used in this chapter.

Term	Definition	Comments
Adventive	Not native (adj.)	More neutral term than *alien* or *exotic*, encompassing both immigrant and introduced species (Frank and McCoy 1990); in botanical literature, can refer to nonnative species only temporarily established (Novak and Mack 2001)
Immigrant	Nonnative species not deliberately or intentionally introduced (n.); pertaining to species not deliberately introduced (adj.)	Accidentally or unintentionally introduced (Sailer 1978, McNeely et al. 2001); includes species arriving on their own (Frank and McCoy 1990; *cf.* Atkinson and Peck 1994): 'true immigrants' (*sensu* Simberloff 2003)
Introduced	Pertaining to nonnative species deliberately or intentionally introduced (adj.)	Sometimes used broadly to refer to all nonnative species (e.g., Simberloff 2003)
Invasive	Pertaining to species that cause socioeconomic or environmental damage or impair human health (adj.)	Variously defined term and subjective, value-based measurement (Hattingh 2001, Ricciardi and Cohen 2007); sometimes applied to any nonnative species
Pathway	Purpose or activity for which adventive species are introduced, either intentionally or unintentionally (n.)	Follows Cowie and Robinson (2003); for alternative uses, see Richardson et al. (2003), Carlton and Ruiz (2005)
Precinctive	Pertaining to a native species known from no other area (adj.)	More restrictive term than indigenous; often misused for endemic (Frank and McCoy 1990)
Vector	Mechanism or vehicle (physical agent) by which adventive species are transported (n.)	Follows Cowie and Robinson (2003), Ruiz and Carlton (2003); 'pathway' often is used to refer to both pathways and vectors (Carlton and Ruiz 2005)

previously described Old World species (Wood 1975, Wheeler and Henry 1992, Booth and Gullan 2006). Certain Eurasian species in North America should be regarded as native to the Pacific Northwest but immigrant in the Northeast (Lindroth 1957, Turnbull 1980, Sailer 1983). An anthocorid bug (*Anthocoris nemoralis*), apparently immigrant in the Northeast, was introduced for biological control in western North America (Horton et al. 2004).

Lindroth (1957) discussed historical, geographic, ecological, biological, and taxonomic criteria useful in evaluating distributional status. His criteria are particularly appropriate for the North Atlantic region (Sadler and Skidmore 1995), including Newfoundland, Canada, which has received numerous western Palearctic insects, often via ships' ballast (Lindroth 1957, Wheeler and Hoebeke 2001, Wheeler et al. 2006). Certain species likely are immigrant in North America even though they do not meet any of Lindroth's (1957) criteria of immigrant status (Turnbull 1979). The ten criteria used to assess the distributional status of a marine crustacean (Chapman and Carlton 1991) are also appropriate for terrestrial insects. The accuracy of criteria used to resolve distributional

status depends on how well the bionomics of the insects in question are known (Turnbull 1980). To address long-standing questions about the origin of certain immigrant pests (Howard 1894), molecular evidence can be used to identify the geographic sources of adventive insects and, in some cases, to detect overlapping or sequential invasions (e.g., Tsutsui et al. 2001, Miller et al. 2005, Austin et al. 2006).

Biogeographers and ecologists often consider a species native if information is insufficient to resolve its distributional status but do so with unwarranted confidence (Carlton 1996). Whitehead and Wheeler (1990) suggested the opposite approach: when in doubt, consider the species adventive ('nonindigenous'). The term *cryptogenic* refers to species that demonstrably are neither native nor adventive (Carlton 1996).

EARLY HISTORY OF ADVENTIVE INSECTS IN NORTH AMERICA

Other organisms accompanied *Homo sapiens* during each major invasion: from Africa, where humans apparently evolved, to Eurasia; thence to Australia, the Americas, and, eventually, to the far reaches of the Pacific (McNeely 2001b, 2005). Lice might have been the first insects to have been transported (Laird 1984). As early as the ninth century, the Norse colonists were responsible for the establishment of European insects in Greenland (Sadler 1991). Insects likely arrived in the New World with landfall by Columbus, who 'mixed, mingled, jumbled, and homogenized the biota of our planet' (Crosby 1994). Insects probably arrived in North America with the *Mayflower*'s landing in 1620 and continued to enter with every ship that brought additional people and supplies from Europe (Sailer 1978, 1983). Earlier, the house fly (*Musca domestica*) might have been brought to tropical latitudes of the Western Hemisphere via canoe or raft by pre-Columbian inhabitants of Central or South America (Legner and McCoy 1966). Outside North America, the Polynesians who colonized Hawaii in prehistoric times might have brought with them a few insects, such as vertebrate ectoparasites, the house fly, and a cockroach (*Balta similis*) (Gagné and Christensen 1985, Beardsley 1991).

Early-arriving insects in the USA and elsewhere mostly were those that could survive a several-month sea voyage under adverse physical conditions: associates of stored products, ectoparasites and blood suckers of humans and their livestock, inhabitants of their excrement, and soil dwellers in dry ballast brought aboard sailing ships (Lindroth 1957, Sailer 1978, Turnbull 1979, Buckland et al. 1995). For many other insect groups, long sea voyages functioned as inadvertent quarantines years before formal quarantines were adopted (Gibbs 1986).

Among early immigrants in the Northeast were the bed bug (*Cimex lectularius*), head (and body) louse (*Pediculus humanus*), and oriental cockroach (*Blatta orientalis*) (Sasscer 1940, Sailer 1978, 1983). Pestiferousness of the stable fly (*Stomoxys calcitrans*) perhaps led to hasty adoption of the Declaration of Independence (Kingsolver et al. 1987). Crop pests generally were not among the early arrivals (Sailer 1978), although archaeological evidence has shown that certain pests were present much earlier than once thought (Bain and LeSage 1998). On the West Coast, where eighteenth-century agriculture was limited to Spanish missions in southern California, various weevils entered with food and seed cereals. Livestock ships brought several species of muscoid flies to California; fur, hide, tallow, and whaling ships allowed additional stored-product insects to enter (Dethier 1976).

Thirteen immigrant insects apparently became established in the USA by 1800 (McGregor et al. 1973), with the total of adventive species about 30 (Sailer 1983); Simberloff's (1986) total of 36 appears to include the mite species noted by Sailer. Numerous other immigrant insects, common in England but not detected in North America until after 1800, probably were present by the eighteenth century (Sasscer 1940, Sailer 1983). Even with establishment of European plant pests such as the codling moth (*Cydia pomonella*), Hessian fly (*Mayetiola destructor*), oystershell scale (*Lepidosaphes ulmi*), and pear sawfly (*Caliroa cerasi*), neither native nor immigrant insects were particularly problematic in the Northeast before 1800 (Dethier 1976, Sailer 1978, 1983). Pest outbreaks did occur, and cessation of burning by the Native Americans favored additional problems from insects (Cronon 1983). Yet, crops in the American colonies mostly were free from the insects that plague modern agriculture (Popham and Hall 1958). The minimal damage from insects was due partly to a lack of extensive and intensive crop production and scarcity of immigrant pests; damage might have been greater because some crop losses went unrecognized (Davis 1952). Drought, however, was the main enemy of colonial farmers (Dethier 1976). Before the nineteenth century, soil exhaustion and limiting socioeconomic conditions also remained more important than insects as deterrents to agriculture (Barnes 1988).

The subsistence-level agriculture of early European settlers, and their cultivation of crops such as corn and squash, did little to disrupt evolutionary relationships between plants and insects (Dethier 1976, Barnes 1988). Increasingly, however, insects emerged as consistent agricultural pests. It is not true, as a politician in a Western state contended, that the USA lacked destructive insects until the country had entomologists (Webster 1892).

Increasing trade with Europe, more rapid means of transportation, planting of additional European crops, and expanding crop acreages favored the arrival and establishment of new insects from Europe. As humans altered ancient relationships between plants and insects, they unwittingly ushered in an era of immigrant pests (Dethier 1976). In roughly 200 years, a 'new ecology' had been created, setting the stage for damage by native insects and the entry of additional European species (Barnes 1988). The slow accumulation of immigrant insects, lasting until about 1860 (Sailer 1978), gave way to a 'continuous, persistent procession' of immigrants (Herrick 1929).

NUMBERS, TAXONOMIC COMPOSITION, AND GEOGRAPHIC ORIGINS OF ADVENTIVE INSECTS

No continent or island is immune to invasion by immigrant organisms. Changes in transport technology and types of commodities transported have affected the predominance of certain groups of insects at different time periods in every world region. Our discussion of the immigrant insect fauna will emphasize the 48 contiguous U.S. states, a principal focus of Sailer's (1978, 1983) seminal work. The specific composition of the U.S. adventive insect fauna is influenced not only by changes in pathways, vectors, and trade routes, but also by the kinds of natural enemies imported for biocontrol, availability of taxonomists who can identify insects in particular families, taxonomic bias in the kinds of insects collected in detection surveys, and changes in quarantine-inspection procedures.

Faunal lists are important but labor intensive and time consuming; thus, few, up-to-date, comprehensive inventories or databases of immigrant insects exist for most world regions. Lists of adventive taxa, however, allow an analysis of the numbers of established species, systematic composition of the most successful groups, and geographic origins, as well as comparisons among biogeographic regions. From data on immigrant insect faunas, we focus on selected countries on different continents, oceanic islands or atolls, and several agriculturally important U.S. states (Table 21.2). The immigrant insect fauna of the USA is one of the best studied and most thoroughly documented (McGregor et al. 1973, Sailer 1978, 1983). Of more than 3500 species of adventive arthropods that reside in the continental USA (Frank and McCoy 1992), more than 2000 insect species are established, representing about 2–3% of the insect fauna (U.S. Congress 1993).

Sailer's (1978, 1983) analysis revealed several major trends among adventive insects in the U.S. fauna. The ubiquitous use of dry ballast during the era of early sailing ships (seventeenth to early nineteenth century) was responsible for an early dominance of Coleoptera. At British ports, beetles and ground-dwelling bugs were common in dry ballast (Lindroth 1957). After the Civil War, Homoptera and Heteroptera arrived with nursery stock and other plant material from western

Europe. Sailer (1978) determined that homopterans contributed the largest number of adventive species, but in his later analysis, hymenopterans, with approximately 390 species (23%), predominated (Sailer 1983). The introduction of numerous parasitic wasps to help control adventive pest arthropods and weeds was thought responsible for the disproportionate increase in hymenopterans. Next in abundance of species were the Coleoptera (372), Homoptera (345), Lepidoptera (134), and Diptera (95). Immigrant arthropods in the USA originate mainly from the western Palearctic (66.2%), followed by the Neotropical (14.3%), and eastern Palearctic and Oriental Regions (13.8%) (Sailer 1983).

Florida has the highest percentage of adventive insects in the conterminous USA (only Hawaii has a higher percentage); about 1000 such species are established, representing about 8% of Florida's insect fauna (Frank and McCoy 1995a). Whereas many of the species entered with commerce (e.g., as stowaways in plant material), others arrived by aerial dispersal from Caribbean islands (Cox 1999). Although Floridian immigrants originate from many world regions, they arrive mainly from the Neotropics and Asia.

Beetles were best represented (~26%) among the 271 species of immigrant insects newly recorded from Florida from 1970 to 1989. Coleoptera were followed by Lepidoptera (~19%), Hymenoptera (~15%), and Homoptera (~13%) (Frank and McCoy 1992). A similar study in Florida for 1986 to 2000 listed 150 adventive insects (Thomas 2006). Unlike Frank and McCoy's (1992) study, Homoptera contributed the largest number of species (~35%), followed by Coleoptera (~26%). The proportion of Coleoptera was similar, but that of Homoptera increased substantially (~13% to ~35%), while that of Lepidoptera declined (~19% to ~3%). From 1970 to 1989 the majority of immigrant species in Florida arrived from the Neotropics (~65%) (Frank and McCoy 1992). In contrast, from 1986 to 2000, the number of Asian immigrants increased substantially (~50%) (Thomas 2006). Between 1994 and 2000, the number of quarantine pest interceptions at Florida ports of entry increased by 162% (Klassen et al. 2002).

Immigrant insects have a long history of crop damage in California, another important agricultural state. Between 1955 and 1988, infestations of 208 immigrant invertebrates were discovered. Homoptera, followed by Coleoptera and Lepidoptera, made up the greatest number of insects; the majority originated

Table 21.2 The adventive insect fauna of selected geographic areas.

Geographic Area	Number of Described Species	Number of Immigrant Species	Percentage of Total Composition	Estimated Rate of Annual Detection	Orders with Most Immigrants (% of Total)	Reference
Conterminous USA	~90,000	>2,000	2–3%	11.0 spp./yr. (1910–1980) 6.2 spp./yr. (1970–1982)	Homoptera (20.5%) Coleoptera (20.3%) Lepidoptera (7.4%)	Hoebeke and Wheeler 1983, Sailer 1983, U.S. Congress 1993, Arnett 2000, Papp 2001
California	~28,000	208	7.4%	6.1 spp./yr. (1955–1988)	Homoptera (33.7%) Coleoptera (12.0%) Lepidoptera (11.5%)	Powell and Hogue 1979, Dowell and Gill 1989
Florida	12,500	945	7.6–10%	12.9 spp./yr. (1971–1991) 7.7 spp./yr. (1970s) 12.0 spp./yr. (1980s)	Coleoptera (33%) Lepidoptera (24.9%) Homoptera (16.3%)	Frank and McCoy 1992, Frank and McCoy 1995b, Frank et al. 1997
Hawaii	7,998	2,598	~32%	12.4 spp./yr. (1912–1935) 14.5 spp./yr. (1937–1961) 18.1 spp./yr. (1962–1976)	Thysanoptera (78.9%) Diptera (44.3%) Homoptera (43.4%)	Beardsley 1962, 1979, 1991; Simberloff 1986; Howarth 1990; Nishida 1994; Eldredge and Miller 1995, 1997, 1998; Miller and Eldredge 1996
Austria	~13,740	212	~1.5%	NA[1]	Coleoptera (1.1%) Lepidoptera (0.6%)	Rabitsch and Essl 2006
Great Britain	21,833	325	1.5%	4.5 spp./yr. (1970–2004)	Homoptera (11.5%) Coleoptera (2.6%)	Williamson and Brown 1986, Smith et al. 2005
Italy	NA	162	NA	4.0 spp./yr. (1975–1995)	Homoptera (64%) Coleoptera (12%) Lepidoptera (7%)	Pellizzari and Dalla Montà 1997, Pellizzari et al. 2005
Japan	29,292	284	9.7%	4.0 spp./yr. (1945–1995)	Coleoptera (32.0%) Homoptera (27.8%) Lepidoptera (12.3%)	Kiritani 1998, 2001; Morimoto and Kiritani 1995
Galápagos Islands	~2,013	463	23%	20.7 spp./yr. (1996–2004)	Hemiptera (25.5%) Coleoptera (24.0%) Diptera (14.3%)	Peck et al. 1998, Causton et al. 2006
New Zealand (Coleoptera only)	5,579	356	6.4%	1.4 spp./yr. (1950–1987) 1.7 spp./yr. (1988–1997)	NA	Klimaszewski and Watt 1997
Tristan da Cunha Islands	152	84[2]	55.3%	NA	Hemiptera (86.7%) Coleoptera (46.5%) Lepidoptera (45.5%)	Holdgate 1960, 1965

[1] Not available.
[2] Includes immigrant ('alien') species and those of doubtful status.

from other regions of North America, followed by the Pacific Region and Europe. Since 1980, the immigration rate from Asia, Australia, Europe, and the Pacific Region increased, especially in Diptera, Hymenoptera, and Homoptera (Dowell and Gill 1989). R.V. Dowell (California Department of Food and Agriculture, Sacramento; cited by Metcalf 1995) compiled a partial list of adventive insects (460 species) established in California between 1600 and 1994.

Hawaii is the most invaded region of the USA because of its particular geography, climate, and history. Approximately 350–400 insects ('original immigrants') colonized the islands before human settlement, probably arriving by ocean or air currents (U.S. Congress 1993). Commerce with the outside world followed the European discovery of the Hawaiian Islands, which allowed additional immigrant insects to enter. Nearly 4600 adventive species have become established, more than half (>2500) of which are arthropods (Nishida 1994, Miller and Eldredge 1996, Eldredge and Miller 1998). More than 98% of pest arthropods are immigrant (Beardsley 1993), and 28% of all Hawaiian insects are adventive (Simberloff 1986), including the entire aphid and ant faunas (Holway et al. 2002, Krushelnycky et al. 2005, Mondor et al. 2007). The rate of establishment of adventive species remains high: an average of 18 new insect species annually from 1937 to 1987 (Beardsley 1979). Since 1965, about 500 immigrant arthropods have become established in the Hawaiian Islands, an annual rate of about 20 species (Beardsley 1979, 1991). With many immigrant pests having become established in Hawaii over the past century, classical biocontrol has been much used (Funasaki et al. 1988).

Canada also is home to numerous adventive insects, with the immigrants of British Columbia, Newfoundland, and Nova Scotia best known. Turnbull (1980) listed 155 immigrant insects in Canada, including human ectoparasites and species associated with dwellings, stored products, cultivated crops and forest trees, and domesticated animals. The Canadian fauna also includes more than 300 species introduced for biological control (Turnbull 1980). The Coleoptera and Hemiptera are reasonably well known in Canada, with recent checklists available for both groups. Of the nearly 7500 coleopteran species recorded in Canada (including Alaska), 469 are considered immigrant ('introduced') in North America (6.2% of the total) (Bousquet 1991). More than 300 species of adventive

Hemiptera have been documented in Canada (and Alaska), representing approximately 7.7% of the nearly 3900 species (Maw et al. 2000); 81 adventive heteropterans, mainly from the western Palearctic, are known from Canada (Scudder and Foottit 2006).

Gillespie's (2001) review of adventive insects detected in British Columbia from the late 1950s to 2000 included information on immigrant lepidopterans such as the codling moth (*Cydia pomonella*), gypsy moth (*Lymantria dispar*), oriental fruit moth (*Grapholita molesta*), and winter moth (*Operophtera brumata*). Among the 48 immigrant lepidopterans are 19 species of Tortricidae (Gillespie and Gillespie 1982). The adventive fauna of British Columbia also includes 42 leafhopper (cicadellid) species (9%) (Maw et al. 2000), 32 plant bug (mirid) species (9%) (Scudder and Foottit 2006), and 10 orthopteroid species (8%) (Scudder and Kevan 1984).

The immigrant fauna of Newfoundland and Nova Scotia has received considerable attention (e.g., Brown 1940, 1950, 1967, Lindroth 1957, Morris 1983). Recent collecting has yielded numerous additional immigrants (e.g., Hoebeke and Wheeler 1996, Majka and Klimaszewski 2004, Wheeler and Hoebeke 2005, Wheeler et al. 2006).

Of 325 adventive invertebrate plant pests that became established in Great Britain from 1787 to 2004, nearly half the species (48.6%) have been recorded since 1970 (Smith et al. 2005). Homopterans (37.1%) and lepidopterans (31.3%) dominate. Of the adventive plant pests, 19% originated from Europe; since 1970, 35.6% of the species originated from continental Europe, 20.3% from North America, and 13.6% from Asia.

At least 500 adventive animal species (~1% of entire fauna) have become established in Austria since 1492 (Essl and Rabitsch 2002, Rabitsch and Essl 2006). The majority (60%) are insects, with Coleoptera and Lepidoptera best represented. About 30% of the Coleoptera are Palearctic (mostly Mediterranean), 23% Oriental, 18% Neotropical, and 7% Nearctic. Nearly 33% of the Lepidoptera originated from the Palearctic, 22% from the Nearctic, 19% from the Oriental, and 10% from the Neotropical Region. The Asian longhorned beetle (*Anoplophora glabripennis*; from China) and western corn rootworm (*Diabrotica virgifera*; from North America) are notable coleopterans detected since 2000.

Insects are the most numerous of all adventive organisms in Switzerland; at least 306 species have become established via human activities (Kenis 2005). Much of this fauna is Mediterranean, although large numbers of tropical or subtropical insects are found in greenhouses (thrips and whiteflies) or are associated with stored products (beetles and moths). Adventive Coleoptera comprise more than 120 species (~40% of the total adventive insect fauna), including the North American Colorado potato beetle (*Leptinotarsa decemlineata*) and western corn rootworm. Fewer than 20 species of adventive Diptera are established in Switzerland (Kenis 2005).

Of insects documented for the Japanese islands, 239 species (0.8%) were considered adventive by Morimoto and Kiritani (1995), but Kiritani (1998) noted that 260 adventives ('exotic species') had been 'introduced' since 1868. Kiritani and Yamamura (2003) increased the number of adventive insect species known from Japan to 415. Coleoptera (31.2%), Homoptera (25.8%), Hymenoptera (11.2%), and Lepidoptera (12.3%) have the largest number of adventive species. Approximately 76% of the immigrant insects are considered pests, whereas only 8% of native Japanese insects are pestiferous (Morimoto and Kiritani 1995). Although the southern islands represent only 1.2% of Japan's land area, about 40% of all immigrant species first became established in these islands (Kiritani 1998).

More than 2000 invertebrate species (mostly insects) in New Zealand are adventive (Brockerhoff and Bain 2000). The ant fauna consists of 28 immigrant and 11 native species (Ward et al. 2006). Several important pests of exotic trees have become widely established (Scott 1984, Charles 1998); additional, potentially important, species are routinely intercepted at ports of entry (Bain 1977, Keall 1981, Ridley et al. 2000). More than 350 adventive beetle species are known from New Zealand (Klimaszewski and Watt 1997). A recent survey for adventive beetles that attack trees or shrubs in New Zealand yielded 51 immigrants in 12 families, most of which were of Australian (58%) and European (25%) origin (Brockerhoff and Bain 2000). Hoare (2001) reported 27 lepidopteran species new to New Zealand since 1988; the majority (67%) represented migrants from Australia, whereas certain others arrived via commerce with Asia or Europe.

The extent of the Australian insect fauna and number of adventive species is unknown (New 1994). Although the Australian Academy of Science estimated that more than 2000 adventive species are established, the actual number might be much greater (Low 2002). More than half the insect orders include an adventive component (New 1994). In the Aphididae, about 80% of some 150 species are immigrant (New 2005b). Adventive insects are most diverse and have had greatest impact in areas most strongly influenced by European settlement ('cultural steppe') (Matthews and Kitching 1984, New 1994).

Analyses and lists of the adventive insects in other countries, islands, and regions include those for Central Europe (Kowarik 2003), France (Martinez and Malausa 1999), Germany (Geiter et al. 2002), Israel (Bytinski-Salz 1966, Roll et al. 2007), Italy (Pellizzari and Dalla Montà 1997, Pellizzari et al. 2005), Kenya (Kedera and Kuria 2003), the Netherlands (van Lenteren et al. 1987), Serbia and Montenegro (Glavendekić et al. 2005), southern oceanic islands (Chown et al. 1998), Spain (Perez Moreno 1999), Tristan da Cunha (Holdgate 1960), and Venezuela (MARN 2001). The U.S. Department of Agriculture's National Agricultural Library created the National Invasive Species Information Center in 2005; its Web site (http://www.invasivespecies info.gov/about.shtml) provides access to sites that contain data on immigrant insects of additional countries and regions.

EFFECTS OF ADVENTIVE INSECTS

Early concerns about the consequences of immigrant insects in the USA involved agriculture and losses to crop production. Direct effects from ectoparasitic and blood-sucking insects would have been apparent to the colonists, but they likely gave no thought to whether the offending species were native. Arthropod-borne diseases were endemic in the colonies from New England to Georgia (Duffy 1953, McNeill 1976, Adler and Wills 2003), and an outbreak of yellow fever in Philadelphia in 1793 was attributed to trade with the West Indies (Inderjit et al. 2005). However, the demonstration that mosquitoes and certain other arthropods transmit the causal organisms of major human diseases was not forthcoming until the late nineteenth or early twentieth century (Mullen and Durden 2002).

Every organism that colonizes a new area can affect native communities and ecosystems (Smith 1933, Lodge 1993a). Most immigrant insects, other than several widespread species of mosquitoes, are terrestrial. Additional immigrant aquatic groups include chironomid midges (Hribar et al. 2008), corixid bugs (Polhemus and Rutter 1997, Polhemus and Golia 2006, Rabitsch 2008), and perhaps a black fly in the Galápagos (Roque-Abelo and Causton 1999).

Not all adventive species produce quick and dramatic effects on native ecosystems (Richardson 2005). Many immigrant insects produce minimal or no observable effects (Williamson 1996, Majka et al. 2006) and may never become problematic; they seemingly become integrated into novel communities as benign or innocuous members of the fauna (Turnbull 1967, Sailer 1978). Although certain immigrants undergo rapid spread, for example, the leucaena psyllid (*Heteropsylla cubana*) (Beardsley 1991) and erythrina gall wasp (*Quadrastichus erythrinae*) (Li et al. 2006), others remain localized. An oak leaf-mining moth (*Phyllonorycter messaniella*) has spread little in Australia since being detected in 1976 and remained innocuous with only occasional outbreaks (New 1994). Two immigrant webspinners (Embiidina) in Australia (New 1994), immigrant beetles in Switzerland in decaying plant material and litter (Kenis 2005), scavenging microlepidopterans in the western USA (Powell 1964), and most immigrant psocopterans in North America (Mockford 1993) likely will remain innocuous and obscure. Some immigrants of cryptic habits remain rarely collected years after their detection, for example, two Palearctic heteropterans in North America: a lace bug (*Kalama tricornis*) (Parshley 1916, Bailey 1951) and leptopodid (*Patapius spinosus*) (Usinger 1941, Lattin 2002). Three immigrant heteropterans (rhyparochromid lygaeoids) that feed on fallen seeds likely would have been predicted to be little-known additions to the fauna of western North America. Instead, these immigrant bugs attracted media attention when they invaded homes, libraries, schools, and businesses, creating anxiety, affecting local economies, and necessitating control measures (Henry and Adamski 1998, Henry 2004). An immigrant thrips (*Cartomothrips* sp.) in California (Arnaud 1983) would have been expected to remain an obscure faunal addition, but it has been said (without evidence, except that it probably feeds on fungi in decaying vegetation) to have potential for affecting ecosystem processes such as decomposition and nutrient cycling (Mooney et al. 1986). Labeling most adventive insects as 'innocuous', however, should be done with caution because their potential for adverse effects might never have been investigated (National Research Council 2002). Studies of immigrant herbivores typically involve their effects on economically important plants rather than possible injury to native species (e.g., Messing et al. 2007). In addition, extended lag times between establishment and explosive population growth are relatively common (Carey 1996, Crooks and Soulé 1999, Loope and Howarth 2003), and competitive exclusion on continental land masses might not take place for long periods (New 1993).

The environmental effects of invasive species sometimes are thought to have received substantial attention only in the twentieth century, mainly after Elton's (1958) book appeared. Yet, Marsh's (1864) book, stressing global anthropogenic disturbances, often is overlooked by current conservationists and invasion biologists. In discussing civilization's effects on the insect fauna of Ohio, Webster (1897) noted vegetational changes such as the reduction of forested areas, apparent disappearance of some native insects, and establishment of European insects. He did not, however, suggest that adventive insects cause environmental changes. We now know that diverse systems are as vulnerable to invasion as those of low diversity, perhaps more so (Levine 2000, D'Antonio et al. 2001; *cf.* Altieri and Nicholls 2002 with respect to agroecosystems); that in some cases, invasive species drive global environmental change (Didham et al. 2005; *cf.* MacDougall and Turkington 2005); and that biotic mixing can involve the introduction of alien alleles and genotypes (Petit 2004).

According to the 'tens rule' (Williamson 1996, Williamson and Fitter 1996), about 10% of the invaders become established, and about 10% of the established species become pests. The rule, despite its statistical regularity, involves considerable variability (Williamson and Fitter 1996, White 1998); it should be considered a working hypothesis (Ehler 1998) and perhaps disregarded for predictive purposes (Suarez et al. 2005). Immigrant insects can become pestiferous at rates almost ten times higher than for native species (Kiritani 2001). Despite impending homogenization of our biota, immigrants are not uniformly dispersed and, as might be predicted, show less cosmopolitanism than plant pathogens (Ezcurra et al. 1978). Even so, immigrant species essentially are everywhere: national parks, nature reserves, and relatively pristine communities (Macdonald et al. 1989, Cole et al. 1992, Pyšek et al. 2003), as well as boreal areas (Simberloff 2004) and Antarctica (Block et al. 1984, Frenot et al. 2005). Fewer immigrant insects have become established in remote areas than along major trade routes and in other areas subject to human disturbance, as is true generally for invasive species (Sala et al. 2000, McNeely 2005; *cf.* Gaston et al. 2003). Urban areas serve as foci of entry for invaders (Frankie et al. 1982, McNeely 1999). Immigrants, including insects (Wheeler and Hoebeke 2001, Majka and Klimaszewski 2004), often are concentrated around shipping and other transport hubs (U.S. Congress 1993, Floerl and Inglis 2005).

The ecological effects of adventive species tend to be severe on old, isolated oceanic islands (Vitousek 1988, Coblentz 1990; *cf.* Simberloff 1995). The economic effects of such species can be particularly devastating to developing nations (Vitousek et al. 1996). Countries that experience the greatest effects from invasive species are heavily tied into systems of global trade (Dalmazzone 2000, McNeely 2006). Invasive species generally have had greater impact in the USA than in continental Europe, where traditionally they have been regarded as a less serious threat (Williamson 1999). Europe, though largely an exporter of species, has experienced recent increases in immigrant species (Pellizzari and Dalla Montà 1997, Essl and Rabitsch 2004, Kenis 2005). More attention, therefore, is being devoted to the problem (Scott 2001, Reinhardt et al. 2003). In addition to the USA (U.S. Congress 1993), countries such as Australia, New Zealand, and South Africa have been substantially affected (Macdonald et al. 1986, McNeely 1999, Pimentel 2002). By 2100, adverse effects from invasive species are expected to be most severe in Mediterranean ecosystems and southern temperate forests (Sala et al. 2000).

The various consequences of invasive species create what Barnard and Waage (2004) termed a 'national, regional, and global development problem'. In contrast to most other human-induced environmental disturbances – pollution and inappropriate use of resources such as poor farming or the draining of wetlands – the invasive species problem is harder to ameliorate and often ecologically permanent (Coblentz 1990). Mooney (2005) created 13 categories for the harmful effects of invasive species, some relevant mainly or solely to plants; at least 6, however, relate to insects: animal disease promoters, crop decimators, forest destroyers, destroyers of homes and gardens, species eliminators, and modifiers of evolution. Whereas invasion biologists and conservationists generally agree that the consequences of invasive species are substantial, they disagree on how best to measure the impact of

invaders (Parker et al. 1999, National Research Council 2002).

We discuss beneficial and detrimental effects of adventive insects here; because of contradictory effects and conflicts of interest, both positive and negative aspects are mentioned for certain species. Although the numerous species thought to have neutral consequences might better be categorized as 'effects unknown' (U.S. Congress 1993), we do not include Unknown as a category.

Adventive insects can be viewed differently in different regions and considered either positive or negative, depending on an observer's perspective. For example, a North American planthopper (*Metcalfa pruinosa*) immigrant in Europe has become a plant pest, but honeybees collect its honeydew in producing a honey that Italian apiarists market as 'Metcalfa honey' (Wilson and Lucchi 2007). Our perceptions of adventive species can be fluid, modified as the result of environmental change, subsequent introductions (U.S. Congress 1993, Simberloff et al. 1997), or as the status of natural enemies introduced for biocontrol changes with changing values (Syrett 2002). The arrival of immigrant fig wasps allowed certain fig species to become weedy in New Zealand and elsewhere (McKey 1989, Kearns et al. 1998). Two immigrant seed wasps detected in Florida in the 1980s, one associated with Brazilian pepper (*Schinus terebinthifolius*) and the other with laurel fig (*Ficus microcarpa*), would have been considered detrimental when these plants were valued as ornamentals but can be viewed as beneficial now that both plants are considered weeds (Nadel et al. 1992, Frank et al. 1997). The presence in Florida of other *Ficus*-associated insects reveals the complexity of evaluating the status of adventive species. The Cuban laurel thrips (*Gynaikothrips ficorum*), upon detection considered a pest of ornamental figs, now is regarded as a beneficial natural enemy of *F. microcarpa*. Another immigrant fig insect, an anthocorid bug (*Montandoniola moraguesi*), until recently would have been viewed as beneficial because it preys on laurel thrips. With reversal in the fig's status, the anthocorid has become an unwanted enemy of a thrips that inflicts severe foliar damage to an undesirable tree (Bennett 1995). In Hawaii, the anthocorid's status changed from a predator introduced to control the Cuban laurel thrips to one that impaired the effectiveness of a thrips introduced for weed biocontrol (Reimer 1988).

Beneficial

Insects are crucial to human existence and ecosystem functions (Waldbauer 2003). The ecological services provided in the USA by mostly native insects is estimated to be nearly $60 billion annually (Losey and Vaughan 2006). Comparable data on adventive species are unavailable, but such insects also contribute important ecological services. Here, we mention immigrant insects involved in several scientific advances and other ways that immigrants might be considered beneficial.

The products of certain insects, such as cochineal, shellac, and silk, are well known and have been used where the insects are not native (Glover 1867, Metcalf and Metcalf 1993, New 1994). For example, the cochineal industry helped save the Canary Islands from starvation after the grape phylloxera (*Daktulosphaira vitifoliae*) devastated the islands' vineyards in the late nineteenth century (Cloudsley-Thompson 1976). Importation of insect galls as drugs and for dyeing and tanning was once important (Fagan 1918). Insects are valuable recyclers of nutrients. Dung beetles (Scarabaeidae) have been introduced into Australia to alleviate problems from slowly decomposing cattle dung. Native scarabs, adapted to feed on the drier dung of precinctive marsupials, feed only to a limited extent on moist cattle dung (Waterhouse 1974, New 1994). African dung beetles also have been introduced into the USA to assist native scarabs (Hoebeke and Beucke 1997). Classical biological control of arthropods and weeds has enjoyed long-term successes (McFadyen 1998, Gurr et al. 2000, Waterhouse and Sands 2001). The use of introduced natural enemies can help reduce pesticide contamination and maintain and manage ecosystem processes (National Research Council 2002, Hoddle 2003, 2004).

Biological control can also help reduce threats from pests of conservation or environmental concern and has been used to conserve endemic plants threatened by immigrant insects (Van Driesche 1994, *cf.* Samways 1997, Louda and Stiling 2004). Several notable campaigns have been carried out despite the difficulty of obtaining financial support (Frank 1998). More than 50 species of parasitoids and predators were introduced into Bermuda from 1946 to 1951 to help reduce infestations of two diaspidid scale insects (mainly *Carulaspis minima*) that were eliminating Bermudian cedar (*Juniperus bermudiana*). This attempt, the first major biocontrol project undertaken expressly for conservation, failed to prevent the death of about 99% of the cedar

forests (Challinor and Wingate 1971, Samways 1994). In the 1990s, a lady beetle (*Hyperaspis pantherina*) introduced into the South Atlantic island of St. Helena saved a precinctive gumwood tree (*Commidendrum robustum*) from extinction by an immigrant ensign scale (*Orthezia insignis*) (Fowler 2004, Wittenberg and Cock 2005).

Adventive insects sometimes can be viewed as ecologically desirable additions to a fauna, such as carabid beetles in impoverished arctic communities (Williamson 1996). In western Canada, immigrant synanthropic carabids were regarded as enriching the fauna; they appear not to threaten native Carabidae because only one native species is strictly synanthropic (Spence and Spence 1988). Immigrant natural enemies arriving with, or separately from, immigrant pests provide fortuitous, though often ineffective, biocontrol (DeBach 1974, Colazza et al. 1996, Nechols 2003). An aquatic moth (*Parapoynx diminutalis*), evaluated for possible use against an aquatic weed (*Hydrilla verticillata*), soon after was detected in Florida as an immigrant (Delfosse et al. 1976). An immigrant parasitoid (*Prospaltella perniciosi*) might provide effective control of the San Jose scale (*Quadraspidiosus perniciosus*) in the USA (Sailer 1972).

Immigrant insects, such as the blow fly (*Chrysomya rufifacies*) in Hawaii, are useful in forensic entomology for establishing postmortem intervals (Goff et al. 1986). Immigrant insects also play a major role in pollination. The alfalfa leafcutter bee (*Megachile rotundata*) is a useful pollinator of alfalfa in the USA (Cane 2003) but is overshadowed in importance by the honeybee, introduced in the early seventeenth century (Sheppard 1989, Horn 2005). Its value to crop pollination is estimated to be $15 to 19 billion annually (Levin 1983, Morse and Calderone 2000), with annual U.S. production of raw honey valued at $150 to 200 million (Flottum 2006, U.S. Department of Agriculture NASS 2006). Societal benefits from the honeybee also include its potential use in monitoring air pollution and hazardous wastes (Shimanuki 1992).

Insects, although important as human food in many parts of the world, are little used in Europe and North America. An increased consumption of insects could promote biodiversity preservation and sustainable agriculture (DeFoliart 1997, Paoletti 2005). In the USA, at least two introduced insects, a belostomatid (*Lethocerus indicus*) and the silkworm (*Bombyx mori*), are sold in Asian food shops (Pemberton 1988, DeFoliart 1999).

Immigrant insects, such as the yellow mealworm (*Tenebrio molitor*), are sold in bait and pet shops as food for insects, birds, and fish (Berenbaum 1989). In nature, adventive species can facilitate native species by providing trophic subsidies (Rodriguez 2006). Immigrant insects provide food for birds, mammals, amphibians, reptiles, and insects, including endangered species (Majka and Shaffer 2008). In the northeastern USA, a European weevil (*Barypeithes pellucidus*) contributes substantially to the diet of a native salamander (*Plethodon cinereus*); other predators, including birds, small mammals, snakes, and invertebrates, also prey on the weevil (Maerz et al. 2005). An immigrant leafhopper (*Opsius stactagalus*) of the adventive saltcedar (*Tamarix ramosissima*) provides food for native birds and riparian herpetofauna in southwestern states (Stevens and Ayers 2002). North American birds prey on egg masses of the introduced gypsy moth (*Lymantria dispar*) (Glen 2004).

Although the prominence of *Drosophila melanogaster* as a research organism does not depend on its immigrant status, the fly has been studied on continents where it is not native. Now almost cosmopolitan, it originally was restricted to the Old World tropics and subtropics (Patterson and Stone 1952). Its arrival in the USA appears undocumented, but it must have been present long before Thomas H. Morgan began his classic studies in genetics in 1909. Experimental work on this immigrant fly has led to four Nobel Prizes (Berenbaum 1997), and it became the third eukaryote whose genome was fully sequenced (DeSalle 2005). An Old World congener (*D. subobscura*) now established in North and South America (Beckenbach and Prevosti 1986) is being used to enhance our understanding of the predictability and rate of evolution in the wild (Huey et al. 2005). Other adventive insects that have advanced our understanding of ecology, evolution, genetics, and physiology include flour beetles (*Tribolium castaneum, T. confusum*) (Sokoloff 1966, Price 1984), a blood-feeding reduviid (*Rhodnius prolixus*) (Wigglesworth 1984), and the yellow mealworm (Schuurman 1937, Costantino and Desharnais 1991). The Madagascan hissing cockroach (*Gromphadorhina portentosa*), which also is used as an experimental animal (Guerra and Mason 2005), is featured in classrooms and insect zoos to introduce children to the pleasures of entomology and pique their interest in insects (Gordon 1996, Rivers 2006); in Florida, it once was a popular pet (Thomas 1995, Simberloff 2003). The popularity of adventive insects as pets can enhance the public's appreciation of insects, but they should be imported legally and not pose a conservation concern (New 2005a). Regardless of how the monarch (*Danaus plexippus*) arrived in Australia, New Zealand, and smaller islands in the Pacific – direct flight, long-distance movement with tropical storms, hitchhiking on ships, or introduction with infested host plants (Zalucki and Clarke 2004) – this butterfly doubtlessly has delighted nature lovers outside North America. The value of insects to human society – in art, decoration, fashion, language, music, spiritual reflection, story, and symbol (Kellert 1996) – might include additional adventive species.

Sterile insect technique, a new principle in population suppression and 'technological milestone in the history of applied entomology' (Perkins 1978), was developed to eradicate the screwworm (*Cochliomyia hominivorax*) from the southern USA (Knipling 1955, Klassen and Curtis 2005). Other immigrants, such as the European corn borer (*Ostrinia nubilalis*) and Hessian fly (*M. destructor*), played key roles in the development of resistant crop varieties in the USA (Painter 1951, Kogan 1982); others, such as the cereal leaf beetle (*Oulema melanopus*), contributed to the modeling of population dynamics and pattern of spread (Kogan 1982, Andow et al. 1990). Identification and synthesis of sex-attractant pheromones of lepidopterans immigrant in North America, including the codling moth (*Cydia pomonella*), European corn borer, and oriental fruit moth (*Grapholita molesta*), helped elucidate chemical communication in insects (Roelofs et al. 1969, 1971, Cardé and Baker 1984) and, in the case of the European corn borer, shed light on evolution of the Insecta (Roelofs et al. 2002). Science and society also have been served by using pheromones of immigrant lepidopterans to monitor pest densities and disrupt mating as an alternative to pesticidal control (Cardé and Baker 1984, Cardé and Minks 1995, Weseloh 2003). Characterization and synthesis of the sex pheromone of the German cockroach (*Blattella germanica*) provided a means for its monitoring and control (Nojima et al. 2005).

Benefits derived from adventive insects can be extended to the red imported fire ant (*Solenopsis invicta*). It not only can be a useful predator under certain conditions (Reagan 1986, Tschinkel 2006, *cf.* Eubanks et al. 2002) and provided insights into the evolution of social organization in insects (Ross and Keller 1995) but also helped shape the career of Edward O. Wilson, indirectly promoting studies in insect biodiversity. Wilson's discovery of the ant at Mobile, Alabama, when he was only 13, represented the earliest U.S. record of the species and his first scientific observation.

At 19, he took a leave of absence from the University of Alabama to study the fire ant's spread and impact. This work for the state conservation department was the first professional position (Wilson 1994) for a biologist who would popularize biodiversity studies and become a leading scientist of the twentieth century.

Detrimental

Invasive species produce adverse socioeconomic, environmental (ecological), and health effects. The problem can be viewed as involving economic as much as ecological issues (Evans 2003). Economic costs can be direct, involving exclusion, eradication, control, and mitigation; or indirect, involving human health or alteration of communities and ecosystems (Perrings et al. 2005, McNeely 2001a). Indirect economic costs, which include ecosystem services (Charles and Dukes 2007), are more difficult to calculate, often are not considered, and can overwhelm direct costs (Ranjan 2006). Seldom considered are both the economic and environmental costs of using pesticides to control (or manage) immigrant pests. Invasive species sometimes act synergistically, their collective effects being greater than those of the species considered individually (Howarth 1985, Simberloff 1997). The red imported fire ant (*Solenopsis invicta*) might facilitate the success of an immigrant mealybug (*Antonina graminis*) that uses the ant's honeydew (Helms and Vinson 2002, 2003). Synergistic effects can make it easier for more species to invade ('invasional meltdown'; Simberloff and Von Holle 1999).

Detecting ecological impacts of invasive species and quantifying their effects on population dynamics of native species are difficult, perhaps more so in the case of insects because of their small size and complex, subtle indirect effects – those involving more than two species (Strauss 1991, White et al. 2006). Indirect effects of adventive insects can include unintended cascading effects unlikely to be predicted by risk analysis or revealed during prerelease screening of biocontrol agents. Quantitative estimates of the probability of indirect effects cannot be made (Simberloff and Alexander 1998). Unintended consequences that affect biodiversity fall within the 'externalities' of economists (McNeely 1999, Perrings et al. 2000).

Harm from invasive insects sometimes is considered only in terms of economic losses to agriculture, forestry, or horticulture: crop damage plus control costs. Costs of control may include only those borne by governments, with costs of private control omitted (U.S. Congress 1993). The calculation of losses may fail to consider that they are dynamic, changing from year to year and by regions (Schwartz and Klassen 1981). Crop losses, coupled with far-reaching societal impacts, can be severe when an invasive insect threatens an industry: for example, grape phylloxera (*Daktulosphaira vitifoliae*) and wine production in Europe (Pouget 1990, Campbell 2005), San Jose scale (*Quadraspidiosus perniciosus*) and deciduous fruit culture in California (Iranzo et al. 2003), and sugarcane delphacid (*Perkinsiella saccharicida*) and sugar production in Hawaii (DeBach 1974).

Economics in relation to invasive insects encompasses more than is treated in most entomological publications, for example, research and management costs (McLeod 2004). Consideration of the consequences of immigrant insects in the USA also should include the federal government's procedures for their exclusion, which can entail substantial annual costs within the country and in the countries of origin (Wallner 1996). How the costs of preventing the introduction of an immigrant species compare with those that would have been caused by that species had it become established represent a 'great unknown' (Cox 1999).

Aggregate costs of invasive species, including insects, seldom have been estimated on a national scale (Reinhardt et al. 2003, Essl and Rabitsch 2004, Colautti et al. 2006). The costs estimated by U.S. Congress (1993) and Pimentel et al. (2001, 2005) are cited frequently. Annual damage to U.S. crops by invasive insects is nearly $16 billion (Pimentel et al. 2001). Because Pimentel et al. (2000, 2001) dealt only with a subset of effects from invasive species, they might have understated the problem (Lodge and Shrader-Frechette 2003; *cf.* Theodoropoulos 2003, p. 116).

Historical data on crop losses from insects in the USA include Walsh's (1868) estimate of $300 million. If immigrant insects are assumed to have caused 40% of the losses, as Pimentel et al. (2001) did, losses in 1868 would have been about $120 million (~$4 billion in 2005 dollars based on Consumer Price Index). Sasscer (1940) estimated annual losses from insects (including costs for maintaining research and quarantine facilities, loss of markets due to quarantines, and processing costs from insect damage) to be about $3 billion annually, with at least half resulting from immigrant species (~20 billion in 2005 dollars).

Estimated losses from immigrants in other countries are Australia: AU\$4.7 billion from insects from 1971 to 1995; British Isles: US\$960 million annually from arthropods; New Zealand: NZ\$437 million annually in crop damage plus control costs for invertebrates, mainly insects; and South Africa: US\$1 billion each year in crop damage plus control costs for arthropods (Pimentel 2002).

Economic losses from individual species can be huge (Table 21.3). In 1927, the U.S. Congress appropriated an unprecedented \$10 million to conduct a clean-up campaign to check further spread of the European corn borer (*Ostrinia nubilalis*) (Worthley 1928). For pests not as well established as the corn borer, actual and predicted costs of eradication, as well as predicted losses, are impressive. Unsuccessful campaigns to eradicate the red imported fire ant from the southeastern USA cost more than \$200 million (Buhs 2004).When the ant was detected in California in 1997, eradication costs were estimated at about \$4 billion to almost \$10 billion (Jetter et al. 2002). Eradication programs often are controversial and unsuccessful; undesirable

consequences include adverse effects on human health, death of wildlife, and reduction of arthropod natural enemies leading to secondary-pest outbreaks (Dreistadt et al. 1990, Buhs 2004). Eradication, however, can provide great financial benefits (Klassen 1989, LeVeen 1989, Myers and Hosking 2002). Annual costs associated with the anticipated arrival of the Russian wheat aphid (*Diuraphis noxia*) in Australia might be as high as several million dollars (New 1994). Full costs to the U.S. bee industry from invasion by the African honeybee (*Apis mellifera scutellata*) are not yet known (Schneider et al. 2004), but prior to its arrival, were estimated at \$26 million to \$58 million for beekeeping and another \$93 million for crop losses due to reduced pollination (Winston 1992).

Losses from the Asian longhorned borer (*Anoplophora glabripennis*) in nine at-risk U.S. cities could range from \$72 million to \$2.3 billion and, if every urban area in the conterminous states became totally infested, might reach nearly \$670 billion (Nowak et al. 2001). The emerald ash borer (*Agrilus planipennis*) poses nearly a \$300 billion threat to U.S. timberlands (Muirhead

Table 21.3 Some economic losses from invasive insects.

Species	Description of Loss[1]	Locality	Reference
Alfalfa weevil (*Hypera postica*)	\$500 million, 1990	USA	Simberloff 2003
Asian papaya fruit fly (*Bactrocera papayae*)	AU\$100 million, 1990s	Queensland, Australia	Clarke et al. 2005
Boll weevil (*Anthonomus grandis*)	\$15 billion, cumulatively since 1893	USA	Cox 1999, Myers and Hosking 2002
Codling moth (*C. pomonella*)	\$10–15 million annually	USA	Gossard 1909
European corn borer (*O. nubilalis*)	\$350 million, 1949	USA	Haeussler 1952
Formosan subterranean termite (*Coptotermes formosanus*)	\$1 billion annually	USA	Pimentel et al. 2000
Hessian fly (*Mayetiola destructor*)	\$40 million annually	USA	Gossard 1909
Mediterranean fruit fly (*Ceratitis capitata*)	\$100 million, eradication 1982–1983; losses to economy exceed projected \$1.4 billion	California, USA	Kim 1983, Kiritani 2001
Melon fly, Oriental fruit fly (*Bactrocera cucurbitae, B. dorsalis*)	\$250 million, eradication	Japan	Kiritani 2001
Mole crickets, *Scapteriscus* spp.	>77 million annually, including control costs	Southeastern USA	Frank 1998
Pink hibiscus mealybug (*Maconellicoccus hirsutus*)	\$125 million annually	Trinidad and Tobago	Ranjan 2006
Russian wheat aphid (*Diuraphis noxia*)	\$500–900 million, through 1990s	USA	Foottit et al. 2006
Sheep blow fly (*L. cuprina*)	AU\$100 million annually	Australia	New 1994
Small hive beetle (*Aethina tumida*)	\$3 million, 1998	USA	Hood 2004

[1]All losses are in U.S. dollars unless otherwise noted; costs have not necessarily been documented by economists.

et al. 2006). Canada could be severely affected by the Asian longhorned borer, as well as the emerald ash borer and brown spruce longhorn beetle (*Tetropium fuscum*) (Colautti et al. 2006). An analysis indicating that the Asian longhorned borer's introduction into Europe would pose a significant threat was in press (MacLeod et al. 2002) when the beetle was detected in Austria (Tomiczek and Krehan 2001).

Immigrant herbivores become problematic by feeding on economically important plants, but they also have indirect effects, such as the transmission of viruses and other phytopathogens. The role of the European elm bark beetle (*Scolytus multistriatus*) in spreading Dutch elm disease in North America is well known (Sinclair and Lyon 2005). With millions of disease-susceptible American elms (*Ulmus americana*) having been planted and a competent immigrant vector already established (a native bark beetle is a less efficient vector (Sinclair 1978b)), conditions were favorable for disease outbreak when the fungal pathogen arrived from Europe. By the mid-1970s, about 56% of urban American elms had died (Owen and Lownsbery 1989). Dutch elm disease has had the greatest societal impact of all insect-related tree diseases of urban areas (Campana 1983); cumulative economic losses have amounted to billions of dollars (Sinclair 1978a). Though often considered an urban problem, this disease also affects plant and animal composition in forests (Sinclair 1978a, Campana 1983). The banded elm bark beetle (*S. schevyrewi*), detected recently in western states, could exacerbate problems from Dutch elm disease in North America (Negrón et al. 2005). A serious problem of eastern North American forests is beech bark disease, which involves American beech (*Fagus grandifolia*), a Palearctic scale insect (*Cryptococcus fagisuga*) detected in Nova Scotia about 1890 (Ehrlich 1934), and nectria fungi (formerly *Nectria* spp. but now placed in other genera (Rossman et al. 1999)). Feeding by the scale insect allows fungi that are unable to infect intact bark to invade injured areas. The disease not only kills beech trees, thereby altering the composition of eastern forests and reducing their commercial and recreational use, but likely also adversely affects birds, small mammals, and arthropods (Sinclair and Lyon 2005, Storer et al. 2005).

Other immigrant insects that transmit phytopathogens are agriculturally and horticulturally important. An example is a Nearctic leafhopper (*Scaphoideus titanus*) that apparently was shipped with grapevine material to the Palearctic Region; it serves as the principal vector of a phytoplasma disease (flavescence dorée) of cultivated grapes in Europe (Lessio and Alma 2004, Bressan et al. 2005). A recent (2000) immigrant, the soybean aphid (*Aphis glycines*), quickly became the most important insect pest of U.S. soybean production (Rodas and O'Neil 2006); this Asian native transmits (or is suspected to transmit) several plant viruses in North America (Heimpel et al. 2004, Damsteegt et al. 2005). Another Old World aphid (*Toxoptera citricida*) transmits the virus that causes citrus tristeza. The disease, though present in Venezuela by 1960, did not threaten the citrus industry until the aphid arrived. By the mid-1980s, the disease had devastated the country's citrus culture (Lee and Rocha-Peña 1992). Whiteflies of the *Bemisia tabaci* species complex, transported with commerce throughout much of the world (Oliveira et al. 2001, Perring 2001), transmit several geminiviruses (Czosnek et al. 2001). In the 1980s, the western flower thrips (*Frankliniella occidentalis*), native to the southwestern USA, assumed near cosmopolitan distribution from global trade in greenhouse plants. Emerging as the main vector of the tospovirus that causes tomato spotted wilt, it induced disease epidemics (Ullman et al. 1997, Morse and Hoddle 2006).

Immigrant insects also transmit pathogens to native uneconomic plants. A recently detected Asian ambrosia beetle (*Xyleborus glabratus*) transmits a fungus responsible for extensive mortality of native red bay (*Persea borbonia*) trees in the southeastern USA (Haack 2006, Mayfield 2006). Immigrant aphids may vector viruses of native Hawaiian plants, including precinctive species (Messing et al. 2007).

Immigrant insects of veterinary importance serve as vectors of disease organisms and otherwise affect productivity or harm domestic and companion animals. Annual losses from long-established species affecting livestock in the USA include nearly $1 billion (Castiglioni and Bicudo 2005) for the horn fly (*Haematobia irritans*) (losses are nearly $70 million in Canada (Colautti et al. 2006)). The stable fly (*Stomoxys calcitrans*), long a pest of cattle in midwestern U.S. feedlots, now affects range cattle. When pest numbers are high, daily decreases in weight gain can be nearly 0.5 lb per head (Hogsette 2003, Campbell 2006). The stable fly and other synanthropic Diptera are nuisance insects that affect the U.S. tourist industry (Merritt et al. 1983). In Australia, annual loss of production and treatment costs for the sheep blow fly (*Lucilia cuprina*), an immigrant ectoparasite responsible for cutaneous myiasis (flystrike) of sheep (Levot 1995), amount to more than AU$160 million (McLeod 1995).

Costs associated with the loss of wildlife as the result of immigrant insects are more difficult to express monetarily than those for domestic animals. Avian malaria, though present in Hawaii, did not seriously affect the native avifauna until a competent vector was in place. Following the establishment of a mosquito (*Culex quinquefasciatus*) in lowland areas of Maui by the early nineteenth century, malaria and avian pox became epidemic, which led to many native birds, especially honeycreepers, becoming endangered or extinct (Warner 1968, Jarvi et al. 2001; *cf.* van Riper et al. 1986). Disease resistance, however, might be evolving in certain Hawaiian forest birds (Woodworth et al. 2005, Strauss et al. 2006). An immigrant muscid fly (*Philornis downsi*) recently was detected on the Galápagos archipelago. This obligate ectoparasite of birds apparently has killed nestlings on the islands and could threaten Darwin's finches (Fessl and Tebbich 2002).

Costs associated with human diseases transmitted by immigrant insects can be estimated (Gratz et al. 2000), as was done in Australia for dengue infections after the yellow fever mosquito (*Aedes aegypti*) became established (Canyon et al. 2002). The impact of invasive insects on human health, however, cannot be expressed adequately in monetary terms. At least five immigrant insects associated with vector-borne diseases helped shape South Carolina's culture and history (Adler and Wills 2003).

Medical effects from invasive insects include mild skin reactions (pruritus, urticaria) from contact with browntail moth or gypsy moth larvae (Allen et al. 1991, Mullen 2002); reactions from exposure to allergens of immigrant cockroaches (Peterson and Shurdut 1999); and life-threatening envenomation and hypersensitive reactions from adventive hymenopterans such as the honeybee, red imported fire ant, and other ant species (Akre and Reed 2002, Klotz et al. 2005, Nelder et al. 2006). Effects on humans are catastrophic when immigrant insects serve as vectors of diseases that cause massive population die-offs (Cartwright 1972, Vitousek et al. 1997). In fourteenth-century Europe, following introductions of the black rat (*Rattus rattus*) and oriental rat flea (*Xenopsylla chaeopis*), about 25 million people were killed by plague in a pandemic often called the Black Death (Cartwright 1972, Cloudsley-Thompson 1976, Laird 1989). A mid-seventeenth-century immigrant to the Western

Hemisphere was the yellow fever mosquito (*Aedes aegypti*), which arrived in the Caribbean with ships bearing Africans for the slave trade and became a notorious vector of viruses that cause dengue and yellow fever (Bryan 1999). Throughout human history, immigrant insects have transmitted agents responsible for major diseases (Cloudsley-Thompson 1976, Lounibos 2002).

With the advent of air travel in the 1920s, airplanes became important transporters of mosquitoes that could serve as disease vectors in new areas (Gratz et al. 2000). Disease outbreaks most often result from independent introductions of vector species and pathogens (Juliano and Lounibos 2005). Mosquito species arriving by ship are more likely to become established than those moved by aircraft (Lounibos 2002).

The establishment of an immigrant mosquito (*Anopheles gambiae* s.l.) in Brazil during the 1930s led to epidemic malaria, imposing great socioeconomic burden on the country (Killeen et al. 2002, Levine et al. 2004). Eradication of the mosquito from northeastern Brazil, rapid and unexpected, ended the severe epidemics (Davis and Garcia 1989). A relatively recent global invader, the Asian tiger mosquito (*Aedes albopictus*), can transmit dengue virus and certain other viral agents of encephalitis (Gratz 2004). Native to the Orient, it has become established on five continents since the late 1970s (Adler and Wills 2003, Aranda et al. 2006). In some regions, the Asian tiger mosquito has displaced an immigrant congener, *A. aegypti* (Juliano 1998, Reitz and Trumble 2002, Juliano et al. 2004), although continental U.S. populations of the latter species had been declining prior to the arrival of *A. albopictus* (Rai 1991). An East Asian mosquito (*Ochlerotatus japonicus*) was first collected in the Western Hemisphere in 1998 in the northeastern USA (Peyton et al. 1999). This public-health threat has spread to the southeastern states (Reeves and Korecki 2004), west coast (Sames and Pehling 2005), and southern Canada (Darsie and Ward 2005), and has become established in Hawaii (Larish and Savage 2005) and continental Europe (Medlock et al. 2005). It is a competent laboratory vector of West Nile virus (and potential vector of others), and the virus has been detected in field-collected specimens (Andreadis et al. 2001, Turell et al. 2001).

The world might be entering another (fourth) transition in the history of human diseases, one characterized by ecological change rather than contact among human populations (Baskin 1999). Insects moved in commerce promise to play crucial roles in additional changes in the patterns of vector-borne diseases.

The toll of vector-borne diseases, in addition to loss of life, impaired health, and socioeconomic consequences, includes environmental effects such as the draining and oiling of U.S. wetlands to reduce mosquito populations and malaria (Adler and Wills 2003). Similarly, wetlands in other countries have long been drained, but the restoration of wetlands or construction of new ones has become more common with realization of the need to conserve biodiversity (Schäfer et al. 2004).

In contrast to long-standing interests in calculating economic losses due to invasive species, ecological costs only recently have begun to be assessed (With 2002). The environmental effects of immigrant insects are difficult to estimate (Simberloff 1996, Binggeli 2003) and perhaps are being overlooked (Kenis 2005), due, in part, to an overemphasis on extinction in the popular press (U.S. Congress 1993). Yet, only a 'small minority' of adventive species appears to be affecting native species (Simberloff and Von Holle 1999). Of 81 adventive heteropterans recorded from Canada, only one species might be causing environmental harm (Scudder and Foottit 2006). Despite their diversity, insects are said not to show 'high potential' for causing environmental harm (Wittenberg 2005). Though adventive insects probably damage the environment less than pathogens, plants, and mammals do (Simberloff 2003), the direct and indirect ecological effects of immigrant insects on eastern North American forests (Liebhold et al. 1995, Cox 1999) alone seem sufficient to negate Wittenberg's (2005) statement. Moreover, immigrant oak-associated herbivores, while not economically important, could adversely affect western oak (*Quercus garryana*) meadows in British Columbia (Gillespie 2001). Simberloff's (2003) comment, therefore, seems more appropriate: 'Relative to the numbers of species introduced, insects rarely cause enormous ecological (as opposed to economic) damage'.

Insects seem not to alter fire regimes as do some invasive plants (D'Antonio and Vitousek 1992, D'Antonio 2000), although an immigrant cerambycid (*Phoracantha semipunctata*) might create a fire hazard in California by killing eucalyptus trees (Dowell and Gill 1989). As underlying mechanisms for adverse effects, ranging from individual to ecosystem levels, competition and predation generally are considered more important than hybridization in insects (Rhymer and Simberloff 1996, National Research Council 2002), with interference competition more easily

demonstrated than resource competition (Simberloff 1997, 2000). Hybridization and introgression, though apparently uncommon in insects (Dowling and Secor 1997), occur in certain species of *Drosophila* (Mallet 2005) and subspecies of the honeybee (Sheppard 1989, Schneider et al. 2006) and between the red imported fire ant (*Solenopsis invicta*) and an immigrant congener (*S. richteri*) in a portion of their U.S. range (Tschinkel 2006). Establishment of the Asian gypsy moth in North America and its possible hybridization with the European form are cause for concern (Cox 2004). Hybridization and genetic disruption between an immigrant and an endemic tiger beetle (*Cicindela* spp.) might be taking place in the Galápagos (Causton et al. 2006). Moreover, multiple immigrations of pest insects enhance genetic diversity (Tschinkel 2006) and potentially create more virulent biotypes (Lattin and Oman 1983, Whitehead and Wheeler 1990).

Environmental effects attributed to invasive insects often are based on anecdotal rather than quantified data; inferences on species interactions may fail to consider alternative hypotheses for explaining the observations (Simberloff 1981). Populations of several native coccinellid beetles appear to have declined after the adventive coccinellids *Coccinella septempunctata* and *Harmonia axyridis* became established in North America (e.g., Wheeler and Hoebeke 1995, Michaud 2002). Declines in native species correlated with the establishment of immigrants do not establish causation (Williamson 1996, Simberloff 1997), and other factors might be involved in the decrease in lady beetle densities (Wheeler and Hoebeke 1995, Day and Tatman 2006). Assessing the proximate and ultimate causes of declines in imperiled native species, which are likely subject to multiple threats, is difficult, as is evaluating the threats and their relative importance (Gurevitch and Padilla 2004). As Tschinkel (2006) emphasized, few studies in which competitive displacement by immigrant ants is claimed actually were designed to measure such an effect. The examples of the environmental effects of adventive insects we give later vary in scientific rigor.

Eurasian phytophagous insects in North America tend to colonize the same genera (and often the same species) they do in the Old World and might not have been able to become established without the presence of their native (or closely related hosts) in the New World (Mattson et al. 1994, Niemelä and Mattson 1996, Frank 2002). Species of *Eucalyptus*, planted in North America since the 1800s, were available for late-twentieth-century colonization by specialized immigrant herbivores (Paine and Millar 2002). Certain immigrants have been found in the Nearctic Region only on Palearctic hosts. Examples include several plant bugs (Miridae) and jumping plant lice (Psyllidae) on European ash (*Fraxinus excelsior*) (Wheeler and Henry 1992, Wheeler and Hoebeke 2004), a psyllid (*Livilla variegata*) on ornamental laburnums (*Laburnum* spp.) (Wheeler and Hoebeke 2005), and a lace bug (*Dictyla echii*) on viper's bugloss (*Echium vulgare*) (Wheeler and Hoebeke 2004). Two Palearctic seed bugs are restricted in North America to cosmopolitan and pantropical cattails (*Typha* spp.) (Wheeler 2002). Even if these specialized phytophages expand their host ranges in North America, they are unlikely to cause environmental harm. In other cases, Eurasian plants serve as alternative hosts of recently established immigrant insects that become crop pests, for instance, the Russian wheat aphid (*Diuraphis noxia*) (Kindler and Springer 1989).

Other immigrant phytophages also are not benign faunal additions. Direct feeding by insects immigrant in Hawaii imperils plants of special concern (Howarth 1985). In the Galápagos Islands, the cottony cushion scale (*Icerya purchasi*) killed endangered plants and, in turn, apparently caused local extirpation of certain host-specific lepidopterans (Causton et al. 2006). A Mexican weevil (*Metamasius callizona*) detected in Florida in 1989 feeds on introduced ornamental bromeliads and kills native epiphytic bromeliads (*Tillandsia* spp.) that are protected by law. Destruction of native bromeliads also destroys the invertebrate inhabitants of water impounded in leaf axils (phytotelmata) on the plants (Frank and Thomas 1994, Frank and Fish 2008).

Immigrant phytophages can threaten not only novel host plants but also their naïve natural enemies. The glassywinged sharpshooter (*Homalodisca vitripennis*), detected in French Polynesia in 1999, developed atypically large populations but did not adversely affect the new hosts on which it fed or affect them indirectly by transmitting the bacterium *Xylella fastidiosa*. Instead, the effects of the leafhopper's arrival were seen at higher trophic levels: a lethal intoxication of its spider predators. The cause of mortality is unknown but might involve the leafhopper's bacterial endosymbionts. By using lethal allelochemicals against spiders, *H. vitripennis* might alter the structure and species composition of food webs in the South Pacific (Suttle and Hoddle 2006).

Ants are among the more spectacular of invasive organisms (Moller 1996); several hundred species have been or are being moved in global trade (McGlynn 1999, Suarez et al. 2005, Ward et al. 2006). Those moving readily in commerce – the so-called tramp species (Passera 1994) – afford opportunities for behavioral, ecological, and evolutionary studies relevant to conservation and agriculture. Immigrant ants not only can reduce biodiversity but also can disrupt the biological control of plant pests (Coppler et al. 2007) and disassemble native ant communities (Sanders et al. 2003). Immigrant ants' competitive displacement of native species often is reported, but, at best, is hard to document (e.g., Krushelnycky et al. 2005). The effects of immigrant ants on ant–plant mutualisms warrant more study (Holway et al. 2002, Ness and Bronstein 2004). Ants' mutualistic tending of homopterans such as aphids and scale insects can protect pest species, increasing their densities and damage and deterring predation by natural enemies (Kaplan and Eubanks 2002, Hill et al. 2003, Jahn et al. 2003). Immigrant ants, in turn, sometimes are replaced by later-arriving ant species (Simberloff 1981, Moller 1996), a phenomenon seen among immigrants in other insect groups and among biocontrol agents (Ehler and Hall 1982, Reitz and Trumble 2002, Snyder and Evans 2006; *cf.* Keller 1984).

The red imported fire ant (*Solenopsis invicta*) adversely affects various invertebrate and vertebrate groups (Porter and Savignano 1990, Vinson 1994, 1997). The recent review of the causes and consequences of ant invasions (Holway et al. 2002), review of the effects of the red imported fire ant on biodiversity (Wojcik et al. 2001, Allen et al. 2004), and critique of purported ecological effects from this fire ant (Tschinkel 2006) provide information and references beyond those we mention here.

Immigrant ants can affect seed dispersal and pollination, processes critical to plant reproductive success. By removing seeds, red imported fire ants are potential threats to spring herbs (e.g., *Trillium* spp.) in deciduous forests of the southeastern USA (Zettler et al. 2001). In the Cape fynbos flora of South Africa, the Argentine ant (*Linepithema humile*) has displaced native ants associated with certain precinctive proteaceous plants (myrmecochores) whose seeds are ant dispersed. Argentine ants are slower to discover the seeds, move them only short distances, and eat the elaiosomes without burying the seeds in subterranean nests, as native ants do. Exposed seeds are vulnerable to predation and desiccation. Plant community composition might change as a result of reduced seedling recruitment (Bond and Slingsby 1984, Giliomee 1986). Because the two native ant species displaced by Argentine ants are more effective dispersers of large-seeded Proteaceae than are the two coexisting native species, the fynbos shrubland community might shift toward smaller-seeded species (Christian 2001). Displacement of native ants in Australia involves interference competition by Argentine ants (Rowles and O'Dowd 2007). Argentine ants also deter insect visitation to flowers of certain fynbos proteas (*Protea nitida*) (Visser et al. 1996), reduce fruit and seed set of a euphorbiaceous shrub (*Euphorbia characias*) in Spain (Blancafort and Gómez 2005) and generally threaten myrmecochory in the Mediterranean biome (Gómez and Oliveras 2003), and pose a threat to precinctive plants in Hawaii by reducing their pollinators and plant reproduction (Loope and Medeiros 1994, 1995, Cox 1999). The longlegged or yellow crazy ant (*Anoplolepis gracilipes*) has had severe direct and indirect effects on Christmas Island, killing an estimated 10–15 million red crabs (*Gecarcoidea natalis*) and eliminating populations of this keystone species that regulates seedling recruitment, composition of seedling species, litter breakdown, and density of litter invertebrates. The crab's elimination has long-term implications for forest composition and structure. The ant's mutualism with honeydew-producing homopterans further disrupts the rainforest ecosystem (O'Dowd et al. 2003, Green et al. 2004). Detected in the Seychelles islands in the 1960s, *A. gracilipes* also has begun to affect biodiversity on Bird Island in the Seychelles, following its discovery in the 1980s (Gerlach 2004). Other immigrant ants (Williams 1994) cause adverse environmental effects, including the bigheaded ant (*Pheidole megacephala*) in Hawaii (Jahn and Beardsley 1994, Asquith 1995) and other Pacific islands (Wetterer 2007), and the little fire ant (*Wasmannia auropunctata*) in the Gálapagos (Lubin 1984).

Nonsocial bees immigrant in North America have not adversely affected native bees (Cane 2003). The introduced honeybee, a social species, by competing for floral resources with native bees and disrupting the pollination of native plants (e.g., Gross and Mackay 1998, Spira 2001, Dupont et al. 2004), might affect native ecosystems. Though harmful effects on native flower visitors have been attributed to honeybees, better experimental data and longer studies generally are needed to support the claims (Butz Huryn 1997, Kearns et al. 1998, Goulson

2003). As principal pollinators of invasive plants, honeybees can also enhance fruit set, thus facilitating invasiveness (Goulson and Derwent 2004). Caution should be exercised before introducing social bees that have become invasive elsewhere, for example, a bumble bee (*Bombus terrestris*) into mainland Australia, when it is highly invasive on the Australian island of Tasmania (Hingston 2006). Immigrant wasps and yellowjackets have been implicated in detrimental ecological effects. Examples include the western yellowjacket (*Vespula pensylvanica*) in Hawaii, which preys on native arthropods, reducing their densities and threatening arthropods of Maui's native ecosystems (Gambino et al. 1990, Asquith 1995); two yellowjackets (*V. germanica, V. vulgaris*) in beech (*Nothofagus*) forests in New Zealand, where they restructure invertebrate communities through predation and competition and compete with the precinctive kaka parrot (*Nestor meridionalis*) by harvesting honeydew from margarodid scale insects (*Ultracoelostoma* spp.), thereby limiting the birds' reproductive success (Beggs and Wilson 1991, Beggs et al. 1998, Beggs and Rees 1999); and a paper wasp (*Polistes versicolor*), which feeds mainly on lepidopteran larvae in the Galápagos and competes for food with native vertebrates such as finches (Causton et al. 2006).

Forest insects continue to be carried to all major continents (Ciesla 1993, Britton and Sun 2002, Haack 2006). Adventive insects that alter forest ecosystems in eastern North America are immigrants except for the European gypsy moth, which was introduced into Massachusetts with the hope of crossing the moth with native silkworms to produce a disease-resistant strain for a U.S. silk industry (Spear 2005). Gypsy moth defoliation of oaks (*Quercus* spp.) and its suppression by applications of the insecticide *Bacillus thuringiensis* have changed forest stand composition; increased nest predation of songbirds; decreased mast (acorn) production, resulting in declines of small mammals and changes in foraging patterns of bear and deer; and decreased lepidopteran populations (Liebhold et al. 1995, Wallner 1996). Cascading effects of this eruptive pest encompass interactions among mast production, mice, deer, and ticks that, in turn, affect the incidence of Lyme disease (Elkinton et al. 1996, Jones et al. 1998, Liebhold et al. 2000). The gypsy moth's sociological impact – on esthetic quality and recreational and residential values – might be even greater than its environmental effects (Liebhold et al. 1995).

An immigrant aphidoid, the balsam woolly adelgid (*Adelges piceae*), affects balsam fir (*Abies balsamea*) forests in the Northeast and has nearly eliminated old-growth Fraser fir (*A. fraseri*) in the spruce–fir ecosystem of the southern Appalachians (Jenkins 2003, Potter et al. 2005). The hemlock woolly adelgid (*A. tsugae*) spread from landscape plantings to native stands of eastern hemlock (*T. canadensis*) in the late 1980s (Hain 2005). This immigrant has caused significant mortality in New England forests, shifting nutrient cycling, composition, and structure and imperiling species that are important culturally, economically, and ecologically (Jenkins et al. 1999, Small et al. 2005, Stadler et al. 2005). The effects of hemlock's decline might extend to long-term effects in headwater stream ecosystems (Snyder et al. 2005). *Adelges tsugae* threatens eastern hemlock and Carolina hemlock (*T. caroliniana*) in the southern Appalachians (Graham et al. 2005).

Although introduced pollinators have not caused substantial ecological harm, scarab beetles released to help remove cattle dung might compete with native beetles (Thomas 2002). In Hawaii, the beetles are eaten by mongooses (*Herpestes javanicus*), perhaps allowing these generalist carnivores to maintain larger-than-normal densities (Howarth 1985). The ill-advised biocontrol release of the mongoose to suppress rat populations in Hawaii, and this carnivore's adverse effects on native birds, is well documented (van Riper and Scott 2001).

That invasion biology and classical biological control are linked has been pointed out by numerous workers (e.g., Ehler 1998, Strong and Pemberton 2000, Fagan et al. 2002). Biological control was once considered to lack environmental risk (DeBach 1974), and as recently as the early 1980s was not discussed among numerous causes of decline in insect populations (Pyle et al. 1981). Evidence for adverse effects of natural enemies, however, had long been available (Howarth 2000) and concern over their unforeseen effects had been expressed at least since the 1890s (Perkins 1897, Spear 2005, p. 260).

During the 1980s, biocontrol began to be criticized by conservationists for its irreversibility and possible adverse effects on nontarget plants and insects (Howarth 1983, 1985). Adverse effects in some cases had been anticipated but considered unimportant because the most vulnerable native plants lacked economic value (McFadyen 1998, Seier 2005). A concern for organisms of no immediate or known human benefit 'provoked a revolution in the field of

biological control that has continued . . . and has yet to be resolved' (Lockwood 2000).

Follett and Duan (2000) reviewed the problem of unintended effects from both biocontrol and conservationist perspectives. Indirect ecological effects of biocontrol were the focus of another edited book (Wajnberg et al. 2001). Louda et al. (2003) gave case histories of problematic biocontrol projects: three dealing with herbivores used to suppress weeds and seven with parasitoids or predators used against other insects. Negative ecological effects of parasitoids generally have been less than for predators (Onstad and McManus 1996). Among the conclusions of Louda et al. (2003) was that North American redistribution of an inadvertently established (immigrant) weevil (*Larinus planus*) to control Canada thistle (*Cirsium arvense*) is having major nontarget impact on a native thistle (*C. undulatum* var. *tracyi*). Effects from releases of a flower-head weevil (*Rhinocyllus conicus*) against carduine thistles were considered severe, especially in relation to densities of the native Platte thistle (*Cirsium canescens*) in western states. *Cactoblastis cactorum*, a pyralid moth released against prickly pear (*Opuntia* spp.) in the West Indies, might enhance the risk of extinction of a rare cactus (*O. corallicola*) in Florida. The moth arrived in Florida via immigration or introduction (Frank et al. 1997, Johnson and Stiling 1998), eventually threatening cacti native to the southwestern states and Mexico (Bloem et al. 2005). Louda et al. (2003) concluded that the tachinid *Compsilura concinnata* used against the gypsy moth could have long-term effects on Nearctic silk moths and might cause local extirpation, and that parasitoids released to control the southern green stink bug (*Nezara viridula*) in Hawaii might be accelerating a decline of koa bug (*Coleotichus blackburniae*) populations that could result in extinction.

Certain nontarget effects from well-screened insects used in biocontrol can be considered trivial from a population perspective (Messing and Wright 2006). 'Spillovers' onto nearby nontarget plants that are associated with weed biocontrol agents at high population densities do not represent host shifts (Blossey et al. 2001); the injury can be considered nontarget feeding rather than impact (van Lenteren et al. 2006). The slight foliar injury on a native willow (*Salix interior*) by adults of leaf beetles (*Galerucella* spp.) used against purple loosestrife (*Lythrum salicaria*) in North America actually had been predicted during prerelease testing and should be regarded as 'verification of science done well' (Wiedenmann 2005).

Most biocontrol projects for insect (Lynch and Thomas 2000, van Lenteren 2006) and weed (Fowler et al. 2000, Gould and DeLoach 2002) suppression are thought to produce slight or inconsequential effects on nontarget organisms, although postrelease monitoring for adverse effects typically has not been done or has been minimal (McFadyen 1998, Hajek 2004). Host-specific species traditionally have been chosen for weed control because of the threat that released herbivores pose to crop plants (Waage 2001, Hajek 2004). Host-range testing of biocontrol agents used against insects has been less rigorous than for weeds (Van Driesche and Hoddle 2000, van Lenteren et al. 2006) and can be constrained by an inadequate ecological and taxonomic knowledge of native insects (Barratt et al. 2003). Behavioral factors can complicate tests for nontarget hosts among insects used in arthropod biocontrol (Messing and Wright 2006), and the complex effects of generalist predators on other species of a community – beneficial or detrimental – are unpredictable (Snyder and Evans 2006). Although predators and parasitoids were not initially subject to as thorough host-range testing as weed agents, the use of generalist parasitoids and predators now is less common (Sands and Van Driesche 2003, Hajek 2004).

Inundative biological control, involving the mass rearing and release of natural enemies, has shown fewer adverse ecological effects than classical biocontrol. Permanent establishment of natural enemies to achieve long-term pest management is not the goal of inundation. Even though inundative biocontrol lacks the irreversibility of classical biocontrol, its use still can produce negative effects on nontarget species and ecosystems. Guidelines have been developed to minimize such risks (van Lenteren et al. 2003).

Biological control, properly conducted and carefully regulated, can be an ally of agriculture and conservation (Hajek 2004, Hoddle 2004, Messing and Wright 2006; cf. Louda and Stiling 2004). Adventive organisms used in classical biological control still add to biotic homogenization (e.g., Louda et al. 1997). Such agents are intentional biotic contaminants (Samways 1988, 1997) whose release has moral implications (Lockwood 2001). Released agents can spread to adjacent regions and neighboring countries (Fowler et al. 2000, Henneman and Memmott 2001, Louda and Stiling 2004). Predicting the impact of candidate biocontrol agents on target species remains problematic (Hopper 2001, Lonsdale et al. 2001). Roitberg

(2000) suggested that biocontrol practitioners incorporate concepts of evolutionary ecology, advocating collaboration with evolutionary biologists who study behavioral plasticity so that variables most likely to determine whether candidate natural enemies would harm nontarget hosts might be identified. Even rigorous host-specificity tests of biocontrol agents (and pest-risk analyses of adventive species) cannot be expected to predict all unintended effects that might disrupt communities and ecosystems (Pemberton 2000, Hoddle 2003). Almost nothing is known about the microsporidia that biocontrol agents of weeds might carry and their potential adverse effects (Samways 1997). Indirect effects essentially are unavoidable in multispecies communities (Holt and Hochberg 2001). Because of documented direct and indirect (including cascading) effects on nontarget organisms, a cautious approach to biocontrol is warranted (Howarth 1991, Follett and Duan 2000, Wajnberg et al. 2001). Classical biological control is a complex discipline that evokes controversy (Osborne and Cuda 2003). Even careful consideration of perceived benefits and risks of a proposed project will not satisfy all those who might be affected: biocontrol specialists, conservationists, regulatory officials, policymakers, and general public.

The stochastic nature of biological systems is exemplified by the recent discovery of human-health implications arising from a seemingly straightforward biocontrol project: release of seed-head flies (Urophora spp.) to suppress spotted knapweed (Centaurea biebersteinii (=maculosa of authors)) in rangelands of western North America. The flies, released in the 1970s, proliferated but did not curtail spread of the weed. Ineffective biocontrol agents such as Urophora (Myers 2000) can become abundant and pose greater risks of nontarget effects than agents that effectively control target organisms (Holt and Hochberg 2001). Although the tephritids have not directly harmed nontarget plants (the host-specific flies have remained on target), their larvae provide a winter food source for deer mice (Peromyscus maniculus) when little other food is available. The mice climb knapweed stalks to forage above the snow cover. Food subsidies thus have allowed densities of deer mice, the primary reservoirs of Sin Nombre hantavirus, to increase as much as threefold. Blood samples from mice showed that seropositive individuals were three times more numerous when flies were present. Elevated densities of seropositive mice might alter hantavirus ecology, increasing the risk of virus infections in humans (Pearson and Callaway 2006).

Spectacular early successes in biological control – suppression of cottony cushion scale in California with importation of the vedalia beetle (*Rodolia cardinalis*) from Australia (Caltagirone and Doutt (1989) and prickly pear cactus (*Opuntia* spp.) by various insects (DeLoach 1997) – gave way to realism: that similar successes would not come as easily. Dunlap (1980) noted that L.O. Howard referred to introducing insects into a new environment as being 'infinitely more complicated than we supposed 20 years ago' (Howard 1930). More than 75 years later, Howard's comment, referring specifically to parasitoids, applies generally to the uncertain behavior of adventive insects in novel environments (e.g., Henry and Wells 2007).

SYSTEMATICS, BIODIVERSITY, AND ADVENTIVE SPECIES

Systematics and taxonomy are fundamental to the study, communication, and identification of agriculturally important pest species (Miller and Rossman 1995). Misidentifications can result in serious miscalculations concerning life-history studies, pest-risk assessments, and biocontrol strategies, as evidenced by species of the moth genus *Copitarsia* (Simmonds and Pogue 2004, Venette and Gould 2006). Numerous pest problems have been solved through a systematic knowledge of organisms that affect agricultural and forest ecosystems (Miller and Rossman 1995, Rossman and Miller 1996). The elucidation of the biology of a pest species for control purposes can be achieved only through accurate identification by taxonomists (Wilson 2000).

The availability of an adequate 'biosystematic service' (Knutson 1989) is needed to deal with the problem of immigrant insects (Dick 1966, Oman 1968). Relatively few nations have biosystematic service centers, and those that do often lack specialists for certain economically important groups. Such gaps in taxonomic coverage (Oman 1968, Wheeler and Nixon 1979) impede the execution of plant-regulatory functions and enforcement of quarantine laws (Knutson 1989, New 1994), although the availability of port identifiers (Shannon 1983) helps compensate for a lack of taxonomic specialists in particular groups. A limited understanding of taxonomy and lack of specialists can lead to catastrophic socioeconomic losses, as happened with Dutch elm disease in North America (Britton and Sun 2002). Accurate identification facilitates determination of an invader's origin, allowing

appropriate areas to be searched for natural enemies that might suppress pest densities by classical biological control (Sabrosky 1955, Delucchi et al. 1976, Danks 1988). Thorough systematic knowledge also is critical to assure accurate identification of natural enemies released by researchers and those sold commercially (e.g., Henry and Wells 2007).

Better support for taxonomy and systematics (Knutson 1989, New 1994) would enhance our ability to identify newly established species that threaten agriculture, forestry, human health, and the environment and determine their areas of origin. It also would enhance our ability to identify insects intercepted in commerce and assist regulatory agencies in determining whether the species are likely to be harmful or innocuous.

CONCLUDING THOUGHTS

Invasive species might soon supplant habitat loss and fragmentation as the principal threats to native biodiversity (Crooks and Soulé 1999) and undoubtedly will continue to provide 'wonder and surprise' (Simberloff 1981) to ecologists who study them. Adventive insects will continue to be redistributed globally given the development of new transportation technologies and emphasis on free trade, coupled with inevitable increases in human migration and tourism. Programs of regulatory enforcement are unlikely to keep pace with increases in global commerce due to liberalization of trade (Jenkins 1996). The use of DNA barcoding eventually may allow immigrant insects to be identified rapidly and accurately. Climate change might affect the abundance, distribution, and phenology of adventive insects (Cannon 1998). The public will remain generally unaware that losses in invertebrate diversity can be detrimental to human well-being (Kellert 1995). Even though it is generally acknowledged that invading insects can affect ecosystem structure and function, more rigorous scientific data are needed to assess their detrimental effects on native biodiversity, as is the case for invasive species in general (Brown and Sax 2007).

As the numbers of immigrant insects continue to increase, so too will opportunities for introducing parasitic and predatory insects to help suppress agricultural pests among the newly established species. Because of the idiosyncratic nature of adventive insects (including biocontrol agents) in new environments, even the most objective and quantitative risk assessments for excluding potential pests, or development of new pest-risk assessment tools, cannot predict with certainty where adventive species might become established or their economic effects, let alone their complex and subtle environmental interactions and consequences. A guilty-until-proven-innocent approach to pest exclusion and use of 'white lists' (e.g., McNeely et al. 2001), however, would represent useful change from current regulatory policy (Ruesink et al. 1995, Simberloff et al. 1997, Simberloff 2005). Messing and Wright (2006) recommended that U.S. policies regulating the introduction of biocontrol agents be made similar to those employed by Australia and New Zealand.

Changes to our first line of defense – attempts at exclusion or prevention of establishment – likely will come slowly. As Van Driesche and Van Driesche (2001) pointed out, Americans tend to view prevention as an unpalatable concept. Attempts to exclude immigrant species conflict with society's emphasis on free trade and travel (Kiritani 2001, Low 2001). Not only will additional immigrant insects continue to become established in the USA and elsewhere, but some species once considered innocuous faunal additions will be revealed as harmful. This prediction follows from the realization that relatively few immigrant insects have received attention from researchers, and with lag times sometimes being protracted, adverse ecological effects can take years to develop and even longer to be detected. Global warming likely will lead to northward spread of immigrant pests in temperate regions (Knight and Wimshurst 2005). Immigrants infused with new genetic material via subsequent introductions may continue to adapt to new environments. Pestiferous immigrants no longer thought to represent a threat might resurge as a result of changes in agricultural practices, climate, and environment.

The invasive species problem is 'a complex social and ethical quandary rather than solely a biological one' (Larson 2007). Invasive species cannot be prevented, but the problem can be minified if attempts at amelioration are viewed as the 'art and science of managing people' (Reaser 2001). We agree that human dimensions of the problem deserve more attention and that effective solutions depend heavily on policymakers appreciating connections between invasive species and global trade, transport, and tourism (McNeely 2001b, 2006). Numerous suggestions for alleviating the invasive-species problem have been made (e.g., Lodge et al. 2006, Nentwig 2007). Recommendations include an obvious need to develop reliable predictive theories of biological invasions; to

be more aware of species that have become invasive elsewhere; and to foster greater international collaboration and cooperation (Clout and De Poorter 2005, Bateman et al. 2007), with continued development of online information networks and less emphasis on political boundaries (McNeely 2001c, McNeely et al. 2001, De Poorter and Clout 2005, Simpson et al. 2006). Greater collaboration among biologists, economists, geographers, psychologists, and sociologists will be particularly crucial in addressing problems (McNeely 2006). Among more innovative suggestions is the development of approaches that would subsidize native species until they are able to adapt to altered environments and coexist with invaders (Schlaepfer et al. 2005). As is the case for most other aspects of invasion biology, researchers, conservationists, policymakers, and the public disagree on how best to deal with adventive organisms. Disparate views have long characterized discussions of adventive species. Before the USA enacted plant-regulatory legislation, a leading federal official once advocated a laissez-faire approach to immigrant insects (Marlatt 1899), which elicited a storm of protest (Wheeler and Nixon 1979). More recently, 'blanket opposition' to adventive organisms has been predicted to become 'more expensive, more irrational, and finally counterproductive as the trickle becomes a flood' (Soulé 1990).

Progress toward documenting the extent of the invasive-species problem and devising solutions has been made in recent years. The Scientific Committee on Problems of the Environment (SCOPE) was a founding partner in the Global Invasive Species Program (GISP). Created in 1997, GISP seeks solutions through new approaches and tools (Mooney 1999, McNeely et al. 2001, Barnard and Waage 2004). Noteworthy U.S. initiatives include creation in 1999 of a National Invasive Species Council. Historical data such as interception records of regulatory agencies (e.g., Worner 2002, McCullough et al. 2006, Ward et al. 2006) are being evaluated to address the lack of information on failed introductions (other than biocontrol agents), a deficiency that Simberloff (1986) pointed out. Other positive signs are increased emphasis on the role of taxonomy in the early detection of immigrants, such as regional workshops for enhancing the identification skills of diagnosticians at land-grant universities and identifiers at U.S. ports of entry (Hodges and Wisler 2005). We also note recent collaboration of the Carnegie Museum of Natural History, traditionally a research institution, with federal and state agencies involved in new-pest detection. This linkage supplements the museum's budget while providing timely identifications of insects taken in traps or surveys in or near ports of entry.

Of the four principal means of dealing with invasive species – exclusion, detection, eradication, and control or management – we feel that detection warrants greater attention. J.W. Beardsley regularly looked for new immigrant insects in Hawaii from 1960 to 1990 (Loope and Howarth 2003). More entomologists familiar with local faunas, and hence more likely to recognize insects that seem out of place (Lutz 1941, p. 6; Hoebeke and Wheeler 1983), are conducting detection surveys. Our own fieldwork in the vicinity of port cities in New England and the Atlantic provinces of Canada (e.g., Hoebeke and Wheeler 1996, Wheeler and Hoebeke 2005), and that by Christopher Majka and colleagues in Atlantic Canada (e.g., Majka and Klimaszewski 2004), attest to the value of detective work in areas vulnerable to entry by immigrant insects. With early detection of immigrants (detecting incipient invasions at low-density populations generally is difficult), more rapid response is possible (Burgess 1959, Oman 1968, Reynolds et al. 1982) and eradication (also usually difficult to achieve) and classical biological control (Ehler 1998) are more likely to succeed. The advantage of early detection, coupled with public involvement (Dick 1966), was demonstrated in Auckland, New Zealand, in 1996; a private citizen gave government scientists a distinctive caterpillar that proved to be the Asian whitespotted tussock moth (*Orgyia thyellina*). This potential pest, though apparently established for more than 1 year, was eradicated (Clout and Lowe 2000). Contact with a local U.S. Department of Agriculture office by a Chicago resident who suspected he had a specimen of the Asian longhorned beetle (*Anoplophora glabripennis*) proved crucial to the city's eradication efforts against the pest (Lingafelter and Hoebeke 2002, Antipin and Dilley 2004).

A review of recent literature on immigrant insects in British Columbia revealed a trend toward reporting the first records of adventives in trade magazines and in-house publications rather than in scientific journals. Outlets for reporting immigrants new to the province might have changed during the 1990s because of inability to pay publication costs for papers in scientific journals, the view that with increasing biotic homogenization the presence of species new to a fauna no longer warrant documentation in journal articles,

a lack of taxonomic specialists capable of identifying immigrant species, and too few entomologists remaining in British Columbia to address new threats to agriculture, forestry, and public health (Gillespie 2001). Britton and Sun (2002) acknowledged that Internet sites can omit relevant references and often are ephemeral. We, therefore, encourage publishing the detection of immigrants in mainstream journals, with accompanying summaries of bionomics in the area where species are native, as well as taxonomic information to facilitate recognition in their new faunas. The availability of at least the approximate time of arrival is important in understanding the long-term effects of invaders (Strayer et al. 2006). We also feel it is useful to follow the spread of immigrants and to document range extensions; such historical records are invaluable in allowing future workers to reconstruct immigration events. Knowledge of the new ranges of transferred species can even enhance our biological understanding of invasive organisms (McGlynn 1999).

A global computerized database of immigrant pests has been envisioned for more than 15 years to complement the Western Hemisphere (formerly North American) Immigrant Arthropod Database (WHIAD), administered by the U.S. Department of Agriculture (Knutson et al. 1990, Kim 1991). Other world regions would benefit from a master list of all adventive species, which would help in inventorying Earth's biota, serve as a database for assessing biotic changes, and facilitate dissemination of information on invasive species (Wonham 2003). Schmitz and Simberloff (2001) proposed a U.S. database administered by a National Center for Biological Invasions. Taking advantage of existing capacities and partnerships (WHIAD was not mentioned), the center would place the administration of rules and regulations pertaining to invasive species under a central agency linked to a major university (perhaps the Institute for Biological Invasions, University of Tennessee, which Simberloff directs; or the Center for Invasive Species Research, University of California, Riverside). Loosely modeled after the Centers for Disease Control and Prevention (Schmitz and Simberloff 2001), the center would be of immeasurable value in dealing promptly and effectively with invasive species.

Creation of a National Center for Biological Invasions might forestall homogenization of the U.S. biota and further erosion in quality of life. Societal effects of immigrants can include loss in the amenity value of ecosystems and reduction in ecotourism, owing to sameness among biotic communities (McLeod 2004, Olden et al. 2005). The harmful effects of immigrant insects might also include development of a biophobic public reluctant to venture outdoors (Soulé 1990) because of the possibility of inhaling small immigrant insects such as whiteflies, and threats from imported fire ants and African honeybees (Vinson 1997, Paine et al. 2003) or mosquito-transmitted diseases. In the event of bioterrorism involving the release of pathogens or other harmful organisms in the USA (Pratt 2004), a rapid and effective response to the threat would be more likely if a national center for invasive species were in place. We feel that congressional action on Schmitz and Simberloff's (2001) proposal, perhaps more than any other initiative, would increase public understanding of the problem, stimulate interest in studying invasive species, improve current programs of pest exclusion and detection, and ensure prompt responses to new invaders. A U.S. center for bioinvasions also could serve as a model for other nations as they try to protect native biodiversity and preserve society's 'sense of place and quality of life' (Olden et al. 2005).

ACKNOWLEDGMENTS

We appreciate the editors' invitation to be part of the book. We are pleased to carry on the analysis of immigrant insects in North America begun by the late Reece Sailer and dedicate our review to Daniel Simberloff for his numerous seminal contributions to invasion ecology. The editors and anonymous referees offered many constructive suggestions for improving the manuscript. AGW thanks interlibrary-loan, reference, and remote-storage librarians at Cooper Library, Clemson University, for their assistance; Tammy Morton and Rachel Rowe for their help with manuscript preparation; and Thomas Henry, Gary Miller, James Stimmel, and Craig Stoops for much-appreciated encouragement.

REFERENCES

Acorn, J. 2007. *Ladybugs of Alberta: Finding the Spots and Connecting the Dots*. University of Alberta Press, Edmonton. 169 pp.

Adler, P. H. and W. Wills. 2003. The history of arthropod-borne human diseases in South Carolina. *American Entomologist* 49: 216–228.

Akre, R. D. and H. C. Reed. 2002. Ants, wasps, and bees (Hymenoptera). Pp. 383–409. *In* G. Mullen and L. Durden (eds). *Medical and Veterinary Entomology*. Academic Press, Amsterdam.

Allen, C. R., D. M. Epperson, and A. S. Garmestani. 2004. Red imported fire ant impacts on wildlife: a decade of research. *American Midland Naturalist* 152: 88–103.

Allen, V. T., O. F. Miller, and W. B. Tyler. 1991. Gypsy moth caterpillar dermatitis – revisited. *Journal of the American Academy of Dermatology* 6: 979–981.

Altieri, M. A. and C. I. Nicholls. 2002. Invasive arthropods and pest outbreaks in the context of the ecology of mechanized agricultural systems. Pp. 1–19. *In* G. J. Hallman and C. P. Schwalbe (eds). *Invasive Arthropods in Agriculture: Problems and Solutions*. Oxford and IBH Publishing, New Delhi.

Andow, D. A., P. M. Kareiva, S. A. Levin, and A. Okubo. 1990. Spread of invading organisms. *Landscape Ecology* 4: 177–188.

Andreadis, T. G., J. F. Anderson, L. E. Munstermann, R. J. Wolfe, and D. A. Florin. 2001. Discovery, distribution, and abundance of the newly introduced mosquito *Ochlerotatus japonicus* (Diptera: Culicidae) in Connecticut, USA. *Journal of Medical Entomology* 38: 774–779.

Antipin, J. and T. Dilley. 2004. *Chicago vs. the Asian longhorned beetle: a portrait of success*. USDA Forest Service MP–1593. 52 pp.

Aranda, C., R. Eritja, and D. Roiz. 2006. First record and establishment of the mosquito *Aedes albopictus* in Spain. *Medical and Veterinary Entomology* 20: 150–152.

Arnaud, P. H., Jr. 1983. The collection of an adventive exotic thrips – *Cartomothrips* sp. (Thysanoptera: Phlaeothripidae) – in California. *Proceedings of the Entomological Society of Washington* 85: 622–624.

Arnett, R. H. 2000. *American Insects: A Handbook of the Insects of America North of Mexico*, Second Edition. CRC Press, Boca Raton, Florida. 1003 pp.

Asquith, A. 1995. Alien species and the extinction crisis of Hawaii's invertebrates. *Endangered Species Update* 12 (6): 8. http://www. hear. org/articles/asquith1995/

Atkinson, T. H. and S. B. Peck. 1994. Annotated checklist of the bark and ambrosia beetles (Coleoptera: Platypodidae and Scolytidae) of tropical southern Florida. *Florida Entomologist* 77: 313–329.

Austin, J. W., A. L. Szalanski, R. H. Scheffrahn, M. T. Messenger, J. A. McKern, and R. E. Gold. 2006. Genetic evidence for two introductions of the Formosan termite, *Coptotermes formosanus* (Isoptera: Rhinotermitidae), to the United States. *Florida Entomologist* 89: 183–193.

Bailey, N. S. 1951. The Tingoidea of New England and their biology. *Entomologica Americana* 31: 1–140.

Bain, A. and L. LeSage. 1998. A late seventeenth century occurrence of *Phyllotreta striolata* (Coleoptera:

Chrysomelidae) in North America. *Canadian Entomologist* 130: 715–719.

Bain, J. 1977. Overseas wood- and bark-boring insects intercepted at New Zealand ports. New Zealand Forest Service Technical Paper 63. 26 pp.

Baker, H. G. and G. L. Stebbins (eds). 1965. *The Genetics of Colonizing Species: Proceedings of the First International Union of Biological Sciences Symposia on General Biology.* Academic Press, New York. 588 pp.

Barnard, P. and J. K. Waage. 2004. *Tackling Biological Invasions Around the World: Regional Responses to the Invasive Alien Species Threat.* Global Invasive Species Programme, Cape Town, South Africa. 40 pp.

Barnes, J. K. 1988. Asa Fitch and the Emergence of American Entomology, with an Entomological Bibliography and a Catalog of Taxonomic Names and Specimens. New York State Museum Bulletin 461. 120 pp.

Barratt, B. I. P., C. B. Phillips, C. M. Ferguson, and S. L. Goldson. 2003. Predicting non-target impacts of parasitoids: Where to go from here? Pp. 378–386. *In 1st International Symposium on Biological Control of Arthropods, Jan. 14–18, 2002.* USDA Forest Service, Morgantown, West Virginia FHTET-03-05.

Baskin, Y. 1999. Winners and losers in a changing world. *BioScience* 48: 788–792.

Bateman, M., C. Brammer, C. Thayer, H. Meissner, and W. Bailey. 2007. Exotic pest information collection and analysis (EPICA)-gathering information on exotic pests from the World Wide Web. *Bulletin OEPP (Organization européenne et méditerranéenne pour la protection des plantes) EPPO (European and Mediterranean Plant Protection Organization)* 37: 404–406.

Bates, M. 1956. Man as an agent in the spread of organisms. Pp. 788–803. *In* W. L. Thomas, Jr. (ed). *Man's Role in Changing the Face of the Earth.* University of Chicago Press, Chicago, Illinois.

Beardsley, J. W. 1962. On accidental immigration and establishment of terrestrial arthropods in Hawaii during recent years. *Proceedings of the Hawaiian Entomological Society* 18: 99–109.

Beardsley, J. W. 1979. New immigrant insects in Hawaii: 1962 through 1976. *Proceedings of the Hawaiian Entomological Society* 23: 35–44.

Beardsley, J. W. 1991. Introduction of arthropod pests into the Hawaiian Islands. *Micronesica Supplement* 3: 1–4.

Beardsley, J. W. 1993. Exotic terrestrial arthropods in the Hawaiian Islands: origins and impacts. *Micronesica Supplement* 4: 11–15.

Beckenbach, A. T. and A. Prevosti. 1986. Colonization of North America by the European species, *Drosophila subobscura* and *D. ambigua. American Midland Naturalist* 115: 10–18.

Beggs, J. R. and J. S. Rees. 1999. Restructuring of Lepidoptera communities by introduced *Vespula* wasps in a New Zealand beech forest. *Oecologia* 119: 565–571.

Beggs, J. R. and P. R. Wilson. 1991. The kaka *Nestor meridionalis*, a New Zealand parrot endangered by introduced wasps and mammals. *Biological Conservation* 56: 23–38.

Beggs, J. R., R. J. Toft, J. P. Malham, J. S. Rees, J. A. V. Tilley, H. Moller, and P. Alspach. 1998. The difficulty of reducing introduced wasp (*Vespula vulgaris*) populations for conservation gains. *New Zealand Journal of Ecology* 22: 55–63.

Bennett, F. D. 1995. *Montandoniola moraguesi* (Hemiptera: Anthocoridae), a new immigrant to Florida: friend or foe? *Vedalia* 2: 3–6.

Berenbaum, M. R. 1989. *Ninety-nine Gnats, Nits, and Nibblers.* University of Illinois Press, Urbana, Illinois. 254 pp.

Berenbaum, M. R. 1997. The fly that changed the course of science. *Wings* 20(2): 7–12.

Binggeli, P. 2003. The costs of biological invasions. *Diversity and Distributions* 9: 331–334.

Blancafort, X. and C. Gómez. 2005. Consequences of the Argentine ant, *Linepithema humile* (Mayr), invasion on pollination of *Euphorbia characias* (L.) (Euphorbiaceae). *Acta Oecologica* 28: 49–55.

Block, W., A. J. Burn, and K. J. Richard. 1984. An insect introduction to the maritime Antarctic. *Biological Journal of the Linnean Society* 23: 33–39.

Bloem, S., R. F. Mizell III, K. A. Bloem, S. D. Hight, and J. E. Carpenter. 2005. Laboratory evaluation of insecticides for control of the invasive *Cactoblastis cactorum* (Lepidoptera: Pyralidae). *Florida Entomologist* 88: 395–400.

Blossey, B., R. Casagrande, L. Tewksbury, D. A. Landis, R. N. Wiedenmann, and D. R. Ellis. 2001. Nontarget feeding of leaf-beetles introduced to control purple loosestrife (*Lythrum salicaria* L.). *Natural Areas Journal* 21: 368–377.

Bond, W. and P. Slingsby. 1984. Collapse of an ant–plant mutualism: the Argentine ant (*Iridomyrmex humilis*) and myrmecochorous Proteaceae. *Ecology* 65: 1031–1037.

Booth, J. M. and P. J. Gullan. 2006. Synonymy of three pestiferous *Matsucoccus* scale insects (Hemiptera: Coccoidea: Matsucoccidae) based on morphological and molecular evidence. *Proceedings of the Entomological Society of Washington* 108: 749–760.

Bousquet, A. 1991. *Checklist of Beetles of Canada and Alaska.* Agriculture Canada Research Branch Publication 1861/E. 430 pp.

Bressan, A., V. Girolami, and E. Boudon-Padieu. 2005. Reduced fitness of the leafhopper vector *Scaphoideus titanus* exposed to Flavescence dorée phytoplasma. *Entomologia Experimentalis et Applicata* 115: 283–290.

Britton, K. O. and J.-H. Sun. 2002. Unwelcome guests: exotic forest pests. *Acta Entomologica Sinica* 45: 121–130.

Brockerhoff, E. G. and J. Bain. 2000. Biosecurity implications of exotic beetles attacking trees and shrubs in New Zealand. *New Zealand Plant Protection* 53: 321–327.

Brown, J. H. and D. F. Sax. 2004. An essay on some topics concerning invasive species. *Austral Ecology* 29: 530–536.

Brown, J. H. and D. F. Sax. 2007. Do biological invasions decrease biodiversity? *Conservation* 8(2): 16–17.

Brown, W. J. 1940. Notes on the American distribution of some species of Coleoptera common to the European and North American continents. *Canadian Entomologist* 72: 65–78.

Brown, W. J. 1950. The extralimital distribution of some species of Coleoptera. *Canadian Entomologist* 82: 197–205.

Brown, W. J. 1967. Notes on the extralimital distribution of some species of Coleoptera. *Canadian Entomologist* 99: 85–93.

Bryan, R. T. 1999. Alien species and emerging infectious diseases: past lessons and future implications. Pp. 163–175. *In* O. T. Sandlund, P. J. Schei, and Å Viken (eds). *Invasive Species and Biodiversity Management*. Kluwer Academic Publishers, Dordrecht.

Buckland, P. C. 1988. North Atlantic faunal connections – introduction or endemic? *Entomologica Scandinavica Supplement* 32: 7–29.

Buckland, P. C., A. C. Ashworth, and D. W. Schwert. 1995. By-Passing Ellis Island: Insect immigration to North America. Pp. 226–244. *In* R. A. Butlin and N. Roberts (eds). *Ecological Relations in Historical Times: Human Impact and Adaptation*. Blackwell, Oxford, United Kingdom.

Buhs, J. B. 2004. *The Fire Ant Wars: Nature, Science, and Public Policy in Twentieth-century America*. University of Chicago Press, Chicago, Illinois. 216 pp.

Burckhardt, D. and R. Mühlethaler. 2003. Exotische Elemente der Schweizer Blattflohfauna (Hemiptera, Psylloidea) mit einer Liste weiterer potentieller Arten. *Mitteilungen der Entomologischen Gesellschaft Basel* 53: 98–110.

Burgess, E. D. 1959. Insect detection. *Bulletin of the Entomological Society of America* 5: 67–68.

Butz Huryn, V. M. 1997. Ecological impacts of introduced honey bees. *Quarterly Review of Biology* 72: 275–297.

Bytinski-Salz, H. 1966. An annotated list of insects and mites introduced into Israel. *Israel Journal of Entomology* 1: 15–48.

Caltagirone, L. E. and R. L. Doutt. 1989. The history of the vedalia beetle importation to California and its impact on the development of biological control. *Annual Review of Entomology* 34: 1–16.

Campana, R. J. 1983. The interface of plant pathology with entomology in the urban environment. Pp. 459–480. *In* G. W. Frankie and C. S. Koehler (eds). *Urban Entomology: Interdisciplinary Perspectives*. Praeger Publishers, New York.

Campbell, C. 2005. *The Botanist and the Vintner: How Wine was Saved for the World*. Algonquin Books, Chapel Hill, North Carolina. 320 pp.

Campbell, J. B. 2006. *A guide for the Control of Flies in Nebraska Feedlots and Dairies*. University of Nebraska Lincoln Extension, Institute of Agriculture and Natural Resources. NebGuide G355 (rev.). 3 pp.

Cane, J. H. 2003. Exotic nonsocial bees (Hymenoptera: Apiformes) in North America: Ecological implications. Pp. 113–126. *In* K. Strickler and J. H. Cane (eds). *For Non-native Crops, Whence Pollinators of the Future? Thomas Say Publications in Entomology: Proceedings*. Entomological Society of America, Lanham, Maryland.

Cannon, R. J. C. 1998. The implications of predicted climate change for insect pests in the UK, with emphasis on nonindigenous species. *Global Change Biology* 4: 785–796.

Canyon, D., R. Speare, I. Naumann, and K. Winkel. 2002. Environmental and economic costs of invertebrate invasions in Australia. Pp. 45–66. *In* D. Pimentel (ed). *Biological Invasions: Economic and Environmental Costs of Alien Plant, Animal, and Microbe Species*. CRC Press, Boca Raton, Florida.

Cardé, R. T. and T. C. Baker. 1984. Sexual communication with pheromones. Pp. 355–383. *In* W. J. Bell and R. T. Cardé (eds). *Chemical Ecology of Insects*. Sinauer Associates, Sunderland, Massachusetts.

Cardé, R. T. and A. K. Minks. 1995. Control of moth pests by mating disruption: successes and constraints. *Annual Review of Entomology* 40: 559–585.

Carey, J. R. 1996. The incipient Mediterranean fruit fly population in California: implications for invasion biology. *Ecology* 77: 1690–1697.

Carlton, J. T. 1996. Biological invasions and cryptogenic species. *Ecology* 77: 1653–1655.

Carlton, J. T. and G. M. Ruiz. 2005. Vector science and integrated vector management in bioinvasion ecology: conceptual frameworks. Pp. 36–58. *In* H. A. Mooney, R. N. Mack, J. A. McNeely, L. E. Neville, P. J. Schei, and J. K. Waage (eds). *Invasive Alien Species: A New Synthesis*. Island Press, Washington, DC.

Cartwright, F. F. 1972. *Disease and History*. Thomas Y. Crowell, New York. 248 pp.

Cassey, P., T. M. Blackburn, R. P. Duncan, and S. L. Chown. 2005. Concerning invasive species: reply to Brown and Sax. *Austral Ecology* 30: 475–480.

Castiglioni, L. and H. E. M. C. Bicudo. 2005. Molecular characterization and relatedness of *Haematobia irritans* (horn fly) populations, by RAPD-PCR. *Genetica* 124: 11–21.

Causton, C. E., S. B. Peck, B. J. Sinclair, L. Roque-Albelo, C. J. Hodgson, and B. Landry. 2006. Alien insects: threats and implications for conservation of Galápagos Islands. *Annals of the Entomological Society of America* 99: 121–143.

Challinor, D. and D. B. Wingate. 1971. The struggle for survival of the Bermuda cedar. *Biological Conservation* 3: 220–222.

Chapman, J. W. and J. T. Carlton. 1991. A test of criteria for introduced species: the global invasion by the isopod *Synidotea laevidorsalis* (Miers, 1881). *Journal of Crustacean Biology* 11: 386–400.

Charles, H. and J. S. Dukes. 2007. Impacts of invasive species on ecosystem services. Pp. 217–237. *In* W. Nentwig (ed.). *Biological Invasions*. Springer, Berlin.

Charles, J. G. 1998. The settlement of fruit crop arthropod pests and their natural enemies in New Zealand: an historical guide to the future. *Biocontrol News and Information* 19: 47N–58N.

Chown, S. L., N. J. M. Gremmen, and K. J. Gaston. 1998. Ecological biogeography of southern ocean islands: species–area relationships, human impacts, and conservation. *American Naturalist* 152: 562–575.

Christian, C. E. 2001. Consequences of a biological invasion reveal the importance of mutualism for plant communities. *Nature* 413: 635–639.

Ciesla, W. M. 1993. Recent introductions of forest pests and their effects: a global overview. *FAO Plant Protection Bulletin* 41: 3–13.

Claassen, P. W. 1933. The influence of civilization on the insect fauna in regions of industrial activity. *Annals of the Entomological Society of America* 26: 503–510.

Clarke, A. R., K. F. Armstrong, A. E. Carmichael, J. R. Milne, S. Raghu, G. K. Roderick, and D. K. Yeates. 2005. Invasive phytophagous pests arising through a recent tropical evolutionary radiation: the *Bactrocera dorsalis* complex of fruit flies. *Annual Review of Entomology* 50: 293–319.

Cloudsley-Thompson, J. L. 1976. *Insects and History*. St. Martin's Press, New York. 242 pp.

Clout, M. N. 1999. Biodiversity conservation and the management of invasive animals in New Zealand. Pp. 349–361. *In* O. T. Sandlund, P. J. Schei, and Å Viken (eds). *Invasive Species and Biodiversity Management*. Kluwer Academic Publishers, Dordrecht.

Clout, M. N. and M. De Poorter. 2005. International initiatives against invasive alien species. *Weed Technology* 19: 523–527.

Clout, M. N. and S. J. Lowe. 2000. Invasive species and environmental changes in New Zealand. Pp. 369–383. *In* H. A. Mooney and R. J. Hobbs (eds). *Invasive Species in a Changing World*. Island Press, Washington, DC.

Coates, P. 2006. *American Perceptions of Immigrant and Invasive Species: Strangers on the Land*. University of California Press, Berkeley. 256 pp.

Coblentz, B. E. 1990. Exotic organisms: a dilemma for conservation biology. *Conservation Biology* 4: 261–265.

Colautti, R. I. and H. J. MacIsaac. 2004. A neutral terminology to define 'invasive' species. *Diversity and Distributions* 10: 135–141.

Colautti, R. I., S. A. Bailey, C. D. A. van Overdijk, K. Amundsen, and H. J. MacIsaac. 2006. Characterised and projected costs of nonindigenous species in Canada. *Biological Invasions* 8: 45–59.

Colazza, S., G. Giangiuliani, and F. Bin. 1996. Fortuitous introduction and successful establishment of *Trichopoda pennipes* F.: adult parasitoid of *Nezara viridula* (L.). *Biological Control* 6: 409–411.

Cole, F. R., A. C. Medeiros, L. L. Loope, and W. W. Zuehlke. 1992. Effects of the Argentine ant on arthropod fauna of Hawaiian high-elevation shrubland. *Ecology* 73: 1313–1322.

Coppler, L. B., J. F. Murphy, and M. D. Eubanks. 2007. Red imported fire ants (Hymenoptera: Formicidae) increase the abundance of aphids in tomato. *Florida Entomologist* 90: 419–425.

Costantino, R. F. and R. A. Desharnais. 1991. *Population Dynamics and the Tribolium Model: Genetics and Demography*. Springer Verlag, New York. 258 pp.

Cowie, R. H. and D. G. Robinson. 2003. Pathways of introduction of nonindigenous land and freshwater snails and slugs. Pp. 93–122. *In* G. M. Ruiz and J. T. Carlton (eds). *Invasive Species: Vectors and Management Strategies*. Island Press, Washington, DC.

Cox, G. W. 1999. *Alien Species in North America and Hawaii: Impacts on Natural Ecosystems*. Island Press, Washington, DC. 387 pp.

Cox, G. W. 2004. *Alien Species and Evolution: The Evolutionary Ecology of Exotic Plants, Animals, Microbes, and Interacting Native Species*. Island Press, Washington, DC. 377 pp.

Cox, J. M. and D. J. Williams. 1981. An account of cassava mealybugs (Hemiptera: Pseudococcidae) with a description of a new species. *Bulletin of Entomological Research* 71: 247–258.

Crooks, J. A. and M. E. Soulé. 1999. Lag times in population explosions of invasive species: causes and implications. Pp. 103–125. *In* O. T. Sandlund, P. J. Schei, and Å. Viken (eds). *Invasive Species and Biodiversity Management*. Kluwer Academic Publishers, New York.

Cronon, W. 1983. *Changes in the Land: Indians, Colonists, and the Ecology of New England*. Hill and Wang, New York. 241 pp.

Crosby, A. W. 1994. *Germs Seeds and Animals: Studies in Ecological History*. M. E. Sharpe, Armonk, New York. 214 pp.

Czosnek, H., S. Morin, G. Rubinstein, V. Fridman, M. Zeidan, and M. Ghanim. 2001. Tomato yellow leaf curl virus: a disease sexually transmitted by whiteflies. Pp. 1–27. *In* K. F. Harris, O. P. Smith, and J. E. Duffus (eds). *Virus–Insect–Plant Interactions*. Academic Press, San Diego, California.

Daehler, C. C. and D. R. Gordon. 1997. To introduce or not to introduce: trade-offs of non-indigenous organisms. *Trends in Ecology and Evolution* 12: 424–425.

Dalmazzone, S. 2000. Economic factors affecting vulnerability to biological invasions. Pp. 17–30. *In* C. Perrings, M. Williamson, and S. Dalmazzone (eds). *The Economics of Biological Invasions*. Edward Elgar, Cheltenham, United Kingdom.

Damsteegt, V., A. Stone, W. Schneider, D. Sherman, F. Gildow, and D. Luster. 2005. The soybean aphid, *Aphis glycines*, a new vector of endemic dwarfing and yellowing isolates of Soybean dwarf luteovirus (Abstr.). *Phytopathology* 95: S22.

Danks, H. V. 1988. Systematics in support of entomology. *Annual Review of Entomology* 33: 271–296.

D'Antonio, C. M. 2000. Fire, plant invasions, and global changes. Pp. 65–93. *In* H. A. Mooney and R. J. Hobbs (eds). *Invasive Species in a Changing World*. Island Press, Washington, DC.

D'Antonio, C. M. and P. M. Vitousek. 1992. Biological invasions by exotic grasses, the grass/fire cycle, and global change. *Annual Review of Ecology and Systematics* 23: 63–87.

D'Antonio, C., L. A. Meyerson, and J. Denslow. 2001. Exotic species and conservation: research needs. Pp. 59–80. *In* M. E. Soulé and G. H. Orians (eds). *Conservation Biology: Research Priorities for the Next Decade.* Island Press, Washington, DC.

Darsie, R. F., Jr. and R. A. Ward. 2005. *Identification and Geographical Distribution of the Mosquitoes of North America, North of Mexico.* University Press of Florida, Gainesville, Florida. 383 pp.

Darwin, C. 1859. *On the Origin of Species by Means of Natural Selection: Or the Preservation of Favoured Races in the Struggle for Life.* J. Murray, London. 502 pp.

Davis, J. J. 1952. Milestones in entomology. Pp. 441–444. *In* F. C. Bishopp (chairman) et al. *Insects: The Yearbook of Agriculture.* U. S. Government Printing Office, Washington, DC.

Davis, J. R. and R. Garcia. 1989. Malaria mosquito in Brazil. Pp. 274–283. *In* D. L. Dahlsten and R. Garcia (eds). *Eradication of Exotic Pests: Analysis with Case Histories.* Yale University Press, New Haven, Connecticut.

Davis, M. A. 2006. Invasion biology 1958–2005: the pursuit of science and conservation. Pp. 35–64. *In* M. W. Cadotte, S. M. McMahon, and T. Fukami (eds). *Conceptual Ecology and Invasions Biology: Reciprocal Approaches to Nature.* Springer, London.

Davis, M. A., K. Thompson, and J. P. Grime. 2001. Charles S. Elton and the dissociation of invasion ecology from the rest of ecology. *Diversity and Distributions* 7: 97–102.

Day, W. H. and K. M. Tatman. 2006. Changes in abundance of native and adventive Coccinellidae (Coleoptera) in alfalfa fields, in northern New Jersey (1993–2004) and Delaware (1999–2004), *U. S. A. Entomological News* 117: 491–502.

DeBach, P. 1974. *Biological Control by Natural Enemies.* Cambridge University Press, London. 323 pp.

DeFoliart, G. R. 1997. An overview of the role of edible insects in preserving biodiversity. *Ecology of Food and Nutrition* 36: 109–132.

DeFoliart, G. R. 1999. Insects as food: why the Western attitude is important. *Annual Review of Entomology* 44: 21–50.

Delfosse ("Del Fosse"), E. S., B. D. Perkins, and K. K. Steward. 1976. A new U.S. record for *Paraponyx diminutalis* (Lepidoptera: Pyralidae), a possible biological control agent for *Hydrilla verticillata. Florida Entomologist* 59: 19–20.

DeLoach, C. J. 1997. Biological control of weeds in the United States and Canada. Pp. 172–194. *In* J. O. Luken and J. W. Thieret (eds). *Assessment and Management of Plant Invasions.* Springer, New York.

Delucchi, V., D. Rosen, and E. I. Schlinger. 1976. Relationship of systematics to biological control. Pp. 81–91. *In* C. B. Huffaker and P. S. Messenger (eds). *Theory and Practice of Biological Control.* Academic Press, New York.

De Poorter, M. and M. Clout. 2005. Biodiversity conservation as part of plant protection: the opportunities and challenges of risk analysis. Pp. 55–60. *In* D. V. Alford and G. F. Backhaus (eds). *Plant Protection and Plant Health in Europe: Introduction and Spread of Invasive Species.* Symposium Proceedings No. 81. British Crop Production Council, Alton, United Kingdom.

DeSalle, R. 2005. Evolutionary developmental biology of the Diptera: the "model clade" approach. Pp. 126–144. *In* D. K. Yeates and B. M. Wiegmann (eds). *The Evolutionary Biology of Flies.* Columbia University Press, New York.

Dethier, V. G. 1976. *Man's Plague: Insects and Agriculture.* Darwin Press, Princeton, New Jersey. 237 pp.

Devine, R. 1998. *Alien Invasion: America's Battle with Non-Native Animals and Plants.* National Geographic Society, Washington, DC. 280 pp.

Deyrup, M. 2007. An acrobat ant, *Crematogastser obscurata* (Hymenoptera: Formicidae), poses an unusual conservation question in the Florida Keys. *Florida Entomologist* 90: 753–754.

Dick, C. V. 1966. Plant protection in the American economy: the front line. *Bulletin of the Entomological Society of America* 12: 40–42.

Didham, R. K., J. M. Tylianakis, M. A. Hutchison, R. M. Ewers, and N. J. Gemmell. 2005. Are invasive species the drivers of ecological change? *Trends in Ecology and Evolution* 20: 470–474.

Dowell, R. V. and R. Gill. 1989. Exotic invertebrates and their effects on California. *Pan-Pacific Entomologist* 65: 132–145.

Dowling, T. E. and C. L. Secor. 1997. The role of hybridization and introgression in the diversification of animals. *Annual Review of Ecology and Systematics* 28: 593–619.

Drake, J. A., H. A. Mooney, F. di Castri, R. H. Groves, F. J. Kruger, M. Rejmánek, and M. Williamson. 1989. *Biological Invasions: A Global Perspective.* John Wiley and Sons, Chichester, United Kingdom. 525 pp.

Dreistadt, S. H., D. L. Dahlsten, and G. W. Frankie. 1990. Urban forests and urban ecology: complex interactions among trees, insects, and people. *BioScience* 40: 192–198.

Duffy, J. 1953. *Epidemics in Colonial America.* Kennikat Press, Port Washington, New York. 274 pp.

Dunlap, T. R. 1980. Farmers, scientists, and insects. *Agricultural History* 54: 93–107.

Dupont, Y. L., D. M. Hansen, A. Valido, and J. M. Olesen. 2004. Impact of introduced honey bees on native pollination interactions of the endemic *Echium wildpretii* (Boraginaceae) on Tenerife, Canary Islands. *Biological Conservation* 118: 301–311.

Ehler, L. E. 1998. Invasion biology and biological control. *Biological Control* 13: 127–133.

Ehler, L. E. and R. W. Hall. 1982. Evidence for competitive exclusion of introduced natural enemies in biological control. *Environmental Entomology* 11: 1–4.

Ehrlich, J. 1934. The beech bark disease: a *Nectria* disease of *Fagus,* following *Cryptococcus fagi* (Baer.). *Canadian Journal of Research* 10: 593–692.

Eldredge, L. G. and S. E. Miller. 1995. How many species are there in Hawaii? *Bishop Museum Occasional Papers* 41: 3–18.

Eldredge, L. G. and S. E. Miller. 1997. Numbers of Hawaiian species: Supplement 2, including a review of freshwater invertebrates. *Bishop Museum Occasional Papers* 48: 3–22.

Eldredge, L. G. and S. E. Miller. 1998. Numbers of Hawaiian species: Supplement 3, with notes on fossil species. *Bishop Museum Occasional Papers* 55: 3–15.

Elkinton, J. S., W. M. Healy, J. P. Buonaccorsi, G. H. Boettner, A. M. Hazzard, H. R. Smith, and A. M. Liebhold. 1996. Interactions among gypsy moths, white-footed mice, and acorns. *Ecology* 77: 2332–2342.

Elton, C. S. 1958. *The Ecology of Invasions by Animals and Plants.* Methuen, London; John Wiley and Sons, New York. 181 pp.

Essl, F. and W. Rabitsch. 2002. *Neobiota in Österreich.* Federal Environment Agency, Vienna. 432 pp.

Essl, F. and W. Rabitsch. 2004. *Austrian action plan on invasive alien species.* Federal Ministry of Agriculture, Forestry, Environment and Water Management, Vienna. 15 pp.

Eubanks, M. D., S. A. Blackwell, C. J. Parrish, Z. D. Delamar, and H. Hull-Sanders. 2002. Intraguild predation of beneficial arthropods by red imported fire ants in cotton. *Environmental Entomology* 31: 1168–1174.

Evans, E. A. 2003. Economic dimensions of invasive species. *Choices* 18(2): 5–9.

Ezcurra, E., E. H. Rapoport, and C. R. Marino. 1978. The geographical distribution of insect pests. *Journal of Biogeography* 5: 149–157.

Fagan, M. M. 1918. The uses of insect galls. *American Naturalist* 52: 155–176.

Fagan, W. F., M. A. Lewis, M. G. Neubert, and P. van den Driessche. 2002. Invasion theory and biological control. *Ecology Letters* 5: 148–157.

Fessl, B. and S. Tebbich. 2002. *Philornis downsi* – a recently discovered parasite on the Galápagos archipelago – a threat for Darwin's finches? *Ibis* 144: 445–451.

Floerl, O. and G. J. Inglis. 2005. Starting the invasion pathway: the interaction between source populations and human transport vectors. *Biological Invasions* 7: 589–606.

Flottum, K. 2006. Honey production – 2005. *Bee Culture* 134: 13–14.

Follett, P. A. and J. J. Duan (eds). 2000. *Nontarget Effects of Biological Control.* Kluwer Academic Publishers, Boston, Massachusetts. 316 pp.

Foottit, R. G., S. E. Halbert, G. L. Miller, E. Maw, and L. M. Russell. 2006. Adventive aphids (Hemiptera: Aphididae) of America north of Mexico. *Proceedings of the Entomological Society of Washington* 108: 583–610.

Fowler, S. V. 2004. Biological control of an exotic scale, *Orthezia insignis* Browne (Homoptera: Ortheziidae), saves the endemic gumwood tree, *Commidendrum robustum* (Roxb.) DC. (Asteraceae) on the island of St. Helena. *Biological Control* 29: 367–374.

Fowler, S. V., P. Syrett, and R. L. Hill. 2000. Success and safety in the biological control of environmental weeds in New Zealand. *Austral Ecology* 25: 553–562.

Frank, J. H. 1998. How risky is biological control? Comment. *Ecology* 79: 1829–1834.

Frank, [J.] H. 2002. Pathways of arrival. Pp. 71–86. *In* G. J. Hallman and C. P. Schwalbe (eds). *Invasive Arthropods in Agriculture: Problems and Solutions.* Oxford and IBH Publishing, New Delhi.

Frank, J. H. and D. Fish. 2008. Potential biodiversity loss in Florida bromeliad phytotelmata due to *Matamasius callizona* (Coleoptera: Dryophthoridae), an invasive species. *Florida Entomologist* 91: 1–8.

Frank, J. H. and E. D. McCoy. 1990. Endemics and epidemics of shibboleths and other things causing chaos. *Florida Entomologist* 73: 1–9.

Frank, J. H. and E. D. McCoy. 1992. The immigration of insects to Florida, with a tabulation of records published since 1970. *Florida Entomologist* 75: 1–28.

Frank, J. H. and E. D. McCoy. 1995a. Precinctive insect species in Florida. *Florida Entomologist* 78: 21–35.

Frank, J. H. and E. D. McCoy. 1995b. Invasive adventive insects and other organisms in Florida. *Florida Entomologist* 78: 1–15.

Frank, J. H. and M. C. Thomas. 1994. *Metamasius callizona* (Chevrolat) (Coleoptera: Curculionidae), an immigrant pest, destroys bromeliads in Florida. *Canadian Entomologist* 126: 673–682.

Frank, J. H., E. D. McCoy, H. G. Hall, G. F. O'Meara, and W. R. Tschinkel. 1997. Immigration and introduction of insects. Pp. 75–99. *In* D. Simberloff, D. C. Schmitz, and T. C. Brown (eds). *Strangers in Paradise: Impact and Management of Nonindigenous Species in Florida.* Island Press, Washington, DC.

Frankie, G. W., R. Gill, C. S. Koehler, D. Dilly, J. O. Washburn, and P. Hamman. 1982. Some considerations for the eradication and management of introduced insect pests in urban environments. Pp. 237–255. *In* S. L. Battenfield (ed). *Proceedings of the Symposium on the Imported Fire Ant, June 7–10, 1982, Atlanta, Georgia.* Organized and managed by Inter-Society Consortium for Plant Protection; sponsored by Environmental Protection Agency/USDA, Animal, Plant Health Inspection Service [Washington, DC].

Frenot, Y., S. L. Chown, J. Whinam, P. M. Selkirk, P. Convey, M. Skotnicki, and D. M. Bergstrom. 2005. Biological invasions in the Antarctic: extent, impacts and implications. *Biological Reviews* 80: 45–72.

Frick, K. E. 1964. *Leucoptera spartifoliella,* an introduced enemy of Scotch broom in the western United States. *Journal of Economic Entomology* 57: 589–591.

Funasaki, G. Y., P.-Y. Lai, L. M. Nakahara, J. W. Beardsley, and A. K. Ota. 1988. A review of biological control introductions in Hawaii: 1890 to 1985. *Proceedings of the Hawaiian Entomological Society* 28: 105–160.

Gagné, R. J. 1995. *Contarinia maculipennis* (Diptera: Cecidomyiidae), a polyphagous pest newly reported for North America. *Bulletin of Entomological Research* 85: 209–214.

Gagné, W. C. and C. C. Christensen. 1985. Conservation status of native terrestrial invertebrates in Hawai'i. Pp. 105–126. *In* C. P. Stone and J. M. Scott (eds). *Hawai'i's Terrestrial Ecosystems: Preservation and Management. Cooperative National Park Resources Studies Unit*, University of Hawaii, Honolulu.

Gambino, P., A. C. Medeiros, and L. L. Loope. 1990. Invasion and colonization of upper elevations on East Maui (Hawaii) by *Vespula pensylvanica* (Hymenoptera: Vespidae). *Annals of the Entomological Society of America* 83: 1088–1095.

Gaston, K. J., A. G. Jones, C. Hänel, and S. L. Chown. 2003. Rates of species introduction to a remote oceanic island. *Proceedings of the Royal Society B Biological Sciences* 270: 1091–1098.

Geiter, O., S. Homma, and R. Kinzelbach. 2002. *Bestandsaufnahme und Bewertung von Neozoen in Deutschland.* Texte des Umweltbundesamtes 25/2002. Umweltbundesamt (Federal Environmental Agency), Berlin, Germany. 293 pp.

Gerlach, J. 2004. Impact of the invasive crazy ant *Anoplolepis gracilipes* on Bird Island, Seychelles. *Journal of Insect Conservation* 8: 15–25.

Gibbs, A. 1986. Microbial invasions. Pp. 115–119. *In* R. H. Groves and J. J. Burdon (eds). *Ecology of Biological Invasions*. Cambridge University Press, Cambridge.

Giliomee, J. H. 1986. Seed dispersal by ants in the Cape flora threatened by *Iridomyrmex humilis* (Hymenoptera: Formicidae). *Entomologia Generalis* 11: 217–219.

Gillespie, D. R. 2001. Arthropod introductions into British Columbia – the past 50 years. *Journal of the Entomological Society of British Columbia* 98: 91–97.

Gillespie, D. R. and B. I. Gillespie. 1982. A list of plant-feeding Lepidoptera introduced into British Columbia. *Journal of the Entomological Society of British Columbia* 79: 37–54.

Glen, D. M. 2004. Birds as predators of lepidopterous larvae. Pp. 89–106. *In* H. van Emden and M. Rothschild (eds). *Insect and Bird Interactions*. Intercept, Andover, United Kingdom.

Glavendekić, M., L. Mihajlović, and R. Petanović. 2005. Introduction and spread of invasive mites and insects in Serbia and Montenegro. Pp. 229–230. *In* D. V. Alford and G. F. Backhaus (eds). *Plant Protection and Health in Europe: Introduction and Spread of Invasive Species*. Symposium Proceedings No. 81. British Crop Production Council, Alton, United Kingdom.

Glover, T. 1867. Report of the Entomologist. Pp. 27–45. *In Report of the Commissioner of Agriculture for the year 1866*. Government Printing Office, Washington, DC.

Goff, M. L., C. B. Odom, and M. Early. 1986. Estimation of postmortem interval by entomological techniques: a case study from Oahu, Hawaii. *Bulletin of the Society of Vector Ecologists* 11: 242–246.

Gómez, C. and J. Oliveras. 2003. Can the Argentine ant (*Linepithema humile* Mayr) replace native ants in myrmecochory? *Acta Oecologica* 24: 47–53.

Gordon, D. G. 1996. *The Compleat Cockroach: A Comprehensive Guide to the Most Despised (and least understood) Creature on Earth.* Ten Speed Press, Berkeley, California. 178 pp.

Gossard, H. A. 1909. Relation of insects to human welfare. *Journal of Economic Entomology* 2: 313–324.

Gould, J. R. and C. J. DeLoach. 2002. Biological control of invasive exotic plant species: protocol, history, and safeguards. Pp. 284–306. *In* B. Tellman (ed). *Invasive Exotic Species in the Sonoran Region.* University of Arizona Press/Arizona-Sonora Desert Museum, Tucson, Arizona.

Goulson, D. 2003. Effects of introduced bees on native ecosystems. *Annual Review of Ecology, Evolution, and Systematics* 34: 1–26.

Goulson, D. and L. C. Derwent. 2004. Synergistic interactions between an exotic honeybee and an exotic weed: pollination of *Lantana camara* in Australia. *Weed Research* 44: 195–202.

Graham, J., G. Walker, R. Williams, Z. Murrell, and A. Rex. 2005. Hemlock ecosystems and spatial patterns of *Adelges tsugae* infestation in northwestern North Carolina. Pp. 293–296. *In* B. Onken and R. Reardon (compilers). *Proceedings of the Third Symposium on Hemlock Woolly Adelgid in the Eastern United States*, February 1–3, 2005, Asheville, North Carolina. U. S. Department of Agriculture, Forest Service FHTET-2005-01.

Gratz, N. G. 2004. Critical review of the vector status of *Aedes albopictus*. *Medical and Veterinary Entomology* 18: 215–227.

Gratz, N. G., R. Steffen, and W. Cocksedge. 2000. Why aircraft disinsection? *Bulletin of the World Health Organization* 78: 995–1004.

Green, D. S. 1984. A proposed origin of the coffee leaf-miner, *Leucoptera coffeella* (Guérin-(Méneville) (Lepidoptera: Lyonetiidae). *Bulletin of the Entomological Society of America* 30(1): 30–31.

Green, P. T., P. S. Lake, and D. J. O'Dowd. 2004. Resistance of island rainforest to invasion by alien plants: influence of microhabitat and herbivory on seedling performance. *Biological Invasions* 6: 1–9.

Gross, C. L. and D. Mackay. 1998. Honeybees reduce fitness in the pioneer shrub *Melastoma affine* (Melastomataceae). *Biological Conservation* 86: 169–178.

Guerra, P. A. and A. C. Mason. 2005. Information on resource quality mediates aggression between male Madagascar hissing cockroaches, *Gromphadorhina portentosa* (Dictyoptera: Blaberidae). *Ethology* 111: 626–637.

Gurevitch, J. and D. K. Padilla. 2004. Are invasive species a major cause of extinctions? *Trends in Ecology and Evolution* 19: 470–474.

Gurr, G. M., N. D. Barlow, J. Memmott, S. D. Wratten, and D. J. Greathead. 2000. A history of methodological, theoretical and empirical approaches to biological control. Pp. 3–37. *In* G. Gurr and S. Wratten (eds). *Biological Control: Measures of Success*. Kluwer Academic Publishers, Dordrecht.

Haack, R. A. 2006. Exotic bark- and wood-boring Coleoptera in the United States: recent establishments and interceptions. *Canadian Journal of Forest Research* 36: 269–288.

Haeussler, G. J. 1952. Losses caused by insects. Pp. 141–146. *In* F. C. Bishopp (chairman), G. J. Haeussler, H. L. Haller, W. L. Popham, B. A. Porter, E. R. Sasscer, and J. S. Wade. *Insects: The Yearbook of Agriculture, 1952*. U. S. Government Printing Office, Washington, DC.

Hain, F. P. 2005. Overview of the Third Hemlock Woolly Adelgid Symposium (including balsam woolly adelgid and elongate hemlock scale). Pp. 3–5. *In* B. Onken and R. Reardon (compilers). *Third Symposium on Hemlock Woolly Adelgid in the Eastern United States, 1–3 February, 2005*, Asheville, North Carolina. U. S. Department of Agriculture, Forest Service FHTET-2005-01.

Hajek, A. E. 2004. *Natural Enemies: An Introduction to Biological Control*. Cambridge University Press, Cambridge. 378 pp.

Hattingh, J. 2001. Human dimensions of invasive alien species in philosophical perspective: towards an ethic of conceptual responsibility. Pp. 183–194. *In* J. A. McNeely (ed). *The Great Reshuffling: Human Dimensions of Invasive Alien Species*. IUCN, Gland, Switzerland; Cambridge, United Kingdom.

Heimpel, G. E., D. W. Ragsdale, R. Venette, K. R. Hopper, R. J. O'Neil, C. E. Rutledge, and Z. Wu. 2004. Prospects for importation biological control of the soybean aphid: anticipating potential costs and benefits. *Annals of the Entomological Society of America* 97: 249–258.

Helms, K. R. and S. B. Vinson. 2002. Widespread association of the invasive ant *Solenopsis invicta* with an invasive mealybug. *Ecology* 83: 2425–2438.

Helms, K. R. and S. B. Vinson. 2003. Apparent facilitation of an invasive mealybug by an invasive ant. *Insectes Sociaux* 50: 403–404.

Henneman, M. L. and J. Memmott. 2001. Infiltration of a Hawaiian community by introduced biological control agents. *Science* 293: 1314–1316.

Henry, C. S. and M. M. Wells. 2007. Can what we don't know about lacewing systematics hurt us? A cautionary tale about mass rearing and release of "*Chrysoperla carnea*" (Neuroptera: Chrysopidae). *American Entomologist* 53: 42–47.

Henry, T. J. 2004. *Raglius alboacuminatus* (Goeze) and *Rhyparochromus vulgaris* (Schilling) (Lygaeoidea: Rhyparochromidae): two Palearctic bugs newly discovered in North America. *Proceedings of the Entomological Society of Washington* 106: 513–522.

Henry, T. J. and D. Adamski. 1998. *Rhyparochromus saturnius* (Rossi) (Heteroptera: Lygaeoidea: Rhyparochromidae), a Palearctic seed bug newly discovered in North America. *Journal of the New York Entomological Society* 106: 132–140.

Herrick, G. W. 1929. The procession of foreign insect pests. *Scientific Monthly* 29: 269–274.

Hill, M., K. Holm, T. Vel, N. J. Shah, and P. Matyot. 2003. Impact of the introduced yellow crazy ant *Anoplolepis gracilipes* on Bird Island, Seychelles. *Biodiversity and Conservation* 12: 1969–1984.

Hingston, A. B. 2006. Is the exotic bumblebee *Bombus terrestris* really invading Tasmanian native vegetation? *Journal of Insect Conservation* 10: 289–293.

Hoare, R. J. B. 2001. Adventive species of Lepidoptera recorded for the first time in New Zealand since 1988. *New Zealand Entomologist* 24: 23–47.

Hoddle, M. S. 2003. Classical biological control of arthropods in the 21st century. Pp. 3–16. *In Proceedings of the 1st International Symposium on Biological Control of Arthropods*, January 14–18, 2002, Honolulu, Hawaii. USDA Forest Service, Morgantown, West Virginia. FHTET-03-05.

Hoddle, M. S. 2004. Restoring balance: using exotic species to control invasive exotic species. *Conservation Biology* 18: 38–49.

Hodges, A. C. and G. C. Wisler. 2005. The importance of taxonomic training to the early detection of exotic pests in the order Hemiptera (Auchenorryca [sic], Sternorrhycha [sic]). *Florida Entomologist* 88: 458–463.

Hoebeke, E. R. and K. Beucke. 1997. Adventive *Onthophagus* (Coleoptera: Scarabaeidae) in North America: geographic ranges, diagnoses, and new distributional records. *Entomological News* 108: 345–362.

Hoebeke, E. R. and A. G. Wheeler, Jr. 1983. Exotic insects reported new to northeastern United States and eastern Canada since 1970. *Journal of the New York Entomological Society* 91: 193–222.

Hoebeke, E. R. and A. G. Wheeler, Jr. 1996. *Meligethes viridescens* (F.) (Coleoptera: Nitidulidae) in Maine, Nova Scotia, and Prince Edward Island: diagnosis, distribution, and bionomics of a Palearctic species new to North America. *Proceedings of the Entomological Society of Washington* 98: 221–227.

Hogsette, J. A. 2003. United States Department of Agriculture-Agricultural Research Service research on veterinary pests. *Pest Management Science* 59: 835–841.

Holdgate, M. W. 1960. The fauna of the mid-Atlantic islands. *Proceedings of the Royal Society of London B Biological Series* 152: 550–567.

Holdgate, M. W. 1965. Part III. The fauna of the Tristan da Cunha Islands. *Proceedings of the Royal Society of London B Biological Series* 249: 361–401.

Holt, R. D. and M. E. Hochberg. 2001. Indirect interactions, community modules and biological control: a theoretical perspective. Pp. 13–37. *In* E. Wajnberg, J. K. Scott, and P. C. Quimby (eds). *Evaluating Indirect Ecological Effects of Biological Control*. CABI Publishing, Wallingford, United Kingdom.

Holway, D. A., L. Lach, A. V. Suarez, N. D. Tsutsui, and T. J. Case. 2002. The causes and consequences of ant invasions. *Annual Review of Ecology and Systematics* 33: 181–233.

Hood, W. M. 2004. The small hive beetle, *Aethina tumida*: a review. *Bee World* 85(3): 51–59.

Hopper, K. R. 2001. Research needs concerning non-target impacts of biological control introductions. Pp. 39–56. *In* E. Wajnberg, J. K. Scott, and P. C. Quimby (eds). *Evaluating indirect ecological effects of biological control.* CABI Publishing, Wallingford, United Kingdom.

Horn, T. 2005. *Bees in America: How the Honey Bee Shaped a Nation.* University Press of Kentucky, Lexington, Kentucky. 333 pp.

Horton, D. R., T. M. Lewis, and D. A. Broers. 2004. Ecological and geographic range expansion of the introduced predator *Anthocoris nemoralis* (Heteroptera: Anthocoridae) in North America: potential for nontarget effects? *American Entomologist* 50: 18–30.

Howard, L. O. 1894. On the geographical distribution of some common scale insects. *Canadian Entomologist* 26: 353–356.

Howard, L. O. 1930. *A History of Applied Entomology (Somewhat Anecdotal).* Smithsonian Miscellaneous Collections Vol. 84. Smithsonian Institution, Washington, DC. 564 pp.

Howarth, F. G. 1983. Classical biocontrol: panacea or Pandora's box. *Proceedings of the Hawaiian Entomological Society* 24: 239–244.

Howarth, F. G. 1985. Impacts of alien land arthropods and mollusks on native plants and animals in Hawai'i. Pp. 149–179. *In* C. P. Stone and J. M. Scott (eds). *Hawai'i's Terrestrial Ecosystems: Preservation and Management.* Cooperative National Park Resources Studies Unit, University of Hawaii, Honolulu.

Howarth, F. G. 1990. Hawaiian terrestrial arthropods: an overview. *Bishop Museum Occasional Papers* 30: 4–26.

Howarth, F. G. 1991. Environmental impacts of classical biological control. *Annual Review of Entomology* 36: 485–510.

Howarth, F. G. 2000. Non-target effects of biological control agents. Pp. 369–403. *In* G. Gurr and S. Wratten (eds). *Biological Control: Measures of Success.* Kluwer Academic Publishers, Dordrecht.

Hribar, L. J., J. H. Epler, J. Martin, and J. E. Sublette. 2008. *Chironomus columbiensis* (Diptera: Chironomidae) new to the fauna of the United States. *Florida Entomologist* 91: 470–471.

Huey, R. B., G. W. Gilchrist, and A. P. Hendry. 2005. Using invasive species to study evolution: case studies with *Drosophila* and salmon. Pp. 139–164. *In* D. F. Sax, J. J. Stachowicz, and S. D. Gaines (eds). *Species Invasions: Insights into Ecology, Evolution, and Biogeography.* Sinauer Associates, Sunderland, Massachusetts.

Inderjit, M. W. Cadotte, and R. I. Colautti. 2005. The ecology of biological invasions: past, present and future. Pp. 19–43. *In* Inderjit (ed). *Invasive Plants: Ecological and Agricultural Aspects.* Birkhäuser Verlag, Basel.

Iranzo, S., A. L. Olmstead, and P. W. Rhode. 2003. Historical perspectives on exotic pests and diseases in California. Pp. 55–67. *In* D. A. Sumner (ed). *Exotic Pests and Diseases: Biology and Economics for Biosecurity.* Iowa State Press, Ames, Iowa.

Jacobsen, R. E. and S. A. Perry. 2007. *Polypedilum nubifer,* a chironomid midge (Diptera: Chironomidae) new to Florida that has nuisance potential. *Florida Entomologist* 90: 264–267.

Jahn, G. C. and J. W. Beardsley. 1994. Big-headed ants, *Pheidole megacephala*: interference with the biological control of gray pineapple mealybugs. Pp. 199–205. *In* D. F. Williams (ed). *Exotic Ants: Biology, Impact, and Control of Introduced Species.* Westview Press, Boulder, Colorado.

Jahn, G. C., J. W. Beardsley, and H. González-Hernández. 2003. A review of the association of ants with mealybug wilt disease of pineapple. *Proceedings of the Hawaiian Entomological Society* 36: 9–28.

Jarvi, S. I., C. T. Atkinson, and R. C. Fleischer. 2001. Immunogenetics and resistance to avian malaria in Hawaiian honeycreepers (Drepanidinae). *Studies in Avian Biology* 22: 254–263.

Jenkins, J. C., J. D. Aber, and C. D. Canham. 1999. Hemlock woolly adelgid impacts on community structure and N cycling rates in eastern hemlock forests. *Canadian Journal of Forest Research* 29: 630–645.

Jenkins, M. A. 2003. Impact of the balsam woolly adelgid (*Adelges piceae* Ratz.) on an *Abies fraseri* (Pursh) Poir. dominated stand near the summit of Mount LeConte, Tennessee. *Castanea* 68: 109–118.

Jenkins, P. T. 1996. Free trade and exotic species introductions. *Conservation Biology* 10: 300–302.

Jetter, K. M., J. Hamilton, and J. H. Klotz. 2002. Red imported fire ants threaten agriculture, wildlife and homes. *California Agriculture* 56: 26–34.

Johnson, D. M. and P. D. Stiling. 1998. Distribution and dispersal of *Cactoblastis cactorum* (Lepidoptera: Pyralidae), an exotic *Opuntia*-feeding moth, in Florida. *Florida Entomologist* 81: 12–22.

Jones, C. G., R. S. Ostfeld, M. P. Richard, E. M. Schauber, and J. O. Wolff. 1998. Chain reactions linking acorns to gypsy moth outbreaks and Lyme disease risk. *Science* 279: 1023–1026.

Juliano, S. A. 1998. Species introduction and replacement among mosquitoes: interspecific resource competition or apparent competition? *Ecology* 79: 255–268.

Juliano, S. A. and L. P. Lounibos. 2005. Ecology of invasive mosquitoes: effects on resident species and on human health. *Ecology Letters* 8: 558–574.

Juliano, S. A., L. P. Lounibos, and G. F. O'Meara. 2004. A field test for competitive effects of *Aedes albopictus* on *A. aegypti* in south Florida: differences between sites of coexistence and exclusion? *Oecologia* 139: 583–593.

Kaplan, I. and M. D. Eubanks. 2002. Disruption of cotton aphid (Homoptera: Aphididae) – natural enemy dynamics by red imported fire ants (Hymenoptera: Formicidae). *Environmental Entomology* 31: 1175–1183.

Keall, J. B. 1981. *Interceptions of Insects, Mites and other Animals Entering New Zealand 1973–1978. Ministry of*

Agriculture and Fisheries, Plant Health Diagnostic Station, Levin. 661 pp.

Kedera, C. and B. Kuria. 2003. Invasive alien species in Kenya: status and management. 7 Pp. *In IPPC Secretariat, Identification of Risks and Management of Invasive Alien Species Using the IPPC Framework. Proceedings of the Workshop on Invasive Alien Species and the International Plant Protection Convention, Braunschweig, Germany, 22–26 September 2003.* FAO, Rome. 301 pp.

Kearns, C. A., D. W. Inouye, and N. M. Waser. 1998. Endangered mutualisms: the conservation of plant–pollinator interactions. *Annual Review of Ecology and Systematics* 29: 83–112.

Keller, M. A. 1984. Reassessing evidence for competitive exclusion of introduced natural enemies. *Environmental Entomology* 13: 192–195.

Kellert, S. R. 1995. Values and perceptions of invertebrates. Pp. 118–128. *In* D. Ehrenfeld (ed). *Readings From Conservation Biology: To Preserve Biodiversity – An Overview.* Society for Conservation Biology [Madison, WI] and Blackwell Science, Cambridge, Massachusetts.

Kellert, S. R. 1996. *The Value of Life: Biological Diversity and Human Society.* Island Press, Washington, DC. 263 pp.

Kenis, M. 2005. Insects-Insecta. Pp. 131–212. *In* R. Wittenberg (ed). *An Inventory of Alien Species and Their Threat to Biodiversity and Economy in Switzerland.* CABI Bioscience Switzerland Centre, Delémont.

Killeen, G. F., U. Fillinger, I. Kiche, L. C. Gouagna, and B. G. J. Knols. 2002. Eradication of *Anopheles gambiae* from Brazil: lessons for malaria control in Africa? *Lancet Infectious Diseases* 2: 618–627.

Kim, K. C. 1983. How to detect and combat exotic pests. Pp. 261–319. *In* C. L. Wilson and C. L. Graham (eds). *Exotic Plant Pests and North American Agriculture.* Academic Press, New York.

Kim, K. C. 1991. Immigrant arthropod pests. *Crop Protection* 10: 4–5.

Kindler, S. D. and T. L. Springer. 1989. Alternate hosts of Russian wheat aphid (Homoptera: Aphididae). *Journal of Economic Entomology* 82: 1358–1362.

Kingsolver, C. H. (chairman) et al. 1987. Pests of Plants and Animals: Their Introduction and Spread. Council for Agricultural Science and Technology Report 12. 40 pp.

Kiritani, K. 1998. Exotic insects in Japan. *Entomological Science* 1: 291–298.

Kiritani, K. 2001. *Invasive Insect Pests and Plant Quarantine in Japan.* Food and Fertilizer Technology Center, Taipei. Bulletin 498. 12 pp.

Kiritani, K. and K. Yamamura. 2003. Exotic insects and their pathways for invasion. Pp. 44–67. *In* G. M. Ruiz and J. T. Carlton (eds). *Invasive Species: Vectors and Management Strategies.* Island Press, Washington, DC.

Klassen, W. 1989. Eradication of introduced arthropod pests: theory and historical practice. *Miscellaneous Publications of the Entomological Society of America* 73: 1–29.

Klassen, W. and C. F. Curtis. 2005. History of the sterile insect technique. Pp. 3–36. *In* V. A. Dyck, J. Hendrichs, and A. S. Robinson (eds). *Sterile Insect Technique: Principles and Practice in Area-Wide Integrated Pest Management.* Springer, Dordrecht.

Klassen, W., C. F. Brodel, and D. A. Fieselmann. 2002. Exotic pests of plants: current and future threats to horticultural production and trade in Florida and the Caribbean Basin. *Micronesica Supplement* 6: 5–27.

Klimaszewski, J. and J. C. Watt. 1997. *Coleoptera: Family-Group Review and Keys to Identification. Fauna of New Zealand No. 37.* Manaaki Whenua Press, Lincoln, Canterbury, New Zealand. 199 pp.

Klimaszewski, J., C. Maus, and A. Gardiner. 2002. The importance of tracking introduced species: new records of atheline rove beetles from South Atlantic Inaccessible Island (Coleoptera, Staphylinidae, Aleocharinae). *Coleopterists Bulletin* 56: 481–490.

Klotz, J. H., R. D. deShazo, J. L. Pinnas, A. M. Frishman, J. O. Schmidt, D. R. Suiter, G. W. Price, and S. A. Klotz. 2005. Adverse reactions to ants other than imported fire ants. *Annals of Allergy, Asthma and Immunology* 95: 418–425.

Knight, B. E. A. and A. A. Wimshurst. 2005. Impact of climate change on the geographical spread of agricultural pests, diseases and weeds. Pp. 241–242. *In* D. V. Alford and G. F. Backhaus (eds). *Plant Protection and Plant Health in Europe: Introduction and Spread of Invasive Species. Symposium Proceedings No. 81.* British Crop Production Council, Alton, United Kingdom.

Knipling, E. F. 1955. Possibilities of insect control or eradication through the use of sexually sterile males. *Journal of Economic Entomology* 48: 459–462.

Knutson, L. 1989. Plant diagnostic problems: insects and mites. Pp. 219–248. *In* R. P. Kahn (ed). *Plant Protection and Quarantine.* Vol. II. Selected Pests and Pathogens of Quarantine Significance. CRC Press, Boca Raton, Florida.

Knutson, L., R. I. Sailer, W. L. Murphy, R. W. Carlson, and J. R. Dogger. 1990. Computerized data base on immigrant arthropods. *Annals of the Entomological Society of America* 83: 1–8.

Kogan, M. 1982. Impact and management of introduced pests in agriculture. Pp. 226–236. *In* S. L. Battenfield (ed). *Proceedings of the Symposium on the Imported Fire Ant, June 7–10, 1982, Atlanta, Georgia. Organized and managed by Inter-Society Consortium for Plant Protection; sponsored by Environmental Protection Agency/USDA,* Animal, Plant Health Inspection Service [Washington, DC].

Kolar, C. S. and D. M. Lodge. 2001. Progress in invasion biology: predicting invaders. *Trends in Ecology and Evolution* 16: 199–204.

Kowarik, I. 2003. *Biologische Invasionen: Neophyten und Neozoen in Mitteleuropa.* Ulmer, Stuttgart. 380 pp.

Krushelnycky, P. D., L. L. Loope, and N. J. Reimer. 2005. The ecology, policy, and management of ants in Hawaii. *Proceedings of the Hawaiian Entomological Society* 37: 1–25.

Laird, M. 1984. Overview and perspectives. Pp. 291–325. *In* M. Laird (ed). *Commerce and the Spread of Pests and Disease Vectors*. Praeger, New York.

Laird, M. 1989. Vector-borne diseases introduced into new areas due to human movements: a historical perspective. Pp. 17–33. *In* M. W. Service (ed). *Demography and Vector-Borne Diseases*. CRC Press, Boca Raton, Florida.

Larish, L. B. and H. M. Savage. 2005. Introduction and establishment of *Aedes* (*Finlaya*) *japonicus japonicus* (Theobald) on the island of Hawaii: implications for arbovirus transmission. *Journal of the American Mosquito Control Association* 21: 318–321.

Larson, B. M. H. 2005. The war of the roses: demilitarizing invasion biology. *Frontiers in Ecology and the Environment* 3: 495–500.

Larson, B. M. H. 2007. An alien approach to invasive species: objectivity and society in invasion biology. *Biological Invasions* 9: 947–956.

Lattin, J. D. 2002. The immigrant leptopodid, *Patapius spinosus* (Rossi), in Oregon (Hemiptera: Heteroptera: Leptopodidae). *Pan-Pacific Entomologist* 78: 62.

Lattin, J. D. and P. Oman. 1983. Where are the exotic insect threats? Pp. 93–137. *In* C. L. Wilson and C. L. Graham (eds). *Exotic Plant Pests and North American Agriculture*. Academic Press, New York.

Lee, R. F. and M. A. Rocha-Peña. 1992. Citrus tristeza virus. Pp. 226–249. *In* J. Kumar, H. S. Chaube, U. S. Singh, and A. N. Mukhopadhyay (eds). *Plant Diseases of International Importance*. Vol. III. Diseases of Fruit Crops. Prentice Hall, Englewood Cliffs, New Jersey.

Legner, E. F. and C. W. McCoy. 1966. The housefly, *Musca domestica* Linnaeus, as an exotic species in the Western Hemisphere incites biological control studies. *Canadian Entomologist* 98: 243–248.

Lenteren, J. C. van, J. Woets, P. Grijpma, S. A. Ulenberg, and O. P. J. M. Minkenberg. 1987. Invasions of pest and beneficial insects in the Netherlands. *Proceedings of the Koninklijke Nederlandse Academie van Wetenschappen* 90C: 51–58.

Lenteren, J. C. van, D. Babendreier, F. Bigler, G. Burgio, H. M. T. Hokkanen, S. Kuske, A. J. M. Loomans, I. Menzler-Hokkanen, P. C. J. van Rijn, M. B. Thomas, M. G. Tommasini, and Q.-Q. Zeng. 2003. Environmental risk assessment of exotic natural enemies used in inundative biological control. *BioControl* 48: 3–38.

Lenteren, J. C. van, J. Bale, F. Bigler, H. M. T. Hokkanen, and A. J. M. Loomans. 2006. Assessing risks of releasing exotic biological control agents of arthropod pests. *Annual Review of Entomology* 51: 609–634.

Lessio, F. and A. Alma. 2004. Seasonal and daily movement of *Scaphoideus titanus* Ball (Homoptera: Cicadellidae). *Environmental Entomology* 33: 1689–1694.

LeVeen, E. P. 1989. Economic evaluation of eradication programs. Pp. 41–56. *In* D. L. Dahlsten and R. Garcia (eds). *Eradication of Exotic Pests: Analysis with Case Histories*. Yale University Press, New Haven, Connecticut.

Levin, M. D. 1983. Value of bee pollination to U. S. agriculture. *Bulletin of the Entomological Society of America* 29(4): 50–51.

Levine, J. M. 2000. Species diversity and biological invasions: relating local process to community pattern. *Science* 288: 852–854.

Levine, R. S., A. T. Peterson, and M. Q. Benedict. 2004. Geographic and ecologic distributions of the *Anopheles gambiae* complex predicted using a genetic algorithm. *America Journal of Tropical Medicine and Hygiene* 70: 105–109.

Levot, G. W. 1995. Resistance and the control of sheep ectoparasites. *International Journal for Parasitology* 25: 1355–1362.

Li, H.-M., H. Xiao, H. Peng, H.-X. Han, and D.-Y. Xue. 2006. Potential global range expansion of a new invasive species, the erythrina gall wasp, *Quadrastichus erythrinae* Kim (Insecta: Hymenoptera: Eulophidae). *Raffles Bulletin of Zoology* 54: 229–234.

Liebhold, A. M., W. L. MacDonald, D. Bergdahl, and V. C. Mastro. 1995. Invasion by exotic forest pests: a threat to forest ecosystems. *Forest Science Monographs* 30: 1–49.

Liebhold, A. [M.], J. Elkinton, D. Williams, and R.-M. Muzika. 2000. What causes outbreaks of the gypsy moth in North America? *Population Ecology* 42: 257–266.

Lindroth, C. H. 1957. *The Faunal Connections between Europe and North America*. Almqvist and Wiksell, Stockholm; John Wiley and Sons, New York. 344 pp.

Lingafelter, S. W. and E. R. Hoebeke. 2002. *Revision of the Genus Anoplophora (Coleoptera: Cerambycidae)*. Entomological Society of Washington, Washington, DC. 236 pp.

Lockwood, J. A. 2000. Nontarget effects of biological control: what are we trying to miss? Pp. 15–30. *In* P. A. Follett and J. J. Duan (eds). *Nontarget effects of biological control*. Kluwer Academic Publishers, Boston, Massachusetts.

Lockwood, J. A. 2001. The ethics of "classical" biological control and the value of place. Pp. 100–119. *In* J. A. Lockwood, F. G. Howarth, and M. F. Purcell (eds). *Balancing Nature: Assessing the Impact of Importing Non-Native Biological Control Agents (an International Perspective)*. Thomas Say Publications in Entomology. Entomological Society of America, Lanham, Maryland.

Lodge, D. M. 1993a. Biological invasions: lessons for ecology. *Trends in Ecology and Evolution* 8: 133–137.

Lodge, D. M. 1993b. Species invasions and deletions: community effects and responses to climate and habitat change. Pp. 367–387. *In* P. M. Kareiva, J. G. Kingsolver, and R. B. Huey (eds). *Biotic Interactions and Global Change*. Sinauer Associates, Sunderland, Massachusetts.

Lodge, D. M. and K. Shrader-Frechette. 2003. Nonindigenous species: ecological explanation, environmental ethics, and public policy. *Conservation Biology* 17: 31–37.

Lodge, D. M., S. L. Williams, H. MacIsaac, K. Hayes, B. Leung, S. Reichard, R. N. Mack, P. B. Moyle, M. Smith, D. A. Andow, J. T. Carlton, and A. McMichael. 2006. Biological invasions: recommendations for U. S. policy and management. Position Paper of the Ecological Society of America. 71 pp. http://www. esa. org/pao/esaPositions/.

Lonsdale, W. M., D. T. Briese, and J. M. Cullen. 2001. Risk analysis and weed biological control. Pp. 185–210. *In* E. Wajnberg, J. K. Scott, and P. C. Quimby (eds). *Evaluating Indirect Ecological Effects of Biological Control.* CABI Publishing, Wallingford, United Kingdom.

Loope, L. L. and F. G. Howarth. 2003. Globalization and pest invasion: where will we be in five years? Pp. 34–39. *In Proceedings of the 1st International Symposium on Biological Control of Arthropods, January 14–18, 2002, Honolulu, Hawaii.* USDA Forest Service, Morgantown, WV. FHTET-03-05.

Loope, L. L. and A. C. Madeiros. 1994. Impacts of biological invasions on the management and recovery of rare plants in Haleakala National Park, Maui, Hawaiian Islands. Pp. 143–158. *In* M. L. Bowles and C. J. Whelan (eds). *Restoration of Endangered Species: Conceptual Issues, Planning, and Implementation.* Cambridge University Press, Cambridge.

Loope, L. L. and A. C. Medeiros. 1995. Haleakala silversword. Pp. 363–364. *In* E. T. LaRoe, G. S. Farris, C. E. Puckett, P. D. Doran, and M. J. Mac (eds). *Our Living Resources: A Report to the Nation on the Distribution, Abundance, and Health of U.S. Plants, Animals, and Ecosystems.* U.S. Department of the Interior-National Biological Service, Washington, DC.

Losey, J. E. and M. Vaughan. 2006. The economic value of ecological services provided by insects. *BioScience* 56: 311–323.

Louda, S. M. and P. Stiling. 2004. The double-edged sword of biological control in conservation and restoration. *Conservation Biology* 18: 50–53.

Louda, S. M., D. Kendall, J. Connor, and D. Simberloff. 1997. Ecological effects of an insect introduced for the biological control of weeds. *Science* 277: 1088–1090.

Louda, S. M., R. W. Pemberton, M. T. Johnson, and P. A. Follett. 2003. Nontarget effects – the Achilles' heel of biological control? Retrospective analyses to reduce risk associated with biocontrol introductions. *Annual Review of Entomology* 48: 365–396.

Lounibos, L. P. 2002. Invasions by insect vectors of human disease. *Annual Review of Entomology* 47: 233–266.

Low, T. 2001. From ecology to politics: the human side of alien invasions. Pp. 35–42. *In* J. A. McNeely (ed). *The Great Reshuffling: Human Dimensions of Invasive Alien Species.* IUCN, Gland, Switzerland; Cambridge, United Kingdom.

Low, T. 2002. *Feral Future: The Untold Story of Australia's Exotic Invaders.* University of Chicago Press, Chicago, Illinois. 394 pp.

Lowenthal, D. 1990. Awareness of human impacts: changing attitudes and emphases. Pp. 121–135. *In* B. L. Turner II (ed). *The Earth as Transformed by Human Action: Global and Regional Changes in the Biosphere Over the Past 300 Years.* Cambridge University Press, Cambridge.

Lubin, Y. D. 1984. Changes in the native fauna of the Galápagos Islands following invasion by the little red fire ant, *Wasmannia auropunctata. Biological Journal of the Linnean Society* 21: 229–242.

Ludsin, S. A. and A. D. Wolfe. 2001. Biological invasion theory: Darwin's contributions from the *Origin of Species. BioScience* 51: 780–789.

Lutz, F. E. 1941. *A Lot of Insects: Entomology in a Suburban Garden.* G. P. Putnam's Sons, New York. 304 pp.

Lynch, L. D. and M. B. Thomas. 2000. Nontarget effects in the biocontrol of insects with insects, nematodes and microbial agents: the evidence. *Biocontrol News and Information* 21: 117N–130N.

McCullough, D. G., T. T. Work, J. F. Cavey, A. M. Liebhold, and D. Marshall. 2006. Interceptions of nonindigenous plant pests at US ports of entry and border crossings over a 17-year period. *Biological Invasions* 8: 611–630.

Macdonald, I. A. W., F. J. Kruger, and A. A. Ferrar (eds). 1986. *The Ecology and Management of Biological Invasions in Southern Africa: Proceedings of the National Synthesis Symposium on the Ecology of Biological Invasions.* Oxford University Press, Cape Town.

Macdonald, I. A. W., L. L. Loope, M. B. Usher, and O. Hamann. 1989. Wildlife conservation and the invasion of nature reserves by introduced species: a global perspective. Pp. 215–255. *In* J. A. Drake, H. A. Mooney, F. di Castri, R. H. Groves, F. J. Kruger, M. Rejmánek, and M. Williamson (eds). *Biological Invasions: A Global Perspective.* John Wiley and Sons, Chichester, United Kingdom.

MacDougall, A. S. and R. Turkington. 2005. Are invasive species the drivers or passengers of change in degraded ecosystems? *Ecology* 86: 42–55.

McFadyen, R. E. C. 1998. Biological control of weeds. *Annual Review of Entomology* 43: 369–393.

McGlynn, T. P. 1999. The worldwide transfer of ants: geographical distribution and ecological invasions. *Journal of Biogeography* 26: 535–548.

McGregor, R. C., R. D. Butler, A. Fox, D. Johnson, C. H. Kingsolver, B. Levy, H. E. Pritchard, and R. I. Sailer. 1973. *The emigrant pests: a report to Dr. Francis J. Mulhern, Administrator, Animal and Plant Health Inspection Service, Hyattsville, MD.* U. S. Department of Agriculture, Animal and Plant Health Inspection Service, Washington, DC. 167 pp.

Mack, R. N., D. Simberloff, W. M. Lonsdale, H. Evans, M. Clout, and F. A. Bazzaz. 2000. Biotic invasions: causes, epidemiology, global consequences, and control. *Ecological Applications* 10: 689–710.

McKey, D. 1989. Population biology of figs: applications for conservation. *Experientia* 45: 661–673.

McKinney, M. L. 2005. Species introduced from nearby sources have a more homogenizing effect than species from distant sources: evidence from plants and fishes in the USA. *Diversity and Distributions* 11: 367–374.

MacLeod, A., H. F. Evans, and R. H. A. Baker. 2002. An analysis of pest risk from an Asian longhorn beetle (*Anoplophora glabripennis*) to hardwood trees in the European community. *Crop Protection* 21: 635–645.

McLeod, R. [H.]. 2004. *Counting the cost: impact of invasive animals in Australia, 2004.* Cooperative Research Centre for Pest Animal Control, Canberra. 70 pp.

McLeod, R. S. 1995. Costs of major parasites to the Australian livestock industries. *International Journal for Parasitology* 25: 1363–1367.

McNeely, J. A. 1999. The great reshuffling: how alien species help feed the global economy. Pp. 11–31. *In* O. T. Sandlund, P. J. Schei, and Å. Viken (eds). *Invasive Species and Biodiversity Management.* Kluwer Academic Publishers, Dordrecht.

McNeely, J. [A.] 2001a. Invasive species: a costly catastrophe for native biodiversity. *Land Use and Water Resources Research* 1(2): 1–10.

McNeely, J. A. 2001b. An introduction to human dimensions of invasive alien species. Pp. 5–20. *In* J. A. McNeely (ed). *The Great Reshuffling: Human Dimensions of Invasive Alien Species.* IUCN, Gland, Switzerland; Cambridge, United Kingdom.

McNeely, J. A. (ed). 2001c. *The Great Reshuffling: Human Dimensions of Invasive Alien Species.* IUCN, Gland, Switzerland; Cambridge, United Kingdom. 242 pp.

McNeely, J. A. 2005. Human dimensions of invasive alien species. Pp. 285–309. *In* H. A. Mooney, R. N. Mack, J. A. McNeely, L. E. Neville, P. J. Schei, and J. K. Waage (eds). *Invasive Alien Species: A New Synthesis.* Island Press, Washington, DC.

McNeely, J. A. 2006. As the world gets smaller, the chances of invasion grow. *Euphytica* 148: 5–15.

McNeely, J. A., H. A. Mooney, L. E. Neville, P. J. Schei, and J. K. Waage (eds). 2001. *A global strategy on invasive alien species.* IUCN, Gland, Switzerland. 50 pp.

McNeill, W. H. 1976. *Plagues and Peoples.* Anchor Press/Doubleday, Garden City, New York. 369 pp.

Maerz, J. C., J. M. Karuzas, D. M. Madison, and B. Blossey. 2005. Introduced invertebrates are important prey for a generalist predator. *Diversity and Distributions* 11: 83–90.

Majka, C. [G.] and J. Klimaszewski. 2004. *Phloeocharis subtilissima* Mannerheim (Staphylinidae: Phloeocharinae and *Cephennium gallicum* Ganglbauer (Scydmaenidae) new to North America: a case study in the introduction of exotic Coleoptera to the port of Halifax, with new records of other species. *Zootaxa* 781: 1–15.

Majka, C. G. and F. Shaffer. 2008. Beetles (Coleoptera) in the diet of Piping Plovers in the Iles de la Madeleine, Québec, Canada. *Wader Study Group Bulletin* 115: 55–61.

Majka, C. G., J. Cook, and S. Westby. 2006. Introduced Carabidae (Coleoptera) from Nova Scotia and Prince Edward Island: new records and ecological perspectives. *Canadian Entomologist* 138: 602–609.

Mallet, J. 2005. Hybridization as an invasion of the genome. *Trends in Ecology and Evolution* 20: 229–237.

Marlatt, C. L. 1899. The laisser-faire philosophy applied to the insect problem. Pp. 5–19 *In Proceedings of the Eleventh Annual Meeting of the Association of Economic Entomologists.* Government Printing Office, Washington, DC.

MARN. 2001. *Informe Sobre Las Especies Exoticas en Venezuela.* Republica Bolivariana de Venezuela Ministerio del Ambiente y de los Recursos Naturales, Caracas. 205 pp.

Marsh, G. P. 1864. *Man and Nature, or, Physical Geography as Modified by Human Action.* Scribner, New York. 560 pp.

Martinez, M. and J. C. Malausa. 1999. Quelques introductions accidentelles d'insectes ravageurs en France (Periode 1950–1999): liste chronologique. Pp. 141–147. *In ANNP 5 eme Conference Internationale sur les Ravageurs en Agriculture,* Montepellier, 7–9 December 1999.

Matthews, E. G. and R. L. Kitching. 1984. *Insect ecology,* Second Edition. University of Queensland Press, St. Lucia. 211 pp.

Mattson, W. J., P. Niemelä, I. Millers, and Y. Inguanzo. 1994. *Immigrant Phytophagous Insects on Woody Plants in the United States and Canada: An Annotated List.* U. S. Department of Agriculture, Forest Service, North Central Forest Experiment Station General Technical Report NC-169. 27 pp.

Maw, H. E. L., R. G. Foottit, K. G. A. Hamilton, and G. G. E. Scudder. 2000. *Checklist of the Hemiptera of Canada and Alaska.* NRC Research Press, Ottawa, Ontario. 220 pp.

Mayfield, A. E. III. 2006. The unusual mortality of red bay. *Florida Arborist* 8(4): 8–9.

Medlock, J. M., K. R. Snow, and S. Leach. 2005. Potential transmission of West Nile virus in the British Isles: an ecological review of candidate mosquito bridge vectors. *Medical and Veterinary Entomology* 19: 2–21.

Merritt, R. W., M. K. Kennedy, and E. F. Gersabeck. 1983. Integrated pest management of nuisance and biting flies in a Michigan resort: dealing with secondary pest outbreaks. Pp. 277–299. *In* G. W. Frankie and C. S. Koehler (eds). *Urban Entomology: Interdisciplinary Perspectives.* Praeger, New York.

Messing, R. H., M. N. Tremblay, E. B. Mondor, R. G. Foottit, and K. S. Pike. 2007. Invasive aphids attack native Hawaiian plants. *Biological Invasions* 9: 601–607.

Messing, R. H. and M. G. Wright. 2006. Biological control of invasive species: solution or pollution? *Frontiers in Ecology and the Environment* 4: 132–140.

Metcalf, R. L. 1995. The need for research on exotic pests in California. Pp. 5–39. *In* J. G. Morse, R. L. Metcalf, J. R. Carey, and R. V. Dowell (eds). *Proceedings: The Medfly in California: Defining Critical Research.* UC Center for Exotic Pest Research Workshop, 9–11 November 1994, Riverside, California.

Metcalf, R. L. and R. A. Metcalf. 1993. *Destructive and Useful Insects: Their Habits and Control,* Fifth Edition. McGraw-Hill, New York. (various pagings).

Michaud, J. P. 2002. Invasion of the Florida citrus ecosystem by *Harmonia axyridis* (Coleoptera: Coccinellidae) and asymmetric competition with a native species, *Cycloneda sanguinea. Environmental Entomology* 31: 827–835.

Miller, D. R. and A. Y. Rossman. 1995. Systematics, biodiversity, and agriculture. *BioScience* 45: 680–686.

Miller, D. R., G. L. Miller, G. S. Hodges, and J. A. Davidson. 2005. Introduced scale insects (Hemiptera: Coccoidea) of the United States and their impact on U. S. agriculture. *Proceedings of the Entomological Society of Washington* 107: 123–158.

Miller, S. E. and L. G. Eldredge. 1996. Numbers of Hawaiian species: supplement 1. *Bishop Museum Occasional Papers* 45: 8–17.

Mockford, E. L. 1993. *North American Psocoptera (Insecta). Flora and Fauna Handbook 10.* Sandhill Crane Press, Gainesville, Florida. 455 pp.

Moller, H. 1996. Lessons for invasion theory from social insects. *Biological Conservation* 78: 125–142.

Mondor, E. B., M. N. Tremblay, and R. H. Messing. 2007. Morphological and ecological traits promoting aphid colonization of the Hawaiian Islands. *Biological Invasions* 9: 87–100.

Mooney, H. A. 1999. The Global Invasive Species Program (GISP). *Biological Invasions* 1: 97–98.

Mooney, H. A. 2005. Invasive alien species: the nature of the problem. Pp. 1–15. *In* H. A. Mooney, R. N. Mack, J. A. McNeely, L. E. Neville, P. J. Schei, and J. K. Waage (eds). *Invasive Alien Species: A New Synthesis.* Island Press, Washington, DC.

Mooney, H. A. and E. E. Cleland. 2001. The evolutionary impact of invasive species. *Proceedings of the National Academy of Sciences* 98: 5446–5451.

Mooney, H. A. and J. A. Drake (eds). 1986. *Ecology of Biological Invasions of North America and Hawaii.* Springer-Verlag, New York. 321 pp.

Mooney, H. A., S. P. Hamburg, and J. A. Drake. 1986. The invasions of plants and animals into California. Pp. 250–272. *In* H. A. Mooney and J. A. Drake (eds). *Ecology of Biological Invasions of North America and Hawaii.* Springer-Verlag, New York.

Morimoto, N. and K. Kiritani. 1995. Fauna of exotic insects in Japan. *Bulletin of the National Institute of Agro-Environmental Sciences* 12: 87–120.

Morris, R. F. 1983. Introduced terrestrial insects. Pp. 551–591. *In* G. R. South (ed). *Biogeography and Ecology of the Island of Newfoundland.* W. Junk, The Hague, Netherlands.

Morrison, L. W. 2002. Long-term impacts of an arthropod-community invasion by the imported fire ant, *Solenopsis invicta. Ecology* 83: 2337–2345.

Morse, J. G. and M. S. Hoddle. 2006. Invasion biology of thrips. *Annual Review of Entomology* 51: 67–89.

Morse, R. A. and N. W. Calderone. 2000. The value of honey bees as pollinators of U. S. crops in 2000. *Bee Culture* 128(3): 15-page supplement between pp. 24–25.

Muirhead, J. R., B. Leung, C. van Overdijk, D. W. Kelly, K. Nandakumar, K. R. Marchant, and H. J. MacIsaac. 2006. Modelling local and long-distance dispersal of invasive emerald ash borer *Agrilus planipennis* (Coleoptera) in North America. *Diversity and Distributions* 12: 71–79.

Mullen, G. R. 2002. Moths and butterflies (*Lepidoptera*). Pp. 363–381. *In* G. Mullen and L. Durden (eds). *Medical and Veterinary Entomology.* Academic Press, Amsterdam.

Mullen, G. and L. Durden (eds). 2002. *Medical and veterinary entomology.* Academic Press, Amsterdam. 597 pp.

Musolin, D. L. 2007. Insects in a warmer world: ecological, physiological and life-history responses of true bugs (Heteroptera) to climate change. *Global Change Biology* 13: 1565–1585.

Musolin, D. L. and K. Fujisaki. 2006. Changes in ranges: trends in distribution of true bugs (Heteroptera) under conditions of the current climate warming. *Russian Entomological Journal* 15: 175–179.

Myers, J. H. 2000. What can we learn from biological control failures? Pp. 151–154. *In* N. R. Spencer (ed). *Proceedings of the X International Symposium on Biological Control of Weeds,* 4–14 July 1999, Montana State University, Bozeman, Montana.

Myers, J. H. and G. Hosking. 2002. Eradication. Pp. 293–307. *In* G. J. Hallman and C. P. Schwalbe (eds). *Invasive Arthropods in Agriculture: Problems and Solutions.* Oxford and IBH Publishing, New Delhi.

Nadel, H., J. H. Frank, and R. J. Knight, Jr. 1992. Escapees and accomplices: the naturalization of exotic *Ficus* and their associated faunas in Florida. *Florida Entomologist* 75: 29–38.

National Research Council. 2002. *Predicting Invasions of Nonindigenous Plants and Plant Pests.* National Academy Press, Washington, DC. 194 pp.

Nechols, J. R. 2003. Biological control of the spherical mealybug on Guam and in the northern Marianas Islands: a classic example of fortuitous biological control. Pp. 324–329. *In Proceedings of the 1st International Symposium on Biological Control of Arthropods, January 14–18, 2002, Honolulu, Hawaii.* USDA Forest Service, Morgantown, West Virginia. FHTET-03-05.

Negrón, J. F., J. J. Witcosky, R. J. Cain, J. R. LaBonte, D. A. Duerr II, S. J. McElwey, J. C. Lee, and S. J. Seybold. 2005. The banded elm bark beetle: a new threat to elms in North America. *American Entomologist* 51: 84–94.

Nelder, M. P., E. S. Paysen, P. A. Zungoli, and E. P. Benson. 2006. Emergence of the introduced ant *Pachycondyla chinensis* (Formicidae: Ponerinae) as a public health threat in the southeastern United States. *Journal of Medical Entomology* 43: 1094–1098.

Nentwig, W. 2007. General conclusion, or what has to be done now? Pp. 419–423. *In* W. Nentwig (ed). *Biological Invasions.* Springer, Berlin.

Ness, J. H. and J. L. Bronstein. 2004. The effects of invasive ants on prospective ant mutualists. *Biological Invasions* 6: 445–461.

New, T. R. 1993. Effects of exotic species on Australian native insects. Pp. 155–169. *In* K. J. Gaston, T. R. New, and M. J. Samways (eds). *Perspectives on Insect Conservation.* Intercept, Andover, United Kingdom.

New, T. R. 1994. *Exotic Insects in Australia*. Gleneagles Publishing, Adelaide. 138 pp.

New, T. R. 2005a. 'Inordinate fondness': a threat to beetles in south east Asia? *Journal of Insect Conservation* 9: 147–150.

New, T. R. 2005b. *Invertebrate Conservation and Agricultural Ecosystems*. Cambridge University Press, Cambridge. 354 pp.

Niemelä, P. and W. J. Mattson. 1996. Invasion of North American forests by European phytophagous insects: legacy of the European crucible? *BioScience* 46: 741–753.

Nishida, G. M. (ed). 1994. Hawaiian Terrestrial Arthropod Checklist. 2nd Ed. Bishop Museum Technical Report 4. 287 pp.

Nojima, S., C. Schal, F. X. Webster, R. G. Santangelo, and W. L. Roelofs. 2005. Identification of the sex pheromone of the German cockroach, *Blatella germanica*. *Science* 307: 1104–1106.

Novak, S. J. and R. N. Mack. 2001. Tracing plant introduction and spread: genetic evidence from *Bromus tectorum* (cheatgrass). *BioScience* 51: 114–122.

Nowak, D. J., J. E. Pasek, R. A. Sequeira, D. E. Crane, and V. C. Mastro. 2001. Potential effect of *Anoplophora glabripennis* (Coleoptera: Cerambycidae) on urban trees in the United States. *Journal of Economic Entomology* 94: 116–122.

Occhipinti-Ambrogi, A. and B. S. Galil. 2004. A uniform terminology on bioinvasions: a chimera or an operative tool? *Marine Pollution Bulletin* 49: 688–694.

O'Dowd, D. J., P. T. Green, and P. S. Lake. 2003. Invasional 'meltdown' on an oceanic island. *Ecology Letters* 6: 812–817.

Olden, J. D., M. E. Douglas, and M. R. Douglas. 2005. The human dimensions of biotic homogenization. *Conservation Biology* 19: 2036–2038.

Oliveira, M. R. V., T. J. Henneberry, and P. Anderson. 2001. History, current status, and collaborative research projects for *Bemisia tabaci*. *Crop Protection* 20: 709–723.

Oman, P. 1968. Prevention, surveillance and management of invading pest insects. *Bulletin of the Entomological Society of America* 14: 98–102.

Onstad, D. W. and M. L. McManus. 1996. Risks of host range expansion by parasites of insects. *BioScience* 46: 430–435.

Osborne, L. S. and J. P. Cuda. 2003. Release of exotic natural enemies for biological control: a case of damned if we do and damned if we don't? *Journal of Land Use and Environmental Law* 18: 399–407.

Owen, D. R. and J. W. Lownsbery. 1989. Dutch elm disease. Pp. 128–146. *In* D. L. Dahlsten and R. Garcia (eds) *Eradication of Exotic Pests: Analysis with Case Histories*. Yale University Press, New Haven, Connecticut.

Paine, T. D. and J. G. Millar. 2002. Insect pests of eucalypts in California: implications of managing invasive species. *Bulletin of Entomological Research* 92: 147–151.

Paine, T. D., K. M. Jetter, K. M. Klonsky, L. G. Bezark, and T. S. Bellows. 2003. Ash whitefly and biological control in the urban environment. Pp. 203–213. *In* D. A. Sumner (ed). *Exotic Pests and Diseases: Biology and Economics for Biosecurity*. Iowa State Press, Ames, Iowa.

Painter, R. H. 1951. *Insect Resistance in Crop Plants*. Macmillan, New York. 520 pp.

Paoletti, M. G. (ed). 2005. *Ecological Implications of Minilivestock: Potential of Insects, Rodents, Frogs, and Snails*. Science Publishers, Enfield, New Hampshire. 648 pp.

Papp, C. S. 2001. *A Comprehensive Guide to North American Insects: With Notes on Other Arthropods of Health Importance*. Gilbert Industries, Jonesboro, Arkansas. 466 pp.

Parker, I. M. 2001. Invasion ecology: echoes of Elton in the twenty-first century. *Conservation Biology* 15: 806–807.

Parker, I. M., D. Simberloff, W. M. Lonsdale, K. Goodell, M. Wonham, P. M. Kareiva, M. H. Williamson, B. Von Holle, P. B. Moyle, J. E. Byers, and L. Goldwasser. 1999. Impact: toward a framework for understanding the ecological effects of invaders. *Biological Invasions* 1: 3–19.

Parshley, H. M. 1916. On some Tingidae from New England. *Psyche* 23: 163–168.

Passera, L. 1994. Characteristics of tramp species. Pp. 23–43. *In* D. F. Williams (ed). *Exotic Ants: Biology, Impact, and Control of Introduced Species*. Westview Press, Boulder, Colorado.

Patterson, J. T. and W. S. Stone. 1952. *Evolution in the genus Drosophila*. Macmillan, New York. 610 pp.

Pauly, P. J. 2002. Fighting the Hessian fly: American and British responses to insect invasion, 1776–1789. *Environmental History* 7: 485–507.

Pearson, D. E. and R. M. Callaway. 2006. Biological agents elevate hantavirus by subsidizing deer mouse populations. *Ecology Letters* 9: 443–450.

Peck, S. B., J. Heraty, B. Landry, and B. J. Sinclair. 1998. Introduced insect fauna of an oceanic archipelago: the Galápagos Islands, Ecuador. *American Entomologist* 44: 218–237.

Pellizzari, G. and L. Dalla Montà. 1997. 1945–1995: fifty years of incidental insect pest introduction to Italy. *Acta Phytopathologica et Entomologica Hungarica* 32: 171–183.

Pellizzari, G., L. Dalla Montà, and V. Vacante. 2005. Alien insect and mite pests introduced to Italy in sixty years (1945–2004). Pp. 275–276. *In* D. V. Alford and G. F Backhaus (eds). *Plant Protection and Plant Health in Europe: Introduction and Spread of Invasive Species*. Symposium Proceedings No. 81. British Crop Production Council, Alton, United Kingdom.

Pemberton, R. W. 1988. The use of the Thai giant waterbug, *Lethocerus indicus* (Hemiptera: Belostomatidae), as human food in California. *Pan-Pacific Entomologist* 64: 81–82.

Pemberton, R. W. 2000. Predictable risk to native plants in weed biological control. *Oecologia* 125: 489–494.

Perez Moreno, I. 1999. Plagas introducidas en Espana peninsular en la segunda mitad del siglo XX. *Aracnet, Boletin Electronico de Entomologia* 4: 1–14. http://entomologia.rediris. es/aracnet/num4/entomap/.

Perkins, J. H. 1978. Edward Fred Knipling's sterile-male technique for control of the screwworm fly. *Environmental Review* 5: 19–37.

Perkins, R. C. L. 1897. The introduction of beneficial insects into the Hawaiian Islands. *Nature* 55: 499–500.

Perring, T. M. 2001. The *Bemisia tabaci* species complex. *Crop Protection* 20: 725–737.

Perrings, C., M. H. Williamson, and S. Dalmazzone (eds). 2000. *The Economics of Biological Invasions*. Edward Elgar, Cheltenham, United Kingdom. 249 pp.

Perrings, C., S. Dalmazzone, and M. Williamson. 2005. The economics of biological invasions. Pp. 16–35. *In* H. A. Mooney, R. N. Mack, J. A. McNeely, L. E. Neville, P. J. Schei, and J. K. Waage (eds). *Invasive Alien Species: A New Synthesis*. Island Press, Washington, DC.

Peterson, R. K. D. and B. A. Shurdut. 1999. Human health risks from cockroaches and cockroach management: a risk analysis approach. *American Entomologist* 45: 142–148.

Petit, R. J. 2004. Biological invasions at the gene level. *Diversity and Distributions* 10: 159–165.

Peyton, E. L., S. R. Campbell, T. M. Candeletti, M. Romanowski, and W. J. Crans. 1999. *Aedes* (*Finlaya*) *japonicus* (Theobald), a new introduction into the United States. *Journal of the American Mosquito Control Association* 15: 238–241.

Pimentel, D. 2002. Introduction: non-native species in the world. Pp. 3–8. *In* D. Pimentel (ed). *Biological Invasions: Economic and Environmental Costs of Alien Plant, Animal, and Microbe Species*. CRC Press, Boca Raton, Florida.

Pimentel, D., L. Lach, R. Zuniga, and D. Morrison. 2000. Environmental and economic costs associated with non-indigenous species in the United States. *BioScience* 50: 53–65.

Pimentel. D., S. McNair, J. Janecka, J. Wightman, C. Simmonds, C. O'Connell, E. Wong, L. Russel, J. Zern, T. Aquino, and T. Tsomondo. 2001. Economic and environmental threats of alien plant, animal, and microbe invasions. *Agriculture, Ecosystems and Environment* 84: 1–20.

Pimentel, D., R. Zuniga, and D. Morrison. 2005. Update on the environmental and economic costs associated with alien-invasive species in the United States. *Ecological Economics* 52: 273–288.

Polhemus, J. T. and V. Golia. 2006. *Micronecta ludibunda* Breddin (Heteroptera: Corixidae: Micronectinae), the second Asian water bug introduced into Florida. *U. S. A. Entomological News* 117: 531–534.

Polhemus, J. T. and R. P. Rutter. 1997. *Synaptonecta issa* (Heteroptera: Corixidae), first New World record of an Asian water bug in Florida. *Entomological News* 108: 300–304.

Popham, W. L. and D. G. Hall. 1958. Insect eradication programs. *Annual Review of Entomology* 3: 335–354.

Porter, S. D. and D. A. Savignano. 1990. Invasion of polygyne fire ants decimates native ants and disrupts arthropod community. *Ecology* 71: 2095–2106.

Potter, K. M., J. Frampton, and J. Sidebottom. 2005. Impacts of balsam woolly adelgid on the southern Appalachian spruce-fir ecosystem and the North Carolina Christmas tree industry. Pp. 25–41. *In* B. Onken and R. Reardon (compilers). *Proceedings of the Third Symposium on Hemlock Woolly Adelgid in the Eastern United States*, February, 1–3, 2005, Asheville, North Carolina. U. S. Department of Agriculture, Forest Service FHTET-2005-01.

Pouget, R. 1990. *Histoire de la Lutte Contre le Phylloxéra de la Vigne en France (1868–1895)*. INRA, Paris. 156 pp.

Powell, J. A. 1964. Two scavenger moths of the genus *Borkhausenia* introduced from Europe to the west coast of North America (Lepidoptera: Oecophoridae). *Pan-Pacific Entomologist* 40: 218–221.

Powell, J. A. and C. L. Hogue. 1979. *California Insects*. University of California Press, Berkeley, California. 388 pp.

Pratt, R. J. 2004. Invasive threats to the American homeland. *Parameters (Journal of the U. S. Army War College)* 33(1): 44–61.

Price, P. W. 1984. *Insect Ecology*. Second Edition. John Wiley and Sons, New York. 607 pp.

Pyle, R., M. Bentzien, and P. Opler. 1981. Insect conservation. *Annual Review of Entomology* 26: 233–258.

Pyšek, P., V. Jarošík, and T. Kučera. 2003. Inclusion of native and alien species in temperate nature reserves: an historical study from Central Europe. *Conservation Biology* 17: 1414–1424.

Rabitsch, W. 2008. Alien true bugs of Europe (Insecta: Hemiptera: Heteroptera). *Zootaxa* 1827: 1–44.

Rabitsch, W. and F. Essl. 2006. Biological invasions in Austria: patterns and case studies. *Biological Invasions* 8: 295–308.

Rai, K. S. 1991. *Aedes albopictus* in the Americas. *Annual Review of Entomology* 36: 459–484.

Ranjan, R. 2006. Economic impacts of pink hibiscus mealybug in Florida and the United States. *Stochastic Environmental Research and Risk Assessment* 20: 353–362.

Reagan, T. E. 1986. Beneficial aspects of the imported fire ant: a field ecology approach. Pp. 58–71. *In* C. S. Lofgren and R. K. Vander Meer (eds). *Fire Ants and Leaf-cutting Ants: Biology and Management*. Westview Press, Boulder, Colorado.

Reaser, J. K. 2001. Invasive alien species prevention and control: the art and science of managing people. Pp. 89–104. *In* J. A. McNeely (ed). *The Great Reshuffling: Human Dimensions of Invasive Alien Species*. IUCN, Gland, Switzerland; Cambridge, United Kingdom.

Reeves, W. K. and J. A. Korecki. 2004. *Ochlerotatus japonicus japonicus* (Theobald) (Diptera: Culicidae), a new invasive mosquito for Georgia and South Carolina. *Proceedings of the Entomological Society of Washington* 106: 233–234.

Reimer, N. J. 1988. Predation on *Liothrips urichi* Karny (Thysanoptera: Phlaeothripidae): a case of biotic interference. *Environmental Entomology* 17: 132–134.

Reitz, S. R. and J. T. Trumble. 2002. Competitive displacement among insects and arachnids. *Annual Review of Entomology* 47: 435–465.

Reinhardt, F., M. Herle, F. Bastiansen, and B. Streit. 2003. *Economic Impact of the Spread of Alien Species in Germany*. Federal Environmental Agency of Germany, Berlin. 190 pp.

Rejmánek, M., D. M. Richardson, M. G. Barbour, M. J. Crawley, G. F. Hrusa, P. B. Moyle, J. M. Randall, D. Simberloff, and M. Williamson. 2002. Biological invasions: politics and the discontinuity of ecological terminology. *Bulletin of the Ecological Society of America* 83: 131–133.

Reynolds, H. T. (chairman), R. Clark, G. Frankie, F. E. Gilstrap, P. J. Hamman, M. Kogan, E. H. Smith, and D. E. Weidhaas. 1982. Management of established vs. introduced pests. Pp. 82–90. *In* S. L. Battenfield (ed). *Proceedings of the Symposium on the Imported Fire Ant*, June 7–10, 1982, Atlanta, Georgia. Organized and Managed by Inter-Society Consortium for Plant Protection; Sponsored by Environmental Protection Agency/USDA, Animal, Plant Health Inspection Service [Washington, DC].

Rhymer, J. M. and D. Simberloff. 1996. Extinction by hybridization and introgression. *Annual Review of Ecology and Systematics* 27: 83–109.

Ricciardi, A. and J. Cohen. 2007. The invasiveness of an introduced species does not predict its impact. *Biological Invasions* 9: 309–315.

Richardson, D. M. 2005. Faster! Faster! Faster! Biological invasions and accelerated evolution. *Diversity and Distributions* 11: 361.

Richardson, D. M., P. Pyšek, M. Rejmánek, M. G. Barbour, F. D. Panetta, and C. J. West. 2000. Naturalization and invasion of alien plants: concepts and definitions. *Diversity and Distributions* 6: 93–107.

Richardson, D. M., J. A. Cambray, R. A. Chapman, W. R. J. Dean, C. L. Griffiths, D. C. Le Maitre, D. J. Newton, and T. J. Winstanley. 2003. Vectors and pathways of biological invasions in South Africa – past, present, and future. Pp. 292–349. *In* G. M. Ruiz and J. T. Carlton (eds). *Invasive Species: Vectors and Management Strategies*. Island Press, Washington, DC.

Ridley, G. S., J. Bain, L. S. Bulman, M. A. Dick, and M. K. Kay. 2000. Threats to New Zealand's indigenous forests from exotic pathogens and pests. *Science for Conservation* 142: 1–67.

Riley, C. V. 1888. The Hessian fly: an imported insect. *Canadian Entomologist* 20: 121–127.

Rivers, D. 2006. Teaching general entomology to disinterested undergraduates. *American Entomologist* 52: 24–28.

Rodas, S. and R. J. O'Neil. 2006. A survey of Indiana soybean producers following the introduction of a new invasive pest, the soybean aphid. *American Entomologist* 52: 146–149.

Rodriguez, L. F. 2006. Can invasive species facilitate native species? Evidence of how, when, and why these impacts occur. *Biological Invasions* 8: 927–939.

Roelofs, W. L., A. Comeau, and R. Selle. 1969. Sex pheromone of the oriental fruit moth. *Nature* 224: 723.

Roelofs, W. [L.], A. Comeau, A. Hill, and G. Milicevic. 1971. Sex attractant of the codling moth: characterization with electroantennogram technique. *Science* 174: 297–299.

Roelofs, W. L., W. Liu, G. Hao, H. Jiao, A. Rooney, and C. E. Linn, Jr. 2002. Evolution of moth sex pheromones

via ancestral genes. *Proceedings of the National Academy of Sciences USA* 99: 13621–13626.

Roitberg, B. D. 2000. Threats, flies, and protocol gaps: can evolutionary ecology save biological control? Pp. 254–265. *In* M. E. Hochberg and A. R. Ives (eds). *Parasitoid Population Biology*. Princeton University Press, Princeton, New Jersey.

Roll, U., T. Dayan, and D. Simberloff. 2007. Non-indigenous insect species in Israel and adjacent areas. *Biological Invasions* 9: 629–643.

Roque-Albelo, L. and C. Causton. 1999. El Niño and introduced insects in the Galápagos Islands: different dispersal strategies, similar effects. *Noticias de Galápagos* 60: 30–36.

Ross, K. G. and L. Keller. 1995. Ecology and evolution of social organization: insights from fire ants and other highly eusocial insects. *Annual Review of Ecology and Systematics* 26: 631–656.

Rossman, A. Y. and D. R. Miller. 1996. Systematics solves problems in agriculture and forestry. *Annals of the Missouri Botanical Garden* 83: 17–28.

Rossman, A. Y., G. J. Samuels, C. T. Rogerson, and R. Lowen. 1999. Genera of Bionectriaceae, Hypocreaceae and Nectriaceae (Hypocreales, Ascomycetes). *Studies in Mycology* 42: 1–260.

Rowles, A. D. and D. J. O'Dowd. 2007. Interference competition by Argentine ants displaces native ants: implications for biotic resistance to invasion. *Biological Invasions* 9: 73–85.

Ruesink, J. L., I. M. Parker, M. J. Groom, and P. M. Kareiva. 1995. Reducing the risks of nonindigenous species introductions: guilty until proven innocent. *BioScience* 45: 465–477.

Ruiz, G. M. and J. T. Carlton (eds.). 2003. Invasive Species: Vectors and Management Strategies. Island Press, Washington, DC. 518 pp.

Sabrosky, C. W. 1955. The interrelations of biological control and taxonomy. *Journal of Economic Entomology* 48: 710–714.

Sadler, J. 1991. Beetles, boats and biogeography: insect invaders of the North Atlantic. *Acta Archaeologica* 61: 199–212.

Sadler, J. P. and P. Skidmore. 1995. Introductions, extinctions or continuity? Faunal change in the North Atlantic islands. Pp. 206–225. *In* R. A. Butlin and N. Roberts (eds). *Ecological Relations in Historical Times: Human Impact and Adaptation*. Blackwell, Oxford, United Kingdom.

Sailer, R. I. 1972. Concepts, principles and potentials of biological control parasites and predators. *Proceedings of the North Central Branch Entomological Society of America* 27: 35–39.

Sailer, R. I. 1978. Our immigrant insect fauna. *Bulletin of the Entomological Society of America* 24: 3–11.

Sailer, R. I. 1983. History of insect introductions. Pp. 15–38. *In* C. L. Wilson and C. L. Graham (eds). *Exotic Plant Pests and North American Agriculture*. Academic Press, New York.

Sala, O. E., F. S. Chapin III, J. J. Armesto, E. Berlow, J. Bloomfield, R. Dirzo, E. Huber-Sanwald, L. F. Huenneke, R. B. Jackson, A. Kinzig, R. Leemans, D. M. Lodge, H. A. Mooney, M. Oesterheld, N. L. Poff, M. T. Sykes, B. H. Walker, M. Walker, and D. H. Wall. 2000. Global biodiversity scenarios for the year 2100. *Science* 287: 1770–1774.

Sames, W. J. and D. Pehling. 2005. Update on *Ochlerotatus japonicus* in the state of Washington. *Journal of the American Mosquito Control Association* 21: 98–99.

Samways, M. J. 1988. Classical biological control and insect conservation: are they compatible? *Environmental Conservation* 15: 349–354, 348.

Samways, M. J. 1994. *Insect Conservation Biology*. Chapman and Hall, London. 358 pp.

Samways, M. J. 1997. Classical biological control and biodiversity conservation: what risks are we prepared to accept? *Biodiversity and Conservation* 6: 1309–1316.

Sanders, N. J., N. J. Gotelli, N. E. Heller, and D. M. Gordon. 2003. Community disassembly by an invasive species. *Proceedings of the National Academy of Sciences USA* 100: 2474–2477.

Sands, D. and R. G. Van Driesche. 2003. Host range testing techniques for parasitoids and predators. Pp. 41–53. *In Proceedings of the 1st International Symposium on Biological Control of Arthropods, January 14–18, 2002, Honolulu, Hawaii*. USDA Forest Service, Morgantown, West Virginia. FHTET-03-05.

Sasscer, E. R. 1940. Undesirable insect aliens. *Journal of Economic Entomology* 33: 1–8.

Sax, D. F., S. D. Gaines, and J. J. Stachowicz. 2005. Introduction. Pp. 1–7. *In* D. F. Sax, J. J. Stachowicz, and S. D. Gaines (eds). *Species Invasions: Insights into Ecology, Evolution, and Biogeography*. Sinauer Associates, Sunderland, Massachusetts.

Schäfer, M. L., J. O. Lundström, M. Pfeffer, E. Lundkvist, and J. Landin. 2004. Biological diversity versus risk for mosquito nuisance and disease transmission in constructed wetlands in southern Sweden. *Medical and Veterinary Entomology* 18: 256–267.

Schlaepfer, M. A., P. W. Sherman, B. Blossey, and M. C. Runge. 2005. Introduced species as evolutionary traps. *Ecology Letters* 8: 241–246.

Schmitz, D. C. and D. Simberloff. 2001. Needed: a national center for biological invasions. *Issues in Science and Technology* 17(4): 57–62.

Schneider, S. S., G. DeGrandi-Hoffman, and D. R. Smith. 2004. The African honey bee: factors contributing to a successful biological invasion. *Annual Review of Entomology* 49: 351–376.

Schneider, S. [S.], G. DeGrandi-Hoffman, D. Smith, and D. Tarpy. 2006. The African honey bee: a case study of a biological invasion. *Bee Culture* 134(4): 21–24.

Schuurman, J. J. 1937. Contributions to the genetics of *Tenebrio molitor* L. *Genetics* 19: 273–355.

Schwartz, P. H. and W. Klassen. 1981. Estimate of losses caused by insects and mites to agricultural crops. Pp. 15–77. *In* D. Pimentel (ed). *CRC Handbook of Pest Management in Agriculture*. Vol. I. CRC Press, Boca Raton, Florida.

Scott, J. K. 2001. Europe gears-up to fight invasive species. *Trends in Ecology and Evolution* 16: 171–172.

Scott, R. R. (ed). 1984. *New Zealand Pests and Beneficial Insects*. Lincoln University College of Agriculture, Canterbury. 373 pp.

Scudder, G. G. E. and R. G. Foottit. 2006. Alien true bugs (Hemiptera: Heteroptera) in Canada: composition and adaptations. *Canadian Entomologist* 138: 24–51.

Scudder, G. G. E. and D. D. McE. Kevan. 1984. A checklist of the orthopteroid insects recorded from British Columbia. *Journal of the Entomological Society of British Columbia* 81: 76–79.

Seier, M.-K. 2005. Exotic beneficials in classical biological control of invasive alien weeds: friends or foes? Pp. 191–194. *In* D. V. Alford and G. F. Backhaus (eds). *Plant Protection and Plant Health in Europe: Introduction and Spread of Invasive Species*. Symposium Proceedings No. 81. British Crop Production Council, Alton, United Kingdom.

Shannon, M. J. 1983. Systematics: a basis for effective regulatory activities. *Bulletin of the Entomological Society of America* 29(3): 47–49.

Sheppard, W. S. 1989. A history of the introduction of honey bee races into the United States. *American Bee Journal* 129: 617–619, 664–667.

Shimanuki, H. 1992. The honey bee deserves to be our national insect. Pp. 34–39. *In* J. Adams (ed). *Insect Potpourri: Adventures in Entomology*. Sandhill Crane Press, Gainesville, Florida.

Simberloff, D. 1981. Community effects of introduced species. Pp. 53–81. *In* M. H. Nitecki (ed). *Biotic Crises in Ecological and Evolutionary Time*. Academic Press, New York.

Simberloff, D. 1986. Introduced insects: a biogeographic and systematic perspective. Pp. 3–26. *In* H. A. Mooney and J. A. Drake (eds). *Ecology of Biological Invasions of North America and Hawaii*. Springer-Verlag, New York.

Simberloff, D. 1995. Why do introduced species appear to devastate islands more than mainland areas? *Pacific Science* 49: 87–97.

Simberloff, D. 1996. Impacts of introduced species in the United States. *Consequences* 2(2): 13–23.

Simberloff, D. 1997. The biology of invasions. Pp. 3–17. *In* D. Simberloff, D. C. Schmitz, and T. C. Brown (eds). *Strangers in Paradise: Impact and Management of Nonindigenous Species in Florida*. Island Press, Washington, DC.

Simberloff, D. 2000. Nonindigenous species – a global threat to biodiversity and stability. Pp. 325–334. *In* P. Raven and T. Williams (eds). *Nature and Human Society: The Quest for a Sustainable World*. National Academy Press, Washington, DC.

Simberloff, D. 2003. Introduced insects. Pp. 597–602. *In* V. H. Resh and R. T. Cardé (eds). *Encyclopedia of Insects*. Academic Press, Amsterdam.

Simberloff, D. 2004. A rising tide of species and literature: a review of some recent books on biological invasions. *BioScience* 54: 247–254.

Simberloff, D. 2005. The politics of assessing risk for biological invasions: the USA as a case study. *Trends in Ecology and Evolution* 20: 216–222.

Simberloff, D. and M. Alexander. 1998. Assessing risks to ecological systems from biological introductions (excluding genetically modified organisms). Pp. 147–176. *In* P. Calow (ed). *Handbook of Environmental Risk Assessment and Management*. Blackwell Science, Oxford, United Kingdom.

Simberloff, D. and B. Von Holle. 1999. Positive interactions of nonindigenous species: invasional meltdown? *Biological Invasions* 1: 21–32.

Simberloff, D., D. C. Schmitz, and T. C. Brown. 1997. Why we should care and what we should do. Pp. 359–367. *In* D. Simberloff, D. C. Schmitz, and T. C. Brown (eds). *Strangers in Paradise: Impact and Management of Nonindigenous Species in Florida*. Island Press, Washington, DC.

Simmonds, R. B. and M. G. Pogue. 2004. Redescription of two often-confused noctuid pests, *Copitarsia decolora* and *Copitarsia incommoda* (Lepidoptera: Noctuidae: Cuculliinae). *Annals of the Entomological Society of America* 97: 1159–1164.

Simpson, A., E. Sellers, A. Grosse, and Y. Xie. 2006. Essential elements of online information networks on invasive alien species. *Biological Invasions* 8: 1579–1587.

Sinclair, W. A. 1978a. Range, suscepts, losses. Pp. 6–8. *In* W. A. Sinclair and R. J. Campana (eds). *Dutch Elm Disease: Perspectives After 60 Years*. Search (Agriculture, Plant Pathology) 8(5).

Sinclair, W. A. 1978b. Epidemiology. Pp. 27–30. *In* W. A. Sinclair and R. J. Campana (eds). *Dutch elm disease: perspectives after 60 years*. Search (Agriculture, Plant Pathology) 8(5).

Sinclair, W. A. and H. H. Lyon. 2005. *Diseases of trees and shrubs*, Second Edition. Cornell University Press, Ithaca, New York. 660 pp.

Small, M. J., C. J. Small, and G. D. Dreyer. 2005. Changes in a hemlock-dominated forest following woolly adelgid infestation in southern New England. *Journal of the Torrey Botanical Society* 132: 458–470.

Smith, H. S. 1933. The influence of civilization on the insect fauna by purposeful introductions. *Annals of the Entomological Society of America* 26: 518–528.

Smith, R. M., R. H. A. Baker, C. P. Malumphy, S. Hockland, R. P. Hammon, J. C. Ostojá-Starzewski, and D. W. Collins. 2005. Non-native invertebrate plant pests established in Great Britain: an assessment of patterns and trends. Pp. 119–124. *In* D. V. Alford and G. F. Backhaus (eds). *Plant Protection and Plant Health in Europe: Introduction and Spread of Invasive Species. Symposium Proceedings No. 81*. British Crop Production Council, Alton, United Kingdom.

Snyder, C. D., J. A. Young, R. M. Ross, and D. R. Smith. 2005. Long-term effects of hemlock forest decline on headwater stream communities. Pp. 42–55. *In* B. Onken and R. Reardon (compilers). *Proceedings of the Third Symposium on Hemlock Woolly Adelgid in the Eastern United States, February 1–3, 2005, Asheville, North Carolina*. U. S. Department of Agriculture Forest Service FHTET-2005-01.

Snyder, W. E. and E. W. Evans. 2006. Ecological effects of invasive arthropod generalist predators. *Annual Review of Ecology, Evolution, and Systematics* 37: 95–122.

Sokoloff, A. 1966. *The genetics of Tribolium and related species. Advances in Genetics Supplement 1*. Academic Press, New York. 212 pp.

Soulé, M. E. 1990. The onslaught of alien species, and other challenges in the coming decades. *Conservation Biology* 4: 233–239.

Southwood, T. R. E. 1960. The flight activity of Heteroptera. *Transactions of the Royal Entomological Society of London* 112: 173–220.

Spear, R. J. 2005. *The Great Gypsy Moth War: The History of the First Campaign in Massachusetts to Eradicate the Gypsy Moth, 1890–1901*. University of Massachusetts Press, Amherst, Massachusetts. 308 pp.

Spence, J. R. and D. H. Spence. 1988. Of ground-beetles and men: introduced species and the synanthropic fauna of western Canada. *Memoirs of the Entomological Society of Canada* 144: 151–168.

Spira, T. P. 2001. Plant–pollinator interactions: a threatened mutualism with implications for the ecology and management of rare plants. *Natural Areas Journal* 21: 78–88.

Stadler, B., T. Müller, D. Orwig, and R. Cobb. 2005. Hemlock woolly adelgid in New England forests: canopy impacts transforming ecosystem processes and landscapes. *Ecosystems* 8: 233–247.

Stevens, L. E. and T. Ayers. 2002. The biodiversity and distribution of exotic vascular plants and animals in the Grand Canyon region. Pp. 241–265. *In* B. Tellman (ed). *Invasive Exotic Species in the Sonoran Region*. University of Arizona Press/Arizona-Sonora Desert Museum, Tucson, Arizona.

Storer, A. J., J. N. Rosemier, B. L. Beachy, and D. J. Flaspohler. 2005. Potential effects of beech bark disease and decline in beech abundance on birds and small mammals. Pp. 72–78. *In* C. A. Evans, J. A. Lucas, and M. J. Twery (eds). *Beech Bark Disease: Proceedings of the Beech Bark Disease Symposium, Saranac Lake, New York, June 16–18, 2004*. U. S. Department of Agriculture Forest Service Northeastern Research Station General Technical Report NE-331.

Strauss, S. Y. 1991. Indirect effects in community ecology: their definition, study and importance. *Trends in Ecology and Evolution* 6: 206–210.

Strauss, S. Y., J. A. Lau, and S. P. Carroll. 2006. Evolutionary responses of natives to introduced species: what do introductions tell us about natural communities? *Ecology Letters* 9: 357–374.

Strayer, D. L., V. T. Eviner, J. M. Jeschke, and M. L. Pace. 2006. Understanding the long-term effects of species invasions. *Trends in Ecology and Evolution* 21: 645–651.

Strong, D. R. and R. W. Pemberton. 2000. Biological control of invading species–risk and reform. *Science* 288: 1969–1970.

Suarez, A. V., D. A. Holway, and P. S. Ward. 2005. The role of opportunity in the unintentional introduction of nonnative ants. *Proceedings of the National Academy of Sciences USA* 102: 17032–17035.

Suttle, K. B. and M. S. Hoddle. 2006. Engineering enemy-free space: an invasive pest that kills its predators. *Biological Invasions* 8: 639–649.

Syrett, P. 2002. New restraints on biological control. Pp. 363–394. *In* G. J. Hallman and C. P. Schwalbe (eds). *Invasive Arthropods in Agriculture: Problems and Solutions.* Oxford and IBH Publishing, New Delhi.

Theodoropoulos, D. I. 2003. *Invasion Biology: Critique of a Pseudoscience.* Avvar Books, Blythe, California. 236 pp.

Thomas, D. B. 2002. Deliberate introductions of exotic insects other than classical biological control agents. Pp. 87–106. *In* G. J. Hallman and C. P. Schwalbe (eds). *Invasive Arthropods in Agriculture: Problems and Solutions.* Oxford and IBH Publishing, New Delhi.

Thomas, D. B., J. E. Eger, W. Jones, and G. Ortega-Leon. 2003. The African cluster bug, *Agnoscelis puberula* (Heteroptera: Pentatomidae), established in the New World. *Florida Entomologist* 86: 151–153.

Thomas, M. C. 1995. Invertebrate pets and the Florida Department of Agriculture and Consumer Services. *Florida Entomologist* 78: 39–44.

Thomas, M. C. 2006. The exotic invasion of Florida: a report on arthropod immigration into the Sunshine State. Florida Department of Agriculture and Consumer Services, Division of Plant Industry, Gainesville. http://doacs. state. fl. us/~pi/enpp/ento/exoticsinjflorida. htm [Accessed 17 October 2006].

Tomiczek, C. and H. Krehan. 2001. Der Asiatische Laubholzbockkäfer – Erstauftreten in Österreich. *Österreichische Forstzeitung* 112: 33.

Tschinkel, W. R. 2006. *The Fire Ants.* Harvard University Press, Cambridge, Massachusetts. 723 pp.

Tsutsui, N. D., A. V. Suarez, D. A. Holway, and T. J. Case. 2001. Relationships among native and introduced populations of the Argentine ant (*Linepithema humile*) and the source of introduced populations. *Molecular Ecology* 10: 2151–2161.

Turell, M. J., M. L. O'Guinn, D. J. Dohm, and J. W. Jones. 2001. Vector competence of North American mosquitoes (Diptera: Culicidae) for West Nile virus. *Journal of Medical Entomology* 38: 130–134.

Turnbull, A. L. 1967. Population dynamics of exotic insects. *Bulletin of the Entomological Society of America* 13: 333–337.

Turnbull, A. L. 1979. Recent changes to the insect fauna of Canada. Pp. 180–194. *In* H. V. Danks (ed). *Canada and Its Insect Fauna.* Memoirs of the Entomological Society of Canada 108.

Turnbull, A. L. 1980. Man and insects: the influence of human settlement on the insect fauna of Canada. *Canadian Entomologist* 112: 1177–1184.

Ullman, D. E., J. L. Sherwood, and T. L. German. 1997. Thrips as vectors of plant pathogens. Pp. 539–565. *In* T. Lewis (ed). *Thrips as Crop Pests.* CAB International, Wallingford, United Kingdom.

U. S. Congress Office of Technology Assessment. 1993. *Harmful Non-Indigenous Species in the United States.* OTA-F-565. U. S. Government Printing Office, Washington, DC. 391 pp.

U. S. Department of Agriculture National Agricultural Statistics Service. 2006. *Honey.* USDA NASS, Washington, DC. 3 pp.

Usinger. R. L. 1941. A remarkable immigrant leptopodid in California. *Bulletin of the Brooklyn Entomological Society* 36: 164–165.

Van Driesche, J. and R. [G.]. Van Driesche. 2001. Guilty until proven innocent: preventing nonnative species invasions. *Conservation in Practice* 2(1): 8–17.

Van Driesche, R. G. 1994. Classical biological control of environmental pests. *Florida Entomologist* 77: 20–33.

Van Driesche, R. G. and M. S. Hoddle. 2000. Classical arthropod biological control: measuring success, step by step. Pp. 39–75. *In* G. Gurr and S. Wratten (eds). *Biological Control: Measures of Success.* Kluwer Academic Publishers, Dordrecht.

van Riper, C. III and J. M. Scott. 2001. Limiting factors affecting Hawaiian native birds. *Studies in Avian Biology* 22: 221–233.

van Riper, C. III, S. G. van Riper, M. L. Goff, and M. Laird. 1986. The epizootiology and ecological significance of malaria in Hawaiian land birds. *Ecological Monographs* 56: 327–344.

Venette, R. C. and J. R. Gould. 2006. A pest risk assessment for *Copitarsia* spp., insects associated with importation of commodities into the United States. *Euphytica* 148: 165–183.

Vinson, S. B. 1994. Impact of the invasion of *Solenopsis invicta* (Buren) on native food webs. Pp. 240–258. *In* D. F. Williams (ed). *Exotic Ants: Biology, Impact, and Control of Introduced Species.* Westview Press, Boulder, Colorado.

Vinson, S. B. 1997. Invasion of the red imported fire ant (Hymenoptera: Formicidae): spread, biology, and impact. *American Entomologist* 43: 23–39.

Visser, D., M. G. Wright, and J. H. Giliomee. 1996. The effect of the Argentine ant, *Linepithema humile* (Mayr) (Hymenoptera: Formicidae), on flower-visiting insects of *Protea nitida* Mill. (Proteaceae). *African Entomology* 4: 285–287.

Vitousek, P. M. 1988. Diversity and biological invasions of oceanic islands. Pp. 181–189. *In* E. O. Wilson (ed). *Biodiversity.* National Academy Press, Washington, DC.

Vitousek, P. M., C. M. D'Antonio, L. L. Loope, and R. Westbrooks. 1996. Biological invasions as global environmental change. *American Scientist* 84: 468–478.

Vitousek, P. M., C. M. D'Antonio, L. L. Loope, M. Rejmánek, and R. Westbrooks. 1997. Introduced species: a significant component of human-caused global change. *New Zealand Journal of Ecology* 21: 1–16.

Vitousek, P. M., L. L. Loope, and C. P. Stone. 1987. Introduced species in Hawaii: biological effects and opportunities for ecological research. *Trends in Ecology and Evolution* 2: 224–227.

Waage, J. K. 2001. Indirect ecological effects in biological control: the challenge and the opportunity. Pp. 1–12. *In* E. Wajnberg, J. K. Scott, and P. C. Quimby (eds). *Evaluating Indirect Ecological Effects of Biological Control.* CABI Publishing, Wallingford, United Kingdom.

Wajnberg, E., J. K. Scott, and P. C. Quimby (eds). 2001. *Evaluating Indirect Ecological Effects of Biological Control.* CABI Publishing, Wallingford, United Kingdom. 261 pp.

Waldbauer, G. 2003. *What Good Are Bugs?: Insects in the Web of Life.* Harvard University Press, Cambridge, Massachusetts. 366 pp.

Wallner, W. E. 1996. Invasive pests ('biological pollutants') and US forests: whose problem, who pays? *Bulletin OEPP (Organisation européenne et méditerranéenne pour la protection des plantes)/EPPO (European and Mediterranean Plant Protection Organization)* 26: 167–180.

Walsh, B. D. 1868. To the agriculturists and horticulturists of the United States. *American Entomologist* 1: 1–2.

Ward, D. F., J. R. Beggs, M. N. Clout, R. J. Harris, and S. O'Connor. 2006. The diversity and origin of exotic ants arriving in New Zealand via human-mediated dispersal. *Diversity and Distributions* 12: 601–609.

Warner, R. E. 1968. The role of introduced diseases in the extinction of the endemic Hawaiian avifauna. *Condor* 70: 101–120.

Waterhouse, D. F. 1974. The biological control of dung. *Scientific American* 230(4): 100–109.

Waterhouse, D. F. and D. P. A. Sands. 2001. *Classical Biological Control of Arthropods in Australia.* CSIRO Entomology; Australian Centre for International Agricultural Research, Canberra. 559 pp.

Webster, F. M. 1892. Early published references to some of our injurious insects. *Insect Life* 4: 262–265.

Webster, F. M. 1897. Biological effects of civilization on the insect fauna of Ohio. Pp. 32–46. *In* Fifth Annual Report of the Ohio State Academy of Science, Columbus, Ohio.

Weseloh, R. M. 2003. People and the gypsy moth: a story of human interactions with an invasive species. *American Entomologist* 49: 180–190.

Wetterer, J. K. 2007. Biology and impacts of Pacific island invasive species. 3. The African big-headed ant, *Pheidole megacephala* (Hymenoptera: Formicidae). *Pacific Science* 61: 437–456.

Wheeler, A. G., Jr. 1999. *Otiorhynchus ovatus, O. rugosostriatus,* and *O. sulcatus* (Coleoptera: Curculionidae): exotic weevils in natural communities, mainly mid-Appalachian shale barrens and outcrops. *Proceedings of the Entomological Society of Washington* 101: 689–692.

Wheeler, A. G., Jr. 2002. *Chilacis typhae* (Perrin) and *Holcocranum saturjae* (Kolenati) (Hemiptera: Lygaeoidea: Artheneidae): updated North American distributions of two Palearctic cattail bugs. *Proceedings of the Entomological Society of Washington* 104: 24–32.

Wheeler, A. G., Jr. and T. J. Henry. 1992. *A Synthesis of the Holarctic Miridae (Heteroptera): Distribution, Biology, and Origin, with Emphasis on North America.* Thomas Say Foundation Monograph 25. Entomological Society of America, Lanham, Maryland. 282 pp.

Wheeler, A. G., Jr. and E. R. Hoebeke. 1995. *Coccinella novemnotata* in northeastern North America: historical occurrence and current status (Coleoptera: Coccinellidae). *Proceedings of the Entomological Society of Washington* 97: 701–716.

Wheeler, A. G., Jr. and E. R. Hoebeke. 2001. A history of adventive insects in North America: their pathways of entry and programs for their detection. Pp. 3–15. *In* Detecting and Monitoring of Invasive Species. Proceedings of the Plant Health Conference 2000, October 24–25, Raleigh, North Carolina. USDA, APHIS, PPQ, Center for Plant Health Science and Technology, Raleigh.

Wheeler, A. G., Jr. and E. R. Hoebeke. 2004. New records of Palearctic Hemiptera (Sternorrhyncha, Cicadomorpha, Heteroptera) in the Canadian maritime provinces. *Proceedings of the Entomological Society of Washington* 106: 298–304.

Wheeler, A. G., Jr. and E. R. Hoebeke. 2005. *Livilla variegata* (Löw) (Hemiptera: Sternorrhyncha: Psyllidae) new to North America, with records of three other Palearctic psyllids new to Newfoundland. *Proceedings of the Entomological Society of Washington* 107: 941–946.

Wheeler, A. G., Jr. and H. F. Nixon. 1979. *Insect Survey and Detection in State Departments of Agriculture. Special Publication.* Pennsylvania Department of Agriculture, Harrisburg. 28 pp.

Wheeler, A. G., Jr., T. J. Henry, and E. R. Hoebeke. 2006. Palearctic plant bugs (Hemiptera, Miridae) in Newfoundland, Canada: first North American records for *Phytocoris longipennis* Flor and *Pilophorus cinnamopterus* (Kirschbaum), new records of eight other species, and review of previously reported species. *In* W. Rabitsch (ed). *Hug the Bug: For Love of True Bugs.* Festschrift zum 70. Geburtstag von Ernst Heiss. Denisia 19: 997–1014.

White, E. M., J. C. Wilson, and A. R. Clarke. 2006. Biotic indirect effects: a neglected concept in invasion biology. *Diversity and Distributions* 12: 443–455.

White, P. S. 1998. Biodiversity and the exotic species threat. Pp. 1–7. *In* K. O. Britton (ed). *Exotic Pests of Eastern Forests: Conference Proceedings, April 8–10, 1997, Nashville, Tennessee.* Tennessee Exotic Pest Plant Council; USDA Forest Service, Atlanta.

Whitehead, D. R. and A. G. Wheeler, Jr. 1990. What is an immigrant arthropod? *Annals of the Entomological Society of America* 83: 9–14.

Wiedenmann, R. N. 2005. Non-target feeding by *Galerucella calmariensis* on sandbar willow (*Salix interior*) in Illinois. *Great Lakes Entomologist* 38: 100–103.

Wigglesworth, V. B. 1984. *Insect Physiology*. Chapman and Hall, London. 191 pp.

Wilcove, D. S., D. Rothstein, J. Dubow, A. Phillips, and E. Losos. 1998. Quantifying threats to imperiled species in the United States. *BioScience* 48: 607–615.

Williams, D. F. 1994. *Exotic Ants: Biology, Impact, and Control of Introduced Species*. Westview Press, Boulder, Colorado. 332 pp.

Williamson, M. [H.]. 1996. *Biological invasions*. Chapman and Hall, London. 244 pp.

Williamson, M. [H.]. 1999. Invasions. *Ecography* 22: 5–12.

Williamson, M. H. and K. C. Brown. 1986. The analysis and modelling of British invasions. *Philosophical Transactions of the Royal Society of London Series B Biological Sciences* 314: 505–522.

Williamson, M. [H.]. and A. Fitter. 1996. The varying success of invaders. *Ecology* 77: 1661–1666.

Wilson, E. O. 1992. *The Diversity of Life*. Harvard University Press, Cambridge, Massachusetts. 424 pp.

Wilson, E. O. 1994. *Naturalist*. Island Press, Washington, DC; Shearwater Books, Covelo, California. 380 pp.

Wilson, M. R. 2000. Loss of taxonomists is a threat to pest control. *Nature* 407: 559.

Wilson, S. W. and A. Lucchi. 2007. Feeding activity of the flatid planthopper *Metcalfa pruinosa* (Hemiptera: Fulgoroidea). *Journal of the Kansas Entomological Society* 80: 175–178.

Winston, M. L. 1992. *Killer Bees: The Africanized Honey Bee in the Americas*. Harvard University Press, Cambridge, Massachusetts. 162 pp.

With, K. A. 2002. The landscape ecology of invasive spread. *Conservation Biology* 16: 1192–1203.

Wittenberg, R. (ed). 2005. *An Inventory of Alien Species and Their Threat to Biodiversity and Economy in Switzerland: Report to the Swiss Agency for Environment, Forests and Landscape*. CABI Bioscience Switzerland Centre, Delémont. 417 pp.

Wittenberg, R. and M. J. W. Cock. 2005. Best practices for the prevention and management of invasive species.

Pp. 209–232. *In* H. A. Mooney, R. N. Mack, J. A. McNeely, L. E. Neville, P. J. Schei, and J. K. Waage (eds). *Invasive Alien Species: A New Synthesis*. Island Press, Washington, DC.

Wojcik, D. P., C. R. Allen, R. J. Brenner, E. A. Forys, D. P. Jouvenaz, and R. S. Lutz. 2001. Red imported fire ants: impact on biodiversity. *American Entomologist* 47: 16–23.

Wonham, M. J. 2003. Ecological gambling: expendable extinctions versus acceptable invasions. Pp. 179–205. *In* P. Kareiva and S. Levin (eds). *The Importance of Species: Perspectives on Expendability and Triage*. Princeton University Press, Princeton, New Jersey.

Wood, S. L. 1975. New synonymy and new species of American bark beetles (Coleoptera: Scolytidae), Part II. *Great Basin Naturalist* 35: 391–401.

Woods, M. and P. V. Moriarty. 2001. Strangers in a strange land: the problem of exotic species. *Environmental Values* 10: 163–191.

Woodworth, B. L., C. T. Atkinson, D. A. LaPointe, P. J. Hart, C. S. Spiegel, E. J. Tweed, C. Henneman, J. LeBrun, T. Denette, R. DeMots, K. L. Kozar, D. Triglia, D. Lease, A. Gregor, T. Smith, and D. Duffy. 2005. Host population persistence in the face of introduced vector-borne diseases: Hawaii amakihi and avian malaria. *Proceedings of the National Academy of Sciences USA* 102: 1531–1536.

Worner, S. P. 2002. Predicting the invasive potential of exotic insects. Pp. 119–137. *In* G. J. Hallman and C. P. Schwalbe (eds). *Invasive Arthropods in Agriculture: Problems and Solutions*. Oxford and IBH Publishing, New Delhi.

Worthley, L. H. 1928. Corn borer control and the ten million dollar clean-up campaign. *Journal of Economic Entomology* 21: 230–234.

Zalucki, M. P. and A. R. Clarke. 2004. Monarchs across the Pacific: the Columbus hypothesis revisited. *Biological Journal of the Linnean Society* 82: 111–121.

Zettler, J. A., T. P. Spira, and C. R. Allen. 2001. Ant–seed mutalisms: can red imported fire ants sour the relationship? *Biological Conservation* 101: 249–253.

Zimmerman, E. C. 1948. *Insects of Hawaii: A Manual of the Insects of the Hawaiian Islands, Including an Enumeration of the Species and Notes on Their Origin, Distribution, Hosts, Parasites, etc. Vol. 1. Introduction*. University of Hawaii Press, Honolulu. 206 pp.

Chapter 22

BIODIVERSITY OF BITING FLIES: IMPLICATIONS FOR HUMANITY

Peter H. Adler

Department of Entomology, Soils, and Plant Sciences, Clemson University, Clemson, South Carolina 29634-0315, USA

Flies are so mighty that they win battles, paralyze our minds, eat our bodies . . .

— *Blaise Pascal (1660)*

Insect Biodiversity: Science and Society, 1st edition. Edited by R. Foottit and P. Adler
© 2009 Blackwell Publishing, ISBN 978-1-4051-5142-9

The anecdotal claims that flies have been responsible for more human deaths and misery than any other group of macroorganisms in recorded history are not easily dismissed. From disseminators of disease to purveyors of famine, flies have marched in lockstep with humanity. They are found in the hieroglyphics of ancient Egypt (Greenberg 1973), and hold a high biblical profile in the fourth of the ten plagues that Moses brought down on the house of Pharaoh (Exodus 8: 21–31). They have driven cultures, economies, land use, and the outcomes of battles (Steiner 1968, Pauly 2002, Adler and Wills 2003). Omnipresent, prolific, and rapid in development, their sudden appearance in rotting organic matter fueled the quaint notion of spontaneous generation. They have tracked humans across the globe, many species now being cosmopolitan. Roughly 6% of immigrant insect species in the 48 contiguous USA are flies (Sailer 1983), including some of history's most destructive species such as the yellow fever mosquito (*Aedes aegypti*) and Hessian fly (*Mayetiola destructor*). Through the transmission of disease agents, flies have been a driving force in human evolution.

Only a few dipteran families – the Syrphidae and perhaps the Tephritidae – hold even a fraction of the allure of butterflies, dragonflies, and tiger beetles. Much of the progress in understanding dipteran biodiversity, therefore, relies on the impetus of socioeconomic and medicoveterinary concerns. Biting flies, because they breach the host's circulatory system, can be important in the transmission of disease-causing agents. The evolutionary antiquity of biting flies (Borkent and Grimaldi 2004) suggests a rich and often tightly evolved system of interactions with hosts and pathogens. Because of their significance to humanity, flies that suck blood and transmit the agents of disease are among the taxonomically best-known Diptera in the world. Their well-developed taxonomy, although still markedly incomplete, suggests that biting flies can provide insight into the biodiversity patterns and trends of other organisms.

In the following treatment, I focus on the biodiversity of biting flies of vertebrate animals, and particularly those that transmit parasites and pathogens. Biting flies are defined here as those species with adults that puncture host skin to obtain a blood meal. Not treated are flies that take blood from invertebrates, flies with blood-feeding larvae (e.g., *Protocalliphora* spp.), flies that scarify wounds and eyes for blood and other secretions (e.g., *Musca*

autumnalis), and nonbiting vectors of disease agents or, borrowing terminology from pollination ecology, the 'mess-and-soil' vectors, such as filth flies (e.g., *Musca domestica*). Also untreated are the dipteran vectors, all nonbiting, of phytopathogens (Harrison et al. 1980). I first consider the size of the world's biting-fly fauna, with a brief summary of the biodiversity and socioeconomic importance of each family. I then examine the philosophy, relevance, status, and future of biodiversity investigations on biting flies in relation to the welfare of humans and domestic animals.

NUMBERS AND ESTIMATES

The order Diptera, representing perhaps 10% of all life forms, comprises more than 150,000 described species worldwide divided among 152 families, many of which are acknowledged to be unnatural, paraphyletic or polyphyletic groups (Thompson 2006, Courtney et al. this volume). How many species of flies inhabit the planet? In the words of E. O. Wilson (1985), referring to all life forms, 'We do not know, not even to the nearest order of magnitude.' Some estimates suggest that nearly half or more of the Nearctic dipteran fauna remains undescribed (Thompson 1990). Estimates of the unknown are far greater for the tropical regions.

The consequences of not knowing how many species – or more importantly, what species – share our planet are disproportionately greater for some dipteran taxa than for others. The relative importance to human welfare of, for instance, the Culicidae far exceeds that of any of the multitude of obscure families of flies such as, say, the Clusiidae. More effort and resources, therefore, have been dedicated to the taxonomy of biting flies than of nonbiting flies. Consequently, the number of described species of biting Diptera should approach the actual number more nearly than that for other flies. The efforts of almost 250 years have brought us to the point where more than 17,500 named species have been placed in 11 families of biting flies, representing approximately 11% of the total described species of Diptera. The biting mouthparts of about one-quarter of these species have been lost or modified over evolutionary time and are unable to cut or penetrate flesh to obtain blood (e.g., some species of Culicidae, Simuliidae, and Tabanidae), or they feed on invertebrates (many Ceratopogonidae). The total number of described species that take vertebrate blood is about 13,225 (Table 22.1). Yet,

Table 22.1 World biodiversity of extant, described biting flies that take blood from vertebrate animals, followed by the number of fly-borne diseases. Family classification follows that of Thompson (2006).

Taxon (English Common Name)[1]	Number of Described Taxa (Genera; Species)[2]	Afrotr.	Austr.	Nearc.	Neotr.	Orient.	Palearc.	References
Athericidae (athericids)[3]	3; 72	13	14	3	19	19	7	James (1968), Nagatomi (1975), Stuckenberg (1980), Webb (1981), Nagatomi and Evenhuis (1989), Rozkošný and Nagatomi (1997)
Ceratopogonidae (biting midges, no-see-ums, punkies)[4]	4; 1619	191	137	155	290	409	437	Borkent and Wirth (1997), Borkent (2007)
Corethrellidae (corethrellids)	1; 97	6	11	7	69	5	2	Borkent (2008)
Culicidae (mosquitoes)[5]	40; 3492	767	746	173	1045	1021	482	Walter Reed Biosystematics Unit (2006), Rueda (2008)
Glossinidae (tsetse)	1; 33	33	0	0	0	0	0	Pont (1980a), Jordan (1993), Gooding and Krafsur (2005)
Hippoboscidae (louse flies, keds, bat flies)[6]	65; 681	134	147	47	226	196	107	Hutson and Oldroyd (1980), Maa and Peterson (1987), Peterson and Wenzel (1987), Wenzel and Peterson (1987), Maa (1989), Guerrero (1997)
Muscidae (horn flies, stable flies, biting muscids)[7]	11; 56	39	5	3	6	14	14	Zumpt (1973), Pont (1980b), Pont and Mihok (2000), A. C. Pont (personal communication)
Psychodidae (sand flies)[8]	7; 921	168	45	15	475	119	140	Lewis (1978), Young and Perkins (1984), Seccombe et al. (1993), Young and Duncan (1994), Ježek (1999), E. A. B. Galati (personal communication)
Rhagionidae (biting snipe flies)[9]	2; 39	0	7	29	0	0	4	Turner (1974), Nagatomi and Evenhuis (1989)

(continued)

Table 22.1 *(continued).*

Taxon (English Common Name)[1]	Number of Described Taxa (Genera; Species)[2]	Afrotr.	Austr.	Nearc.	Neotr.	Orient.	Palearc.	References
Simuliidae (black flies)[10]	26; 1956	216	250	231	356	322	671	Takaoka (2003), Adler et al. (2004), Adler and Crosskey (2008)
Tabanidae (horse flies, deer flies)[11]	127; 4259	736	462	326	1149	939	656	Chainey and Oldroyd (1980), Fairchild and Burger (1994), Burger (1995), J. F. Burger (personal communication)
Totals	287; 13,225	2303	1824	989	3635	3044	2520	
Diseases of humans	45	18	8	15	20	16	8	Table 22.2
Diseases of domestic animals	36	23	15	20	17	17	16	Table 22.2

[1] Taxa with biting mouthparts, which are not known to attack vertebrates, are not included. The family Chironomidae includes three species with mandibulate mouthparts but with unknown (if any) hosts (Cranston et al. 2002). For the carnid bird flies of the genus *Carnus*, whether or not the feeding behavior involves piercing and sucking of blood from birds remains controversial (Grimaldi 1997).

[2] Subspecies are included for those families in which they are recognized (except the Culicidae, which at the end of 2006 included 100 subspecies), with the view that closer scrutiny might confirm many as valid species. Subspecies typically account for less than 6% of the total species count for each family, except in the Glossinidae where they account for about 26%. The total number of species does not equal the sum of regional faunas because of representation of some species in more than one region.

[3] Includes only genera with representatives known to take vertebrate blood (*Atrichops, Dasyomma*, and *Suragina*).

[4] Includes only members of the family that feed on vertebrate blood: *Austroconops, Culicoides, Leptoconops*, and *Forcipomyia* (*Lasiohelea*). Species numbers for each region are based on type localities given by Borkent (2007); the 234 species with China as the type locality are divided equally between the Oriental and Palearctic Regions.

[5] Includes all species of the family, except 88 species of *Toxorhynchites* with nonbiting mouthparts and 12 species of *Malaya* that feed on regurgitated stomach contents of ants; the number of vertebrate blood-feeding mosquitoes, nonetheless, might be overstated because hosts of many species remain unknown.

[6] The family Hippoboscidae is often presented as three separate families: Hippoboscidae *sensu stricto* (21 genera; 200 species), Nycteribiidae (13 genera, 260 species), and Streblidae (31 genera, 221 species).

[7] Includes only those species with biting mouthparts (subfamily Stomoxyinae plus one species, *M. crassirostris*, in subfamily Muscinae).

[8] Includes only those taxa in subfamilies with biting mouthparts (subfamilies Phlebotominae (890 species) and Sycoracinae (31 species; Ježek 1999, Bravo personal communication)); 6–25 genera of Phlebotominae are recognized, with most of the discrepancy in numbers being in the New World (Galati 1995); 5.4% of Phlebotominae are subspecies, not including nominotypical subspecies (E. A. B. Galati, personal communication).

[9] Includes only those genera with representatives known to take vertebrate blood (*Spaniopsis* and *Symphoromyia*).

[10] Includes all members of the family except 50 species (25 in the Nearctic Region and at least 32 in the Palearctic Region) that do not have mouthparts adapted for cutting host tissue.

[11] Includes all members of the family except *Goniops* (1 Nearctic species), *Stonemyia* (8 Nearctic and 2 Palearctic species), the subfamily Scepsidinae (6 Afrotropical and 1 Neotropical species), and the tribe Mycteromyiini (23 Neotropical species), which do not take vertebrate blood; additional species probably do not feed on vertebrate blood, but they are poorly documented.

counting species is akin to running in place; no sooner are the species counted than new names are added and older ones synonymized or resurrected. In the hyperactive 5-year period from 2000 to 2004, for example, 179 new species of black flies were described, 28 names were synonymized, and 16 names were recovered to validity from synonymy. In a similar 5-year time frame, 1999–2003, 242 new species of ceratopogonids (including 40 vertebrate-blood feeders) were described, increasing the number of species for the entire family by 4.6%.

The discovery and recognition of species varies over time, often depending on significant events or revolutions in approaches and methodologies. For the North American Simuliidae, for example, significant advancements include the first use of male genitalia (1911) and later female genitalia (1927), exploration of western North America (1930s and 1940s), and the introduction of cytogenetic techniques (1956). Molecular techniques represent the current revolution, with DNA barcoding (2003) among the most touted, and can be expected to reveal a great deal of genetically hidden biodiversity.

New species of biting flies continue to be discovered in poorly sampled, as well as more thoroughly sampled, geographic areas. For most families, the greatest potential for new species lies in untapped areas of the tropical rain forests. Seventy-one new species of black flies, for example, recently were described from three of the five major islands of Indonesia (Takaoka 2003), and more can be expected as the high mountains and surrounding tropical forests of areas such as Irian Jaya reluctantly yield their species. Additional biodiversity hot spots are in politically unstable areas of the world that will prove hazardous for decades, littered with political unrest, armed insurgents, land mines, and government restrictions on free-range collecting. Some of the potentially species-rich but politically troublesome areas include Cambodia, Colombia, Myanmar, and northern India and Pakistan.

An unknown number of species remains hidden within described morphospecies (species recognized on anatomical criteria), where prospecting in the genome is likely to reveal a wealth of additional biodiversity. Cytogenetic studies, and more recently molecular studies, of black flies and mosquitoes have shown that morphologically similar species, known as sibling species, or cryptic species, are commonplace among biting flies. About one-fourth of the 255 species of black flies in North America were discovered chromosomally (Adler

et al. 2004). In Latin America, the black fly *Simulium metallicum* is a complex of more than seven sibling species (Conn et al. 1989). At least 40 of the 484 species in the mosquito genus *Anopheles* are unnamed members of species complexes (Harbach 2004). The *Anopheles punctulatus* complex in Papua New Guinea, for example, consists of at least six sibling species (Foley et al. 1993), whereas the New World *An. crucians* complex includes five sibling species (Wilkerson et al. 2004). Nearly 30 years of chromosomal studies of African vectors have shown that *Anopheles gambiae* consists of seven or more sibling species (Coluzzi et al. 2002), and *Simulium damnosum* consists of 55 sibling species and cytoforms (Post et al. 2007), making it the largest species complex of any blood-feeding organisms. Hidden more deeply than sibling species are homosequential sibling species, which have the same banding sequences in their polytene chromosomes and virtually identical external morphologies (Bedo 1979). Their discovery is facilitated by ecological differences and cytological evidence such as heterozygote deficiencies of polymorphic inversions. Because of the difficulty in uncovering homosequential sibling species, no statements yet can be made about their prevalence. Molecular techniques hold the potential to reveal hidden species that are reproductively isolated but lack diagnostic cytological and morphological markers.

A complete biodiversity catalog for extant biting flies must include not only all species arranged phylogenetically, but also all life stages, both sexes, and ultimately a comprehensive genetic library for each species. Although the number of blood-sucking species currently stands at roughly 13,225, the actual structural and ecological diversity is about five times greater (i.e., ca. 66,085 biodiversity units), taking into account the differences among eggs, larvae, pupae, males, and females. In many cases, however, only one or two life stages or a single sex are known. Species-level taxonomy of larvae and pupae is well developed for the Culicidae, Glossinidae, Simuliidae, and a few regional groups, but poorly known for the remaining families. The larvae and pupae of North American simuliids, for example, are described for 93% and 98% of the species, respectively (Adler et al. 2004), whereas the immatures of only 13.8% (819 species) of all ceratopogonids have been described (A. Borkent, personal communication). The eggs have been described for not more than 10% of the world's species of biting flies, perhaps largely a matter of choice, for they often can be extracted from females that are captured gravid or are allowed to mature their eggs after capture. The job of collecting, rearing, curating, describing, naming, and publishing the remaining biodiversity of biting Diptera is daunting.

OVERVIEW OF BITING FLIES AND DISEASES

Of the roughly 13,225 described species of flies that acquire blood from vertebrate hosts, about 2% serve as vectors of the agents of at least 45 diseases of humans, while less than 1% currently are known to transmit the agents of nearly 40 significant diseases of domestic animals (Table 22.2). The following accounts summarize the taxonomic status, bionomics, vector-borne diseases, and socioeconomic importance of the 11 families of flies with biting representatives (Fig. 22.1). The families are listed alphabetically; the sequence would differ if the families were arranged according to their effect on society, with mosquitoes at the top.

Athericidae   The athericids have one of the lowest profiles of the families with species that feed on vertebrate blood. The larvae are predaceous and live in streams. They are poorly known taxonomically (Webb 1995). Females of three genera take vertebrate blood. Those of *Atrichops* feed on frogs and those of *Dasyomma* and *Suragina* on mammals, including humans, cattle, and horses, with one record from an owl (Nagatomi and Soroida 1985). The athericids have not earned pest status, and are not yet known to transmit pathogens or parasites.

Ceratopogonidae   The family Ceratopogonidae consists of 5923 described, extant species in 110 genera (Borkent 2007). It is the most species-rich family of biting flies, a status related to the small body size, diverse larval habitats, breadth of larval and adult feeding behaviors, worldwide distribution, and evolutionary age, the oldest fossils being from the Lower Cretaceous. Only three genera (*Austroconops*, *Culicoides*, and *Leptoconops*) and the subgenus *Forcipomyia* (*Lasiohelea*), comprising about 1619 species (Borkent 2007), take vertebrate blood. Hosts of the vertebrate feeders include amphibians, birds, fish, mammals, and reptiles. Females of some small genera do not bite, whereas those in the vast majority (107) of genera feed on invertebrates (Downes and Wirth 1981). Larvae develop in an impressive range of habitats including coastal

marshes, streams, ponds, rotting plants, manure, and phytotelmata (water held by plants). They feed on algae, bacteria, detritus, and small invertebrates. The taxonomy of the immature stages, with a few regional exceptions (e.g., Murphree and Mullen 1991), is poorly developed.

In addition to biting, female ceratopogonids transmit 66 viruses, 15 species of protozoans, and 26 species of filarial nematodes (Borkent 2005). Two significant human diseases are caused by ceratopogonid-borne agents. Mansonelliasis is caused by three species of filarial nematodes transmitted by biting midges in the Caribbean, northern South America, and Africa. Oropouche fever, caused by a virus, strikes in the Caribbean and the Amazon region. More than 50 viruses have been isolated from females of the genus *Culicoides* (Mellor et al. 2000). The most prominent ceratopogonid-borne disease of domestic animals is bluetongue, an economically costly, international disease of livestock. African horse sickness, one of the most lethal infectious diseases of horses, is a ceratopogonid-borne viral disease that can cause equid mortality rates in excess of 90% (Mellor et al. 2000).

Corethrellidae   The Corethrellidae are best represented in the Neotropical Region where nearly three-quarters of the described species occur. About 100 species are currently recognized (Borkent 2008), with estimates of more than twice that number still awaiting discovery (A. Borkent, personal communication). Larvae are predators and live in lentic habitats such as pools alongside streams and in phytotelmata. Females feed on the blood of male frogs that they locate by following the calls (Bernal et al. 2006). Some species transmit trypanosomes to frogs (Johnson et al. 1993).

Culicidae   The medicoveterinary importance of the Culicidae has made them one of the taxonomically best-known families of arthropods, with 3492 species and 100 subspecies. They are the only family of insects for which two international journals have been devoted (*Journal of the American Mosquito Control Association* and the now defunct *Mosquito Systematics*). The larvae and pupae are known for most species, and the eggs have been described for a greater percentage of species than for any other group of biting flies.

Mosquitoes inhabit all regions of the world except Antarctica, and are among the most abundant and

Table 22.2 Biting fly-borne diseases of humans and domestic animals of the world; dipteran families and diseases within each family are listed alphabetically.

| Taxon | Diseases[1] | | References |
	Humans	Domestic Animals[2]	
Athericidae	None known	None known	
Ceratopogonidae	Mansonelliasis, Oropouche fever	African horse sickness, Akabane viral disease?, avian trypanosomiasis, bluetongue disease, [bovine ephemeral fever?], bovine onchocerciasis, [epizootic hemorrhagic disease], [equine encephalosis], equine onchocerciasis, leucocytozoonosis, [Palyam viral disease?], [additional viral diseases]	Mellor et al. (2000), Mullen (2002)
Corethrellidae	None known	None known	
Culicidae	[Anthrax], Bancroftian filariasis (elephantiasis), Barmah forest viral disease, brugian filariasis, [Cache Valley viral disease], [California encephalitis], Chikungunya fever, dengue, dirofilariasis, eastern equine encephalitis, [Ilheus fever], [Jamestown Canyon viral disease], Japanese encephalitis, La Crosse encephalitis, malaria, Mayaro fever, Murray Valley encephalitis, O'nyong-nyong fever, Rift Valley fever, [Rocio encephalitis], Ross River fever, Semliki Forest encephalitis, Sindbis fever, [snowshoe hare viral disease], St. Louis encephalitis, Tahyna fever, [tularemia], Venezuelan equine encephalitis, Wesselsbron disease, western equine encephalitis, West Nile encephalitis, yellow fever, [additional viral diseases]	Avian malaria, avian pox (e.g., fowlpox), canine dirofilariasis (dog heartworm disease), eastern equine encephalitis, Japanese encephalitis, [various onchocerciases], Rift Valley fever, Venezuelan equine encephalitis, Wesselsbron disease, western equine encephalitis, [additional viral diseases]	Harwood and James (1979), Eldridge et al. (2000), Foster and Walker (2002)
Glossinidae	African trypanosomiasis (sleeping sickness)	Nagana	Krinsky (2002)
Hippoboscidae	[Deer ked dermatitis]	Avian trypanosomiasis [filariasis of canids]	Baker (1967), Lloyd (2002), Dehio et al. (2004)
Muscidae	[Anthrax], [cutaneous streptothrichosis]	[Anthrax], [bovine leukosis], [cutaneous streptothrichosis], [equine infectious anemia], habronemiasis (summer sores), stephanofilariasis	Zumpt (1973), Harwood and James (1979), Moon (2002)

(continued)

Table 22.2 *(continued)*.

Taxon	Diseases[1] Humans	Diseases[1] Domestic Animals[2]	References
Psychodidae	Bartonellosis (Oroya fever and verruga peruana), Chandipura virus disease, Changuinola fever, leishmaniasis (cutaneous and visceral forms), sand fly fever (papatasi fever), vesicular stomatitis?	Leishmaniasis (canine and feline), vesicular stomatitis?	Rutledge and Gupta (2002)
Rhagionidae	None known	None known	
Simuliidae	Mansonelliasis, onchocerciasis (river blindness), [tularemia]	Avian trypanosomiasis, bovine onchocerciasis, leucocytozoonosis, vesicular stomatitis	Crosskey (1990), Adler and McCreadie (2002), Mead et al. (2004)
Tabanidae	[Anthrax], loiasis, tularemia	Anaplasmosis, [anthrax], [besnoitiosis], [bovine leukosis], elaeophorosis, equine infectious anemia (swamp fever), [hog cholera], mal de caderas, surra, [additional trypanosomiases]	Mullens (2002)

[1] Diseases in square brackets indicate that the families under which they are listed probably play a minor epidemiological role; other vectors or means of transmission are more common. Many additional disease-causing agents, particularly viruses, have been isolated from biting flies, but the role of the flies in transmission of the agents remains largely unknown.
[2] Livestock, poultry, cats, and dogs.

conspicuous organisms in areas such as the tundra and boreal forest, although species richness is greatest at tropical latitudes. The larvae and pupae are found in nearly all still-water aquatic habitats, the most common including salt marshes, rain pools, tree holes, artificial containers, pitcher plants, lakes, and swamps. Larvae use a variety of modes, such as filtering, to feed on microorganisms, small macroinvertebrates, and detritus. Females of the majority of species take blood, feeding principally on birds and mammals, but also on amphibians, reptiles, fish, and insects (Rueda 2008).

Familiar to nearly everyone, mosquitoes can become such a nuisance that they disrupt outdoor activity, destroy personal composure, and impede economic development (Laird et al. 1982). Nuisance problems drive mosquito abatement programs in the USA, involving hundreds of millions of dollars annually. Threats of disease, such as West Nile encephalitis, further swell these budgets.

Mosquitoes are the terrorists of the arthropodan world. The legion of protozoans, viruses, and other disease agents that they transmit make them the greatest global threat to humanity of any group of organisms other than humans and microbes. They transmit a disproportionate number of pathogens, relative to other biting flies. About half of the more than 500 known arboviruses have been isolated from mosquitoes and more than 100 infect humans (Karabatsos 1985). The principal mosquito-borne diseases of humans include dengue, arboviral encephalitis, lymphatic filariasis, malaria, and yellow fever. Human malaria, caused by any of four species of *Plasmodium* protists transmitted by about 70 species of anopheline mosquitoes, is a devastating disease, with up to half a billion new cases and more than 1 million deaths annually in tropical and subtropical regions (Foster and Walker 2002). The World Health Organization (WHO) estimates that human lymphatic filariasis, caused by three species of nematodes transmitted

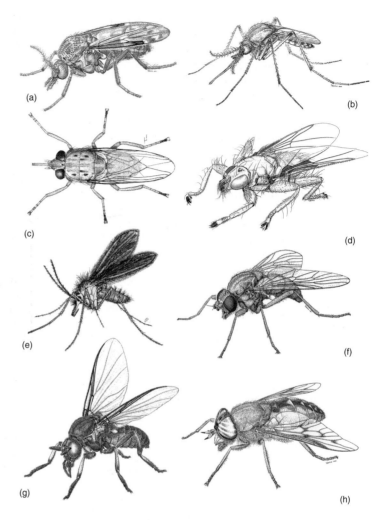

Fig. 22.1 Representative females of the major groups of biting flies. (a) No-see-um (Ceratopogonidae), (b) Mosquito (Culicidae), (c) Tsetse (Glossinidae), (d) Louse fly (Hippoboscidae), (e) Sand fly (Psychodidae, Phlebotominae), (f) Biting snipe fly (Rhagionidae), (g) Black fly (Simuliidae), (h) Horse fly (Tabanidae). (Source: The figures a, b, d, f, and h from Agriculture and Agri-Food Canada used with permission of the Minister of Public Works and Government Services; c and e from Jordan (1993) and Lane (1993), respectively, © The Natural History Museum, London and used with permission; g, from Adler et al. (2004) used by permission of the publisher, Cornell University Press.

by more than 60 species of mosquitoes, accounts for about 128 million active infections. These statistics belie the immeasurable socioeconomic burden and losses to future generations through the winnowing of human potential. Many of the mosquito-borne human-disease agents, such as the encephalitides, afflict domestic animals, which also suffer their own unique complement of mosquito-borne diseases.

Glossinidae Thirty-one species and subspecies of tsetse are restricted to Africa south of the Sahara, with an additional two species in southwestern Saudi Arabia. Four Oligocene fossil species from the Florissant shales of Colorado suggest that tsetse vexed

the American megafauna during the Age of Great Mammals. The family is well known taxonomically. The last new species was named more than 20 years ago, and the adults and about half of the puparia of all species and subspecies have been described (Jordan 1993).

Females have a low lifetime fecundity, maturing one egg at a time and nurturing the larva in the uterus. The final (third) instar is deposited on the ground and within a few hours transforms into a puparium in the soil. Both males and females take blood. Hosts include reptiles (e.g., crocodiles), large mammals (e.g., buffalo, giraffe, human, rhinoceros, and suids), and occasionally birds (e.g., ostrich).

Despite transmitting only about seven recognized species of trypanosomes to mammals (Jones 2001), tsetse have had a powerful influence in Africa. The trypanosomiases – sleeping sickness in humans and nagana in livestock – have caused enormous human suffering, inhibited economic development, and restricted animal agriculture, while also serving as the single greatest conservator of the African savannah by restricting the encroachment of humans and livestock (Harwood and James 1979, Krinsky 2002).

Hippoboscidae The 680-plus species of the worldwide family Hippoboscidae are sometimes divided among three discrete families, the Hippoboscidae *sensu stricto*, Nycteribiidae, and Streblidae. The latter two taxa, comprising slightly more than 70% of the Hippoboscidae *sensu lato*, are obligate ectoparasites of bats and have little relevance to the health and welfare of humans or domestic animals. About 75% of the Hippoboscidae *sensu stricto* are ectoparasites of birds, with the remainder on mammals other than bats. Species richness is greatest in the Neotropics.

All known species of the family Hippoboscidae are larviparous, nourishing a single larva, which forms a puparium upon deposition. Both sexes are obligate blood feeders. Among the bird and mammal feeders, each species is exclusively mammalophilic or ornithophilic. No species habitually feeds on humans, but some attack domestic animals. Premier among them is the sheep ked (*Melophagus ovinus*), an economically important pest of sheep that causes damage to the pelts. A few species of hippoboscids transmit trypanosomes and apicomplexan parasites to domestic pigeons, and some transmit parasites and pathogens to wildlife (Lloyd 2002). The role of hippoboscids in transmitting these agents to birds and mammals is probably underappreciated.

Muscidae The biting muscids include members of the subfamily Stomoxyinae and one species (*Musca crassirostris*) in the subfamily Muscinae. The number of described morphospecies has plateaued at 56; only one new species has been described in the past 30 years (Pont and Mihok 2000). About 70% of the species are found in the Afrotropical Region. The larvae and puparia of about 15% and 6%, respectively, of the biting species are described. The larvae typically develop in dung associated with hosts of the adult flies. Larvae of the stable fly (*Stomoxys calcitrans*) also develop in decomposing vegetation, especially grass cuttings.

The males and females of all biting muscids feed on mammal blood, although some probably feed from sores and host exudates rather than pierce the hide (Zumpt 1973). Humans can be annoyed by some species, especially the stable fly. Ungulates are hosts of most species. The two principal livestock pests are the widely distributed horn fly (*Haematobia irritans*) and the cosmopolitan stable fly. Both species cause anxiety, decreased immunological response, reduced milk production, and weight loss (Moon 2002). The stomoxyines, however, are not considered major vectors of pathogens or parasites to livestock. Some pathogens can be transmitted mechanically (i.e., by contamination of the mouthparts), but the frequency of transmission is not known (Zumpt 1973). The horn fly and stable fly transmit spirurid nematodes that cause cutaneous diseases known as habronemiasis in horses and stephanofilariasis in cattle, respectively.

Psychodidae The nearly 1000 species of biting psychodids, or sand flies, are restricted to two subfamilies: the Phlebotominae (sometimes considered a separate family) and the Sycoracinae. These subfamilies are widespread on all continents except Antarctica, although more than half of the species are Neotropical. Taxonomic knowledge of the immature stages, with some exceptions (e.g., Hanson 1968), is restricted to species of concern to human health. Sand flies develop in damp terrestrial environments, where the larvae feed on decaying organic matter.

Female phlebotomines feed on amphibians (toads), birds, mammals, and reptiles, and include the major vectors (*Lutzomyia* and *Phlebotomus*) of human-disease agents. Females of the species-poor, seldom-encountered sycoracines feed on frogs. Most sand flies have broad host ranges (Rutledge and Gupta 2002), which in some cases might reflect undiscovered sibling species. Biting by sand flies can be annoying, but their significance as medically important insects is related to the bacterial, protozoan, and viral agents they transmit. Foremost among the human diseases are the cutaneous and visceral forms of leishmaniasis, caused by about 23 protozoan species of *Leishmania* transmitted by about 80 species of sand flies; approximately 12 million people are afflicted worldwide (Lawyer and Perkins 2000). WHO estimated 59,000 human deaths from leishmaniasis in 2002. Phlebotomine-borne viruses cause the influenza-like illness known as sand fly fever, as well as several more geographically local human diseases. Bartonellosis is a human disease caused by phlebotomine-borne bacteria in Colombia, Ecuador, and Peru. Sand flies are of little veterinary importance, although they might be involved in transmission of the causal agents of leishmaniasis to dogs, the latter serving as a reservoir for human infection (Ostfeld et al. 2004).

Rhagionidae Of the 21 genera of Rhagionidae, only the Australian *Spaniopsis* and the Holarctic *Symphoromyia* include species that feed on vertebrate blood. The immature stages live in moist soil. Larval feeding habits are sketchy but apparently involve detrital feeding and possibly facultative predation. The larvae, as well as the eggs and pupae, are poorly known. The actual number of species of *Symphoromyia* is considered much greater than the 32 currently recognized (Turner 1974). Molecular work is needed to resolve existing species issues and reveal sibling species.

The females of *Spaniopsis* feed on humans and probably other mammals. Those of *Symphoromyia* acquire blood from a wide range of hosts, including cattle, deer, horses, and humans, and are significant pests in some areas of western North America and Central Asia (Turner 1974). No animal pathogens or parasites are known to be transmitted by the females of *Spaniopsis* or *Symphoromyia*.

Simuliidae The family Simuliidae is distributed worldwide between 55°S and 73°N, being absent only from some oceanic islands and areas without flowing water. Species richness is greatest in the Palearctic Region, a real reflection of the number of species, but also an artifact of oversplitting taxa based on structural minutiae without benefit of cytotaxonomic or other corroboration. The family is one of the more completely known among biting flies, largely because of the routine practice of describing the larva, pupa, male, and female of new species; the extensive use of cytotaxonomy; and the well-defined habitat of the larvae and pupae, which facilitates the collection of immatures and association of life stages.

The larvae require flowing water and are often the numerically dominant macroinvertebrates in lotic habitats (Adler and McCreadie 1997, Malmqvist et al. 2004). The females of all but about 2.5% of the world's 2015 species take blood (Adler and Crosskey 2008). Hosts are restricted to birds and mammals. Black flies are major biting and nuisance pests and are the only insects that have routinely killed animals by exsanguination and toxic shock from injected salivary components (Crosskey 1990). No species is strictly anthropophilic, but many take blood from humans (Adler and McCreadie 2002, Adler et al. 2004).

Two species of filarial nematodes cause diseases in humans. *Onchocerca volvulus* causes human onchocerciasis, which afflicts about 18 million people in equatorial Africa, southern Yemen, and localized areas of Central and South America. *Mansonella ozzardi* causes mansonelliasis, a marginally pathogenic disease in the rain forests of Brazil, Colombia, Guyana, Venezuela, and southern Panama (Adler and McCreadie 2002). Domestic fowl are plagued by simuliid-borne *Leucocytozoon* protists, whereas cattle are infected by the more benign filarial worm *Onchocerca lienalis*.

Tabanidae The large size of tabanids, often eliminating the need for a microscope in routine identification, has appealed to taxonomists and promoted the primarily morphological approach to their taxonomy. Morphotaxonomy, nonetheless, often suggests the presence of sibling species, confirmed in some instances by additional techniques (Sakolsky et al. 1999). The larvae and pupae require diligent searching to locate and are moderately well known only in the Nearctic Region (e.g., Burger 1977).

Tabanids are present throughout the world, primarily below the tree line. The larvae are generalist predators and inhabit wet soil and debris in bogs, marshes, ponds, and stream margins, although some species live in drier soil and tree holes. Females feed chiefly on mammals and occasionally amphibians, birds, and reptiles. An unknown number of the 4300 described species do not take blood, including species of *Goniops*, *Stonemyia*, the subfamily Scepsidinae, and the tribe Mycteromyiini. Humans and livestock are annoyed by both deer flies and horse flies.

The only human disease that is consistently tabanid borne (via *Chrysops*) is loiasis, caused by a filarial nematode known as the African eyeworm (*Loa loa*). Tabanids have the potential to transmit disease organisms mechanically (Mullens 2002). The bacterium that causes tularemia, for example, can be transmitted mechanically, although the proportion of human and livestock cases attributable to infection through tabanids is unknown.

RATIONALE FOR BIODIVERSITY STUDIES OF BLOOD-SUCKING FLIES

A major objective of biodiversity studies is to document all species on Earth. New biological riches (e.g., foods and medicinal compounds), aesthetics, the impending loss of species, and matters of conservation typically justify biodiversity studies for most organisms. But beyond the unifying goal of documenting species, the chief motives driving biodiversity studies of biting flies are likely to diverge from those for other organisms. For biting flies, threats to human and animal welfare, such as economic losses and disease epidemics caused by pests and vectors, motivate biodiversity research, with the ultimate goal of eliminating the risk.

Yet, among biting flies, only one of the four life stages, and often only the female, directly creates risk. The male and three pre-adult stages, particularly the larval stage, often perform key functional roles in the environment (Malmqvist et al. 2004). Larval black flies, for example, have been termed 'ecosystem engineers' because they play a vital role in processing dissolved and particulate matter into larger fecal pellets used as food by other aquatic organisms and as a possible source of fertilizer for flood-plain vegetation (Wotton et al. 1998, Malmqvist et al. 2004). A trade-off thus exists between reducing the troublesome adult stage and conserving the beneficial immature stages. Ironically, the larval stage often is targeted for control of the adult pest or vector. Complicating matters further, adult biting flies, too, can play an integral role in the environment as pollinators, food for predators, and regulators of wildlife populations (Malmqvist et al. 2004).

One would find it difficult to argue for conservation of the yellow fever mosquito, a vector not only of the virus of its namesake but also, *inter alia*, Chikungunya virus, dengue viruses, and Ross River virus. The intrinsic difficulties of eradicating a vector, the unintended consequences of doing so, and the desire to maintain all life forms can be addressed by breaking the cycle of disease while conserving the vector. The transmitted causal agent can be targeted with drugs

given to the vertebrate hosts, and vector populations can be suppressed to levels simply sufficient to disrupt the transmission cycle. Well into the 1930s, malaria was one of the major socioeconomic burdens in much of the USA, afflicting 6–7 million people annually and reducing the South's industrial potential by about one third (Russell 1959, Desowitz 1998). DDT, antimalarial drugs, habitat alteration (e.g., oiling and draining of wetlands), and improved standards of living eventually broke the cycle (Adler and Wills 2003). The principal vectors, members of the *Anopheles quadrimaculatus* complex, remain common in the formerly malarious regions of the southern USA. The *Plasmodium* species responsible for human malaria, however, are essentially gone from the areas where they were unintentionally introduced by the slave trade.

The dilemma of conserving or eliminating biting flies is epitomized by the black fly *Simulium ochraceum* (formerly *S. bipunctatum*) in the Galápagos Islands. *S. ochraceum* was discovered in the archipelago (on San Cristóbal) in 1989 (Abedraabo et al. 1993) and has since become a serious biting pest of humans, causing some residents to abandon their farms. Because of limited breeding sites, the black flies could be eradicated from the archipelago by applying the larva-specific biopesticide *Bacillus thuringiensis* var. *israelensis*. The critical issue is whether the species is a recent introduction, as some authors (Abedraabo et al. 1993) maintain, or has been on the archipelago longer. The distinction is significant because if it has long been present, it might have become a keystone species, serving as the basis of the food web for other aquatic invertebrates; in this scenario, its eradication could disrupt the island's flowing-water ecosystem. *S. ochraceum* is actually a species complex of at least two sibling species in Central and South America, one or more of which transmits the causal agent of human onchocerciasis (Hirai et al. 1994). Still unknown is which member of the complex is present in the Galápagos Islands and to what extent it has diverged from mainland populations. Cytogenetic and molecular techniques are now being used to determine the identity of the insular sibling species, as well as its mainland source, extent of divergence, and length of time in the Galápagos. Similar quandaries will continue to arise as land use changes and as the interface between civilization and raw nature expands. Persistent incursions into undeveloped areas of the tropics, for example, expose people to new and greater risks of vector-borne diseases (Basáñez et al. 2000), while development projects (e.g., irrigation) increase the prevalence of diseases such as leishmaniasis (Lane 1993).

Regardless of the motivation for biodiversity studies of biting flies, the deep understanding that has accrued from such studies has fostered the use of biting flies as research models in areas applicable to other organisms. Biting flies have provided models for studies of community ecology (McCreadie and Adler 1998), genomic mapping (Holt et al. 2002), host-symbiont evolution (Chen et al. 1999), filter feeding (Merritt et al. 1996), island biogeography (Craig 2003), speciation (Rothfels 1989), transgenic arthropods (Alphey et al. 2002), and a bounty of other relevant topics.

BIODIVERSITY EXPLORATION

Progress in documenting the biodiversity of biting flies has been uneven among families. This imbalance stems not only from the differential importance of the various families to society, but also from the largely independent research efforts for each family, embracing different terminologies, techniques, and philosophies. These independent trajectories are maintained by specialists who typically confine their work to a single group. Some progress, nonetheless, has been made in homologizing structures and, hence, terminology across families. The *Manual of Nearctic Diptera*, for instance, provides a *lingua franca* of morphological terms for larvae and adults, although rarefied family-specific terms persist, especially for genitalia and pupae. Tools and approaches are applied variously among families. Polytene chromosomes have played an integral role in the taxonomy of some groups but not others; yet even among the groups for which cytotaxonomy has been practiced routinely (Culicidae and Simuliidae), the terminology is family specific. The Culicidae and Simuliidae are among the top three families of Diptera with the most DNA-barcoded specimens, while other families of blood-sucking flies have far fewer barcoded representatives or none (Barcode of Life Data Systems 2008). Although barcoding appears to be the wave of the near future in species identification, its controversial nature not withstanding (Besansky et al. 2003), molecular taxonomy currently is applied regularly only in certain groups of mosquitoes and sand flies that transmit human-disease agents. The criteria used to define species also vary among groups. Morphological criteria are used to define most species, although

cytological and molecular criteria also are used in the Culicidae and Simuliidae. Subspecies are recognized in some families (e.g., Culicidae, Glossinidae) but not others (e.g., Corethrellidae and Simuliidae).

In some groups of flies, certain taxonomic methods have been more profitable than others. A particular technique, for example, might be taxon specific or even stage specific. In black flies, a group in which speciation likely has been driven by chromosomal rearrangements, cytological approaches have been instrumental in the discovery and resolution of species (Rothfels 1989). The Simuliidae provide the lesson that significant biodiversity lies hidden, not only as sibling species, but also as homosequential sibling species (Bedo 1979). Although limited prospecting has been done at the molecular level among black flies (Krüger et al. 2006), no *bona fide* species has yet been discovered with molecular approaches, and until recently (Duncan et al. 2004), molecular techniques have failed to discriminate some of the most similar species. This situation is destined to change as the field of molecular biology advances. Though chromosomes currently provide the most powerful tool for revealing hidden species in the Simuliidae, they are useful, with notable exceptions (Bedo 1976), only for middle- to late-instar larvae.

Among mosquitoes, cytogenetic techniques have been invaluable in revealing sibling species, although not all mosquitoes are readily amenable to band-by-band chromosomal analysis. While the polytene chromosomes of anophelines are often workable, those of many culicines prove difficult, although the limits of the technique can be pushed (McAbee et al. 2007), for example by using Malpighian tubules rather than the standard tissues such as larval salivary glands (Zambetaki et al. 1998, Campos et al. 2003). Molecular techniques are now used routinely to reveal and discriminate sibling species of mosquito vectors (Kengne et al. 2003, Benet et al. 2004, Wilkerson et al. 2004) and have largely supplanted cytogenetic approaches. DNA barcoding will play an increasingly frequent role in mosquito taxonomy (Cywinska et al. 2006, Kumar et al. 2007).

For sand flies, polytene chromosomes have been the subject of few studies (White and Killick-Kendrick 1975). Molecular techniques, however, increasingly are being brought to bear on species problems of phlebotomines. These studies have been restricted largely to major vector complexes associated with human leishmaniasis. For several decades, the Neotropical sand fly *Lutzomyia longipalpis* has been suspected of being a species complex. Isozymes, molecular data, and pheromonal chemistry support this notion and have revealed at least four sibling species (Arrivillaga et al. 2003, Hamilton et al. 2004). Similarly, the Mediterranean *Phlebotomus perniciosus* complex consists of sibling species (Pesson et al. 2004). Establishing the vectorial abilities of each of these sibling species should help explain observed differences in the distribution and clinical manifestations of leishmaniasis, and ultimately improve vector-management strategies (Lanzaro and Warburg 1995).

Tsetse have a rich history of classical genetic studies (Gooding and Krafsur 2005), and are now frequent subjects of molecular investigation. No unequivocal cases of sibling species have been discovered among tsetse. Yet, hybridization between some subspecies results in male sterility of the offspring (Gooding 1985), and heterozygote deficiencies and linkage disequilibria occur within other subspecies, indicating population substructuring (Luna et al. 2001) that might reflect sibling species or incipient species. Of particular interest with regard to hidden biodiversity of tsetse is the potential variation in the ability to transmit trypanosomes.

For the remaining families, morphological approaches to species questions continue to dominate despite the option of using molecular and other techniques. The few studies that have explored the possibility of concealed biodiversity in the Tabanidae and Ceratopogonidae support the trend of sibling-species richness. Enzyme electrophoresis established the validity of two sibling species of the salt marsh greenheads *Tabanus conterminus* and *T. nigrovittatus* (Jacobson et al. 1981), and cuticular hydrocarbon analysis later corroborated the existence of the two species and revealed a probable third species (Sutton and Carlson 1997). Isozyme studies demonstrated that the ceratopogonid *Culicoides variipennis* is a complex of three reproductively isolated species (Holbrook et al. 2000). Molecular investigations of the economically important Ceratopogonidae are picking up pace (Mathieu et al. 2007) and can be predicted to yield additional cryptic biodiversity.

With time, biodiversity prospecting in biting flies will come to rely heavily – some believe exclusively – on molecular techniques. This suborganismal approach might uncover the genetic basis for reproductive isolation. Molecular, chromosomal, and morphological approaches, however, reflect hierarchical scales of investigation; information at one scale might not be accessible at another scale. A pluralistic approach, zooming in and out on the biodiversity scale, although not always producing concordance (della Torre et al. 2001, 2002), is likely to be the most illuminating strategy. Despite the value of a synthetic approach, the trend is swinging toward an exclusively molecular approach, propelled by the increasing ease, speed, and accessibility of molecular techniques, while cytogenetic and morphological techniques remain relatively difficult and time consuming.

Laboratory aspects of revealing and describing the biodiversity of biting flies are often corroborated, informed, or initiated when biological discrepancies (e.g., in seasonality or host usage) are observed in natural populations. Future studies of biodiversity must integrate the study of dead specimens with information from populations in nature. For a complete appreciation of biting-fly biodiversity, scientists must strive for an understanding of living organisms, including habitat preferences, reproductive behaviors, and interactions with other organisms.

SOCIETAL CONSEQUENCES OF DISREGARDING BIODIVERSITY

Accurate identification is critical to all scientific research and its application, perhaps no more so than when dealing with biting pests and vectors. Yet misidentification is rampant in the literature on biting flies, owing in part to the similarity among so many species, the dwindling numbers of specialists, and the lack of mandatory taxonomy courses. More than 600 misidentifications have been documented in the 2200 or so articles and theses on North American black flies (Adler et al. 2004). Misidentifications can cost lives, disrupt economies, and waste money. The misidentification of *Anopheles varuna*, a nonvector of malarial agents, misdirected efforts to control the actual vector in central Vietnam (Van Bortel et al. 2001). Misidentification of another nonvector, *An. filipinae*, in the Philippines set the stage for similarly unproductive management efforts (Foley et al. 1996). Targeting the wrong species also could exacerbate the pest or disease problem by opening habitat for development of the real villain. The success of antimalarial programs is inversely correlated with the biodiversity of the vectorial system (Coluzzi 1984). Failure to recognize all entities of the system compromises success. Efforts to manage pests and vectors biologically, for example by introducing sterile males, genetically engineered individuals, or host-specific natural enemies, must ensure compatibility with the targeted populations.

At special risk for misidentification are the morphologically similar, if not identical, sibling species and the more covert homosequential sibling species. Given their frequency and the difficulty in recognizing them, the chance of misidentification is considerable. The different vectorial capacities of sibling species underscore the importance of precise identification. The demonstration that the biting midge *C. variipennis* is actually three distinct species, only one of which (*C. sonorensis*) is a principal vector of bluetongue virus in North American ruminants, implies that vector-competence genes have a low probability of flowing from *C. sonorensis* to *C. variipennis sensu stricto* (Holbrook et al. 2000). The significance for North American agriculture is that livestock and germplasm can be moved more freely from bluetongue-free areas of the USA to bluetongue-free countries. At least half of the major vectors of human malarial agents belong to complexes of sibling species (Krzywinski and Besansky 2003). The seven or more sibling species in the *An. gambiae* complex differ in their ability to transmit human malarial agents, ranging from nonvector to highly competent vectors (Coluzzi 1984). Only a few of the ten or more sibling species of African biting midges in the *Culicoides imicola* complex have been implicated in the transmission of African horse sickness virus (Meiswinkel 1998, Mellor et al. 2000). Of the 30 sibling species and cytoforms in the *S. damnosum* complex in northeastern Africa and East Africa, only 6 have been incriminated as vectors of the filarial parasite responsible for human onchocerciasis (Post et al. 2007). Further, the vector species of the *S. damnosum* complex show differential susceptibilities and resistance to insecticides, requiring accurate identification to monitor the development and spread of resistance genes (Kurtak 1990).

To avoid misidentifications, as well as reveal and delineate hidden biodiversity, clarify pest status and epidemiological complexity, and generally keep biodiversity discoveries flowing, a synthetic approach is required. Molecular techniques are now part of the toolkit that includes the tried-and-true methods of classical systematics. Molecular methods also provide a reliable means of linking immature stages with adults (including type specimens), as well as identifying specimens in poor condition or larvae too young for morphological or chromosomal identifications. No single method, however, is likely to answer all questions

regarding species. Even the powerful triumvirate of cytotaxonomy, morphotaxonomy, and molecular taxonomy can fail to resolve or provide concordance for certain critical species issues among disease-bearing flies such as the *An. gambiae* complex (della Torre et al. 2001, 2002). Additional technologies, such as cuticular hydrocarbon analysis, have been used in most groups of biting flies and generally, though not always, corroborate species limits based on other criteria (Carlson et al. 1997). The resolution of five sibling species in the North American *An. quadrimaculatus* complex, using biochemical, chromosomal, ecological, hybridization, molecular, and morphological techniques (Reinert et al. 1997), provides a blueprint for multifaceted approaches. An integrated strategy also will be needed to relate the vectorial capacity of the 5 siblings to the 15 or more disease agents transmitted by the members of the complex.

New approaches and techniques bring new problems, some immediately apparent, others latent. A new problem of the molecular age is that some sequences placed in public databases are linked with misidentified species; this problem has surfaced for mosquitoes (Krzywinski and Besansky 2003). To avoid these mistakes, molecular biologists must work collaboratively with morphotaxonomists and cytotaxonomists. More importantly, future biodiversity researchers must be *au fait* with each of these techniques and capable of applying them to solve the thornier problems. They must 'be emboldened to venture across the dissolving disciplinary barriers' (Kafatos and Eisner 2003).

PRESENT AND FUTURE CONCERNS

To further guard against misidentifications and facilitate future discoveries of hidden biodiversity, voucher specimens must be deposited in accessible museums. Vouchers provide some measure of immortality to an author's work by permitting subsequent verification or recalibration should hidden diversity be discovered. They also provide a historical basis for evaluating genetic and morphological changes over time. The notorious cattle pest of the Canadian prairies, responsible for the deaths of more than 3500 head of cattle since 1886, was long known as *S. arcticum*. In reality, it was an entirely different and undescribed species, subsequently named *S. vampirum* (Adler et al. 2004). So abundant at times that it formed a loud, buzzing fog, the species later proved difficult to study taxonomically because limited specimens had been deposited in museums before it was eradicated from much of its former habitat; evidently, it was so abundant that previous researchers saw little need to deposit material in museums. Future researchers hoping to track the genetic changes from a time of superabundance to virtual extermination (bottleneck) will be challenged to find adequate historical material of this species. More generally, future workers are not likely to view present vouchering practices approvingly. Of 84 articles on biting flies published in the *Journal of Medical Entomology* in 2003, only 5% mentioned the deposition of voucher specimens – the other 95% representing an enormous potential loss of future information.

Voucher specimens and type series can include not only actual specimens but also chromosome preparations, preferably with photographs that ensure permanency should the chromosome preparations prove unstable over time (Adler and Kim 1985). Images of biochemical and molecular gels similarly can be deposited as complementary, but not exclusive, items to actual specimens or specimen parts. So that representative material might be available in depositories for future scrutiny and analysis, the practice of grinding an entire organism for molecular analysis, or grinding a portion and discarding the remainder, must be avoided.

At a time when the need for collections and storage of biological material is critical, administrators are calling for the downsizing, sale, or abandonment of collections. Such myopic calls are epitomized by the suggestion of a former administrator at one North American institution to replace the specimens in 'space-consuming cabinets' with two-dimensional photographs that could be stored in compact files. Solutions are not easy, but none are satisfactory if they do not allow for storage of actual specimens. No scientist has the prescience to predict what information will be required from museum specimens, which otherwise would be unavailable from photographs, CD ROMs, and other ersatz items. Nor can the techniques be foreseen that one day might be used to extract information from real specimens. With continued destruction of habitat and concomitant extinctions, all we ever will know of some species will be the information associated with real specimens housed in collections (Wheeler 2004).

The greatest need facing future biodiversity studies of biting flies is a pervasive appreciation of the value of the work and the importance of organismal education. All other needs flow logically from this perspective.

Currently, an appreciation is held by only a small cadre of scientists and lay people. A certain irony in acknowledging the value of biodiversity issues comes from the WHO Onchocerciasis Control Programme in Africa. The cytotaxonomists whose discoveries of sibling species in the *S. damnosum* complex revolutionized the playing field by allowing more precise targeting of the vectors of *O. volvulus*, nonetheless, repeatedly had to justify their efforts to identify the vector species in control programs (Meredith 1988). A majority of students now graduate from college without an adequate understanding of the biodiversity crisis or how biting flies and vector-borne diseases might affect their lives.

Yet another irony, and a dangerous pitfall, is success itself. Effective control can become the enemy of continued biodiversity exploration. The heady success of DDT in controlling malaria, beginning in the late 1940s, led to a false belief that mosquito biologists were no longer needed; a generation or more went untrained, by which time resistance to DDT, as well as to certain antimalarial drugs, had become entrenched. The unprecedented success of *B. thuringiensis* var. *israelensis* against black flies has dampened not only studies of natural enemies (Adler et al. 2004), but also studies of the family in general.

'Every insect taxon needs a fanatical specialist who will make a life of passionate proprietary concern for his own personal charges' (Lloyd 2003). To improve on James Lloyd's sentiment, we need only pluralize 'specialist' and adjust the corresponding orthography. Yet, the number of trained specialists continues to dwindle while the hackneyed plea goes out for more taxonomic expertise. In nearly 250 years of describing and naming biting flies, about 415 people have authored or coauthored formal species names for the world's black flies, averaging roughly 5.0 valid species per author. About 50 authors have named biting muscids, for an average of 1.1 valid species and subspecies per author. For tsetse and corethrellids, 24 authors each have described species and subspecies, or about 1.4 and 4.0 valid taxa, respectively, per author. Perhaps with more sophisticated techniques, in the context of rapid information processing, the task of describing all species might be completed with fewer workers, although this view seems naively optimistic. Current papers dealing with complex species problems often carry up to seven or more authors (e.g., della Torre et al. 2002, Arrivillaga et al. 2003, Pesson et al. 2004).

Cataloging biodiversity must not remain the static end game. The dynamic, coevolutionary system of vector, vertebrate host, and disease agent requires vigilant monitoring to protect human and animal welfare. The rapidly changing nature of the world's ecological landscape exerts intense and swift selection on pests and vectors, the development of pesticide resistance being a textbook example. Two sibling species in the *S. damnosum* complex, *S. damnosum sensu stricto* and *S. sirbanum*, are now hybridizing at increased rates as a result of habitat disturbance and resistance stemming from long-term, repeated insecticide applications in the Onchocerciasis Control Programme area of Africa (Boakye et al. 2000). The danger is that genetic material could be transferred from one species to the other, giving rise to new variants or even new species, and compromising vector-management programs. This scenario has been described for mosquitoes in the *Culex pipiens* complex, which are vectors of West Nile virus. Molecular data suggest that, in the USA, hybridization between the bird-feeding *Culex pipiens* and the mammal-feeding *Culex molestus* (recognized by many workers as a physiological form of *Cx. pipiens*) has produced efficient vectors that are able to transfer the virus from birds to humans (Fonseca et al. 2004). Because some species of the *An. gambiae* complex feed exclusively on humans and breed in artificial containers, which have been available for fewer than 10,000 years, at least some radiation within the complex might have occurred in recent times (Coluzzi et al. 2002). Ongoing speciation within sibling species of the *An. gambiae* complex complicates the control of malaria in Africa by extending the transmission potential of the vectors in space and time (della Torre et al. 2002).

As a final caveat, new disease concerns are likely to emerge from previously unappreciated or *de novo* vector–pathogen relationships. Climate change, for example, has driven the major Old World vector of bluetongue virus into new areas of Europe, and bluetongue virus has expanded its range even further, putatively through the involvement of novel *Culicoides* vectors (Purse et al. 2005). The rapid spread of *Aedes albopictus* over large portions of the globe is fueling unexpected outbreaks of diseases, such as chikungunya, with increased potency (Enserink 2008). Future biodiversity efforts, therefore, must continue to link the species of disease-causing agents with their dipteran hosts. The total number of zoopathogens transmitted by biting flies remains unknown, not only because the taxonomy of some microorganisms is poorly developed or confused, but also because many hosts of minimal economic importance (i.e., wildlife)

have not been surveyed for pathogens, or the pathogens have not been associated with the vectors. About 82 species of *Leucocytozoon* are known from 54 families of birds (Peirce 2005), but only about 13 species have been linked with vectors (all Simuliidae, except one Ceratopogonidae) (Adler and McCreadie 2002). Species resolution of disease-causing organisms, such as protozoans, has been difficult because of a dearth of morphological characters. Based on the prevalence of sibling species among vectors, a working hypothesis is that transmitted microorganisms also consist of sibling species. Molecular evidence supports this hypothesis for the genus *Leucocytozoon* (Sehgal et al. 2006, Hellgren et al. 2008). A corollary hypothesis is that the vector specificity and vertebrate-host specificity of transmitted disease agents is greater than currently realized. Molecular approaches should be useful in testing these ideas. They have proven useful in groups such as the tsetse-transmitted *Trypanosoma* in which species in some subgenera have been oversplit, while species richness in other subgenera has been underestimated (Gibson 2003).

CONCLUSIONS

More than 17,500 species of biting flies have been described, of which about 75% take blood from vertebrates. The Neotropics are richest in species, although new species continue to be discovered in biodiversity hot spots and inadequately sampled areas throughout the world. Still more biodiversity remains to be harvested from the genome of currently known morphospecies. Structurally similar sibling species are revealed routinely when appropriate techniques are applied, with some morphospecies consisting of as many as 50 sibling species.

What do biodiversity studies of biting flies reveal about other groups of organisms? More than anything, they suggest that estimates of the total number of life forms are too low. Most estimates of life's richness are based on known numbers of morphospecies, failing to account for sibling species and homosequential sibling species that, with closer scrutiny or more sophisticated techniques, await discovery. The plurality of approaches that has been used to uncover hidden biodiversity in particular groups of biting flies provides a template for biodiversity investigations of other organisms.

As a final thought, we must appreciate that discovering, describing, and cataloging the rich biodiversity of biting flies is not an endpoint, but rather a means of addressing a multiplicity of questions directly relevant to the future of humanity. Biting flies, as pests and vectors of disease agents, have profoundly influenced human history and continue to exact a brutal toll. They are responsible for more than 1.25 million human deaths annually and an incalculable socioeconomic burden. Threats to human and animal welfare arise not only from well-known pests and vectors, but also from undiscovered species, recent invasive species, novel vector-microorganism associations, and environmental dynamics that foster hybridization and exchange of genomic material between vector taxa. The degree to which risks are eliminated is directly correlated with our understanding of the biodiversity of the incriminated biting flies. At odds with the destruction levied by adult biting flies is the key ecological role of the immature stages. Safeguarding the future of humanity while conserving the ecological benefits of biting flies hinges on continued investments in biodiversity research.

ACKNOWLEDGMENTS

I thank A. Borkent for insights into the number of corethrellid and ceratopogonid species and early access to his world ceratopogonid catalogue; F. Bravo, J. F. Burger, and E. A. B. Galati for information on the number of sycoracine, tabanid, and phlebotomine species, respectively; and A. Borkent, R. W. Crosskey, A. C. Pont, and W. K. Reeves for thoughtful reviews of the manuscript.

REFERENCES

Abedraabo, S., F. Le Pont, A. J. Shelley, and J. Mouchet. 1993. Introduction et acclimatation d'une simulie anthropophile dans l'ile San Cristobal, archipel des Galapagos (Diptera, Simulidae [sic]). *Bulletin de la Société Entomologique de France* 98: 108.

Adler, P. H. and R. W. Crosskey. 2008. World Blackflies (Diptera: Simuliidae): A Fully Revised Edition of the Taxonomic and Geographical Inventory. http://entweb .clemson.edu/biomia/pdfs/blackflyinventory.pdf [Accessed 31 March 2008].

Adler, P. H. and K. C. Kim. 1985. Taxonomy of black fly sibling species: two new species in the *Prosimulium mixtum* group (Diptera: Simuliidae). *Annals of the Entomological Society of America* 78: 41–49.

Adler, P. H. and J. W. McCreadie. 1997. The hidden ecology of black flies: sibling species and ecological scale. *American Entomologist* 43: 153–161.

Adler, P. H. and J. W. McCreadie. 2002. Black flies (Simuliidae). Pp. 185–202. *In* G. Mullen and L. Durden (eds). *Medical and Veterinary Entomology*. Academic Press, New York.

Adler, P. H. and W. Wills. 2003. Legacy of death: the history of arthropod-borne human diseases in South Carolina. *American Entomologist* 49: 216–228.

Adler, P. H., D. C. Currie, and D. M. Wood. 2004. *The Black Flies (Simuliidae) of North America*. Cornell University Press, Ithaca, NY. 941 pp + 24 plates.

Alphey, L., C. B. Beard, P. Billingsley, M. Coetzee, A. Crisanti, C. Curtis, P. Eggleston, C. Godfray, J. Hemingway, M. Jacobs-Lorena, A. A. James, F. C. Kafatos, L. G. Mukwaya, M. Paton, J. R. Powell, W. Schneider, T. W. Scott, B. Sina, R. Sinden, S. Sinkins, A. Spielman, Y. Touré, and F. H. Collins. 2002. Malaria control with genetically manipulated insect vectors. *Science* 298: 119–121.

Arrivillaga, J., J.-P. Mutebi, H. Piñango, D. Norris, B. Alexander, M. D. Feliciangeli, and G. C. Lanzaro. 2003. The taxonomic status of genetically divergent populations of *Lutzomyia longipalpis* (Diptera: Psychodidae) based on the distribution of mitochondrial and isozyme variation. *Journal of Medical Entomology* 40: 615–627.

Baker, J. R. 1967. A review of the role played by the Hippoboscidae (Diptera) as vectors of endoparasites. *Journal of Parasitology* 53: 412–418.

Barcode of Life Data Systems. 2008. http://www .barcodinglife.org/views/login.php [Accessed 20 January 2008].

Basáñez, M. G., L. Yarzábal, H. L. Frontado, and N. J. Villamizar. 2000. *Onchocerca-Simulium* complexes in Venezuela: can human onchocerciasis spread outside its present endemic area? *Parasitology* 120: 143–160.

Bedo, D. G. 1976. Polytene chromosomes in pupal and adult blackflies (Diptera: Simuliidae). *Chromosoma* 57: 387–396.

Bedo, D. G. 1979. Cytogenetics and evolution of *Simulium ornatipes* Skuse (Diptera: Simuliidae). II. Temporal variation in chromosomal polymorphisms and homosequential sibling species. *Evolution* 33: 296–308.

Benet, A., A. Mai, F. Bockarie, M. Lagog, P. Zimmerman, M. P. Alpers, J. C. Reeder, and M. S. J. Bockarie. 2004. Polymerase chain reaction diagnosis and the changing pattern of vector ecology and malaria transmission dynamics in Papua New Guinea. *American Journal of Tropical Medicine and Hygiene* 71: 277–284.

Bernal, X. E., A. S. Rand, and M. J. Ryan. 2006. Acoustic preferences and localization performance of blood-sucking flies (*Corethrella* Coquillett) to túngara frog calls. *Behavioral Ecology* 17: 709–715.

Besansky, N. J., D. W. Severson, and M. T. Ferdig. 2003. DNA barcoding of parasites and invertebrate disease vectors: what you don't know can hurt you. *Trends in Parasitology* 19: 545–546.

Boakye, D. A., C. Back, and P. M. Brakefield. 2000. Evidence of multiple mating and hybridization in *Simulium damnosum* s. l. (Diptera: Simuliidae) in nature. *Journal of Medical Entomology* 37: 29–34.

Borkent, A. 2005. The biting midges, the Ceratopogonidae (Diptera). Pp. 113–126. *In* W. C. Marquardt (ed). *Biology of Disease Vectors*, Second Edition. Elsevier Academic Press, San Diego, CA.

Borkent, A. 2007. World Species of Biting Midges (Diptera: Ceratopogonidae). http://www.inhs.uiuc.edu/cee/FLYTREE/CeratopogonidaeCatalog.pdf [Accessed 12 January 2007].

Borkent, A. 2008. The frog-biting midges of the world (Corethrellidae: Diptera). *Zootaxa* 1804: 1–456.

Borkent, A. and D. A. Grimaldi. 2004. The earliest fossil mosquito (Diptera: Culicidae), in mid-Cretaceous Burmese amber. *Annals of the Entomological Society of America* 97: 882–888.

Borkent, A. and W. W. Wirth. 1997. World species of biting midges (Diptera: Ceratopogonidae). *Bulletin of the American Museum of Natural History* 233: 1–257.

Burger, J. F. 1977. The biosystematics of immature Arizona Tabanidae (Diptera). *Transactions of the American Entomological Society* 103: 145–258.

Burger, J. F. 1995. Catalog of Tabanidae (Diptera) of North America north of Mexico. *Contributions on Entomology, International* 1: 1–100.

Campos, J., C. F. S. Andrade, and S. M. Recco-Pimentel. 2003. A technique for preparing polytene chromosomes from *Aedes aegypti* (Diptera, Culicinae). *Memórias do Instituto Oswaldo Cruz* 98: 387–390.

Carlson, D. A., J. F. Reinert, U. R. Bernier, B. D. Sutton, and J. A. Seawright. 1997. Analysis of the cuticular hydrocarbons among species of the *Anopheles quadrimaculatus* complex (Diptera: Culicidae). *Journal of the American Mosquito Control Association* 13(Supplement): 103–111.

Chainey, J. E. and H. Oldroyd. 1980. Family Tabanidae. Pp. 275–308. *In* R. W. Crosskey (ed). *Catalogue of the Diptera of the Afrotropical Region*. British Museum (Natural History), London.

Chen, X., S. Li, and S. Aksoy. 1999. Concordant evolution of a symbiont with its host insect species: molecular phylogeny of genus *Glossina* and its bacteriome-associated endosymbiont, *Wigglesworthia glossinidia*. *Journal of Molecular Evolution* 48: 49–58.

Coluzzi, M. 1984. Heterogeneities of the malaria vectorial system in tropical Africa and their significance in malaria epidemiology and control. *Bulletin of the World Health Organization* 62 (Supplement): 107–113.

Coluzzi, M., A. Sabatini, A. della Torre, M. A. Di Deco, and V. Petrarca. 2002. A polytene chromosome analysis of the *Anopheles gambiae* complex. *Science* 298: 1415–1418.

Conn, J., K. H. Rothfels, W. S. Procunier, and H. Hirai. 1989. The *Simulium metallicum* species complex (Diptera: Simuliidae) in Latin America: a cytological study. *Canadian Journal of Zoology* 67: 1217–1245.

Craig, D. A. 2003. Geomorphology, development of running water habitats, and evolution of black flies on Polynesian Islands. *BioScience* 53: 1079–1093.

Cranston, P. S., D. H. D. Edward, and L. G. Cook. 2002. New status, species, distribution records and phylogeny for Australian mandibulate Chironomidae (Diptera). *Australian Journal of Entomology* 41: 357–366.

Crosskey, R. W. 1990. *The Natural History of Blackflies*. John Wiley and Sons Ltd., Chichester. 711 pp.

Cywinska, A., F. F. Hunter, and P. D. N. Hebert. 2006. Identifying Canadian mosquito species through DNA barcodes. *Medical and Veterinary Entomology* 20: 413–426.

Dehio, C., U. Sauder, and R. Hiestand. 2004. Isolation of *Bartonella schoenbuchensis* from *Lipoptena cervi*, a blood-sucking arthropod causing deer-ked dermatitis. *Journal of Clinical Microbiology* 42: 5320–5323.

della Torre, A., C. Fanello, M. Akogbeto, J. Dossou-yovo, G. Favia, V. Petrarca, and M. Coluzzi. 2001. Molecular evidence of incipient speciation within *Anopheles gambiae* s. s. in West Africa. *Insect Molecular Biology* 10: 9–18.

della Torre, A., C. Costantini, N. J. Besansky, A. Caccone, V. Petrarca, J. R. Powell, and M. Coluzzi. 2002. Speciation within *Anopheles gambiae* – the glass is half full. *Science* 298: 115–117.

Desowitz, R. 1998. *Tropical Diseases from 50,000 BC to 2500 AD*. Flamingo, London. 256 pp.

Downes, J. A. and W. W. Wirth. 1981. Ceratopogonidae. Pp. 393–421. *In* J. F. McAlpine, B. V. Peterson, G. E. Shewell, H. J. Teskey, J. R. Vockeroth, and D. M. Wood (eds). *Manual of Nearctic Diptera*, Volume 1. Monograph No. 27. Research Branch, Agriculture Canada, Ottawa.

Duncan, G., P. H. Adler, K. P. Pruess, and T. O. Powers. 2004. Molecular differentiation of two sibling species of the black fly *Simulium vittatum* (Diptera: Simuliidae), based on randomly amplified polymorphic DNA. *Genome* 47: 373–379.

Eldridge, B. F., T. W. Scott, J. A. Day, and W. J. Tabachnick. 2000. Arbovirus diseases. Pp. 415–460. *In* B. F. Eldridge and J. D. Edman (eds). *Medical Entomology*. Kluwer Academic Publishers, Netherlands.

Enserink, M. 2008. Chikungunya: no longer a Third World disease. *Science* 318: 1860–1861.

Fairchild, G. B. and J. F. Burger. 1994. A catalog of the Tabanidae (Diptera) of the Americas south of the United States. *Memoirs of the American Entomological Institute* 55: 1–249.

Foley, D. H., R. Paru, H. Dagoro, and J. H. Bryan. 1993. Allozyme analysis reveals six species within the *Anopheles punctulatus* complex of mosquitoes in Papua New Guinea. *Medical and Veterinary Entomology* 7: 37–48.

Foley, D. H., N. W. Beebe, E. Torres, and A. Saul. 1996. Misidentification of a Philippine malaria vector revealed by allozyme and ribosomal DNA markers. *American Journal of Tropical Medicine and Hygiene* 54: 46–48.

Fonseca, D. M., N. Keyghobadi, C. A. Malcolm, C. Mehmet, F. Schaffner, M. Mogi, R. C. Fleischer, and R. C. Wilkerson. 2004. Emerging vectors in the *Culex pipiens* complex. *Science* 303: 1535–1538.

Foster, W. A. and E. D. Walker. 2002. Mosquitoes (Culicidae). Pp. 203–262. *In* G. Mullen and L. Durden (eds). *Medical and Veterinary Entomology*. Academic Press, New York.

Galati, E. A. B. 1995. Phylogenetic systematics of the Phlebotominae (Diptera, Psychodidae) with emphasis on American groups. *Boletín de la Dirección de Malariología y Saneamiento Ambiental* 35(Supplement 1): 133–142.

Gibson, W. 2003. Species concepts for trypanosomes: from morphological to molecular definitions? *Kinetoplastid Biology and Disease* 2: 10.

Gooding, R. H. 1985. Electrophoretic and hybridization comparison of *Glossina morsitans morsitans*, *G. m. centralis*, and *G. m. submorsitans* (Diptera: Glossinidae). *Canadian Journal of Zoology* 63: 2694–2702.

Gooding, R. H. and E. S. Krafsur. 2005. Tsetse genetics: contributions to biology, systematics, and control of tsetse flies. *Annual Review of Entomology* 50: 101–123.

Greenberg, B. 1973. *Flies and Disease, Volume 2. Biology and Disease Transmission*. Princeton University Press, Princeton, NJ. 447 pp.

Grimaldi, D. 1997. The bird flies, genus *Carnus*: species revision, generic relationships, and a fossil *Meoneura* in amber (Diptera: Carnidae). *American Museum Novitates* 3190: 1–30.

Guerrero, R. 1997. Catalogo de los Streblidae (Diptera: Pupipara) parasitos de murcielagos (Mammalia: Chiroptera) del Nuevo Mundo. VII. Lista de especies, hospedadores y paises. *Acta Biologica Venezuelica* 17: 9–24.

Hamilton, J. G. C., R. P. Brazil, and R. Maingon. 2004. A fourth chemotype of *Lutzomyia longipalpis* (Diptera: Psychodidae) from Jaibas, Minas Gerais State, Brazil. *Journal of Medical Entomology* 41: 1021–1026.

Hanson, W. J. 1968. The immature stages of the subfamily Phlebotominae in Panama (Diptera: Psychodidae), Ph. D. thesis. University of Kansas, Lawrence. 104 pp. + 27 plates.

Harbach, R. E. 2004. The classification of genus *Anopheles* (Diptera: Culicidae): a working hypothesis of phylogenetic relationships. *Bulletin of Entomological Research* 95: 537–553.

Harrison, M. D., J. W. Brewer, and L. D. Merrill. 1980. Insect involvement in the transmission of bacterial pathogens. Pp. 201–292. *In* K. F. Harris and K. Maramorosch (eds). *Vectors of Plant Pathogens*. Academic Press, New York.

Harwood, R. F. and M. T. James. 1979. *Entomology in Human and Animal Health*. Seventh Edition. Macmillan Publishing, New York. 548 pp.

Hellgren, O., S. Bensch, and B. Malmqvist. 2008. Bird hosts, blood parasites and their vectors – associations uncovered by molecular analyses of blackfly blood meals. *Molecular Ecology* 17: 1605–1613.

Hirai, H., W. S. Procunier, J. O. Ochoa, and K. Uemoto. 1994. A cytogenetic analysis of the *Simulium ochraceum* species complex (Diptera: Simuliidae) in Central America. *Genome* 37: 36–53.

Holbrook, F. R., W. J. Tabachnick, E. T. Schmidtmann, C. N. McKinnon, R. J. Bobian, and W. L. Grogan. 2000. Sympatry in the *Culicoides variipennis* complex (Diptera: Ceratopogonidae): a taxonomic reassessment. *Journal of Medical Entomology* 37: 65–76.

Holt, R. A. et al. (123 authors). 2002. The genome sequence of the malaria mosquito *Anopheles gambiae*. *Science* 298: 129–149.

Hutson, A. M. and H. Oldroyd. 1980. Family Hippoboscidae. Pp. 766–771. *In* R. W. Crosskey (ed). *Catalogue of the Diptera of the Afrotropical Region*. British Museum (Natural History), London.

Jacobson, N. R., E. J. Hansens, R. C. Vrijenhoek, D. L. Swofford, and S. H. Berlocher. 1981. Electrophoretic detection of a sibling species of the salt marsh greenhead, *Tabanus nigrovittatus*. *Annals of the Entomological Society of America* 74: 602–605.

James, M. T. 1968. Family Rhagionidae (Leptidae). Pp. 29.1–29.12. *In* N. Papavero (ed). *A Catalogue of the Diptera of the Americas South of the United States*. São Paulo, Brazil.

Ježek, J. 1999. Comments on the correct grammatic gender of *Sycorax* Curt. and *Philosepedon* Eat. (Diptera: Psychodidae) with world catalogue. *Dipterologica Bohemoslovaca* 9: 83–87.

Johnson, R. N., D. G. Young, and J. F. Butler. 1993. Trypanosome transmission by *Corethrella wirthi* (Diptera: Corethrellidae) to the green treefrog, *Hyla cinerea* (Anura: Hylidae). *Journal of Medical Entomology* 30: 918–921.

Jones, T. W. 2001. Animal trypanosomiasis. Pp. 33–46. *In* M. W. Service (ed). *Encyclopedia of Arthropod-Transmitted Infections of Man and Domesticated Animals*. CABI Publishing, New York.

Jordan, A. M. 1993. Tsetse-flies (Glossinidae). Pp. 333–388. *In* R. P. Lane and R. W. Crosskey (eds). *Medical Insects and Arachnids*. Chapman and Hall, London.

Kafatos, F. C. and T. Eisner. 2003. Unification in the century of biology. *Science* 303: 1257.

Karabatsos, N. (ed). 1985. *International Catalogue of Arboviruses Including Certain other Viruses of Vertebrates*. Third Edition. American Society of Tropical Medicine and Hygiene, San Antonio, TX.

Kengne, P., P. Awono-Ambene, C. Antonio-Nkondjio, F. Simard, and D. Fontenille. 2003. Molecular identification of the *Anopheles nili* group of African malaria vectors. *Medical and Veterinary Entomology* 17: 67–74.

Krinsky, W. L. 2002. Tsetse flies (Glossinidae). Pp. 303–316. *In* G. Mullen and L. Durden (eds). *Medical and Veterinary Entomology*. Academic Press, New York.

Krüger, A., A. K. Kalinga, A. M. Kibweja, A. Mwaikonyole, and B. T. A. Maegga. 2006. Cytogenetic and PCR-based identification of *S. damnosum* 'Nkusi J' as the anthropophilic blackfly in the Uluguru onchocerciasis focus in Tanzania. *Tropical Medicine and International Health* 11: 1066–1074.

Krzywinski, J. and J. J. Besansky. 2003. Molecular systematics of *Anopheles*: from subgenera to subpopulations. *Annual Review of Entomology* 48: 111–139.

Kumar, N. P., A. R. Rajavel, R. Natarajan, and P. Jambulingam. 2007. DNA barcodes can distinguish species of Indian mosquitoes (Diptera: Culicidae). *Journal of Medical Entomology* 44: 1–7.

Kurtak, D. C. 1990. Maintenance of effective control of *Simulium damnosum* in the face of insecticide resistance. *Acta Leidensia* 59: 95–112.

Laird, M., A. Aubin, P. Belton, M. M. Chance, F. J. H. Fredeen, W. O. Haufe, H. B. N. Hynes, D. J. Lewis, I. S. Lindsay, D. M. McLean, G. A. Surgeoner, D. M. Wood, and M. D. Sutton. 1982. *Biting Flies in Canada: Health Effects and Economic Consequences.*. National Research Council of Canada Publication No. 19248. Pp. 1–157.

Lane, R. P. 1993. Sandflies (Phlebotominae). Pp. 78–119. *In* R. P. Lane and R. W. Crosskey (eds). *Medical Insects and Arachnids*. Chapman and Hall, London.

Lanzaro, G. C. and A. Warburg. 1995. Genetic variability in phlebotomine sandflies: possible implications for leishmaniasis epidemiology. *Parasitology Today* 11: 151–154.

Lawyer, P. G. and P. V. Perkins. 2000. Leishmaniasis and trypanosomiasis. Pp. 231–298. *In* B. F. Eldridge and J. D. Edman (eds). *Medical Entomology*. Kluwer Academic Publishers, Netherlands.

Lewis, D. J. 1978. The phlebotomine sandflies (Diptera: Psychodidae) of the Oriental Region. *Bulletin of the British Museum of Natural History (Entomology)* 37: 217–343.

Lloyd, J. E. 2002. Louse flies, keds, and related flies (Hippoboscidae). Pp. 349–362. *In* G. Mullen and L. Durden (eds). *Medical and Veterinary Entomology*. Academic Press, New York.

Lloyd, J. E. 2003. On research and entomological education VI: firefly species and lists, old and new. *Florida Entomologist* 86: 99–113.

Luna, C., M. Bonizzoni, Q. Cheng, A. S. Robinson, S. Aksoy, and L. Zheng. 2001. Microsatellite polymorphism in tsetse flies (Diptera: Glossinidae). *Journal of Medical Entomology* 38: 376–381.

Maa, T. C. 1989. Family Hippoboscidae, Family Nycteribiidae, Family Streblidae. Pp. 785–789, 790–794, 795–796. *In* N. L. Evenhuis (ed). *Catalog of the Diptera of the Australasian and Oceanian Regions*. Bishop Museum Special Publication 86. Bishop Museum Press, Honolulu, Hawaii and E. J. Brill, Leiden, The Netherlands.

Maa, T. C. and B. V. Peterson. 1987. Hippoboscidae. Pp. 1271–1281. *In* J. F. McAlpine, B. V. Peterson, G. E. Shewell, H. J. Teskey, J. R. Vockeroth, and D. M. Wood (eds). *Manual of Nearctic Diptera*, Volume 2. Monograph No. 28. Research Branch, Agriculture Canada, Ottawa.

Malmqvist, B., P. H. Adler, K. Kuusela, R. W. Merritt, and R. S. Wotton. 2004. Black flies in the boreal biome, key

organisms in both terrestrial and aquatic environments: a review. *Écoscience* 11: 187–200.

Mathieu, B., A. Perrin, T. Baldet, J.-C. Delécolle, E. Albina, and C. Cêtre-Sossah. 2007. Molecular identification of Western European species of *obsoletus* complex (Diptera: Ceratopogonidae) by an internal transcribed spacer-1 rDNA multiplex polymerase chain reaction assay. *Journal of Medical Entomology* 44: 1019–1025.

McAbee, R. D., J. A. Christiansen, and A. J. Cornel. 2007. A detailed larval salivary gland polytene chromosome photomap for *Culex quinquefasciatus* (Diptera: Culicidae) from Johannesburg, South Africa. *Journal of Medical Entomology* 44: 229–237.

McCreadie, J. W. and P. H. Adler. 1998. Scale, time, space, and predictability: species distributions of preimaginal black flies (Diptera: Simuliidae). *Oecologia* 114: 79–92.

Mead, D. G., E. W. Gray, R. Noblet, M. D. Murphy, E. W. Howerth, and D. E. Stallknecht. 2004. Biological transmission of vesicular stomatitis virus (New Jersey serotype) by *Simulium vittatum* (Diptera: Simuliidae) to domestic swine (*Sus scrofa*). *Journal of Medical Entomology* 41: 78–82.

Meiswinkel, R. 1998. The ten species in the *Culicoides imicola* Kieffer complex: an update (Ceratopogonidae). *Abstracts of the Fourth International Congress of Dipterology*. Pp. 144–145. Oxford, UK.

Mellor, P. S., J. Boorman, and M. Baylis. 2000. *Culicoides* biting midges: their role as arbovirus vectors. *Annual Review of Entomology* 45: 307–340.

Meredith, S. E. O. 1988. The role of systematics in black fly population management. Pp. 53–61. *In* K. C. Kim and R. W. Merritt (eds). *Black Flies: Ecology, Population Management, and Annotated World List*. Pennsylvania State University Press, University Park, PA.

Merritt, R. W., D. A. Craig, R. S. Wotton, and E. D. Walker. 1996. Feeding behavior of aquatic insects: case studies on black fly and mosquito larvae. *Invertebrate Biology* 115: 206–217.

Moon, R. D. 2002. Muscid flies (Muscidae). Pp. 279–301. *In* G. Mullen and L. Durden (eds). *Medical and Veterinary Entomology*. Academic Press, New York.

Mullen, G. R. 2002. Biting midges (Ceratopogonidae). Pp. 163–183. *In* G. Mullen and L. Durden (eds). *Medical and Veterinary Entomology*. Academic Press, New York.

Mullens, B. A. 2002. Horse flies and deer flies (Tabanidae). Pp. 263–277. *In* G. Mullen and L. Durden (eds). *Medical and Veterinary Entomology*. Academic Press, New York.

Murphree, C. S. and G. R. Mullen. 1991. Comparative larval morphology of the genus *Culicoides* Latreille (Diptera: Ceratopogonidae) in North America with a key to species. *Bulletin of the Society for Vector Ecology* 16: 269–399.

Nagatomi, A. 1975. Family Rhagionidae (Leptidae). Pp. 82–90. *In* M. D. Delfinado and D. E. Hardy (eds). *A Catalog of the Diptera of the Oriental Region*, Volume 2. University Press of Hawaii, Honolulu, Hawaii.

Nagatomi, A. and K. Soroida. 1985. The structure of the mouthparts of the orthorrhaphous Brachycera (Diptera) with special reference to blood-sucking. *Beiträge zur Entomologie* 35: 263–368.

Nagatomi, A. and N. L. Evenhuis. 1989. Family Rhagionidae. Pp. 296–298. *In* N. L. Evenhuis (ed). *Catalog of the Diptera of the Australasian and Oceanian Regions*. Bishop Museum Special Publication 86. Bishop Museum Press, Honolulu, Hawaii and E. J. Brill, Leiden, The Netherlands.

Ostfeld, R. S., P. Roy, W. Haumaier, L. Canter, F. Keesing, and E. D. Rowton. 2004. Sand fly (*Lutzomyia vexator*) (Diptera: Psychodidae) populations in Upstate New York: abundance, microhabitat, and phenology. *Journal of Medical Entomology* 41: 774–778.

Pauly, P. J. 2002. Fighting the Hessian fly: American and British responses to insect invasion, 1776–1789. *Environmental History* 7: 377–400.

Peirce, M. A. 2005. A checklist of the valid avian species of *Babesia* (Apicomplexa: Piroplasmorida), *Haemoproteus*, *Leucocytozoon* (Apicomplexa: Haemosporida), and *Hepatozoon* (Apicomplexa: Haemogregarinidae). *Journal of Natural History* 39: 3621–3632.

Pesson, B., J. S. Ready, I. Benabdennbi, J. Martín-Sánchez, S. Esseghir, M. Cadi-Soussi, F. Morillas-Marquez, and P. D. Ready. 2004. Sandflies of the *Phlebotomus perniciosus* complex: mitochondrial introgression and a new sibling species of *P. longicuspis* in the Moroccan Rif. *Medical and Veterinary Entomology* 18: 25–37.

Peterson, B. V. and R. L. Wenzel. 1987. Nycteribiidae. Pp. 1283–1291. *In* J. F. McAlpine, B. V. Peterson, G. E. Shewell, H. J. Teskey, J. R. Vockeroth, and D. M. Wood (eds). *Manual of Nearctic Diptera*, Volume 2. Monograph No. 28. Research Branch, Agriculture Canada, Ottawa.

Pont, A. C. 1980a. Family Glossinidae. Pp. 762–765. *In* R. W. Crosskey (ed). *Catalogue of the Diptera of the Afrotropical Region*. British Museum (Natural History), London.

Pont, A. C. 1980b. Family Muscidae. Pp. 721–761. *In* R. W. Crosskey (ed). *Catalogue of the Diptera of the Afrotropical Region*. British Museum (Natural History), London.

Pont, A. C. and S. Mihok. 2000. A new species of *Haematobosca* Bezzi from Kenya (Diptera, Muscidae). *Studia Dipterologica* 7: 25–32.

Post, R. J., M. Mustapha, and A. Krueger. 2007. Taxonomy and inventory of the cytospecies and cytotypes of the *Simulium damnosum* complex (Diptera: Simuliidae) in relation to onchocerciasis. *Tropical Medicine and International Health* 12: 1342–1353.

Purse, B. V., P. S. Mellor, D. J. Rogers, A. R. Samuel, P. P. C. Mertens, and M. Baylis. 2005. Climate change and the recent emergence of bluetongue in Europe. *Nature Reviews Microbiology* 3: 171–181.

Reinert, J. F., P. E. Kaiser, and J. A. Seawright. 1997. Analysis of the *Anopheles* (*Anopheles*) *quadrimaculatus* complex of sibling species (Diptera: Culicidae) using morphological, cytological, molecular, genetic, biochemical, and ecological techniques in an integrated approach. *Journal of the American Mosquito Control Association* 13 (Supplement): 1–102.

Rothfels, K. 1989. Speciation in black flies. *Genome* 32: 500–509.

Rozkošný, R. and A. Nagatomi. 1997. Family Athericidae. Pp. 439–446. *In* L. Papp and B. Darvas (eds). *Contributions to a Manual of Palaearctic Diptera*, Volume 2. Science Herald, Budapest.

Rueda, L. M. 2008. Global diversity of mosquitoes (Insecta: Diptera: Culicidae) in freshwater. *Hydrobiologia* 595: 477–487.

Russell, P. F. 1959. Insects and the epidemiology of malaria. *Annual Review of Entomology* 4: 415–434.

Rutledge, L. C. and R. K. Gupta. 2002. Moth flies and sand flies (Psychodidae). Pp. 147–161. *In* G. Mullen and L. Durden (eds). *Medical and Veterinary Entomology*. Academic Press, New York.

Sailer, R. I. 1983. History of insect introductions. Pp. 15–38. *In* C. L. Wilson and C. L. Graham (eds). *Exotic Plant Pests and North American Agriculture*. Academic Press, New York.

Sakolsky, G., D. A. Carlson, B. D. Sutton, C.-M. Yin, and J. G. Stoffalano, Jr. 1999. Detection of cryptic species in the *Tabanus nigrovittatum* (Diptera: Tabanidae) complex in Massachusetts using cuticular hydrocarbon and morphometric analyses. *Journal of Medical Entomology* 36: 610–613.

Seccombe, A. K., P. D. Ready, and L. M. Huddleston. 1993. *A Catalogue of Old World Phlebotomine Sandflies (Diptera, Psychodidae, Phlebotominae)*. Occasional Papers on Systematic Entomology. Natural History Museum, London 8: 1–57.

Sehgal, R. N. M., A. C. Hull, N. L. Anderson, G. Valkiūnas, M. J. Markovets, S. Kawamura, and L. A. Tell. 2006. Evidence for cryptic speciation of *Leucocytozoon* spp. (Haemosporida, Leucocytozoidae) in diurnal raptors. *Journal of Parasitology* 92: 375–379.

Steiner, P. E. 1968. *Disease in the Civil War: Natural Biological Warfare in 1861–1865*. Charles C. Thomas, Springfield, IL. 243 pp.

Stuckenberg, B. R. 1980. Family Athericidae. Pp. 312–313. *In* R. W. Crosskey (ed). *Catalogue of the Diptera of the Afrotropical Region*. British Museum (Natural History), London.

Sutton, B. D. and D. A. Carlson. 1997. Cuticular hydrocarbon variation in the Tabanidae (Diptera): *Tabanus nigrovittatus* complex of the North American Atlantic Coast. *Annals of the Entomological Society of America* 90: 542–549.

Takaoka, H. 2003. *The Black Flies (Diptera: Simuliidae) of Sulawesi, Maluku and Irian Jaya*. Kyushu University Press, Fukuoka, Japan. 581 pp.

Thompson, F. C. 1990. Biosystematic information: dipterists ride the third wave. Pp. 179–201. *In* M. Kosztarab and C. W. Schaefer (eds). *Systematics of the North American Insects and Arachnids: Status and Needs*. Virginia Agricultural Experiment Station Information Series 90-1. Blacksburg, VA.

Thompson, F. C. (ed). 2006. Biosystematic Database of World Diptera. http://www.sel.barc.usda.gov/Diptera/names/famlistt.htm [Accessed 5 November 2006].

Turner, W. J. 1974. A review of the genus *Symphoromyia* Frauenfeld (Diptera: Rhagionidae). I. Introduction. Subgenera and species-groups. Review of biology. *Canadian Entomologist* 106: 851–868.

Van Bortel, W., R. E. Harbach, H. D. Trung, P. Roelants, T. Backeljau, and M. Coosemans. 2001. Confirmation of *Anopheles varuna* in Vietnam, previously misidentified and mistargeted as the malaria vector *Anopheles minimus*. *American Journal of Tropical Medicine and Hygiene* 65: 729–732.

Walter Reed Biosystematics Unit. 2006. 2001 Systematic Catalog of Culicidae. http://www.mosquitocatalog.org/main.asp [Accessed 3 December 2006].

Webb, D. W. 1981. Athericidae. Pp. 479–482. *In* J. F. McAlpine, B. V. Peterson, G. E. Shewell, H. J. Teskey, J. R. Vockeroth, and D. M. Wood (eds). *Manual of Nearctic Diptera*, Volume 1. Monograph No. 27. Research Branch, Agriculture Canada, Ottawa.

Webb, D. W. 1995. The immature stages of *Suragina concinna* (Williston) (Diptera: Athericidae). *Journal of the Kansas Entomological Society* 67: 421–425.

Wenzel, R. L. and B. V. Peterson. 1987. Streblidae. Pp. 1293–1301. *In* J. F. McAlpine, B. V. Peterson, G. E. Shewell, H. J. Teskey, J. R. Vockeroth, and D. M. Wood (eds). *Manual of Nearctic Diptera*, Volume 2. Monograph No. 28. Research Branch, Agriculture Canada, Ottawa.

Wheeler, Q. D. 2004. Taxonomic triage and the poverty of phylogeny. *Philosophical Transactions of the Royal Society, Series B, Biological Sciences* 359: 571–583.

White, G. B. and R. Killick-Kendrick. 1975. Polytene chromosomes of the sandfly *Lutzomyia longipalpis* and the cytogenetics of Psychodidae in relation to other Diptera. *Journal of Entomology A* 50: 187–196.

Wilkerson, R. C., J. F. Reinert, and C. Li. 2004. Ribosomal DNA ITS2 sequences differentiate six species in the *Anopheles crucians* complex (Diptera: Simuliidae). *Journal of Medical Entomology* 41: 392–401.

Wilson, E. O. 1985. The biological diversity crisis: a challenge to science. *Issues in Science and Technology* 1985(Fall)2: 20–29.

Wotton, R. S., B. Malmqvist, T. Muotka, and K. Larsson. 1998. Fecal pellets from a dense aggregation of suspension-feeders in a stream: an example of ecosystem engineering. *Limnology and Oceanography* 43: 719–725.

Young, D. G. and M. A. Duncan. 1994. Guide to the identification and geographic distribution of *Lutzomyia* sand flies in Mexico, the West Indies, Central and South America (Diptera: Psychodidae). *Memoirs of the American Entomological Institute* 54: 1–881.

Young, D. G. and P. V. Perkins. 1984. Phlebotomine sand flies of North America (Diptera: Psychodidae). *Mosquito News* 44: 263–304.

Zambetaki, A., N. Pasteur, and P. Mavragani-Tsipidou. 1998. Cytogenetic analysis of Malpighian tubule polytene chromosomes of *Culex pipiens* (Diptera: Culicidae). *Genome* 41: 751–755.

Zumpt, F. 1973. *The Stomoxyine Biting Flies of the World*. Gustav Fischer Verlag, Suttgart. 175 pp.

Chapter 23

RECONCILING ETHICAL AND SCIENTIFIC ISSUES FOR INSECT CONSERVATION

Michael J. Samways

Department of Conservation Ecology and Entomology, and Centre for Invasion Biology, Stellenbosch University, Matieland 7602, South Africa

Insect Biodiversity: Science and Society, 1st edition. Edited by R. Foottit and P. Adler
© 2009 Blackwell Publishing, ISBN 978-1-4051-5142-9

Insects are the most speciose animal group on earth, with 1,004,898 described species (introduction to this volume). If an estimated 99% of all species of organisms that have ever existed on earth have gone extinct, we may ask why we should concern ourselves with conservation when, on balance, extinction is perhaps the norm.

Paleontological evidence suggests that insect biodiversity has increased, but not steadily, over the last 350 million years (Labandeira and Sepkoski 1993). Despite comparatively high loss of some specialist species at the end of the Cretaceous, insects as a group, nevertheless, survived into the Tertiary, and indeed some have thrived and diversified (Labandeira et al. 2002). The Isoptera, Lepidoptera, and Mantodea are among those groups that have shown relatively modern radiation. Later, during the Quaternary (the last 2.4 million years), even with the advance and retreat of the glaciers, insect populations moved north and then south, in the Northern Hemisphere, with alternating warm and cold events. The geographic range shifts across latitudes were associated with little species extinction (Coope 1995). These population shifts apparently have been a regular feature of insect assemblages for at least 110,000 years (Ponel et al. 2003, Samways et al. 2006), and probably much longer.

With the appearance of tool-wielding humans and agricultural settlement, the landscape began to change rapidly, and insects faced a new challenge. From about 6000 years ago, the landscape became gradually and increasingly transformed (Steadman 1995, Burney and Flannery 2005), so that today, insects are experiencing new types of influences on their habitats, such as landscape fragmentation and draining of marshland habitats. These adverse effects have increased gradually in both extent and intensity, particularly over the last century.

Mawdsley and Stork (1995) estimate that between 100,000 and 500,000 insect species could go extinct over the next 300 years, while McKinney (1999) suggests that a quarter of all insect species are under threat of imminent extinction. These extinctions are primarily due to human transformation of the landscape and destruction of their habitats as a result of direct competition with humans for space and resources. Many of the effects are additive, one upon another, leading to particularly harsh and threatening conditions for habitat specialists (Travis 2003). Insects appear to be especially sensitive to these modern compound effects – they are declining faster than birds or vascular plants (Thomas

et al. 2004) (Fig. 23.1), at least in Britain. Of concern is that the British biota is essentially post-glacial and with very few narrow-range specialists. Because the tropics have many geographically restricted, specialist species, the situation might be even more acute in areas of the world where a host of sensitive specialists are living

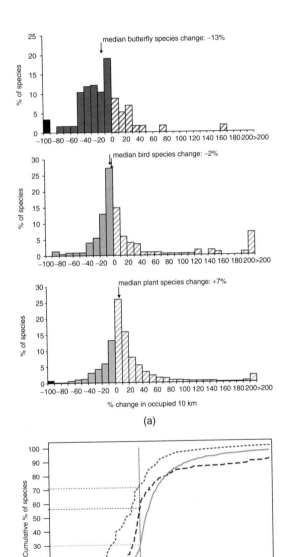

(a)

(b)

in small geographic areas and subject to rapid decline when the footprint of humankind coincides with their restricted range (Samways 2006).

A further consideration is that coextinctions also are beginning to occur and are associated with the enormous numbers of mutualistic or other coevolutionary interactions between different insects and other organisms. This coextinction appears to be the case when the relationships are particularly close. In both Britain and the Netherlands, pollinator declines are most frequent in habitat and flower specialists, in univoltine species, and in nonmigrants (Biesmeijer et al. 2006). The decline of pollinators leaves certain plants as the living dead, with the dependent herbivores that live on these plants effectively adrift on a doomed raft (Vamosi et al. 2006). In many parts of the world, all sorts of curious and specific interactions are under threat. The rather uncharismatic and even more uncharmingly named pygmy hog louse (*Haematopinus oliveri*) is under threat because its own special host, the northern Indian pygmy hog is threatened.

For the first time in the history of the earth, an insect-extinction spasm is taking place that is being caused by just one other species. Humans are steadily and surely pulling the thread out of the jersey of life and, to use Al Gore's expression, we are not facing 'the inconvenient truth' that we are rapidly unraveling ecosystems across the globe. The point is that this is also the first time that a supposedly conscious and sentient being is causing such extinction.

To address the opening point: as we are the conscious cause of such extinction, we surely have a moral responsibility to do something about it. Although not all insects will be lost, and indeed some will benefit from our influences, many unique evolutionary characters and even ecologically keystone species will be extinguished.

Fig. 23.1 Changes in the number of 10^2km $(10 \times 10$ km) in Britain occupied by native butterfly, bird, and plant species between the two censuses (20–40 years apart) of each taxon. (a) Frequency distributions: median butterfly species (top, $n = 58$) < median bird species (middle, $n = 201, P < 0.001$) < median plant native species (bottom, $n = 1253, P < 0.001$). Black, extinct species; solid color, declining species; hatched color, increasing species. (b) Cumulative frequency distributions (butterfly declines (......) > bird declines (– – –) > plant declines (——); Kolmogorov–Smirnov tests, $P < 0.001$). (From Thomas et al. 2004).

This issue is not just simply a moral one, but a practical one, too, with aspects of our survival also being in jeopardy. We must deal with this crisis, because 'business as usual' is not an option (Millenium Ecosystem Assessment 2005). How we deal with it, in turn, depends at the outset on the values we hold. I will now explore this question of values a little more, as it is the platform upon which our action must take place. I will then explore some practical options, bearing in mind that insect conservation is not a separate entity but rather intertwined with biodiversity conservation as a whole.

VALUING NATURE

Types of value

At the morally most simplistic and superficial level, organisms, including insects, are there for our benefit. This level is also one of practicality and the language of most real-life conservation. Conservation becomes ever more practical as the pressures mount. The human increase and the conversion of natural resources is, in turn, putting more pressure on insects. This practical level is about utilitarian or instrumental value. Such value can be consumptive or nonconsumptive. A consumptive value might relate, say, to pollinators because they are needed to produce some particular human food. An example of nonconsumptive utilitarian value is the viewing of insects for enjoyment at specific sites (Hill and Twist 1998). Utilitarian value, particularly when it is consumptive, must have sustainability, with due cognizance of the Precautionary Principle (Fauna and Flora International 2006). This principle recognizes that we must be sensitive to the complexity and levels of current biodiversity; yet, also, we must be cautious in using it because we do not know the extent to which using it will adversely affect those natural resources.

At the morally deeper level, beyond utilitarian value, is intrinsic value. While utilitarian value is about how we might benefit from nature, intrinsic value is about organisms having a share in the survival stakes in their own right. Intrinsic value also has come into prominence in one sector, which relates to the 'IUCN Red List of Threatened Species' (www.redlist.org). The Red List is proving to have enormous benefit in conservation planning (Baillie et al. 2004, Rodrigues et al. 2006). The significance of the intrinsic value of insects is that each species has equal space on the Red List. Thus, a bee, bird, or buffalo is given equal exposure, which is a major step forward for insect conservation, as individual insect species are given higher prominence than normally would be the case. Let us now explore some of these values more extensively.

Sensitive use of ecosystem services

The utilitarian level of practicality is currently the most persuasive for policy makers. A recent emphasis, for example, is to value biodiversity in terms of services (Millenium Ecosystem Assessment 2005) and economics (Bishop et al. 2007). Such a utilitarian approach emphasizes that we need insects because they supply goods and services of benefit to us, either directly or indirectly through sustenance of the natural world on which we depend. In purely monetary terms, these instrumental values are staggering. Losey and Vaughan (2006) estimate that 'wild' insects, which control pests, pollinate flowers, bury dung, and provide nutrition for other wildlife, are worth $57 billion per year in the USA alone. These small creatures are clearly providing big services. They are also an essential part of the fabric of healthy ecosystems as we know them (Coleman and Hendrix 2000).

This utilitarian approach separates humans from the rest of nature. It is a value system based on *us* using *them* (other organisms). With the connectedness of nature in mind, this is an untenable way of valuing nature, including insects, in the long term, unless we temper it with careful stewardship of ecosystems. Nevertheless, this utilitarian approach is the only conceivable one for the near future, both for local and international conservation policy and management.

This utilitarian approach, which does not necessarily exclude intrinsic value, is being explored under the European Union's Coordination Action Project 'Rationalising Biodiversity Conservation in Dynamic Ecosystems' (www.RUBICODE.net). The project is reviewing and developing concepts of dynamic ecosystems and the services that they provide. Those components of biodiversity that provide specific services to society are being defined and evaluated to increase our understanding of the value of biodiversity services, as well as the cost of losing them. This approach will give decision makers a more rational base and will help in understanding the need for adequate conservation policies, which are essential to halting biodiversity loss.

One way of viewing dynamic ecosystems and the services they provide is based on valuing a population through the concept of the Service Providing Unit (SPU) (Luck et al. 2003). An SPU provides a recognized service at some temporal or spatial scale. For example, if the service required is the European-wide sequestration of carbon, then the SPU would be the biodiversity of Europe. At a lower level, if the service required is pollination of particular plants, then the SPU would be the species and abundance of the pollinators. If the service is conservation of a rare herbivorous insect for its own sake, the SPU would be the population of its host plant(s) present in suitable habitat in the area over which special conservation measures are required. In turn, this service provision can occur at multiple levels, and will depend on social institutions as well as scientific issues.

SPUs often comprise more than one species, but their contribution to any given service might not be equal. They also can contribute to more than one service or be adverse or antagonistic to another service (termed Service Antagonising Units (SAUs)) (www.RUBICODE.net). Wild flowers, for example, not only can support crop pollinators and biocontrol agents, but also harbor SAUs in the form of pests. Quantifying the potential positive effects of biodiversity on a service should involve the subtraction of the effects of SAUs. One person's SPU, however, might be another's SAU, and so the concept varies according to various economic, cultural, and esthetic contexts in which it is measured. For example, while a certain biological control agent might be an SPU for an agriculturalist, it could be an SAU for a conservationist who is preserving a rare and threatened insect species that happens to be a nontarget for the biological control agent.

The SPU/SAU concept provides a framework for linking changes in key characteristics of populations with implications for service provision. The concept is also easily extendable to include other levels of organization (e.g., functional groups). Identifying quantitative links between components of ecosystems and service provisions is crucial to guiding the management of services. This quantitative information is of most value to policy makers and land managers, as it facilitates specific rather than vague management guidelines, which ensure the sustainability of nature's services. This relationship is illustrated by the use of economic thresholds in pest management (Pringle 2006), where the exact levels of the pest (in this case an SAU) and the natural enemy (the SPU) determine the service provision of the agricultural production patch.

Common good approaches

To value nature in context, we need to bring in common good approaches (Harrison and Burgess 2000). These authors show how farmers and residents contest scientific approaches to valuing nature when adjudicating conflicts over protected natural areas. For effective conservation, the knowledge base for the goals and practices of nature conservation must be widened. A way to tackle this challenge is to develop a common good approach based on ethical and moral concerns about nature. The important feature is to translate these concerns into practice by expressing the values through a social and political process of consensus building. The aim is to build coalitions and common thought between different interests through a process of debate and systematic analysis of values. Different perspectives on the utilitarian values of nature are thus reconciled. The point for insect conservation is that such a process of making decisions at the outset truly benefits conservation of, for example, the agri-environment, an arena where biodiversity conservation is at a critical stage (Conrad et al. 2004, Perrings et al. 2006).

Another dimension to this common good approach is beginning to surface. It hinges on our losing touch with nature (Stokes 2006), aptly called by Miller (2005) the 'extinction of experience'. His concern is that we are beginning to forget just how much we rely on nature. Because insects are so crucial to so many essential services, yet far from most peoples' thoughts, we, as a collective consciousness, are oblivious to what we are doing to the rich tapestry of species that grace and fine-tune ecosystems. Lyons et al. (2005) convincingly illustrated that less common species, which by reasonable extension include the majority of insects, make significant ecosystem contributions. Furthermore, many prejudices against insects exist, with mosquitoes and other flies being seen by a large sector of the public to represent insects at large. Entomologists, like other invertebrate conservationists, have a long way to go before the human world recognizes the general utilitarian, let alone intrinsic value, of insects as a whole.

Additionally, what we (the current adult population) value is not necessarily what our children value, either now or in the future when they are adults. To give one example, children (and the elderly...the wise?) are particularly fascinated by dragonflies, and much more so than economically active (too busy?) adults (Suh and Samways 2001). In response, Palmer and Finlay (2003), speaking no less than on behalf of the World Bank, recommend from Baha'i scriptures 'Train your children from the earliest days to be infinitely tender and loving to animals'. This generational and cultural sensitivity is just what the World Bank considers as valuing and investing in the future.

Intrinsic value and conservation action

Cincotta and Engelman (2000) suggest that our world is in crisis but not doomed. This view is not blind optimism, as biotic recoveries from prehistorical mass extinction events have shown. However, it does not mean that we simply abandon any effort to help nature survive in as much of its entirety as possible. We must apply our moral conscience to the full. Yet it need not be a penance. It makes for much more positive action and creativity when we are joyous about these tasks. This is the language of deep ecology, which provides not only a positive foundation and sense of wisdom, but also a course of action (Naess 1989).

With this intrinsic value approach, humans and nature are inseparable. A course of action then arises when all critics are recognized, alongside a more simplistic lifestyle, where harmony with nature is the goal. This is not a vague notion, but builds on sound ecological knowledge, including identifying which of our actions are beneficial for nature and which are harmful. Yet these terms are presumptuous: 'we know best for nature'. Arguably, a better way of expressing this sentiment, and interpreting it in terms of quantitative biodiversity, is to say that our actions should maintain ecological integrity (compositional and structural diversity) and encourage ecological health (functional diversity) (Rapport et al. 1998). In turn, a combination of these two, integrity and health, begets ecosystem resilience (Peterson et al. 1998). They also make available many more evolutionary opportunities than would an impoverished system. Thus, scientific values easily can be reconciled with those of intrinsic value.

This science and faith in ecosystem health, integrity, and resilience goes to the heart of insect conservation simply because the insect world is so vast and complex, with such an unimaginable number of biotic interactions, that we are obliged to employ the Precautionary Principle, whereby we maintain all the parts and their function as best we can. Translated into practical terms, landscape conservation over large areas is critically important for conserving insects.

With the landscape-scale approach, we let nature 'know best' by employing some basic, interrelated tenets that are beginning to emerge (Samways 2007a). Three important ones are (1) maintaining *large* landscape patches, (2) encouraging patch *quality*, and (3) *reducing* patch *isolation*. With this landscape-scale approach (which does not exclude species-level approaches as a fine-tuning), we emerge with what is, in effect, an Earth harmony, where all living things, no matter how small, have the right to live. This approach, according to Johnson (1991), is also the only morally acceptable way forward.

This landscape way of thinking inherently also incorporates a sense of *place* (Lockwood 2001) and, additionally, a sense of *change* across place. This change across place is, in ecological terms, change across space (i.e., beta and gamma diversity), and is where deep ecology departs from spirituality, as it favors location over omnipotence. Moving this thought into the realm of insect conservation means maintaining population levels of insect species *in situ* across the multitude of habitats across the globe, that is, giving habitat heterogeneity an important place on the conservation agenda.

Reconciling values

We can argue for reconciliation between utilitarian and intrinsic value from yet another perspective. In their extremes, utilitarian and intrinsic value approaches could be considered confrontational because, according to Norton (2000), they share four questionable assumptions and obstacles: (1) a mutual exclusion of each other, (2) an entity, not process, orientation, (3) moral monism, and (4) placeless evaluation. What Norton (2000) posits here are the two extremes, starting with mutual exclusion. He also draws attention to *things*, that is, *organisms* that supposedly have either utilitarian *or* intrinsic value. Reconciliation comes when one takes an ecological (short-term), as well as an evolutionary (long-term), dynamic view at large spatial scales, giving prominence to ecological integrity and ecosystem health and resilience. The obstacle is only apparent when utilitarianism and intrinsic values are each taken in their extreme, monistic sense.

Intrinsic value has to be seen alongside a deep ecological sense of place before reconciliation can begin. Norton (2000) summarizes this by suggesting an alternative value system that recognizes a continuum of ways in which we value nature. Such a spectrum gives value to all the natural interactions in their natural place in a pluralistic way. He calls this a Universal Earth Ethic, which means all places on Earth. This is not, however, a melting down into a common currency of homogenous biota, which is happening, for example, with the vast interchange of invasive alien organisms. The importance of such an Earth ethic is that it values nature for the creativity of its processes.

Rolston (2000), another influential environmental ethicist, also envisions an Earth ethic, with a blending of anthropocentric and biocentric values. This melding is for the patent reason that we and this planet share entwined destinies. This wider Earth ethic neither excludes a land ethic, nor does it ignore differences across the planet. It simply recognizes that we must think globally as well as locally. Even among insects, some interactions are local, such as a parasitoid and its host, while others are global and even affect planetary function. This point is brought home by the calculations of Bignell et al. (1997) that global gas production by termites in tropical forests represents 1.5% of carbon dioxide and a massive 15% of all methane. Plausibly, if we upset termites too much, we could be tinkering with planetary function. When primary forest in Indonesia is converted to cassava fields, for example, termite species diversity drops from 34 to 1 (Jones et al. 2003).

INSECTS AND ECOSYSTEMS

Interactions and multiple effects

Insects become inseparable from the rest of nature when deeper values are considered. That insects are open systems – yet there are so many of them, both as individuals and as species – inevitably means that they are functionally integral to ecosystem processes as we know them, clearly illustrated by the gaseous termites of Bignell et al. (1997). Viewed in another way, an enormous number of niches were open and selection pressures were such that insects diversified enormously. Then, as a group, they became fundamental to the connectance within most terrestrial ecosystems. If we compare the total biomass of vascular plants to that of animals, the ratio is about 99.999 : 0.0001, while the total number of vascular plant species to animal species (and mostly insects) is virtually an exact reversal, with a ratio of about 0.026 : 99.974 (Samways 1993).

Among plants, functional diversity, as measured by the value and spectrum of species' traits, rather than by simply species numbers, strongly determines ecosystem functioning (Diaz and Cabido 2001). However, a basic difference exists between plants and insects in terms of presence. Plants are fixed on the spot, but insect populations often blink on and off like lights across the landscape, as populations survive and thrive and then locally disappear (Dempster 1989). Insects, thus, are an intrinsic part of the ecological tissue of the landscape, yet dynamic in terms of their presence and influence and hence in their connectance relationships.

What we see as a land mosaic is not necessarily how insects perceive and react to it (Haslett 2001). This dynamic population response of insects to the world around them was particularly evident in the Quaternary when geographic ranges of species shifted back and forth, tracking optimal temperatures. This response emphasizes that the 'sense of place', in terms of evolutionary conservation, is an artifact of time.

While rapid evolutionary change in insects does occur (Mavárez et al. 2006), in response either to these historical climate-driven events or even to current anthropogenic landscape changes (Williams 2002), in general, the modern mix of pressures on many species will be too much for them to survive. This situation is largely because many effects are synergistic, as with the joint effect of fragmentation and global climate change,

described by Travis (2003) as a 'deadly anthropogenic cocktail'. Habitat loss and global climate change seem to be responsible for huge declines of some British butterflies (Warren et al. 2001) (Fig. 23.2) and probably also of some British moths (Conrad et al. 2004). Reserve networks in Europe also are beginning to appear as if they will be unable to cope with the dynamic geographic range shifts necessary for the long-term survival of many insect species (Kuchlein and Ellis 1997).

The crucial question, then, hinges on whether there will be decoupling of interactions (e.g., pollination, herbivory, and parasitism), with cascade effects as the climate changes across a world dominated by human landscapes. These mosaics inhibit movement of many species and prevent them from finding suitable source-habitat conditions. This predicament, combined with deterioration of patch quality, is causing considerable local extinction, especially among habitat specialists (Kotze et al. 2003, Valladares et al. 2006), leading to ecosystem discontinuities (Samways 1996).

Insects and food webs

Because insects function at various trophic levels, multiple effects involving insects are likely to occur in food webs. Dunne et al. (2002) have shown that food-web structure mediates dramatic biodiversity loss, including

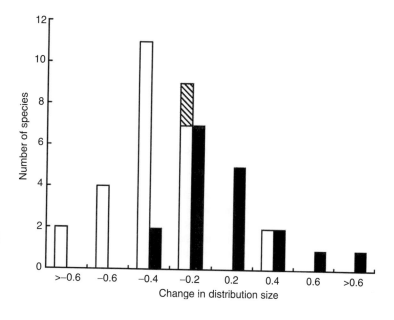

Fig. 23.2 Proportional changes in geographic distribution sizes of British butterflies between 1970–1982 and 1995–1999. Sedentary habitat specialists (white) have declined the most, while mobile specialists (hatched) have also declined. Wider, countryside species (black) have not fared so badly, with several increasing their ranges. (From Warren et al. 2001).

secondary and cascading extinctions. Their findings emphasize how removal of a *number* of species affects ecosystems differently depending on trophic *functions* of the species removed. They found that food webs are more robust to random removal of species than to selective removal of species that have the most trophic links to other species. The implication is that an increase in the robustness of a food web through improved connectivity occurs within it, although it is apparently independent of species richness per se.

Removal of ecologically highly connected species can have a devastating cascading effect. Conversely, but not always, removal of species with few trophic connections generally has a much lesser effect. The upshot is that to maintain food-web stability, the diversity of highly connected species must be maintained. Removal of only 5–10% of these highly connected species can lead to major ecosystem change (Sole and Goodwin (2000). Loss of only a few important predators of grazers can have a disproportionately large effect on ecosystem diversity, which may involve subtle effects with time delays (Duffy 2003). Evidence also is accumulating that changes in biodiversity can be both the cause and the result of changes in productivity, as well as in stability. This two-way effect creates feedback loops and other effects that influence how communities respond to biodiversity loss. Food webs mediate these interactions, with consumers modifying, dampening, and even reversing these biodiversity–productivity linkages (Worm and Duffy 2003).

Trophic cascades can occur even across ecosystems. When fish consume large numbers of dragonfly larvae, pressure is reduced on pollinators, which are normally eaten by adult dragonflies (Knight et al. 2005). As a result, plants near ponds with fish receive higher levels of pollination than plants away from ponds (Fig. 23.3).

Importance of maintaining landscape connectance

Evidence is beginning to point toward the importance of conserving whole landscapes, with all levels and types of connectance intact. Such connectance should include the myriad of species, including insects, which collectively make up a major component, even though individually they might appear ecologically somewhat redundant. Evidence from plant communities suggests that the collective effect of rare species increases community resistance to invasion by aliens and minimizes any effect (Lyons and Schwartz 2001), while maintaining rare species in the community also helps maintain ecosystem function (Lyons et al. 2005).

With the increased stresses and gradual loss of rare species and specialists, some catastrophic regime shifts are likely, where pressures build to reach a point where a radical change to a new state occurs (Scheffer and Carpenter 2003). Transformation of formerly extensive ecosystems into remnant patches, therefore, does not leave these fragments as simply smaller reflections of the whole. Each patch will gradually go on its own trajectory, sometimes catastrophically.

These changes will come about because the stability of food webs can be subtle. Although local species richness can affect ecosystem functioning, such as productivity and stability, generally species diversity positively influences this functioning – but not always. The differences seem to arise depending on the spatial scale under consideration. At the small, local scale, an increase might occur to a point where all available niches are filled, but as species diversity increases further, through high immigration of sink-population competitors, functioning decreases. In contrast, at the larger spatial scale, regional species complement each other, with the result that ecosystem functioning increases with an increase in species diversity (Bond and Chase 2002). This relationship emphasizes the importance of maintaining large complementary patches (Valladares et al. 2006) (Fig. 23.4) and networks (Samways 2007b) across wide areas to maximize insect and other biotic diversity.

TWO CHALLENGES

The taxonomic challenge

One of the greatest tasks for insect conservation is identifying the focal species in any particular study. The lack of taxonomic information on the very things we are trying to conserve is the taxonomic challenge. How do we deal with this? If landscape conservation is an umbrella for such a wide range of insects and all their interactions, we may even ask whether it matters to know all their names.

When deciding which landscapes have value (which might not simply be rarity value), we need to know the components that make up the landscapes to assess their value. One of the components – and not the only one, of course – is the actual species that live there

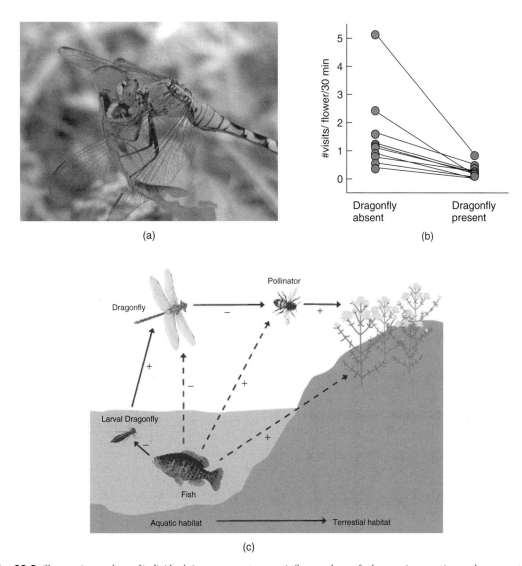

(a)

(b)

(c)

Fig. 23.3 Changes in numbers of individuals in one ecosystem can influence those of other species, even in another ecosystem. Here, fish population levels influence dragonfly population levels, which, in turn, affect pollinator levels, so affecting levels of pollination and seed set. (a) Dragonfly (female *Erythemis simplicicollis*) consuming a bee-fly pollinator (*Bombylius* sp. (Diptera: Bombyliidae). *(See color plate)*. (b) Comparison of pollinator visitation rates in pairs of large-mesh cages placed around an experimental shrub, one with a dragonfly in it, and another as a control. Overall, visitation was much lower in the cage with the dragonfly than in the control. (c) An interaction web where fish can facilitate plant reproduction. Solid arrows indicate direct interactions; dashed arrows denote indirect interactions. The sign refers to the expected direction of the direct or indirect effect. (From Knight et al. 2005).

and are something tangible into which we can get our conservation teeth (New 1999). Explicit population models suggest that prediction of the effects of fragmentation requires a good understanding of the biology and habitat use of the species in question, and

that the uniqueness of species and the landscape in which they live confound simple analysis (Wiegand et al. 2005). Yet, for practical large-scale conservation, and given the regular shortage of resources, we cannot know all the species, even in a small area, especially in

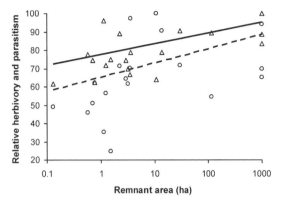

Fig. 23.4 Two processes at different trophic levels (herbivory (circles) and parasitism (triangles)) decline equally with increased fragmentation and smaller fragment size, as shown here for Argentinian woodlands. (From Valladares et al. 2006).

the species-rich lower latitudes and other biodiversity hotspots.

This taxonomic challenge is further highlighted because some, and probably many, putative species are species complexes (Hebert et al. 2004). To get a measure of the value of a landscape, we choose a subset of taxa. This subset can be certain taxa that are well known, or a size class, particular functional types, and even Red Listed species, which can be used effectively to stand in for the rest of the insects, and indeed much of the rest of biodiversity. In other words, we select surrogates and even icons. Use of icons is widespread, and important even in mammal conservation, where charismatic species get preference over less glamorous ones.

A whole range of possible surrogates exists, the choice of which, in part, depends on the conservation question being posed. In most cases – and this is a harsh reality in all realms of invertebrate conservation – the final choice depends on the taxonomic knowledge available. Identified species are so much more valuable for quality-landscape conservation than are unidentified species. This issue of using named species is critical when, for example, selecting landscapes that have rare and endemic taxa that might be threatened and irreplaceable.

The challenge of complementary surrogates

Landscape planners have been exploring ways to prioritize areas of conservation value so that conservation management can be directed first to physical areas of high biodiversity value and those that are under most threat. Evidence is accumulating that the best way forward is to use surrogates for the landscape, in addition to surrogates for species, in this planning process. This approach, in turn, enables selection of reserve areas. When such thorough and insightful selection is done, about half the land area needs to be conserved if regional biodiversity is to be maintained (Reyers et al. 2002). This situation is not generally tenable if we wish to conserve all biodiversity, given the human demand for land.

We are facing a situation where we have to consider some sort of landscape and species triage, whereby the land and its organisms will benefit most from our intervention when it is focused on the points for which we can obtain maximum return for minimum resources. The alarming feature of this approach is that even if we select optimally, those reserves of today will not necessarily be optimally sited for the future, given global climate change.

SYNTHESIZING DEEPER VALUES AND PRACTICAL ISSUES

From our discussion on conservation value, we found that conservation of whole landscapes across the whole of the globe is the moral way forward. This perspective involves a distinct sense of place. Some sort of reconciliation of human needs with this tenet is required, but the point is that we can do much better than we are if we simply put our minds to some of the challenges and engage in landscape conservation, complemented by single-species actions where the need arises. Some suggestions are given by Samways (2007a, 2007b). Similarly, an Earth ethic (Norton 2000, Rolston 2000) also purports to maintain the differences between landscapes. In short, we aim, through a sense of moral concern and joy for nature, to conserve as much of the world's original integrity as possible, even if it has to involve some triage. The quantitative ecological approach is leading to the same conclusion. To maintain ecological health and resilience, we need to conserve the range of connectance, even involving rare species, throughout a range of landscapes.

A dark cloud facing us is that much of the human effect has momentum that is not stoppable for a long time, with the inevitability of extinction debt (Tilman et al. 1994). While, in the past, insect populations could shift across natural landscapes on a regional scale unimpeded by anthropogenic obstacles, the situation today is not the same, with the myriad of human-transformed land mosaics. To maintain any semblance of the past, we need to think of ecological health and vigor in terms of massive linkages between large, quality-landscape patches. This approach has been forcefully emphasized by Erwin (1991) in a landmark but neglected paper. For many parts of the world it is too late to develop these huge networks proactively, but we can engage restoration triage (Samways 2000), and we can introduce ecological networks at the landscape level (Hilty et al. 2006, Samways 2007b). Whatever path we choose, for terrestrial ecosystems at least, the collective moral, human consciousness must consider insects as part of the initiative.

SUMMARY

Despite various global catastrophes over the last 350 million years, insects have increased in species diversity. Yet today, they face a human meteoric impact estimated to eliminate perhaps a quarter of them. We need to be concerned because insects play enormous roles in ecosystem services and have great monetary value. Besides, we have a moral duty to conserve them, as they are our planetary partners. As insect biodiversity is so vast and essentially unknowable, we need to adopt a precautionary approach whereby we conserve as much natural and near-natural land as possible. We also need to maintain diversity of landscapes across the globe, where this landscape diversity becomes an umbrella for a huge range of insect species. Where insects per se are not the focal issue, and where the concern is for ecosystem health, integrity, and resilience, we still need to maintain as much species diversity and food-web connectance as possible. This approach will, in turn, benefit an enormous range of insects. While keystone species, whether insects or not, play the major role in shaping ecosystems as we know them, rare species *en masse* also play a role. Thus, all efforts now should be directed to reduce homogenization of the world and to radically stem biodiversity loss. These efforts need to take place across the whole globe as part of an Earth ethic for the survival of all.

ACKNOWLEDGMENTS

Tim New and G. G. E. Scudder provided excellent critical comment. This work was supported by the RUBICODE Coordination Action Project, funded under the Sixth Framework Programme of the European Commission (Contract number 036890).

REFERENCES

Baillie, J. E. M., C. Hilton-Taylor, and S. Stuart. (eds). 2004. *2004 IUCN Red List of Threatened Species. A Global Assessment*. IUCN, Cambridge, United Kingdom.

Biesmeijer, J. C., S. P. M. Roberts, M. Reemer, R. Ohlemüller, M. Edwards, T. Peeters, A. P. Schaffers, S. G. Potts, R. Kleukers, C. D. Thomas, J. Settele, and W. E. Kunin. 2006. Parallel declines in pollinators and insect-pollinated plants in Britain and the Netherlands. *Science* 313: 351–354.

Bignell, D. E., P. Eggleton, L. Nunes, and K. L. Thomas. 1997. Termites as mediators of carbon fluxes in tropical forests. Pp. 109–134. *In* A. D. Watt, N. E. Stork, and M. D. Hunter (eds). *Forests and Insects*. Chapman and Hall, London.

Bishop, J., S. Kapila, F. Hicks, and P. Mitchell. 2007. *Building Biodiversity Business: Report of a Scoping Study*. Shell International Limited and the World Conservation Union: London, United Kingdom.

Bond, E. M. and J. M. Chase. 2002. Biodiversity and ecosystem functioning at local and regional scales. *Ecology Letters* 5: 467–470.

Burney, D. A. and T. F. Flannery. 2005. Fifty millennia of catastrophic extinctions after human contact. *Trends in Ecology and Evolution* 20: 395–401.

Cincotta, R. P. and R. Engelman. 2000. *Nature's Place: Human Population and the Future of Biological Diversity*. Population Action International. Washington, DC.

Coleman, D. C. and P. F. Hendrix. (eds). 2000. *Invertebrates as Webmasters in Ecosystems*. CAB International, Wallingford, Untied Kingdom.

Conrad, K. F., I. P. Woiwod, M. Parsons, R. Fox, and M. S. Warren. 2004. Long-term population trends in widespread British moths. *Journal of Insect Conservation* 8: 119–136.

Coope, G. R. 1995. Insect faunas in ice age environments: why so little extinction? Pp. 55–74. *In* J. H. Lawton and R. M. May (eds). *Extinction Rates*. Oxford University Press, Oxford, United Kingdom.

Dempster, J. P. 1989. Insect introductions: natural dispersal and population persistence in insects. *Entomologist* 108: 5–13.

Diaz, S. and M. Cabido. 2001. Vive la difference: plant functional diversity matters to ecosystem processes. *Trends in Ecology and Evolution* 16: 646–655.

Duffy, J. E. 2003. Biodiversity loss, trophic skew and ecosystem functioning. *Ecology Letters* 6: 680–687.

Dunne, J. A., R. J. Williams, and N. D. Martinez. 2002. Network structure and biodiversity loss in food webs: robustness increases with connectance. *Ecology Letters* 5: 558–567.

Erwin, T. L. 1991. An evolutionary basis for conservation strategies. *Science* 253: 750–752.

Fauna and Flora International. 2006. *Guidelines for Applying the Precautionary Principle to Biodiversity Conservation and Natural Resource Management*. Flora and Fauna International, Cambridge, United Kingdom.

Harrison, C. and J. Burgess. 2000. Valuing nature in context: the contribution of common-good approaches. *Biodiversity and Conservation* 9: 1115–1130.

Haslett, J. R. 2001. Biodiversity conservation of Diptera in heterogenous land mosaics: a fly's eye view. *Journal of Insect Conservation* 5: 71–75.

Hebert, P. D. N., E. H. Penton, J. M. Burns, D. H. Janzen, and W. Hallwachs. 2004. Ten species in one: DNA barcoding reveals cryptic species in the neotropical skipper butterfly *Astaptes fulgerator*. *Proceedings of the National Academy of Sciences USA* 101: 14812–14817.

Hill, P. and C. Twist. 1998. *Butterflies and Dragonflies: A Site Guide*, Second Edition. Arlequin, Chelmsford, United Kingdom.

Hilty, J. A., W. Z. Lidicker, Jr., and A. M. Merenlender. 2006. *Corridor Ecology*. Island Press, Washington, DC.

Johnson, L. E. 1991. *A Morally Deep World*. Cambridge University Press, Cambridge.

Jones, D. T., F. X. Susilo, D. E. Bignell, S. Hardiwinoto, A. N. Gillison, and P. Eggleton. 2003. Termite assemblage collapse along a land-use intensification gradient in lowland central Sumatra, Indonesia. *Journal of Applied Ecology* 40: 380–391.

Knight, T. M., M. W. McCoy, J. M. Chase, K. A. McCoy, and R. D. Holt. 2005. Trophic cascades across ecosystems. *Nature* 437: 880–883.

Kotze, D. J., J. Niemelä, R. B. O'Hara, and H. Turin. 2003. Testing abundance-range size relationships in European carabid beetles (Coleoptera, Carabidae). *Ecography* 26: 553–566.

Kuchlein, J. H. and W. N. Ellis. 1997. Climate-induced changes in the microlepidoptera fauna of The Netherlands and implications for nature conservation. *Journal of Insect Conservation* 1: 73–80.

Labandeira, C. C., K. R. Johnson, and P. Wilf. 2002. Impact of the terminal Cretaceous event on plant–insect associations. *Proceedings of the National Academy of Sciences USA* 99: 2061–2066.

Labandeira, C. C. and J. J. Sepkoski, Jr. 1993. Insect diversity and the fossil record. *Science* 261: 310–315.

Lockwood, J. A. 2001. The ethics 'Classical' of biological control and the value of *Place*. Pp. 100–119. *In* J. A. Lockwood, F. G. Howarth, and M. F. Purcell (eds). *Importing Non-Native Biological Control Agents (An International Perspective)*. Entomological Society of America, Lanham, Maryland.

Losey, J. E. and M. Vaughan. 2006. The economic value of ecological services provided by insects. *BioScience* 56: 311–323.

Luck, G. W., G. C. Daily, and P. R. Ehrlich. 2003. Population diversity and ecosystem services. *Trends in Ecology and Evolution* 18: 331–336.

Lyons, K. G., C. A. Brigham, B. H. Traut, and M. W. Schwartz. 2005. Rare species and ecosystem functioning. *Conservation Biology* 19: 1019–1024.

Lyons, K. G. and M. W. Schwartz. 2001. Rare species loss alters ecosystem function – invasion resistance. *Ecology Letters* 4: 358–365.

Mavárez, J., C. A. Salazar, E. Bermingham, C. Salcedo, C. D. Jiggins, and M. Linares. 2006. Speciation by hybridization in *Heliconius* butterflies. *Nature* 441: 868–871.

Mawdsley, N. A. and N. E. Stork. 1995. Species extinctions in insects: ecological and biogeographical considerations. Pp. 321–369. *In* R. Harrington and N. E. Stork (eds). *Insects in a Changing Environment*. Academic Press, London.

McKinney, M. L. 1999. High rates of extinction and threat in poorly studied taxa. *Conservation Biology* 13: 1273–1281.

Millenium Ecosystem Assessment. 2005. *Ecosystems and Human Well-Being: Biodiversity Synthesis*. World Resources Institute, Washington, DC.

Miller, J. R. 2005. Biodiversity conservation and the extinction of experience. *Trends in Ecology and Evolution* 20: 430–434.

Naess, A. 1989. *Ecology, Community and Lifestyle*. Cambridge University Press, Cambridge.

New, T. R. 1999. Descriptive taxonomy as a facilitating discipline in invertebrate conservation. Pp. 154–158. *In* W. Ponder and D. Lunney (eds). *The Other 99%: The Conservation and Biodiversity of Invertebrates*. Royal Zoological Society of New South Wales, Mosman, Australia.

Norton, B. G. 2000. Biodiversity and environmental values: in search of a universal earth ethic. *Biodiversity and Conservation* 9: 1029–1044.

Palmer, M. and V. Finlay. 2003. *Faith in Conservation*. The World Bank, Washington, DC.

Perrings, C., L. Jackson, K. Bawa, L. Brussard, S. Brush, T. Gavin, R. Papa, U. Pascual, and P. De Ruiter. 2006. Biodiversity in agricultural landscapes: saving natural capital without losing interest. *Conservation Biology* 20: 263–264.

Peterson, G., C. R. Allen, and C. S. Holling. 1998. Ecological resilience, biodiversity, and scale. *Ecosystems* 1: 6–18.

Ponel, P. Orgeas J., M. J. Samways, V. Andrieu-Ponel, L. De Beaulieu, M. Reille, P. Roche, and T. Tatoni. 2003. 110 000 years of Quaternary beetle diversity change. *Biodiversity and Conservation* 12: 2077–2089.

Pringle, K. L. 2006. The use of economic thresholds in pest management: apples in South Africa. *South African Journal of Science* 102: 201–204.

Rapport, D., R. Costanza, P. R. Epstein, C. Gaudet, and R. Levins. 1998. *Ecosystem Health*. Blackwell Science, Oxford.

Reyers, B., K. J. Wessels, and A. S. van Jaarsveld. 2002. An assessment of biodiversity surrogacy options in the Limpopo Province of South Africa. *African Zoology* 37: 185–195.

Rodrigues, A. S. L., J. D. Pilgrim, J. F. Lamoreux, M. Hoffman, and T. M. Brooks. 2006. The value of the IUCN Red List for conservation. *Trends in Ecology and Evolution* 21: 71–76.

Rolston, H., III. 2000. The land ethic at the turn of the millennium. *Biodiversity and Conservation* 9: 1045–1058.

Samways, M. J. 1993. Insects in biodiversity conservation: some perspectives and directives. *Biodiversity and Conservation* 2: 258–282.

Samways, M. J. 1996. Insects on the brink of a major discontinuity. *Biodiversity and Conservation* 5: 1047–1058.

Samways, M. J. 2000. A conceptual model of restoration triage based on experiences from three remote oceanic islands. *Biodiversity and Conservation* 9: 1073–1083.

Samways, M. J. 2006. Insect extinctions and insect survival. *Conservation Biology* 20: 245–246.

Samways, M. J. 2007a. Insect conservation: a synthetic management approach. *Annual Review of Entomology* 52: 465–487.

Samways, M. J. 2007b. Implementing ecological networks for conserving insects and other biodiversity. Pp. 127–143. *In* A. Stewart, O. Lewis, and T. R. New (eds). *Insect Conservation Biology*. CABI, Wallingford, Oxon, United Kingdom.

Samways, M. J., P. Ponel, and V. Andrieu-Ponel. 2006. Palaeobiodiversity emphasizes the importance of conserving landscape heterogeneity and connectivity. *Journal of Insect Conservation* 10: 215–218.

Scheffer, M. and S. R. Carpenter. 2003. Catastrophic regime shifts in ecosystems: linking theory to observation. *Trends in Ecology and Evolution* 18: 648–656.

Sole, R. and B. Goodwin. 2000. *Signs of Life*. Basic Books, New York.

Steadman, D. W. 1995. Prehistoric extinctions of Pacific island birds: biodiversity meets zooarchaeology. *Science* 267: 1123–1131.

Stokes, D. L. 2006. Conservators of experience. *BioScience* 56: 6–7.

Suh, A. N. and M. J. Samways. 2001. Development of a dragonfly awareness trail in an African botanical garden. *Biological Conservation* 100: 345–353.

Thomas, J. A., M. G. Telfer, D. B. Roy, C. D. Preston, J. J. D. Greenwood, J. Asher, R. Fox, R. T. Clarke, and J. H. Lawton. 2004. Comparative losses of British butterflies, birds and plants and the global extinction crisis. *Science* 303: 1879–1881.

Tilman, D., R. M. May, C. L. Lehman, and M. A. Nowak. 1994. Habitat destruction and the extinction debt. *Nature* 371: 65–66.

Travis, J. M. J. 2003. Climate change and habitat destruction: a deadly anthropogenic cocktail. *Proceedings of the Royal Society of London, B, Biological Sciences* 270: 467–473.

Valladares, G., A. Salvo, and L. Cagnolo. 2006. Habitat fragmentation affects on trophic processes of insect-plant food webs. *Conservation Biology* 20: 212–217.

Vamosi, J. C., T. M. Knight, J. A. Steets, S. J. Mazer, M. Burd, and T.-L. Ashman. 2006. Pollination decays in biodiversity hotspots. *Proceedings of the National Academy of Sciences USA* 103: 956–961.

Warren, M. S., J. K. Hill, J. A. Thomas, J. Asher, R. Fox, B. Huntly, D. B. Roy, M. G. Telfer, S. Jeffcoate, P. Harding, G. Jeffcoate, S. G. Willis, J. N. Greatorex-Davies, D. Moss, and C. D. Thomas. 2001. Rapid responses of British butterflies to opposing forces of climate and habitat change. *Nature* 414: 65–69.

Wiegand, T., E. Revilla, and K. A. Maloney. 2005. Effects of habitat loss and fragmentation on population dynamics. *Conservation Biology* 19: 108–121.

Williams, B. L. 2002. Conservation genetics, extinction, and taxonomic status: a case history of the regal fritillary. *Conservation Biology* 16: 148–157.

Worm, B. and J. E. Duffy. 2003. Biodiversity, productivity and stability in real food webs. *Trends in Ecology and Evolution* 18: 628–632.

Chapter 24

TAXONOMY AND MANAGEMENT OF INSECT BIODIVERSITY

Ke Chung Kim

Frost Entomological Museum, Department of Entomology, Pennsylvania State University, University Park, Pennsylvania 16802 USA

Insect Biodiversity: Science and Society, 1st edition. Edited by R. Foottit and P. Adler
© 2009 Blackwell Publishing, ISBN 978-1-4051-5142-9

Insects are incredibly diverse, abundant, and spectacular in color, form, and function. They are found in nearly every conceivable habitat throughout the world. Their lives are intertwined with the lives of humans, and they make up a major portion of global biodiversity. They feed on plants and animals, through which considerable economic losses are incurred, and transmit parasites to humans, causing diseases such as malaria, onchocerciasis, and plague. Above all, insects are important ecological partners in sustaining our life-support system (Coleman and Hendrix 2000). Biodiversity, the totality of living things and their variations, is the essence of life on Earth. All species of animals, fungi, microorganisms, and plants are elements of our life-support system, of which insects represent the most diverse and important partners (Wilson 1992, Kim and Weaver 1994, Kim 2001a).

As the most successful group of animals, insects and arachnids make up more than 60% of global biodiversity and are major players and important ecological partners in ecosystem function and humanity's existence (Kim 1993b, Coleman and Hendrix 2000). Despite centuries of extensive studies, global biodiversity is still poorly known. Only 1.75 million species have been described of what currently is estimated to be a total of 13.6 million species (10 million is commonly used). Our knowledge base, however, has been skewed toward vertebrates (>90% described) and plants (>84% described), leaving behind small animals such as insects, arachnids, and lower plants (algae, ferns, mosses, and others). Only 11.9% of the insects and 10% of the arachnids have been described, along with 4.8% of the fungi and 0.4% of the bacteria (Heywood and Watson 1995). Our current knowledge of global biodiversity – with a current human population of 6.7 billion – is not too much better than what it was a century before when the human population was barely 1.65 billion. Today's known biodiversity is far too small, merely 20% or less of the extant total. At the same time, we barely know what biodiversity exists in our own backyards. 'Backyard biodiversity', defined as 'local biodiversity in and near human habitation', refers to the natural resources and capital for ecosystem services at the grassroots level. Backyard biodiversity, the foundation for local sustainable development, is poorly explored and documented for almost all places in the world (Kim and Byrne 2006).

Insect biodiversity is important in managing ecosystems. Insects, with their combination of unique biological traits and ecological roles, are valuable indicator organisms for assessing, monitoring, and managing biodiversity in natural ecosystems and combating the invasion and spread of nonindigenous insects (Kim and Wheeler 1991, Kim 1993a, 1993b, US Congress 1993, Büchs 2003). Today's biodiversity is the culmination of long-term ecological and evolutionary processes going back more than 410 million years to the Devonian period (Geological History 2008). Biodiversity is constantly changing in species composition and abundance. As a result, no one community contains the same biodiversity as any other unit, even in rather homogenous forests, due to site-specific characteristics of soil, topography, vegetation, weather, and other environmental factors. Because biodiversity conservation and resource management are targeted for specific local units, biodiversity assessment must be designed specifically for the particular site. Yet, practically few locally defined sites exist, whose ecosystem services are influenced by human activities, and whose local or 'backyard' biodiversity has been inventoried for natural resources and capital (Kim and Byrne 2006).

In this chapter, I discuss the state of insect biodiversity in the context of insects as ecological partners for sustaining our lives. I also discuss the need for taxonomy in biodiversity assessment and monitoring, the problems associated with taxonomic bottlenecks, new opportunities for taxonomy in biodiversity science, and the application of insect biodiversity science and taxonomy to ecosystem management, pest management, and conservation.

INSECT BIODIVERSITY

Honey bees, American cockroaches, and Japanese beetles are distinct species of insects, but they are not biodiversity – they are units of biodiversity. Biodiversity is not something we can touch, although we see and feel the parts of a community. Biodiversity is commonly defined as 'the totality of variety of species of plants, animals, fungi, and microbes, the genetic variation within them, the ecological roles they play, and their interrelationships in biological communities in which they occur' (Kim 2001a), although it is defined somewhat differently by leading scientists (Takacs 1996). The essence of biodiversity, however, is perceived differently by people of different backgrounds.

Our lives depend completely on biodiversity; dynamic ecosystem processes are sustained by the interactions of all species of animals, fungi, microbes, and plants (Kinzig et al. 2001). From an economic perspective, biodiversity includes the natural resources and capital assets that provide the basic resources for all organisms and human enterprises (Baskin 1997, Kim 2001a). Biodiversity also is the endpoint of anthropocentric influences on the biosphere, which is continually infused with chemicals, many of which are newly created by humans. A person strolling through a flower garden, a honey bee visiting a flower, a cow grazing in a pasture, or a woodpecker pecking on a tree are all parts of the interconnected whole of our living world. Our survival depends on this intricate web of life, the network of interacting organisms. The world's biodiversity supplies our basic necessities such as food, clean air and water, fuel, fibers, building materials, medicines, and natural areas where we enjoy recreational activities (Kim 2001a)

The 1,004,898 species of insects (introduction to this volume) make up 58% of all described species, with an estimated range of 5 million (Grimaldi and Engel 2005) to 8 million (Heywood and Watson 1995) extant insect species. Insects and terrestrial arachnids are also cosmopolitan animals closely associated with human enterprises. Because humans are exploring and exploiting every corner of the world, we should expect to find and document many new species and new distributions. Insects continue to exploit new habitats and expand their ranges, producing lineages of great ecological and behavioral diversity and making their study – entomology – one of the most fascinating and important sciences.

Natural biodiversity in wild lands such as the Amazon, northern Alaska, and Siberia continue to be threatened by human enterprises such as clearing, logging, gas and oil drilling, mining, housing development, and urban sprawl. As a result, biodiversity continues to decline, and likely will continue as the human population grows and economic development expands in developing nations such as China and India.

As our ancestors moved from a hunter-gatherer to an agrarian lifestyle, human settlement and farming offered new territory and habitats for insects (Ponting 1991). Many of these insects readily invaded human habitation. The close relationship between humans and insects, known as 'synanthropy' (Harwood and James 1979), can also be referred to as 'human insect biodiversity'. In other words, insects have successfully evolved with humans in agricultural societies.

Insects are not merely pests of our crops or vectors of disease agents such as malarial parasites. They are also

the most successful group of organisms in terms of the numbers of species and individuals. They are small and agile with short life cycles, enabling long-distance dispersal and invasion of new territories and habitats (Kim 1983, 1993a). They have exploited ecological opportunities to occupy unique niches and microhabitats (Kim 1993a, 1993b) and have become major players in maintaining ecosystems and providing diverse ecosystem services (Coleman and Hendrix 2000).

BIODIVERSITY LOSS AND HUMANITY

Earth is rapidly becoming a network of human ecosystems, with urban and suburban communities transforming natural environments into human habitats. This process continues everywhere humans live at nature's expense. The destruction of natural habitats and loss of species draw a different group of insects and other invertebrates, forming secondary human biodiversity that includes pests and imported exotic species of plants and animals. Today's cities and suburbs are being connected or merged with other newly developed municipalities, forming megacities (Cohen 2006, Lee 2007). Secondary biodiversity of synanthropic arthropods often contributes to economic loss or causes diseases in humans and animals. In most developing and underdeveloped countries in Africa, South America, and Southeast Asia, people still live in natural ecosystems with bustling biodiversity (Kim 1993b, 1994, UNDP et al. 2000, Sodhi et al. 2004). These lands, with their rich backyard biodiversity, are exploited for local economic development, becoming agricultural and industrial lands. In tropical regions, people make their living without knowing what they have in their backyard beyond some species of herbal plants and common animals. In this setting, biodiversity represents a basic economic resource and natural capital for human survival, but it is not considered important for the economy or land-use planning. This paradigm should be corrected for sustainable development; the assessment of backyard biodiversity is the first step in managing biological resources and ecosystems (UNDP et al. 2000).

Extinction is a natural process. Today's global biodiversity represents less than 1% of all species that ever existed. Throughout the history of life, billions of species, perhaps 5–50 billion, arose and became extinct. In light of the average lifespan of a species, estimated from fossils, one species in every 4 million likely died out without involving humans (Raup 1991). Today's extinction is caused primarily by humans, particularly during the last two centuries. In recent human history, global biodiversity has been destroyed at an unprecedented rate because of rapid human population growth and economic development that increases urbanization and land conversion. The loss of biodiversity becomes the primary concern for our own sustainability. Humans, the greatest force shaping the evolution of global ecosystems, continue to convert natural ecosystems to human habitat (Palumbi 2001). Without environmental safeguards, this process will continue to destroy habitats and their native species, perhaps as many as 1000 per year, throughout the world (Myers 1979, Wilson 1985, 1992, Kim and Weaver 1994), but this problem has not been recognized by political leaders or the public in most countries.

The great challenge is to estimate and predict anthropogenic extinctions in a void of knowledge about global biodiversity, especially for small organisms that make up the core of this biodiversity. The 1995 Global Biodiversity Assessment shows that 484 animal and 654 plant species have become extinct since 1600. The World Conservation Monitoring Centre listed 602 species of vertebrates, 582 invertebrates, and 2632 plants as endangered and 5366 animals and 26,106 plants as threatened or of special concern (Heywood and Watson 1995). For vertebrates and plants, between 5% and 20% of biodiversity for some taxa is threatened with extinction. If this estimate is applied to insects, the expected number of species that would be lost in the next 25 years comes to a whopping 51,250 species at 5% and 2,005,000 species at 20%. This loss of global biodiversity is caused by two mechanisms: extirpation (local population loss) and extinction (permanent species loss from the planet). Accurate estimation of the actual numbers lost remains difficult because our knowledge base for global biodiversity is inadequate and backyard biodiversity is practically unknown (Heywood and Watson 1995, MEA 2005).

BIODIVERSITY AND TAXONOMY

Biodiversity does not refer to a single species, such as human (*Homo sapiens* Linnaeus) or European corn borer (*Ostrinia nubilalis* (Hübner)), but rather to the totality of all species in a defined area. The importance of biodiversity lies not in numbers but in the interactions of species in a specific habitat, community, or ecosystem. On the other hand, taxonomy (= alpha and beta taxonomy of Mayr's 'systematics') is the science of discovery, documentation, and organization of organisms (Mayr and Ashlock 1991). Good taxonomy for a specific taxon is built by taxonomists who study the taxon at the species level, for example, sucking lice (Anoplura). Biodiverse taxa such as Insecta pose great challenges to scientists in community ecology and natural resource management interested in designing biodiversity studies and processing large field samples because species identification of many taxa is difficult, even for general taxonomists, often requiring identification by specialists (Kim and Byrne 2006).

Today's knowledge of global biodiversity represents the culmination of research by taxonomists and natural historians of more than 250 years – since the publication of the tenth edition of 'Systema Naturae' by Carl von Linnaeus (1758), the starting point for zoological nomenclature, following the publication of 'Species Plantarum' in 1753, the starting point for plant nomenclature (Mayr and Ashlock 1991). Taxonomists have been highly motivated, determined, and dedicated scientists who continued to add new species and information to global biodiversity by discovering, naming, and describing unknown organisms one by one from the thicket of unknowns. In other words, discovery and documentation of biodiversity began with backyard efforts and expanded to regional and global efforts through western expeditions around the world and taxon-based surveys by natural historians and taxonomists from the late nineteenth century to the present.

Taxonomy is the oldest discipline in biology, with Carl von Linnaeus (1707–1778) bringing a hierarchical order to natural history and establishing the basis for biological nomenclature and classification that guides the science of life to this day (Blunt 2001, Nature 2007, Warne 2007). Linnaeus's followers, natural historians and taxonomists, have explored nature and methodically documented biodiversity worldwide through the outset of the New Millennium. The culmination of these efforts is what we call 'global biodiversity'. Yet, today's taxonomy barely scratches the surface of extant global biodiversity (Heywood and Watson 1995). Taxonomy with evolutionary and speciation perspectives is known as 'systematics', the foundation of biology, fundamental to all branches of biological and environmental sciences since the New Systematics (Huxley 1940). Systematics encompasses three aspects: (1) alpha taxonomy, the most basic aspect, involving biodiversity exploration, description and documentation of extant species, and development of taxonomic tools, (2) beta taxonomy involving the synthesis of taxonomic data and development of biological classifications, and (3) gamma taxonomy involving the study of speciation and phylogeny of organisms, which is increasingly fragmented under the umbrella of evolutionary biology (Mayr and Ashlock 1991). Taxonomy is fundamental to all biological sciences. Taxonomic data are applied to all aspects of agriculture, conservation, ecology, fisheries, forestry, and environmental studies (Kim and Byrne 2006). Yet, we are facing a continued decline of taxonomy worldwide and are now left with a small pool of taxonomists and trained parataxonomists (a term initially used for trained lay professionals of Costa Rica's INBio) who are explicitly trained professionals that explore, collect, sort, document, and identify specimens, usually to order, family, or generic levels. At the same time, demands for taxonomic services are growing in biodiversity science, conservation, and natural resource management (Kim and Byrne 2006).

Sucking lice (Anoplura) are obligate, permanent ectoparasites of selected groups of mammals, for which biodiversity can be estimated accurately because the survival of sucking lice depends on the host's survival (Kim et al. 1990). If the host species becomes extinct, sucking lice on that host also become extinct (Stork and Lyal 1993). The world fauna of sucking lice is estimated at 1540 species, of which 532 from more than 828 species of mammals were listed in 'Sucking Lice of the World' (Durden and Musser 1994) (Kim et al. 1990, Kim 2007). Thus, about 1831 species of extant mammalian host species are expected to harbor 1008 new species of sucking lice. Even a well-studied regional fauna, such as North American insects, is barely 50% known, with only a small number of species described in the larval stage (Kosztarab and Schaefer 1990a). North American biodiversity of insects was estimated to be about 200,000 species: the total number of known species (90,968), plus the estimated total number of undescribed species (98,257–98,767). The task to describe North American insects, including the immature stages and both sexes, would involve about 1,200,000 descriptions for 200,000 species (Kosztarab and Schaefer 1990b).

While rapid advances have been made in molecular biology, genetics, theoretical ecology, and other branches of biology over the last quarter century, alpha taxonomy has declined precipitously. Financial support and training for taxonomists has been eroded by declining prestige and job opportunities. The most troubling aspects of this erosion are the rapid decline in human resources, namely practicing taxonomic specialists at the Ph.D. level and parataxonomists with undergraduate degrees and 1 or 2 years of postgraduate training in taxonomic services. Some important taxa have a shortage or absence of taxonomists. Even if the level of human resources at the 1990 level were maintained, the number of taxonomists still would be too low to facilitate the advancement of alpha taxonomy for the roughly 8 million species requiring description. The moribund state of insect taxonomy has negatively affected the advancement of community ecology, ecosystem management, and conservation (Gotelli 2004), forcing ecologists and conservation biologists to come up with taxonomic surrogates (e.g., genus taxa or family taxa instead of species) and 'morphospecies' without generic names (Krell 2004, Kim and Byrne 2006, Bertrand et al. 2006, Biaggini et al. 2007).

Taxonomy is no longer solely the taxonomists' domain but has become an important scientific tool for applied biologists and environmental scientists, particularly those working at the species level. A dichotomy exists in the perspective of taxonomy. Systematists view taxonomy as a research discipline, whereas all other scientists consider it a service discipline (Donoghue and Alverson 2000, Brooks and McLennan 2002). In reality, it is both. In recent years, demand for taxonomic services has increased, while the capacity for taxonomic services has declined. These contradictory trends have stymied the advancement of community ecology and nature conservation. Taxonomic keys that can be used by people without much training in insect identification are increasingly demanded by the conservation community. Because of this trend, professional concern over misidentification has been raised (Ehrlich 2005, Kim and Byrne 2006).

A shortage of taxonomists is now felt at all levels because of the decline of job and training opportunities and lack of taxonomic instruction at educational institutions (Wheeler et al. 2004, Kim and Byrne 2006). The interest in taxonomy slowly is shifting its center of gravity, from Europe and North America where it was historically centered for research and training, to developing regions of the world where hotspots of biodiversity exist (Kim and Byrne 2006). Global climate change, along with habitat destruction and pollution, directly affects biodiversity at various spatial and temporal scales, making the study of biodiversity an urgent challenge to the scientific community everywhere. The scientific community must take these challenges seriously to build new infrastructure and provide taxonomic services through which a new generation of applied taxonomists can be trained. A self-supporting infrastructure would promote biodiversity science that trains undergraduate students and attracts young scientists who enjoy working with insects. Backyard biodiversity, thus, could be explored and documented for local sustainable development at the grassroots level, enriching and expanding the knowledge base of global biodiversity (Kim and Byrne 2006).

BIODIVERSITY INVENTORY AND ECOLOGY

Biodiversity can be discussed in abstract or theoretical terms, without referring to spatial or geographic units. We can talk about biodiversity and ecosystem functions because they are interdependent and related, without regard to specific locations or scales. Yet, biodiversity is meaningless when it is not defined by spatial or geographic coordinates or at specific biological scales because species composition and assemblage patterns are site specific. We can understand patterns of what and how different species or guilds function and interact in a community by using numerical indices of diversity, richness, and evenness. Such empirical observations appear obvious to most eyes for large organisms such as primates, seals, and trees. But such observations and statistics do not provide information about what species are present and what they do in sustaining ecosystem functions at a specific site.

The task to undertake exploration and documentation of biodiversity throughout the world is urgent. In my view, we should increase our knowledge base of global biodiversity to at least 50–60% of the extant biodiversity on the planet by 2020. However, with what is left of the contemporary taxonomic regiment, this challenge appears impossible. We need new approaches to global biodiversity inventories. Although individual taxonomists continue to discover and describe new species, efforts to explore and document backyard biodiversity at the grassroots level everywhere in the world are needed. The resulting database is fundamental to building a local economy based on local natural resources and capital.

Ecology, the study of the interactions among organisms and their physical environment, is a science of uncovering the cause of patterns in natural and managed ecosystems (Tilman and Lehman 2001). Biodiversity, from an ecological perspective, is commonly expressed in terms of the number of species and their relative abundance in a locality (e.g., the Shannon index), without much reference to species composition. Tilman and Lehman (2001) recognized two different perspectives of biodiversity: (1) *species diversity*, the number of species, or species richness versus evenness, in a habitat, and (2) *functional diversity*, the range of species traits in an area. As ecological models usually focus on quantitative indices to study productivity, resource dynamics, and stability in the context of ecosystem function (Kinzig et al. 2001), taxonomic diversity and species composition of a community are considered unimportant and are often ignored. However, the goals of biodiversity science include assessing taxonomic diversity and species composition of a community because interacting species can be affected by specific stresses. In this context, we must know which species of insects are involved.

Conservation and ecosystem management need specific data on species composition because community assemblages differ in habitat characteristics and show different responses to extinction thresholds (Lande 1987, Bascompte and Solé 1996, 1998, Lin et al. 2005, Lin and Liu 2006). There is a void in surveying and sampling methodologies in insect biodiversity inventories (Mahan et al. 1998, Boone et al. 2005). As a result, most natural resource agencies, such as national and state parks, lack baseline, site-specific biodiversity data that should be a basic guide for all other resource management programs. Biodiversity inventory data should be available for ecosystem management of public lands. The species composition and assemblage of every community and ecosystem are site specific (Hector and Bagchi 2007). In other words, the management of communities and ecosystems requires more than just knowing general patterns. It needs to focus on building the baseline data on species composition, assemblage structure, and interactions of guilds (Mahan et al. 1998, Boone et al. 2005, Kim and Byrne 2006), particularly for insects and other invertebrates that make up the bulk of every community and ecosystem (Coleman and Hendrix 2000).

Every species is unique; thus, species are not interchangeable (Nee and May 1997, Brooks and McLennan

2002). Site-specific biodiversity is unique because of different physical architectures and ecological processes at each site, although general patterns of distribution and species richness might be similar among habitats. Local species assemblages are influenced and shaped by local and regional processes (Ricklefs 1987, 2004, Cornell and Karlson 1997, Gaston and Blackburn 2000, He et al. 2005, Shurin and Srivastava 2005). Biodiversity involves the total composition of resident species in a spatially defined area. Every species matters in conservation until we know more about global biodiversity and what the species are doing to sustain our life-support system. We simply do not know enough about biodiversity to make a final call about the destiny of certain species. Ecological processes at both local and regional scales include competition, disturbance factors, immigration, mutualism, parasitism, and predation. Species of a guild in one community might not play the same ecological role as related species of the same guild in a different community (Kim 1993b, Coleman and Hendrix 2000, Hector and Bagchi 2007). This trend is particularly evident when insect biodiversity is involved in any community or ecosystem (Coleman and Hendrix 2000, Wardle 2002, Schmitz 2007).

To manage natural resources, restore habitats, or conserve species of concern, the biodiversity of specific areas or targeted habitats needs to be assessed, requiring a comprehensive but efficient means to inventory the organisms and what they do in the system (Kinzig et al. 2001, Loreau et al. 2001, Loreau et al. 2002, Hooper et al. 2005, Kim and Byrne 2006). Without knowing which species are represented or which species previously recorded are now missing, biodiversity assessment would be meaningless and could not contribute much to conservation, restoration, or management of ecosystems. To advance biodiversity science, therefore, biodiversity assessment and appropriate measures of biodiversity change become important.

Ecosystem management requires biodiversity assessment with social and humanistic factors considered, which includes documentation of endangered or threatened species for ultimate mitigation (Kim 1993b). Assessing and conserving biodiversity invariably involve serious taxonomic issues that include both taxonomic inflation (Bertrand and Härlin 2006, Padial and de la Riva 2006) and taxonomic surrogacy (e.g., so-called morphospecies, albeit not in a taxonomic sense) (Oliver and Beattie 1996, Derraik

et al. 2002, Bertrand et al. 2006, Biagfini et al. 2007). These approaches are taken by applied scientists to support their immediate problem solving and do not add much understanding to biodiversity patterns (Krell 2004, Bertrand et al. 2006). They actually make taxonomic problems in biodiversity conservation more complicated, without adding much useful information.

Considering the heterogeneity and diversity of ecological niches, no simple survey technique can cover all species in a community. Many survey techniques used in taxon-based surveys and ecological research are suitable only for targeted taxa or specific research objectives. Biodiversity inventories must be designed for specific objectives. If an inventory is to assess the biodiversity of an area, a survey strategy must consider quantitative and qualitative information on the resident biodiversity over all seasons. The inventory design must include a sampling scheme, techniques, and protocols for capturing representatives of all resident species at the site (Mahan et al. 1998, Kim 2001b, 2006). Because quantitative techniques often miss species that are seasonal, periodic, or influenced by local weather conditions, taxon-based collecting or specific surveys can be supplemented with surveys to capture unique or specialized species that do not appear in standard samples or general surveys.

BACKYARD BIODIVERSITY AND SUSTAINABILITY

Since our ancestors became settlers with farming technology, natural ecosystems have been transformed into cultivated lands, with selected species of plants and animals domesticated for food and other human amenities. With the development of farming communities, small towns began to flourish and eventually expanded and merged with others into larger cities (Ponting 1991). This process invariably destroyed biodiversity and habitats of resident species in the transformed land. Modern development likewise transforms nature into farmland or human habitation. In the process, primary (or natural) biodiversity is destroyed. Anthropocentric transformation of natural lands then acquires secondary (or human-based) biodiversity that usually has limited species richness.

The concept of backyard biodiversity highlights the importance of appreciating local biodiversity on a scale at which human activities determine local ecosystem services. Backyard biodiversity promotes local conservation efforts by providing the basic natural knowledge for local leadership (Schwartz et al. 2002, Mascia et al. 2003, Berkes 2004). Backyard biodiversity also encompasses the organisms inhabiting private properties, neighborhoods, and local municipalities. All these perspectives are relevant to local needs, cultures, and land-use regulations (Center for Wildlife Law 1996, Farber et al. 2006). Because global biodiversity hotspots (Mittermeier et al. 2000, Mittermeier 2004) are located in underdeveloped or developing countries of the world, backyard biodiversity will promote local conservation efforts as a part of sustainable economic development.

Since the Rio Declaration of Biological Diversity and Sustainable Development in 1992, conservation plans and policies for biodiversity and ecosystems have been developed at global, national, and regional levels. However, biodiversity conservation and ecosystem management must occur at a local scale, where factors influencing realistic conservation practices are related to cultural and economic interests of the local people. Knowing the backyard biodiversity provides the necessary knowledge base for improving economic well-being. Comprehensive backyard biodiversity databases provide the baseline information about local natural resources and capital for ecosystem services (Lundmark 2003). A backyard biodiversity database provides a scientific basis for sustainable economic development, and collectively, the global database of all backyard biodiversity from around the world will help develop a comprehensive plan for sustainable economic development at all geographic levels, from local to global.

TAXONOMIC BOTTLENECKS IN MANAGING INSECT BIODIVERSITY

Biodiversity assessment requires taxonomic services. The assessment process involves (1) inventory of an area, which yields a large collection of specimens, (2) field-collection management, which requires preparation, sorting, labeling and management of specimens, (3) taxonomic services, which include identification of sorted specimens, and (4) building a biodiversity database for the inventory site (Kim 1993b, Mahan et al. 1998, Kim and Byrne 2006).

A biodiversity inventory program faces major scientific and technical challenges that require resolution. Taxonomic infrastructure must be built for each country or a core of regional setups. Biodiversity inventories

generate great numbers of specimens, from several thousand in small projects to a million or more in large, multiyear projects (Mahan et al. 1998, Kim 2001b, 2006), which must be processed and prepared for species identification and which invariably include new species that must be described. Species identification and taxonomic processing of field samples involve a good number of trained taxonomic technicians (e.g., parataxonomists) and taxonomic scientists. New technologies, such as DNA barcoding, have been developed for identification (Cameron et al. 2006, Hickerson et al. 2006). Yet, all of this requires a stable infrastructure with a regular source of funding to retain well-trained parataxonomists and a network of taxonomic specialists and scientists who provide identifications. This inventory process is time consuming and labor intensive for technicians or parataxonomists and requires the close attention of taxonomic specialists, often discouraging ecologists and conservation biologists from undertaking biodiversity assessments involving insects. As a result, practically no resource management units in federal and state public lands (e.g., national parks) have developed biodiversity assessments, and little advancement has been made in community ecology and conservation biology (Gotelli 2004, Kim and Byrne 2006, Rohr et al. 2006).

Taxonomic bottlenecks in biodiversity research have been recognized and their resolution advocated (Gotelli 2004, Boone et al. 2005). Taxonomic bottlenecks have serious consequences for applied science, conservation biology, and natural resource and ecosystem management (Kim and Byrne 2006). Historically, species identification was provided gratis by taxonomic specialists and parataxonomists of federal (e.g., USDA/ARS Systematic Entomology Laboratory) and state agencies associated with agriculture (e.g., Florida State Collection of Arthropods and taxonomists in land-grant institutions). This practice has virtually ceased. With the decline of taxonomy worldwide and shortage of taxonomic specialists, taxonomic services are increasingly costly and difficult to obtain.

Backyard biodiversity must be documented so that the resulting database becomes the basis on which ecosystem management and sustainable development can be applied to local human-dominated ecosystems. New strategies for taxonomic services and biodiversity education must be developed. Even if education programs were instituted now, at least 5 years would be needed before taxonomic specialists were available for taxonomic services. The Integrated Biodiversity

Assessment Center (IBAC), an infrastructure for providing taxonomic services, would be networked from the grassroots to national level to share specific taxonomic expertise and informatics throughout the world (Kim and Byrne 2006).

The IBAC could provide a global infrastructure for taxonomic services and training of applied taxonomists. Each IBAC, with several permanently employed taxonomists, should be based on a native systematics collection or at least associated with established collections for reference and voucher deposition. IBACs would provide taxonomic services, including (1) planning and execution of exploratory field sampling and collection, (2) sorting and preparation of field samples for identification and management, (3) species identifications, (4) individualized biodiversity databases, and (5) long-term storage and management of voucher collections and archival field collections. Once established with government funds and grants, IBACs should be self-supporting from the fees for services rendered. Taxonomic services provided by IBACs would be tailored for specific needs of users in research, conservation, detection of exotic and invasive species, and ecosystem management, and could provide diverse training programs such as annual seminars, workshops for specific regional taxa, and internships for undergraduate and graduate students in applied taxonomy (Kim 2006, Kim and Byrne 2006). IBACs would promote the advancement of biodiversity science, which in turn would increase the knowledge base of global biodiversity from approximately 1.8% to as much as 50% by 2020. Local IBACs and a network of IBACs globally will facilitate the process of building taxonomic human resources and training future biodiversity scientists. At the grassroots level, local IBAC associates could be trained as parataxonomists familiar with local backyard biodiversity. They then could lead biodiversity-related activities such as integrated pest management, sustainable agriculture, and monitoring of nonindigenous invasive pests (Kim and Byrne 2006). A centralized international IBAC database, based on the merger of backyard biodiversity databases through a network of global IBACs, would be a *de facto* global biodiversity database, providing the scientific basis for a broad range of civic and public works readily available to users in diverse scientific and technical communities including systematics, ecology, conservation biology, environmental technology, land-use planning, and drug prospecting. The globally networked backyard biodiversity inventory would not only produce local biodiversity databases, but also facilitate the discovery and documentation of new species.

ADVANCING THE SCIENCE OF INSECT BIODIVERSITY

Massive industrial and technological advancements during the last century brought about great economic development and affluence. This anthropocentric success came with enormous environmental baggage. The process continues but at a far faster pace throughout the world. In the last part of the twentieth century we identified the largest human footprint of environmental decline – global warming and climate change – created by human abuse of the environment. Humanity is now at the crossroads to control the causes and effects of human endeavors, such as development, expansion, material wastes, and pollution.

Global climate change affects all organisms, including insects and humans, and could cause devastating destruction to human infrastructure and the economy. Its effect on biodiversity, particularly insect biodiversity, would be serious and long lasting because the effects would involve changes in physiologies, populations, and life histories of organisms, shifts in distributions and geographic ranges of species, and changes in species composition and the structure and function of ecosystems (Canadell and Noble 2007). The effects of these changes might show up, for example, in the seasonal patterns of events such as the timing of migration and reproductive cycles (McCarty 2001). Global climate change affects population growth and ecological roles of insects, particularly of common herbivores of our cropping systems and those species that transmit pathogens of humans and animals (Kim and McPheron 1993, McCarty 2001, Easterbrook 2007, IPCC 2007, Botkin et al. 2007). Ecosystem processes such as carbon cycling and storage that are already stressed, along with those species already endangered or threatened, likely would feel the brunt of global warming (Sala et al. 2000, Parmesan and Galbraith 2004). Climate changes due to global warming, such as the frequency and strength of hurricanes and droughts, would elevate the risk of extinction or extirpation (ENS 2006a, 2006b).

In addition to global warming, today's environmental concerns such as biodiversity loss, land transformation, habitat loss, pollution, and changes in ecosystem services are caused by human activities (MEA 2005).

Global warming and climate change represent the foremost manifestation of the anthropocentric effects of development and pollution. Our basic approaches to economic development have not changed much through the dawn of the new millennium, despite global environmental movements persistently pursued by the United Nations (UN General Assembly, UNEP, UNESCO, UNDAP, and others), including ambitious targets for 2010 to significantly reduce the rate of biodiversity loss to achieve the 2015 targets of the Millennium Development Goals (CBD 2006a, 2006b).

We must systematically inventory backyard biodiversity throughout the world as a part of sustainable economic development. These inventories are fundamental to a scientific basis for land-use planning, conservation and management of lands, natural resource management, and sustainable economic development. They provide the information for combating immigrant pests and invasive species that are major factors in current biodiversity loss. Our knowledge base for biodiversity, particularly of insects, would contribute to protection and management of backyard biodiversity. The pursuit of sustainability requires new conceptual and practical approaches for which biodiversity science, an integration of biodiversity and human sustainability based on taxonomy, ecology, and conservation science (Kim and Byrne 2006), provides a means to meet those challenges.

We face a paradox involving the need for taxonomic services on the one hand and taxonomic demise on the other (Kate and Kress 2005, Pegg 2006, Stribling 2006, Olden and Rooney 2006, Henningsen 2007, Walter and Winterton 2007). Concerns over the scientific and social effects of taxonomic decline have been expressed in major scientific journals, without much heed (Savage 1995, Wheeler 2004, Wheeler et al. 2004). In recent years, however, renewed pronouncements on the needs for taxonomy and a reversal of taxonomic decline have appeared in popular magazines and news media. They differ from past efforts because the writers represent broader scientific disciplines, such as conservation biology and ecology (Ehrlich and Wilson 1991, Kim 1993, Godfray and Knapp 2002, Gotelli 2004, Ehrlich 2005, Kim and Byrne 2006). Yet, the core issues in the taxonomic domain have not shifted to meet the applied needs, and science policy makers have not seen an accountability of taxonomy for biology and environmental sciences. Global climate change, along with habitat destruction and pollution, directly affects biodiversity at various spatial and temporal scales,

making the study of biodiversity urgent. All of these issues represent important challenges to science policy makers and the scientific community at large.

The historic decline of taxonomy and the increasing shortage of taxonomic specialists make it difficult to obtain accurate identification of scientifically and economically important species. At the same time, no other means are available to have insects identified to lower taxonomic units because trained and knowledgeable parataxonomists are scarce and no private infrastructure for taxonomic services exist (Gotelli 2004, Kim and Byrne 2006). New generations of biologists and environmental scientists bypassed taxonomic training and natural history because many institutions of higher learning reduced curricular requirements in these subject areas as new discoveries in molecular biology shifted the interests of young scientists away from systematics. Major curricular shifts in undergraduate and graduate education produced biologists who are knowledgeable about molecular biology, genetics, and perhaps phylogenetics but who have little understanding of species concepts, taxonomy, or basic methods for classification and identification of organisms. As a result, today's generation of biologists is ill prepared to conduct the alpha-taxonomic tasks required.

REFERENCES

Bascompte, J. and R. V. Solé. 1996. Habitat fragmentation and extinction thresholds in spatially explicit models. *Journal of Animal Ecology* 65: 465–473.

Bascompte, J. and R. V. Solé. 1998. Effects of habitat destruction in a prey-predator metapopulation model. *Journal of Theoretical Biology* 195: 383–393.

Baskin, Y. 1997. *The Work of Nature – How the Diversity of life Sustains Us. A Project of SCOPE: The Scientific Committee on Problems of the Environment*. Island Press, Washington, DC. 263 pp.

Berkes, F. 2004. Rethinking community-based conservation. *Conservation Biology* 18: 621–630.

Bertrand, Y. and M. Härlin. 2006. Stability and universality in the application of taxon names in phylogenetic nomenclature. *Systematic Biology* 55: 848–858.

Bertrand, Y., F. Pleijel and G. W. Rouse. 2006. Taxonomic surrogacy in biodiversity assessments, and the meaning of Linnean ranks. *Systematics and Biodiversity* 4: 149–159.

Biaggini, M., R. Consorti, L. Dapporto, M. Dellacasa, E. Paggetti and C. Dorti. 2007. The taxonomic level order as a possible tool for rapid assessment of arthropod diversity in agricultural landscapes. *Agriculture, Ecosystems & Environment* 122: 183–191.

Blunt, W. 2001. *Linnaeus: the Compleat Naturalist.* Princeton University Press. Princeton, New Jersey. 264 pp.

Boone, J. H., C. G. Mahan and K. C. Kim. 2005. *Biodiversity Inventory: Approaches, Analysis, and Synthesis.* Technical Report NPS/NER/NRTR-2005/015. US Department of Interior, National Park Service, Northeast Region, Philadelphia, Pennsylvania.

Botkin, D. B., H. Saxe, M. B. Araújo, R. Betts, R. H. W. Bradshaw, T. Cedhagen, P. Chesson, T. P. Dawson, J. R. Etterson, D. P. Faith, S. Ferrier, A. Guisan, A. S. Hansen, D. W. Hilbert, C. Loehle, C. Margules, M. New, M. J. Sobel and D. R. B. Stockwell. 2007. Forecasting the effects of global warming on biodiversity. *BioScience* 57: 227–236.

Brooks, D. R. and D. A. McLennan. 2002. *The Nature of Diversity: An Evolutionary Voyage of Discovery.* University of Chicago Press, Chicago, Illinois. 668 pp.

Büchs, W. (ed.). 2003. *Biotic Indicators for Biodiversity and Sustainable Agriculture.* Elsevier, Amsterdam, Netherlands. 550 pp.

Cameron, S., D. Rubinoff and K. Will. 2006. Who will actually use DNA barcoding and what will it cost? *Systematic Biology* 55: 844–847.

Canadell, J. C. and I. Noble. 2007. Changing metabolism of terrestrial ecosystems under global change. *Ecological Applications* 10: 1551–1552.

CBD (Secretariat of the Convention on Biological Diversity). 2006a. *Sustaining Life on Earth: How the Convention on Biological Diversity Promotes Nature and Human Well-being.* CBD, Montreal, Quebec.

CBD. 2006b. *Global Biodiversity Outlook 2.* CBD, Montreal, Quebec.

Center for Wildlife Law. 1996. *Saving Biodiversity: A Status Report on State Laws, Policies and Programs.* Defenders of Wildlife. Washington, DC.

Cohen, J. E. 2006. Human population: the next half century. Pp. 13–21. In D. Kennedy and the Editors of Science. *Science Magazine's State of the Planet 2006–2007, AAAS.* Island Press, Washington, DC.

Coleman, D. C. and P. F. Hendrix (eds). 2000. *Invertebrates as Webmasters in Ecosystems.* CABI Publishing, Oxford, United Kingdom. 336 pp.

Cornell, H. V. and R. H. Karlson. 1997. Local and regional processes as controls of species richness. Pp. 250–263. In D. Tilman and P. Kareiva (eds). *Spatial Ecology.* Princeton University Press, Princeton, New Jersey.

Derraik, J. G. B., G. P. Closs, K. J. M. Dickinson, P. Sirvid, B. I. P. Barratt and B. H. Patrick. 2002. Arthropod morphospecies versus taxonomic species: a case study with Araneae, Coleoptera, and Lepidoptera. *Conservation Biology* 16: 1015–1023.

Donoghue, M. J. and W. W. Alverson. 2000. A new age of discovery. *Annals of the Missouri Botanical Garden* 87: 110–126.

Durden L. A. and G. G. Musser. 1994. The sucking lice (Insecta, Anoplura) of the world: taxonomic checklist with records of mammalian hosts and geographical distributions. *Bulletin of the American Museum of Natural History* 218: 1–90.

Easterbrook, G. 2007. Global warming: who loses – and who wins? *The Atlantic April* 2007: 52–64.

Ehrlich, P. R. 2005. Twenty-first century systematics and the human predicament. *Proceedings of the California Academy of Sciences* 56 (Supplement 1) (12): 130–148.

Ehrlich, P. R. and E. O. Wilson. 1991. Biodiversity studies: science and policy. *Science* 253: 758–762.

ENS. 2006a. Global warming could spread extreme drought. Environmental News Service October 5, 2006, 3 pages. http://www.ens-newswire.com/ens/oct2006/2006-10-05-01.asp [Accessed July 8, 2008].

ENS. 2006b. Scientists predict future of weather extremes. Environmental News Service October 20, 2006, 2 pp. http://www.ens-newswire.com/ens/oct2006/2006-10-20-03.asp [Accessed July 8, 2008].

Farber, S., R. Costanza, D. L. Childers, J. Erickson, K. Gross, M. Grove, C. S. Hopkinson, J. Kahn, S. Pincetl, A. Troy, P. Warren and M. Wilson. 2006. Linking ecology and economics for ecosystem management. *BioScience* 56: 121–133.

Gaston, K. J. and T. M. Blackburn. 2000. *Pattern and Process in Macroecology.* Blackwell Science, Oxford, United Kingdom. 377 pp.

Geological History. 2008. Geological History across Geological Time: http://www.fossilmuseum.net/Geological History .htm [Accessed July 8, 2008].

Godfray, H. C. J and S. Knapp. 2004. Introduction: taxonomy for the twenty-first century. *Philosophical Transactions of the Royal Society B Biological Sciences* 395: 559–569.

Gotelli, N. J. 2004. A taxonomic wish-list for community ecology. *Philosophical Transactions of the Royal Society B Biological Sciences* 359: 585–597.

Grimaldi, D. and M. S. Engel. 2005. *Evolution of the Insects.* Cambridge University Press, Cambridge, United Kingdom. 755 pp.

Harwood, R. F. and M. T. James. 1979. *Entomology in Human and Animal Health.* 7th edition. Macmillan Publishing, New York. 548 pp.

He, F., K. J Gaston, E. F. Connor and D. S. Srivastava. 2005. The local-regional relationships: immigration, extinction, and scale. *Ecology* 86: 360–365.

Hector, A. and R. Bagchi. 2007. Biodiversity and ecosystem multifunctionalisty. *Nature* 448: 188–190.

Henningsen, C. 2007. Managing ecological investment risk. Environmental News Service ENS August 2007: 1–5; http://www.ens-newswire.com/ens/aug2007/2007-08-13-03.asp [Accessed July 8, 2008].

Heywood, V. H. and R. T. Watson (eds). 1995. *Global Biodiversity Assessment. United Nations Environment Programme.* Cambridge University Press, Cambridge, United Kingdom.

Hickerson, M. J., C. P. Meyer and C. Moritz. 2006. DNA Barcoding will often fail to discover new animal species over broad parameter space. *Systematic Biology* 55: 729–739.

Hooper, D. U., F. S. Chapin, III, J. J. Ewel, A. Hector, P. Inchausti, S. Lavorel, J. H. Lawton, D. M. Lodge, M. Loreau, S. Naeem, B. Schmid, H. Setälä, A. J. Symstad, J. Vandermeer and D. A. Wardle. 2005. Effects of biodiversity on ecosystem functioning: a consensus of current knowledge. *Ecological Monographs* 75: 2–35.

Huxley, J. S. (ed). 1940. *The New Systematics*. Clarendon Press, Oxford, United Kingdom. viii + 583 pp.

IPCC (Intergovernmental Panel on Climate Change). 2007. IPCC Fourth Assessment Report. UNEP and WMO.

Kate, K. T. and W. G. Kress. 2005. The Convention on Biological Diversity. *Tropinet* 16 (2): 1–3 (Supplement to Biotropica 37, 2).

Kim, K. C. 1983. How to detect and combat exotic pests. Pp. 261–319. *In* C. W. Wilson and C. L. Graham (eds.). *Exotic Plant Pests and North American Agriculture*. Academic Press, New York.

Kim, K. C. 1993a. 1. Insect pests and evolution. Pp. 3–26. *In* K. C. Kim and B. A. McPheron (eds.). *Evolution of Insect Pests: Patterns of Variation*. John Wiley and Sons, New York.

Kim, K. C. 1993b. Biodiversity, conservation, and inventory: why insects matter. *Biodiversity and Conservation* 2: 191–214.

Kim, K. C. 2001a. *Biodiversity, Our Living World: Your Life Depends on It!* Pennsylvania State University, College of Agricultural Sciences, Cooperative Extension and Center for BioDiversity Research, Environmental Resources Research Institute, University Park, Pennsylvania. 16 pp.

Kim, K. C. 2001b. *Interim Technical Report: Inventory of Invertebrates at Gettysburg National Military Park and Eisenhower National Historic Sites, with Special Reference to Forest Removal*. Center for BioDiversity Research, Environmental Resources Research Institute, Frost Entomological Museum, Department of Entomology, Pennsylvania State University, University Park, Pennsylvania.

Kim, K. C. 2006. *Biodiversity Inventory and Assessment of the National Guard Training Center, Fort Indiantown Gap (FIG-NGTC), Pennsylvania (Contract: PMDVA Task 03/PAA2001033; 4000000661/#359494; 2003–2006)*. Final Report to Fort Indiantown Gap-National Guard Training Center, Pennsylvania Department of Military and Veterans Affairs, Environmental Division, Building 11–19, Annville, PA 17003-5002. Institutes of the Environment and Energy, Pennsylvania.

Kim, K. C. 2007. Blood-sucking lice (Anoplura of small mammals: true parasites). Pp. 141–160. *In* S. Morand, B. R. Krasnov and R. Poulin (eds). *Micromammals and Macroparasites*. Springer, Tokyo, Japan.

Kim, K. C. and B. A. McPheron. 1993. *Evolution of Insect Pests: Patterns of Variation*. John Wiley & Sons, New York. 479 pp.

Kim, K. C. and L. B. Byrne. 2006. Biodiversity loss and the taxonomic bottleneck: emerging biodiversity science. *Ecological Research* 21: 794–810.

Kim, K. C. and R. D. Weaver. 1994. *Biodiversity and Landscapes: A Paradox of Humanity*. Cambridge University Press, New York. 431 pp.

Kim, K. C. and A. G. Wheeler, Jr. 1991. *Pathways and Consequences of the Introduction of Non-indigenous Insects and Arachinds in the United States*. Contractor Report for the United States Congress Office of Technology Assessment, December 1991.

Kim, K. C., K. C. Emerson and R. Traub. 1990. Diversity of parasitic insects: Anoplura, Mallophaga, and Siphonaptera. Pp. 91–103. *In* M. Kosztarab and C. W. Schaefer (eds.). *Systematics of the North American Insects and Arachnids: Status and Needs*. Virginia Agricultural Experiment Station Information Series 90-1. Virginia Polytechnic Institute and State University, Blacksburg, Virginia.

Kinzig, A. P., S. W. Pacala and D. Tilman. 2001. *The Functional Consequences of Biodiversity: Empirical Progress and Theoretical Extensions*. Princeton University Press, Princeton, New Jersey. 365 pp.

Kosztarab, M. and C. W. Schaefer (eds). 1990a. *Systematics of the North American Insects and Arachnids: Status and Needs*. Virginia Agricultural Experiment Station Information Series 90-1. Virginia Polytechnic Institute and State University, Blacksburg, Virginia. 247 pp.

Kosztarab, M, and C. W. Schaefer. 1990b. Conclusions, Pp. 241–247. *In* M. Kosztarab and C. W. Schaefer (eds). *Systematics of the North American Insects and Arachnids: Status and Needs*. Virginia Agricultural Experiment Station Information Series 90-1. Virginia Polytechnic Institute and State University, Blacksburg, Virginia. 247 pp.

Krell, F.-T. 2004. Parataxonomy vs. taxonomy in biodiversity studies – pitfalls and applicability of 'morphospecies' sorting. *Biodiversity and Conservation* 13: 795–812.

Lande, R. 1987. Extinction thresholds in demographic models of territorial populations. *American Naturalist* 130: 624–635.

Lee, K. N. 2007. An urbanizing world. Pp. 3–25. *In 2007 State of the World: Our Urban Future. A Worldwatch Institute Report on Progress Toward a Sustainable Society*. W. W. Norton & Company, New York.

Lin, Z. S., X. Z. Qi and B. L. Li. 2005. Can best competitors avoid extinction as habitat destruction? *Ecological Modeling* 182: 107–112.

Lin, Z. S. and H. Y. Liu. 2006. How species diversity responds to different kinds of human-caused habitat destruction. *Ecological Research* 21: 100–106.

Linnaeus, C. 1758. *Systema Naturae per Regna Tria Naturae, Secundum Classes, Ordines, Genera, Species cum Characteribu, Differniis Synonymis, Locis, Editio Decima, Reformata*. Tomus 1, Lauarentii Salvii. Homiae.

Loreau, M., S. Naeem, P. Inchausti, J. Bengtsson, J. P. Grine, A. Hector, D. U. Hooper, M. A. Huston, D. Raffaelli,

B. Schmid, D. Tilman and D. A. Wardle. 2001. Biodiversity and ecosystem functioning: current knowledge and future challenges. *Science* 294: 804–808.

Loreau, M., S. Naeem and P. Inhausti (eds.). 2002. *Biodiversity and Ecosystem Functioning: Synthesis and Perspectives.* Oxford University Press, New York.

Lundmark, C. 2003. BioBlitz: getting into backyard biodiversity. *BioScience* 54: 329.

Mahan, C. G., K. Sullivan, K. C. Kim, R. H. Yahner and M. Abrams. 1998. *Ecosystem Profile Assessment of Biodiversity; Sampling Protocols and Procedures.* Final Report, USDI, National Park Service, Mid-Atlantic Region.

Mayr, E. and P. D. Ashlock. 1991. *Principles of Systematic Zoology.* 2nd edition. McGraw-Hill College, New York.

Mascia, M. B., J. P. Brosius, T. A. Dobson, B. C. Borbes, L. Horowitz, M. A. McKean and N. J. Turner. 2003. Conservation and the social sciences. *Conservation Biology* 17: 649–650.

McCarty, J. P. 2001. Ecological consequences of recent climate change. *Conservation Biology* 15: 320–331.

MEA: Millennium Ecosystem Assessment. 2005. *Ecosystems and Human Well-being: Synthesis Report.* Island Press, Washington, DC.

Mittermeier, R. A. 2004. *Hotspots Revisited.* CEMEC, Mexico City. 390 pp.

Mittermeier, R. A., N. Myers, P. C. Gill and C. G. Mittermeier. 2000. *Hotspots: Earths Biologically Richest and Most Endangered Terrestrial Ecoregions.* CEMEX, Mexico City. 430 pp.

Myers, N. 1979. *The Sinking Ark: a New Look at the Problem of Disappearing Species.* Pergamon Press, Oxford, United Kingdom. 240 pp.

Nature. 2007. Linnaeus at 300. Nature 446/15 March 2007: 231 (Editorial), 247–256 (News Features), 259–262 (Commentary), 268 (Essay).

Nee, S. and R. M. May. 1997. Extinction and the loss of evolutionary history. *Science* 278: 692–694.

Olden, J. D. and T. P. Rooney. 2006. On defining and quantifying biotic homogenization. *Global Ecology and Biogeography* 15: 113–120.

Oliver, I. and A. J. Beattie. 1996. Invertebrate morphospecies as surrogates for species: a case study. *Conservation Biology* 10: 99–109.

Padial, J. M. and I. de la Riva. 2006. Taxonomic inflation and the stability of species lists: the perils of ostrich's behavior. *Systematic Biology* 55: 859–867.

Palumbi, S. R. 2001. Humans are the world's greatest evolutionary force. *Science* 293: 1786–1790.

Parmesan C. and H. Galbraith. 2004. *Observed Impacts of Global Climate Change in the U. S.* Pew Center on Global Climate Change, Washington, DC.

Pegg, J. R. 2006. Global Ecological Assessment Call for Humanity to Value Nature. Environmental News Service, ENS, Latest Environmental Information. January 23, 2006: 1–5 http://www.ens-newswire.com/ens/Jan2006/2006-01-23-10.asp [Accessed July 8, 2008].

Ponting, C. 1991. *A Green History of the World: the Environment and the Collapse of Great Civilizations.* Penguin Books, New York. 432 pp.

Raup, D. M. 1991. *Extinction: Bad Genes or Bad Luck?* W.W. Norton, New York. 228 pp.

Ricklefs, R. E. 1987. Community diversity: relative roles of local and regional processes. *Science* 235: 167–171.

Ricklefs, R. E. 2004. A comprehensive framework for global patterns in biodiversity. *Ecology Letters* 7: 1–15.

Rohr, J. R., C. B. Mahan and K. C. Kim. 2006. Developing a monitoring program for invertebrates: guidelines and a case study. *Conservation Biology* 21: 422–435.

Sala, O. E., F. S. Chapin, J. J. Armesto, E. Berlow, J. Bloomfield, R. Dirzo, E. Huber-Sanwald, L. F. Huenneke, R. B. Jackson, A. Kinzig, R. Leemans, D. M. Lodge, H. A. Mooney, M. Oesterheld, N. L. Poff, M. T. Sykes, B. H. Walker, M. Walker and D. H. Wall. 2000. Global biodiversity scenarios for the year 2100. *Science* 289: 1770–1774.

Savage, J. M. 1995. Systematics and the biodiversity crisis. *Bioscience* 45: 673–679.

Schmitz, O. J. 2007. *Ecology and Ecosystem Conservation.* Island Press, Washington, DC. 184 pp.

Schwartz, M. W., N. L. Jurjavcic and J. M. O'Brian. 2002. Conservation's disenfranchised urban poor. *BioScience* 52: 601–606.

Shurin, J. B. and D. S. Srivastava. 2005. New perspectives on local and regional diversity: beyond saturation. Pp. 399–417. *In* M. Holyoak, M. A. Leibold and R. D. Holt (eds.). *Metacommunities.* University of Chicago Press, Chicago, Illinois.

Sohdi, N. S., L. P. Koh, B. W. Brook and P. K. L. Ng. 2004. Southeast Asian biodiversity: an impending disaster. *Trends in Ecology and Evolution* 19: 654–660.

Stork, N. E. and C. H. C. Lyal. 1993. Extinction or co-extinction rates. *Nature* 366: 307.

Stribling, J. B. 2006. Viewpoint: environmental protection using DNA barcodes or taxa? *BioScience* 56: 878.

Takacs, D. 1996. *The Idea of Biodiversity and Philosophies of Paradise.* Johns Hopkins University Press, Baltimore, Maryland. 393 pp.

Tilman, D. and C. Lehman. 2001. Biodiversity, composition, and ecosystem. Pp. 9–41. *In* A. P. Kinzig, S. W. Pacala and D. Tilman (eds.). *The Functional Consequences of Biodiversity: Empirical Progress and Theoretical Extensions.* Princeton University Press, Princeton, New Jersey.

UNDP, UNEP, WB, WRI. 2000. *World Resources 2000–2001: People and Ecosystems – The Fraying Web of Life.* United Nations Development Programme (UNDP), United Nations Environmental Programme (UNEP), World Bank (WB), and World Watch Institute (WRI). Elsevier Science, Amsterdam.

U. S. Congress, Office of Technology Assessment. 1993. *Harmful Non-Indigenous Species in the United States.* OTA-F-565.

Washington, D C: U.S. Government Printing Office, September 1993. 391 pp.

Walter, D. V. and S. Winterton. 2007. Keys and the crisis in taxonomy: extinction or reinvention. *Annual Review of Entomology* 52: 193–208.

Wardle, D. A. 2002. *Communities and Ecosystems: Linking the Aboveground and Belowground Components*. Princeton University Press, Princeton, New Jersey. 400 pp.

Warne, K. 2007. Tribute: organization man. *Smithsonian* (magazine) May 2007: 105–111.

Wheeler, Q. D. 2004. Taxonomic triage and the poverty of phylogeny. *Philosophical Transactions of the Royal Society B Biological Sciences* 359: 571–583.

Wheeler, Q. D., P. H. Raven and E. O. Wilson. 2004. Taxonomy impediment or expedient? *Science* 303: 285.

Wilson, E. O. 1985. The biological diversity crisis: A challenge to science. *Issues in Science and Technology* 2: 20–29.

Wilson, E. O. 1992. *The Diversity of Life*. Belknap Press of Harvard University Press, Cambridge, Massachusetts. 424 pp.

Chapter 25

INSECT BIODIVERSITY – MILLIONS AND MILLIONS

May Berenbaum

Department of Entomology, 320 Morrill Hall, University of Illinois, 505 S. Goodwin, Urbana, IL 61801-3795

Insect Biodiversity: Science and Society, 1st edition. Edited by R. Foottit and P. Adler
© 2009 Blackwell Publishing, ISBN 978-1-4051-5142-9

R elatively few scientific truths persist across centuries; new technologies, new hypotheses, and new information mean that most scientific conclusions are constantly modified, updated, or even sometimes discarded. One conclusion, however, seems robust even as new information has accumulated across the centuries – namely, that there are a lot of insects. Ever since Linnaeus imposed order on the classification of living organisms, six-legged animals with exoskeletons have had an edge over other groups. Of the 4203 species of animals described by Linnaeus, insects constituted 2102, over half of the known animal diversity at the time. Species descriptions proceeded apace once Linnaeus provided a flexible framework for naming and classifying them, but as species piled up across all taxa they piled up fastest for insects. Between 1758 and 1800, 58,833 insect species were described and between 1800 and 1850, 363,588 species were added to the pantheon of known species. As other paradigms in biology came and went, the notion that insects are the dominant life form on the planet remained unchanged by new data.

Relatively early in the history of insect taxonomy, even before Linnaeus imposed order and structure on biological diversity, the question arose as to just how many insects there might be on the planet. Virtually every chronicler of insect life dating back to the seventeenth century has felt obligated to venture a guess. Sir John Ray, the great British naturalist who authored the 'Wisdom of God in the Works of His Creation' in 1691, attempted a projection based on his own experiences in his homeland:

'Supposing, then, there be a thousand several sorts of insects in this island and the sea near it, if the same proportion holds between the insects native of England and those of the rest of the world . . . , the species of insects in the whole earth (land and water) will amount to 10,000: and I do believe they rather exceed than fall short of this sum' (cited in Westwood 1833).

Ray later revised his own estimate upward by twofold, based on his discoveries of new species of English moths and butterflies, but in 1815 William Kirby and William Spence, authors of *Introduction to Entomology*, the first textbook of entomology, still regarded the estimate of 20,000, 'which in his time was reckoned a magnificent idea' as 'beggarly', by nineteenth-century standards. Their own assessment was influenced by the fact that their contemporary DeCandolle had estimated the number of plant species at 110,000 to 120,000. Observing that each British plant was typically associated with at least six insect species, Kirby and Spence reasoned, 'if we reckon the phanerogamous vegetables of the globe, in round numbers at 100,000 species, the number of insects would amount to 600,000'. By 1833, John Obadiah Westwood, a curator at Oxford and thus familiar with the 'immense influx of novelties which has been poured into our museums and cabinets since the days of Linnaeus . . . especially in the insect tribes' attempted to estimate the 'numerical extent of this department of nature'. He arrived at his estimate, a more conservative 400,000, by comparing the rate at which new species were described to the number already known across different key groups. Illustrative of the expansion of knowledge were the carabid genera *Carabus* and *Cicindela*; within 70 years, for example, the 50 species collectively described in the genera *Carabus* and *Cicindela* by Linnaeus had expanded to more than 2000 species, a 40-fold increase.

Such estimates opened a can of (very diverse) worms and kicked off a discussion among insect systematists that continues to this day. Howard (1932) provided a summary of the controversy through the end of the nineteenth century. In 1883, Dr. David Sharp, a British coleopterist, hazarded his guess: 'As the result of a moderate estimate it appears probable that the number of species of true insect existing at present on our globe is somewhere between five hundred thousand and one million', adding that 'the number probably exceeds the higher of these figures and will come in near to two million'. Thomas de Grey, Lord Walsingham, President of the Entomological Society of London from 1889 to 1890, chose in his presidential address to endorse Sharp's estimate. Across the Atlantic, however, Charles Valentine Riley arrived at a different conclusion: 'after considering the fact that the species already collected are mostly from the temperate regions of the globe and that many portions of the world are as yet unexplored by collectors of insects, and further that the species in many groups of insects are apparently unknown – that the estimate of 2,000,000 species in the world, made by Dr. Sharp and Lord Walsingham, was extremely low and that it probably represented not more than one fifth of the species that actually exist'. And, with that, Riley boldly elevated the estimate by an order of magnitude to 10,000,000.

Such a high estimate did not sit well with the entomological establishment of the era; according to Howard, 'it met with no favorable comment. Everyone thought it was too high'. In reviewing the history, Howard speculated that at the time few people other than astronomers and geologists had any grasp of a figure as large as 10 million.

Having reviewed the various estimates, Howard himself tried another approach. He took the question to a meeting of the Entomological Society of Washington, attended by a number of experts in many of the larger insect taxa. There, he asked everyone present to estimate the proportion described within the groups they knew best, figuring that 'because of the character of the men who took part in this discussion it was probably the most authoritative expression of opinions that could very well be had'. Summing the individual estimates led Howard to suggest that an estimate of 4,000,000 was reasonable.

Character notwithstanding, debate did not end with the meeting of the Entomological Society of Washington. Metcalf (1940) picked up the gauntlet soon thereafter. Eschewing any effort to count the total number of insect individuals (because 'no one has been foolhardy enough to attempt to make a world census of insect individuals'), he attempted, as his predecessors had, to estimate the number of species. At the time, textbooks pegged the number at between 250,000 and 1,000,000. As the basis for his estimate, he examined the changing ratios of genus to species over time and revised the estimate upward to 1,500,000. He was realistic enough to recognize the limitations on his estimate, however, and philosophized about the futility of such efforts:

'At least they have done this; they have occupied my thoughts on a hazy Indian summer day; their calculations have kept an otherwise idle adding machine busy; and last by no means least important I hope they have stimulated your thoughts in this field. . . . they will make those of us who profess to be systematists more systematic as we go about our daily business of describing new genera and species, so that those who come after us will not have to do too much counting and recounting'.

Despite the good intentions, counting and recounting continued. Sabrosky (1952, 1953) pointed out that 6000 to 7000 species of insects were being described every year. He engaged in a different kind of calculation to convey the magnitude of the number of species already described, estimating that, 'If the names were printed one to a line in an eight-page, eight-column newspaper of average size, without headlines and pictures, more than eight weeks, including Sundays, would be needed to print only

the names of the insects that are already known in the world' (Sabrosky 1953). Sabrosky was among the first to suggest that synonymies might inflate the count of described insects and accordingly elevate estimates of total insect diversity. Subsequently, he proposed a one-time census of all animal species, to commemorate the two-hundredth anniversary of 'Systema Naturae' (Sabrosky et al. 1953).

As it happens, 1958 came and went without that definitive census and, after an extended pause, discussions of total insect biodiversity resumed once more. What apparently precipitated the next round of pitched discussion was an estimate based on a new approach. Terry Erwin (1982) evaluated the number of beetle species found in association with one species of tropical tree, *Luehea seemannii*, exhaustively collecting them by fogging the canopy with an insecticide. After identifying the species known to science, he could estimate the ratio of described to undescribed species of Coleoptera. He then estimated the degree of host specificity of each feeding guild represented in the collection, calculated the ratio of coleopterans to other arthropods, guessed at the relative ratio of canopy to ground-dwelling arthropods, and multiplied by the estimated number of tropical tree species in the world, to obtain an estimate of world insect diversity of approximately 30 million (Erwin 1988).

Although Erwin's estimates were advanced as a testable hypothesis, rather than a definitive number (Erwin 1991), the sheer magnitude of the number commanded attention. Not everyone embraced the new methodology and a flurry of publications followed; in the context of growing concerns about biodiversity losses, estimates of total insect diversity had gained new importance. Robert May (1988, 1990) took another tack, extrapolating based on the general relationship between size and species diversity across all organisms. Basically, over the range of body lengths from several meters to 1 cm, for each order of magnitude reduction in length (or 100-fold reduction in body weight), the number of species increases 100-fold. Problematically, the relationship is less clear for taxa less than 1 cm, which encompasses the majority of insects, but extrapolating the relationship leads to an estimate of approximately 10 million species of terrestrial animals (May 1988), of which approximately three-quarters are insects.

Yet another approach was taken by Hodkinson and Casson (1990), who enumerated the hemipteran species represented in an exhaustive collection of Sulawesi arthropods and determined that approximately 62.5% of the 1690 species collected were undescribed. Assuming that the same proportion of the world's bugs were undescribed, they arrived at an estimate of 184,000–193,000 species of bugs worldwide, and assuming that bugs comprise approximately 7.5 to 10% of the world's insects, they produced an estimate of 1.84–2.57 million. Gaston (1991a, 1991b) returned to the approach used by Westwood (1833) and Howard (1932) in earlier centuries and polled expert taxonomists for their best estimate of what proportion of their taxon of interest was undescribed; most estimates were optimistic (i.e., that the difference was less than fivefold), leading him to an estimate of 5 million (given that the number of described species was in the neighborhood of 1 million). Concerns about synonymies reared their ugly head once again, as did concerns about sampling, leading Stork (1993) to revise the Erwin estimates downward to between 5 and 15 million. By 2002, sufficient data existed from a variety of biotic diversity inventories to allow Ødegaard et al. (2000) to correct earlier estimates based on the degree of host-plant specialization of phytophagous insects associated with a restricted number of plant species and extrapolating to a larger number of plants (as did Kirby and Spence almost two centuries earlier), arriving at a modified estimate of 5 to 10 million species.

So systematists continue to debate the issue without any hope of consensus. Even the number of insects already described, a number that would seem to be robust and beyond dispute, is contentious. On one hand, there are problems with synonymies; Alroy (2002) has estimated that '24–31% of currently accepted names eventually will prove invalid' due to synonymies or *nomina dubia* (that is, '30% of named species are illusions created by unsettled taxonomy'). On the other hand, new molecular approaches have revealed substantially greater species diversity than was hitherto suspected to exist. Smith et al. (2007) used DNA barcoding to evaluate 16 morphospecies of apparently generalist tropical parasitoid flies in the family Tachinidae and found that the 16 generalist species apparently represent 9 generalist species and 73 specialist lineages. Given that parasitoids are thought to represent 20% of all insect species, gross underestimates of parasitoid species diversity may mean that global species richness of insects also may be grossly underestimated.

That there has been a debate raging for more than four centuries has gone largely unnoticed by the general public. Indeed, it is unlikely that there is even a

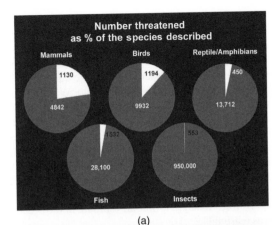

(a)

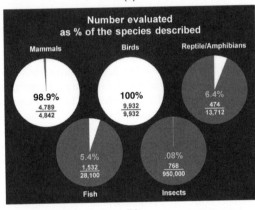

(b)

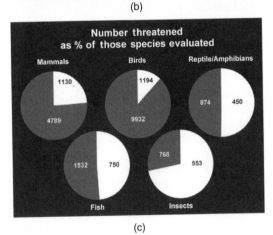

(c)

Fig. 25.1 Taxa threatened (in white) with extinction worldwide. (a) Number threatened as a percentage of species described. (b) Number evaluated as percentage of species described. (c) Number threatened as percentage of species evaluated (data from www.redlist.org, 2004)

vague notion among the general public of the extent of arthropod diversity. The level of appreciation of insect diversity rarely extends beyond the ordinal level and at times fails to achieve even that degree of differentiation. This unfortunate state of affairs is not new; Dan Beard, for example, in his 1900 book *Outdoor Games for All Seasons: The American Boy's Book of Sport*, bemoaned the fact that, 'With the exception of butterflies, the general public class the whole insect world under two heads – worms and bugs – and regard them with unqualified disgust. But this is only a sign of universal ignorance'.

In attempting to characterize public attitudes toward invertebrates, Kellert (1993) administered a questionnaire to determine the 'level of knowledge' of invertebrates in general. Individuals surveyed included randomly selected residents of the New Haven, Connecticut, area, along with subsamples of farmers, conservation organization members, and scientists. Of the various categories of knowledge investigated, the general public 'revealed the least knowledge of taxonomic differences among invertebrates . . . and – taxonomically – toward bees, cockroaches, grasshoppers, termites and beetles'. Specifically, 'only a minority had much concept of the overall number of insect species', fewer than one-quarter knew that spiders are not insects, and only 11% recognized that cockroaches are not beetles. Taxonomic confusion extended beyond the phylum Arthropoda – a majority of respondents thought that 'snails are more closely related to turtles than to spiders' and that 'the snail darter [a fish] is an endangered butterfly'.

Even relatively sophisticated members of the public appear content with their taxonomic ignorance. Snopes.com, a Web site devoted to mythbusting and correcting popular misconceptions on a wide range of topics, including scientific ones, claims that the statement 'The food colorants cochineal and carmine are made from ground beetles' is 'True' (http://www .snopes.com/food/ingredient/bugjuice.asp), despite the fact that cochineal and carmine are derived not from any of the 350,000 or so species in the order Coleoptera but rather from *Dactylopius coccus*, a scale insect in the order Hemiptera. Oddly enough, the species is accurately named on the Web site but is described as 'a beetle that inhabits a type of cactus called Opuntia'. When an entomologist pointed out their error in an email message, he was curtly informed that the authoritative reference for calling

cochineal scale a beetle was Webster's Dictionary, which defines 'beetle' not only as a member of the order Coleoptera, but anything even vaguely beetle-like ('any of various insects resembling a beetle', http://www.m-w.com/dictionary/beetles) (S. Bambara, personal communication).

So, while biologists passionately debate the number of species awaiting discovery, the general public appears blissfully unaware of and indifferent to this discussion. In fact, the business of differentiating among insect species has long been regarded as a pursuit of dubious value. Kirby and Spence (1815), in the introduction to their entomology text, lamented the fact that, in the eighteenth century, the will of one Lady Glanville was 'attempted to be set aside on the ground of lunacy, evinced by no other act than her fondness for collecting insects; and [Sir John] Ray had to appear at Exeter on the trial as a witness of her sanity'. What the public rarely recognizes, however, is that differentiating among and inventorying insect species is not now, nor has it ever been, an irrelevant or trivial pursuit. Insect species, as numerous and as seemingly similar as they may appear, are generally not ecologically interchangeable, and failing to recognize that fact has had tremendous economic and public health consequences over the centuries.

Examples of the importance of differentiating among arthropod species are legion. In agriculture, identifying pest species correctly is often key to understanding their life histories and developing approaches for managing them. The varroa mite, for example, is a devastating parasite of the European honeybee *Apis mellifera*. When mites first appeared attacking honeybees in North America in the 1980s, they were assumed to be *Varroa jacobsoni*, a species native to Indonesia and Malaysia that had hitherto been thought to attack only *Apis cerana*, the Eastern honeybee. However, Anderson and Trueman (2000) conducted morphological and molecular studies and determined that the mite attacking bees in North America is a distinct species, which they named *Varroa destructor*. Unlike *A. jacobsoni*, *A. destructor* infests *A. cerana* throughout much of mainland Asia and is also capable of parasitizing *A. mellifera* throughout the world to devastating effect (National Academy of Sciences 2006).

No less important than identifying pest species to control them is identifying and appreciating the diversity of potential biological control agents that can be used in pest-management programs. Many programs have failed or experienced decades-long delays simply because the diversity of potential control agents was not fully recognized (Caltagirone 1981). California red scale *Aonidiella aurantii*, for example, is an important pest of citrus that was accidentally introduced into California in the nineteenth century, most likely from Southeast Asia. For close to 60 years, biocontrol efforts ignored ectoparasitoid wasps in the genus *Aphytis* as potential control agents because *Aphytis chrysomphali* was already present in the state, having been accidentally introduced at the turn of the twentieth century, and apparently had little impact on the pest. Thus, dismissing ectoparasitoids as ineffective, entomologists concentrated on potential predators and endoparasitoids, without much success. Eventually, taxonomic study of the genus revealed a complex of species, including two, *A. lingnanensis* and *A. melinus*, which, once introduced, proved to be significantly superior biocontrol agents for the scale (Price 1997).

Just as biocontrol agents are not interchangeable, neither are insect-pollinating agents. Establishing a fig industry in California, today second only to Turkey in production of figs worldwide, was stymied for a decade in the late nineteenth century until entomologists recognized that one particular agaonid fig wasp species, one of hundreds in the genus, had to be imported to pollinate the trees (Swingle 1908). Similarly, cacao cultivation in Africa, outside its area of indigeneity in Mexico, was not profitable until the specific pollinators – midges in the genus *Forcipomyia* – were imported (Young 1982).

There have been significant public health consequences of the failure to recognize and differentiate among insect species. *Anopheles gambiae*, for example, was long regarded as the most important vector of malaria in Africa south of the Sahara. The species *Anopheles gambiae*, however, turned out to be in reality a complex of seven essentially morphologically identical species, some of which are efficient vectors and others are not vectors (White 1974, Hunt et al. 1998). These species also differ in the degree to which they are resistant to insecticides, which has major implications for control efforts (Davidson 1964). Effective management of vectors of malaria, a disease that kills 2 to 3 million people worldwide annually, requires precise identification of species (Gentile et al. 2002).

Even such unglamorous ecosystem services as waste disposal depend on a diversity of noninterchangeable arthropods. Introduction of placental mammals such as cattle and sheep into Australia led to monumental problems with dung accumulation; Australian dung beetles, adapted to using dung of marsupial mammals, could not process the dung of the introduced placental livestock species. The accumulated dung threatened the livestock industry by taking substantial amounts of pastureland out of commission but also led to population explosions of *Musca vetustissima*, the bush fly, whose larval stages thrived in the dung of the introduced species. Ultimately, over 50 species of dung beetles, with different habitat requirements, food preferences, and phenologies, were imported to manage the dung problem (Doube et al. 1991).

More than 5000 species of scarabaeine dung beetles have been described to date (Hanski and Cambefort 1991), but how many species remain to be described is an open question. The general public likely would consider an inventory of the dung beetles of the world to be a scientific enterprise of little import. Typically, arguments made for inventorying biodiversity are based on utilitarian grounds – that hitherto undescribed species may contribute valuable new pharmaceuticals or provide useful genes for bioengineering. The vast majority, however, contribute in smaller ways that might not become apparent until they are no longer abundant enough to make that contribution.

May (1990) has argued that there is a compelling argument, beyond utilitarian concerns, for cataloguing biodiversity – the 'same reasons that compel us to reach out toward understanding the origins and eventual fate of the universe or the structure of the elementary particles that it is built from'. Cosmology, however, has long enjoyed greater popularity with the general public than have coleopterans and it is unlikely that appealing to the public thirst for pure knowledge will soon pay dividends in the form of increased appreciation for insect biodiversity (or even a greater understanding of the ordinal limits of the group). This has long been true; even Kirby and Spence (1815) bemoaned the fact that their detailed illustrations were viewed as amusing diversions to inspire ladies' needlework.

Unfortunately, although the long-term future of atomic particles seems assured and immutable, the same cannot be said for the Earth's insect biodiversity. While there may not be a consensus on how many insect species remain to be described, an examination of known numbers is an argument for stepping up the inventory effort. According to the Red List (http://www.redlist.org/), 9932 species of birds have been described, of which 100% have been evaluated as to their ability to survive; of the 4842 species of mammals that have been described, 4782, almost 97%, have been evaluated (Fig. 25.1). By contrast, of the 1,004,898 or so species of insects that have been described (introduction to this volume), only 768 – about 0.08% – have been evaluated. Of the mammal species that have been evaluated, 1130 of 4782, or 23.6%, are threatened; of the 9932 bird species, 1194, or 12%, are threatened. By contrast, of the 768 insect species that have been evaluated, 563, or 73%, are threatened.

Insects, then, despite their almost ungraspably large numbers, are at disproportionate risk of extinction. How disproportionate is an open question in that, of all the major animal taxa on the planet, they are the least well-characterized group. To characterize even the majority of insect species would be a massive undertaking, but the fact that the rate at which species are being described has steadily increased since Linnaeus's day suggests that it is not an impossible one. If there is general consensus that conserving biodiversity is a good thing, then a necessary first step is to inventory that which is to be conserved. It would be a shame that one of the most durable and time-tested biological observations – that there are a lot of insects – may cease to be true in the foreseeable future.

ACKNOWLEDGMENTS

I thank Robert Foottit not only for inviting me to contribute this chapter but also for patiently waiting for me to complete it well after the deadline had passed and I thank Peter Adler for his patience in waiting for seriously late revisions. My UIUC colleague Jim Whitfield graciously provided helpful comments on the manuscript on short notice. This manuscript was supported in part by an NSF OPUS award.

REFERENCES

Alroy, J. 2002. How many named species are valid? *Proceedings of the National Academy of Sciences USA* 99: 3706–3711.

Anderson, D. and J. W. H. Trueman. 2000. *Varroa jacobsoni* (Acari: Varroidae) is more than one species. *Experimental and Applied Acarology* 24: 165–189.

Beard, D. 1900. *Outdoor Games for All Seasons: The American Boy's Book of Sport*. Charles Scribner's Sons, New York. 496 pp.

Caltagirone L. 1981. Landmark examples in classical biological control. *Annual Review of Entomology* 26: 213–222.

Davidson, G. 1964. The five mating types of the *Anopheles gambiae* complex. *Rivista di Malariologia* 13: 167–183.

Doube, B. M., A. Macqueen, T. J. Ridsill-Smith, and T. A. Weir. 1991. Native and introduced dung beetles in Australia. Pp. 255–278. *In* I. Hanski and Y. Cambeford (eds). *Dung Beetle Ecology*, Princeton University Press, New Jersey.

Erwin, T. 1982. Tropical forests: their richness in Coleoptera and other arthropod species. *Coleopterists Bulletin* 36: 74–75.

Erwin, T. 1988. Tropical forest canopies: the heart of biotic diversity. Pp. 123–129. *In* E. O. Wilson and F. M. Peter (eds). *Biodiversity*. National Academy Press, Washington, DC.

Erwin, T. L. 1991. How many species are there?: Revisited. *Conservation Biology* 5: 330–333.

Gaston, K. J. 1991a. The magnitude of global insect species richness. *Conservation Biology* 5: 283–296.

Gaston, K. J. 1991b. Estimates of the near-imponderable: a reply to Erwin. *Conservation Biology* 5: 564–566.

Gentile, G., A. della Torre, B. Maegga, J. R. Powell, and A. Caccone. 2002. Genetic differentiation in the African malaria vector, *Anopheles gambiae* s.s., and the problem of taxonomic status. *Genetics* 161: 1561–1578.

Hanski, I. and Y. Cambefort (eds). 1991. *Dung Beetle Ecology*. Princeton University Press, Princeton, New Jersey. 520 pp.

Hodkinson, I. D. and D. Casson. 1990. A lesser predilection for bugs: Hemiptera (Insecta) diversity in tropical rain forests. *Biological Journal of the Linnaean Society* 43: 101–109.

Howard, L. O. 1932. *The Insect Menace*. The Century Company, New York. 347 pp.

Hunt, R. H., M. Coetzee, and M. Fettene. 1998. The *Anopheles gambiae* complex: a new species from Ethiopia. *Transactions of the Royal Society of Tropical Medicine and Hygiene* 92: 231–235.

Kellert, S. R. 1993. Values and perceptions of invertebrates. *Conservation Biology* 7: 845–855.

Kirby, W. and W. Spence. 1815. *An Introduction to Entomology: Or Elements of the Natural History of Insects*. Longman, London.

May, R. L. 1988. How many species are there in the world? *Science* 241: 1441–1449.

May, R. L. 1990. How many species? *Philosophical Transactions of the Royal Society of London, Series B, Biological Sciences* 330: 293–304.

Metcalf, Z. P. 1940. How many insects are there in the world? *Entomological News* 51: 219–222.

National Academy of Sciences. 2006. *The Status of Pollinators in North America*. National Academy Press, Washington, DC. 312 pp.

Ødegaard, F., O. H. Diserud, S. Engen, and K. Aagaard. 2000. The magnitude of local host specificity for phytophagous insects and its implications for estimates of global species richness. *Conservation Biology* 14: 1182–1186.

Price, P. 1997. *Insect Ecology*. John Wiley and Sons, New York. 874 pp.

Sabrosky, C. W. 1952. How many insects are there? Pp. 1–7. *In* A. Stefferud (ed). *Insects: The Yearbook of Agriculture.* United States Department of Agriculture, Washington, DC.

Sabrosky, C. W. 1953. How many insects are there? *Systematic Zoology* 2: 31–36.

Smith, M. A., D. M. Wood, D. H. Janzen, W. Hallwachs, and P. D. N. Hebert. 2007. DNA barcodes affirm that 16 species of apparently generalist tropical parasitoid flies (Diptera, Tachinidae) are not all generalists. *Proceedings of the National Academy of Sciences USA* 104: 4967–4972.

Stork, N. E. 1993. How many species are there? *Biodiversity and Conservation* 2: 215–232.

Swingle, W. T. 1908. The fig in California. *Papers and Discussions Presented before the 34th and 35th Fruit-Growers' Convention, California.* State Commission on Horticulture. Sacramento, California.

Westwood, J. O. 1833. On the probable number of species of insects in the creation; together with descriptions of several minute Hymenoptera. *Magazine of Natural History and Journal of Zoology, Botany, Mineralogy, Geology, and Meteorology* 6: 116–123.

White, G. B. 1974. *Anopheles gambiae* complex and disease transmission in Africa. *Transactions of the Royal Society of Tropical Medicine and Hygiene* 68: 278–295.

Young, A. M. 1982. Effects of shade cover and availability of midge breeding sites on pollinating midge populations and fruit set in two cocoa farms. *Journal of Applied Ecology* 19: 47–63.

INDEX OF ARTHROPOD TAXA ARRANGED BY ORDER AND FAMILY

Supraordinal taxa and informal clades are listed as primary entries coordinate with order. Taxa and informal clades between order and family level are treated as secondary entries coordinate with family within the relevant order. Family-group names below family level are treated as tertiary entries coordinate with genus under the appropriate family. Page numbers in bold face indicate in table entries, and numbers in italic face indicate entries on figures and in figure captions.

ALPHABETIC INDEX TO ARTHROPOD TAXA

Page numbers in bold face indicate entries in tables; page numbers in italic face indicate entries on figures or in figure captions

INDEX OF NON-ARTHROPOD ORGANISMS

Implicit references to pathogenic organisms through the associated disease are included.

SUBJECT INDEX

A

abundance, 8, 59, 95, 189, 199, 203, 448, 497, 537–538, 566
Access to Biological Collections Data (ABCD), 437–438
acoustic communication, 168, 391
adaptive radiation (see radiations)
adhesion to surfaces, 284
adventive species, 4, 75–76, 80, 88, 91, 93, 94–95, 96, 98, 149, 176–177, 198, 199, 200, 233, 237, 242, 244, 249, 271, 281, 282, 290, 317, 361, 399, 421, 423, 446, 469, 476–500, 524, 540, 562, 569, 570
aesthetics, 534
aestivation, 85, 170
Africa, 11, 70–80, 108, 114, 129, 150, 196, 197, 228, 234, 235, 237, 238, 242, 244, 274, 277, 314, 316, 317, 327, 336, 337, 338, 340, 341, 343, 418, 454, 457, 478, 527, 528, 530, 532, 533, 537, 538, 539, 563, 580
African Fruit Fly Initiative (AFFI), 75
African horse sickness, 197, 528, 529, 537
Afrotropical Region, 70–80, 108, 130, 245, 247, 277, 327, 338, 340, 342, 343, 346, 525–526, 532
agamospecies, 385, 387, 388, 391–392
aging, 202
agriculture, 37, 75–77, 148, 178, 195, 279–280, 312, 317, 453, 458, 476, 479, 483, 495–497, 537, 579
agroecosystems, 73, 94, 234, 273, 305, 449, 452
Akabane viral disease, 529
Alaska, 37, 563
Albania, 286
alcoholism, 202
Algeria, 286, 287
alien species (see adventive species)

All Taxa Biological Inventory (ATBI), 350
allelochemicals, 493
allergic reactions, 177, 195, 198, 491
allochronic speciation (see speciation)
allopatry, 385–386, 387, 401
allozymes, 389, 401, 403, 404, 406, 422
alpine insects, 108, 128, 140–141
Altai Mountains, 120, 135, 139, 140, 141, 148
amateurs, 371
Amazon Region, 2, 4, 50–65, 528, 563
amber, 91
American blight, 469
American Samoa, 237
amplified fragment length polymorphisms (AFLPs), 404, 406
anaphylactic shock from insect toxins, 310
anaplasmosis, 199, 530
Anatolia, 114, 131, 142, 147, 150
Andrews Experimental Forest (Oregon), 8
Angola, 70, 80
anholocyclic populations, 467
Animal Base, 440
Antarctica, 85, 129, 186, 425, 484
anthrax, 529, 530
anthropogenic pressures, 108, 476, 484, 553, 564
anticancer properties (see medicinal properties of insects)
antifreeze proteins, 16, 278
antimicrobial properties, 278
Antipodes Islands, 89
antivenins, 310
Aotearoa, 85
aphrodisiacs, 283
apiculture, 317
aposematism, 236, 244, 248, 283

Appalachian Mountains, 177, 335, 495
apterous insects (see flightless insects)
aquaculture, 189
aquatic insects, 90, 92, 93, 97, 98, 129, 130, 142, 144, 152, 166–179, 186, 188, 189, 203, 228, 229–231, 249, 272, 273, 274, 276, 425–426, 484, 534
arboviruses, 530
Arctic, 36, 37, 43, 120, 122, 124, 128, 129, 139, 141–142, 143, 150, 425, 486
Argentina, 90, 91, 236, 242, 288, 335, 336, 341 556
aridization, 96, 124, 139, 146
Aristotle, 382
Arizona, 36, 37, 43, 343, 348
Armenia, 114, 145, 147, 286
Arrente people, 85
art, insects in, 304, 350, 487
asexual reproduction (see parthenogenesis)
Asia, 76, 234, 237, 242, 244, 247, 271, 281, 342, 343, 346, 347, 348, 390, 391, 418, 457, 480, 482, 483, 579
Auckland Islands, 89
Australasian Region, 78, 84–98, 169, 243, 248, 327, 339, 341, 344, 348, 525–526
Australia, 9, 10, 11, 12, 52, 78, 84, 85, 88–98, 129, 139, 175, 192, 193, 197, 198, 199, 224, 234, 236, 237, 241, 242, 243, 245, 247, 266, 272, 273, 277, 281, 282, 309, 311, 313, 317, 318, 335, 336, 337, 338, 339, 340, 341, 342, 343, 346, 348, 390, 391, 421, 422, 478, 482, 483, 484, 485, 486, 487, 489, 491, 494, 497, 498, 580
Australian Biological Resources Study (ABRS), 97

623